An Introduction to
CELL AND MOLECULAR BIOLOGY

STEPHEN L. WOLFE
University of California at Davis

Wadsworth Publishing Company
I(T)P™ *An International Thomson Publishing Company*
Belmont • Albany • Bonn • Boston • Cincinnati • Detroit • London
Madrid • Melbourne • Mexico City • New York • Paris • San Francisco
Singapore • Tokyo • Toronto • Washington

To Andrea, Eric, and Christopher

BIOLOGY PUBLISHER: Jack C. Carey

DEVELOPMENTAL EDITOR: Mary Arbogast

EDITORIAL ASSISTANT: Kristin Milotich

PRODUCTION EDITOR:The Book Company

PRINT BUYER: Diana Spence

PERMISSIONS EDITOR: Peggy Meehan

COPY EDITOR: Ellen Silge, Geetline Editorial Service

TECHNICAL ILLUSTRATORS: Darwen and Vally Hennings, John and Judy Waller, Illustrious, Inc., Academy ArtWorks, Inc., Precision Graphics

COVER DESIGNER: Gary Head

COVER ART: Tomo Narashima, based on a design by John and Judy Waller

COMPOSITOR: Thompson Type

COLOR SEPARATOR: H&S Graphics, Inc.

PRINTER: Banta Company, Menasha

I(T)P The ITP logo is a trademark under license

Printed in the United States of America
1 2 3 4 5 6 7 8 9 10—01 00 99 98 97 96 95

For more information, contact Wadsworth Publishing Company.

Wadsworth Publishing Company
10 Davis Drive
Belmont, California 94002
USA

International Thomson Publishing Europe
Berkshire House 168-173
High Holborn
London, WC1V 7AA
England

Thomas Nelson Australia
102 Dodds Street
South Melbourne 3205
Victoria, Australia

Nelson Canada
1120 Birchmount Road
Scarborough, Ontario
Canada M1K 5G4

International Thomson Editores
Campos Eliscos 385, Piso 7
Col. Polanco
11560 México D.F. México

International Thomson Publishing GmbH
Königswinterer Strasse 418
53227 Bonn
Germany

International Thomson Publishing Asia
221 Henderson Road
#05-10 Henderson Building
Singapore 0315

International Thomson Publishing Japan
Hirakawacho Kyowa Building, 3F
2-2-1 Hirakawacho
Chiyoda-ku, Tokyo 102
Japan

Library of Congress Cataloging-in-Publication Data

Wolfe, Stephen L.
    An introduction to cell and molecular biology / Stephen L. Wolfe
        p.   cm.
    Includes bibliographical reference and index
    ISBN 0–534–23778–9
    1. Cytology.   2. Molecular biology.   I. Title
QH581.2.W644 1995
574.87--dc20   94–42680
                                CIP

PREFACE

An Introduction to Cell and Molecular Biology is a condensation of *Molecular and Cellular Biology*. It is intended primarily for undergraduate courses in molecular and cellular biology, particularly those given in the first two or three years of college. It can also be used in more advanced courses when a shorter text is of central importance.

Although generally shortened and condensed, the emphasis on experimental studies is retained and a number of topics have been expanded to reflect new developments of central importance that have appeared since publication of the longer text. And, for those topics included, new developments have been integrated in all areas of the book. Several sections have been revised and updated in accordance with many helpful suggestions received from those using the longer book.

Like the longer book, the aim of this text is to integrate molecular biology, biochemistry, and cell biology into a unified course of study. Until the longer book appeared, molecular biology texts concentrated on gene structure and activity to the near exclusion of more traditional biochemistry and cell biology. This reflects a past tendency to throw out the baby with the bath water—to regard many of the findings and conclusions from traditional biochemical, cell biological and genetic investigations as unreliable and uninteresting because the research was conducted before the molecular revolution. However, the emphasis in molecular biology has shifted from a concentration on genes for their own sake to the application of molecular genetic studies to all areas of cell biology and biochemistry. As a result, more traditional areas—ion transport and the control of cell division and the cell cycle come to mind as two examples—have become the subjects of highly productive investigations in which the powerful methods of molecular biology are producing many new findings by expanding traditional research and pushing these areas to the forefront.

Like the longer text, this book reflects these developments by offering an integrated and balanced view of contemporary molecular biology, biochemistry and cell biology. The central topics of molecular biology are included, among them DNA structure, gene structure and regulation and the molecular methods for studying genes. These topics are integrated with a description of the biochemistry of gene transcription, along with the cell biology of the nucleus, including the structure of nucleosomes, chromatin, the nucleolus, and the nuclear envelope; the changes occurring in nucleosomes and chromatin during the shift between inactivity and gene transcription, the role of the nuclear envelope in transport between nucleus and cytoplasm, and the routes and mechanisms by which proteins enter nuclear structures. The organization of genes into genomes is also included in this integration. Similarly, the coverage of cell division includes an integrated view of the molecular biology and biochemistry of cell cycle genes and their regulation, DNA replication, the changes in chromosome structure that accompany cell division, the molecules, forces, and structures separating the chromosomes during division, and the molecular biology, biochemistry, and cell biology of genetic recombination. All these topics are put in place by considering them against the background of traditional descriptions of mitosis and meiosis and their stages.

Other more traditional topics of cell biology and biochemistry that are integrated with molecular biology and emphasized in this text are membrane structure and transport; the activities of mitochondria and chloroplasts in cellular energy metabolism, and the mechanisms governing genetic inheritance in these organelles; protein synthesis and the modification and distribution of newly synthesized proteins by the endoplasmic reticulum and Golgi complex; the roles of microtubules and microfilaments in motility, and with intermediate filaments, in cytoskeletal support; and the structure and function of the cell surface and extracellular matrix, along with the cellular regulatory mechanisms linked to receptors at the cell surface.

These topics are distributed between the chapters of the text, which cover subjects that are central to most or all courses and emphasize eukaryotic cells, and chapter Supplements that make parallels with prokaryotes and add information that is more specialized. In areas where research with prokaryotes has provided the primary and fundamental information of the field, as in the area of DNA replication, prokaryotic and eukaryotic findings are integrated in the main chapter. Also included in the chapters are Information Boxes, presenting short but essential items of background information. By selecting among the material presented in the chapters and supplements, an instructor can tailor the text to suit the aims of a course and its students.

Diagrams and micrographs are used generously to illustrate words with pictures. The many light and electron micrographs, in particular, emphasize that molecular and biochemical processes have a structural basis and help relate these processes to cell biology. In addition, each chapter opens with a major piece of art that focuses interest and illustrates a central topic of the chapter.

The findings, conclusions, and principles described in the text are presented in terms of experimental evidence drawn from work with prokaryotes, fungi, animals, and plants. Controversies as well as conclusions are presented in order to show that the body of scientific information in molecular and cellular biology is not fixed, and that many significant questions remain to be answered. This experimental foundation is bolstered by essays contributed by original investigators, presenting their experiences in the conception and execution of experiments that have produced key contributions in molecular biology, biochemistry, and cell biology. The essays add a personal element to the information presented in the book, and emphasize that this information is the result of experimental work by individual scientists. The essayists include both established researchers, who are essentially household names among biologists, and relative newcomers, who are likely to be among the next generation of famous names. Hopefully, in addition to piquing students' interests, the essays may help turn them toward a career in research.

This book could not have been written without the help of colleagues and friends who reviewed the text and offered many suggestions for improving its accuracy and content. I am also deeply indebted to the many investigators who generously supplied micrographs, diagrams, and tables from their published research. I must also acknowledge my debt to my biology editor, Jack Carey, who provided encouragement and guidance in the preparation of the text, and to George Calmenson of the Book Company, who very capably coordinated the many details attending to production of the book with Wadsworth Publishing. I am also indebted to Darwen and Valley Hennings, and John and Judy Waller, who prepared original art and redrew many of the diagrams sent by other authors. Darwen Hennings in particular was instrumental in the design and preparation of the more complex illustrations in the book, including the majority of the chapter openers.

REVIEWERS

David Asai, *Purdue University*
Aimee Bakken, *University of Washington*
Stephen C. Benson, *California State University, Hayward*
Colin F. Blake, *Oxford University*
Douglas C. Braaten, *Washington University*
B. R. Brinkley, *University of Alabama*
Iain L. Cartwright, *University of Cincinnati*
G. P. Chapman, *University of London*
Susan Conrad, *Michigan State University*
Barbara Danowski, *Union College*
David T. Denhardt, *University of Western Ontario*
William Dentler, *University of Kansas*
William Dunphy, *University of California, San Diego*
Douglas Easton, *State University College of Buffalo*
D. J. Finnegan, *University of Edinburgh*
Christine H. Foyer, *The University of Sheffield*
Carl Frankel, *Pennsylvania State University*
Michael Freeling, *University of California, Berkeley*
David Fromson, *California State University, Fullerton*
Carol Guzé, *California State University, Dominguez Hills*
Leah T. Haimo, *University of California, Riverside*
Elizabeth D. Hay, *Harvard University*
John Hershey, *University of California, Davis*
Peter Heywood, *Brown University*
Ricky R. Hirschhorn, *University of Kentucky*
J. Kenneth Hoober, *Temple University*
Yasuo Hotta, *Nagoya University*
Lon Kaufman, *University of Illinois*
Charles C. Lambert, *California State University, Fullerton*
Elias Lazarides, *California Institute of Technology*
Mary Lee Ledbetter, *College of the Holy Cross*
John Lye, *University of Colorado*
Thomas Lynch, *University of Arkansas, Little Rock*
Robert MacDonald, *Northwestern University*
S. K. Malhotra, *University of Alberta*
Peter B. Moens, *York University*
Stephen M. Mount, *Columbia University*
Dominic Poccia, *Amherst College*
P. Elaine Roberts, *Colorado State University*
Thomas Roberts, *Florida State University*
Enrique Rodriguez-Boulan, *Cornell University*
Richard G. Rose, *West Valley College*
Jonathan A. Rothblatt, *Dartmouth College*
John Ruffolo, *Indiana State University*
Robert Savage, *Swarthmore College*
David Saxon, *Morehead State University*
R. L. Seale, *Scripps Clinic Research Foundation*
Barbara Sears, *Michigan State University*
Joel Sheffield, *Temple University*
Sheldon S. Shen, *Iowa State University*
Roger Sloboda, *Dartmouth College*
Gary Stein, *University of Florida*
Herbert Stern, *University of California, San Diego*
J. William Straus, *Vassar College*
Albert P. Torzilli, *George Mason University*
David H. Vickers, *University of Central Florida*
Fred Warner, *Syracuse University*
Ruth Welti, *Kansas State University*
Larry Williams, *Kansas State University*
John G. Wise, *University of California, Los Angeles*
William L. Wissinger, *St. Bonaventure University*
Debra J. Wolgemuth, *Columbia University*
D. L. Worcester, *University of Missouri*
Charles F. Yocum, *University of Michigan, Ann Arbor*

THE EXPERIMENTAL PROCESS ESSAYISTS

S. J. Singer, *How Do Ions Get Across Membranes?* (Ch. 5)

Lowell E. Hokin, *The Road to the Phosphoinositide-Generated Second Messengers* (Ch. 6)

Peter Goodwin, *Cell/Cell Communication in Plants: An Open and Shut Case* (Ch. 7)

Richard E. McCarty, *Divide and Conquer: Deciphering the Functions of a Polypeptide of the Chloroplast ATP Synthase* (Ch. 8)

Marshall D. Hatch, *The C_4 Pathway: A Surprise Option for Photosynthetic Carbon Assimilation* (Ch. 9)

Michael P. Sheetz, *Identification of Microtubule Motors Involved in Axonal Transport* (Ch. 10)

Yoshie Harada and Toshio Yanagida, *Do the Two Heads of the Myosin Molecule Function Independently or Cooperatively?* (Ch. 11)

Elaine V. Fuchs, *Of Mice and Men: Genetic Skin Diseases Arising from Defects in Keratin Filaments* (Ch. 12)

Alexander Rich, *The Discovery of Left-Handed Z-DNA* (Ch. 13)

Alina C. Lopo, *When Sperm Meets Egg: Initiation Factors and Translational Control at Fertilization* (Ch. 16)

George C. Prendergast, *Sequential Clues to the DNA Binding Specificity of Myc* (Ch. 17)

Gordon Ada, *Instruction or Selection: The Role of Antigen in Antibody Production* (Ch. 19)

Vishwanath R. Lingappa, *Redirecting Protein Traffic from Cytoplasm to the Secretory Pathway: A Test of the Signal Hypothesis* (Ch. 20)

Peter E. Thorsness, *Escape of DNA from Mitochondria to the Nucleus During Evolution and in Real Time* (Ch. 21)

Tim Hunt, *Finding Cyclins and After* (Ch. 22)

Carol W. Greider, *Telomerase: The Search for a Hypothetical Enzyme* (Ch. 23)

W. Zacheus Cande, *Spindle Elongation* in vitro: *The Relative Roles of Microtubule Sliding and Tubulin Polymerization* (Ch. 24)

Herbert Stern, *m-rec: An Enzyme That Affects Genetic Recombination in Meiotic Cells* (Ch. 25)

Noam Lahav and Sherwood Chang, *How Were Peptides Formed in the Prebiotic Era?* (Ch. 26)

BRIEF CONTENTS

DETAILED CONTENTS

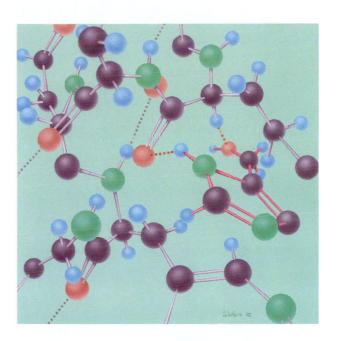

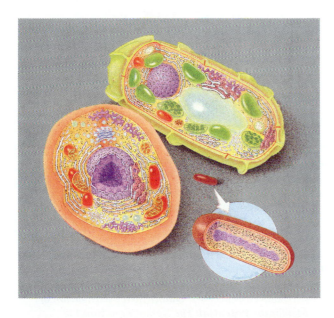

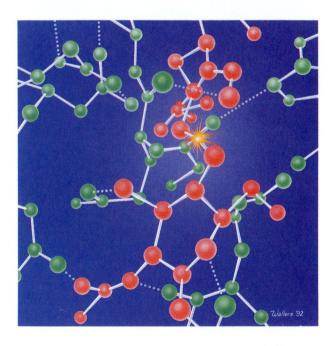

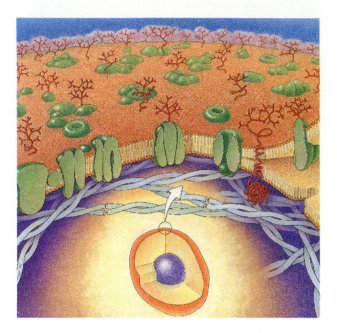

CHAPTER 5

THE FUNCTIONS OF MEMBRANES IN TRANSPORT

CHAPTER 4

THE STRUCTURE OF CELLULAR MEMBRANES

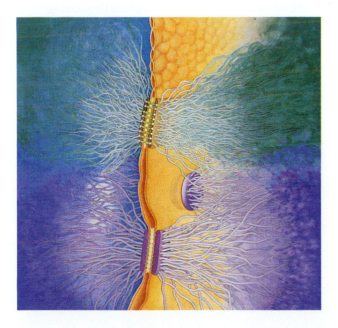

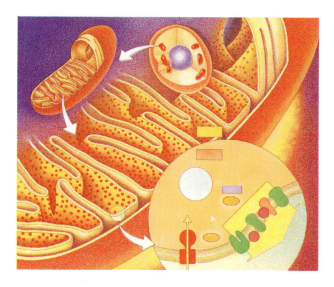

CHAPTER 9

PHOTOSYNTHESIS AND THE CHLOROPLAST

CHAPTER 10

MICROTUBULES AND MICROTUBULE-BASED CELL MOTILITY

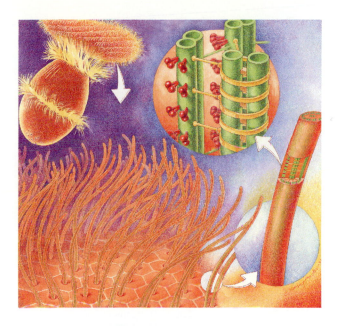

CHAPTER 12

MICROTUBULES, MICROFILAMENTS, AND INTERMEDIATE FILAMENTS IN THE CYTOSKELETON

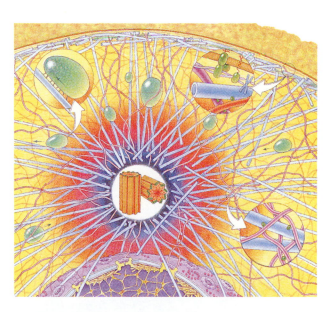

CHAPTER 11

MICROFILAMENTS AND MICROFILAMENT-GENERATED CELL MOTILITY

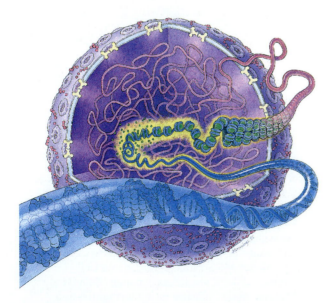

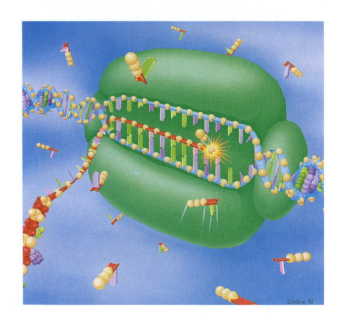

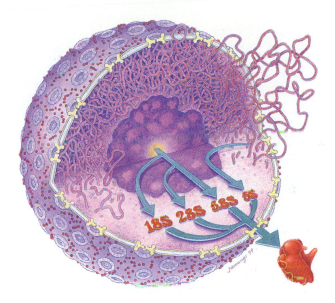

CHAPTER 16

INTERACTION OF mRNA, rRNA, AND tRNA IN PROTEIN SYNTHESIS

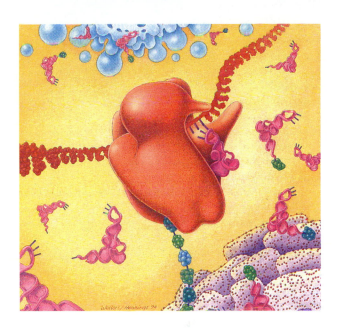

CHAPTER 17

TRANSCRIPTIONAL AND TRANSLATIONAL REGULATION

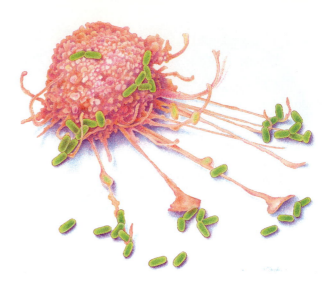

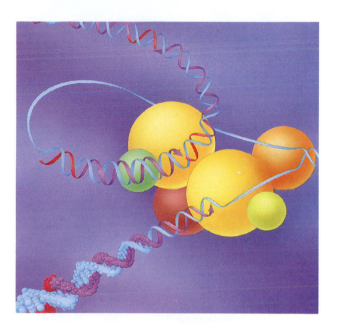

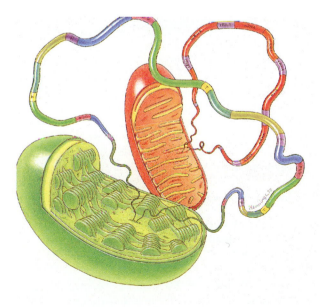

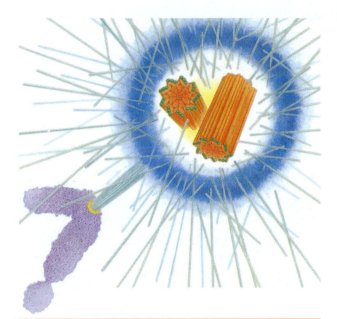

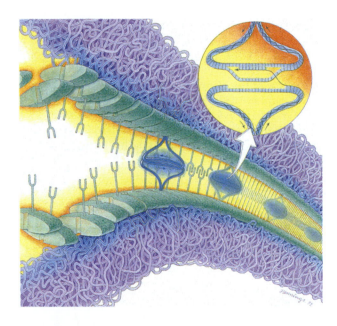

A STUDENT'S GUIDE TO THIS BOOK

DNA replication and repair . . . the genetic code and its exceptions . . . cell cycle regulation and cancer . . . the extracellular matrix . . . these are just a few of the many exciting new areas of research you'll read about in this book. Up-to-date topics like these will keep you on the cutting edge of knowledge in a field where the knowledge base is in a constant state of change.

You'll find that studying about the recent advances in cell and molecular biology is fascinating--especially with the right tools to guide you. On the next few pages I will tell you about the tools you'll find in this book and how they can help you study more effectively.

Stephen Wolfe

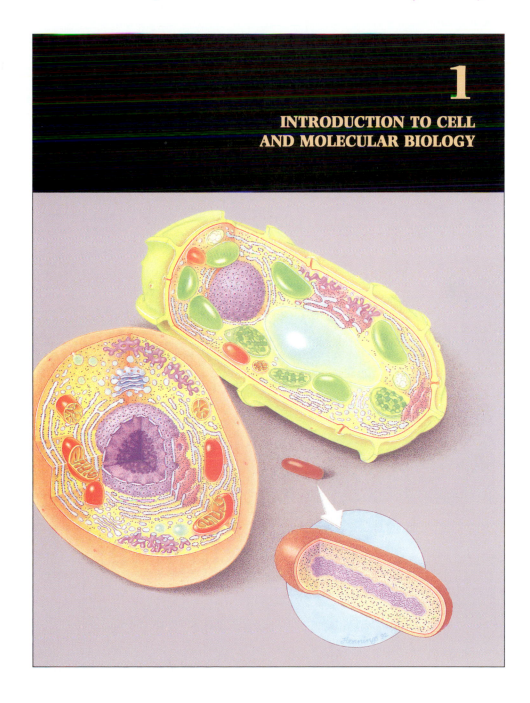

1

INTRODUCTION TO CELL AND MOLECULAR BIOLOGY

Because of the rapid changes in the field of cell and molecular biology, currency is vital in a textbook. I've included the latest information on topics such as protein synthesis, structure and function of the nuclear envelope, and membrane transport. See, for example, the pages shown here on gene regulation.

This domain contains a zinc-fold structure that interacts with the DNA. The other includes a region capable of directly or indirectly activating transcription by RNA polymerase II. The activating region may bind either RNA polymerase II or one of the transcription factors (see p. 412) required for transcriptional initiation by the polymerase enzyme. Division into two structural and functional regions—one recognizing and binding DNA and the other activating transcription—is typical of many eukaryotic regulatory proteins.

The way GAL4 activates transcription is still unknown. Ptashne proposed that GAL4, after binding to the upstream site in the DNA (Fig. 17-7a), interacts with a protein that binds to the promoter region. The protein interacting with GAL4 may be either TFIID, a transcription factor that binds to the TATA box (see p. 409), or RNA polymerase II. (TFIID is shown in Fig. 17-7 as the protein interacting with GAL4.) Interaction between GAL4 and the protein brings the promoter and enhancerlike regions together, forming a *loop* in the DNA (Fig. 17-7b). The combination promotes tight binding of RNA polymerase II to the promoter (as shown in Fig. 17-7c), and transcription begins.

GAL80 acts as a negative regulatory protein controlling the activity of GAL4. When no galactose is present in the cytoplasm of the yeast cell, GAL80 binds to GAL4 (Fig. 17-7d). Although GAL4 can still recognize and bind the enhancerlike region upstream

Figure 17-7 Regulation by GAL4/GAL80. After recognizing and binding a *cis* element upstream of the promoter (a), GAL4 interacts with a protein that binds to the promoter (b), here shown as TFIID, a transcription factor that binds to the TATA box. The interaction brings the promoter and enhancerlike regions together, forming a loop in the DNA. (c) This conformation induces tight binding of RNA polymerase II to the promoter, and transcription begins. (d) When no galactose is present in the cytoplasm, GAL80, a negative regulatory protein, binds to GAL4. Although GAL4 can still recognize and bind the enhancerlike region upstream of galactose genes when GAL80 is attached, its ability to activate transcription is eliminated. (e) In the presence of galactose or some of its metabolic derivatives, GAL80 undergoes a conformational change that releases it from GAL4. The release frees GAL4 from inhibition and allows initiation of transcription to proceed.

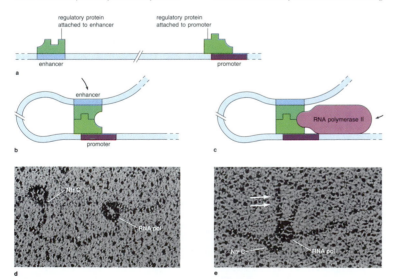

How enhancers work is one of the more intriguing questions of molecular biology. According to the *loop model*, proposed by H. Echols, R. Tjian, and M. Ptashne and their colleagues, a regulatory protein binds the enhancer as one of the initial steps in initiation (Fig. 14-11a). The protein bound at the enhancer then interacts with other proteins bound at the promoter (Fig. 14-11b), throwing the length of DNA between the enhancer and the promoter into a loop. The combined regulatory proteins, according to the model, take on a conformation that binds RNA polymerase II more avidly than the promoter complex alone

(Fig. 14-11c) and thereby enhances the rate of transcription. Among the several lines of evidence supporting the loop model are electron micrographs directly showing loop formation by interaction of proteins at the enhancer and promoter sites (Fig. 14-11d and e).

Sequence Elements in the Transcribed Region: Exons and Introns Other than the initiator and terminator codons for protein synthesis, the sequences of primary interest in the coding segment of mRNA genes are those defining exons and introns. These sequences have been identified by several methods, including

Figure 14-11 The loop model for the action of enhancers in increasing gene transcription (see text). (a) The regulatory protein bound to the enhancer. (b) Interaction between the regulatory protein and a protein bound to the promoter creates a loop in the DNA. (c) The combination sets up a conformation with high binding affinity for RNA polymerase II, which adds to the complex and initiates transcription. (d) The enhancer/promoter region of the *glnA* operon of *E. coli*, showing the regulatory protein NtrC bound to the enhancer and RNA polymerase (RNA pol) bound, presumably loosely, to the promoter. The two proteins can be distinguished because NtrC is more lightly coated by the tungsten metal used to shadow the preparation (see Appendix p. 790). (e) Formation of a loop between the regions bound by NtrC and RNA polymerase. Presumably the interaction leads to tight binding by the enzyme and initiation of transcription. ([d and e] courtesy of H. Echols, from *J. Biolog. Chem.* 265:14699 [1990].)

EXPERIMENTAL ESSAYS

Two-page essays from original investigators discuss the design and execution of research studies that have made key contributions to the fields of molecular and cellular biology and biochemistry. The essays (by both established researchers and relative newcomers in biology) show science in action and scientists as real people, actively engaged on the forefront of change.

The Experimental Process

Redirecting Protein Traffic from Cytoplasm to the Secretory Pathway: A Test of the Signal Hypothesis

Vishwanath R. Lingappa

VISHWANATH R. LINGAPPA received the B.A. degree from Swarthmore College in 1975, after which he pursued graduate research with Gunter Blobel at The Rockefeller University. He received the Ph.D. degree in 1979 and the M.D. degree from Cornell University Medical College in 1980. After residency in internal medicine at the University of California Hospitals in San Francisco, he joined the UCSF faculty where he is Professor of Physiology and Medicine. A Board Certified internist, Dr. Lingappa is engaged in medical student teaching, research in molecular mechanisms of protein trafficking, and general internal medicine practice in San Francisco.

For several years after it was first proposed, the signal hypothesis[1] (see p. 579) was met with skepticism from some quarters. Some critics doubted that a small, discrete sequence could be the sole determinant of chain targeting to, and translocation across, the ER membrane, especially since considerable variation in amino acid sequence occurs between signal sequences even within a single species. Rather, they favored notions such as (1) that global features of protein folding distinguished secretory from cytosolic proteins, or (2) that other specific regions of secretory proteins in addition to the signal sequence might be involved in translocation, or (3) that a specific sequence in cytosolic proteins prevented their translocation. Others wondered if cell-free systems, from which most of the data in favor of the signal hypothesis had been derived, were an adequate reflection of events occurring in vivo.

In planning experiments it is often useful to consider the possible outcomes in a preliminary "thought experiment" before actually investing a major amount of time and effort. We considered the following possibilities:

1. The signal hypothesis is incorrect or overly simplistic. Hence a signal sequence is either unnecessary or, if necessary, may not be sufficient for translocation of a normally cytosolic protein across the ER membrane. In either case, the chimeric protein would remain in the cytoplasm.

2. The signal hypothesis is correct. Hence a signal sequence, being both necessary and sufficient for translocation across ER membranes, would redirect an otherwise cytosolic protein into the secretory pathway.

3. Finally, regardless of whether the hypothesis is correct or incorrect, the experimental approach might be flawed, for either systematic or fortuitous reasons. For example, juxtaposition of the specific sequences from the two proteins composing the chimera might cause the chain to fold into an insoluble aggregate and hence be scored as nontranslocating. A different chimera might have given a different result, and the one we chose to study may have been an (unfortunate) exception to the general rule.

Experimental science is an inductive rather than a deductive process. Thus, a negative result (e.g., lack of translocation of a particular chimeric protein) is relatively uninformative because it does not distinguish between possibilities (1) and (3). On the other hand, a positive re-

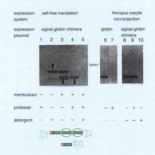

	expression system:	cell-free translation					*Xenopus* oocyte microinjection			
	expression plasmid:	signal-globin chimera				globin	globin	signal-globin chimera		
		1	2	3	4	5	6	7	8	9	10
membranes:		−	−	+	+			−	+	+	+
protease:		+	−	−	+			−	+	−	+
detergent:		−	−	−	+			−	−	−	+

Figure A   Signal globin expression and translocation in a cell-free translation system (with cartoon interpretation) and in *Xenopus* oocytes.

p. 420) at the start of the globin coding region made deletions of the upstream lactamase codons beyond the signal sequence technically quite easy. Thus, for the initial experiments we chose to use the beta lactamase signal sequence. These DNA molecules were constructed behind an active promoter (see p. 409). This allowed us to make mRNA from the cloned DNA in a test tube and then translate this mRNA in a cell-free translation system.

In the absence of added ER membranes, the full chimera was synthesized, which was larger than authentic globin by the size of the signal sequence (Fig. A, lane 2, downward pointing arrowhead). When proteases were added to this translation product it was totally digested,

bin without a signal sequence was shown not to be translocated even when membrane vesicles were present during its synthesis.[2]

To test whether these results applied to events occurring *in vivo*, we used microinjection to introduce the mRNA for authentic globin or the signal-globin chimera into the cytoplasm of *Xenopus* oocytes. Large living cells with an active secretory pathway. mRNA so introduced is translated by ribosomes in the oocyte cytoplasm. Radiolabeled amino acids included in the medium which bathes the oocytes are taken up and incorporated into the newly synthesized proteins. Homogenization of the oocytes converts the ER into vesicles, and analysis of localization can be carried out either by fractionation of the homogenate into cytosol and vesicles or by protease digestion, as previously described. Authentic-sized globin was synthesized from both mRNAs (lanes 6 and 8). However, globin synthesized from the chimeric mRNA was protected from added protease (lane 9) unless detergent was added (lane 10), while that synthesized from globin mRNA not encoding a signal sequence was not protected from added proteases (lane 7). Moreover, fractionation demonstrated that authentic globin synthesized from globin mRNA was entirely in the cytosol while authentic-sized globin synthesized from the chimeric mRNA was entirely in the vesicles' (not shown). These data suggested that, in the case of the chimera, authentic globin was generated as a result of cleavage of the signal sequence from the growing chain as it translocated to the ER lumen. Thus, both *in vitro* and *in vivo*, the beta-lactamase signal sequence was sufficient to direct chimeric alpha globin out of the cytoplasm and into the secretory pathway.[2,3]

An ironic twist to this story was that, some time later, an insect globin gene was cloned and found to have a remarkable structure. Globin is found in the cytoplasm of red blood cells in most animal species. However insects lack red blood cells. Instead, they secrete their globin into a fluid, called hemolymph, which bathes the cells. Thus insect globin, whose mature sequence is quite conserved with other animal globins, is naturally secreted rather than a cytoplasmic protein. When the mRNA

The Experimental Process

When Sperm Meets Egg: Initiation Factors and Translational Control at Fertilization

Alina C. Lopo

ALINA C. LOPO's interests have focused on the cellular and molecular details of early development for most of her research career. After obtaining B.S. and M.S. degrees from the University of Miami, she completed doctoral research at the University of California at Davis and San Diego, then postdoctoral work at the U.C. San Francisco and U.C. Davis Schools of Medicine. In 1985 she joined the faculty of the Division of Biomedical Sciences at U.C. Riverside, where most of the work described here was carried out. She recently received an M.D. at the UCLA School of Medicine and is now in the residency training program at the UCLA Medical Center. In her spare time, she is a wife and mother.

A key consequence of fertilization is the dramatic activation of the metabolic machinery of the dormant egg. One of the many cellular processes that is rapidly and efficiently turned on at fertilization is protein synthesis. This translational activation is a fundamental step in the activation of development, since without the full activation of protein synthesis, development cannot be completed successfully. The mechanisms regulating translational activation at fertilization have been the subject of extensive investigation by cell, molecular, and developmental biologists since egg activation was first described by Tore Hultin in 1950. A puzzling aspect of this phenomenon is that the unfertilized egg contains all the necessary molecules to carry out protein synthesis. Although progress had been made in understanding how the ionic changes at fertilization influence translational activation, until relatively recently the macromolecular changes were poorly understood.

In the early 1980s work by several laboratories demonstrated that the regulation of protein synthesis at fertilization was, as in most other eukaryotic cells, primarily by regulation of the initiation phase. Working with John Hershey, we showed that possibly the earliest change in initiation was somehow related to the activity of an initiation factor, eIF4E,[1] which binds the 5' cap structure of mRNAs and greatly speeds attachment of mRNA to the small ribosomal subunit. In the preceding few years several groups, including Hershey's, had shown that, in rabbit reticulocyte lysate and in mammalian cells in culture, turning protein synthesis on or off correlated well with the phosphorylation or dephosphorylation of some initiation factors. This led us to ask whether similar changes took place in the egg following fertilization.

We focused our initial studies on the smallest subunit of eIF4F, the so-called cap-binding protein, or eIF4E. The reasoning behind this centered on the knowledge that the presence of the 5' or m7G cap on mRNA is an absolute requirement for efficient protein synthesis in

eukaryotes. Previous studies suggested that the bulk of mRNAs stored in the egg during oogenesis were capped, so we thought the key to translational activation may reside in the level of activity of the cap-binding protein. If this protein were somehow activated at fertilization, then perhaps mRNA could enter the initiation pathway. At approximately the same time we were working on this problem, Roger Duncan and John Hershey obtained evidence that, at least in HeLa cells, eIF4E appeared to be dephosphorylated when protein synthesis was repressed following heat-shock treatment. This reinforced our belief that we were on the right track, that the change was biologically likely.

We purified the eIF4E protein from unfertilized eggs of the sea urchin using m7G affinity chromatography.[2] We used sea urchin eggs because it is easy and inexpensive to obtain very large quantities of pure unfertilized eggs. The large quantities were necessary because, based on evidence from mammalian cells in culture, eIF4E is of relatively low abundance in cells. The sea urchin egg eIF4E turned out to be quite similar in molecular weight and amino acid composition to the mammalian factor, which was being characterized by Bob Rhoads' group at the University of Kentucky and Nahum Sonenberg at McGill University. This was not surprising, because protein synthesis is a highly conserved cellular process.

We had improved our yields of pure sea urchin egg eIF4E approximately ten-fold, and we were soon ready to start making rabbit antibodies against the factor. Our plan was this: first, raise rabbit antibodies against sea urchin eIF4E. Then, prepare electrophoretic gels of homogenates from unfertilized eggs and from eggs fertilized and homogenized at 1-minute intervals up to 15 minutes after fertilization. Use each of these gels to prepare Western blots (see p. 799) and use our antibody to probe the blots and ask how many forms of the eIF4E were present. Since phosphate groups are charged, phosphorylation or dephosphorylation of a protein will change its charge. This would be observable as a change in position of a band in the gels.

Although this plan is quite straightforward conceptually, it took six months of very long hours and hard work to perfect the procedure and obtain results. Finally, we consistently obtained the following results: In the unfertilized egg, all of the eIF4E detectable by our method was in one band on the gel. However, no earlier than 4 and no later than 5 minutes after fertilization, a large fraction of the detectable eIF4E was converted to a more acidic form, as might be expected if it is phosphorylated.

While these results indicated we were on the right track, they did not conclusively prove that the protein was phosphorylated. There were other possible explanations. For example, acetylation of the protein would produce a change of similar magnitude and direction. There

certainly is precedent for the covalent modification of proteins by acetylation, although it is less common than phosphorylation. Another even less likely possibility was that the protein was glycosylated. This, too, could have yielded a comparable change in mobility on the gel. Thus, we needed to establish the exact nature of the change before we could draw any meaningful conclusions from our data.

There were two ways we could go about demonstrating that the covalent modification was due to addition of a phosphate group. One way was to reverse the change *in vitro*, using an enzyme known to be specific for the removal of phosphate groups from proteins. Examples of such enzymes are alkaline phosphatase and acid phosphatase. While this works well in some systems, in our system we encountered enough problems early on to lead us to look for a more workable alternative. This was to give the unfertilized egg radioactive phosphate, then fertilize the eggs and show that, 4 to 5 minutes after fertilization, the eIF4E contained radioactivity. While relatively straightforward in concept, this experiment proved problematic in its execution for one reason: Unfertilized eggs are exceedingly impermeable to many substances, especially phosphate. However, it was essential that we get the radioactive phosphate into the egg prior to fertilization, since we had to show that the unfertilized egg eIF4E had no phosphate. To circumvent the problem, we preloaded the unfertilized eggs with a high level of radioactive phosphate by incubating a tiny amount of eggs for six hours with a very large amount of radioactive phosphate. By increasing the gradient in this way, we were able to load the phosphate pools in the eggs with radioactive material.

What we did next required mostly timing and speed. A labeled, unfertilized egg sample was homogenized and passed over the m7G affinity column. The eIF4E would attach to the beads, while all other proteins would go through. We removed the eIF4E bound to the column by adding a large excess of m7GTP to the column. An identical batch of radioactively preloaded unfertilized eggs was then fertilized, and homogenized at 5 minutes after fertilization. This fertilized sample was treated in a similar fashion, by passing over the m7G affinity column. The two samples containing eIF4E were then analyzed by slab gel electrophoresis, and autoradiograms were prepared by exposing the gels to X-ray film. Much to our delight, when we examined the autoradiograms, we saw a radioactively labeled band right where eIF4E should be in the fertilized egg sample, but not in the unfertilized sample.

This showed that the mobility change we had observed was due to phosphate, but it raised another question: was eIF4E unlabeled in the fertilized egg because we were not getting radioactive phosphate into the ATP

pools of the egg? The enzymes that catalyze the reaction of protein phosphorylation use the gamma-phosphate on ATP, not free phosphate. Since the unfertilized egg is metabolically quiescent, it was possible that we were getting hot phosphate into the egg but not into the ATP.

The question, an important one to answer, turned out to be easy to address. We simply ran a slab gel of the material that had not bound to the m7G affinity column and prepared an autoradiogram from it. This material contained all the proteins in the unfertilized egg except eIF4E. The autoradiogram showed that many proteins were radioactively labeled in the unfertilized egg. This established that we were getting the radioactive phosphate into the ATP pools in the egg, and that protein kinases were capable of using this phosphate, even in the relatively quiescent egg, to phosphorylate proteins.

A final issue that concerned us was whether the eIF4E phosphorylation was unique to *Strongylocentrotus purpuratus*, the species we used in our work, or whether this was a more general phenomenon. We examined two other species of sea urchin that were evolutionarily distant from *S. purpuratus*, *Lytechinus pictus* and *Arbacia punctulata*. In all cases we were able to demonstrate phosphorylation of the eIF4E concomitant with the activation of protein synthesis.

Additional controls further strengthened our conclusions and provided more information on the possible triggers for phosphorylation of eIF4E. Looking back, it is interesting to note that about 75% of the time spent on this project was devoted to carrying out the experimental controls necessary to establish that we were not dealing with an artifact.

What did our results tell us? At the time we carried out this work, it was the first report of a specific biochemical change in the translational machinery that correlated with the activation of protein synthesis. However, it is important to keep in mind that like all scientific endeavor, our results raised as many, if not more questions than were answered. For example, what enzyme phosphorylates the eIF4E? How is it activated at fertilization? And, most important, does phosphorylation of eIF4E allow it to participate more efficiently in binding of the 5' cap, and if so, how?

References

[1]Lopo, A. C.; MacMillan, S.; and Hershey, J. W. B. *Biochemistry* 27:351 (1988).

[2]Lopo, A. C., and Hershey, J. W. B. *J. Cell Biol.* 101:221a (1985).

Where possible, I use **examples from human biology** (especially medicine) to show you how molecular and cellular biology and biochemistry impact your own life so dramatically. For example, as shown here, the regulation of cell division is illustrated by a description of the cell cycle controls that go awry in the transformation of normal cells into cancer cells.

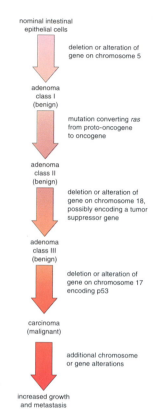

nominal intestinal epithelial cells

→ deletion or alteration of gene on chromosome 5

adenoma class I (benign)

→ mutation converting *ras* from proto-oncogene to oncogene

adenoma class II (benign)

→ deletion or alteration of gene on chromosome 18, possibly encoding a tumor suppressor gene

adenoma class III (benign)

→ deletion or alteration of gene on chromosome 17 encoding p53

carcinoma (malignant)

→ additional chromosome or gene alterations

increased growth and metastasis

Figure 22-11 One of several possible sequences in the multistep progression of colorectal cancer (see text).

Information Box 10-1

Dynein Arms and the Immotile Cilia Syndrome

In 1933 a German physician, M. Kartagener, noted an odd coincidence between three inherited human abnormalities: acute bronchitis, sinusitis, and reversal of the position of the heart from the left to the right side of the body. Later it was noted that patients with these conditions frequently also had chronic headaches and, if male, were sterile. The complete syndrome is inherited with a frequency of 1 in 15,000 persons.

The pieces of this medical puzzle were put together in 1975 by B. A. Afzelius and his colleagues at the Wenner-Gren Institute in Stockholm. Afzelius discovered that sterile males with Kartagener's syndrome had immotile flagella in sperm and other body cells. In the flagella the 9 + 2 system lacked dynein arms (see the figure in this box). In some cases the spoke heads or inner sheath and one or both central singlets were also missing.

Afzelius's discovery explained the connection between male sterility, bronchitis, sinusitis, and the headaches. The cilia in cells lining the respiratory system were paralyzed and unable to maintain the flow of mucus that normally removes irritating and infectious matter from the lungs and respiratory tract. This deficiency explains the

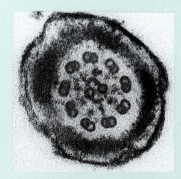

sinusitis and bronchitis. Presumably an insufficient flow of fluids in the ventricles of the brain, normally maintained by ciliated cells lining these cavities, produced the headaches. The connection between nonmotile cilia and the reversed position taken by the heart during embryonic development remains unexplained.

Information Box 18-2

DNA Fingerprinting

A forensic technique detecting small, individual differences in DNA sequence elements, developed in 1985 by A. J. Jeffreys, V. Wilson, and S. V. Thien, is gradually gaining acceptance as a method for identifying persons responsible for crimes such as murder and rape. The technique, *DNA fingerprinting*, has also been employed to determine paternity. The method depends on groups of repeated sequences called *hypervariable elements*. Although different in each individual, hypervariable elements have *core* sequences that are similar enough from one person to the next to be represented by an average or consensus sequence. The repeated DNA sequences are frequently preserved well enough in semen or bloodstains, even in stains that have been dried for some time, to allow their extraction and comparison with those of a suspect or victim.

In the technique (see part **a** of the figure in this box) a DNA sample obtained from fresh or dried semen or blood, after being increased in quantity by the polymerase chain reaction (see Supplement 14-1), is digested with a restriction endonuclease, yielding fragment classes of different sizes. The fragments are then run on an electrophoretic gel and extracted by Southern blotting (see Appendix pp. 796 and 798). The bands containing hypervariable sequences of interest are identified by hybridizing them with a radioactive probe consisting of DNA with the consensus sequence for the hypervariable elements.

These samples are compared with samples from the victim or suspect. In a rape case, for example, a DNA fingerprint made from the semen or blood of the suspect could be compared with a fingerprint made from a semen sample obtained from the vagina or clothing of the victim. In a murder case a DNA fingerprint made from bloodstains found on the suspect's person, clothing, or other articles could be compared with the DNA fingerprint of the victim. If comparisons show the DNA finger-

prints obtained from these sources to be the same or closely similar, the results can be taken as valid evidence for or against conviction.

DNA fingerprints can also be used as an indicator of paternity because the DNA fingerprints of a parent and offspring are much more similar than those of unrelated persons. Similar DNA fingerprints, when taken along with other tests such as blood typing, can provide a good indication of whether a child has been fathered or mothered by a given person (part **b** of the figure in this box shows a fingerprint made to determine paternity).

To be dependable, DNA fingerprints must be made and compared with great care. Degradation of DNA in samples or contamination by bacteria can produce extra bands that invite errors in identification. The polymerase chain reaction can also produce extra bands (as in the figure in this box). Because of these difficulties, many scientists have proposed that laboratories and individual technicians preparing and comparing DNA fingerprints should be tested regularly and licensed only if competence is demonstrated.

Courts have experienced some operational problems with admission of evidence from DNA fingerprinting. Some experts estimate that the chance of an innocent person producing an incriminating DNA fingerprint is as small as one in hundreds of millions; others maintain that the chance may be as great as 1 in 50. Opinions this diverse have provided defense attorneys with ammunition for casting doubt on the validity of the technique. Some civil libertarians have also expressed concern that DNA fingerprinting constitutes an invasion of privacy. Further experience with the method, agreement on standards to be used for comparison of fingerprints, establishment of rules governing submission to the test, and licensing of technicians and laboratories promise to make the technique more useful as a source of definitive legal evidence.

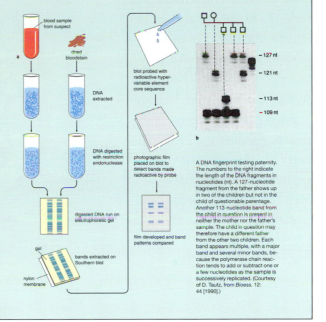

A DNA fingerprint testing paternity. The numbers to the right indicate the length of the DNA fragments in nucleotides (nt). A 127-nucleotide fragment from the father shows up in two of the children but not in the child of questionable parentage. Another 113-nucleotide band from the child in question is present in neither the mother nor the father's sample. The child in question may therefore have a different father from the other two children. Each band appears multiple, with a major band and several minor bands, because the polymerase chain reaction tends to add or subtract one or a few nucleotides as the sample is successively replicated. (Courtesy of D. Tautz, from *Bioess.* 12: 44 [1990].)

Information Boxes in many chapters present short yet essential items of background information to help you put what you're learning into context.

Supplement 19-1
AIDS (Acquired Immune Deficiency Syndrome)

AIDS appears to be a relatively recent human affliction. The first cases were reported in 1981, and the disease was probably established in the human population not too long before this time, perhaps some 30 to 40 years. The earliest known blood samples containing antibodies against the AIDS virus (the *human immunodeficiency virus* or *HIV*) were taken from African patients in Zaire in 1959. HIV-1, the virus responsible for the most common form of AIDS in the industrial world, was isolated and identified in 1983 in the laboratories of R. C. Gallo in the United States and L. Montagnier in France.

As of mid-1993 the AIDS virus was estimated to have infected 14 million people worldwide and to infect one new person each 15 seconds. By the year 2000 the total may reach 40 million. In North America more than 1 million people have been infected with the virus. The disease is most prevalent in sub-Saharan Africa, where 8 million people have been infected. In some cities of this region, one out of every three adults carries the virus. Over 2 million individuals worldwide have developed full-blown AIDS as a result of HIV infection; most of these individuals are dead.

HIV is a retrovirus (see p. 536). Like all retroviruses, its genome is encoded in a pair of RNA molecules when the virus is in the free, infective form (Fig. 19-14). In the free virus the nucleic acid core is surrounded by a protein coat, covered on its exterior surface by a lipid bilayer derived from a host cell. Extending from the lipid bilayer are *spike proteins* consisting of a basal glycoprotein, *gp41*, which anchors the spike in the membrane, and a surface glycoprotein, *gp120*, which extends from the basal unit.

The gp120 glycoprotein has the central role in HIV infection. This glycoprotein is recognized and bound by the CD4 receptor on the surfaces of helper T cells (see Fig. 19-12b). Some other cell types without the CD4 receptor, including macrophages, regulatory cells of the intestinal lining (chromaffin cells), glial cells of the brain (see p. 348), and cells of the duodenum, colon, and rectum, also apparently have receptors able to bind gp120. Once bound to a receptor by gp120, the viral particles enter the cell, probably by fusion of the viral membrane coat with the plasma membrane. Fusion introduces the naked viral particle, free of its boundary membrane, directly into the cytoplasm of the cell. At this point the virus loses its protein coat and its RNA core is copied into DNA by a reverse transcriptase (see p. 523) included among the proteins of the infecting viral particle.

One or more of the DNA copies of the viral genome then insert into the host cell genome. In this form the virus is protected from attack by the immune system, even though the infected individual may mount an impressive immune reaction against proteins of the virus. The integrated DNA contains the *gag*, *env*, and *pol* genes typical of retroviruses, which encode coat proteins and enzymes required for duplication, excision, and insertion of the virus. The viral genome also includes several genes that encode regulatory proteins. The regulatory proteins, unusually extensive for a retrovirus, allow the integrated retroviral DNA to exist in three primary states: (1) inactivity; (2) slow activity in which only a few active, infective particles are produced; or (3) rapid, intensive production of infective viral particles. Rapid viral production is often accompanied by lysis and death of the host cell.

The primary detrimental effect of HIV infection is a gradual but steady destruction of helper T cells. Elimination of the helper Ts disables the responses of macrophages and B cells in the immune response. As the infection progresses, the immune system becomes less and less effective, until at final stages the infected person is left essentially defenseless against infective microorganisms and viruses. It is still uncertain how HIV infection leads to T-cell destruction. The virus may directly kill T cells or may stimulate an immune response that kills the cells (or both).

Symptoms and Progress of AIDS

Many of the pathogens infecting individuals with AIDS are common organisms of the environment that rarely cause problems for persons with healthy immune systems. For example, people with advanced AIDS become especially susceptible to infection by a fungus, *Pneumocystis carinii*, that causes an otherwise rare form of pneumonia. Many AIDS patients also develop an otherwise rare skin cancer known as *Kaposi's sarcoma*, in which reddish-purple lesions appear on the skin.

AIDS progresses through a number of stages that are identified by the presence of antibodies to HIV, various symptoms, and the number of surviving helper T cells in the blood. Soon after the initial infection, usually within two to five weeks, symptoms resembling flu or mononucleosis may appear, including fatigue, fever, a rash, and headaches. These symptoms ordinarily last for only a few days to two weeks. During this initial stage, antibodies against HIV usually

---

END OF CHAPTER SUPPLEMENTS

At the end of most chapters you'll find one or more **Supplements**--additional sections that add important specialized or peripheral information. They are not central to understanding the material, but they will greatly enhance your knowledge.

Visual aids can help you better understand the more difficult concepts in cellular and molecular biology. It often helps to see a representation of a written explanation or description. To this end, I've used diagrams or micrographs wherever possible to enhance the presentation. All of the illustrations are in full color and have been carefully and accurately rendered.

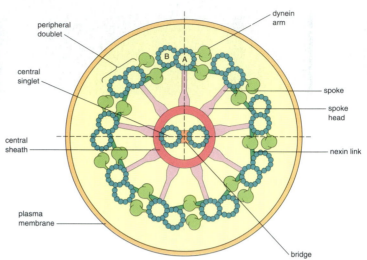

Figure 10-14 Microtubules and connecting elements in the 9 + 2 system of a flagellum. The complete A subtubules of a doublet usually have walls made up of a circle of 13 protofilaments; the B subtubules consist of a partial circle usually containing 10 to 11 protofilaments. The direction pointed by the dynein arms allows the point of view with respect to the tip or base of the flagellum to be determined. If the arms extend from the A subtubules in a clockwise direction, the flagellum is being viewed from base to tip. Compare with Figure 10-6.

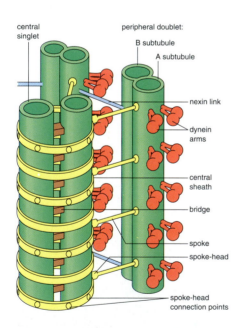

Figure 10-15 Arrangement of arms and linking elements in the 9 + 2 system. Depending on the species, the spokes are spaced evenly in groups of two or three along the A subtubules of the doublets. Entire spoke groups, containing two or three spokes per repeating unit, repeat at a 96-nm interval. The sheath elements contacted by the spoke heads are spaced at intervals of 16 nm. Dynein arms occur in pairs at intervals of 24 nm along the A subtubule; nexin links are spaced at intervals of 96 nm.

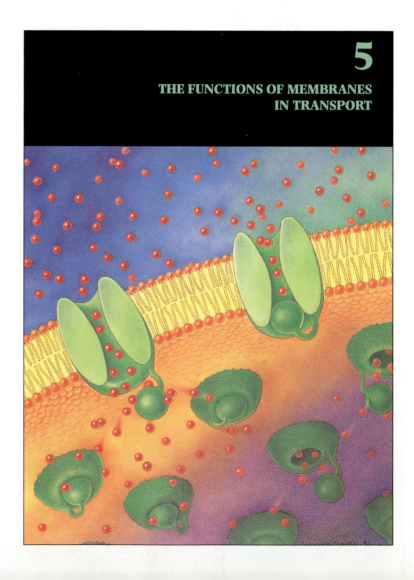

5

THE FUNCTIONS OF MEMBRANES IN TRANSPORT

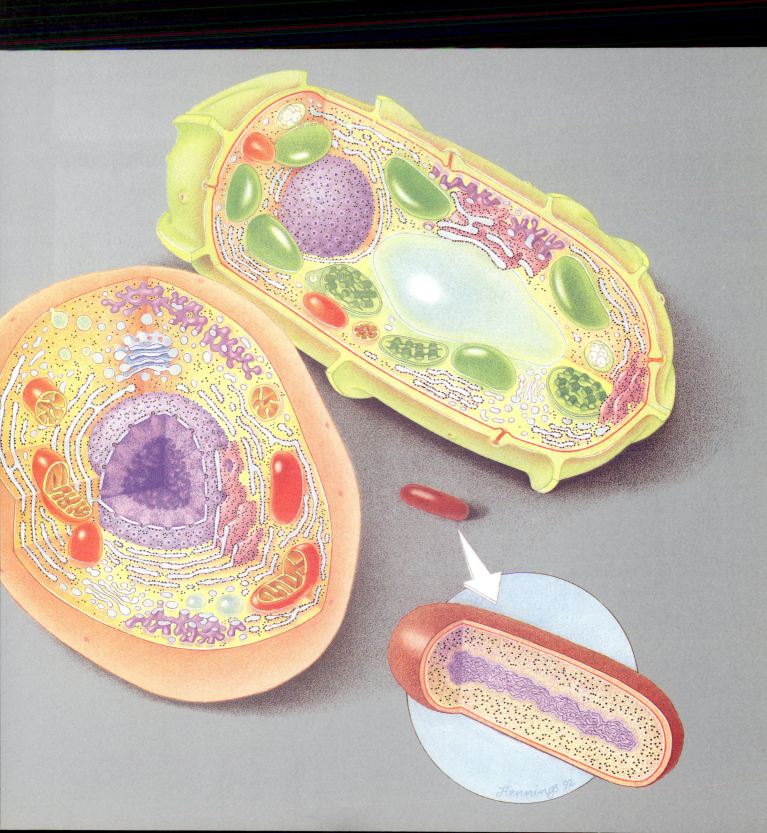

C ells are the fundamental structural and functional units of all living organisms. They contain highly organized molecular and biochemical systems that can store information and translate it into cellular molecules. Cells use energy sources to power their activities, are capable of movement, and can compensate for environmental changes by altering their internal biochemical reactions. Cells can also duplicate and pass on their hereditary information as well as their major biochemical and molecular systems. All these activities are packed into units that are the ultimate in miniaturization—most cells are microscopic, either invisible or barely visible to the naked eye.

In some organisms, such as bacteria and protozoa, cells and individuals are one and the same. Each cell of these organisms is functionally independent and capable of carrying out all the activities of life. In more complex multicellular organisms, major life activities are divided among groups of specialized cells. However, even the cells of multicellular organisms are potentially capable of independent activity.

If cells are broken, the quality of life is lost: Their interior structures, although capable of limited biochemical or molecular activities if placed under the proper conditions, are unable to grow, reproduce, or respond to outside stimuli in a coordinated, potentially independent fashion. Therefore life as we know it does not exist in units more simple than individual cells.

CELL STRUCTURE

Cells occur in highly varied forms in different plants, animals, and microorganisms (Figs. 1-1 to 1-3). They may exist singly, as in bacteria and protozoa, or packed together by the millions or billions as in larger plants and animals. In size, cells range from bacteria, which with diameters of about 0.5 micrometer (μm) are barely visible in the light microscope, to units as large as a hen's egg: The yolk of a hen's egg is a single cell, several centimeters in diameter. (Table 1-1 explains the micrometer and other units of measurement used in cell and molecular biology.) In multicellular animals, cells range from about 10 to 30 μm in diameter; plant cells range from diameters of about 10 μm to several hundred micrometers. Cells may be roughly spherical in shape, flattened, cuboidal, or columnar; some, like the nerve cells of larger animals,

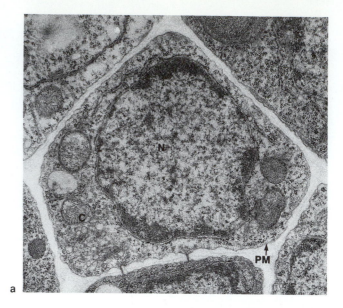

a

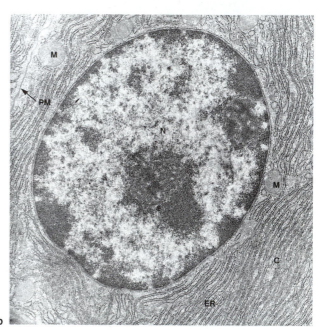

b

Figure 1-1 Electron micrographs of plant and animal cells. N, nucleus; C, cytoplasm; PM, plasma membrane. **(a)** An embryonic plant cell of *Sorghum bicolor*, a type of grass. × 12,000. (Courtesy of Chin Ho Lin.) **(b)** A cell from the pancreas of the rat. × 11,500. (Photograph by the author.)

may carry long extensions that are of microscopic diameter but can be meters in length.

In spite of their varied sizes, shapes, and activities, all cells are divided into two major internal regions, the *nuclear region* and the *cytoplasm*, that reflect a fundamental division of labor. The nuclear region contains *deoxyribonucleic acid* (*DNA*) molecules that store the hereditary information required for cell

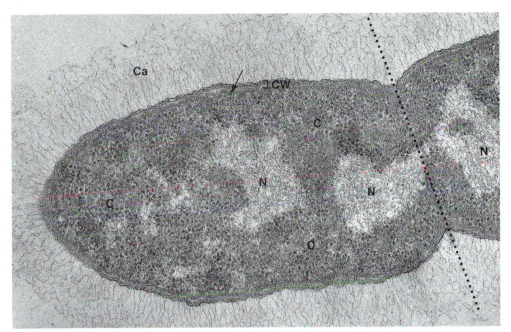

Figure 1-2 A prokaryotic cell, the bacterium *Klebsiella pneumoniae*. The nucleoid (N), the prokaryote equivalent of a nucleus, occupies the center of the cell. The cytoplasm (C) surrounding the nucleoid is packed with ribosomes. The cell is surrounded by a cell wall (CW); the plasma membrane (arrow) lies just beneath the cell wall. The capsule (Ca) is visible just outside the cell wall. The cell is dividing; the plane of division is shown by the dotted line. × 52,000. (Courtesy of E. N. Schmid, from *J. Ultrastr. Res.* 75:41[1981].)

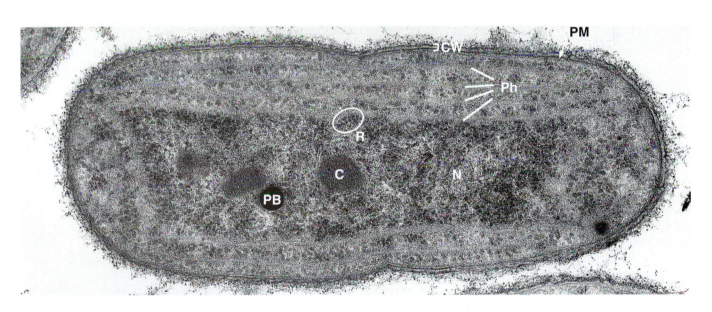

Figure 1-3 A cyanobacterium, *Agmenellum quadruplicatum*. N, nucleoid; R, ribosomes; PM, plasma membrane; Ph, photosynthetic membranes; PB, polyphosphate body; C, carboxysome, a body containing semicrystalline deposits of the enzyme active in carbon dioxide fixation; CW, cell wall. × 60,000. (Courtesy of S. A. Nierzwicki-Bauer, D. L. Balkwill, and S. E. Stevens, from *J. Ultrastr. Res.* 84:73[1983].)

growth and reproduction. The nuclear region also contains enzymes that can copy and duplicate this information. The cytoplasm makes proteins according to directions copied in the nucleus and also synthesizes most of the other molecules required for growth and reproduction (some are made in the nuclear region). The cytoplasm carries out several additional vital functions. Among the most important are conversion of fuel substances into forms of chemical energy that can be used by cells, conversion of light to chemical energy in photosynthetic cells, conduction of stimulatory signals from the outside into the cell interior, movement of materials to and from cells, and cell motility.

Cells are maintained as distinct environments and collections of matter by exceedingly thin layers of

Table 1-1 Units of Measure Used in Cell and Molecular Biology

Unit	Angstroms	Equivalent in: Nanometers	Equivalent in: Micrometers	Equivalent in: Millimeters
Angstrom (A)	1	0.1	0.0001	0.0000001
Nanometer (nm)	10	1	0.001	0.000001
Micrometer (µm)	10,000	1000	1	0.001
Millimeter (mm)	10,000,000	1,000,000	1000	1

Units on a log scale:

1 A	10 A	100 A	1000 A	10,000 A	10,000,000 A
	1 nm	10 nm	100 nm	1000 nm	1,000,000 nm
				1 µm	1000 µm
					1 mm

lipid and protein molecules called *membranes*. These molecular layers, not much more than 7 to 8 nanometers (nm) in thickness, set up continuous outer boundaries that separate the cell contents from the exterior. In many cells, membranes also divide the cell interior into compartments with their own distinct environments and collections of molecules. The lipid part of membranes, consisting of a double layer (a *bilayer*) of molecules (Fig. 1-4), provides a structural framework and sets up a barrier to the passage of water-soluble substances. Membrane proteins, which are suspended in the lipid bilayer individually or in groups, form channels that allow selected water-soluble molecules to pass from one side of the membrane to the other. The selectivity of the protein channels allows membranes to control the movement of molecules in and out of cells and between membrane-bound compartments within cells. Most other functions of membranes, such as recognition and binding of molecules at the membrane surfaces, are also carried out by proteins. Thus the structural framework of membranes depends primarily on lipids, and the functions of membranes primarily on proteins.

The cells of all organisms fall into one of two major divisions according to the number and arrangement of cellular membranes and the complexity of the nuclear region. The smaller and more primitive division, the *prokaryotes* (from *pro* = before and *karyon* = nucleus), includes only two major groups—bacteria and *cyanobacteria* (or *blue-green algae*). Prokaryotic membranes are limited to the surface or *plasma membrane* and a relatively simple collection of inner membranes in the cytoplasm. The nuclear region of prokaryotes, the *nucleoid*, has no boundary membrane separating it from the surrounding cytoplasm.

The second major division of living organisms, the *eukaryotes* (from *eu* = true, and *karyon* = nucleus),

phospholipid molecule

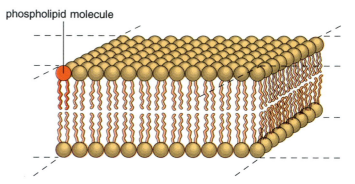

Figure 1-4 The arrangement of phospholipid molecules in a layer two molecules in thickness, the bilayer. The bilayer provides the framework of all biological membranes.

includes all the remaining organisms of the earth—animals, plants, fungi, and protozoa. A plasma membrane covers the surface of the cells of these organisms as in the prokaryotes. In addition, several distinct internal membrane systems divide eukaryotic cells into interior compartments with specialized functions. Among these interior compartments is the nucleus, set off from the surrounding cytoplasm by a double system of membranes.

PROKARYOTIC CELLS

Prokaryotic cells (see Figs. 1-2 and 1-3) are comparatively small, usually not much more than a few micrometers long and a micrometer or slightly less in width. (Figure 1-5 compares the dimensions of several molecules and cell structures in prokaryotes and eukaryotes.) In almost all prokaryotes the boundary membrane of the cell, the plasma membrane, is surrounded by a rigid external layer of material, the *cell*

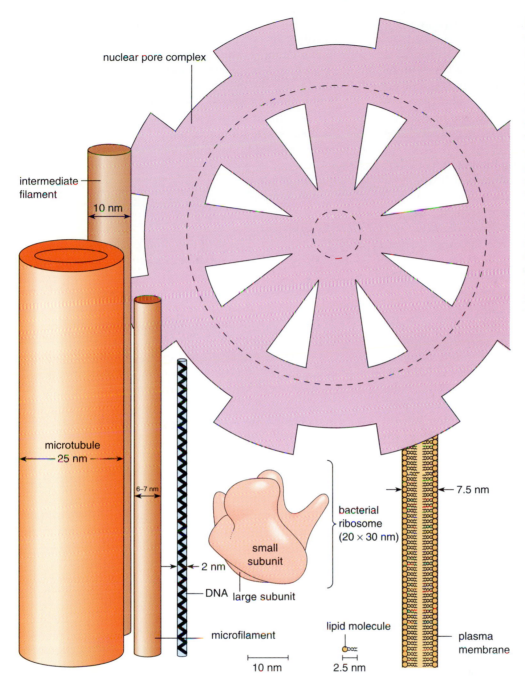

nuclear pore complex

intermediate filament

10 nm

microtubule
25 nm

6–7 nm

small subunit

2 nm

DNA large subunit

microfilament

bacterial ribosome (20 × 30 nm)

7.5 nm

lipid molecule

plasma membrane

10 nm

2.5 nm

Figure 1-5 The dimensions of some cell structures in prokaryotes and eukaryotes. In the diagram of a nuclear envelope pore complex the annulus that fills the pore is shown in outline as it might appear from the outside of the nuclear envelope. The smaller dotted circle at the center is the approximate location of the channel extending through the annulus; the larger dotted circle is the diameter of the opening in the nuclear envelope membranes that is filled by the annulus. (The diagram shown for the pore complex is based on a model developed by C. W. Akey in *J. Cell Biol.* 109:955[1989].)

wall. The cell wall may range in thickness from 15 to 100 nm or more and may itself be coated with a thick, jellylike *capsule* (as in Fig. 1-2). The cell wall provides rigidity to prokaryotic cells and, with the capsule, protects the cell within the wall.

The plasma membrane carries out a variety of vital functions in prokaryotes. Most important of these is *transport*, the movement of substances in and out of cells. In addition, most of the molecular systems that break down fuel substances to release energy for cell activities are housed in the prokaryotic plasma membrane. In photosynthetic bacteria and cyanobacteria the molecules absorbing light and converting it

to chemical energy are also associated with the plasma membrane and its interior extensions or derivatives. In addition, the plasma membrane contains proteins that act as *receptors*, recognizing and binding specific molecules that penetrate through the wall from the surrounding medium. Binding the external molecules triggers internal reactions that allow prokaryotic cells to respond to their environment. The prokaryotic plasma membrane may also play a part in replication and division of the nuclear material.

The nucleoid of a prokaryotic cell, suspended in the cytoplasm without boundary membranes, appears as a structure of irregular outline that contains masses of very fine fibers 3 to 5 nm in thickness (see Figs. 1-2 and 1-3). When isolated, the nucleoid of bacteria proves to contain a single DNA molecule in the form of a closed circle. In different bacteria the circles range from about 250 μm to a maximum of about 1500 μm of included DNA. Associated with prokaryotic DNA are protein molecules that hold the nucleic acid in its folded form and regulate its activity in cell synthesis and duplication.

The cytoplasm surrounding the prokaryotic nucleoid usually appears densely stained in electron micrographs. Most of this density is due to the presence of large numbers of *ribosomes*, small, roughly spherical particles about 20 to 30 nm in diameter (see Fig. 1-2). In bacteria, ribosomes contain more than 50 different proteins in combination with several types of *ribonucleic acid* (*RNA*). These small but complex spherical bodies are the sites where amino acids are assembled into proteins.

Other structures may be present in the cytoplasm of more complex prokaryotes. In some bacteria and cyanobacteria the cytoplasm contains numerous bag-like, closed sacs, collectively called *vesicles* or *vacuoles*, with walls formed by a single, continuous membrane. Molecules carrying out photosynthesis are associated with some of these internal sacs in the photosynthetic bacteria and cyanobacteria. Deposits of lipids, polysaccharides, or inorganic phosphates, appearing as small, very dense spherical bodies scattered in the cytoplasm, are also present in some prokaryotic cells (as in Fig. 1-3).

Many types of bacteria are capable of rapid movement generated by the action of long, threadlike protein fibers called *flagella* (singular = *flagellum*) that extend from the cell surface (Fig. 1-6). Bacterial flagella, which produce motion by rotating like a propeller (see page 304), are fundamentally different from the much larger and more complex flagella of eukaryotic cells (see below).

The apparent simplicity of prokaryotic cells is deceptive. Most bacteria and cyanobacteria can use a wide variety of substances as energy sources and are able to synthesize all their required organic molecules from simple starting substances such as water, carbon dioxide, and inorganic sources of nitrogen, phosphorus, and sulfur. In many respects, in fact, prokaryotes are more versatile than eukaryotes in their biochemical activities.

EUKARYOTIC CELLS

The complex membrane systems of eukaryotic cells (Figs. 1-7 and 1-8) cover the cell surface, define the nuclear region as a true *nucleus*, and separate the cytoplasm into distinct compartments called *organelles*. The eukaryotic plasma membrane, like its prokaryotic counterpart, carries out a variety of functions. Most significant of these is transport, provided by proteins forming channels in the membrane. Eukaryotic plasma membranes also contain a variety of proteins acting as receptors, which can recognize and bind specific molecules from the surrounding medium. Most receptors are linked to internal reaction systems that are triggered when the receptor binds a target molecule at the cell surface. Many internal responses that coordinate the activities of individual cells in animals, for example, are triggered through receptors that recognize and bind "signal" molecules such as hormones at the cell surface. Additional proteins in eukaryotic plasma membranes can recognize and adhere to specific molecules on the surfaces of other cells; this ability is critical to the development and maintenance of tissues and organs in multicellular animals. Other

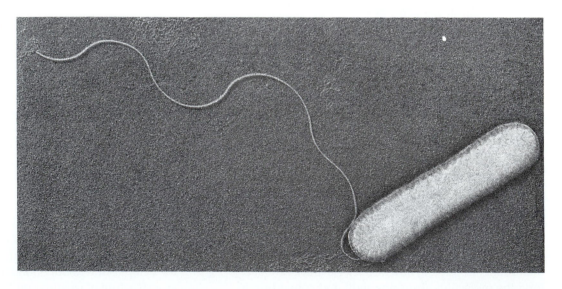

Figure 1-6 A bacterial cell with a single flagellum. × 30,000. (Courtesy of J. Pangborn.)

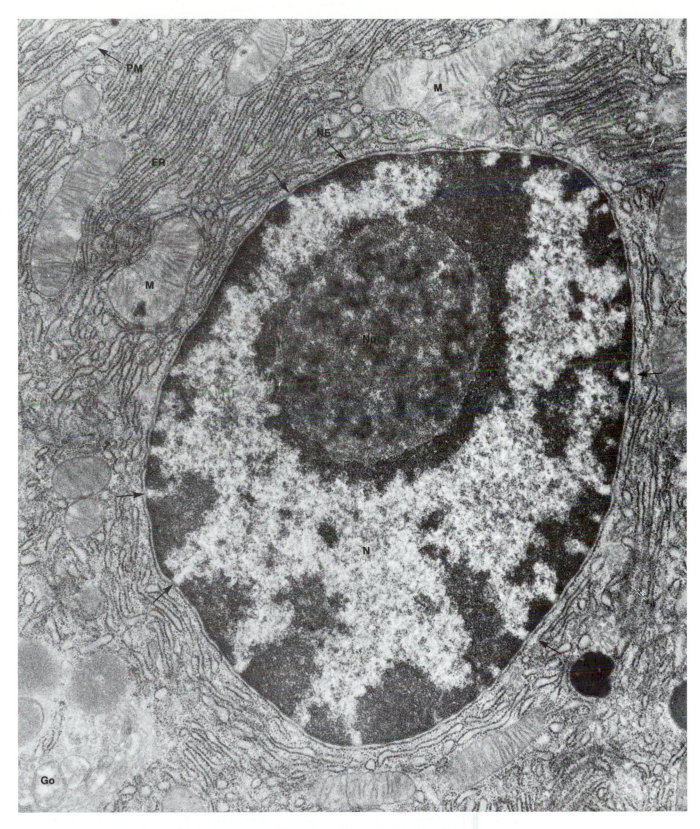

Figure 1-7   A eukaryotic cell from rat pancreas. N, nucleus; Nu, nucleolus; M, mitochondrion; ER, endoplasmic reticulum; Go, Golgi complex; PM, plasma membrane; NE, nuclear envelope; arrows, pore complexes. × 24,000. (Photograph by the author.)

Figure 1-8 Eukaryotic cell structure (see text).

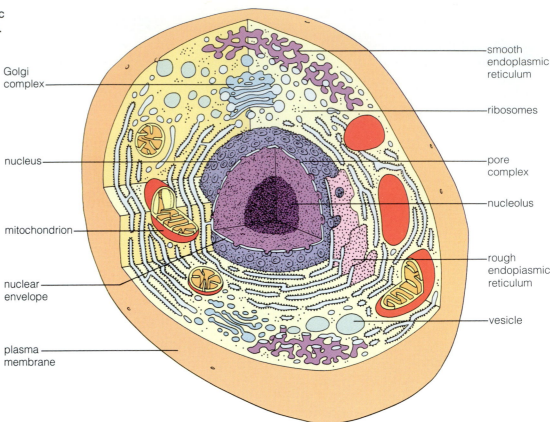

Golgi complex

nucleus

mitochondrion

nuclear envelope

plasma membrane

smooth endoplasmic reticulum

ribosomes

pore complex

nucleolus

rough endopiasmic reticulum

vesicle

proteins form markers that, in effect, identify cells as part of the same individual or as foreign. In contrast to prokaryotes, the plasma membranes of eukaryotes do not contain molecules that break down fuel substances to provide cellular energy or convert light to chemical energy. Membranes carrying out these activities are located in internal organelles in eukaryotic cells.

The Nucleus

The nucleus is separated from the cytoplasm by the *nuclear envelope*, a membrane system that consists of two concentric membranes, one layered just inside the other. The outer membrane faces the cytoplasm, and the inner membrane, separated from the outer membrane by a narrow space, faces the nuclear interior. The nuclear envelope is perforated by *pores* that form openings some 70 to 90 nm wide through both membranes (arrows, Fig. 1-7; see also Figs. 13-15 to 13-21). The pores are filled by a ringlike mass of proteins, the *annulus*, which controls movement of larger molecules such as RNA and proteins through the nuclear envelope.

Most of the space inside the nucleus is occupied by masses of very fine, irregularly folded *chromatin fibers*, about 10 to 30 nm in thickness, which contain

the nuclear DNA. In chromatin, DNA is associated with two major types of proteins, the *histone* and *nonhistone chromosomal proteins*. The histones are primarily structural molecules that pack DNA into chromatin fibers. The nonhistones include proteins that carry out one of the most important cellular functions, the regulation of gene activity.

Instead of being concentrated in a single circular molecule as in bacteria, the hereditary information of a eukaryotic nucleus is subdivided among several to many linear DNA molecules. Each individual DNA molecule, with its associated histone and nonhistone proteins, is a *chromosome*.

Eukaryotic nuclei contain much greater quantities of DNA than prokaryotic nucleoids. The entire complement of 46 chromosomes in a human cell, for example, includes a total DNA length of about a meter. The nuclei of some eukaryotes contain even more— a single cell nucleus of an amphibian such as a frog or salamander, for example, contains about 10 meters of DNA!

Suspended within the chromatin of the nucleus are one or more irregularly shaped masses of small fibers and granules, the *nucleoli* (singular = *nucleolus*). The nucleolus is a subpart of the chromatin specialized for assembly of ribosomal subunits. The RNA of ribosomes is synthesized from genes in the nucleolus;

ribosomal proteins are synthesized in the cytoplasm and assembled with ribosomal RNA into ribosomes in the nucleolus. The fibers and granules visible in the nucleolus are structural forms taken by successive stages in this assembly. No membranes separate nucleoli from the surrounding chromatin.

The nucleus is the ultimate control center for cell activities. Within the chromatin the information required to synthesize cellular proteins is coded into the DNA. Each DNA segment containing the information for making a protein constitutes a *gene*. The information in a protein-encoding gene is copied into a *messenger RNA (mRNA)* molecule that moves to the cytoplasm through the pores of the nuclear envelope. In the cytoplasm, mRNA molecules are used by ribosomes as directions for the assembly of proteins. The DNA of the entire nucleus contains the codes for many thousands of different proteins.

Other DNA regions store the information for making additional RNA types that carry out accessory roles in protein synthesis and other functions in the nucleus and cytoplasm. *Ribosomal RNA (rRNA)*, as previously noted, is encoded in DNA regions forming parts of the nucleolus. Another RNA, *transfer RNA (tRNA)*, binds to amino acids during protein synthesis and provides a link between the information coded into nucleic acids and the amino acid sequence of proteins. Other RNAs aid in reactions processing mRNA, rRNA, and tRNA molecules from initial to finished forms. Each DNA segment encoding an rRNA, tRNA, or other accessory RNA type is also known as a gene. The process by which any of the genes is copied into an mRNA, rRNA, tRNA, or other RNA equivalent is called *transcription*.

A second major function of the nucleus involves duplication of the chromatin as a part of cell reproduction. Just before cell division, all the components of chromatin, including both DNA and chromosomal proteins, are precisely doubled. The duplication of DNA in chromatin is known as *replication*. During cell division the two copies of each duplicated chromosome are precisely separated so that the two cells resulting from the division each receive a complete set of chromosomes and genes.

The Cytoplasm

Eukaryotic cytoplasm is packed with ribosomes, vesicles, and a variety of membrane-bound organelles. The boundary membranes of the organelles set them off as distinct chemical and molecular environments, specialized to carry out different functions.

Ribosomes and Protein Synthesis The reactions assembling proteins take place entirely in the cyto-plasm. Following their transcription in the nucleus, mRNAs and tRNAs pass through the nuclear envelope and enter the cytoplasm as individual molecules; rRNAs enter in the form of ribosomal subunits. In the cytoplasm the ribosomal subunits join by twos with mRNA molecules to form complete ribosomes active in protein synthesis. Eukaryotic ribosomes, about 25 to 35 nm in diameter, are somewhat larger than the ribosomes of prokaryotes and contain more protein and RNA molecules.

In protein synthesis, called *translation*, a ribosome moves along an mRNA molecule, reading the code for protein assembly as it goes. As it moves, the ribosome assembles amino acids into a gradually lengthening protein chain. At the end of the coded message, translation stops, the ribosomal subunits separate and detach from the mRNA, and the completed protein is released.

Transfer RNA molecules function as the "dictionary" in the translation mechanism. Each of the 20 amino acids used in protein synthesis is linked to a specific kind of tRNA. The tRNA is capable of recognizing and binding the nucleic acid code word (called a *codon*) specifying its attached amino acid in an mRNA molecule.

The interaction of tRNAs with mRNA codons takes place on ribosomes (Fig. 1-9). As a ribosome encounters an mRNA codon specifying a given amino acid (codon X in Fig. 1-9a), the tRNA carrying that amino acid recognizes and binds to the codon (Fig. 1-9b). This binding places the amino acid in its correct location in the growing protein chain. The ribosome then moves to the next mRNA codon (codon Y in Fig. 1-9c), causing the next tRNA-amino acid complex specified by the code to bind (Fig. 1-9c and d). As each successive amino acid arrives at the ribosome, it is split from its tRNA and linked into the gradually lengthening protein chain. The process repeats until the ribosome reaches the end of the message and completes assembly of the protein. Prokaryotic ribosomes carry out protein synthesis in essentially the same way.

Ribosomes, Endoplasmic Reticulum, and the Golgi Complex Eukaryotic ribosomes that are active in protein synthesis may be either freely suspended in the cytoplasm or attached to the surface of membranous sacs (Fig. 1-10). The freely suspended ribosomes make proteins that primarily go into solution in the cytoplasm or form important cytoplasmic structural or motile elements. The ribosomes attached to membranous sacs assemble proteins that become a part of membranes or are packaged into vesicles for storage in the cytoplasm or export to the cell exterior.

The membranous sacs with their attached ribosomes form an extended, interconnected system of

Figure 1-9 The overall mechanism by which amino acids are assembled into proteins on ribosomes (see text).

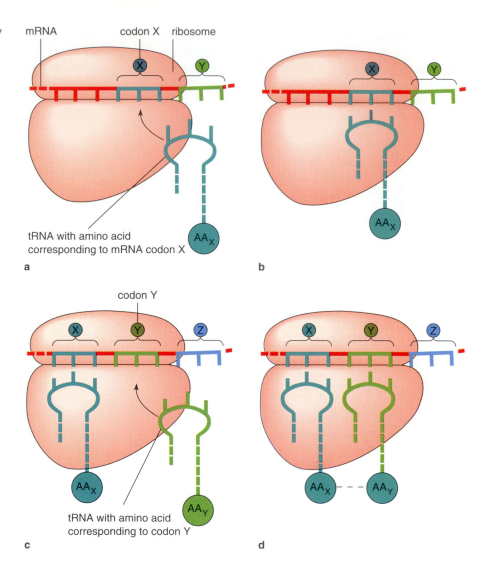

mRNA codon X ribosome

tRNA with amino acid corresponding to mRNA codon X

a

b

codon Y

tRNA with amino acid corresponding to codon Y

c

d

cytoplasmic channels known as the *rough endoplasmic reticulum*, or *rough ER* (Fig. 1-7 and 1-10; see also Figs. 20-1 to 20-3). Proteins synthesized on the ribosomes penetrate into the rough ER membranes and form a part of the membrane structure or pass entirely through the membranes to enter the enclosed ER channels. The proteins are subsequently processed into chemically altered forms and distributed to various sites in the cytoplasm or released to the cell exterior.

The processing and sorting of proteins made in the rough ER occur primarily in another system of membranous sacs, the *Golgi complex*. This system, named for its discoverer, Camillo Golgi, often appears as a closely stacked pile of flattened sacs (Fig. 1-11; see also Fig. 20-4). Small vesicles containing newly synthesized proteins pinch off from the rough ER and fuse with the Golgi membranes. Within the Golgi complex the proteins are modified by the attachment of other chemical groups, which, depending on the protein type, may include small groups such as sul-

fates or larger chemical structures such as lipid or sugar units. Following modification the proteins are sorted into small vesicles that pinch off from the Golgi membranes. The protein-containing sacs may either remain suspended in the cytoplasm as *storage vesicles* or form *secretory vesicles*, which release their contents to the cell exterior.

Release of the contents of secretory vesicles to the cell exterior occurs by a process known as *exocytosis*. In exocytosis a secretory vesicle moves to the plasma membrane (Fig. 1-12a); as contact is made, the vesicle membrane fuses with the plasma membrane (Fig. 1-12b). The fusion makes the secretory vesicle membrane continuous with the plasma membrane and spills the vesicle contents to the cell exterior (Fig. 1-12c). The proteins and lipids of the vesicle membrane become a temporary or permanent part of the plasma membrane.

Eukaryotic cells regularly take up particulate matter or large molecules such as proteins by a mechanism called *endocytosis*, which essentially reverses

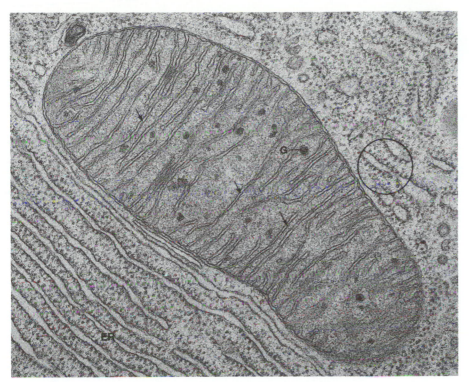

Figure 1-10 A mitochondrion from bat pancreas, surrounded by cytoplasm containing rough endoplasmic reticulum (rough ER). Cristae (arrows) extend into the interior of the mitochondrion as folds from the inner boundary membrane. The darkly stained granules (G) are believed to be lipid deposits. A segment of the rough ER is circled; ribosomes (arrows within circle) are clearly visible on the surfaces of the ER membranes. × 50,000. (Courtesy of K. R. Porter.)

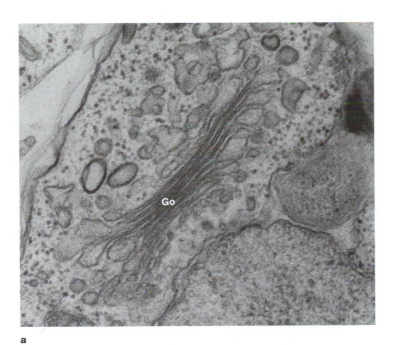

a

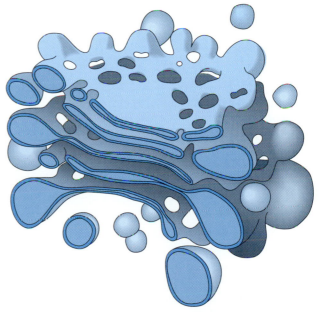

b

Figure 1-11 The Golgi complex. **(a)** A Golgi complex (Go) in a plant cell. (Courtesy of W. A. Jensen.) **(b)** The Golgi complex in three dimensions.

exocytosis. In endocytosis a molecule or particle is attached to the external surface of the cell, usually by receptor molecules forming part of the plasma membrane (Fig. 1-12d). The plasma membrane then invaginates as a pocket (Fig. 1-12e) that pinches off (Fig. 1-12f) and sinks into deeper layers of the cytoplasm as an *endocytotic vesicle*. Once in the cytoplasm, the vesicle contents are routed to various locations in the cell. One of the major destinations is the Golgi

Exocytosis

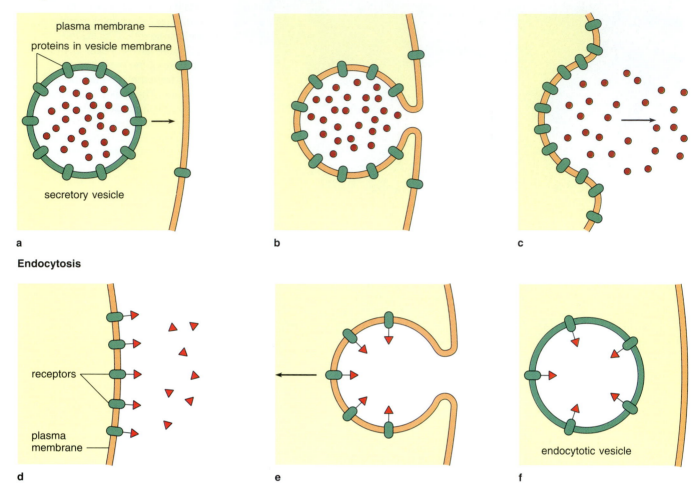

Endocytosis

Figure 1-12 Exocytosis (**a** to **c**) and endocytosis (**d** to **f**). (See text.)

complex, where proteins taken in by endocytosis are sorted and placed into vesicles for routing to other locations. The Golgi complex thus serves as a major sorting station for materials traveling both into and out of the cell, as well as a site modifying proteins newly synthesized in the rough ER.

Not all the interconnected membranous vesicles collectively identified as endoplasmic reticulum are associated with ribosomes. The ribosome-free membranes, known as *smooth endoplasmic reticulum* (or *smooth ER*; see Fig. 20-1), have various functions in the cytoplasm. One is the breakdown of fats as a first step in using them as an energy source. Some segments of smooth ER have also been identified with lipid syn-

thesis or with reactions that break down toxic substances absorbed by cells.

Lysosomes One type of protein-containing sac produced by the activity of the rough ER and Golgi complex, the *lysosome* (see Fig. 20-19), is especially important in animal cells. Lysosomes contain enzymes that, in aggregate, are capable of breaking down all the major classes of biological molecules. In many cell types, molecules taken in by endocytosis are frequently delivered to lysosomes, where they are digested by the lysosomal enzymes. The lysosomal enzymes may also be released to the cell exterior by exocytosis to break down molecules on the outside. In

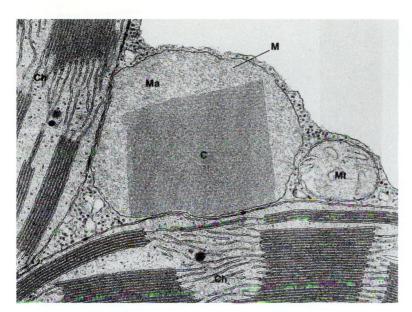

Figure 1-13 A microbody (M) in the cytoplasm of a tobacco leaf cell. Ma, matrix of the microbody; C, crystalline core, Ch, chloroplast; Mt, mitochondrion. × 49,000. (Courtesy of S. E. Frederick, from *Ann. N. Y. Acad. Sci.* 386:228[1982].)

some situations, lysosomes rupture to release their enzymes inside the cell, where the enzymes break down the cell itself and cause cell death. Self-destruction of this type may be a part of pathological conditions or may be programmed as part of normal developmental processes. Loss of the tail as tadpoles develop into adult frogs and disappearance of the webbing between fingers and toes in human development, for example, are examples of programmed cell death in which lysosomal enzymes are released directly into cells.

Mitochondria Most of the chemical energy required for the activities of ribosomes, the ER, the Golgi complex, and other functions of eukaryotic cells is derived from reactions taking place in a membrane-bound cytoplasmic organelle known as the *mitochondrion* (plural = *mitochondria*; from *mitos* = thread and *chondros* = grain). The name refers to the fact that mitochondria may be long and filamentous, or compact and granulelike, or may change between these forms.

Mitochondria are surrounded by two separate membrane systems, one enclosed within the other (see Fig. 1-10 and Fig. 8-2). The *outer boundary membrane* is smooth and continuous. The *inner boundary membrane* extends into the mitochondrial interior in numerous folds or tubular projections called *cristae* (singular = *crista*). The innermost mitochondrial compartment, surrounded by both membranes, is the *matrix*.

Mitochondria are frequently called the "powerhouses" of the cell because they carry out most of the oxidative reactions that release energy for cellular activities. Fuel for these reactions is provided by break-down products of all major cellular molecules, including carbohydrates, fats, proteins, and nucleic acids. The oxidative reactions of mitochondria are distributed between the cristae membranes and the matrix.

Mitochondria are partially autonomous; they contain their own DNA and ribosomes, enzymes, and other factors required for transcription and protein synthesis, all concentrated in the matrix. Many features of mitochondrial DNA, transcription, and translation resemble the equivalent systems of bacteria. The similarities to bacterial systems indicate that mitochondria probably evolved from bacteria that became established as symbionts in cell lines destined to form the eukaryotes.

Microbodies Eukaryotic oxidative and other reactions are also localized in another group of organelles collectively known as *microbodies* (Fig. 1-13). Microbodies are relatively simple structures enclosed by a single boundary membrane; inside a microbody is a *matrix* consisting primarily of a solution or suspension of enzymes. Often a crystalline protein *core* is visible in the matrix (as in Fig. 1-13). Rather than directly providing chemical energy for cellular activities, microbodies carry out reactions that link major oxidative pathways occurring elsewhere in the cytoplasm. Typically microbodies make hydrogen peroxide (H_2O_2) as a final product of their oxidative reactions and break down this toxic substance using *catalase*, an enzyme that converts H_2O_2 to water and oxygen. Because of this characteristic reaction, microbodies are frequently termed *peroxisomes*. In plants a group of microbodies called *glyoxisomes* converts fats to sugars.

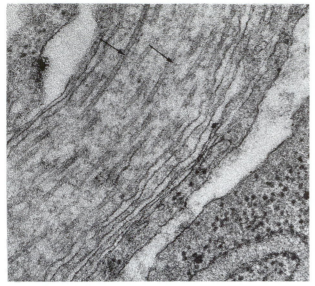

a

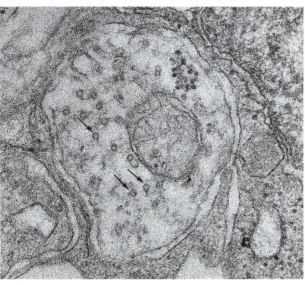

b

Figure 1-14 Microtubules (arrows) in longitudinal section **(a)** and cross section **(b)**. × 65,000. (Courtesy of M. P. Daniels and the New York Academy of Sciences.)

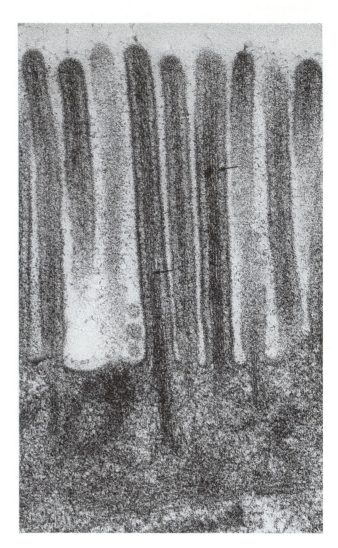

Figure 1-15 Microfilaments (arrows) inside microvilli, fingerlike projections extending from the surface of a chick intestinal cell. × 95,000. (Courtesy of C. Chambers.)

Microtubules, Microfilaments, and Cell Movement
Almost all cell movements in eukaryotes are generated by either of two cytoplasmic structures—*microtubules* and *microfilaments*—acting singly or in coordination. Microtubules (Fig. 1-14; see also Fig. 1-5) are fine, tubelike structures about 25 nm in diameter. Microfilaments (Fig. 1-15; see also Fig. 1-5) are thin fibers 5 to 7 nm in diameter. Each structure is assembled from subunits of a different protein—microtubules from *tubulin* and microfilaments from *actin*.

Both microtubules and microfilaments produce motion through the activity of protein crossbridges that are capable of converting chemical energy to me-chanical energy. One end of a crossbridge is firmly attached to the surface of a microtubule or microfilament. The opposite end has a reactive site that may attach to another microtubule or microfilament or to other cell structures. The crossbridge makes an attachment at its reactive end, forcefully swivels a short distance, and then releases, working much like the oar of a boat (see Figs. 10-13 and 11-9). Each of the two motile elements has distinct proteins forming the swiveling crossbridges.

Some cellular movements, such as the beating of flagella, depend exclusively on the activity of microtubules. Microfilaments are solely responsible for

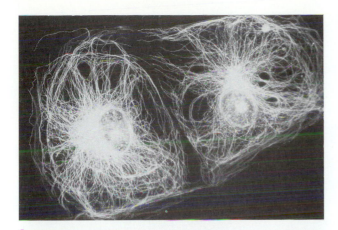

a

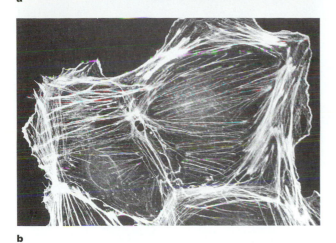

b

Figure 1-16 The cytoskeleton in cultured rat kangaroo cells. The staining technique used to prepare these micrographs makes the cytoskeletal elements appear white against a dark background. **(a)** The microtubule network of the cytoskeleton; **(b)** the microfilament network. × 550. (Reprinted from M. Osborn, W. W. Franke, and K. Weber, *Proc. Nat. Acad. Sci.* 74:2490[1977].)

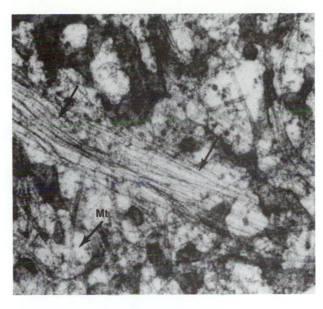

Mt

Figure 1-17 Intermediate filaments (arrows) in the cytoplasm of cultured rat kangaroo cell. Microtubules are also visible. × 71,000. (Courtesy of K. McDonald.)

other types of movements, including the active, flowing motion of cytoplasm called *cytoplasmic streaming*, ameboid motion, and contraction of muscle cells. Cell division in animals involves the coordinated activities of both microtubules and microfilaments—the chromosomes are divided by microtubules and the cytoplasm by microfilaments.

The Cytoskeleton In addition to their functions in cell motility, both microtubules and microfilaments form supportive networks in the cytoplasm of eukaryotic cells (Fig. 1-16). In some eukaryotic cells, most notably in animals, another set of structural elements, the

intermediate filaments (Fig. 1-17; see also Figs. 12-1 and 12-2) also provides support to the cytoplasm. Intermediate filaments, with diameters averaging about 10 nm, assemble from a diverse but interrelated group of proteins that is completely distinct from the tubulins and actins.

The supportive networks set up by microtubules, microfilaments, and intermediate filaments are known collectively as the *cytoskeleton*. Different cytoskeletal systems support the plasma membrane, cell extensions, regions of the cytoplasm between the plasma membrane and the nucleus, and the nuclear envelope.

Specialized Cytoplasmic Structures of Plant Cells

All the nuclear and cytoplasmic structures described up to this point, with the exception of lysosomes and possibly intermediate filaments, occur in plant as well as animal cells. Plant cells also have several organelles and components not found in animal cells (Figs. 1-18 and 1-19). The most conspicuous of these are *plastids*; large, specialized membrane-bound sacs called *central vacuoles*; and the *cell wall*.

Plastids are a family of organelles with various functions. In green plant tissues the characteristic plastid is the *chloroplast* (Fig. 1-20; see also Figs. 9-2 and 9-3), an organelle built up from three membrane systems. A smooth *outer boundary membrane* completely covers the surface of the organelles. A highly folded and convoluted *inner boundary membrane* lines

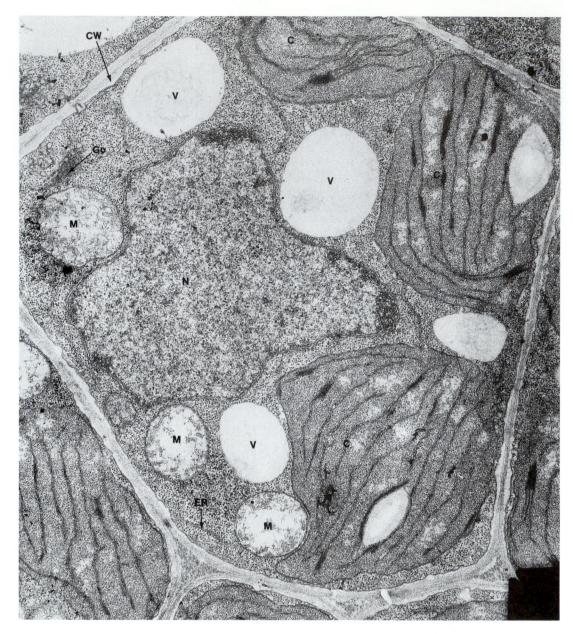

Figure 1-18 A plant cell from a bean seedling (*Phaseolus vulgaris*), showing the major structures occurring in plant but not animal cells: the cell wall (CW), chloroplasts (C), and a series of large vacuoles (V). Structures common to both plant and animal eukaryotes are also visible: the nucleus (N), surrounded by the nuclear envelope; mitochondria (M); endoplasmic reticulum (ER); and the Golgi complex (Go). In this embryonic plant cell the vacuoles are small in size and have not yet coalesced into a large central vacuole. (Courtesy of Chin Ho Lin.)

the outer membrane. These two boundary membranes enclose an inner compartment, the *stroma*, equivalent in location to the mitochondrial matrix. Within the stroma is the third membrane system, consisting of flattened, closed sacs called *thylakoids*.

Thylakoid membranes house photosynthetic pigments and molecules that absorb light energy and convert it to chemical energy. The chemical energy is used by enzyme systems suspended in the stroma to drive the assembly of carbohydrates and other com-

plex organic molecules from water, carbon dioxide, and other simple inorganic precursors.

The chloroplast stroma, like the mitochondrial matrix, also contains DNA, ribosomes, and all the enzymes and other factors required for transcription and protein synthesis. As in mitochondria, the DNA and other elements are believed to be derived from the biochemical systems of ancient prokaryotes probably resembling cyanobacteria. The ancient prokaryotes became established as permanent residents and

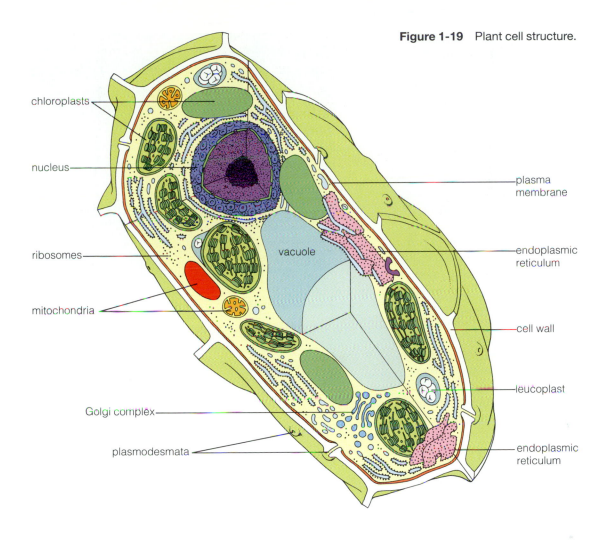

Figure 1-19 Plant cell structure.

chloroplasts

nucleus

ribosomes

mitochondria

Golgi complex

plasmodesmata

plasma membrane

endoplasmic reticulum

cell wall

leucoplast

endoplasmic reticulum

vacuole

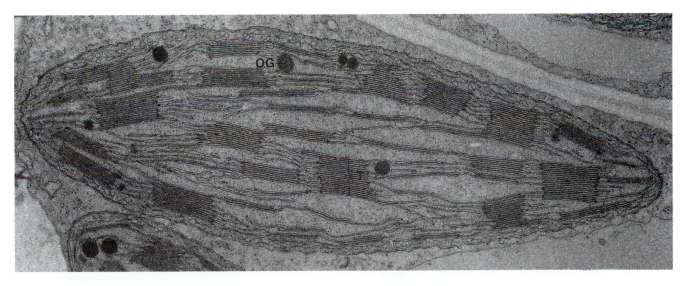

Figure 1-20 Chloroplast in a tobacco leaf cell. The reactions converting light into chemical energy are concentrated in the thylakoid membranes (T). Lipid deposits called osmiophilic granules (OG) are commonly observed inside chloroplasts. (Courtesy of W. M. Laetsch.)

OG

evolved into chloroplasts in the cytoplasm of cell lines destined to found the modern eukaryotic algae and plants.

Central vacuoles are identified as separate, distinct organelles of plant cells because they reach much larger dimensions than the vesicles or vacuoles of animal cells and carry out functions that are unique to plants. Often most of the volume of a mature plant cell, as much as 90% or more, is occupied by one or more large vacuoles consisting of a single, continuous membrane enclosing a fluid-filled space.

Central vacuoles support plant cells and tissues through the development of osmotic pressure (see p. 119), which presses the cells against the surrounding cell walls. In plants with little or no woody tissue, stems and leaves are supported primarily by pressure developed in the vacuoles, which may be as much as 20 times greater than atmospheric pressure.

Various substances are held in solution in central vacuoles, including organic and inorganic salts, organic acids, sugars, storage proteins, and pigments. Frequently the colors of flowers are produced by pigments concentrated in vacuoles. Also present in some central vacuoles are enzymes capable of breaking down many biological molecules, giving these vacuoles some of the properties of lysosomes. Molecules providing chemical defenses against pathogenic organisms also occur in some plant vacuoles.

Cell walls are layered structures located outside the plasma membrane. The *primary cell wall* of an embryonic plant cell is flexible and capable of extension during growth, but the wall of a mature cell, called a *secondary cell wall*, is usually rigid and inextensible. Both primary and secondary walls contain fibers of *cellulose* (see Figs. 2-8 and 7-12), a long, chainlike molecule consisting of repeating subunits derived from the sugar glucose. The cellulose fibers, which give tensile strength to cell walls, are embedded in a network of proteins and highly branched carbohydrates. At intervals, cell walls in higher plants are perforated by minute channels, the *plasmodesmata* (singular = *plasmodesma*; see Figs. 1-19 and 7-20), which contain cytoplasmic bridges that directly connect adjacent cells.

PROKARYOTIC AND EUKARYOTIC CELL STRUCTURE: A SUMMARY

Prokaryotes are relatively small cells surrounded by a plasma membrane and, in most groups, enclosed by a cell wall. The prokaryotic nuclear material, contained in the nucleoid, is suspended directly in the cytoplasm with no boundary membranes. No nucleolus is present in the nucleoid. Prokaryotic cytoplasm consists primarily of masses of ribosomes, along with

Table 1-2	Major Cell Structures of Prokaryotes and Eukaryotes		
Structure	Prokaryotes	Plants	Animals
Cell wall	+	+	−
Chloroplasts	−	+	−
Endoplasmic reticulum	−	+	+
Golgi complex	−	+	+
Intermediate filaments	−	?	+
Microbodies	−	+	+
Microfilaments	−	+	+
Microtubules	−	+	+
Mitochondria	−	+	+
Nuclear envelope	−	+	+
Nucleoid	+	−	−
Nucleolus	−	+	+
Nucleus	−	+	+
Plasma membrane	+	+	+
Ribosomes	+	+	+

nonmembranous inclusions such as lipid or phosphate deposits. Although small membranous vesicles may occur in the cytoplasm of the more complex prokaryotes, none of the discrete membrane-bound organelles of eukaryotes, such as mitochondria or chloroplasts, are present.

Eukaryotic cells are surrounded by a plasma membrane and divided into specialized compartments by internal membranes. The nucleus is separated from the cytoplasm by a double layer of membranes, the nuclear envelope. The nucleus contains chromatin fibers, consisting of DNA in association with the histone and nonhistone proteins. Within the nucleus is the nucleolus, a segment of the chromatin specialized for rRNA transcription and assembly of ribosomal subunits. The cytoplasm contains numerous ribosomes and membrane systems forming the endoplasmic reticulum, Golgi complex, mitochondria, microbodies, and a variety of vesicles and vacuoles. Both microtubules and microfilaments are present in eukaryotic cells as motile elements. Microtubules and microfilaments, along with intermediate filaments in animal cells, also form cytoskeletal systems that support internal cellular regions. Plant cells are surrounded by a cell wall and contain the structures listed above (with the possible exception of intermediate filaments), as well as chloroplasts and large central vacuoles. Table 1-2 summarizes the major structures of prokaryotic and eukaryotic cells.

VIRUSES

Although cells are the smallest functional units of living organisms, simpler units can invade cells and subvert their synthetic machinery. These are the *viruses*, which in the free, infective form consist of little more than a coat of protein surrounding one or more nucleic acid molecules, all concentrated into a structure that in most cases is smaller than a ribosome. The nucleic acid of a virus, when inserted into a host cell, can convert the synthetic machinery of its host to synthesize and release viruses of the same kind. Viruses are important in their own right as the vectors of diseases in humans, domestic animals, and plants. They are also of great value in research because their patterns of infection reflect the inner workings of their host cells.

Viral Structure

The nucleic acid of a free virus particle, either DNA or RNA, is the *core* of the virus. The protein layer surrounding the core is the *capsid* or *coat*. Some viruses are simple particles consisting of a core with a single nucleic acid molecule and a capsid assembled from protein molecules of a single type. Other, more complex viruses have cores containing several nucleic acid molecules and capsids made up of several to many different proteins—in some, as many as 50 or more. Some viruses, such as those causing AIDS, influenza, and herpes in humans, are surrounded by a membrane derived from the plasma membranes of their host cells. Viruses of this type are known as *enveloped viruses* (see Fig. 1-23).

The DNA or RNA molecules storing viral genetic information may contain from as few as three or four to as many as several hundred genes. At a minimum these genes encode enzymes required to duplicate the nucleic acid core and protein coat of the virus. They may also encode recognition proteins that become implanted in the protein coat. Segments of the recognition proteins recognize and bind surface molecules of the host cell and promote entry of the particle or its nucleic acid into the host.

Depending on the virus, the protein coat may be rodlike or spherical. Many viruses infecting plant cells are rodlike (Fig. 1-21). The coat of the much studied tobacco mosaic virus (TMV), for example, is a rod-shaped structure about 15 by 300 nm long, assembled from more than 2000 identical protein units. As in all rodlike viruses, the protein subunits of the TMV coat are assembled in a spiral or helix (Fig. 1-22). The RNA molecule of this virus winds into a matching spiral inside the protein helix.

The spherical viruses (Fig. 1-23) include types that infect animals, plants, and bacteria. Rather than being perfectly spherical, the coats of these viruses are built up from a number of flat, triangular facets forming a polyhedron (Fig. 1-24). In some spherical

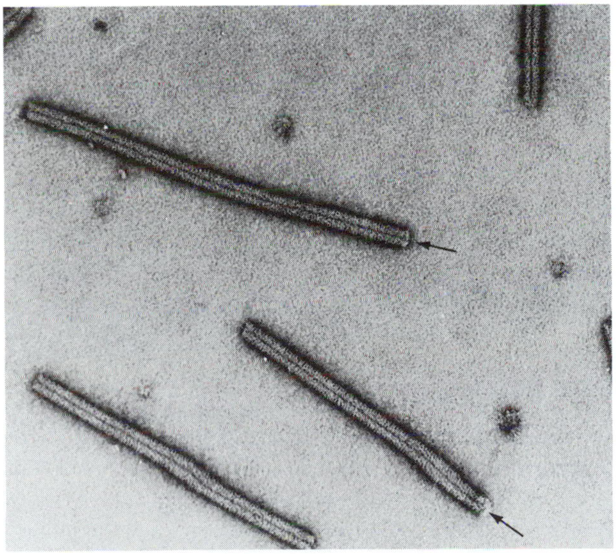

Figure 1-21 The beet necrotic yellow virus, an example of a rodlike virus. The viral particles have been prepared for electron microscopy by negative staining, in which the stain covers the surfaces and penetrates into interior compartments of the specimen. In this micrograph the stain has penetrated into the central channel (arrows) containing the nucleic acid core of the virus. × 80,000. (Courtesy of A. C. Steven and Academic Press, from *Virology* 113:428[1981].)

Figure 1-22 The arrangement of the nucleic acid core (in red) and coat protein molecules of the tobacco mosaic virus. (Courtesy of D. L. D. Caspar, from *Science* 227: 773[1985]. Copyright 1985 by the AAAS.)

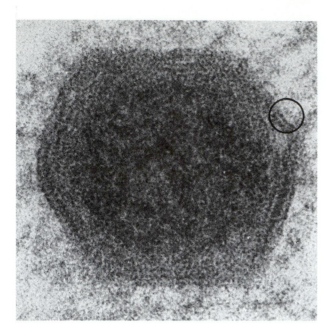

Figure 1-23 A spherical virus, in which the coat is actually a polyhedron built up from flat, triangular facets. This virus, which causes lymphocytosis in fishes, is covered by a membrane (circled) derived from a host cell. × 260,000. (Courtesy of L. Berthiaume and Academic Press, from *Virology* 135:10[1984].)

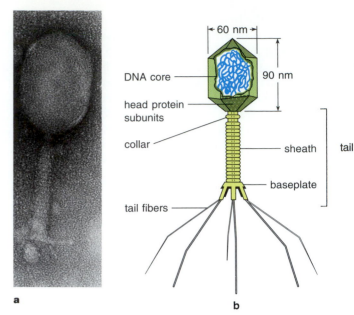

Figure 1-25 Tailed bacteriophages. **(a)** A tailed bacteriophage that infects *E. coli* cells. × 340,000. (Courtesy of the Perkin-Elmer Corporation.) **(b)** The structures of a tailed bacteriophage. The tail fibers function in host cell recognition and binding.

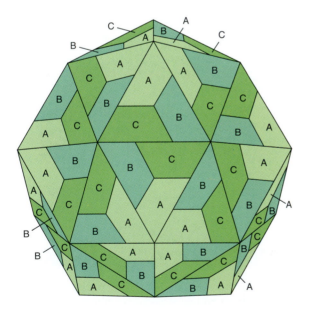

Figure 1-24 The arrangement of facets in the polyhedral coat of the southern bean mosaic virus. In this virus the subunits A, B, and C of each facet are the same protein arranged in slightly different folding patterns. Some polyhedral viruses have more than three subunits in a triangular facet.

viruses, *spike proteins* providing host recognition extend from the points or corners where the facets fit together. Some spherical viruses are covered by a membranous envelope derived from their host cells (as in Fig. 1-23).

Among the most complex spherical viruses are the *bacteriophages* infecting bacteria. Some bacteriophages have a *tail* (Fig. 1-25) that functions in host cell recognition and attachment and in injection of the nucleic acid core.

Viral Infective Cycles

In the free form, viral particles are incapable of independent movement. They are carried about by random molecular collisions until contact is made with the surface of a host cell. A series of events then inserts either the entire virus or its nucleic acid core into the host cell. Once inside the host, the genes encoded in the viral nucleic acid direct the host cell to make and release additional viral particles of the same type. In most cases the cycle of viral infection and release damages or kills the host cell. Typically viruses have a narrow host range—they are capable of attaching to, and infecting, only the cells of a single species or group of closely related species.

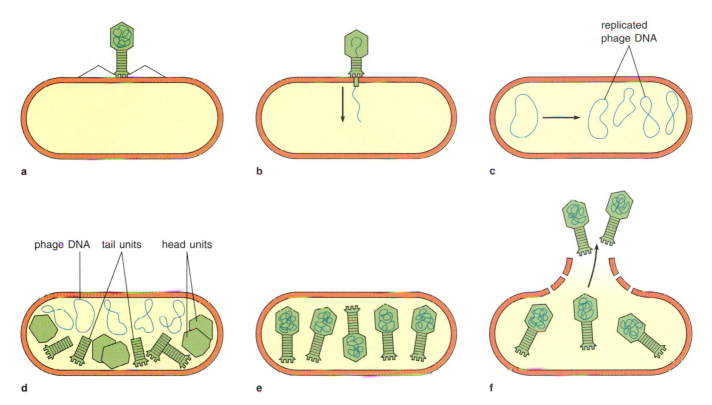

Figure 1-26 Life cycle of a T-even bacteriophage. **(a)** Attachment of bacteriophage to a bacterial cell; **(b)** injection of viral DNA; **(c)** replication of viral DNA; **(d)** synthesis of head and tail proteins; **(e)** assembly of DNA, head, and tail units into finished virus particles; **(f)** release of completed bacteriophage particles by rupture of the infected cell.

Infection of Bacterial Cells In infection of a bacterial cell by a tailed bacteriophage, random collision of a viral particle with the host leads to tight binding between the viral tail fibers and molecules of the bacterial cell wall (Fig. 1-26a). The head and tail sheath of the virus then contract and inject the DNA core into the cell (Fig. 1-26b). The proteins of the virus remain outside.

Once inside the bacterial cell, a part of the bacteriophage DNA is immediately transcribed into mRNAs by bacterial enzymes. These first mRNAs direct the bacterial ribosomes to synthesize several viral proteins, among them enzymes necessary for replication of the viral DNA. The bacterial cell, now converted to the production of viral DNA, continues replication until as many as a thousand new bacteriophage DNA molecules are made (Fig. 1-26c). At some time after viral DNA replication begins, other segments of the viral DNA are transcribed into mRNAs. These late viral messengers direct the bacterial cell to synthesize viral coat proteins and assemble them into heads and tails (Fig. 1-26d). As the coats assemble, the newly synthesized bacteriophage DNA packs into the heads, and the heads and tails link into completed particles (Fig. 1-26e). After particle assembly is complete, a final viral protein is synthesized that opens breaks in the bacterial cell wall, rupturing the cell (Fig. 1-26f). The rupture releases the newly completed viral particles to the surrounding medium. Chance collisions may lead to another cycle of infection and release of virus particles if a host cell is encountered.

Infection of Eukaryotic Cells Infection of eukaryotic cells follows a similar pattern, except that both the core and capsid enter the host cell. In infection of animal cells, viral particles without membrane envelopes are bound tightly to the outer surface of the plasma membrane by their recognition proteins. The binding stimulates the host cell to take in the viral particles by endocytosis. Enveloped viruses enter by a fusion of their surface membranes with the host cell plasma membrane (Fig. 1-27a to c). The fusion inserts the virus inside the host cell with its protein coat but without the surface membrane. In plants, viral particles usually enter initially through cellular regions exposed by wounds or abrasions in the cell wall. Within the plant, infective particles are passed from infected to healthy cells via plasmodesmata and the vascular system.

Once inside the eukaryotic host cell, the viral protein coat ruptures or disassembles to release the

Figure 1-27 Entry (**a** to **c**) and release (**d** to **f**) of enveloped viruses (see text).

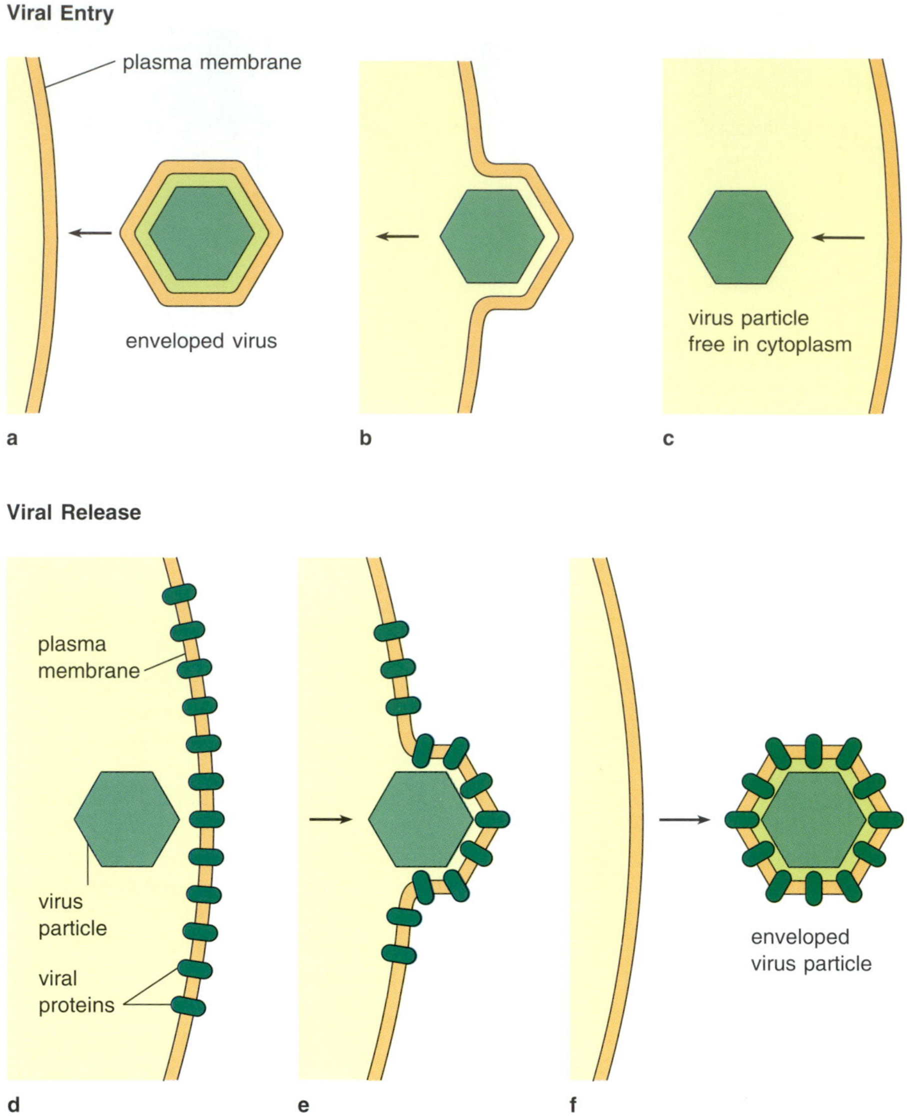

Viral Entry

plasma membrane

enveloped virus

a

virus particle free in cytoplasm

b

c

Viral Release

plasma membrane

virus particle

viral proteins

enveloped virus particle

d

e

f

nucleic acid core. The nucleic acid directs the assembly of additional viral nucleic acids and coat proteins as in the bacterial viruses. When synthesis of these components is complete, the new viral particles are assembled and released from the infected cell.

The enveloped viruses (see Fig. 1-23) derive their surface membranes from the host cell plasma membrane during release (Figs. 1-27 and 1-28). After assembly in the host cell cytoplasm a new viral particle attaches to the inside surface of the plasma membrane (see Fig. 1-27d). A pocket containing the particle then pushes out as a membrane-covered bud from the host cell surface (Fig. 1-27e). The bud then pinches off (Fig. 1-27f) to release the virus in free form. This pattern of viral release, since it does not break the plasma membrane, may proceed without rupture of the host cell.

Many viruses with a DNA core enter a stage in which the viral DNA inserts into the host cell DNA. During much of the time the integrated viral DNA may remain inactive. In this form, called the *lysogenic* phase of the infective cycle, the inactive viral DNA is replicated and passed on with the host cell DNA to all descendants of the infected cell. At some later time, frequently during environmental conditions that are stressful to cells containing the integrated virus, the viral DNA becomes active. Among the first viral proteins synthesized in response are enzymes that cut the viral DNA from the host cell DNA. The viral DNA then directs its replication and the production of coat proteins. This active stage, culminating in the release of infective viral particles, is called the *lytic* phase of the infective cycle. Among the viruses entering lysogenic phases in which their DNA is integrated into the

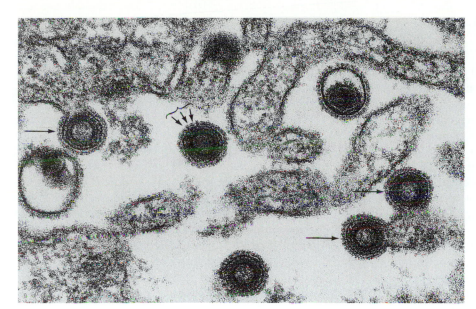

Figure 1-28 Particles of a mouse mammary tumor virus (arrows) acquiring membranous coats during release from a host cell. Viral proteins (bracketed arrows) are closely packed in the membranous coats. × 140,000. (Courtesy of N. H. Sarkar.)

host cell DNA are some of the bacteriophages and the herpesviruses infecting humans.

Origins and Significance of Viruses

Viral particles in free form carry out none of the activities of life and are inert except for the capacity to attach to their host cells. Many viruses in the free form can be purified and crystallized and stored indefinitely without change or damage. Thus a free viral particle is probably best classified as nonliving matter.

Although the origin of viruses is obscure, they may represent fragments of a nucleoid or chromosome derived from a once-living cell or an RNA copy of such a fragment, surrounded by a layer of protein with protective and cell-recognition functions. The information encoded in the core of the virus is reduced to a set of directions for maintenance of the virus in a host cell and production of more viral particles of the same kind.

The scientific, medical, and economic importance of viruses can hardly be overestimated. The molecular revolution in biology has primary roots in experiments carried out with viruses, particularly the bacteriophages. These experiments were among the first to reveal how genes are structured, regulated, and duplicated.

Viruses are responsible for human disabilities ranging from relatively mild irritations such as the common cold, through more serious infections such as influenza, chicken pox, and polio, to deadly diseases such as AIDS and some forms of cancer. Viruses also cause diseases of domestic animals and infect plants in both agricultural and natural environments. With a few notable exceptions, such as polio and influenza, which are controllable by vaccines, most viral diseases are difficult or impossible to treat and must simply be allowed to run their course. Antibiotics, which typically work by interfering with bacterial metabolism, are of little or no use except as a treatment for secondary bacterial infections that frequently accompany viral diseases. Some viruses, such as the herpesviruses responsible for oral and genital sores, produce infections that last the lifetime of the individual.

Although vaccines developed against the viruses causing polio and smallpox have remained highly effective, some infective viruses mutate rapidly to forms for which a previously successful vaccine is ineffective. The influenza virus, for example, regularly appears in new forms that cannot be treated by previously effective vaccines. The fact that the virus responsible for AIDS also mutates rapidly will undoubtedly cause problems for the treatment of this disease by vaccines.

Some viruses that usually cause only moderately incapacitating diseases can also mutate into more deadly forms. The influenza virus, for example, infects hundreds of millions of people annually but normally causes relatively few deaths, primarily among infants and the aged. However, in 1918 the virus mutated into a highly dangerous form that killed between 20 and 30 million people of all ages in Europe and America alone, so many that few families escaped without at least one close relative dying from influenza.

HISTORICAL ORIGINS OF CELL AND MOLECULAR BIOLOGY

Cell and molecular biology developed gradually from the first description of cells in the seventeenth century. During the earliest period, investigations in cell biology consisted almost entirely of morphological descriptions of cell structure, using the primitive light microscopes available at the time. In the eighteenth and nineteenth centuries the study of cell chemistry and physiology began, largely as an effort that proceeded independently from morphological studies. Study of cell structure, cell chemistry, and physiology continued as separate fields of experimentation until the beginning of the twentieth century, when the rapidly developing field of biochemistry began to influence cell biology. At the same time, genetics became established as a new field of study.

The integration of cell biology, genetics, physiology, and biochemistry began in earnest in the 1930s, and research in cell biology started its shift from primarily morphological investigations to biochemical and molecular studies of cell function. In recent years the biochemical and molecular approach has dominated the study of cells.

Throughout the progress of cell and molecular biology run two threads common to all scientific endeavor: Major discoveries follow closely on the heels of the development of new techniques, and as advances appear, the research leads to more questions than answers. At each level of advancement, however, the questions become more penetrating and sophisticated and lead to solutions of more fundamental and far-reaching importance.

The Discovery of Cells

The earliest developments in cell biology were closely tied to the invention and gradual improvement of the light microscope. In 1665, soon after the microscope was invented, the English scientist Robert Hooke published the first descriptions of cells. Hooke reported seeing in the woody tissues of plants small, compartmentlike units that he named "pores" or "cells." It is frequently claimed that Hooke did not actually observe living cells because in some of the tissues he examined, as in cork, the cells were simply empty spaces outlined by residual cell walls. But in other plant tissues, such as the "inner pulp or pith of an Elder, or almost any other tree," Hooke "plainly enough discover'd these cells or Pores fill'd with juices." Thus he actually saw living cells.

Hooke's observations of cells were extended by several investigators toward the end of the 1600s, most notably by Anton van Leeuwenhoek, a Dutch amateur microscopist. Leeuwenhoek made remarkably accurate observations of the microscopic structure of protozoa, blood, sperm cells, and a variety of other "animalcules," as he called them. He reported his findings to the British Royal Society in some 200 letters written over a period of about 50 years.

After Leeuwenhoek's death in the early 1700s, cell biology entered a period of relative quiescence that lasted until well into the nineteenth century. This lag was largely because of imperfections in the lenses of the early light microscopes, which were serious enough to prevent investigators from seeing the inner details of cells. The correction of some of the more serious imperfections in the early nineteenth century led to a burst of new discoveries in cell biology, among them the finding that both plant and animal tissues are composed of cells.

The early observations of Hooke and others, made primarily in plant tissues, gave the impression that cells were units of living matter outlined by conspicuous cell walls. Because the cells of animal tissues lack distinct walls, structures equivalent to plant cells could not at first be seen in animals. As a result, the basic similarities in microscopic structure between plants and animals escaped notice.

The parallels in structure between animal and plant tissues were drawn in 1839, when the German biologist Theodor Schwann observed that animal cartilage contains a microscopic structure that "exactly resembles . . . [the] cellular tissues of plants." Schwann's work was aided by the fact that extracellular material in cartilage occupies a position analogous to the cell wall in plants and clearly outlines the cells. This enabled Schwann to recognize the cellular nature of animal tissue for the first time. Schwann also remarked on the presence of nuclei in the cartilage cells, but he was not as impressed by the cell contents as by the structures he thought were walls.

Continued improvements in light optics soon allowed microscopists to recognize the much thinner boundaries of cells in other types of animal tissues. As the details of structure in the cell interior became discernible, interest gradually shifted from the walls to the contents, and the term *cell* began to take on its modern connotations.

Development of the Cell Theory

In 1833 Robert Brown published a paper in England describing the microscopic structure of the reproductive organs of plants. In his paper, Brown drew attention to the nucleus as a constant feature of plant cells. Brown's study established that nucleated cells are the units of living tissue in plants and laid the foundations for the concept that cells with nuclei are the fundamental units of all living organisms. This idea is part of what is now known as the *cell theory*.

The cell theory was developed in part by Theodor Schwann and an eminent German botanist, Matthias Schleiden. During his work with cartilage in the 1830s, Schwann's attention was drawn to the cell nucleus by Schleiden, who had developed a series of hypotheses emphasizing the importance of the nucleus in cell reproduction. Although Schleiden's ideas about cell reproduction were erroneous, his preoccupation with the cell nucleus led Schwann to recognize the universality of nucleated cells as the structural units of living matter. This work led to two of the three postulates of the cell theory: that all living organisms are composed of one or more nucleated cells, and that cells are the minimum functional units of living organisms. Historians usually attribute these two hypotheses jointly to Schwann and Schleiden.

Their conclusions were soon supplemented by a third postulate that completed the cell theory. This idea was developed by scientists investigating cell origins, who observed that cells arise in both plants and animals by the division of a parent cell into two daughter cells. By 1855 this work had progressed far enough for Rudolf Virchow, a German pathologist, to affirm that all cells arise only from preexisting cells by a process of division. Virchow's famous statement of this concept, "*Omnis cellula e cellula*," completed the cell theory:

1. All living organisms are composed of nucleated cells.

2. Cells are the functional units of life.

3. Cells arise only from preexisting cells by a process of division.

Further work established that the nucleus is the repository of hereditary information and that the essential feature of cell division is transmission of hereditary information from parent to daughter cells. The physical continuity of the nuclear material through cycles of division was confirmed by the German scientists Eduard Strasburger and Walther Flemming, who discovered that the chromatin is transformed into compact rodlets, the chromosomes, during cell division. When fully formed, chromosomes are clearly seen to be double. As division progresses, the two halves of each chromosome are separated and placed in separate daughter cells, where they form the daughter nuclei. Later work by Flemming and others showed that the number of chromosomes remains constant for all members of a species.

With the discovery that chromosomes are duplicated and passed on in constant numbers during cell division, enthusiasm grew for the hypothesis that cell heredity is probably controlled by the cell nucleus. In 1884 Strasburger declared that the physical basis of heredity resides in the nucleus; in 1885 the German zoologist August Weismann concluded that "the complex mechanism for cell division exists practically for the sole purpose of dividing the chromatin, and . . . thus the [chromatin] is without doubt the most important part of the nucleus."

Some years before these developments took place, an Austrian monk, Gregor Mendel, had studied the inheritance of traits such as seed shape and color in garden peas. Through his analysis of the distribution of these traits in parents and offspring, Mendel discovered genes and their patterns of inheritance. Among other observations, Mendel noted that genes occur in pairs in an individual, that one member of each pair is derived from the male parent and the other from the female parent of the individual, and that the pairs separate independently in the formation of gametes and are reunited in fertilization.

The significance of Mendel's work was not appreciated at the time of its publication in 1865. One reason for its lack of impact was that the biologists of the time were unaccustomed to thinking about cellular processes in the abstract, mathematical terms used by Mendel. Another was that investigations had not progressed far enough to allow Mendel's mathematical conclusions to be related to physical units in the cell. Mendel's conclusions also ran counter to the beliefs of many prominent biologists of the time, who held that inheritance depended on an averaging or mixing of parental traits rather than being determined by discrete hereditary units. As a result, Mendel's findings lay unnoticed until just after the turn of the century, when the same conclusions were reached independently by a Dutch plant physiologist, Hugo de Vries, and the German and Austrian botanists Carl Correns and Erich von Tschermak. These investigators, on searching the earlier literature, were surprised to find that they had been scooped by an obscure monk some 40 years earlier!

By the time Mendel's results were rediscovered, the behavior of chromosomes in cell division was sufficiently well known for the correlation between Mendel's genes and chromosomes to become apparent. In 1903 the American cell biologists W. A. Cannon, Edmund Wilson, and Walter Sutton pointed out the precise equivalence between the patterns of inheritance of genes and chromosomes in organisms that reproduce sexually, that is, by the union of eggs and sperm. Both genes and chromosomes occur in pairs in which one member of the pair is inherited from the male and the other from the female parent of an individual. Both gene and chromosome pairs are separated in the formation of gametes and reunited in fertilization.

In total these findings and conclusions established the concept that heredity is controlled by discrete physical units, the genes; that genes are carried on the chromosomes; and therefore that the nucleus and its contents store and transmit hereditary information. Research with genes and their patterns of transmission expanded greatly during the early decades of this century, particularly in the laboratory of Thomas Hunt Morgan at Columbia University. Morgan's work with the fruit fly, *Drosophila melanogaster*, led directly to an understanding of the linear order of genes on chromosomes and how genes are mixed into new combinations during the formation of gametes in sexually reproducing organisms.

Early Chemical Investigations

The application of physics and chemistry to cell biology began in 1772 when the English chemist Joseph Priestley discovered that green plants release oxygen when they are exposed to light. At almost the same time, Antoine Lavoisier, a Frenchman who is considered the father of modern chemistry, recognized that "respiration is a . . . combustion, slow, it is true, but otherwise perfectly similar to that of charcoal." The fundamental importance of these discoveries was not appreciated at the time because it was still widely believed that the substances and processes in living organisms were different from those of the inorganic world. Therefore the chemical and physical techniques used to study inanimate objects were not believed to be applicable to life.

Significant movement from these attitudes came in 1828, when the German chemist Friedrich Wöhler achieved the first artificial synthesis of organic molecules from inorganic precursors. Wöhler converted the inorganic chemical ammonium cyanate to urea, an organic substance commonly excreted by animals. Wöhler also synthesized oxalic acid, an organic chemical found in plant tissues. Many additional organic molecules were subsequently synthesized by others, and it gradually became clear that the same chemical elements occur in both living and inanimate objects and are governed by the same chemical and physical laws.

By the end of the nineteenth century, investigators had isolated, identified, and synthesized many organic substances found in plants and animals. Most successful in this work was the German organic chemist Emil Fischer, who extracted, degraded, and resynthesized many substances from living organisms and laid the foundation for the chemical description of amino acids, proteins, fats, and sugars. In 1902

Fischer and another German chemist, F. Hofmeister, independently described the structure and formation of the peptide bond, the chemical linkage that ties amino acids together in proteins. The structure of nucleotides, the chemical building blocks of the nucleic acids, was also worked out by the turn of the century.

This chemical work was complemented in the nineteenth century by the first functional biochemical studies. In the 1850s the French scientist Louis Pasteur began his efforts to determine the cause of fermentation. Pasteur found that fermentation occurs in nature only if particular living microorganisms are present— if the microorganisms are eliminated or killed, sugar is not converted into alcohol. Pasteur's work also confirmed that microorganisms and cells in general can arise only from preexisting cells and eliminated the possibility of spontaneous generation—the production of living cells from nonliving matter.

The catalytic nature of fermentation was discovered through the research of Eduard Buchner in Germany. In 1897 Buchner was attempting to isolate and preserve yeast extracts for medicinal purposes. To ground and pressed yeast cells, Buchner added a sugar solution as a preservative. To his surprise, the sugar was rapidly fermented by the cell-free extract. Buchner subsequently demonstrated, in yeast extracts from which all intact cells were carefully removed, that the fermentation was catalyzed by protein-based catalysts. The protein-based catalysts were given the name *enzymes*, coined from a Greek word that means "in yeast."

This work culminated in general acceptance of the conclusion that living organisms contain the same elements as nonliving matter and are motivated by no vital forces other than enzymes and the reactions they catalyze. Living systems could now be studied by chemical and physical techniques with the confidence that their activities follow the same chemical and physical laws as nonliving systems.

Integration of Chemical and Morphological Studies

Significantly, investigations into the chemical nature of the nucleus were among the first efforts to integrate chemical and morphological approaches to the study of cellular life. This work stemmed directly from the growing realization in the latter half of the nineteenth century that the nucleus is of central importance in cell function and heredity.

Johann Friedrich Miescher, a Swiss physician and physiological chemist, became interested in the chemical composition of cell nuclei and developed a method for isolating nuclei in quantity for analysis. From his preparations he isolated a previously unknown substance with properties then considered unusual for organic matter, including high phosphorus content and a strongly acidic reaction. Announcing his discovery in 1871, Miescher called the new substance *nuclein*. Soon afterward, Flemming concluded that if the chromatin of the nucleus and nuclein were not one and the same substance, "one carries the other." Nuclein was later called a nucleic acid.

A method for purifying nucleic acids was worked out by R. Altman in 1889, and their chemical subunits, the nucleotides, were identified. One type of nucleic acid, DNA, was subsequently established to be characteristic of all cell nuclei. Work in the 1920s and 1930s confirmed that DNA is located in the chromosomes, and many cell biologists began to suspect direct involvement of this substance in heredity. Finally, in the 1940s and 1950s, a series of experiments with bacteria by Oswald Avery and his colleagues, and with viruses by Alfred D. Hershey and Martha Chase, confirmed that DNA is the hereditary molecule.

Avery and his coworkers Colin MacLeod and MacLyn McCarty at the Rockefeller Institute worked with two forms of a bacterium, *Streptococcus pneumoniae*, that causes pneumonia in humans and mice. Cells of a virulent, infective form of the bacterium are surrounded by capsules and produce smooth, gel-like colonies when grown in culture dishes. Cells of a nonvirulent form have no capsules and form colonies that appear lusterless or rough. It had been found previously that killed cells of the virulent form could transform uncapsulated, nonvirulent cells into fully infective cells with capsules. The change was permanent and was passed on to descendants of the transformed cells.

Avery and his colleagues were interested in identifying the agent in killed virulent cells that could transform nonvirulent cells to the virulent form. They treated extracts of killed cells with enzymes that catalyze the breakdown of DNA, RNA, or proteins and exposed nonvirulent cells to the treated extracts (Fig. 1-29). Only the enzyme breaking down DNA destroyed the capacity of the extract to transform nonvirulent cells into the infective form. From these findings, Avery, MacLeod, and McCarty proposed in 1944 that DNA is the substance responsible for transforming the noninfective cells. Because DNA could carry genetic information in the experimental situation, it was considered likely to be the normal carrier of genetic information in the cell nucleus.

Their conclusion was directly supported and extended by the experiments of Hershey and Chase in 1952 with bacterial viruses at the Carnegie Laboratory

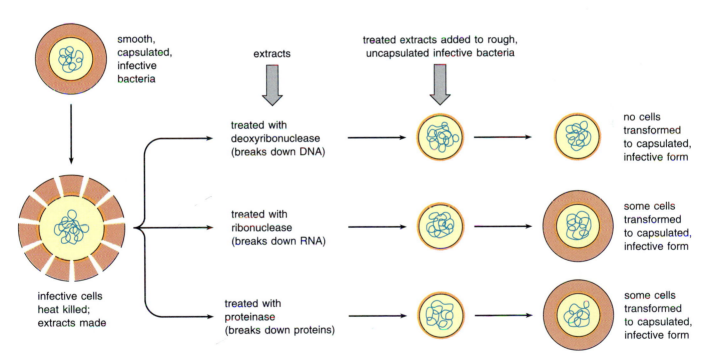

Figure 1-29 The experiment by Avery, MacLeod, and McCarty demonstrating that DNA is able to transform nonvirulent *Streptococcus pneumoniae* into the virulent, infective form. Heat-killed extracts of the infective form were treated with either deoxyribonuclease, ribonuclease, or a proteinase. Only deoxyribonuclease, which breaks down DNA, destroyed the ability of the extracts to transform nonvirulent cells into the virulent form. On this basis, Avery and his colleagues concluded that the transforming agent was DNA.

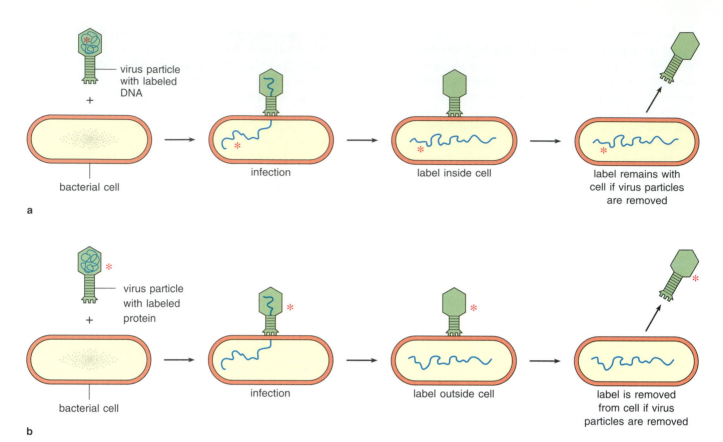

Figure 1-30 The Hershey and Chase experiment. Viral particles were labeled with radioactive phosphorus or sulfur by growing infected cells in the presence of compounds containing either of these substances. The radioactive phosphorus labeled the DNA, but not the protein of the particles; the sulfur labeled the protein, but not the DNA. **(a)** When viral particles containing labeled DNA were used to infect bacteria, the radioactivity entered the cells. **(b)** When particles containing labeled protein were used for infection, the radioactivity remained outside the cells and could be removed by shaking. The experiment demonstrated that the DNA enters the cell and delivers the genetic information necessary to convert the cell to production of new viral particles.

of Genetics. It was known that the viruses, which consist only of a DNA core surrounded by a protein coat, cause bacterial cells to cease production of their own molecules and to make instead the DNA and protein of new viral particles of the same kind as the infecting virus. It was assumed that during the infection of a bacterium, either the DNA or the protein of the virus enters the bacterial cell to alter its genetic and synthetic activity.

Hershey and Chase used radioactive isotopes of phosphorus and sulfur to label differently the protein or DNA of the virus. This was possible because the protein of the virus contains sulfur but no phosphorus, and DNA contains phosphorus but no sulfur. The labeled viruses were used to infect bacterial cells. Radioactivity was found inside the infected cells only if the virus contained labeled DNA (Fig. 1-30a). If the proteins of the infecting virus were labeled, the radio-

activity remained outside the cell and could be dislodged by shaking (Fig. 1-30b). The experiments showed that only the DNA of the virus entered the bacterial cells and therefore that it must contain the genetic information that converts the cells to production of viral DNA and protein. The Hershey and Chase and Avery experiments, taken together, established that DNA is the hereditary molecule.

Insights into the type of genetic information encoded in DNA were obtained during the same period by George Beadle and Edward Tatum. Previous work indicated that genes control individual steps in biochemical reactions. Beadle and Tatum set out to identify the basis for this control by investigating mutants in the bread mold *Neurospora crassa*. They found that mutations in individual genes frequently caused the mold to grow poorly. Growth could be restored to normal levels by the addition of single substances, such as individual amino acids, to the growth medium. Beadle and Tatum concluded that a mutant gene encodes a faulty form of an enzyme necessary to produce a substance needed for normal growth. On this basis they proposed that each gene identified in their study codes for a single enzyme. This proposal became famous as the "one gene–one enzyme" hypothesis.

Research with cytoplasmic organelles led to similar integrations of morphology and biochemistry, beginning just before the turn of the century with investigations into the functions of chloroplasts and mitochondria. Work in this area proceeded slowly until the 1930s, when Albert Claude developed a technique for isolating and purifying cell parts by cell fractionation and centrifugation. In centrifugation, centrifugal force developed in tubes held in a spinning rotor is used to separate structures or molecules into distinct groups according to size or density (for details, see p. 795). Claude and his associates George Palade, Keith Porter, and Christian de Duve were able to separate ribosomes, mitochondria, lysosomes, peroxisomes, and the Golgi complex into distinct fractions by this method, which allowed biochemical analysis of the fractions. By the 1950s it had even become possible to break mitochondria and chloroplasts into separate outer membrane, inner membrane, and matrix or stroma fractions.

Application of the electron microscope to biological materials, which began in the late 1940s, greatly extended the visibility of details in cells. Membranes and their arrangement could be directly observed, and features as small as ribosomes and even individual protein molecules could be seen in the microscope. Palade and Porter pioneered application of the electron microscope to cell biology and worked out the ultrastructure of many of the organelles studied biochemically in cell fractions.

As this work progressed, research with cell extracts, particularly from *E. coli*, yeast, and green algae, gradually identified the molecules and enzymes participating in individual steps of cellular oxidation and photosynthesis. Fractionation of eukaryotic cells by centrifugation soon allowed these reaction sequences to be placed in their correct locations in mitochondria and chloroplasts or in the cytoplasm surrounding these organelles. Similar work with ribosomes led to development of cell-free systems able to carry out all of the reactions of protein synthesis.

In more recent years, centrifugation has been supplemented by gel electrophoresis, which uses an electrical field instead of centrifugal force to drive molecules through highly viscous plastic gels (for details, see p. 796). Even small molecules are separated readily and rapidly according to size, density, and

Figure 1-31   James D. Watson (left) and Francis H. C. Crick, demonstrating their model for DNA structure deduced from X-ray diffraction data obtained by Maurice Wilkins and Rosalind Franklin. (Courtesy of Cold Spring Harbor Laboratory Archives.)

charge by the technique. Application of gel electrophoresis has allowed many thousands of protein, nucleic acid, and other cellular molecules to be individually purified and biochemically characterized.

The Molecular Revolution

The experiments of Avery and Hershey and Chase sparked an intensive effort to work out the structure of DNA. Although its chemical components had been identified, the three-dimensional structure of DNA was still unknown when the molecule was established as the hereditary material. An intense competition among scientists culminated in the discovery of DNA structure at Cambridge University in 1953 by an American, James D. Watson, and an Englishman, Francis H. C. Crick. Their model for DNA structure (Fig. 1-31) was based primarily on data obtained by X-ray diffraction (see p. 802) of DNA samples by Maurice Wilkins and Rosalind Franklin.

The DNA structure worked out by Watson and Crick, perhaps the most significant single accomplishment in the history of biology, sparked an effort to determine the molecular structure of genes and their modes of action. This effort was greatly facilitated by rapid methods for nucleic acid sequencing developed by F. Sanger, A. M. Maxam, and W. Gilbert (see p. 423).

Using these methods, many genes, and in some organisms the entire DNA complement, have been completely sequenced. Through the gene sequences the amino acid sequences of many proteins have been deduced. The comparisons of gene and protein structure in normal and mutant forms made possible by these accomplishments have provided fundamental insights into the molecular functions of genes and proteins and how genes are regulated and controlled. The aims of this approach have been furthered by the development of genetic engineering, which allows the genes of living organisms to be altered in specific, predetermined ways. Experimentally induced mutations have revealed the functions not only of specific regions of genes and their controls but also of proteins encoded in the genes.

The molecular approach produced an explosion in our knowledge of cell structure and function and a revolution in the methods and attainments of almost all the related fields of biology: genetics, biochemistry, developmental biology, physiology, medicine, and even taxonomy, systematics, evolution, and behavior. The importance of the molecular approach has made an understanding of cell and molecular biology a prerequisite for basic comprehension and advanced study in any field of biology. It has also created a new responsibility for nonscientist citizens and their governments, who must now understand and cope with the achievements and conclusions of molecular biology to vote and govern wisely and effectively.

The contemporary field of cell biology has thus developed through an integration of structural and biochemical studies, recently revolutionized by molecular studies of gene structure and function and of the primary encoded products of the genes—the proteins. The unity of cellular structure and function revealed by the molecular revolution and its accomplishments are the subjects of this book.

Suggestions for Further Reading

Allen, M. M. 1984. Cyanobacterial cell inclusions. *Ann. Rev. Microbiol.* 38:1–25.

Bracegirdle, B. 1989. Microscopy and comprehension: The development of understanding of the nature of the cell. *Trends Biochem. Sci.* 14:464–468.

Brock, T. D. 1988. The bacterial nucleus: A history. *Microbiol. Rev.* 52:397–411.

Claude, A. 1975. The coming of age of the cell. *Science* 189:433–435.

De Duve, C. 1975. Exploring cells with a centrifuge. *Science* 189:186–194.

De Duve, C., and Beaufay, H. 1981. A short history of tissue fractionation. *J. Cell Biol.* 91:293s–299s.

Fruton, J. S. 1976. The emergence of biochemistry. *Science* 192:327–334.

Hughes, A. 1959. *A History of Cytology.* New York: Abelard-Schuman.

May, Y., Pattison, B. W. J., and Pattison, J. R., eds. 1984. *The Microbe, 1984, Part 1: Viruses* (Symposium of the Society for General Microbiology).

Quastel, J. H. 1984. The development of biochemistry in the 20th century. *Canad. J. Cell Biol.* 62:1103–1110.

Schenk, F. 1988. Early nucleic acid chemistry. *Trends Biochem. Sci.* 13:67–69.

Watson, J. D. 1968. *The Double Helix.* New York: Atheneum.

Review Questions

1. Define and outline the overall functions of the nuclear region and cytoplasm.

2. What are the major differences between eukaryotic and prokaryotic cells? What organisms fall into the two divisions?

3. What cytoplasmic structures occur in prokaryotic cells?

4. Compare the structures of prokaryotic nucleoids and eukaryotic nuclei.

5. What major differences exist in the DNA of eukaryotic and prokaryotic cells?

6. What is an organelle?

7. Define transcription and translation.

8. Outline the major roles of mRNA, rRNA, and tRNA in protein synthesis.

9. Outline the structure and primary functions of mitochondria, the endoplasmic reticulum, the Golgi complex, lysosomes, and microbodies.

10. Contrast microtubules and microfilaments, and outline their major cellular roles.

11. What are intermediate filaments?

12. What cellular and extracellular structures occur in plant but not animal cells? Briefly outline the roles of these structures.

13. Outline the structure of a rod-shaped and spherical virus particle.

14. How do coated viruses obtain their surface membrane?

15. Describe the pattern of infection followed by bacterial viruses. What does this pattern suggest about the living or nonliving nature of viruses?

16. Contrast the lytic and lysogenic phases of viral infective cycles.

17. List the major hypotheses forming parts of the cell theory. Why are the first and second hypotheses listed as separate ideas?

18. What was the significance of Wöhler's synthesis of an organic molecule to the development of the chemical approach to cell biology?

19. Why was Mendel's discovery of genes largely unappreciated by biological scientists when he published his results?

20. What parallels between the behavior of genes and chromosomes convinced investigators that genes are carried on the chromosomes?

21. What experiments established that DNA is the hereditary molecule? Outline the steps carried out in these experiments.

22. What major techniques and discoveries led to the molecular revolution in biology?

2

CHEMICAL BONDS AND BIOLOGICAL MOLECULES

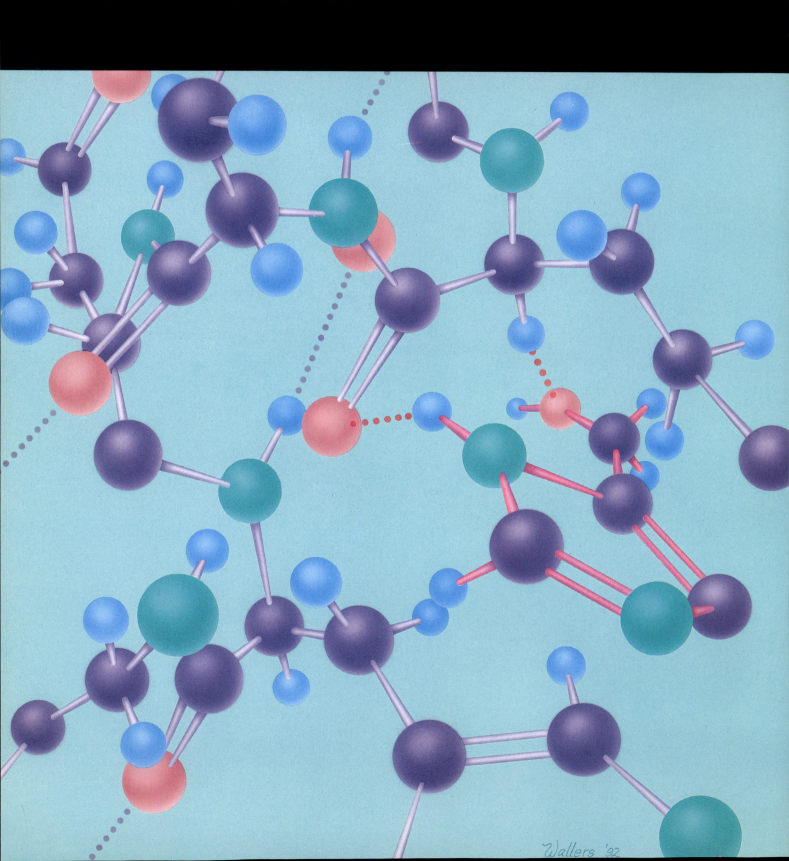

Wallers '92

- *Ionic bonds* ▪ *Covalent bonds* ▪ *Polarity*
- *Hydrogen bonds* ▪ *van der Waals forces*
- *Functional groups* ▪ *Carbohydrates* ▪ *Lipids*
- *Glycolipids* ▪ *Proteins* ▪ *Glycoproteins*
- *Lipoproteins* ▪ *Nucleotides* ▪ *Nucleic acids*

Cell biology has been transformed into a science that is primarily biochemical and molecular. Molecular biology is by nature chemical and physical in its approach. As a result, it is impossible to understand the conclusions of cell and molecular biology without an introduction to the forces holding atoms together in molecules, molecular structure, chemical reactions, and the major types of biological molecules carrying out the activities of life.

Although many molecules of living systems are highly complex, most fall into one of four classes—*carbohydrates*, *lipids*, *proteins*, and *nucleic acids*—or contain subunits belonging to one or more of these classes. Along with water, these four classes form almost all the matter of living organisms. Molecules of these substances are held together and interact through chemical bonds and forces of several different types: *ionic*, *covalent*, and *hydrogen bonds*; *nonpolar associations*; and *van der Waals forces*.

For those already familiar with chemical bonds and the major classes of biological molecules, this chapter will serve for reference and review. For others the chapter provides a foundation for the discussions of molecular structures, functions, and interactions in this book.

CHEMICAL BONDS IMPORTANT IN BIOLOGICAL MOLECULES

Molecules are formed from atoms linked together in definite numbers and ratios by chemical bonds. The chemical properties of an atom are determined largely by the number of electrons in its outermost "shell" or energy level. Most significant is the difference between the number in the outermost level and the stable numbers of either two electrons for atoms with only one energy level (hydrogen and helium) or eight electrons for larger atoms. Helium, with two electrons in its single level, and atoms such as neon and argon, with eight electrons in their outer levels, are stable and essentially inert chemically. Atoms with outer levels containing electrons near these numbers tend to gain or lose electrons to approximate these stable configurations. For example, sodium has two electrons in its first energy level, eight in the second, and

one in the third and outermost level. This atom readily loses its single outer electron to leave a stable second energy level with eight electrons. Chlorine, with seven electrons in its outermost energy level, tends to attract an electron from another atom to attain the stable number of eight. Atoms differing from the stable configuration by more than one electron are inclined to *share* electrons with other atoms rather than to gain or lose electrons completely. Among the atoms forming biological molecules, electron sharing is most characteristic of carbon, which has four electrons in its outer energy level and thus falls at the midpoint between the tendency to gain or lose electrons. Oxygen, with six electrons in its outer level, nitrogen, with five electrons in its outer level, and hydrogen, with one electron in its single energy level, also share electrons readily. The relative tendency to gain, share, or lose electrons underlies the chemical bonds and forces holding the atoms of molecules together.

Ionic Bonds

Ionic bonds (also called *electrostatic bonds*) form between atoms that tend to gain or to lose electrons completely. Under the correct conditions, for example, a sodium atom (Fig. 2-1) readily loses the single electron in its outermost energy level to a chlorine atom:

$$Na^{.} + .\overset{..}{\underset{..}{Cl}}: \qquad Na^{+} :\overset{..}{\underset{..}{Cl}}:^{-}$$

(The dots in the diagram represent the electrons in the outermost energy level.) After the transfer the sodium atom, now with 11 protons in its nucleus and 10 electrons in surrounding orbitals, carries a single positive charge. The chlorine atom, now 17 protons and 18 electrons, including a complete outer energy level with 8 electrons, carries a single negative charge. In this charged condition the sodium and chlorine atoms are called *ions* and are identified as Na^+ and Cl^-. The difference in charge sets up an attraction—the ionic or electrostatic bond—that holds the ions together in solid NaCl.

Many other atoms that differ from stable outer energy levels by one electron, including hydrogen, can gain or lose electrons completely to form ions and ionic bonds. A number of atoms with outer energy levels differing from the stable number of eight by two or three electrons, particularly metallic atoms such as calcium, magnesium, and iron, also lose their electrons readily to form positively charged ions and ionic bonds.

Ionic bonds have several features with great significance for their activities in biological and other molecules. They exert an attractive force over distances greater than any other chemical bond. The attractive force extends in all directions, so that ionic

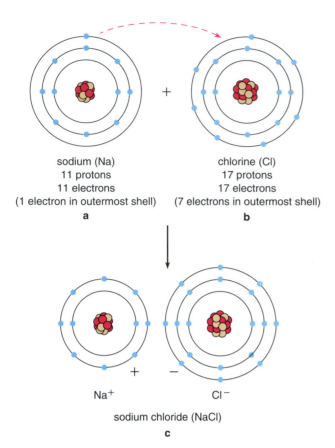

sodium (Na)
11 protons
11 electrons
(1 electron in outermost shell)
a

chlorine (Cl)
17 protons
17 electrons
(7 electrons in outermost shell)
b

Na⁺

Cl⁻

sodium chloride (NaCl)
c

Figure 2-1 Formation of an ionic or electrostatic bond. **(a)** Sodium, with one electron in its outermost energy level, readily loses an electron to attain a stable state in which its second energy level, with eight electrons, becomes the outer level. **(b)** Chlorine, with seven electrons in its outer energy level, readily gains an electron to attain the stable number of eight. **(c)** Transfer of an electron from sodium to chlorine creates stable outer energy levels in both atoms and creates the ions Na⁺ and Cl⁻. The attractive force holding the oppositely charged ions together is the ionic or electrostatic bond.

bonds are not limited to discrete angles as are, for example, covalent bonds. In solids or crystals containing no charged particles other than ions, the strength of ionic bonds is very high. The strength of a chemical bond can be measured as the energy required to break it; ionic bonds between ions in solids typically require about 80 kilocalories of energy per mole (kcal/mol)[1] for breakage (Table 2-1).

[1]A *mole* of a substance is defined as 6.022×10^{23} molecules (Avogadro's number) of that substance. A *gram molecular weight* is the weight in grams of one mole of a substance. Molecular weights are often given in *daltons*; one dalton is approximately equal to the weight of 6.022×10^{23} hydrogen atoms and exactly equal to 1.000 on the scale of atomic weights. A *calorie* is the amount of heat required to raise 1 gram of water 1°C (from 14.5° to 15.5°C at atmospheric pressure). The *Calorie* familiar to dieters, written with a capital *C*, is equal to 1 kcal or 1000 calories.

Table 2-1 Energy Required to Break Chemical Bonds or Attractions

Bond or Attraction	Approximate Energy (kcal/mol)
Ionic (in solids)	80
Ionic (in water)	1–3
Covalent	
O—H	110
C—H	100
N—H	93
C—C	83
S—H	80
S—S	50
Hydrogen	2–5
van der Waals (at optimum)	1

The presence of other charged particles reduces the strength of ionic attractions. When other charged particles are present, they are attracted by opposite charges and form a neutralizing coat surrounding the ions. This coat weakens the attraction between ions, making them easier to separate. Water molecules are particularly effective in this role because they have relatively positive and negative ends (see below). When surrounded by a surface layer of water molecules, only 1 to 3 kcal/mol are required to break ionic bonds. As a consequence, ions are easily separated in water to enter solution as *free ions*. (If negatively charged, free ions are identified as *anions*; if positively charged, as *cations*.)

Ionic bonds are common among the forces holding ions, atoms, and molecules together in biological systems. Many systems take advantage of the fact that ionic bonds vary in strength depending on the presence or absence of water or other polar or charged substances. In some, ionic bonds form in locations that exclude water or other charged particles, setting up strong and stable attractions that are not easily disturbed. For example, metallic ions are held stably by strong ionic bonds in interior regions of some enzymes and other proteins such as chlorophyll and hemoglobin. In other systems, particularly at molecular surfaces exposed to water molecules, ionic bonds are relatively weak, allowing ionic attractions to be set up quickly or broken. For example, many enzymes bind and release molecules by forming and breaking relatively weak ionic bonds as part of their activity in speeding biological reactions.

Covalent Bonds

Covalent bonds form when atoms share electrons rather than giving them up completely. The formation

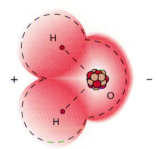

Figure 2-2 Polarity in the water molecule, created by unequal electron sharing in the covalent bonds between the two hydrogens and oxygen. Regions of deepest red indicate the most frequent locations of the shared electrons. At any instant the shared electrons are likely to be closer to the oxygen nucleus, making this end of the molecule relatively negative and the hydrogen end relatively positive. The orbitals occupied by the electrons are more complex than the spherical forms shown here.

of molecular hydrogen, H_2, by two hydrogen atoms is the simplest example of the sharing mechanism. If two hydrogen atoms approach closely enough, the single electron of each atom may join in a new, combined two-electron orbital that surrounds both nuclei. The new orbital fills the energy level, so the hydrogen atoms tend to remain linked together. The linking force set up by the shared electrons is the covalent bond.

The energy required to break covalent bonds varies considerably, from about 50 to 110 kcal/mol, depending on the atoms sharing electrons and their positions in a molecule (see Table 2-1). However, at even the smallest values, covalent bonds are relatively stable. In molecular diagrams a covalent bond is designated by a pair of dots or a single line representing the shared electrons:

$$H. + \cdot H \longrightarrow H \!:\! H \text{ or } H\!-\!H$$

Carbon, with four unpaired outer electrons, forms four separate covalent bonds to complete its outermost shell. The formation of methane (CH_4) by electron sharing between one carbon and four hydrogen atoms illustrates this process:

$$H\cdot + \cdot \overset{\displaystyle \cdot}{\underset{\displaystyle \cdot}{C}} \cdot + \cdot H \longrightarrow H \!:\! \overset{\displaystyle \cdot\cdot}{\underset{\displaystyle \cdot\cdot}{C}} \!:\! H \quad \text{or} \quad H\!-\!\overset{\displaystyle H}{\underset{\displaystyle H}{C}}\!-\!H$$

Because carbon forms four separate covalent bonds, carbon atoms link readily into chains or branched structures. Such structures form the backbones of an almost unlimited variety of molecules. Figures 2-6

and 2-16 show some examples of molecules with carbon-chain backbones. (Double bonds in the figures indicate that the atoms in the linkage share two pairs of electrons.) The other atoms in these molecules—oxygen, hydrogen, nitrogen, and sulfur—also share electrons readily to form covalent linkages and are commonly found with carbon in biological molecules. In these linkages, oxygen typically forms two covalent bonds, hydrogen one, nitrogen three, and sulfur two.

Unequal Electron Sharing and Polarity   Electrons are not always shared equally in a covalent bond. In traveling their orbitals the electrons may pass near one of the two atomic nuclei more frequently. The atom retaining the electrons a greater part of the time tends to carry a relatively negative charge, and the atom deprived of electrons becomes relatively positive. Depending on their positions, the positive and negative atoms may give an entire molecule relatively plus and minus ends. A molecule of this type, with relatively positive and negative ends resulting from unequal electron sharing, is said to be *polar*. A molecule without relatively plus and minus ends is *nonpolar*.

Water is the primary biological example of a polar molecule. The shared electrons in a water molecule are more likely at any instant to be nearest the oxygen nucleus, giving the oxygen a relatively negative charge and the hydrogens a relatively positive charge. The asymmetrical location of the hydrogen atoms at one side of the oxygen atom gives the entire water molecule plus and minus ends and makes it strongly polar (Fig. 2-2).

Oxygen, nitrogen, and sulfur also share electrons unequally when linked with hydrogen. Therefore, the presence of —OH, —NH, or —SH groups tends to make regions containing them polar. Electrons in the double bond between carbon and oxygen in the C=O (carbonyl) group are also shared unequally to some extent and may make regions of a molecule containing this group slightly polar. Carbon and hydrogen share electrons equally, so regions containing only C—H linkages are typically nonpolar.

Polar and Nonpolar Associations   Polar molecules tend to attract and align themselves with other polar or charged ions and molecules. The polar environments created by this association tend to exclude nonpolar substances. The exclusion of nonpolar molecules by water is believed to depend on a lattice of attractions between water molecules (the attractions are hydrogen bonds; see Fig. 2-4 and below). The lattice is stable enough to resist disruption by other molecules unless the invading molecule also contains polar or charged regions that can form competing attractions with water molecules. The competing attractions open the water lattice, creating a cavity into which the polar molecule can move. At temperatures at which water is liquid, kinetic movements are easily sufficient to move polar molecules into the water lattice.

Nonpolar molecules, in contrast, have no groups with attractions able to disturb the water lattice. As a result, nonpolar molecules are unable to enter the lattice and are excluded. When present in quantity, the excluded nonpolar molecules tend to clump together in arrangements that reduce the surface area exposed to water. This clumping tendency of nonpolar molecules in the presence of water is termed a *nonpolar association*.

The exclusion of nonpolar molecules by water can be illustrated easily by shaking a vessel containing water and a nonpolar substance such as cooking oil. No matter how vigorously the vessel is shaken, the oil will not dissolve in the water; when suspended in the water, droplets of the oil assume a spherical form that exposes the least surface area to the polar surroundings. If the vessel is placed at rest, the oil and water quickly separate into two layers representing distinct polar and nonpolar environments. Because of these reactions, water and other polar substances are identified as *hydrophilic* (from *hydro* = water and *philic* = preferring) and nonpolar substances as *hydrophobic* (from *phobic* = avoiding).

Polar and nonpolar molecules and environments are critical to the organization of cells. Cellular membranes are maintained as stable boundaries primarily by polar and nonpolar associations. The watery medium on either side of cellular membranes is strongly polar, while the lipid molecules making up cellular membranes have regions that are completely nonpolar. Exclusion by water molecules holds the nonpolar segments of the membrane molecules in a stable, thin film and creates a barrier that effectively prevents polar substances from passing through the nonpolar membrane interior.

The tendency of polar and nonpolar groups to associate in separate regions also contributes to the three-dimensional folding patterns taken on by biological molecules. In proteins, for example, many of the amino acids contain polar or nonpolar regions that form mutually exclusive associations. In a water solution these associations induce a folding pattern in which the nonpolar amino acids are pushed to the interior of the molecule, and the polar ones are held at the surface. Proteins suspended in a nonpolar environment, as in the interior of a membrane, fold in an opposite pattern, with the nonpolar amino acids at the surface and polar groups inside.

The strong polarity and small size of water molecules give water one of its most significant and fundamental properties, its ability to act as a *solvent* for polar and charged substances. Water molecules readily penetrate and coat the surfaces of other polar molecules and ions, producing a surface *hydration layer* that reduces the attraction between the molecules or ions in a solid and promotes their separation and entry into solution. Once the polar molecules or ions are in solution, the hydration layer prevents them from reassociating.

Unequal Electron Sharing, Hydrogen Bonds, and the Properties of Water

Hydrogen atoms, made relatively positive by unequal electron sharing with oxygen, nitrogen, or sulfur, are attracted to other atoms made relatively negative by unequal electron sharing (Fig. 2-3). This attractive force, highly important in biological molecules and their interactions, is the *hydrogen bond* (shown by a dotted line in Fig. 2-3). Hydrogen bonds may be set up between a hydrogen atom and relatively negative atoms—most commonly oxygen, nitrogen or sulfur—in different parts of the same molecule or between different molecules.

Only about 2 to 5 kcal/mol are required to break a hydrogen bond, making this attractive force relatively weak (see Table 2-1). Although individually weak, opportunities for hydrogen bonding are frequently so abundant that many such links may form in or between molecules. When numerous, hydrogen bonds are collectively strong and lend stability to the three-dimensional structure of complex molecules such as nucleic acids and proteins. The many hydrogen bonds stabilizing protein structures are shown as dotted lines in Figures 2-21 and 2-22.

The relatively weak attractive force of hydrogen bonds makes them much easier to break than covalent bonds, particularly at elevated temperatures. Hydrogen bonds separate readily in most biological molecules at temperatures above 50° to 60°C and become practically nonexistent at 100°C. Disruption of hydrogen bonds at elevated temperatures is the primary reason why proteins such as enzymes are inactivated by heat.

Water molecules readily link together via hydrogen bonds. The asymmetric structure of a water molecule places the oxygen and hydrogens in positions that allow formation of multiple hydrogen bonds with neighboring water molecules (Fig. 2-4). In liquid water each water molecule averages about three hydrogen bonds with its neighbors. The bonds in liquid water are not static; they form, break, and reform constantly at rates of billions of times per molecule each second. Despite this constant breaking and reforming, the average hydrogen-bonding level of three for each molecule imparts an unusually stable structure to liquid water.

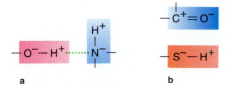

a   b

Figure 2-3   Hydrogen bonds. **(a)** A hydrogen bond (dotted line) between the hydrogen of an —OH group and a nearby nitrogen nucleus that also shares electrons unequally with another hydrogen. Hydrogen bonds are strongest when all three atoms involved in the attraction are arranged in a straight line as shown. **(b)** Other chemical groups commonly forming hydrogen bonds in biological molecules. Regions of deepest color indicate the most likely locations of electrons.

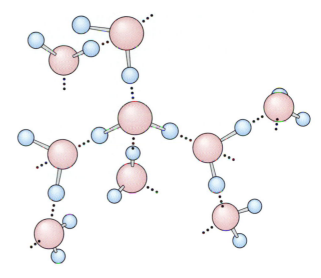

Figure 2-4   Hydrogen bonds (dotted lines) forming a lattice of attractions between water molecules. The bonds are formed through an attraction between the relatively negative oxygen atoms and relatively positive hydrogen atoms of adjacent molecules. In solid water (ice) the lattice is quite rigid; in liquid water the lattice is less extensive, and the bonds are rapidly formed and broken. The bonds are disrupted completely as water is heated to the boiling point.

We have already noted that the lattice in liquid water is strong enough to resist invasion by nonpolar substances. The lattice also gives water other properties that make it unique among inorganic substances and ideal as the medium suspending the molecules and reactions of life. The stability of the lattice makes water molecules highly *cohesive*. This resistance to separation gives water a relatively high surface tension and allows water columns, as in the conductive vessels of plants, to resist breakage. The cohesiveness produced by the hydrogen bond lattice also gives water a relatively high boiling point and stability as a liquid over the wide temperature range of 0° to 100°C. Related substances that do not form an extended hydrogen-bonded lattice, such as H_2S, have much lower boiling points. The properties of these related substances indicate that water would boil at −81°C without its hydrogen bond lattice. It is difficult to imagine how life could exist if water boiled at this temperature.

Also related to the stability conferred by the hydrogen bond lattice is the relatively high *specific heat* of water—that is, the amount of heat required to raise the temperature of water by 1°C (defined as the calorie; see footnote 1 on p. 34). As heat flows into water, much of it is absorbed in the breakage of hydrogen bonds. As a consequence, the temperature of water, reflected in the average kinetic motion of its molecules (see p. 71), rises relatively slowly as heat is added. The high specific heat of water creates a temperature "buffer"—it allows cells to absorb and release large quantities of heat without undergoing extreme changes in temperature.

Water molecules also readily form hydrogen bonds with other substances such as proteins, creating a structured water layer over the surfaces of such molecules. This structured surface layer contributes strongly to water's solvent properties, and is critical to the form and function of complex molecules such as proteins and multimolecular assemblies such as membranes.

The various charged, polar, and nonpolar attractions and repulsions set up by water molecules provide much of the force holding biological molecules in their three-dimensional folding patterns. For proteins in particular, these folding patterns are vital to their biological functions.

van der Waals Forces

The electrons surrounding covalently bonded atoms are responsible for another set of forces important in biological molecules, the *van der Waals forces*. Unlike ionic, covalent, and hydrogen bonds, van der Waals forces can be either attractive or repulsive depending on the distance separating atoms.

Attractive van der Waals forces result from momentary, chance inequalities in the distribution of electrons in any covalent bond, whether sharing is equal or unequal. These momentary inequalities make the atoms sharing a covalent bond relatively negative or positive, in a charge pattern that fluctuates rapidly between the two ends of a bond. As a result, the ends of the bond switch rapidly between relatively positive and negative charges. These rapidly fluctuating charges can set up opposite fluctuations in nearby covalent bonds, establishing a weak attractive force between the atoms held by the bonds. The attraction, though weak, becomes progressively stronger as atoms approach more closely, up to the point at which their outermost electron orbitals begin to overlap. As atoms reach this point, the electrons in the overlapping orbitals strongly repel each other, producing a repulsive force that tends to push the atoms apart. This force, which is relatively strong, is the repulsive phase of van der Waals forces.

The net effect of van der Waals attractions and repulsions is a tendency for different molecules, or the atoms within a molecule, to take up optimum positions that represent the closest packing (due to attractive van der Waals forces) that does not violate the minimum space requirement of individual atoms (due to repulsive van der Waals forces).

The minimum space occupied by an atom, represented by the distance from the atomic nucleus at which van der Waals forces switch from attraction to repulsion, can be expressed as a radius extending from an atom's center. This radius establishes the diameter of the spheres representing atoms in a *space-filling model* such as those shown in Figure 2-9.

An attractive van der Waals force at the optimum separation between two atoms requires only about 1 kcal/mol for breakage (see Table 2-1). At this level the force is only about one-half to one-fifth as strong as a hydrogen bond. However, the collective effects of van der Waals forces can be enough to stabilize the three-dimensional shape of a molecule or a combination between interacting molecules. The van der Waals attractions are also significant as forces holding together nonpolar substances excluded by water. In many intra- and intermolecular reactions the collective van der Waals attractions add stabilizing forces to those provided by hydrogen bonds and ionic attractions.

Covalent linkages and their patterns of electron sharing are thus responsible for several interatomic and intermolecular attractions and repulsions. Besides holding atoms together in molecular structures, the patterns of electron sharing in covalent bonds may produce polarity, polar and nonpolar associations, hydrogen bonds, and van der Waals forces. Along with ionic bonds, these bonding forces are critical to the structure and interactions of biological molecules.

FUNCTIONAL GROUPS IN BIOLOGICAL MOLECULES

Although covalent bonds are relatively stable, some arrangements of atoms place bonds in positions in which they are disturbed more readily. The atoms of such groups can enter into reactions in which their existing covalent bonds break and new ones form. The reactive arrangements, called *functional groups*, serve as centers active in converting one type of organic subunit to another or in forming linkages that bind subunits into larger molecular assemblies. Although the complete list of such groups is long, several stand out as frequent reactive subparts of biological molecules: *hydroxyl* ($-OH$), *carbonyl* ($-\overset{|}{C}=O$), *carboxyl* ($-COOH$), *amino* ($-NH_2$), *phosphate* ($-PO_4$), and *sulfhydryl* ($-SH$) groups (Table 2-2).

Depending on their positions, these functional groups take on characteristic patterns of interaction, giving the organic molecules containing them properties of *alcohols, aldehydes, ketones,* or *acids.* Each figures importantly in the molecular interactions that take place in cells and often serves as a building block for the assembly of carbohydrates, lipids, proteins, and nucleic acids.

The Hydroxyl Group and Alcohols

Hydroxyl ($-OH$) groups are split readily from their molecules or enter into interactions that link subunits into larger molecular assemblies. Among the most frequent of these interactions are a *condensation*, in which the elements of a water molecule are released as subunits assemble, and a *hydrolysis*, in which the elements of a water molecule are incorporated as subunits split from a larger molecule. (Condensation and hydrolysis are discussed further in Information Box 2-1.) Hydroxyl groups contribute to the polarity of molecules because of unequal electron sharing between the oxygen and hydrogen atoms in the structure.

A hydroxyl group forms the reactive part of the alcohol group, which also includes a hydrogen linked to the same carbon atom:

$$\diagdown \overset{\displaystyle\diagup \text{H}}{\underset{\displaystyle \diagdown \text{OH}}{\text{C}}}$$

The alcohol group is readily oxidized[2] to form aldehydes or ketones and, because of its hydroxyl segment, easily forms linkages. Alcohol groups are

---

[2]An *oxidation* is a chemical reaction in which electrons are removed from a molecule or reactive group. The opposite reaction, in which electrons are added to a molecule or reactive group, is a *reduction.* Because electrons have an associated energy, oxidations remove energy from the molecule oxidized, and reductions add energy to the molecule reduced. (For further details, see Chapter 9 and Information Box 9-1.)

Table 2-2 Common Reactive Groups of Organic Molecules

Functional Group	Structural Formula	Reactivity
Hydroxyl	—OH	Strongly polar; highly reactive with other groups in formation of covalent bonds, particularly in condensations.
Alcohol	$\overset{\displaystyle H}{\underset{\displaystyle |}{-\overset{|}{C}-OH}}$	Strongly polar; readily oxidized to aldehydes and ketones; also reacts with organic acids through —OH group to form many biological substances; part of many carbohydrate molecules.
Carbonyl	$-\overset{|}{C}{=}O$	Weakly polar; highly reactive; enters into linkages, particularly with bases; primary reactive group of aldehydes and ketones.
Aldehyde	$-\overset{\displaystyle H}{\underset{}{\overset{|}{C}}}{=}O$	Similar in reactivity because of carbonyl group but aldehydes more reactive than ketones; oxidized to form acids or reduced to form alcohols; components of carbohydrates, fats, and intermediates formed in synthesis and oxidative breakdown of fats and carbohydrates.
Ketone	$\overset{|}{\underset{|}{C}}{=}O$	
Carboxyl	$-C\overset{\displaystyle O}{\underset{\displaystyle OH}{\big\langle}}$	Strongly polar; ionizes in solution to release H^+ and thus acts as acid; part of amino acids, fatty acids, and many additional organic acids that form fuel substances and intermediates of a variety of biological reactions.
Amino	$-N\overset{\displaystyle H}{\underset{\displaystyle H}{\big\langle}}$	Polar; basic because it combines with H^+ in solution to produce $-NH_3^+$ group, thereby reducing H^+ concentration; enters reactions with other groups forming covalent linkages; component of amino acids and other biological molecules.
Phosphate	$-O-\overset{\displaystyle O^-}{\underset{\displaystyle O}{\overset{|}{\underset{\|}{P}}}}-O^-$	Acidic and polar; forms linking bridge between major organic groups in complex biological molecules; part of molecules storing and releasing chemical energy; serves as control element regulating molecular activity.
Sulfhydryl	—S—H	Readily oxidized; reactive group that forms linkages between major organic groups; two —SH groups enter into disulfide linkages that stabilize biological molecules including proteins.

important reactive groups in many organic molecules including carbohydrates.

The Carbonyl Group, Aldehydes, and Ketones

The oxygen atom of the carbonyl ($-\overset{|}{C}{=}O$) group is highly susceptible to chemical attack by other reactive groups, particularly bases. (Information Box 2-2 defines acids, bases, and pH.) The carbonyl group can be oxidized or reduced, depending on the presence of other active groups. Although only slightly polar, the carbonyl group may form a part of other functional groups, such as the carboxyl (—COOH) group, that are strongly polar.

Carbonyl groups figure as part of aldehydes and ketones, which are particularly important as building blocks of carbohydrates and fats. In an aldehyde the carbonyl group occurs with a hydrogen at the end of a carbon chain:

$$-C\overset{\displaystyle O}{\underset{\displaystyle H}{\big\langle}}$$

In a ketone the carbonyl group occurs in the interior of a carbon chain:

$$-\overset{\displaystyle }{\underset{\displaystyle O}{\overset{|}{\underset{\|}{C}}}}-$$

Although it reacts similarly in aldehydes and ketones, the carbonyl oxygen is more exposed in the aldehyde group, making this the more reactive of the two.

Aldehydes and ketones serve as potential linkage sites and can be oxidized to form acids or reduced to form alcohols. Their susceptibility to oxidation, particularly in aldehydes, makes them important intermediates in cellular energy metabolism as well as

Condensation and Hydrolysis

Many types of biological reactions involve addition or removal of a molecule of water. When two glucose molecules interact to form a disaccharide, for example, a molecule of water is assembled as an additional product. (See part **a** of the figure in this box; the atoms contributing to the formation of water are boxed in blue.) In this type of reaction, called a *condensation*, the components of a molecule of water, —H and —OH, are split from reacting groups of the combining molecules.

Most biological reactions in which complex molecules are assembled from smaller subunits are condensations. For example, the reactions assembling proteins from

amino acids and nucleic acids from nucleotides are condensations.

The disassembly of biological molecules into subunits usually involves *hydrolysis*, a reaction that is the reverse of condensation. In this process the components of a molecule of water are added as a covalent bond is broken (see part **b** of the figure in this box). The reactions of digestion, which break down proteins, fats, carbohydrates, and nucleic acids into smaller chemical subunits that can be absorbed in the small intestine, are typical hydrolysis reactions.

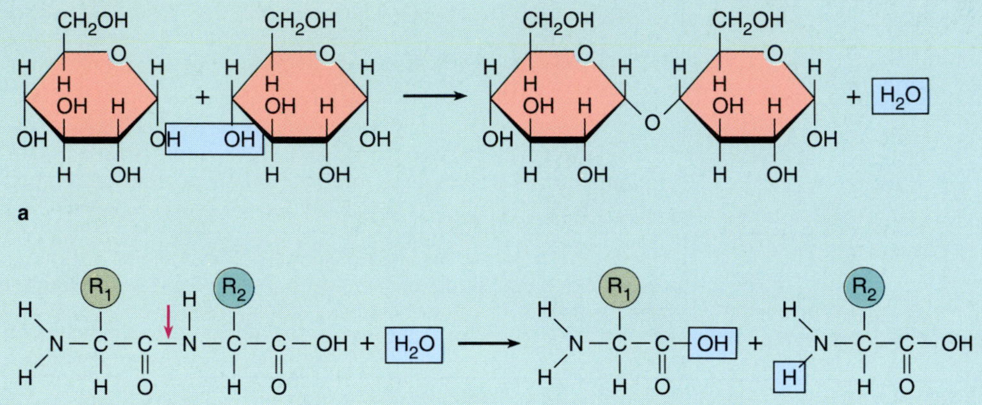

a

b

building blocks for reactions assembling macromolecules.

The Carboxyl Group and Organic Acids

Carbonyl and hydroxyl groups combine to form the carboxyl group, the characteristic functional group of organic acids:

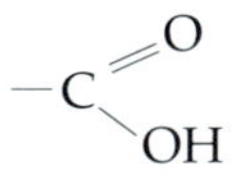

$$-C\overset{\displaystyle O}{\underset{\displaystyle OH}{}}$$

The reactivity of the carboxyl group depends primarily on its hydroxyl segment, which readily dissociates to release its hydrogen as H^+:

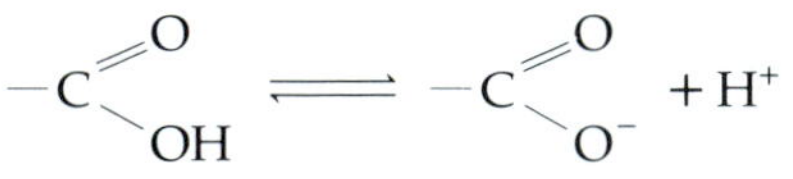

$$-C\overset{O}{\underset{OH}{}} \rightleftharpoons -C\overset{O}{\underset{O^-}{}} + H^+$$

The strongly polar carboxyl group is characteristic of amino acids, fatty acids, and a long list of other organic acids. Many organic acids, such as citric, pyruvic, and acetic acid, are oxidized readily and serve as important intermediates in energy metabolism. Carboxyl groups also enter into linkages combining organic subunits into larger assemblies.

The Amino Group

The amino ($—NH_2$) group is significant both for its chemical reactivity and its ability to act as a base in organic molecules. It is particularly important in amino acids, which contain an amino group at one end and a carboxyl group at the other. A condensation reaction between the amino group of one amino acid and the carboxyl group of another is responsible for the formation of a *peptide linkage*, which ties amino acids into the backbone chains of proteins (see Fig. 2-19).

Acids, Bases, and pH

Many inorganic and organic molecules act as either acids or bases in water solution. Acids and bases are substances that affect the concentration of hydrogen ions (H^+) in water. Water always contains H^+ because a proportion of the molecules in liquid water dissociates to produce H^+ and OH^-:

$$H_2O \rightleftharpoons H^+ + OH^-$$

When dissolved in water, acids release additional hydrogen ions, increasing the H^+ concentration. Bases bind hydrogen ions, or release additional hydroxyl ions when dissolved in water, thereby reducing the H^+ concentration.

The relative concentrations of H^+ and OH^- in a water solution determine the *acidity* of the solution. The degree of acidity alters the chemical reactivity of many organic and inorganic substances dissolved in water and modifies the folding conformations of protein molecules.

Acidity is expressed quantitatively as *pH*, on a number scale ranging from 0 to 14 (see the figure in this box). The number scale reflects a constant relationship between the relative H^+ and OH^- concentration in water solutions. At a temperature of 25°C the product of the H^+ and OH^- concentrations has a constant value of 1×10^{-14}:

$$(\text{concentration of } H^+) \times (\text{concentration of } OH^-) = 1 \times 10^{-14}$$

These concentrations are given in moles per liter, abbreviated *M*. The pH of a solution is defined as the negative logarithm (to the base 10) of the H^+ concentration, or:

$$pH = -\log_{10} [H^+]$$

where the brackets [ ] indicate concentration in moles per liter. For example, if H^+ is present at a concentration of 0.0000001 M (1×10^{-7} M), the $\log_{10}$ of this concentration is -7. The negative of the logarithm -7 is 7. Thus, a water solution with H^+ at a concentration of 1×10^{-7} M has a pH of 7.

At pH 7 the concentrations of H^+ and OH^- are equal:

$$(1 \times 10^{-7} M\ H^+)(1 \times 10^{-7} M\ OH^-) = 1 \times 10^{-14}$$

At this concentration the solution is said to be neutral. Solutions with pH higher than 7 have OH^- in excess and are basic or alkaline; solutions with pH less than 7 have H^+ in excess and are acid.

The product of the H^+ and OH^- concentrations remains equal to 1×10^{-14}. Thus, if H^+ concentration rises from 1×10^{-7} to 1×10^{-5}, the OH^- concentration falls so that the product remains equal to 1×10^{-14}:

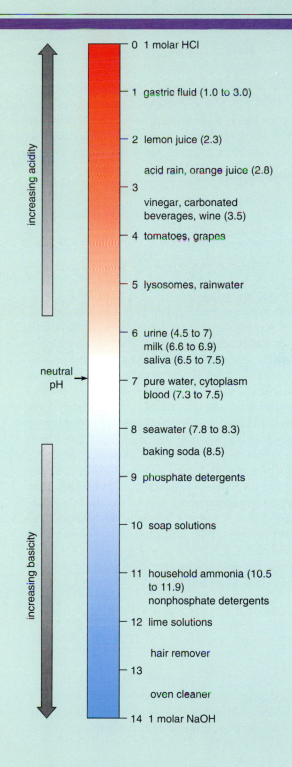

increasing acidity

neutral
pH →

increasing basicity

- 0 1 molar HCl
- 1 gastric fluid (1.0 to 3.0)
- 2 lemon juice (2.3)

    acid rain, orange juice (2.8)
- 3

    vinegar, carbonated beverages, wine (3.5)
- 4 tomatoes, grapes

- 5 lysosomes, rainwater

- 6 urine (4.5 to 7)
    milk (6.6 to 6.9)
    saliva (6.5 to 7.5)
- 7 pure water, cytoplasm blood (7.3 to 7.5)

- 8 seawater (7.8 to 8.3)

    baking soda (8.5)
- 9 phosphate detergents

- 10 soap solutions

- 11 household ammonia (10.5 to 11.9)
    nonphosphate detergents
- 12 lime solutions

    hair remover
- 13

    oven cleaner
- 14 1 molar NaOH

Because pH values are logs to the base 10 of the concentration, a change of one pH unit represents a concentration difference of 10 times. Thus, a solution of pH 8 has 10 times as many OH^- ions as a solution at pH 7. Cells typically have a neutral cytoplasmic pH, or a slightly basic pH in the range 7.0 to 7.2.

An amino group reacts as a base because of its ability to take up a hydrogen ion:

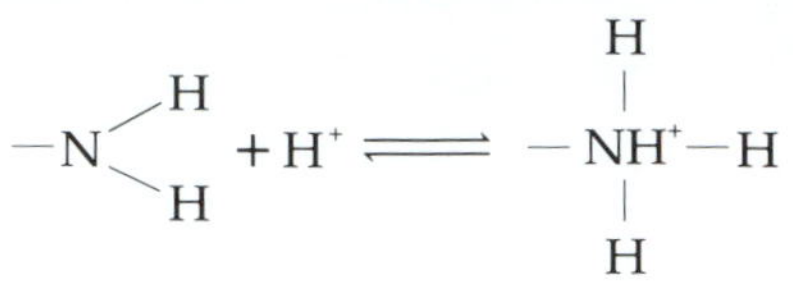

$$-N\diagdown{\raise4pt\hbox{H}}\atop{\lower4pt\hbox{H}} + H^+ \rightleftharpoons -NH^+-H$$

Removal of the H^+ effectively lowers the acidity of the surrounding medium.

The Phosphate Group

The phosphate ($-PO_4$) group:

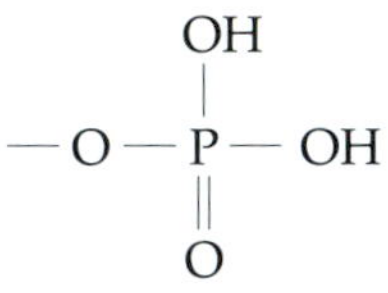

$$-O-\underset{\underset{O}{\parallel}}{\overset{\overset{OH}{|}}{P}}-OH$$

plays several important roles in biological molecules. One of the most significant depends on its ability to form chemical bridges linking organic building blocks into larger molecules. These bridges form because the oxygens held by single bonds are available to form covalent bonds on either side of the central phosphorus atom. Among the most important of the phosphate-bridged superstructures are the nucleic acids DNA and RNA (see Fig. 2-28), and the phospholipids (see Fig. 2-11), which provide the structural framework of biological membranes. Phosphate groups are also important in molecules that conserve and release energy and in metabolic regulation. The activity of many proteins, for example, is turned on or off by the addition or removal of phosphate groups.

Phosphate groups readily dissociate to release H^+, thus acting as an acid:

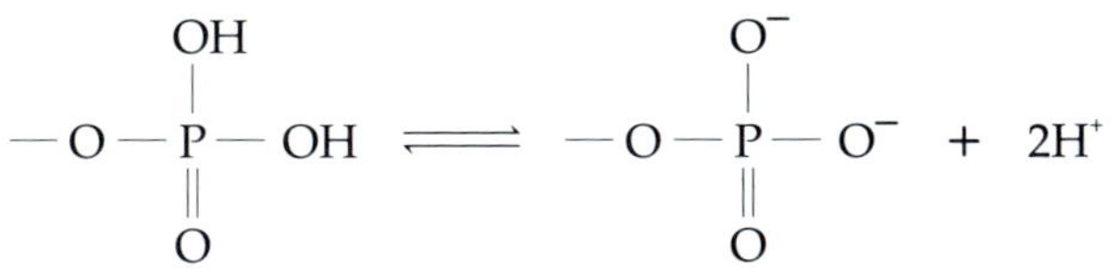

$$-O-\underset{\underset{O}{\parallel}}{\overset{\overset{OH}{|}}{P}}-OH \rightleftharpoons -O-\underset{\underset{O}{\parallel}}{\overset{\overset{O^-}{|}}{P}}-O^- + 2H^+$$

The strongly acidic reaction of the nucleic acids, for example, results from the release of H^+ from the many phosphate groups linking the subunits of these molecules.

The Sulfhydryl Group

The sulfhydryl ($-SH$) group is easily oxidized or converted into a covalent linkage. One of the most important of these linkages occurs when two sulfhydryl groups interact within a protein to form a *disulfide* ($-S-S-$) linkage (see Fig. 2-17):

$$-SH + HS- \longrightarrow -S-S-$$
$$+ 2H^+ + 2\ electrons$$

The reaction is an oxidation, in which two electrons and two hydrogen nuclei, or *protons*, are removed from the interacting $-SH$ groups. The disulfide linkage is significant in the structure of many proteins, in which it forms a molecular "button" that holds subparts of a protein together.

The functional groups and organic building blocks containing them combine in various ways to produce the four major classes of biological molecules—the carbohydrates, lipids, proteins, and nucleic acids. The particular combinations of functional groups and building blocks give the classes individual chemical properties that allow them to serve vital roles in living systems: carbohydrates and lipids primarily as structural and fuel molecules; proteins as enzymes, structural, transport, motile, and recognition molecules, among other functions; and nucleic acids as informational molecules.

CARBOHYDRATES

Carbohydrates serve many functions. Along with fats, they provide the primary molecules oxidized to provide chemical energy for cell activities. Carbohydrate fuels are stored in cells as *starches*, molecules consisting of long chains of repeating carbohydrate units linked end to end. Molecules assembled from chains of carbohydrate subunits also form structural molecules such as *cellulose*, one of the primary constituents of plant cell walls. Carbohydrates link with lipids to form *glycolipids* and with proteins to form *glycoproteins*. Both glycolipids and glycoproteins are found in quantity at cell surfaces, where they act in roles such as reception, cell-cell recognition, and adhesion. Glycoproteins are also abundant among the molecules released by cells as secretions.

Carbohydrate Structure

Carbohydrates are named for their characteristic content of carbon, hydrogen, and oxygen atoms, which occur in a 1C:2H:1O ratio, or very near this ratio. Short chains containing from three to seven carbons form the *monosaccharides*, the individual building blocks of carbohydrates. Of these, *trioses* (three carbons), *pentoses* (five carbons), and *hexoses* (six carbons) are most common in cells. (Table 2-3 lists a number of monosaccharides and gives some of their functions in living organisms.)

All monosaccharides can occur in linear form. In this form (Fig. 2-5) each carbon atom in the chain, except one, carries an $-OH$ group. The remaining carbon carries a $-C=O$ (carbonyl) group. In mono-

Table 2-3	Carbohydrate Units Found in Nature		
Number of Carbons	Type	Examples	Major Activity
3	Triose	Glyceraldehyde, dihydroxyacetone	Intermediates in energy-yielding reactions and photosynthesis.
4	Tetrose	Erythrose	Intermediate in photosynthesis.
5	Pentose	Ribose, deoxyribose, ribulose	Intermediates in photosynthesis; components of molecules carrying energy; components of the informational nucleic acids DNA and RNA; structural molecules in cell walls of plants.
6	Hexose	Glucose, fructose, galactose, mannose	Fuel substances; products of photosynthesis; building blocks of starches, cellulose, and carbohydrate units of glycolipids and glycoproteins.
7	Heptose	Sedoheptulose	Intermediate in photosynthesis.

saccharides all the other available binding sites of the carbons are occupied by hydrogen atoms. The carbonyl oxygen of a linear sugar may be located at the end of the carbon chain as an aldehyde group or inside the chain as a ketone group (see Table 2-2).

Monosaccharides with five or more carbons can form a ring as well as a linear configuration (Fig. 2-6). The rings form through a reaction between two functional groups in the same molecule. In the six-carbon monosaccharide glucose, for example, a covalent bond can form through a reaction between the aldehyde at the 1-carbon and the hydroxyl at the 5-carbon. This reaction produces either of the two closely related *glucopyranose* ring structures shown in Figure 2-6a and b. The aldehyde at the 1-carbon can also react with the hydroxyl at the 4-carbon to produce a *glucofuranose* ring (Fig. 2-6c). The glucopyranose ring is the most common form of glucose in living cells.

Monosaccharide rings such as that of glucose are frequently depicted as a *Haworth projection* (Fig. 2-6d), a diagram that suggests the three-dimensional orientation of the attached —H, —OH, and —CH$_2$OH groups in relation to the ring. However, the glucopyranose ring does not actually lie in a flat plane as suggested by this projection. Instead, the ends of the ring are bent up or down, most frequently in the *"chair" conformation* shown in Figure 2-6e. Moreover, the side groups attached to the ring extend at the various angles shown in Figure 2-6e, not at the right angles depicted in the Haworth projection. A space-filling model of the glucopyranose ring, showing the van der Waals radii of its atoms, appears in Figure 2-6f.

The glucopyranose ring occurs in two forms that differ only in the orientation of the —OH group at the 1-carbon (see Fig. 2-6b). In a Haworth projection the —OH group "points" downward, in the *alpha* (α) form of the sugar, as in α-glucose. The other form, in

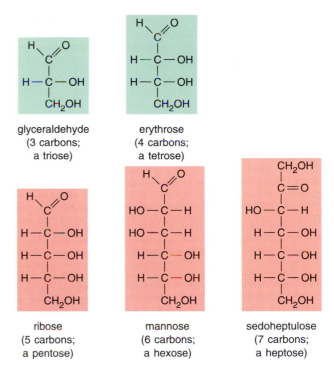

glyceraldehyde
(3 carbons; a triose)

erythrose
(4 carbons; a tetrose)

ribose
(5 carbons; a pentose)

mannose
(6 carbons; a hexose)

sedoheptulose
(7 carbons; a heptose)

which the —OH group points upward from the ring, is the *beta* (β) form of the sugar, as in β-glucose.

Although the difference between the two ring forms might seem trivial, it has great significance for the chemical properties of polysaccharides assembled from monosaccharide rings. For example, starches, which are assembled from α-glucose units, are soluble and easily digested. Cellulose, synthesized from β-glucose units, is insoluble and cannot be digested as a food source by most animals.

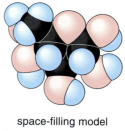

α-glucose

β-glucose

glucopyranose rings

glucofuranose ring

c

Haworth projection

d

"chair" conformation

e

space-filling model

f

a

b

Figure 2-6 Various states and representations of the glucose molecule. **(a)** Glucose in linear form. The carbons of glucose and other monosaccharides are numbered in order with the carbon of the aldehyde (—CHO) group as number 1. **(b)** Pyranose ring formation by glucose, in which the aldehyde group at the 1-carbon reacts with the hydroxyl group at the 5-carbon. The reaction produces the two alternate conformations of the glucopyranose ring, α and β, determined by the orientation of the hydroxyl (—OH) group linked to the 1-carbon (shaded). **(c)** The glucofuranose ring, formed by reaction of the aldehyde at the 1-carbon with the hydroxyl group at the 4-carbon. The glucofuranose ring, which is relatively rare in nature, also exists in two forms equivalent to α- and β-glucopyranose. **(d)** A Haworth projection of glucose, in which the *C*'s designating carbons are omitted and the covalent bonds of the ring are shown as thicker lines along one side. This depiction indicates that the ring lies in a plane perpendicular to the page, with the thickest edge closest to the viewer. The various groups attached to the ring are considered to extend at right angles to the ring, in a plane parallel to the surface of the page. **(e)** The "chair" conformation, a more accurate depiction of the most common three-dimensional arrangement of the glucose ring. The side groups extend at various angles from the ring as shown. **(f)** A space-filling model of glucose, showing the volumes and arrangement of the atoms according to their van der Waals radii. Carbon atoms are shown in black, oxygen in red, and hydrogen in white.

The orientation of other groups attached to the carbon chain of carbohydrates also affects the properties and interactions of monosaccharides. The alternative forms of a molecule in which atoms or groups attached to the carbon chain point in different directions are termed *stereoisomers*. By convention, stereoisomers are identified as D- or L- depending on the directions pointed by atoms attached to the carbon chain (see Information Box 2-3).

With several exceptions, the different stereoisomers of the monosaccharides have identical chemical and physical properties. Most important of the exceptions for biological substances is that stereoisomers react at significantly different rates with the D- and L-forms of other stereoisomers. Because the amino acid subunits of enzymes are stereoisomers, enzymes typically react much more efficiently with one of the two stereoisomers of monosaccharides. In most cases this is the D-form of the monosaccharide, which accounts for the fact that D-forms are much more common than L-forms among cellular carbohydrates.

Linkage of Monosaccharides into Disaccharides and Polysaccharides

Monosaccharides in the ring form can link together by twos to form *disaccharides* or in greater numbers to

Stereoisomers

Many internal carbon atoms of monosaccharides and other organic molecules are *asymmetric*—each of their four covalent bonds links to a different atom or chemical group. The middle carbon of the triose *glyceraldehyde*, for example, is asymmetric because it shares electrons in covalent bonds with — H, —OH, —CHO, and —CH$_2$OH:

D-glyceraldehyde   L-glyceraldehyde

The groups attached to asymmetric carbon atoms can take up either of two fixed positions with respect to other carbon atoms in a carbon chain. In glyceraldehyde (see above) the —OH group can extend either to the left or to the right of the carbon chain with reference to the —CHO and —CH$_2$OH groups. These two possible forms are called *stereoisomers*. By convention the stereoisomer in which the —OH extends to the right is called D-*glyceraldehyde* (from the Latin *dexter* = right), and the stereoisomer in which the —OH extends to the left is L-*glyceraldehyde* (from *laevus* = left).

Monosaccharides typically exist in different stereoisomer forms. The longer-chain carbohydrates have several asymmetric carbons, giving more than two possible types, and so naming them becomes complicated. However, by convention, the longer chains are designated as D- or L-isomers according to the direction pointed by the —OH group on the next-to-last carbon at the end of the chain opposite the carbonyl (—C=O) oxygen:

D-glucose   L-glucose

Many other kinds of biological molecules are stereoisomers, including the amino acids. These are named by comparing them to the "handedness" of glyceraldehyde. The amino acids, for example, are named as D- and L-forms by noting the direction pointed by the NH$_2$ group attached to the central asymmetric carbon:

D-alanine   L-alanine

The different stereoisomers of a molecule generally have the same chemical and physical properties. However, they frequently interact at significantly different rates with other substances that are also stereoisomers or contain subunits that are stereoisomers, such as enzymes. The rates are often different enough to limit use effectively to one stereoisomer, as in the exclusive use of L-isomers of the amino acids in protein synthesis.

---

form *polysaccharides*. In some polysaccharides, such as cellulose, the chain is unbranched; in others, such as glycogen, the chain is branched. Some polysaccharides, such as cellulose, contain only a single type of monosaccharide building block. Others, including the polysaccharide units of glycolipids and glycoproteins, are built up from a variety of different units.

Linkage of two glucose molecules to form the disaccharide *maltose* illustrates the general pattern of disaccharide and polysaccharide assembly (Fig. 2-7). In maltose the linkage, called a *glycosidic* bond, extends between the 1-carbon of the first glucose unit and the 4-carbon of the second. Bonds of this type, which commonly link monosaccharides, are known as 1→4 glycosidic linkages; 1→2, 1→3, and 1→6 linkages are also common. The linkages are designated as alpha (α) or beta (β) depending on the orientation of the —OH group at the 1-carbon forming the bond. In maltose the —OH group is in the α position. Therefore, the link between the two glucose subunits of maltose is an α(1→4) linkage.

Cellulose (Fig. 2-8a) is formed by a series of reactions of the type shown in Figure 2-7, except that the glycosidic bonds holding the glucose subunits together are β(1→4) linkages. The glucose subunits of plant starch molecules (Fig. 2-8b) are held together by α(1→4) linkages. Glycogen, the primary storage polysaccharide of animal cells, also contains glucose

Figure 2-7 Combination of two glucose molecules to form the disaccharide maltose. In the reaction the elements of a water molecule are removed from the monosaccharides (shown in blue).

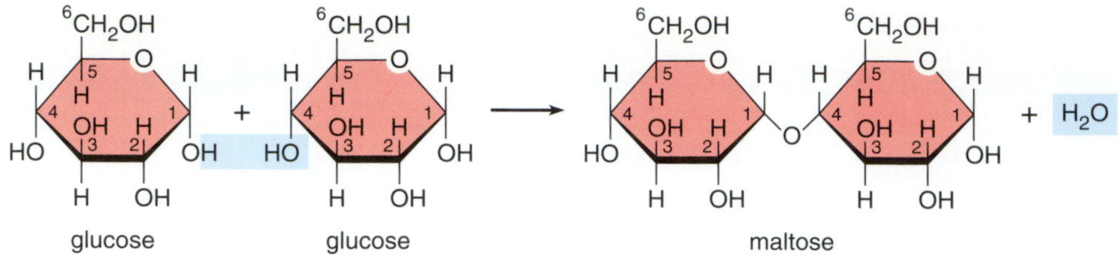

subunits held in chains by α(1→4) linkages. Glycogen molecules are branched structures in which side branches are linked to main chains by α(1→6) linkages (shown in blue in Fig. 2-8c). As noted, α-linkages are characteristic of sugars and starches that can be metabolized and used as an energy source by higher organisms; β-linkages occur in structural molecules such as cellulose that cannot be digested by most eukaryotes.

Linkage of glucose units into chains in cellulose or starch molecules illustrates the general pattern of a *polymerization* reaction. In a polymerization, multiple identical or nearly identical subunits, called *monomers*, join like links in a chain to form a *polymer*. Thus in cellulose formation the individual glucose molecules are monomers and the finished cellulose molecule is the polymer. The importance of polymerization reactions can hardly be overemphasized, because they underlie the assembly of most biological macromolecules—monosaccharides polymerize into polysaccharides, amino acid monomers into proteins, and nucleotide monomers into nucleic acid polymers. Polymerization of monomers into polymers also occurs in the assembly of several major cell structures, including the assembly of tubulin into microtubules and actin into microfilaments (see pp. 283 and 310 for details).

Much of the earth's organic matter consists of (1) carbohydrates in the form of monosaccharides, (2) polysaccharides containing glucose units as the sole building block, or (3) polysaccharides containing glucose in combination with other hexoses. One of the glucose-containing polysaccharides, cellulose, is probably the most abundant organic molecule. Most polysaccharides serve as a reservoir of stored chemical energy or provide structural support for cells. Other polysaccharides are linked to proteins in glycoproteins, which take part in an almost endless variety of reactions and function as enzymes, recognition, adhesion, receptor, and structural molecules. Figure 4-13 shows the branched or antennalike polysaccharide structure typical of many membrane glycoproteins. Another carbohydrate-protein combination forms *proteoglycans* (see Fig. 7-5), a complex molecular component of animal extracellular materials such as cartilage. Similar carbohydrate "antennas" attached to lipids in glycolipids (see Fig. 2-15) occur commonly in cellular membranes, including the plasma membrane and the membranes of chloroplasts and mitochondria.

LIPIDS

Lipids are a diverse group of biological substances made up primarily or exclusively of nonpolar groups. As a result of their nonpolar character, lipids typically dissolve more readily in nonpolar solvents such as acetone, ether, chloroform, and benzene than in water. This solubility characteristic is of extreme importance in cells because lipids tend to associate into nonpolar groups and barriers, as in the cell membranes that form boundaries between and within cells.

Besides having important roles in membranes, lipids are stored and used in cells as an energy source. Other lipids form parts of cellular regulatory mechanisms. Lipids link covalently with carbohydrates to form *glycolipids* and with proteins to form *lipoproteins*. Both of these substances play a wide variety of structural and functional roles in cells.

The lipid molecules carrying out these roles occur in one of three major forms: (1) *neutral lipids*, (2) *phospholipids*, and (3) *steroids*. Each of these lipid types is present in quantity in different cells and tissues.

Neutral Lipids

Neutral lipids, commonly found in cells as storage *fats* and *oils*, are so called because at cellular pH they bear no charged groups. Generally, they are completely nonpolar, with no affinity for water or aqueous solutions. Almost all neutral lipids are a combination of *fatty acids* with the alcohol *glycerol*.

A fatty acid is a long, unbranched chain of carbon atoms with attached hydrogens and other groups (Fig. 2-9a and b). A carboxyl (—COOH) group at one end gives the molecule its acidic properties. Most naturally occurring fatty acids contain an even number

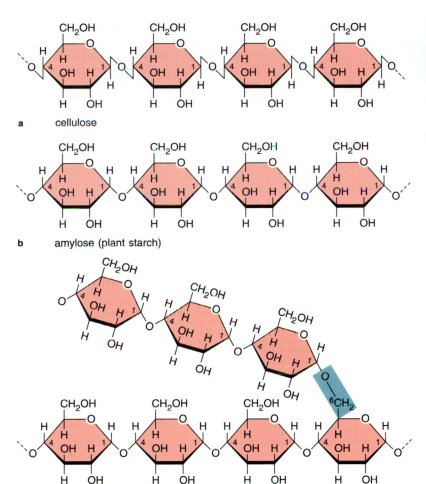

a cellulose

b amylose (plant starch)

c glycogen

Figure 2-8 Three common polysaccharides. **(a)** Cellulose, formed by end-to-end binding of β-glucose molecules in β(1→4) linkages. **(b)** A plant starch, formed by end-to-end binding α-glucose molecules in α(1→4) linkages. **(c)** Glycogen, a branched polysaccharide formed by α-glucose units joined by α(1→4) and α(1→6) linkages; the α(1→6) linkage, which occurs at branches, is shown in blue in **(c)**.

of carbon atoms in their backbone chains. Although a few with odd numbers are found in all organisms, these make up only a minor fraction of the total.

The backbone chains of fatty acids vary in length from as few as 4 to 24 or more carbons. Most fatty acids linked into neutral lipids have even-numbered chains with 14 to 22 carbons; those with either 16 or 18 carbons occur most frequently. The polar —COOH group is enough to make the shortest fatty acid chains water soluble. As chain length increases, the fatty acid types become progressively less water soluble and take on oily or fatty characteristics.

Carbon atoms within the fatty acid chain and at the end opposite the —COOH group bind hydrogen atoms. This combination of carbon and hydrogen atoms forms what is known as a *hydrocarbon* chain (the prefix *hydro* refers to hydrogen, not to water). If the carbons of a fatty acid chain bind the maximum possible number of hydrogen atoms, the fatty acid is *saturated* (Fig. 2-9a). If the number of hydrogen atoms bound by the carbons is less than the possible maximum, the fatty acid is *unsaturated* (Fig. 2-9b). At points

where hydrogen atoms are missing from adjacent carbon atoms, the carbons share a double instead of a single bond. If double bonds occur at multiple sites (up to a maximum of about six), the fatty acid is *polyunsaturated*. Unsaturated fatty acids have lower melting points than saturated fatty acids and are more abundant in living organisms. (Table 2-4 lists common saturated and unsaturated fatty acids.)

The carbon chain of a fully saturated fatty acid is more or less straight, without major bends or kinks (see Fig. 2-9a). An unsaturated fatty acid may take one of two forms at a double bond. In the *cis* form the chain bends at an angle of about 30°, producing a definite kink (see Fig. 2-9b). In the *trans* form the chain is doubly bent so that it continues in the same direction, without a pronounced kink, after the double bond (Fig. 2-9c). The kink of the *cis* form affects the packing of unsaturated fatty acid chains in cell structures, making the chains more disordered and consequently more fluid at biological temperatures. Unsaturated fatty acids in the kinked, *cis* form are much more common in cells than the *trans* form.

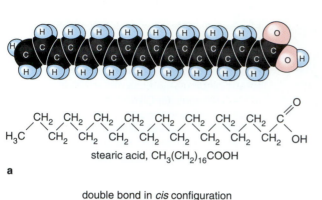

stearic acid, $CH_3(CH_2)_{16}COOH$

a

double bond in *cis* configuration

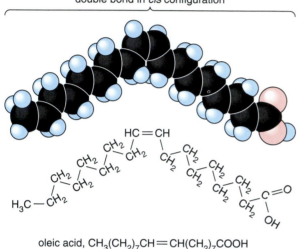

$HC=CH$

oleic acid, $CH_3(CH_2)_7CH=CH(CH_2)_7COOH$

b

double bond in
trans configuration

c

glycerol, $HOCH_2CH(OH)CH_2OH$

d

◀ **Figure 2-9** The components of a neutral lipid or triglyceride, which include fatty acids and glycerol. **(a)** Stearic acid, a saturated fatty acid. **(b)** Oleic acid, an unsaturated fatty acid. The space-filling molecular model shows the "kink" introduced in the fatty acid chain by a double bond in the *cis* configuration. **(c)** The *trans* configuration of a double bond in a fatty acid, in which the chain is not conspicuously kinked. **(d)** Glycerol. The shaded —OH groups of glycerol react with the —COOH groups of fatty acids to form triglycerides.

The other major component of neutral lipids, glycerol, has three —OH groups at which fatty acids may attach (blocked in color in Fig. 2-9d). If a fatty acid binds to each of the three sites, the resulting compound is known as a *triglyceride* or *triglycerol*. Most lipids stored as a fuel reserve in living systems are triglycerides.

In the formation of triglycerides each of the three —OH sites of a glycerol molecule reacts with the —COOH group of a fatty acid (Fig. 2-10). The additions are condensations in which one water molecule is released as each linkage forms. (See Information Box 2-2; fatty acids and other organic subunits such as amino acids are called *residues* after their linkage into larger molecules.) The combination eliminates the polar —OH groups of glycerol and the —COOH groups of the fatty acids, forming the completely nonpolar triglyceride product.

The fatty acid residues linked to glycerol may be different or the same. As a result, many varieties of the basic triglyceride format are possible. In fishes, for example, at least 20 different fatty acids may link in various combinations with glycerol, giving more than 1500 possible triglycerides. Different species usually have distinctive combinations of fatty acid residues in their triglycerides, and the combinations may change depending on diet.

Like individual fatty acids, triglycerides generally become less fluid as the length of their fatty acid

Table 2-4	Saturated and Unsaturated Fatty Acids		
Saturated Fatty Acids		**Unsaturated Fatty Acids**	
Butyric acid	$CH_3(CH_2)_2CO_2H$	Crotonic acid	$CH_3CH=CHCO_2H$
Caproic acid	$CH_3(CH_2)_4CO_2H$	Palmitoleic acid*	$CH_3(CH_2)_5CH=CH(CH_2)_7CO_2H$
Caprylic acid	$CH_3(CH_2)_6CO_2H$	Oleic acid*	$CH_3(CH_2)_7CH=CH(CH_2)_7CO_2H$
Capric acid	$CH_3(CH_2)_8CO_2H$	Linoleic acid*	$CH_3(CH_2)_3(CH_2CH=CH)_2(CH_2)_7CO_2H$
Lauric acid	$CH_3(CH_2)_{10}CO_2H$	Linolenic acid	$CH_3(CH_2CH=CH)_3(CH_2)_7CO_2H$
Myristic acid	$CH_3(CH_2)_{12}CO_2H$	Arachidonic acid	$CH_3(CH_2)_3(CH_2CH=CH)_4(CH_2)_3CO_2H$
Palmitic acid	$CH_3(CH_2)_{14}CO_2H$	Nervonic acid	$CH_3(CH_2)_7CH=CH(CH_2)_{13}CO_2H$
Stearic acid	$CH_3(CH_2)_{16}CO_2H$		
Arachidic acid	$CH_3(CH_2)_{18}CO_2H$		
Lignoceric acid	$CH_3(CH_2)_{22}CO_2H$		

* Occur as major fatty acids in human storage fats.

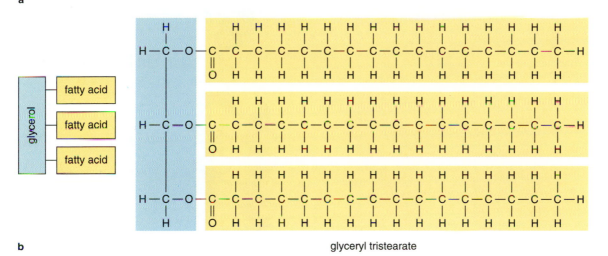

Figure 2-10 Triglyceride formation. **(a)** Combination of glycerol with three fatty acids to form a triglyceride. The R groups represent the carbon chains of the fatty acids. The components of a water molecule (shown in blue) are removed from the glycerol and fatty acids in each of the three bonds formed. **(b)** A triglyceride formed by combination of glycerol with three fatty acids.

glyceryl tristearate

chains increases. The degree of saturation of the fatty acid chains also affects the fluidity of triglycerides— the more saturated, the less fluid the triglyceride. Triglycerides that are liquid at biological temperatures are called oils; those that are semisolid or solid are called fats.

Most of the oils and fats inside cells are stored as energy reserves. In animals a large proportion of the stored triglycerides have saturated fatty acid chains, producing semisolid or solid deposits of fat. In plants, triglycerides are typically unsaturated and stored as oils. Plant oils are converted commercially to fats by adding hydrogens to increase the degree of saturation, as in the conversion of vegetable oils to margarine.

Fatty acids combine with long-chain alcohols to form *waxes*. These substances occur on the exterior surfaces of plant cell walls, where they form a protective coating that resists water loss or invasion by infective agents.

Phospholipids

The primary lipids of membranes are *phospholipids*, a group of phosphate-containing molecules with structures related to the triglycerides. In the most common phospholipids, called *phosphoglycerides*, glycerol forms the backbone of the molecule as in triglycerides, but only two of its binding sites link to fatty acid residues (Fig. 2-11). The third site links instead to a bridging phosphate group. The carbons linked to fatty acid residues are the 1- and 2-carbons; the carbon linked to the phosphate group is the 3-carbon. The other end of the phosphate group links to a nitrogen-containing alcohol or another organic subunit such as the amino acids serine or threonine (Fig. 2-12).

The phosphoglyceride diagramed in Figure 2-11, *phosphatidyl choline*, is a major lipid component of cellular membranes. Because different fatty acids may bind at the 1- and 2-carbons of the glycerol in phosphoglycerides, phosphatidyl choline, rather than being a single molecular type, is a family of closely related molecules differing in their fatty acid combinations.

Several other types of phospholipids (see pp. 95–98) occur in cellular membranes in addition to the phosphoglycerides. Although these types differ from the phosphoglycerides to a greater or lesser extent, all contain a phosphate group linking a polar unit to a segment containing nonpolar hydrocarbon chains. Depending on the alcohol or other group present, the polar segment may carry a net charge at cellular pH.

The combination of a polar unit with nonpolar hydrocarbon chains gives a phospholipid dual solubility properties; that is, one end of the molecule is hydrophilic, the other hydrophobic. Molecules of this

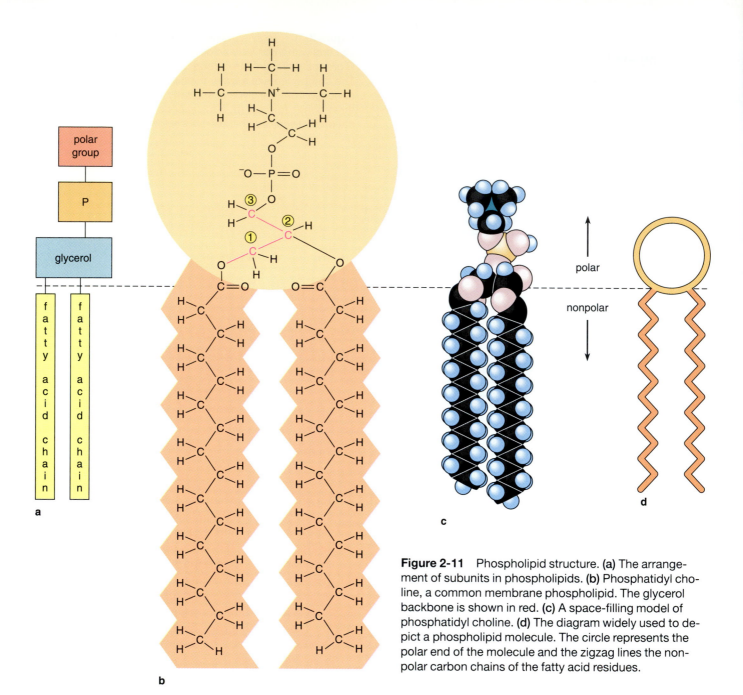

Figure 2-11 Phospholipid structure. **(a)** The arrangement of subunits in phospholipids. **(b)** Phosphatidyl choline, a common membrane phospholipid. The glycerol backbone is shown in red. **(c)** A space-filling model of phosphatidyl choline. **(d)** The diagram widely used to depict a phospholipid molecule. The circle represents the polar end of the molecule and the zigzag lines the nonpolar carbon chains of the fatty acid residues.

type, with dual solubility properties, are termed *amphipathic* or *amphiphilic*.

Phospholipids take up arrangements in polar or nonpolar environments that satisfy their amphipathic properties. When introduced into the interface formed when a nonpolar solvent such as benzene is layered on top of water, phospholipids orient so that their nonpolar hydrocarbon chains extend into the benzene and their polar groups extend into the water (Fig. 2-13a). Phospholipids placed under the surface of water, or any strongly polar liquid, satisfy their dual solubility requirements by forming a *bilayer*, a film just two molecules thick. In a bilayer the phospholipid molecules orient so that the hydrocarbon chains associate in a nonpolar, hydrophobic region in the interior of the layer. The polar phosphate-alcohol groups face the surrounding water (Fig. 2-13b).

A phospholipid bilayer is stable in water because the hydrophilic and hydrophobic parts of its molecules are suspended in environments with like solubility properties. Displacement of the phospholipids would expose their nonpolar regions to the surrounding aqueous medium or bury their polar segments into the nonpolar interior of the bilayer. Phospholipid

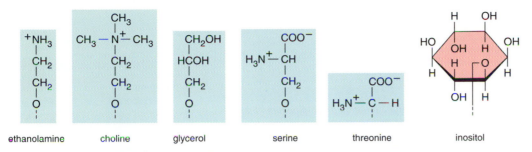

ethanolamine   choline   glycerol   serine   threonine   inositol

Figure 2-12 Organic subunits commonly linked to glycerol by a phosphate group in phospholipids. The site at which the subunit links to glycerol via a phosphate group is indicated by the dashed line.

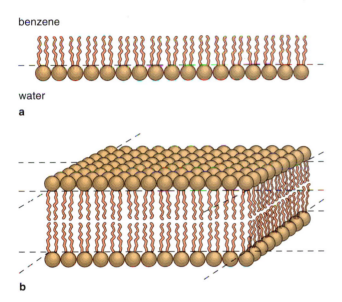

Figure 2-13 Arrangements satisfying the dual solubility property of phospholipids. **(a)** Phospholipid molecules introduced into the interface between a nonpolar solvent such as benzine and water orient with their nonpolar tails extending into the benzene and their polar heads facing the water. **(b)** Phospholipid molecules completely surrounded by water assemble into bilayers with their nonpolar tails associated together in the bilayer interior and their polar heads facing the surrounding water molecules.

bilayers, held together by such polar and nonpolar associations, are the primary structural framework of cellular membranes. (Chapter 4 discusses additional phospholipid types and the arrangement of phospholipids in cellular membranes.)

Steroids

Steroids are lipids based on a framework of four carbon rings (Fig. 2-14a). The members of this group differ in the position and number of double bonds in the rings and in the side groups attached to the rings. The

most abundant group of steroids, the *sterols*, has a hydroxyl group linked to one end and a complex, nonpolar carbon chain at the opposite end of the ring structure (blocked groups in Fig. 2-14b). Although sterols are almost completely hydrophobic, the single hydroxyl group gives one end of the molecules a slightly polar, hydrophilic character. As a result, the sterols are slightly amphipathic and tend to orient in arrangements that satisfy their dual solubility properties.

Cholesterol (Fig. 2-14b and c) is an important component of the plasma membrane in all animal cells; similar sterols occur in plant plasma membranes. When present in membranes, cholesterol loosens the packing of membrane phospholipids. Among other effects, its interference with phospholipid packing maintains the fluidity of membranes at low temperatures. Deposits derived from cholesterol also form a major part of the material collecting inside arteries in the disease *atherosclerosis* (hardening of the arteries).

Other steroids are important as *hormones* in animals. Although steroid hormones occur in only trace amounts in animal tissues, they have regulatory effects far out of proportion to their concentrations. The male and female sex hormones of humans and other animals are steroids; so are the hormones of the adrenal cortex that regulate cell growth and activity (details of steroid hormone regulation are given in Chapter 17). Some toxic steroids occur in the venoms of toads and other animals.

Combinations Between Lipids and Carbohydrates: Glycolipids

Glycolipids are amphipathic lipids with one or more carbohydrate groups linked to their polar segments. The carbohydrate groups may be relatively simple and include only one or two monosaccharide building blocks. Or, they may be complex, branched structures

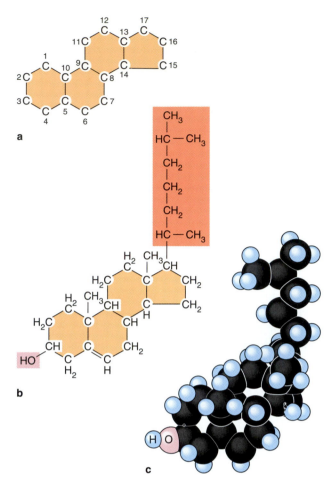

Figure 2-14 Steroids. **(a)** The typical arrangement of four carbon rings in a steroid molecule. **(b)** and **(c)** A sterol, cholesterol. Sterols have a long, nonpolar side chain linked to the 17-carbon at one end and a single —OH group at the 3-carbon at the other end (boxed in red in **b**). The —OH group makes the end that carries it slightly polar. The remaining parts of the molecule are nonpolar.

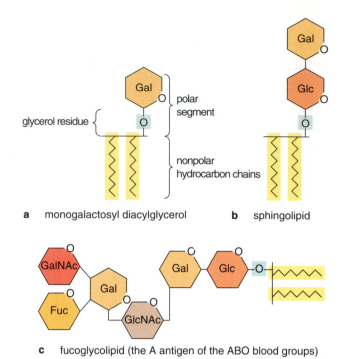

a monogalactosyl diacylglycerol   **b** sphingolipid

c fucoglycolipid (the A antigen of the ABO blood groups)

Figure 2-15 Some representative glycolipids in diagrammatic form. Monogalactosyl diacylglycerol and the sphingolipids occur in both plant and animal membranes; sphingolipids are especially abundant in brain and nerve cells in higher vertebrates. Glc, glucose; Gal, galactose; Fuc, fucose; GlcNAc, glucosamine; GalNAc, galactosamine. The structures of these sugars appear in Figure 4-4.

containing a variety of carbohydrate units linked in different patterns. Figure 2-15 diagrams several representative glycolipids.

Glycolipids occur in the highest proportions in bacterial plasma membranes and in the thylakoid membranes of chloroplasts. Animal plasma membranes, depending on the species and cell type, contain glycolipids in varying amounts. The plasma membranes of most mammalian cells contain low concentrations of glycolipids, with carbohydrate units ranging from very simple structures with only a few sugar units to complex "antennae" built up from as many as 30 different sugars. A few mammalian cell types, including cells of the brain and intestinal lining, contain glycolipids in relatively high concentrations.

Glycolipids lend varied structural and functional properties to membranes. Because their carbohydrate groups have many sites capable of forming hydrogen bonds, glycolipids reinforce membrane structure and occur in quantity in membranes subject to disruptive physical stress or chemical agents. For example, glycolipids in cells lining the intestinal cavity of mammals stabilize plasma membranes against the effect of bile salts, which act as detergents (see p. 101) and would otherwise disrupt membrane structure. Glycolipids also reduce the rate at which ions penetrate membranes. This effect probably accounts for their elevated concentrations in the plasma membranes of bacteria living in highly saline environments. Glycolipids also function as recognition markers at the surfaces of some mammalian cells. The human ABO blood groups, for example, depend in part on small differences in the carbohydrate groups linked to membrane glycolipids (Fig. 2-15c; carbohydrate groups linked to membrane proteins also determine the ABO antigens; see p. 153).

PROTEINS

Proteins carry out many vital functions in living organisms. As structural molecules, proteins provide much of the cytoskeletal framework of cells; as enzymes, they act as biological catalysts that speed the rate of cellular reactions; as motile molecules, they impart movement to cells and cell structures. They stabilize and control the activity of the two nucleic acids DNA and RNA, including the DNA of genes; form active parts of ribosomes; and provide the transport and recognition functions of cellular membranes. Proteins are also secreted to the cell exterior. Some secreted proteins, such as the collagen of animal cells and the wall proteins of plant cells, form parts of extracellular structures. Others, including peptide hormones and growth factors, transmit control signals from one cell to another; still others act as antibodies or digestive enzymes. Many toxins or venoms are also based on secreted proteins.

The protein molecules carrying out these diverse functions are fundamentally similar in structure. All consist of one or more long, unbranched chains of amino acid subunits. Although the most common proteins contain from 50 to 1000 amino acids, some are found in nature with as few as 3 to 10 or more than 50,000 amino acids.

Proteins are assembled initially in cells by linking 20 different amino acids in various numbers and combinations. After synthesis, one or more of the original 20 types may be chemically modified to other forms, so that as many as 150 different amino acids may occur in finished proteins. The potential variety of proteins made possible by differences in number, length, and chemical modification of amino acids is practically infinite. For any given protein the amino acid sequence determines its structure, properties, and chemical activity.

The Amino Acids

With one exception, the 20 amino acids initially assembled into proteins are based on the same structural plan (Fig. 2-16). These amino acids have a central carbon atom, the *alpha* (α) carbon, to which are attached an amino ($-NH_2$) and a carboxyl ($-COOH$) group. One of the two remaining bonds of the central carbon is linked to a hydrogen atom, giving the

$$NH_2 - \overset{\displaystyle |}{\underset{\displaystyle |}{C}} - COOH$$
$$H$$

structure common to these amino acids. The fourth bond of the α-carbon may be attached to any one of

19 different side chains, ranging in complexity from a single hydrogen atom in the simplest amino acid, *glycine*, to longer carbon chains or rings in other amino acids. Some of the more complex side chains contain oxygen, nitrogen, or sulfur in addition to carbon and hydrogen. The remaining amino acid, *proline*, an exception to this general plan, is based on a ring structure that includes the central carbon atom (see Fig. 2-16). However, proline has a $-COOH$ group at one side and an *imino* ($-NH-$) group forming part of the ring at the other, which reacts in essentially the same pattern as the $-NH_2$ groups of the other amino acids in the formation of proteins.

The types and configurations of atoms in the side chains give the individual amino acids distinct chemical properties. Some of the side chains include functional groups, such as $-NH_2$, $-OH$, $-COOH$, and $-SH$, which contribute to their chemical properties. These functional groups may interact with atoms, molecules, and ions located outside the protein or elsewhere on the amino acid chains of the same protein. Many side groups have atoms in positions that can form hydrogen bonds. The various side groups differ in polarity, and some of the polar ones are charged or capable of acting as acids or bases at cellular pH (see Information Box 2-2).

The $-SH$ group of the amino acid cysteine is particularly important in molecular interactions. The $-SH$ group of a cysteine residue can react with the $-SH$ group of another cysteine on the same or a different amino acid chain to form an $-S-S-$ (disulfide) linkage that helps maintain proteins in folded form (Fig. 2-17).

In all the amino acids except glycine, the α-carbon is linked to four different chemical groups. It is thus asymmetric (see Information Box 2-3) and has D- and L-stereoisomers. The amino acids used by ribosomes to assemble proteins are exclusively L-isomers.

Modified amino acids are common in natural proteins. For example, many proline residues in collagen, one of the most abundant animal proteins, are modified to *hydroxyproline* by the addition of an $-OH$ group. Figure 2-18 shows hydroxyproline and several other modified amino acids common in proteins.

Amino acids readily link covalently into chains of two or more units. The covalent link, the *peptide bond* or *peptide linkage*, is produced by a condensation reaction between the amino group of one amino acid and the carboxyl group of a second (Fig. 2-19). Because the product in Figure 2-19, a *dipeptide*, retains an amino group at one end and a carboxyl group at the other, both ends are potentially available for the formation of additional peptide linkages. In cells, additional amino acids are added only to the $-COOH$ end. This means that a finished protein has an $-NH_2$ exposed

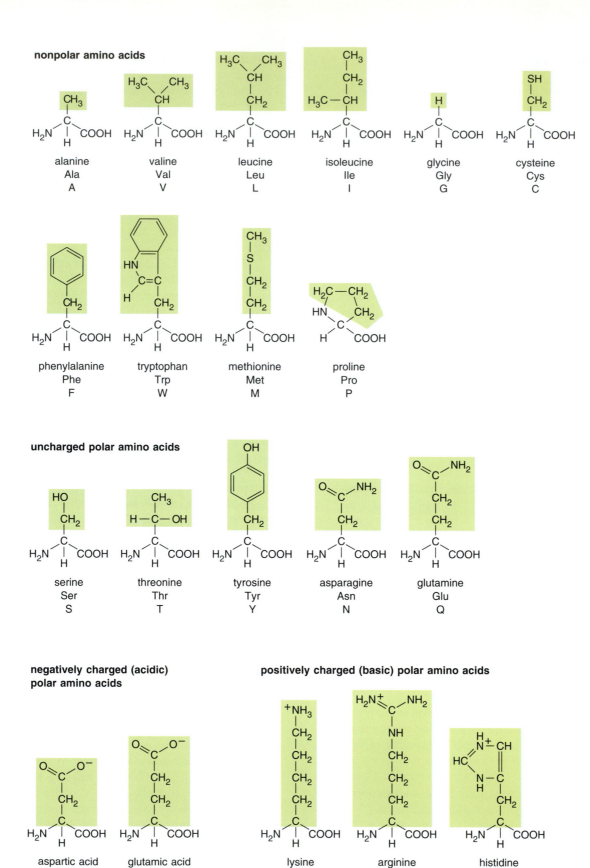

Figure 2-16 The 20 amino acids used by ribosomes in protein synthesis. The side chains of each amino acid are boxed in green. One amino acid, proline, differs from the common structural plan (see text). Three-letter and one-letter abbreviations used for the amino acids are included below each structural diagram.

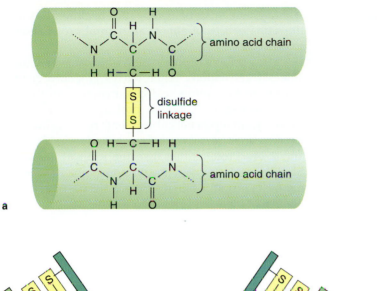

Figure 2-17 Disulfide linkages. **(a)** The atoms involved in a disulfide linkage, formed by a reaction between the sulfhydryl (—SH) groups of two cysteine residues in different parts of an amino acid chain. The reaction is an oxidation in which two electrons and two hydrogens are removed from the reacting groups. **(b)** The pattern of disulfide linkages within and between the four polypeptide subunits of an antibody molecule.

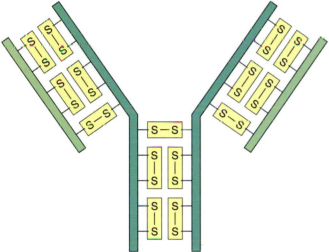

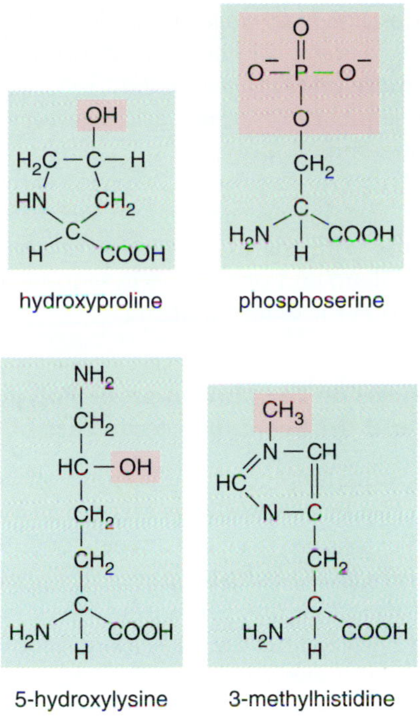

hydroxyproline

phosphoserine

5-hydroxylysine

3-methylhistidine

Figure 2-18 Some modified amino acids appearing in proteins. Modified amino acids are produced in cells by chemical conversion of the 20 amino acids used in initial assembly of proteins. The modifying groups are boxed in red.

Figure 2-19 Reaction of the carboxyl group of one amino acid with the amino group of a second amino acid to form a peptide linkage. The reaction is a condensation in which the components of a water molecule are removed from the reacting groups.

side chain   amino group

side chain   carboxyl group   side chain

side chain   peptide linkage

side chain   + H_2O

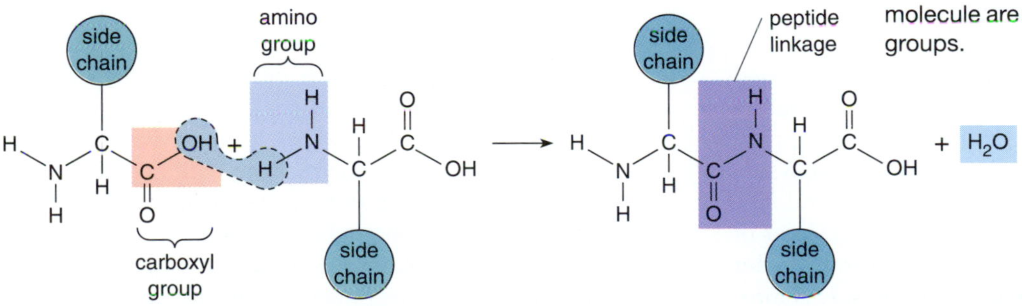

at its oldest, or *N-terminal end* and a —COOH at its newest, or *C-terminal end*. An amino acid chain of three or more units is a *polypeptide*.

The sequence of amino acid residues in a completely assembled polypeptide chain is called the *primary structure* of a protein. By virtue of the direction in which proteins are assembled, the primary structure begins with the first amino acid residue at the N-terminal end and progresses in sequence to the last amino acid at the C-terminal end.

Restrictions on the Movement of Polypeptide Chains and the Secondary Structure of Proteins

In a peptide linkage the electrons shared in the O=C—N part of the linkage take up an average distribution extending from the oxygen, over the carbon, to the nitrogen. This distribution, designated by the dashed lines in Figure 2-20, introduces a partial double bond between the C and N atoms, converting the linkage to the form O═C═N. This arrangement greatly increases the energy required to rotate the atoms around the C—N bond. As a consequence, the total structure, including all the atoms in Figure 2-20, tends to remain fixed in a single plane.

The resistance to twisting around the C—N bond has a significant effect on the possible folding conformations of a polypeptide chain. Because of this resistance, twists of the backbone chain are restricted to the single bonds on either side of the α-carbons, the carbons carrying the amino acid side groups. Even the degree of rotation around these bonds is limited, because in some positions the side groups approach too closely.

By analyzing the likelihood of different degrees of rotation around the bonds on either side of the α-carbons and the possibilities for formation of stabilizing hydrogen bonds, protein chemists have determined two ordered arrangements that are particularly stable—the *alpha helix* and the *beta strand*. Hydrogen bonding and van der Waals forces set up in or around these arrangements make the backbone chain relatively stiff and resistant to bending. A third, less regular arrangement, the *random coil*, provides flexible regions allowing amino acid chains to bend and fold. The number and positions of these arrangements determine the *secondary structure* of a polypeptide chain.

The Alpha Helix In the alpha helix, first proposed by Linus Pauling and Robert Corey, the alternating α-carbons and peptide linkages follow a zigzag path tracing a regular spiral (Fig. 2-21). The amino acid side chains extend outward from the axis of the spiral. The structure is stabilized by a regularly spaced series of hydrogen bonds (dotted lines in Fig. 2-21) and van der Waals forces set up between atoms in the side chains.

Although an alpha helix may turn in either direction, the alpha helices of almost all natural proteins wind to the right.

Most proteins contain alpha helical segments in at least some regions. Where such segments occur, the portion of the amino acid chain containing them is rigid and rodlike. Globular proteins usually contain several short alpha helical segments running in different directions (shown as coils or cylinders in Figs. 2-23 and 2-24). Proteins that take the form of elongated fibers typically contain one or more alpha helical segments that run the length of the molecule with few or no breaks.

The Beta Strand Pauling and Corey proposed a second regular arrangement of amino acid chains that they termed the beta strand (Fig. 2-22a). In a beta strand the amino acid chain also follows a zigzag path but is extended rather than wound into a coil.

No stabilizing hydrogen bonds form within a single length of beta strand. However, the —NH and C=O groups of a beta strand may set up hydrogen bonds with adjacent beta strands or alpha helical segments. Beta strands aligned in side-by-side register by hydrogen bonds form a *beta sheet* or *pleated sheet* (Fig. 2-22b). The polypeptide chains in a pleated sheet may run in opposite directions, as in Figure 2-22b, or in the same direction. In many proteins the entire beta sheet twists to some extent rather than lying in a perfectly flat plane (Fig. 2-22c; individual beta strands are frequently depicted in protein diagrams by an arrow as in this figure, with the arrow pointing toward the C-terminal end of the polypeptide chain). In some proteins the entire beta sheet winds into a cylinder, forming a *beta barrel*, typically stabilized by a ring of alpha helices spaced around the barrel structure (visible in the triosephosphate isomerase diagrams in Fig. 2-23). Beta barrels occur in some enzymes and are common in the coat proteins of icosahedral viruses. The beta sheet in any of these forms allows even greater opportunities for hydrogen bonding than the alpha helix, making beta sheets highly rigid and stable structures.

Beta strands, sheets, and barrels occur in many natural proteins, usually in combination with alpha-helical segments. Only a few proteins contain only beta strands or sheets. One of these is the silk protein secreted by silkworms, in which the beta-sheet structure contributes to the unusually high tensile strength of silk fibers.

The Random Coil In the random coil the amino acid chain takes an irregular configuration with no tendency to form alpha helices or beta strands. Proline residues often contribute to random coil structures because their ring form does not fit into an alpha helix

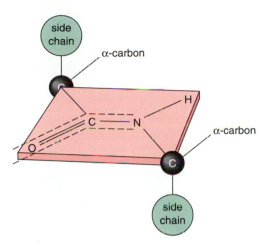

Figure 2-20 Spatial arrangement of the atoms in a peptide linkage. The dashed lines show the average distribution of electrons shared among the oxygen, carbon, and nitrogen atoms of the linkage. The distribution has the effect of introducing a partial double bond between the carbon and nitrogen atoms, preventing rotation of the atoms around the bonds and holding the entire linkage in a flat plane. Only the α-carbons at the corners of the plane can rotate. (Redrawn by permission from R. E. Dickerson and I. Geis, *The Structure and Action of Proteins*, Harper & Row, 1969. All rights reserved.)

or beta sheet, and no sites are available for formation of stabilizing hydrogen bonds. Other factors, such as polar or nonpolar associations, attractions between charged regions of amino acid side groups, or reactions with substances outside the protein, may also disturb the more regular conformations and introduce segments of random coil.

The term "random coil" is something of a misnomer, because an irregularly bent or folded region in an amino acid chain is usually not completely random in form. Instead, its amino acid sequence and segments of amino acid chain located nearby limit its conformation.

Segments of random coil are as significant to protein structure as the more regular conformations. They provide opportunities for the amino acid chain to bend back on itself, thereby allowing proteins that contain alpha-helical or beta-strand segments to fold into compact, globular forms. More sharply bent segments of random coil often occur at the surfaces of proteins, linking segments of alpha helix or beta strand located more deeply in the protein. These surface loops are especially important in membrane proteins, which frequently include a series of alpha-helical segments that extend back and forth across the membrane, connected at their ends by loops of random coil that project from the polar

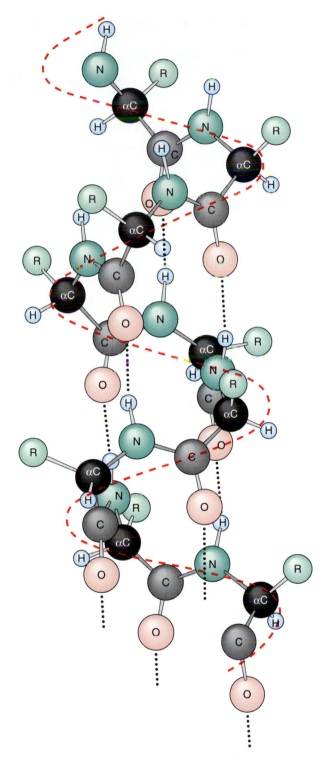

Figure 2-21 The alpha helix. The backbone of the amino acid chain is held in a spiral (red dashed line) by hydrogen bonds (short dotted lines) formed at regular intervals. The spheres labeled R represent amino acid side chains. (Redrawn from "Proteins" by Paul Doty, *Scientific American* 197:173[1957]. Copyright 1957 by Scientific American, Inc. All rights reserved.)

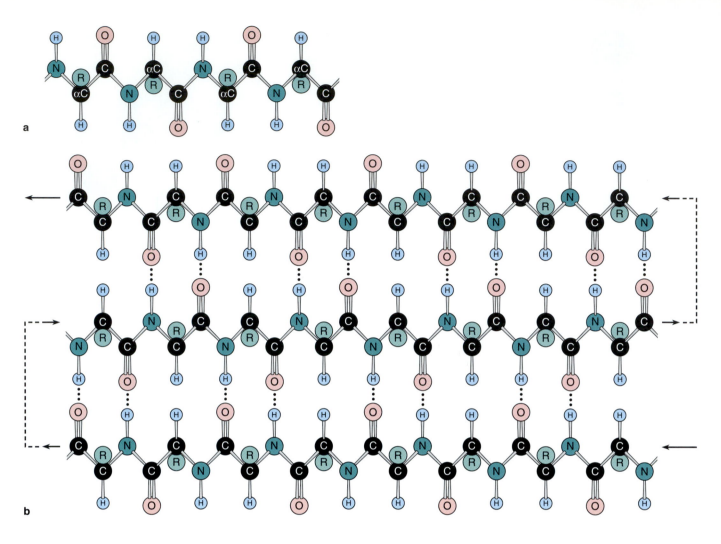

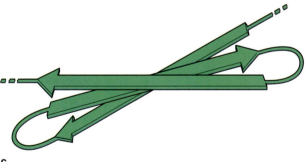

Figure 2-22 The beta strand. **(a)** Amino acids arranged in a beta strand. The spheres labeled R represent the amino acid side chains. **(b)** Side-by-side arrangement of beta strands in a beta sheet. Hydrogen bonds (dotted lines) between the strands stabilize the sheet. **(c)** Twisted arrangement of beta strands in a beta sheet.

membrane surfaces (as in the protein diagramed in Fig. 4-8). Segments of random coil also commonly act as "hinges" that allow subparts of proteins to move with respect to each other.

Tertiary Structure: The Three-Dimensional Arrangement of Polypeptide Chains

At a given pH and temperature the amino acid sequence of a polypeptide chain primarily determines its content of alpha helical, beta strand, and random coil segments. The amino acid sequence also determines the number and position of disulfide linkages between cysteine residues, the location of proline residues, and the position at which other substances, such as metallic ions, lipids, and carbohydrates, may bind. Also important, and equally dependent on the amino acid sequence, are the positions of hydrogen bonds, attractions between positively and negatively charged side groups, van der Waals forces, and polar and nonpolar associations. The net effect of these factors is to establish a distinct three-dimensional shape

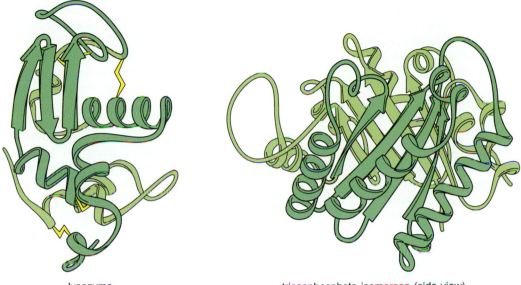

lysozyme   triosephosphate isomerase (side view)

Figure 2-23 Tertiary structure of two proteins, lysozyme and triosephosphate isomerase, depicted in a diagrammatic form in which alpha helices are shown as coils or cylinders, beta strands as arrows, and random coils as ribbons. The triosephosphate isomerase molecule contains a beta barrel consisting of a sheet of parallel beta strands wound into a cylinder. Alpha-helical segments are typically associated with beta strands wound into a beta barrel, as in triosephosphate isomerase. (Redrawn from originals courtesy of J. S. Richardson, from *Adv. Prot. Chem.* 34:167[1981].)

for each protein of a unique sequence. This three-dimensional shape is the *tertiary structure* of a protein. (Fig. 2-23 shows the tertiary structure of several proteins.)

The three-dimensional shape of a protein represents a compromise between many opposing factors, such as regions at the surface or interior occupied by polar and nonpolar amino acid side groups, maximum possible packing to the limits of van der Waals radii, and opportunities for hydrogen bonding. On the average, globular proteins fold to about 75% of the minimum volume expected from their van der Waals radii. This degree of packing makes a protein a solid structure, equivalent in rigidity to most organic solids.

The compromise represented by the three-dimensional folding pattern is a state in which the smallest possible amount of energy is required to maintain the folded structure. Departures from this *minimum energy state* require an input of energy from the surrounding medium, such as an increase in temperature. Because the minimum energy state is a balance of opposing factors, the three-dimensional shape of many proteins remains sensitive to disturbance and may change readily.

In addition to temperature changes, alterations in pH or salt concentration of the medium may contribute to changes in protein shape. Molecular collisions

or binding ions or chemical groups may also cause changes in tertiary structure. These changes may involve alterations in the angle of rotation of atoms around a single bond, movements of segments of the amino acid chain, or vibrational or "breathing" movements of the entire protein. Any of these structural alterations is known as a *conformational change*.

The flexibility of protein structure contributes to the function of many proteins, particularly those involved in enzymatic catalysis, transport, and motility. For example, in muscle contraction a segment of the protein *myosin* undergoes conformational changes that cycle it through oarlike motions (see Fig. 11-9). The oarlike movement, pushing against other proteins forming parts of muscle, is responsible for the voluntary and involuntary muscular movements of animals.

Conformational changes can be detrimental under extreme conditions. Extreme changes in temperature or pH can seriously disrupt the internal bonds and attractions that hold proteins in their tertiary conformations. Excessive heat increases the motions of amino acid side groups until the relatively weak attractions of hydrogen bonds and van der Waals forces can no longer hold them in place. The resultant unfolding is one reason why few living organisms can tolerate temperatures above 45° to 55°C. Changes in pH alter the charge of amino acid side groups, destroying or weakening ionic bonds that hold polypeptide chains in their three-dimensional form. If unfolding is so extensive that functional activity is destroyed, the protein is said to be *denatured*.

Some proteins can return to their native folded form if the temperature or pH returns to natural values, so that denaturation is reversible. In others, denaturation is permanent. The most familiar example of permanent denaturation is a cooked egg white, in which the protein albumin, originally in a clear solution, is converted into an insoluble white mass.

Disulfide linkages are particularly important as a factor limiting the degree of protein denaturation by heat and other factors. These linkages prevent amino acid chains from completely unfolding and frequently hold them in intermediate conformations that greatly facilitate refolding to a native conformation. The extra stability conferred by disulfide linkages probably accounts for their characteristic occurrence in proteins that are exposed to the more disruptive conditions in the cell exterior, such as antibodies and the collagen proteins forming parts of extracellular supports of some types of animal cells.

The tertiary structure of proteins frequently contains large blocks of amino acids folded into distinct subregions called *domains* (Fig. 2-24). Domains are usually separated by a *cleft* or *crevice* spanned by a segment of random coil. In many enzymatic proteins the regions catalyzing chemical reactions are located in clefts or crevices between domains. The clefts and crevices provide unique environments containing charged, acidic, or basic amino acid side groups, or patterns of hydrogen bonding. These environments promote reactions that would proceed only very slowly in free and open solutions.

In proteins with multiple functions, such as an enzyme capable of catalyzing several different reactions, individual functions are often associated with different domains (as in the protein segment diagramed in Fig. 2-24). Thus protein domains often represent subdivisions of function as well as structure. Multifunctional proteins that have some of their functions in common frequently share domains associated with the similar functions. The shared domains suggest that these related but different proteins may have evolved through a mechanism that mixed existing domains into new combinations (see p. 411).

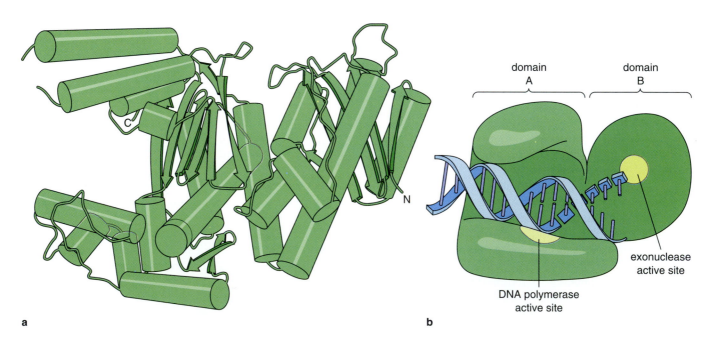

Figure 2-24 Two domains in part of the *E. coli* DNA polymerase enzyme. **(a)** The domains, with alpha helices as cylinders and beta strands as arrows. **(b)** A three-dimensional diagram from the same view showing only the domain surfaces. The part of the enzyme shown catalyzes both the polymerization of DNA from nucleotide subunits (called *DNA polymerase activity*) and the reverse reaction, which removes nucleotides from the end of a nucleotide chain (called *DNA exonuclease activity*). (Redrawn from originals courtesy of T. A. Steitz, from *Trends Biochem. Sci.* 12:288[1987].)

Associations of Multiple Polypeptide Chains into Quarternary Structures

Many complex proteins, such as hemoglobin and antibody molecules, are built up from several polypeptide chains. The same bonds and forces folding individual amino acid chains into tertiary structures hold the chains together. (Figure 2-17b shows the disulfide linkages stabilizing the quaternary structure of antibody molecules.) Hemoglobin and antibody molecules both consist of 4 individual polypeptide chains; some proteins have as many as 10 or 12. The pattern in which the individual polypeptide chains of a multichain protein are held together is called its *quaternary structure.*

Fibrous proteins frequently contain specialized structures conferring high elasticity and tensile strength that are intimate combinations of tertiary and quaternary structure. One of these is the *coiled coil*, a structure in which two long alpha-helical segments wind around each other in a double spiral (see Fig. 2-25a and Information Box 2-4). Coiled coils occur, for example, in several of the fibrous proteins of muscle tissue. A *triple helix* of three alpha-helical polypeptide segments makes up much of the structure of collagen, the primary fibrous protein of the extracellular matrix in animals (Fig. 2-25b; for further details, see Fig. 7-1 and p. 179).

Proteins thus have four levels of structure. The sequence of amino acids in a protein constitutes its primary structure. The arrangement of segments of the amino acid chain into alpha helix, beta strand, or random coil is its secondary structure. Folding of the chain upon itself produces its tertiary or three-dimensional structure. Quaternary structure refers to the number and arrangement of individual polypeptide chains within a complex protein containing more than one chain.

a coiled coil in a myosin segment

b triple helix in a collagen segment

Figure 2-25 Coiled structures of fibrous proteins. **(a)** The coiled-coil structure in a segment of the myosin molecule, in which two polypeptide chains wind in a double helix. **(b)** The triple helix of polypeptide chains in a collagen molecule.

Many types of proteins associate noncovalently by twos as *dimers*, by threes as *trimers*, by fours as *tetramers*, and so on. Many of the transport proteins of the plasma membrane, for example, take the form of dimers or tetramers. In these cases the activity of the protein is dependent on the association; the proteins are inactive, or largely so, in single form.

Combinations of Proteins with Other Substances

The many reactive groups on the side chains of amino acid residues allow proteins to enter a wide range of biochemical interactions with other substances. Covalent linkage of proteins with carbohydrates or lipids produces composite molecules with great importance to cellular functions. Glycoproteins—formed by covalent binding of proteins to carbohydrate units—function as enzymes, antibodies, recognition and receptor molecules at the cell surface, and as parts of extracellular supports such as collagen. All the known proteins located at the cell surface or in the spaces between cells, in fact, are glycoproteins. The attached carbohydrate groups, because of their many sites capable of forming hydrogen bonds, increase conformational stability. The hydrogen bonding capacity also gives carbohydrate groups the ability to take up and partially immobilize water molecules in large quantity. This property makes some glycoproteins highly effective as extracellular cushions or lubricants. The carbohydrate groups increase the resistance of glycoproteins to enzymatic attack; in some glycoproteins, carbohydrates provide groups recognized by receptors at the cell surface.

Lipoproteins—formed by covalent linkage of proteins to lipid units, including individual fatty acids and entire phospholipids—are widely distributed among prokaryotic and eukaryotic cells. Some lipoproteins also have covalent linkages to glycolipids, producing a complex, three-way structure that includes covalently linked protein, lipid, and carbohydrate units. In most of these complexes the lipid units link to either the N- or C-terminal end rather than to residues in the interior of the amino acid chain. Most lipoproteins occur in cellular membranes, particularly in the plasma membrane, where the lipid segment provides an "anchor" tying the protein to the membrane.

Many proteins interact with inorganic ions such as sodium, potassium, zinc, copper, iron, iodine, magnesium, and calcium, through electrostatic attractions set up by charged amino acid residues. The presence of one or more of these ions is necessary for full activity of many proteins, particularly enzymes. Individual amino acid side chains also covalently bind

The Coiled Coil

The coiled coil, which occurs in fibrous regions of proteins such as the long, tail-like extension of the myosin molecule, consists of two alpha-helical amino acid chains twisted into a double spiral. This highly stable structure depends on a regularly repeated sequence of amino acids in the form *a-b-c-d-e-f-g*, in which amino acids *a* and *d* are nonpolar and *b, c, e, f* and *g* are polar (see part **a** of the figure in this box). Among the polar amino acids in the sequence, *g* is positively charged, and *e* is negatively charged.

The twist of the coiled coil brings the hydrophobic

amino acid groups together in the region of contact between the two alpha helices, creating a hydrophobic association that stabilizes the structure (see part **b** of the figure in this box). The association is further stabilized by attractions between the positively and negatively charged amino acid groups in the coil. The outer surfaces of the coiled coil, formed by the amino acids at the *b, c, e, f,* and *g* positions in the repeated sequences, are hydrophilic and stabilize the coil in its interactions with the surrounding aqueous medium.

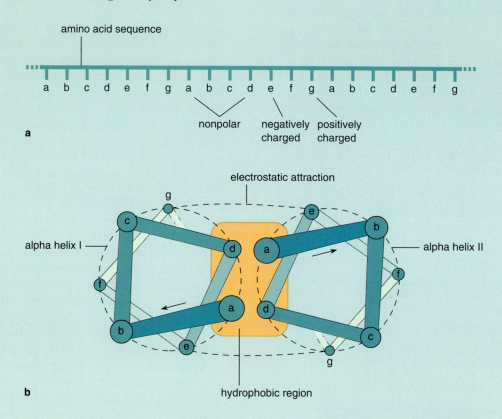

inorganic and organic groups, including hydroxyl, methyl (—CH₃), acetyl (—COCH₃), and phosphate groups. Addition of these groups alters amino acids to other forms and often modifies or regulates activity of the entire protein. Addition and removal of calcium ions or phosphate groups are particularly important as interactions modifying the activity of proteins and form part of pathways regulating cellular activities as diverse as cell development, differentiation and division, transport of substances across membranes, and muscle contraction. Proteins in addition combine ei-

ther covalently or noncovalently with larger, more complex organic structures (as, for example, the electron carriers NAD and FAD shown in Figs. 8-7 and 8-12). In most instances the organic structure adds a vital function to the protein, such as the ability to accept and release electrons in oxidative reactions.

Linked ions or other substances that contribute to the function of an enzyme or other protein are called *cofactors*. Large, complex organic structures are called *coenzymes*; if linked to the enzyme or protein by covalent bonds, they are often termed *prosthetic groups*.

Many complex coenzymes or prosthetic groups of enzymatic proteins are derived from vitamins.

NUCLEOTIDES AND NUCLEIC ACIDS

The two nucleic acids *deoxyribonucleic acid* (*DNA*) and *ribonucleic acid* (*RNA*) are the informational molecules of all living organisms. Besides storing or transmitting information, RNA forms structural and functional parts of units such as the ribosome and in some systems has a catalytic function as *ribozymes*. Both DNA and RNA are long polymers assembled from repeating subunits, the *nucleotides*. The sequence of nucleotides in informational nucleic acid molecules makes up a code that stores and transmits the directions required for assembling all types of proteins.

Individual nucleotides, in addition to providing the building blocks of nucleic acids, carry out a variety of biological functions. Many nucleotides or molecules built on nucleotides transport chemical energy in the form of phosphate groups or electrons from one reaction system to another. (The prosthetic groups NAD and FAD shown in Figs. 8-7 and 8-12, for example, are nucleotides.) Others carry metabolites such as acetyl groups between reactions. Still other nucleotides in cyclic form (see Fig. 6-5) are important in cell regulation.

The Nucleotides

A nucleotide (Fig. 2-26) consists of three covalently-linked parts: (1) a nitrogen-containing base, (2) a five-carbon sugar, and (3) one or more phosphate groups. The nitrogenous bases of naturally occurring nucleotides, the *pyrimidines* and *purines* (Fig. 2-27), are ring structures containing both carbon and nitrogen atoms. Pyrimidines contain one carbon-nitrogen ring, and purines contain two rings. Three pyrimidine bases, *uracil* (*U*), *thymine* (*T*), and *cytosine* (*C*), and two purine bases, *adenine* (*A*) and *guanine* (*G*), are assembled into nucleic acids in cells.

The nitrogenous bases link covalently in nucleotides to one of two five-carbon sugars, either *ribose* or *deoxyribose*. The two sugars differ only in the chemical group bound to their 2'-carbon (boxed in blue in Figure 2-26; the carbons in the sugars are written with primes—1', 2', 3', and so on—to distinguish them from the carbons of the bases, which are written without primes). Ribose has an —OH group at this position and deoxyribose a single hydrogen. A chain of one, two, or three phosphates links to the ribose or deoxyribose sugar at its 5'-carbon to complete the mono-, di-, or triphosphate form of a nucleotide.

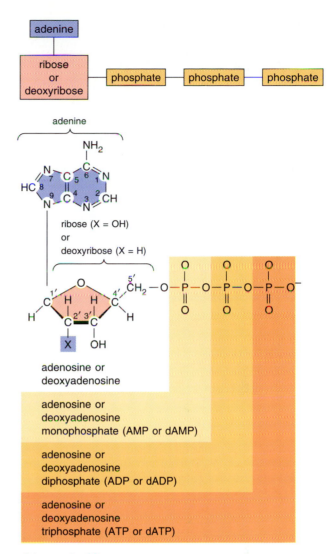

Other nucleotides:
containing guanine: guanosine or deoxyguanosine mono-, di-, or triphosphate
containing cytosine: cytidine or deoxycytidine mono-, di-, or triphosphate
containing thymine: thymidine mono-, di-, or triphosphate*
containing uracil: uridine mono-, di-, or triphosphate*

*Thymidine occurs only in DNA in the deoxyribose form; uridine occurs only in RNA in the ribose form.

Figure 2-26 Structural plan of the nucleotides (see text).

The names used for nucleotides can be confusing. The term nucleo*tide* refers to a complete unit containing all three subunits: a nitrogenous base, a five-carbon sugar, and one or more phosphates. A unit consisting of only the base and sugar without phosphates is called a nucleo*side* and is named according to its nitrogenous base. For example, the base-sugar complex containing adenine and ribose is called *adenosine*; if deoxyribose is the sugar in the complex, it is

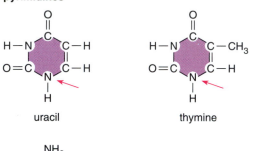

purines

adenine

guanine

pyrimidines

uracil

thymine

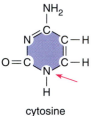

cytosine

Figure 2-27 The purine and pyrimidine bases of nucleic acids and nucleotides. The red arrows indicate where the base links to ribose or deoxyribose sugars in the formation of nucleotides.

called *deoxyadenosine*. To number the phosphate groups, nucleotides are usually named by considering them as nucleosides with added phosphates. The adenine-ribose complex with one phosphate, for example, is called *adenosine monophosphate* (*AMP*); with two phosphates, *adenosine diphosphate* (*ADP*); with three, *adenosine triphosphate* (*ATP*). All the remaining nucleoside mono-, di-, and triphosphates use equivalent names and abbreviated forms (see Fig. 2-26). In the abbreviations a lower case *d* prefix indicates that the deoxyribose form of the five-carbon sugar is present, as in dATP or dGTP.

DNA and RNA

The two nucleic acids DNA and RNA (Fig. 2-28) consist of nucleotides held in chains by a bridging phosphate group that extends between the 5′-carbon of one sugar and the 3′-carbon of the next sugar in line. This arrangement produces a backbone chain of alternating sugar and phosphate groups.

The nucleotides of DNA chains contain the sugar deoxyribose and one of the four bases A, T, G, or C. The nucleotides of RNA chains contain the sugar ribose and one of the four bases A, U, G, or C. Thus DNA and RNA differ in the sugar present, either ribose or deoxyribose, and the presence of either T in DNA or U in RNA. (Thymine and uracil differ only in a methyl group linked to the ring in T but absent in U; see Fig. 2-27.)

Fully processed and finished forms of both DNA and RNA usually contain a number of *modified bases* formed by chemical alteration of the original nucleotides to other types. (Fig. 14-4 shows some of the modified bases.) Modified bases are particularly common in some types of RNA. As many as 10 to 15% of the bases in tRNA molecules, for example, may be modified to other forms.

DNA exists in cells as a *double helix* containing two intertwined chains of nucleotides (Fig. 2-29). In the DNA double helix, discovered by J. D. Watson and F. H. C. Crick in 1953, the sugar-phosphate backbones of the two chains twist together in a right-handed direction to form the double spiral. The backbone chains, which are located at the surface of the double helix, are separated across the helix by a regular space that is filled in by the nitrogenous bases. The bases extend inward from the sugars toward the axis of the helix as *base pairs*, which stack in flat planes roughly perpendicular to the long axis of the helix. (Figure 2-29a shows these planes as seen from the side; Figure 2-30 shows the base pairs as seen from one end of the molecule.) Each complete turn of the double helix includes about 10 base pairs.

The space separating the sugar-phosphate backbones of a DNA double helix is just wide enough to accommodate a purine-pyrimidine base pair. Purine-purine pairs are too wide, and pyrimidine-pyrimidine pairs too narrow to fit this space exactly. The shapes of the bases and the locations of groups capable of forming stabilizing hydrogen bonds impose further restrictions on base pairing. (Three hydrogen bonds can form between guanine and cytosine, and two between adenine and thymine.) The shapes of the bases and the sites available for hydrogen bonding effectively restrict the pairing to the G-C and A-T pairs shown in Figure 2-30.

The pairing restrictions also mean that the sequence of one nucleotide chain fixes the sequence of its partner in the double helix as a *complementary copy* in which C on one side is paired with a G on the opposite side, and an A is paired with a T. This relationship is critical to the processes of DNA replication and RNA transcription, in which one nucleotide chain is used as a *template* for assembly of a complementary copy.

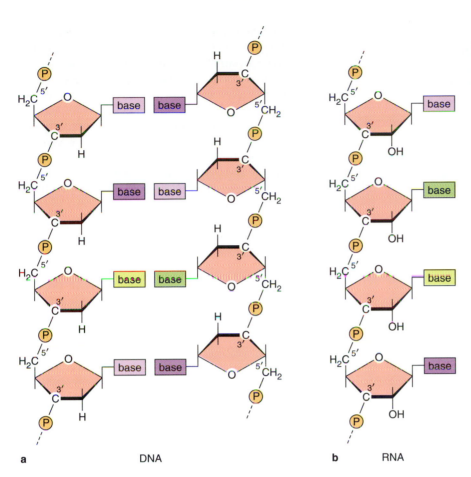

a   DNA

b   RNA

Figure 2-28 Linkage of nucleotides to form the nucleic acids DNA **(a)** and RNA **(b)**. In DNA, any of the four bases adenine (A), thymine (T), cytosine (C), or guanine (G) may be bound at the positions marked *base*. In RNA, A, G, C, or uracil (U) may occur at these sites. P, phosphate group.

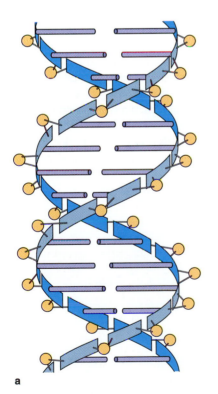

a

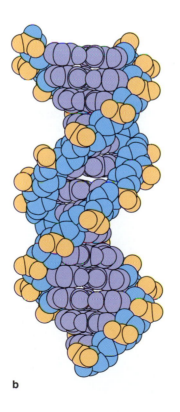

b

Figure 2-29 The DNA double helix. **(a)** The arrangement of sugars, phosphate groups, and bases in the DNA double helix. **(b)** Space-filling model of the DNA double helix. The paired bases, which lie in flat planes, are seen on edge in this view. ([b] redrawn from originals courtesy of W. Saenger and Springer-Verlag, from *Principles of Nucleic Acid Structure*, 1984.)

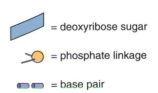

= deoxyribose sugar

= phosphate linkage

= base pair

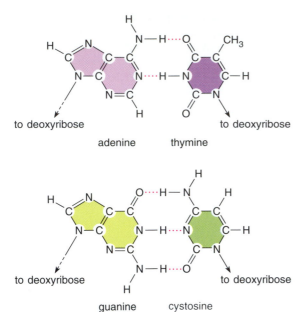

Figure 2-30 The A-T (adenine-thymine) and G-C (guanine-cytosine) base pairs of DNA, as seen from one end of a DNA molecule. Dotted lines designate hydrogen bonds.

Another significant feature of a DNA double helix, also determined by the pattern of hydrogen bonding and fit of the base pairs, is that its two chains are *antiparallel*—they run in opposite directions. This arrangement is most easily understood if the two chains are unwound and laid out flat, as illustrated in Figure 2-28a. The phosphate linkages in the chain on the left, if traced from the bottom to the top, extend from the 5′-carbon of the sugar below to the 3′-carbon of the sugar above each linkage. On the other chain the 5′→3′ linkages run in the opposite direction, from top to bottom. This feature of DNA structure has great significance for both DNA replication and RNA transcription, since a new DNA or RNA chain being copied must run in the direction opposite to its template (see Fig. 14-2).

Three primary intermolecular forces hold the DNA double helix together. One is hydrogen bonding between base pairs in the interior of the molecule. Cumulatively, the hydrogen bonds form a stable structure if the helix includes at least 10 base pairs. Attractive van der Waals forces between the closely packed atoms of the double helix, particularly between atoms in the tightly stacked base pairs, provide the second stabilizing force. The third results from hydrophobic associations among base pairs in the interior of the helix. The nitrogenous bases, which are primarily nonpolar, pack tightly enough to exclude water and form a stable, primarily nonpolar environment in the helix interior.

RNA exists largely as single, rather than double, nucleotide chains in living cells. However, segments of RNA molecules may pair temporarily in double-helical form or may fold back on themselves to set up extensive double-helical regions (see Fig. 15-2). These fold-back double helices and their arrangement are often more important to RNA functions than nucleotide sequence, particularly in noncoding RNAs such as those of ribosomes. RNA molecules also wind temporarily with DNA to form hybrid double helices consisting of one DNA and one RNA chain. Such hybrid helices are set up, for example, during transcription of RNA molecules from DNA templates (see Fig. 14-3).

Like the three-dimensional structures of proteins, the arrangement of DNA and RNA double helices is sensitive to disturbance by conditions in the surrounding medium. Particularly important as a disturbing factor is elevated temperature. As temperature increases, the hydrogen bonds and other forces holding the backbones together are increasingly disturbed until the two chains come apart and unwind. This denaturation, or *melting*, as it is called, takes place in most DNA molecules at temperatures of about 50° to 60°C. In many cases, DNA or RNA molecules can wind back into native form if temperatures return to normal values. Unwinding and rewinding DNA and RNA molecules by adjustments in temperature is often used as an experimental technique in cell and molecular biology (for details, see Information Box 13-1).

For Further Information

Carbohydrates
   in cell walls, *Ch. 7*
   formation in photosynthesis, *Ch. 9*
   oxidation, *Ch. 8*
   as subunits of membrane glycoproteins and glycolipids, *Chs. 4 and 6*

Suggestions for Further Reading

Adams, R. L. P., Burdon, R. H., Campbell, A. M., Leader, D. P., and Smellie, R. M. S. 1981. *The Biochemistry of the Nucleic Acids*, 9th ed. New York: Chapman and Hall.

Barford, D. 1992. Molecular mechanisms for the control of enzymic activity by protein phosphorylation. *Biochim. Biophys. Acta* 1133:55–62.

Blake, C. C. F. 1984. Protein structure. *Trends Biochem. Sci.* 10:421–425.

Brandon, C., and Tooze, J. 1991. *Introduction to Protein Structure*. New York: Garland.

Brown, T. L., and Lemay, H. E. 1985. *Chemistry: The Central Science*, 3rd. ed. New York: Prentice-Hall.

Creighton, T. E. 1984. *Proteins, Structures and Molecular Properties*. New York: Freeman.

Creighton, T. E. 1988. Disulfide bonds and protein stability. *Bioess.* 8:57–63.

Dickerson, R. E., and Geis, I. 1969. *The Structure and Action of Proteins*. New York: Harper & Row.

Dickerson, R. E. 1983. The DNA double helix and how it is read. *Sci. Amer.* 249:94–111 (October).

Doolittle, R. F. 1985. Proteins. *Sci. Amer.* 253:88–99 (October).

Eisenberg, D., and Kautzmann, T. 1969. *The Structure and Properties of Water*. New York: Oxford University Press.

Felsenfeld, G. 1985. DNA. *Sci. Amer.* 253:58–66 (October).

Ginsburg, V., and Robbins, P., eds. 1984. *Biology of Carbohydrates*. New York: Wiley.

Goodsell, D. S., and Olson, A. J. 1993. Soluble proteins: Size, shape, and functions. *Trends Biochem. Sci.* 18:65–70.

Jeffry, G. A., and Saenger, W. 1991. *Hydrogen Bonding in Biological Structures*. New York: Springer-Verlag.

Karplus, M., and McCammon, J. A. M. 1986. The dynamics of proteins. *Sci. Amer.* 254:42–51 (April).

Lehninger, A. L. 1982. *Principles of Biochemistry*. New York: Worth.

Olson, A. J., and Goodsell, D. S. 1992. Visualizing biological molecules. *Sci. Amer.* 267:76–81 (November).

Opdenakker, G., Rudd, P. M., Ponting, C. P., and Dwek, R. A. 1993. Concepts and principles of glycobiology. *FASEB J.* 7:1330–1337.

Rich, A., and Kim, S.-H. 1978. The three-dimensional structure of a transfer RNA. *Sci. Amer.* 238:52–62 (January).

Saenger, W. 1984. *Principles of Nucleic Acid Structure*. New York: Springer-Verlag.

Schultz, G. E., and Schirmer, R. H. 1979. *Principles of Protein Structure*. New York: Springer-Verlag.

Sharon, N. 1980. Carbohydrates. *Sci. Amer.* 243:90–114 (November).

Stillinger, F. H. 1980. Water revisited. *Science* 209:451–457.

Stryer, L. 1988. *Biochemistry*, 3rd ed. New York: Worth.

Williams, R. J. P. 1993. Are enzymes mechanical devices? *Trends Biochem. Sci.* 18:115–117.

Yaron, A., and Naider, F. 1993. Proline-dependent structure and biological properties of peptides and proteins. *Crit. Rev. Biochem. Molec. Biol.* 28:31–81.

Review Questions

1. What determines the chemical activity of an atom?

2. What is an ionic or electrostatic bond? How is this bond formed?

3. What is a covalent bond? How are covalent bonds formed? How are covalent bonds represented in molecular diagrams?

4. What are polar and nonpolar molecules? What conditions produce polarity?

5. What do the terms *hydrophobic* and *hydrophilic* mean? How is the property of being hydrophobic or hydrophilic related to polarity?

6. How is polarity important in molecular and cellular structure?

7. What is a hydrogen bond? What conditions are necessary for formation of a hydrogen bond? In what ways are hydrogen bonds related to polarity?

8. What are van der Waals forces? How can these forces be either attractive or repulsive?

9. Diagram a hydroxyl, carboxyl, aldehyde, carbonyl, amino, sulfhydryl, and phosphate group. What are the overall chemical properties of each group?

10. What structural features do carbohydrates have in common? What is a monosaccharide? A disaccharide? A polysaccharide?

11. What is a polymerization? List several polymerization reactions of importance in biological systems.

12. What is the difference between the α- and β-forms of glucose? Does this difference have any biological significance?

13. What are isomers? Do the different isometric forms of molecules such as sugars and amino acids have any biological significance?

14. What are the primary differences between cellulose, plant starches, and glycogen?

15. Define a lipid.

16. Diagram a saturated and unsaturated fatty acid. What are the characteristics of cellular fatty acids?

17. What is the structural plan of a neutral lipid? Why are neutral lipids nonpolar?

18. Diagram a phospholipid. Why do phospholipids have dual solubility properties?

19. What is a bilayer? How is bilayer formation related to the properties of phospholipids? Could a neutral lipid form a bilayer? Why are bilayers stable?

20. What is a steroid? A sterol? In what way are the solubility properties of steroids and phospholipids similar?

21. What is a glycolipid? Where do glycolipids occur in cells?

22. What functions do proteins carry out in cells?

23. What is the structural plan of an amino acid? List the acidic, basic, polar, and nonpolar amino acids. How does proline differ from the other amino acids?

24. How do amino acids link together to form proteins? Where does this reaction take place in cells?

25. Define the primary, secondary, tertiary, and quaternary structure of proteins. What determines the three-dimensional structure of proteins?

26. What is an alpha helix? A beta strand? A random coil? How do these folding arrangements affect protein structure?

27. What is a protein domain? What is the significance of domains for protein structure and function?

28. What are conformational changes in proteins? How are they related to the functions of proteins? What is denaturation?

29. What are glycoproteins and lipoproteins? Coenzymes? Cofactors? Prosthetic groups?

30. What are the organic subunits of nucleotides? How are nucleotides named? What is a purine? A pyrimidine?

31. What is the difference between ATP and dATP?

32. List the structural similarities and differences between DNA and RNA.

33. What is the DNA double helix? What are base pairs? Can RNA exist in double helical form? What does the term *antiparallel* mean with reference to nucleic acids?

34. List the primary functions of nucleotides and nucleic acids.

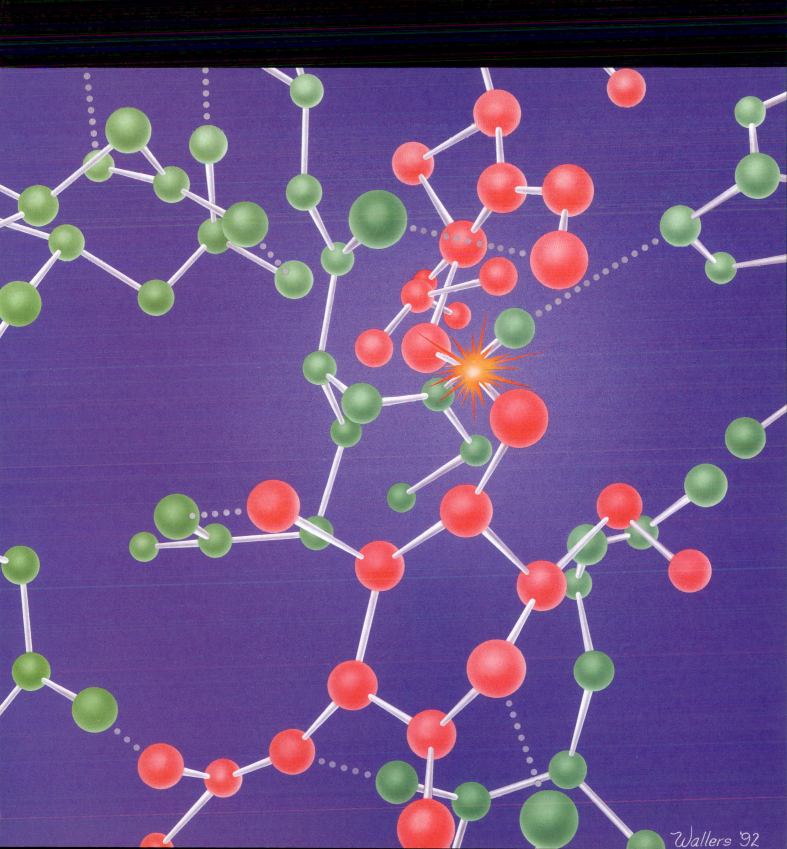

Wallers '92

At the cellular level, literally thousands of biochemical reactions accomplish the special activities we associate with life, among them growth, reproduction, movement, and the ability to respond to stimuli. The progress of these cellular reactions depends on enzymes: Enzymes speed the rate of chemical interactions that otherwise would take place far too slowly at the temperatures characteristic of cellular life.

The timing and location of the reactions occurring inside cells are also precisely regulated. This is accomplished by controls of the time and place at which enzymes are newly synthesized, or by regulation of enzymes already present. As a consequence of controlled synthesis and regulation of enzymes, the right enzymes appear in active form at the right place and time to increase the rate of the biochemical interactions necessary for a particular life activity.

The chemical reactions making up the activities of life are not unique to cells or to the living state. As long as the necessary physical and chemical conditions are supplied, biochemical reactions can proceed in a test tube with no cells present. Not even the enzymes normally catalyzing the reactions are absolutely required: The role of enzymes is simply *to increase the rate of reactions that could still proceed, however slowly, without enzymes.* This is true because all reactions obey the same chemical and physical laws operating anywhere in the universe, whether inside cells or in the outside world.

Therefore, understanding whether biological reactions will take place and predicting their direction depend on a knowledge of the basic chemical and physical laws that govern chemical interactions. Understanding the rate at which such reactions take place inside cells depends on an understanding of enzymes and how they work. These fundamental relationships are the subjects of this chapter.

CHEMICAL AND PHYSICAL LAWS GOVERNING REACTIONS

The possibility and direction of reactions of any kind depend on physical laws deduced through the study of *thermodynamics.* Thermodynamics examines chemical and physical changes and considers their relationship to energy.

Reactions that are possible are called *spontaneous* reactions in thermodynamics. In this usage "spontaneous" does not imply that the change is instantaneous. In thermodynamics, spontaneous reactions are simply reactions that will "go"; they may take place at any rate from unmeasurably slow to practically instantaneous.

Groups of reacting molecules are called *systems* in thermodynamics. A system can be defined to include any reacting molecules of interest. Everything outside the defined system is called the *surroundings.* In undergoing any type of change, such as a chemical reaction, systems go from an *initial state* to a *final state.* If the total amount of matter in the system remains the same during the change, the system is *closed.* If matter is transferred between the system and its surroundings during the change, the system is *open.* Living organisms are open systems because they constantly exchange matter with their surroundings. However, many of the individual reactions within living organisms may operate as closed systems.

Spontaneous Reactions and the Laws of Thermodynamics

Two fundamental laws, the *first* and *second laws of thermodynamics,* have been deduced from the study of energy changes in reacting systems. Taken together, the laws can be used to predict the rate and direction of all chemical and physical reactions, including those taking place in living organisms.

The First Law of Thermodynamics The first law of thermodynamics was developed partly through a study of the many unsuccessful attempts to construct perpetual-motion machines. All such machines, no matter how ingenious, eventually run down and stop. The initial push of energy used to start the machines is inevitably lost to the surroundings as heat resulting from friction between the working parts. To keep the machines running, the lost energy must constantly be replaced by energy added from the surroundings.

Careful measurement of the heat flowing from such machines shows that as they come to a stop, the total amount of heat lost is exactly equivalent to the amount of energy used to start or maintain their motion. These and similar observations in other reacting systems led to the deduction of the first law, which states that in any nonnuclear process involving an energy change (as in the change from mechanical to heat energy in a machine), the *total amount of energy in a system and its surroundings remains constant.* In other words, in such nonnuclear changes *energy can neither be created nor destroyed.*

Because the systems of interest in cell and molecular biology consist of molecules taking part in chemical reactions, it is useful to consider what the total amount of energy means when applied to collections of molecules. One part of the energy content of molecules at temperatures above absolute zero ($-273°C$) is *kinetic energy*, reflected in constant rotation, vibration, and lateral movement of the molecules. The second part of the energy content, *potential energy*, depends on energy contained in the arrangement of atoms and chemical bonds in the molecules.

Kinetic and potential energy are interchangeable. For example, an unlit match at room temperature contains considerable potential energy but only moderate kinetic energy. The potential energy content reflects the complex arrangements of atoms and their bonds in the wood of the match and the phosphorus at its tip. Once the match is struck, much of the potential energy is transformed into kinetic energy, reflected in increased motion of molecules in the match and its immediate surroundings. Some energy also flows from the burning match as heat and light.

The energy content of the reactants in a chemical interaction is usually larger than that of the products. This can be determined directly by burning equal molar quantities of reactants and products separately in bomb calorimeters and comparing the heat given off. Such observations show that spontaneous reactions generally progress to a state in which the reactants have *minimum energy content*. The difference between the energy content of reactants and products is lost to the surroundings as heat. Reactions that release heat to the surroundings as they proceed are *exothermic*.

Although most spontaneous reactions are exothermic and proceed to a state of minimum energy content, there are exceptions. For example, as ice melts spontaneously at temperatures above 0°C, the energy content of the meltwater is higher than that of the ice. In this case the energy added as ice melts is absorbed from the surroundings. Reactions that absorb heat from the surroundings as they proceed are called *endothermic* reactions.

The Second Law of Thermodynamics   The exceptional behavior of melting ice shows that knowing the energy content of the reactants and products is not enough by itself to predict whether a process or chemical reaction will take place spontaneously. The missing element was supplied through thermodynamic investigation into a phenomenon that is part of our common experience. As any type of change takes place, the things involved generally tend to get out of order rather than spontaneously assuming more ordered arrangements (your room is probably the best example of this phenomenon). The scientific study of this phenomenon led to deduction of the second law

of thermodynamics: In any process involving a spontaneous change from an initial to a final state, *the total disorder of the system plus its surroundings always increases*. In thermodynamics, disorder is termed *entropy*. If the system and its surroundings are defined as the entire universe, the second law means that as changes take place anywhere in the universe, the total disorder or entropy of the universe constantly increases.

Using the First and Second Laws to Predict the Rate and Direction of Chemical Reactions

Combining the first and second laws allows us to predict with confidence whether a reaction should be possible and spontaneous. The reactions most likely to proceed spontaneously are those in which *the system goes toward a condition of minimum energy and maximum entropy* (*maximum disorder*) as it goes from the initial to final states.

Energy released by reactions moving spontaneously to the final state is *free energy*, meaning energy that is available to do work. In living organisms the primary work accomplished by free energy is the chemical and physical work involved in activities such as the synthesis of cellular molecules, movement, and reproduction.

It is possible for reactions to proceed spontaneously when only one of the energy and entropy conditions is met. A reaction may take place in which the entropy as well as the energy content of a system decreases as it goes to the final state. Or, conversely, although entropy increases, the energy content of the system may rise between the initial and final state. In such reactions the change in energy content and entropy in the system act as opposing tendencies, one tending to favor the reaction and the other to oppose it. In these interactions the *balance* of the opposing tendencies determines whether the reaction will proceed spontaneously.

For example, when ice melts, the energy content of the system increases, which is an unfavorable condition for a spontaneous reaction. However, there is a very large increase in entropy, reflecting the change in the water molecules from the highly ordered arrangement in ice crystals to the more random and disordered condition in liquid water. The balance between the two opposing factors is favorable enough to make the change proceed spontaneously. This change does not violate the first law of thermodynamics. Because the required energy is absorbed from the surroundings, the total energy content of the system and its surroundings remains the same.

The reverse process, the conversion of liquid water to ice, provides a convenient example of a change in which a large loss in energy content in the system compensates for a decrease in entropy. At temperatures below 0°C, water freezes spontaneously, even though the arrangement of water molecules in ice crystals is more ordered than in liquid water. Although entropy decreases in this reaction, the very large drop in energy content as water freezes is sufficient to balance the entropy change.

Many biological systems appear to follow a pathway in which both laws of thermodynamics are violated: Energy content increases and entropy decreases. For example, as a fertilized egg develops spontaneously into an adult animal, it synthesizes more and more organic molecules from less complex substances, greatly increasing its energy content and decreasing its entropy. However, the increase in energy content and decrease in entropy are illusionary in this and similar biological situations because the system is incompletely defined—not all the reactants and products have been included.

In order to include all reactants and products, the system in the initial state—fertilization—must include the complex organic molecules such as carbohydrates and fats that the developing animal uses as energy sources. At the final state, when development is complete, the system must include the waste products produced by the animal, all simple substances such as water and carbon dioxide that are much less complex than the organic molecules used as fuels. When these necessary reactants and waste products are included, the total change satisfies both criteria—the system moves toward both minimum energy and maximum entropy.

Reversible Reactions

In many chemical reactions the conversion of reactants to products does not proceed entirely to completion because other factors, concentration of reactants and products or pressure, become critical. With respect to concentration differences, the minimum energy and maximum entropy state is one in which the reactants and products are present in equal concentrations in the initial and final states. Pushing the system toward all products and no reactants in the final state represents a very high degree of order (greatly reduced entropy) with respect to concentrations. The entropy reduction required for a reaction to go completely in the direction of products is often so high that it cannot be balanced by the energy and entropy changes in the reaction itself.

Reactions of this type proceed to a point at which the tendency of the reaction to go toward products with minimum energy and maximum entropy is exactly balanced by the tendency to move backward toward the most disordered state with respect to concentration differences, in which there are equal numbers of reactants and products. The balance achieved is called the *equilibrium point* for the reaction. The balance is not static; at the equilibrium point equal numbers of molecules change constantly from products to reactants, and from reactants to products. In other words, at the equilibrium point, *the rates of the forward and backward reactions are equal.* If more reactants are added, the reaction proceeds toward products until the equilibrium is reestablished. Conversely, if more products are added, the reaction moves toward reactants until the system again balances. Such reactions are termed *reversible* and written with a double arrow:

$$\text{reactants} \rightleftharpoons \text{products} \qquad (3\text{-}1)$$

Most biological reactions are reversible.

The equilibrium point of a reversible reaction depends primarily on the difference in energy content and entropy between the reactants and products. If this difference is large, so that the products contain considerably less energy and much greater entropy than the reactants, the equilibrium point lies far to the right, and most of the reaction mixture consists of products. If the energy and entropy difference between reactants and products is small, the equilibrium point lies farther to the left, and a greater proportion of reactants is present in the equilibrium mixture.

Pressure changes have similar effects on reversible reactions. Increases in pressure as reactants go to products work as a condition opposing the tendency of a reaction to proceed in a forward direction according to energy and entropy changes. In such reactions the tendency to go forward is balanced by increased pressure at the equilibrium point.

The ability of reversible reactions to proceed in either direction allows them to be written reversibly, so that the names given to reactants and products are interchangeable. For example, in one important biological reaction the amino acid glutamine is converted to glutamic acid by the removal of an amino group:

$$\text{glutamine} + H_2O \longrightarrow \text{glutamic acid} + NH_3 \quad (3\text{-}2)$$

The reaction proceeds far in the direction of products and releases free energy.

Because the reaction is reversible, it can be written in the opposite direction:

$$\text{glutamic acid} + NH_3 \longrightarrow \text{glutamine} + H_2O \quad (3\text{-}3)$$

The reaction will actually proceed spontaneously in this direction if the solution initially contains only glutamic acid and NH_3. However, very little glutamine and water are expected as products because the equilibrium point, when the reaction is written in this way, now lies far to the left.

How Living Organisms Push Reversible Reactions Uphill

Cells carry out many reactions such as Reaction 3-3 even though the equilibrium point lies far to the left. How do cells circumvent this problem, so that the products can be made in useful quantities? The answer is a biological "trick" in which the desired uphill reaction is joined, or *coupled*, with another reaction that proceeds far downhill. The total change for the combined reaction produces products with minimum energy content and maximum entropy, so that the equilibrium point now lies far to the right.

One of the most interesting features of the mechanism coupling uphill and downhill reactions in living cells is that the primary coupling agent, the nucleotide *ATP* (*adenosine triphosphate*; see Fig. 3-1a and p. 63), is the same in all organisms from bacteria to humans. ATP consists of a nitrogenous base, *adenine*, linked to a five-carbon sugar, *ribose*. The ribose sugar links in turn to a chain of three phosphate groups.

Much of the energy available in the ATP molecule is associated with the arrangement of the three phosphate groups. The oxygens of the phosphate groups carry a strongly negative charge at cellular ranges of acidity. The three phosphate groups line up side by side in ATP, in highly ordered positions that bring the strongly negative charges close together. Removing one or two of the phosphates produces a large drop in energy content and greatly increases the entropy. As a consequence, free energy is released in large quantity:

$$ATP + H_2O \longrightarrow ADP + P_i + \text{free energy} \quad (3\text{-}4)$$

$$ATP + H_2O \longrightarrow AMP + 2P_i + \text{free energy} \quad (3\text{-}5)$$

The products of these reactions are *adenosine diphosphate* (ADP) and *adenosine monophosphate* (AMP; see Fig. 3-1b). P_i represents the phosphate group removed in the reactions (i = inorganic).

Cells use ATP breakdown to drive uphill reactions by coupling Reaction 3-4 or 3-5 to the reactions requiring energy. Coupling is accomplished not simply by running the energy-requiring and energy-releasing reactions simultaneously, but through a temporary chemical attachment of ATP or its breakdown products to intermediates in the reaction series, or to the enzyme catalyzing the reaction.

For example, the reaction synthesizing glutamine from glutamic acid runs far to the right if Reactions 3-3 and 3-4 are coupled. In this coupling the phosphate group removed from ATP is chemically attached to glutamic acid, in an initial downhill reac-

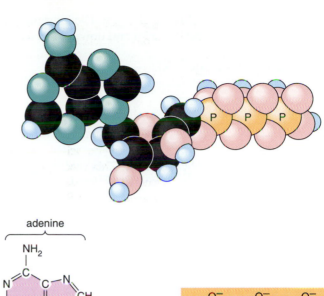

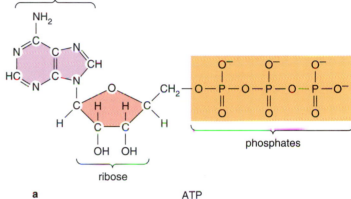

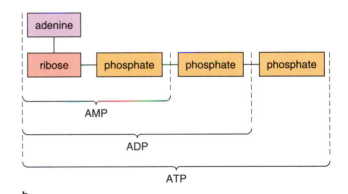

Figure 3-1 Adenosine triphosphate (ATP) and its derivatives adenosine diphosphate (ADP) and adenosine monophosphate (AMP). **(a)** Space-filling model and chemical structure of the ATP molecule. P, the phosphorus atoms of the three phosphate groups. **(b)** Sequential removal of the phosphate groups produces ADP and AMP.

tion that forms an intermediate substance, glutamyl phosphate:

$$\text{glutamic acid} + ATP \longrightarrow \text{glutamyl phosphate} +$$
$$ADP + \text{free energy} \quad (3\text{-}6)$$

The glutamyl phosphate, which remains attached to the enzyme catalyzing the reaction, retains much of the energy released by ATP breakdown. The intermediate then reacts with NH_3 in a second downhill step:

$$\text{glutamyl phosphate} + NH_3 \longrightarrow \text{glutamine} + P_i + \text{free energy} \quad (3\text{-}7)$$

Combining 3-6 and 3-7 and canceling like terms gives the overall reaction:

$$\text{glutamic acid} + NH_3 + \text{ATP} \longrightarrow \text{glutamine} + \text{ADP} + P_i + \text{free energy} \quad (3\text{-}8)$$

The end products—glutamine, ADP, and phosphate—have considerably less energy content and much greater entropy than the reactants, which include glutamic acid, ATP, and NH_3. As a result, the total reaction runs spontaneously far to the right and releases free energy.

Cells use energy released by cellular oxidations to replenish the ATP broken down in coupled reactions such as 3-8. For example, large quantities of free energy (686,000 cal/mol) are released if glucose is oxidized by burning it in air:

$$C_6H_{12}O_6 + 6O_2 \longrightarrow 6CO_2 + 6H_2O +$$
$$\text{free energy} \quad (3\text{-}9)$$

In cells, glucose oxidation is coupled to ATP synthesis:

$$C_6H_{12}O_6 + 38\text{ADP} + 38P_i + 6O_2 \longrightarrow$$
$$6CO_2 + 6H_2O + 38\text{ATP} + \text{free energy} \quad (3\text{-}10)$$

The coupled reaction is still spontaneous, but it releases less free energy (about 260,000 cal/mol). The difference in free energy released by Reactions 3-9 and 3-10 represents energy packed into ATP. Reaction 3-10 is accomplished through many stepwise reactions in which ATP or its breakdown products are temporarily attached to intermediates or enzymes catalyzing individual steps (details of these reactions are presented in Chapter 8).

The effect of adding or removing phosphate groups in the ATP/ADP/AMP system is much like compressing or releasing a spring. Adding phosphate groups, up to a limit of three, compresses the spring and stores energy. Removing one or both terminal phosphates releases the spring and makes energy available for useful work.

The Standard Free Energy Change

The amount of energy released by ATP breakdown, or required for the synthesis of glutamine from glutamic acid, is often shown in biological reactions in calories

per mole (defined in footnote on p. 34), as determined from reactions run under standard conditions. (Standard conditions include a temperature of 25°C, constant pressure at 1 atmosphere, and the reaction initiated with the reactants and products in equal quantities of 1 mole.) Under these conditions the reaction will proceed from the initial conditions to equilibrium and give off or absorb free energy as it proceeds. The free energy released or absorbed under these conditions is termed the *standard free energy change*, designated $\Delta G°$ (the G is in honor of J. W. Gibbs, an American physicist who was a major contributor to the science of thermodynamics). If standard conditions used to measure the free energy change include maintenance of the reaction at neutral pH (pH = 7; see Information Box 2-2), the standard free energy change is written as $\Delta G°'$, in which the prime (') indicates standard pH.

The standard free energy change $\Delta G°$ or $\Delta G°'$ indicates directly whether a reaction is spontaneous. It also provides a way to compare different reactions on the same scale and to estimate in relative terms the amount of energy released or required in running them (standard free energy changes for a number of reactions of biological interest are given in Table 3-1).

For example, running Reaction 3-3 to the right requires an input of 3400 calories for each mole of glutamic acid fully converted to glutamine under standard conditions. By convention, the standard free energy change is written as a positive number if a reaction requires the addition of energy:

$$\text{glutamic acid} + NH_3 \longrightarrow \text{glutamine} + H_2O \quad (3\text{-}11)$$
$$\Delta G° = +3400 \text{ cal/mol}$$

A positive value for the standard free energy change indicates that a reaction is nonspontaneous.

Running Reaction 3-4 to the right releases 7000 calories for each mole of ATP converted into ADP + P_i. The standard free energy change is written as a negative number if a reaction releases free energy:

$$\text{ATP} + H_2O \longrightarrow \text{ADP} + P_i \quad (3\text{-}12)$$
$$\Delta G° = -7000 \text{ cal/mol}$$

A negative value for the standard free energy change indicates that a reaction is spontaneous.

This information can be used to estimate the free energy released when Reactions 3-11 and 3-12 are coupled to drive the synthesis of glutamine under standard conditions. Adding +3400 cal/mole and −7000 cal/mole gives an expected free energy change of −3600 calories for each mole of glutamic acid converted to glutamine by the coupled reaction:

Table 3-1 Standard Free Energy Changes at pH7 and 25°C for Some Chemical Reactions of Biological Interest

Reaction	$\Delta G'$ (cal/mol)
Oxidation	
glucose + $6O_2$ → $6CO_2$ + $6H_2O$	−686,000
lactic acid + $3O_2$ → $3CO_2$ + $3H_2O$	−326,000
palmitic acid + $23O_2$ → $16CO_2$ + $16H_2O$	−2,338,000
Hydrolysis	
sucrose + H_2O → glucose + fructose	−5,500
glucose-6-phosphate + H_2O → glucose + H_3PO_4	−3,300
glycylglycine + H_2O → 2 glycine	−4,600
Rearrangement	
glucose-1-phosphate → glucose-6-phosphate	−1,745
fructose-6-phosphate → glucose-6-phosphate	−400
Ionization	
CH_3COOH + H_2O → H_3O^+ + CH_3COO^-	+6,310
Elimination	
malate → fumarate + H_2O	+750

SOURCE: From *Bioenergetics*, 2nd ed., by Albert L. Lehninger (Menlo Park, CA: Benjamin/Cummings Publishing Company, 1971), p. 32. Reprinted by permission.
NOTE: $\Delta G' = \Delta G°$ at pH 7.

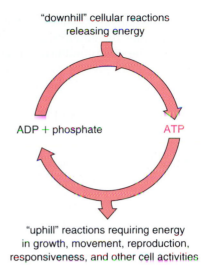

Figure 3-2 ATP and ADP cycle between energy-releasing and energy-requiring activities in the cell. Reactions that release energy are indirectly coupled to the synthesis of ATP from ADP and phosphate (top half of diagram). Reactions that require energy are made favorable by coupling them to the breakdown of ATP to ADP and phosphate, which releases energy for the reactions to proceed (bottom half of diagram).

$$\text{glutamic acid} + NH_3 + ATP \longrightarrow$$
$$\text{glutamine} + ADP + P_i \quad (3\text{-}13)$$
$$\Delta G° = -3600 \text{ cal/mol}$$

Because the value for the standard free energy change is negative, the reaction is expected to be spontaneous and release energy.

It should be kept in mind that $\Delta G°$ is valuable only for making comparisons between reactions. The actual energy released or required for a reaction taking place inside cells depends on the initial concentrations of reactants and products, temperature, pressure, and pH, which under biological conditions are likely to differ from the standards used in calculating $\Delta G°$. In particular, the breakdown of ATP to ADP + P_i is likely to release considerably more than 7000 calories per mole at the ATP under cellular conditions.

In coupled reactions an uphill reaction requiring energy is linked to the breakdown of ATP to produce an overall reaction that is spontaneous and releases energy. Almost all the uphill chemical, electrical, and mechanical work of cells is made energetically favorable in this way, including the energy-requiring activities of growth, reproduction, movement, and response to stimuli (Fig. 3-2). The ATP broken down to drive these reactions is replenished through energy released by cellular oxidations.

THE ROLE OF ENZYMES IN BIOLOGICAL REACTIONS

Enzymes and Enzymatic Catalysis

The laws of thermodynamics apply to chemical reactions anywhere in the universe. No reactions can violate the rule that they must proceed to a level of minimum energy and maximum entropy or to a favorable balance between the two factors. Then what effects do enzymes have on biochemical reactions? The answer is that enzymes simply increase the *rate* at which spontaneous reactions take place (rate = number of reactant molecules converted to products per unit time). Enzymes cannot make a reaction occur that would not proceed spontaneously without the enzyme.

The same principles apply to reversible reactions. Enzymes do not alter the equilibrium point of a reversible reaction. The same equilibrium is established whether an enzyme is present or not—an enzyme simply increases the rate at which a reaction reaches equilibrium.

Although spontaneous, most biological reactions would proceed so slowly without enzymes that their rate would be essentially zero at the temperatures characteristic of living organisms. For example, the oxidation of glucose to CO_2 and H_2O is spontaneous and proceeds almost completely in the direction of the products. However, without enzymes, glucose oxidation occurs so slowly at physiological temperatures that the rate is essentially unmeasurable. In cells, enzymes speed the oxidation of glucose in a number of substeps. Even though the overall pathway is complex, glucose oxidation requires only seconds or minutes at the relatively low temperatures characteristic of living organisms. Enzymes thus facilitate and speed reactions within the narrow range of temperatures that can be tolerated by living organisms.

The increases in rate achieved by enzymes, depending on the enzyme and reaction, range from a minimum of about a million to as much as a trillion times faster than the uncatalyzed reaction at equivalent concentrations and temperatures (Table 3-2). In some instances the rates are as much as a billion times faster than even the rates achieved by inorganic catalysts used in the chemical industry.

Enzymes and Activation Energy

How do enzymes accomplish these feats? Their most basic and fundamental effect is to reduce the *activation energy* required for a chemical interaction to proceed. Activation energy is the energy barrier over which the molecules in a system must be raised for a reaction to take place (Fig. 3-3a). This condition is analogous to a rock resting in a depression at the top of a hill (Fig. 3-3b). As long as the rock remains undisturbed, it will not spontaneously roll downhill, even though the total "reaction"—the progression of the rock downward—is energetically favorable. In this physical example the activation energy would be the effort required to raise the rock over the lip of the depression.

The requirement for activation energy raises the question of why reactions proceed spontaneously at all. What is the source of the energy required to push molecules over the energy barrier? Spontaneous movement over the barrier occurs because molecules, unlike the rock in Figure 3-3b, are in constant motion at temperatures above absolute zero. Although the average amount of movement, or kinetic energy, is below the amount required for activation, some mo-

Table 3-2	Enzymatic Increases in Reaction Rates
Enzyme	Increase in Reaction Rate over Uncatalyzed Rate
Carbonic anhydrase	1.1×10^8
Creatine kinase	4×10^8
Hexokinase	8×10^{10}
Lysozyme	1×10^7 to 1×10^9
Phosphorylase	9×10^{11}
Serine proteases	1×10^5 to 1×10^{10}
Urease	1×10^{14}

lecular collisions may raise a number of molecules to the energy level required for the reaction to proceed. The higher the activating barrier, the fewer the molecules that will proceed over the energy barrier per unit time.

The probability that molecules in the reacting system will go over the barrier increases with increasing temperature. As the temperature rises, both the speed of individual molecules and the frequency of collisions increase, making it more likely that sufficiently forceful collisions will occur at the correct angle and place. Once large numbers of molecules begin to pass over the barrier, the reaction is frequently self-sustaining—the molecules being converted to products release enough heat to push the remaining reactants over the barrier. Glucose oxidation may be pushed into such a self-sustaining reaction, for example, by igniting glucose over an open flame.

Ignition is obviously not a satisfactory approach for pushing reacting molecules over the activation barrier in biological systems. Rather than igniting the reactants, enzymes speed reactions by *lowering the activation energy required for a reaction to proceed* (Fig. 3-4). Lowering the activation energy increases the possibility that molecules will gain enough energy to pass over the energy barrier at normal temperatures.

Characteristics of Protein-Based Enzymes

Enzymes are protein molecules that are tailored to recognize and bind specific reactants and speed their conversion into products. These proteins are responsible for increasing the rates of practically all of the many thousands of reactions taking place inside cells.

All enzymatic proteins have several characteristics in common (Table 3-3). Enzymes combine briefly with the reacting molecules during catalysis and are released unchanged when the reaction is complete. Depending on the enzyme, the combination may occur through the temporary formation of any type of

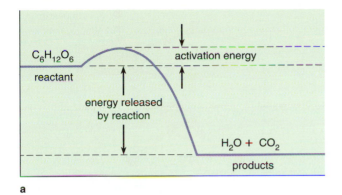

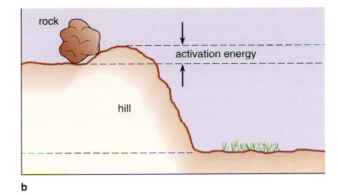

a   b

Figure 3-3 Activation energy. **(a)** The activation energy for the oxidation of glucose, an energy barrier over which glucose molecules must be raised before they can be oxidized to H_2O and CO_2. **(b)** An analogous physical situation in which a rock is poised in a depression at the top of a hill. The rock will not move downward unless activating energy is added by raising it over the lip of the depression.

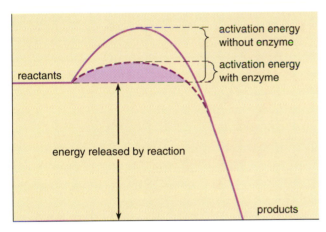

Figure 3-4 Enzymes increase the rate of a spontaneous reaction by reducing the activation energy. This reduction allows biological reactions to proceed rapidly at the relatively low temperatures that can be tolerated by living organisms. Enzymes combine briefly with the reactants as a part of the mechanism reducing the activation energy.

Table 3-3 Characteristics of Enzymatic Proteins
1. Enzymes combine briefly with reactants during an enzyme-catalyzed reaction.
2. Enzymes are released unchanged after catalyzing the conversion of reactants to products.
3. Enzymes are specific in their activity: each enzyme catalyzes the reaction of a single type of molecule or a group of closely related molecules.
4. Enzymes are saturated by high substrate concentrations.
5. Many enzymes contain nonprotein groups called cofactors, which contribute to their activity. Inorganic cofactors are all metallic ions. Organic cofactors, called coenzymes, are complex groups derived from vitamins.

chemical bond, association, or attraction—ionic, covalent, or hydrogen bonds, polar or nonpolar associations, or van der Waals forces.

Because enzymes are released unchanged after a reaction, a single enzyme molecule may combine repeatedly with reactants and release products. The rate of combination and release, known as the *turnover number*, lies near 1000 per second for most enzymes. However, some enzymes have turnover numbers as small as 100 per second or as large as 10 million per second. As a result of enzyme turnover, a relatively small number of enzyme molecules can catalyze a large number of reactant molecules.

Enzymes catalyze only a single type of biochemical reaction and combine with only a single type of molecule or a group of closely related molecules. This characteristic of enzymatic proteins is known as *specificity*. The specific molecule or molecular group whose reaction is catalyzed is known as the *substrate* of the enzyme.

The part of an enzyme that combines with a substrate molecule is the *active site*. In most enzymes the active site is located in a cavity or pocket on the enzyme surface, frequently within a cleft marking the boundary between two or more major domains. Within the cleft or pocket, amino acid side groups are

Figure 3-5 The cleft forming the active site of the lysozyme enzyme, and the fit of the enzyme's substrate (in red) into the cleft, as deduced from X-ray diffraction (see p. 802). The substrate is a polysaccharide consisting of a chain of six-carbon sugars. The reaction catalyzed by the enzyme breaks the sugar chain by hydrolysis at the site indicated by the arrow. (Redrawn from an original courtesy of D. C. Phillips, from *Biochem. J.* 193: 553[1981].)

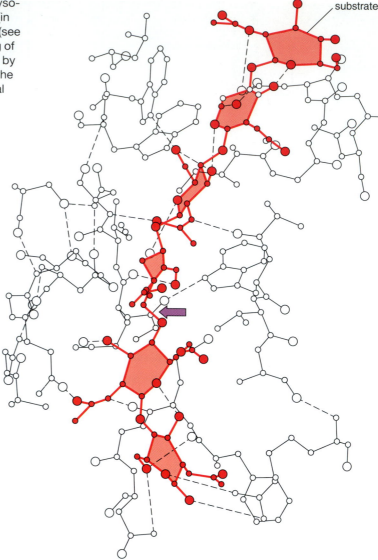

substrate

situated to fit and bind parts of substrate molecules that are critical to the reaction catalyzed by the enzyme (as, for example, in the active site shown in Fig. 3-5). The active site also separates substrate molecules from the surrounding solution and places them in environments with unique characteristics, including partial or complete exclusion of water.

Thousands of different enzymes have been detected and described. They vary from relatively small molecules with single polypeptide chains containing as few as 100 amino acids, to large complexes with several to many polypeptide chains totaling thousands of amino acids. Many enzymes include an inorganic ion or a nonprotein organic group that contributes to their catalytic function. These groups, called *cofactors*, may link to the enzyme by covalent, ionic, or hydrogen bonds or other attractions.

The inorganic cofactors, all metallic ions, include iron, copper, magnesium, zinc, potassium, manga-

nese, molybdenum, and cobalt. When present as cofactors, these ions contribute directly to the reduction of activation energy by the enzyme. The organic cofactors, called *coenzymes*, are all complex chemical groups of various kinds, many of them derived in higher animals from vitamins (Table 3-4; Figs. 8-7 and 8-10 show representative coenzymes). Coenzymes frequently act as carriers of chemical groups, atoms, or electrons removed from substrates during reactions. When tightly linked to an enzyme by covalent bonds, coenzymes are often called *prosthetic groups*.

How Enzymes Lower the Energy of Activation

The mechanisms by which enzymes lower the energy of activation are still not totally understood. However, the mechanisms are believed to be directly or indirectly related to achievement of what is known as the *transition state* for a reaction. During any chemical in-

Table 3-4	Some Coenzymes and Their Vitamin Sources		
	Coenzyme	Vitamin Source	Units Carried
NAD (nicotinamide adenine dinucleotide)		Nicotinic acid	Electrons and hydrogen
CoA (coenzyme A)		Pantothenic acid	Acetyl groups
FMN (flavin mononucleotide)		Riboflavin	Electrons and hydrogen
FAD (flavin adenine dinucleotide)		Riboflavin	Electrons and hydrogen
TPP (thiamin pyrophosphate)		Thiamin	Aldehyde groups

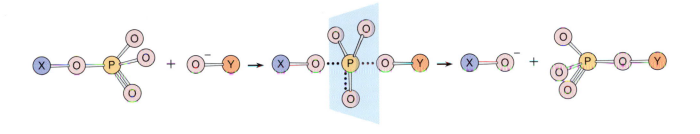

transition state

Figure 3-6 Formation of the transition state in transfer of a phosphate group between the two molecules X and Y. The transition state is unstable and easily pushed in the direction of either reactants or products.

teraction the reactants briefly enter a state in which old chemical bonds are incompletely broken and new ones are incompletely formed. In this transition state, electron orbitals assume intermediate positions between their locations in the reactants and their positions in the products. The transition state is highly unstable and can easily move in either direction with little change in energy—forward toward products or backward toward reactants. In effect, achievement of the transition state places a reacting system in a poised and precariously balanced position at the top of the activation energy barrier.

For example, in the transfer of a phosphate group from one molecule to another, a transition state is set up in which both molecules (shown as X and Y in Fig. 3-6) link to the phosphate group for a fraction of a second via transitory bonds (dotted lines in Fig. 3-6). The oxygens attached to the phosphate groups move to a position intermediate between the positions they take when the phosphate group links to either X or Y. This unstable state can change readily in the direction of either products or unchanged reactants.

Supposedly, an enzyme's active site matches the transition state for the reaction it catalyzes. By binding the reactants the enzyme "warps" them toward the transition state, thereby facilitating their conversion to reactants.

Testing the Transition State Hypothesis The idea that pushing molecules toward the transition state under-

lies enzymatic catalysis was first proposed by Linus Pauling in 1946. Some years after Pauling advanced his hypothesis, W. P. Jencks realized that the hypothesis could be tested with antibodies tailored to fit the transition state. Antibodies are protein molecules generated in higher animals in response to exposure to a foreign substance called an *antigen*. The antibody molecules contain a binding site tailored exactly to fit the antigen. Combination of an antibody with its antigen leads to removal of the invading substance from the body (see Chapter 19 for details). If an antibody could be made to bind the transition state for a reaction, Jencks reasoned, it should be able to act as a catalyst as well as an antibody.

To test this hypothesis, Jencks proposed that *transition state analogs* be injected into animals as antigens. These analogs are molecules that happen to have a stable structure that resembles the transition state for a particular reacting system. The antibodies developed in response should fit the analogs exactly and be able to push the analogous reactants into the transition state. In 1986 two groups working independently, those of R. A. Lerner and P. G. Schultz, successfully tested Jencks's hypothesis by showing that antibodies can indeed be induced to act as enzymes by modeling them to the transition state of a reacting system.

For example, the Lerner group studied the enzymatic catalysis of a typical biological reaction called an *acyl transfer* (Fig. 3-7a). The transition state for this reaction is mimicked by a group of stable, unrelated

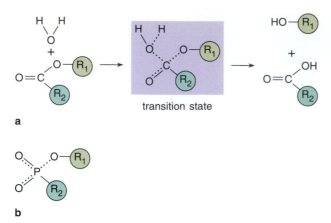

Figure 3-7 The acyl transfer reaction used by R. A. Lerner and his coworkers to test the ability of antibodies tailored to fit the transition state to act as enzymes (see text). **(a)** Formation of the transition state in an acyl transfer reaction. R_1 and R_2 designate organic groups. **(b)** A phosphonate ester, a stable structure that closely resembles the transition state of acyl transfer reactions.

molecules called *phosphonate esters* (Fig. 3-7b). By injecting phosphonate esters into test animals, the Lerner group was able to induce formation of antibodies tailored to fit the transition state for the acyl transfer reaction. The resulting antibodies, as predicted by the Pauling-Jencks hypotheses, were able to act as enzymes (or *abzymes*, as they are now called) and speed the rate of the acyl transfer reaction.

The results of these experiments directly support the proposal that achievement of the transition state is an important part of enzymatic catalysis. The technique of converting antibodies to enzymes by tailoring them to fit the transition state also opens possibilities for the production of "designer enzymes"—artificial enzymes designed specifically to catalyze a reaction desired in research, medicine, or industry.

Enzymatic Mechanisms Contributing to the Transition State   A number of mechanisms operating singly or together at the active site probably contribute to formation of the transition state or help to push reacting molecules over the activation barrier once they are in the transition state. One is *bringing reacting molecules into close proximity*. Many reactions involve combination or interaction of two or more reactant molecules. For the reaction to take place, the substrate molecules must collide. The required collisions may be rare among reactant molecules suspended in free solution, particularly if the substrates are present in low concentrations. Binding at the active site of an enzyme brings the reactants close together, raising their effective concentration in the active site to many times the concentration in the surrounding solution.

A second contributing mechanism is *orienting reactants in positions favoring their interaction*. Binding at the active site may bring substrate molecules into an arrangement in which they can collide at exactly the correct positions and angles required for achievement of the transition state. Orientation of reactant molecules at the active site may also restrict the movement of parts of the reactants around single bonds, thereby confining them to the positions most favorable for reaction.

The third contributing mechanism is *exposing reactant molecules to altered environments that promote their interaction*. Some reactions, for example, take place more readily in nonpolar environments. Active sites may create such an environment by binding reactants so closely that water molecules are excluded. Another important environmental change is creation of acidic or basic conditions by groups in the active site that release or take up H^+.

Conformational changes, involving movements of amino acid chains in or near the active site, may contribute to one or more of these mechanisms. For example, conformational changes in response to substrate binding may produce forces in the active site warping the substrate toward the transition state. In this sense, enzymatic activity may have a mechanical as well as chemical basis.

The Active Site, Transition State, and Reversible Reactions

The contemporary view of the active site as fitting most tightly to the transition state explains how enzymes can catalyze reversible reactions in either direction. The active site, while binding most strongly to the transition state, probably also has conformations that fit the reactants and the products. This is not as difficult to achieve as it might at first appear because the reactants, transition state, and products of enzymatically catalyzed reactions generally have similar molecular structures. Whether reactants or products bind to the active site depends on their relative concentrations in the solution surrounding the enzyme. If the reactants are present in highest concentration, collisions between the enzyme and reactant molecules are more likely and the reactants bind most frequently. If the products are present in highest concentration, the reverse is true. Once the molecules bind to the enzyme, they are pushed toward the tightly bound but unstable transition state, from which they can easily move in either direction, toward reactants or products.

Factors Affecting Enzyme Activity

A number of external factors affect the activity of enzymes in speeding conversion of reactants to products. These factors, including variations in the concentration of substrate molecules, temperature, and pH, speed or slow enzymatic activity in highly characteristic patterns.

Substrate Concentration   Enzymes react distinctively to alterations in the concentration of reacting molecules. At very low substrate concentrations, collisions between enzyme and substrate molecules are infrequent and the reaction proceeds slowly. As the substrate concentration increases, the reaction rate initially increases proportionately as collisions between enzyme molecules and the reactants become more frequent (Fig. 3-8). When the enzymes begin to approach the maximum rate at which they can combine with reactants and release products, the effects of increasing substrate concentration diminish. At the point at which the enzymes are cycling as rapidly as possible, further increases in substrate concentration have no effect on the reaction rate. At this point the enzyme is *saturated* and the reaction remains at the saturation level represented by the horizontal dashed line in Figure 3-8.

The characteristic saturation curve in Figure 3-8 provides a valuable biochemical tool for determining whether a given reaction or process is speeded by an enzyme. To determine whether enzymatic catalysis is involved, the concentration of reactants is increased experimentally and the rate of the reaction followed. If the reaction reaches a point at which further increases in reactants have no effect in increasing its rate, indications are good that the reaction is catalyzed by an enzyme. Uncatalyzed reactions, in contrast, increase in rate almost indefinitely as the concentration of reactants increases. (Supplement 3-1 presents a quantitative description of the responses of enzymes to increasing substrate concentrations, the *Michaelis–Menten equation*.)

Enzyme Inhibition   The fact that enzymes combine briefly with their reactants makes them susceptible to *inhibition* by unreactive molecules that resemble the substrate. The inhibiting molecules can combine with the active site of the enzyme but tend to remain bound without change, blocking access by the normal substrate. As a result, the rate of the reaction slows. If the concentration of the inhibitor becomes high enough, the reaction may stop completely. Inhibition of this type is called *competitive* because the inhibitor competes with the normal substrate for binding to the active site.

Some inhibitors interfere with enzyme-catalyzed reactions by combining with enzymes at locations outside the active site. These inhibitors, rather than reducing accessibility of the active site to the substrate, cause changes in folding conformation that re-

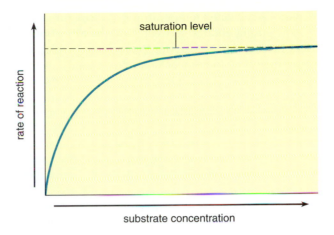

Figure 3-8   The effect of increases in substrate concentration on the rate of an enzyme-catalyzed reaction. At saturation (horizontal dashed line), further increases in substrate concentration do not increase the rate of the reaction.

duce the ability of the enzyme to lower the activation energy. Because such inhibitors do not directly compete for binding to the active site, their pattern of inhibition is called *noncompetitive*. (Supplement 3-1 shows how the Michaelis–Menten equation can be used to determine whether inhibition is competitive or noncompetitive.)

Some poisons or toxins exert their damaging effects by acting as enzyme inhibitors. For example, the action of cyanide and carbon monoxide as poisons depends on their ability to inhibit enzymes important in the utilization of oxygen in cellular respiration. Poisons and toxins typically act irreversibly by combining so strongly with enzymes, either covalently or noncovalently, that the inhibition is essentially permanent. Some irreversible poisons, rather than combining with the enzyme, destroy enzyme activity by chemically modifying critical amino acid side groups.

Temperature and pH   Both temperature and pH can profoundly affect catalysis by altering the folding conformation of enzymatic proteins. Proteins are held in their three-dimensional conformations by interactions between amino acid side groups (see p. 58). The pattern and strength of three of these interactions—hydrogen bonds, van der Waals forces, and ionic bonds—are highly sensitive to changes in the conditions of the solution surrounding the enzyme: hydrogen bonds and van der Waals forces to elevated temperatures, and ionic bonds primarily to alterations in pH.

As the temperature rises, the kinetic motions of the amino acid chains forming an enzyme molecule increase; the strength and frequency of collisions between enzymes and surrounding molecules also increase. At elevated temperatures these disturbances become strong enough to overcome the attraction of hydrogen bonds and van der Waals forces, which are individually relatively weak. As these bonds and attractions break, an enzyme gradually unfolds and loses its native three-dimensional conformation.

These changes affect enzymatic activity in a characteristic way (Fig. 3-9). Over the range from 0° to about 40°C, increases in temperature have little significant effect on enzyme structure. Over this range, enzyme activity follows the course of all chemical reactions: Each 10°C rise in temperature approximately doubles the reaction rate. This effect results from increases in the force and frequency of collisions between enzymes and reactant molecules, reflecting the heightened kinetic motion of all molecules in the solution. As the temperature rises above 40°C, the increase in rate begins to fall off as collisions become violent enough to break hydrogen bonds and unfold the enzymes. For most enzymes the drop in activity becomes steep at 55°C and falls to zero at about 60°C, when the disturbance in bonding causes the enzyme to unfold into a completely inactive, denatured form (see also p. 59). Once complete, denaturation totally counteracts the positive effects of increased kinetic motion at elevated temperatures.

As a result of the two opposing effects of a rise in temperature, all enzymes have an optimum temperature at which kinetic motion is greatest but no significant unfolding of the enzyme has yet occurred. For most enzymes this optimum lies between 40° and 50°C. Some enzymes have lower temperature optima. The enzymes of maize pollen, for example, have optima near 30°C and become inactive above 32°C. As a result, extremely hot weather seriously inhibits fertilization in maize. Among the enzymes with the lowest known temperature optima are those of arctic snow fleas, which are most active at −10°C. At the other extreme, a few organisms, such as the bacteria living in hot springs, possess enzymes with structures so resistant to disturbance that they remain active at temperatures of 85°C or more.

Changes in pH affect enzyme structure and activity primarily by altering charged amino acid side groups. These groups change between charged and uncharged forms at a characteristic pH. The alterations affect the strength of ionic bonds holding enzymes in their final three-dimensional shape, and in some cases also affect the functions of charged groups in the active site.

Charge alterations due to changes in pH also affect enzyme activity enzymes in characteristic patterns (Fig. 3-10). Typically, each enzyme has a pH optimum at which its conformation and the charge of its amino acid side groups are "correct" and the en-

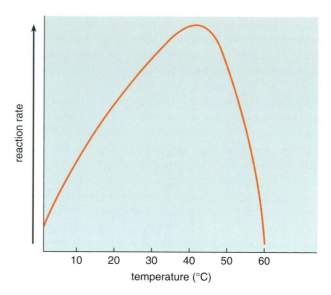

Figure 3-9 The effect of temperature on enzymatic activity. As the temperature rises, the rate of the catalyzed reaction increases proportionally until the temperature reaches the point at which the enzyme begins to denature. The rate drops off steeply as denaturation progresses and becomes complete.

zyme is most efficient in speeding the rate of its specific biochemical reaction. On either side of this pH optimum the rate of the catalyzed reaction falls because alterations in the charged groups change the folding conformation of the enzyme and the activity of charged amino acid residues in the active site. The effects become more extreme at pH values farther from the optimum, until the rate drops to zero.

The optimum activity of most enzymes lies in the vicinity of neutrality, near pH 7 (Fig. 3-10a). A relatively few enzymes have pH optima at significantly higher or lower values (Fig. 3-10b and c). For example, pepsin, an enzyme that speeds the breakdown of proteins in the stomach, has its optimum activity at about pH 3, near the level of acidity characteristic of the stomach contents. Lysosomal enzymes are also most active at acid pH, at about pH 5.

Alterations in pH act as regulatory mechanisms in some cell types. For example, the enzymes catalyzing partial breakdown of cell wall molecules in plants are active in acid pH and almost totally inactive at neutral pH. As part of the regulatory changes leading to cell growth, plant cells pump H^+ from the cytoplasm into the cell wall, leading to a drop in wall pH. This change is believed by some investigators to activate the wall enzymes, leading to partial breakdown of the cell wall to allow expansion of the cell (for details, see Chapter 8).

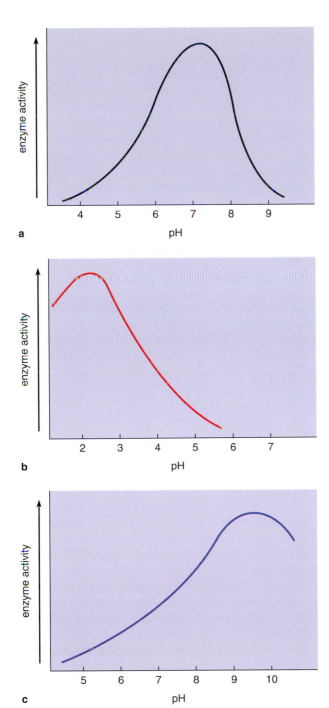

a

b

c

Figure 3-10 The effects of pH changes on enzyme activity. Enzymes typically have an optimum activity with respect to pH. At the pH optimum the charge of amino acid side groups produces a folding conformation and pattern of charges in the active site that is most favorable for speeding the rate of a reaction. Changes in pH to values lower or higher than the optimum alter the conformation and cause the reaction rate to drop. At extreme pH values the rate drops to zero. **(a)** The effect of pH changes on enzymes with maximum activity at a neutral pH. **(b)** The effect of pH changes on enzymes with maximum activity at an acidic pH. **(c)** The effect of pH changes on enzymes with maximum activity at an alkaline pH.

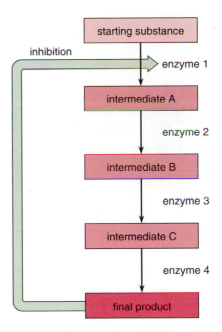

Figure 3-11 End-product or feedback inhibition. The final product of a multistep pathway acts as an inhibitor of the enzyme catalyzing an early step in the pathway. If the final product accumulates in excess, inhibition by the product of the enzyme catalyzing the early step turns off the entire pathway (see also Fig. 3-15).

Enzyme Regulation

The activity of many enzymes can be adjusted upward or downward to meet the needs of the cell for product molecules. We have seen how some enzymes are regulated by alterations in pH. Enzymes are also controlled by several other pathways, most notably by *inhibition, allosteric regulation,* and *covalent modification.*

Inhibition Enzyme inhibition regulates many cellular enzymes. Some inhibitors acting as enzyme controls are substances that compete with substrate molecules for binding to the active site; others work noncompetitively by binding to sites elsewhere on the enzyme.

Inhibitors acting as regulators typically combine reversibly with an enzyme, so that changes in their concentrations result in upward or downward adjustments in the amount of the inhibitor combined with the enzyme. As a consequence, changes in the concentration of the inhibitors can produce precise and sensitive adjustments in enzyme activity.

Frequently the substance acting as an inhibitor is a product of the enzyme-catalyzed reactions that it regulates. As an excess of the product accumulates, the enzymatic reaction producing it is automatically slowed, producing *end-product* or *feedback* inhibition (Fig. 3-11).

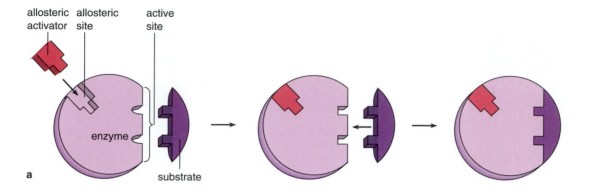

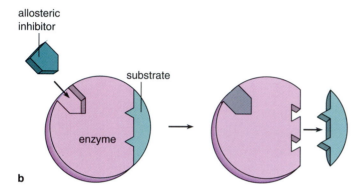

Figure 3-12 Allosteric regulation. **(a)** Activation of an allosteric enzyme by an activator. Combination of the activator with the allosteric site causes a conformational change in the active site enhancing substrate binding. **(b)** Allosteric inhibition. Combination of the inhibitor with the allosteric site induces a conformational change that reduces the affinity of the active site for the substrate.

Allosteric Regulation   Allosteric regulation (from *allo* = different and *steric* = structure or state), first proposed in 1965 by Jaques Monod and his colleagues J. P. Changeaux and J. Wyman, occurs by reversible combination of substances with sites on the enzyme other than the active site. The pattern is thus superficially similar to noncompetitive inhibition, but with several important differences.

The name *allosteric* refers to the fact that enzymes regulated allosterically can exist in at least two different states controlled by regulatory substances. In the *R state* (R = relaxed) the enzyme has high affinity for substrate molecules. In the *T state* (T = tense) the enzyme has low or no affinity for the substrate. The site at which the regulating substance binds is called the *allosteric* or *regulatory* site. Binding with regulatory substances may induce either the R or T states: Binding an *allosteric inhibitor* induces conversion of an allosteric enzyme from the R to the T state, and binding an *allosteric activator* induces conversion from the T to the R states. Combination with the activator or inhibitor at the allosteric site is considered to cause a conformational change in the active site that decreases or increases the ability of the enzyme to combine with the substrate (Fig. 3-12).

The pattern of allosteric regulation has several characteristics that sets it apart from noncompetitive inhibition. One of the most significant is the fact that

allosteric enzymes may be either inhibited or activated by a regulator substance. The reaction rate for most enzymes that are regulated allosterically, when plotted against substrate concentration, produces a *sigmoid*, or *S*-shaped, curve (Fig. 3-13) rather than the hyperbola typical of nonallosteric enzymes (compare with Fig. 3-8). The sigmoid curve results from a stimulatory effect of the substrate itself on the enzyme. Allosteric enzymes are typically large, complex molecules assembled from two or more subunits; combination of one subunit with a substrate molecule induces a conformational change in other subunits of the enzyme that increases their ability to react with additional substrate molecules. As substrate concentration increases from zero, this effect produces the especially rapid rise of the first segment of the sigmoid curve. (Because of the sigmoid curve produced by a plot of reaction rate against substrate concentration, allosteric enzymes are not described accurately by the Michaelis–Menten relationship.) Some allosteric enzymes have the unusual property of structural instability in the cold, so that activity tapers off steeply at both low and high temperatures.

The molecular mechanism underlying allosteric regulation has been worked out for a few enzymes. *Phosphofructokinase*, for example, investigated by P. R. Evans and his colleagues, is an allosteric enzyme that in active form adds a phosphate group to a molecule

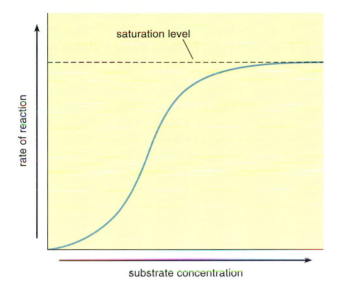

Figure 3-13 The sigmoid, or *S*-shaped, increase in reaction rate produced by most allosteric enzymes as substrate concentration increases.

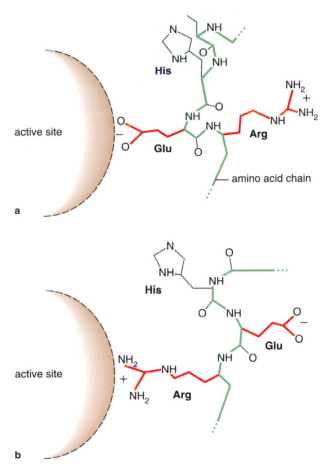

Figure 3-14 The molecular basis for allosteric regulation of the phosphofructokinase enzyme. **(a)** The arrangement of the amino acid chain in the active state, in which an arginine residue occupies a critical location. **(b)** Change to the inactive state results from a rotation of the amino acid chain that exchanges a glutamine for the arginine at the critical position. (Redrawn from originals courtesy of P. R. Evans; reprinted by permission from *Nature* 343: 140[1990]. Copyright 1990 Macmillan Magazines Ltd.)

forming part of a major oxidative pathway (see p. 213). The nucleotides ADP or GDP act as allosteric activators for the enzyme; a product of one of the final reactions of the oxidative pathway, *phosphoenolpyruvate* (*PEP*), acts as an allosteric inhibitor.

The Evans laboratory investigated conformational changes in the enzyme by X-ray diffraction studies of its R and T states. In the active R state the positively charged side group of an arginine residue, necessary for binding the substrate molecules, is located in the active site (Fig. 3-14a). In the inactive T state, the amino acid chain rotates, swiveling the arginine residue out of the active site and inserting a negatively charged glutamine residue (Fig. 3-14b). The glutamine residue blocks binding of the substrate molecule. When ADP or GDP is bound to the allosteric site, the arginine swivels into the active site, activating the enzyme. If PEP binds to the allosteric site, the arginine swivels out and glutamine swivels in, inactivating the enzyme.

Allosteric regulation controls many biochemical pathways, particularly through feedback inhibition. The reaction series converting threonine to isoleucine (Fig. 3-15) is a classic example of allosteric regulation by feedback inhibition. Five enzymes acting in sequence catalyze the pathway. The final product of the sequence, isoleucine, is an allosteric inhibitor of the first enzyme of the pathway, threonine deaminase. As

the pathway runs, any isoleucine molecules made in excess of cell requirements combine reversibly with threonine deaminase at a location outside the active site. The combination converts threonine deaminase to the T state and inhibits its ability to combine with threonine, the substrate for the first reaction in the pathway.

Regulation by Covalent Modification Many key enzymes are regulated by covalent binding of ions or chemical groups to the enzyme structure. The additions, which may include calcium ions, phosphate, methyl ($-CH_3$), or acetyl ($-COCH_3$) groups or derivatives of nucleotides, induce conformational

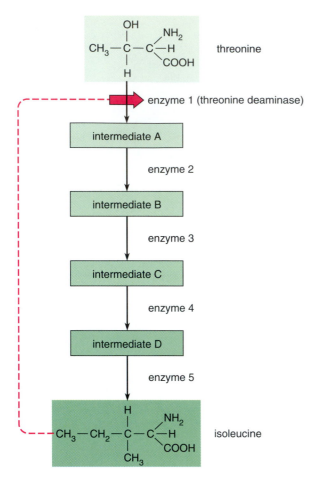

Figure 3-15 Allosteric regulation in the pathway producing isoleucine from threonine. Five enzymes catalyze successive steps in the pathway. If the end product of the pathway, isoleucine, is present in excess, it acts as an allosteric inhibitor of the enzyme catalyzing the first step in the sequence.

changes in the enzyme that adjust its activity upward or downward.

Among these modifications, reversible addition of phosphate groups occurs most frequently as a cellular control regulating enzyme activity. The phosphate groups are added to enzymatic proteins by other enzymes known as *protein kinases*; removal occurs through the activity of *phosphatases*.

For example, two enzymes involved in glucose metabolism, *phosphorylase* and *glycogen synthase*, are regulated by addition and removal of phosphate groups. Phosphorylase becomes active when phosphorylated and inactive when dephosphorylated. Glycogen synthase responds in the reverse pattern: The enzyme is inactive when phosphorylated and active when dephosphorylated. The addition or removal of phosphate groups, catalyzed by protein kinases and phosphatases specific in their activity for either of the two enzymes, provides a precise control

of the availability of glucose for cellular oxidations (for details, see p. 160). Phosphorylation and dephosphorylation also occur as an additional mechanism regulating some allosteric enzymes.

The enzymes regulated allosterically or by covalent modification frequently catalyze key reactions that form one of the initial steps in complex, multistep pathways involving the sequential activity of several enzymes, as in the threonine-isoleucine pathway. By speeding or slowing the initial step, the control regulates the entire pathway. Among the pathways regulated in this way are reactions synthesizing products required by an organism and complex reaction series accomplishing a major function such as oxidation of fuel substances.

RNA-Based Catalysts: Ribozymes

Until the early 1980s, the statement "all enzymes are proteins" was apparently one of the unshakable truths of biology. As such, it formed the basis for test questions posed to many generations of biology students. The statement had to be revised in 1981, when Thomas R. Cech discovered a group of RNA molecules capable of acting as biological catalysts. Since Cech's initial discovery the list of RNA molecules with this ability has grown to the point that *RNA-based catalysts* or *ribozymes*, as they are known, are now accepted as a part of the biochemical machinery of all living organisms.

The RNA-based catalyst initially discovered in the Cech laboratory is a ribosomal RNA (rRNA) molecule of the protozoan *Tetrahymena* (Fig. 3-16). The rRNA is initially transcribed in the form of a precursor that contains an extra segment 419 nucleotides in length (Fig. 3-16a). The segment must be removed for the rRNA to become fully mature and active. Cech and his colleagues discovered that the RNA precursor is capable of self-catalyzing removal of the extra segment, including cutting out the 419 nucleotides as a unit and splicing the free ends into a continuous, mature molecule. No proteins are necessary for the catalysis to proceed.

The reaction takes place through breakage of the phosphate bonds at the beginning and end of the 419-nucleotide segment. (Fig. 3-16b to d outlines the progress of the reaction; for further details, see p. 435 and Fig. 15-7a.) Normally the phosphate linkages binding DNA or RNA nucleotides into nucleic acids are highly stable and undergo spontaneous breakage only at a very slow rate. In the self-catalyzed reaction removing the 419-nucleotide insert, the *Tetrahymena* rRNA folds into a form that exposes phosphate bonds in the two locations destined to break to chemical attack by hydroxyl (—OH) groups within the RNA structures,

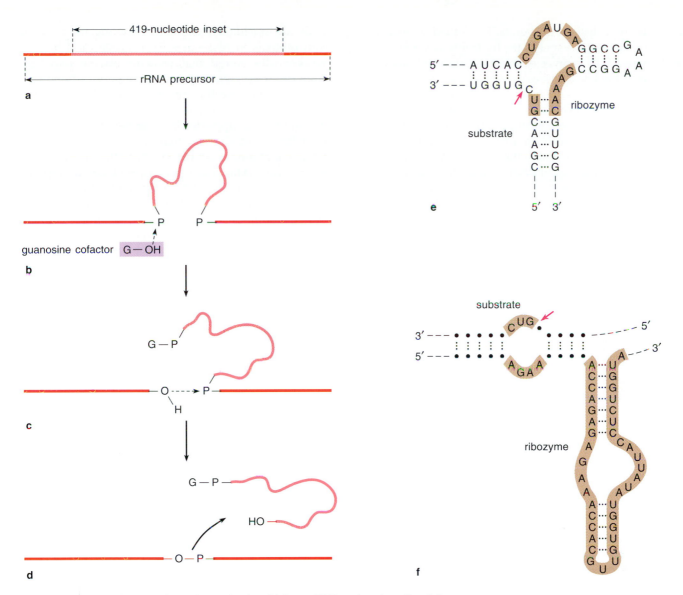

Figure 3-16 An RNA-catalyzed reaction in which an rRNA molecule self-catalyzes removal of an internal surplus segment. **(a)** The rRNA molecule with the 419-nucleotide surplus segment. **(b)** The rRNA folded in the conformation that catalyzes removal of the insert. The strain induced by the folding pushes the phosphate linkages at the ends of the insert toward the transition state for their hydrolysis, in a reaction equivalent to that of the phosphate transfer shown in Figure 3-6. In this state the phosphate bond at the beginning of the insert is susceptible to cleavage by the oxygen in an —OH group of a guanosine nucleotide, which acts as a cofactor for the cleavage (boxed in purple in *b*). **(c)** Cleavage of the phosphate bond leaves the —OH group attached to the rRNA chain and the guanosine linked to the insert. The —OH attached to the cleavage site now attacks the second unstable phosphate linkage at the end of the insert (*c*, dotted arrow). **(d)** Cleavage of this bond removes the insert and links the rRNA molecule into a single, continuous chain through a second phosphate transfer. Examples of **(e)**, a "hammerhead," and **(f)**, a "hairpin," ribozyme, so called because their sequences can be folded in two dimensions in these shapes. The three-dimensional structures of the ribozymes are unknown. Both ribozyme types occur in some plant viruses; a hairpin ribozyme has also been reported in the newt. The red arrow shows the site at which the substrate is broken. The substrate RNA may be part of the same RNA chain or a separate molecule. Invariant parts of the sequence are boxed in blue. Parts of the sequence that vary in different examples of the ribozymes are in black in **(e)** and shown as dots in **(f)**.

leading to hydrolysis of the bonds. The folded state of the *Tetrahymena* rRNA speeds the hydrolysis by a factor of about a billion times over the uncatalyzed rate.

Most of the known ribozymes speed cutting and splicing reactions in which surplus segments are removed from tRNA, rRNA, and the RNAs of a group of plant viruses. In all these examples, hydroxyl groups of RNA molecules are brought by conformational changes into positions where their reactivity leads to hydrolysis and breakage of RNA chains. Metallic ions, particularly Mg^{2+}, Mn^{2+}, and Ca^{2+}, typically act as cofactors in these reactions (Fig. 3-16e and f show two of these ribozyme types).

In the self-cutting and splicing reactions of the pattern shown in Figure 3-16, the catalytic RNA may be said to fail to fulfill the complete requirements for a catalyst because it is changed as the reaction proceeds. However, several ribozymes catalyze reactions from which they emerge unchanged. For example, one RNA-based catalyst, *ribonuclease P*, investigated by Sidney Altman and his colleagues, speeds a reaction in which a surplus segment is removed from a different molecule, a tRNA. Ribonuclease P is unchanged by the reaction and retains full capability to catalyze further reactions. Thus, this RNA fulfills all the requirements of a biological catalyst. Ribonuclease P does contain a protein component, which was long thought to be the part of the molecule with enzymatic activity. However, detailed investigation of this enzyme in the Altman laboratory revealed that the RNA portion of the enzyme alone is capable of catalyzing the tRNA-processing reaction. The protein segment by itself is without enzymatic activity. (The protein, however, enhances the activity of the RNA segment; for details of RNAse P and its role in tRNA processing, see p. 445.) Cech and Altman received the Nobel Prize in 1989 for their discovery of RNA-based catalysts.

For a time it was thought that ribozymes were confined to catalyzing reactions in which RNA molecules were split or spliced together. However, recent investigations have revealed that RNA-based catalysts may speed the reactions assembling amino acids into proteins in ribosomes. H. F. Noller and his coworkers discovered that proteins can be removed from ribosomes without destroying peptide bond formation, indicating that ribozyme activity may catalyze this fundamental reaction of protein synthesis. This finding is supported by recent evidence from the Cech laboratory showing that a ribozyme can catalyze breakage of amino acids from their tRNA carriers in the test tube. This reaction is the first step in peptide bond formation (for details, see Chapter 16).

The existence of RNA-based catalysts provides a possible solution to a long-standing paradox about the evolution of life. Scientists have argued for many years whether proteins or nucleic acids came first in evolution. The problem is paradoxical because it is difficult to understand how DNA could carry out its functions without the enzymatic proteins required for transcription and replication, or how the enzymes could exist without the information encoded in the nucleic acids necessary for their synthesis. RNA-based catalysts offer a way around this dilemma because the earliest forms of life may conceivably have been based entirely on RNA molecules able to act as both informational molecules and catalysts. The earliest forms of life may therefore have inhabited an "RNA world" with no initial requirements for either DNA or proteins. (Chapter 26 provides further details of the proposed RNA world and the evolution of cellular life.)

This book includes many examples of energy changes, the role of ATP in cells, and enzymes. In many respects the story of cellular life is the story of enzymes, so much so that cells have been described simply as "bags of enzymes." Although this is an obvious oversimplification, each of the thousands of individual biochemical reactions coordinated to achieve cellular life is speeded by its own enzyme, tailored by evolution to fit its normal substrates with specificity approaching perfection.

For Further Information

Suggestions for Further Reading

von Ahsen, U., and Schroeder, R. 1993. RNA as a catalyst: Natural and designed ribozymes. *Bioess.* 15:299–307.

Altman, S. 1990. Enzymatic cleavage of RNA by RNA. *Biosci. Rep.* 10:317–337.

Aragon, J. J., and Sols, A. 1991. Regulation of enzyme activity in the cell: Effect of enzyme concentration. *FASEB J.* 5:2945–2950.

Barford, D. 1991. Molecular mechanisms for the control of enzyme activity by protein phosphorylation. *Biochim. Biophys. Acta* 1153:55–62.

Benkovic, S. J. 1992. Catalytic antibodies. *Ann. Rev. Biochem.* 61:29–54.

Castanotto, D., Rossi, J. J., and Deshler, J. O. 1992. Biological and functional aspects of catalytic RNAs. *Crit. Rev. Eukaryotic Gene Express.* 2:331–357.

Cech, T. R. 1986. RNA as an enzyme. *Sci. Amer.* 255:64–74 (November).

Cech, T. R. 1987. The chemistry of self-splicing RNA and RNA enzymes. *Science* 236:1532–1539.

Cech, T. R. 1990. Self-splicing and enzymatic activity of an intervening sequence RNA from *Tetrahymena. Biosci. Rep.* 10:239–261.

Cedergren, R. 1990. RNA—the catalyst. *Biochem. Cell Biol.* 68:903–906.

Changeaux, J.-P. 1993. Allosteric proteins: From regulatory enzymes to receptors. *Bioss.* 11:625–634.

Fersht, A. 1985. *Enzyme Structure and Mechanisms*, 2nd ed. San Francisco: Freeman.

Klotz, I. M. 1967. *Energy Changes in Biochemical Reactions.* New York: Academic Press.

Knowles, J. R. 1987. Tinkering with enzymes: What are we learning? *Science* 236:1252–1258.

Knowles, J. R. 1991. Enzyme catalysis: Not different, just better. *Nature* 350:121–124.

Koshland, D. E., Jr. 1984. Control of enzyme activity and metabolic pathways. *Trends Biochem. Sci.* 9:155–159.

Kraut, J. 1988. How do enzymes work? *Science* 242:533–540.

Lehninger, A. L. 1971. *Bioenergetics*, 2nd ed. Menlo Park, Calif.: Benjamin-Cummings.

Lerner, R. A., and Tramontano, A. 1988. Catalytic antibodies. *Sci. Amer.* 258:58–70 (March).

Lerner, R. A., Benkovic, S. J., and Schultz, P. G. 1991. At the crossroads of chemistry and immunology: Catalytic antibodies. *Science* 252:659–667.

Lolis, E., and Petsko, G. A. 1990. Transition state analogs in protein crystallography: Probes of the structural source of enzyme catalysis. *Ann. Rev. Biochem.* 59:597–630.

Long, D. M., and Uhlenbeck, O. C. 1993. Self-cleaving catalytic RNA. *FASEB J.* 7:25–30.

Pace, N. R. 1992. New horizons for RNA catalysis. *Science* 256:1402–1403.

Perutz, M. F. 1988. Allosteric enzymes. Control by phosphorylation. *Nature* 336:202–203.

Price, N. P., and Stevens, L. 1982. *Fundamentals of Enzymology.* New York: Oxford Scientific.

Pyle, A. M. 1993. Ribozymes: A distinct class of metalloenzymes. *Science* 261:709–714.

Racker, E. 1976. *A New Look at Mechanisms in Bioenergetics.* New York: Academic Press.

Schirmer, T., and Evans, P. R. 1990. Structural basis of the allosteric behavior of phosphofructokinase. *Nature* 343:140–145.

Stryer, L. 1988. *Biochemistry*, 3rd ed. New York: Freeman.

Symons, R. H. 1992. Small catalytic RNAs. *Ann. Rev. Biochem.* 61:641–671.

Review Questions

1. State the first and second laws of thermodynamics. Give examples from your own experience that illustrate operation of the laws.

2. What does the word *spontaneous* mean in thermodynamics?

3. Define a system, the surroundings, and the initial and final states in the thermodynamic sense.

4. What does energy content mean with reference to a collection of molecules? Define kinetic and potential energy.

5. What is entropy? How do energy content and entropy interact in determining whether chemical reactions will proceed to completion?

6. What is a reversible reaction? What happens at the equilibrium point of a reversible reaction? What happens if more reactants are added to a reaction at equilibrium? If more products are added?

7. How do biological organisms run reactions that require energy? What does coupling mean?

8. It is sometimes claimed that organisms violate the second law of thermodynamics because they become more complex as they develop spontaneously from a seed or fertilized egg to an adult. Why is this statement incorrect?

9. How does ATP act as the primary agent coupling uphill and downhill reactions in living organisms?

10. Draw the structures of ATP, ADP, and AMP.

11. What are calories? Moles? Calories per mole?

12. What effects do enzymes have on spontaneous reactions?

13. What are the characteristics of protein-based enzymes? What is the active site of an enzyme?

14. What is the activation energy of a reaction? What effects do enzymes have on the activation energy?

15. Define the transition state and outline its importance to enzyme-catalyzed reactions. What mechanisms are believed to push molecules toward the transition state at the active site of enzymes? How does attainment of the transition state affect the activation energy for a reaction?

16. Why is it likely that the active site of an enzyme can fit both the reactants and products of a reaction?

17. Outline one experiment demonstrating that the active site of enzymes also binds the transition state of the reactants.

18. What effects do changes in temperature have on enzyme activity? Why?

19. What effects do changes in pH have on enzyme activity? Why?

20. Define enzyme saturation and inhibition. What is the probable molecular basis for these effects on enzymes?

21. What mechanisms regulate the active site? What is allosteric inhibition? End-product or feedback inhibition?

22. What is an RNA-based catalyst? What types of reactions are catalyzed by these "ribozymes"?

Supplement 3-1
A Quantitative Treatment of Enzyme Saturation and Inhibition: The Michaelis-Menten Equation

An equation derived in 1913 by Leonor Michaelis and Maud L. Menten describes quantitively the rate or velocity of an enzyme-catalyzed reaction, as plotted in Figure 3-8:

$$\text{velocity} = V = V_{max} \frac{[S]}{[S] + K_M} \qquad (3\text{-}14)$$

in which V_{max} is the maximum velocity of the reaction when the enzyme is saturated (horizontal dotted line in Fig. 3-8), [S] is the substrate concentration in moles, and K_M is a constant that reflects the affinity of an enzyme for its substrate (that is, the strength by which the enzyme binds to its substrate). Generally, the lower the value of K_M, called the *Michaelis constant*, the greater the affinity of the enzyme for its substrate. (Some representative values for K_M appear in Table 3-5.) The value of V_{max} is related to the turnover number for the enzyme being studied (turnover number = $V_{max}/[S]$).

If the substrate concentration is adjusted so that it equals K_M, K_M can be substituted for [S] in Equation 3-14:

$$V = V_{max} \frac{K_M}{K_M + K_M} = V_{max} \cdot \frac{1}{2} \qquad (3\text{-}15)$$

Since this algebraic manipulation shows that $V = 1/2 V_{max}$ when $K_M = [S]$, the reverse statement is also true: $K_M = [S]$ when $V = 1/2 V_{max}$. Therefore, K_M *is equal to the substrate concentration when an enzyme-catalyzed reaction is running at half of its maximum velocity.* Figure 3-17 graphically depicts the relationships between V, V_{max}, [S], and K_M.

In practice, it is difficult to use experimental results to generate a saturation curve of the type shown in Figures 3-8 and 3-17 because many points must be plotted to draw the curve of this type (a hyperbolic curve) accurately. It is also often difficult to determine the saturation level, or V_{max}, directly from experiments. To solve these problems a conversion called the *Lineweaver-Burk plot* is often used. For this plot, reciprocals are taken of both sides of Equation 3-14:

$$\frac{1}{V} = \frac{K_M + [S]}{V_{max}[S]}$$

$$\frac{1}{V} = \frac{K_M}{V_{max}[S]} + \frac{[S]}{V_{max}[S]}$$

$$\frac{1}{V} = \frac{K_M}{V_{max}} \cdot \frac{1}{[S]} + \frac{1}{V_{max}} \qquad (3\text{-}16)$$

Equation 3-16 has the form of a linear equation with slope equivalent to K_M/V_{max} and the intercept of the y-axis equivalent to $1/V_{max}$. This mathematical technique converts the hyperbola of Figure 3-17 into a straight line that can be plotted accurately from a relatively few experimental points (Fig. 3-18). Since the y-intercept is equivalent to $1/V_{max}$, and the x-intercept to $-1/K_M$, the values for V_{max} and K_M can be determined quickly and accurately.

The Michaelis-Menten equation and its Lineweaver-Burk conversion accurately describe the behavior of many enzymes. The Lineweaver-Burk plot is especially effective as a means for determining whether inhibition is competitive or noncompetitive. Competitive inhibitors operate by combining with the active

Table 3-5 The K_M Values of Some Enzymes	
Enzyme	K_M Value
Aminotransferase	0.0009
Carbonic anhydrase	0.008
Catalase	0.025
Chymotrypsin	0.005
Glutamic acid dehydrogenase	0.00012
Hexokinase	0.00015
Pyruvic acid carboxylase	0.004

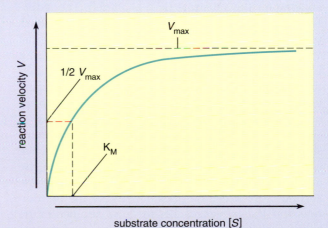

Figure 3-17 Enzyme saturation in terms of reaction velocity V, substrate concentration [S], maximum velocity V_{max}, and the Michaelis constant K_M (see text).

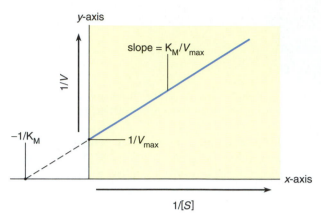

Figure 3-18 A plot of the reciprocals of the velocity ($1/V$) and of the substrate concentration ($1/[S]$). This plot, the Lineweaver-Burk plot, produces a straight line with the y-intercept equivalent to $1/V_{max}$ and the x-intercept equal to $-1/K_M$. The slope of the line is equivalent to K_M/V_{max} (see text).

site of the enzyme. Because the combination is reversible and concentration-dependent, the amount of inhibitor bound to the enzyme can be reduced by adding more substrate (this has the effect of reducing the relative concentration of the inhibitor molecules). At very high substrate concentrations, practically all the enzyme molecules form enzyme-substrate complexes rather than enzyme-inhibitor complexes. As a result, V_{max} can eventually reach its normal level as more and more substrate is added, and the value obtained for $1/V_{max}$ in the Lineweaver-Burk plot (the y-intercept) remains unchanged (Fig. 3-19). At substrate concentrations below the levels required to reach V_{max} in the presence of the inhibitor, the velocity of the reaction will be slower for a given substrate concentration because of interference with the active site by the inhibiting molecule. In other words, the affinity of the enzyme for the substrate (K_M) is altered. The reduction in velocity has the effect in the Lineweaver-Burk plot of increasing the slope of the line $1/V$ against $1/[S]$ and altering the x-intercept ($-1/K_M$). Thus, a Lineweaver-Burk plot of a reaction taking place in the presence of a competitive inhibitor has the same y-intercept ($1/V_{max}$) but an altered slope and x-intercept ($-1/K_M$).

In noncompetitive inhibition, since the inhibitor slows the enzyme by combining outside the active site, adding more substrate does not reduce the amount of inhibitor bound to the enzyme. Therefore, adding more substrate does not restore the reaction rate, and V_{max} never reaches the level attained in the absence of the inhibitor. Since the ability of the enzyme to combine with the substrate (enzyme affinity) remains the same, however, K_M has its usual value. Thus, a Lineweaver-Burk plot for a reaction slowed

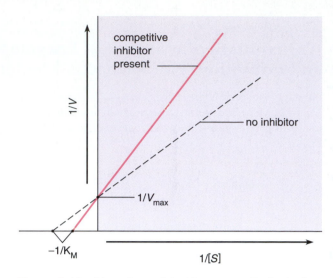

Figure 3-19 Alterations of the Lineweaver-Burk plot for an enzyme-catalyzed reaction as a result of competitive inhibition. The plot has the same y-intercept ($1/V_{max}$) as a reaction with no inhibition but an altered slope and x-intercept ($-1/K_M$).

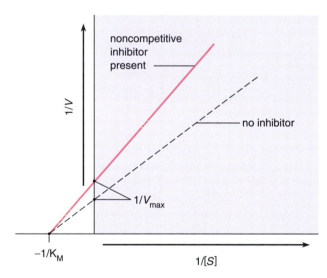

Figure 3-20 Alterations of the Lineweaver-Burk plot for an enzyme-catalyzed reaction as a result of noncompetitive inhibition. The plot has the same x-intercept ($-1/K_M$) as a reaction taking place without inhibition but an altered y-intercept ($1/V_{max}$) and slope.

by noncompetitive inhibition will show the same x-intercept ($-1/K_M$) but an altered y-intercept ($1/V_{max}$) and slope (Fig. 3-20). In summary:

competitive inhibitors: increase in K_M, no change in V_{max}

noncompetitive inhibitors: reduction in V_{max}, no change in K_M

Membranes organize and maintain all cells as separate and distinct molecular environments. The plasma membrane, the outermost membrane of both prokaryotic and eukaryotic cells, keeps the cell contents from mixing freely with molecules outside the cell and serves as the primary zone of contact between the cell and its external environment. In eukaryotic cells, internal membranes also divide the cell interior into compartments with distinct biochemical environments and specialized functions. Regions within the narrow membrane interior provide hydrophobic environments in which certain reactions take place almost exclusively. Among the reactions of the membrane interior are some of the most vital interactions providing cellular energy, including major pathways in mitochondria and chloroplasts.

These fundamental roles depend on the basic structural plan of membranes. The contemporary view of this structure is based on the *fluid mosaic model*, a hypothesis advanced in 1972 by S. J. Singer and G. L. Nicolson. According to the model (see Fig. 4-12) membrane structure is based on a double layer of lipid molecules, the *bilayer* (see Fig. 4-8). The lipids of the bilayer are amphipathic (see p. 50) molecules with both hydrophobic and hydrophilic segments. Their hydrophobic portions, concentrated in the membrane interior, create a barrier that prevents free movement of polar molecules across the membrane. Their hydrophilic portions face the watery, polar environments at the membrane surfaces. Under physiological conditions the bilayer is in a fluid state—that is, the lipid molecules forming the bilayer are free to exchange places and rotate in place. Membrane proteins float individually in the lipid bilayer, suspended like "icebergs in the sea." The segregation of the membrane into separate regions, some occupied by proteins and others by lipids, is the *mosaic* part of the fluid mosaic model.

The proteins embedded in the lipid framework provide membranes with a variety of specialized functions. Some plasma membrane proteins transport selected polar molecules and ions between the cell interior and exterior, thereby maintaining the internal balance of molecules required for cellular life. Similar proteins transport polar molecules through the boundary membranes of internal organelles. Other proteins of the plasma membrane are markers, which identify the cell as part of the same individual or as foreign, or receptors, which can recognize and bind molecules originating as chemical signals from other cells. (Peptide hormones, for example, are among the signal molecules bound by receptors in the plasma membrane.) Some membrane proteins promote cell associations by recognizing and binding chemical marker groups on other cells. Still other proteins give membranes the capacity to carry out important reactions of respiration and photosynthesis. Each membrane type, including the plasma membranes of different cells and the membranes of cell organelles, has a characteristic group of proteins that is responsible for its specialized functions.

This chapter describes the structure of biological membranes, including the lipid bilayer and its associated proteins. The functions of membranes in transport are covered in Chapter 5. Other roles of membranes are the subjects of later chapters in this book.

LIPID AND PROTEIN MOLECULES OF BIOLOGICAL MEMBRANES

All known biological membranes contain both lipid and protein molecules. The relative amounts of the two molecular types vary from the extremes of myelin (a membrane found in nerve tissue), with about 80% lipids and 20% proteins by weight, to the inner membrane of mitochondria, which contains more than 75% proteins (Table 4-1). Some bacterial membranes, such as those of photosynthetic bacteria, also contain about 75% proteins. Most plant and animal membranes fall midway between these extremes,

Table 4-1	Amounts of Lipid, Protein, and Carbohydrate in Different Biological Membranes		
	Approximate Percent of Dry Mass		
Membrane	Protein	Lipid	Carbohydrate
Plasma membranes			
erythrocyte	49	43	8
nerve myelin	18	79	3
liver cell	54	36	10
Nuclear envelope	66	32	2
Endoplasmic reticulum	62	27	10
Golgi complex	64	26	10
Mitochondrion			
outer membrane	55	45	Trace
inner membrane	78	22	——
Chloroplast	70	30	——

Adapted from G. Guidotti; reproduced, with permission, from *Ann. Rev. Biochem.* 41:731 (1972). © 1972 by Annual Reviews, Inc., and R. Lotan and G. L. Nicolson, *Advanced Cell Biology* (Schwartz, L. M., and Azar, M. M., eds.), New York: Van Nostrand Reinhold, 1981.

with lipids making up from 30% to 50% of the total and proteins the remainder.

Carbohydrates also occur in membranes as chemical groups in glycolipids and glycoproteins (see p. 61). In eukaryotes, membrane carbohydrates are detected in greatest abundance in plasma membranes, where they are linked almost exclusively to the ends of membrane glycolipids and glycoproteins facing the cell exterior. When present, carbohydrates make up from 1% to 10% of total membrane components by dry weight.

Membrane Lipids

The lipid part of membranes, which provides the bilayer framework, is highly varied. Depending on the species and cell type, various plant and animal membranes may include *phosphoglycerides, sphingolipids,* and *sterols* as major classes. The phosphoglycerides and some sphingolipids contain phosphate groups and are collectively called *phospholipids* (see also p. 49). Many of the lipids are linked to one or more carbohydrate groups to form glycolipids.

Phosphoglycerides Phosphoglycerides, the most abundant and characteristic membrane lipids, are based on a glycerol framework. To this framework are bound two fatty acid chains and a phosphate group (see Fig. 2-11 and p. 49 for details). The phosphate group acts as a linker binding one of a group of alcohols or amino acids, including choline, ethanolamine, glycerol, inositol, serine or threonine, to the lipid structure (see Fig. 2-12).

The most significant feature of the phosphoglycerides from the standpoint of membrane structure is that they are amphipathic: One end of the molecule containing the hydrocarbon chains is nonpolar and hydrophobic, while the end containing the phosphate group is polar and hydrophilic (see Fig. 2-11a). The polar end of a phosphoglyceride extends approximately to the level of the bonds connecting the fatty acid residues to glycerol (above the dashed line in Fig. 2-11; the circle in Fig. 2-11d represents the polar portion of the molecule). This end of a phospholipid molecule readily associates with water molecules if exposed to an aqueous medium. In contrast, the nonpolar hydrocarbon chains (below the dashed line in Figure 2-11; represented by the zigzag lines in Fig. 2-11d) "prefer" a nonpolar environment from which water is excluded. Almost all membrane lipids and proteins possess this dual solubility property (see p. 49), which is of fundamental importance to their interactions in membranes.

Membranes contain several major types of phosphoglycerides. The particular combinations present appear to be important to function because the proportions of the various phosphoglycerides are maintained in a given membrane type. This maintenance may reflect the fact that membrane proteins such as enzymes and transporters appear to work at optimum levels only when they are associated with certain phosphoglycerides. The maintenance of phospholipid proportions also indicates that there are biochemical mechanisms that detect changes in phosphoglyceride content and make compensatory adjustments as required.

Membrane Glycolipids Some membrane glycolipids are based on a glycerol framework with fatty acyl chains attached to the 1- and 2-carbons as in phosphoglycerides. However, the carbohydrate group of these glycolipids is linked directly to the remaining glycerol site, with no bridging phosphate group (Fig. 4-1). These glycolipids are the primary carbohydrate-linked lipids of plants and photosynthetic protists; they also occur as a major glycolipid type in bacteria.

Although glycolipids based on glycerol also occur in animals, the primary animal glycolipids are based on sphingolipids. The various sphingolipids are

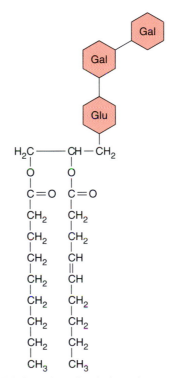

glycolipid based on diacyl glycerol

Figure 4-1 A glycolipid based on a glycerol residue with fatty acyl chains at the 1- and 2-carbons, and a carbohydrate group at the 3-carbon. Glycolipids of this type are common in plants and bacteria but not in animals. Gal, galactose; Glu, glucose.

built on a backbone structure known as *sphingosine* (Fig. 4-2a), consisting of an amino acid residue, serine, linked to a fatty acid chain. The serine residue has two additional reactive sites, an amino (—NH_2) and a hydroxyl (—OH) group (boxed in Fig. 4-2a). In sphingolipids the amino group links to a second fatty acid chain (as in Fig. 4-2b and c). The hydroxyl may be unbound, as in *ceramides* (Fig. 4-2b), or linked by a phosphate group to one of the same series of polar alcohols occurring in phosphoglycerides, as in *sphingomyelin*, a phosphate-containing sphingolipid with choline as the polar alcohol (Fig. 4-2c). The hydroxyl may also link directly to a carbohydrate group to form a *glycosphingolipid* (as in Fig. 4-3; Fig. 4-4 shows the carbohydrate groups common in membrane glycolipids).

The nonpolar hydrocarbon chains at one end of a sphingolipid or glycosphingolipid give that end of the molecule a hydrophobic character, and the —OH or the various polar groups linked at the 3-carbon site make the other end hydrophilic. These lipids are therefore amphipathic molecules like the phosphoglycerides and fit into membrane structure in the same way.

Glycosphingolipids occur in plasma membranes throughout the plant and animal kingdoms. They are especially abundant in the plasma membranes of animal nerve and brain cells. In plasma membranes, glycosphingolipids orient with their polar carbohydrate groups extending from the outer membrane surface. These carbohydrate groups form stable, interlocked networks of hydrogen bonds extending over the membrane surface. Such networks are highly developed on the plasma membranes of cells exposed to physical or chemical stress, such as the cells lining the small intestine in mammals. These characteristics suggest that one of the functions of glycosphingolipids is membrane stabilization.

Several glycosphingolipids of the ganglioside group (see Fig. 4-3b) have been identified as sites recognized by antibodies in immune reactions or as antigens responsible for blood group interactions. Some glycosphingolipids at the cell surface also act as binding sites for substances taken up by cells, including some peptide hormones and the cholera and tetanus toxins. Several viruses and bacteria also use glycosphingolipids of animal plasma membranes as recog-

Figure 4-2 Sphingolipids (see text).

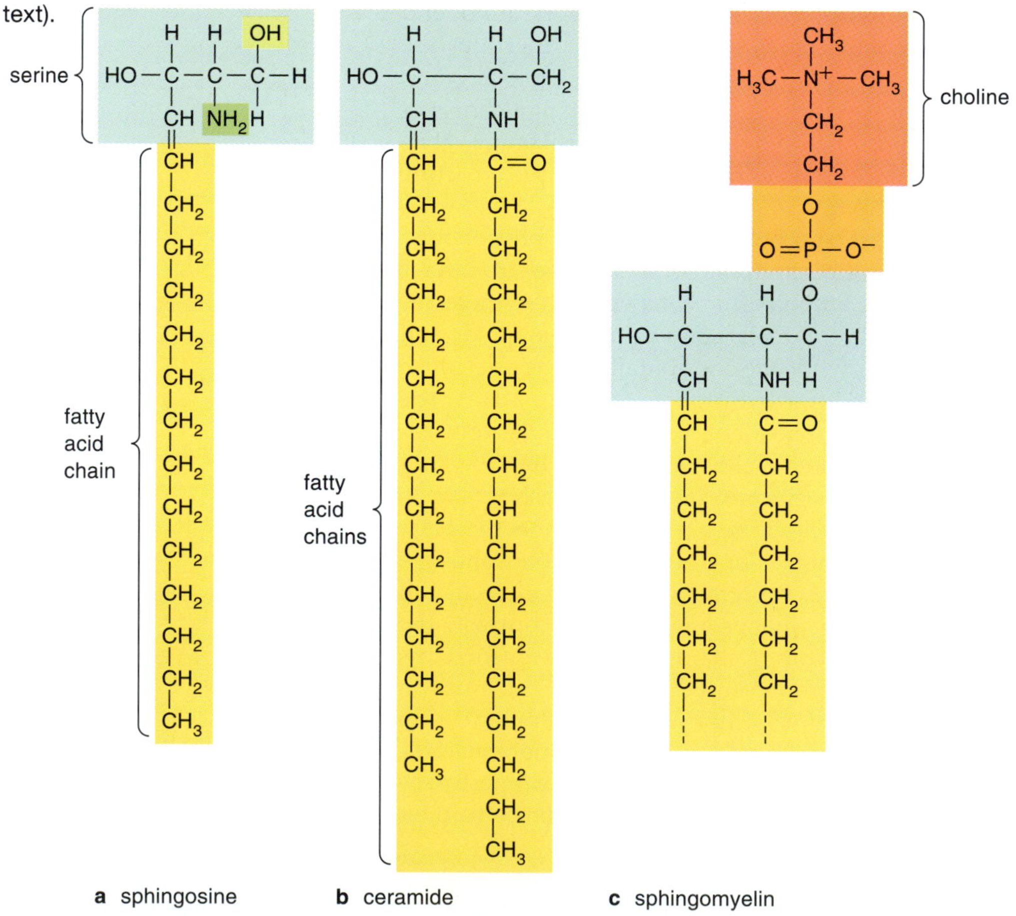

a sphingosine   b ceramide   c sphingomyelin

nition and attachment sites. Breakdown products of sphingolipids have been implicated in metabolic regulation, including systems that regulate the activity of surface receptors, blood platelets, and growth factors.

Deficiencies in the metabolism of glycosphingolipids cause several important human disorders. One example is Tay-Sachs disease, in which normal breakdown and turnover of these molecules are deficient. Glycosphingolipids accumulate as a result and interfere with nerve and brain function, leading to paralysis and severe mental impairment. Glycosphingolipids of the plasma membrane may also change radically—old ones may disappear or new ones may appear—when normal cells are transformed into

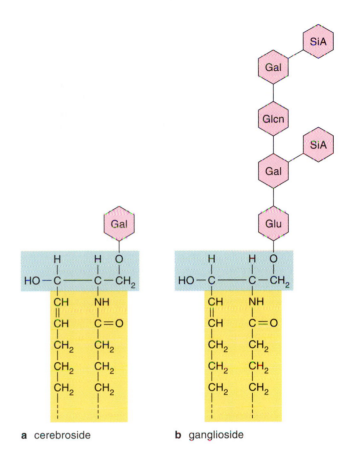

Figure 4-3 Glycosphingolipids, formed by the addition ▶ of carbohydrate units to a sphingolipid. Neutral glycolipids formed by linkage of a single sugar unit such as a galactose to the sphingosine unit are called *cerebrosides* (a); charged glycolipids with more complex straight or branched sugar chains containing sialic acid residues are termed *gangliosides* (b). The sialic acid residues (see Fig. 4-4) give gangliosides a negative (acidic) charge at cellular pH. The names given to cerebrosides and gangliosides reflect their abundance in the plasma membranes of nerve and brain cells of animals. However, they are not restricted to this location and are found in membranes throughout the plant and animal kingdoms. Glu, glucose; Gal, galactose; GlcN, glucosamine; SiA, sialic acid.

a cerebroside   b ganglioside

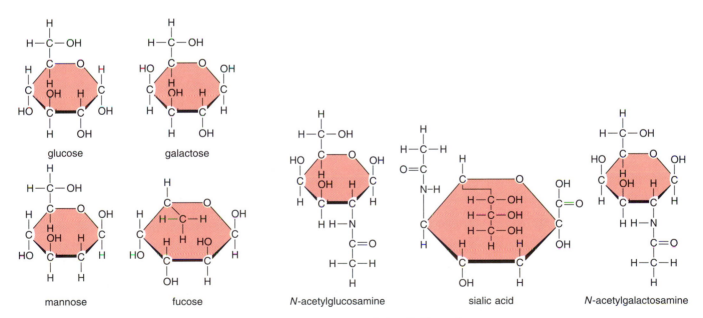

Figure 4-4 Sugars commonly found in the carbohydrate groups of glycolipids and glycoproteins.

cancer cells. Programmed changes in glycosphingolipid types at cell surfaces are also noted during embryonic development in animals.

Sterols The sterols (see Fig. 2-14 and p. 51), based on a framework of four carbon rings, are found in both plant and animal membranes. The arrangement of the four rings makes sterols rigid, flattened structures; only the side chain extending from one end of the ring structure is flexible. In contrast to phospholipids and sphingolipids, polar groups are very limited in sterols. The polar segment of cholesterol, for example, is limited to a single hydroxyl group at one end of the molecule, which is characteristic of the sterols (see Fig. 2-14c).

Cholesterol is the predominant sterol of animal cell membranes. Plasma membranes in animals contain almost as much cholesterol as phospholipids. Cholesterol also occurs in reduced quantities in some internal cellular membranes. Plant membranes contain small amounts of cholesterol and larger quantities of related sterols called *phytosterols*. Sterols of any kind are rare or completely absent in prokaryotes.

A primary effect of cholesterol and related sterols on biological membranes is to disturb the close packing of hydrocarbon chains of membrane phospholipids. As a result of this disturbance, membranes containing cholesterol remain fluid at low temperatures rather than "freezing" into nonfluid forms. Cholesterol also appears to increase both the flexibility and mechanical stability of membranes.

Membrane Proteins

The proteins of membranes occur in types and numbers that reflect the functional activity of membranes. Most membranes contain from 10 to 50 different major protein types, with molecular weights ranging from as little as 10,000 to more than 250,000. The relatively simple plasma membranes of red blood cells, with about 11 different major protein types, lie near one end of this distribution; the plasma membranes of HeLa cells (a cultured line of human cells), with more than 50 major protein types, lie near the other extreme. Many additional protein types probably occur in amounts too small to be readily detected in these and other membrane types. An individual protein type, depending on the membrane, may occur in copies numbering from tens to hundreds of thousands.

The Structure of Membrane Proteins Membrane proteins contain about the same proportion of hydrophobic amino acids as the soluble proteins of the internal cytoplasm—slightly less than 50%. However, the distribution of hydrophobic and hydrophilic amino acids in membrane and soluble proteins is quite different.

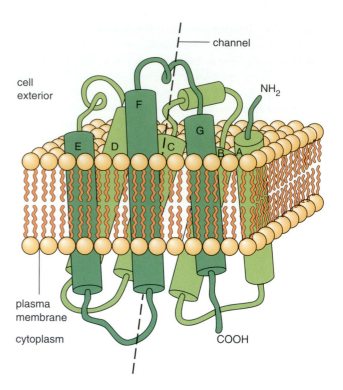

Figure 4-5 Bacteriorhodopsin, a protein absorbing light energy in plasma membranes of a group of photosynthetic bacteria (the halobacteria). The molecule contains seven alpha-helical transmembrane segments (cylinders A to G) that span the membrane. Each alpha-helical segment is connected to the next by a loop of hydrophilic amino acids that extends into or through the polar membrane surfaces. The three-dimensional arrangement of the alpha helices was deduced from electron diffraction patterns by R. Henderson and coworkers.

The hydrophobic amino acids of membrane proteins are frequently clustered in segments containing about 20 to 25 residues, separated by stretches of hydrophilic amino acids of varying length. In most soluble proteins, in contrast, hydrophobic residues tend to be scattered among hydrophilic amino acids without clustering. In membrane proteins the hydrophobic blocks of membrane proteins form segments, usually wound into an alpha helix, that are long enough to span the thickness of the membrane. These membrane-spanning segments, numbering from 1 to 20 or more in different membrane proteins, form anchors that hold the proteins in stable alignment in the lipid bilayer (as in the membrane protein shown in Fig. 4-5). The transmembrane segments typically extend back and forth across the membrane, each connected to the next by a short stretch of hydrophilic amino acids that makes a loop at the membrane surface. (Figure 4-6 summarizes various patterns in which the hydrophobic and hydrophilic segments of membrane proteins are distributed in membranes.)

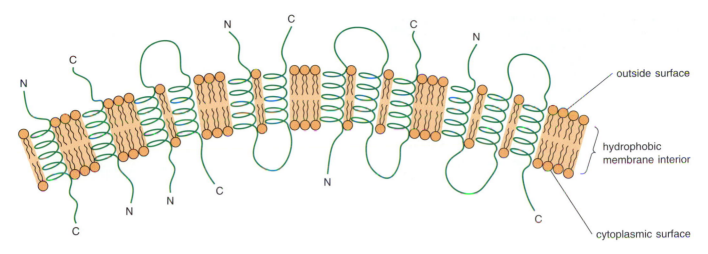

Figure 4-6 The arrangements of membrane-spanning hydrophobic segments and surface hydrophilic segments observed in different membrane proteins. N and C indicate the N- and C-terminal ends of the amino acid chains.

Some membrane proteins contain transmembrane segments in which one to three hydrophobic amino acids alternate in a repeating pattern with similar numbers of hydrophilic residues. When wound into an alpha helix, this arrangement places the hydrophobic residues on one side of the helix and the hydrophilic ones on the other side (Fig. 4-7a). Such alpha helices therefore have a polar and nonpolar face. These helices occur only in membrane proteins with multiple transmembrane segments. Evidently in these proteins the segments align so that the hydrophobic sides of the amphipathic helices face the membrane interior, and the hydrophilic sides cluster around the central axis of the protein (Fig. 4-7b). This arrangement creates a polar channel that extends through the protein from one membrane surface to the other. Such arrangements are believed to be typical of transport proteins, which conduct charged and polar molecules across the membrane via the polar channel.

The structure of membrane proteins thus places hydrophobic amino acids in regions facing the nonpolar membrane interior and positions hydrophilic amino acids in the inside of the protein or at the polar membrane surfaces. This segregation produces proteins with solubility properties similar to membrane lipids, that is, with distinctly polar and nonpolar regions.

Glycoproteins Proteins with covalently attached carbohydrate groups occur in many types of cellular membranes. The carbohydrate groups, which include essentially the same monosaccharides as those of glycolipids (see Fig. 4-4), link into straight or branched chains containing from 2 to 60 residues. (Figure 4-8

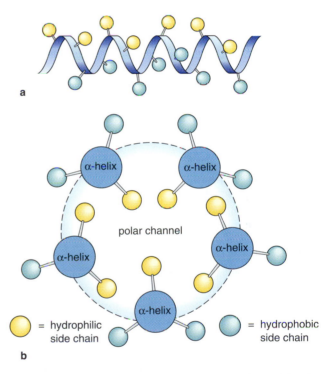

Figure 4-7 A frequent arrangement of amino acid residues in membrane-spanning alpha helical segments, in which hydrophobic amino acid side groups are located on one side of the helix and hydrophilic groups on the other (a). In proteins with multiple membrane-spanning segments, alpha helices of this type are often oriented with their hydrophilic side groups clustered inside the protein, forming a polar channel through the membrane (b).

Figure 4-8 Typical carbohydrate groups linked to membrane proteins in glycoproteins. The carbohydrate groups shown are linked to **(a)** glycophorin and **(b)** human fibroblast surface glycoprotein. Gal, galactose; Man, mannose; GlcN, glucosamine; SiA, sialic acid.

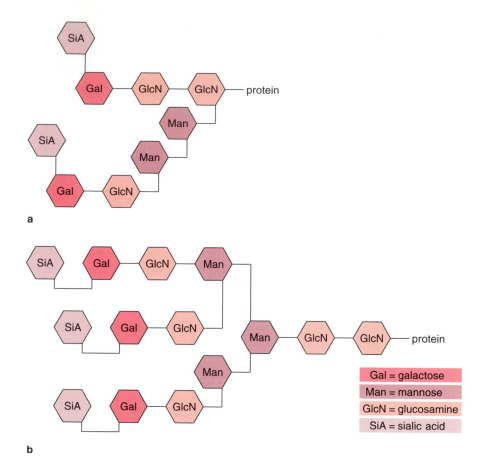

| Gal = galactose |
| Man = mannose |
| GlcN = glucosamine |
| SiA = sialic acid |

shows several representative carbohydrate units of membrane glycoproteins.)

Glycoproteins are most abundant in plasma membranes, where their carbohydrate groups occur almost exclusively on the outer membrane surface. Along with the carbohydrate groups of glycolipids, they give the cell surface what is often described as a "sugar coating" or *glycocalyx* (from the Greek *glykys* = sweet, and *calyx* = cup or vessel). In animals, glycoproteins are predominant in the glycocalyx; in plants, glycolipids make up most of the membrane coat. Relatively small amounts of glycoproteins also occur in internal membranes such as those of the endoplasmic reticulum (ER), Golgi complex, and nuclear envelope.

The functions of the carbohydrate groups of membrane glycoproteins are not completely understood. However, they are considered likely to add stability to membranes by forming a network interlinked by hydrogen bonds. The carbohydrate groups probably also act as parts of the recognition sites of membrane receptors involved in binding extracellular signal molecules and cell-to-cell adhesion. Like those of glycolipids, the carbohydrate groups of membrane proteins are also used as recognition and binding sites by infective bacteria and viruses.

LIPIDS AND PROTEINS IN MEMBRANE STRUCTURE

Singer and Nicolson's fluid mosaic model broke new ground with its central proposition that proteins are suspended individually in a fluid membrane bilayer. The burst of research triggered by the hypothesis fully supported its basic proposals about membrane structure and provided additions that have clarified and greatly extended the model.

Membrane Lipids and Bilayer Structure

The bilayer proposed for membranes in the fluid mosaic model was adapted from earlier models. It had been known for some time that, when exposed to a watery medium, phospholipids spontaneously assemble into the bilayer arrangement (Fig. 4-9). Within a bilayer, individual molecules pack with their long axes at right angles to the plane of the bilayer, in an orientation that satisfies their dual solubility properties. The polar head groups face the surrounding aqueous medium, and the nonpolar hydrocarbon chains associate end to end in the membrane interior, in a region that excludes water.

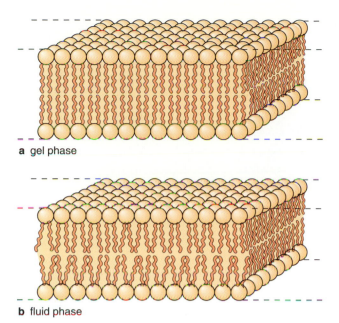

a gel phase

b fluid phase

Figure 4-9 Lipid bilayer structure, in which a double layer of phospholipid molecules is oriented so that the polar head groups face the surrounding aqueous medium and the nonpolar fatty acid chains associate in the nonpolar membrane interior. **(a)** The more highly ordered arrangement of phospholipid molecules in a bilayer at temperatures below the phase transition, in which the bilayer is "frozen" into a semisolid gel. **(b)** The less ordered arrangement above the phase transition, in which the bilayer "melts" into a fluid state in which individual molecules are free to flex, rotate, and exchange places.

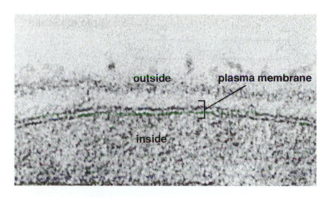

Figure 4-10 A plasma membrane at high magnification, showing the typical trilaminar or "railroad track" image seen in thin-sectioned membranes. × 240,000. (Courtesy of R. B. Park.)

Bilayers are stable and resist disturbances that would expose their hydrophobic interior region to surrounding water molecules. However, membranes disperse almost instantaneously if exposed to a nonpolar environment or to detergents, which can form a hydrophilic coat around the hydrophobic portions of membrane lipids and proteins in water solutions.

A variety of evidence supports the conclusion that membrane lipids actually occur in bilayers. One major line of evidence comes from experiments that compared natural membranes with artificial bilayers, made by painting a drop of phospholipid, dissolved in a nonpolar solvent, over a narrow opening in a teflon plate. When held under a water surface, the droplet gradually thins out until a film two molecules thick—a bilayer—stretches across the opening. Such artificial bilayers are more manageable experimentally than natural membranes because they can be assembled from single lipids or a completely defined lipid mixture.

In the electron microscope, thin-sectioned natural membranes and artificial phospholipid bilayers have equivalent dimensions, with the thickness expected for a double layer of lipid molecules—about 7 to 8 nm. Moreover, artificial bilayers and many cross-sectioned membranes appear in electron micrographs as two dark surface lines separated by a less dense interzone (Fig. 4-10). This image is consistent with a bilayer structure if the dark lines represent the polar membrane surfaces, and the less dense interzone the nonpolar membrane interior.

Another line of supporting evidence comes from comparisons of artificial bilayers and natural membranes prepared for electron microscopy by the freeze-fracture technique. In this method, artificial bilayers or tissue samples are rapidly frozen by placing them in liquid nitrogen. The frozen specimen is then fractured or split by striking it with a sharp knife edge. Because the nonpolar interior of artificial or natural bilayers produces weakly frozen faults in the preparations, which split open more readily than the surrounding polar regions, a fracture tends to follow planes along membrane interiors, revealing bilayers and splitting them into inner and outer halves (Fig. 4-11).

The bilayer arrangement for membrane lipids was first proposed in the 1920s by E. Gorter and F. Grendel, from experiments carried out with mammalian red blood cells (erythrocytes). Because these cells contain no internal organelles or nuclei, all the lipid in the preparations could be assumed to originate from the plasma membranes. Gorter and Grendel exposed the erythrocytes to distilled water, which causes the cells to burst. The treatment released the cell contents, consisting of little more than a solution of hemoglobin, and left empty plasma membranes called erythrocyte "ghosts." By measuring individual, flattened

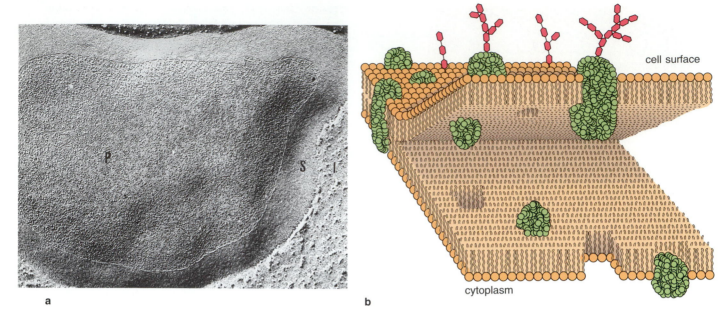

a
b

Figure 4-11 Membrane structures visible in freeze-fracture preparations. **(a)** A fracture exposing the bilayer interior of a human erythrocyte plasma membrane. The particles are membrane proteins embedded within the membrane bilayer. P, membrane particles in bilayer; S, external membrane surface; I, ice crystals surrounding the specimen. × 38,000. (Courtesy of T. W. Tillack, from *J. Cell Biol.* 45:649 [1970], by permission of the Rockefeller University Press.) **(b)** The path followed by a fracture that splits a membrane into bilayer halves.

ghosts under a light microscope, Gorter and Grendel were able to estimate the total area of a single erythrocyte plasma membrane. They then measured the total lipid quantity and number of cells in a preparation of erythrocyte ghosts. Comparing the total lipid quantity with the total surface area of the ghosts indicated that enough lipid was present to make a layer two molecules in thickness around each cell. Using this information, Gorter and Grendel made the first proposal that the lipid molecules of cell membranes occur in a double layer, or bilayer.

Bilayer Characteristics   Much has been learned about bilayer characteristics from the study of artificial bilayers. The artificial films allow study of membrane proteins as well as lipids, because in many instances it is possible to extract individual proteins from natural membranes and insert them singly or in combinations into artificial phospholipid bilayers.

Studies of the physical properties of artificial bilayers have revealed two fundamental characteristics that apply to both natural and artificial membranes: (1) Bilayers undergo what is known as a *phase transition*, in which they "melt" or "freeze" above or below certain temperatures; and (2) at temperatures above the phase transition, phospholipid bilayers are in a highly fluid state. Natural membranes must be held above the phase transition to remain fully functional.

The phase transition of a phospholipid bilayer results from a change in the packing and mobility of phospholipids induced by changes in temperature. At low temperatures the hydrocarbon chains of bilayer phospholipids are tightly packed and restricted in movement (see Fig. 4-9a). This tightly packed state is termed the *gel* or *crystalline* phase of a bilayer. As the temperature rises, a level is reached at which both the space separating adjacent phospholipid molecules and the movements of their hydrocarbon chains increase abruptly. At this temperature the phospholipid bilayer undergoes the phase transition by "melting" and becoming fluid (Fig. 4-9b). The increase in fluidity includes rotation of entire phospholipid molecules around their long axis, flexing of hydrocarbon chains, and movement of molecules from one position to another in the bilayer.

The temperature at which the phase transition occurs depends primarily on the length and degree of saturation of the hydrocarbon chains of bilayer phospholipids. Generally, the longer the hydrocarbon chains, the higher the temperature at which the bilayer melts. The sharp bends introduced in hydrocarbon chains by double bonds in the *cis* configuration, which interfere with tight packing of the chains (see Fig. 2-9), reduce the temperature at which the phase transition takes place and allow the bilayer to remain fluid at lower temperatures.

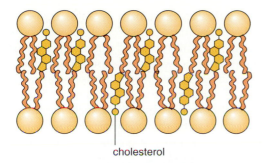

cholesterol

Figure 4-12 The position taken by cholesterol in bilayers. The hydrophilic —OH group, represented by the small circle at one end of the molecule, lies at the level of the bonds linking glycerol to the fatty acyl chains of the membrane phospholipids. The remainder of the cholesterol molecule, which is completely hydrophobic, lies parallel to the fatty acyl chains in the nonpolar membrane interior.

In artificial bilayers the motions of phospholipid molecules at temperatures above the phase transition are largely restricted to movement within the same bilayer half. Depending on the phospholipid type, the time required for movement of half of the phospolipid quantity from one bilayer half to the other, called *flip-flop*, is on the order of hours to days. In natural membranes, in contrast, flip-flop occurs much more rapidly, with half-times often on the order of seconds. These rapid rates result from the activity of proteins specialized for this function. If introduced into artificial bilayers, the proteins increase flip-flop to rates comparable to those of natural membranes (see p. 113).

Cholesterol and the Phase Transition Adding cholesterol to artificial phospholipid bilayers has a marked effect on the phase transition. When added in small amounts, up to about 20% of the total bilayer lipid, cholesterol widens the temperature range over which the phase transition takes place. At levels above about 20% it eliminates the gel phase entirely at most temperatures likely to be encountered by living organisms (bilayers can accommodate cholesterol up to a 1:1 molar ratio with phospholipid). At the same time, cholesterol restricts rapid movement of the hydrocarbon chains of membrane phospholipids, which has the effect of reducing membrane fluidity at elevated temperatures. Cholesterol therefore has the dual effect of increasing membrane fluidity at low temperatures and reducing it at elevated temperatures. For these reasons, cholesterol is said to be a "plasticizer" of natural membranes.

The plasticizing effects of cholesterol are believed to be due to the position it takes in bilayers and to its interaction with surrounding bilayer lipids. X-ray dif-

fraction (see Appendix p. 802) indicates that cholesterol lines up between the phospholipid molecules of a bilayer half, with the long axis of the cholesterol molecule parallel to the hydrocarbon chains of the phospholipids (Fig. 4-12). The cholesterol molecule orients in the bilayer with its single polar —OH group at the level of the bonds linking fatty acyl chains to glycerol. The remainder of the rigid, hydrophobic cholesterol ring structure extends into the nonpolar membrane interior. In this position the cholesterol molecule interferes with tight packing of the hydrocarbon chains of nearby phospholipids and inhibits formation of the gel phase. Above the phase transition, cholesterol interferes with the more extreme movements of the surrounding phospholipid molecules and reduces fluidity.

Cholesterol has other effects in phospholipid bilayers. It forces the polar head groups of phospholipid molecules farther apart, increasing access to the bilayer interior by molecules at the membrane surface. Through this effect, cholesterol may aid attachment to the membrane surface of molecules such as proteins that bind by extending hydrophobic groups into the nonpolar membrane interior. Even though it increases the separation of phospholipid molecules in the bilayer, cholesterol helps to seal the nonpolar part of bilayers against the penetration of water, ions, and small polar molecules such as glucose. This effect may result from the ability of cholesterol to fit into the spaces between hydrocarbon chains of membrane phospholipids, eliminating faults through which ions and small polar molecules may pass.

Regulation of the Phase Transition in Living Organisms
Many organisms adapt to colder temperatures by decreasing the length or increasing the degree of saturation of the fatty acid chains in membrane phospholipids or by increasing cholesterol content. These alterations interfere with packing of the hydrocarbon chains in membrane phospholipids, thereby lowering the temperature at which transition to the gel state occurs.

Many eukaryotic organisms, including algae, higher plants, protozoa, and animals, use one or more of these alterations to keep membranes fluid at low temperatures. In animals in which body temperature fluctuates with environmental temperature, the proportions of both cholesterol and double bonds in membrane phospholipids are increased at lower temperatures as a means of regulating membrane fluidity. Similar changes are noted as mammals enter hibernation. In nonhibernating mammals the phase transition from fluid to gel takes place at about 15°C. As mammals enter hibernation, their body temperature may fall to as low as 5°C. As body temperature falls, the proportion of double bonds in membrane phospholipids increases, keeping the membranes fluid at the reduced temperatures.

The ability to keep membranes fluid at low temperatures is particularly significant to the nervous system of hibernating mammals. Retention of membrane fluidity allows cells of the nervous system to remain active in transmitting nerve impulses. A hibernating animal, although sluggish, can still maintain its basic body functions and respond to external stimuli.

Higher plants with resistance to chilling also contain membrane lipids with greater proportions of double bonds. Because resistance to chilling is a critical factor in survival and growth of crop plants in colder regions, there is considerable interest in the possibility of adding cold resistance by genetic engineering.

Prokaryotes are also able to regulate the phase transition to keep membranes fluid at lower temperatures. In most bacteria, reduced temperatures induce synthesis of an enzyme that adds double bonds to fatty acid chains. As a consequence, more fatty acids containing double bonds are made in the cytoplasm and introduced into membrane lipids. The human intestinal bacterium *E. coli* is among the many bacteria using this mechanism to regulate the phase transition. In some bacteria the phase transition is pushed downward by decreases in fatty acid chain length.

Bilayers, Membrane Proteins, and the Fluid Mosaic Model

The fluid mosaic model reaffirmed earlier proposals that membrane lipids occur in bilayers, and emphasized the fluid nature of the bilayer. The primary and revolutionary proposals of the model, however, concerned the distribution of membrane proteins. Until the model was advanced, membrane proteins were generally thought to be extended in exceedingly thin layers over bilayer surfaces. Singer and Nicolson's model proposed instead that proteins are suspended individually as globular units within or on the membrane bilayer (Fig. 4-13).

The fluid mosaic model defined two major classes of membrane proteins—*integral* and *peripheral*—according to their mode of association with membrane bilayers (Table 4-2). Integral proteins are deeply embedded in the bilayer and held in place by nonpolar interactions with membrane lipids. Suspension in the bilayer depends on the dual solubility properties of membrane proteins. At polar membrane surfaces, protein molecules fold to expose only hydrophilic amino acid side chains. Nonpolar side chains are exposed on the protein surfaces facing the hydrophobic membrane interior, held in this position by their association with the nonpolar hydrocarbon chains of membrane lipids.

Integral proteins remain in stable suspension in the bilayer because, as with membrane lipids, any change in orientation would expose their hydrophobic regions to the watery surroundings. Because of their intimate association with the nonpolar membrane interior, integral membrane proteins can be removed from membranes only by agents, such as detergents or nonpolar solvents, that disperse the bilayer. Within these limitations, integral proteins are potentially free to displace phospholipid molecules and move laterally through the fluid bilayer.

Peripheral proteins are hydrophilic molecules that bind noncovalently to polar membrane surfaces. Because of their hydrophilic nature and polar associations, peripheral membrane proteins can be removed by relatively mild treatments that do not disrupt the bilayer, such as adjustments in the salt concentration or pH of the suspending medium.

The hydrophilic carbohydrate groups of membrane glycoproteins and glycolipids are considered in the fluid mosaic model to extend into the polar environment at membrane surfaces (see Fig. 4-13). In plasma membranes, in accordance with observations that sugar groups are restricted to the outside surfaces of cells, these carbohydrate groups are thought to extend only from the exterior surface of membranes. The carbohydrate groups are probably distributed so thickly over the exterior surface of plasma membranes that little of the phospholipid bilayer is exposed directly to the surrounding aqueous medium.

Some peripheral proteins are anchored at one end to membranes by a fatty acid chain, phospholipid, or other nonpolar hydrocarbon group. The anchoring group links covalently to an amino acid side chain at the N-terminal end of the protein. The anchor, often called a "greasy foot" or "greasy finger," holds the peripheral protein in association with the membrane. Several cytoskeletal proteins, and other proteins taking part in pathways transmitting extracellular signals from the plasma membrane to the cell interior, are anchored in this way.

The Singer–Nicolson model originally proposed that integral proteins are completely free to move in the fluid bilayer. It has become apparent, however, that free movement may be restricted by linkage of some integral proteins into aggregates or by combinations between integral and peripheral proteins. In the latter case, peripheral proteins may join in a cytoskeletal lattice or network at the membrane surface that holds many of the integral proteins in fixed position. Recent evidence has shown that such anchoring networks of peripheral proteins occur widely in cellular membranes, particularly in plasma membranes. Cytoskeletal networks immobilizing membrane proteins occur on the inner membrane surface facing the cytoplasm (for details, see Chapter 12). Linkages fix-

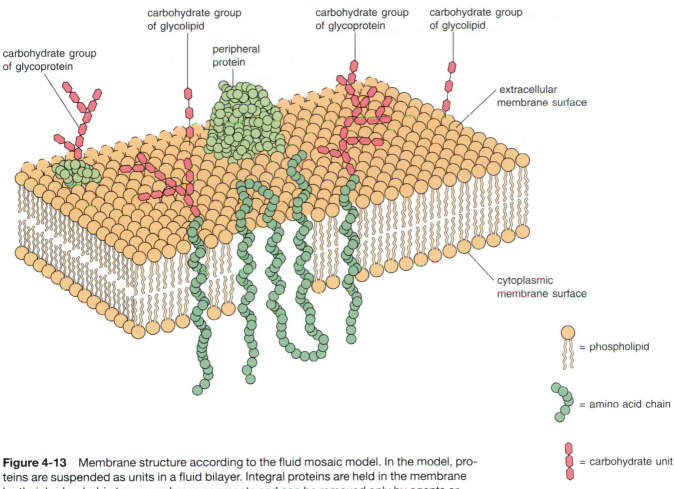

carbohydrate group
of glycoprotein

carbohydrate group
of glycolipid

peripheral
protein

carbohydrate group
of glycoprotein

carbohydrate group
of glycolipid

carbohydrate group
of glycoprotein

extracellular
membrane surface

cytoplasmic
membrane surface

= phospholipid

= amino acid chain

= carbohydrate unit

Figure 4-13 Membrane structure according to the fluid mosaic model. In the model, proteins are suspended as units in a fluid bilayer. Integral proteins are held in the membrane
by their hydrophobic transmembrane segments and can be removed only by agents or
treatments that disrupt the bilayer. Peripheral proteins, which are hydrophilic, are attached
to the bilayer surfaces and can be removed by less drastic techniques. The diagram shows
a plasma membrane as proposed in the model, with carbohydrate groups of membrane
glycoproteins and glycolipids facing the cell exterior.

Table 4-2 Characteristics of Integral and Peripheral Membrane Proteins		
Characteristic	Integral Protein	Peripheral Protein
Location in membrane	Buried in hydrophobic membrane interior	Bound to membrane surface
Requirements for release from membrane	Released only by agents that disrupt membrane bilayer, such as detergents	Released by treatments that leave bilayer intact, such as increases in salt concentration
Association with lipids when released	Usually associated with lipids	Not associated with lipids
Solubility	Usually insoluble in aqueous media	Usually soluble in aqueous media

ing integral proteins in place may also be formed with
molecules of the extracellular matrix (see Chapter 7),
or by interactions among carbohydrate groups of
membrane glycoplipids and glycoproteins at the membrane surface.

Another major feature of the fluid mosaic model
proposes that the arrangement of phospholipids and
both integral and peripheral proteins is *asymmetric* in
biological membranes. For membrane lipids, asymmetry means that the inner and outer bilayer halves

contain different proportions of various lipid types. For integral membrane proteins, asymmetry means that these molecules have specific orientations with respect to the two membrane surfaces and do not tumble or rotate to change their orientation. Such movements would be energetically unfavorable because rotation would expose the hydrophobic portions of the proteins to polar membrane surfaces, and polar regions of the proteins, including any attached carbohydrate groups, to the nonpolar membrane interior. (The Experimental Process Essay by S. J. Singer on p. 124 describes his classic experiment eliminating the rotating carrier model.) The restriction of peripheral proteins to one bilayer surface depends on their hydrophilic nature, which prevents them from crossing from one side to the other.

Evidence Supporting the Fluid Mosaic Model

Research carried out since its proposal has fully supported the fluid mosaic model, and it is now accepted as an accurate depiction of the fundamental structure of biological membranes. The supporting evidence has confirmed that bilayers are fluid at physiological temperatures; that proteins are suspended as individual units in membranes, with at least some able to move laterally through the fluid bilayer; and that both lipids and proteins are asymmetrically arranged in membranes.

Evidence That Membranes Are Fluid Evidence that membrane bilayers are fluid comes from several sources. A method using *photobleaching* provides a graphic demonstration of the fluidity of both natural and artificial bilayers. In this technique, fluorescent dyes are attached to membrane lipids or proteins. The dye markers are then bleached by a laser beam in one region of the bilayer, in a spot only a few micrometers in diameter. The bleached spot, initially sharply defined on the membrane, gradually spreads and fades as the marked molecules diffuse and exchange places with unbleached molecules from the surroundings. The time required for the spot to fade provides a measure of fluidity, which for both natural and artificial bilayers turns out to be approximately the same as for olive oil or light machine oil. Observations of the movements of membrane proteins (see below) also clearly support the conclusion that membrane bilayers are in a fluid state at temperatures above the phase transition.

One curious aspect of membrane fluidity is its possible relationship to substances used as anesthetics. Many of these substances, which interrupt consciousness or the sensation of pain, touch, temperature, and other sensory inputs, are nonpolar. In fact, the more soluble anesthetics are in lipid solvents, the greater

their anesthetic effects. Although their exact effects on transmission of nerve impulses are unknown, anesthetics alter the phase transition of lipid bilayers, increase the rate at which substances pass through membranes, and disrupt membrane organization. They may also increase membrane fluidity. There are also indications that some anesthetics act directly on membrane proteins, particularly those forming ion channels in nerve cell plasma membranes. These alternations may so alter the arrangement of receptors and other membrane proteins in nerve synapses (see p. 142) that transmission of nerve impulses stops in anesthetized regions, thus impairing or interrupting the flow of information to or within the brain.

Evidence That Proteins Are Suspended Within Membrane Bilayers The best evidence that proteins are suspended in biological membranes comes from electron microscopy of freeze-fracture preparations. When a fracture follows the interior of a membrane, splitting the bilayer into inner and outer halves, globular particles the size of protein molecules are clearly seen embedded in the bilayer (as in Fig. 4-11a).

The particles exposed in fractured membranes have been identified as proteins by various methods. For example, myelin membranes, known to contain relatively few proteins, show almost no globular particles in freeze-fracture preparations. Other membranes known to contain proteins in greater quantity, such as the plasma membrane shown in Figure 4-11a, are crowded with particles when freeze-fractured.

Direct support that the particles are proteins comes from freeze-fracture studies of artificial bilayers carried out by D. W. Deamer, D. Branton, and others. Artificial bilayers created by suspending pure phospholipids in water show smooth interior surfaces without particulate units when freeze-fractured (Fig. 4-14a). When proteins are added to the artificial bilayers, globular particles closely resembling those in natural membranes appear in the freeze-fracture preparations (Fig. 4-14b).

Other experiments have shown parallels in the behavior of membrane proteins and the particles in freeze-fractured membranes. By various treatments such as changes in pH of the surrounding solution, proteins in living membranes can be induced to pack together in clumps. Freeze-fracture preparations of these membranes show that the membrane particles are also clumped together. As a group, these experiments establish that the particles in freeze-fractured membranes are proteins. Their distribution clearly supports the idea that proteins are suspended within the bilayers of biological membranes.

Several lines of experimental evidence have established that most integral proteins extend entirely through the bilayer and protrude from both mem-

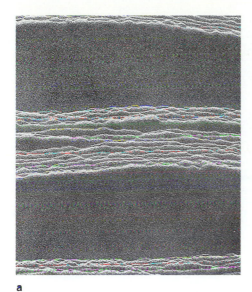 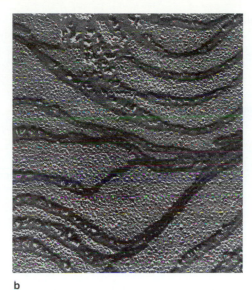

a   b

Figure 4-14 Artificial bilayers demonstrating the appearance of membrane proteins in freeze-fracture preparations. **(a)** Freeze-fracture preparation of layered artificial bilayers without added proteins; none of the layers shows evidence of embedded particles. **(b)** Artificial bilayers to which proteins have been added. Particles similar to those seen in freeze-fractured natural membranes are now visible. (Courtesy of D. W. Deamer.)

brane surfaces. In some of the experiments a radioactive label or chemical group was attached to parts of proteins exposed at either membrane surface. Other experiments detected segments of proteins that extend from membrane surfaces by combining them with fluorescent antibodies (see p. 786) or digesting them with proteolytic enzymes.

For example, E. Rechstein and A. Blostein added radioactive iodine to protein segments exposed at the surfaces of membranous vesicles (Fig. 4-15). Their experiment took advantage of the fact that membrane fragments can be induced to seal into right-side-out or inside-out vesicles by adjustments in the salt concentration and pH of the suspending solution. By this means they could identify the parts of proteins that extend from the inside and outside surfaces of the membrane vesicles. Their experiments, and the results of similar tests using other techniques, confirm that almost all integral proteins extend entirely through membranes and have segments exposed on both membrane surfaces. The same experimental approaches also provide a major part of the evidence that integral proteins take an asymmetric orientation in membranes (see below).

Evidence That Membrane Proteins Can Move Laterally in the Bilayer Evidence that many integral proteins are free to move laterally in cellular membranes comes from a classic experiment carried out in 1970 by L. D. Frye and M. A. Edidin. These investigators worked with mouse and human cells, using a technique that fuses the plasma membranes of separate cells (see Information Box 4-1). The fusing technique involved exposure to the Sendai virus, which alters cells so that their plasma membranes flow together upon collision, joining the cells into a larger structure

with two or more nuclei and a composite cytoplasm and plasma membrane.

Frye and Edidin used the cell fusion technique in combination with antibodies that bind specifically to plasma membrane proteins of either the mouse or human cell type. The antimouse antibodies were attached to molecules that fluoresce green under ultraviolet light, and the antihuman antibodies to molecules that fluoresce red (Fig. 4-16a; fluorescence microscopy is described in the Appendix on p. 786). The mouse and human cells were then fused, forming cells with a composite mouse-human plasma membrane (Fig. 4-16b). At first the fluorescent colors were in separate regions on the surfaces of the fused cells, with half of the membranes fluorescing red and half green. After 40 minutes at 37°C the fluorescent colors were completely intermixed on 90% of the cells, indicating the mouse and human proteins had become uniformly distributed throughout the composite membranes (Fig. 4-16c). Lowering the temperature gradually slowed intermixing; at 15°C the membrane proteins of the two species remained separate. From these results, Frye and Edidin concluded that intermixing resulted from lateral movement of the mouse and human proteins through the fluid membrane bilayer. The interruption of diffusion at 15°C reflected transition to the gel phase, in which the membrane lipids and proteins were immobilized.

The time required for intermixing of the mouse and human membrane proteins gives a measure of the rate at which proteins can move through bilayers. From this data, Edidin calculated that the mouse and human proteins move about 100 times more slowly than membrane lipids. Other experiments using similar or different methods show protein movement over a range of from 10 to 1000 times slower than

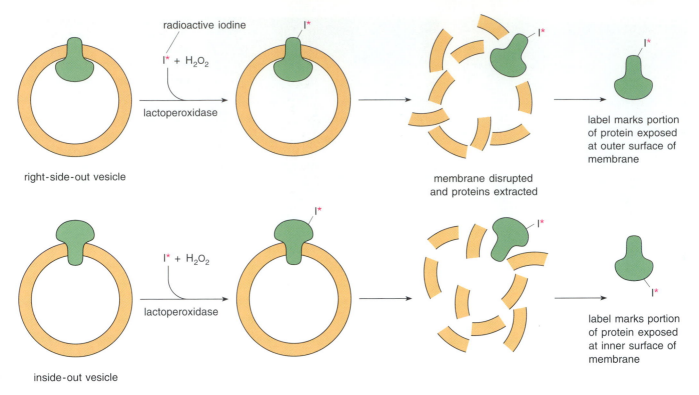

radioactive iodine

I^*

$I^* + H_2O_2$

lactoperoxidase

right-side-out vesicle

membrane disrupted
and proteins extracted

label marks portion
of protein exposed
at outer surface of
membrane

$I^* + H_2O_2$

lactoperoxidase

inside-out vesicle

label marks portion
of protein exposed
at inner surface of
membrane

Figure 4-15 Rationale for Rechstein and Blostein's experiment demonstrating that integral proteins extend entirely through membranes. They used the enzyme lactoperoxidase to attach a radioactive iodine label to the exposed portions of the proteins. (Lactoperoxidase catalyzes the covalent linkage of halogens to proteins.)

membrane lipids.[1] The slower diffusion rates for membrane proteins probably reflect several factors, including their larger size, interactions between charged and other groups on protein and membrane surfaces, and linkage of membrane proteins to peripheral proteins, particularly those of the cytoskeleton.

Some additional experiments have shown that certain membrane proteins, rather than diffusing passively in the membrane bilayer, are actively moved by systems that require energy. This active movement, called *capping*, has been demonstrated with fluorescent antibodies that bind to specific proteins of the plasma membrane of cells such as lymphocytes (white blood cells). When antibodies are first added, the membrane proteins binding them are distributed randomly over the entire cell surface (Fig. 4-17a).

Within minutes, however, the marked proteins are swept to one end of the cell, where they form a dense cap (Fig. 4-17b). The cap may remain in position or be taken into the cell interior by endocytosis (see p. 593). Inhibitors of reactions supplying ATP arrest the capping reaction, indicating that it is an active process that requires expenditure of cellular energy. (For further details of the capping reaction, which depends on the activity of microfilaments, see Chapter 11.)

Some membrane proteins and glycoproteins also prove to be more or less immobile in biological membranes. These molecules are held in place by linkages to cytoskeletal networks or extracellular materials, or by interactions between the glycoprotein carbohydrate groups extending from the membrane surface. For example, *band 3 protein*, an integral protein of erythrocyte plasma membranes, is held essentially immobile through its linkage to cytoskeletal proteins that form a network underneath the plasma membrane (see p. 351).

Evidence That Membranes Are Asymmetric An extensive series of experiments has demonstrated that both lipids and proteins are distributed asymmetrically in natural membranes. A group of enzymes that hydrolyze phospholipids, the *phospholipases*, has been used

[1] The rate at which molecules move through bilayers is described quantitatively by the *diffusion coefficient*. For lipid molecules, the diffusion coefficient is surprisingly high, about 10^{-8} cm^2/sec. At this rate, a phospholipid molecule could move the entire length of a small cell such as a bacterium in about one second! The rates observed for proteins reflect diffusion coefficients ranging from about 10^{-9} to 10^{-11} cm^2/sec, from 10 to 1000 times slower than membrane lipids.

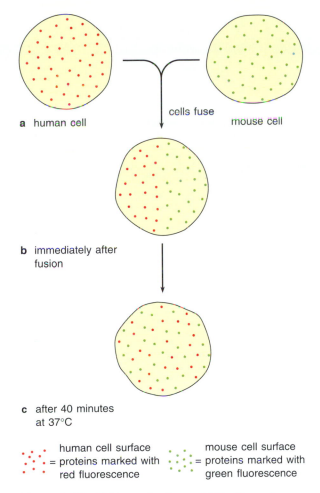

a human cell   cells fuse   mouse cell

b immediately after
   fusion

c after 40 minutes
   at 37°C

human cell surface
= proteins marked with
red fluorescence

mouse cell surface
= proteins marked with
green fluorescence

Figure 4-16 Evidence for the mobility of membrane proteins in the combined plasma membranes of fused mouse-human cells. **(a)** The cells before fusion. **(b)** Immediately after fusion, membrane proteins from the mouse and human cells are segregated into distinct regions in the plasma membrane. **(c)** After 40 minutes at 37°C the mouse and human membrane proteins are completely intermixed in the fused cells.

widely in this research. Using these enzymes, L. L. M. van Deenen and his coworkers found that 76% of the total membrane phospholipids containing choline and 20% of those containing ethanolamine were hydrolyzed in intact erythrocytes (Fig. 4-18). These results indicate that these phospholipids occur in these proportions in the outside half of the erythrocyte bilayer. If so, the inner half of the bilayer contains 24% of the choline-containing and 80% of the ethanolamine-containing phospholipids. No serine-containing phospholipids were hydrolyzed by the enzymes as long as the erythrocytes were intact. However, if the plasma membranes were broken, exposing the inner membrane surface to the enzymes, all the phospholipids containing serine were hydrolyzed, indicating that this phospholipid type is confined to the inner bilayer half.

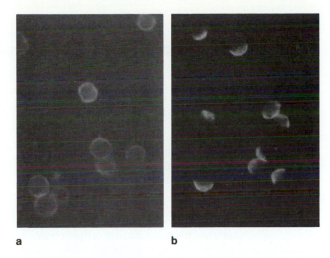

a   **b**

Figure 4-17 Capping in lymphocytes (white blood cells). **(a)** Binding of fluorescent antibodies (light areas) to proteins distributed over the entire cell surface. **(b)** Formation of the cap, in which the marked proteins are actively swept to one end of the cell. (Courtesy of G. M. Edelman.)

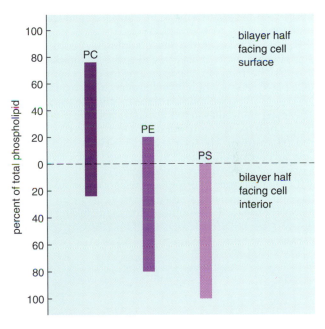

Figure 4-18 The asymmetric distribution of phosphatidyl choline (PC), phosphatidyl ethanolamine (PE), and phosphatidyl serine (PS) in erythrocyte membranes as revealed by the experiments of L. L. M. van Deenen and his coworkers.

Equivalent results have been obtained with the plasma membranes of various other cell types. Although most studies have concentrated on plasma membranes, a few studies of organelle membranes, such as those of mitochondria, the Golgi complex, and

Membrane Fusion

Some biological membranes commonly undergo *fusion,* a process in which the bilayers of two separate membranes join to form a single, continuous membrane. The fusion inserts the lipids and proteins of one membrane into another, allowing the membrane components to mix freely. Membrane fusion occurs regularly when vesicles pass from the ER to the Golgi complex, when vesicles connect with the plasma membrane during secretion, when vesicles enclosing materials from outside the cell join with lysosomes, and when sperm and egg plasma membranes fuse during fertilization. During secretion, for example, cytoplasmic vesicles containing substances to be released to the cell exterior contact the plasma membrane on its cytoplasmic side. On contact the vesicle and plasma membrane bilayers join and flow together, adding the lipids and proteins of the vesicle membrane to the plasma membrane and releasing the vesicle contents to the cell exterior. Membrane fusion also is a part of pathological processes such as cell infection by enveloped viruses such as the Sendai and influenza types. (Enveloped viruses are those surrounded by a surface membrane coat; see p. 19.)

For fusion to occur, the phospholipid bilayers of the joining membranes must come into direct contact. At least two barriers must be overcome for direct contact to take place. One is a film of water molecules covering polar groups of phospholipids at the membrane surface. The water molecules in this *hydration layer* are more highly ordered than surrounding water molecules and resist disruption. The second barrier is electrostatic repulsion by phospholipid head groups, most of which carry a negative charge under physiological conditions.

The mechanisms overcoming the barriers to fusion have been extensively studied in connection with the enveloped viruses. Some of these viruses, including the herpes and Sendai viruses, attach and fuse directly with the plasma membrane. The attachment takes place by means of "spike" glycoproteins that extend from the viral membrane coat (see part **a** of the figure in this box). The spike glycoproteins, which resemble segments of molecules normally reacting with the cell surface, are bound by the cell's receptors. On binding to the cell surface the viral proteins undergo a conformational change that exposes a hydrophobic segment (part **b** of the figure). This segment probably inserts into the hydrophobic interior of the plasma membrane and promotes fusion of the host cell and viral membrane bilayers. Presumably the hydrophobic segment of the viral spike protein provides a surface along which lipids of the bilayer halves can flow in order to fuse. Proteins able to promote membrane fusion by this or a related mechanism are known as *fusogens.*

Fusion of cellular membranes in activities such as secretion also appears to be promoted by fusion proteins. The fusion proteins involved in cellular membrane fusions, like those of enveloped viruses, have hydrophobic segments that are exposed by the activating sequence, leading to attachment of vesicles and fusion with target membranes. In cellular systems, fusion proteins act in cooperation with other proteins in a combination that J. E. Rothman, one of the primary investigators in this field, calls a *fusion machine* (see also pp. 142 and 578). The proteins of a fusion machine work in a sequence that in many systems is triggered by local increases in Ca^{2+} concentration that typically occur just before membrane fusion takes place.

Ca^{2+} may also exert an effect directly on membranes, because the ion alone can promote fusion between artificial phospholipid films, with no proteins in evidence. The Ca^{2+} may neutralize the negative charges at the membrane surfaces or, as some investigators have suggested, may aggregate membrane phospholipids into "islands" that create faults in the bilayer. The faults may open regions that lack the hydration and electrostatic barriers and promote membrane approach and contact.

---

the ER, indicate that phospholipid distribution in all cellular membranes is probably asymmetric.

The functional significance of lipid asymmetry is not completely understood. However, the asymmetric distribution of lipids between the two bilayer halves may maintain necessary differences in the charge, fluidity, or phase transition of the two membrane surfaces or may be associated in some way with the asymmetric orientation of membrane proteins.

One of the experiments demonstrating protein asymmetry came from the Singer laboratory. Nicolson and Singer developed a glycoprotein marker for elec-

tron microscopy by linking ferritin, a large, electron-dense molecule, to *concanavalin A (con A),* a plant protein that binds to the carbohydrate groups of some glycoproteins. The ferritin-labeled con A was found to bind only to outer surfaces of erythrocyte membranes and never to inner surfaces.

The asymmetric arrangement of membrane proteins is central to their function in membranes. In transport proteins that use energy to move substances across the membrane it ensures that transported molecules move preferentially in only one direction. The absorption of light in photosynthesis and its conver-

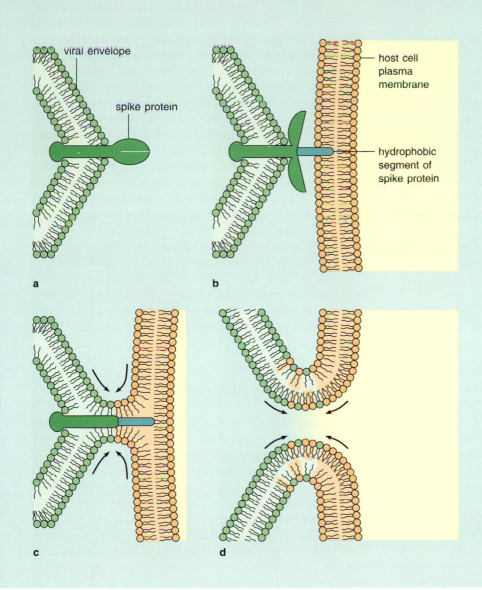

viral envelope

spike protein

host cell plasma membrane

hydrophobic segment of spike protein

a

b

c

d

sion to chemical energy in chloroplasts, and the synthesis of ATP in both chloroplasts and mitochondria are equally dependent on membrane "sidedness" and the asymmetric arrangement of the proteins carrying out these activities. Moreover, the asymmetric orientation of glycoproteins on the outside surfaces of plasma membranes provides the cell with receptors and with recognition and adhesion groups directed toward its surroundings.

Some proteins are restricted to certain regions of cellular membranes as well as being arranged asymmetrically. For example, in cells lining the mammalian intestinal tract the enzyme alkaline phosphatase and a membrane channel that admits sodium ions occur only in the part of the plasma membrane facing the intestinal cavity. Other proteins of the intestinal cells, such as a surface recognition marker and a different sodium transporter, one that uses energy to pump sodium ions out of the cell, are restricted to the part of the plasma membrane facing the circulatory system.

The unequal distribution of plasma membrane proteins in the intestinal cells is maintained by linkage of proteins to the cytoskeleton underlying the membrane and by a type of cell-to-cell junction (a *tight* or sealing junction; see p. 170). In the junction, plasma membranes of adjacent cells fuse so closely that membrane proteins cannot diffuse through the junction region. Because the junction extends in a beltlike band entirely around the cell, it forms a complete barrier to diffusion of membrane proteins between the sides of the cell facing the intestinal cavity and the bloodstream. Destruction of the junction allows some proteins of the two membrane regions to intermix, as expected if the junction helps maintain the different membrane domains.

The unequal lateral distribution of proteins creates subregions of the membrane that are specialized for different functions. In the cells lining the intestinal tract, for example, the sodium channels facing the intestinal cavity specialize the plasma membrane in this region for free entry of sodium ions from the intestinal contents. The location of active sodium pumps on the side facing the bloodstream specializes the plasma membrane in this region for removal of sodium ions from the cytoplasm. As a consequence of these specializations, sodium ions flow through the cells from the intestinal lumen to the bloodstream.

Developments Leading to the Fluid Mosaic Model

Until the fluid mosaic model revolutionized membrane concepts, it was generally believed that proteins were distributed over membrane surfaces and extended into layers only one amino acid in thickness on both sides of the bilayer. In 1966, J. Lenard and S. J. Singer showed that as much as 30% of the amino acid chains of membrane proteins are twisted into an alpha helix. This made it unlikely that membrane proteins could be spread over the membrane surface in completely extended form. Singer noted that the total content of alpha helix in membrane proteins, in fact, is typical of polypeptides with a spherical rather than a flattened shape.

Singer also pointed out that unfolding proteins into fully extended form on membrane surfaces would inevitably expose nonpolar amino acid side chains to the polar environment of the surrounding aqueous medium. He realized that this condition is thermodynamically unlikely, since large amounts of energy must be expended to maintain polar or nonpolar groups in a position that exposes them to the opposite environment. Instead, proteins tend to fold into minimum energy states, with hydrophilic and hydrophobic groups located in separate regions facing environments of like polarity. Singer calculated that the energy required to maintain proteins in the fully extended form would be so great that membranes would be highly unstable, rather than the stable structures they are observed to be.

Some clues were available as to the actual arrangement of protein molecules in membranes. Freeze-fractured membrane preparations revealed that large numbers of particles with the dimensions expected for protein molecules are suspended in the membrane interior. These findings suggested that proteins are embedded in membranes as globular units rather than extended on membrane surfaces. In 1972, Singer and Nicolson combined all the available evidence and arguments into their hypothesis for membrane structure, the fluid mosaic model. Their model retained the phospholipid bilayer advanced in earlier models as the basic structure underlying biological membranes and proposed that the bilayer is fluid. Their proposal that integral membrane proteins are suspended in globular, particulate form in the lipid bilayer, maintained in this position by polar and nonpolar associations, explained how proteins can exist in membranes in a stable, minimum-energy state. They added the idea that both membrane proteins and lipids are distributed asymmetrically in membranes, and they proposed further that membrane proteins occur in two types—integral proteins suspended in the bilayer and peripheral proteins attached to membrane surfaces.

MEMBRANE BIOGENESIS

Membranes are dynamic structures that rapidly assemble and disassemble. C. E. Bracker has calculated that a growing fungal cell adds new membranes at the rate of 32 square micrometers per minute. In dividing cells that are roughly spherical in shape, the surface area of the plasma membrane increases by about 1.6 times during the growth phase between divisions. Where are the new lipids and proteins assembled, and how are they inserted in membranes?

Lipid Origins

Using radioactively labeled precursors as markers in autoradiography experiments (see Appendix p. 791), D. J. Morré showed in 1970 that in both plant and animal cells, radioactive lipids first show up in the endoplasmic reticulum (ER). The label then moves from the ER to the Golgi complex, and later to the plasma membrane. These observations indicate that membrane lipids are synthesized in the ER and are first inserted into membrane bilayers in this location. The enzymes required for membrane lipid synthesis have also been identified in ER membranes isolated by cell fractionation (see Appendix p. 794), primarily in the smooth ER.

Newly synthesized phospholipids enter ER membranes at the bilayer half facing the surrounding cytoplasm. They are then flip-flopped to the opposite bilayer half by enzymes called *flippases*, which make up part of the integral proteins of the ER membranes. Although phospholipid transfer by the flippases is readily detected, the mechanisms by which the enzymes move phospholipid molecules between bilayer halves remains unknown. Presumably the flippases undergo conformational changes that bury polar phospholipid groups in the active site during their transfer across the nonpolar bilayer interior.

J. M. Backer and E. A. Dawidowicz successfully removed a flippase from rat ER membranes and inserted it into artificial phospholipid membranes. The ER flippase was capable of rapidly transferring phospholipids between bilayer halves in the artificial membranes, with no requirement for an energy input. Whether all flippases are similarly energy-independent remains unknown.

The newly synthesized and inserted phospholipids then move from the ER membranes through the Golgi complex to the plasma membrane. Movement from the ER to the Golgi complex occurs via membranous vesicles that pinch off from the ER and fuse with the Golgi (*transition vesicles*; see Fig. 4-19); similarly, movement from the Golgi complex to the plasma membrane occurs via secretory vesicles that pinch off from the Golgi complex, move through the intervening cytoplasm, and fuse with the plasma membrane. (Supplement 4-1 describes possibilities for the molecular mechanisms underlying membrane fusion; see Chapter 20 and Fig. 20-5 for further details of vesicle movement between the ER, Golgi complex, and the plasma membrane.) Flow from the rough ER to the nuclear envelope is also possible through the direct connections that occur between these membrane systems (see p. 574). The fluidity of the membrane bilayer provides the basis for the flow of phospholipids through these permanently or intermittently connected membrane systems. Cholesterol follows a similar pathway except that flippases may not be required to move it between bilayer halves—because it is almost entirely nonpolar, cholesterol is able to change sides readily in bilayers.

How do newly synthesized phospholipids move from the ER to the membranes of mitochondria and chloroplasts? These organelles, which rarely show connections to other cellular membranes, contain enzymes only for synthesis of a few minor membrane

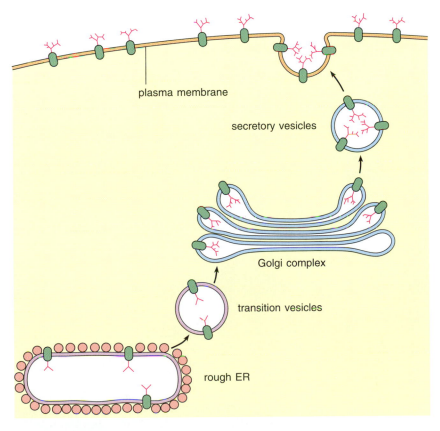

plasma membrane

secretory vesicles

Golgi complex

transition vesicles

rough ER

— carbohydrate group
— polypeptide } membrane glycoprotein

Figure 4-19 Orientation of glycoproteins during synthesis and insertion in the plasma membrane. The central stem structures of carbohydrate groups are added to glycoproteins in the rough ER, and branches are added in the Golgi complex. The carbohydrate groups face the interior of ER, Golgi complex, and secretory vesicles; as the vesicles fuse with the plasma membrane, the carbohydrate groups are placed on the side facing the cell exterior.

lipids; most of their membrane lipids apparently originate from their sites of final assembly in the ER.

Research by K. W. A. Wirtz and D. B. Zilbersmit and others established that phospholipid flow between the ER and totally disconnected organelles such as mitochondria occurs by means of soluble cytoplasmic proteins called *lipid transfer* or *exchange proteins*. Wirtz and Zilbersmit isolated a group of lipid transfer proteins that promote exchange of phospholipid molecules between isolated ER and mitochondria and, within mitochondria, between the inner and outer mitochondrial membranes. Others detected lipid transfer proteins that move phospholipids between the ER and the plasma membrane or between the ER and chloroplast. Proteins with these functions have been detected in every cell type in which they have been investigated, including those of animals, plants, yeast, and bacteria. Many of the transfer proteins are identical in structure in these groups, indicating that their evolutionary origins are very ancient.

The mechanism by which the lipid exchange proteins move hydrophobic lipids through the aqueous cytoplasm is unknown. Presumably they undergo a conformational change on binding a lipid that buries the hydrophobic portions of the lipid molecule in the protein interior, so that only hydrophilic portions, if any, face the surrounding polar medium. On contacting a target membrane, another conformational change releases the lipid from the protein interior and introduces it into the outer bilayer half. Transfer to the opposite bilayer half is catalyzed by flippases forming part of the target membranes. (Fig. 4-20 summarizes the origins and distribution of membrane lipids.)

Origins of Membrane Proteins

Experiments with radioactively labeled amino acids have demonstrated that the ER is also the assembly site for integral proteins of membranes directly or indirectly connected to this system. L. G. Caro and G. E. Palade were the first to show that initial assembly of labeled amino acids into proteins occurs in the rough ER—that is, in ER membranes with attached ribosomes. The label subsequently moves to the Golgi complex and plasma membrane (see Fig. 20-6). Equivalent experiments by Morré with plant tissues showed the same pattern.

These results indicate that proteins, like membrane lipids, are first incorporated into cellular membranes in the ER. The new membrane proteins then flow through the Golgi complex to reach the plasma membrane by the same routes as membrane lipids.

Figure 4-20 The pathway followed by newly synthesized membrane lipids. Lipids are assembled in the ER on the bilayer half facing the surrounding cytoplasm. Flippases catalyze flip-flop of lipids to the opposite bilayer half. From the ER the lipids move to the nuclear envelope by direct membranous connections, or to the Golgi complex, lysosomes, and plasma membrane via membranous vesicles that travel between these structures. Mitochondria, chloroplasts, and other cytoplasmic organelles not connected to the ER→Golgi→ plasma membrane distribution system receive membrane lipids by means of lipid transfer proteins, which are capable of moving lipids from the ER to these structures through the aqueous cytoplasmic medium.

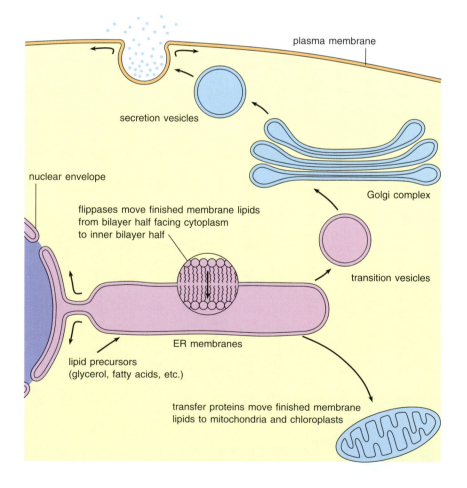

plasma membrane

secretion vesicles

Golgi complex

nuclear envelope

flippases move finished membrane lipids from bilayer half facing cytoplasm to inner bilayer half

transition vesicles

ER membranes

lipid precursors (glycerol, fatty acids, etc.)

transfer proteins move finished membrane lipids to mitochondria and chloroplasts

Flow through ER membranes to the nuclear envelope also occurs.

Proteins are inserted into ER membranes by a process known as the *signal mechanism*. The messenger RNAs for these proteins include a segment that, when translated into a polypeptide chain, forms a *signal* at the beginning of the protein. The signal causes attachment of the polypeptide chain, and the ribosome assembling it, to the ER. After attachment to the ER the signal, which contains a membrane-spanning hydrophobic segment, promotes entry of the polypeptide into the membrane bilayer. After the signal has performed its functions, it is enzymatically clipped from most proteins entering the ER pathway. (The signal mechanisms and distribution of proteins from the ER to membranes and other locations are described in detail in Chapter 20.)

Exposing cells to carbohydrate precursors, such as labeled glucose, shows that assembly of the carbohydrate groups of glycoproteins is initiated in the ER and completed in the Golgi complex. The assembly takes place on the ends of newly synthesized membrane proteins that project into the interior of the ER sacs. The carbohydrate groups are completed as the proteins move through the Golgi complex. As in the ER, the finished carbohydrate units face the inside of vesicles pinching off from the Golgi. Fusion of the vesicles with the plasma membrane places the carbohydrates on the exterior surface of the cell (see Fig. 4-19; see also Supplement 20-2).

The proteins of mitochondrial and chloroplast membranes, and those of other cytoplasmic organelles not directly or indirectly connected to the ER and Golgi complex, originate from ribosomes freely suspended in the cytoplasm or attached to the surfaces or the organelles. These proteins reach their target organelles by a process analogous to the signal mechanism. In this case the signal recognizes and binds specifically to the organelles rather than to the ER, causing insertion of the newly synthesized proteins into the organelle membranes. Many of these proteins initially take on a conformation that exposes polar amino acid side groups at their surfaces, making them soluble in the cytoplasm. On reaching their target organelle membrane they undergo a conformational change that promotes insertion into the membrane, with nonpolar regions facing the membrane interior and polar regions at the membrane surfaces. A group of proteins called *chaperones* maintains the organelle membrane proteins in conformations facilitating their transfer through the cytoplasmic solution and insertion in target membranes (for details, see Ch. 20).

For Further Information

Addition of proteins during membrane synthesis, *Ch. 20*
Carbohydrate groups of glycoproteins and glycolipids, synthesis, *Ch. 20*
Cell-cell recognition and adhesion, *Ch. 6*
Cytoskeleton, *Ch. 12*
Extracellular materials, *Ch. 7*
Glycoproteins and glycolipids in cell surfaces, *Ch. 6*
Membrane function in
    cell division, *Chs. 24 and 25*
    cell junctions, *Ch. 6*
    cell surfaces, *Chs. 6 and 12*
    chloroplasts, *Ch. 9*
    cytoskeleton, *Ch. 12*
    ER and Golgi complex, *Ch. 20*
    exocytosis and endocytosis, *Ch. 20*
    mitochondria, *Ch. 8*
    nerve conduction, *Ch. 5*
    nuclear envelope, *Ch. 13*
    transport, *Ch. 5*
Protein distribution to cellular membranes, *Ch. 20*
Signal mechanism, *Ch. 20*
Receptors and intercellular communication, *Ch. 6*

Suggestions for Further Reading

Aloia, R. C., and Raison, J. K. 1989. Membrane function in mammalian hibernation. *Biochim. Biophys. Acta* 988: 123–146.

Bretscher, M. S. 1985. The molecules of the cell membrane. *Sci. Amer.* 253:100–109 (October).

Burgoyne, R. D. 1990. Secretory vesicle-associated proteins and their role in exocytosis. *Ann. Rev. Physiol.* 52: 647–659.

Curatolo, W. 1987. Glycolipid function. *Biochim. Biophys. Acta* 906:137–160.

England, P. T. 1993. The structure and biosynthesis of glycosyl phosphatidylinositol protein anchors. *Ann. Rev. Biochem.* 62:121–138.

Gennis, R. B. 1989. *Biomembranes: Molecular Structure and Function.* New York: Springer-Verlag.

Hakomori, S.-I. 1986. Glycosphingolipids. *Sci. Amer.* 254: 45–53 (June).

Harwood, J. L., and Walton, T. J., eds. 1988. *Plant Membranes: Structure, Assembly, and Function.* London: Biochemical Society.

Lipowsky, R. 1991. The conformation of membranes. *Nature* 349:475–481.

Nelson, W. J. 1992. Regulation of cell surface polarity from bacteria to mammals. *Science* 258:948–955.

Rademacher, T. W., Parekh, R. B., and Dwek, R. A. 1988. Glycobiology. *Ann. Rev. Biochem.* 57:785–838.

Rodrigues-Boulan, E., and Powell, S. K. 1992. Polarity of epithelial and neuronal cells. *Ann. Rev. Cell Biol.* 8:395–427.

Singer, S. J. 1990. The structure and insertion of integral proteins in membranes. *Ann. Rev. Cell Biol.* 6:247–229.

Singer, S. J. 1992. The structure and function of membranes—a personal memoir. *J. Membr. Biol.* 129:3–12.

Singer, S. J., and Nicolson, G. L. 1972. The fluid mosaic model of the structure of cell membranes. *Science* 173:720–731.

Traxler, B., Boyd, D., and Beckwith, J. 1993. The topological analysis of integral cytoplasmic membrane proteins. *J. Membr. Biol.* 132:1–11.

Unwin, N., and Henderson, R. 1984. The structure of proteins in biological membranes. *Sci. Amer.* 250:78–94 (February).

White, J. M. 1992. Membrane fusion. *Science* 258:917–924.

Wirtz, K. W. A. 1991. Phospholipid transfer proteins. *Ann. Rev. Biochem.* 60:73–99.

Yeagle, P. L. 1985. Cholesterol and the cell membrane. *Biochim. Biophys. Acta* 822:267–287.

Review Questions

1. What are the proportions of lipid and protein molecules in biological membranes?

2. How are major functions divided between lipids and proteins in biological membranes? List major functions of cellular membranes, and indicate whether the functions are carried out by membrane lipids or by proteins.

3. Outline the structures of phosphoglycerides, sphingolipids, sterols, cholesterol, glycolipids, and glycosphingolipids. What major functions do these substances carry out in membranes?

4. Outline the structural characteristics of membrane proteins and glycoproteins. What are membrane-spanning segments, and how are they formed?

5. Describe the structure of a lipid bilayer. What forces hold a bilayer together?

6. What is the phase transition of a bilayer? What physical events are responsible for the phase transition? What is bilayer fluidity? How do living organisms regulate the phase transition?

7. What is lipid flip-flop? Does flip-flop occur readily in artificial lipid bilayers? In natural membranes? What is responsible for the differences in flip-flop in these systems?

8. Outline the structure of membranes according to the fluid mosaic model. What is the "fluid" part of the model? What is the "mosaic" part of the model? What is membrane asymmetry? What are integral and peripheral proteins?

9. What evidence supports the proposal that bilayers are actually present in biological membranes?

10. What evidence supports the proposal that bilayers are fluid?

11. What evidence supports the idea that proteins are suspended in membranes as individual, particulate units? What holds proteins in suspension in lipid bilayers?

12. What evidence demonstrates that proteins and lipids are distributed asymmetrically in membranes? What positions do the carbohydrate portions of glycolipids and glycoproteins take in membranes?

13. What evidence indicates that proteins are mobile in membrane bilayers? What is capping?

14. What new ideas were advanced in the fluid mosaic model? What observations led Singer and Nicolson to reject the proposal that membrane proteins are spread in extended form on membrane surfaces?

15. How are lipids and proteins synthesized and inserted in membranes?

16. How do membrane lipids and proteins move from their sites of initial assembly in the ER to the Golgi complex, plasma membrane, and nuclear envelope? How do proteins enter the membranes of mitochondria and chloroplasts?

17. What are flippases? Lipid exchange proteins? Chaperones?

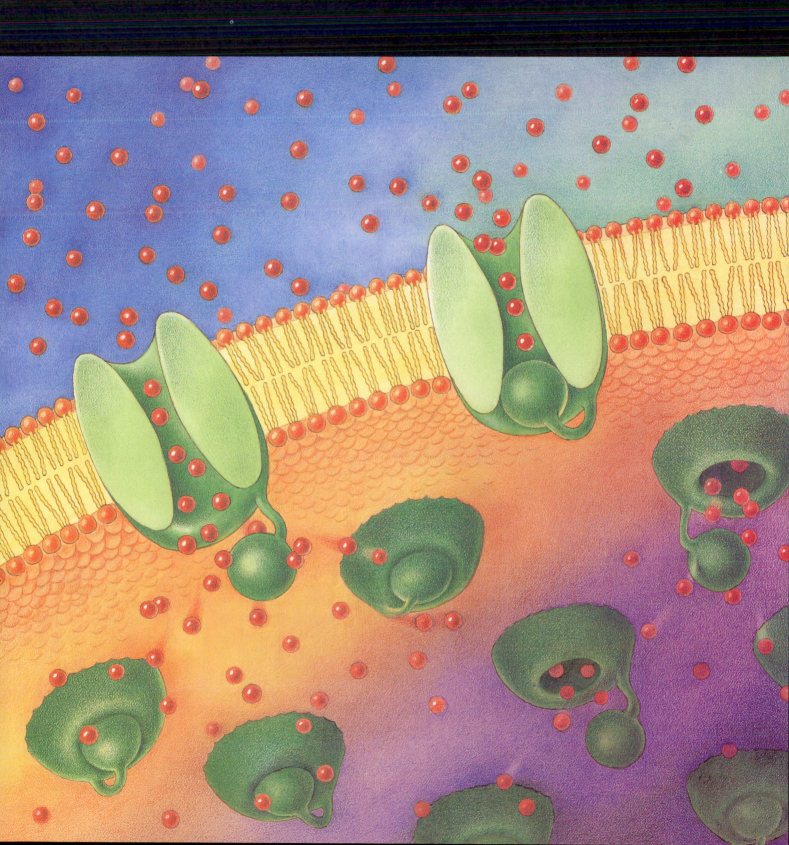

The concentrations of molecules and ions inside cells are maintained by membranes, which regulate the passage of all substances moving in and out of cells and among their membrane-bound interior compartments. As part of this maintenance, which is vital to cellular life, ions and molecules move continuously in both directions across cellular membranes. Ions move continuously back and forth; metabolites, including all necessary fuel substances and raw materials, enter the cell from the outside, and waste materials and cell secretions exit in the opposite direction. Molecules and ions also flow continuously in both directions across the boundary membranes of organelles in the cell interior.

Two primary mechanisms underlie this vital ionic and molecular transport. One, *passive transport*, results from differences in the concentration of substances inside and outside cells. If ions or molecules are more concentrated outside, the direction of movement is from outside to inside. If concentration is higher inside, movement is in the opposite direction. The difference between inside and outside concentrations, which provides the energy source driving movement in either direction, is called a *concentration gradient*. Because passive transport depends on concentration gradients, it requires no expenditure of cellular energy.

In the second mechanism, *active transport*, ions or molecules move *against* their concentration gradients. Unlike passive transport, active transport requires cells to expend energy; it stops if the reactions supplying energy for cellular activities are experimentally inhibited.

The rates of transport reflect the properties of both the lipid and protein parts of membranes. Transport through the lipid component of membranes, which is strictly passive, is influenced by the hydrophobic, nonpolar character of this component. In general, nonpolar molecules diffuse passively through the lipid part of membranes much more readily than polar or charged substances. The only exception to this pattern is water, which passes rapidly through the lipid component even though it is strongly polar.

Transport by membrane proteins primarily involves movement of ions and polar molecules. Some of this transport is passive, driven by concentration gradients. Other protein-mediated movement is active and requires expenditure of cellular energy. Transport by membrane proteins has the additional characteristic of *specificity*: In general, membrane proteins transport only certain types or classes of ions and polar molecules.

For most cells the great majority of transported substances are hydrophilic molecules and ions that move through membranes via transport proteins. Relatively few nonpolar molecules pass through the lipid part of membranes; the nonpolar component acts primarily as a seal that prevents passage of hydrophilic substances for which no selective protein carriers exist in a particular membrane type.

Because ions carry a charge, the passive and active movement of these substances can produce significant electrical effects on either side of a membrane. These electrical effects provide the basis for many important cellular functions, including the activities of nerve and brain cells in communication.

This chapter describes the functions of membrane proteins and lipids in passive and active transport. Also discussed are the electrical effects of ion transport and the utilization of these effects in nerve conduction.

PASSIVE TRANSPORT

Diffusion as the Basis for Passive Transport

Passive transport is a form of *diffusion*, the net movement of molecules from one region to another. Diffusion is a physical process that depends on the constant kinetic motion of molecules at temperatures above absolute zero ($-273°C$; see also p. 71). In this movement, molecules travel in straight lines until they collide with other molecules. After colliding the molecules rebound and move in a direction and speed that depend on the angles and force of the collisions.

Kinetic motion may result in diffusion if two regions in communication have different initial concentrations of molecules (concentration = number of molecules per unit volume). Consider two spaces of equal volume that are initially separated by a barrier that molecules cannot pass. The absolute temperature of the spaces is the same, but one of them contains more molecules. The space with higher concentration of molecules contains a greater amount of energy because it includes more moving particles.

If the barrier between the two compartments is removed, the movement and collisions on the more concentrated side propel the molecules into the less concentrated side. Molecules also move from the less to the more concentrated side, but over any interval of time there are more collisions and movement from the more concentrated side. As a result, there is a net movement of molecules from the side of greater concentration to the side of lesser concentration. This net movement is diffusion in response to a concentration difference; it continues until the molecules are evenly distributed throughout the available space. Molecules continue to move from one space to the other even after the concentration is the same on both sides. However, no *net* increase on either side is expected after the distribution becomes uniform.

As molecules diffuse to the evenly distributed state, they release energy to their surroundings. This energy is *free energy*, and can accomplish work.[1] Among the work accomplished in cells by the free energy released in diffusion is ATP synthesis (for details, see Ch. 8).

The movement of charged particles such as ions in response to concentration gradients is modified by electrical attractions and repulsions. In response to electrical charge, net movement tends to proceed toward a condition in which all parts of a space contain the same number of positive and negative charges and the space is thus electrically neutral. The free energy made available by movements in response to electrical charge is called *electrical potential* or *voltage* and can do work. The work of the nervous system, for example, is powered by free energy released by gradients of electrical charge.

---

[1]The free energy released by the movement of a mole of molecules down a concentration gradient can be calculated from a modification of the equation used for free energy changes in chemical reactions:

$$\text{free energy} = \Delta G = -RT\ln C_1/C_2 \qquad (5\text{-}1)$$

in which R is the gas constant, T is the absolute temperature, and C_1 and C_2 are the high and low concentrations of the gradient. At a temperature of 25°C (absolute temperature = 298 Kelvin) and concentration differences of 0.1 M and 0.0001,

$$\begin{aligned}
\Delta G &= -[1.98 \text{ cal}/(\text{degree}\cdot\text{mole})(298)(\ln 0.1/0.0001) \\
&= -(590)(\ln 1000) \\
&= -(590)(6.908) \\
&= -4075 \text{ cal}/\text{mol}
\end{aligned}$$

The free energy released as the mole of molecules travels down the gradient, −4075 cal/mole, exceeds the energy available from hydrolysis of a mole of ATP at 25°C, about −3000 cal/mole. The concentration difference used for this calculation, which amounts to 1000 times, would not be unusual for biological systems.

The Effects of Semipermeable Membranes on Diffusion

An artificial or natural membrane placed between two regions containing molecules at different concentrations has no effect on the final outcome of diffusion if all the molecules or ions can pass equally through the membrane. However, the net movement may be altered, sometimes in unexpected ways, if some molecules or ions pass through the membrane less readily than others or are excluded entirely. Membranes having this effect, including all biological membranes, are said to be *semipermeable*.

Semipermeable Membranes and Osmosis One of the most fundamental effects of semipermeable membranes on diffusion is *osmosis*, diagrammed in Figure 5-1. The figure shows two spaces of equal volume separated by a semipermeable membrane, in which one side contains pure water and the other a solution of protein molecules in water. The pores in the semipermeable membrane admit water molecules but are too small to allow the protein molecules to pass.

It is obvious that a concentration gradient exists for the protein molecules; however, no net movement of the protein can occur because of the semipermeable barrier. It is less obvious that there is also a concentration gradient for water molecules. This gradient exists because some of the available space on one side is taken up by the protein molecules. Therefore there are fewer water molecules per unit volume on the side

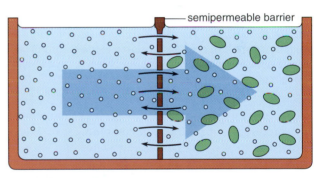

○ = water molecule
🟢 = protein molecule

Figure 5-1 Osmotic flow of water in a system in which a semipermeable barrier separates two compartments of equal volume, with pure water on the left and a solution of protein molecules in water on the right (see text). The small black arrows show water molecules moving in both directions across the semipermeable barrier; however, because the proteins take up some of the available space on the right side, there are fewer water molecules per unit volume on this side. In response, a net movement of water molecules occurs from left to right (large arrow).

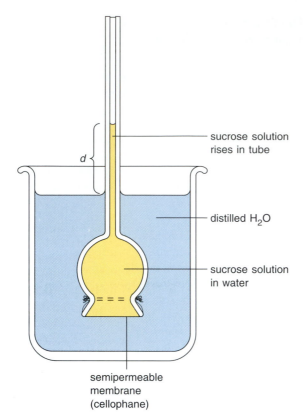

sucrose solution rises in tube

d

distilled H_2O

sucrose solution in water

semipermeable membrane (cellophane)

Figure 5-2 An apparatus demonstrating osmotic flow of water. The beaker contains pure water. Suspended in the beaker is an inverted thistle tube containing a solution of sucrose in water. The bottom of the thistle tube is covered by a cellophane film that allows water molecules, but not sucrose molecules, to pass in either direction. The level of the solution in the tube will rise because of the net movement of water molecules from the beaker into the thistle tube. Osmotic flow continues until the weight of the water column (labeled d in the diagram) develops sufficient pressure to counterbalance the osmotic movement of water molecules into the tube.

containing the protein. In response, a net movement of water occurs from the side containing only water molecules to the side containing water and protein molecules. Osmosis is the net diffusion of water molecules in response to a gradient of this type. Because osmosis occurs in response to a concentration gradient, it releases free energy and can accomplish work.

Osmosis is often demonstrated with an apparatus that consists of an inverted thistle tube closed at its lower end by a sheet of cellophane (Fig. 5-2). Inside the tube is a sucrose solution in water; the tube is suspended in a beaker of distilled water. The cellophane film acts as a semipermeable barrier because its pores are large enough to allow water molecules but not sucrose to pass in either direction. A concentration gradient exists for water molecules in this system because some of the available space for water is taken

up by the sucrose molecules in the tube. After a short time the solution in the tube rises as water passes across the cellophane film from the beaker in response to its concentration gradient.

The level of the solution continues to rise until the pressure created by the weight of the raised solution exactly balances the tendency of water molecules to move from the beaker into the tube. At this point the system is in balance, and although water molecules still move in both directions across the cellophane membrane, no further net movement of water occurs. The work accomplished by osmosis in the apparatus is equivalent to the height reached by the solution in the tube. The pressure required to counterbalance exactly the tendency of water molecules to move into the tube is the *osmotic pressure* of the solution in the tube.

The relationship between osmotic pressure and the concentration of a nonpermeable solute in a situation such as the apparatus in Figure 5-2 is described by a simple relationship, the *van't Hoff equation*. For a system in which a semipermeable membrane separates pure water from a dilute solution of a solute that cannot pass across the membrane,

$$\text{osmotic pressure} = \pi = RTC \qquad (5\text{-}2)$$

in which R is the gas constant, T the absolute temperature, and C the concentration of the solute in moles (or the total concentration of all nonpermeable solutes on the other side). Conversely, the osmotic pressure developed by a solution can be used to determine the concentrations of a solute in moles or even to approximate its molecular weight.

Osmosis in Living Cells Cells act like osmotic devices similar to the apparatus shown in Figure 5-2 because they contain solutions of proteins and other molecules that are retained inside by a membrane impermeable to them but freely permeable to water. The resulting osmotic movement of water produces a force that operates continuously in living cells. This force may be used as an energy source for some of the activities of life, or it may be a disturbance that must be counteracted for survival.

The root hair cells of most land plants, for example, contain proteins and other molecules that cannot move across the plasma membrane but are surrounded by almost pure water. As a result, water flows into the root cells by osmosis. The osmotic pressure developed by this flow contributes part of the force required to raise water into the stems and leaves of the plant. In stems and leaves, osmotic pressure pushes the cells tightly against their walls and supports softer tissues against the force of gravity. Bacterial and cyanobacterial cells are also kept tightly pressed against their walls by osmotic pressure.

Freshwater unicellular organisms lacking rigid walls, such as protozoa, must expend considerable energy to excrete the water continuously entering by osmosis. Organisms living in surroundings containing highly concentrated salt solutions have opposite problems and must continuously expend energy to replace water lost by osmosis. Within the bodies of many-celled animals, ions, proteins, and other molecules occur in significant concentrations in the extracellular fluids as well as inside cells, so that the concentration of water inside and outside cells is more closely balanced. However, the concentration difference is great enough that animal cells must still expend considerable energy to counteract the effects of the inward movement of water by osmosis (in animal cells, osmotic pressure is reduced by active transport of Na^+ out of cells; see below).

Semipermeable Membranes and Movement of Charged Particles The presence of a semipermeable barrier can have unexpected effects on the movement of charged particles if not all the particles can move across the barrier with equal ease. Consider two water-filled spaces separated by a semipermeable barrier, with one side containing a solution of negatively charged proteins that cannot move across the barrier. If a salt such as NaCl is added to either side, the sodium and chloride ions released as the salt dissolves will tend to diffuse until their concentrations are the same on both sides of the barrier. However, their distribution cannot become completely uniform because such a distribution would produce more total negative charges on the side containing both Cl^- and proteins. As a result, the buildup of a negative charge on the side containing the proteins opposes the tendency of the system to run toward an even distribution of particles. In this case neither uniform concentrations nor electrical neutrality can be attained, and net movement of the ions stops when the charge difference between the two sides becomes large enough to exactly balance movement in response to concentration differences. At this point the system reaches an equilibrium in which both a charge difference and concentration difference remain for the ions on either side of the barrier.

Biological Membranes and Passive Transport: Facilitated Diffusion

The effects of biological membranes as semipermeable barriers are complicated by the fact that they have polar and nonpolar regions. Polarity significantly affects the ability of molecules to diffuse pas-sively through biological membranes. The ability of membrane proteins to specifically admit some polar molecules and exclude others also greatly modifies the responses of biological membranes to concentration gradients.

The effects of biological membranes on diffusion were first studied at the turn of the century by Ernst Overton, who observed that, for a given molecular size, molecules soluble in lipid solvents penetrated into cells more rapidly than the water-soluble molecules he tested. This behavior contrasted with the passage of molecules across artificial barriers such as cellophane, in which the relative degree of water solubility had no noticeable effect. Overton's work provided the first clue that cells are surrounded by a lipidlike surface layer.

Later work confirmed that many substances penetrate across biological membranes according to lipid solubility. However, certain polar molecules such as glucose and amino acids were found to enter many cells much more rapidly than expected according to their solubility in lipids. Ions such as Na^+ were also found to penetrate rapidly across some natural membranes in spite of their hydrophilic nature. This passive transport of substances at rates higher than predicted from their lipid solubility is termed *facilitated diffusion*.

Proteins and Facilitated Diffusion Proteins were directly implicated in facilitated diffusion by comparisons of natural membranes with artificial phospholipid bilayers (see p. 101). All molecules except water pass through pure phospholipid bilayers strictly according to lipid solubility and molecular size. (The exceptional behavior of water is discussed further below.) Ions are essentially excluded from passage, and only a very few polar substances occurring naturally in cells, such as ethyl alcohol, urea, and glycerol, are soluble enough in phospholipid bilayers to penetrate to any extent. The situation is different if membrane proteins are added to the bilayers. Addition of membrane proteins frequently allows polar and charged substances to penetrate at rates comparable to those of natural membranes.

Cells control the particular group of polar and charged molecules and ions passing through their membranes by regulating the types of transport proteins that are synthesized in the ER, modified in the Golgi complex, and placed in cellular membranes (see p. 577). As a result, each cell type has its own spectrum of plasma membrane proteins, passing a characteristic group of hydrophilic substances by facilitated diffusion (or by active transport; see below). The membranes surrounding interior compartments such as those of mitochondria and chloroplasts also have

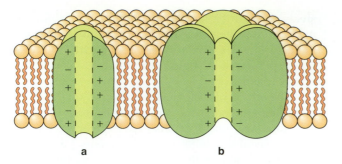

Figure 5-3 Formation of a polar or charged channel through the interior of a single membrane protein **(a)** or by alignment of several membrane proteins **(b)**.

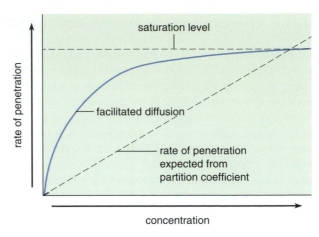

Figure 5-4 Relationship between the concentration and rate at which polar substances enter cells by facilitated diffusion. At low concentrations, polar substances penetrate through natural membranes by facilitated diffusion (solid line) at rates much higher than expected from their lipid solubility (dashed line). At increasingly higher concentrations, penetration gradually drops off until it reaches a maximum rate that does not respond to further increases in concentration. At this concentration the facilitated diffusion transporters are saturated.

transport proteins that allow the inward and outward flow of specific polar and charged substances.

The transport proteins of natural membranes are integral membrane proteins that completely span the lipid bilayer (Fig. 5-3). The hydrophilic channels in these proteins are set up by an arrangement of amino acid chains that opens the channels and lines their sides with polar groups (see also Fig. 4-7). The channels may be formed by a polar opening through a single protein (Fig. 5-3a) or by several proteins that combine to form a channel between them (Fig. 5-3b).

The Characteristics of Facilitated Diffusion Facilitated diffusion has several important characteristics (Table 5-1). Although membrane proteins may greatly enhance the diffusion of polar molecules, the energy required for facilitated diffusion is provided only by a favorable concentration gradient, and the process stops if the gradient falls to zero. Because facilitated diffusion depends entirely on favorable concentration gradients, it is passive and requires no direct expenditure of cellular energy.

Comparisons of the rate of facilitated diffusion and the concentrations of the transported molecules reveal another fundamental characteristic of the mechanism (Fig. 5-4). At successively higher concentrations the degree of enhancement drops off until at some point the mechanism becomes *saturated*: Further increases in concentration cause no further rise in the rate of penetration (solid blue line in Fig. 5-4). This saturation is in sharp contrast to the behavior of nonpolar molecules, which penetrate across the phospholipid portion of membranes by simple diffusion according to lipid solubility and molecular size. For these molecules the rate of penetration across membranes is directly proportional to concentration, with no marked dropoff at high concentrations (dashed line in Fig. 5-4).

The saturation of facilitated diffusion at elevated concentrations closely resembles the behavior of en-

zymes in catalyzing biochemical reactions (see pp. 81 and Fig. 3-8). Enzymes also become saturated as substrate concentration increases, and the rate of the reaction levels off. The similarity in behavior between the two systems indicates that facilitated diffusion is carried out by membrane proteins with properties similar to those of enzymes.

Another similarity between enzymes and the proteins carrying out facilitated diffusion is the property of specificity: Each molecule transported by facilitated diffusion is carried by a separate protein specific only for that substance or a group of closely related substances. For example, the protein facilitating the transport of glucose will also transport the closely related monosaccharides mannose, galactose, xylose, and arabinose. However, it will carry only the naturally occurring D isomers of these sugars and not the L isomers (see Chapter 2). The specificity of facilitated diffusion further resembles the specificity of enzymes in that substances with structures closely related to the normally transported molecule can inhibit the rate of transport of that molecule.

Like enzymes, the proteins carrying out facilitated diffusion have been demonstrated to combine briefly with their transported substances. The facilitated diffusion transporters combine with molecules or ions on the side of the membrane of higher concentration and release them on the side of lower concentration. The

Table 5-1   Characteristics of Transport Mechanisms

Characteristic	Simple Diffusion	Facilitated Diffusion	Active Transport
Membrane component responsible for transport	Lipids	Proteins	Proteins
Binding of transported substance	No	Yes	Yes
Energy source	Concentration gradients	Concentration gradients	ATP hydrolysis or concentration gradients
Direction of transport	With gradient of transported substance	With gradient of transported substance	Against gradient of transported substance
Specificity for molecules or molecular classes	Nonspecific	Specific	Specific
Saturation at high concentrations of transported molecules	No	Yes	Yes

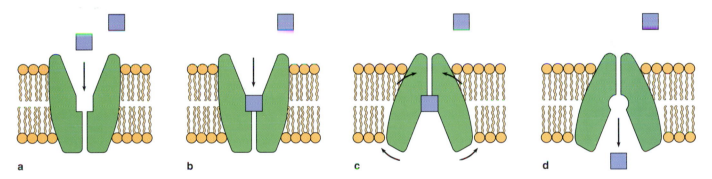

a   b   c   d

Figure 5-5   Facilitated diffusion according to the alternating conformation model (see text).

transporters have also been demonstrated to undergo conformational changes in restricted regions as facilitated diffusion proceeds. These changes alternate the protein between conformations in which the binding site for the transported molecule faces either the inside or outside surface of the membrane. In some systems the binding sites have been shown to have significantly different affinities for the transported substance in the two conformations.

How Facilitated Diffusion Transporters Are Believed to Operate   The characteristics of facilitated diffusion transporters (summarized in Table 5-1) and the general properties of integral membrane proteins have provided the basis for a hypothesis, the *alternating conformation* model, proposed by S. J. Singer and others for the mechanism by which transport proteins carry out facilitated diffusion (Fig. 5-5; see also Singer's Experimental Process Essay on p. 124). According to the model, a transport protein shifts between two conformations that direct the binding site of the protein alternately toward the inside and outside of

the membrane. In the initial step of the process the protein is folded so that the site binding the transported molecule is exposed toward the membrane surface facing the region of higher concentration (Fig. 5-5a). In this position the binding site is in a *high affinity state*, in which it binds strongly to the transported molecule. The transported substance binds to the transporter through random collisions resulting from constant molecular movements (Fig. 5-5b). Binding causes a change in the folding conformation of the transporter, so that it shifts to the alternate conformation in which the binding site is exposed to the membrane surface facing the region of lower concentration (Fig. 5-5c). The conformational change also alters the active site to a *low affinity state*, in which it binds the transported molecule relatively weakly. As a result, the transported substance is readily released on the low concentration side of the membrane (Fig. 5-5d). The release returns the protein to its original folding conformation, with the binding site facing the opposite membrane surface again (as in Fig. 5-5a), ready to initiate another cycle of facilitated diffusion.

The Experimental Process

How Do Ions Get across Membranes?

S. J. Singer

S. JONATHAN SINGER is a University Professor in the Department of Biology at the University of California at San Diego. His early scientific training was in physical chemistry, and he obtained A.B. and A.M. degrees from Columbia University and a Ph.D. from the Polytechnic Institute of Brooklyn. After postdoctoral training with Dr. Linus Pauling at Caltech, Dr. Singer was on the faculty of chemistry at Yale University before joining the newly developing U.C. campus at San Diego in 1961. He is a member of the National Academy of Sciences and a former Research Professor of the American Cancer Society. In 1991, he received the E. B. Wilson Award of the American Society for Cell Biology for his pioneering work on the fluid mosaic model of membrane structure.

The transport of ions and small hydrophilic molecules (such as sugars and amino acids) across otherwise impermeable and hydrophobic membrane lipid bilayers is a critical phenomenon in cell biology. Such transport is involved in the regulation of cell metabolism and in all signal transmission processes, including transmission in the nervous system. Until the early 1970s, a prevalent view of the molecular mechanism of transport was that of the "rotating carrier" (Fig. A). The ion or hydrophilic molecule (the ligand) was supposed to be bound to a specific site on a transport protein molecule, which then rotated and diffused across the bilayer to release the ligand on the other side. However, the same thermodynamic arguments that we used in proposing the fluid mosaic model for the molecular organization of the proteins and lipids

of membranes indicated that the rotation of a membrane protein molecule across a bilayer was not at all likely to occur, because such rotation would necessitate the transient insertion of some of the ionic amino acid residues of the protein molecule into the hydrophobic interior of the bilayer, an event that would be energetically quite unfavorable. We suggested instead[1] that transport proteins in membranes would have three main structural properties (see Fig. 5-3 on p. 122):

1. They would be specific aggregates of a small number (two to six) of identical or similar polypeptide chains traversing the bilayer.

2. Down the central axis through the subunit aggregate a narrow transmembrane channel would form that would be lined by ionic and hydrophilic amino acid residues of the protein subunits, and this channel would be filled with water molecules.

3. A specific binding site for the ligand to be transported would be present on one or more of the subunits within the channel, such that a *quarternary rearrangement* of the subunits, requiring only small amounts of energy, would transfer the ligand from one side of the membrane to the other (Fig. B).

This was at a time when the structure of no transport protein was known. We decided to do an experiment that might rule out the rotating carrier model. The idea of the experiment was that if a large protein molecule, such as an antibody molecule, was attached to the exposed surface of a transport protein in an intact membrane, it should strongly inhibit any rotation of the transport protein across the membrane, and therefore greatly reduce

Figure A The rotating carrier mechanism, the prevalent hypothesis of membrane transport until the early 1970s. According to this hypothesis, a transport protein molecules rotates to carry a ligand across the bilayer, where it is released.

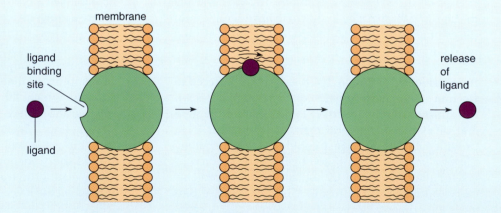

the rate of ligand transport through the membrane. However, if transport occurred by way of a protein aggregate and a mechanism such as pictured in Figure B, little if any effect of the bound antibody on the transport rate might be expected.

We chose for our experimental system the transport of Ca^{2+} across the membranes of the sarcoplasmic reticulum (SR) of muscle cells. This transport is a critical part of the process that controls the Ca^{2+} concentration in the muscle cell cytoplasm and thereby regulates muscle contractility. The Ca^{2+} transport is carried out by a membrane protein, Ca^{2+}-ATPase, which is an enzyme that uses the energy of ATP hydrolysis to power Ca^{2+} import into the lumen of the SR. Vesicles of the SR, whose membranes consist of over 80% just the one protein, the Ca^{2+}-ATPase, can be isolated from the muscle. We could have produced an antibody to the Ca^{2+}-ATPase, but we reasoned that, with the methods available at the time, it would be uncertain that an antibody that did not inactivate transport was really directed to the Ca^{2+}-ATPase. We therefore chose instead to modify the Ca^{2+}-ATPase in the intact SR vesicles to contain an exposed covalently bound 2,4-dinitrophenyl (DNP) group, for which highly purified specific antibodies (anti-DNP) were already available in our laboratory.

The experimental procedure[2] involved, first, the chemical modification of the intact SR vesicles using the enzyme liver transglutaminase and [^{3}H]-DNP-cadaverine to catalyze the attachment of a DNP-cadaverine molecule by its amino group in amide linkage to some carboxyl group on the SR membrane protein. Radioactive counting of the attached [^{3}H] allowed us to show that, depending upon the particular reaction conditions used, between 0.55 and 0.80 mol of DNP group was covalently bound per mol of Ca^{2+}-ATPase. Second, we showed that upon the addition of excess anti-DNP to the modified SR vesicles, $0.55 \pm .05$ mol of antibody was bound per mol of covalently-bound DNP group, probably indicating that each molecule of the antibody, bearing two anti-DNP binding sites, was bound to two DNP groups simultaneously. Third, we showed that such antibody binding to the DNP-modified SR vesicles had no significant effect either on the Ca^{2+}-ATPase enzyme activity or on the rate of radioactive [$^{45}Ca^{2+}$] transport into the vesicles.

These experiments, and similar ones carried out independently on other transport systems by J. Kyte,[3] and by A. Martonosi and F. Fortier,[4] indicated that "rotating carrier mechanisms for protein-mediated membrane transport should be laid to rest." Since 1976, when these experiments were performed, many transport proteins have been isolated and structurally characterized, and in

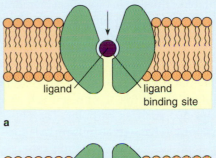

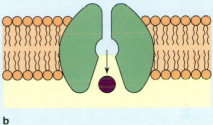

Figure B Active transport of a molecule through a membrane protein channel. A molecule interacts with the active site of a protein subunit (a). This triggers an energy-yielding enzyme reaction that produces a shift in the subunit conformation (b), "squeezing" the molecule through the membrane.

many cases their amino acid sequences have been determined from the analysis of their isolated cDNAs. From these and other results, the general features of the model proposed in Figure B (see also Fig. 5-3) have been confirmed. (For an example, see ref. 5 concerning the acetylcholine receptor, its structure and mechanism of transport.)

References

[1]Singer, S. J. In *Structure and function of biological membranes*, Chapter 4. Ed. L. I. Rothfield. New York: Academic Press, pp. 145–222 (1971).

[2]Dutton, A. H.; Rees, E. D.; and Singer, S. J. *Proc. Natl. Acad. Sci. USA* 73:1532 (1976).

[3]Kyte, J. *J. Biol. Chem.* 249:3652 (1974).

[4]Martonosi, A., and Fortier, F. *Biochem. Biophys. Res. Commun.* 60: 382 (1974).

[5]Unwin, N.; Toyoshima, C.; and Kubalek, E. *J. Cell. Biol.* 107:1123 (1988).

The energy required for the conformational changes is provided by the favorable concentration gradient.

Transporter Types Operating in Facilitated Diffusion

Several different types of transporter proteins carry out facilitated diffusion in biological membranes. One type, found primarily in animal cells, is specialized for facilitated transport of relatively small organic molecules such as glucose and other hexose sugars (glucose is also actively transported; see below). A second animal transport protein, called an *anion carrier*, facilitates the diffusion of negatively charged inorganic groups such as carbonate or phosphate. A third type forms channels admitting ions such as Na^+, K^+, Ca^{2+}, and Cl^-. Most of the ion channels, which occur in both plant and animal cells, are *gated*; that is, they are controllable and can exist in open, closed, or intermediate states. Gated ion channels are particularly important in animals because they provide the basis for such fundamental processes as conduction of impulses along and between nerve cells.

All these transporter proteins share several structural characteristics. All are integral membrane proteins containing several to many alpha-helical segments that zigzag back and forth across the membrane. The transporters of plasma membranes are glycoproteins with carbohydrate groups directed toward the cell exterior.

A Carrier of Small Organic Molecules: The Glucose Transporter The plasma membranes of most vertebrate cell types contain proteins transporting glucose and other six-carbon sugars by facilitated diffusion. Mammals have at least six different forms of these hexose transporters, encoded in as many separate genes. All consist of a single protein with 12 alpha-helical segments capable of spanning the membrane, indicating that the protein chain probably zigzags back and forth across the plasma membrane 12 times, with the membrane-spanning segments connected by loops extending into the cytoplasm or the extracellular space. Transmembrane segments 7, 8, and 11 form alpha helices that are polar on one side and nonpolar on the other (see Fig. 4-7); these segments combine to form a polar pore extending through the glucose transporter.

Facilitated diffusion of glucose is possible for most animal cells because glucose is about seven times more concentrated in the circulation than inside cells. The concentration difference is maintained partly by a biochemical "trick" in which glucose is phosphorylated to glucose 6-phosphate after it enters the cytoplasm. Because glucose 6-phosphate cannot pass backward through the glucose transport carrier

(phosphorylated molecules generally are unable to pass through biological membranes), the phosphorylation effectively removes the glucose entering cells from the concentration inside. The phosphorylation trick is also used by cells to effectively lower the inside concentrations of other molecules moved inward by facilitated diffusion. (Glucose is also transported actively in some cells; see below.)

Transport proteins related to the mammalian glucose transporter occur in many other animals, if not all animal species, and in fungi and bacteria. The wide distribution of these related hexose transporters suggest that they are members of an ancient family of transport proteins descended from a single ancestral type.

Some facilitated glucose transporters, such as those in muscle and fat cells, are regulated by the hormone *insulin*. When secreted into the circulation in response to elevated blood glucose levels, the hormone stimulates muscle and fat cells to insert more glucose transporters in the plasma membrane. The glucose transporters of most other body cells in mammals, such as those of the brain and liver, are unaffected by insulin; they are present in plasma membranes in the same numbers whether the hormone is present in the bloodstream or not. In one common form of diabetes, mutations alter the sequences of transmembrane segments of the glucose transporter, inhibiting or blocking its ability to transport glucose molecules into cells. Glucose accumulates in the bloodstream as a result.

The transporter carrying glucose by facilitated diffusion has been extracted, purified, and inserted in artificial phospholipid bilayers. When the transporter is added, artificial bilayers are able to carry out facilitated diffusion of glucose by essentially the same patterns as natural membranes.

Channels Passively Transporting Ions Proteins forming channels for passive transport of ions occur in all eukaryotes. Almost all these channels conduct positively charged ions, in most either Na^+, K^+, or Ca^{2+}, at rates approaching a million particles per second when fully open. A channel conducting Cl^- has also been detected and is believed to be widely distributed.

The mechanism by which ion channels admit one ion selectively and restrict the passage of others is not completely understood. Part of the selectivity may depend on binding sites in the channel that recognize and bind only a single ion or group of related ions. The shape of the polar channel formed by ion transporters may also contribute to the selectivity. The ion channels investigated in detail have proved to be shaped like a double funnel, with a wider opening at either membrane surface leading to a narrow, constricted region at the channel midpoint (see Fig. 5-8).

Presumably the narrow funnel neck operates as a "selective filter" that allows one ion type to pass readily and excludes others to a greater or lesser extent.

Exclusion is believed to depend on the fact that the selective site in the channel neck binds the transported ion much more strongly than other ions. Once the selected ion binds, it blocks the narrow channel neck to other ions and can be displaced only by another ion that binds equally strongly—that is, another ion of the same type. Successive binding and displacement by the selected ion, occurring millions of times per second, moves the ion through the channel and excludes others. Differences in the binding sites are small; for example, a single mutation causing substitution of one amino acid in the channel wall can change selectivity from Na^+ to Ca^{2+}.

Most ion channels are opened or closed by gates. Gating is rapid; ion flow can change from rates of millions of particles per second to essentially zero in a matter of milliseconds. The gating can also be modulated—that is, regulated upward or downward—through adjustment of positions between fully open and fully closed states. Some of the modulation is exerted by long-term controls that regulate channel function. In most cases, long-term control modulating gated channels is imposed by addition or removal of phosphate groups.

Three types of gated ion channels have been detected in animal cells. One, the *voltage-gated channel*, opens or closes in response to changes in the voltage difference across the membrane housing the channel. In most cases the voltage changes result from alterations in the distribution of ions or charged molecules on either side of the membrane. The second type, the *ligand-gated* channel (from *ligare* = to bind or tie), opens or closes in response to binding specific control molecules. Many hormones, and molecules released by nerve cells as *neurotransmitters* (see below), act as control molecules that open or close ligand-gated channels. The third channel type, the *mechanosensitive* or *stretch-gated* channel, opens or closes in response to mechanical stresses on the membrane.

In general, gated channels occur in multiple, closely related forms rather than as a single channel type. Mammalian brain cells, for example, contain at least three different but closely related forms of voltage-gated Na^+ channels, encoded in different but related genes. Many other cell types have distinct Na^+ channels with slight but significant differences in their gating properties. In all, more than 30 types of K^+ channels have been identified, some gated by voltage changes and some by ligands.

Ion-conducting channels have been studied extensively by inserting them into artificial phospholipid films after isolation and purification. Voltage-gated channels, in particular, have also been studied by

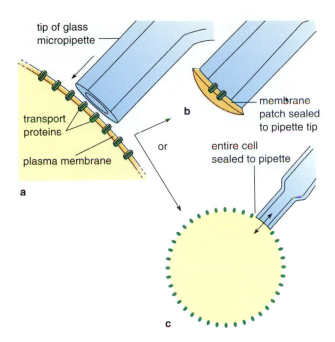

Figure 5-6 The patch clamp technique for attachment of membrane segments to the tip of a glass micropipette. **(a)** The pipette is touched to a cell surface to seal the membrane to the pipette; variations in the technique can produce a small membrane patch attached to the pipette tip **(b)** or an entire cell **(c)**. After attachment the effects of solutions introduced on either side of the attached membrane can be followed; electrodes can also be introduced on either side of the membrane.

the *patch clamp technique*, developed in the mid-1970s by the German scientists E. Neher and B. Sakmann (Fig. 5-6). In this method a very fine glass micropipette is touched to the cell surface (Fig. 5-6a). The plasma membrane forms a tight seal around the micropipette, so that ions can move in or out of the pipette only through channels in the membrane. The technique can be varied so that the membrane sealed to the pipette includes only a small patch (Fig. 5-6b) containing only a few or even a single ion channel, or the entire plasma membrane (Fig. 5-6c). Segments of artificial phospholipid films with or without added membrane proteins can also be picked up and sealed to the end of the pipette, which can be made less than 5 μm in diameter. The solutions inside and outside the pipette can be varied at will and electrodes introduced on either side. The technique allows ion flows through single channel types, or even a single channel, to be recorded and varied experimentally. Neher and Sakmann received the Nobel Prize in 1991 for their work, which revolutionized membrane studies in cell physiology.

Voltage-Gated Channels Voltage-gated channels occur characteristically in cells that respond to stimuli by undergoing rapid changes in plasma membrane voltage. Cells with this property are termed *excitable*. The voltage-gated channels are opened by a relatively small voltage disturbance at some point on the plasma membrane. On opening they admit very large inflows of the ion they transport, usually either Na^+, K^+, or Ca^{2+}. The inflows greatly magnify the voltage change and extend it over the cell surface.

Voltage-gated Na^+, K^+, and Ca^{2+} channels occur in essentially all animal cells. Voltage-gated Na^+ and K^+ channels are responsible for much of the voltage change involved in conduction of impulses along nerve and muscle cells (for details, see p. 139). The importance of Na^+ channels in excitable cells is underscored by the fact that many neurotoxins in naturally occurring poisons and venoms target voltage-gated Na^+ or K^+ channels in nerve and muscle. When bound, the neurotoxins lock the channels in an open or closed state and so seriously disturb the functions of the nervous and muscular systems that their effects are usually fatal.

Voltage-gated Ca^{2+} channels contribute to generation of impulses in some nerve cells. In some cell types, voltage-gated Ca^{2+} channels provide important links between nerve impulses and the regulation of major cell functions. Typically Ca^{2+} released into the cytoplasm through the opened channels combines with one or more Ca^+-dependent proteins that require calcium ions for conversion to active form. In active form they work as control proteins that trigger a specific cell activity. One of the primary control proteins activated by combination with Ca^{2+} is *calmodulin*, which regulates many cytoplasmic functions (see p. 164 and Information Box 6-2). In muscle cells, for example, voltage changes induced by the arrival of a nerve impulse at the cell surface open voltage-gated Ca^{2+} channels, releasing Ca^{2+} into the muscle cell cytoplasm. The Ca^{2+} combines with and activates a Ca^+-dependent protein that controls muscle contraction (for details, see p. 324). In secretory cells, Ca^{2+} released into the cytoplasm activates proteins that trigger fusion of secretory vesicles with the plasma membrane, thereby releasing the secretory product to the cell exterior. A similar secretory mechanism triggered by the opening of voltage-gated Ca^{2+} channels takes part in the transmission of impulses across nerve synapses (see below).

Voltage-gated channels conducting K^+ and Ca^{2+} have also been detected in fungi and higher plants. In one plant, *Chara*, a voltage-gated K^+ channel is associated with an excitable response. Voltage-gated Ca^{2+} channels are also suspected to occur in higher plants.

Molecular studies show that the voltage-gated channels share a basic structural unit including six alpha-helical segments that span the membrane (Fig. 5-7). Each channel consists of four of these units, giving a total of 24 transmembrane helices. In the K^+ channel the units are separate polypeptides; in Na^+ and Ca^{2+} channels the four units are linked into a single large polypeptide. Several lines of evidence indicate that transmembrane helices 5 and 6 are the pore-forming part of the channels. Among the most significant experiments were those of H. A. Hartmann and A. M. Brown and their coworkers, in which substitutions of amino acid segments in this region were shown to be most effective in altering ion flow through K^+ channels. Similar experiments have implicated a segment of transmembrane helix 4 as the part of the channels that senses voltage changes. Conformational changes in this helix in response to a voltage change may trigger opening of the gate.

Ligand- and Stretch-Gated Channels Many epithelia lining body cavities or ducts such as those of kidney tubules, colon, lungs, trachea, and sweat glands have ligand-gated channels, most of them conducting Na^+. Entry of ions through these channels is regulated by hormones such as aldosterone and vasopressin, neurotransmitters of various kinds (see below), and by cytoplasmic ions such as Ca^{2+}. In higher plants, Ca^{2+}, K^+, H^+, and Cl^- act as ligand-gating channels conducting the same or different ions. These ligands, on binding the channel, induce a conformational change that opens the channel gate. In addition to Na^+, ligand-gated channels conducting K^+ and Ca^{2+} have also been detected in animals. Ligand-gated Ca^{2+} channels, for example, occur on arterial smooth muscle cells. Other ligand-gated Ca^{2+} channels, located in both the plasma membrane and internal cell membranes, are important in cellular regulatory pathways, particularly those triggered by the activation of receptors at the cell surface (for details, see Chapter 6).

A recently discovered gated Cl^- channel is implicated in the hereditary disease *cystic fibrosis*, the most common potentially fatal hereditary disease of humans. In this disease, regulation of the channel is faulty; altered operation of the channel in epithelial cells releases unusually high levels of Na^+ and Cl^- in sweat and, through unknown processes, causes accumulation of mucus in the respiratory tract and failure of exocrine secretion in glands such as the pancreas. Blockage of airways by mucus leads to chronic lung infections that, with other effects of the Cl^- transport deficiency, are fatal by about age 20 for half the persons born with the disease.

The Cl^- channel involved in cystic fibrosis is called the *CFTR* protein (CFTR = Cystic Fibrosis Transmembrane conductance Regulator). The CFTR is encoded in the *CF* gene, discovered in the laboratories of L.-C. Tsui, F. S. Collins, and J. R. Riordan. Although CFTR operates as a gated ion channel, it appears to be related to a group of ATP-driven active

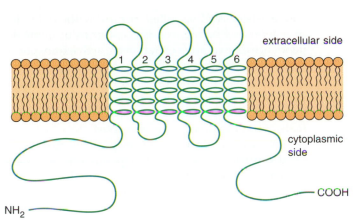

Figure 5-7 The six transmembrane segments forming the structural unit of a voltage-gated ion channel. Each channel consists of four of these units. In the K^+ channel the units are separate polypeptides; in Na^+ and Ca^{2+} channels the four units are linked end to end into a single large polypeptide. Of the six transmembrane segments in each unit, 5 and 6 are believed to surround the channel pore; number 4, in which every third residue is a positively charged lysine, contains the voltage-sensing site.

transport pumps that occurs in both bacteria and eukaryotes (see p. 148). The CFTR channel actually binds ATP; however, ATP appears to have a regulatory function rather than providing energy for transport—binding the nucleotide opens the channel. In this way, CFTR acts as a ligand-gated channel. Mutations in *CF* are common—about 1 in 25 persons of Northern European extraction are unaffected carriers, and 1 in 2500 are homozygous recessives born with the disease. About 60% of the individuals with cystic fibrosis have a three-nucleotide deletion in the *CF* gene that is reflected in the absence of a single phenylalanine residue from CFTR.

Stretch-gated channels undergo conformational changes opening or closing the channel gate when the membrane containing them is subjected to a force that distorts the membrane. Mammalian muscle cells, for example, contain Na^+ channels that open in response to stretching of the muscle cell plasma membranes. Stretch-gated Ca^{2+} channels have been detected in fungi and animals and are suspected to occur in higher plants. In yeast the channels may be part of pathways regulating responses to osmotic pressure and growth by budding. In animals, stretch-gated Ca^{2+} channels have been detected in cells of the epithelia lining blood vessels, where their activity may be linked to blood pressure regulation, and in the hair cells of the inner ear.

How Gated Channels May Work   Several experiments with voltage-gated channels have provided clues to the possible molecular structure and function of this and possibly other gated channels. In 1977, C. M. Armstrong and F. Bezanilla found that the ability of voltage-gated channels to close could be blocked by a proteinase introduced on the cytoplasmic side of the membrane. From this finding they proposed that the gate consists of a segment of the protein that extends from the cytoplasmic side of the channel, in a position

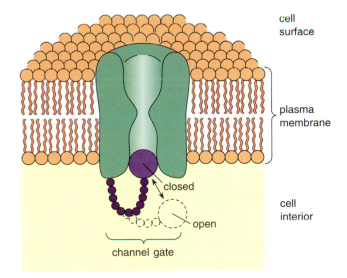

Figure 5-8   Structure of a gated ion channel. The narrow midpoint in the channel may operate as a selective filter that binds and passes only ions of the type transported by the channel. Conformational changes triggered by voltage changes, ligand binding, or mechanical stress open the gate to admit ion flow. In most channels the gate is on the cytoplasmic side of the membrane, as shown in this diagram.

that makes it susceptible to removal by the proteinase. According to their hypothesis (Fig. 5-8), the extended amino acid segment folds into a ball, suspended by an extended chain of amino acids, that fits into the opening of the ion channel on its cytoplasmic side. Gating is accomplished by conformational changes that swing the ball toward or away from the channel opening.

Recent experiments by R. W. Aldrich and his coworkers lend support to the ball-and-chain hypothesis. By using induced mutations to change individual amino acids in a voltage-gated Na^+ channel, the Aldrich group found that alterations in the last 20 amino acids at the N-terminal end of the protein could destroy channel gating. This segment may form the ball at the end of the chain. Deletions of amino acids in a segment preceding the terminal group, instead of destroying gating, speeded the rate at which the channel closed. Presumably this segment is the chain; shortening the chain would reduce the time required for the ball to swing into the channel mouth.

In another series of experiments, Aldrich and his colleagues investigated a naturally occurring *Drosophila* mutation, called *shaker*, in which a voltage-gated K^+ channel is defective because of deletions in the channel protein at the N-terminal end. To test whether an artificial ball could close the mutant channels, they synthesized a short peptide with the same amino acid sequence as the presumed ball of a normal Na^+ channel. Addition of a solution of the artificial peptide enabled the mutant channels to close, in keeping with the ball-and-chain hypothesis.

Passive transport, involving simple and facilitated diffusion, accounts for the movement of a wide variety of substances through cellular membranes. By this mechanism, cells are able to absorb many of the hydrophobic and hydrophilic molecules required for their biological reactions and to release waste materials or secreted products to the outside. Among the most important substances entering cells by facilitated diffusion are ions and fuel molecules such as sugars (ions and sugars also enter by active transport). Passive transport, because it is driven by concentration gradients, takes place without an expenditure of cellular energy. It greatly enhances the transport of key substances that would otherwise penetrate cells too slowly to support cellular life.

ACTIVE TRANSPORT

Many substances are pushed across membranes against gradients of chemical or electrical potential by active transport "pumps." Among the substances transported actively are Na^+, H^+, Ca^{2+}, and other ions; glucose and other six-carbon sugars; and amino acids. In some cases the pumps moving these substances are powerful enough to continue pushing their transported molecules against concentration differences as great as a million times.

Characteristics of Active Transport

The most fundamental characteristic of active transport is its energy requirement. For most active transport, energy is derived directly or indirectly from hydrolysis of ATP, supplied in turn by the oxidation of cellular fuel substances. Because of its ultimate dependence on oxidative metabolism, active transport is sensitive to metabolic poisons and, in contrast to simple or facilitated diffusion, slows or stops if cellular reactions supplying ATP are inhibited. Much of the total energy expended by cells, particularly in animals, is used to drive active transport.

Other features of active transport resemble facilitated diffusion. The process depends on membrane proteins and ceases if these proteins are denatured or removed. Like the proteins carrying out facilitated diffusion, active transport proteins can be saturated by high concentrations of the transported substance, and they are specific: Only a certain molecule or closely related group of molecules is moved across membranes by a given active transport system. In summary, active transport:

1. Goes against gradients of chemical or electrical potential

2. Requires energy and is sensitive to metabolic poisons

3. Depends on the presence and activity of membrane proteins

4. Can be saturated

5. Is specific for certain substances or closely related groups of substances

During active transport the pump proteins combine briefly with the transported substance, as in facilitated diffusion and enzymatic catalysis. (Table 5-1 compares the characteristics of simple diffusion, facilitated diffusion, and active transport.)

Active transport proteins exhibit many of the same structural characteristics as facilitated diffusion transporters. They have active sites that bind the transported molecules. They also alternate between at least two major conformational changes in which the active sites are directed to either side of the membrane during the pumping cycle. Depending on the side faced, the active sites bind the transported molecules with high or low affinity.

Active transport takes place in eukaryotes by two primary mechanisms—one directly dependent on ATP hydrolysis and one indirectly dependent. In directly dependent mechanisms, ATP hydrolysis and active movement of the transported substance are carried out by the same transporter protein, and the energy released is used directly by the transporter to push substances across the membrane. All the known pumps working in this way transport positively charged ions across cellular membranes. Indirectly dependent mechanisms, in contrast, use the concentration gradient of an ion, built up by other, directly ATP-dependent transporters, as the energy source for active transport of a different substance. Indirect pumps move a variety of organic substances as well as either positively or negatively charged ions across membranes.

Active Transport Directly Dependent on ATP

Active transport pumps directly dependent on ATP occur in two primary types. One type temporarily binds a phosphate group removed from ATP during the pumping cycle. Because they bind phosphates, these direct carriers are known as *P-type pumps*. The second type, the *V-type pumps* (*V* stands for vesicle), are so called because they occur in the membranes of vesicles such as those of lysosomes and the Golgi complex. Although V-type pumps break down ATP as their energy source for active transport, they do not bind an ATP-derived phosphate during their pumping action.

P-type Pumps Three P-type pumps, all moving positively charged ions against their concentration gradients, occur commonly in eukaryotes. One, the *H^+-ATPase* pump (ATPase refers to *adenosine triphosphatase*, the enzymatic activity that breaks down ATP), moves H^+ out of cells. The second, the *Ca^{2+}-ATPase* pump, pushes Ca^{2+} out of cells or into cytoplasmic vesicles such as those of the ER. The third, the *Na^+/K^+-ATPase*, moves Na^+ outward and K^+ inward across the plasma membrane in the same pumping action.

All these pumps are transmembrane proteins in which one polypeptide carries out all functions of the pump including ATP breakdown, phosphate binding, and movement of the pumped ion across the membrane (Fig. 5-9a). The pumps contain 10 transmembrane helices; about 80% of their mass consists of three extensive domains extending into the cytoplasm. One of these domains couples energy released by ATP hydrolysis to active transport; the second hydrolyzes ATP and binds the phosphate group released by ATP breakdown. The remaining domain binds calmodulin in combination with Ca^{2+}, which stimulates pumping action. This domain is also modified by phosphorylation. All the helices show similarities in their amino acid sequences, indicating that they probably evolved from a single ancestral type. The Na^+/K^+-ATPase pump contains an auxiliary polypeptide subunit that may regulate its activity (Fig. 5-9b). Because related pumps exist in bacteria (see Supplement 5-1), the ancestral gene must have very ancient origins indeed.

Characteristics of the P-type Pumps H^+-ATPase pumps are widely distributed among fungi, algae, and plants, and also occur in prokaryotes. They are located in plasma membranes, where they push H^+ from the cytoplasm to the cell exterior. The H^+ gradient created, high in the cell exterior and low in the cytoplasm, provides the primary energy source for indirect active transport in fungi and plants (see below). H^+-ATPase pumps also occur in the membrane surrounding the large central vacuole of plant cells, where they increase the acidity of the vacuole by pumping H^+ inward. P-type H^+ pumps are more limited in distribution in animals. However, a P-type H^+ pump is responsible for the movement of H^+ from stomach-lining cells into the gastric juice. The stomach contents may become as acidic as pH 0.8 as a result of this movement.

Ca^{2+}-ATPase pumps are widely distributed among animals and have also been detected in plants, a protozoan, and a prokaryote. Most animal cells have multiple types of Ca^{2+}-ATPase pumps. The pumps occurring in the plasma membrane and ER membranes of muscle cells, for example, are of different but closely related types. The eukaryotic Ca^{2+}-ATPase pumps push Ca^{2+} from the cytoplasm to the cell exterior and into the vesicles of the ER. As a result, the Ca^{2+} concentration is typically high outside cells and in ER vesicles and very low in the cytoplasmic solution. This Ca^{2+} gradient is used as a regulatory control of activities as diverse as muscle contraction, secretion, and microtubule assembly. An activity such as muscle contraction is at rest at the low Ca^{2+} concentrations maintained by the Ca^{2+}-ATPase pumps; a sudden, controlled release of Ca^{2+} into the cytoplasm, by facilitated diffusion through voltage- or ligand-gated channels in the plasma membrane and ER, starts the activity (see, for example, p. 324). As noted, these systems work through Ca^{2+}-dependent control proteins that are activated by combination with calcium ions.

The Na^+/K^+-ATPase pump moves Na^+ outward and K^+ inward across the plasma membrane. Three Na^+ are pumped out and two K^+ are pumped into the cell by each cycle of the pump. Because the Na^+/K^+-ATPase pump moves three positively charged ions outward for every two moved inward, positive charges accumulate in excess outside the membrane. As a result, the inside of the cell becomes negatively charged with respect to the outside, and a measurable difference in voltage develops across the plasma membrane. (Other factors, such as the passive distribution of ions across the plasma membrane, also contribute to the voltage difference.)

The distribution of ions set up by the Na^+/K^+-ATPase pump has several effects of fundamental

Figure 5-9 (a) The possible arrangement of transmembrane helices and domains in P-type active transport pumps, in which 10 alpha helices zigzag back and forth across the membrane. The transducing domain couples the energy released by ATP hydrolysis to ion transport; the ATP-binding and -hydrolysis domain is hinged so that the phosphorylation site is brought near the ATPase site during ATP breakdown, permitting transfer of a phosphate group from ATP to the site. The Ca^{2+}/calmodulin-binding domain regulates pump activity. Binding calmodulin activated by combination with Ca^{2+} stimulates pump activity. This domain also contains sites that are phosphorylated to regulate pump activity. The entire pump structure totals slightly more than 100,000 daltons. (b) Structure of the Na^+/K^+-ATPase pump. The additional β-subunit may regulate the pumping activity of the α-subunit, which is similar in amino acid sequence and structure to other P-type pumps.

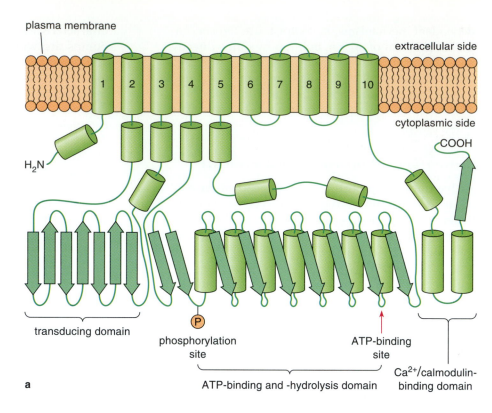

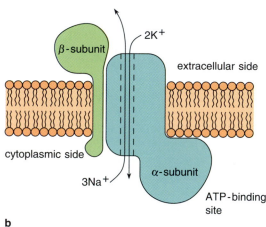

importance to eukaryotic cells. The voltage difference set up across the plasma membrane by the pump and other factors forms the basis for conduction of electrical impulses by nerve and muscle cells. In addition, the high outside/low inside Na^+ gradient supplies the energy source for indirect active transport of sugars and amino acids (see p. 136). The ion distribution set up by the Na^+/K^+-ATPase pump, particularly the removal of Na^+ from the cytoplasm, also controls the osmotic pressure of eukaryotic cells. This has been clearly demonstrated by the use of *ouabain*, an inhibitor that specifically binds to the Na^+/K^+-ATPase pump and stops its action. When exposed to ouabain, many types of animal cells swell and burst because of inward osmotic movement of water in response to the increased cytoplasmic Na^+ concentration.

In spite of its importance in animal cells, the Na^+/K^+-ATPase pump evidently has no equivalent in plants, fungi, or bacteria. Ouabain, for example, has no effect on Na^+ transport in these organisms. In plants, fungi, and bacteria, Na^+ concentrations are controlled by the combined activity of the H^+-ATPase pump and an indirect active transporter that uses energy stored in the H^+ gradient to drive Na^+ outward.

Some investigators have proposed that the Na^+/K^+-ATPase pump is an animal adaptation that made cellular life possible without cell walls. Even with the combined activity of the H^+-ATPase pump and indirect Na^+ transport, plant and bacterial cells swell and burst in most natural environments if their cell walls are ruptured or removed. The Na^+/K^+-ATPase pump, evolved as part of adaptations leading to the appearance of animal cells, keeps internal Na^+ concentrations low enough to prevent cells without walls from bursting from osmotic pressure. Some idea of the importance of the Na^+/K^+-ATPase pump can be gained from the fact that about 30% of the total energy budget of most animal cells is used to operate this pump alone.

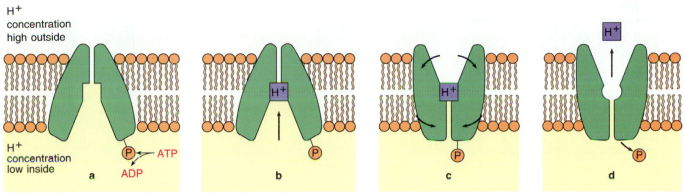

H⁺ concentration high outside

H⁺ concentration low inside

Figure 5-10 Active transport in a P-type active transport pump (see text).

In mammals Na^+/K^+-ATPase pumping activity is regulated by an extensive variety of hormones and growth factors. In some of these pathways, the pump is regulated by addition of phosphate groups (see p. 164).

How P-type Pumps May Operate Figure 5-10 illustrates how P-type pumps are believed to operate, using the H^+-ATPase transporter as an example. The transporter is initially folded so that its H^+ binding site faces the inside of the cell (H^+ is pumped from inside to outside in most systems). The pump first binds and hydrolyzes ATP; the phosphate removed from ATP is bound covalently to the transporter (Fig. 5-10a). Much of the energy released by ATP breakdown is retained in the complex formed between the transporter protein and the phosphate, which can be regarded as a "high-energy" complex. Attaching the phosphate also converts the H^+-binding site to the high-affinity state in which it readily binds H^+ colliding with the transporter on the cytoplasmic side of the membrane (Fig. 5-10b). In response to binding H^+, the transporter protein undergoes a conformational change that exposes the H^+-binding site to the opposite side of the membrane (Fig. 5-10c). At the same time, the conformational change alters the H^+-binding site to a low-affinity state in which its binding is so weak that H^+ is released to the outside even when the concentration of the ion is relatively high (Fig. 5-10d). The phosphate group is also released at the inside of the membrane at this time. Release of the H^+ and phosphate induces a final conformational change that returns the protein to the initial conformation shown in Figure 5-10a. The pumping cycle is now ready to begin again. (Fig. 5-11 shows how the more complex Na^+/K^+-ATPase pumping cycle may operate.)

One of the most fundamental and far-reaching discoveries made through the study of P-type pumps is that *they can be made to synthesize ATP by forcing them to run in reverse.* For example, the Na^+/K^+-ATPase pump normally moves Na^+ out of animal cells. I. M. Glynn and his coworkers showed that if the external Na^+ concentration is raised to extremely high levels, Na^+ is pushed backward through the pump by the gradient, forcing the pump to run in reverse. Under these conditions the pump protein *adds phosphate groups to ADP molecules to form ATP.* Similar observations have been made by others studying H^+ and Ca^{2+} active transport pumps. These observations are among the most significant findings in cellular energetics because they demonstrate that *a concentration gradient can be used by membrane proteins as an energy source for synthesizing ATP.* This principle, discussed at length in Chapters 8 and 9, underlies the primary mechanism synthesizing ATP in all eukaryotic and most prokaryotic organisms.

V-type Pumps V-type pumps occur in the membranes of internal vesicles such as endocytotic vesicles, the Golgi complex, lysosomes, and secretory vesicles in all eukaryotic cells examined to date. In plants and lower eukaryotes such as yeasts, pumps of this type also occur in the boundary membrane of the large central vacuole. In all these locations, V-type pumps push H^+ from the cytoplasm into the space enclosed by the vesicle membrane. As a consequence, these pumps lower the pH of the vesicle contents and contribute toward maintenance of neutral pH in the cytoplasm.

V-type pumps are large, complex structures with two segments, one buried in the membrane and one extending into the cytoplasm on the outer side of the vesicle membrane (Fig. 5-12). The buried segment forms a channel conducting H^+ through the membrane. The cytoplasmic segment carries out ATP hydrolysis and perhaps the major activity pumping H^+ through the channel.

The amino acid sequences of two of the polypeptides of the cytoplasmic segment (the A and B subunits in Fig. 5-12) are highly conserved in animals,

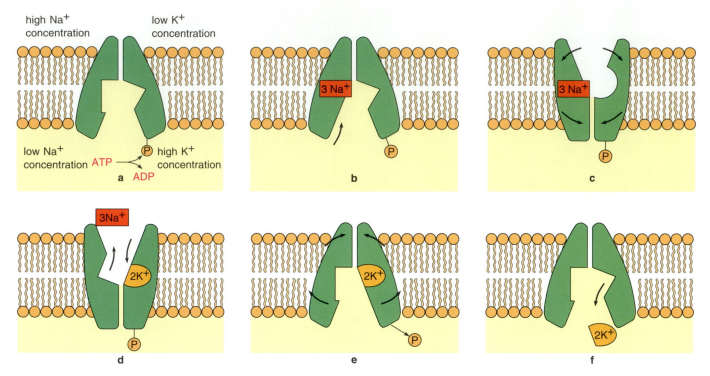

Figure 5-11 A possible mechanism for operation of the Na$^+$/K$^+$-ATPase pump. In the first step of the cycle **(a)** the pump binds and hydrolyzes ATP; the phosphate group is transferred from ATP to the transporter in the process. The phosphorylated pump may be considered as a high-energy complex. **(b)** Binding the phosphate converts the Na$^+$-binding site to its high-affinity state, in which it avidly binds Na$^+$ on the cytoplasmic side of the membrane. **(c)** Binding Na$^+$ triggers a conformational change that exposes the Na$^+$-binding site to the medium outside the cell. **(d)** As the conformational change becomes complete, the Na$^+$-binding site is altered to a low-affinity state, releasing Na$^+$ to the outside medium. At the same time, the K$^+$-binding site is activated so that it now binds K$^+$ at the outside surface with high affinity. **(e)** K$^+$ binding induces a conformational change that exposes the K$^+$-binding site to the cell interior and releases the phosphate group. At the same time, the K$^+$-binding site is converted to a low-affinity state in which K$^+$ is released **(f)**. Completion of this step returns the transporter to the initial state, in which it is ready to repeat the cycle. The transporter operates in such a way that three Na$^+$ are moved outside and two K$^+$ inside for each turn of the cycle.

plants, and fungi. A V-type pump with structure closely related to eukaryotic V-types was recently discovered in a bacterial group, the *archaebacteria*, indicating very ancient evolutionary origins for the pumps in this group.

V-type pumps are remarkably similar to a group of vital ATPases occurring in the membranes of bacteria, cyanobacteria, mitochondria, and chloroplasts (compare Figs. 5-12 and 8-23). These ATPases, known as the F_oF_1 *ATPases* (see pp. 230 and 258), work in the opposite direction from the V-type H$^+$ pumps—they use an H$^+$ gradient as the energy source for synthesizing ATP from ADP and phosphate. The F_oF_1 ATPases have membrane and cytoplasmic segments with functions and polypeptide subunits equivalent to those of

V-type pumps. These similarities indicate that V-type active transport pumps, which break down ATP as an energy source to pump H$^+$, and the F_oF_1 ATPase enzymes, which use an H$^+$ gradient as an energy source for ATP synthesis, evolved from a single ancestral type.

Indirect Active Transport Pumps

Indirect active transport pumps use a ion concentration gradient maintained by another, direct active transport pump as their energy source. Because molecules carried in or out of cells by indirect active transport always move across membranes in conjunction

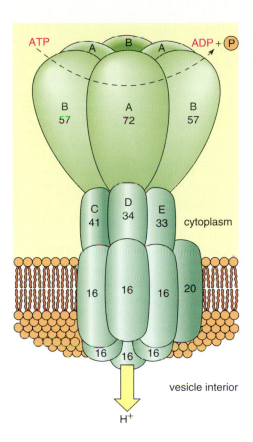

ATP   ADP + P

cytoplasm

vesicle interior

H⁺

Figure 5-12 Structure of a V-type active transport pump (the numbers give the molecular weights of the subunits in thousands). A membrane channel conducting H^+ from the cytoplasm into an organelle or vacuole is formed from six copies of a 16,000-dalton polypeptide subunit in association with one copy of a 20,000-dalton polypeptide subunit. One 38,000- and one 100,000-dalton polypeptide (not shown) also occur in association with the membrane channel. The large cytoplasmic segment consists of at least five different subunits, with two of them, the A and B subunits, present in three copies each. The catalytic activity associated with ATP binding and hydrolysis is believed to be associated with the B subunit.

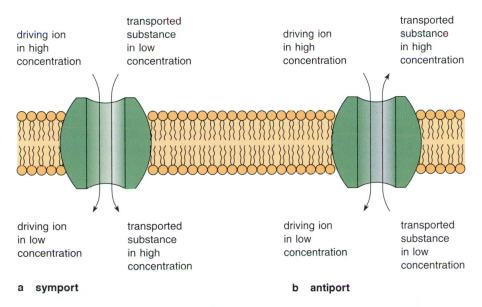

| driving ion in high concentration | transported substance in low concentration | | driving ion in high concentration | transported substance in high concentration |

| driving ion in low concentration | transported substance in high concentration | | driving ion in low concentration | transported substance in low concentration |

a symport   **b antiport**

Figure 5-13 Indirect active transport by symport and antiport, in which a favorable concentration gradient of an ion is used as the energy source for active transport of another substance. **(a)** In symport the transported substance moves in the same direction as the driving ion. **(b)** In antiport the transported substance moves in the direction opposite to the gradient of the driving ion.

with an ion supplying the driving force, this pathway of active transport is also termed *cotransport*.

Characteristics of Indirect Active Transport Indirect active transport takes place in either of two patterns, *symport* and *antiport* (Fig. 5-13). In *symport* the cotransported substance moves in the same direction as the

driving ion. Among the most important metabolites moved actively into cells by symport are sugars and amino acids. Ions are also transported inward by this mechanism in many cell types. In *antiport* the cotransported substance moves in the direction opposite to the driving ion. This pattern also pumps sugars, amino acids, and ions.

Most of the transport proteins in this group have 12 transmembrane helices, with a large hydrophilic domain extending into the cytoplasm between helices 6 and 7. The group is highly varied in type and activity; more than 50 known symport or antiport carriers, all related in structure, transport sugars of various kinds, including pentoses, hexoses, and disaccharides, in animals, plants, fungi, and bacteria.

The driving force for indirect active transport in animal cells is the high outside/low inside Na^+ gradient set up by the Na^+/K^+-ATPase pump. The equivalent energy source for cotransport in plants and fungi is the high outside/low inside H^+ gradient set up by the H^+-ATPase pump. Bacteria also carry out H^+-driven cotransport, but in bacteria the high outside/low inside H^+ gradient driving the transport is set up by oxidative metabolism instead of active transport (see Supplement 5-1).

In eukaryotes practically all the organic substances actively transported into cells are moved by cotransport. Indirect carriers cotransporting all the amino acids, either singly or in related groups, have been detected in animals, plants, fungi, and bacteria. In vertebrates, indirect pumps transport glucose and other six-carbon sugars in cells of the intestine and kidney. In animals indirect active transport of H^+ works along with direct H^+ transport by V-type pumps to eliminate excess H^+ produced by metabolic reactions. (The equivalent function in plants is served by direct H^+-ATPase pumps working in conjunction with V-type pumps.) There is also some evidence that Mg^{2+} is pumped out of many cell types by an Na^+-driven antiport system.

Plants have an interesting cotransport system in which sugars and amino acids, produced by photosynthesis in leaves and circulating through vascular tissue, are taken up by cells by H^+-driven symport across the plasma membrane. Once in the cells, the sugars (primarily sucrose) and amino acids are concentrated in the large central vacuole by H^+-driven antiport carriers in the vacuole membrane. The H^+ gradient, high inside the vacuole, is set up by H^+-ATPase or V-type active transport carriers working in the vacuole membrane. The amino acids moving from leaf cells to the rest of the plant are the primary form in which fixed nitrogen is circulated in plants.

How Indirect Pumps May Work   Figure 5-14 shows the mechanism by which an indirect pump of animal cells may use an Na^- gradient as the energy source for the active transport (symport) of glucose. The mechanism assumes that Na^+ exists at a high outside/low inside gradient established by the Na^+/K^+-ATPase pump. In the first step in the pumping cycle the transport protein is folded so that sites tailored to fit Na^+ and glucose are directed toward the outside surface of the membrane (Fig. 5-14a). In this step the glucose binding site is considered to be in a high-affinity state. Binding Na^+ and glucose (Fig. 5-14b) triggers a conformational change in the transporter that closes access to the outside and directs the Na^+ and glucose binding sites toward the inside membrane surface (Fig. 5-14c). Energy for the conformational change is derived from movement of Na^+ down its concentration gradient. In this conformation the binding sites are altered to a low-affinity state in which glucose and Na^+ are released simultaneously to the inside of the membrane (Fig. 5-14d). In response the transporter returns to the original conformation (Fig. 5-14a), ready to enter another cycle.

In the transport mechanism, Na^+ moves passively in the direction of its concentration gradient, and glucose molecules are moved actively against their gradient. Antiport works by a similar mechanism except that binding and release of the driving ion and transported substances occur on opposite sides of the membrane.

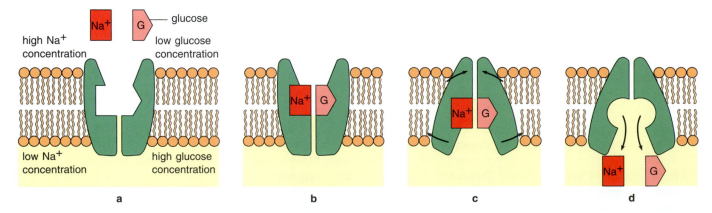

Figure 5-14   Indirect active transport of glucose by symport (see text).

The distinguishing feature of active transport is movement of substances against their concentration gradients. Energy for this movement is provided directly or indirectly through ATP hydrolysis. Other features of active transport—dependence on membrane proteins, specificity, and saturation—are shared with passive transport.

In direct active transport, which pumps positively charged ions across cellular membranes, ATP is hydrolyzed as the immediate energy source. Direct pumps occur in two major types: the P-type pumps, which combine with a phosphate group removed from ATP during the pumping cycle, and V-type pumps, which also hydrolyze ATP but operate without binding an ATP-derived phosphate group. P-type pumps move H^+, Na^+, K^+, and Ca^{2+} against their concentration gradients; all known V-type pumps move H^+ against its concentration gradient.

Indirect active transport pumps use H^+ or Na^+ gradients set up by direct active transport as their energy source. A substance transported by this mechanism may move in the same direction as the driving H^+ or Na^+ gradient (symport) or in the opposite direction (antiport). Various ions and all the organic molecules actively transported into eukaryotic cells, including glucose and amino acids, are pumped by indirect active transport.

ACTIVE TRANSPORT, MEMBRANE POTENTIALS, AND NERVE CONDUCTION

The Na^+/K^+-ATPase active transport pump, by moving three Na^+ out and only two K^+ inward in each pumping cycle, contributes significantly to a voltage difference across the plasma membrane in which the cytoplasm is negative with respect to the extracellular medium. The voltage difference forms the basis for the activity of nerve, muscle, and other excitable cells in animals.

The Resting and Action Potentials

In nerve and muscle cells the membrane voltage remains constant when the cells are not conducting electrical impulses. This steady membrane voltage of nonconducting nerve and muscle cells is called the *resting potential*. Depending on the cell type, the resting potential varies from about 20 to 100 millivolts (1 millivolt = 1/1000 volt; Information Box 5-1 on p. 138 describes the *Nernst equation*, used to calculate the resting potential from differences in ion concentration on either side of a membrane).

When triggered by an appropriate stimulus, the voltage difference across the plasma membrane changes abruptly in excitable cells. The abrupt change results from the activity of voltage-gated Na^+ and K^+ channels, which open in response to the stimulus and allow the ions to flow across the plasma membrane in response to their concentration gradients. The resulting disturbance of the resting potential, which takes a characteristic form in nerve and muscle cells, is called the *action potential*. The voltage changes characteristic of the action potential can be measured directly by placing minute electrical contacts (electrodes) inside nerve cells (*neurons*).

Neurons typically have an enlarged *cell body* that contains the nucleus and most of the cytoplasmic organelles such as the ER and mitochondria (Fig. 5-15). Branching extensions called *dendrites* and a long process called the *axon* extend from the cell body. Measurements of the resting and action potential can be made directly by placing one electrode inside an axon and

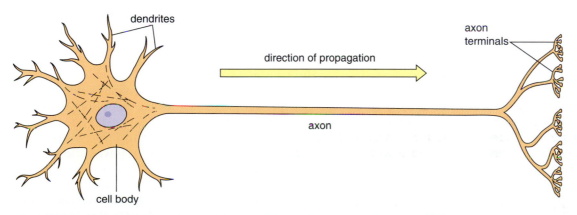

Figure 5-15 Structure of a neuron. The cell body contains the nucleus and major cytoplasmic organelles. The structural arrangement shown, with short dendrites and a long axon, is typical of motor neurons, the nerve cells that conduct impulses from the central nervous system to muscle cells.

Quantitative Estimation of Membrane Potential: The Nernst Equation

The magnitude of the voltage difference or electrical potential across the plasma membrane can be calculated from the inside and outside concentrations of the contributing ions. The potential difference due to any single ion can be calculated from the *Nernst equation:*

$$\text{potential difference} = E = \frac{RT}{z\mathscr{F}}\ln\frac{C_o}{C_i} \qquad (5\text{-}3)$$

in which R is the gas constant, T is the absolute temperature, z is the charge carried by the ion, $\mathscr{F}$ is the Faraday constant (23,062 cal/mol·V, or 96,000 coulombs/mol·V), and C_o and C_i are the outside and inside concentrations of the ion in moles.

For K^+ ions, for example, which in squid axons are maintained at outside and inside concentrations of 20 and 400 millimoles, substitution in the Nernst equation gives:

$$\text{voltage difference due to } K^+ = E_k = \frac{RT}{z\mathscr{F}}\ln\frac{20}{400}$$

Since $\log_{10} = 2.3 \ln$, and $RT/\mathscr{F}$ at 18°C = 25 and z = 1,

$$E_k = 2.3 \times 25\log_{10}\frac{1}{20} = -75 \text{ mV}$$

The result, −75 mV, gives a reasonably good approximation of the −60 mV actually measured for the membrane potential of resting squid axons.

---

another electrode in the solution surrounding the axon. For this work, investigators frequently take advantage of giant neurons of the squid, which have axons that are about 1 mm in diameter.

The voltage changes measured as an action potential develops take the form shown in Figure 5-16. A giant squid axon has a resting potential of −60 mV (negative inside). As an action potential begins, the inside potential rises rapidly and reverses, so that the inside of the membrane becomes positive with respect to the outside. At the peak of the voltage change, called the *spike,* the inside of the membrane reaches about +40 mV. Full development of the spike from the resting potential takes less than 1 millisecond. The inside potential then falls at a slower rate, dropping slightly below the resting value before stabilizing again at the resting potential. The entire change, from initiation of the spike to the return to the resting value, takes only a little more than 5 milliseconds in the squid giant axon. Action potentials take a similar form in neurons of other types except for differences in the negative value of the resting potential, absolute negative and positive limits of the response, and the time required to complete the voltage change.

The Role of Na⁺ and K⁺ in the Action Potential   The ion movements producing the action potential were worked out by A. L. Hodgkin and A. L. Huxley in 1952. In their work the flow of current that produces the action potential was of major interest, since current is a measure of the flow of charges through a conductor or across a membrane. The charges flowing in this case are the ions moving to produce the action potential.

Figure 5-17 shows the changes in current that Hodgkin and Huxley detected during generation of an action potential. Immediately after a stimulus generating an action potential, a positive current flows rapidly inward across the plasma membrane of an axon (Fig. 5-17a). The inward flow, lasting for less than 1 millisecond, is followed by a change to a slower outward flow of positive current.

The contributions of Na^+ and K^+ to these current changes were sorted out by further experimentation. By replacing Na^+ with choline in the solution outside the axon, any effects due to Na^+ could be eliminated. (Choline is a positively charged organic molecule that does not penetrate the membrane.) Under these conditions the initial inward flow of positive charges did not occur, and only the outward flow was recorded (Fig. 5-17b). Therefore the initial inward flow of current could be attributed to sudden movement of Na^+ across the membrane from outside to inside. The outward current flow, which begins immediately after the stimulus and builds up relatively slowly, was identified with movement of K^+ from inside to outside by the use of labeled potassium (^{42}K). During buildup of the outward current flow, radioactivity due to ^{42}K

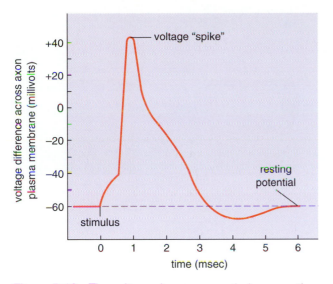

Figure 5-16 The voltage changes recorded as an action potential develops in a segment of a squid nerve axon (see text).

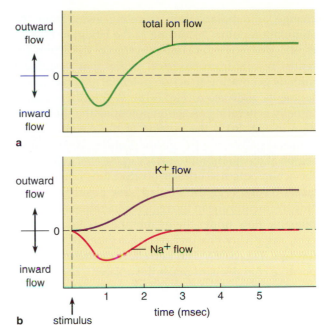

Figure 5-17 Ion flows across the plasma membrane, recorded as changes in current as an action potential develops in a squid nerve axon. **(a)** Total changes noted during an action potential. An inward flow of positive charges is followed by an outward flow of positive charges. **(b)** Current flow when Na^+ is eliminated from the solution surrounding an axon (purple curve). The first rapid inward flow is absent, indicating that the initial inward flow noted in **(a)** is due to Na^+ (red curve).

built up outside the membrane at the same rate as the outward flow of charges. This accumulation directly linked the outward flow to K^+ movement.

From these observations Hodgkin and Huxley concluded that the action potential results from a sudden, explosive inflow of Na^+, lasting for less than 1 millisecond. This flow changes the potential inside the axon from negative to positive, and the membrane is said to be *depolarized*. At the same time a slower outflow of K^+ gradually builds up; within 2 milliseconds after the inward Na^+ flow reaches its peak, the outward movement of K^+ restores membrane potential to the resting value. Hodgkin and Huxley received the Nobel Prize in 1963 for their work with nerve conduction.

Voltage-Gated Na^+ and K^+ Channels and the Action Potential   The ion flows of the action potential are now known to depend on voltage gating of the Na^+ and K^+ channels. In resting neurons, both the Na^+ and K^+ channels in the plasma membrane are closed. The channels are gated open if the voltage difference across the plasma membrane is sufficiently reduced. The Na^+ channels open rapidly, allowing Na^+ to flow massively from outside to inside in response to the high outside/low inside concentration gradient, while the K^+ channels open more slowly. The sudden inward flow of positive charges as the Na^+ channels open causes the inside of the membrane to become positive, producing the spike of the action potential. The Na^+ channels close again, about 1 millisecond after opening; by this time the K^+ channels have opened enough to release K^+ from inside the membrane to the outside in response to the high inside/

low outside concentration gradient of this ion. The outward flow of positive charges gradually increases as the K^+ channels open more fully; within about 2 milliseconds enough K^+ has moved outside to compensate for the Na^+ ions moving inward, and the inside of the membrane becomes negative again. As the inside of the membrane becomes fully negative, the K^+ channels close and the voltage difference stabilizes at the resting potential.

Dependence of the action potential on an inward flow of Na^+ has been found to be characteristic of most excitable cells in animals. The voltage-gated Na^+ channel appears to work in closely similar patterns in various neurons; the differences in firing patterns and duration of the action potential of different cell types and organisms seem to depend primarily on variations in the K^+ channel.

Movement of Impulses Along an Axon: Propagation of the Action Potential

Once an action potential develops, it passes along the surface of a nerve or muscle cell as a wave of electrical change or depolarization traveling away from the stimulation point. According to the Hodgkin-Huxley model, this movement or *propagation* of the action potential results from electrical effects that a segment initially generating an action potential has on adjacent segments of an axon (Fig. 5-18). Within the segment initially generating an action potential (the shaded region in Fig. 5-18) the outside of the membrane becomes negative and the inside positive. The negative region at the membrane surface in the area generating the action potential attracts positive ions from the adjacent resting regions. Similarly, the positive ions inside the membrane in the region generating the action potential flow under the membrane toward the adjacent, negatively charged resting regions of the axon (arrows, Fig. 5-18). The net effect of these flows of positive charges is to make the adjacent, resting segments of the axon membrane less positive on the outside and more positive on the inside, or, in other words, to reduce the potential difference across the membrane in the adjacent regions.

The reduction in potential produced by these ion flows is easily large enough to open the Na^+ and K^+ channels in the adjacent membrane segments, and the action potential develops in these regions. In this way, each segment of the axon stimulates the next segment to "fire," and the action potential moves rapidly along the axon as a nerve impulse.

The impulse does not reverse at any point because of a phenomenon noted by Hodgkin and Huxley in their original experiments. For a very brief interval after the action potential reaches its peak, known as the *refractory period*, a second action potential cannot be generated by the same region of an axon. Evidently during the refractory period, which occurs while K^+ outflow restores the resting potential, Na^+ channels are held shut and regions that have just fired are unable to open and generate a second action potential. The refractory period prevents a nerve impulse from reversing because, by the time the Na^+ channels in a recently fired region can open again, the action potential has moved too far away for its electrical disturbances to open the Na^+ gates in the region.

An action potential normally develops through the activity of ligand-gated channels in the dendrites of a neuron (see below). The action potential is then propagated over the neuron through the action of voltage-gated channels and moves along the axon in a one-way direction toward the axon terminal. However, if an axon is artificially stimulated at any point between its ends, two action potentials are generated that separate and move in opposite directions. Because neurons are normally stimulated only by ligand-gated channels in the dendrites, bidirectional conduction from a point along an axon does not occur.

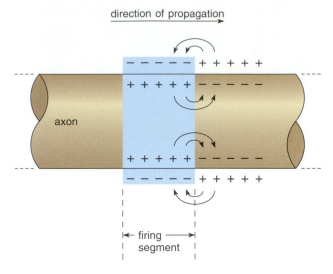

direction of propagation

axon

firing segment

Figure 5-18 Movement or propagation of an action potential by flow of charges between a firing segment (shaded area) and an adjacent unfired region of an axon (see text). Each firing segment induces the next to fire, causing the action potential to move directionally along the axon.

Some evidence for generation and propagation of action potentials has also been obtained in higher plants able to respond to stimuli. In the sensitive plant *Mimosa*, for example, which closes its leaves in response to touch, action potentials are generated and propagated by voltage-sensitive K^+ and Cl^- channels.

Saltatory Conduction in Myelinated Neurons

In axons firing by the pattern shown in Figure 5-18, impulses are conducted by a smooth and continuous flow of the action potential over the axon membrane surface, much like a burning fuse. The rate of conduction is proportional to the diameter of the axon, so that increases in conduction rate can be achieved by increases in axon diameter. The giant axons of squids and some other invertebrates represent the maximum development of this mechanism for increasing conduction rate.

Conduction rate is increased in most vertebrate neurons by a different but related mechanism. The axons of these cells are surrounded by layers of tightly packed *myelin* membranes, which act as an electrical insulator (Figs. 5-19 and 5-20). The covering is complete except for regularly spaced interruptions called *nodes*, which expose the axon membrane directly to the surrounding extracellular fluids.

Na^+ channels are restricted almost entirely to the segments of plasma membrane exposed at the nodes. Because of the insulating myelin sheath, current is conducted electrically through the axon from one

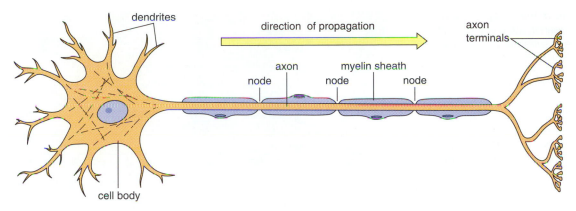

Figure 5-19 A myelinated neuron. The myelin sheath, which acts as an electrical insulator, covers the axon except for regularly spaced openings called nodes. Impulses jump from node to node by electrical conduction, greatly increasing their rate of travel along the axon.

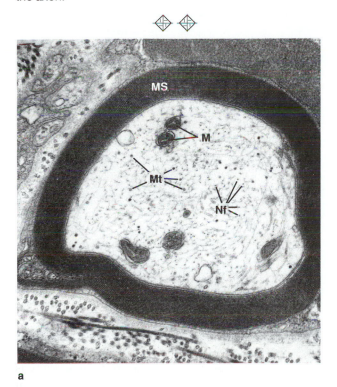

a

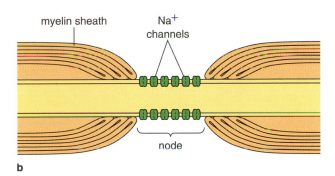

b

Figure 5-20 Myelinated axons and nodes. **(a)** A myelinated axon of the newt in cross section. The myelin sheath (MS) appears as successive layers of membranes surrounding the axon. M, mitochondria; Mt, microtubules; Nf, neurofilaments. × 27,000. (Courtesy of T. L. Lentz, from *J. Cell Biol.* 52:719 [1972], by permission of the Rockefeller University Press.) **(b)** A myelinated axon in longitudinal section showing how the sheath is interrupted at the node. Na⁺ channels are concentrated in the segment of the axon exposed at the node.

node to the next, in a jumping or *saltatory* fashion. Application of a stimulus at a node at one end of the axon develops an action potential at this node; the voltage change is conducted electrically and almost instantaneously through the myelinated portion of the axon to the next node. Arrival of the electrical change stimulates the next node to fire an action potential, producing a current that fires the next node, and so on. The result is very rapid travel of the impulse along the axon as the action potential skips from one node to the next, at rates two to five times faster than steady conduction through unmyelinated axons.

Reversal is prevented by the same mechanism operating in an unmyelinated neuron. After a node fires, it takes 2–3 milliseconds to recover its resting potential. By that time the impulse has skipped too far along the axon for its electrical current to stimulate a previously fired and recovered node.

Conduction Between Nerve Cells via the Synapse

Although a few axons make a direct, membrane-to-membrane contact with the dendrite of another neuron, the axon tips of most neurons are separated from

Figure 5-21 Structures of the synapse. **(a)** Thin-sectioned synapse as seen in the electron microscope. Large numbers of synaptic vesicles (arrows) are visible inside the axon terminal. SC, synaptic cleft. × 103,000. (Courtesy of C. Sotelo, from *International Cell Biology*, eds. B. R. Brinkely and K. R. Porter, 1977, by permission of the Rockefeller University Press.) **(b)** Major structures of a synapse. The synaptic vesicles contain neurotransmitter molecules that are released into the synaptic cleft as an action potential arrives at the axon terminal. The neurotransmitter molecules diffuse across the space between the sending and receiving cells and bind to ligand-gated channels in the receiving cell plasma membrane. The binding opens the ligand-gated channels and initiates an action potential in the receiving cell.

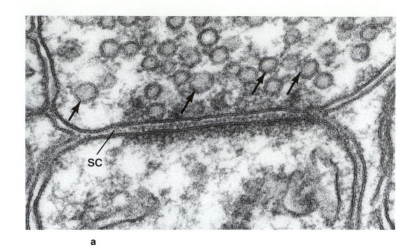

a

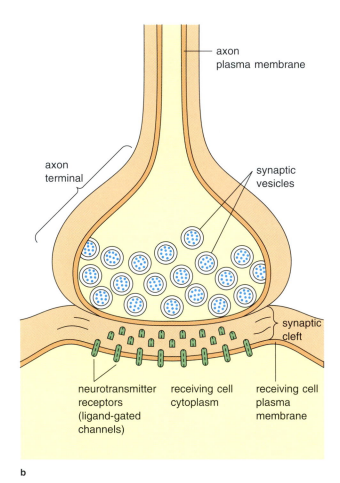

b

the next neuron in line, or from a responding muscle cell, by a narrow gap called the *synapse* (Fig. 5-21). As a result of the gap, an action potential arriving at an axon terminal fails to leap the intervening distance and stops at this point. The impulse crosses the gap by the release of neurotransmitters and their action on ligand-gated channels in the cell on the receiving side of the synapse.

The neurotransmitters, which include *acetylcholine*, various amino acids, or other substances, are stored in *synaptic vesicles* within the cytoplasm of the axon terminal (see Fig. 5-21). Arrival of an impulse causes the vesicles to fuse with the plasma membrane of the axon terminal and releases the neurotransmitter into the synaptic cleft. The neurotransmitter molecules diffuse across the synapse to make contact with the plasma membrane of the receiving cell. There they are bound by ligand-gated Na^+ or K^+ channels, which open in response, and Na^+ or K^+ flows inward or outward along their concentration gradients. If the electrical disturbance created by the ion flow is large enough, voltage-gated Na^+ and K^+ channels open in membrane regions adjacent to the synapse, and a new action potential is triggered in the receiving cell. Although the mechanism may seem cumbersome, the entire process of synaptic transmission is almost as rapid as propagation of a nerve impulse along an axon.

Recovery of the synapse after action potentials cease arriving at the axon terminal involves several steps. First, the voltage-gated Ca^{2+} channels in the axon terminal snap shut and any excess Ca^{2+} in the axon cytoplasm is quickly removed by Ca^{2+}-ATPase pumps. This inhibits vesicle fusion with the plasma membrane and release of neurotransmitter molecules

into the synaptic cleft. Any neurotransmitter molecules remaining in the cleft are then quickly removed by enzymatic breakdown, diffusion away from the cleft region, or resorption into the axon terminal. These mechanisms remove the neurotransmitter from the cleft within several microseconds and stop generation of action potentials in the receiving cell.

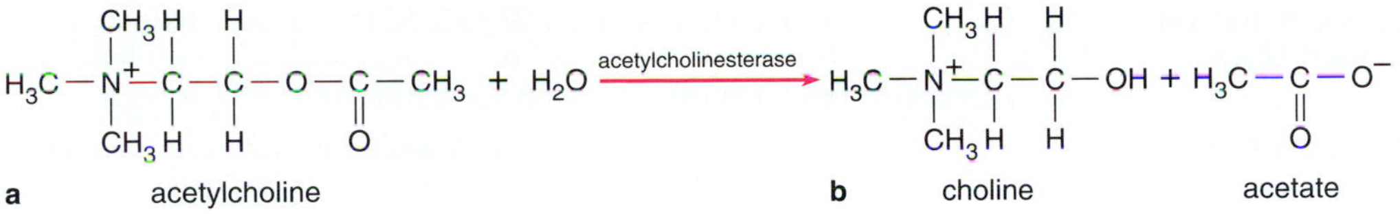

Figure 5-22 Acetylcholine and the reaction deactivating it by enzymatic breakdown. **(a)** The acetylcholine molecule. **(b)** The reaction hydrolyzing acetylcholine, catalyzed by the enzyme acetylcholin-esterase.

One of the best-studied neurotransmitters is acetylcholine (Fig. 5-22a), released in the synapses connecting neurons to muscle cells in vertebrates. The ligand-gated channel binding the neurotransmitter on the receiving end of the synapse, the *acetylcholine receptor*, is a complex protein containing four different polypeptides encoded in four separate but related genes. Binding acetylcholine opens the receptor channel to the flow of positively charged ions, primarily Na^+, which initiates the action potential on the receiving side of the synapse. Acetylcholine released into the synaptic cleft is hydrolyzed into choline and acetate (Fig. 5-22b) by *acetylcholinesterase*, an enzyme that is constantly active in the cleft. The constant hydrolysis of acetylcholine by the enzyme quickly clears the synapse of the neurotransmitter when action potentials cease to arrive at the synapse.

Acetylcholine and its receptor are central to nerve-muscle coordination in vertebrates. Their central importance is underscored by the fact that several highly dangerous venoms and toxins target the acetylcholine system. When introduced into the body, these poisons paralyze skeletal muscles and leave the animal helpless. *Curare*, extracted from the bark of several South American plants, was used by the native Indians as an arrow poison. This substance competes directly with acetylcholine for its binding site on the receptor. *Bungarotoxin*, a component of the venom of a Southeast Asian snake, has a similar effect. Other agents, such as the drug *procaine*, combine noncompetitively (see p. 83) by binding with other sites on the acetylcholine receptor to block its action.

In neurons, gated ion flows produce the action potential and, through propagation of the action potential, generate nerve impulses. The flow of ions through the gated channels occurs by facilitated diffusion along concentration gradients; the Na^+ and K^+ gradients responsible for the flow are set up by the Na^+/K^+-ATPase active transport pump. The impulses, passing from cell to cell via synapses, are the primary route of communication between excitable cells. Transmission across synapses depends on the combined activity of voltage- and ligand-gated channels on either side of a synapse. All the functions of animal nervous and muscular systems, including those of your eyes and brain in reading and comprehending the words on this page, depend on these nerve impulses; in turn, the impulses result from the activities of passive and active membrane transporters, embedded as integral proteins in the plasma membranes of neurons and muscle cells.

For Further Information

ATP synthesis in mitochondria and chloroplasts, *Chs. 8 and 9*
F_oF_1 ATPase in chloroplasts and cyanobacteria, *Ch. 9*
F_oF_1 ATPase in mitochondria and bacteria, *Ch. 8*
Membrane structure, *Ch. 4*
Muscle contraction, *Ch. 11*
Transport in mitochondria and chloroplasts, *Chs. 8 and 9*
Transport through the nuclear envelope, *Ch. 13*

Suggestions for Further Reading

Baldwin, S. A., and Henderson, P. J. F. 1989. Homologies between sugar transporters from eukaryotes and prokaryotes. *Ann. Rev. Physiol.* 51:459–471.

Bloom, F. E. 1988. Neurotransmitters: Past, present, and future directions. *FASEB J.* 2:32–41.

Briskin, D. P. 1990. The plasma membrane H^+-ATPase of higher plant cells: Biochemical and transport functions. *Biochim. Biophys. Acta* 1019:95–109.

Bush, D. R. 1993. Proton-coupled sugar and amino acid transporters in plants. *Ann. Rev. Plant Physiol. Plant Molec. Biol.* 44:513–542.

Carafoli, E., and Chiesi, M. 1992. Calcium pumps in the plasma and intracellular membranes. *Curr. Top. Cellular Regulat.* 32:209–224.

Catterall, W. A. 1991. Functional subunit structure of voltage-gated calcium channels. *Science* 253:1499–1500.

Changeaux, J.-P. 1993. Chemical signalling in the brain. *Sci. Amer.* 269:58–62 (November).

Cohen, S. A., and Barchi, R. L. 1993. Voltage-dependent Na^+ channels. *Internat. Rev. Cytol.* 137C:55–103.

Collins, F. S. 1992. Cystic fibrosis: Molecular biology and therapeutic implications. *Science* 256:774–779.

DeCamilli, P., and Jahn, R. 1990. Pathways to regulated exocytosis in neurons. *Ann. Rev. Physiol.* 52:625–645.

Devillers-Thiérry, A., Galzi, J. L., Eiselé, J. L., Bertrand, S., and Changeaux, J. P. 1993. Functional architecture of the nicotinic acetylcholine receptor: Prototype of ligand-gated ion channels. *J. Membr. Biol.* 136:97–112.

Doige, C. A., and Ames, G. F.-L. 1993. ATP-dependent transport systems in bacteria and humans: Relevance to cystic fibrosis and multidrug resistance. *Ann. Rev. Microbiol.* 47:291–319.

Frace, A. M., and Gargus, J. J. 1991. Molecular biology of membrane transport proteins. *Curr. Top. Membr. Transport* 39:3–36.

French, A. S. 1992. Mechanotransduction. *Ann. Rev. Physiol.* 54:135–152.

Gennis, R. B. 1989. *Biomembranes: Molecular Structure and Function.* New York: Springer-Verlag.

Gluck, S. L. 1993. The vacuolar H^+-ATPases: Versatile proton pumps participating in constitutive and specialized functions of eukaryotic cells. *Internat. Rev. Cytol.* 137C:105–137.

Gould, G. W., and Bell, G. I. 1990. Facilitative glucose transporters: An expanding family. *Trends Biochem. Sci.* 15:18–23.

Henderson, P. J. F. 1990. Proton-linked sugar transport systems in bacteria. *J. Bioenerget. Biomembr.* 22:571–592.

Higgins, C. F. 1992. ABC transporters: From microorganisms to man. *Ann. Rev. Cell Biol.* 8:67–113.

Jan, L. Y., and Jan, Y. N. 1992. Structural elements involved in specific K^+ channel function. *Ann. Rev. Physiol.* 54:537–555.

Linehard, G. E., Slot, J. W., James, D. E., and Muekler, M. M. 1992. How cells absorb glucose. *Sci. Amer.* 266:86–91 (January).

Llinas, R. R. 1988. The intrinsic electrophysiological properties of mammalian neurons: insights into central nervous system function. *Science* 242:1654–1663.

McCleskey, E. W., Womack, M. D., and Fieber, L. A. 1993. Structural properties of voltage-dependent Ca^{2+} channels. *Internat. Rev. Cytol.* 137C:39–54.

McIntosh, I., and Cutting, G. R. 1992. Cystic fibrosis transmembrane conductance regulator and the etiology and pathogenesis of cystic fibrosis. *FASEB J.* 6:2775–2782.

Montal, M. 1990. Molecular anatomy and molecular design of channel proteins. *FASEB J.* 4:2623–2695.

Morell, P., and Norton, W. T. 1980. Myelin. *Sci. Amer.* 242:88–118 (May).

Neher, E., and Sakmann, B. 1992. The patch clamp technique. *Sci. Amer.* 266:44–51.

Nelson, N. 1992. Structural conservation and functional diversity of V-ATPases. *J. Bioenerget. Biomembr.* 24:407–414.

Nikaido, H. 1992. Porins and specific channels of bacterial outer membranes. *Molec. Microbiol.* 6:435–442.

Nikaido, H., and Saier, M. H., Jr. 1992. Transport proteins in bacteria: Common themes in their design. *Science* 258:936–942.

Silverman, M. 1994. Structure and function of hexose transporters. *Ann. Rev. Biochem.* 60:757–794.

Skou, J. C., and Esman, M. 1992. The Na,K-ATPase. *J. Bioenerget. Biomembr.* 24:249–261.

Smith, P. R., and Benos, D. J. 1991. Epithelial Na^+ channels. *Ann. Rev. Physiol.* 53:509-530.

Stein, W. D. 1990. *Channels, Carriers, and Pumps: An Introduction to Membrane Transport.* New York: Academic Press.

Strehler, E. E. 1991. Recent advances in the molecular characterization of plasma membrane Ca^{2+} pumps. *J. Membr. Biol.* 123:93–103.

Review Questions

1. What is a concentration gradient? What is diffusion? What causes diffusion?

2. What is a semipermeable membrane? What is the relationship of osmosis to semipermeable membranes? What conditions are necessary for osmosis to occur? What provides the energy for osmosis?

3. What is osmotic pressure? How is osmotic pressure related to cellular life? Compare a cell with the osmosis apparatus shown in Figure 5-2.

4. What is passive transport? What is the relationship between passive transport and lipid solubility? What is the relationship between passive transport and the fluid mosaic structure of cellular membranes?

5. What is facilitated diffusion? What is the relationship between membrane proteins and facilitated diffusion? How does facilitated diffusion resemble the activity of enzymes?

6. Outline the mechanism by which facilitated diffusion is believed to work.

7. What are ion channels? What is gating? What are voltage-gated channels? Ligand-gated channels? Stretch-

gated channels? What functions do gated channels carry out in cells?

8. Compare facilitated diffusion and active transport. What supplies the energy for active transport? What is the difference between direct and indirect active transport?

9. How are direct active transport pumps believed to work? What are P-type active transport pumps? V-type active transport pumps? Give examples of both types and outline their importance in living cells. Compare the structure of P-type and V-type pumps and the F_oF_1 ATPase. In what organisms do these structures occur?

10. How are indirect active transport pumps believed to work? What kinds of substances are moved across membranes by indirect active transport?

11. What happens if a direct active transport protein is forced to run in reverse? What significance does this have for cellular energy metabolism?

12. How can transport produce a voltage difference across membranes?

13. Outline the structure of a neuron.

14. What is resting potential? Action potential? What ion movements are responsible for generating the action potential? What is the role of gated membrane channels in generation of the action potential? What experimental evidence supports the conclusion that these ion movements actually produce the action potential?

15. What is propagation of the action potential? What is responsible for this propagation? What is myelin? A node? Outline the difference between continuous and saltatory conduction.

16. What are refractory periods? What is the relationship of refractory periods to propagation of the action potential in continuous and saltatory conduction?

Transport in Bacteria

Bacteria possess complete systems transporting ions and metabolites by a combination of passive and active transport. Passive transport of polar substances depends on facilitated diffusion catalyzed by membrane proteins, as it does in eukaryotes; however, bacteria have relatively few facilitated diffusion transporters because ions and metabolites rarely occur in bacterial environments in concentrations high enough to drive diffusion into the cell. As a consequence, most ions and metabolites are moved across bacterial plasma membranes by active transport.

Direct active transport in bacteria is driven by mechanisms more diverse than those of eukaryotes. Some active transport is carried out by pumps using ATP as a direct energy source. Other pumps appear to be driven by phosphate-bond energy in forms such as polyphosphates (inorganic phosphates linked into chains) or by a combination of ATP hydrolysis and a driving H^+ gradient—that is, by a combination of direct and indirect active transport. In a few photosynthetic bacteria, active transport is driven by the energy of absorbed light.

Indirect active transport in bacteria is driven primarily by an H^+ gradient. However, in contrast to fungi and plants, which also use an H^+ gradient to drive cotransport, the high outside/low inside H^+ gradient of bacteria is established primarily through active H^+ transport by membrane proteins taking part in oxidative reactions. These proteins use the energy of electrons removed from substances undergoing oxidation to drive H^+ from the cytoplasm to the cell exterior. Similar electron-driven transporters establish an H^+ gradient in mitochondria and chloroplasts. In these organelles the H^+ gradient, in turn, is used to drive ATP synthesis through activity of the F_oF_1 ATPase (see Chapters 8 and 9 for details).

A few bacterial cotransport systems use a high outside/low inside Na^+ gradient as their energy source. These Na^+-dependent transporters are actually indirectly dependent on the H^+ gradient set up by electron transport. This is because the Na^+ gradient itself is created by an indirect transport pump that uses the H^+ gradient to move Na^+ outward across the plasma membrane.

Bacterial transport has been studied most extensively in *E. coli*. Transport in this bacterium is complicated by the fact that, as in other gram-negative bacteria, the cytoplasm is surrounded by two boundary membranes (see p. 199 and Fig. 7-21b). The inner membrane, the plasma membrane, lies just under the cell wall. This membrane is structurally and functionally equivalent to the plasma membranes of other cells. The outer membrane, which lies just outside the cell wall, is also based on a bilayer but contains unusual molecules found nowhere else among living organisms (see p. 200). (For further details of the outer membrane of gram-negative bacteria, see Chapter 7.)

The outer membrane effectively seals gram-negative bacteria against entry of most molecules from the extracellular medium except for those that move passively through channel-forming proteins called *porins* (see Fig. 5-23), which restrict passage of molecules above a certain size limit. Once through the outer membrane, substances cross the intervening cell wall region (called the *periplasmic space*) and enter the cytoplasm through specific passive and active transport proteins embedded in the plasma membrane. Thus there are two major sets of transport proteins in gram-negative bacteria: the porins of the outer membrane and the passive and active transporters of the plasma membrane. Gram-positive bacteria, which lack the outer membrane, have only plasma membrane-associated transport systems.

Porins and Transport Through the Outer Membrane of Gram-Negative Bacteria

Porins allow hydrophilic molecules below a certain size to pass through the outer membrane of gram-negative bacteria. Entry through the porins of *E. coli*, for example, is limited to ions and molecules with molecular weights below about 600 to 650. Some porins are selective and admit only certain types or classes of molecules; others act simply as molecular sieves that admit essentially any substance small enough to pass through their channels. In either event the sizes admitted exclude viruses and potentially damaging molecules of larger dimensions such as antibiotics and enzymatic proteins.

Bacterial porins are distinguished by a pore structure based on a ring of beta strands (p. 56) rather than alpha helices as in most transport proteins. Nonselective porins consist of a trimer of three identical polypeptide subunits, each with a pore formed by a circle of 16 beta strands. The pore channel is shaped like a double funnel, with a diameter of about 1 to 1.2 nm at its narrowest region. Selective porins have a similar structure except for a site in the pore channel that recognizes and binds the molecule or group of molecules transported.

E. coli normally has only a single nonselective porin type in its outer membrane. Each cell may have as many as 100,000 or more copies of the porin molecule per cell, however, making it one of the most abundant proteins of the bacterium. Stressful environmental conditions such as high salinity or limitations in the supply of inorganic phosphates may induce other nonselective or selective porin types to appear in *E. coli*.

Many of the selective porins, which preferentially admit substances such as lactose- or maltose-containing carbohydrates, ribose, or short peptides through the outer membrane, work in association with active transport pumps in the underlying plasma membrane. These porins are encoded by a separate gene family with no sequence similarities to the nonselective porins. At least 20 different active transport-associated porin types have been identified in *E. coli*, each one feeding a specific active transport pump that moves the molecules admitted by the porin through the plasma membrane.

Porins with characteristics similar to the sieving porins of gram-negative bacteria also occur in the outer membranes of mitochondria and chloroplasts (see pp. 236 and 268). These porins, also based on a beta-strand structure, admit molecules of larger dimensions than bacterial porins—up to about 6000 daltons for the mitochondrial porin and as much as 10,000 to 13,000 daltons for the chloroplast porin. At these dimensions the mitochondrial and chloroplast porins have channels that are large enough to allow free passage of all the metabolites entering and leaving the organelles but still small enough to exclude larger molecules such as enzymes.

Transport Through the Bacterial Plasma Membrane

Two major systems transport substances across bacterial plasma membranes. One system, found primarily in gram-negative bacteria, includes the active transport pumps working in conjunction with selective porins. These pumps, built up from multiple polypeptide subunits, transport only the organic substances admitted by the selective porins. The second system, characteristic of all bacteria, transports both ions and organic substances. In general the second class, which includes both passive and active transporters, is formed from single polypeptides unrelated in structure to the multiple-subunit carriers of the first class.

Pumps Occurring in Conjunction with Selective Porins The pumping systems associated with selective porins in gram-negative bacteria, which transport amino acids, sugars, and short polysaccharides and peptides into the cell, contain several components

(Fig. 5-23). One component is the selective porin itself. The second component is a soluble *periplasmic protein* located in the periplasmic space between the outer and plasma membranes. Each transport system has its own periplasmic protein, which combines with the substance admitted by the porin associated with the system and greatly increases the efficiency of its transport. The final component of the system is the active transport pump embedded in the plasma membrane. Most of the active transport pumps of these systems, termed *periplasmic protein-related transporters*, consist of four polypeptide subunits. Two of the subunits are transmembrane proteins, each with six transmembrane segments, and two are hydrophilic subunits located on the cytoplasmic side of the plasma membrane. In various members of this group the two membrane subunits or all four subunits may be linked into a single polypeptide. Each of the cytoplasmic segments has a site binding and hydrolyzing ATP. The membrane subunits, which determine the substance transported, show little similarity in amino acid sequence among different members of this group. The cytoplasmic subunits hydrolyzing ATP, in contrast, are highly conserved in structure among different transporters. The closely related amino acid sequence segment of the cytoplasmic subunit is often termed the *ATP-binding cassette (ABC)*, and the carriers of the group are called *ABC transporters*.

One of the best-characterized periplasmic protein-dependent transporters is the maltose-transporting system of *E. coli*. The porin of the maltose system, encoded in the *lamB* gene, is highly selective for maltose and small polysaccharides with up to six or seven maltose units. The periplasmic protein for the maltose system, encoded in the *malE* gene, is a 370-amino acid polypeptide with one binding site that specifically recognizes maltose sugars. When bound to a sugar molecule, the MalE protein diffuses across the periplasmic space and in some as yet unknown manner greatly speeds uptake of the sugar by the active transport pump in the plasma membrane. Presumably, combination with the sugar activates a binding site on the periplasmic protein that specifically recognizes the active transport pump. Binding between the periplasmic protein and the active transport pump releases the sugar in a position that greatly favors its entry into the pump channel. Experimental elimination of the periplasmic protein markedly slows active transport of maltose sugars, confirming its importance to the transport system.

The active transport protein, like other active transport pumps, undergoes conformational changes

Figure 5-23 The periplasmic active transport system of *E. coli* and other gram-negative bacteria, including a selective porin in the outer membrane, the periplasmic protein, and the active transport pump of the inner membrane. The system transporting maltose sugars in *E. coli* is shown.

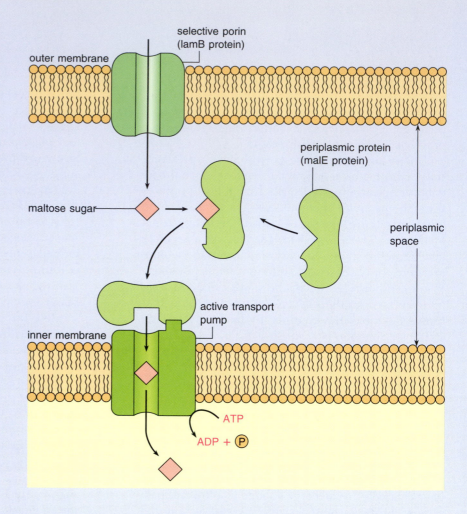

that alternately expose the maltose-binding site on the outside and inside surfaces of the plasma membrane as the pumping cycle turns (as in Fig. 5-10). Energy for the pumping cycle is derived from hydrolysis of ATP or other phosphates. If this energy source is eliminated in experimental systems, pumping stops. Other porin-associated active transport systems work similarly to the maltose system in *E. coli*.

Members of the ABC group have recently been discovered in gram-positive bacteria. Although no porins occur in these systems, proteins equivalent to the periplasmic carriers are present. In gram-positive bacteria the periplasmic proteins stimulating transport of selected molecules are anchored at one end to the outer surface of the plasma membrane by a "greasy foot" (see p. 104).

Several active transport pumps recently discovered in humans, yeast, *Drosophila*, a plant chloroplast, and the protozoan responsible for malaria have structural relationships to the bacterial ABC pumps. Each of these pumps, although structured from a single protein, has two ABC subunits with amino acid sequences corresponding to the ABC subunits of peri-

plasmic protein-dependent pumps of bacteria. One of these eukaryotic active transport pumps, the *MDR pump* (MDR = *MultiDrug Resistance*), appears in increased numbers in the plasma membranes of human cells exposed to a wide variety of toxic substances. Unfortunately the toxic substances removed from the cytoplasm by the pump include drugs used to treat cancer. The MDR pump frequently nullifies the effect of anticancer drugs by transporting them rapidly out of tumor cells. A *CQR pump* (CQR = *ChloroQuinone Resistance*) recently discovered in the malaria parasite has a similar function, allowing the parasite to pump out the chloroquine drug used to treat malaria.

Another human transporter related to the same family of active transport pumps is the CFTR protein encoded in the *CF* gene, which in mutant form is responsible for defects in chloride transport associated with cystic fibrosis (see p. 128). Although related transporters are active pumps, CFTR apparently operates as an ATP-gated channel passively admitting Cl^-.

Other Bacterial Transporters The remaining bacterial transport systems occur in the plasma membranes of both gram-negative and gram-positive bacteria. The active transporters of this group are single polypeptides that may use phosphate bond energy, H^+ or Na^+ gradients, or a combination of both gradients and phosphate hydrolysis to drive substances across the plasma membrane. The various pumps may transport ions or organic substances. Although ion transport may be direct or indirect, most carriers of organic molecules are cotransporters that use the H^+ ion gradient as their energy source.

Ion Transport Active transport systems have been detected in *E. coli* and other bacteria that push K^+ and Na^+ across the plasma membrane. K^+ is concentrated inside bacterial cells by a transporter with clear structural and functional affinities to the P-type pumps of eukaryotes. This K^+-ATPase directly uses ATP as its energy source and binds a phosphate group derived from ATP during the pumping cycle. The K^+ driven inward by the bacterial K^+-ATPase provides potassium necessary for metabolic reactions and maintains internal osmotic pressure at positive values.

Na$^+$ may be pushed inward or outward depending on conditions. Active inward movement is driven by an Na^+-ATPase pump that directly uses ATP as its energy source; active outward movement takes place by an antiporter driven by the H^+ gradient.

Bacteria also have a series of ATP-driven active transport pumps that specifically transport toxic ions out of the cell. Pumps in this group, collectively known as *resistance ATPases*, expel such toxic ions as arsenate, arsenite, and antimony ions. The resistance ATPases, which provide part of the environmental defenses of both gram-positive and gram-negative bacteria, form a family of structurally related pumps that has no eukaryotic counterparts and shows no sequence relationships to other classes of ATP-driven active transport pumps in bacteria.

Transport of Organic Substances Inward movement of organic molecules, primarily sugars and amino acids, can take place either passively or actively in bacteria. Passive transporters of these substances can be detected in the plasma membrane when their substrates exist in plentiful supply in the medium. The active transporters of organic molecules are cotransporters; although some are Na^+-dependent, most use the H^+ gradient established by electron transport as their energy source. In *E. coli*, cotransporters collectively able to concentrate all 20 amino acids in the cytoplasm have been detected. Some of these pumps are specific for single amino acids; others carry groups of amino acids with related structures.

H^+-driven pumps for sugars such as lactose, arabinose, galactose, and xylose also occur in *E. coli*. All these sugar transporters have similar amino acid sequences capable of forming 12 alpha-helical transmembrane segments. As noted, the sugar transporters of *E. coli* also share sequence homologies with the protein moving glucose into erythrocytes and other mammalian cells by facilitated diffusion (see p. 126).

In many bacterial transport systems, sugars are phosphorylated as they enter the cytoplasm. As in eukaryotic systems (see p. 126), this phosphorylation effectively removes sugar from the concentration gradient because the phosphorylated types cannot pass through the membrane.

Light-Driven Active Transport in Photosynthetic Bacteria The plasma membranes of one bacterial group, the purple photosynthetic bacteria (see p. 271) possess an active transport pump that uses light energy to push H^+ from the cytoplasm to the cell exterior. The protein, *bacteriorhodopsin* (see Fig. 4-5), contains a light-absorbing organic group identical to *rhodopsin*, the characteristic visual pigment of animal cells. Absorption of light by rhodopsin induces a conformational change in the protein that results in expulsion of a hydrogen ion across the plasma membrane. The bacteriorhodopsin system supplements the H^+ gradient established by oxidative electron transport in the purple bacteria, especially under conditions in which the oxygen necessary for electron transport is in limited supply.

Bacteriorhodopsin has been successfully isolated and placed in operating condition in artificial phospholipid films. These isolated bacteriorhodopsin systems have been critical to experiments demonstrating that an H^+ gradient, set up normally by electron transport in bacteria, mitochondria, and chloroplasts, can be used as a direct energy source for ATP synthesis (see p. 233 for details).

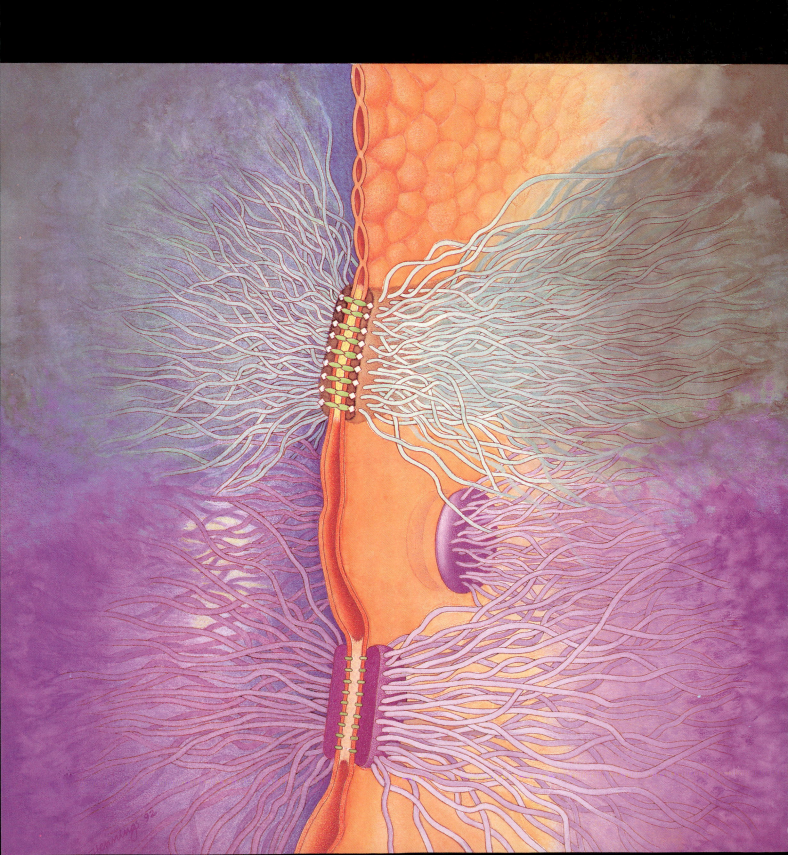

C ells make contact with the outside world through the cell surface, which includes the plasma membrane, its lipid and protein molecules, and parts of these molecules that extend from the membrane into the extracellular medium. Through the activities of this contact layer, cells recognize other cells as part of the same individual or as foreign, send and receive chemical and physical signals, and adhere to other cells or to extracellular materials.

The cell surface is specialized in various ways to carry out these tasks. Cell-to-cell recognition is based on glycolipid and glycoprotein molecules of the plasma membrane that, in effect, provide cells with a name and address. These molecules function in development, maintenance of cell associations in adults, and immune recognition and rejection.

Intercellular signaling is based on other membrane glycoproteins acting as *receptors* that can recognize and bind a signal molecule such as a peptide hormone or neurotransmitter. Binding the signal triggers internal responses that range from adjustments in metabolic reactions to changes in the rate of secretion and cell division. The signals, surface receptors, and internal response mechanisms are the primary elements coordinating cell growth and activity in animals.

Other glycoproteins are specialized for cell adhesion. Some of this function depends on the interaction of surface molecules on adjacent cells or between cells and extracellular materials. Other adhesions depend on highly organized arrangements of membrane proteins, glycoproteins, and lipids into *cell junctions* of various kinds. The junctions anchor cells to each other, seal cell boundaries to the flow of ions and molecules, or form channels for direct transport and communication between cells.

The cell surface of eukaryotes and its role in recognition, intercellular signaling, and adhesion, including the various types of cell junctions, are the subjects of this chapter. Surface receptors also act in linkage of cells to the extracellular matrix and in receptor-mediated endocytosis, in which molecules specifically recognized by receptors are bound to the cell surface and taken into the cytoplasm in vesicles. These processes are covered in later chapters—linkage of cells to the extracellular matrix in Chapter 7 and receptor-mediated endocytosis in Chapter 20.

CELL RECOGNITION

The cell surfaces of many organisms, particularly in the animal kingdom, contain glycolipid and glycoprotein molecules that serve as sites of recognition (Information Box 6-1 outlines the structure of cell surface glycoproteins). Some of these surface molecules identify cells as belonging to a single individual or to a particular type of tissue. Other surface glycoproteins function as receptors that have the capacity to recognize other cells and to identify them as part of the same or a foreign individual. In vertebrates, recognition of surface molecules as foreign triggers an immune response that normally destroys the invading cell. Rejection of organ transplants in mammals, for example, results from recognition of foreign surface molecules by cells of the immune system, followed by an immune response that kills the transplanted cells. The reactions of humans to blood transfusions also depend on recognition of cell surface molecules.

Major Histocompatibility Complex (MHC) Molecules

The group of molecules primarily responsible for recognition of mammalian cells as part of self or foreign is termed the *major histocompatibility complex* (*MHC*; Fig. 6-1). Although now applied to the cell surface molecules themselves, the name originally referred to the large and complex group of genes encoding these molecules. MHC molecules occur in two different types, *class I* and *class II MHC* (see Fig. 6-1). In mammals, class I MHC molecules occur on the surfaces of all body cells except those of the immune system. Cells producing an immune response—primarily leukocytes (white blood cells) of various kinds—have class II MHC proteins on their surfaces. Both MHC types contain constant and variable sequence regions. The constant regions are closely similar or identical between different individuals of a species. The variable regions differ significantly in amino acid sequence, so that no two individuals (except identical twins) are even remotely likely to have the same MHC proteins. The differences arise from random combinations of a large number of alleles (see p. 522) of the MHC genes.

The primary role of the class I MHC proteins is "presentation" of foreign molecules (antigens) to cells of the immune system (see p. 560 for details). In this presentation the antigen, usually a polypeptide, is taken into a body cell bearing class I MHC molecules and broken into fragments. The fragments then appear on the cell surface in combination with class I MHC molecules. Recognition of the class I MHC-antigen combination stimulates a *T-cell*, a specialized leucocyte bearing class II MHC molecules, to initiate an immune response against the antigen. If the cell

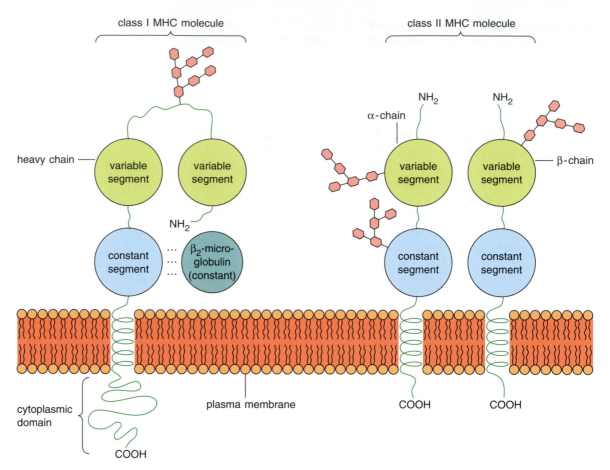

Figure 6-1 The class I and class II MHC surface molecules of mammalian cells (see text). Both glycoproteins consist of two polypeptide chains. The extracellular portions of each molecule consist of two constant and two variable domains. The constant domains are similar in amino acid sequence to the constant domains of other members of the immunoglobulin superfamily. The variations in amino acid sequence responsible for differences in MHC molecules between different individuals are concentrated in the variable domains. Class I MHC consist of a heavy chain forming three of the four major domains of the molecule, with an additional small polypeptide, β_2-*microglobin*, that completes the structure. Class II molecules are built up from two polypeptides of approximately equal size, both of which contain one constant and one variable domain.

bearing the class I MHC molecule is from the same individual, the immune response is mounted only against the antigen (unless the antigen is derived from a virus infecting the cell). If the cell bearing the class I MHC-antigen combination is from a different individual, as it might be as a result of an organ or tissue transplant, both the antigen and MHC are recognized as foreign and an immune response is mounted against the cell as well as the antigen. The combination of a class I MHC molecule with an antigen derived from an infecting virus is also likely to lead to destruction of the cell bearing the marker, even if it is from the same individual.

The class I and II MHC molecules and antibodies are related in amino acid sequence and structure (compare Figs. 6-1 and 19-1), indicating that all evolved from common ancestral genes. Several other cell surface glycoproteins, including some of the molecules involved in cell-cell adhesions, are also structurally related to antibodies and form part of a large group of cell surface glycoproteins known as the *immunoglobulin superfamily*. (Further details of the immunoglobulin superfamily are presented in Chapter 19.)

Blood Groups and Other Surface Markers

The glycolipid and glycoprotein markers responsible for blood groups in humans and other vertebrates provide another important example of surface recog-

Glycolipids and Glycoproteins of the Cell Surface

Most glycolipids and glycoproteins active in recognition, reception, and adhesion consist of a basal phospholipid or protein structure embedded within the plasma membrane and one or more carbohydrate chains that extend from the cell surface. (For details of glycolipid and glycoprotein structure, see Chapters 2 and 5.) The polypeptide units of the glycoproteins typically have domains that protrude from both the exterior and cytoplasmic sides of the plasma membrane, connected by one or more alpha-helical transmembrane segments.

The monosaccharides commonly linked into membrane glycolipids and glycoproteins (see Fig. 4-4) include glucose, galactose, mannose, and fucose, the *amino sugars* glucosamine and galactosamine, and *sialic acid*. The two amino sugars consist of a glucose or galactose unit in which one or more of the —OH groups has been replaced by an amino (—NH$_2$) group. Individual amino sugars differ in the positions of the amino groups and side groups at other points on the molecules. Sialic acid is based on a seven-carbon framework with a —COOH group attached to the 1-carbon (see Fig. 4-4), which gives the sugar its acidic properties. Various side groups attached to the carbons at other points form more than 20

different types of sialic acid. Different membrane glycolipids and glycoproteins may bear single or multiple carbohydrate segments, in straight or branched chains ranging in length from a single monosaccharide unit to chains of 70 or more.

Although the functions of the complex carbohydrate groups of membrane glycoproteins and glycolipids are incompletely understood, they probably provide part of the specific structure allowing these molecules to act in recognition, reception, and adhesion. In addition, hydrogen bonds set up between carbohydrate groups probably stabilize the membrane surface.

Added to the carbohydrate variability are differences in the amino acid sequences in the protein portion of cell surface glycoproteins. These two sources of variability provide so many possible combinations that the potential number of different membrane glycoproteins is essentially infinite. Thus the carbohydrate structures of membrane glycolipids and both the carbohydrate and protein segments of membrane glycoproteins easily provide the variability required for the diverse functions of the cell surface.

nition groups. *Glycophorin*, a surface glycoprotein of red blood cells (erythrocytes), carries the markers responsible for the MN blood groups of humans; evidently both the protein and carbohydrate portions of glycophorin contribute to the MN markers.

The ABO blood groups, in contrast, depend on small differences in carbohydrate structures attached to both glycolipids or glycoproteins in erythrocyte plasma membranes. Persons with type A blood have the amino sugar N-acetylgalactosamine at one position in the carbohydrate complex of these cell surface molecules. (Fig. 2-15c shows this form of the carbohydrate structure.) Persons with type B blood have a galactose sugar unit at the same position. Individuals with type AB blood have carbohydrate groups of both types on their red blood cells. In persons with type O blood the corresponding position in the carbohydrate group is empty and contains neither sugar. The same ABO group of cell surface markers also occurs on epithelial cells.

Similar surface recognition groups are involved in sperm-egg interactions in animals and in recognition of mating types between reproductive cells in al-

gae and fungi. Surface recognition groups in higher plants are involved in graft acceptance or rejection and in the recognition of symbiotic bacteria such as *Rhizobium*, which stimulates the formation of root nodules in peas, clover, and other legumes.

SURFACE RECEPTORS AND INTERCELLULAR SIGNALING

The surface receptors involved in eukaryotic cell-to-cell signaling are membrane glycoproteins that can recognize and bind signal molecules circulating in the extracellular medium. The signal molecules, including peptide hormones, growth factors, and neurotransmitters, are released by one group of cells and bound by receptors of another cell group. Binding by the receptors triggers a complex internal response in the cells receiving the signal, such as an increase or decrease in the rate of transport, secretion, oxidative metabolism, initiation of cell division, or cell movement. (Supplement 6-1 describes the sensory-response mechanisms of bacteria.)

The Characteristics of Receptor-Response Signaling Pathways

The surface receptors binding extracellular signal molecules have all been identified as membrane-spanning glycoproteins that have segments extending from both sides of the plasma membrane. The segment extending from the outer membrane surface contains the site that recognizes and binds a hormone or neurotransmitter acting as an extracellular signal. In response to binding, the receptor undergoes a conformational change that activates a site at the end of the protein extending into the cytoplasm. The cytoplasmic site usually catalyzes or initiates a reaction that serves as the first step in a series of interactions leading to the cellular response.

The cell surface may contain from hundreds to thousands of individual receptor molecules. Receptors for different peptide hormones, for example, may number from 500 to as many as 100,000 or more per cell, with 10,000 to 20,000 per cell being typical. Different cell types contain distinct combinations of receptors, allowing them to react individually to the group of hormones and neurotransmitters circulating in the bloodstream or extracellular fluids. The combination of surface receptors on particular cell types is not fixed, but changes as cells develop. Changes in receptors also frequently occur as normal cells are transformed into cancer cells.

Receptor-response mechanisms initiated at the cell surface have several characteristics that give important clues to their operation. First, the peptide hormone, growth factor, or neurotransmitter acting as an extracellular signal does not penetrate into the cell to induce a response; binding to a surface receptor is sufficient. Second, no response is produced if the signal molecule is directly injected into the cytoplasm. Thus, involvement of the receptor is necessary for a response to occur. Third, the receptor remains in position in the plasma membrane while the cellular response is initiated; it does not enter the cell to initiate the response.

These characteristics indicate that the extracellular signal has no function other than binding to the receptor at the cell surface. This conclusion is confirmed by the observation that binding unrelated molecules to a receptor can often induce a full cellular response in the absence of the hormone or growth factor normally bound. One series of experiments demonstrating this feature used *lectins*, a group of glycoproteins that can bind to a wide spectrum of cell surface molecules. Frequently lectins binding to a surface receptor trigger the same internal response as the corresponding extracellular signal molecule. A full cellular response can also be produced by antibodies developed against receptors.

Other important characteristics reflect operation of the internal response mechanisms triggered by the receptors. Although cells may have many different receptor types, stimulated by a variety of signal molecules, their internal response mechanisms follow a relatively few pathways that share several features in common.

One shared feature is the means by which internal cellular responses are produced and controlled. This is through *protein kinases*, enzymes that attach phosphate groups derived from ATP to specific target proteins. Addition of the phosphate groups inhibits or stimulates the activity of the target proteins in carrying out a cellular reaction. Among the target proteins are: (1) enzymes carrying out critical steps in metabolic pathways, (2) transport proteins such as ion channels, (3) ribosomal proteins, (4) proteins regulating gene activity, and (5) the receptors themselves. The type of cellular response produced depends on the types of protein kinases present in a cell, controlled ultimately by the systems regulating gene transcription in the cell nucleus.

In a relatively few systems the protein kinase activity is part of the receptor itself, built into the receptor segment extending into the cytoplasm. More commonly the protein kinases are separate molecules working in the final steps of a reaction sequence activated by the receptor. These enzymes may be associated with membranes or suspended in solution in the cytoplasm or nucleus.

In all receptor-response pathways the phosphate groups added by protein kinases can be removed by the activity of a varied group of enzymes known collectively as *protein phosphatases*. These enzymes counterbalance the activity of protein kinases by continually removing phosphate groups from target proteins. The initial discovery of control of protein phosphorylation by protein kinases, and evidence showing that protein phosphatases reverse these phosphorylations, was developed by two scientists, Edwin Krebs and Edmond Fischer, in experiments begun in the 1950s. Krebs and Fischer received the Nobel Prize in 1992 for their pioneering research.

As receptor-response mechanisms run their course, the receptors and their bound signal molecules are removed from the cell surface by endocytosis. Both the receptor and its bound signal molecule may be degraded in lysosomes. Or the receptors may be separated from the signals and recycled to the cell surface for reuse. (See Fig. 6-7 for details of these processes.)

Receptors with Integral Protein Kinase Activity

Although relatively few receptors have integral protein kinase sites, the cellular responses controlled by these receptors are among the most fundamental processes of animal cells. One receptor of this group recognizes and binds the peptide hormone *insulin*. This hormone controls glucose uptake and the rate of many metabolic reactions in its target cells. Other receptors of this group bind *epidermal growth factor* (*EGF*) or *platelet-derived growth factor* (*PDGF*), both of which regulate cell growth and division in higher animals.

All receptors in this group are related proteins with similarities in amino acid sequence and structure (Fig. 6-2). The extracellular portion of the receptor contains the binding site for the extracellular signal. The cytoplasmic end of the receptor contains the protein kinase site. The extracellular and cytoplasmic ends are connected by a single alpha helix that spans the membrane. The protein kinase site on the cytoplasmic side of the membrane is inactive when the extracellular site is unbound (Fig. 6-2a). Binding a hormone or growth factor at the cell surface induces a conformational change in the receptor that is transmitted in some unknown way through the alpha helical segment and activates the protein kinase site on the cytoplasmic end (Fig. 6-2b). Generally the receptors in this group link by twos, or *dimerize*, as they combine with signal molecules. (The insulin receptor is a permanent dimer.) An interaction between the two transmembrane helices of the receptor dimer, such as a relative twist or pistonlike movement, may contribute to transmission of the signal through the plasma membrane.

The activated protein kinase site on the cytoplasmic end of the receptor, which adds phosphate groups to tyrosine residues, may phosphorylate the cytoplasmic end of the receptor itself or add phosphate groups directly to tyrosines of target proteins in the cytoplasm. Self-phosphyrolation of the receptor, or *autophosphorylation* as it is called, provides the first step in a pathway of reactions leading to activation of other protein kinases in the cytoplasm. The proteins targeted for phosphorylation, either directly by the receptor protein kinase or by other kinases activated indirectly by pathways linked to the receptor, are usually enzymes controlling key steps in metabolic pathways or proteins controlling gene activity in the nucleus. Addition of the phosphate groups may either stimulate or inhibit activity of the target proteins.

The indirect pathway initiated by autophosphorylation of the receptor, recently pieced together for the EGF receptor by a large group of investigators including J. Downward, G. Rubin, J. Schlessinger, R. Weinberg, M. Wigler, and others, involves a protein knowns as *Ras*. This protein acts as a critical on/off switch in the pathway. Ras binds and breaks down the nucleoside triphosphate GTP (see Fig. 2-26). When GTP is bound, Ras is turned on; in this state, Ras acti-

vates protein kinases leading to phosphorylation of target proteins. The bound GTP, however, is quickly hydrolyzed by Ras itself to GDP; when GDP is bound, Ras is turned off and the protein kinases following Ras in the pathway are inactivated. It remains locked in the off state until the GDP is released, opening its binding site to occupancy by another molecule of GTP.

Ras occupies the third position in a series of steps leading from the autophosphorylated receptor (Fig. 6-2c). In the first step the phosphate groups bound to the receptor are recognized and bound by an *adaptor* protein. The combination of adaptor and autophosphorylated receptor activates a second protein known as the *guanine nucleotide release factor* (*GRF*; step 2 in the pathway). When activated, GRF removes GDP from Ras, enabling Ras to bind GTP and cycle to the on state (step 3). In the on state, Ras activates one or more protein kinases leading to phosphorylation of target proteins (step 4).

As long as the receptor remains autophosphorylated (or, in other words, as long as the receptor is bound to the extracellular signal molecule), GRF is active and continually removes GDP from Ras. This allows Ras to bind another GTP and keeps the pathway switched on. However, Ras's ability to hydrolyze its bound GTP to GDP quickly switches it, as well as the linked protein kinases, to the off state if the receptor becomes inactive or is removed from the cell surface by endocytosis. Ras's built-in GTPase activity thus automatically shuts down the pathway if the receptor becomes inactive.

Proteins related to Ras have been detected in eukaryotes from yeast to humans and higher plants. Human Ras, in fact, will work successfully as a substitute in mutant yeast cells with faulty Ras proteins. The presence of Ras in yeast and higher plants such as maize and the discovery of relatives of the EGF, PDGF, and insulin receptors in organisms such as *Drosophilia* indicate that this receptor family has roots expanding back to the origins of eukaryotic cells. Another indicator of the importance of the receptors and pathways in this family is that many genes promoting cancer (*oncogenes*; see p. 641) are altered forms of genes encoding either the EGF or PDGF receptors. The faulty receptors encoded by the oncogenes have protein kinase sites that are continually active, whether the growth factor is bound or not. Mutant forms of Ras with defective GTPase activity that keeps the protein in the on state also figure importantly in many types of cancer. Ras, in fact, gets its name from a tumor type (*rat* sarcoma) in which the protein was first detected in mutant form.

Hereditary defects in the insulin receptor are responsible for some forms of diabetes, a disease in which glucose cannot be taken up in sufficient quantity by body cells and accumulates in the bloodstream.

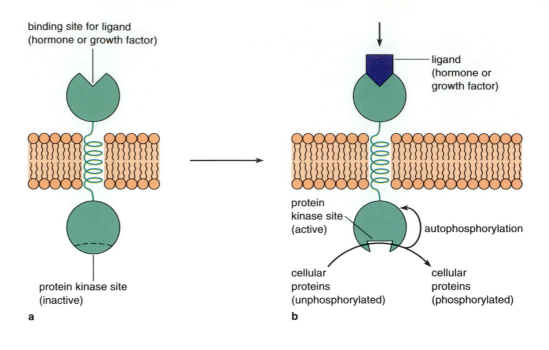

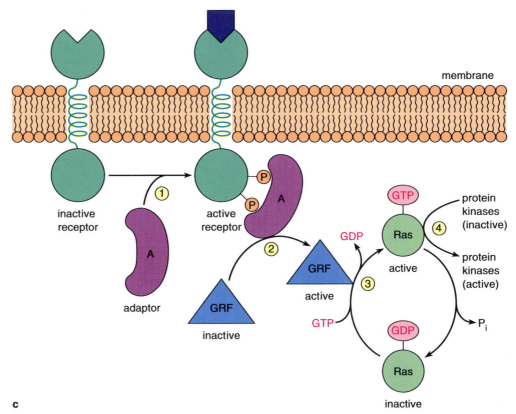

Figure 6-2 Receptors with integral protein kinase activity. **(a)** When the extracellular site is unbound, the protein kinase site on the cytoplasmic side of the membrane is inactive. **(b)** Binding a hormone or growth factor at the cell surface induces a conformational change that is transmitted through the transmembrane segment and activates the protein kinase site, which adds phosphate groups to tyrosine residues in the target proteins. **(c)** Pathway linking receptors in this group indirectly to other protein kinases through the Ras protein (see text). GRF, guanine nucleotide release factor; P$_i$, inorganic phosphate.

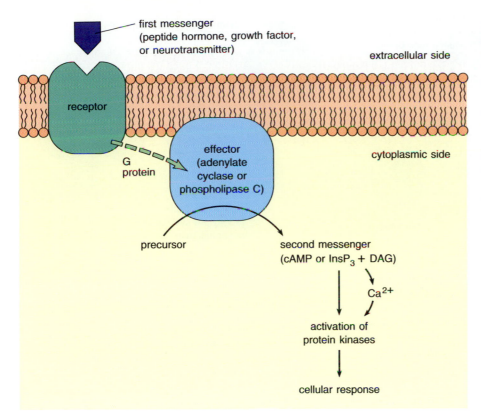

first messenger
(peptide hormone, growth factor,
or neurotransmitter)

receptor

extracellular side

G
protein

effector
(adenylate
cyclase or
phospholipase C)

cytoplasmic side

precursor

second messenger
(cAMP or InsP$_3$ + DAG)

Ca^{2+}

activation of
protein kinases

cellular response

Figure 6-3 Receptor-response systems in which the effector is separate from the receptor. Binding the first messenger activates the receptor. By a series of steps involving G proteins the receptor activates the effector. The effector is an enzyme, either adenylate cyclase or phospholipase C, that converts precursor substances into the second messengers cyclic AMP (cAMP) or inositol triphosphate (InsP$_3$) and diacyl glycerol (DAG). Ca^{2+}, released into the cytoplasm by InsP$_3$, acts as a supplementary second messenger. The second messengers, in turn, activate one or more protein kinases that add phosphate groups to enzymes or other molecules. The phosphorylations, by promoting or inhibiting enzymatic and other activities, produce a cellular response.

Usually, relatively minor defects in the receptor impair but do not completely interfere with insulin binding or kinase activity. In rarer instances a mutation eliminates most of the protein kinase domain from the receptor, so that the enzymatic segment is inactive even though the receptor binds insulin. In other forms of diabetes, insufficient quantities of insulin are released to the body circulation because insulin production by the pancreas is faulty.

Receptors with Separate Protein Kinase Activity

Receptors without built-in protein kinase activity are much more varied and numerous than integral receptors. The receptors in this group are complex proteins with seven transmembrane segments. Binding an extracellular signal molecule, termed the *first messenger* in these pathways, activates an enzymatic site on the cytoplasmic end of the receptors. The cytoplasmic site catalyzes the first step in a series of separate reactions leading to activation of one or more protein kinases (Fig. 6-3).

The primary reaction series triggered by the activated receptors follow one of two distinct but similar pathways. In both, the first step catalyzed by the receptor is activation of a *G protein*. In turn the G protein activates an enzyme known as an *effector*. The role of the effector is to generate one or more *second mes-*

sengers, small molecules that activate protein kinases. The protein kinases activated by the pathways add phosphate groups to serine or threonine residues in their target proteins. The overall pattern of these steps was first proposed by E. W. Sutherland, who received the Nobel Prize in 1971 for his elucidation of receptor-response mechanisms.

A great variety of protein kinases activated by the pathways have been identified in eukaryotes, so many that in humans, for example, as much as 1% of the DNA is estimated to encode different versions of these important enzymes. Although highly diverse in structure and targets, all share a similar core segment of about 260 amino acid residues that forms the active site transferring phosphate groups from ATP to target proteins.

G Proteins of the Two Major Pathways The G proteins, first detected by M. Rodbell and his colleagues, serve as the critical link between activated receptors and the effectors generating second messengers. So called because they bind the nucleotides GTP and GDP (see p. 63), G proteins consist of three polypeptides called the α-, β-, and γ-*chains*. All three chains are associated with the cytoplasmic side of the plasma membrane as peripheral proteins.

Figure 6-4 shows how G proteins, which show structural and functional similarities to Ras, set up

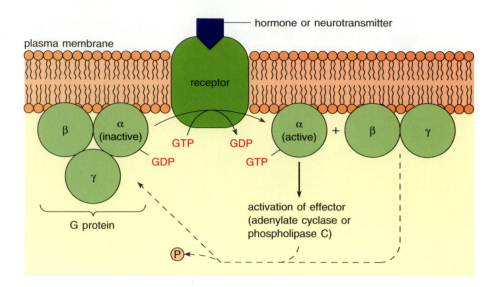

Figure 6-4 The G-protein cycle in second-messenger pathways. The receptor, when bound to the first messenger, activates the α-chain of the G protein by catalyzing the exchange of GTP for GDP on the α-chain. The activated α-chain then releases from the β- and γ-subunits of the G-protein and, in turn, activates the effector. Hydrolysis of GTP to GDP, catalyzed by an enzymatic site forming part of the α-chain, inactivates the α-chain. The inactivated α-chain rejoins the complex with the β- and γ-chains to complete the cycle.

functional links between receptors and effectors in the pathways. The α-chain contains the site binding GTP or GDP and also acts as the part of the G protein directly activating the effector. When GDP is bound, the α-chain has strong affinity for the β- and γ-chains. As a result, the three subunits of the G protein associate tightly and the α-chain is inactive. When the receptor binds a first messenger, the enzymatic site on its cytoplasmic side catalyzes the exchange of GTP for the GDP bound to the α-chain of the G protein. GTP activates the α-chain and causes its release from the β- and γ-chains. In this state the α-chain activates the effector.

The α-chain, like Ras, can hydrolyze as well as bind GTP. As a result, the GTP bound to the α-subunit is hydrolyzed to GDP very quickly after the GDP-GTP exchange catalyzed by the receptor. This converts the α-subunit back to the conformation binding the β- and γ-subunits and inactivates the G protein and the effector. As long as the receptor remains active, its catalytic site catalyzes another exchange of GTP for GDP. This switches the α-chain back to its active form, and the effector is activated. However, if the receptor becomes inactive, automatic conversion of the α-chain to the inactive form quickly shuts down the pathway. As in the Ras-based pathway, the ability of the α-chain to hydrolyze its own GTP thus supplies an automated "off switch" that stops the response if the receptor becomes inactive. Each of the two major receptor-response pathways has its own set of distinct G proteins.

G proteins have been detected in both vertebrate and invertebrate animals and in higher plants. This wide distribution indicates that they arose early in the evolution of multicellular life, probably as an adaptation promoting the communication between cells necessary for coordination of cellular activity.

Effectors and Second Messengers of the Two Major Pathways Although the two major receptor-response pathways both generate second messengers to activate their protein kinases, the effectors and second messengers are different. The second messenger of one pathway is *cyclic AMP* (*cAMP*), generated from ATP by the effector *adenylate cyclase* (Fig. 6-5). cAMP is a relatively small, water-soluble substance that can diffuse rapidly through the cytoplasm to activate a series of protein kinases. The other pathway generates two second messengers by breakdown of a membrane phospholipid, *phosphatidyl inositol* (see Fig. 6-6 and p. 95), in a reaction catalyzed by the effector *phospholipase C*. One of the two second messengers produced by phosphatidyl inositol breakdown, *inositol triphosphate* (*InsP₃*), is another relatively small, water-soluble molecule consisting of the inositol unit with three added phosphate groups. The other second messenger, *diacylglycerol* (*DAG*), produced from the remainder of the membrane phospholipid, consists of a glycerol unit with two fatty acid residues. DAG has polar and nonpolar segments and remains suspended in the plasma membrane bilayer. Ca^{2+}, liberated through the activity of $InsP_3$, acts as a supplementary second messenger of this pathway. The two pathways are identified in this chapter by their primary second messengers as the *cAMP* and *InsP₃/DAG* pathways.

The second messengers of both systems are rapidly eliminated once formed. cAMP is quickly degraded by an enzyme, *cyclic nucleotide phosphodiesterase* (see Fig. 6-5b). $InsP_3$ is cycled rapidly through reactions that reassemble phosphatidyl inositol. DAG is degraded by lipases into glycerol and fatty acids or recycled into membrane lipids. Ca^{2+} is rapidly cleared from the cytoplasm by Ca^{2+}-ATPase active transport pumps in the plasma membrane and endoplasmic reticulum (ER). The rapid elimination of the second mes-

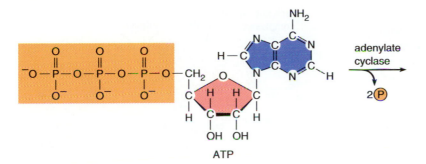

ATP

Figure 6-5 Synthesis and breakdown of cAMP. **(a)** The effector, adenylate cyclase, catalyzes conversion of ATP into the second messenger cAMP. **(b)** Breakdown of cAMP, catalyzed by cAMP phosphodiesterase. The breakdown produces 5'-AMP.

a

cyclic AMP (cAMP)   →   adenosine 5'-monophosphate

b

sengers provides another off switch for the pathways by ensuring that stimulation of protein kinases ceases almost instantly if the receptor becomes inactive. Still another off switch is provided by *serine/tyrosine protein phosphatases*, which remove the phosphate groups added to proteins by the protein kinases.

Activity of the receptor-response pathways is also terminated by endocytosis of receptors and their bound extracellular signals. Depending on the pathway, the receptors may be degraded in lysosomes after entry by endocytosis, or may be removed from their extracellular signals and recycled to the cell surface (Fig. 6-7).

Operation of the cAMP and InsP₃/DAG Receptor-Response Pathways

The cAMP Pathway A long and highly diverse list of receptors triggers the cAMP pathway in animals and primitive eukaryotes such as yeast. Among the many hormones or neurotransmitters acting as first messengers for this pathway in mammals and other higher vertebrates are *adrenaline (epinephrine), adrenocorticotrophic hormone (ACTH), glucagon, luteinizing hormone, parathormone,* and *acetylcholine.* The receptors binding these signals control such varied cellular processes as the uptake and oxidation of glucose, glycogen breakdown or synthesis, bone resorption, progesterone and thyroid hormone secretion, ion transport by voltage-gated Na^+, K^+, or Ca^{2+} channels, and transport of amino acids into the cell.

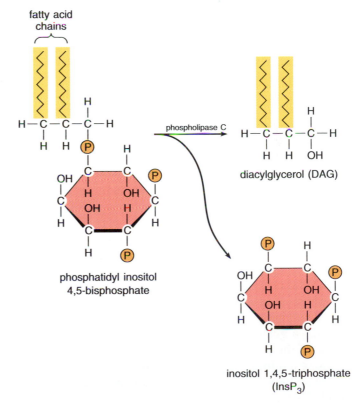

phosphatidyl inositol 4,5-bisphosphate

diacylglycerol (DAG)

inositol 1,4,5-triphosphate (InsP₃)

Figure 6-6 Synthesis of the second messengers InsP₃ and DAG by breakdown of phosphatidylinositol 4,5-bisphosphate, a phospholipid of the plasma membrane. The reaction is catalyzed by the effector phospholipase C.

The action of glucagon provides a well-studied example of a cAMP-based response pathway (Fig. 6-8). Glucagon is a peptide hormone secreted by the pancreas when blood sugar falls to low levels. Binding of the hormone by a receptor on a target cell in the liver (step 1 in Fig. 6-8) activates the receptor, which triggers the series of reactions leading through the G proteins and adenylate cyclase to generation of the second messenger cAMP (steps 2 to 4 in Fig. 6-8).

The protein kinases activated by cAMP make up the *protein kinase A* family (A = cAMP). Two enzymes are primary targets of the protein kinases A activated in the glucagon-triggered pathway. One, *phosphorylase kinase*, is activated by the added phosphate groups (step 6). Phosphorylase kinase is also a protein kinase (but not a protein kinase A). It adds a phosphate group to *glycogen phosphorylase* (step 7), an enzyme that catalyzes the breakdown of glycogen into glucose units (step 8). The added phosphate groups activate the enzyme; the glucose units liberated by glycogen breakdown pass from the liver cells into the bloodstream, raising glucose levels in the circulation. The second target protein of the protein kinase A activated by cAMP in the pathway is *glycogen synthase*, an enzyme that in active form catalyzes assembly of glycogen from glucose units. Phosphorylation by protein kinase A inactivates glycogen synthase (step 9), ensuring that the glucose units liberated by glycogen phosphorylase are not converted back into glycogen. Many other receptor-response pathways using cAMP as a second messenger work similarly, with the phosphorylations catalyzed by protein kinases A either activating or inhibiting their target proteins.

Whether cells respond to a hormone such as glucagon or a specific neurotransmitter or whether the pathway is stimulatory or inhibitory depends on the receptor type and the set of G proteins in the cell. Liver cells, for example, make glucagon receptors and insert them in their plasma membranes as part of the developmental pathways producing mature cells of this type. Because they have glucagon receptors, they are able to respond to this hormone when it appears in the blood circulation. Other cells, such as those of neural tissues, do not produce glucagon receptors and thus do not bind or respond to this hormone. In a cell carrying a given receptor the type of G protein activated by the receptor determines initial steps of the response pathway. Some G proteins, for example, stimulate membrane K^+ or Ca^+ channels to open instead of activating adenylate cyclase. Some G proteins may actually link the same receptor to cAMP-based pathways in one cell and to the $InsP_3$/DAG pathway in another cell, so that the same hormone may stimulate entirely different pathways in the two cell types. For example, receptors for glutamate, used as a neurotransmitter in the brain, may be linked to either re-

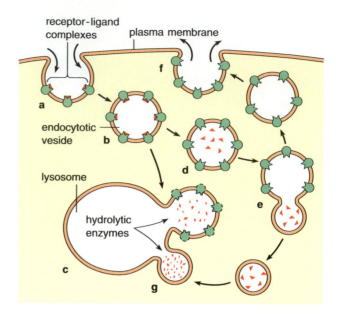

Figure 6-7 Elimination or recycling of receptor-ligand complexes by endocytosis. **(a)** Initially, the receptor-ligand complexes are collected into pockets that form in the plasma membrane. **(b)** The pockets then invaginate and pinch off as endocytotic vesicles that sink into the underlying cytoplasm. **(c)** After enclosure in the vesicles the receptor and the hormone or neurotransmitter are degraded by hydrolytic enzymes introduced into the vesicles from lysosomes (see p. 597). In some systems the receptors release the signal molecules into the vesicle interior **(d)**; following the release the empty receptors and ligands are sorted into separate vesicles **(e)**. The vesicles containing receptors return to the plasma membrane, where fusion of the vesicle with the membrane returns the receptors to the cell surface **(f)**. The vesicles containing the hormone or neurotransmitter fuse with lysosomes, in which the signal molecules are degraded **(g)**.

sponse pathway in this way. Most of this selectivity depends on the presence of different G_α subunits in the G protein, although in some systems the β and γ subunits may also stimulate or inhibit effectors.

In the cells responding to first messengers triggering cAMP pathways the proteins phosphorylated in the final steps depend on the types of protein kinase A present. As in the presence or absence of a given receptor on the cell surface, the kinds of protein kinase A vary from cell to cell and represent part of the development producing mature cell types. Thus whether cells respond to a hormone depends on the presence or absence of a receptor for the hormone at the cell surface. If a receptor is present, the character of the internal response depends on the G protein and protein kinase A types present in the cytoplasm and on the group of target proteins specifically activated or inactivated by the protein kinases.

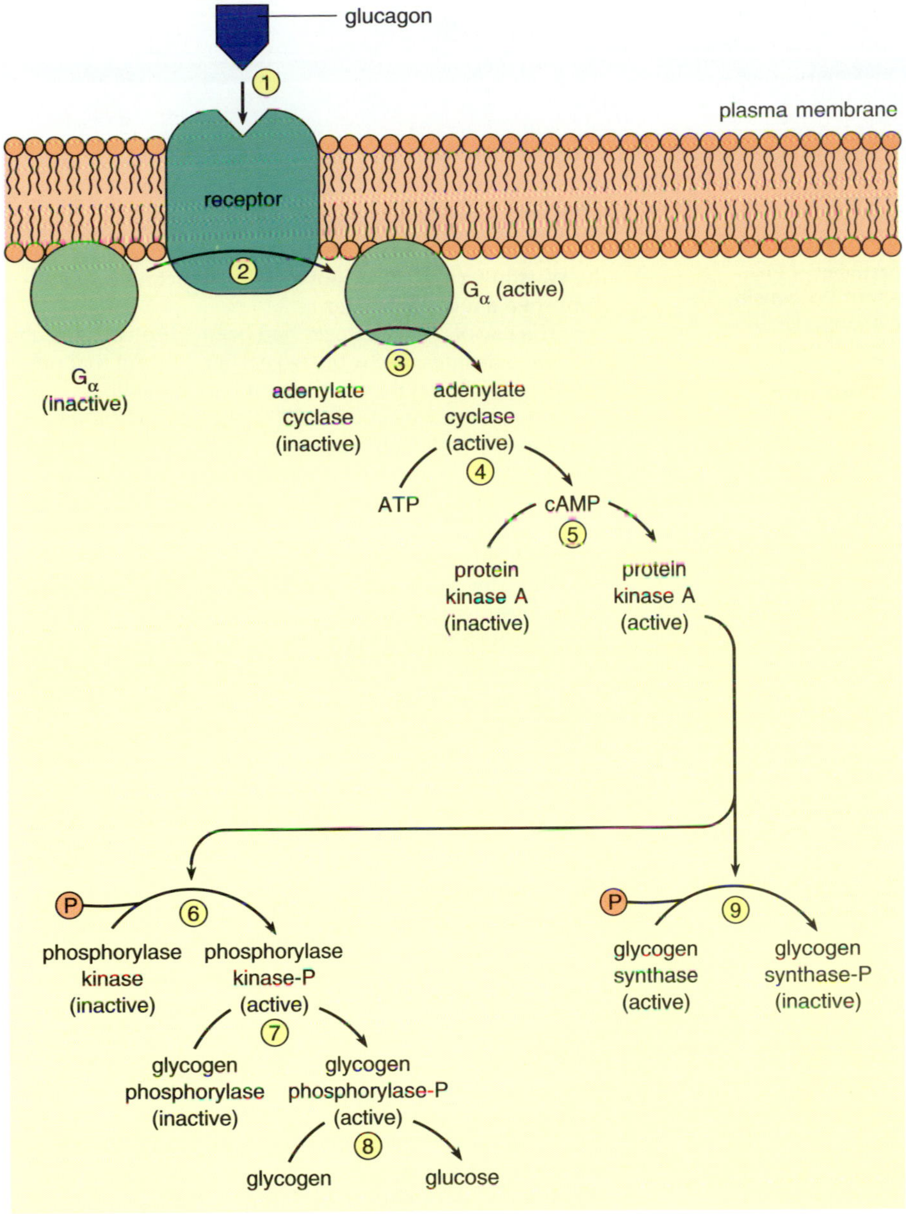

Figure 6-8   The cAMP-based pathway triggered in liver cells by the peptide hormone glucagon (see text).

G proteins, and the effector of the cAMP pathway, adenylate cyclase, occur in higher plants. However, the protein kinases through which cAMP exerts its effects in animal cells have not been discovered in plants; neither has any link been detected between cAMP concentrations and the known physiological reactions of plants. Therefore, although at least some elements of the cAMP-based pathway exist in plants, the degree to which this regulatory system works in plants is presently unknown.

A recently discovered group of receptors acts much like those using cAMP as second messenger, except that their cytoplasmic extension contains an enzymatic site with *guanylate cyclase* activity. On binding a ligand the guanylate cyclase site of these receptors converts cytoplasmic GTP to *cyclic GMP (cGMP)*,

a substance identical to cAMP except that guanine substitutes for adenine in its structure. Among other effects, cGMP acts in opposition to cAMP-based control pathways by activating a cGMP-dependent phosphodiesterase that breaks down cAMP (see below).

The InsP₃/DAG Pathway   The InsP$_3$/DAG pathway is universally distributed among eukaryotic organisms, including both vertebrate and invertebrate animals, fungi, and plants. The system, first detected by M. N. Hokin, L. E. Hokin, and R. Michell (see the Experimental Process essay by L. E. Hokin on p. 162), controls responses as varied as secretion of hormones and neurotransmitters by gland and nerve cells, cell division, early events in fertilization, sugar and ion transport, movements such as contraction of smooth

The Experimental Process

The Road to the Phosphoinositide-Generated Second Messengers

Lowell E. Hokin

LOWELL E. HOKIN is Professor and Chairman of the Department of Pharmacology, University of Wisconsin Medical School, Madison. He received his M.D. in 1948 at the University of Louisville School of Medicine and his Ph.D. under Hans Krebs at the University of Sheffield, U.K., in 1952. After postdoctoral work, he was Assistant Professor of Pharmacology at McGill University, moving to the University of Wisconsin in 1957. He was appointed to the Chairmanship of Pharmacology in 1968. His current emphasis is the mechanism of action at the molecular level of the antimanic drug lithium. His current interests are reading, swimming, and downhill skiing.

Neurotransmitters, growth factors, and hormones generally activate intracellular processes by combining with cell surface receptors. This interaction in turn activates membrane enzymes, usually via a GTP binding protein, to produce "second messengers" that diffuse to an intracellular target. In recent years, it has become clear that on binding of a variety of agonists (or ligands) to their receptors, inositol lipids in cell membranes, called phosphoinositides, are cleaved to inositol phosphates and diacylglycerol or DAG, which is the lipid backbone of phosphoinositides.

This is the story of the discovery of the phosphoinositide effect in the early 1950s, a "discovery before its time." Twenty years elapsed after our initial demonstration of agonist-stimulated phospatidylinositol (PI) turnover before Bob Michell was to rekindle interest in this effect—an effect that has answered (and posed!) so many questions about signal transduction mechanisms.

The story really begins in 1949, when I arrived at Hans Krebs' laboratory in Sheffield to study for a Ph.D. in biochemistry. Krebs' policy was to have a student formulate his own problem, pursue it independently with a minimum of supervision, and publish by himself. With some background in gastroenterology, which I had received in the laboratory of Warren S. Rehm while attending medical school, I looked into whether I could use pancreatic tissue as a model to study the synthesis and secretion of proteins, using amylase as a measure of these two processes.

Pigeon pancreas slices proved to be an excellent system for this purpose. Near the end of my doctoral studies in Krebs' laboratory, I became interested in the possible involvement of RNA in protein synthesis. This was based on some cytological studies in the late forties by Caspersson and Brachet showing that RNA levels in a variety of dividing and nondividing cells were correlated with protein synthetic activity. RNA levels were particularly high in the pancreas. (My interest in this problem antedated the Watson-Crick DNA double helix, which was to be published a few years later.)

The use of ^{32}P as a tracer had been adopted by Krebs and his associates a year or two earlier. I found that on stimulation of enzyme secretion in pancreas slices with carbachol, there was about a 100% increase in the incorporation of [^{32}P]orthophosphate into RNA. Near the end of this work, I began to suspect that an alkaline hydrolytic product of phospholipids was contaminating the then crude RNA fraction, and that this might be responsible for the stimulation of ^{32}P incorporation into RNA.

This idea came to me shortly after I had completed my requirements for the Ph.D. and was about to set sail with Mabel Hokin for Halifax, Nova Scotia. I did manage to do one experiment, but I did not have time to go through the necessary procedures to rid the lipid fraction of contaminating [^{32}P]orthophosphate and to count the radioactivity. I literally took with me on the boat a rack of about a dozen rather large test tubes containing the ethanol-ether extracts. (This would be prohibited with the strict regulations governing radiation safety now in place.) When we arrived at J. H. Quastel's laboratory at McGill, we counted the purified total lipid fractions and found an enormous increase in specific radioactivity in the lipids from pancreas slices that had been stimulated with carbachol. The 1953 *Journal of Biological Chemistry* paper arising from this work is often incorrectly quoted as demonstrating the stimulated turnover of PI. We did not actually show this until 1955, when techniques for measuring radioactivity in individual phospholipids from small samples of tissue became available.

The discovery of the PI effect launched Mabel Hokin and me on a 15-year quest for its explanation, as well as a pursuit of its biochemical mechanism. Mabel Hokin continued this search after I had left the field around 1965. I returned in the early 1980s. We made several key observations in the fifties and early sixties concerning the PI effect, including: (1) the effect occurred with many agonists and in many cell types; (2) the effect for the most part was due to a turnover and not a net synthesis of PI and phosphatidic acid (PA), with DAG being partly or completely conserved; and (3) omission of Ca^{2+} blocked secretion but did not inhibit PI turnover to any great extent. This was part of the underpinnings of Michell's hypothesis relating PI turnover to Ca^{2+} gating.

With the aid of kinetic studies on the turnover of PI and PA in the avian salt gland on cholinergic stimulation (followed by atropine clamping), we were able to present the first version of the PI cycle. The main points of this cycle were that on cholinergic stimulation of salt gland slices phospholipase C catalyzed the breakdown of PI to DAG and inositol-1-phosphate (IP). Diacylglycerol kinase then formed PA. On quenching with atropine, the resting steady-state level of PI was restored at the expense of PA by the sequential actions of PA-cytidyl transferase and PI synthase.

Mabel Hokin made some key observations in the 1970s further supporting the PI cycle, including the findings that on stimulation in pancreas, there was a loss in the mass of PI and a rise in the mass of PA. Also, on stimulation, the fatty acid composition of PA approached more closely that of PI. These data gave additional strong support to the view that PA was derived from PI via DAG during stimulated turnover of PI and PA. She further found an increase in mass of DAG in pancreas on agonist stimulation. At about the same time, Michell showed independently a fall in mass in PI in the parotid gland on stimulation. Mabel Hokin also confirmed the salt gland kinetic data in the pancreas. Her recalculation of some old measurements of PI in the salt gland (expressed as percentages of controls rather than as absolute values) showed a significant loss in mass in PI in this tissue as well. The original PI cycle has, of course, been modified to incorporate the polyphosphoinositides. There does appear, however, to be direct breakdown of PI as well, which appears to be an additional source of second messenger DAG.

The discovery of the PI effect in the early 1950s was indeed a "discovery before its time." Very little was known about membrane structure. The classic experiments from the laboratories of Douglas and Katz pointing to the importance of Ca^{2+} in stimulus-response coupling were not carried out until the early 1960s. We did not know for certain the structure of the "phosphoinositide" that showed increased turnover in pancreas and brain until around 1958, when we showed it was PI. The existence of the intracellular "trigger pool" of Ca^{2+}, which rapidly releases part of its stores of Ca^{2+} on agonist stimulation, was not known until the work of Schulz and others in the late 1970s. The technique of cell permeabilization to allow entry of phosphorylated compounds into the cell was not developed until the late 1970s and early 1980s. This, of course, permitted Streb, Irvine, Berridge, and Schulz (1983) to demonstrate the release of nonmitochondrial intracellular stores of Ca^{2+} by $Ins(1,4,5)P_3$.

Apparently, there was some derision of the PI effect in the early years. At a meeting a few years ago, John Fain said that when he arrived at Berridge's laboratory in the late 1970s, the common attitude toward the PI response (presumably not that of Berridge) was that it was "Hokin's hokum and Michell's folly."

We are currently interested in the effects of the antimanic drug lithium on $Ins(1,4,5)P_3$ levels in brain. We have found that in species ranging from the mouse to the monkey, lithium elevates levels of $Ins(1,4,5)P_3$ in cerebral cortex slices. The $Ins(1,4,5)P_3$ levels can be measured by prelabeling the slices with [^{3}H]inositol. After incubation with lithium, the slices are extracted with perchloric acid, the inositol phosphates are separated by high-performance liquid chromatography, and their radioactivity is determined. Another method is an $Ins(1,4,5)P_3$ receptor binding assay. Extracts of brain cortex slices incubated without [^{3}H]inositol are mixed with standard [^{3}H]$Ins(1,4,5)P_3$, the $Ins(1,4,5)P_3$ receptor, and the unknown sample. The $Ins(1,4,5)P_3$ in the unknown sample will displace [^{3}H]$Ins(1,4,5)P_3$ from the receptor, and the level of $Ins(1,4,5)P_3$ in the sample can be calculated.

Naturally, I am elated and I know I speak for Mabel Hokin as well, when I contemplate that what we started 40 years ago has led to the discovery of two phosphoinositide-derived second messengers—$Ins(1,3,5)P_3$ and DAG (by Nishizuka). Many additional inositol phosphate compounds formed on stimulation have been found. It appears that there may be more phosphoinositide-derived second messengers. For example, $Ins(1,3,4,5)P_3$ also appears to be involved in Ca^{2+} movements. Provocative studies on the involvement of the phosphoinositide system in fertilization, growth, and oncogene action are emerging. Functions for phosphoinositides other than as second messenger generators are beginning to surface. For example, PI is an important anchor for many proteins on the cell surface.

References

Berridge, M. J., and Irvine, R. F. *Nature* 341:197–205 (1989).

Rana, R. S., and Hokin, L. E. *Physiol. Rev.* 70:115–64 (1990).

Farago, A., and Nishizuka, Y. *FEBS Lett.* 268:350–54 (1990).

Bansal, V. S., and Majerus, P. W. *Annu. Rev. Cell. Biol.* 6:41–67 (1990).

Lee, C. H.; Dixon, J. F.; Reichman, M.; Moummi, C.; Los, G.; and Hokin, L. E. *Biochem. J.* 282:377–85 (1992).

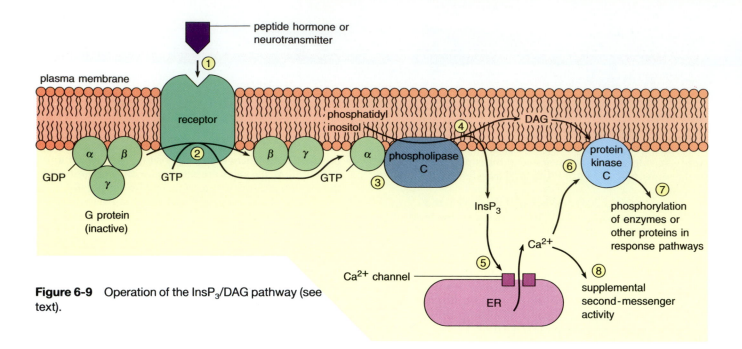

Figure 6-9 Operation of the InsP₃/DAG pathway (see text).

muscle, and glucose metabolism. The list of more than 40 known peptide hormones and neurotransmitters acting as first messengers for the InsP₃/DAG pathway includes *vasopressin, angiotensin,* and *norepinephrine.*

Figure 6-9 illustrates operation of the InsP₃/DAG pathway. Binding of a hormone or neurotransmitter activates the receptor (step 1 in Fig. 6-9), in turn leading to sequential activation of G proteins and phospholipase C and production of the InsP₃ and DAG second messengers (steps 2 to 4). The water-soluble InsP₃ messenger is released from the membrane and diffuses through the cytoplasm to the ER. In this location it binds to the *InsP₃ receptor,* a ligand-gated Ca^{2+}-channel (see p. 127) for which InsP₃ is the ligand. InsP₃ binding opens the channel, releasing Ca^{2+} from the ER into the surrounding cytoplasm (step 5). One function of the Ca^{2+} released into the cytoplasm by InsP₃ is as a coactivator with DAG of protein kinases of the pathway (step 6). The released Ca^{2+}, by directly or indirectly regulating a variety of cellular mechanisms, also works as a supplemental second messenger for the InsP₃/DAG pathway (step 8).

The other second messenger, DAG, remains suspended in the plasma membrane. Its primary role is to complete activation of the protein kinases associated with the pathway (step 6 in Fig. 6-9). This occurs as the protein kinases, partially activated by combination with Ca^{2+} in the cytoplasm, collide with DAG located at the inner side of the plasma membrane. The completely activated protein kinases phosphorylate target proteins controlled by the InsP₃/DAG pathway (step 7).

The protein kinases of the InsP₃/DAG pathway are all members of the *protein kinase C* family (C =

Ca^{2+}-activated). These kinases, like those of the cAMP pathways, add phosphates to serine and threonine residues in their target proteins. Among the targets of protein kinases C are chromosomal proteins; transport proteins of the plasma membrane including those for Ca^{2+}, Na^+, K^+, and glucose; contractile and cytoskeletal proteins; proteins regulating secretion and endocytosis, including many involved in the secretion of neurotransmitters; and a variety of enzymes catalyzing oxidative and other metabolic reactions. Many of the surface receptors stimulating either the InsP₃/DAG or cAMP pathways are also targets of protein kinases C activated in the InsP₃/DAG pathway.

The phosphate groups added by protein kinases C, like those added by the protein kinases A of cAMP-based pathways, may either stimulate or inhibit the activity of target proteins. One of the many proteins inhibited through phosphorylation by a Ca^{2+}-dependent protein kinase is adenylate cyclase, the enzyme converting ATP to cAMP in cAMP-based pathways. The phosphate groups added by the protein kinase C enzymes activated in the InsP₃/DAG pathway are removed by a series of serine/threonine protein phosphatases.

The Ca^{2+} released by InsP₃ may operate directly as a cellular regulator, as it does in activating protein kinases C or regulating ion and other channels in the plasma membrane. Alternatively the released Ca^{2+} may operate indirectly, by combining with a control protein known as *calmodulin* (see Information Box 6-2). Calmodulin is converted from inactive to active form by binding Ca^{2+}. In its active form the Ca^{2+}/calmodulin complex binds to a variety of enzymes and other proteins that depend on the complex for their activity.

Calmodulin

Calmodulin occurs in all eukaryotic cells as a Ca^{2+}-activated regulatory protein. The protein, which is highly conserved in sequence and structure among eukaryotic groups as diverse as animals, plants, fungi, and protozoa, is a relatively small molecule. The calmodulin of bovine brain tissue, for example, contains 148 amino acids with a total molecular weight approximating 16,700.

Calmodulin has two globular domains that can each bind two calcium ions, giving a total Ca^{2+} binding capacity of four ions (see the figure in this box). Ca^{2+} binding produces extensive conformational changes that convert calmodulin to its active form, in which it binds strongly to target proteins of various kinds. Some of the targets are protein kinases that, in turn, activate or deactivate their targets by adding phosphate groups. Others, such as the Ca^{2+}-ATPase active transport pump and some of the proteins regulating microtubule assembly, are regulated directly by combination with the activated Ca^{2+}/calmodulin complex.

The molecules activated by combination with the Ca^{2+}/calmodulin complex are highly diversified in amino acid sequence and structure. All, however, share a structural element about 20 amino acids in length that contains pos-

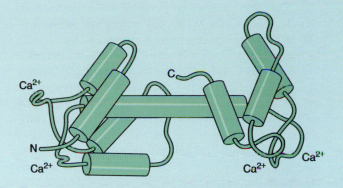

itive residues—arginine, lysine, or histidine—alternating with hydrophobic residues. This combination, which may wind into an amphipathic alpha helix, is apparently the sequence element recognized and bound by calmodulin. The specific amino acid sequence of the target element does not appear to be as significant for its recognition by calmodulin as its structural arrangement of alternating positive and hydrophobic residues.

Among the enzymes activated by the Ca^{2+}/calmodulin complex is another group of protein kinases with target proteins that play key roles in cellular metabolism, protein synthesis, and secretion. The assembly of microtubules and the contraction of striated muscle are also regulated by this complex or closely related proteins (see pp. 287 and 321).

Most of the elements of the $InsP_3$/DAG pathway, including G proteins, $InsP_3$ itself, calmodulin, and a group of protein kinases and other proteins activated by Ca^{2+} have been identified in plants. Injection of GTP into plant cells stimulates $InsP_3$ release; injection of $InsP_3$ or Ca^{2+} produces responses such as alterations in ion transport and photosynthesis. In plants, $InsP_3$ and Ca^{2+} release into the cytoplasm is triggered by nonprotein hormones such as auxin and abscisic acid and by environmental conditions including changes in light intensity, osmotic stress, and water loss. Responses to these stimuli appear to be triggered by elements in the cytoplasm rather than receptors embedded in the plasma membrane as in animals. For example, responses to light intensity are triggered by light-sensitive cytoplasmic pigments called *phytochromes* in plants.

Pathogens, Poisons, Cancer, and the cAMP and $InsP_3$/DAG Pathways The important cAMP and $InsP_3$ pathways are targets for toxins secreted by pathogenic bacteria and other poisons. G proteins of the cAMP pathway are targeted by no less than three known bacterial toxins—the *cholera* toxin (secreted by the bacterium *Vibrio cholerae*), the *pertussis* toxin (produced by *Bordetella pertussis*, the bacterium causing whooping cough), and a toxin produced by pathogenic forms of *E. coli*. All three modify the G proteins to make them permanently active. The activity leads to overproduction of cAMP, causing cAMP-based pathways to run at uncontrolled, high levels.

Many oncogenes code for mutated receptors or other elements of the $InsP_3$/DAG pathway. The effect of the mutations is to activate the $InsP_3$/DAG pathway permanently, producing a response that leads directly or indirectly to the rapid and uncontrolled cell division characteristic of cancer cells. One group of plant substances promoting tumor growth, the *phorbol esters*, depends for its effects on a molecular resemblance to DAG. The phorbol esters imitate DAG in activating protein kinase C, thus increasing the rate at which cancer cells divide.

CELL ADHESION

The capacity for cell adhesion, which underlies maintenance of body form and structure in animals, extends throughout the animal kingdom, from sponges to the most advanced vertebrate and invertebrate species. The glycoproteins responsible for cell adhesion, called *CAMs* (for *cell adhesion molecules*), work as a specialized class of receptors forming cell surface attachments. Although many cell types react internally to binding by CAMs at their surfaces through changes such as alterations to the cytoskeleton, the responses do not usually involve complex reaction series such as those triggered in the cAMP and $InsP_3$/DAG pathways. In many cases the responses are inhibitory—cell growth and division are retarded or stop, for example, in any animal cells as they form surface attachments via CAMs.

Cell attachments by CAMs are normally permanent once embryonic development is complete. However, they may change in highly programmed patterns while development is in progress. Embryonic cells regularly make initial attachments that are broken and remade as individual cells or tissues change position in the developing embryo. Some cells, such as erythrocytes, lose their attachments when mature and become freely suspended in a fluid medium. A return to the unattached state characteristic of embryonic cells often occurs in cancer, when malignant cells break their attachments in a tumor and move to new locations in the body.

CAMs have been extensively studied in mammals and other vertebrates by G. M. Edelman, S. D. Marlin, M. Takeichi, T. A. Springer, and others. They occur in several families, each containing a group of molecules of related types. All are integral glycoproteins that span the plasma membrane, with domains extending from both the outside and cytoplasmic sides of the membrane. They may bind CAMs of the same type on other cells (called *homotypic* binding) or may bind to different CAMs (called *heterotypic binding*). Multiple connections, involving several to many different CAM types, are typical of most body cells. Antibodies developed against CAMs block their binding and lead to partial or complete disassembly of tissues, confirming their importance in the cell-cell adhesions maintaining body structures in intact form. In addition to direct cell-to-cell adhesions by CAMs, animal cells are also held in place through linkages formed by receptors that recognize and bind extracellular materials such as collagen (described in Chapter 7).

One major CAM family, the *Ig-CAMs*, contains molecules with surface domains related to antibodies, making them members of the immunoglobin super-family. The Ig-CAMs may bind homotypically or heterotypically. Among the more than 30 Ig-CAMs discovered so far are *C-CAM*, found on the surfaces of liver cells, *N-CAM*, the neural adhesion molecule, and *Ng-CAM*, the neuron-glial adhesion molecule. (Glial cells form a supportive tissue surrounding neurons.) Another Ig-CAM, I-CAM, occurs on many cell types, including leucocytes. I-CAMs on leucocytes lead to temporary adhesions of these cells to epithelia as part of responses to wounding or infections.

A second major group of CAMs, the *cadherins*, binds homotypically; all the members of this family are Ca^{2+}-dependent—their binding may join or release depending on Ca^{2+} concentration in the extracellular medium. Among the members of this family are L-CAM, found on liver and other cells, and P-cadherin, found on placental and epithelial cells. A third important family, the *LEC-CAMs*, occurs on leucocytes and other cells of the circulatory system.

The connections made by CAMs have been extensively studied in nerve tissues. N-CAM molecules on the surface of a neuron bind directly to N-CAMs on other neurons. The N-CAM interaction is one of the associations setting up and maintaining nerve tissues in organized arrays. The Ng-CAMs on neurons bind to Ng-CAM receptors on glial cells. The Ng-CAM-receptor linkages effectively fix neurons and glial cells in place in nerve and brain tissues. In adult tissues, Ng-CAMs are distributed on axons and dendrites but not on cell bodies of nerve cells (see Fig. 5-15), so that the neuron-glial interactions are restricted to these regions. The association between neurons and glial cells that forms myelin sheaths around axons (see p. 140) is a primary example of an adhesion maintained by bonds between Ng-CAMs and Ng-CAM receptors.

Cell-cell connections by CAMs figure importantly in embryonic development. N-CAM and L-CAM appear as early as the blastoderm stage in chick embryos. At this stage, L-CAM occurs in cells in all three primary germ layers; N-CAM is generally distributed among ectoderm and mesoderm. Later in development, N-CAMs become concentrated in cells forming parts or derivatives of the nervous system. L-CAMs become concentrated on liver and other cells of mesodermal origin. Ng-CAMs appear much later in development as glial cells differentiate and take up their positions in the nervous system.

CAMs on some cell types may change extensively as development proceeds. This is evident during the migrations of neural crest cells, a cell type that migrates from the developing central nervous system. These cells display N-CAMs and adhere to their neighbors in the nerve tube early in development. As mesodermal structures begin to differentiate,

N-CAMs disappear from their surfaces and the neural crest cells break loose from their adhesions in the nerve tube. They migrate to other locations in the embryo, where they develop into a wide variety of structures. N-CAMs and connections to neurons reappear as the neural crest cells reach their destinations. As these and many other cell types take up their final positions in embryos, their cell-cell adhesions are reinforced by adhesive junctions of various kinds (see below).

In adults, changes in cell adhesions remain important in processes such as wound healing and the action of leucocytes. Cell-cell adhesions and the CAMs displayed at cell surfaces may change radically as cells are transformed from normal to cancer types. Typically cancer cells lose their normal adhesions, allowing them to break loose from their tissues of origin and migrate to other regions of the body.

Some pathogens use cell adhesion molecules as attachment sites during infection. The *rhinovirus* causing the common cold, for example, has a coat protein that is recognized and bound by I-CAM. The poliovirus also attaches to a cell adhesion molecule.

An interesting experiment by S. D. Marlin and his colleagues shows promise of leading to the long-sought cure for the common cold. These investigators found that I-CAM molecules, made soluble by removal of a segment anchoring them to the plasma membrane, effectively tie up the recognition sites on free rhinovirus particles. The particles blocked in this way are unable to attach and infect host cells. Whether the soluble I-CAMs, which could be produced in commercial quantities in genetically engineered bacteria, can be applied to prevent cold infections without serious side effects remains to be seen.

CELL JUNCTIONS

In many animal tissues the cell-cell adhesions set up during embryonic development by individual cell adhesion molecules are quickly followed by formation of intercellular junctions. Cell junctions occur in three major types, each with different functions. *Adhesive junctions* hold cells together, acting as intercellular "buttons" or "zippers" that help maintain cells in their fixed positions in tissues. *Tight junctions* close the space between cells to diffusion, forming a sort of dam that prevents flow of molecules and ions through the extracellular space. *Gap* or *communicating junctions* set up open channels in the plasma membranes of adjacent cells that allow ions and small molecules to flow directly from one cell to the other. Junctions of all three types occur widely in the animal kingdom.

The molecular components of junctions have been studied primarily by breaking apart cell types with large numbers of junctions, such as epithelial cells of various kinds, and purifying membrane fractions by centrifugation. Proteins are extracted from the fractions and purified by gel electrophoresis or other means. Once provisionally identified as part of a given junction type, the proteins are used as antigens to develop antibodies in test animals. The resulting antibodies, linked to a fluorescent dye or heavy metal marker, are then used as probes in light or electron microscope preparations of tissue containing the junction. Regions of the junction combining with the antibody are assumed to contain the protein of interest. Genes for some junction proteins have been isolated and cloned, allowing the amino acid sequence and secondary structure of the proteins to be deduced. (The various methods used in these studies are described in the Appendix.)

Adhesive Junctions

Adhesive junctions occur in two primary forms, *desmosomes* and *adherens junctions*. Although similar in overall function, the two types link to different cytoskeletal elements—desmosomes to intermediate filaments, and adherens junctions to microfilaments.

Desmosomes First described in 1954 by K. R. Porter, a *desmosome* (from *desmos*, meaning "bond") is usually a circular or elliptical structure (Fig. 6-10). In a desmosome a thick layer of dense material, the *plaque*, lies just under the plasma membrane on either side of the junction. Intermediate filaments extend from the plaques for some distance into the underlying cytoplasm. Depending on the cell, the intermediate filaments may be of the cytokeratin or vimentin type (see pp. 343 and 344). Called *tonofilaments* when they are connected to desmosomes, intermediate filaments evidently serve as cytoplasmic anchors for the junction. The extracellular space in the region of a desmosome is filled with faint filaments and granular material.

Half desmosomes, or *hemidesmosomes* (Fig. 6-11) occur where cells are anchored to extracellular materials instead of other cells, as to the layers of collagen underlying some epithelia. Hemidesmosomes also appear along the plasma membrane in cells grown in tissue culture, where they anchor cells to the glass or plastic substrate on which they are grown (Appendix p. 792 describes cell culture techniques).

The proteins of desmosomes have been isolated from rat, bovine, and other animal tissues by W. W. Franke, M. S. Steinberg, and others. Among the several major proteins consistently found in desmosomes in different animals and tissues, two of the largest,

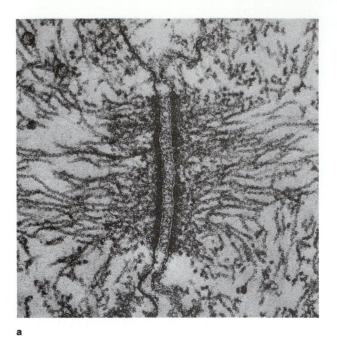

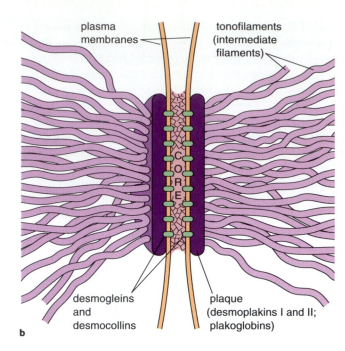

plasma membranes

tonofilaments (intermediate filaments)

C O R E

desmogleins and desmocollins

plaque (desmoplakins I and II; plakoglobins)

a

b

Figure 6-10 Desmosomes. **(a)** A desmosome binding adjacent cells in the skin of a salamander. × 22,000. (Courtesy of D. E. Kelly, from *J. Cell Biol.* 28:51 [1966], by permission of The Rockefeller University Press.) **(b)** Desmosome structure. The plaques contain at least three major proteins, desmoplakins I and II and plakoglobin. The desmogleins and desmocollin probably serve as cell-cell adhesion molecules in the region of the junction.

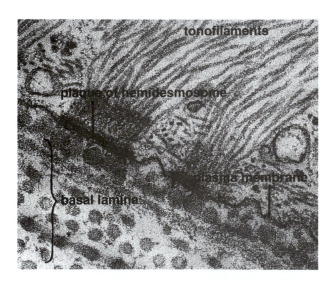

tonofilaments

plaque of hemidesmosome

plasma membrane

basal lamina

Figure 6-11 Hemidesmosomes anchoring cells of the skin to the basal lamina in a salamander. × 68,000. (Courtesy of D. E. Kelly, from *J. Cell Biol.* 28:51 [1966], by permission of the Rockefeller University Press.)

desmoplakins I and *II*, are related glycoproteins associated with the plaques. Another protein, *plakoglobin*, also occurs in the plaques. Although the desmoplakins are located in regions where intermediate filaments connect to the plaques, these molecules do not appear to bind directly to intermediate filaments. In-

stead, these connections may be made by interlinkers, among them *desmocalmin* and *keratocalmin*, that form ties between plaque proteins and intermediate filaments. Two additional glycoproteins, *desmoglein* and the *desmocollins*, both members of the cadherin family, are integral membrane glycoproteins that may set up cell-cell adhesions in the region of the junction. The cytoplasmic domains of desmoglein and the desmocollins probably contribute to the mass of the plaques.

Although there are similar desmoplakins in hemidesmosomes, desmogleins cannot be detected. Instead of desmogleins, the principal adhesion molecule of hemidesmosomes appears to be a member of the *integrin* family, a group of transmembrane glycoproteins that bind cells to the extracellular matrix (see p. 187). These differences indicate that hemidesmosomes are not simply half desmosomes as the name and electron micrographs such as Figure 6-11 suggest, but are molecularly distinct structures.

If living cells held together by desmosomes are digested briefly by dilute solutions of a proteinase such as trypsin, only the extracellular filamentous material is attacked, allowing the cells to separate without apparent damage. This method is routinely used to disperse cells linked together by desmosomes, such as tissue culture cells or the cells of early embryos. Reducing extracellular Ca^{2+} concentration to very low levels also causes desmosomes to separate. The sensitivity of desmosomes to Ca^{2+} concentration may depend in part on the desmogleins and desmocollins, which require Ca^{2+} for their binding activity.

Desmosomes are present between many cell types in multicellular animals. They are especially abundant in tissues normally subjected to shearing forces or lateral stress, such as skin and cardiac muscle cells and the epithelia lining the internal surfaces of body cavities. Desmosomes appear early in embryonic development and contribute to the maintenance of cell position from the earliest stages. They also form readily between cells maintained in cultures. Desmosomes frequently disappear when cells are transformed into cancerous types. This change undoubtedly contributes to the tendency of cancer cells to break loose from tumors and move to new locations in the body (see p. 653).

Adherens Junctions   Adherens junctions bind cells in certain animal tissues such as heart muscle and the thin cell layers covering organs and lining body cavities (Fig. 6-12). As in desmosomes, the regions just under the plasma membrane on either side of an adherens junction are thickened into plaques. However, the plaques of adherens junctions appear as loosely structured mats rather than solidly dense layers like the plaques of desmosomes. The anchoring fibers radiating from the plaques into the underlying cytoplasm are actin microfilaments. A faintly visible filamentous network fills the extracellular space in the region of an adherens junction, as in desmosomes.

Adherens junctions occur in forms ranging from buttonlike structures to beltlike bands that completely encircle the cell. The presence of microfilaments in adherens junctions suggests that these junctions, in addition to their adhesive function, may form part of actin-based motile systems in some cells. The extracellular connections holding plasma membranes together in the region of an adherens junction, like those of desmosomes, are calcium dependent; extreme reductions in extracellular Ca^{2+} concentrations release the connections.

Adherens junctions also occur in a "hemi" form consisting of an apparent half junction. These structures occur on the plasma membranes of cells linked to extracellular structures and also at the margins of cells bound to the surfaces of culture vessels.

Isolation of the molecular components of adherens junctions reveals that transmembrane proteins of the cadherin family, such as *E-cadherin*, form major

a

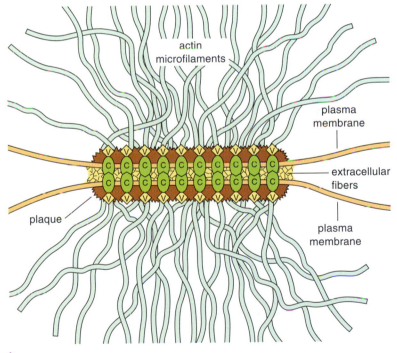

b

Figure 6-12   Adherens junctions. **(a)** An adherens junction (bracket) between intestinal cells. A portion of a desmosome (arrow) is also visible just below the adherens junction. × 99,000. (Courtesy of T. S. LeCount.) **(b)** Possible structural elements of an adherens junction. C, a cadherin linking the cell surfaces in the region of the junction; V, vinculin or other molecules linking the receptors and plasma membrane to microfilaments of the cytoskeleton.

Figure 6-13 Tight junctions. **(a)** A tight junction between cells of a capillary in muscle tissue. Arrows mark the sites of fusion between the two plasma membranes. × 190,000. (Courtesy of J. E. Rash, from *Cell* 28: 441[1982]; copyright Cell Press.) **(b)** The possible arrangement of membrane proteins (P) and bilayer halves forming the fused ridges of plasma membranes sealed by tight junctions. The fusions block the movement of substances through the intercellular space; the tightly packed rows of membrane proteins within the ridges block the lateral flow of other membrane proteins past the junction. The flow of lipid molecules is blocked in the outer bilayer half by the junction.

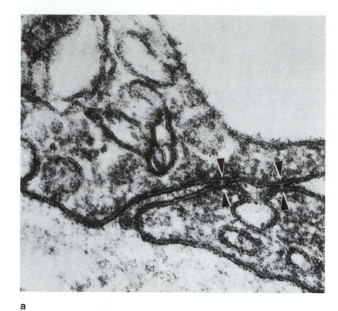

a

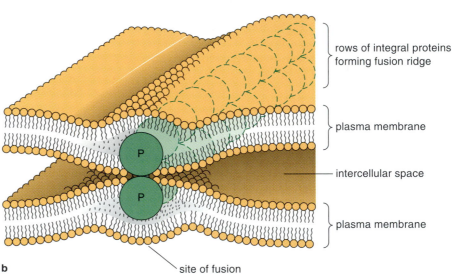

rows of integral proteins forming fusion ridge

plasma membrane

intercellular space

plasma membrane

b   site of fusion

components. The portions of these cadherins that extend outward from the plasma membrane probably form the cell-cell adhesions holding adherens junctions together. The cytoplasmic segments of these molecules contribute to the plaques and organize a group of proteins in the plaque regions, including α-*actinin*, plakoglobin, the *catenins*, and *vinculin*. Some of the plaque proteins, including α-actinin and vinculin, set up links between the plaques and actin microfilaments. Some adherins junctions have links to the extracellular matrix via integrins.

Adherens junctions often occur along with desmosomes and tight junctions in cell layers lining body cavities such as the digestive tract. Adherens junctions also link cardiac muscle cells. In these cells the actin microfilaments of the adherens junctions are continuous with the internal microfilaments of the

heart muscle cells, which are responsible for the contractions producing the heartbeat (see p. 330). These connections directly implicate adherens junctions in the motile systems of these cells.

Tight Junctions

Tight junctions (Fig. 6-13), discovered in 1963 by M. G. Farquhar and G. E. Palade, form a continuous belt around cells lining body cavities, as in the layer of cells lining the digestive tract. The junctions seal the cells together and prevent fluids from leaking between the cavity and the body cells underlying the cavity.

In the region of a tight junction the outer bilayer halves of adjacent plasma membranes fuse along sharply defined ridges (arrows, Fig. 6-13a). The fused ridges eliminate the extracellular space and bar pas-

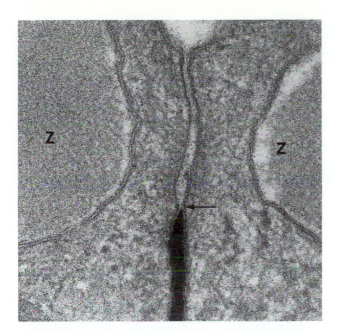

Figure 6-14 Penetration of a stain visible in the electron microscope (lanthanum hydroxide) through the intercellular space only as far as the ridge of a tight junction (arrow) between two pancreatic cells. Z, zymogen granule, containing a protein secreted by pancreatic cells. × 100,000. (Courtesy of D. S. Friend, from *J. Cell Biol.* 53:758 [1972] by copyright permission of the Rockefeller University Press.)

sage to extracellular fluids and dissolved ions and molecules. The degree to which tight junctions close off the extracellular space depends on the number of parallel ridges—the more ridges, the more perfect the seal. The molecular structure of tight junctions remains unknown. However, they may be formed by tightly packed rows of integral membrane proteins in the outer bilayer halves (Fig. 6-13b).

Several proteins have been tentatively identified in tight junctions, including *ZO-1, ZO-2* (ZO = *zonula occludens*, another name for tight junctions), *7H6*, and *cingulin*. All these proteins have been localized in an ill-defined, plaquelike region in the cytoplasm underlying tight junctions. There is some evidence that cadherins may also form part of tight junctions. The identity of the integral membrane protein forming the tight junction ridges remains unknown.

The sealing function of tight junctions has been convincingly demonstrated by experiments in which a staining solution is introduced into the extracellular space on one side of the junction. Lanthanum hydroxide, for example, can be seen under the electron microscope to penetrate through extracellular spaces only as far as the fused ridges of tight junctions (Fig. 6-14).

The ridges of tight junctions also block the lateral movement of proteins and lipids within the membrane bilayers. The movement of integral membrane proteins is completely blocked by the ridges. The blockage of lipid movement is confined to the outer bilayer half of the membrane; lipids in the bilayer half facing the underlying cytoplasm can still move laterally.

The ability of tight junctions to block the movement of membrane proteins has been most clearly demonstrated in the epithelial cells lining body cavities. These cells have different distributions of plasma membrane molecules such as surface receptors, transport proteins, enzymes, and lipid molecules on either side of their tight junction belts (see also p. 111). In cells lining the intestinal tract, for example, Na^+-ATPase active transport pumps and type I MHC molecules are restricted to the side of the cells facing the body interior; other integral membrane proteins, such as Na^+ channels and alkaline phosphatase enzymes, are confined to the side facing the intestinal cavity. Breakage of tight junctions at the margins of these cells allows many of these proteins to flow to all parts of the cell surface, confirming the importance of these junctions in maintaining the proteins in their normal distributions.

Although tight junctions have been found in both vertebrates and invertebrates, they are much more characteristic of vertebrates. They are particularly abundant in the epithelia lining the digestive system, including those lining the cavities and ducts of the digestive tract. Tight junctions are also common between the cells of capillary walls in certain locations such as the brain and testis, where they prevent the blood from mixing with the extracellular fluids. In some locations, such as the lining of the urinary bladder, tight junctions maintain very large differences in ionic concentrations between the contents of the body cavity and the surrounding tissues.

Gap Junctions

The third type of cell junction, the gap junction, provides open passages through which ions and small molecules can pass directly between cells. First described in 1967 by J. P. Revel and M. J. Karnovsky, these junctions appear as regions where the plasma membranes of adjacent cells are aligned and separated by a very regular 2 to 3 nanometer space (Fig. 6-15a). Negatively stained gap junctions show masses of hollow cylinders that are aligned perpendicular to the membrane surface (Fig. 6-15b). In vertebrates the narrow channel through the center of each cylinder is about 1.5 nm in diameter at its narrowest point, easily large enough to allow the free passage of ions and small molecules. D. A. Goodenough, N. B. Gilula, and others later showed that the cylinders in the adjacent plasma membranes meet end to end, forming an open

Figure 6-15 Gap junctions. **(a)** A gap junction (bracket) between chick intestinal cells. × 92,000. (Courtesy of C. Chambers.) Inset to **(a)**, an isolated segment of a gap junction prepared by negative staining. The junction segment is seen as if the viewer is looking directly at the membrane surface, showing a tightly packed mass of hollow cylinders. The stain has penetrated into the narrow channel in the center of each cylinder. Inset × 277,000. (Courtesy of D. A. Goodenough, from *J. Cell Biol.* 68:220[1976], by permission of the Rockefeller University Press.) **(b)** Gap junction structure. The cylinders or connexons (brackets) of the adjacent membranes align to form continuous channels placing the cytoplasm of the two cells sharing the junction in direct communication. Each connexon is formed from a circle of four or six subunits of the protein connexin. **(c)** The dimensions of a connexon channel. A caplike gate at the cytoplasmic end of the channel may open and close to regulate movement of substances through the channel. **(d)** Each connexin protein unit has four alpha helices that extend across the plasma membrane, with a sizable domain extending into the cytoplasm at the C-terminal end.

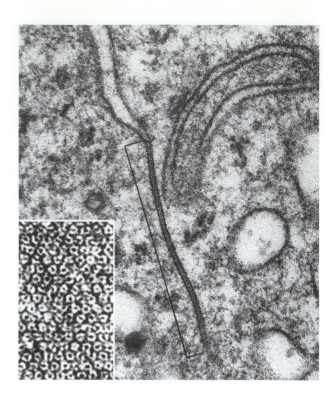

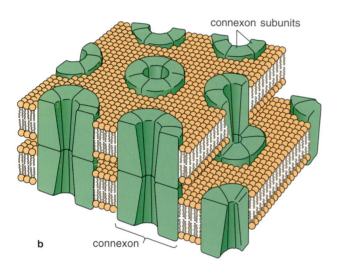

b   connexon subunits / connexon

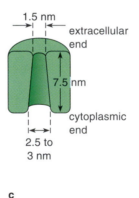

c   1.5 nm / extracellular end / 7.5 nm / cytoplasmic end / 2.5 to 3 nm

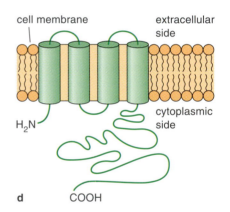

d   cell membrane / extracellular side / cytoplasmic side / H₂N / COOH

pathway directly between the cells but sealing the junction from the surrounding extracellular fluids (Fig. 6-15c). The number of cylinders in a single gap junction may range from a few to many tens of thousands.

Each gap junction cylinder, termed a *connexon*, consists of a circle of four or six units of a protein called *connexin*. The gap junctions of a given tissue may contain one or a mixture of two or more connexin types. Introduction of isolated connexin proteins into artificial lipid bilayers produces functional gap junctions, confirming that this protein is the primary unit of gap junction structure and indicating further that the connexin proteins have the capacity to self-assemble into connexons.

The Function of Gap Junctions Various experiments have shown that ions and small molecules can pass freely through gap junctions. In one of these experiments, W. R. Loewenstein injected a dye into one cell of a row of insect cells connected by gap junctions and followed movement of the dye in the light microscope. The dye, a 300-dalton hydrophilic molecule, moved rapidly from one cell to the next, much faster than expected if the molecule had traveled between the cells by penetrating through plasma membranes. In similar experiments, Loewenstein and others showed that molecules with molecular weights up to about 1200 can pass freely between cells connected by gap junctions in mammalian tissues such as rat liver, human mammary, and calf lens. Molecular weights in this range mean that the connexon channels are large enough to permit passage of monosaccharides and amino acids and second messengers such as cAMP and $InsP_3$.

An experiment by N. B. Gilula and his coworkers with cultured cells confirmed that second messengers can move directly through gap junctions. Ovarian cells, which have no receptors for norepinephrine and do not normally respond to this hormone, were placed in mixed cultures with heart muscle cells, which have norepinephrine receptors and respond fully to stimulation by the hormone. The cultured ovarian and heart muscle cells frequently developed connections by gap junctions. Stimulation of heart muscle cells with norepinephrine in the cultures also produced a hormone response in the ovarian cells. The chemical signal passing through the gap junctions from the muscle to the ovarian cells was probably cAMP, which acts as the second messenger for the pathway stimulated by norepinephrine in heart muscle cells.

Regulation of Gap Junctions The channels of gap junctions can open and close, probably through conformational changes in the connexin proteins that form the channel walls. The channels are thus functionally similar to voltage- and ligand-gated channels.

Gap junctions have been found to respond to changes in the cytoplasmic concentrations of either Ca^{2+} or H^+. Gap junctions are open at the low Ca^{2+} concentrations typical of "resting" eukaryotic cytoplasm. As Ca^{2+} concentration rises, the channels narrow until closing fully at about 10^{-5} M Ca^{2+}. Regulation by Ca^{2+} may be direct or through calmodulin. Increases in H^+ concentration, as reflected in a drop in cytoplasmic pH from 7 to 6.8 or below, also close the gap junction channels. Other stimuli, such as changes in voltage across the plasma membrane, can close the connexon channels in some cells; in others, phosphorylation by a protein kinase A induces changes that open the channels more fully. Regulation of gap junctions by Ca^{2+} and protein kinase A links control of these structures to both the $InsP_3$/DAG and cAMP pathways and suggests that in many cell types, conduction through these junctions may be regulated upward or downward by signal molecules binding to receptors at the cell surface.

Besides functioning as a regulatory mechanism, closure by elevated Ca^{2+} concentrations probably provides a failsafe function, sealing off broken cells from a tissue interconnected by gap junctions. Because external Ca^{2+} concentrations are high in animals (see p. 131), any cells with broken plasma membranes immediately undergo an increase in cytoplasmic Ca^{2+}. This increase closes the gap junction channels between the broken cell and its neighbors and eliminates leakage of ions and small molecules from undamaged neighboring cells through the broken cell. Otherwise, a leak in one cell in a tissue interconnected by gap junctions could lead to critical ionic and molecular imbalances in all the cells of the tissue.

Occurrence and Distribution of Gap Junctions Gap junctions have been observed between cells in animals from coelenterates to mammals. In vertebrates they occur between cells within practically all body tissues, but not between the cells of different tissues. Only a few cell types, including those of skeletal muscle, blood, and nerve tissues, have cells regularly lacking gap junctions. Gap junctions are particularly important in the uterus, in which the direct connections integrate smooth muscle cells into a single functional unit. Although gap junctions are absent from most nerve cells, a few types, such as some of the sensory neurons in the cerebral cortex and in dental pulp, are directly connected by these junctions. In these tissues the junctions provide almost instantaneous electrical communication that proceeds much more rapidly than conduction across synapses (see p. 141).

Gap junctions appear between cells early in embryonic development. In early mammalian embryos, gap junctions are present by the eight-cell stage and soon interconnect all cells in the developing embryo. As development proceeds, the connections are generally maintained among the cells of the same tissue, but not between the cells of different tissues. The cells of some tissues, such as skeletal muscle and most nervous tissue, lose their gap junctions completely as they differentiate.

Changes may occur in gap-junction numbers in adults. In mammals, gap junctions between muscle cells of the uterus increase greatly in females about to give birth. The added gap junctions, probably developed in response to hormonal stimulation, contribute to the synchronization and strength of uterine contractions during labor. The number of gap junctions between uterine cells falls again soon after delivery. In the uterine and other cells the removed gap junctions are taken in by endocytosis and degraded in lysosomes.

Gap junctions are thus implicated in nutrient exchange, cell regulation, conduction of electrical impulses, development and differentiation, and synchronization and integration of the activities of cells within organs such as the uterus. These junctions thus serve along with intracellular signals from the nervous and endocrine systems as a primary means of cell coordination in animals.

The surface specializations discussed in this chapter serve a variety of functions in eukaryotic cells. In animals, some glycoproteins of the cell surface act as markers that allow recognition of cells as part of self or foreign. Other surface glycoproteins act as receptors that bind to extracellular signal molecules and stimulate internal responses ranging from increases or decreases in metabolic rates to cell division. Still other surface molecules are responsible for cell adhesions. In animals the initial adhesions established by interactions between individual glycoproteins are reinforced in some tissues by adhesive junctions, including desmosomes and adherens junctions. Where necessary, sealing or tight junctions close off the space between cells to diffusion of molecules and ions. Communicating or gap junctions open channels that permit ions and small molecules to diffuse directly between the cytoplasm of adjacent cells with related functions.

For Further Information

Suggestions for Further Reading

Albeda, S. M., and Buck, C. A. 1990. Integrins and other cell adhesion molecules. *FASEB J.* 4:2868–2880.

Bansal, S., and Majeras, P. W. 1990. Phosphatidylinositol-derived precursors and signals. *Ann. Rev. Cell Biol.* 6:41–67.

Berridge, M. J. 1993. Inositol triphosphate and calcium signalling. *Nature* 361:315–325.

Beyer, E. C. 1993. Gap junctions. *Internat. Rev. Cytol.* 137C: 1–37.

Bourne, H. R., Sanders, D. A., and McCormick, F. 1991. The GTPase superfamily: Conserved structure and molecular mechanism. *Nature* 349:117–127.

Boyer, E. C., Paul, D. L., and Goodenough, D. A. 1990. Connexin family of gap junction proteins. *J. Membr. Biol.* 116:187–194.

Brown, A. M., and Birnbaumer, L. 1990. Ionic channels and their regulation by G proteins. *Ann. Rev. Physiol.* 52: 197–213.

Cadena, D. L. and Gill, G. N. 1992. Receptor tyrosine kinases. *FASEB J.* 6:2332–2337.

Carpenter, G., and Cohen, S. 1990. Epidermal growth factors. *J. Biolog. Chem.* 265:7709–7712.

Charbonneau, H., and Tonks, N. K. 1992. 1002 protein phosphatases? *Ann. Rev. Cell Biol.* 8:463–493.

Citi, S. 1993. The molecular organization of tight junctions. *J. Cell Biol.* 121:485–489.

Czech, M. P. 1989. Signal transmission by the insulin-like growth factors. *Cell* 59:235–238.

Edelman, G. M., and Crossin, K. L. 1991. Cell adhesion molecules: Implications for a molecular histology. *Ann. Rev. Biochem.* 60:155–190.

Fantl, W. J., Johnson, D. E., and Williams, L. T. 1993. Signalling by receptor tyrosine kinases. *Ann. Rev. Biochem.* 62:453–481.

Geiger, B., and Ayalon, O. 1992. Cadherins. *Ann. Rev. Cell Biol.* 8:307–332.

Geiger, B., and Ginsberg, D. 1991. The cytoplasmic domain of adherens-type junctions. *Cell Motil. Cytoskel.* 20: 1–6.

Hynes, R. O. 1992. Integrins: Versatility, modulation, and signalling in cell adhesion. *Cell* 69:11–25.

Legan, P. K., Collins, J. E., and Garrod, D. R. 1992. The molecular biology of desmosomes and hemidesmosomes: What's in a name? *Bioess.* 14:385–392.

Levitzki, A. 1988. From epinephrine to cAMP. *Science* 241: 800–806.

Linder, M. E., and Gilman, A. E. 1992. G proteins. *Sci. Amer.* 267:56–65 (July).

McClay, D. R., and Ettensohn, C. A. 1987. Cell adhesion in morphogenesis. *Ann. Rev. Cell Biol.* 3:319–345.

Milner, C. M., and Campbell, R. D. 1992. Genes, genes, and more genes in the human major histocompatibility complex. *Bioess.* 14:565–571.

Muto, S. 1992. Intracellular Ca^{2+} messenger system in plants. *Internat. Rev. Cytol.* 142:305–345.

Neer, E. J., and Clapham, D. E. 1988. Roles of G protein subunits in transmembrane signalling. *Nature* 333:129–139.

Parkinson, J. S. 1993. Signal transduction systems of bacteria. *Cell* 73:857–871.

Rademacher, T. W., Parekh, R. B., and Dwek, R. A. 1988. Glycobiology. *Ann. Rev. Biochem.* 57:785–838.

Rasmussen, H. 1989. The cycling of calcium as an intracellular messenger. *Sci. Amer.* 261:66–73 (October).

Rasmussen, H., and Rasmussen, J. E. 1990. Calcium as intracellular messenger: From simplicity to complexity. *Curr. Top. Cellular Regulat.* 31:1–109.

Rosen, O. M. 1987. After insulin binds. *Science* 237: 1452–1458.

Satoh, T., Nakafuku, M., and Kaziro, Y. 1992. Function of Ras as a molecular switch in signal transduction. *J. Biolog. Chem.* 267:24149–24152.

Schwarz, M. A., Owaribe, K., Kartenbeck, J., and Franke, W. W. 1990. Desmosomes and hemidesmosomes: Constitutive molecular components. *Annu. Rev. Cell Biol.* 6:461–491.

Sharon, N., and Lis, H. 1993. Carbohydrates in cell recognition. *Sci. Amer.* 268:82–89 (January).

Simon, M. I., Strathmann, M. P., and Gautam, N. 1991. Diversity of G proteins in signal transduction. *Science* 252:802–808.

Singer, S. J. 1992. Intercellular communication and cell-cell adhesion. *Science* 255:1671–1677.

Taylor, S. S., Buechler, J. A., and Yonemoto, W. 1990. cAMP-dependent protein kinase: Framework for a diverse family of regulatory enzymes. *Ann. Rev. Biochem.* 59:971–1005.

White, M. F. 1991. Structure and function of tyrosine kinase receptors. *J. Bioenerget. Biomembr.* 23:63–82.

Yamamoto, F. I., Clausen, H., White, T., Marken, J., and Hakamore, S. 1990. Molecular genetic basis of the histo-blood group ABO systems. *Nature* 345:229–233.

Review Questions

1. Outline the structure of membrane glycolipids and glycoproteins. How are these molecules embedded in the fluid mosaic structure of the plasma membrane?

2. How is the capacity for cell-to-cell recognition related to membrane surface glycolipids and glycoproteins?

3. What are MHC molecules? What are the major classes of these molecules? How are these molecules involved in the rejection of organ transplants? In the immune response to viral infections? What is the relationship between MHC molecules and antibodies?

4. Outline the molecular basis for the ABO blood groups of humans.

5. What is a cell surface receptor? What evidence indicates that receptors transmit a signal to the cytoplasm? What types of molecules are bound by receptors?

6. What are protein kinases? What is the relationship of protein kinases to surface receptors? What two major types of receptors exist with reference to protein kinase activity?

7. What are first messengers? Second messengers? G proteins? Effectors? How do these elements act in receptor-response mechanisms?

8. Outline the role of G proteins in receptor-response pathways.

9. What two types of effectors operate in receptor-response pathways? What second messengers are products of the enzymatic activity of these effectors? What activates the effectors?

10. Outline the reaction steps taking place in a pathway using cAMP as second messenger. What types of first messengers stimulate operation of cAMP-based pathways?

11. Outline the reaction steps taking place in a pathway using $InsP_3$ and DAG as second messengers. What types of first messengers stimulate $InsP_3$/DAG-based pathways? What is the role of Ca^{2+} ions in these pathways?

12. Compare the activities of the protein kinases activated in cAMP- and $InsP_3$/DAG-based pathways.

13. What is the relationship between the receptor-response pathways and cancer? Between substances acting as toxins or poisons?

14. What is the importance of cell adhesion in animals? What kinds of molecules are responsible for cell adhesion?

15. Outline the role of cell adhesion molecules in development. How is cell adhesion related to surface reception?

16. Outline the major types of junctions formed between animal cells. What are the major functions of each junction type?

17. What is a desmosome? An adherens junction? In what ways are these two junction types similar, and in what ways are they different?

18. Where do desmosomes and adherens junctions occur in animals? What is the relationship between adhesive junctions and cancer?

19. What is a tight junction? What evidence supports the idea that tight junctions close intercellular spaces to the flow of molecules?

20. Outline the possible structure of a tight junction. Where do tight junctions occur in animals?

21. Describe the structure of a gap junction. What evidence indicates that gap junctions provide open channels for the flow of molecules and ions between cells?

22. How are gap junctions regulated? What keeps damaged cells from causing leakage from other cells in tissues interconnected by gap junctions?

23. Where do gap junctions occur in animals? List some major body functions that depend on cell communication via gap junctions.

Supplement 6-1

Bacterial Receptor-Response Mechanisms

Bacteria also have systems that function in reception of external stimuli. This reception depends on receptor molecules that bind external molecules acting as signals, as in eukaryotic systems. Cellular responses triggered by binding the signal adjust functions as diverse as transport, nitrogen metabolism, sporulation, and *chemotaxis*, a motile response in which cells swim toward or away from attractive or repellent substances in the medium.

More than 20 different receptors triggering internal responses have been detected in gram-negative bacteria such as *E. coli* and *Salmonella*, in which receptor mechanisms have been best studied. The systems detect environmental changes in concentration and availability of nutrients such as sugars and amino acids, the presence of toxic substances such as acids, alcohols, or heavy metals, and changes in environmental conditions such as alterations in light, osmotic pressure, temperature, and electrical fields.

The receptor mechanisms are based on proteins consisting of a *sensor* and a *regulator* (Fig. 6-16). The sensor in some systems is a transmembrane protein that combines with a molecule originating from outside the cell that acts as a sensory signal or detects change in an environmental condition. In others the sensor is cytoplasmic and is activated indirectly by additional membrane proteins that combine directly with a signal molecule or detect a condition. The second member of the pair, the regulator, is cytoplasmic in all known systems. Additional cytoplasmic components working in a cascade activated by the regulator are present in some receptor-response mechanisms.

Although much of the sensor protein varies in amino acid sequence between different sensor-regulator pairs, all sensors share a homologous, 200-amino acid domain at their C-terminal ends. In sensors suspended in the plasma membrane the variable element, which combines with molecules acting as sensory signals, extends from the outside membrane surface. The homologous domain extends into the cytoplasm on the other side of the membrane. The regulator proteins also have homologous and variable sequence elements. The domain shared by regulators includes a sequence element of approximately 120 amino acids at the N-terminal end.

Binding a molecule acting as a sensory signal induces a conformational change in the sensor that is transmitted through the protein's structure to its C-terminal domain. The conformational change activates a self-phosphorylation in the C-terminal domain in which a phosphate group is transferred from ATP to a histidine residue in the protein. The sensor then transfers the phosphate group to an aspartic acid residue in the N-terminal domain of the regulator. This phosphorylation activates the regulator, which triggers an internal response. For many of the bacterial systems the activated regulator promotes transcription of one or more genes by combining directly with control regions in the DNA.

The sensor-regulator pair responding to changes in osmotic pressure in *E. coli* is one of the better understood bacterial systems (Fig. 6-17). Under external conditions producing low osmotic pressure a sensor protein located in the plasma membrane, *EnvZ*, is activated only at relatively low levels. In this condition the transfer of phosphate groups to the regulator protein, *OmpR*, is limited, and the regulator combines with elements in the DNA that activate transcription of *ompF*, a gene encoding the porin *OmpF*. The porins, which are channel-like proteins inserted in the outer membrane of gram-negative bacteria, control the upper limit of molecular sizes able to penetrate through the outer membrane to reach the plasma membrane (see p. 146). When the concentrations of ions and molecules in medium produces high osmotic pressure inside the cell, EnvZ is highly activated and heavily phosphorylates OmpR. In the highly phosphorylated form, OmpR combines with sites in the DNA regulating a gene encoding a different porin, *OmpC*. The presence of either OmpF or OmpC as the predominant outer membrane protein makes compensating adjustments in the osmotic pressure of the cell.

Because each of the response pathways has its own set of sensors and receptors, the bacterial systems differ fundamentally from the receptor mechanisms of eukaryotes, in which a wide variety of cellular responses is controlled through a small number of second messengers common to many pathways. There are in fact no known bacterial equivalents to the second messengers of eukaryotes; while cAMP occurs in bacteria, this cyclic nucleotide functions as a part of pathways regulating genetic activity that are not a direct part of the sensor-regulator systems (see p. 517). There are, however, a few eukaryotic systems equivalent to the bacterial pathways that use proteins with sequence structure and functions related to the bacterial sensor-regulator proteins. These include a system in higher plants responding to ethylene, which operates as a plant hormone controlling growth and the ripening of fruit, and a system regulating growth in yeast.

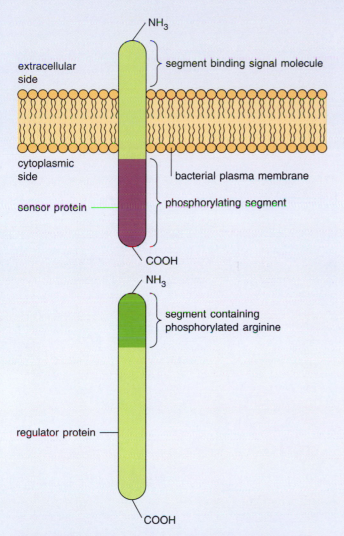

Figure 6-16 Sensor-regulator pairs in bacteria. The N-terminal domain of the sensor, which varies in sequence in the sensors of different pairs, contains the site responding to signals. The C-terminal domain shared by different sensors is shown in purple; this domain is self-phosphorylated at a histidine residue when the sensor detects its signal. The homologous N-terminal domain shared by different regulators (shown in dark green) contains an aspartic acid residue that is phosphorylated by transfer of a phosphate group from the sensor; phosphorylation activates the regulator, which triggers an internal response. In many of the sensor-regulator systems the regulator directly controls transcription of one or more genes.

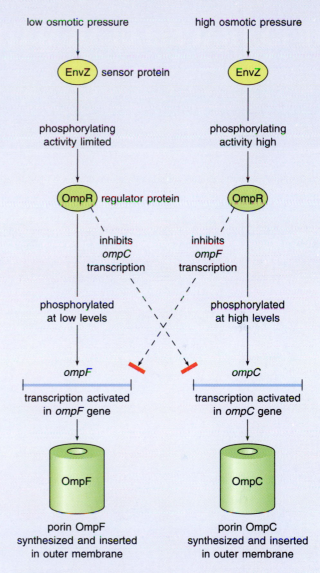

Figure 6-17 The sensor-regulator pair responding to changes in osmotic pressure in *E. coli* (see text).

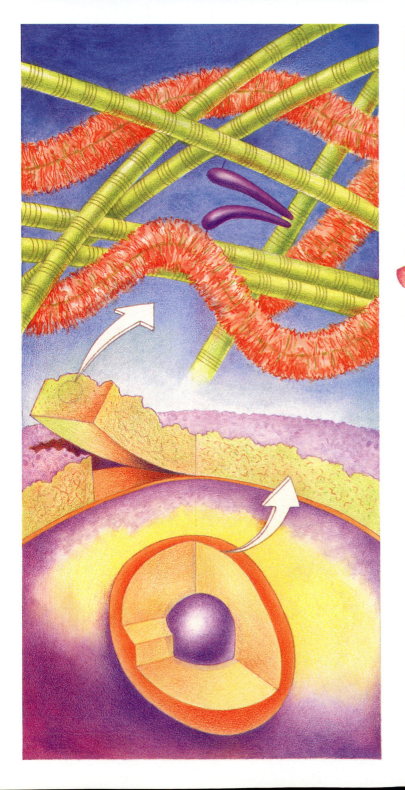

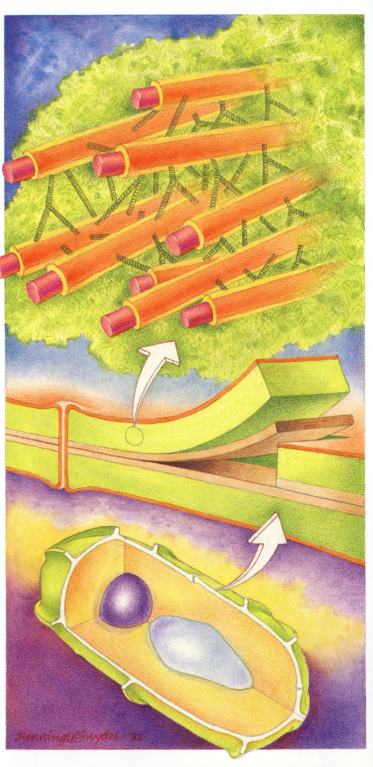

Most cells form extracellular structures, collectively called the *extracellular matrix*, that carry out a variety of functions. In animals the extracellular matrix ranges from the tough, elastic structures of tendons, cartilage, and bone to the clear, crystal-like cornea of the eye. In plants and prokaryotes, extracellular materials form the prominent cell walls of these organisms.

The primary function of the extracellular matrix is support. However, it has a variety of additional functions, particularly in animals. The patterns and types of extracellular materials laid down in animals also regulate cell division, adhesion, cell motility and migration, and differentiation during embryonic development. A meshlike extracellular layer acts as a filter that excludes larger molecules such as proteins from passing through cell layers in the walls of capillaries and kidney glomeruli. Extracellular material also figures importantly in the reactions to wounding and disease. It is altered in cancer cells, which frequently lose the connections anchoring them to the extracellular matrix.

In plants, algae, fungi, and prokaryotes, extracellular materials of cell walls may carry sites that recognize and bind molecules from the surrounding medium. Plant cell walls also function in cell-to-cell communication. In addition, the external walls in all these forms provide protection from mechanical damage and attack by infective organisms and viruses.

In spite of their diverse types and functions, the extracellular structures of eukaryotes share two common structural elements, *fibers* and a surrounding *network*. Fibers are long, semicrystalline elements that provide resistance to stretching and other tensile forces. The network, a more or less elastic, interlocked assembly of branched molecules, holds the fibers in place and resists compression by trapping and retarding the flow of water molecules. The degree of hardness and elasticity of extracellular structures depend on variations in the kinds of fiber and network molecules, the degree of crosslinking, the amount of trapped water, and the types of added substances such as the calcium phosphate crystals of bone or the lignin of plant cell walls.

This chapter describes the structure and function of the extracellular matrix in animals and plants. The cell walls of bacteria are discussed in Supplement 7-1.

THE ANIMAL EXTRACELLULAR MATRIX

The extracellular matrix of animal cells contains glycoprotein molecules called *collagens* as the primary fiber. The main components of the surrounding network are the *proteoglycans*, a unique group of glycoproteins of unusually high carbohydrate content.

The Collagens

Collagens are semicrystalline, fibrous molecules that hold cells in place, provide tensile strength and elasticity to the extracellular matrix, and serve other functions related to cell motility and development. They occur in all animal phyla from the sponges to the chordates. In vertebrates the collagens of tendons, cartilage, and bone are the most abundant proteins of the body, making up about half of the total body proteins by weight. Among invertebrates, collagen fibers are particularly abundant in the extracellular material of sponges and the thick surface cuticle of nematode and annelid worms.

Collagen Structure   The collagens are insoluble glycoproteins characterized by a high content of glycine and two modified amino acids, hydroxylysine and hydroxyproline (see Fig. 2-18). Almost all the carbohydrate portion of the collagens is built up from two sugars, glucose and galactose (see Fig. 4-4). These sugars link in short two-unit chains attached to the protein component of the molecules. About 10% of the total weight of collagens is carbohydrate.

Individual collagen molecules are built up from three alpha-helical polypeptide chains known as α-*chains* (Fig. 7-1a). Three of the α-chains twist together into a right-handed triple helix that forms about 95% of the rigid, rodlike central portion of the molecule (Fig. 7-1b and c). This helical structure is made possible by the regularly spaced glycine residues, which are small enough to fit along the central axis of the helix (shown in green in Fig. 7-1b). Depending on the collagen type, the triple helix may be almost continuous and completely uninterrupted, or it may contain segments in which the triple helix gives way to less ordered conformations. The less structured segments act as hinges that give the central region greater flexibility.

At either end of the central triple helix the amino acid chains fold into a caplike, globular segment. The entire structure is held together by hydrogen bonds between the α-chains in both the triple helix and

Figure 7-1 Collagen structure.
(a) An α-chain, a single polypeptide subunit of a collagen molecule. In most of the chain, glycine is repeated at every third position in the amino acid sequence in the pattern (Gly-X-Y)$_n$, in which X and Y are other amino acids, most commonly proline or lysine in either unmodified or modified form. The chain winds into a shallow left-handed helix, the *polyproline helix*. The two hydrogens associated with the glycine residues are shown in green; other hydrogens capable of forming hydrogen bonds are shaded. **(b)** A collagen molecule, a triple helix of three α-chains wound together in a right-hand helix. The spacing of glycine residues (in green) places them in the axis of the triple helix, where their small size allows close packing of the three α-chains. **(c)** A collagen molecule in diagrammatic form. The carbohydrate groups of collagen consist almost completely of short sugar-galactose-glucose chains. The hydroxyl (—OH) groups on hydroxyproline and hydroxylysine residues exposed at the surface form hydrogen bonds to join collagen molecules into collagen fibers. ([a] and [b] redrawn from originals courtesy of R. D. B. Fraser, from *J. Molec. Biol.* 129:463[1979].)

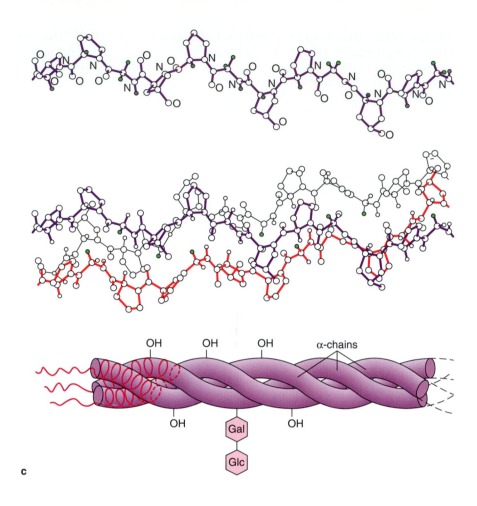

globular end caps. The —OH groups added to the modified forms of proline and lysine are particularly important as hydrogen-bonding sites that stabilize the triple helix.

As many as 25 different kinds of α-chains, all closely related but of distinct amino acid sequence, occur in vertebrates. These α-chains twist by threes in different combinations to form at least 16 distinct types of collagen molecules, designated types I to XVI (Table 7-1). The molecules associate through both noncovalent and covalent bonding to produce collagen fibers of various structures and dimensions. Most of these fibers are assembled primarily from one collagen type, with smaller amounts of other types in fibers in some cellular locations.

Type I, II, and III collagen molecules make up the main fibers of animal extracellular structures. These molecules, in which the central triple helix is almost uninterrupted and contains about 1000 amino acids, are the most fibrous and least flexible of the collagen types. In the remaining collagens the central triple helix has more interruptions and is more flexible. All these more flexible collagens, with the exception of type IV, are minor constituents of supportive structures.

Type I collagen molecules, the primary constituents of bone, skin, and tendons, form about 90% of the body's collagen. In bone, collagen fibers are structured from essentially pure type I molecules; in skin and tendons, collagen fibers contain type I in combination with lesser amounts of type III and V molecules. Fibers containing type I and smaller amounts of type V occur in the cornea. Type II collagen makes up the major fibers of cartilage.

Collagen fibers are arranged in rigid plates in bones, in parallel bundles in tendons, and in a dense meshwork in cartilage. In the cornea, collagen occurs in clear, crystalline layers with each layer overlapping the next at regular 90° angles (as in Fig. 7-2a); a similar pattern occurs in the skin of some animals. In all these locations, collagen fibers combine with greater or lesser quantities of other molecules forming the surrounding network of the extracellular matrix (see below).

Collagen fibers containing type I, II, III, V, and XI molecules show a regular pattern of cross-striations in the electron microscope (Figs. 7-2 and 7-3). This pattern is believed to reflect a packing arrangement in which the triple helices of individual collagen molecules line up in overlapping, parallel rows (see Fig. 7-3). A single

Table 7-1 Collagen Types and Their Occurrence

Type	Fiber Length (approximate)	Occurrence
I	300 nm	Skin, bones, teeth, tendons, ligaments, fascia, cornea
II	300 nm	Cartilage, notochord, intervertebral discs, vitreous humor of eye
III	300 nm	Skin, tendons, blood vessel walls, uterine walls; not in bone
IV	390 nm	Basal laminae, kidney glomerulus, lens capsule of eye
V	300 nm	Cornea, skin, bones, and tendons
VI	105 nm	All connective tissues, including both cartilagenous and noncartilagenous; may connect between extracellular matrix and cell surfaces
VII	450 nm	Basal lamina of skin; interface between stroma and epithelia in oral and vaginal mucosa, tongue, cornea, sclera, amnion
VIII	150 nm	Minor constituent of cartilage, sclera, basal laminae
IX	200 nm	Minor constituent of cartilage; also occurs in mixed fibers with type II
X	150 nm	Minor constituent of cartilage; also in cartilagenous regions undergoing transition to bone and in basal laminae
XI	Unknown	Minor constituent of cartilage
XII	Unknown	Minor constituent of tendons; associates with type I collagen
XIII	Unknown	Endothelial tissues (identified from gene sequence)
XIV	Unknown	Fetal skin and tendon
XV	Unknown	Identified from gene sequence
XVI	Unknown	Identified from gene sequence

collagen fiber may contain more than 10,000 collagen molecules per micrometer of length.

Type IV collagen molecules assemble into very fine, unstriated fibers in which either end can link to points near the tips of other type IV fibers. This linkage pattern enables type IV collagens to set up a fine, sheetlike layer, the *basal lamina*, which underlies the skin and epithelia covering internal organs and lining body cavities. In addition, basal laminae surround muscle cells, fat cells, and the Schwann cells forming myelin sheaths around nerve axons (see p. 140). Basal laminae covering cells in the glomeruli of the kidneys and the walls of blood capillaries work as molecular filters that block the movement of proteins but allow smaller molecules to pass. Basal laminae also serve as tracks along which cells migrate during embryonic development, and set up barriers to cell migration between major embryonic structures. Type VIII and X collagens also contribute to some of these basal laminae.

The remaining vertebrate collagen types form collagen fibers of small diameter that do not appear to be striated. Most of these collagens form parts of fibers containing the major I, II, and III types or serve as elements linking structures formed by the major col-

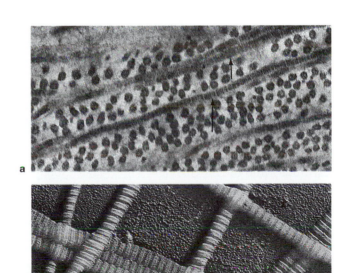

Figure 7-2 (a) Collagen fibers (arrows) in the extracellular matrix of the salamander. × 64,000. Successive fiber layers are arranged at 90° angles to each other. (Courtesy of R. Tucker.) (b) Collagen fibers isolated from bovine skin and prepared for electron microscopy by shadowing. × 30,000. (Courtesy of J. Gross.)

Figure 7-3 The arrangement of collagen molecules believed to produce the striations seen in the electron microscope. **(a)** Positively stained collagen fiber, revealing a pattern of cross-striations that repeats every 67 nm. The pattern and repeat length are believed to reflect a regular packing arrangement in which individual collagen molecules line up in staggered, parallel rows **(b)**. (Micrograph courtesy of R. Bruns.)

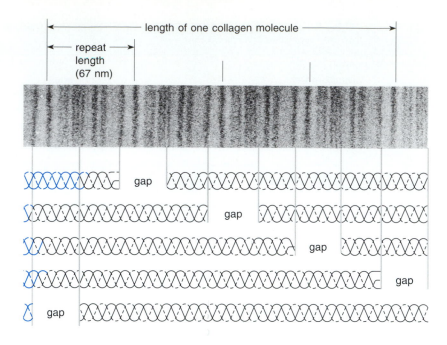

lagens to other elements in the extracellular matrix or to cell surfaces. Types IX, XII, and XIV form a group known as the *FACITS* (*f*ibril-*a*ssociated *c*ollagens with *i*nterrupted *h*elices); these associate in various minor proportions with type I and II collagens in extracellular structures. Types XIII, XV, and XVI are known primarily from cloned gene sequences and have not as yet been identified in the extracellular matrix. Type XVI, however, shows structural similarities to the FACITS and may be part of this group. Relatively little is known about the structure of invertebrate collagens except that their fibers are nonstriated and appear to resemble the flexible, fine fibers formed by type IV collagen in vertebrates.

Collagen Assembly The α-chains of collagen molecules are assembled on ribosomes attached to membranes of the ER. As they are synthesized, they penetrate individually into the ER channels (Fig. 7-4a). Carbohydrate units are added during penetration. Further steps in collagen synthesis are best known for types I, II, and III. Initially the polypeptides of these collagens contain extra amino acid sequences in the globular regions at both ends. These collagen precursors, called *pro-α-chains*, wind together by threes to form soluble *procollagen* molecules while still in the ER channels. The newly assembled procollagens then move to the Golgi complex (Fig. 7-4b), where they are packaged into secretion vesicles (Fig. 7-4c). Fusion of the vesicles with the plasma membrane (Fig. 7-4d) releases the procollagens to the cell exterior. In the extracellular region, processing enzymes remove the extra sequences at both ends of the molecules (Fig. 7-4e). This removal reduces the size of the globular end segments and converts the molecules into their insol-

uble finished form, in which they assemble into fibers (Fig. 7-4f).

Less is known about the synthesis of other collagens. At least two, types IV and VI, appear to be secreted in finished form, with no requirement for further processing. Collagens are synthesized and secreted primarily by cell types known as *fibroblasts* in connective tissues throughout the body, *chondroblasts* in cartilage, and *osteoblasts* in bone.

The collagens are altered by mutations that cause a variety of human disabilities ranging from mild to severe. Excessive collagen synthesis is responsible for various kinds of *fibrosis*, a condition particularly important in lung and liver diseases. Excessive collagen deposition has also been implicated in hardening of the arteries (atherosclerosis). At the other end of the spectrum, insufficient deposition or mutations in type III collagen are responsible for diseases such as *Ehlers-Danlos* or "*rubber man*" *syndrome*, in which the skin, tendons, and internal viscera are easily stretched and weak, producing loose joints, hyperextensive limbs, and susceptibility to rupture of internal organs and major arteries. In *osteogenesis imperfecta* or "*brittlebone*" *syndrome*, mutations produce type I procollagen molecules that fail partially or completely to assemble into triple helices. The bones are so fragile that literally hundreds of fractures may occur during the lifetime of an afflicted individual.

The Network Surrounding Collagen Fibers: Proteoglycans

A highly diversified group of glycoproteins, the *proteoglycans*, sets up the network surrounding collagen fibers in the extracellular matrix. These glycoproteins are distinguished by an unusually high carbohydrate content, which may approach as much as 90% to 95% of their total dry weight.

Proteoglycan Structure The proteoglycans are built around a linear polypeptide *core* to which long carbohydrate chains are linked, radiating outward from the core like the bristles of a bottlebrush (Fig. 7-5a). Depending on the proteoglycan type, from as few as 1 or 2 to as many as 200 carbohydrate chains of different types and lengths may link covalently along the core.

The primary carbohydrate groups of the proteoglycans, the *glycosaminoglycans*, or *GAGs* (Fig. 7-6), consist of unbranched chains assembled from hundreds of repeating two-sugar units. One of two amino sugars, *glucosamine* or *galactosamine*, is always present as one of the two sugars in a repeating unit. In the GAGs attached to core proteins of proteoglycans, many of the amino sugars are modified by the addition of sulfate groups. A small number of more highly diversified carbohydrate groups resembling the branched or unbranched structures of other glycoproteins may also attach to the polypeptide core. Depending on the length of the protein core, proteoglycans range from relatively small types occurring in bone and cornea, with total molecular weights of about 250,000, to the very large molecules characteristic of cartilage, which may approach 3,000,000 daltons.

In most extracellular structures, proteoglycans occur in the form shown in Figure 7-5a. In some structures, primarily in cartilage, proteoglycans combine further with an unsulfated GAG, *hyaluronic acid* (see Fig. 7-6), to form even larger complexes (see Fig. 7-5b). A single hylauronic acid molecule, which may be many nanometers in length, forms the axis of one of these hyaluronic acid–proteoglycan complexes. Linked noncovalently along the hyaluronic acid axis are proteoglycans, each having attached GAGs. The total molecular weight of a single hyaluronic acid–proteoglycan complex, which may contain hundreds of attached proteoglycans, can easily range in the hundreds of millions.

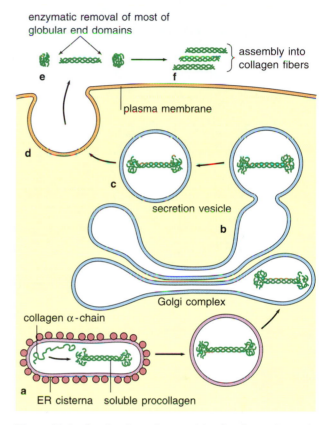

Figure 7-4 Synthesis and assembly of collagen types I, II, and III in vertebrates (see text).

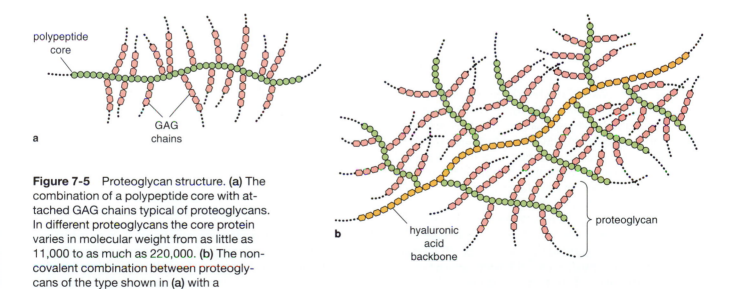

Figure 7-5 Proteoglycan structure. **(a)** The combination of a polypeptide core with attached GAG chains typical of proteoglycans. In different proteoglycans the core protein varies in molecular weight from as little as 11,000 to as much as 220,000. **(b)** The noncovalent combination between proteoglycans of the type shown in **(a)** with a hyaluronic acid chain.

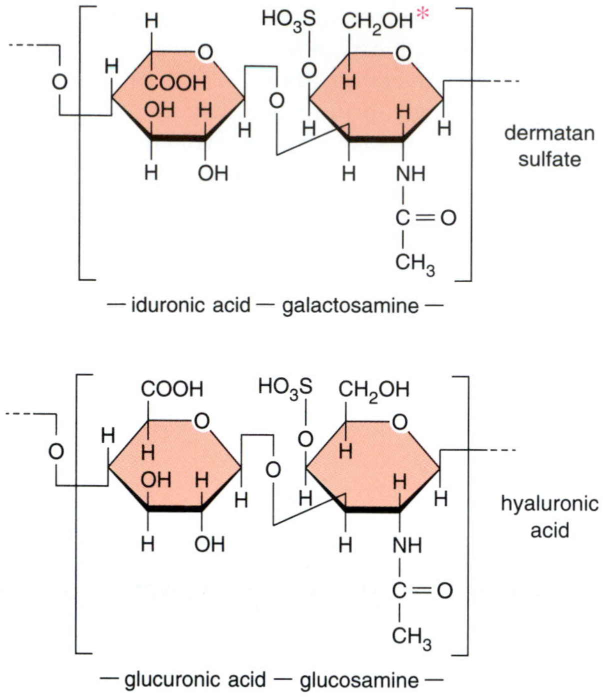

Figure 7-6 Representative GAG types. One part of each repeating two-sugar unit is always an amino sugar, either glucosamine or galactosamine. In all GAGs except hyaluronic acid the glucosamine or galactosamine residues are bound to sulfate groups as shown. An additional sulfate may occur at the asterisk in dermatan sulfate, a GAG occurring in tendons, fibrillar connective tissue, skin, and cornea. The many carboxyl and sulfate groups in GAGs make them highly acidic at physiological pH. Hyaluronic acid remains unsulfated.

These complexes are among the largest known biological molecules.

Functions and Occurrence of the Proteoglycans Networks formed by proteoglycans or hyaluronic acid–proteoglycan complexes trap and impede the flow of water molecules. Depending on the structure and composition of the networks, they may collect as much as 50 times their dry weight in water. The trapped water makes the networks resist compression and spring back to shape readily if deformed. Resistance to deformation is especially marked in the hyaluronic acid-proteoglycan network of cartilage, which is normally compressed to a fraction of its fully hydrated forms. The resistance to further compression contributes much of the rigidity and elasticity characteristic of cartilage.

In vertebrates, proteoglycans occur universally with collagens in the extracellular matrix (see Fig. 7-10). They are most abundant in cartilage, which consists of about 50% hyaluronic acid–proteoglycan complexes by dry weight. In bone, tendons, and in the extracellular structures of smooth muscle and aorta, proteoglycans occur primarily without attachment to hyaluronic acid chains. The fluid surrounding the egg in mammalian ovarian follicles contains proteoglycans in single form. These proteoglycans evidently increase the viscosity of the follicular fluid and help maintain the expanded volume of the follicles. Proteoglycans have been detected in at least one invertebrate, in *Drosophila*.

The degree of elasticity and the hardness of the animal extracellular matrix depend on the relative proportions and types of collagen fibers and proteoglycans and on the presence of other substances added to the complex. Cartilage, which contains proteoglycans in high proportions, is relatively soft. Proteoglycans are also abundant in basal laminae, in which the collagen-proteoglycan combination produces a fine and delicate, meshlike layer. Tendons, which contain almost pure collagen fibers, are tough and elastic.

In bone the proteoglycan network is impregnated with crystals of $Ca_{10}(PO_4)_6(OH)_2$—calcium hydroxyapatite—along with calcium carbonate and citrate. The combination of collagen fibers, proteoglycans, and minerals produces a total structure with extreme density and hardness, but also elasticity and high tensile strength. The mineral crystals embedded in the extracellular matrix of bone, which may make up from 45% to 85% of the bone tissue by weight, may be arranged in highly ordered woven, parallel, or layered structures.

In addition to their primary role as the network structures surrounding collagen fibers, proteoglycans also serve as sites for temporary or permanent cell adhesions. Cells use the temporary adhesions as anchors to push or pull against during ameboid movements (see p. 334 for details). Temporary adhesions to proteoglycans are especially important during embryonic development, when individual cells and cell layers migrate through the embryo. During these movements, some migrating cells follow tracks marked by changes in the types and kinds of proteoglycans covering stationary cells. Proteoglycans also appear to be necessary in the extracellular medium for many cell types to differentiate normally during embryonic development. During development of cancer, extensive changes in the types and kinds of proteoglycans are observed as normal cells change into malignant ones.

Molecules Interlinking and Binding the Extracellular Matrix to the Cell Surface

The collagen-proteoglycan network is interlinked and reinforced by a large and complex group of molecules. All these molecules contain multiple binding sites that can recognize more than one element of the extracellular matrix, allowing them to form extensive networks among the various collagens and proteoglycans, as well as with each other. Among the most important and frequent of the interlinking molecules are *fibronectin* and *laminin*. Fibronectin occurs in many, if not all extracellular structures, while laminin is apparently restricted to basal laminae.

Both fibronectin and laminin, other interlinking elements, and many collagens are also recognized and bound by cell surface receptors. Most of these molecules, which are membrane-spanning glycoproteins with both extracellular and cytoplasmic domains, are members of a large family of receptors known as the *integrins*. The integrins tie cells to the extracellular matrix and also form connections on their cytoplasmic extensions with elements of the cytoskeleton. Connections made by the integrins accomplish more than simply tying cells to the intracellular matrix; in many systems, linkage by the integrins triggers internal cellular responses via the cAMP-, InsP$_3$/DAG, or other pathways (see Chapter 6 for details of these pathways). The responses triggered by the attachments in-

clude adjustments in processes as fundamental as growth, cell division, and differentiation. The molecular steps linking these responses to the integrins remain unknown.

Connections from the extracellular matrix to the cell surface are also made by some proteoglycans that anchor directly in the plasma membrane via nonpolar regions of their core proteins or a nonpolar segment forming a "greasy foot" (see p. 104). The entire pattern of interlinks, membrane anchorages, cell surface receptors, and linkages to the cytoskeleton ties cells and the tissues containing them in position in animal body structures.

Fibronectin   Fibronectin (from *fibre* = fiber and *nectare* = to bind or connect) is a large glycoprotein assembled from two fibrous polypeptides of closely similar structure (Fig. 7-7), tied at one end into a V-shaped arrangement by disulfide linkages. Fibronectin has binding sites for cell surface receptors, collagens I, III, and IV, GAGs, hyaluronic acid, and other fibronectin molecules. These varied binding

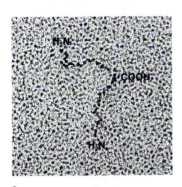

Figure 7-7   Fibronectin. **(a)** Fibronectin in the electron microscope. **(b)** Distribution of domains and binding sites along the fibronectin molecule. The molecule is a dimer consisting of two nearly identical polypeptide chains connected at their C-terminal ends by disulfide linkages. Each chain is arranged as a series of rigid domains (cylinders in the diagram) connected by short, more flexible segments (shown as lines connecting cylinders). The rigid segments carry binding sites for elements at the cell surface or in the extracellular material. The cell surface binding site contains the short amino acid sequence RGDS (arginine-glycine-aspartic acid-serine), which is recognized by cell surface receptors binding fibronectin. Other sites bind heparin, heparan sulfate, collagen, hyaluronic acid, other fibronectin molecules, and gangliosides (a type of membrane sphingolipid; see p. 95). (Micrograph × 280,000; courtesy of H. Furthmayr, from J. Engel et al., *J. Mol. Biol.* 150:97[1981].)

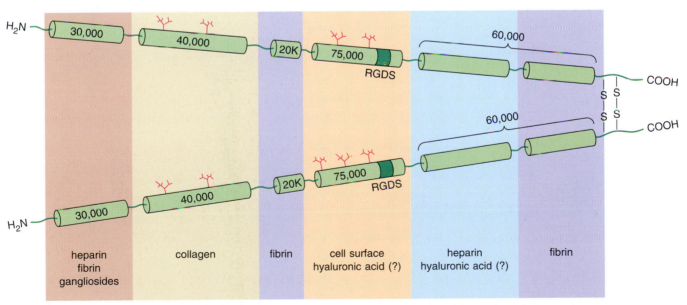

sites set up extensive interlinks that stabilize extracellular material and attach it to cell surfaces. Fibronectin occurs as a family of molecules of related structure.

Fibronectin has been implicated in functions other than linking extracellular material. Many cultured cell types, including fibroblasts and fat storage cells, can take on shapes approaching their appearance in their normal body locations only if surfaces of the culture vessel are coated with fibronectin. Many types of cancer cells lose their ability to synthesize fibronectin. At the same time, they lose their normal cell shape, surface morphology, cytoskeletal organization, and attachments to the extracellular matrix. Supplying these cells with fibronectin often restores the normal morphology and organization of the cell surface and cytoskeleton and reattaches the cells to the extracellular network.

Fibronectin can also be identified in the "tracks" of extracellular material followed by embryonic cells in their migrations through the embryo during development. Antibodies against fibronectin inhibit the migrations of embryonic cells, including the movement of neural crest cells from their original location in the nerve tube to other parts of the embryo. Similarly, the

extension of nerve cells into axons and dendrites during development is arrested if fibronectin is eliminated from the extracellular material. These discoveries link fibronectin to development and motility as well as structural support.

Fibronectin has been studied most extensively in mammals, where it is especially abundant in basal laminae, in the loose connective tissue underneath the skin and between organs of the body, and in the capsular material surrounding nerves and muscles. Fibronectin appears to be generally distributed among vertebrates and has been detected in many invertebrates including *Drosophila*. Several pathogenic bacteria, among them *Staphylococcus* and *Streptococcus*, have surface receptors that can recognize and bind firmly to extracellular structures containing fibronectin. The firm binding aids the infective process by stabilizing attachment of these bacteria to cell surfaces and extracellular materials.

Laminin Laminin, a large, cross-shaped glycoprotein assembled from three polypeptides, has binding sites for type IV collagen, cell surface receptors, GAGs, and other laminin molecules (Fig. 7-8). Laminin is re-

Figure 7-8 Laminin. (a) Laminin in the electron microscope. (b) Laminin is built up from three polypeptide chains: one α-chain and two closely related β-chains, β_1 and β_2, bound together by disulfide linkages at the intersection of the short arms. Some segments of the long arm of the cross wind into a three-stranded coiled-coil structure (not shown). Besides binding at the sites shown, laminin molecules can self-associate into aggregates through laminin-laminin binding sites at the tips of the short arms. (Micrograph $\times$ 367,000; courtesy of H. Furthmayr, from J. Engel and H. Furthmayr, *Meth. Enzymol.* 145:3[1987].)

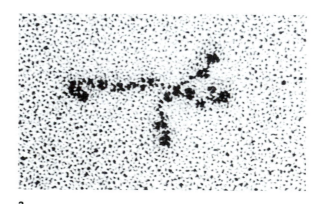

a

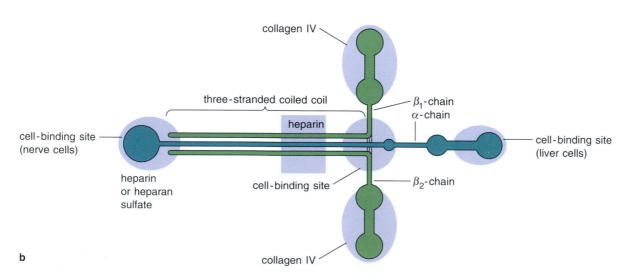

b

stricted primarily to basal laminae, in which it is the most abundant interlinking glycoprotein.

Laminin is widely distributed among vertebrates and has been detected in invertebrates such as insects and sea urchins. It appears as early as the 16-cell stage in some animal embryos, in which it shares many of the functions of fibronectin in the stimulation of cell adhesion, motility, and development. Laminin promotes growth of nerve axons in embryos and also appears to be a significant element in nerve regeneration after injuries in adults. Thus laminin, like fibronectin, has many functions in addition to its role in crosslinking and anchoring extracellular structures to the cell surface.

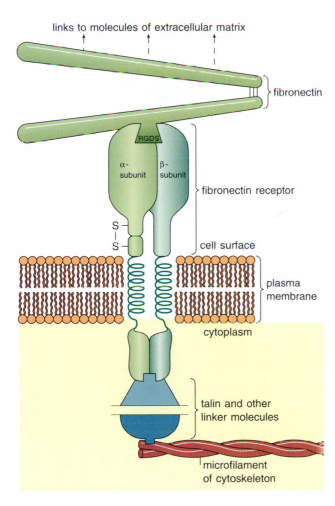

Figure 7-9 The fibronectin receptor, a member of the integrin family of cell surface receptors. Like other members of this family, the fibronectin receptor consists of two transmembrane polypeptides, the α- and β-subunits. At the cell surface the receptor links to fibronectin, which in turn forms linkages to collagen and other molecules of the extracellular matrix. On the cytoplasmic side of the plasma membrane the receptor forms connections via talin and other linkers to actin microfilaments of the cytoskeleton.

The Integrins At least 17 different but related integrins have been discovered, each with different binding specificities for elements of the extracellular matrix. All contain one each of two polypeptides, the α- and β-*subunits* (see Fig. 7-9). The particular structural element of the extracellular matrix bound by an integrin, such as fibronectin, laminin, or collagen, is determined by the combination of α- and β-subunits in the receptor. At least 14 α- and 8 β-subunits join in different combinations in integrins.

Many of the extracellular interlinkers recognized and bound by integrins share the amino acid sequence element RGD (arginine-glycine-aspartic acid in the one-letter symbols for the amino acids). Polypeptides containing the RGD sequence element alone can compete with linkers such as fibronectin or laminin for binding to integrins. Other parts of the interlinkers also contribute recognition elements; these elements may allow integrins to bind individual interlinker types with greater specificity.

The integrin binding fibronectin, for example (Fig. 7-9), has a site that recognizes and binds fibronectin at the cell surface. Fibronectin, in turn, links to collagen and other elements of the extracellular matrix. At the inner surface of the plasma membrane the fibronectin receptor forms linkages via proteins such as *talin* and *vinculin* (see Table 11-1) to microfilaments forming parts of the cytoskeleton. As many as 500,000 fibronectin receptors may occur on a single cell. At least some major laminin receptors have similar structure and properties; others appear to differ significantly from the integrin family.

Mature, fully differentiated cells often possess multiple integrin receptors linking them to more than one molecular type of the extracellular matrix. Some receptors in the integrin group recognize and bind directly to collagen.

In embryos, attachments to the extracellular matrix change in programmed patterns as development proceeds. For example, neural crest cells have no associations with elements of the extracellular matrix until their migrations begin. As these cells break loose from their cell-cell attachments in the neural tube, linkages form with strands of fibronectin that extend throughout the embryo. The fibronectin attachments are made and broken repeatedly as the neural crest cells follow the strands of extracellular material, which apparently provide at least part of the information directing the cells to their final destinations in the embryo. Once the neural crest cells reach their final locations, attachments to fibronectin may persist or may be replaced by other cell-matrix or cell-cell adhesions as the neural cells differentiate into mature forms. Attachments made to the extracellular matrix are normally permanent once cells mature in their final locations in the body.

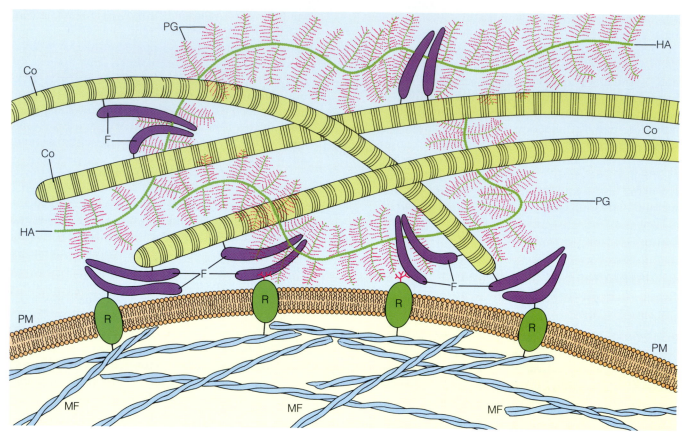

Figure 7-10 Crosslinking between collagen fibers (Co), proteoglycans (PG), hyaluronic acid (HA), and fibronectin (F) in the extracellular matrix of animal cells. Fibronectin also ties the network to surface receptors (R) in the plasma membrane (PM); the surface receptors are transmembrane glycoproteins that bind at their cytoplasmic ends to microfilaments (MF) of the cytoskeleton.

Collagens and proteoglycans form an extensive fiber-network complex that is interlinked and attached to cell surface receptors either directly or by fibronectin, laminin, and other interlinking molecules. Many cell surface receptors in the integrin family tie the extracellular matrix to the cytoskeleton on the cytoplasmic side of the plasma membrane, thereby linking the entire system into a framework that extends from the inside to the outside of the cell (Fig. 7-10).

Through the connections and reinforcements provided by bones and tendons, animals are able to withstand the force of gravity, translate muscle contractions into body movements, and absorb the shocks of locomotion without injury. The extracellular matrix forms a part of the protective surface skin of animals and ties the skin to the underlying muscles, bones, cartilage, and organs. Besides providing structural support, the extracellular matrix acts as a molecular filter in capillaries and kidney glomeruli, provides sites for cell adhesion, and promotes cell motility. The latter function is especially important in embryos, in which cell and tissue movements follow tracks set up by the extracellular matrix. The extracellular material also plays a role in wound healing and the behavioral and structural changes that occur in transformation of normal to cancer cells. These manifold structural and functional activities reflect the importance of the extracellular matrix as one of the fundamental evolutionary adaptations making multicellular life possible in animals.

THE EXTRACELLULAR STRUCTURES OF PLANTS: CELL WALLS

The walls of plant cells provide support for plant tissues, protect cells from mechanical injury, and act as barriers to infection by bacteria and fungi. Only a very few cell types, in tissues such as reproductive tissue and endosperm in some species, are wall-free in higher plants. Even these cells are without walls for only a brief period during development.

When first laid down during cell division, plant cell walls are relatively soft, flexible, and extensible.

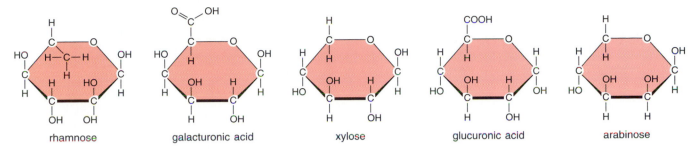

rhamnose   galacturonic acid   xylose   glucuronic acid   arabinose

Figure 7-11 Carbohydrate subunits commonly found in plant cell walls in addition to the glucose, galactose, mannose, and fucose sugars shown in Figure 4-4.

These *primary cell walls* elongate as cells enlarge during maturation of plant tissues. During and after cell elongation, additional cell wall material is laid down to produce the much thicker and more rigid *secondary cell walls* of mature plant cells. Primary cell walls are relatively simple structures consisting of a single layer of more or less homogeneous wall material. Secondary walls are multilayered, complex, and occur in an almost endless variety of structural forms.

The walls of plant cells are often thought of as dead extracellular material that confines the living material of cells, with functions essentially limited to skeletal support. However, plant cell walls are no more dead or alive than the extracellular matrix of animals, such as the material making up most of the mass of bones, tendons, and cartilage. Plant cell walls are actually an extension of the cytoplasm outside the plasma membrane, with its own complement of enzymes and other proteins, which changes in composition and function depending on development and conditions such as mechanical stress, wounding, and attack by pathogens. Cell walls extend and grow, and they may break down and reform during processes such as wound healing, leaf abscission, adventitious bud formation, and fruit ripening.

Molecular Constituents of Plant Cell Walls

The majority of the varied fiber and network molecules in plant cell walls are polysaccharides assembled from different combinations of only nine monosaccharide subunits, linked in straight or branched chains. Seven of these, *glucose, galactose, mannose, fucose, rhamnose, galacturonic acid,* and *glucuronic acid,* are six-carbon sugars. The remaining two, *xylose* and *arabinose,* are five-carbon sugars (see Figs. 4-4 and 7-11).

Proteins, usually forming about 5% to 10% of the network, are present as either structural elements or enzymes involved in wall synthesis. Most of the structural proteins are glycoproteins.

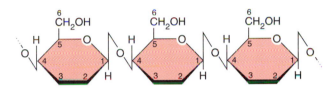

a

b

Figure 7-12 Cellulose molecules and microfibrils. **(a)** Cellulose, a polymer of glucose units bound into a linear chain by β(1→4) linkages. **(b)** Combination of cellulose chains to form a cellulose microfibril. Except for their parallel alignment the precise arrangement of cellulose chains within microfibrils is not known.

Cell Wall Fibers *Cellulose* is the predominant polysaccharide fiber of higher plant cell walls. It is formed entirely from hundreds to thousands of glucose subunits linked together in unbranched, linear chains by β(1→4) bonds (Fig. 7-12a; see p. 45 for a description of these and other bonds linking sugars into polysaccharides). Individual cellulose molecules combine in ordered, parallel rows to form *microfibrils* ranging in diameter from 10 to as much as 85 nm (Fig. 7-12b). X-ray diffraction studies (see Appendix p. 802) suggest that some 60 to 70 individual cellulose molecules

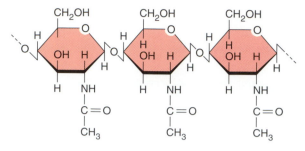

Figure 7-13 Chitin, assembled from repeating acetylglucosamine units joined by β(1→4) linkages. Chitin is a primary cell wall constituent of fungi.

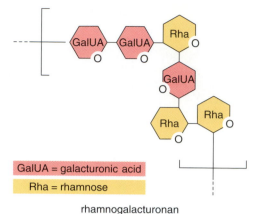

GalUA = galacturonic acid
Rha = rhamnose

rhamnogalacturonan

Figure 7-14 Structure proposed by P. Albersheim for a fragment of rhamnogalacturonan, a pectin from sycamore cell walls. The fragment has a backbone structure of galacturonic acid and rhamnose units. Attached to the repeating units of the backbone in this and other pectins are extensive side branches containing arabinose, galactose, and other sugar units. A complete pectin molecule consists of hundreds of the units (700 for the rhamnogalacturonan fragment shown) plus added side chains.

line up side by side to form a cellulose microfibril, which approximates a crystalline structure. Within a microfibril, individual cellulose molecules are held in a parallel, unbent form by hydrogen bonds between adjacent glucose chains. Because of difficulties in interpreting X-ray patterns obtained from cellulose, the pattern of hydrogen bonds and arrangement of molecules within cellulose microfibrils are only partially understood.

Cellulose makes up from 20% to 50% of the dry weight and as much as 15% of the volume of the wall material in most plant cells. Wall fibers of other types also occur in the walls of some higher plant cells. For example, *callose*, assembled from glucose units in β(1→3) instead of β(1→4) links, occurs in regions where cell walls have been damaged by wounding. Callose is also found as a regular wall constituent in yeasts and other fungi. Algal cell walls contain polymers of xylose, mannose, or glucose units in pure form or mixed polymers made up from any two of these three monosaccharides. *Chitin*, a polysaccharide built up from acetylglucosamine units joined by β(1→4) linkages (Fig. 7-13), is a primary cell wall constituent of fungi. Chitin is also the primary structural fiber of the external skeletons of insects, crustaceans, and other arthropods.

Network Molecules of Plant Cell Walls The network surrounding cellulose fibers in plant cell walls is made up primarily of two types of polysaccharide molecules, *pectins* and *hemicelluloses*. Pectins are a varied group of branched polysaccharides built on backbones containing only galacturonic acid units or mixtures of galacturonic acid and rhamnose. Carbohydrate side chains attached to the backbones may contain glucose, galactose, fucose, xylose, arabinose, and other sugars. (Fig. 7-14 shows a possible structure for a pectin fragment from dicotyledonous plants.) Pectins carry multiple negative charges contributed by the —COOH groups of their many galacturonic acid units. As a result, pectins are highly acidic at the

pH ranges characteristic of cell walls. Their acidic reaction may be important in the formation of electrostatic attractions with *extensins*, positively charged glycoproteins of cell walls (see below). Interactions with Ca^{2+} are also considered likely to play an important role in the reactions linking pectins into cell wall networks.

The highly branched structure of pectins traps water molecules and enables these molecules to form stable gels. (Extracted pectins are used to produce the gel-like consistency of fruit jams and jellies.) The pectin gels, which range from almost waterlike to highly rigid, probably have the greatest influence among the various polysaccharides in establishing the physical consistency of plant cell walls. Indications are that pectins vary greatly in types and structure between different plant species, and even in different regions of cell walls in the same plant. In dicots, in which pectins have been most extensively studied, they may make up 25% to 30% of the dry weight of cell walls. Hemicelluloses are built on backbones of either glucose or xylose units. In dicots the primary hemicellulose is *xyloglucan*, structured from a glucose backbone with xylose side chains (Fig. 7-15). Other sugars, including galactose, fucose, and arabinose, occur in smaller amounts in the side chains. Like pectins, hemicelluloses trap water molecules and contribute to the gel-like characteristics of the cell wall network. Hemicelluloses may be neutral or acidic at wall pH. In cell walls, hemicelluloses are hydrogen-bonded as

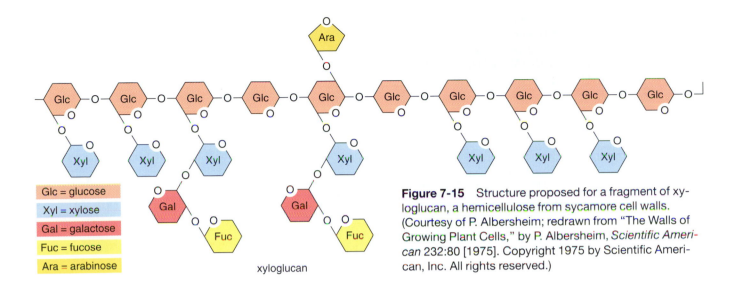

Glc = glucose
Xyl = xylose
Gal = galactose
Fuc = fucose
Ara = arabinose

xyloglucan

Figure 7-15 Structure proposed for a fragment of xyloglucan, a hemicellulose from sycamore cell walls. (Courtesy of P. Albersheim; redrawn from "The Walls of Growing Plant Cells," by P. Albersheim, *Scientific American* 232:80 [1975]. Copyright 1975 by Scientific American, Inc. All rights reserved.)

a sheath closely surrounding cellulose microfibrils (see Fig. 7-17). Hemicelluloses make up about 20% of the dry weight of cell walls in angiosperms.

Cell Wall Proteins Cell walls contain structural glycoproteins and a varied complement of enzymes. Four types of structural glycoproteins occur universally in higher plant cell walls: *extensins, glycine-rich proteins, proline-rich proteins*, and *arabinogalactan proteins*. All the structural proteins, except the glycine-rich proteins, contain hydroxyproline as a common modified amino acid (the same modified amino acid occurs in quantity in collagens; see p. 179). Among the enzymes are a long list of transferases shifting sugar groups from precursors to matrix polysaccharides, hydrolytic enzymes that remove sugar groups and open breaks in matrix polysaccharides, peroxidases, phosphatases, proteases, oxidases, and reductases. Some of the enzymes that can hydrolyze molecules in the cell walls of invading fungi and bacteria play a defensive role in protecting plants from infection.

Extensins are primary structural glycoproteins of cell walls (Fig. 7-16). In addition to a high proportion of hydroxyproline, serine and lysine also occur in significant amounts; the lysines give the extensins a strongly positive charge. The positive charge suggests that the extensins may interact through ionic attractions with negatively charged matrix components of cell walls, particularly the pectins. The extensins also interlink among themselves through covalent bonds between tyrosine residues, and possibly also with pectins. Extensins contain carbohydrates in very high proportions, as much as 65% or more by weight. The carbohydrate groups are primarily short, linear chains attached at frequent intervals to the polypeptide backbone.

Although the name suggests that the extensins are flexible molecules, the opposite actually appears to be the case. They occur in highest proportions in the rigid walls of tissues that act as skeletal supports, and at lowest levels in the pliant walls of growing cells. One cell wall type containing extensin (the sclerenchyma fibers of leaves or stems), in which the glycoprotein occurs in highest concentration, falls between wrought iron and steel in tensile strength!

Extensins may also be involved in wound healing, developmental changes, and defense against pathogens as well as structural support. They are widely distributed in lower plants as well; some algae, such as *Chlamydomonas*, contain extensins as the primary wall constituent.

Glycine-rich glycoproteins occur in the plant vascular system and are also present in regions undergoing wound repair. Proline-rich glycoproteins also occur in quantity in the vascular system, particularly in vessel walls. Both glycoprotein types probably undergo conversion to highly hydrophobic residues in walls undergoing lignification (see below). The remaining glycoprotein group, the arabinogalactan proteins, are often located in the region just outside the plasma membrane and have been suggested as functioning in signal reception. Other proposed functions for these proteins are as glues and lubricants. It should be mentioned that the functions of all the cell wall glycoproteins, including the extensins, remain incompletely understood.

Other cell wall glycoproteins are identified as *lectins*, soluble molecules with multiple binding sites that link to carbohydrate groups with varying degrees of specificity. Lectins may contribute to the extracellular network by crosslinking carbohydrate units of

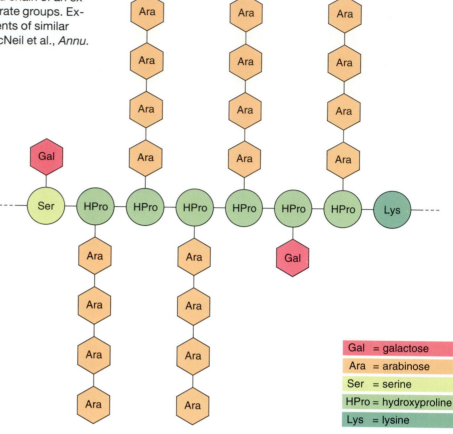

Figure 7-16 Segment of the amino acid chain of an extensin molecule with attached carbohydrate groups. Extensins consist of many repeated segments of similar structure. (From data presented in M. McNeil et al., *Annu. Rev. Biochem.* 53:625 [1984].)

Gal	= galactose
Ara	= arabinose
Ser	= serine
HPro	= hydroxyproline
Lys	= lysine

matrix polysaccharides. They also form part of the defensive armor of plants; lectins recognize and combine with chitins and other wall elements of invading fungi. They also serve as part of the mechanism recognizing the symbiotic *Rhizobium* bacteria that fix nitrogen in the roots of leguminous plants. Lectins are secreted into plant vacuoles as well as the cell wall.

Plant lectins have proved useful as *mitogens* for animal cells—that is, molecules that stimulate cell division. Many animal cell surface receptors are recognized and bound by plant lectins; some respond by triggering rapid cell division. A standard method for growing leucocyte cultures, for example, involves exposing freshly drawn blood to phytohemagglutinin, a plant lectin extracted from red kidney beans. The lectin combines with surface receptors on the leucocytes, stimulating them to divide and producing a rich culture of white blood cells.

Lignin Cell wall networks may be secondarily impregnated with varying amounts of *lignin* (from *lignum* = wood), a dense, insoluble substance formed by complex alcohols linked covalently into branched networks. During lignification, water is removed, wall glycoproteins are converted into an insoluble residue, and the wall material becomes a dense, hydrophobic

mass. This lignified, residual wall material, present in quantity in woody tissues, is probably responsible for much of the common misconception that plant cell walls are dead material. Minerals may also be deposited in cell walls, as in the silica impregnating cell walls in grasses.

Lignin is so resistant to solution or breakdown that its precise structure has so far defied chemical analysis. Much of the density and strength of wood, particularly the hardwoods, can be attributed to large quantities of lignin in the cell walls forming the woody tissue. The walls of different cell types and species, and even different regions of the walls of individual cells, contain distinct lignin types.

Cell Wall Structure

Cell wall molecules interlink through covalent, ionic, and hydrogen bonds into an immensely complex "supermolecule" that surrounds individual cells and extends the length and breadth of the plant. The structure is presently understood only in broad outline; even the individual pectin, hemicellulose, extensin, and lignin molecules are known only as fragments of the types shown in Figures 7-14 and 7-15, released from cell walls by chemical or enzymatic digestion.

P. Albersheim and his coworkers and other investigators have analyzed cell wall structure by digesting walls with enzymes and noting the linkages retained between fragments released by this treatment. The results of this work have led to several tentative models for cell wall structure. (Figure 7-17 summarizes major features of these models in simplified form.) The models agree that cellulose microfibrils are embedded in the cell wall network, immediately surrounded by a hydrogen-bonded sheath of hemicellulose. The cellulose-hemicellulose complexes are linked by hydrogen bonds to the rest of the network, which is formed by pectins of various kinds. The cellulose-hemicellulose-pectin structure may be reinforced by extensin and other glycoprotein molecules linked into an arrangement resembling chicken wire. The entire structure, held together by numerous covalent and noncovalent linkages, has great resistance to stretching and compression. At the same time, it is open enough to admit water, ions, and small molecules.

The proportions of cellulose, pectins, hemicellulose, and other components vary considerably between primary and secondary cell walls, between layers within walls and cell walls in different parts of the same plant, and between different species. This variability indicates that in different cell wall types there are likely to be extensive differences in the kinds of individual network molecules present and their patterns of linkage.

Cell Wall Synthesis

Primary cell walls contain cellulose microfibrils in a loose, meshlike array embedded in a relatively soft, gel-like matrix. As cells mature, they may elongate to as much as 10 to 100 times their embryonic length. During this elongation the cellulose meshwork is first loosened and then stretched. Force for stretching is supplied by osmotic pressure built up in the cytoplasm of the cell. Throughout the period of stretching and elongation, new cellulose fibers are laid down just outside the plasma membrane, gradually thickening the wall as it extends. These newly added microfibrils are laid down in many cell wall types in beltlike, parallel courses in an arrangement that is more ordered than the irregular meshwork characteristic of primary walls.

The new cellulose molecules and microfilbrils laid down during wall growth are evidently assembled from precursors at the plasma membrane. Electron microscopy of freeze-fracture preparations (see p. 791) by R. M. Brown shows that cellulose microfibrils are assembled in groovelike depressions in the plasma membrane (Fig. 7-18). Fractures splitting the membrane into bilayer halves show protein-sized

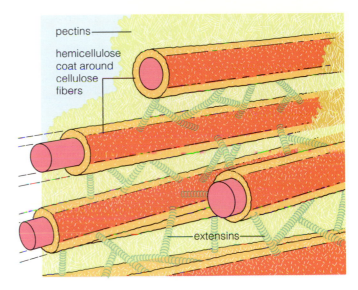

Figure 7-17 The possible arrangement of cellulose microfibrils, hemicellulose, pectins, and extensins in cell wall structure.

Figure 7-18 Freeze-fracture preparation of a plasma membrane in regions synthesizing cellulose microfibrils. New microfibrils are believed to form in the grooves. The rows of particles (arrows) are considered to be the enzyme complexes synthesizing cellulose. From the alga *Oocystis*; × 64,000. (Courtesy of R. M. Brown, Jr.)

particles embedded within the membrane at the ends of the grooves, arranged either in linear rows (as in Fig. 7-18) or in small circles called *rosettes*. The linear rows or rosettes are believed to be arrangements of the enzymes that synthesize cellulose microfibrils from glucose (Fig. 7-19).

Microtubules appear to be involved in some way in orienting newly synthesized cellulose microfibrils.

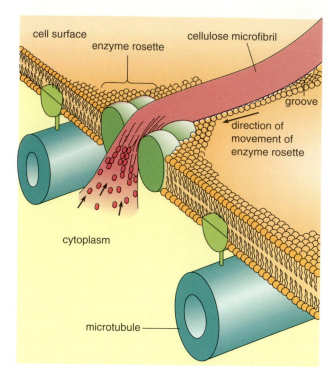

Figure 7-19 Possible relationship of an enzyme rosette and plasma membrane groove in generation of cellulose microfibrils. Each rosette, consisting of a circle of six protein subunits arranged in a hexagon, is thought to cast a microfibril about 5 nm in diameter. Larger microfibrils are assembled by groups of rosettes that may number from two or three to hundreds. As they are assembled, the 5-nm microfibrils fuse into larger microfibrils. The rosette may be guided by parallel tracks of microtubules as shown.

Many investigators have noted a layer of microtubules in the cytoplasm just under growing cell walls. In cells in which microfibrils are laid down in parallel arrays the microtubules just inside the plasma membrane frequently extend in the same direction. Interference with these microtubules usually, but not always, causes disarray of the cellulose microfibrils in the walls. This was first noted by P. B. Green, who studied the effects of colchicine on cells of the alga *Nitella*. (Colchicine is a drug that interferes with microtubule assembly—see p. 288.) Growing *Nitella* cells normally deposit microfibrils in parallel arrays at an angle to the long axis of the cell. When colchicine is added during cell wall growth, the microfibrils are laid down in random patterns. With some exceptions, other experiments carried out with microtubule inhibitors have had the same result: Interference with microtubules destroys cellulose microfibril orientation but does not inhibit cellulose synthesis.

How microtubules determine the orientation of cellulose microfibrils remains unknown. One possibil-

ity is that microtubules bind to integral membrane proteins on either side of the enzyme rosettes, establishing boundaries or "fences" (as shown in Fig. 7-19) that confine the movement of the enzymes synthesizing microfibrils to linear pathways.

Synthesis of the network molecules holding cellulose fibers in place in cell walls has been traced by experiments using labeled glucose. The label is first incorporated into precursors of pectins and hemicelluloses in the Golgi complex. From this location the molecules, still in the form of unfinished intermediates, move to the plasma membrane in vesicles. As they reach the cell border, the vesicles fuse with the plasma membrane and release their contents to the cell exterior. There enzymes located in the cell wall complete synthesis of the pectins and hemicelluloses and set up covalent bonds linking them into the wall. Few of the steps in these reactions are known.

Plasmodesmata: Communicating Junctions of Plant Cells

Both primary and secondary plant cell walls retain minute openings, the *plasmodesmata* (singular = *plasmodesma*), through which the cytoplasm of adjacent cells apparently remains in open communication (Fig. 7-20). A plasmodesma is usually roughly cylindrical in shape, with the cylinder narrowed in diameter at both ends. The plasma membranes of adjacent cells completely line the cylinder walls, so that there is no break in the membranes between cells. A narrow, tubelike extension of the ER runs through the central axis of the plasmodesma channel. Surrounding the central tubule is a sleevelike extension of the cytoplasm. The structure of a plasmodesma thus apparently sets up continuities between the plasma membranes, soluble cytoplasm, ER membranes, and ER cisternae of adjacent plant cells. Plasmodesmata are found in all plant groups from bryophytes to angiosperms. Equivalent structures have also been observed in some fungi and multicellular algae.

The experiments linking plasmodesmata to direct communication between plant cells closely resemble those linking gap junctions to direct cytoplasmic communication in animal cells. For example, R. M. Spanswick and J. W. F. Costerton placed electrodes in adjacent plant cells and found that electrical resistance between the cells is about 50 times lower than that expected if the cells were separated by continuous plasma membranes. Experiments by R. L. Overall and B. E. S. Gunning showed that the flow of current reflecting ion movements between adjacent cells is directly proportional to the number of plasmodesmata linking the cells. P. B. Goodwin and his colleagues found that fluorescent dyes with molecular dimensions too large to pass directly across membranes

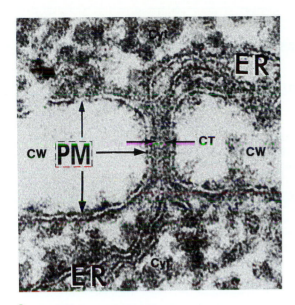

a

Figure 7-20 Plasmodesmata. **(a)** A plasmodesma in the wall separating two onion root cells. × 240,000. (Courtesy of B. E. S. Gunning and Springer-Verlag, from *Protoplasma* 111:134 [1982].) **(b)** Plasmodesma structure. A central tubule, evidently derived from cisternae of the ER on either side of the cell wall, lies in the channel of a plasmodesma. Channel diameter at the narrowest point is about 25 to 30 nm. CW, cell wall; PM, plasma membrane; ER, endoplasmic reticulum; CT, central tubule or desmotubule of plasmodesma; Cyt, cytoplasm underlying the wall and held in communication by the plasmodesma.

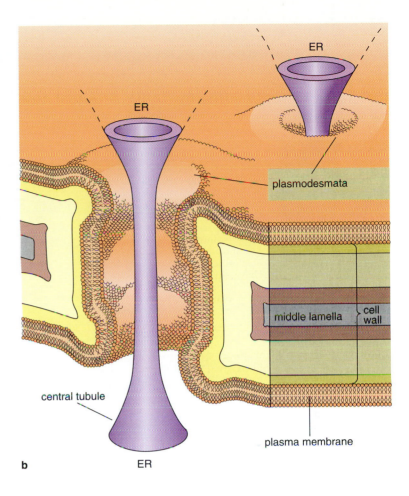

b

move readily from one cell to its neighbors if injected; the upper limit for free movement is about 700 daltons. Goodwin also found that injected Ca^{2+} drastically reduced the flow of substances between adjacent plant cells, indicating that movement through plasmodesmata is regulated (see the Experimental Process essay by Goodwin on p. 196).

Plasmodesmata are thus equivalent in function to the gap junctions of animal cells. B. Epel, M. Schindler, and others have found that antibodies developed against the connexin proteins of gap junctions (see p. 172) cross react with plasmodesmata, indicating that related proteins may occur in the two structures.

Plasmodesmata develop during deposition of the primary cell wall between dividing cells, at sites where spindle microtubules and strands of the ER are trapped (see p. 719 for details). As later growth converts primary cell walls to secondary form, plasmodesmata persist in their original numbers, which in mature walls may vary from 1 to 140 per square micrometer. At these levels a mature plant cell may have

as many as 100 to 100,000 plasmodesmata connecting it to its neighbors.

Cell walls protect and bind cells together in plants, give permanent shape to plant cells, resist the internal forces developed by osmosis, and protect cells from infection. The primary fiber of plant cell walls is cellulose; pectins, hemicellulose, and the extensins are the primary network molecules. New cellulose microfibrils are laid down by enzyme complexes embedded in the plasma membrane, possibly guided by microtubules on the cytoplasmic side of the membrane. Plasmodesmata persist in cell walls as sites where the cytoplasm of adjacent cells is continuous, providing a function in plants equivalent to the gap junctions of animal cells. The extracellular materials of animals, plants, and other eukaryotes are unusual among biological structures in that they are assembled, in all their complexity, primarily by enzymatic systems located outside the cell. (The extracellular materials of prokaryotes are described in Supplement 8-1.)

Cell/Cell Communication in Plants: An Open and Shut Case

Peter Goodwin

PETER GOODWIN is Reader in Horticulture in the School of Crop Sciences, University of Sydney, NSW. He obtained his Bachelor and Master of Science in Agriculture Degrees at Sydney University, and then went on to the University of Nottingham for his Ph.D. He was employed at the Scottish Horticultural Research Institute and at the Research School of Biological Sciences, Australian National University, before moving to his present position. As well as interests in the developmental biology of plants, he has been involved in the selection and breeding of native Australian plants.

The work described in this paper sprang from what, in the late 1970s, was our almost total ignorance of the actual properties of the cell-to-cell pathway via plasmodesmata in plants. The state of knowledge in this now distant era was reviewed by various authors in Gunning and Robards (1976).[1] In general, while a large conjectural framework had been built up over the previous century, the actual characteristics of the system were unknown. Cells of higher plants had been shown to be connected by a relatively low electrical resistance pathways—for example by Spanswick in 1972[2]—but this technique was too insensitive to give any detailed understanding of the physiology of plasmodesmata.

Meanwhile, progress in the understanding of cell-to-cell communication in animal cells, although begun much later than the plant studies, had proceeded much more rapidly. A key advance was in the development of a graded series of fluorescent probes to examine the size and permeability of the intercellular pores, the gap junctions. Application of this technique to plant cells showed that the molecular exclusion limit is similar in both plants and animals.[3]

If the cytoplasm of all plant cells is linked by plasmodesmata, which allow molecules of the order of 3 nm (700 daltons) to move from cell to cell, then most small metabolite molecules should be able to move from cell to cell, so complicating the normal processes of tissue differentiation. In addition, when some cells interconnected by plasmodesmata in a tissue are damaged—a very common occurrence—then the same small metabolites should leak from the uninjured to the damaged cells, making the tissue vulnerable to metabolite leakage through damage to only a fraction of its cells. The obvious suggestion is that perhaps the plasmodesmata can shut when they are exposed to the "outside" or to a damaged cell. In addition, the same valving perhaps operates during differentiation.

At this moment an able student, Michael Erwee, arrived and was given the challenge of discovering whether plasmodesmata can be caused to shut, if this is reversible, and what is the key signal which causes closure and opening. The obvious candidate was the cytoplasmic calcium ion concentration, which is far higher "outside" than "inside" the cytoplasm, and which was already known to be involved in many processes in the cellular biology of both animals and plants. Of special relevance was the demonstration by Rose, Simpson, and Loewenstein in 1977[4] that calcium ions regulate gap junction permeability. In addition, a number of treatments likely to increase cytoplasmic calcium ion concentrations were known to decrease electrical coupling.

The approach used combined trials with various inhibitors, which would be expected to increase cytoplasmic calcium ion concentrations, and also the direct injection of various ions. The work employed leaves of the simple water plant *Egeria densa.* The leaves have only two layers of cells, so that all cells can be easily seen under the microscope. In addition, one can avoid a major possible artifact in fluorescent probe movement, where the probe moves from cell to cell in the cell wall. In *Egeria* the cells are all exposed to water, so that probes show very limited movement in the cell wall—they diffuse out into the solution. In addition, because of the large, clearly defined cells, it is easy to tell if the probe is in the cell wall, cytoplasm, vacuole, or nucleus.

The next possible source of artifact was that the probes might simply leak across the plasmalemma (plasma membrane). The tissues were soaked in solutions of the probes and washed. No detectable dye entry occurred within 1 hour. The next step was to inject the probes. Those of molecular weight up to 665 daltons moved rapidly from cell to cell. At the same time it was possible to check for another possible source of artifacts—that the probes damage the cells and then leak to successive cells. As judged by a continuation of cytoplasmic streaming, the cells were not damaged by the probes. The largest probe dye able to move, fluorescein glutamyl glutamic acid $(F(GLU)_2)$, was chosen to assess the effects of various ions and inhibitors, since its movement would be sensitive to even a small restriction in the pores. The problem was how to inject both the divalent ions and the probes. In fact, a simple solution presented itself: to inject the divalent ions and the probes from the same electrode, with the divalent cations injected using a positive current, and the probes (which are anions) injected using a negative current. This overcame the possibility that the different molecules would be injected into different parts of the cell, or that the use of two electrodes would cause major damage to the cell.

It was found that Ca^{2+}, Mg^{2+}, and Sr^{2+} injected a short time (0–1 min) before the probe inhibited its movement. Perhaps any cation would have the same effect? Not K^+ or Na^+. Was only the largest mobile probe affected? No, the movement of all probes tested was inhibited, suggesting that the plasmodesmata shut, rather than showing a small fall in the size of molecule able to pass. The inhibition of movement by the divalent ions was an all-or-none effect. Either the dye moved as well as in the control (in about a third of cases) or it did not move at all. Presumably, in a third of cases the cation either did not shut the plasmodesmata, or shut them too slowly. The inhibiting effect of the cations was greater if there was a 1 minute delay between injection of cation and probe, suggesting that shutting is time dependent, i.e., being greater after 1 minute.

The next question was, is this a reversible process? Since the cell is very competent to reduce divalent ion levels, one might expect the plasmodesmata to eventually re-open. This was found: if there was a delay of 5–30 minutes between divalent cation and dye injection, then the dye moved from cell to cell, showing that the calcium-1 or magnesium-mediated closing of the plasmodesmata is a reversible process. It also showed that the injection of the divalent ions had not caused some type of permanent damage to the cell or the plasmodesmata.

There is always uncertainty in the use of inhibitors, as they invariably have multiple effects. However, a number of chemicals likely to increase the concentration of cytoplasmic calcium ions, notably the mitochondrial uncoupler carbonyl cyanide p-trifluoromethoxy-phenyl hydrazone, the inhibitor of mitochondrial Ca^{2+} uptake trifluralin, and the ionophore A23187, all inhibited probe movement.

The significance of the work was that it showed for the first time that plasmodesmata are highly dynamic structures, able to respond rapidly to changes in the cellular environment. The work, of course, raised as many questions as it answered. How do the divalent ions work? Do other gating molecules exist? Is the observed response the normal system? Does it occur in other species?

The whole field of cell-to-cell communication has expanded since these early studies. For a recent review see Robards and Lucas (1990).[5] Of special interest has been the demonstration of changes in cell/cell communication with development,[6] and the viral regulation of plasmodesmata.[7] Unexpectedly, in view of the work on *Egeria*, it has been shown that plasmodesmata can stay open in isolated bundle sheath cell strands,[8] a system offering considerable potential for the direct study of plasmodesmatal function.

References

[1]Gunning, B. E. S., and Robards, A. W. *Intercellular communications in plants. studies on plasmodesmata.* Berlin: Springer (1976).

[2]Spanswick, R. M. *Planta* 102:215–27 (1972).

[3]Goodwin, P. B. *Planta* 157:124 (1983); Tucker, E. B. *Protoplasma* 113:193 (1982).

[4]Rose, B.; Simpson, I.; and Loewenstein, W. R. *Nature* 267:625 (1977).

[5]Robards, A. W., and Lucas, W. J. *Annu. Rev. Plant Phys. Plant Mol. Biol.* 41:369 (1990).

[6]Erwee, M. G., and Goodwin, P. B. *Planta* 163:9 (1985).

[7]Wolf, S.; Deom, C. M.; Beachy, R. N.; and Lucas, W. J. *Science* 246:377 (1989).

[8]Burnell, J. N. *J. Exp. Bot.* 39:1575 (1988); Weiner, H.; Burnell, J. N.; Woodrow, I. E.; Heldt, H. W.; and Hatch, M. D. *Plant Physiol.* 88:815 (1988).

For Further Information

Suggestions for Further Reading

Akiyama, S. K., Nagata, K., and Yamada, K. M. 1990. Cell surface receptors for extracellular matrix components. *Biochim. Biophys. Acta* 1031:91–110.

Albeda, S. M., and Buck, C. A. 1990. Integrins and other cell adhesion molecules. *FASEB J.* 4:2868–2880.

Beck, K., Hunter, I., and Engel, J. 1990. Structure and function of a laminin: Anatomy of a multidomain glycoprotein. *FASEB J.* 4:148–160.

Beveridge, T. J., and Graham, L. L. 1991. Surface layers of bacteria. *Microbiol. Rev.* 55:684–705.

Bolwell, G. P. 1993. Dynamic aspects of the plant extracellular matrix. *Internat. Rev. Cytol.* 146:261–324.

Boulnois, G. J., and Jann, K. 1989. Bacterial polysaccharide capsule synthesis, export, and evolution of structural diversity. *Molec. Microbiol.* 3:1819–1823.

Brush, S. G. 1992. Bacterial endotoxins. *Sci. Amer.* 157: 54–61 (August).

Caplan, A. I. 1984. Cartilage. *Sci. Amer.* 251:84-94 (October).

Chrispeels, M. J., and Raikhel, N. V. 1991. Lectins, lectin genes, and their role in plant defense. *Plant Cell* 3:1–9.

Delmer, D. P. 1987. Cellulose biosynthesis. *Ann. Rev. Plant Physiol.* 38:259–290.

Ekblom, P., Vestveber, D., and Kemler, R. 1986. Cell-matrix interactions and cell adhesion during development. *Ann. Rev. Cell Biol.* 2:28–47.

Goetnick, P. F. 1991. Proteoglycans in development. *Curr. Top. Devel. Biol.* 25:111–131.

Hayashi, S., and Wu, H. C. 1990. Lipoproteins in bacteria. *J. Bioenerget. Biomembr.* 22:451–470.

Hayashi, T. 1989. Zyloglucans in the primary cell wall. *Ann. Rev. Plant Phys. Plant Molec. Biol.* 40:139–168.

Hultgren, S. J., Abraham, S., Caparon, M., and Falk, P. 1993. Pilus and nonpilus bacterial adhesins: Assembly and function in cell recognition. *Cell* 73:887–901.

Hynes, R. O. 1986. Fibronectins. *Sci. Amer.* 254:42–51 (June).

Hynes, R. O. 1992. Integrins: Versatility, modulation, and signalling in cell adhesion. *Cell* 69:11–25.

Juliano, R. L., and Haskill, S. 1993. Signal transduction from the extracellular matrix. *J. Cell Biol.* 120:577–585.

Kjellen, L. and Lindahl, U. 1991. Proteoglycans: Structures and interactions. *Ann. Rev. Biochem.* 60:443–475.

Knox, P. 1990. Emerging patterns of organization at the plant cell surface. *J. Cell Sci.* 96:557–561.

Kucharz, E. J. 1991. *The Collagens: Biochemistry and Pathophysiology.* New York: Springer-Verlag.

Kuivaneiemi, H., Tromp, G., and Prockop, D. 1991. Mutations in collagen genes: Causes of rare and some common diseases in humans. *FASEB J.* 5:2052–2060.

Lee, B., D'Alessio, M., and Ramirez, F. 1991. Modifications in the organization and expression of collagen genes associated with skeletal disorders. *Crit. Rev. Eukary. Gene Express.* 1:173–187.

Lewis, N. G., and Yamamoto, E. 1990. Lignin: Occurrence, biogenesis, and biodegradation. *Ann. Rev. Plant Physiol. Plant Molec. Biol.* 41:455–496.

Mecham, R. P. 1991. Receptors for laminin on mammalian cells. *FASEB J.* 5:2538–2546.

Rest, M. van der, and Garrone, R. 1991. Collagen family of proteins. *FASEB J.* 5:2814–2823.

Robards, A. W., and Lucas, W. J. 1990. Plasmodesmata. *Ann. Rev. Plant Physiol. Plant Molec. Biol.* 41:369–419.

Scott, J. E. 1992. Supramolecular organization of extracellular matrix of glycosaminoglycans, in vitro and in the tissues. *FASEB J.* 6:2639–2645.

Showalter, A. M. 1993. Structure and function of plant cell wall proteins. *Plant Cell* 5:9–23.

Varner, J. E., and Lin, L.-S. 1989. Plant cell wall architecture. *Cell* 56:231–239.

Vuorio, E., and de Crombrugghe, B. 1990. The family of collagen genes. *Ann. Rev. Biochem.* 59:837–872.

Yurcheno, P. D., and Schittny, J. C. 1990. Molecular architecture of basement membranes. *FASEB J.* 4:1577–1590.

Review Questions

1. Outline the structure of the collagens. List each collagen type and its overall functions in the extracellular matrix.

2. How do collagen polypeptides combine to form collagen molecules? How do collagen molecules combine to form collagen fibers? Where and how are collagen molecules and fibers synthesized?

3. What are proteoglycans? GAGs? How do proteins and GAGs combine in proteoglycan structure? How do proteoglycans and hyaluronic acid combine into molecular superstructures?

4. In what structures do proteoglycans occur in the animal extracellular matrix? What functions do proteoglycans carry out in these locations?

5. What are integrins? Outline the structures of fibronectin and laminin. What is the relationship between these molecules and the extracellular matrix? The cytoskeleton?

6. What changes occur in the extracellular matrix as cells change into cancerous forms?

7. What are basal laminae? What functions does this extracellular structure have in animal cells?

8. Outline the structure of cellulose. How are cellulose molecules believed to combine to form cellulose microfibrils?

9. What are pectins? Hemicelluloses? Extensins? Lignins? How do these molecules function in cell wall structure?

10. How are new cellulose microfibrils laid down in growing cell walls?

11. What evidence indicates that microtubules are involved in the deposition of cellulose microfibrils? How might microtubules determine the direction in which cellulose microfibrils are laid down in cell walls?

12. What are plasmodesmata? In what ways are plasmodesmata similar to the gap junctions of animal cells? What evidence indicates that plasmodesmata provide direct cytoplasmic connections between plant cells?

Supplement 7-1
Cell Walls in Prokaryotes

Like the walls of plant cells, prokaryotic cell walls are assembled from complex polysaccharides linked with other substances into a "supermolecule" that completely surrounds the cell. The walls of prokaryotes support and give shape to cells and prevent them from bursting as a result of osmotic pressure. Prokaryotic walls also protect cells from infection by viruses and provide sites for cell recognition and adhesion. In addition, the cell wall serves as a partial permeability barrier and houses some molecules involved in molecular transport. Because cell walls are so important to the viability of prokaryotes, chemical interference with the walls or wall synthesis is an effective and much used way to control the growth of disease-causing bacteria.

Cell Wall Structure in Gram-Positive and Gram-Negative Bacteria

Almost all bacteria fall into two major groups that are distinguished by the *gram stain*, devised in the 1800s by Hans Christian Gram. In the technique, cells are stained with a blue dye and then exposed to a decolorizing solvent. Gram-positive cells are not decolorized and retain a deep blue color; gram-negative cells are decolorized.

The staining difference reflects clear distinctions in wall structure. Walls in gram-positive bacteria (Fig. 7-21a) consist of a single, relatively thick layer of polysaccharide wall material that is not differentiated into sublayers. Gram-negative bacteria have walls containing two distinct layers (Fig. 7-21b). The most unusual of these layers is an extra boundary membrane, the *outer membrane*, lying outside the plasma membrane. The region between the plasma and outer membranes, the *periplasmic space*, is filled with a thin layer of polysaccharide wall material. The difference in reaction to the gram stain, which is not understood, may depend on the relatively thick layer of wall material in gram-positive bacteria, which may retard removal of the stain from cells more effectively than the very thin walls of gram-negative bacteria.

One or more structures may extend outward from cell walls of both gram-positive and gram-negative bacteria. Many bacteria have an external *capsule* or *slime layer* surrounding the cell wall (Fig. 7-22; see also Fig. 1-2). The capsule, which may be from tens of nanometers to several micrometers in thickness, shows great structural and chemical diversity in various species of bacteria and cyanobacteria. The wall may also bear motile *flagella* consisting of long,

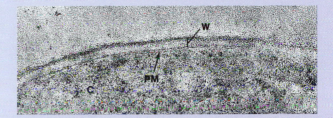

a

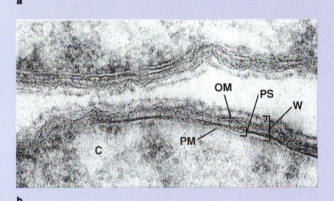

b

Figure 7-21 Wall structure in bacteria. **(a)** The single-layered wall characteristic of gram-positive bacteria. × 103,000. **(b)** The multilayered wall of gram-negative bacteria. × 112,000. W, wall; PM, plasma membrane; OM, outer membrane; C, cytoplasm. (Courtesy of J. W. Costerton, reproduced, with permission, from *Ann. Rev. Microbiol.* 33:459 [1979]. Copyright 1979 by Annual Reviews, Inc.)

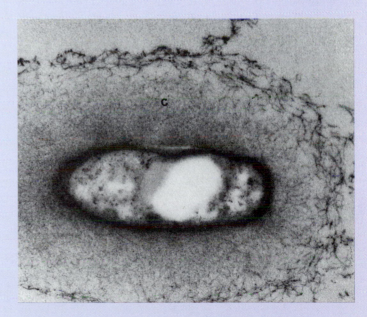

Figure 7-22 The capsule (C) surrounding the cell wall in the soil bacterium *Rhizobium*. × 30,000. (Courtesy of W. J. Brill, from *J. Bact.* 137:1362 [1979].)

filamentous protein fibers (see Fig. 1-6 and pp. 304–307) and shorter, nonmotile rods of protein called *pili* (singular = *pilus*). The pili, which stick out like rigid hairs from the surface, are involved in cell recognition and adhesion.

Molecular Constituents of Prokaryotic Cell Walls

Chemical analysis of bacterial walls reveals a variety of substances, some with no counterparts in other living organisms. Gram-positive and gram-negative bacterial walls share a unique group of molecules, the *peptidoglycans* (Figs. 7-23 and 7-24), that provide rigidity and strength to bacterial walls. Both gram-positive and gram-negative bacterial cell walls also contain enzymes concerned with wall synthesis, transport (see p. 199), and breakdown of organic molecules absorbed by the cell.

Other molecules occur only in one of the two bacterial types. Almost all gram-positive bacterial walls contain *teichoic acids* (see Fig. 7-25) as a wall constituent. These molecules, among other functions, may serve as recognition and binding sites in gram-positive cell walls. Gram-negative bacterial walls lack teichoic acids but contain several unique molecular types: *lipopolysaccharides* (see Fig. 7-26), which occur as a part of the outer membrane; and *lipoproteins*, which link the outer membrane to the underlying peptidoglycan layer. Gram-negative outer membranes additionally contain *porins*, channel-forming proteins that extend entirely through the outer membrane. Porins provide openings that allow small polar molecules such as sugars and amino acids to penetrate from the medium into the wall and eventually to the plasma membrane. (See p. 146 for details; Table 7-2 summarizes the major chemical constituents of gram-positive and gram-negative bacterial walls.)

Peptidoglycans The primary structural molecules of prokaryotic cell walls, the peptidoglycans, are assembled on a backbone of repeating two-sugar units with attached short peptide chains (see Figs. 7-23 and 7-24). Adjacent backbones are crosslinked by a single peptide bond or a short sequence of amino acids. The polysaccharide backbones and peptide crosslinks make peptidoglycans the equivalent of both the fiber and network molecules of eukaryotic extracellular structures. With few exceptions the same fundamental peptidoglycan structure occurs in all gram-negative and gram-positive bacteria and in the cyanobacteria.

Because peptidoglycan backbone chains are fairly rigid and too long to lie at right angles to the membrane, they are believed to occur in layers parallel to the plasma membrane. If the parallel arrangement is actually the case, there is room for about 20 to 40 layers of peptidoglycan chains in the relatively thick walls of gram-positive bacteria. Gram-negative bacteria have space between the plasma and outer membranes for only a very thin layer, in some species probably only one molecule thick. The thin peptidoglycan layer of gram-negative bacteria is believed to form a stretched, tightly fitted sheath around the cell. The peptidoglycan layer in the gram-negative bacteria is open enough to allow molecules as large as proteins to diffuse between the plasma and outer membranes.

The peptidoglycan layer grows by deposition of new wall material inside the old. Although the process is not understood, it is considered likely that new peptidoglycans are initially connected in loose, looplike chains underlying the existing wall. Wall enzymes then open linkages in the short peptide chains of the old peptidoglycan sheath, allowing the old subunits to slide past each other as the cell wall expands and stretches to accommodate the newly added chains. Force for the expansion is supplied by osmotic

Table 7-2 Proteoglycans and Their Distribution		
Type	Molecular Weight of Core Proteins	Distribution
Chondroitin sulfate	200,000 to 300,000 in cartilage 45,000 in bone 45,000 and 500,000 in skin	Cartilage, bone, skin; small amounts in basal laminae
Keratan sulfate	45,000	Bone, cornea
Dermatan sulfate	40,000 to 50,000	Tendon, fibrillar connective tissue; small amounts in skin and cornea
Heparan sulfate	30,000 to 200,000	Basal laminae, cornea, nerve tissue
Hyaluronic acid– proteoglycan complexes	Hundreds of thousands to millions	Cartilage; minor amounts in bone, tendon, aorta, smooth muscle

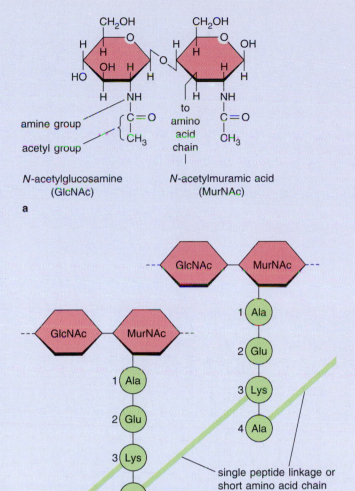

amine group
acetyl group

N-acetylglucosamine
(GlcNAc)

to amino acid chain

N-acetylmuramic acid
(MurNAc)

a

1 Ala
2 Glu
3 Lys
4 Ala

single peptide linkage or short amino acid chain

b

Figure 7-23 The structural units of peptidoglycans. **(a)** The repeating two-sugar unit that forms the backbone of peptidoglycans, composed of the amino sugars acetylglucosamine and acetylmuramic acid. Attached to each acetylmuramic acid residue is a short sequence of amino acids, usually four as in **(b)**. These short peptide chains connect the backbones of adjacent peptidoglycan chains through a linkage between the amino acid at position 3 in one repeating unit and the amino acid at position 4 in a repeating unit of an adjacent chain, as in **(b)**. The linkage tying adjacent backbone chains may be a single peptide bond (gram-negative bacteria) or a short chain of amino acids (gram-positive bacteria).

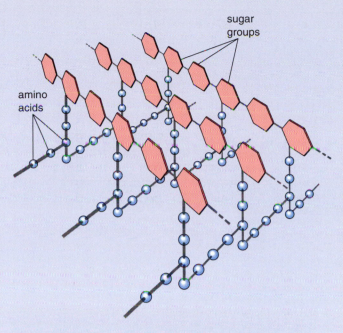

Figure 7-24 Three-dimensional structure of a peptidoglycan "supermolecule." The hexagons represent sugar units, and the spheres amino acids. The peptidoglycan shown is the type occurring in gram-positive bacteria, in which the interchain linkages consist of a short chain of amino acids.

pressure developed in the cell, as it is in plants. Almost all the enzyme systems required for synthesis of the molecules in the peptidoglycan sheath are contained in the plasma membrane.

The walls of cyanobacteria (blue-green algae) resemble those of the gram-negative bacteria in most characteristics. Each of the complex layers of gram-negative bacterial walls has a counterpart in the cyanobacteria, including an outer membrane. For unknown reasons, cyanobacteria are gram positive in spite of their chemical and morphological similarities to gram-negative bacteria.

Penicillin interferes with wall synthesis in both bacteria and cyanobacteria by inhibiting the enzymes that close links between peptide side chains in peptidoglycans after cell wall growth is complete. The open links weaken the wall, causing the cells to swell and burst as a result of osmotic pressure.

Teichoic Acids The teichoic acids of gram-positive bacteria are assembled from repeating alcohol units linked together by phosphate groups (Fig. 7-25). Often one or more amino acid or glucose units are linked to the alcohols of the backbone (as shown in Fig. 7-25b and c).

Cell wall fragments frequently reveal teichoic acid chains linked covalently at one end to the peptidoglycan backbones, indicating that they tie into the cell wall structure in this position. The other end may link to the plasma membrane or extend outward from the wall surface into the surrounding medium.

The function of teichoic acids is not fully known. However, the closely spaced phosphate groups form

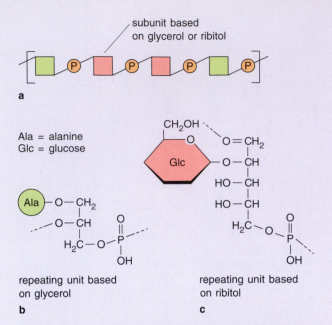

a

subunit based on glycerol or ribitol

Ala = alanine
Glc = glucose

repeating unit based on glycerol

b

repeating unit based on ribitol

c

Figure 7-25 Teichoic acids. **(a)** The backbone structure of teichoic acids consists of a repeating series of subunits, usually glycerol or ribitol, linked by phosphate groups. Typical repeating groups, based on glycerol and ribitol, are shown in **(b)** and **(c)**. Frequently the amino acid alanine or more complex chains of several amino acids are linked to one or more of the available —OH groups on glycerol or ribitol as shown in **(b)**. In other variations, glucose may be attached at an available —OH group as shown in **(c)**. Teichoic acid polymers may contain from 5 or so to more than 30 repeating glycerol or ribitol subunits.

powerful attractants for positively charged ions, particularly divalent cations such as Ca^{2+} and Mg^{2+}. Through this attraction teichoic acids are believed to maintain the concentrations of divalent cations necessary for maximum activity of many wall enzymes. In addition, teichoic acids also provide major recognition sites at the cell surface. Many viruses infecting gram-positive bacteria recognize and bind teichoic acid chains as their means of attachment to the cell. The teichoic acids also form the primary groups recognized and bound by antibodies in humans and other vertebrates.

Lipopolysaccharides The lipid bilayer forming the outer membrane of gram-negative prokaryotes characteristically contains lipopolysaccharides (Fig. 7-26). These unique molecules are constructed from three major subunits. A basal subunit, called the *lipid A subunit*, consists of a two-sugar unit to which are connected several nonpolar fatty acid chains. This subunit is anchored in the outer membrane. Connected to the lipid A subunit is a long, hydrophilic carbohydrate

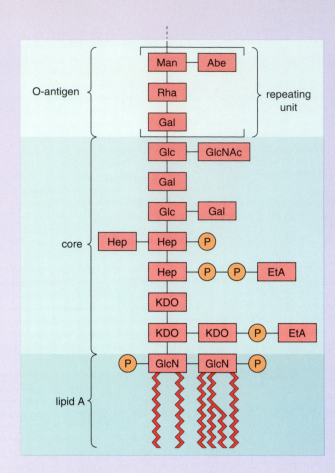

Abe	=	abequose
Rha	=	rhamnose
Man	=	mannose
Gal	=	galactose
GlcNAc	=	*N*-acetyl glucosamine
Glc	=	glucose
Hep	=	heptulose
EtA	=	ethanolamine
KDO	=	2-keto-3-deoxyoctonate
GlcN	=	glucosamine

Figure 7-26 Lipopolysaccharides. The basal unit, lipid A, contains a short backbone constructed from a pair of glucosamine subunits linked end to end. Connected to the backbone are several nonpolar fatty acid chains, frequently lauric or palmitic acid. Next to lipid A is the core subunit, a branched polysaccharide chain containing the common six-carbon sugars glucose, galactose, and glucosamine and several seven- and eight-carbon sugars, such as Hep and KDO, found nowhere else among living organisms. Extending outward from the core is the O-antigen, consisting of a chain of as many as 40 or more repeating units of six-carbon hexose sugars.

"tail" consisting of two major subunits, the *core* and *O-antigen*. Within the O-antigen are repeating units of three to six sugars. The sugars in these subunits and the pattern of the repeats are highly variable among different gram negative-bacteria, and even among different genetic lines of the same species. The O-antigen

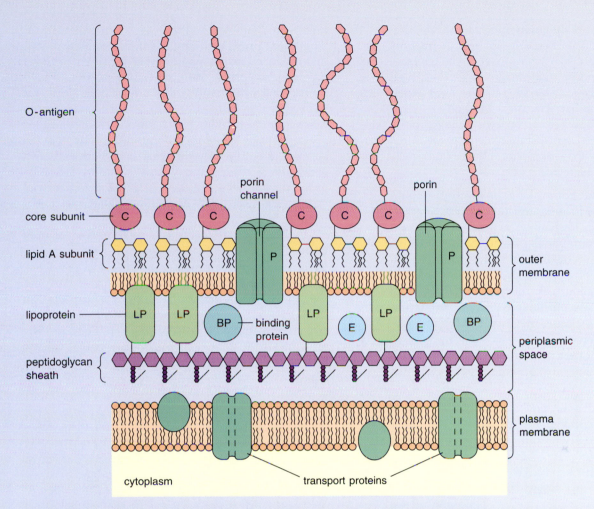

O-antigen

core subunit

lipid A subunit

porin channel

porin

outer membrane

lipoprotein

LP   LP   BP — binding protein   LP   E   LP   E   BP

periplasmic space

peptidoglycan sheath

plasma membrane

cytoplasm   transport proteins

Figure 7-27 Structural units of gram-negative cell walls. C, core subunit of lipopolysaccharide; P, porin; LP, lipoprotein; BP, binding protein; E, hydrolytic enzyme. The binding proteins specifically bind substances, primarily sugars and amino acids, and transfer them to active transport systems of the plasma membrane (see p. 147).

is so called because it provides the primary group recognized by antibodies in gram-negative cell walls. Infecting viruses also use the O-antigen as a recognition site for attachment.

Lipopolysaccharides are confined to the outermost bilayer half of the outer membrane, positioned with their basal units buried in the nonpolar bilayer interior and the O-antigen chains extending outward as much as 30 nanometers from the wall surface (see Fig. 7-27). The lipopolysaccharide layer in the outer membrane is much more resistant than phospholipids to disruption by detergents and lipid solvents, which probably accounts for the ability of *E. coli* to tolerate the strongly detergent effect of bile salts in the human digestive tract. They form such an effective barrier to diffusion that essentially no molecules can pass through the outer membrane except those admitted by the porins.

The extreme variability of the O-antigen segment of outer membrane lipopolysaccharides can present severe problems for the immune response. *E. coli*, for example, is normally a harmless gram-negative inhabitant of the human large intestine. Occasionally *E. coli* invades other regions of the body, particularly the urogenital tract, bloodstream, or the fluid within the brain and spinal cord, where it may become a dangerous pathogen. In these locations *E. coli* infections are often very persistent because of variability in the O-antigen. The O-antigen mutates so rapidly from generation to generation that, in effect, the bacterium keeps ahead of the body's antibody defenses by continually developing new types that are not recognized by previously developed antibodies.

Many of the toxic effects of some invading bacteria, including *Salmonella*, *E. coli*, *Chlamydia*, *Vibrio cholerae*, and the *Pertussis* bacterium responsible for whooping cough, are due to lipopolysaccharide fragments of cell walls, particularly the lipid A fraction. Their primary effect is on macrophages, which release substances in response that can produce shock, high fever, dangerously low blood pressure, and blood clotting. When the bacteria infect the body in large numbers, these responses can be lethal. The lipopolysaccharide fragments producing the effects are called *endotoxins* because they form part of the bacterial cells rather than being secreted.

Lipoproteins The lipoproteins of the outer membrane consist of a protein unit to which is attached one or more fatty acid chains. The nonpolar fatty acid chains anchor lipoproteins in the lipid bilayer of the outer membrane, in this case in the inner bilayer half facing the underlying peptidoglycan sheath (see Fig. 7-27). In *E. coli* about one-third of the lipoprotein molecules form covalent links with peptidoglycans, firmly anchoring the outer membrane to the peptidoglycan sheath. There are as many as 70,000 lipoprotein molecules in the outer membrane of an *E. coli* cell, making it one of the most abundant proteins of the organism.

The cell wall of a gram-positive bacterium consists primarily of a peptidoglycan sheath, with teichoic acid molecules extending from the sheath into the region outside the wall. The gram-negative wall has two distinct layers—the peptidoglycan sheath and, outside it, the outer membrane. The outer membrane, with its porins, lipopolysaccharides, and lipoproteins, sets up an effective barrier to essentially all substances except ions and small molecules admitted by the porins. The barrier protects the cell from chemical attack by larger molecules such as enzymes and antibiotics. The outer membrane is also less susceptible than the underlying plasma membrane to the disruptive effects of agents such as bile and detergents.

Phospholipids are also present in gram-negative outer membranes, in quantities amounting to about half of the lipopolysaccharide content by weight. They are confined with the lipoproteins almost entirely to the inner bilayer half of the outer membrane; the outer bilayer half contains essentially only lipopolysaccharides. Most of the protein complement of the outer membrane is supplied by porins; enzymatic activity is limited to a few proteins able to catalyze breakdown of lipids and proteins, and an enzyme associated with capsule assembly. Most enzymes required for synthesis of molecules in the outer membrane are concentrated in the plasma membrane. (Figure 7-27 summarizes outer membrane structure.)

Pili and Bacterial Adhesions

Pili (singular = *pilus*; pili are also called *fimbriae*) are hollow, cylindrical protein fibers about 5 to 7 nanometers in diameter and 100 to 200 nanometers in length that project entirely through the cell wall and into the surrounding medium (Fig. 7-28). The walls of the cylinder consist of linear chains of protein subunits called *pilins*, arranged in a tight helix. In many bacteria a thin fibrous protein called an *adhesin* extends from the tip of the pilus. Pili and their adhesions are capable of binding carbohydrate groups on the surfaces of both prokaryotic and eukaryotic cells and also serve as recognition groups bound by receptors on eukaryotic cells and the coat proteins of bacterial viruses. Pili occur in both gram-positive and gram-negative bacteria.

Individual cells may occur without pili or may possess from 1 to 300 of the same or different types. In some bacteria the presence or absence of pili determines the "sex" of conjugating cells. During conjugation, in which a cytoplasmic bridge is set up between mating pairs, bacterial cells with pili act as donor cells that transfer DNA to cells lacking pili. Because cells with pili are the donors, they are considered as "male" or + cells in the change. Receiving cells, considered as "female" or − in the exchange, possess a surface receptor that can recognize and bind the pilin protein.

Pili figure importantly in promoting bacterial infections. For example, in the bacterium that causes gonorrhea, *Neisseria gonorrhoeae*, pili are strongly bound by epithelia cells lining the urinary tract, so strongly that they are not readily dislodged by urine flow. The strong adherence and other factors, including constant mutation of the pili proteins that allow the cells to escape immune reactions, make gonorrheal infections highly persistent unless treated by antibiotics. Adhesins on the pili of *E. coli* also make this bacterium a persistent pathogen if it is introduced into body regions such as the urinary tract.

Bacterial adhesions in infections are not limited to pili. Adhesins are also distributed in other regions of bacterial cell walls; many infecting bacteria also contain other surface proteins that have binding affinity for receptors on animal cell surfaces or for elements of the extracellular matrix, including fibronectin, laminin, collagen, and the proteoglycans. *Staphylococcus* and *Streptococcus*, for example, have proteins binding fibronectin and collagen; *E. coli* has groups binding these proteins and laminin.

The Capsule

The polysaccharides of the capsule, which occurs in gram-positive and gram-negative bacteria and cyanobacteria, vary in chemical structure between different species. However, within a given species the capsule is of uniform structure and usually contains a single polysaccharide type assembled primarily from glucose, galactose, fucose, mannose, or glucuronic acid

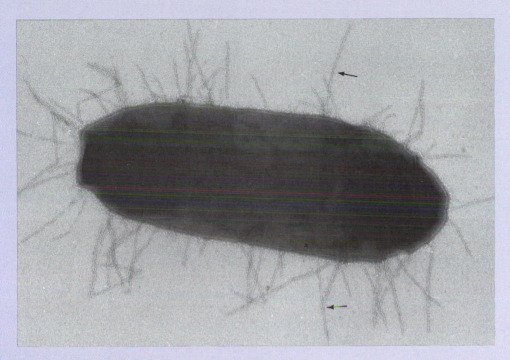

Figure 7-28 Pili (arrows) extending from the surface of an *E. coli* cell prepared for electron microscopy by negative staining. (Courtesy of A. Gbarah and N. Sharon.)

subunits. Generally two or more of these sugars link in various combinations to form a tetrasaccharide unit. The unit repeats in end-to-end linkages to form the capsule polysaccharide.

Capsule polysaccharides absorb many times their weight in water molecules, forming a highly hydrated layer of slimelike material coating the surface of the wall. The capsule provides protection against attack by bacterial viruses. In many bacterial species the capsule's adhesive properties attach cells to each other in colonies or to a substrate.

The capsule figures importantly in bacterial diseases of both plants and animals. The thick layer of capsular material impedes diffusion and attachment of antibodies to the bacterial cell surface. The capsule also interferes with the ability of phagocytes such as white blood cells to attach and engulf bacterial cells by endocytosis (see p. 593). The capsule is so effective in these protective roles that the difference between the virulent and noninfective forms of many disease-causing bacteria simply reflects the presence or absence of the capsule. For example, *Pneumococcus* is nonvirulent and can easily be eliminated by natural

defense mechanisms if it is injected into mice or other mammals as an unencapsulated mutant. The normal, encapsulated form, however, is a highly pathogenic bacterium that avoids destruction by the immune system and causes severe pneumonia in humans and other mammals.

The capsular material in some cases contributes directly to the pathogenic effects of bacterial infections. Some species of *Pseudomonas* bacteria infecting plants, for example, produce capsular material that greatly increases the viscosity of fluids in the vascular system of the plants. The highly viscous fluid blocks the vascular system to such an extent that water cannot flow from roots to stems and leaves, causing severe wilting. A *Pseudomonas* species that infects humans suffering from cystic fibrosis causes an analogous condition. The bacteria accumulate in the air passages of the lungs, where their capsular material makes the mucus so viscous that the lungs cannot be cleared. Bacterial capsules are also a major component of dental plaque, the slimy material that collects on the surfaces of teeth and figures as a major cause of tooth decay.

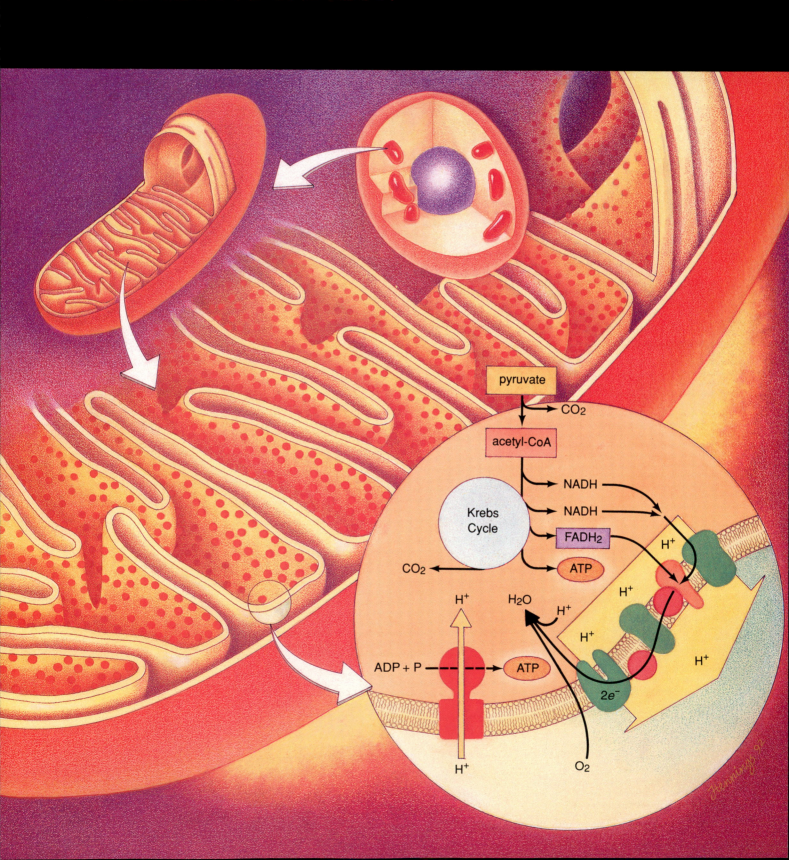

T he energy required for life is derived from oxidation of complex organic fuels such as carbohydrates and fats. Energy from this source is converted into the chemical energy of ATP (adenosine triphosphate; see p. 73), the "dollar" of the cellular energy economy. The oxidative reactions releasing energy drive the synthesis of ATP from ADP and inorganic phosphate. Reactions that require energy obtain it by splitting ATP into adenosine diphosphate (ADP) or adenosine monophosphate (AMP) and inorganic phosphate. ATP thus cycles between reactions that release energy and those that require energy.

ATP moves through the ADP/ATP cycle in impressively large quantities in living organisms. For example, in humans at rest an ATP quantity equivalent to one-half the body weight passes through the ADP/ATP cycle each day. Moderate exercise increases the quantity to several times the body weight each day.

This chapter discusses the reactions that synthesize ATP, which proceed by one or both of two mechanisms in living organisms. One, *oxidative phosphorylation*, is common to eukaryotic cells and many prokaryotes. This mechanism, which requires oxygen

to proceed, is highly efficient in converting the potential energy of fuel substances into energy bound into ATP. The second mechanism, *substrate-level phosphorylation*, is common to all organisms. Although less efficient, substrate-level phosphorylation is the primary source of ATP for cells living temporarily or permanently without oxygen.

ATP PRODUCTION BY OXIDATIVE PHOSPHORYLATION: AN OVERVIEW

Oxidative phosphorylation can be conveniently divided into three overall parts, which are common to all living organisms possessing this pathway (Fig. 8-1):

1. A series of reactions providing a *source of electrons at elevated energy levels*

2. An *electron transport system*, embedded in a membrane, that uses the energy of the electrons provided in part 1 to build an H^+ gradient across the membrane

3. An *ATP-synthesizing enzyme*, also embedded in the membrane, that uses the H^+ gradient produced in part 2 as an energy source to phosphorylate ADP to ATP

The Source of Electrons at Elevated Energy Levels

Cellular oxidations, in which electrons are removed from organic molecules, are the immediate source of

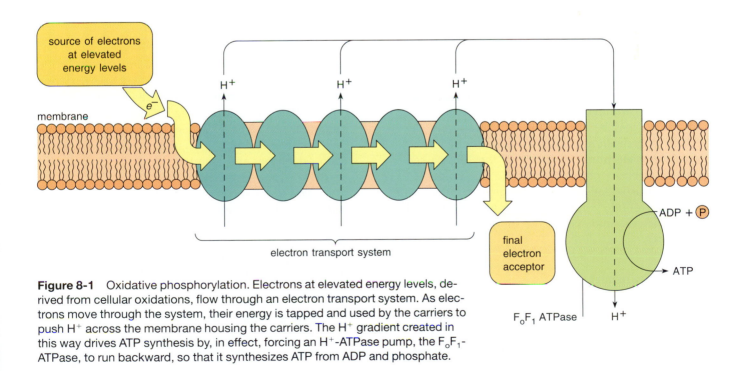

Figure 8-1   Oxidative phosphorylation. Electrons at elevated energy levels, derived from cellular oxidations, flow through an electron transport system. As electrons move through the system, their energy is tapped and used by the carriers to push H^+ across the membrane housing the carriers. The H^+ gradient created in this way drives ATP synthesis by, in effect, forcing an H^+-ATPase pump, the F_oF_1-ATPase, to run backward, so that it synthesizes ATP from ADP and phosphate.

Oxidation and Reduction

Many substances, both organic and inorganic, can gain or lose electrons. Removal of electrons is called *oxidation*, and acceptance of electrons is termed *reduction*. A substance from which electrons are removed is *oxidized*; an accepting substance is *reduced*. When any substance is oxidized, another molecule is usually reduced by combining with the removed electrons. Therefore each oxidation is usually accompanied by a simultaneous reduction. Frequently one or two hydrogen ions (protons) as well as electrons are removed from a molecule during an oxidation. The molecules acting as acceptors for the removed electrons may also combine with one or both of these hydrogens. In general, the energy content of a given substance is greater in its reduced state than in its oxidized state because of the energy of the added electrons.

The amount of energy associated with electrons removed in an oxidation depends on the orbitals they occupied in the oxidized molecule. The energy of the removed electrons is measured and expressed as a relative *potential* or *voltage* by comparison with an arbitrary standard, the energy of electrons removed from hydrogen in the reaction $H_2 \longrightarrow 2H^+ + 2e^-$ (see the diagram in this box). The potential assigned to these electrons is 0.00 V. All other potentials, called *redox* or *reduction-oxidation potentials*, are measured and assigned a value with respect to this standard. (The table in this box lists redox potentials of electrons removed from some important intermediates in cellular oxidations.)

Although all electrons carry a negative charge, their relative voltage may be greater or less than the 0.00 standard. The voltage of electrons with energy greater than the standard is given a minus value; the voltage of electrons with energy lesser than the standard is written as a positive value. Therefore an electron with a potential of -0.2 V, for example, has greater energy than one with a potential of $+0.2$ V; the difference in voltage between these electrons totals 0.4 V.

Electrons at elevated energy levels, as defined in this chapter, are those that have relatively high voltage or electrical potential. Electrons at elevated energy levels travel at higher velocities and at shorter wavelengths than electrons at lower energy levels. Electrons at the most elevated energy levels normally encountered in living systems have a potential of about -0.6 V; the lowest usable energies are at about $+0.8$ V.

In general, electrons are transferred from one substance to another located lower on the redox scale; that is, the donor is higher on the scale than the acceptor. As electrons are transferred from a donor to an acceptor, the amount of free energy released is equivalent to the distance between the donor and acceptor on the redox scale (or the difference in their redox potentials). In the series of electron carriers that are alternately oxidized and reduced as electrons flow through electron transport systems of mitochondria, chloroplasts, and prokaryotes, much of the free energy released during redox reactions is used to build an H^+ gradient and, ultimately, to synthesize ATP.

---

electrons at elevated energy levels. (Oxidation and reduction are explained in Information Box 8-1.) Although any of the four major classes of biological molecules—carbohydrates, lipids, proteins, and nucleic acids—may be oxidized as electron sources, the primary sources in most cells are carbohydrates and lipids. After electrons are removed, they are passed on to the electron transport system setting up the H^+ gradient. The oxidative reactions take place in the soluble background cytoplasm of both prokaryotic and eukaryotic cells and within the mitochondria of all eukaryotes.

Electron Transport

The system transporting electrons removed in cellular oxidations consists of a series of carrier molecules embedded in a membrane (see Fig. 8-1). Each electron carrier in the series accepts and releases electrons at lower energy levels than the carriers preceding it in the series. The energy lost by the electrons as they pass from one carrier to the next is released as free energy—that is, energy that can do work (see p. 71). Some of the free energy is used by the carriers to pump H^+ from one side of the membrane to the other. This active transport establishes and maintains the H^+ gradient used to drive ATP synthesis.

Most of the molecules transporting electrons are integral membrane proteins combined with a nonprotein organic group. The organic group is the part of the carrier that picks up and releases electrons to accomplish electron transport. The activity that uses energy released by electrons to push H^+ across the membrane is concentrated in the protein part of the carrier molecules.

After traveling through an electron transport system, electrons are delivered to a final electron acceptor, which may be either an organic or inorganic

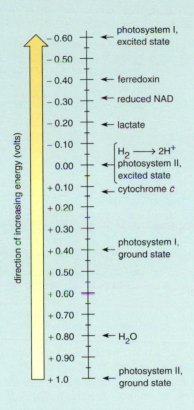

direction of increasing energy (volts)	
− 0.60	← photosystem I, excited state
− 0.50	
− 0.40	← ferredoxin
− 0.30	← reduced NAD
− 0.20	← lactate
− 0.10	
0.00	← $H_2 \longrightarrow 2H^+$ photosystem II, excited state
+ 0.10	← cytochrome c
+ 0.20	
+ 0.30	
+ 0.40	← photosystem I, ground state
+ 0.50	
+ 0.60	
+ 0.70	
+ 0.80	← H_2O
+ 0.90	
+ 1.0	← photosystem II, ground state

Reduction Potentials of Biological Oxidations (at pH ≈ 7.0 and temperature between 25° to 30°C)

Oxidation	Redox Potentials (in volts)
Acetaldehyde → acetate + $2H^+$ + $2e^-$	−0.58
Isocitrate → α-ketoglutarate + CO_2 + $2H^+$ + $2e^-$	−0.38
β-Hydroxybutyrate → acetoacetate + $2H^+$ + $2e^-$	−0.346
NADH + H^+ → NAD^+ + $2H^+$ + $2e^-$	−0.320
NADPH + H^+ → $NADP^+$ + $2H^+$ + $2e^-$	−0.324
Ethanol → acetaldehyde + $2H^+$ + $2e^-$	−0.197
Lactate → pyruvate + $2H^+$ + $2e^-$	−0.185
Malate → oxaloacetate + $2H^+$ + $2e^-$	−0.166
Succinate → fumarate + $2H^+$ + $2e^-$	−0.031
Coenzyme Q_{red} → coenzyme Q_{ox} + $2H^+$ + $2e^-$	+0.10
2 cytochrome $b_{K(red)}$ → 2 cytochrome $b_{K(ox)}$ + $2e^-$	+0.030
2 cytochrome c_{red} → 2 cytochrome c_{ox} + $2e^-$	+0.254
2 cytochrome $a_{3(red)}$ → 2 cytochrome $a_{3(ox)}$ + $2e^-$	+0.385
H_2O → $\frac{1}{2} O_2$ + $2H^+$ + $2e^-$	+0.816

SOURCE: From A. L. Lehninger, *Biochemistry*, 2nd ed. Worth Publishers, New York, 1975.

substance. In oxidative phosphorylation the final electron acceptor is oxygen. The oxygen combines with hydrogen as it accepts electrons, producing water in the final reaction of electron transport.

ATP Synthesis

The molecules directly using the H^+ gradient built up by electron transport can be considered to be H^+-ATPase pumps (see p. 131) that are forced to operate in reverse. If operating in the forward direction, the pumps would use energy released by ATP hydrolysis to drive H^+ across a membrane. However, in oxidative phosphorylation the pumps are driven in reverse by the H^+ gradient produced by electron transport. When running in reverse, the pumps, instead of breaking down ATP, use the energy of the H^+ gradient to drive synthesis of ATP from ADP and phosphate. This ability to reverse is not unique to the

H^+-ATPase pumps operating in ATP synthesis; many active transport pumps directly using ATP as an energy source can be forced to run in reverse and synthesize ATP (see p. 133).

In oxidative phosphorylation, both the electron transport system and the H^+-ATPase pumps are embedded in the same membrane (see Fig. 8-1). The H^+-ATPase pumps synthesizing ATP in response to very high H^+ gradients are called F_oF_1 *ATPases*. These pumps occur in the plasma membrane and its derivatives in prokaryotes and in the innermost membranes of mitochondria and chloroplasts in eukaryotes.

The overall mechanism linking an H^+ gradient to ATP synthesis was first proposed in 1961 by Peter Mitchell, in a model he called the *chemiosmotic hypothesis*. Since then, all the experimental tests of Mitchell's model have supported its basic tenets. Mitchell received the Nobel Prize in chemistry in 1975 for the chemiosmotic hypothesis and his work supporting it.

Substrate level phosphorylation differs fundamentally from oxidative phosphorylation. In this mechanism, phosphate groups are transferred from an organic molecule directly to ADP. Removal of the phosphate group from the organic donor releases free energy; some of this energy is captured in attachment of the phosphate group to ADP to form ATP. No electron carriers, H^+ gradient, or F_oF_1-ATPase are involved in this relatively simple process. Substrate-level phosphorylation is the primary source of ATP for organisms that do not possess an oxidative electron transport system. It becomes the primary source for cells in other organisms, including some cells of humans, if oxygen is unavailable as a final acceptor for electron transport pathways.

MITOCHONDRIAL STRUCTURE AND OCCURRENCE

Most of the reaction systems leading to ATP synthesis are housed inside mitochondria. These include almost all the reactions removing electrons from sugars and other molecules, the electron transport system, and the F_oF_1-ATPase directly synthesizing ATP.

Mitochondrial Structure

In most cells, mitochondria appear as spherical or elongated bodies about 0.5 μm in diameter and 1 to 2 μm in length, about the size of a bacterium. They may be much larger in some cells—up to several micrometers in diameter, and as long as 10 μm.

Two separate membrane systems, the *outer* and *inner boundary membranes*, define mitochondrial structure (Fig. 8-2; see also Fig. 1-10). The outer boundary membrane forms a single, continuous surface layer around the organelle. The inner boundary membrane is thrown into folds or tubular extensions, the *cristae* (singular = *crista*), which project into the mitochondrial interior and greatly increase the surface area of this membrane.

The outer and inner boundary membranes separate the mitochondrial interior into two distinct regions (see Fig. 8-2b): The *intermembrane compartment*, between the inner and outer membranes, and the *matrix*, the innermost compartment enclosed by the inner boundary membrane. The matrix contains granules of various sizes and, in some cases, fibers or crystals. The most common large granule is the *intramitochondrial granule*, a dense, spherical particle that may store Ca^{2+} and other ions (visible in Fig. 1-13). Other smaller granules, the *mitochondrial ribosomes*, occur in the matrix in large numbers (see Figs. 8-2 and 21-12a). In structure and function these ribosomes

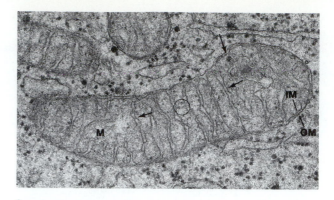

a

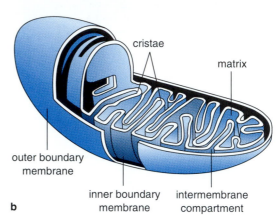

b

Figure 8-2 Mitochondrial structure. **(a)** A mitochondrion from a chick intestinal cell. The outer boundary membrane (OM) is smooth and covers the entire mitochondrion; the inner boundary membrane (IM) folds into cristae (arrows) that extend into the mitochondrial interior. The matrix (M), the innermost mitochondrial compartment, contains proteins and other molecules in solution, including the enzymes and intermediates of pyruvate oxidation and the citric acid cycle. Mitochondrial ribosomes (circle) and DNA are also present in the matrix. × 45,000. (Courtesy of J. Mais.) **(b)** The membranes and compartments of mitochondria.

resemble prokaryotic ribosomes much more closely than those in the eukaryotic cytoplasm outside mitochondria. Also embedded within the matrix are scattered deposits of DNA, also quite similar in structure and properties to bacterial DNA. (Chapter 21 gives details of the structure and function of mitochondrial DNA and ribosomes and their possible relationships to bacterial systems.) Glycogen deposits may appear in the mitochondrial matrix in some cell types.

Isolation of Mitochondrial Components

The outer and inner mitochondrial membranes react differently to disturbances such as exposure to detergents or high-frequency sound waves (sonication), al-

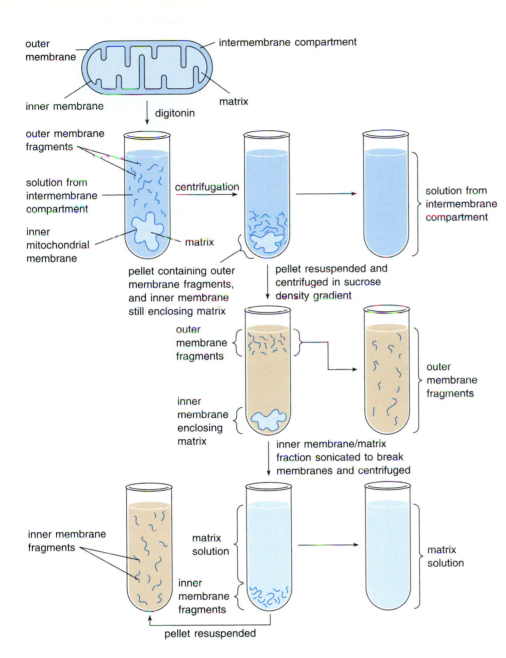

Figure 8-3 Fractionation of mitochondrial membranes and compartments. Mild detergents such as bile salts or digitonin break but do not completely disrupt the outer membrane. These agents release the solution in the intermembrane compartment but leave the inner membranes and the enclosed matrix intact. The preparation is centrifuged, which pellets the membranes and leaves the solution from the intermembrane compartment in the supernatant; the inner membranes still enclose the matrix in the pellet. The pellet is resuspended, and the outer membrane fragments and inner membrane/matrix fraction are separated by density-gradient centrifugation (see p. 795; the inner membrane/matrix fraction is significantly more dense than the outer membrane fragments). The inner membrane, still enclosing the matrix, is then broken by homogenization or sonication and isolated from the matrix by a final centrifugation.

lowing the two membranes and the solutions in the intermembrane compartment and matrix to be separated and individually purified (Fig. 8-3). The four fractions—outer boundary membrane, intermembrane compartment, inner boundary membrane, and matrix—retain much of their native activity and can be analyzed biochemically when separately purified in this way.

Analyses of the fractions reveal that the outer boundary membrane is relatively rich in cholesterol as compared to the inner membrane, which is virtually cholesterol-free. The inner boundary membrane contains high levels of *cardiolipin* (Fig. 8-4), an unusual phospholipid type present in very low concentrations or virtually absent from the outer membrane. The

contrasts in cholesterol and cardiolipin content between the two membranes probably account for the greater susceptibility of the outer membrane to detergents and sonication. (The distinct structure of the two membranes may reflect differences in their evolutionary origins—see p. 776 for details.)

The outer boundary membrane contains several enzymes associated with initial breakdown of fats and amino acids. Other outer membrane proteins are *porins* (see p. 146), transport molecules that form sieve-like channels through the membrane. They admit small molecules such as amino acids but bar the passage of enzymes and other proteins. The inner membrane fraction contains the F_o/F_1-ATPase, the carriers and complexes of the electron transport

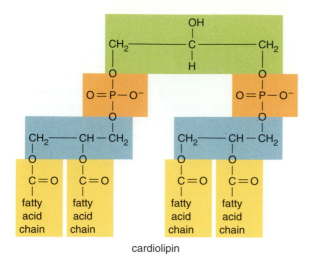

cardiolipin

Figure 8-4 Cardiolipin, an unusual phospholipid that forms part of the lipid framework of the inner mitochondrial membrane. Cardiolipin is absent or present at only very low concentrations in the outer boundary membrane of mitochondria and other cellular membranes of eukaryotes. It is common, however, in bacterial membranes.

system, and proteins involved in the transport of metabolites across the inner boundary membrane. The matrix contains the enzymes and reactants of two major systems removing electrons from fuel substances (pyruvic acid oxidation and the citric acid cycle; see below).

Occurrence and Distribution of Mitochondria

Almost all eukaryotic cells contain mitochondria. The exceptions are a few metabolically inert types such as the red blood cells of higher animals, which contain no cytoplasmic organelles of any kind when fully mature, and some fungi and protozoa that do not carry out oxidative phosphorylation. A few algae and some protozoan cells contain only a single mitochondrion. Most cells, however, contain mitochondria numbering in the tens to hundreds. Some cell types may contain many more; liver cells in higher animals, for example, contain from slightly less than a thousand to more than 2500 mitochondria. A few very large cells, such as the egg cells of various animals, may have hundreds of thousands of mitochondria. In cells in which mitochondria are especially abundant they take up a significant portion of the cell volume. In heart muscle cells, for example, mitochondria may occupy as much as 30% of the total cell volume. Light microscopy of living cells shows that mitochondria may grow in length, branch, and divide. The physical or chemical mechanisms underlying these movements are unknown.

Mitochondria do not occur in bacteria and cyanobacteria. In these prokaryotes, oxidative reactions are distributed between the cytoplasm and the plasma membrane. (The oxidative systems of the bacteria and cyanobacteria are described in Supplement 8-1.)

THE OXIDATIVE REACTIONS SUPPLYING ELECTRONS

Before being oxidized, the primary substances used as electron sources are split into their component parts. Complex carbohydrates such as starches are split into individual monosaccharides, and fats are broken into glycerol and fatty acids. Proteins to be oxidized as an electron source are hydrolyzed into individual amino acids.

Oxidation of these products proceeds in a series of reactions that may include one or both of two major stages, the first occurring outside mitochondria and the second inside. In the first stage, *glycolysis* (from *glykys* = sweet and *lysis* = breakdown), fuels are partially oxidized and converted to three-carbon segments. These short carbon chains are then completely oxidized to carbon dioxide and water in the second stage inside mitochondria, which includes two reaction sequences, *pyruvate oxidation* and the *citric acid cycle*. A relatively small amount of ATP is produced by substrate-level phosphorylation during both stages. However, the major synthesis of ATP occurs as the electrons removed during these oxidative stages pass through the electron transport system.

Monosaccharides and many other carbohydrate units follow a main line of oxidation through both stages. Glycerol enters the glycolytic pathway in the first stage, and fatty acids enter in the second stage. Amino acids are converted by *deamination* (removal of the amino, or —NH_2, group) into molecules that, depending on the product, may enter the pathway in either stage.

The First Major Stage: Glycolysis

Glycolysis uses glucose as an input and yields electrons and the three-carbon substance *pyruvic acid* or *pyruvate*.[1] This three-carbon molecule is the primary fuel substance for the second stage of oxidative reac-

---

[1] Pyruvic *acid* refers to the undissociated form of the acid, and pyru*vate* to the dissociated form:

$$CH_3CH_2COOH \rightleftharpoons CH_3CH_2COO^- + H^+$$

pyruvic acid   pyruvate

Because organic acids typically dissociate under physiological conditions, the *-ate* form is used in the text and diagrams in this book.

tions inside mitochondria. Glycolysis also produces a small quantity of ATP by substrate-level phosphorylation. Although the amount of ATP made in glycolysis is relatively small, it can be vitally important to survival when oxygen supplies are low.

An Overview of Glycolysis Glycolysis has two major parts (Fig. 8-5). In the first part (red, "uphill" portion in Fig. 8-5), driven by ATP hydrolysis, six-carbon derivatives of glucose are raised to an energy level high enough for entrance into the second part of the glycolytic sequence. In the second part of the glycolytic sequence (blue, downhill portion in Fig. 8-5), electrons are removed from the derivatives of glucose, and the energy expended in the first part is recovered with a net gain in ATP by substrate-level phosphorylation. In the process the glucose derivatives are split into two molecules of pyruvate. Each step in the pathway is catalyzed by a specific enzyme. The various enzymes, reactants, and products of glycolysis are suspended in the cytoplasmic solution outside mitochondria.

Glycolytic Reactions The first reactions of glycolysis (reactions 1 through 3 in Fig. 8-6) use two molecules of ATP to convert glucose into a highly reactive six-carbon, two-phosphate sugar. These initial reactions go uphill in terms of the energy content of the products and proceed only because they are coupled to ATP breakdown.

The first reaction transfers a phosphate group from ATP to glucose, producing *glucose 6-phosphate*. The product of the first reaction is rearranged (reaction 2) and then phosphorylated (reaction 3) at the expense of a second molecule of ATP, producing *fructose 1,6-bisphosphate*. The characteristics of this reaction illustrate one of the several mechanisms regulating glycolysis. The enzyme catalyzing the reaction, *phosphofructokinase*, is allosterically inhibited (see p. 84) by high concentrations of ATP. If ATP is present in elevated concentrations in the cytoplasm, inhibition of phosphofructokinase slows or stops the subsequent reactions of glycolysis. Phosphoenolpyruvate, the product of a reaction later in glycolysis (reaction 9; see below), also inhibits the enzyme. Because the glycolytic sequence produces a three-carbon fuel substance for the oxidations taking place in mitochondria, slowing glycolysis can retard the oxidative reactions driving ATP synthesis inside mitochondria. A surplus of other products or intermediates of oxidative metabolism, such as NADH or citrate (see below), also inhibits phosphofructokinase. If energy-requiring activities take place in the cell, resulting in extensive conversion of ATP to ADP, the reduction in ATP concentration relieves inhibition of phosphofructokinase. In addition, ADP acts as an allosteric activator of the enzyme. The rates of glycolysis and ATP production therefore

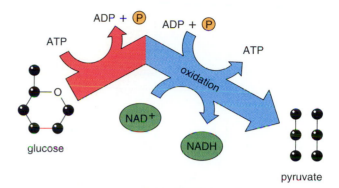

Figure 8-5 The overall reactions of glycolysis. The reactions split glucose (six carbons) into pyruvate (three carbons) and yield ATP and NADH. The reactions take place in two major parts. In the first part (in red), six-carbon molecules derived from glucose are raised to higher energy levels at the expense of ATP. In the second part the products of the first part are oxidized and split into pyruvate. The second part yields NADH and a net gain in ATP.

increase proportionately as ADP concentration rises. Regulation of phosphofructokinase is probably the most sensitive and significant of the several controls of glycolysis because it is directly keyed to the relative concentrations of ATP and ADP and thus to the rate at which cells use energy for their activities.

The ATP used in the initial steps of glycolysis is recovered with a net gain in the remaining reactions of the pathway (reactions 4 through 10). In reaction 4, fructose 1,6-bisphosphate is broken into two different three-carbon sugars, each with one phosphate group. Only one of these sugars, *3-phosphoglyceraldehyde (3PGAL)*, directly enters the next step in glycolysis. However, as this three-carbon sugar is used, a rearranging enzyme converts the second three-carbon sugar, *dihydroxyacetone phosphate*, into 3PGAL (reaction 5). As a result, both products of reaction 4 are used in the remainder of the sequence.

At the next step (reaction 6), two electrons and two hydrogens are removed from 3PGAL. Some of the energy released is trapped by the addition of an inorganic phosphate group from the medium (not from ATP) to form a three-carbon, two-phosphate product, *1,3-bisphosphoglycerate*. The electrons removed in reaction 6 have a relatively high energy level and are accepted by a carrier molecule, *nicotinamide adenine dinucleotide*[2] (*NAD*; Fig. 8-7). NAD is one of a group of nucleotide-based carriers that shuttle electrons between major reaction systems. One of the two hydrogens removed from 3PGAL in reaction 6 binds to

[2]The oxidized form of this electron carrier is symbolized as NAD^+; the reduced form as NADH. A similar convention is used for several other nucleotide-based electron carriers.

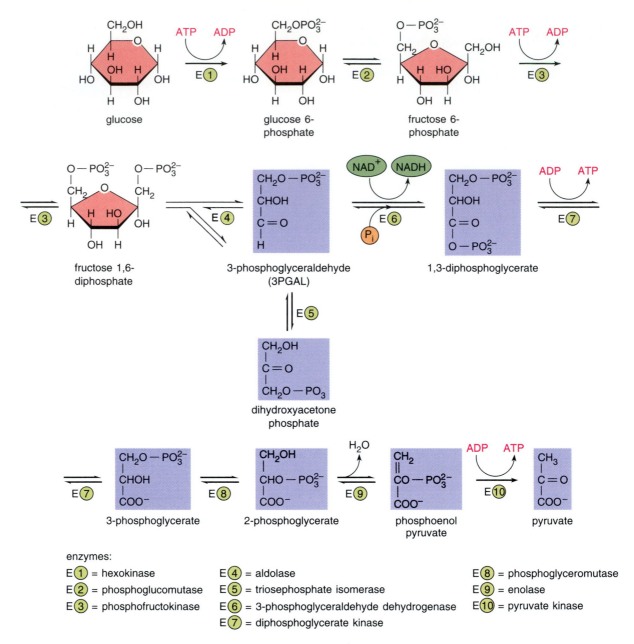

enzymes:

E① = hexokinase   E④ = aldolase   E⑧ = phosphoglyceromutase
E② = phosphoglucomutase E⑤ = triosephosphate isomerase   E⑨ = enolase
E③ = phosphofructokinase E⑥ = 3-phosphoglyceraldehyde dehydrogenase   E⑩ = pyruvate kinase
                          E⑦ = diphosphoglycerate kinase

Figure 8-6 The reactions and enzymes of glycolysis (see text).

NAD⁺ in the position shown in Figure 8-7 to form NADH. The second is released to the medium as H^+. The reduced NAD (or NADH) formed at this step can be considered a "high-energy" substance by virtue of the electrons it gains in reaction 6. When oxygen is abundant, the electrons carried by NADH eventually enter the electron transport system in mitochondria, where much of their energy is captured in the conversion of ADP to ATP.

The 1,3-bisphosphoglycerate product of reaction 6 may also be regarded as a high-energy substance because removal of the two phosphates releases large amounts of free energy, which is captured in substrate-level phosphorylation of ADP to ATP. In the first of these reactions (reaction 7), 1,3-bisphosphoglycerate is hydrolyzed to *3-phosphoglycerate*; the phosphate group removed in the reaction is added directly to ADP to form ATP. The 3-phosphoglycerate enters the final series of reactions, which produces another ATP and pyruvate. Reaction 8 is a rearrangement that shifts the phosphate group from the 3- to the 2-carbon, producing *2-phosphoglycerate*. Reaction 9 is an internal oxidation in which electrons are removed from one part of 2-phosphoglycerate and delivered to another part at lower energy levels; most of the energy lost by the electrons is retained in the product of the reaction, *phosphoenolpyruvate*. The shift greatly increases the energy made available when its phosphate group is re-

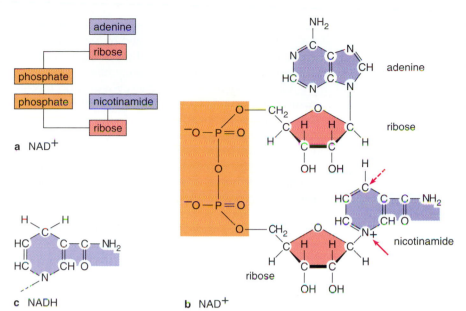

Figure 8-7 NAD (nicotinamide adenine dinucleotide), an electron carrier reduced in glycolysis and other cellular reactions. The molecule consists of two nucleotides linked end to end. (**a** and **b**) The oxidized form of the carrier, designated NAD$^+$. One electron is added at each of the two positions marked by an arrow as the carrier is reduced; a hydrogen is also added at the position marked by the dashed arrow. (**c**) The reduced form of the nicotinamide portion of the carrier, designated NADH.

moved in the next reaction. The final reaction of the pathway (reaction 10) removes the phosphate group from phosphoenolpyruvate and yields pyruvate, the end product of glycolysis. The phosphate group is transferred directly to ADP to form ATP in the second substrate-level phosphorylation of the pathway.

The substrate-level phosphorylations of the second part of glycolysis provide a net gain in ATP because each glucose molecule entering the pathway produces two molecules of 3PGAL. As these two 3PGALs are oxidized to two molecules of pyruvate, a total of 4ATP is produced. Two of these ATPs pay back the 2ATP required to convert glucose to fructose 1,6-bisphosphate in the initial steps of glycolysis (reactions 1 through 3). The remaining 2ATP are a net gain for each molecule of glucose oxidized to two molecules of pyruvate. The second segment of glycolysis (reactions 4 through 10) also produces 2NADH, which represent a significant reservoir of energy that can also be captured in the conversion of ADP to ATP. The total reactants and products of glycolysis are therefore:

$$\text{glucose} + 2\text{ ADP} + 2\text{ P}_i + \text{NAD}^+ \longrightarrow$$
$$2\text{ pyruvate} + 2\text{ NADH} + 2\text{ ATP} \quad \text{(8-1)}$$

where P_i indicates inorganic phosphate.

Important Variations of the Glycolytic Pathway A wide variety of substances can be used as inputs for glycolysis. Animal glycogen and plant starch are both polysaccharides built up from glucose chains (see Fig. 2-9). Glycogen, which is stored in the liver and striated muscle cells of humans and other vertebrates, is hydrolyzed to yield glucose 6-phosphate, which enters the glycolytic pathway as an input into reac-

tion 2. The phosphate added to the 6-carbon of the glucose units removed from glycogen is derived from inorganic supplies rather than ATP. (Glycogen breakdown is regulated by a cAMP-based second messenger pathway; see p. 160.) Plant starch is hydrolyzed to individual glucose molecules, which enter as inputs into the first reaction of the glycolytic pathway. The enzymes necessary for complete hydrolysis of plant starch, *amylase* and *maltase*, occur as digestive enzymes in humans and other animals.

3PGAL also serves as an important entry point for glycolysis. Besides taking part in the glycolytic sequence, it is formed as the first net product of photosynthesis (see pp. 262–265 for details). In plants the 3PGAL produced in photosynthesis may serve as a cellular fuel by entering glycolysis directly in reaction 6. The glycerol released by fat breakdown is also converted indirectly into 3PGAL and enters cellular oxidations in reaction 6.

The most significant variations in glycolysis from the standpoint of the cellular energy economy occur at the end of the pathway. These variations enable glycolysis to operate as the primary source of ATP when oxygen supplies are too low for ATP production in mitochondria.

In one of the most important of these variations, the end product of glycolysis, pyruvate, is converted into lactate:

$$\text{pyruvate} + \text{NADH} \longrightarrow \text{lactate} + \text{NAD}^+ \quad \text{(8-2)}$$

In the reaction, pyruvate acts as an acceptor for the electrons carried by NADH. The NAD$^+$ is then free to cycle back to the oxidative reaction of glycolysis (reaction 6) to accept further electrons from 3PGAL.

Because NAD^+ is constantly regenerated by lactate formation, glycolysis can continue to run, with net production of ATP.

Lactate formation by this variation of glycolysis occurs in the muscle cells of animals, including humans, if intensive activity is carried out before increases in breathing and heart rate are able to meet the demand for oxygen in the muscle tissue. The lactate accumulating under these conditions serves as a temporary storage site for electrons that can later be passed on to the mitochondrial electron transport system when the oxygen content of the muscle cells returns to normal levels. This occurs by a sequence that essentially reverses reaction 8-2 and regenerates pyruvate and NADH.

Another important variation occurs in microorganisms such as yeasts. In this modification, pyruvate accepts electrons from NADH and is converted by additional reactions into *ethyl alcohol* (two carbons) and CO_2:

$$\text{pyruvate} + \text{NADH} \longrightarrow$$
$$\text{ethyl alcohol} + CO_2 + NAD^+ \quad (8\text{-}3)$$

This variation has been of central importance to human activities since the earliest days of recorded history. It provides alcohol for brewing and other industries and CO_2 for raising dough in baking.

When electrons removed from 3PGAL in glycolysis are delivered to an organic molecule by NADH, as they are in lactate or ethyl alcohol formation, the total pathway is termed a *fermentation*. Various fermentations, producing a wide range of products, generate ATP in different fungi and bacteria.

Some bacterial and fungal species, the *strict anaerobes*, can produce ATP only by fermentations. Others can switch between fermentations and full oxidation pathways, depending on the oxygen supply. Cells in this category are *facultative anaerobes*. Some cells, the *strict aerobes*, are unable to live solely by fermentation at any time. Many cells of higher organisms, including vertebrate muscle cells, are facultative and can switch between fermentation and complete oxidation depending on the availability of oxygen. Others, such as the brain cells of humans and other vertebrates, are strict aerobes and cannot survive unless oxygen is available.

Glycolysis can also be reversed in effect, so that glucose or glucose 6-phosphate can be synthesized from pyruvate. Other, noncarbohydrate substances such as fats and amino acids can also be converted into glucose after conversion into pyruvate or intermediates of the glycolytic pathway. Synthesis of glucose from noncarbohydrate precursors by this route is called *gluconeogenesis*. Some of the reactions of glycolysis (reactions 1, 3, and 10 in Figure 8-4) are essentially irreversible when catalyzed by their glycolytic enzymes; in gluconeogenesis these steps are circumvented by the use of different enzymes.

Gluconeogenesis is an energy-requiring mechanism that proceeds only through conversion of phosphate groups removed from ATP and other organic molecules to inorganic form. Although requiring an energy input to proceed, gluconeogenesis is a vital source of glucose when its supplies are limited. For example, in humans and other vertebrates the reserves of glycogen stored in the liver are depleted by only a few hours of fasting. Under these conditions, liver cells compensate for glycogen depletion by converting breakdown products of fats and proteins into glucose through gluconeogenesis. Maintenance of blood glucose by gluconeogenesis is vital to survival because brain cells can continue to function only if glucose remains available at normal concentrations in the blood.

In addition to reversal, glycolysis can be circumvented by a reaction sequence that oxidizes glucose 6-phosphate into a five-carbon sugar, *ribose 5-phosphate*. This sequence, called the *pentose phosphate pathway*, is an important source of five-carbon sugars for synthesis of nucleotides and other substances. (Details of the pathway are given in Information Box 8-2.)

Discovery of the Glycolytic Pathway Research leading to the discovery of glycolysis began in the late 1800s. In Germany, Eduard and Hans Buchner found that alcoholic fermentation could be carried out by nonliving extracts of yeast cells. Through research spurred by this discovery the first enzymes of any kind were discovered and described. Intensive studies of alcoholic fermentation continued until well into the 1930s, when Gustave Embden assembled the known reactions into a provisional reaction sequence that explained glycolysis. Other investigators, most notably Otto Meyerhoff, Carl Cori, and Otto Warburg, fitted final pieces into the puzzle. By 1940 the reactions of glycolysis were known essentially as they are today.

More recent research has revealed that different organisms, and in some cases different tissues of the same organism, have subtle variations in the pathway. The differences, involving mechanisms such as substrate affinity or the effects of inhibitors, depend on *isoenzymes*—slightly different versions of the glycolytic enzymes that catalyze the same reaction but have distinct amino acid sequences and properties. In spite of these differences, the enzymes catalyzing glycolysis, which occur in most prokaryotes and all eukaryotes, are among the most highly conserved enzymes known. The reaction sequence was probably present in essentially the same form as today

The Pentose Phosphate Pathway

Not all the glucose metabolized in cellular oxidations is oxidized by the glycolytic pathway. Certain inhibitors, such as fluoride or iodoacetate, block the glycolytic pathway but do not completely halt glucose oxidation. This finding led to discovery of the *pentose phosphate pathway*, in which glucose is oxidized and converted into a five-carbon sugar (a pentose). $NADP^+$ serves as electron acceptor for the pathway.

The metabolism of glucose via the pentose phosphate pathway begins with glucose 6-phosphate, which is oxidized to *6-phosphogluconate* in a two-step process (see reactions 1 and 2 of the figure in this box). The oxidation takes place in the first step; $NADP^+$ acts as the acceptor for the electrons. After addition of the elements of a molecule of water in the second step, the product, 6-phosphogluconate, enters the second oxidation of the pathway (reaction 3 of the figure). $NADP^+$ is also the electron acceptor for this oxidation; as part of the reaction, the carbon chain is shortened to yield the five-carbon sugar *ribulose 5-phosphate*. The carbon removed is released as CO_2. Ribulose 5-phosphate is rearranged to *ribose 5-phosphate* in the final step of the pathway (reaction 4 of the figure).

The two major products of this pathway, ribose 5-phosphate and NADPH, are both significant in cellular metabolism. The pentose is used as the starting point for the synthesis of a variety of three-, four-, five-, six-, and seven-carbon sugars in both plants and animals. Five-carbon sugars form part of the structure of nucleotides, including those of ATP, FAD, FMN, and coenzyme A, and of the nucleic acids DNA and RNA. A derivative of ribose 5-phosphate (*ribulose 1,5-bisphosphate*; see p. 261) combines with atmospheric CO_2 as the first step in the photosynthetic reactions producing sugars. Ribose 5-phosphate may also enter a complex series of reactions with its own three-, seven-, and four-carbon derivatives

to regenerate glucose 6-phosphate, which may enter the pentose phosphate pathway for another series of oxidations. In this way, all the carbons of an original glucose molecule may eventually be oxidized to CO_2 by the pentose phosphate pathway, with further generation of NADPH.

Because NADPH is important in synthetic reactions in which a reduction is required, the pentose phosphate pathway may be considered as an alternative to glycolysis that generates reducing power in the cell in the form of NADPH. The NADH produced in glycolysis, in contrast, drives ATP synthesis by delivering electrons removed from glucose to the mitochondrial electron transport system.

Whether glucose is oxidized via glycolysis or the pentose phosphate pathway depends on the enzyme most active in conversion of glucose 6-phosphate, the product of the first reaction in the glycolytic pathway. If NADPH is present in high concentrations in the cytoplasm, the first enzyme of the pentose phosphate pathway (*glucose 6-phosphate dehydrogenase*) is inhibited, and glucose 6-phosphate is converted to fructose 6-phosphate by phosphoglucomutase as it normally would be in the glycolytic pathway. If NADPH is present in low concentrations and the levels of $NADP^+$ are high, glucose 6-phosphate dehydrogenase is stimulated and glucose 6-phosphate is oxidized by the pentose phosphate pathway. The effects of the relative concentrations of $NADP^+$ and NADPH on the enzyme tie the pentose phosphate pathway to the cell's needs for reducing power in the form of NADPH. The pathway operates only when NADPH is required by synthetic reactions somewhere else in the cell. Because the pentose phosphate pathway diverts glucose 6-phosphate from glycolysis when cells need reducing power, it is sometimes called the pentose phosphate *shunt*.

glucose 6-phosphate → (1) glucose 6-phosphate dehydrogenase [$NADP^+$ → NADPH] → 6-phosphoglucono-δ-lactose → (2) lactonase [H_2O → H^+] → 6-phosphogluconate → (3) 6-phosphogluconate dehydrogenase [$NADP^+$ → NADPH, CO_2] → ribulose 5-phosphate → (4) phosphopentose isomerase → ribose 5-phosphate when eukaryotic cells first evolved from prokaryotes some 1800 million years ago. The relatively small differences noted between the glycolytic enzymes of organisms as distantly related as plants and animals indicate that only 5% of the amino acid residues have changed in the enzymes during each 100 million years of evolution.

The Second Major Stage: Pyruvate Oxidation and the Citric Acid Cycle

In the second major stage of cellular oxidations the pyruvate produced by glycolysis is oxidized inside mitochondria. The mitochondrial oxidations occur in two overall parts. In the first, *pyruvate oxidation*, pyruvate is oxidized into two-carbon *acetyl* (CH_3—$\overset{\displaystyle \|}{\underset{\displaystyle O}{C}}$—)

units and CO_2. In the second part, the *citric acid cycle*, the acetyl units produced by pyruvate oxidation are completely oxidized to CO_2. Because oxygen is normally the final acceptor for the electrons removed in pyruvate oxidation and the citric acid cycle, the oxidative activities of mitochondria are frequently termed *respiration*. This term is distinct from the common physiological meaning of "breathing." However, the primary use of the oxygen we inhale in physiological respiration is actually as the final electron acceptor for the electrons removed in mitochondrial oxidations.

Part 1 of Mitochondrial Respiration: Pyruvate Oxidation In pyruvate oxidation a molecule of pyruvate is converted into an acetyl unit in a cyclic series of reactions that removes two electrons, two hydrogens, and one carbon in the form of CO_2 (Figure 8-8 shows the overall cycle; details of the cycle are given in Fig. 8-9). The two electrons and one of the hydrogens removed are accepted by NAD^+, which is converted to NADH. The acetyl unit is transferred to *coenzyme A* (Fig. 8-10), a molecule that transfers acetyl units between major reaction systems:

$$\text{pyruvate} + \text{coenzyme A} + NAD^+ \longrightarrow$$
$$\text{acetyl coenzyme A} + NADH + H^+ + CO_2 \quad (8\text{-}4)$$

Coenzyme A, like ATP and NAD, is based on a nucleotide (compare Figs. 2-26, 8-7, and 8-10). The CO_2 and the remaining hydrogen are released to enter the surrounding medium. Because each molecule of glucose entering glycolysis produces two molecules of pyruvate, all the reactants and products in reaction 8-4 are multiplied by a factor of 2 if pyruvate oxidation is considered as a continuation of glycolysis.

Both organic products of pyruvate oxidation are important elements in cellular energy metabolism. The acetyl units carried by coenzyme A serve as the

Figure 8-8 Overall reactants and products of pyruvate oxidation (see text).

immediate fuel for the citric acid cycle. The electrons carried by NADH represent potential energy that is eventually tapped to drive ATP synthesis.

Part 2 of Mitochondrial Respiration: The Citric Acid Cycle In the citric acid cycle (Fig. 8-11; also called the *Krebs cycle* in honor of Hans Krebs, who was the first to piece together the reactions of the series) there is a continuous input of two-carbon acetyl units carried by coenzyme A, and ADP, NAD^+, and another oxidized electron carrier, *flavin adenine dinucleotide* (*FAD*; Fig. 8-12). There is a continuous output of "empty" coenzyme A, and CO_2, ATP, NADH, and $FADH_2$ (the reduced form of FAD).

Only one molecule of ATP (or GTP; see below) is formed as a direct product of the citric acid cycle. Most of the energy released by the several oxidations of the cycle is trapped in the electrons carried by NADH and $FADH_2$. These electrons enter the electron transport system to drive synthesis of most of the ATP produced in cellular oxidations.

In the first of the individual reactions of the citric acid cycle (reaction 1 in Fig. 8-13) the two-carbon acetyl group carried by coenzyme A is transferred to *oxaloacetate* (four carbons) to form *citrate* (six carbons), the acid for which the cycle is named. The reaction is catalyzed by a specific enzyme, as are all the reactions of the cycle. (The enzymes are listed individually in Fig. 8-13.) Coenzyme A, relieved of its acetyl unit by reaction 1, is free to enter another cycle of pyruvate oxidation.

The enzyme catalyzing the first reaction of the cycle, *citrate synthase*, is one of the major regulators of the citric acid cycle. The enzyme is inhibited if succinyl-coenzyme A, an intermediate product of a reaction later in the cycle (reaction 5), accumulates. The enzyme is also inhibited by high concentrations of ATP. These inhibitors effectively slow or stop the cycle if ATP supplies are abundant or if later steps in the cycle are slowed (usually also due to inhibitions imposed by elevated ATP concentrations; see below).

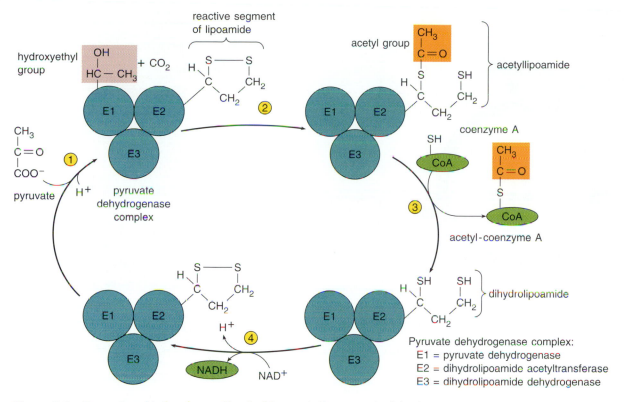

Figure 8-9 Pyruvate oxidation. In reaction 1 of the cycle the pyruvate dehydrogenase complex, which consists of the three enzymes E1, E2, and E3, binds a molecule of pyruvate. After binding, E1 catalyzes removal of one carbon from pyruvate; the carbon is released as CO_2. The two-carbon segment remains attached to the prosthetic group (thiamin pyrophosphate, or TPP) of E1 as a hydroxyethyl group. In reaction 2, E1 oxidizes the hydroxyethyl group, forming an acetyl group. As part of this reaction the acetyl group is transferred to lipoamide, the prosthetic group of E2. The transfer converts lipoamide to acetyllipoamide. In reaction 3, E2 catalyzes transfer of the acetyl group to coenzyme A, forming acetyl-coenzyme A. The transfer converts the prosthetic group of E2 to dihydrolipoamide, a fully reduced form that retains both electrons removed from pyruvate during its oxidation. These electrons are transferred to NAD^+ in reaction 4 of the cycle, catalyzed by E3. The electron transfer converts dihydrolipoamide back to lipoamide, and the pyruvate dehydrogenase complex is ready to repeat the cycle.

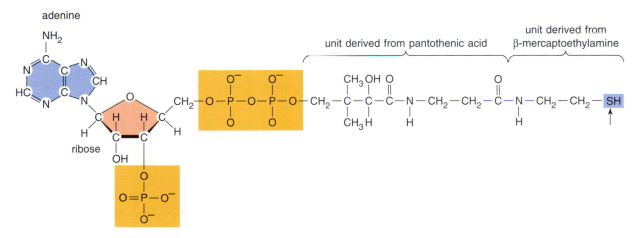

Figure 8-10 Coenzyme A, a nucleotide-based molecule that carries acetyl groups. Coenzyme A also carries other substances between oxidative reactions in cells, including fatty acid chains and succinate. The groups carried attach to coenzyme A at the —SH group marked by the arrow.

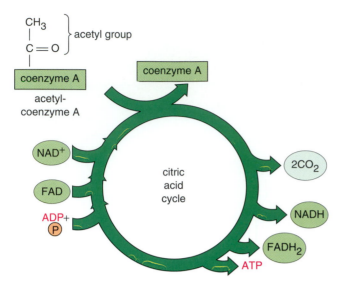

Figure 8-11 Overall reactions of the citric acid cycle. The cycle oxidizes two-carbon acetyl groups to two molecules of CO_2. Acetyl-coenzyme A, NAD^+, FAD, and ADP enter the cycle; coenzyme A, NADH, $FADH_2$, ATP, and CO_2 are released as products. Under certain conditions one of the oxidations of the cycle may use $NADP^+$ instead of NAD^+ as electron acceptor.

The citrate formed in the first reaction is rearranged in reactions 2 and 3 into *isocitrate*, which becomes the reactant for the first oxidation in the cycle. In this oxidation (reaction 4), two electrons and two hydrogens are removed from isocitrate, forming *α-ketoglutarate*. The carbon chain is also shortened at this step, with the release of one molecule of CO_2.

NAD^+ is the usual acceptor for the electrons removed in this oxidation. However, depending on the relative concentrations of ATP, ADP, and NADH, a closely related molecule, *nicotinamide adenine dinucleotide phosphate* (*NADP*; Fig. 8-14), may accept the electrons. Which of the two electron acceptors is reduced illustrates yet another of the many controls regulating cellular oxidations. Reaction 4 is catalyzed by either of two enzymes that are identical in activity except that one uses NAD^+ and the other $NADP^+$ as electron acceptor. If ADP concentrations are high and ATP and NADH are limited in concentration, the NADP-specific enzyme is inhibited, and the NAD^+-specific enzyme is most active (this enzyme requires ADP as a cofactor; see p. 78). If the concentration of ATP or NADH is high, the NAD^+-specific enzyme is inhibited, and $NADP^+$ is favored as electron acceptor.

These controls are significant because NADH normally delivers its electrons to the electron transport system of mitochondria, which drives the synthesis of ATP, whereas NADPH acts as an electron donor for synthetic reactions that require a reduction. Under the usual conditions, in which cellular activity demands a

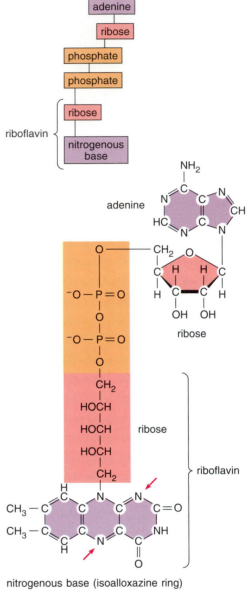

Figure 8-12 FAD (flavin adenine dinucleotide), an electron carrier that accepts electrons removed from the succinate oxidized in the citric acid cycle. FAD adds a hydrogen and an electron at two positions (arrows) during reduction to form $FADH_2$.

more or less continuous supply of ATP, the enzyme using NAD^+ as an electron acceptor predominates at this step, leading to ATP synthesis in response.

The second oxidation of the cycle (reaction 5) converts α-ketoglutarate into *succinate*. The reaction occurs by means of a subcycle (not shown in Fig. 8-13) that is similar to pyruvate oxidation. During the subcycle, two electrons and two hydrogens are removed from the five-carbon α-ketoglutarate entering the cycle, and one carbon is removed and released as CO_2. NAD^+ acts as electron acceptor for the reaction. Some

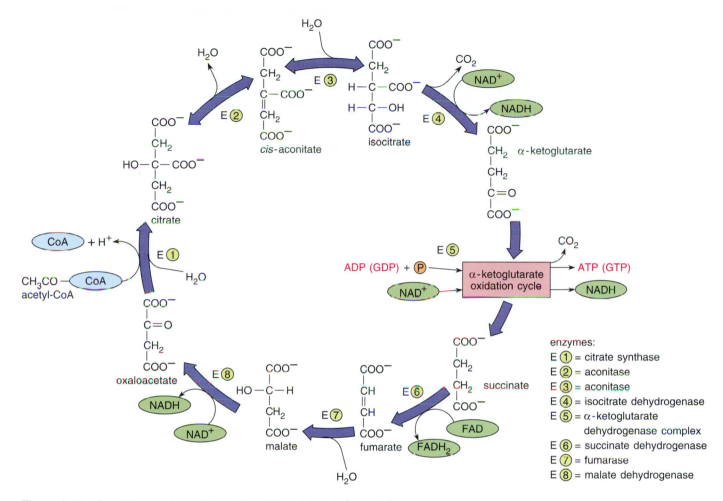

Figure 8-13 Reactions and enzymes of the citric acid cycle (see text).

of the energy released in reaction 5 is captured in substrate-level synthesis of a molecule of GTP (in animals) or ATP (in bacteria). This is the only GTP or ATP originating directly from the citric acid cycle. The net products of reaction 5 are therefore a four-carbon product (succinate) and one molecule each of CO_2, NADH, and ATP or GTP.

Succinate is oxidized at the next step of the cycle (reaction 6) by the enzyme *succinate dehydrogenase*. This enzyme is unique among the various enzymes of the citric acid cycle because it is tightly bound to the inner mitochondrial membrane. In this location it forms part of one of the protein complexes of the electron transport system, as does the electron acceptor for reaction 6, FAD. FAD goes to its reduced form, $FADH_2$, when it accepts two electrons and two hydrogens in reaction 6.

The four-carbon product of reaction 6, *fumarate*, is rearranged (reaction 7) with the addition of a molecule of water to form another four-carbon substance, *malate*. This acid is oxidized to oxaloacetate in the final reaction of the citric acid cycle (reaction 8). NAD⁺ acts as electron acceptor for the reaction. The molecule of

oxaloacetate replaces the one used in reaction 1, and the cycle is ready to turn again.

In a complete turn of the citric acid cycle, one two-carbon acetyl unit is consumed and two molecules of CO_2 are released. Electrons are removed at each of four oxidations in the cycle. Assuming that the first oxidation of the cycle uses NAD⁺ rather than NADP⁺ as electron acceptor, each of three of the oxidations produce one NADH. The remaining oxidation produces one $FADH_2$. One molecule of ATP (or GTP) is generated by substrate-level phosphorylation. The oxaloacetate used in the initial reaction of the citric acid cycle is replaced in the final reaction:

$$acetyl\ unit + 3NAD^+ + FAD + ADP + P_i \longrightarrow$$
$$2CO_2 + 3NADH + FADH_2 + ATP \quad (8\text{-}5)$$

Work with isolated mitochondrial fractions shows that all the reactions of pyruvic acid oxidation and the citric acid cycle, with one exception, are suspended in solution in the mitochondrial matrix. The exception is the enzyme catalyzing oxidation of succinate (reaction 5 in Fig. 8-13), which is tightly bound

Figure 8-14 NADP (nicotinamide adenine dinucleotide phosphate), an alternate electron acceptor for reaction 4 of the citric acid cycle. NADP is identical to NAD except for an additional phosphate group (shaded). The oxidized form of the carrier, designated as NADP$^+$, is shown. The carrier adds a hydrogen and two electrons at the same positions as NAD$^+$ (see Fig. 8-5) when reduced to NADPH. NADPH carries electrons to cellular reactions requiring a reduction rather than to systems synthesizing ATP.

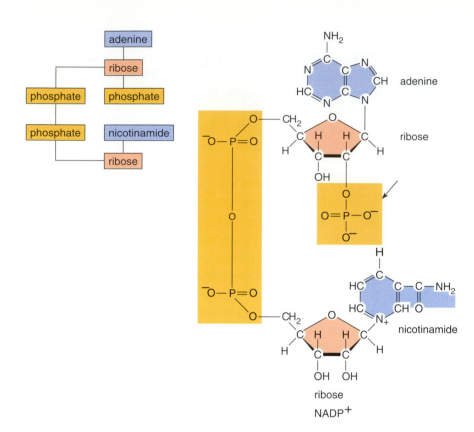

with its electron acceptor, FAD, to the inner mitochondrial membrane as part of the mitochondrial electron transport system.

Hans Krebs worked out the majority of the reactions in the citric acid cycle over a five-year period beginning in 1932. He tested various compounds thought to be important in cellular metabolism and discovered that a number of acids, including citrate, succinate, fumarate, and acetate, are rapidly oxidized in slices of liver and kidney tissue. Concurrent contributions were made by other scientists, most notably Albert Szent-Gyorgyi, Carl Martius, and Franz Knoop. Martius and Knoop pieced together the reaction series from citrate to succinate, and Szent-Gyorgyi proposed the series leading from succinate through fumarate and malate to oxaloacetate. Finally, Krebs discovered that oxaloacetate and a product derived from pyruvate are combined to form citrate. This discovery was the key enabling Krebs to assemble the series of oxidative reactions into a cycle. He was awarded the Nobel Prize in 1937 for his discoveries in cellular oxidation.

The Products of Glucose Oxidation

We can now sum all the products of complete oxidation of glucose to CO_2, from glycolysis through the citric acid cycle. For each molecule of glucose entering the sequence, the citric acid cycle turns twice. Glycolysis and pyruvate oxidation together yield 4NADH, 2ATP, and 2CO_2 (remember that two molecules of pyruvate are oxidized for each glucose molecule entering the sequence). Adding to this the products of two turns of the citric acid cycle:

$$\text{glucose} + 4\text{ADP} + 4\text{P}_i + 10\text{NAD}^+ + 2\text{FAD} \longrightarrow$$
$$4\text{ATP} + 10\text{NADH} + 2\text{FADH}_2 + 6\text{CO}_2 \quad (8\text{-}6)$$

At this point, relatively little ATP has been produced. However, the electrons carried by the 10NADH and 2FADH$_2$ contain much of the energy obtained from oxidation of glucose to CO_2. The free energy released as these electrons travel through the mitochondrial electron transport system provides the energy driving most of the ATP synthesis carried out in the cell.

Funneling Fats and Proteins into the Oxidative Pathways

Oxidation of fats and proteins proceeds after initial hydrolysis of these substances into smaller molecules—fats into glycerol and fatty acids and proteins into amino acids. Glycerol is converted in several steps into 3PGAL and enters glycolysis in this form. Fatty acids are oxidized in a preliminary series of reactions in mitochondria and split into two-carbon acetyl units attached to coenzyme A, which then enter

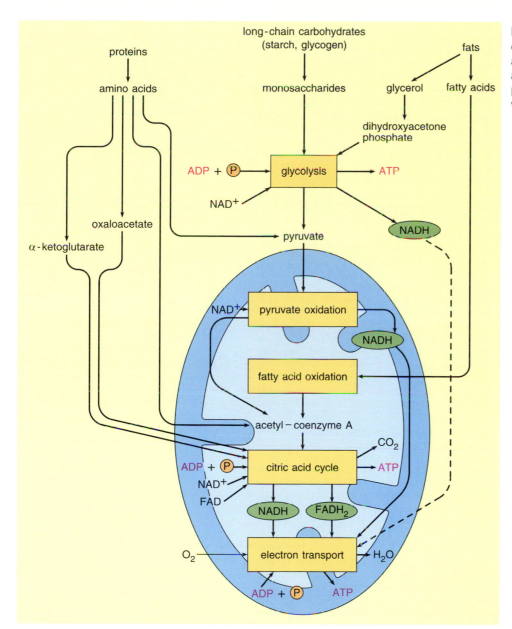

Figure 8-15 Major pathways oxidizing carbohydrates, fats, and proteins. Coenzyme A forms a central channel that funnels the products of many oxidative pathways into the citric acid cycle.

the citric acid cycle. (The reactions converting fatty acids to acetyl units, called *fatty acid oxidation*, are described in Supplement 8-2.) Amino acids derived from protein hydrolysis, during or after deamination, are converted into products that enter the main line of carbohydrate oxidation as either pyruvate, acetyl units carried by coenzyme A, or intermediates of the citric acid cycle.

Figure 8-15 summarizes the cellular pathways that oxidize carbohydrates, fats, and proteins. The figure shows the central role of coenzyme A as a carrier funneling acetyl units originating from different pathways into the citric acid cycle. The reactions of mitochondria are coordinated with those of another class of oxidative organelles called *microbodies* or *peroxisomes*, which provide essential links in the breakdown

of carbohydrates, fats, proteins, and nucleic acids. (Supplement 8-3 outlines the activities of microbodies in these reactions.)

THE MITOCHONDRIAL ELECTRON TRANSPORT SYSTEM

The electron transport system of mitochondria contains four large protein-based complexes that work in sequence to deliver electrons carried by NADH and FADH$_2$ to oxygen (Fig. 8-16). Each complex contains (1) a major enzyme that catalyzes electron transfer, (2) nonprotein organic groups (prosthetic groups; see p. 78) that directly accept and release electrons carried

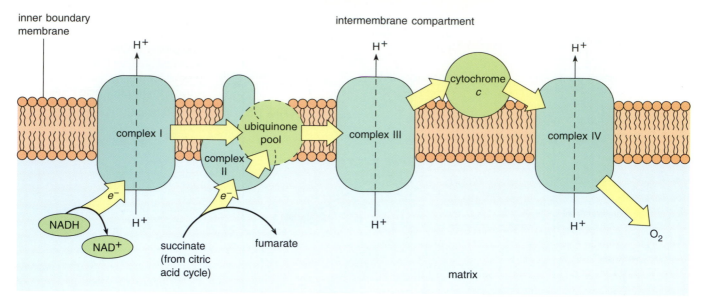

Figure 8-16 The mitochondrial electron transport system, which includes four major complexes and two smaller electron carriers (ubiquinone Q and cytochrome c) that act as shuttles between major complexes. Electron flow is shown in yellow arrows; hydrogen movement by dashed arrows. Passage of electrons through complexes I, III, and IV results in pumping of H⁺ from the matrix to the intermembrane compartment.

complex I = NADH-ubiquinone oxidoreductase
complex II = succinate-ubiquinone oxidoreductase
complex III = ubiquinone-cytochrome c oxidoreductase
complex IV = cytochrome c-O₂ oxidoreductase

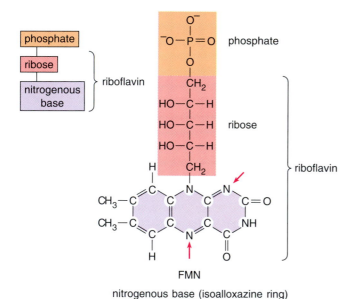

Figure 8-17 FMN (flavin mononucleotide). FMN binds a hydrogen and an electron at each of the two points marked by arrows when going to the reduced state as FMNH₂.

by the complex, and (3) structural proteins that may contribute to movement of H⁺ across the membrane. The system also contains two small, highly mobile carriers that shuttle electrons between major complexes. One of these mobile carriers is *cytochrome c*, a relatively small protein with its own prosthetic group; the other is a nonprotein organic molecule, *ubiquinone* (also called *coenzyme Q*), which directly picks up and releases electrons. Three of the major complexes (complexes I, III, and IV in Fig. 8-16) move H⁺ across the membrane in response to electron flow.

Prosthetic Groups and Their Sequence in Electron Transport

With one exception (ubiquinone), all the organic groups directly carrying electrons in the mitochondrial system are prosthetic groups tightly linked to proteins. These prosthetic groups and ubiquinone cycle between the reduced and oxidized states as they accept and release electrons.

Four major types of prosthetic groups serve as electron carriers in the chain. One type consists of either of the two nucleotide-based carriers FAD (see Fig. 8-12) or *flavin mononucleotide* (*FMN*; Fig. 8-17). The segment alternately oxidized and reduced in FAD and FMN is derived from *riboflavin*, a vitamin of the

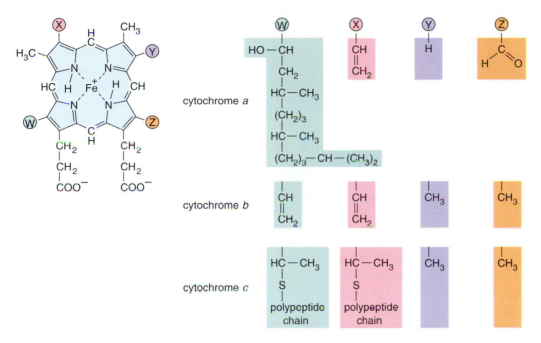

Figure 8-18 The heme group, the prosthetic group of the cytochromes. The iron atom in the center of the ring alternates between the Fe^{2+} and Fe^{3+} forms during reduction and oxidation. In the various cytochromes, different side groups attach to the ring structure (the groups linked to the ring in cytochromes *a*, *b*, and *c* are shown). Similar rings occur as central structures in the chlorophylls, hemoglobins, and other important biological molecules.

B group. Like FAD, FMN carries electrons in pairs; it also binds two hydrogens to form $FMNH_2$ during reduction. The linkage between FAD or FMN and their attached proteins forms a combination known as a *flavoprotein.*

The prosthetic groups of the second major carrier type, the *cytochromes*, consist of a ring structure containing a central iron atom (Fig. 8-18). The ring structure alone is termed a *tetrapyrrole* ring; with the central iron atom included it is called a *heme* group. The heme groups of mitochondrial cytochromes, including *a*, a_3, *b*, *c*, and c_1, differ in minor substitutions in the side groups of the tetrapyrrole ring and in their manner of attachment to proteins. Each cytochrome alternates between oxidized and reduced forms through gain or loss of a single electron by the central iron atom, which can exist in the Fe^{2+} (reduced) or Fe^{3+} (oxidized) state. The cytochromes do not combine with hydrogen during conversion to the reduced state.

Cytochrome *c* is the only cytochrome linked to a peripheral protein (see p. 104) of the inner mitochondrial membrane. The remaining cytochromes—*a*, a_3, *b*, and c_1—are linked to integral proteins that are deeply embedded in the inner mitochondrial membrane.

Iron-sulfur centers form the third major type of prosthetic group in the mitochondrial electron transport system. These centers contain iron and sulfur atoms in covalently linked groups, most commonly with either two iron and two sulfur atoms (2Fe/2S) or four iron and four sulfur atoms (4Fe/4S; Fig. 8-19). Some S atoms are part of cysteine residues in the polypeptide chain of an iron-sulfur protein. The iron atoms of the Fe/S centers change between Fe^{2+} and Fe^{3+} states as the centers pick up and release single electrons. No hydrogens are bound by the Fe/S carriers as they cycle to the reduced state.

The combination between the iron-sulfur centers and their protein complement forms a class of molecules known as *iron-sulfur proteins*. The iron-sulfur proteins are easily the most numerous electron carriers in the electron transport system—at least 13 distinct Fe/S centers have been identified in mitochondria.

The fourth type of prosthetic group is the *copper center*. These groups consist of individual copper atoms linked into a protein structure, often to the side chains of histidine, cysteine, or methionine residues. The copper atoms alternate between the Cu^{1+} and Cu^{2+} states as the centers are reduced and oxidized.

Ubiquinone, the only unattached carrier of the mitochondrial electron transport system (Fig. 8-20), contains a ring structure called a *quinone*. Linked to the quinone ring is a side chain containing repeats of a nonpolar group. Ubiquinone carries both electrons and hydrogens and has two levels of reduction in

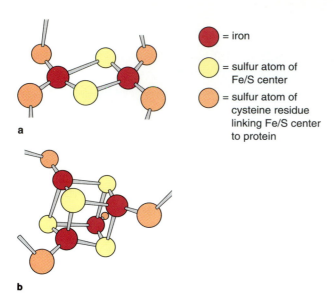

a

b

Figure 8-19 The two types of Fe/S centers occurring as prosthetic groups in iron-sulfur proteins. **(a)** A 2Fe/2S center; **(b)** a 4Fe/4S center. The iron atoms in the structures alternate between the Fe^{2+} and Fe^{3+} states during cycles of reduction and oxidation.

= iron

= sulfur atom of Fe/S center

= sulfur atom of cysteine residue linking Fe/S center to protein

which it may accept or release either one electron and hydrogen (Fig. 8-20a and b) or electrons and hydrogens in pairs (Fig. 8-20a through c). Ubiquinone occurs primarily in solution in the nonpolar membrane interior, forming a mobile "pool" that transfers electrons between the major complexes of the electron transport system (see Fig. 8-16 and below).

While most of the mitochondrial quinones are suspended in the membrane interior, a number are tightly bound to proteins in the complexes carrying out mito-

chondrial electron transport. Within the complexes they may participate in electron transfers between other prosthetic groups. Because the quinones link to hydrogens as well as to electrons in the reduced state, a part of their function may be to deliver protons to the sites pumping H^+ in the carrier complexes.

Many different biochemical techniques have been used to place the various electron carriers in a tentative sequence in the mitochondrial electron transport chain (see Information Box 8-3). Combining the results of these approaches has established the overall sequence shown in Figure 8-21. Note that electrons have two routes of entry into the sequence, one at FMN and one at FAD, which join at ubiquinone. The figure is a simplified pathway that describes only the major, overall sequence of electron flow. Within the carriers of the electron transport system, electrons probably follow complex routes among and between prosthetic groups that act in coordination rather than in the directly sequential pathway shown in the figure.

Organization of Prosthetic Groups with Proteins in the Electron Transport System

The organization of prosthetic groups in mitochondrial carrier complexes was revealed by isolation of proteins from inner boundary membranes. This research, pioneered in the laboratories of Y. Hatefi, E. Racker, Y. Kagawa, and D. E. Green, showed that the prosthetic groups shown in Figure 8-21 (except for ubiquinone) combine with proteins into the four major transmembrane complexes shown in Figure 8-16. Each complex has been purified and placed in artificial phospholipid membranes in active form.

The four complexes include *complex I*, conducting electrons from NADH to a ubiquinone pool; *com-*

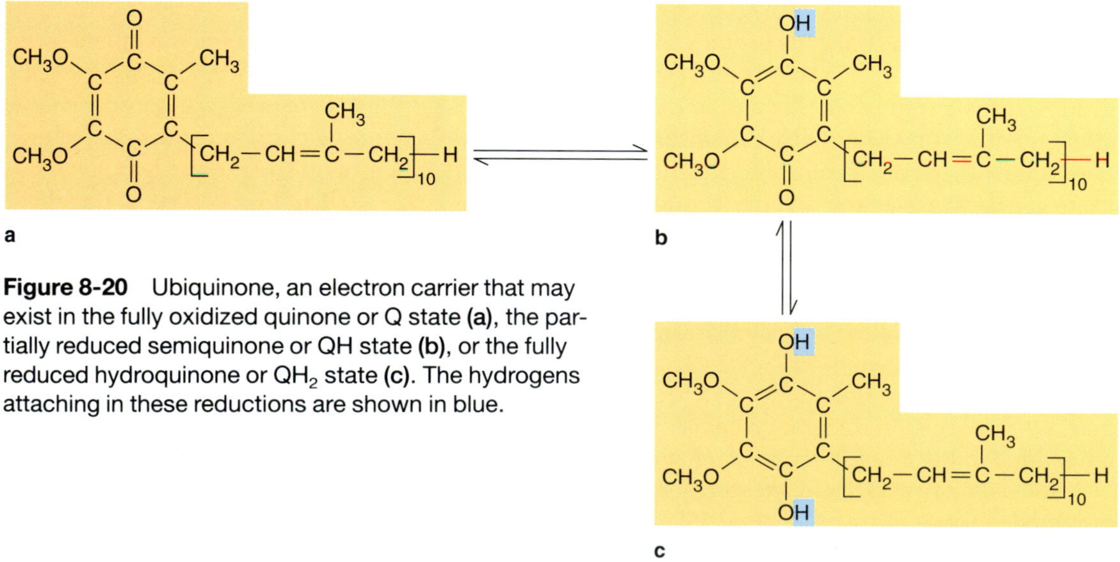

a

b

c

Figure 8-20 Ubiquinone, an electron carrier that may exist in the fully oxidized quinone or Q state **(a)**, the partially reduced semiquinone or QH state **(b)**, or the fully reduced hydroquinone or QH_2 state **(c)**. The hydrogens attaching in these reductions are shown in blue.

Placing Electron Carriers in Sequence

Part of the research placing the carriers of the mitochondrial electron transport system in sequence was based on measurements of the potential of electrons removed from purified prosthetic groups. By arranging the carriers in order of decreasing voltage, their sequence in mitochondrial electron transport could be approximated. This method introduces some error because the potentials measured under standard conditions, when the carriers are in isolated form, differ to an unknown extent from the potentials in the natural situation in the inner boundary membrane.

A more direct and accurate method for sequencing the carriers is based on the fact that most carrier molecules absorb light at characteristic wavelengths. The wavelengths absorbed change as the carriers cycle between oxidized and reduced states. The primary advantage of the method is that the absorption patterns of the carriers can be detected, and their level of oxidation or reduction determined, in intact mitochondria.

These light-absorption patterns are used in different ways to place the carriers in sequence. One of the most

effective, used by the pioneer of this method of approach, B. Chance, employs inhibitors known to block the activity of specific electron carriers (see the diagram in this box). Carriers falling after the inhibited carrier in the sequence gradually become oxidized by removal of electrons and show the characteristic light-absorption patterns of their oxidized forms. Carriers located before the blocked carrier remain in reduced form and show absorption patterns characteristic for this state.

For example, the drug *antimycin A* blocks release of electrons by cytochrome *b*. In mitochondria exposed to antimycin A, NAD^+ and the flavoproteins remain in the reduced state, and cytochromes *a* and *c* become oxidized. This indicates that NAD and the flavoproteins fall before cytochrome *b*, while cytochromes *a* and *c* fall after it in the sequence. Using inhibitors for different electron carriers and noting which are held in the reduced or oxidized state in mitochondria exposed to the inhibitors allowed researchers to place many of the carriers in overall sequence in the mitochondrial system.

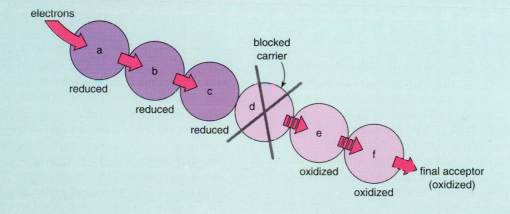

plex II, carrying electrons from the oxidation of succinate in the citric acid cycle to ubiquinone; *complex III*, conducting electrons from the ubiquinone pool to cytochrome *c*; and *complex IV*, conducting electrons from cytochrome *c* to oxygen.

Complex I contains an enzyme, *NADH dehydrogenase*, that catalyzes the initial step in electron transport, oxidation of NADH. Electrons removed from NADH are transferred to FMN, bound to the complex as a prosthetic group. The transfer forms $FMNH_2$, which is subsequently oxidized as electrons move from complex I to the ubiquinone pool. Also present in complex I are as many as nine iron-sulfur centers

and two tightly bound quinones. The iron-sulfur centers, and possibly the quinones, participate in the oxidation of $FMNH_2$, in movement of electrons from complex I to the ubiquinone pool, and possibly also in the initial reduction of FMN, in a complex interaction that is not yet understood in detail. The oxidation of reduced NAD occurs on the segment of complex I that faces the mitochondrial matrix.

Complex I contains a very large number of proteins in eukaryotes—more than 30 in lower eukaryotes and higher plants and more than 40 in vertebrates (41 different polypeptide subunits have been detected in beef heart mitochondrial complex I). As yet, little is

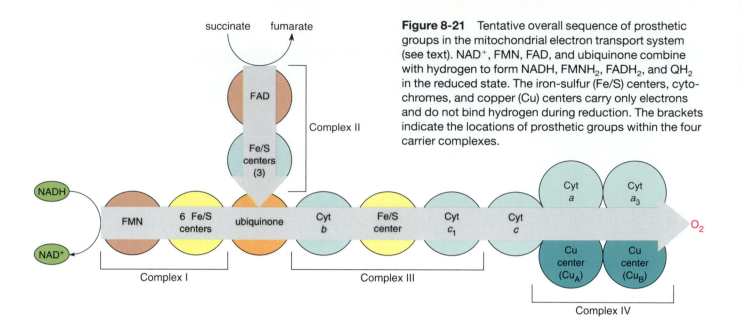

Figure 8-21 Tentative overall sequence of prosthetic groups in the mitochondrial electron transport system (see text). NAD$^+$, FMN, FAD, and ubiquinone combine with hydrogen to form NADH, FMNH$_2$, FADH$_2$, and QH$_2$ in the reduced state. The iron-sulfur (Fe/S) centers, cytochromes, and copper (Cu) centers carry only electrons and do not bind hydrogen during reduction. The brackets indicate the locations of prosthetic groups within the four carrier complexes.

known about the arrangement and associations of these proteins within the complex. A number are associated with NADH dehydrogenase and FMN. Some carry the Fe/S centers; others may act as structural subunits or as part of the mechanism pumping H$^+$ across the membrane. Of the proteins, 7 are encoded in mitochondrial DNA and the remainder in the cell nucleus in most organisms.

Complex II contains succinate dehydrogenase, catalyzing reaction 6 in the citric acid cycle. Because the enzyme is part of complex II, succinate must collide with the membrane surface facing the matrix for oxidation in the citric acid cycle. The acceptor for the electrons removed in the oxidation, FAD, is bound as a prosthetic group to the complex. At least three iron-sulfur centers are also present. Electrons are believed to pass from FAD to the iron-sulfur centers within the complex and then from the complex to the ubiquinone pool. Complex II contains three polypeptide subunits in addition to the succinate dehydrogenase enzyme. All four subunits are encoded in the cell nucleus.

Complex III contains *cytochrome* c *reductase*, the enzyme catalyzing transfer of electrons from the ubiquinone pool to cytochrome c. Some of the remaining peptides are associated with the prosthetic groups of the complex, which include two cytochrome b heme groups, one cytochrome c_1, and one iron-sulfur center. Two ubiquinone molecules bound to complex III may participate in movement of electrons within the complex. Complex III picks up electrons from the ubiquinone pool and delivers them to cytochrome c. The complex, which plays a central role in the electron transport systems of mitochondria, chloroplasts, and

most aerobic bacteria, has been highly conserved in evolution. Complex III contains as many as 11 polypeptides. One of these, the protein component of cytochrome b, is encoded in mitochondrial DNA.

The final complex of the electron transport system, complex IV, conducts electrons from cytochrome c to oxygen. The transfer is catalyzed by the major enzyme of this complex, *cytochrome* c *oxidase*. The four prosthetic groups of complex IV, which include cytochrome a, cytochrome a_3, and two copper centers identified as Cu_A and Cu_B, are linked to three polypeptide subunits designated *I*, *II*, and *III*. These polypeptides, which are central to the electron transfer and pumping activity of the complex, occur in all eukaryotes and in most bacteria with equivalents to mitochondrial complex IV. Cytochromes a, a_3, and Cu_B are linked to polypeptide subunit I; subunit II contains Cu_A. Additional subunits with auxiliary functions occur in eukaryotes, including 6 in yeast and up to 10 in mammals, for a maximum total of 13. The three central I, II, and III polypeptides are encoded in mitochondrial DNA; the auxiliary subunits are encoded in the cell nucleus.

Recent experiments indicate that cytochrome a_3 and Cu_B form a *binuclear center* within complex IV that transfers electrons to oxygen and triggers the conformational changes pumping H$^+$. Electrons flow from cytochrome c through Cu_A and cytochrome a to reach the binuclear center. Two H$^+$ are taken up from the matrix to form H$_2$O as oxygen is reduced at the binuclear center.

In its function as the electron carrier delivering electrons to O$_2$ in mitochondria, complex IV accounts

Table 8-1   Complexes of the Mitochondrial Electron Transport System

Complex	Total Molecular Weight	Polypeptide Subunits	Prosthetic Groups	H^+ Transport
I	700,000–850,000	32–41	1 FMN 6 to 9 Fe/S centers 2 bound quinones	Yes
II	97,000	4	1 FAD 3 Fe/S centers	No
III	280,000–300,000	11	2 cytochrome b 1 cytochrome c, 1 Fe/S center 2 bound quinones	Yes
IV	200,000	9–13	1 cytochrome a 1 cytochrome a_3 2 Cu centers	Yes

for 90% of the oxygen consumed by living organisms. The activity of cyanide as a poison depends on its ability to block transfer of electrons from complex IV to oxygen. (Table 8-1 summarizes the components of the four electron transport complexes.)

When isolated and added to artificial phospholipid bilayers, complexes I, III, and IV pump H^+ across the bilayer as they cycle between reduced and oxidized states. Only complex II fails to pump H^+ when it is inserted into artificial phospholipid films and supplied with a suitable electron donor and acceptor. Evidently complex II functions solely as an entry point for the electrons removed from succinate, which contain insufficient energy to reduce NAD^+.

In the random collisions that accomplish electron transport, molecules of the ubiquinone pool, moving more rapidly because of their relatively small size, probably act as shuttles transferring electrons from complexes I and II to complex III. Similarly, cytochrome c molecules, also moving more rapidly because of their relatively small size, shuttle electrons from complex III to complex IV. (Cytochrome c is bound loosely to the surface of the inner mitochondrial membrane facing the intermembrane space.)

H^+ Pumping by the Electron Transport System

The mitochondrial electron transport system pumps H^+ from the matrix into the intermembrane compartment as electrons flow through the carriers from NADH or $FADH_2$ to oxygen. Measurements of the number of H^+ moved across membranes by isolated complexes and by the intact system indicate that four

H^+ are pushed across the inner boundary membrane as a pair of electrons moves through complex I, and two H^+ as an electron pair moves through complexes III or IV. The entire system therefore adds at least eight H^+ to the gradient for each pair of electrons traveling from NADH to oxygen. Electron pairs moving from $FADH_2$ to oxygen, since they bypass complex I, account for a total of four H^+ added to the concentration gradient. Two more H^+ are removed from the mitochondrial matrix for each pair of electrons delivered to oxygen as part of the reaction producing water, contributing to the total H^+ gradient created by mitochondrial electron transport.

Although the available evidence clearly supports Mitchell's proposal that electron transport creates an H^+ gradient across the inner mitochondrial membrane, the mechanism by which the major complexes add to the H^+ gradient as they cycle between reduced and oxidized states is not completely known. However, the weight of evidence favors the *conformational hypothesis* proposed by P. D. Boyer and others. According to this hypothesis, reduction of a complex induces a conformational change in one or more of its polypeptides. The change activates a site that binds H^+ with high affinity on the matrix side of the membrane (Fig. 8-22a and b). Subsequent oxidation of the carrier changes its conformation so that the binding site now faces the opposite side of the membrane (Fig. 8-22c). In this state the affinity of the binding site for H^+ is reduced, causing release of H^+ to the intermembrane compartment (Fig. 8-22d). The complexes thus work much like active transport proteins (see p. 131) except that their energy source is in electron

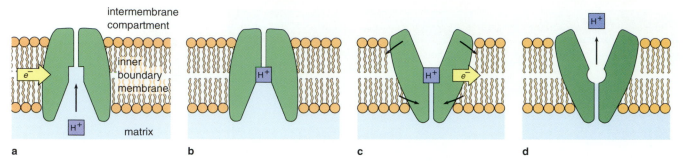

Figure 8-22 The conformational model for H$^+$ pumping by mitochondrial electron transport complexes (see text).

transfers rather than ATP. Quinones bound to some of the complexes may participate in delivery of H$^+$ to the pumping centers.

The H$^+$ gradient set up by the electron transport system provides energy for the synthesis of ATP. Measurements of the quantities of ATP synthesized in response to electron flow through the electron transport system, made by A. L. Lehninger and others, established that the H$^+$ gradient produced by travel of one electron pair through the entire pathway from NAD to oxygen is sufficient to drive synthesis of as much as three ATP. The shorter pathway followed by electrons removed from succinate, from the FAD of complex II to oxygen, results in the synthesis of as much as two ATP per electron pair. (Information Box 8-4 uses these conclusions to estimate total ATP production from complete oxidation of glucose to CO_2 and H_2O.)

Discovery of Mitochondrial Electron Transport

Research into mitochondrial electron transport began in the 1920s, when Otto Warburg discovered an iron-containing enzyme that catalyzes reduction of oxygen to water in aerobic cells. In 1925, D. Keilen identified the enzyme as a cytochrome. Keilen also discovered other cytochromes, which he proposed act in chains to transfer electrons in stepwise fashion from donor molecules to oxygen. His concept was extended by the research of D. W. Green and his coworkers, who found that ubiquinone, iron-sulfur proteins, and copper centers also act as carriers in the chain. The work of B. Chance was critical in the assignment of the carriers to a tentative sequence; Green's group and the laboratory of Y. Hatefi were instrumental in isolation of the carriers and description of the major carrier complexes. Intensive research continues today in mitochondrial electron transport, particularly into the structure of electron carrier complexes and the mechanism by which they pump H$^+$ across the inner mitochondrial membrane.

THE MITOCHONDRIAL F$_o$F$_1$-ATPase AND ATP SYNTHESIS

The H$^+$ gradient developed through electron transport is the energy source for the F$_o$F$_1$-ATPase synthesizing ATP from ADP and phosphate. Like the carrier complexes of the electron transport system, the F$_o$/F$_1$-ATPase contains multiple polypeptide subunits, as many as 24 in eukaryotes. These polypeptides are organized into a complex that completely spans the inner mitochondrial membrane.

Structure of the F$_o$F$_1$-ATPase

The name given to the F$_o$F$_1$-ATPase is based on the fact that it can be separated into two subparts, the F$_o$ and F$_1$ subunits (the o in the F$_o$F$_1$ ATPase is the lowercase letter o, not a zero; see below). The F$_1$ subunit consists of a group of polypeptides that can be removed from inner mitochondrial membranes in stable, soluble form by mild treatments such as adjustments in salt concentration of the isolating medium. The F$_o$ subunit, in contrast, can be removed only by relatively harsh treatments such as the use of detergents that destroy the membrane. The polypeptides of the F$_o$ subunit are therefore suspended in the nonpolar membrane interior.

The two subunits can be seen in the electron microscope as a lollipop-shaped particle with a clearly defined *headpiece*, connected to a *basal unit* by a *stalk* (Fig. 8-23a). When F$_1$ subunits are removed from inner mitochondrial membranes, the headpiece and stalk disappear from preparations made for electron microscopy, showing that the headpiece and stalk are contained in the F$_1$ subunit. The segment remaining in the membrane after removal of the headpiece and stalk is the F$_o$ subunit.

The active site of ATP synthesis was identified with the F$_1$ subunit primarily in the laboratory of E. Racker. Racker's group isolated inner mitochondrial membranes in the form of sealed vesicles that, under

Total ATP Production from Glucose Oxidation

There is no fixed ratio between the number of hydrogens pumped by the electron transport system and the amount of ATP synthesized by the F_oF_1-ATPase. This is because the H^+ gradient established by electron transport is an energy source that may be utilized for functions other than ATP synthesis, most notably for active transport of substances in or out of mitochondria. If mitochondria used the H^+ gradient solely for ATP synthesis, electron flow through the mitochondrial electron transport system from NADH to oxygen would result in the synthesis of three ATP per electron pair according to Lehninger's results. Electron flow over the shorter route from $FADH_2$ to oxygen would drive the synthesis of two ATP per electron pair.

Under this assumption, complete oxidation of one molecule of glucose through glycolysis, pyruvate oxidation, and the citric acid cycle would produce a total of 4ATP, 10NADH, 2FADH$_2$, and 6CO$_2$ (from reaction 8-6). The 10 electron pairs carried by 10NADH, traversing the electron transport system, could drive the synthesis of a maximum of $10 \times 3 = 30$ molecules of ATP. The 2 electron pairs carried by the 2FADH$_2$ could drive the synthesis of

only $2 \times 2 = 4$ molecules of ATP because they enter the sequence of carriers farther along in the chain. This gives a grand total of 34 ATP from electron transport. This 34 ATP, when added to the 4 ATP produced by substrate-level phosphorylation in glycolysis and the citric acid cycle, gives a grand total of 38 ATP for each molecule of glucose completely oxidized to CO_2 and H_2O.

This total assumes that the more efficient of the two mechanisms shuttling electrons from NADH outside to inside mitochondria is operating (see Fig. 8-24a), and that the two NADH produced in glycolysis induce synthesis of two NADH inside. If the less efficient shuttle is operating (see Fig. 8-24b), the two NADH from glycolysis result in production of two FADH$_2$ inside mitochondria. These enter the electron transport system farther along in the chain and drive synthesis of 2 ATP for each molecule of reduced FAD. In this case, complete oxidation of glucose results in a total of 36 ATP instead of 38. Which shuttle predominates depends on the particular species and cell types involved. Heart and liver cells in mammals, for example, possess the more efficient shuttle.

---

the electron microscope, clearly showed headpieces and stalks. When placed in a solution setting up an inside/outside H^+ gradient, these vesicles were fully capable of ATP synthesis. Removal of the headpieces and stalks from the vesicles stopped ATP synthesis. The removed headpieces and stalks, when separately purified, could catalyze the reversible reaction synthesizing ATP from ADP + phosphate. Adding the headpieces and stalks back to the vesicles restored their ability to synthesize ATP. Racker called the purified headpiece-stalk particles *coupling factor 1*, later shortened to the "F_1" subunit of the ATPase complex.

F_o channels inserted into membranes introduce an H^+ leak and are incapable of synthesizing ATP unless capped by F_1 subunits. The F_o subunit is therefore considered to supply an H^+-conducting channel for the F_oF_1-ATPase. The F_o subunit gets its name from *oligomycin*, a drug that inhibits the ATPase complex. Racker's group found that of the two ATPase subunits, oligomycin binds to the portion firmly attached to the membrane. This subunit was subsequently termed the *oligomycin-sensitive factor* or F_o.

The F_oF_1-ATPase of *E. coli*, the best studied of the enzymes in this group, contains eight subunits (Fig. 8-23b). The F_1 headpiece and stalk are built up

from five subunits, α, β, γ, δ, and ϵ, in the combination $\alpha_3\beta_3\gamma\delta\epsilon$. The β subunits contain the sites converting ADP to ATP. These sites are believed to work in cooperative fashion, so that at any instant each site is at a different stage of the reactions binding ADP and inorganic phosphate, converting them to ATP, and releasing the product. The F_o basal unit is assembled from three subunits, a, b, and c, in the combination ab_2c_{6-12}. The channel conducting H^+ through the membrane is probably formed by a circle of 6 to 12 c subunits. The total molecular weight of the *E. coli* F_oF_1-ATPase is about 529,000, assuming 10 copies of the c subunit.

Mitochondria and chloroplasts both contain F_1 subunits with polypeptides equivalent to the $\alpha_3\beta_3\gamma\delta\epsilon$ polypeptides of bacterial F_1. The β subunit is 70% homologous in amino acid sequence in organisms from bacteria to humans. The F_o subunit is more variable. While both mitochondria and chloroplasts contain equivalents of the bacterial a, b, and c polypeptides, chloroplasts contain one additional F_o polypeptide; mitochondrial F_o subunits may contain as many as seven additional polypeptides.

In 1983, J. E. Walker and M. Futai and their colleagues successfully isolated the gene complex coding

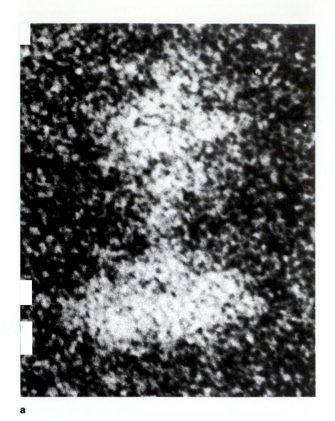

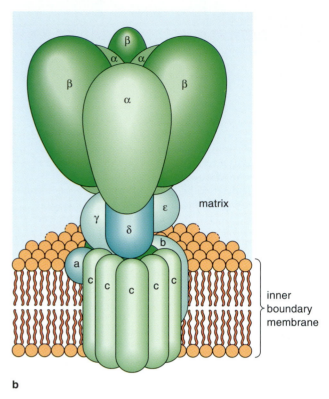

a

b

Figure 8-23 The F_oF_1-ATPase complex. **(a)** The complex isolated from a mitochondrion and prepared for electron microscopy by negative staining. × 4,000,000. (Courtesy of P. L. Pedersen, The Johns Hopkins University, from *J. Biolog. Chem.* 254:11170[1979].) **(b)** Possible arrangement of the polypeptide subunits of the *E. coli* F_oF_1-ATPase. The β subunits contain the sites converting ADP to ATP. The α subunits promote ATP synthesis by the β subunits; the minimum combination of subunits able to work fully as an ATPase appears to be the $\alpha_3\beta_3\gamma$ combination in *E. coli*. The γ, δ, and ε polypeptides of the stalk are necessary for F_oF_1 assembly; ε in addition inhibits ATP hydrolysis by β subunits, so that the complex cannot work backward like an ATP-driven H^+ pump. A circle of 6 to 12 c subunits probably forms the channel conducting H^+ through the inner mitochondrial membrane. The a and b polypeptides may stabilize the channel; the b polypeptide, which occurs in two copies in the bacterial F_o subunit, is also considered likely to bind the channel to the stalk portion of the F_1 unit.

for the entire group of polypeptides forming the F_oF_1-ATPase in *E. coli*. S. D. Dunn and L. A. Heppell, as well as E. Schneider and K. Altendorf, successfully reassembled the *E. coli* F_oF_1-ATPase in functional form from its polypeptide subunits. The Experimental Process essay by R. E. McCarty on page 234 describes a major approach used to assign functions to subunits of the chloroplast ATPase.

The F_oF_1-ATPase is a member of a family of H^+-ATPase pumps that includes the V-type pumps located in vesicle membranes in the cytoplasm of eukaryotic cells (see p. 133). Each pump in this family includes headpieces, stalks, and membrane segments of similar structure, assembled from related proteins.

The Mechanism of ATP Synthesis by the F_oF_1-ATPase

Chapter 5 describes the active transport pumps that directly hydrolyze ATP as an energy source to move ions and other substances across membranes against their concentration gradients (see p. 131ff). These active transport pumps, which have enzymatic activity as ATPases, remove a phosphate group from ATP when working in the forward direction:

$$ATP + H_2O \longrightarrow ADP + P_i \qquad (8\text{-}7)$$

Many of the ATP-dependent active transport pumps, especially those transporting ions, can be made to run in reverse if the concentration of the ion is raised to very high levels on the side of the membrane toward which it is normally pumped. Under these conditions the ion moves backward through the membrane channel, and the active transport pump *adds phosphate groups to ADP to form ATP.*

This is likely to be what happens in the F_oF_1-ATPase of mitochondrial membranes. The very high H^+ gradient set up by the electron transport system "overpowers" the F_oF_1-ATPase, which, if working in the forward direction, would hydrolyze ATP to pump H^+ in the same direction as the electron transport system. (The bacterial F_oF_1-ATPase is fully reversible and works in both the forward and reverse directions; see Supplement 8-1.) When pushed in the reverse direction by the high H^+ gradient, the F_oF_1-ATPase instead synthesizes ATP from ADP and phosphate:

$$ADP + P_i \longrightarrow ATP + H_2O \qquad (8\text{-}8)$$

The free energy required to run this reaction to the right, in the uphill direction, is supplied by the H^+ gradient. The V-type ATPases evidently work entirely in the forward direction as H^+-ATPase pumps.

The details of the molecular mechanism by which the F_oF_1-ATPase converts the energy of an H^+ gradient into ATP synthesis remain unknown. It is considered likely that movement of H^+ through the channel of the F_o subunit induces a conformational change that is transmitted through the stalk and activates ATP synthesis by the headpiece. This view is supported by extensive conformational changes that can actually be detected in the F_oF_1-ATPase as ATP synthesis proceeds. In the reaction sequence synthesizing ATP, ADP and inorganic phosphate bind to the β segments of the F_1 subunit and ATP is formed, with no requirement for an energy input by H^+ movement through the F_o subunit. Evidently the coordinated activity of the ADP/ATP binding sites on the three β segments favors progress of the reaction sequence to this point. However, the ATP formed by the reaction cannot be released from the β segments until H^+ flow through the F_o channel transmits energy to the F_1 subunit.

Evidence that an H^+ Gradient Drives ATP Synthesis

Support for Mitchell's hypothesis that the H^+ gradient established by electron transport drives ATP synthesis comes from a number of sources. The first is the simple observation that closed, intact vesicles are necessary for ATP synthesis to occur in inner membrane preparations isolated from mitochondria. If the vesicle membranes are broken or made leaky, which destroys any possibility of an H^+ gradient, ATP synthesis stops. The second comes from observations of the effect of *uncouplers* of oxidative phosphorylation. These substances are so called because they separate electron transport from ATP synthesis when they are added to mitochondria or mitochondrial vesicles. Under the action of an uncoupler, electron transport runs continuously with no relationship to ATP synthesis. With few exceptions the uncouplers, such as *DNP (2,4-dinitrophenol)*, are agents that make membranes leaky to H^+. Destruction of the H^+ gradient removes the driving force for ATP synthesis, in agreement with Mitchell's chemiosmotic hypothesis.

More direct evidence comes from experiments in which an imposed pH gradient results in ATP synthesis. The first experiment was carried out in 1966 by A. T. Jagendorf and E. Uribe, who used chloroplasts instead of mitochondria for several reasons. The chloroplast electron transport system is similar to the mitochondrial system (see pp. 253–256), and ATP synthesis in chloroplasts is coupled to electron transport in the same way as in mitochondria. The primary difference in ATP synthesis is that the energy of the electrons carried by the chloroplast electron transport system originates from absorbed light rather than from oxidations. By placing isolated chloroplasts in darkness, Jagendorf and Uribe eliminated electron transport as a source of energy for ATP synthesis. They next created a surplus of H^+ inside the chloroplasts by placing them in a medium containing an acid that could penetrate inside. The chloroplasts were then transferred to a second medium at lower H^+ concentration. As H^+ moved from inside to outside the chloroplasts in response to the gradient, ATP was synthesized. (In chloroplasts the H^+ gradient normally established by electron transport is opposite in direction to mitochondria—H^+ is concentrated inside the organelle by electron transport rather than being pushed to the outside as it is in mitochondria; see Fig. 8-16). Since electron transport was eliminated as an energy source, the observed ATP synthesis could be ascribed directly to the artificially created pH gradient. Similar experiments were subsequently carried out with mitochondria by Mitchell and his coworkers and others, with essentially the same results.

Another photosynthetic system has provided an elegant demonstration that an H^+ gradient can provide the energy required for ATP synthesis. W. Stoeckenius discovered that the plasma membrane of *Halobacterium* contains a pigment that resembles the visual pigments found in the retina of animal eyes. This pigment, combined with protein in a molecule called *bacteriorhodopsin* (see Fig. 4-5), acts as a light-driven H^+ pump that uses the energy of absorbed light to transport H^+ from one side of the membrane

Divide and Conquer: Deciphering the Functions of a Polypeptide of the Chloroplast ATP Synthase

Richard E. McCarty

RICHARD E. MCCARTY received an A.B. degree in Biology and a Ph.D. degree in Biochemistry from The Johns Hopkins University in Baltimore, Maryland. After postdoctoral training at the Public Health Research Institute of the City of New York, he joined the faculty of Cornell University in Ithaca, New York, in 1966. He was Chairman of the Section of Biochemistry, Molecular, and Cell Biology from 1981 to 1985 and Director of the Biotechnology Program from 1988 to 1990. In 1990, Dr. McCarty returned to Johns Hopkins as Professor and Chairman of the Department of Biology. His research is in the area of photosynthesis, with an emphasis on chloroplast membrane proteins.

ATP is required for photosynthesis. The green energy-converting membranes of chloroplasts, known as thylakoid membranes, carry out the synthesis of ATP from ADP and inorganic phosphate. This energy-requiring reaction is driven by the energy-liberating flow of protons down their electrochemical gradient. The electrochemical proton gradient is generated across the thylakoid membrane by light-dependent electron flow in much the same way that electron transport powers proton gradient formation in mitochondria.

The enzyme that couples the synthesis of ATP to proton movement across thylakoid membranes is known as the ATP synthase and also as H^+-ATPase. An ATPase is an enzyme that catalyzes the hydrolysis of ATP, and ATP hydrolysis is the reverse of ATP synthesis. Under special conditions, the ATP synthase can act as an ATPase.

The ATP synthase is remarkable in its structure and complexity. The enzyme is made up of nine different polypeptides ranging in molecular weight from 56,000 to 8,000 and a total of about 20 polypeptide chains. The synthase is thus a rather large protein complex with a molecular weight of approximately 550,000. The enzyme consists of two parts, CF_o and CF_1. CF_o is a collection of rather hydrophobic membrane proteins that anchor CF_1 to the thylakoid membrane. CF_o also functions in the rapid transport of protons across the membrane. CF_1 (molecular weight, 400,000) bears the catalytic sites of the ATP synthase and is comprised of a total of nine polypeptide chains and five different kinds of polypeptides. Unlike CF_o, which can be removed from the thylakoid membrane only by destroying the membrane with detergents, CF_1 may be readily removed from membranes and purified. Once removed from the membrane CF_1 is soluble in aqueous solution.

The polypeptide composition of CF_1 is highly unusual. There are three copies of the larger α and β subunits, but only one copy of the three smaller CF_1 subunits (γ, δ, and ϵ). This composition means that the enzyme is structurally asymmetric. The β subunits, which likely bear the catalytic centers, for example, cannot be in equivalent environments.

An intriguing feature of the ATP synthase is its complexity. What are the functions of the individual polypeptide subunits? Which polypeptides are required for catalytic activity? Which may be involved in proton transport? Which are required for CF_1-CF_o interactions?

A classical biochemical approach to the study of complex systems is to dissect the system into its components and put it all back together again. This approach is sometimes called "resolution and reconstruction analysis," but I prefer "divide and conquer." The idea is quite simple. Find ways to remove polypeptides selectively from the ATP synthase (or parts thereof) and test the depleted enzyme for function. If the depleted ATP synthase is defective, readdition of the polypeptide removed from the enzyme should restore normal function.

The divide-and-conquer approach will be illustrated by the selective removal of the smallest CF_1 subunit, ϵ (molecular weight, 14,700). CF_1 may be isolated and purified in relatively large amounts (200–400 mg) from only 2 kg of a convenient, inexpensive source—spinach leaves purchased at a local market. The chloroplast thylakoid membrane is by far the most prevalent membrane in green leaves and, thus, is very easy to prepare. The leaves are ground in a blender with the buffered sucrose solution and, after filtration to remove debris, the homogenate is centrifuged. Thylakoid membranes are large and relatively dense. Thus, they sediment much more readily than all of the other membranes in the homogenate. To remove contaminating soluble proteins, the thylakoid membranes are extensively washed by repeated resuspension in fresh buffer and centrifugation to pellet the membranes.

Dilution of the washed membranes into a medium of very low salt concentration that also contains a reagent that binds Mg^{2+} causes CF_1 to pop off the thylakoid membranes. The membranes are removed and the protein solution, which is at least 50% CF_1, is concentrated. Further purification is achieved by ion exchange chromatography.

A fascinating feature of the chloroplast ATP synthase

is that it is virtually inactive in the dark, but very active in the light. In the light the enzyme makes ATP. In the dark, the enzyme should, in principle, catalyze the reverse reaction, ATP hydrolysis. It does not. Elaborate mechanisms have evolved that prevent ATP hydrolysis from occurring. If the ATP synthase were active in the dark in ATP hydrolysis, it would deplete the ATP pools in both the chloroplast and cytoplasm of leaf cells and strongly inhibit metabolism. In the light, the enzyme operates as an ATP synthase not as an ATPase because the electrochemical proton gradient drives the reaction in the synthetic direction.

We were interested in finding out more about the ways in which the ATP synthase is switched on in the light and off in the dark. That is, how is the enzyme interconverted between inactive and active forms? One question of interest was, which of the polypeptides of the ATP synthase are involved in this regulation of activity?

We were fortunate in that CF_1 in solution has properties in common with membrane-bound CF_1. CF_1 has low ATPase activity. Several treatments, including the incubation of CF_1 with certain detergents or organic solvents, increase the ATPase activity markedly. Could this activation be the result of either the denaturation or release of an inhibitory subunit?

We decided to test this idea. CF_1 was bound to a column of cellulose to which positively charged diethylaminoethyl groups were attached. CF_1 is negatively charged and sticks tightly to the column at low salt concentrations. The column was then eluted with a buffered solution containing 20% (by volume) ethanol and 30% (by volume) glycerol. This concentration of ethanol was chosen since work in other laboratories had established that 20% ethanol gave the maximum stimulation of the ATPase activity of CF_1.

We were delighted to find[1] that the ethanol-glycerol buffer caused the elution of the ϵ subunit of CF_1. If ethanol was omitted from the elution buffer, no ϵ was eluted. CF_1 now depleted in ϵ ($CF_1(-\epsilon)$) could then be washed from the column by elution with a buffer of high ionic strength.

The $CF_1(-\epsilon)$ had very high ATPase activity. Also, the addition of small amounts of purified ϵ to the $CF_1(-\epsilon)$ strongly inhibited the ATPase. This experiment shows rather conclusively that the ϵ subunit is an inhibitor of the enzyme. Having divided the enzyme, we attempted to conquer the system. Our ability to remove ϵ selectively allowed us to determine the binding stoichiometry of ϵ (one ϵ per CF_1) and to gain information about the position of ϵ within the CF_1 structure.

It was also now possible to show that the ϵ subunit is absolutely required for ATP synthesis. Thylakoid membranes from which CF_1 had been removed cannot catalyze ATP synthesis. Adding CF_1 to the depleted membranes restores ATP formation. $CF_1(-\epsilon)$ binds to the membranes (really to CF_o) equally as well as CF_1, but does not restore ATP synthesis.

It was a surprise that an inhibitor of the ATP synthase is required for ATP synthesis. Logic dictates that an inhibitor cannot block only one direction of a reversible reaction. This, however, appeared to be the case. The ϵ subunit seemed to inhibit ATP hydrolysis, but not ATP synthesis. We are still uncertain as to the exact mechanism of this unusual regulation. However, polyclonal antibodies prepared against the ϵ polypeptide have given some clues. The incubation of thylakoid membranes with an anti-ϵ serum in the dark had no effect on either ATP synthesis or hydrolysis. If, however, the membranes were incubated in the light with the antiserum, ATP synthesis was inhibited. The antibodies actually removed part of the ϵ subunit from the CF_1 in the membrane.[2]

This result suggests that illumination of thylakoid membranes causes changes in the structure of CF_1 that expose the ϵ subunit to the antibodies. The effect of light is indirect. Very likely, the enzyme senses the electrochemical proton gradient generated by light-dependent electron flow and alters its structure. In particular, the interactions between the inhibitory ϵ subunit and other CF_1 polypeptides are changed. It seems plausible that these changes overcome the inhibitory effects of ϵ. CF_1 is simply not the same enzyme in the dark as it is in the light.

The divide-and-conquer approach has given new insights into the functions of the ϵ subunit of CF_1. In the future we will attempt to use recombinant DNA methods to engineer mutant forms of ϵ. In this way we will be able to exploit our reconstitution systems to define further the roles of the ϵ subunit in the regulation of ATP synthesis in chloroplasts.

References

[1] Richter, M. L.; Patrie, W. P.; and McCarty, R. E. *J. Biol. Chem.* 259: 7371 (1984).

[2] Richter, M. L., and McCarty, R. E. *J. Biol. Chem.* 262:15037 (1987).

to the other. In 1974, Racker and Stoeckenius inserted bacteriorhodopsin molecules into artificial phospholipid vesicles along with mitochondrial F_oF_1-ATPase. When exposed to light, the vesicles were able to synthesize ATP. This simple system, containing nothing more than a closed membrane, a molecule establishing an H^+ gradient, and the F_oF_1-ATPase, provides conclusive evidence that H^+ gradients supply the energy for ATP synthesis.

MITOCHONDRIAL TRANSPORT

The integration of reactions inside mitochondria with the surrounding cytoplasm requires transport of an extensive group of substances across the mitochondrial boundary membranes. Molecules entering oxidative reactions or ATP synthesis inside mitochondria, including pyruvate, various intermediates of the citric acid cycle, ADP, and inorganic phosphate ions, are transported inward. Amino acids, fatty acids, and a variety of positively charged ions are also transported into mitochondria. ATP is among the most important mitochondrial products that must be transported outward. Several mechanisms also shuttle electrons from the NADH reduced in glycolysis to either NAD^+ or FAD inside mitochondria.

Transport between mitochondria and the surrounding cytoplasm is complicated by the fact that, although the outer membrane is freely permeable to essentially any molecules of molecular weight less than about 6000, the lipid portion of the inner mitochondrial membrane is highly impermeable to essentially all substances except H_2O, O_2, and CO_2. The unusual lipid present in the inner mitochondrial membrane, cardiolipin, is believed to contribute to its impermeability. The problem is circumvented in mitochondria, as it is elsewhere in the cell, by transport proteins that specifically move the required substances across the inner boundary membrane.

Some of the transporters of the inner membrane operate by facilitated diffusion, in response to favorable concentration gradients between the cytoplasm and the mitochondrial matrix. Others operate as active transport pumps, powered by the H^+ gradient set up across the inner mitochondrial membrane by electron transport. More of the energy of the H^+ gradient, in fact, may be consumed in this active transport than in ATP synthesis.

The permeability of the outer mitochondrial membrane to molecules with molecular weights below about 6000 depends on the presence of porinlike proteins (see p. 146), which form more or less open channels in this membrane. The mitochondrial porin has a channel wide enough to admit all the metabolites transported by the inner membrane, including ADP and ATP, NAD, pyruvate, intermediates of the citric acid cycle, fatty acids, and amino acids. The outer membrane thus effectively provides an open "window" for the substances passing between the cytoplasm and the mitochondrial matrix through the inner membrane. However, the window is too small to allow larger molecules, such as enzymes catalyzing unrelated activities in the cytoplasm, to pass.

Among the more important active transport carriers of the inner boundary membrane is the *ADP/ATP exchange carrier*, which exchanges ATP synthesized inside mitochondria for ADP on the outside on a one-for-one basis. The exchange is driven by the H^+ gradient, at the rate of one H^+ for each ADP/ATP exchanged by the carrier.

Another carrier related in structure to the ADP/ATP exchange carrier, the *phosphate carrier*, transports the inorganic phosphate groups needed for ATP synthesis into mitochondria. The phosphate carrier is also an active transport pump driven by the mitochondrial H^+ gradient. Operation of the two carriers, the ADP/ATP exchange carrier and the phosphate carrier, provides raw materials needed for ATP synthesis inside mitochondria. The gradient in phosphate groups set up by the carrier is also employed as an energy source for active transport of some metabolites.

Separate carriers in the inner membrane transport metabolites such as pyruvate, intermediates of the citric acid cycle, fatty acids, and amino acids into mitochondria. Many of the carriers for intermediates of the citric acid cycle are active transport pumps that use the phosphate gradient set up by the phosphate carrier as an energy source. These carriers exchange one molecule of an intermediate such as oxaloacetate or malate, moved from the outside to the inside, for an inorganic phosphate group moved outward. Still other carriers work by exchanging one of the metabolites already transported inside for a different one outside. Citrate, for example, enters mitochondria in exchange for a molecule of malate moved outward.

Amino acids enter mitochondria by systems similar to the carriers transporting pyruvate and intermediates of the citric acid cycle. Some, depending on the amino acid, operate as facilitated diffusion carriers driven by favorable concentration gradients; others work as active transport pumps driven by the mitochondrial H^+ gradient. Some work as carriers that exchange an amino acid outside for one already inside. Fatty acids move inside mitochondria via a shuttle mechanism involving *carnitine*, a small, nonpolar molecule that moves back and forth across the inner mitochondrial membrane.

The electrons carried by NADH outside mitochondria are transferred inside by unusual and complex

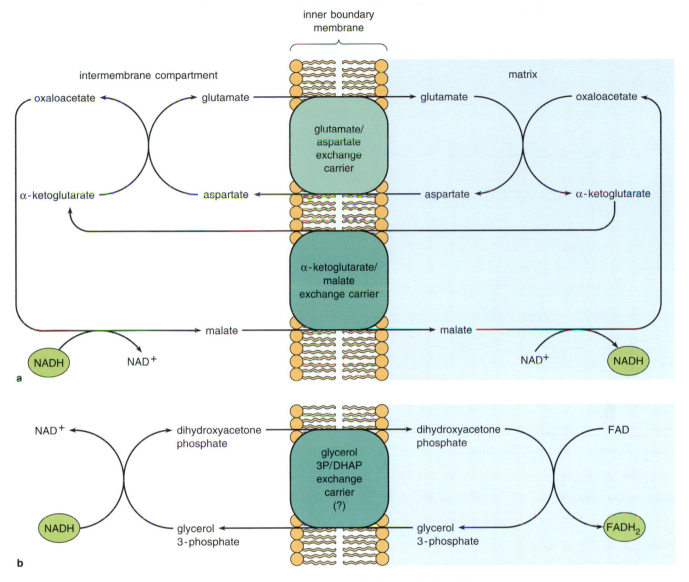

Figure 8-24 Shuttles transferring electrons from cytoplasmic NADH into mitochondria. The more efficient shuttle shown in **(a)** conducts electrons from NADH outside to NAD$^+$ inside, with no net loss in energy. The less efficient shuttle shown in **(b)** transfers electrons from cytoplasmic NADH to FAD inside mitochondria.

shuttles that employ the exchange carriers moving amino acids and intermediates of the citric acid cycle (Fig. 8-24). In principle the shuttles act by transferring electrons from NADH to an organic molecule on the outside of the mitochondrial membrane. The reduced molecule, after transport into the mitochondrion, is oxidized on the inside by transfer of electrons to NAD$^+$ or FAD on the inside. The shuttle shown in Figure 8-24a transfers electrons from NADH reduced in glycolysis outside mitochondria to NAD$^+$ inside with no significant energy loss. A less efficient shuttle transferring electrons from NADH outside to FAD inside is shown in Figure 8-24b.

Positively charged ions, among them Na$^+$, K$^+$, Ca^{2+}, and Mg^{2+}, are pumped into mitochondria by active transport carriers. Of these ion carriers, the one moving Ca^{2+}, which is powered by ATP hydrolysis rather than the H$^+$ gradient, is critical to the many cellular mechanisms regulated by this ion (see Chapter 5 for details).

Eukaryotic cells employ two major pathways for ATP synthesis driven by oxidative metabolism. In one pathway, substrate-level phosphorylation, ATP is synthesized by direct transfer of phosphate groups from an organic molecule to ADP. ATP synthesis takes place by this pathway in glycolysis and the citric acid cycle. In the second pathway, oxidative phosphorylation, electrons removed in oxidative reactions pass through an electron transport system, which taps off their energy to establish an H^+ gradient. The H^+ gradient is used in turn as the energy source for ATP synthesis by an F_oF_1-ATPase. Oxidative phosphorylation, which takes place inside mitochondria, is responsible for most of the ATP produced in eukaryotes. The equivalent oxidative systems of prokaryotes are described in Supplement 8-1.

For Further Information

ATP-linked active transport, *Ch. 5*
DNA, ribosomes, and protein synthesis in mitochondria, *Ch. 21*
Electron transport and ATP synthesis in photosynthesis, *Ch. 9*
Enzymes and enzymatic catalysis, *Ch. 3*
Evolutionary origins of mitochondria, *Ch. 26*
Facilitated diffusion, *Ch. 5*
H^+-linked active transport, *Ch. 5*
Porins in prokaryotes, *Ch. 5*
    in chloroplasts, *Ch. 9*
Signals sorting proteins to mitochondria, *Ch. 20*

Suggestions for Further Reading

Anraku, Y. 1988. Bacterial electron transport chains. *Ann. Rev. Biochem.* 57:101–132.

van den Bosch, H., Schutgens, R. B. H., Wanders, R. J. A., and Tager, J. M. 1992. Biochemistry of peroxisomes. *Ann. Rev. Biochem.* 61:157–97.

Capaldi, R. A. 1990. Structure and function of cytochrome *c* oxidase. *Ann. Rev. Biochem.* 59:569–596.

Cooper, C. E., Nicholls, P., and Freedman, J. A. 1991. Cytochrome *c* oxidase: Structure, function, and membrane topology of the polypeptide subunits. *Biochem. Cell Biol.* 69:586–607.

deDuve, C. 1983. Microbodies in the living cell. *Sci. Amer.* 248:74–84 (May).

Fearnley, I. M., and Walker, J. E. 1993. Conservation of sequence of subunits of mitochondrial complex I and their relations with other proteins. *Biochim. Biophys. Acta* 1140:105–134.

Fothergill-Gilmore, L. A., and Michels, P. A. M. 1993. Evolution of glycolysis. *Prog. Biophys. Molec. Biol.* 59:105–235.

Futai, M., Noumi, T., and Maeda, M. 1989. ATP synthase (H^+-ATPase): Results by combined biochemical and molecular biology approaches. *Ann. Rev. Biochem.* 58:111–136.

Glaser, E., and Norling, B. 1991. Chloroplast and plant mitochondrial ATP synthases. *Curr. Top. Bioenerget.* 16:223–263.

Hers, G. A., and Hue, L. 1983. Gluconeogenesis and related aspects of glycolysis. *Ann. Rev. Biochem.* 52:617–653.

Journal of Bioenergetics and Biomembranes, vol. 24, part 5: Series of reviews on the F_oF_1-ATPase.

Journal of Bioenergetics and Biomembranes, vol. 25, part 3: Series of reviews on the mitochondrial electron transport complex III.

Journal of Bioenergetics and Biomembranes, vol. 25, part 4: Series of reviews on electron transport complex I.

Journal of Bioenergetics and Biomembranes, vol. 25, part 5: Series of reviews on mitochondrial transport.

Kuan, J., and Saier, M. H. 1993. The mitochondrial carrier family of transport proteins: Structural, functional, and evolutionary relationships. *Crit. Rev. Biochem. Molec. Biol.* 28:209–233.

Mannella, C. A., and Tedeschi, H. 1992. The emerging picture of mitochondrial membrane channels. *J. Bioenerget. Biomembr.* 24:3–5.

Masters, C. J., Reid, S., and Don, M. 1987. Glycolysis—new concepts in an old pathway. *Molec. Cellular Biochem.* 76:3–14.

Mitchell, P. 1991. Foundations of vectorial metabolism and osmochemistry. *Biosci. Rep.* 11:297–346.

Munn, E. A. 1974. *The Structure of Mitochondria*. New York: Academic Press.

Nelson, N. 1992. Evolution of organellar proton-ATPases. *Biochim. Biophys. Acta* 1100:109–124.

Palmieri, F., Bisaccia, F., Iacobazzi, V., Indiveri, C., and Zara, V. 1992. Mitochondrial substrate carriers. *Biochim. Biophys. Acta* 1101:223–227.

Patel, M. S., and Roche, T. E. 1990. Molecular biology and biochemistry of pyruvate dehydrogenase complexes. *FASEB J.* 4:3224–3233.

Pedersen, P. L., and Amzel, L. M. 1993. ATP synthases: Structure, reaction center, mechanism, and regulation of one of nature's most unique machines. *J. Biolog. Chem.* 268:9937–9940.

Pilkis, S. J., El-Maghrabi, M. R., and Claus, T. H. 1988. Hormonal regulation of hepatic gluconeogenesis and glycolysis. *Ann. Rev. Biochem.* 57:755–783.

Senior, A. E. 1990. The proton-translocating ATPase of *E. coli*. *Ann. Rev. Biophys. Biophys. Chem.* 19:7–41.

Stryer, L. 1988. *Biochemistry*, 3rd ed. San Francisco: Freeman.

Tolbert, N. E. 1981. Metabolic pathways in peroxisomes and glyoxisomes. *Ann. Rev. Biochem.* 50:133–157.

Tzagoloff, A. 1982. *Mitochondria*. New York: Plenum.

Wallace, D. C. 1992. Diseases of the mitochondrial DNA. *Ann. Rev. Biochem.* 61:1175–1212.

Wilson, M. T., and Bickar, D. 1991. Cytochrome oxidase as a proton pump. *J. Bioenerget. Biomembr.* 23:755–771.

Zaar, K. 1992. Structure and function of peroxisomes in the mammalian kidney. *Eu. J. Cell Biol.* 59:233–254.

Review Questions

1. Outline the structure and function of the membranes and compartments of mitochondria.

2. Define oxidation and reduction. What are redox potentials?

3. What are the sources of electrons in cells? How is the energy of these electrons tapped off in electron transport systems? How is this energy used to drive the synthesis of ATP?

4. What is substrate-level phosphorylation? Oxidative phosphorylation?

5. Outline the sequence of events in glycolysis. What are the net inputs and outputs of the glycolytic sequence? Where does glycolysis take place?

6. What is fermentation? What advantages does the ability to carry out fermentations have for cells? Can cells live by fermentations alone? Do human cells carry out fermentations?

7. What substances can act as inputs to glycolysis? What is gluconeogenesis?

8. What are aerobes? Anaerobes? Facultative anaerobes? Strict anaerobes?

9. Outline the chemical inputs and outputs of pyruvate oxidation.

10. Outline the chemical inputs and outputs of the citric acid cycle. Where do the citric acid cycle and pyruvate oxidation take place?

11. Compare the structure and function of ATP, NAD, NADP, FAD, and coenzyme A.

12. What is a heme group? The prosthetic group of a flavoprotein? An iron-sulfur center? A quinone? How are these prosthetic groups placed in a tentative sequence in electron transport systems?

13. Outline the structure of complexes I, II, III, and IV of the mitochondrial electron transport system. Where are these complexes located in mitochondria?

14. How are electron transport complexes believed to set up an H^+ gradient in response to electron flow?

15. Outline the structure of the F_oF_1-ATPase of mitochondria. Where is the F_oF_1-ATPase located in mitochondria? How is the F_oF_1-ATPase believed to use an H^+ gradient as its energy source for the synthesis of ATP?

16. How many molecules of ATP are synthesized through complete oxidation of a molecule of glucose to CO_2 and H_2O? (Consider that the H^+ gradient set up by electron transport is used at maximum efficiency for the synthesis of ATP.)

17. Consider that the maximum efficiency of glucose oxidation to CO_2 and H_2O is 39%. What proportion of this efficiency is contributed by glycolysis? By pyruvate oxidation? By the citric acid cycle? (Assume that all electrons accepted by NAD or FAD are delivered to oxygen.)

18. Why do electrons carried by NAD potentially result in the synthesis of greater amounts of ATP than those carried by FAD?

19. Outline major evidence supporting the chemiosmotic hypothesis.

20. How does ATP made inside mitochondria get outside? How do electrons carried by NAD outside mitochondria get inside? What are the energy sources for active transport of substances in or out of mitochondria? What kinds of molecules are transported through mitochondrial membranes? What are mitochondrial porins?

21. What are microbodies? What enzymatic reactions are common to most microbodies? What important links do microbodies supply in fatty acid oxidation? In gluconeogenesis?

Supplement 8-1

Oxidative Metabolism in Prokaryotes

Most bacteria oxidize fuels by essentially the same reactions as eukaryotes. Glycolysis takes place in the cytoplasmic solution as it does in eukaryotic cells and follows essentially the same routes, except that in some bacteria the initial reactions leading to 3-phosphoglyceraldehyde (reactions 1 to 4 in Fig. 8-6) differ to a greater or lesser extent. The final products of fermentations are also more varied in bacteria. Pyruvate oxidation and the citric acid cycle also occur by the same pathways in bacteria and eukaryotes. However, because bacteria have no mitochondria, the enzymes and intermediates of the citric acid cycle are located in the cytoplasmic solution. The enzymes carrying out pyruvate oxidation and the electron transport system are associated with the plasma membrane in bacteria.

There are few organic molecules that cannot be oxidized by at least one bacterial species. Some bacteria oxidize inorganic substances such as iron (as Fe^{2+}), H_2S, or hydrogen (as H_2) as electron sources. In many of these oxidative pathways the electrons removed from the oxidized substances are passed directly to the electron transport system with no intermediate steps in the reaction pathway. The wide variety of oxidative reactions in bacteria is another reflection of the biochemical versatility of these minute but highly successful organisms.

Electron Transport and the F_oF_1-ATPase in Bacteria

Although the electron transport systems of many bacteria include the same prosthetic groups present in mitochondria, including FMN, FAD, iron-sulfur groups, quinones, and cytochromes, different bacteria vary considerably in the types within these groups. For example, while most bacteria utilize cytochromes as electron carriers, unique types such as cytochrome o and d may be present in their electron transport systems.

The bacterial electron carriers are organized with proteins into carrier complexes as in mitochondria (Fig. 8-25). Some bacteria contain a complex similar to mitochondrial complex I, with FMN and Fe/S carriers. Some of the complex I–like bacterial carriers, however, may contain FAD instead of FMN, or neither of these prosthetic groups. Carriers may also be present with similarities to complex II (with FAD and Fe/S carriers), complex III (with carriers similar to cytochrome b_1), and complex IV (with cytochromes o, a, a_2, a_3, or d in various combinations, along with Cu centers). Typically these carriers are simpler in polypeptide composition than their mitochondrial equivalents—bacterial complexes equivalent to mitochondrial complex IV, for example, may contain as few as two or three polypeptides. Like the mitochondrial carrier complexes, most bacterial electron transport complexes are capable of pumping H^+ as they cycle between oxidized and reduced states.

The bacterial carrier complexes are tightly bound to the plasma membrane. In this location the carrier complexes move H^+ from the cytoplasm to the outside of the plasma membrane as they cycle between oxidized and reduced states.

In some bacteria, equivalents of one or more of the mitochondrial electron transport complexes may be missing; a few bacterial electron transport systems contain only a single carrier complex. In species with more extensive electron transport systems the pathways of electron flow are frequently more branched than the system in eukaryotic mitochondria. That is, they have supplemental electron inputs and exits in addition to or in place of the eukaryotic routes.

Some bacteria differ in the substance used as final electron acceptor. While aerobic bacteria by definition use oxygen as final electron acceptor, some anaerobic species use inorganic substances such as sulfur, sulfate, or nitrate or organic molecules such as fumarate in this role instead. A few anaerobic bacteria entirely lack a functional electron transport system.

The most striking similarity between bacterial and eukaryotic oxidative mechanisms is the form taken by the F_oF_1-ATPase. The polypeptides of the headpiece and stalk and the a, b, and c polypeptides of the basal unit of the bacterial F_oF_1-ATPase have similar counterparts in the mitochondrial F_oF_1-ATPase.

In mitochondria an inhibitor protein linked to the ATPase complex prevents the complex from operating as an H^+-ATPase active transport pump; that is, as a pump pushing H^+ to the outside of the plasma membrane at the expense of ATP. In many bacteria, however, no equivalent inhibitor is present and the enzyme is fully reversible. In these bacteria the enzyme synthesizes ATP when electron transport maintains a sufficiently strong H^+ gradient and reverses, using ATP energy to push H^+ across the plasma membrane, when electron transport and the H^+ gradient fall off. This sustains the H^+ gradient necessary to power H^+-linked active transport and flagellar motility, both of which depend directly on the H^+ gradient in bacteria. (The role of H^+ in powering flagellar movement is described in Supplement 10-1.) In bacteria lacking a functional electron transport system the H^+ gradient necessary to power transport and flagellar motion is produced entirely by the F_oF_1-ATPase complex at the expense of ATP. In this case

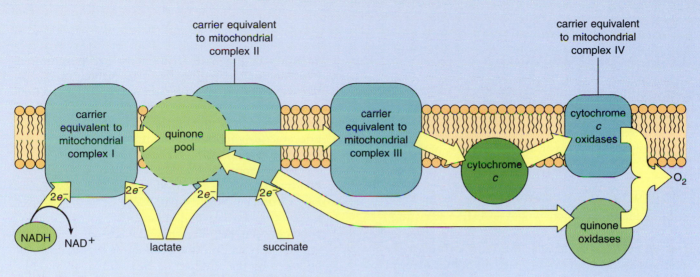

Figure 8-25 Pathways of electron transport in bacteria. Carriers equivalent to mitochondrial complexes I and II are not always present in bacterial systems. Carrier I, when present, contains Fe/S centers; however, FAD or FMN may be present or absent. Complex II, when present, closely resembles mitochondrial complex II. A quinone pool and a carrier equivalent to mitochondrial complex III are usually present in bacterial systems. Like its mitochondrial equivalent, the bacterial complex III usually contains cytochromes b and c_1 and an Fe/S center. As equivalents of mitochondrial complex IV, bacterial systems frequently contain oxidases that deliver electrons from cytochrome c or directly from the quinone pool to oxygen.

the required ATP is supplied primarily by anaerobic fermentations.

Oxidation and ATP Synthesis in Cynaobacteria

Relatively little is known about oxidative metabolism in the cyanobacteria. When placed in the dark, many cyanobacteria are unable to oxidize fuel substances such as glucose. Those that can, do so only very slowly and are limited to a very few molecules that can be used as fuels. These include little more than glucose and fructose and, in a few species, ribose or glycerol.

The limited oxidative reactions of cyanobacteria evidently proceed primarily or exclusively by a reaction series in which glucose is oxidized by the pentose phosphate pathway (see Information Box 8-2). Electrons removed during oxidations in the pathway are accepted by $NADP^+$, which in cyanobacteria may deliver them in turn to the electron transport system. Through this means the oxidations carried out in the pentose-phosphate pathway may lead to ATP synthesis. Because enzymes of the glycolytic pathway can be detected in some cyanobacteria, these organisms are probably also capable of glycolysis. Several key enzymes of the citric acid cycle are either rare or unde-

tectable in cyanobacteria, indicating that they cannot carry out the complete cycle.

The limited oxidative capabilities of cyanobacteria probably reflect the fact that these prokaryotes depend primarily on photosynthesis for their ATP requirements. All cyanobacteria contain a membrane-linked electron transport system as part of their photosynthetic apparatus, including essentially the same carriers as the system operating in photosynthesis in higher plants (see p. 276). The electron transport systems of cyanobacteria may be located in the plasma membrane as in bacteria, or in membranous vesicles suspended in the cytoplasm (see Figs. 1-3 and 9-26).

Recently the genes encoding the F_oF_1-ATPase of two cyanobacteria, *Anabaena* and *Synechococcus*, were isolated and sequenced by A. L. Cozens and J. E. Walker, S. E. Curtis, and others. The amino acid sequences deduced for the polypeptides of the cyanobacterial F_oF_1-ATPase indicate that its structure is closely similar to those of bacteria, mitochondria, and chloroplasts. Of the various F_oF_1-ATPases, the cyanobacterial enzyme is most closely related to the equivalent ATPase of chloroplasts, known as the CF_oCF_1-*ATPase* (see p. 248). This similarity provides another of the many functional and structural relationships between cyanobacteria and chloroplasts, strengthening the conclusion that chloroplasts are evolutionary descendants of cyanobacteria (see Chapter 26 for details).

Supplement 8-2

Fatty Acid Oxidation

The fatty acid chains produced by fat breakdown enter mitochondria via a carnitine carrier that delivers them to the mitochondrial interior as fatty acyl-coenzyme A complexes. Attachment of the fatty acid chain to coenzyme A is driven by breakdown of ATP to AMP:

fatty acid + ATP + coenzyme A ⟶

$\quad$ fatty acyl-coenzyme A + AMP + 2P$_i$ $\quad$ (8-9)

Once in the mitochondrial matrix, fatty acyl-coenzyme A serves as the input for the reactions of the fatty acid oxidation pathway.

The pathway shortens the fatty acid chain two carbons at a time, yielding units of acetyl-coenzyme A

as products (Fig. 8-26). If the fatty acid is saturated (with no double bonds between its carbons) and contains an even number of carbon atoms, the reaction series resembles the segment of the citric acid cycle between succinate and oxaloacetate; that is, an oxidation, an addition of the elements of a molecule of water, and a second oxidation occur in sequence.

In reaction 1 of the pathway, two electrons and two hydrogens are removed from a fatty acyl-coenzyme A unit entering the pathway. FAD is the electron acceptor for the reaction, which introduces a double bond between the two carbon atoms involved (dotted arrow). In reaction 2 the product of the first step is hydrated by the addition of the elements of an H_2O molecule, eliminating the double bond and preparing the group for the second oxidation. The product is oxidized in reaction 3, in which another electron pair

Figure 8-26 The pathway of fatty acid oxidation inside mitochondria (see text).

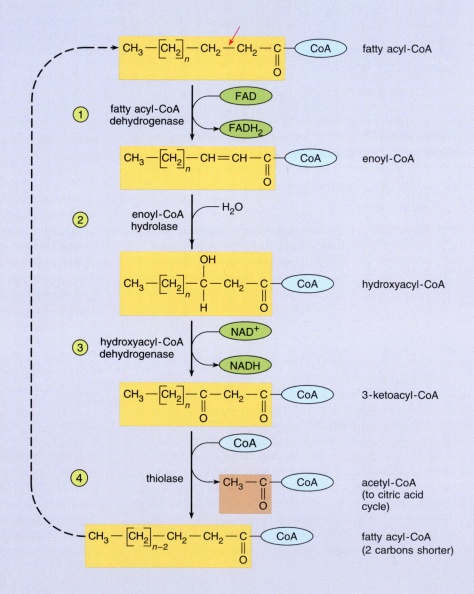

and two hydrogens are removed. NAD^+ is the electron acceptor for this oxidation.

Much of the energy released in the second oxidation is conserved in the structure of the product, *3-ketoacyl coenzyme A*. Most of this energy is transferred to a second molecule of coenzyme A in reaction 4. In this step the last two carbons of the chain are split off and attached to coenzyme A, forming acetyl-coenzyme A. The other product of the reaction is the remaining fatty acid chain attached to the second molecule of coenzyme A. This product is identical to the fatty acyl-coenzyme A complex entering the first reaction of the series except that it is now two carbons shorter.

The shortened fatty acyl-coenzyme A complex then repeats the four steps, which split off another two carbons and produce another molecule of acetyl-coenzyme A. The reactions repeat until the entire fatty acid chain is split into two-carbon acetyl units attached to coenzyme A. The acetyl-coenzyme A produced by the pathway directly enters the citric acid cycle like acetyl-coenzyme A units originating from pyruvic acid oxidation.

Unsaturated fatty acids with even numbers of carbon atoms are oxidized by essentially the same sequence. However, wherever the fatty acyl groups contain a double bond, the sequence skips the first step and proceeds directly to the second, missing the initial oxidation. As a result, unsaturated fatty acids reduce one FAD less for each double bond in their carbon chains.

Odd-numbered fatty acid chains are oxidized into two-carbon units by the same sequences as even-numbered fatty acids until the last cycle of the pathway. At this point the final turn of the cycle splits the remaining five-carbon fatty acid into one molecule of acetyl-coenzyme A and a three-carbon segment (a *proprionyl* group), also attached to coenzyme A. The acetyl-coenzyme A enters the citric acid cycle as usual. The proprionyl-coenzyme A is converted to succinyl-coenzyme A by addition of CO_2 (a carboxylation):

$$\text{proprionyl-coenzyme A} + CO_2 \longrightarrow$$
$$\text{succinyl-coenzyme A} \quad (8\text{-}10)$$

The succinyl-coenzyme A then enters the citric acid cycle directly at reaction 5 in Figure 8-13, as a part of the complex side cycle that yields one ATP by substrate-level phosphorylation.

The energy potentially available from fats is comparatively high—about twice the energy yield of carbohydrates by weight. This explains why fats are such an excellent source of energy in the diet and why they are used so extensively as a low-weight form of stored energy in animals. Storage of the equivalent amount of energy as carbohydrates rather than fats would add more than 100 pounds to the weight of an average man or woman!

Supplement 8-3
Microbodies (Peroxisomes)

Microbodies (see Fig. 1-13) are roughly spherical organelles somewhat smaller than mitochondria, about 0.1 to 1.5 μm in diameter. They are enclosed by a single membrane; frequently the material inside the membrane, called the *matrix*, contains a *core* that shows a regular lattice or crystalline structure (visible in Fig. 1-13). Biochemical analysis of microbody fractions by C. deDuve and his coworkers in the 1960s revealed that the organelles typically contain enzymes that oxidize organic substances and use oxygen directly as an electron acceptor. In the reactions catalyzed by these oxidases, oxygen combines with hydrogen to produce hydrogen peroxide (H_2O_2):

$$RH_2 + O_2 \longrightarrow R + H_2O_2 \quad (8\text{-}11)$$

R indicates an organic molecule. H_2O_2 is a toxic substance that is quickly converted into water and oxygen by *catalase*, an enzyme universally present in microbodies:

$$2H_2O_2 \longrightarrow H_2O + O_2 \quad (8\text{-}12)$$

On this basis, deDuve named the organelles *peroxisomes*, a term that is also widely used for microbodies.

Microbodies are present in the cytoplasm of all eukaryotic cells, including those of animals, plants, fungi, and protozoa. They are particularly abundant in the liver and kidney cells of vertebrates, including humans and other mammals, and in photosynthetic cells of plants.

There is great diversity in the biochemical reactions and pathways of microbodies in different cell types and species. Among the most important pathways in microbodies are fatty acid oxidation and the *glyoxylate cycle*. Many microbodies also contain enzymes that can convert products of the glyoxylate cycle into amino acids. The microbodies of leaf cells and other plant tissues active in photosynthesis contain another oxidative sequence, the *glycolate pathway*, which eliminates glycolate, a potentially toxic substance that may be indirectly produced by the reactions of photosynthesis (see Supplement 9-2).

Although many biochemical pathways of microbodies are oxidative, no ATP is produced directly within these organelles. However, some of the electrons removed in the oxidative pathways are accepted by NAD^+ or FAD. These electrons may eventually be carried into mitochondria by transport shuttles to power ATP synthesis.

Microbodies and Fatty Acid Oxidation

The fatty acid oxidation pathway in microbodies is almost identical to the mitochondrial pathway described in Supplement 8-2. Only the first reaction in Figure 8-26 differs; in microbodies the oxidase catalyzing this step transfers electrons directly to oxygen instead of FAD, yielding H_2O_2 as a product. Since the electrons removed at this step are delivered directly to oxygen, all the free energy released at this step is lost as heat. The NAD^+ or FAD accepting electrons at later steps in the peroxisome pathway may eventually deliver its electrons to the electron transport system of mitochondria by means of various shuttles.

The microbody pathway increases the cell's capacity to oxidize fats, particularly those containing long-chain fatty acids. Fatty acid chains longer than 10 to 12 carbons are oxidized relatively slowly by mitochondria. In microbodies, in contrast, fatty acid chains as long as 22 carbons are rapidly oxidized. The long chains are clipped two carbons at a time by the peroxisomal pathway until they reach the shorter 10–12 carbon length that can be oxidized with greatest efficiency by mitochondria; at this point in many species the shortened fatty acid chain is then transferred from microbodies to mitochondria for further oxidation.

Fatty acid oxidation in microbodies is particularly important in oily seeds such as the peanut and castor bean, where the reactions provide an important link in pathways converting stored fats into glucose and other carbohydrates via the glyoxylate cycle. In humans, defects in microbody formation and function underlie a group of diseases characterized by faulty metabolism of long-chain fatty acids. In *adrenoleukodystrophy*, for example, β-oxidation of long-chain fatty acids is impaired, leading to progressive demyelinization of the central nervous system, neurological degeneration, impairment of adrenal function, and death within a few years of onset of the disease, usually in childhood. The mutated gene responsible for the disease is carried on the X chromosome (the movie *Lorenzo's Oil* is about the efforts of parents to alleviate

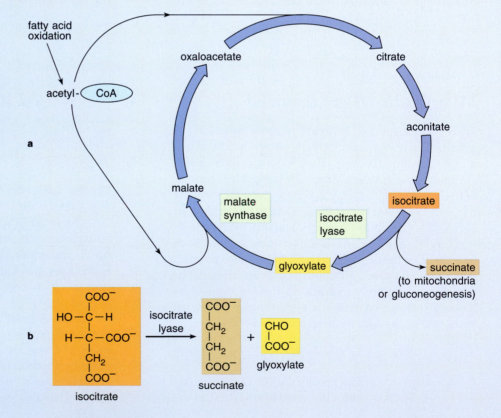

Figure 8-27 The glyoxylate cycle. **(a)** In the cycle, acetyl-coenzyme A units produced by fatty acid oxidation combine with oxaloacetate to form citrate as in the citric acid cycle. Isocitrate is then produced by interconversion between citrate and aconitate, also as in the citric acid cycle. From this point on the glyoxylate cycle diverges from the citric acid cycle through the action of two enzymes that are unique to microbodies. One, *isocitrate lyase*, splits isocitrate into succinate and glyoxylate **(b)**. Glyoxylate then interacts with a second molecule of acetyl coenzyme A to produce malate. This reaction is catalyzed by the second enzyme unique to glyoxisomes, *malate synthase*. Malate is oxidized to oxaloacetate by the same enzyme oxidizing this substance in the citric acid cycle, malate dehydrogenase, and the cycle is ready to turn again.

the symptoms of a child with adrenoleukodystrophy by changes in diet).

Microbodies and the Glyoxylate Cycle: Glyoxisomes

The glyoxylate cycle, which occurs typically in the microbodies of plant seeds, represents a partial circuit around the citric acid cycle in which some reactions are included and others bypassed. (Fig. 8-27 outlines the cycle.) Each turn of the cycle converts two molecules of acetyl-coenzyme A into one molecule of succinate. The succinate may diffuse from microbodies to mitochondria, where it can enter the citric acid cycle. Alternatively, it may indirectly enter gluconeogenesis (see p. 216), leading to generation of glucose from the fat originally entering oxidation in the microbody. Through this pathway the stored fat of the seeds contributes to products formed from glucose, such as the cellulose of new cell walls. Microbodies containing the glyoxylate cycle are frequently termed *glyoxisomes*.

Other Biochemical Pathways in Microbodies

Microbodies in different cell types may have other activities. Some contain enzymes that convert amino acids to sugars or vice versa by catalyzing transfer of amino groups, thereby providing important links between the metabolism of carbohydrates and proteins. In some animals, microbodies oxidize urate, a product of nucleic acid and protein breakdown, into allantoin:

$$\text{urate} + O_2 \longrightarrow \text{allantoin} + H_2O_2 \quad (8\text{-}13)$$

The crystalline core present in some peroxisomes is probably a deposit of urate oxidase, the enzyme catalyzing this reaction. The allantoin product is either excreted directly, as it is in most animals, or converted into urea and then excreted, as it is in fishes and amphibians.

Enzymes in the microbodies of some fungi allow these organisms to metabolize the group of organic molecules known as *alkanes* (methane, ethane, propane, butane, and longer-chain members of this series). Alkanes are primary constituents of crude oil; the ability of fungi to metabolize these substances makes them useful as a biological resource for the cleanup of oil spills.

Microbodies and the Cellular Energy Economy

The H_2O_2-producing oxidations of microbodies seem wasteful to the cellular energy economy because they do not directly conserve energy in the form of ATP. Why have microbodies persisted as ubiquitous eukaryotic cellular organelles in spite of their apparently futile oxidative pathway? The answer to this question must lie in the biochemical versatility of microbodies, which provides necessary links between the metabolism of carbohydrates, lipids, proteins, and nucleic acids in the cytoplasm. Without microbodies these linking pathways might not otherwise exist in eukaryotic cells.

PHOTOSYNTHESIS AND THE CHLOROPLAST

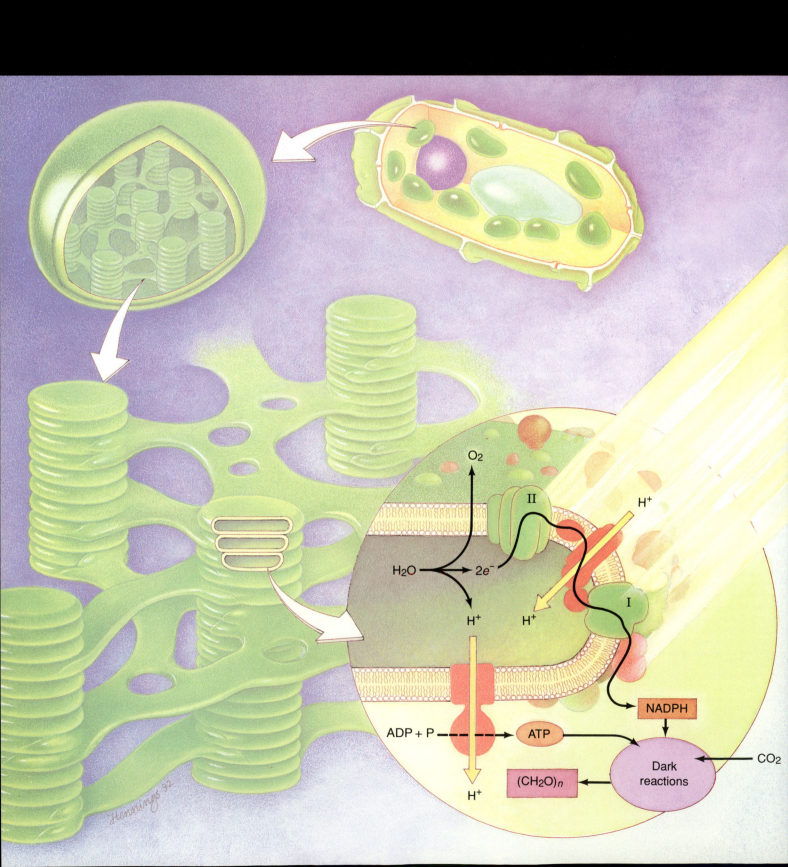

Photosynthesis, carried out by green plants, eukaryotic algae, cyanobacteria, and photosynthetic bacteria, uses the energy of sunlight to convert or "fix" inorganic substances into organic molecules. In most photosynthetic organisms, organic molecules are assembled from raw materials no more complex than water, carbon dioxide, and a supply of inorganic minerals.

Photosynthetic organisms typically make more organic molecules than they require for their own activities. Much of the surplus is used as a fuel source by animals and other organisms that live by eating plants. These plant-eating forms are consumed in turn by other organisms, and so on down the food chain until the last of the organic matter assembled by photosynthesis is completely oxidized to carbon dioxide and water.

Because the reactions capturing light energy provide the first step in this extended pathway of energy flow, photosynthesis is the vital link between the energy of sunlight and the vast majority of living organisms. Without the activity of photosynthetic eukaryotes and prokaryotes in converting light to chemical energy, most of the earth's creatures, including the human population, would soon cease to exist.

This chapter describes the capture and utilization of light energy in photosynthesis, emphasizing photosynthetic pathways in eukaryotic plants. The photosynthetic mechanisms of prokaryotes are outlined in Supplement 9-1.

THE PATHWAYS OF PHOTOSYNTHESIS: AN OVERVIEW

Photosynthesis proceeds by two major steps in which electrons play a primary role. In the first step the energy of sunlight is absorbed and used to push electrons to an elevated energy level (see Information Box 8-1). In cyanobacteria, eukaryotic algae, and higher plants, water is the source of electrons for this process. After being elevated in energy level, the electrons are used

in the second step of photosynthesis as an indirect energy source for ATP synthesis, and also for synthetic reactions in which reductions are required. (See Information Box 8-1 for a definition of oxidation and reduction.) In the cyanobacteria, algae, and plants the primary substance reduced by the addition of electrons in photosynthesis is carbon dioxide. The reduction, which also adds hydrogens, converts carbon dioxide into organic substances. Oxygen, derived from the H_2O used in step 1, is released to the environment as an important by-product of photosynthesis. In eukaryotes, both major steps of photosynthesis take place inside chloroplasts.

The First Major Step of Photosynthesis: The Light Reactions

The reactions of the first major step of photosynthesis follow the same fundamental pathway as the energy-trapping mechanisms of mitochondria. These mechanisms, outlined in Chapter 8, include: (1) provision of electrons at elevated energy levels; (2) an electron transport system, embedded in a membrane, that taps energy from the electrons and uses it to build an H^+ gradient across the membrane; and (3) an ATPase that uses the H^+ gradient as an energy source for ATP synthesis (see Fig. 8-1). In operation of these mechanisms in chloroplasts, only part of the energy of the electrons is tapped off and used to drive ATP synthesis. The remainder is used as reducing power for the second overall step in photosynthesis, reduction of CO_2. Because the mechanisms of the first major step are dependent on light and stop if the source of light is interrupted, they are called the *light* or *light-dependent reactions* of photosynthesis.

The Source of Electrons in Water   The electrons used in eukaryotic and cyanobacterial photosynthesis are removed from water in the reaction:

$$H_2O \longrightarrow 2H^+ + 2e^- + \tfrac{1}{2}O_2 \qquad (9\text{-}1)$$

The electrons removed from water, which are initially at energy levels too low to enter the electron transport system directly, are elevated in energy level by absorbed light energy. In all photosynthetic organisms, prokaryotes as well as eukaryotes, the primary molecules pushing electrons to higher energy levels by absorbing light are *chlorophylls* (see Fig. 9-4).

Electrons pushed to an elevated energy level by light absorption in chlorophyll are initially highly unstable. They are converted to stable form by transfer from chlorophyll to a molecule called a *primary acceptor. Once the electrons are trapped in stable orbitals of the primary acceptor, light energy has been captured as chemical energy.*

The provision of electrons for eukaryotic and cyanobacterial photosynthesis thus takes place in three steps: (1) removal of electrons at low energy levels from an inorganic donor, water; (2) elevation of the electrons to higher energy levels by the absorption of light energy; and (3) transfer of the electrons to stable orbitals in a primary acceptor. The molecules carrying out these reactions are embedded in the internal membranes of chloroplasts.

Electron Transport and ATP Synthesis in the Light Reactions   Like its mitochondrial counterpart, the electron transport system of photosynthesis consists of a series of carriers that accept and release electrons at successively lower energy levels. At the end of the pathway, electrons are delivered to a final acceptor. As electrons move through the carrier sequence, some of the released energy is used by the carrier molecules to push H^+ across the membrane housing the electron transport system. The resulting H^+ gradient provides the energy source driving ATP synthesis.

Photosynthetic electron transport differs from the mitochondrial system in that the electrons delivered to the final acceptor at the end of the pathway are still at relatively high energy levels. Further, the final electron acceptor in photosynthetic electron transport in chloroplasts is an organic substance, *NADP (nicotinamide adenine dinucleotide phosphate* (see Figs. 9-7 and 8-14). In contrast, the final electron acceptor in mitochondrial electron transport is an inorganic substance, O_2. The NADPH produced at the end of photosynthetic electron transport carries electrons to the reactions converting CO_2 to organic molecules in the second major step of photosynthesis.

Utilization of the H^+ gradient for ATP synthesis proceeds in chloroplasts by exactly the same mechanism as in mitochondria (see Fig. 8-1). The chloroplast ATPase, called the CF_oCF_1-*ATPase*, is driven by the H^+ gradient to assemble ATP from ADP and phosphate. The CF_oCF_1-ATPase, embedded in the same membrane as the photosynthetic electron transport system, contains molecular subunits similar to those of the mitochondrial F_oF_1-ATPase, arranged in the same headpiece-stalk-basal unit structure that extends from the membranes as a lollipop-shaped particle (see p. 230 and Figs. 8-23 and 9-12).

The light reactions of photosynthesis thus use energy derived from light to produce two substances, ATP and NADPH, that serve as reactants for the second major step. The oxygen released when water is split in the initial step of the light reactions (reaction 9-1) is also an important by-product of photosynthesis.

The Second Major Step in Photosynthesis: The Dark Reactions

In the second major step of photosynthesis, CO_2 is reduced and converted into organic substances. Since the reactions of the second major step depend only on a supply of ATP, NADPH, and CO_2 and do not directly require light, they are called the *dark* or *light-independent reactions* of photosynthesis. In the dark reactions, electrons carried by NADPH are used directly to reduce CO_2 into carbohydrates and other organic molecules. The ATP produced in the light reactions also supplies energy for the dark reactions.

Figure 9-1 summarizes the interrelationships of the light and dark reactions. Note from the figure that the ATP and NADPH produced by the light reactions, along with CO_2, are the reactants of the dark reactions. The ADP, inorganic phosphate, and $NADP^+$ produced by the dark reactions, along with H_2O, are the reactants for the light reactions. The light and dark reactions thus form a cycle in which the net inputs are H_2O and CO_2 and the net outputs are organic molecules and O_2:

$$H_2O + CO_2 \longrightarrow \text{organic molecules} + O_2 \qquad (9\text{-}2)$$

For sugars the organic molecules can be considered to be units of carbohydrate (CH_2O), so that this reaction can be written in balanced form as:

$$H_2O + CO_2 \longrightarrow (CH_2O) + O_2 \qquad (9\text{-}3)$$

Location of the Light and Dark Reactions in Eukaryotes: The Chloroplast

Chloroplast Structure   Chloroplasts are lens-shaped, membranous bodies that range from about 3 to 10 μm

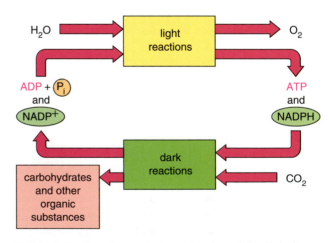

Figure 9-1   Relationships between the light and dark reactions of photosynthesis. The two reaction pathways are linked by ATP and NADP (see text).

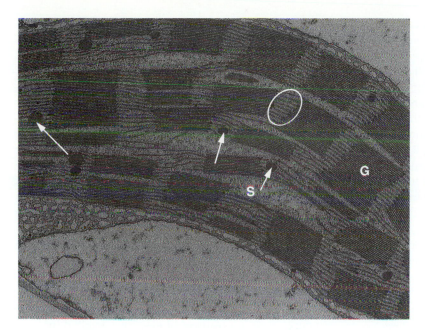

Figure 9-2 A chloroplast from maize. The chloroplast is surrounded by a smooth outer boundary membrane and a much folded and convoluted inner boundary membrane. Thylakoids stacked into grana (G) are visible inside the chloroplast. Adjacent grana are connected by stromal lamellae (circled). The thylakoid membranes are suspended in the stroma (S), a solution of proteins and other molecules. The stroma also contains numerous osmiophilic granules (arrows). × 24,500. (Courtesy of L. K. Shumway, Washington State University.)

in length and 1 to 5 μm in diameter (Fig. 9-2). Chloroplasts are surrounded by two continuous membranes, the *outer* and *inner boundary membranes*, separated by an *intermembrane compartment*. The outer boundary membrane is smooth; frequently the inner boundary membrane folds extensively or extends into tubelike forms (as in Fig. 9-2). The two boundary membranes enclose an inner compartment, the *stroma*, analogous in location to the mitochondrial matrix. Within the stroma is a third membrane system consisting of small, flattened sacs, the *thylakoids*. Chloroplasts thus have three separate internal compartments set off by membranes: the intermembrane compartment between the boundary membranes, the stroma, and the compartments enclosed within the thylakoids.

In green algae and higher plants, thylakoids in most chloroplasts are arranged in stacks called *grana* (singular = *granum*; see Figs. 9-2 and 9-3). Chloroplasts may contain from a few to 40 to 60 grana, each formed from stacks of 2 or 3 to as many as 100 individual thylakoids. The thylakoids of grana stacks are interconnected by flattened, tubular membranes called *stromal lamellae* (circled in Fig. 9-2). Stromal lamellae probably connect the thylakoid compartments into a single, continuous compartment within the stroma (diagramed in Fig. 9-3).

The chloroplast stroma may contain a variety of inclusions, among them *osmiophilic granules* or *droplets* (visible in Fig. 9-2) and, in some chloroplasts, large *starch granules*. The osmiophilic granules are probably collections of lipids suspended in the stroma. The chloroplast stroma also contains DNA, ribosomes, and all the factors required for DNA replication, RNA

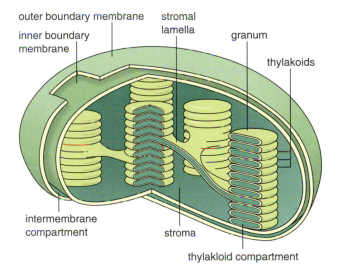

Figure 9-3 The membranes and compartments of chloroplasts (see text).

transcription, and protein synthesis. (The functions of chloroplast DNA and ribosomes in replication, transcription, and protein synthesis are discussed in Chapters 21 and 23.)

Localization of Molecules and Functions in Chloroplasts
Localizing the light and dark reactions in specific parts of chloroplasts was first accomplished by R. B. Park and N. G. Pon, who used a centrifuge to separate broken chloroplasts into thylakoid and stroma fractions. These and other investigators using the same

techniques found that the thylakoids, which are released intact when boundary membranes are broken, contain the light reactions—the mechanism splitting water, the pigments capturing and converting light to chemical energy, all the molecules and complexes transporting electrons and reducing $NADP^+$, and the CF_oCF_1-ATPase. The stroma, which remains in suspension as a colorless solution of proteins, contains the major enzymes of CO_2 fixation and the dark reactions. The outer boundary membranes contain proteins that transport substances between the stroma and the surrounding cytoplasm and enzymes that catalyze synthesis of fatty acids and chloroplast lipids, including carotenoids and quinones.

Usually chloroplasts are considerably larger than mitochondria in the same plant cell and occur in smaller numbers. The number may vary from a single chloroplast, as in *Micromonas* and several other green algae, to several hundred in the cells of most higher plants. In higher plant cells active in photosynthesis, chloroplasts may take up as much as 25% of the total cell volume.

THE LIGHT REACTIONS OF PHOTOSYNTHESIS

Light and Light Absorption

Visible light is a form of radiant energy at wavelengths ranging from about 400 nm, seen as blue light, to 680 nm, seen as red. Although radiated in apparently continuous beams, the energy of light actually flows in discrete units called *photons*. Photons contain an amount of energy that is inversely proportional to their wavelength. The shorter the wavelength, the greater the energy of a photon.

Molecules such as chlorophyll appear colored or pigmented because they absorb light at certain wavelengths and transmit other wavelengths. The color of a pigment is produced by the transmitted light. Chlorophyll, for example, absorbs blue and red light and transmits intermediate wavelengths that are seen in combination as green.

Light is absorbed in a pigment molecule by electrons occupying certain orbitals. In darkness or when exposed to light at wavelengths not absorbed by the molecule, these electrons occupy orbitals at a relatively low energy level known as the *ground state*. If an electron absorbs the energy of a photon, it moves to a new orbital at a higher energy level. In the new orbital, which is more distant from the atomic nucleus, the electron is said to be in an *excited state*. The difference in energy level between the ground state and excited state is exactly equivalent to the energy contained in the photon of light absorbed.

An electron can remain in the highly unstable excited state for only a billionth of a second or less. Return to the ground state may occur by one of several pathways. The excited electron may simply drop back to its ground state, releasing all its absorbed energy as heat or as a combination of heat and light. Two other possible fates for an excited electron have greater significance for photosynthesis. If a molecule containing an excited electron is situated close enough to another molecule, the excited orbital may extend far enough to overlap possible orbitals in the second molecule. The excited electron may then transfer to stable ground-state orbital in the second molecule. If the transfer takes place, the energy of the electron is effectively trapped as chemical energy. Any difference in energy level between the excited orbital in the light-absorbing molecule and the stable ground-state orbital in the acceptor is released as heat.

In photosynthesis, light is converted to chemical energy through transfer of electrons, excited by light absorption in chlorophyll, to stable orbitals in primary acceptor molecules. The transfer occurs within a few trillionths of a second after the chlorophyll is excited by the absorption of light energy.

A third possible fate for excitation energy is transmission to a nearby molecule by a mechanism called *inductive resonance*. On absorbing light, a pigment molecule sets up an electromagnetic field because of the rapid vibration of its excited electrons. Electrons in equivalent orbitals of an adjacent pigment molecule, lying within the vibrational field created by the excited molecule, may be induced to vibrate or resonate at the same frequency. If this occurs, the electrons in the second molecule absorb the excitation energy of the first. As the excitation energy is transferred to the second molecule, electrons in this molecule rise to an excited state, and the electrons in the first molecule return to the ground state. Energy transfer by inductive resonance occurs in billionths of a second with relatively little energy loss. It is important in the light reactions as a mechanism transferring absorbed light energy from pigment to pigment until the energy reaches chlorophyll molecules that pass electrons to stable orbitals in acceptor molecules.

The Structure of Chlorophyll and Other Light-Absorbing Molecules

Chlorophylls are the molecules directly involved in light absorption and electron transfer. Other pigments, the *carotenoids*, act as *accessory pigments* that absorb light energy and pass it on to the chlorophylls. Both chlorophylls and carotenoids are lipid molecules bound to thylakoid membranes and stromal lamellae in chloroplasts.

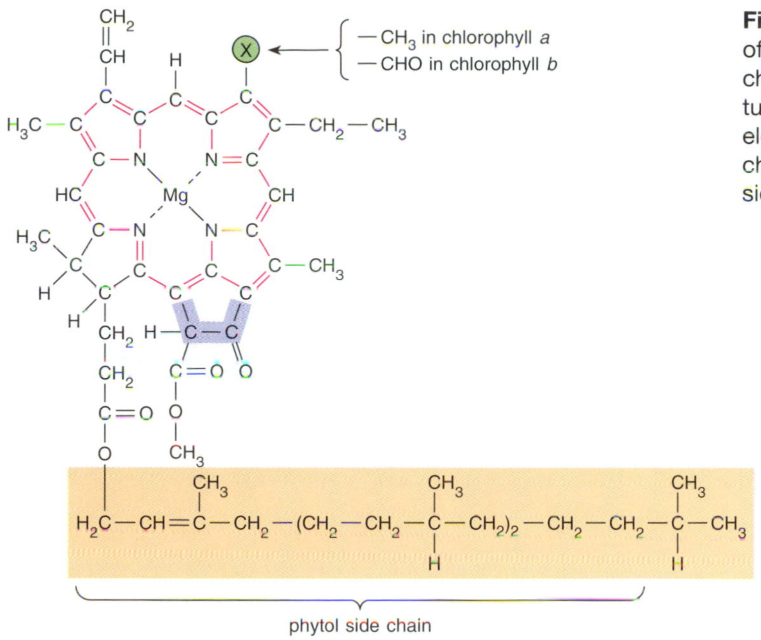

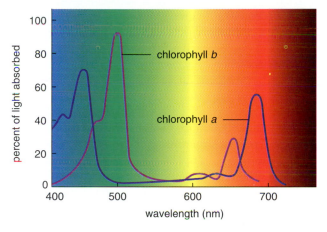

Figure 9-4 Chlorophylls *a* and *b*. The tetrapyrrole ring of the chlorophylls is similar to the ring structure of cytochromes and hemoglobins except for the additional structure shown in blue. The distribution of light-absorbing electrons in chlorophyll is shown in red. The nonpolar character of the chlorophylls is due primarily to the phytol side chain, which is strongly hydrophobic.

The Chlorophylls The chlorophylls are a family of closely related molecules based on a tetrapyrrole ring (Fig. 9-4). Their ring structure is similar to those of the cytochromes and hemoglobin except for the presence of an extra subunit (outlined in blue in Fig. 9-4). A magnesium atom is bound at the center of the chlorophyll ring. Attached to the ring is a long, hydrophobic side chain that gives the two major chlorophylls found in higher plants, *chlorophyll* a and *chlorophyll* b, their nonpolar solubility characteristics. These chlorophylls differ only in the side groups attached to one carbon of the tetrapyrrole ring. Of the two types, chlorophyll *a* plays the central role in the conversion of light to chemical energy in all photosynthetic eukaryotes and also in cyanobacteria; chlorophyll *b* is one of several pigments that pass the energy of absorbed light to chlorophyll *a*. The closely related chlorophyll *c* passes absorbed energy to chlorophyll *a* in brown algae, diatoms, and dinoflagellates.

The chlorophylls contain many electrons capable of moving to excited orbitals by absorbing light. These electrons are distributed in a network of alternating single and double bonds extending around the tetrapyrrole ring (shown in red in Fig. 9-4). The electrons in the network each absorb light at different wavelengths. The combination of wavelengths broadens the distribution absorbed by chlorophyll into a broad curve called its *absorption spectrum* (Fig. 9-5). Each chlorophyll type has a distinct absorption spectrum.

The absorption spectra of the chlorophylls are modified by association of the pigments with other molecules in chloroplast membranes, particularly with proteins. For example, chlorophyll *a*, when extracted

Figure 9-5 Absorption of light at different wavelengths by chlorophylls *a* and *b*. The positions of the peaks change to some extent depending on combination of chlorophylls with proteins.

and purified, absorbs red light most strongly at 665 nm in an acetone solution. In chloroplast membranes, where it is closely associated with other molecules, chlorophyll *a* may absorb light most strongly at other wavelengths, such as 660, 670, 680, and 700 nm. Some of the associations, particularly those producing absorption peaks in chlorophyll *a* at 680 and 700 nm, figure importantly in the light reactions of photosynthesis.

The Carotenoids The carotenoids (Fig. 9-6) are a separate family of light-absorbing lipids built upon a long

Figure 9-6 Carotenoids, which consist of **(a)** the carotenes, which are pure hydrocarbons, and **(b)** the xanthophylls (also called carotenols), which contain oxygen atoms in their terminal structures.

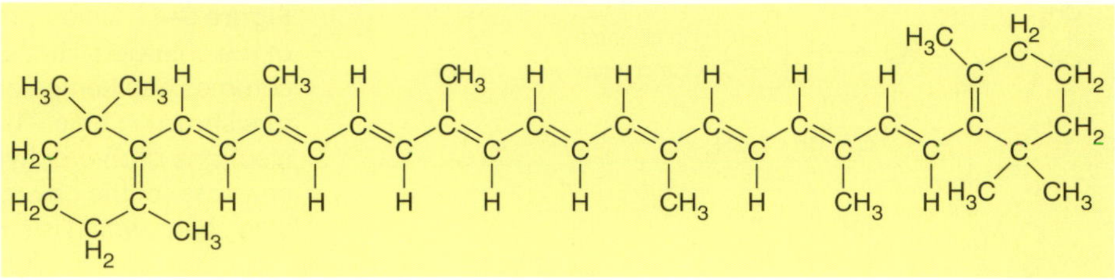

a   β-carotene

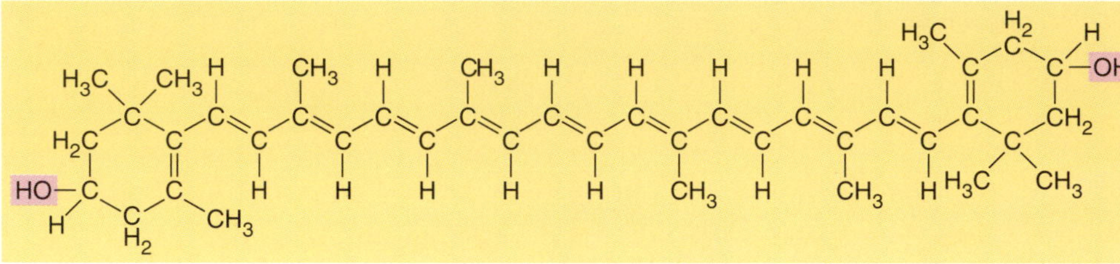

b   lutein (a xanthophyll)

carbon backbone containing 40 carbon atoms. Various substitutions in side groups attached to the backbone give rise to different carotenoid pigments. Two types of carotenoids occur in all green plants. The *carotenes* are pure hydrocarbons that contain no oxygen atoms; chief in abundance among these pigments in higher plants is β-*carotene* (Fig. 9-6a). The second major carotenoid type, the *xanthophylls* (also called the *carotenols*), are closely similar except for the presence of oxygen atoms in their terminal structures (Fig. 9-6b). Carotenoids of both types have multiple light-absorbing electrons associated with the alternating single and double bonds of the backbone chain. These electrons absorb light at blue-green wavelengths from 400 to 550 nm, which are only weakly absorbed by chlorophylls, and transmit other wavelengths in combinations that appear yellow, orange, red, or brown.

Light energy absorbed by the carotenoids and by chlorophyll *b* is eventually transferred by inductive resonance to the chlorophyll *a* molecules involved in transforming light into chemical energy. The net effect of the entire combination of chlorophylls and carotenoids is to broaden the spectrum of wavelengths used efficiently as energy sources for photosynthesis.

Organization of Photosynthetic Pigments in Chloroplasts

The carotenoid and chlorophyll molecules are organized with proteins and other molecules into complexes that play central roles in the light reactions. Two of these complexes, called *photosystems I and II* (see Fig. 9-10), absorb light and convert it to chemical energy in eukaryotic plants. The two photosystems have been isolated and purified from chloroplast

membranes; each contains from about 50 to 100 chlorophyll *a* molecules and a smaller number of carotenes. These pigment molecules are combined with polypeptides—as many as 20 in both photosystems—that carry out structural, enzymatic, regulatory, and other functions in the photosystems. One or two chlorophyll *a* molecules within the photosystems form a *reaction center* in which the central event of the light reactions, transfer of excited electrons to stable orbitals in a primary acceptor, takes place. The photosystems also contain a sequence of carriers involved in the transport of electrons away from the primary acceptor. Photosystem II, in addition, contains the mechanism splitting water (reaction 9-1).

The internal membranes of chloroplasts also contain pigment-protein assemblies called *light-harvesting complexes* (*LHCs*; see Fig. 9-10), which act as accessory light-gathering "antennas." LHCs have no reaction centers or primary acceptors, and they do not convert light to chemical energy. Instead, they pass on the energy of absorbed light to the two photosystems. Most eukaryotes have two types of LHCs: *LHC-I*, associated with photosystem I, and *LHC-II*, associated with photosystem II.

Photosystem I   Photosystem I (see Fig. 9-10) contains a group of about 110 chlorophyll *a* molecules and 16 β-carotenes combined with a group of proteins into a light-absorbing assembly known as the *core antenna*. Absorbed light energy is passed, probably by inductive resonance, from the core antenna to the reaction center, which in photosystem I consists of a specialized form of chlorophyll *a* called *P700* (P = pigment). P700 is given this name because its light absorption changes sharply at a wavelength of 700 nm as it passes electrons to the primary acceptor of the photosystem.

Eleven integral and peripheral polypeptide subunits are central to photosystem I structure (see Fig. 9-10). At least 10 additional minor proteins occur in eukaryotic photosystem I complexes, giving a total molecular weight of about 340,000. Two of the major subunits, A and B, are large, integral membrane proteins that house the core antenna chlorophylls and the reaction center.

Photosystem II   The core antenna of photosystem II (see Fig. 9-10) includes about 40 chlorophyll a molecules and a much smaller number of β-carotenes. The reaction center of this photosystem contains a specialized form of chlorophyll a, *P680*, which undergoes a conspicuous change in light absorption at 680 nm as electrons pass to the primary acceptor. Another pair of chlorophyll molecules lacking Mg^{2+} (*pheophytins*; see below) participates in transfer of electrons from the reaction center to the primary acceptor.

At least 12 integral and peripheral proteins are central to the photosystem II complex. Another 10 or so are minor components. Two of the major integral proteins, *D1* and *D2*, link the reaction center, primary acceptor, and all the major electron carriers of the complex; the integral proteins *CP43* and *CP47* bind the chlorophyll a and b molecules of the core antenna. A cytochrome of unknown function, b_{559}, is also present in photosystem II. Although this cytochrome may participate in electron transfers within the complex, its function does not appear to be essential.

The water-splitting mechanism associated with photosystem II includes three peripheral polypeptides and four manganese atoms arranged in a *manganese center*. One Ca^{2+} and one Cl^- work as essential cofactors for the manganese center. By a process that is still only incompletely understood, the manganese center promotes the oxidation of water (reaction 9-1) as the first step in the light reactions.

The Light-Harvesting Complexes (LHCs)   The two LHC antennas (see Fig. 9-10) contain both chlorophylls a and b. LHC-I antennas contain about 80 to 120 chlorophyll molecules in a chlorophyll a:chlorophyll b ratio of about 3:4. LHC-II antennas contain about 50 chlorophyll a and b molecules in an approximate 1:1 ratio. In each antenna type the chlorophylls are associated with about three or four polypeptides. About half of the chlorophyll a, and three-fourths of the chlorophyll b of higher plant chloroplasts, is concentrated in LHC-II antennas. LHC-II antennas also contain a few carotenoid molecules, including both β-carotene and essentially all the xanthophylls of higher plant chloroplasts.

The LHCs are large particles that completely span the photosynthetic membranes. LHC-I forms a more or less permanent complex with photosystem I. LHC-II, however, dissociates from photosystem II under certain conditions, so that the accessory antenna and photosystem may occur as either a tight complex or separately in chloroplast membranes. Separation of LHC-II from photosystem II, which greatly reduces the light-harvesting efficiency of photosystem II, provides an important mechanism balancing the light reactions (see below). The energy of photons absorbed within the LHCs is probably passed on to the photosystems by inductive resonance.

The Electron Transport System Linking the Photosystems

The two photosystems are linked in chloroplasts by an electron transport system that conducts electrons from one photosystem to the next and delivers electrons to $NADP^+$ at the end of the pathway. Most of the molecules oxidized and reduced in the photosynthetic system, like the electron carriers of mitochondria (see p. 223), consist of nonprotein prosthetic groups organized with proteins into large complexes. Some of the prosthetic groups are linked into the photosystems. Most of the prosthetic groups of the chloroplast system are the same types active in mitochondrial electron transport—flavoproteins, cytochromes, Fe/S centers, copper centers, and quinones (see pp. 224–226 for a description of these electron carriers).

The same battery of techniques used to place mitochondrial carriers in sequence (see Information Box 8-3) has been used to assign tentative positions to the photosystems and electron carriers of chloroplasts. Among the inhibitors used to block steps in the pathway are various herbicides that target the photosystems or carriers. These techniques have revealed that the photosynthetic carriers are arranged in a Z *pathway* (Fig. 9-7), first advanced as a possibility by R. Hill and F. Bendall. The first leg of the "Z" is the flow of electrons from water through photosystem II.[1] The electrons then flow through a series of carriers that connect photosystems II and I; this series forms the diagonal of the Z. As they pass through this series, electrons lose energy; some of the released energy is

---

[1]Photosystems I and II are numbered in the order in which they were first discovered. However, in the overall pattern of electron flow in the light reactions, photosystem II occurs before photosystem I in the pathway. Thus the identifiers given to the two photosystems are reversed in the sense of electron flow. Tradition being what it is in science, the pathway of electron flow in photosynthesis is far more likely to reverse than is the nomenclature.

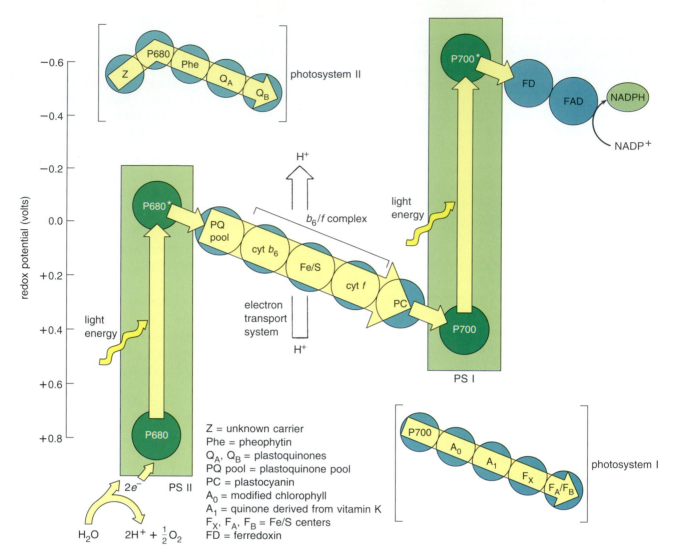

Figure 9-7 Electron flow through photosystems and prosthetic groups in the Z pathway. The first leg of the Z is the flow of electrons from water through photosystem II. The diagonal is formed by a series of carriers that connects photosystems II and I; some of the energy released by the electrons in this passage is used to pump H⁺ across the membrane housing the carriers. The electrons next move through photosystem I and are finally transferred to a short transport chain leading to NADP⁺. The asterisks indicate the excited forms of P680 and P700 in photosystems II and I.

used to pump H^+ across the membrane housing the carriers. The electrons then pass through photosystem I and are transferred to a short transport chain leading to the final acceptor of the chloroplast system, $NADP^+$.

Individual Carriers and Their Activities in the Z Pathway
The tentative positions of individual prosthetic groups acting as carriers within the Z pathway are shown in Figure 9-7. Electrons entering the Z pathway are removed from water at low energy levels by the water-splitting complex of photosystem II and are delivered to the reaction center of photosystem II by an electron carrier identified as Z. This carrier is the side chain of a

tyrosine residue within the D1 polypeptide of photosystem II. Z picks up electrons from the water-splitting reaction and releases them to a P680 chlorophyll at the reaction center. P680 accepts single electrons and raises them to excited orbitals through the absorption of light energy.

After excitation, electrons are released to the primary acceptor of photosystem II, Q_A. This substance is a *plastoquinone*, a quinone type typical of chloroplasts (Fig. 9-8). From the primary acceptor, electrons flow to a second plastoquinone, Q_B, the final electron carrier in photosystem II. The H^+ removed from water by photosystem II is released to add to the H^+

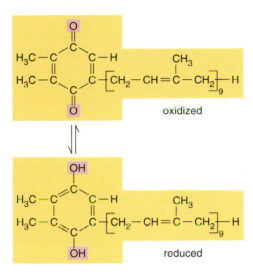

Oxidized:

$H_3C-C \cdots C-H$

$H_3C-C \cdots C-[CH_2-CH=C-CH_2]_9-H$ with CH_3

oxidized

reduced:

$H_3C-C \cdots C-H$

$H_3C-C \cdots C-[CH_2-CH=C-CH_2]_9-H$ with CH_3

reduced

Figure 9-8 Plastoquinone, a carrier that transports electrons within the photosystem II complex and in the electron transport system linking the photosystems in the Z pathway. A related but different quinone derived from vitamin K_1 carries electrons within photosystem I.

gradient produced by electron transport in chloroplasts (see below).

The transfer of electrons from P680 to Q_B in photosystem II is believed to follow a three-dimensional molecular pathway closely similar to the system operating in a purple photosynthetic bacterium, *Rhodopseudomonas* (Fig. 9-9). The pathway was worked out by H. Michel, J. Deisenhofer, and R. Huber, who crystallized the photosystem of the bacterium and analyzed the crystals by X-ray diffraction. Michel, Deisenhofer, and Huber received a Nobel Prize in 1988 for their analysis.

Electrons are transferred from the plastoquinone carrier of photosystem II (Q_B in Fig. 9-7) to a pool of plastoquinone molecules that forms the first carrier of the transport system linking the two photosystems (see Fig. 9-10). The pool, in which individual plastoquinone molecules are unassociated with proteins and free to diffuse within the membrane interior, is analogous to the ubiquinone pool of the mitochondrial electron transport system. Electron transfer to the pool probably occurs simply by detachment and entry into the pool of a reduced quinone Q_B molecule from photosystem II. The reduced plastoquinone diffusing into the pool is replaced in photosystem II by an oxidized plastoquinone from the pool.

From the plastoquinone pool, electrons flow through three carriers—cytochrome b_6, an Fe/S protein, and cytochrome f, combined with proteins into a membrane-spanning structure known as the b_6/f complex. This complex is closely similar to complex III of the mitochondrial system. (The cytochrome f of the

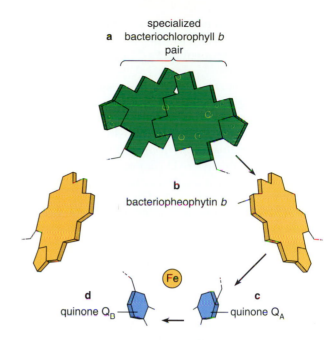

Figure 9-9 The three-dimensional arrangement of the reaction center and internal electron carriers of the *Rhodopseudomonas viridis* photosystem, as deduced from X-ray diffraction by Michel, Deisenhofer, and Huber. After excitation in the reaction center, which consists of a pair of bacteriochlorophyll b molecules, electrons flow to bacteriopheophytin b and through the two quinones Q_A and Q_B in sequence. The reaction center, primary acceptor, and plastoquinones of eukaryotic photosystem II probably take on the same three-dimensional arrangement.

b_6/f complex, although given the "f" identifier, is a c-type cytochrome closely related to the cytochrome c_1 of mitochondrial complex III.)

From the b_6/f complex, electrons pass to the final carrier linking the photosystems, *plastocyanin*. This carrier, a protein containing a copper atom that varies between the Cu^+ and Cu^{2+} states during alternate cycles of reduction and oxidation, shuttles electrons from the b_6/f complex to photosystem I. Plastocyanin thus occupies a position in the chloroplast system equivalent to cytochrome c of the mitochondrial electron transport system. (In many cyanobacteria and some eukaryotic algae, cytochrome c substitutes for plastocyanin at this point in the Z pathway.) The pathway from the plastoquinone pool to plastocyanin is thus essentially the same as the mitochondrial pathway from ubiquinone to cytochrome c (compare Figs. 9-10 and 8-16).

From plastocyanin, electrons flow to the P700 chlorophyll at the reaction center of photosystem I. After excitation by light absorption, electrons are transferred from P700 to the primary acceptor of this photosystem, a modified form of chlorophyll known as A_0. From A_0, electrons flow to A_1, a quinone derived from vitamin K. The electrons then flow through a chain of three sequential Fe/S centers within the photosystem.

Electrons next pass from photosystem I at very high energy levels to *ferredoxin*, an Fe/S protein that acts as a separate, highly mobile electron carrier in the chloroplast system. From ferredoxin, electrons flow along the short final chain from FAD to NADP to complete the Z pathway. FAD (flavin adenine dinucleotide: see Fig. 8-12) is the prosthetic group of an enzyme called *ferredoxin-NADP oxidoreductase*. This oxidoreductase, although a separate protein of the chloroplast electron transport system, evidently remains closely associated with photosystem I.

Electrons receive two boosts in energy as they move through the Z pathway, one in photosystem II and one in photosystem I (see Fig. 9-7). The two consecutive boosts raise the electrons to energy levels high enough to reduce NADP. Along the way some energy is tapped from the electrons and used to produce an H^+ gradient. The gradient is established primarily through the activity of the b_6/f complex, which, like complex III, its structural counterpart in mitochondria, acts as an electron-driven pump actively transporting H^+. In chloroplasts, H^+ is pumped from the stroma into the thylakoid compartments.

The pattern of electron flow in the Z pathway from H_2O to NADP is frequently called *noncyclic photosynthesis* because electrons are removed from water and travel in a one-way direction to NADP. Figure 9-10 summarizes the organization of individual electron carriers within the photosystems and carrier complexes of chloroplasts. All the components of the Z pathway are located on or within the thylakoid membranes or stromal lamellae of chloroplasts.

The water-splitting reaction carried out by photosystem II, developed in cyanobacteria some 2 to 3 billion years ago, is one of the most significant and fundamentally important of all biological interactions. The reaction enabled photosynthetic organisms to use one of the most abundant substances on earth—water—as an electron source. The reaction is also responsible for the oxygen present in the atmosphere of our planet. The appearance of oxygen in the atmosphere, released by the water-splitting reactions of photosynthesis, paved the way for the evolution of aerobic organisms, in which oxygen serves as the final acceptor for electrons removed in cellular oxidations.

A variety of experiments have verified that the two photosystems are actually linked into the Z pathway proposed by Hill and Bendall. For example, L. N. M. Duysens showed that when chloroplasts are irradiated with red light at wavelengths longer than 680 nm (expected primarily to excite photosystem I but not II), cytochrome *f* becomes oxidized. At shorter wavelengths expected primarily to excite photosystem II, cytochrome c_1 is reduced. This is the expected result if cytochrome *f* occurs in an electron transport system linking the two photosystems as proposed by the Z pathway. Equivalent experiments have supported other segments of the Z pathway.

Cyclic Electron Flow Within the Z Pathway

Electrons may flow cyclically within the Z pathway through a closed circuit around photosystem I (Fig. 9-11). In this pathway, electrons pass from ferredoxin back to the b_6/f complex instead of traveling to $NADP^+$. From the b_6/f complex, electrons flow to plastocyanin, through photosystem I, and back to ferredoxin. This cycle pumps additional H^+ each time electrons flow through the b_6/f complex. The resulting H^+ gradient is probably sufficient to drive the synthesis of one ATP molecule for each electron pair traveling the circuit around photosystem I. The net result of cyclic electron flow is conversion of light to chemical energy in the form of ATP without reduction of $NADP^+$.

The degree to which cyclic electron flow operates in eukaryotic photosynthesis is unknown. However, it might allow varying requirements for ATP and NADPH in the dark reactions to be precisely met. That is, cyclic photosynthesis might supplement the noncyclic pathway under conditions in which the dark reactions require significantly higher amounts of ATP than NADPH.

H^+ Pumping by the Z Pathway

Although the b_6/f complex is the primary H^+ pump of the Z pathway, the H^+ gradient is increased at two other sites as well. One is the water-splitting reaction (reaction 9-1) of photosystem II, which takes place on the side of thylakoid membranes facing the thylakoid compartment. The reaction releases two H^+ into the thylakoid compartment for each electron pair entering the Z pathway. The second takes place as electrons flow from ferredoxin to NADP via the ferredoxin-NADP oxidoreductase carrier. As $NADP^+$ is reduced to NADPH, it removes one H^+ from the stroma. Removal of this H^+ adds to the gradient created by electron transport.

As many as four H^+ per electron pair are pumped by the b_6/f complex as it cycles between oxidized and reduced forms. Recently E. C. Hart and his colleagues inserted isolated and purified b_6/f complexes into artificial lipid vesicles (see p. 101). When supplied with reduced plastoquinone as an electron donor and oxidized plastocyanin as an acceptor, the reconstituted

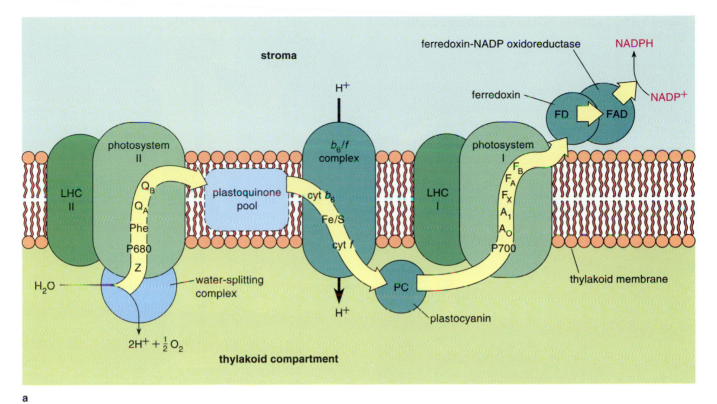

a

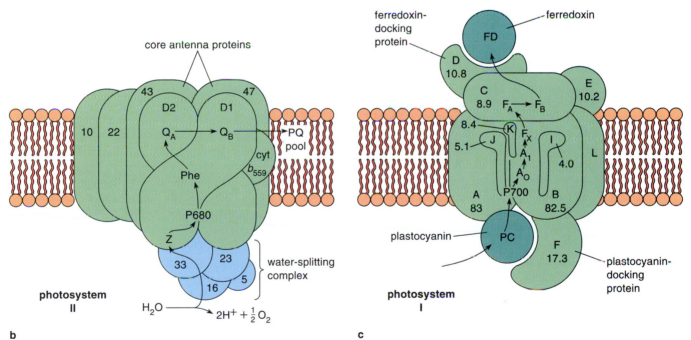

b

c

system pumped H^+ from the medium into the vesicle interiors as the b_6/f complex cycled between the oxidized and reduced states.

Using the H^+ Gradient to Synthesize ATP

The chloroplast CF_oCF_1-ATPase uses the H^+ gradient established by the Z pathway as the energy source for

Figure 9-10 The chloroplast electron transport system. (a) Locations of the two photosystems and electron transport carriers in thylakoid membranes. (b) Major polypeptide subunits in photosystem II. (c) Major polypeptides of the photosystem I complex. Polypeptides A and B, in addition to binding the reaction center and many of the electron carriers of photosystem I, also house the chlorophyll molecules of the core antenna. The numbers give molecular weights in thousands.

Figure 9-11 Cyclic electron flow around photosystem I (see text).

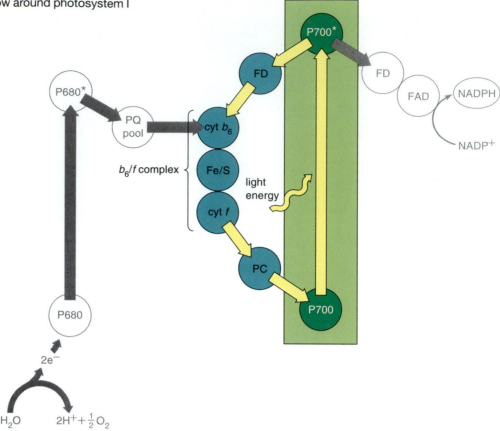

ATP synthesis. The polypeptides of the chloroplast ATPase, which are similar to those of the mitochondrial and bacterial F_oF_1-ATPases (see Fig. 8-23), are arranged into a spherical headpiece carried on a stalk (the CF_1 subunit), anchored to a basal subunit buried in the membrane (the CF_o subunit). The headpieces and stalks of the CF_oCF_1-ATPase can be seen as "lollipop" structures extending from the outer surfaces of photosynthetic membranes prepared for electron microscopy by negative staining (Fig. 9-12).

As in the mitochondrial F_oF_1-ATPase, the active site of the enzymatic mechanism converting ADP to ATP is concentrated in the CF_1 head subunit (see p. 230). The CF_o basal subunit provides the membrane channel conducting H^+ through the CF_oCF_1-ATPase. Conformational changes resulting from the H^+ flow are transmitted through the stalk to the headpiece; in some unknown way the conformational changes drive synthesis of ATP from ADP and inorganic phosphate (see also p. 232).

The chloroplast CF_oCF_1-ATPase is essentially irreversible, even more so than the mitochondrial F_oF_1-ATPase. The irreversibility of the chloroplast CF_oCF_1-ATPase may be connected with the fact that no electron transport occurs and no H^+ gradient is set up across thylakoid membranes in the dark. Because there is no H^+ gradient in darkness, no opposing force acts as a brake to prevent the CF_oCF_1-ATPase from working as an active transport pump (in mitochondria, electron transport is essentially continuous so that an H^+ gradient always exists across the membrane housing the F_oF_1-ATPase). An easily reversible CF_oCF_1-ATPase, unopposed by an H^+ gradient, might run backward in the dark, hydrolyzing all available ATP to push H^+ across the photosynthetic membranes. The irreversibility of the chloroplast CF_oCF_1-ATPase evidently depends on the ϵ-subunit, which inhibits ATP breakdown by the complex (see the Experimental Process essay by R. E. McCarty on p. 234).

Organization of Light Reaction Components in Thylakoid Membranes

The Z pathway is organized into four large complexes and three individual electron carriers (see Fig. 9-10). Photosystem I and II, each combined with its LHC antenna, form two of the large complexes. The remaining two complexes consist of the b_6/f complex and the ferredoxin-NADP oxidoreductase. Tests for the organization of membrane components show that the two photosystems and the b_6/f complex extend entirely through the thylakoid membranes. The water-

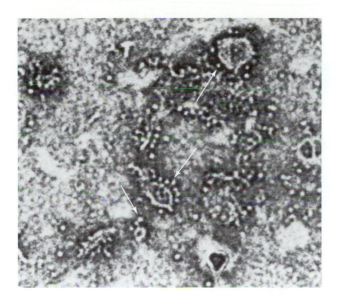

Figure 9-12 Isolated thylakoid membranes prepared for electron microscopy by negative staining. The head-pieces and stalks of the CF_oCF_1-ATPase can be seen extending from the membrane surfaces (arrows). (Courtesy of E. Racker, from *J. Biolog. Chem.* 248:8281[1973]).

splitting complex is located on the part of photosystem II extending into the thylakoid compartment. The three individual carriers of the Z pathway—plastoquinone, plastocyanin, and ferredoxin—occur as single molecules in the system. Plastocyanin is located on the membrane surface facing the thylakoid compartments, and ferredoxin and ferredoxin-NADP oxidoreductase on the side facing the stroma. Because of this orientation, NADP reduction occurs in the stroma surrounding thylakoid membranes.

The locations of these Z-pathway components within thylakoid membranes have been analyzed by the same techniques used for other cellular membranes (see p. 108ff). These techniques include exposure of inside or outside membrane surfaces to radioactive or other marker groups, digestive enzymes, antibodies developed against the Z-pathway components, or artificial electron donors or acceptors. For example, artificial electron donors that can substitute for water work only on inside-out and not right-side-out vesicles produced from thylakoids, demonstrating that the water-splitting segment of photosystem II faces the thylakoid compartments.

Although there are enough photosystem complexes and electron transport elements to form several hundred Z pathways in each thylakoid, the light reactions do not seem to be organized into rigid Z assemblies. With certain exceptions the components are evidently free to move through or over the fluid thylakoid membrane and interact by random collisions.

The exceptions were discovered by B. Andersson and J. M. Anderson and others by an application of the freeze-fracture technique for electron microscopy (see p. 791), used in combination with mutants that lack one or more of the Z-pathway components. These techniques allowed the particles visible in freeze-fracture preparations to be identified by correlating the absence of a particle of characteristic size and shape with a mutant lacking a particular Z-pathway component.

The most striking restriction to free mobility revealed by these techniques involves photosystems I and II. Photosystem I is apparently confined almost entirely to stromal lamellae and regions where outer thylakoid surfaces directly face the stroma (Fig. 9-13). Most photosystem II complexes lie in regions of thylakoid membranes that are fused to other thylakoid membranes within stacks. The remaining components of the Z pathway appear to be distributed more or less uniformly throughout thylakoid membranes and stromal lamellae.

Segregation of photosystems I and II in distinct regions of thylakoid membranes means that noncyclic electron transport from photosystem II to I in the Z pathway must proceed via mobile electron carriers that can shuttle between the stacked and unstacked regions of thylakoids. The most likely candidates for this shuttle are plastoquinone and plastocyanin, both relatively small molecules that can diffuse rapidly through or along thylakoid membranes. In their role as shuttles, reduced plastoquinone molecules would transport electrons from photosystem II to b_6/f complexes in stacked thylakoid regions. Mobile plastocyanin carriers would make the necessary connections between the b_6/f complexes and photosystem I in unstacked regions. The shuttle activity of plastoquinone and plastocyanin molecules probably proceeds simply through random motions and collisions, aided by the relatively small size of the two shuttle molecules and the highly fluid nature of the thylakoid bilayer.

Photosystem II complexes are evidently held in the fused regions of stacked thylakoid membranes by interactions between the LHC-II antennas of adjacent membranes. Among the evidence supporting this conclusion are observations of mutants lacking the LHC-II complex. In these mutants, thylakoids do not fuse into grana stacks.

The "lollipops" formed by the CF_oCF_1-ATPase can be seen in electron microscope preparations to extend from unfused segments of thylakoid membranes into the stroma (see Fig. 9-13). Extension of the ATPase headpieces into the stroma and the direction of H^+ pumping in chloroplasts makes thylakoid compartments the functional equivalents of the intermembrane compartment in mitochondria (Fig. 9-14).

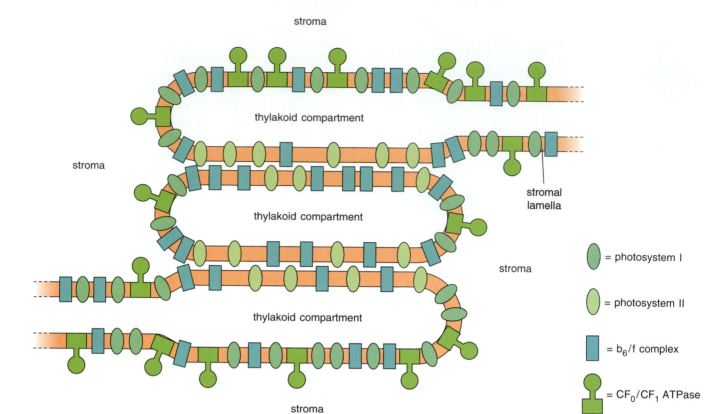

Figure 9-13 Distribution of major Z-pathway components in photosynthetic membranes.

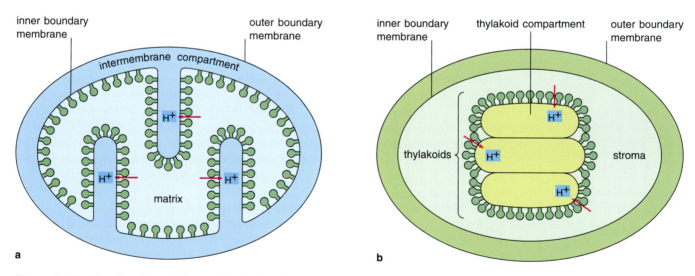

Figure 9-14 The direction of H⁺ pumping by the electron transport systems of mitochondria (a) and chloroplasts (b). H⁺ is transported into the intermembrane compartment in mitochondria and into the thylakoid compartments in chloroplasts.

The Light Reactions and the Chemiosmotic Mechanism

In 1954, D. I. Arnon and his colleagues first demonstrated the connection between electron transport and ATP synthesis in chloroplasts. They supplied isolated spinach chloroplasts with ADP and inorganic phosphate and detected ATP synthesis in the light. Since the quantity of ATP synthesized was correlated with the amount of electron flow in response to light, Arnon and his coworkers proposed that ATP synthesis in the isolated chloroplasts was coupled in some way to electron transport. This light-driven ATP synthesis was termed *photophosphorylation*.

The nature of the connection between electron transport and ATP synthesis was not understood until Mitchell advanced his chemiosmotic hypothesis in 1962 (see pp. 232–236). The applicability of the chemiosmotic hypothesis to chloroplasts is supported by several major lines of evidence:

1. Photophosphorylation depends on the presence of closed membranous vesicles.

2. Electron transport through photosystems II and I, or cyclic transport around photosystem I, results in accumulation of H^+ inside isolated thylakoids or inside vesicles derived from thylakoids, as predicted by the chemiosmotic hypothesis.

3. Destruction of the H^+ gradient by agents that increase permeability of thylakoid membranes to H^+ stops ATP synthesis.

4. Imposition of an artificial H^+ gradient across closed thylakoid membranes stimulates ATP synthesis.

The fourth line of evidence, stemming from experiments conducted by A. T. Jagendorf and E. Uribe in 1966, is among the strongest in favor of the chemiosmotic hypothesis. (For details of Jagendorf and Uribe's experiments, see p. 233.)

The light reactions absorb light energy and convert it into chemical energy in the form of ATP and electrons held in stable orbitals at elevated energy levels. The electrons are derived from water in a reaction that releases oxygen as an important by-product. As electrons pass along the electron transport system toward their final acceptor, $NADP^+$, some of their energy is tapped off and used to establish an H^+ gradient across thylakoid membranes. The H^+ gradient serves as the immediate energy source for ATP synthesis by the chloroplast CF_o/CF_1-ATPase. The primary products of the light reactions, ATP and NADPH, serve as the sources of energy and reducing power for the dark reactions of photosynthesis.

THE DARK REACTIONS

In the dark reactions, chemical energy produced in the light reactions is used to convert carbon dioxide into carbohydrates and other organic products. Although commonly called the dark reactions because they do not depend directly on light, the reactions of this series actually take place primarily in the daytime, when ATP and NADPH are available from the light reactions.

Tracing the Pathways of the Dark Reactions

The first significant progress in unraveling the dark reactions was made in the 1940s, when radioactive compounds became available to biochemists. One substance in particular, CO_2 labeled with the radioactive carbon isotope ^{14}C, was critical to this research.

In work beginning in 1945, M. Calvin, A. A. Benson, and their colleagues used the newly available ^{14}C isotope to trace the pathways of the dark reactions. Calvin and his colleagues allowed photosynthesis to proceed in a green alga, *Chlorella*, in the presence of CO_2 containing ^{14}C. At various times after exposure to radioactive CO_2, cells were removed and placed in hot alcohol, which killed the cells and stopped the dark reactions instantly. Extracts of carbohydrates and other substances were made and separated by two-dimensional paper chromatography (Fig. 9-15). The substances in the radioactive spots separated by the chromatography were then identified chemically.

If the carbohydrate extracts were made within a few seconds after exposure of *Chlorella* to labeled CO_2, most of the radioactivity could be identified with the three-carbon substance *3-phosphoglycerate* (*3PGA*). The rapidity with which the labeled CO_2 was incorporated into 3PGA indicated that it is one of the earliest products of photosynthesis. If extracts were made after longer periods, radioactivity showed up in the three-carbon sugar *3-phosphoglyceraldehyde* (*3PGAL*) and in more complex substances, including a variety of six-carbon sugars, sucrose, and starch.

In other experiments, Calvin and his colleagues reduced the amount of CO_2 available to the *Chlorella* cells so that photosynthesis was delayed even though adequate light was supplied. Under these conditions a five-carbon sugar, *ribulose 1,5-bisphosphate* (*RuBP*), accumulated in the chloroplasts. This result suggested that RuBP is the first substance to react with CO_2 in the dark reactions and accumulates if CO_2 is in short supply. By similar methods, most of the remaining intermediate compounds between CO_2 and six-carbon sugars were identified.

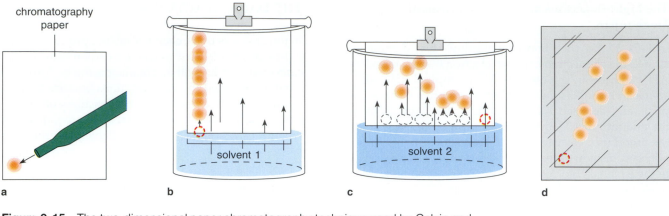

Figure 9-15 The two-dimensional paper chromatography technique used by Calvin and his colleagues to trace the dark reactions in *Chlorella*. **(a)** A drop of the alcohol extract of *Chlorella* cells, containing labeled and unlabeled compounds, is placed at one corner of the chromatography paper and dried. **(b)** The paper is then placed with its edge touching the solvent used to produce separation in the first direction (a water solution of butyl alcohol and propionic acid was used). The compounds in the extract separate and are carried upward by the migrating solvent at rates that vary according to molecular weight and solubility. (The location of the original drop of alcohol extract is shown as a red dashed circle in **b** to **e**.) **(c)** The paper is dried, turned 90°, and placed in the second solvent (a water solution of phenol was used). The compounds migrate upward at rates that are different from their mobility in the first solvent. This step separates compounds that migrated similarly in the first solvent. The final positions allow the compounds to be identified by comparison with previously established standards. **(d)** The dried paper is covered with a sheet of photographic film. Any radioactive compounds expose spots on the film. **(e)** The film is developed and compared with the paper chromatograph to identify radioactive compounds.

Using this information, Calvin, with his colleagues Benson and J. A. Bassham, pieced together the dark reactions of photosynthesis. The cycle is now called the C_3 *cycle* (because the first product of the cycle, 3PGA, is a three-carbon molecule) or the *Calvin–Benson cycle*. Calvin was awarded the Nobel Prize in 1961 for his brilliant work deducing the reactions of the cycle.

The C_3 Cycle

The cycle worked out by Calvin and his colleagues uses CO_2, ATP, and NADPH as net reactants and releases ADP, $NADP^+$, and 3PGAL as net products (Figs. 9-16 and 9-17). In the first reaction of the cycle (reaction 1 in Fig. 9-17), CO_2 combines directly with RuBP. This reaction, catalyzed by the enzyme *ribulose 1,5-bisphosphate carboxylase* (*RuBP carboxylase*), produces two molecules of 3PGA, the substance detected by the labeling experiments as the first product of the dark reactions. One of the 3PGAs contains the newly incorporated CO_2 in the position marked by an asterisk in Figure 9-17. The reaction requires no input of energy because the two three-carbon products exist at a much lower energy level than RuBP, which can be

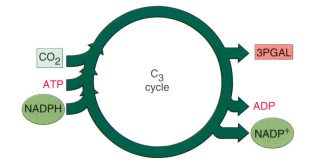

Figure 9-16 Overall reactants and products of the C_3 cycle.

considered a high-energy substance. The first reaction of the C_3 cycle accomplishes carbon *fixation*: conversion of the carbon of an inorganic molecule, CO_2, into the carbon of an organic substance.

The next two reactions of the cycle, which proceed at the expense of ATP and NADPH, yield the net carbohydrate product of the cycle, 3PGAL. In the first of these reactions (reaction 2), a phosphate group derived from ATP is added to each of the two 3PGA molecules produced by reaction 1. The products of the second reaction each accept an H^+ and two electrons from NADPH in the next reaction (reaction 3). One phosphate is removed from each of the reactants at the same time, yielding two molecules of 3PGAL for each molecule of CO_2 added to RuBP in reaction 1. Some of the 3PGAL produced at this step is used to regenerate the RuBP used in the first step in the cycle, and some is released as a product of the cycle.

Regeneration of RuBP occurs through a series of reactions (reaction series 4, not shown in Fig. 9-17) that yields as an initial product the five-carbon, one-phosphate sugar *ribulose 5-phosphate*. Another phosphate is removed from ATP and added to this product in reaction 5 to produce RuBP. This reaction replaces the RuBP used in reaction 1, and the cycle is ready to turn again.

Three turns of the C_3 cycle are required to yield one molecule of 3PGAL as a surplus. In three turns of the cycle through reaction 3 in Figure 9-17, six molecules of 3PGAL are formed (for a total of 18 carbons). Five of these molecules (totaling 15 carbons) enter the complex series that regenerates the three RuBP molecules used in the three turns. The remaining molecule of 3PGAL is surplus and can enter reaction pathways

yielding glucose, sucrose, starch, and a variety of other complex organic substances. The C_3 cycle can be considered as a "breeder" reaction because it makes a greater quantity of one of its own intermediates (3PGAL) than it uses.

A single turn of the cycle thus takes up one molecule of CO_2 and yields one unit of carbohydrate (CH_2O) as a product of the cycle. Three turns are required to form one surplus molecule of 3PGAL; six turns are required to make enough (CH_2O) units to synthesize a six-carbon sugar such as glucose.

For each turn of the cycle a total two ATP and two NADPH are used in reactions 2 and 3, and one additional ATP is used in reaction 5, for a total of three ATP and two NADH for each turn. Although one of the phosphates derived from ATP is attached to 3PGAL, this phosphate is eventually released as 3PGAL is converted into other substances. As net reactants and products, one complete turn of the cycle therefore includes:

$$CO_2 + 3ATP + 2NADPH \longrightarrow$$
$$(CH_2O) + 2NADP^+ + 3ADP + 3P_i \quad (9\text{-}4)$$

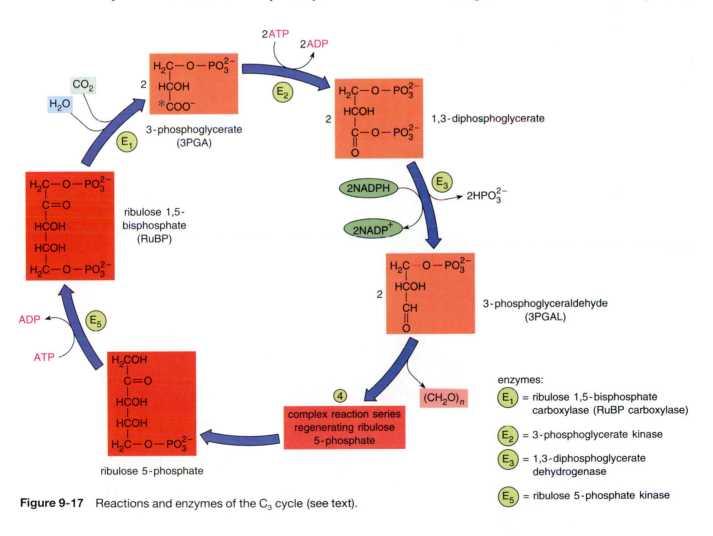

Figure 9-17 Reactions and enzymes of the C_3 cycle (see text).

The cycle is about 80% efficient in conserving the energy of the three ATP and two NADPH in the (CH_2O) carbohydrate product.

The Key Enzyme of the C_3 Cycle: RuBP Carboxylase
RuBP carboxylase, the enzyme catalyzing the first reaction of the C_3 cycle, is unique to photosynthetic organisms, including photosynthetic bacteria, cyanobacteria, algae, and plants. There are so many copies of the enzyme in chloroplasts that RuBP carboxylase may make up as much as 50% or more of the total protein of plant leaves. As such, it is probably the world's most abundant single protein. It is estimated to exist in a worldwide amount of some 40 million tons, equivalent to about 20 pounds for every human on earth! In this quantity the enzyme fixes about 100 billion tons of CO_2 into carbohydrates annually. Its abundance is probably an adaptation that compensates for the fact that the enzyme catalyzes CO_2 fixation relatively slowly, at rates of only about three molecules of CO_2 per enzyme per second.

RuBP carboxylase is assembled from two polypeptides. One, the *large subunit*, is a 56,000-dalton polypeptide encoded in DNA molecules included in the chloroplast (see below). The second, the *small subunit*, is a 14,000-dalton polypeptide encoded in the DNA of the plant cell nucleus. An active RuBP carboxylase enzyme is an octamer assembled from eight each of the large and small subunits, forming a complex with a total molecular weight of more than 550,000.

The position of RuBP carboxylase as the initial enzyme of the C_3 cycle makes it a key regulatory site. Among the several controls regulating RuBP carboxylase are two based on the amounts of ATP and NADPH available from the light reactions. NADPH is an allosteric regulator (see p. 84) for RuBP carboxylase. Unless NADPH is available to the allosteric site on the enzyme, RuBP carboxylase remains inactive. Unless the light reactions run, too little NADPH is present in the stroma to combine with the allosteric sites, and the C_3 pathway remains inactive. The control based on ATP/ADP concentrations is similar. As ATP concentrations rise and ADP or phosphate concentrations fall due to increases in the rate of the light reactions, the activity of RuBP carboxylase is stimulated. High concentrations of ADP or phosphate relative to ATP have the opposite effect, inhibiting the enzyme. Similar controls involving availability of ATP and NADPH also regulate the enzymes catalyzing reactions 2 and 3 in Figure 9-17 and other reactions of the C_3 cycle.

The controls linking activity of the C_3 cycle to the light reactions ensure that the dark reactions operate primarily or exclusively in the daytime, when ATP and NADPH are available for CO_2 fixation. Thus the term "light-independent reactions" often applied to the dark reactions is a misnomer in this regard—although the dark reactions can take place in darkness, the stringent requirements for activation of key enzymes of the C_3 pathway actually limit the dark reactions primarily to the daytime, when the light reactions are operating.

The controls limiting operation of the dark reactions at night probably evolved as a means to conserve ATP and oxidizable fuel molecules. In darkness, when the light reactions are inactive, plants become the bioenergetic equivalents of animals—plants must oxidize organic molecules as the energy source for ATP synthesis. Shutdown of the dark reactions at night conserves ATP and ensures that oxidizable fuels do not enter the C_3 pathway.

Another series of evolutionary adaptations compensates for a major defect in the activity of RuBP carboxylase that can greatly reduce the efficiency of photosynthesis. The defect stems from the fact that RuBP carboxylase can also work as an *oxygenase*, in a reaction that converts RuBP into 3-phosphoglycerate and phosphoglycolate (see also Fig. 9-28):

$$RuBP + O_2 \longrightarrow$$
$$\text{3-phosphoglycerate + phosphoglycolate} \quad (9\text{-}5)$$

The phosphogylcolate is subsequently dephosphorylated to glycolate, a toxic substance that is oxidized inside microbodies by the glycolate pathway. Among other effects, glycolate oxidation may convert CO_2 fixed in the dark reactions back to inorganic form in a process known as *photorespiration*. (Details of the glycolate cycle and photorespiration are presented in Supplement 9-2.)

Formation of Complex Carbohydrates and Other Substances from 3PGAL 3PGAL is the starting point for synthesis of a variety of more complex carbohydrates and polysaccharides. It may either remain inside chloroplasts for conversion into starch and other complex substances, or it may enter the surrounding cytoplasm to act as the input for various synthetic pathways. Alternatively, 3PGAL entering the cytoplasm may be oxidized in the second half of glycolysis, producing pyruvate that subsequently enters mitochondria for complete oxidation to CO_2 and H_2O (see p. 218 for details).

Sucrose, formed from a combination of glucose and fructose units, is the main form in which the products of photosynthesis circulate from cell to cell within most higher plants. Nutrients are stored in most higher plants as sucrose or starch or a combination of the two in proportions that depend on the plant species.

In addition to starch and other carbohydrates, chloroplasts can also synthesize amino acids, fatty acids, and lipids. The lipids synthesized inside chloroplasts include chlorophylls and carotenoids, which are assembled from simple precursors originating from the C_3 cycle. Protein synthesis, occurring on ribosomes suspended in the chloroplast stroma (see below), can also be detected inside chloroplasts, along with linkage of nucleotides into DNA and RNA. Chloroplasts, in fact, have synthetic capacity practically equivalent to entire cells.

The C_4 Cycle

During the 1960s, several groups working independently, including M. D. Hatch and C. R. Slack and H. P. Kortschak and his colleagues, discovered that certain plants carry out a series of reactions, the C_4 cycle, that circumvents problems caused by the oxygenase activity of RuBP carboxylase. (Hatch's Experimental Process essay on p. 266 describes his experiments leading to discovery of the C_4 pathway.)

The C_4 cycle (Fig. 9-18) uses a series of enzymes that initially combines CO_2 with a three-carbon molecule, *phosphoenolpyruvate*, eventually producing *malate*, a four-carbon acid. (The C_4 cycle derives its name from this four-carbon product.) An enzyme critical to the C_4 pathway, *phosphoenolpyruvate carboxylase*, catalyzes the step fixing CO_2. The C_4 cycle occurs instead of the C_3 cycle in cells in which oxygen is abundant. In these cells the relatively high oxygen content would stimulate the oxygenase activity of RuBP carboxylase if the C_3 cycle were operating. The malate product of the C_4 cycle then diffuses to deeper cell layers, where oxygen concentrations are lower. Here malate is oxidized to pyruvate, a three-carbon mole-

cule, with release of CO_2. The CO_2 enters the C_3 cycle in these cells, in which the relatively low oxygen concentration defeats the oxygenase activity of RuBP carboxylase. The pyruvate diffuses back to the first cell layer to enter another turn of the C_4 cycle. As a part of the reentry, pyruvate is converted to phosphoenolpyruvate at the expense of one molecule of ATP converted to AMP. Because two phosphates are removed from ATP for each turn of the cycle, making a molecule of glucose by the combined activities of the C4 and C3 pathways costs an additional 12 ATP/glucose. The C_4 cycle thus adds a significant increase in the amount of ATP required to synthesize sugars and other organic molecules in the dark reactions.

In spite of the price paid in ATP, the C_4 pathway provides an overall increase in photosynthetic efficiency at elevated temperatures, at which plants possessing only the C_3 pathway are penalized by the greatly increased oxygenase activity of RuBP carboxylase (the oxygenase activity of the enzyme increases under these conditions because the solubility of oxygen in plant tissues is higher at elevated temperatures). The crossover point seems to lie at about 30°C. Above this temperature, C_4 plants become significantly more efficient than C_3-limited plants; below 30°C, the additional ATP used by C_4 plants evidently makes C_3-limited species more efficient in spite of photorespiration.

The 30°C crossover point places C_4 plants at an advantage in the tropics and in temperate regions with elevated summer temperatures, as in the central United States. Several important tropical and temperate crop plants benefit from the C_4 adaptation, including sugar cane, corn, sorghum, and some pasture grasses. A group of highly successful weed pests, including bermuda and crab grass, are also C_4 plants.

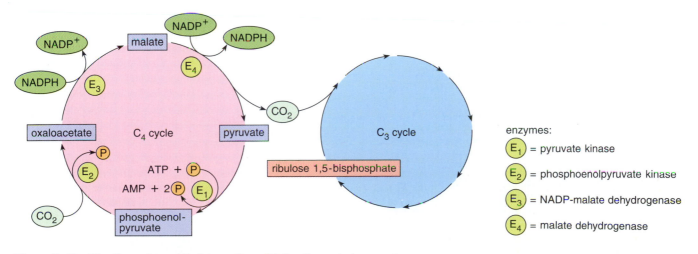

Figure 9-18 The C_4 cycle and its integration with the C_3 cycle (see text).

The C_4 Pathway: A Surprise Option for Photosynthetic Carbon Assimilation

Marshall D. Hatch

MARSHALL D. HATCH completed his undergraduate studies at Sydney University in 1954 and was awarded a Ph.D. from the same University in 1959. After two years post doctoral studies at the University of California, Davis, he joined the laboratory of the Colonial Sugar Refining Company in Brisbane, Australia, to study aspects of the biochemistry and physiology of sugarcane. In 1970 he moved to the Division of Plant Industry, CSIRO, in Canberra, where he is currently a Chief Research Scientist. Since 1965 he has worked almost exclusively on aspects of the mechanism and function of C_4 photosynthesis. He was elected to the Australian Academy of Science in 1975, the Royal Society in 1980, and as a Foreign Associate of the National Academy of Science of the United States in 1990. He was awarded the Rank Prize in 1981 and the International Prize for Biology in 1991.

By 1960, Melvin Calvin and colleagues Benson, Bassham, and many others had worked out the essential features of the path of CO_2 assimilation now known as the Calvin cycle, or C_3 pathway. The critical first step in this process is the assimilation of CO_2 by a reaction with the 5-carbon sugar ribulose 1,5-bisphosphate giving two molecules of the 3-carbon compound 3-phosphoglyceric acid. This compound is then transformed through many steps to give sucrose, starch, and various other metabolites.

The general rule in biochemistry is for uniformity rather than diversity so that biochemical mechanisms for particular processes are usually the same both within and between the different classes of living organisms. Therefore, there was no reason to expect that a photosynthetic process differing from the Calvin cycle might occur in plants. However, it is interesting to note, in retrospect, that there was in fact a considerable amount of physiological and anatomical evidence available at the time which pointed to that possibility.

In the early 1960s Dr. Roger Slack and I were working in Brisbane, Australia, in the research laboratory of the Colonial Sugar Refining Company on various aspects of sugarcane metabolism. We were aware, through private communication, of some unusual results of colleagues working on sugarcane in Hawaii. In these studies, initiated in the late 1950s, Hugo Kortschak and collaborators looked at the radioactive products formed when sugarcane leaves were allowed to assimilate radioactive carbon dioxide ($^{14}CO_2$) in the light. What they noted was that after short periods in $^{14}CO_2$ more radioactivity appeared in the 4-carbon (C_4) acids malate and aspartate than in the C_3 acid 3-phosphoglyceric acid. Of course the latter compound should be the first product labeled with radioactivity if the normal Calvin cycle was operating alone.

There were various possible explanations of this result, which Dr. Slack and I had often discussed. However,

to cut a long story short, when Kortschak and colleagues finally published their results in 1965[1] Dr. Slack and I set about trying to repeat these observations and resolve which of these options might be correct. To complete the background history, there is one other interesting twist to this story. Three or four years after starting our study on this process we discovered a report published in 1960 in an obscure Proceedings of a Russian agricultural institute. This report, by Russian scientist Yuri Karpilov, clearly described the predominant radioactive labeling of malate and aspartate in maize (corn) leaves exposed to $^{14}CO_2$ in the light. This observation was not followed up by Karpilov until several years after our first publication, however.

Our initial experiments were designed to trace the exact fate of carbon assimilated during photosynthesis. Following the general procedure developed in Calvin's studies, we exposed sugarcane leaves to $^{14}CO_2$ under natural conditions in the light and then stopped photosynthesis after various periods up to 90 seconds by boiling or freezing leaves in killing mixtures containing ethanol. The leaf contents were then extracted and the compounds containing radioactivity were identified by paper chromatography. Later, these compounds were purified and degraded to find out which particular carbons were labeled with radioactivity. These studies confirmed the original observation of Kortschak and colleagues that in shorter times most of the radioactive carbon incorporated from $^{14}CO_2$ appears in malate and aspartate. In addition, we showed that the first product formed is actually an unstable dicarboxylic acid, oxaloacetic acid, and that malate and aspartate were formed from oxaloacetate.

We went on to show that this radioactivity appears exclusively in the C-4 carboxyl of these C_4 acids. So called pulse-chase experiments were also performed where the fate of ^{14}C incorporated in a short pulse in $^{14}CO_2$ is followed in a subsequent "chase" period after leaves are transferred back to air containing normal non-radioactive CO_2. The results of these pulse-chase experiments appearing in our first publication on this subject[2] are shown in Fig. A. It can be inferred from this data that the C-4 carboxyl of C_4 acids is transferred to the C-1 carboxyl of 3-phosphoglycerate, which is then metabolized to hexose phosphates and finally sucrose and starch. The pattern of radioactive labeling of 3-phosphoglycerate and sugar molecules was consistent with incorporation via a reaction similar to the one catalysed by the Calvin cycle carboxylase, ribulose 1,5-bisphosphate carboxylase. Several control experiments were conducted to make sure that the results we got were due to photosynthesis and were not an artifact of the way leaves were killed or extracted.

Our conclusions were summarized in a scheme published in the original paper[2] and reproduced here (Fig. B). This proposes a novel path of carbon assimilation differ-

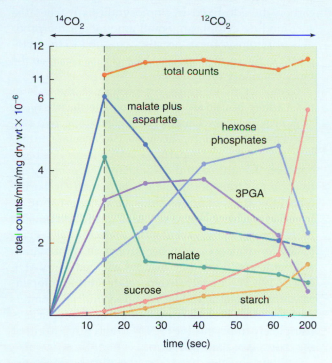

Figure A Changes in radioactive carbon (^{14}C) content in photosynthetic intermediates and products in sugarcane leaves following a 15-second pulse in $^{14}CO_2$ and then a "chase" in normal ($^{12}CO_2$) air. (Data from Hatch and Slack [1966].[2])

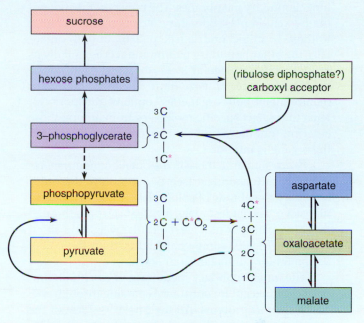

Figure B Proposal for the pathway of photosynthetic CO_2 fixation in sugarcane leaves as it was presented in our final paper on C_4 photosynthesis.[2]

ing substantially from the Calvin cycle. The simplest interpretation of the data was that CO_2 is initially assimilated by a carboxylation reaction involving a 3-carbon acceptor, possibly pyruvate or phosphoenolpyruvate, giving rise to oxaloacetate and then other C_4 dicarboxylic acids as first products. The carbon assimilated into the C-4 carboxyl is apparently transferred to appear in 3-phosphoglycerate and then into sugar phosphates, sucrose, and starch. The pattern of slower secondary incorporation of radioactive carbon into the other carbons (C-1, C-2, and C-3) of C_4 acids was consistent with it originating by exchange between 3-phosphoglycerate and phosphoenolpyruvate. We named this process the C_4 dicarboxylic acid pathway of photosynthesis, since abbreviated to the C_4 pathway. The Calvin cycle is now commonly referred to as the C_3 pathway on the grounds that its first product is a 3-carbon compound.

In the following three or four years our studies elaborated on these findings and showed that the pathway operated in a number of other higher plant species. We identified several of the key enzymes involved. For instance, we showed that phosphoenolpyruvate carboxylase was the enzyme responsible for the primary carboxylation reaction giving oxaloacetate. Amongst the others were two previously undiscovered enzymes, pyruvate P_i dikinase (which converts pyruvate to phos-

phoenolpyruvate) and NADP-malate dehydrogenase (which reduces oxaloacetate to malate). It was also shown that the carboxylation reaction leading to C_4 acid formation occurs in mesophyll cells while the decarboxylation reaction leading to transfer of carbon to 3-phosphoglycerate occurs in adjacent bundle sheath cells. We reviewed these early studies in 1970.[3]

In the intervening time the full complexity of this process has been revealed. As outlined in a recent review,[4] there are three distinct variants of C_4 photosynthesis and these options are species-specific. We now know that the function of the reactions unique to C_4 photosynthesis is to concentrate CO_2 in bundle sheath cells.[4] The primary purpose is to suppress the oxygenase reaction catalysed by Rubisco, and hence to eliminate the process known as photorespiration.[4] This endows C_4 plants with advantages in terms of photosynthetic capacity and water use efficiency. C_4 plants are at a particular advantage in drier and especially hotter locations. The recognition of the C_4 process and its physiological significance has been a powerful influence in understanding the factors affecting plant dry matter production and growth.

References

[1]Kortschak, H. P.; Hartt, C. E.; and Burr, G. O. *Plant Physiol.* 40: 209 (1965).

[2]Hatch, M. D., and Slack, C. R. *Biochem. J.* 101:103 (1966).

[3]Hatch, M. D., and Slack, C. R. *Annu. Rev. Plant Physiol.* 21:141 (1970).

[4]Hatch, M. D. *Biochim. Biophys. Acta* 895:81 (1987).

CHLOROPLAST TRANSPORT

All substances entering and leaving chloroplasts must cross the double barrier presented by the outer and inner boundary membranes. The transport functions of the two chloroplast boundary membranes are similar to those of mitochondria (see p. 236). In both organelles the outer boundary membrane acts essentially as a nonspecific molecular sieve that admits all molecules below a certain size limit. The inner boundary membrane of both organelles contains transport systems that select specific molecules from among those entering the intermembrane space for transport into the organelle interior.

Porins similar to those in the outer boundary membrane of mitochondria (see p. 236) allow the outer boundary membrane of chloroplasts to act as a molecular sieve. The largest molecules that can pass through the chloroplast porins have molecular weights of about 10,000 to 13,000, somewhat higher than the 6,000-dalton limit of mitochondrial porins.

The inner boundary membrane is essentially impermeable except for substances passing through specific transport proteins. All the carrier systems of the inner boundary membrane appear to be driven by favorable concentration gradients rather than by active transport. This is in distinct contrast to mitochondria, in which much of the transport is active, driven directly or indirectly by the H^+ gradient established by electron transport.

Inorganic phosphate is required in quantity by chloroplasts because the products of photosynthesis are released to the surrounding cytoplasm primarily in the form of phosphorylated molecules such as 3PGAL. Phosphate is transported by a *phosphate exchange carrier*, a protein that exchanges inorganic phosphate from the outside for a three-carbon phosphorylated sugar made on the inside on a one-for-one basis. If phosphate ions are unavailable for the exchange, the products of photosynthesis cannot be exported and both the light and dark reactions are strongly inhibited. The phosphate exchange carrier therefore provides the means by which most of the carbon dioxide fixed in photosynthesis reaches the surrounding cytoplasm. Small amounts of the products of photosynthesis are also exported to the cytoplasm in the form of glucose by a *glucose carrier* in the inner boundary membrane.

Another major transport protein of the chloroplast inner boundary membrane, the *dicarboxylate exchange carrier*, exchanges acids containing two carboxyl groups—malate, succinate, oxaloacetate, fumarate, glutamate, and aspartate—on a one-for-one basis between the stroma and cytoplasm. The primary activity of this transport protein is apparently to participate in *shuttles* of various kinds. One of these shuttles transfers electrons between NADP inside and outside the chloroplast by exchanging malate in the cytoplasm for oxaloacetate in the stroma, in combination with side reactions that oxidize and reduce NADP (Fig. 9-19). The malate/oxaloacetate shuttle compensates for the fact that, like its mitochondrial counterpart, the inner boundary membrane of chloroplasts is completely impermeable to NADP and NAD in either the reduced or oxidized form.

Other shuttles based on the dicarboxylate exchange carrier participate in the movement of NH_4^+ or, in C_4 plants, of bound CO_2 into chloroplasts. In the latter shuttle, CO_2 bound into malate enters chloroplasts in exchange for another dicarboxylate; once inside, malate is oxidized to pyruvate, delivering CO_2 to the chloroplast interior. The pyruvate is subsequently exported to the cytoplasm for completion of the C_4 cycle by another carrier protein specialized for this activity.

Chloroplasts in most plants appear to be highly impermeable to either ADP or ATP. If so, ATP for cytoplasmic activities cannot be obtained directly from

Figure 9-19 The dicarboxylate exchange carrier, a shuttle that exchanges acids containing two carboxyl groups on a one-for-one basis between the chloroplast interior and the cytoplasm. The shuttle can, in effect, move high-energy electrons from NADPH inside the chloroplast to $NADP^+$ in the cytoplasm.

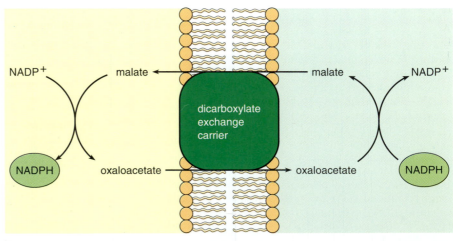

b

the light reactions in these plants. However, since 3PGAL can enter the cytoplasm via the phosphate exchange carrier for oxidation, the reactions inside chloroplasts may lead indirectly to synthesis of cytoplasmic ATP through glycolysis, pyruvate oxidation, and the citric acid cycle.

Another transport protein of interest is the *glycolate carrier*. This carrier exports glycolate produced through the oxygenase activity of RuBP carboxylase.

The reactions of photosynthesis enable eukaryotic plants to use the energy of sunlight to drive assembly of organic molecules from simple inorganic precursors. Some of the organic molecules are used by plants themselves as fuels, particularly during periods of reduced light or at night when photosynthesis is inactive. The complex molecules synthesized by plants also form the primary energy source for animals, fungi, and other organisms that live by eating plants. The photosynthetic systems of plants operate with such efficiency that not only plants themselves but most living organisms are supported by the light energy captured and converted into chemical energy inside chloroplasts. (Photosynthesis in prokaryotes is outlined in Supplement 9-1.)

For Further Information

DNA, ribosomes, and protein synthesis in chloroplasts, *Ch. 21*
Electron transport and ATP synthesis in mitochondria, *Ch. 8*
Evolutionary origins of chloroplasts, *Ch. 26*
Mechanisms delivering proteins to chloroplasts, *Ch. 20*
Membrane structure, *Ch. 4*
Microbodies (Peroxisomes), *Supplement 8-3*
Mitchell's chemiosmotic hypothesis, *Ch. 8*
Porins in mitochondria, *Ch. 8*
  in bacteria, *Ch. 5*
Transport, active and passive, *Ch. 5*

Suggestions for Further Reading

Anderson, J. M. 1992. Cytochrome b_6f complex: Dynamic molecular organization, function, and acclimation. *Photosynth. Res.* 34:341–357.

Andersson, B., and Styring, S. 1991. Photosystem I: molecular organization, function, and acclimation. *Curr. Top. Bioenerget.* 16:2–81.

Arnon, D. I. 1991. Photosynthetic electron transport: Emergence of a concept, 1949–1959. *Photosynth. Res.* 29:117–131.

Babcock, G. T. 1993. Proteins, radicals, isotopes, and mutants in photosynthetic oxygen evolution. *Proc. Nat. Acad. Sci.* 90:10893–10895.

Beck, W. F., and de Paula, J. C. 1989. Mechanism of photosynthetic water oxidation. *Ann. Rev. Biophys. Biophys. Chem.* 18:25–46.

Bendich, A., and Olson, J. A. 1989. Biological actions of carotenoids. *FASEB J.* 3:1927–1932.

Chitnis, P. R., and Thornber, J. P. 1988. The major light-harvesting complex of photosystem II: aspects of its molecular and cell biology. *Photosynth. Res.* 16:41–63.

Cramer, W. A., Furbacher, P. N., Szczepaniek, A., and Tae, G.-S. 1991. Electron transport between photosystem II and photosystem I. *Curr. Top. Bioenerget.* 16:180–222.

Deisenhofer, J., and Michel, H. 1991. Structures of bacterial reaction centers. *Ann. Rev. Cell Biol.* 7:1–23.

Douce, R., and Joyard, J. 1990. Biochemistry and function of the plastid envelope. *Annu Rev. Cell Biol.* 6:173–216.

Edwards, G., and Walker, D. A. 1983. C_3, C_4. Berkeley: University of California Press.

Feber, G., Allen, J. P., Okamura, M. Y., and Rees, D. C. 1989. Structure and function of bacterial photosynthetic reaction centers. *Nature* 339:111–116.

Flügge, U.-I., and Heldt, H. W. 1991. Metabolite translocators of the chloroplast envelope. *Ann. Rev. Plant Physiol. Plant Molec. Biol.* 42:129–144.

Ghanotakis, F., and Yocum, C. F. 1990. Photosystem II and the oxygen-evolving complex. *Ann. Rev. Plant Physiol. Plant Molec. Biol.* 41:255–276.

Glaser, E., and Norling, B. 1991. Chloroplast and plant mitochondrial ATP synthesis. *Curr. Top. Bioenerget.* 16:223–263.

Golbeck, J. H. 1992. Structure and function of photosystem I. *Ann. Rev. Plant Physiol. Plant Molec. Biol.* 43:293–324.

Golbeck, J. H. 1993. Shared thematic elements in photochemical reaction centers. *Proc. Nat. Acad. Sci.* 90:1642–1646.

Govindjee and Coleman, W. J. 1990. How plants make O_2. *Sci. Amer.* 262:50–58 (February).

Grossman, A. R., Schaefer, M. R., Chiang, G. G., and Collier, J. L. 1993. The phycobilisome, a light-harvesting complex responsive to environmental conditions. *Microbiol. Rev.* 57:725–749.

Hatch, M. D. 1987. C_4 photosynthesis: A unique blend of modified biochemistry, anatomy, and ultrastructure. *Biochim. Biophys. Acta* 895:81–106.

Hoober, J. K. 1984. *Chloroplasts.* New York: Plenum.

Hope, A. B. 1993. The chloroplast cytochrome *bf complex*: A critical focus on function. *Biochim. Biophys. Acta* 1143: 1–22.

Journal of Bioenergetics and Biomembranes, vol. 24, part 5: Series of reviews on F_oF_1-ATPase.

Khorana, H. G. 1988. Bacteriorhodopsin, a membrane protein that uses light to translocate protons. *J. Biolog. Chem.* 263:7439–7442.

Knaff, D. B. 1990. The cytochrome bc_1 complex of photosynthetic bacteria. *Trends Biochem. Sci.* 15:289–290.

Melis, A. 1991. Dynamics of photosynthetic membrane structure and function. *Biochim. Biophys. Acta* 1058: 87–106.

Nelson, N. 1992. Evolution of organellar proton-ATPases. *Biochim. Biophys. Acta* 1100:109–124.

Nelson, T., and Langdale, J. A. 1992. Developmental genetics of C_4 photosynthesis. *Ann. Rev. Plant Physiol. Plant Molec. Biol.* 43:25–47.

Oesterhelt, D., and Tittor, J. 1989. Two pumps, one principle: Light-driven ion transport in halobacteria. *Trends Biochem. Sci.* 14:57–61.

Ogren, W. L. 1984. Photorespiration: Pathways, regulation, and modification. *Ann. Rev. Plant Physiol.* 35: 415–442.

Pedersen, P. L., and Amzel, L. M. 1993. ATP synthases: Structure, reaction center, mechanism, and regulation of one of nature's most unique machines. *J. Biolog. Chem.* 268:9937–9940.

Portis, A. R. 1992. Regulation of ribulose 1,5-bisphosphate carboxylase/oxygenase activity. *Ann. Rev. Plant Physiol. Plant Molec. Biol.* 43:415–437.

Spreitzer, R. J. 1993. Genetic dissection of rubisco structure and function. *Ann. Rev. Plant Physiol. Plant Molec. Biol.* 43:411–434.

Witt, H. T. 1991. Functional mechanism of water splitting photosynthesis. *Photosynth. Res.* 29:55–77.

Youvan, D. C., and Marrs, B. L. 1987. Molecular mechanisms of photosynthesis. *Sci. Amer.* 256:42–48.

Review Questions

1. What structures are visible in chloroplasts in the electron microscope? What is a thylakoid? What is the relationship between thylakoids, grana, and stromal lamellae?

2. What substances link the light and dark reactions of photosynthesis?

3. What is a photon? What is the relationship between the wavelength and energy of a photon?

4. What makes some molecules appear pigmented or colored? What happens when light is absorbed by a pigmented molecule? What molecules absorb light in chloroplasts? Outline their structures. What wavelengths do they absorb? Why do leaves appear green?

5. What is the excited state of an electron? The ground state? What is a reaction center? How is light energy passed to a reaction center? How is light energy trapped in a reaction center? What is chlorophyll P700 and P680?

6. Outline the structure of photosystems I and II. Trace the flow of electrons within the photosystems. What are light-harvesting complexes, and what is their relationship to the photosystems?

7. What kinds of carriers are present in the Z pathway? Describe two methods by which the carriers of the Z pathway were sequenced.

8. Trace the electron flow through the Z pathway. What is cyclic photosynthesis? Noncyclic photosynthesis?

9. Outline the structure of the CF_oCF_1-ATPase.

10. How are the carriers and complexes of the Z pathway organized in thylakoid membranes? How do electrons flow between different elements of the system?

11. How is the organization of the Z pathway in thylakoids related to H^+ pumping and the H^+ gradient produced by electron transport?

12. What experiments support the proposal that Mitchell's chemiosmotic mechanism operates in photosynthesis?

13. How were the dark reactions traced? How are ATP and NADPH used in the dark reactions? Where does CO_2 enter the dark reactions?

14. Outline the reactions of the C_3 cycle.

15. Describe the characteristics of RuBP carboxylase. What is photorespiration? How is photorespiration related to the characteristics and activity of RuBP carboxylase?

16. What is the C_4 cycle? How is the C_4 cycle related to photorespiration and the properties of RuBP carboxylase?

17. What major types of molecules are synthesized in photosynthesis?

18. Where are the light and dark reactions located in chloroplast structures?

Supplement 9-1
Photosynthesis in Prokaryotes

The two groups of photosynthetic prokaryotes, the bacteria and cyanobacteria, differ fundamentally in the pathways of their photosynthetic reactions. The photosynthetic bacteria, the *green* and *purple bacteria*, have comparatively primitive systems that cannot use H_2O as an electron donor and do not evolve oxygen in photosynthesis. Nevertheless, in terms of their structure and internal pathways of electron transport, the photosystems of bacteria may resemble either photosystem I or II of eukaryotic plants.

The cyanobacteria, although typically prokaryotic in cellular organization, have two photosystems equivalent to eukaryotic photosystems I and II and carry out photosynthesis by essentially the same mechanisms as the eukaryotic plants. Cyanobacteria can use H_2O as an electron donor and evolve oxygen in photosynthesis.

All photosynthetic prokaryotes fix CO_2 through the activity of RuBP carboxylase working in a C_3 cycle as in the eukaryotic algae and higher plants. Many of these prokaryotes also possess enzymes of the C_4 pathway, indicating that they may circumvent photorespiration by evolutionary strategies similar to those of eukaryotes. In addition to the C_3 and C_4 pathways, some photosynthetic bacteria are able to fix CO_2 by reactions that essentially reverse the CO_2-generating reactions of pyruvate oxidation or the citric acid cycle (see p. 218 for details of these reactions).

The Photosynthetic Bacteria

The purple and green photosynthetic bacteria comprise less than 50 known species. Purple bacteria occur in two recognized subgroups, the *purple sulfur bacteria*, which use sulfur-containing compounds such as H_2S as electron donors for noncyclic photosynthesis, and the *purple nonsulfur bacteria*, which use organic substances such as malate and succinate as electron donors. Different species among the green bacteria may use either inorganic sulfur-containing compounds or nonsulfur organic molecules as electron donors for noncyclic photosynthesis. However, the different green bacterial types have not been formally classified as sulfur and nonsulfur subgroups.

The pigments, light-harvesting antennas, photosystems, electron carriers, and the F_oF_1-ATPase carrying out the light reactions in photosynthetic bacteria are embedded in the plasma membrane or in tube- or saclike membranes in the cytoplasm (Fig. 9-20).

Bacterial Photosynthetic Pigments   Both purple and green photosynthetic bacteria use a distinct family of chlorophylls, the *bacteriochlorophylls* (Fig. 9-21) as primary light-absorbing pigments. The bacteriochlorophylls, like the chlorophylls of eukaryotic plants, are built on a tetrapyrrole ring containing a central magnesium atom. They differ only in minor substitutions in side groups attached to the ring.

Bacteriochlorophylls absorb light most strongly in the near-ultraviolet and far-red region of the spectrum and transmit most of the visible wavelengths. As a result, they contribute little color to either bacterial group. The distinctive colors of the green and purple photosynthetic bacteria come from different carotenoids occurring as accessory pigments.

The bacteriochlorophylls and carotenoids of photosynthetic bacteria are organized with proteins into photosystems and light-harvesting complexes as in the eukaryotic plants. Within the photosystems, specialized bacteriochlorophylls form a reaction center that raises electrons to excited energy levels and passes them on to primary acceptors. The primary acceptors are followed by short sequences of electron carriers within the photosystems that, in different purple and green bacteria, may resemble the systems of either eukaryotic photosystem I or II. Because the photosystems of bacteria cannot be reduced by the very low-energy electrons released from water, electron donors with higher reducing power must be used. These donors, including H_2S and other sulfur-containing compounds and complex organic molecules such as malate and succinate, differ in the major green and purple bacterial groups.

Electron Transport in Bacterial Photosynthesis   Electrons flow both cyclically and noncyclically in the electron transport systems of purple and green bacteria. Although the carriers active in bacterial electron transport include most of the major types active in eukaryotic photosynthesis—quinones, cytochromes, and Fe/S proteins—there is great variety in the types and combinations of these carriers. Many different *b*- and *c*-type cytochromes, most of them distinct from their eukaryotic counterparts, carry electrons in the bacterial systems; even quinones occur in four or more types. Complicating the situation is the fact that some carriers transport electrons in both photosynthetic and respiratory pathways.

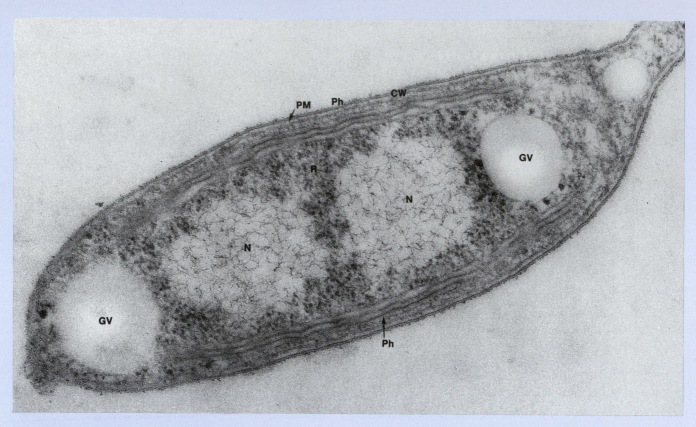

Figure 9-20 A photosynthetic bacterium. Membranes carrying out photosynthesis (Ph) are visible in the cytoplasm. N, nucleoid; PM, plasma membrane; R, ribosomes; GV, gas vacuole; CW, cell wall. (Courtesy of W. C. Trentini and the American Society for Microbiology, from *J. Bact.* 93:1699 [1967].)

◄ **Figure 9-21** Bacteriochlorophylls *a* and *b*. An additional —H occurs at the 4-carbon in bacteriochlorophyll *a*. One form of bacteriochlorophyll *a* is linked to a different hydrophobic, long-chain alcohol instead of the phytol chain. Other side-group substitutions produce bacteriochlorophylls *c*, *d*, and *e*. Purple bacteria contain bacteriochlorophylls *a* and *b*. Bacteriochlorophylls *c*, *d*, and *e* predominate in the green bacteria in combination with small quantities of bacteriochlorophyll *a*.

Cyclic electron flow builds up an H^+ gradient that is used as an energy source for ATP synthesis. Noncyclic flow leads from an electron donor directly or indirectly to NAD^+ rather than $NADP^+$ as in cyanobacteria and eukaryotic plants. After reduction, NADH may donate its electrons to synthetic reactions requiring a reduction, such as those of the dark reactions fixing CO_2, or to an electron transport chain producing an H^+ gradient for ATP synthesis.

The Light Reactions in the Purple Bacteria The single photosystem of purple bacteria is built around a reaction center containing either bacteriochlorophyll *a* or *b* and a short series of electron carriers closely resembling those of photosystem II of eukaryotic plants (see Fig. 9-9). The similarities extend to two polypeptides of the bacterial reaction center, which are related in amino acid sequence to two polypeptides at the reaction center of eukaryotic photosystem II. The similarities in structure and function suggest that cyanobacterial and eukaryotic photosystem II may have evolved from the purple bacterial photosystem by development of the polypeptides carrying out the water-splitting reaction (see p. 253).

The reaction center of purple bacteria consists of a specialized pair of bacteriochlorophyll *b* molecules. After excitation in the reaction center (see Fig. 9-9), electrons flow to *bacteriopheophytin* b, which resembles bacteriochlorophyll *b* without the central magnesium atom. From bacteriopheophytin *b*, electrons flow through two quinones, Q_A and Q_B. At this point the electrons pass from the photosystem to carriers of the electron transport system. In most purple bacteria the bacteriochlorophyll molecules at the reaction center undergo a conspicuous change in absorption at a wavelength of 870 or 960 nm, depending on the species, as they undergo cycles of oxidation and reduction in connection with the excitation of electrons. The reaction center bacteriochlorophylls of these bacteria are identified accordingly as *P870* or *P960*.

Electrons may flow cyclically or noncyclically around the single photosystem of the purple sulfur bacteria. In cyclic electron transport (Fig. 9-22a), electrons released from the photosystem enter a quinone pool. Although variations are noted in different purple bacteria, in many species electrons are later transferred from the quinone pool to a b/c_1 complex (see pp. 228 and 255). Like its chloroplast and mitochondrial counterparts, the bacterial b/c_1 complex contains a *b*-type and *c*-type cytochrome, linked with an iron-sulfur protein and a group of polypeptides. Electron flow through the bacterial b/c_1 complex pumps H^+ across the bacterial membrane, building an H^+ gradient linked to electron transport as in eukaryotic systems.

In most purple bacteria, electrons flow from the b/c_1 complex to another *c*-type cytochrome, cytochrome c_2, a peripheral membrane protein that takes a position in bacterial cyclic electron flow equivalent to

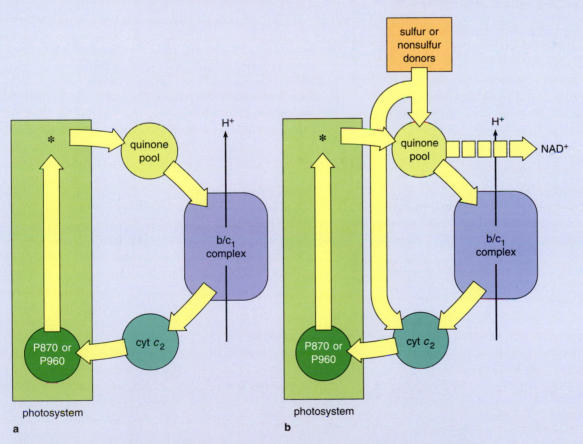

Figure 9-22 Photosynthetic electron transport in the purple bacteria. **(a)** Cyclic electron transport; **(b)** noncyclic electron transport. Electron donors for noncyclic photosynthesis may be sulfur-containing compounds such as H_2S or nonsulfur organic substances such as succinate. The asterisk indicates the excited form of the photosystem.

plastocyanin in eukaryotic plants. From cytochrome c_2, electrons return at lower energy levels to the reaction center of the single photosystem. After another energy boost through light absorption, they may repeat the cyclic pathway.

In noncyclic flow (Fig. 9-22b), electrons derived from various sulfur or nonsulfur donors, depending on their energy level, may be passed by a carrier (usually a cytochrome) to the photosystem and then to the quinone pool, or they may directly enter the quinone pool. In either case, electrons in the quinone pool initially contain too little energy to directly reduce NAD^+. Some electrons in the pool, however, evidently receive an additional energy boost from the membrane potential built up by cyclic electron transport. This possibility is supported by the observation that uncoupling agents, which make the membrane "leaky" to H^+ and destroy the H^+ gradient, completely block NAD^+ reduction.

Noncyclic flow results in one-way transfer of electrons from donor substances to NAD^+. The NADH produced by the reduction, like the NADPH produced in eukaryotic systems, provides a source of electrons for reductions elsewhere in the cell, as in the dark reactions fixing CO_2 into carbohydrates.

The electrons carried by NADH can also enter electron transport linked to ATP synthesis. The same F_oF_1-ATPase active in oxidative phosphorylation (see p. 240) uses the H^+ gradient built up by photosynthetic electron transport as an energy source for ATP synthesis in the purple photosynthetic bacteria. The ATP produced by the bacterial F_oF_1-ATPase, along with the NADH formed by noncyclic photosynthesis, provides the energy and reducing power for CO_2 fixation in the dark reactions.

The Light Reactions in Green Bacteria   The photosynthetic systems of green photosynthetic bacteria, although less well known than those of purple bacteria, appear to fall into two groups. One group is anaerobic and possesses a photosystem resembling photosystem II of the eukaryotic plants. The second group is aerobic, with a photosystem similar to eukaryotic photosystem I.

Electron transport around the photosystem of anaerobic green bacteria, which contains quinones as internal electron carriers like those of eukaryotic photosystem II, appears to be primarily cyclic (Fig. 9-23). Within the photosystem of anaerobic green bacteria, bacteriochlorophyll *a* molecules forming the reaction center change in absorbance at a wavelength of 870 nm and are identified as *P870*. Following excitation in the reaction center, electrons flow through a series of internal carriers including bacteriopheophytin and iron-associated quinones as in the purple

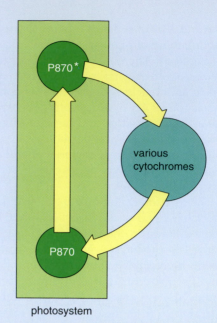

Figure 9-23   Photosynthetic electron flow in anaerobic green bacteria, which progresses primarily or exclusively by a cyclic pathway (see text).

bacteria or eukaryotic photosystem II. Only a few cytochromes occur in the electron transport system of these bacteria. The connection between this electron flow and ATP synthesis or reduction of NAD^+ remains to be established.

The photosystem of the aerobic green bacteria contains specialized bacteriochlorophyll *a* molecules absorbing light at 840 nm. These molecules, identified as *P840*, pass electrons to a primary acceptor and a chain of Fe/S centers, in a pattern similar to eukaryotic photosystem I. The electron transport system conducting excited electrons from the photosystem varies greatly among different aerobic green bacteria but often contains a b/c_1 complex, ferredoxin, and the ferredoxin-NAD oxidoreductase complex.

Figure 9-24 shows patterns by which aerobic green bacteria may accomplish either cyclic or noncyclic electron flow. Because the b/c_1 complex is present in the system, H^+ is pumped across the membrane housing the system each time electrons flow through this carrier in either cyclic or noncyclic flow around the photosystem.

In cyclic flow (Fig. 9-24a), electrons pass to ferredoxin after excitation and movement through the internal carriers of the photosystem. The direct reduction of ferredoxin as the first carrier reflects the fact that electrons are excited to significantly higher energy levels by the photosystem in these bacteria.

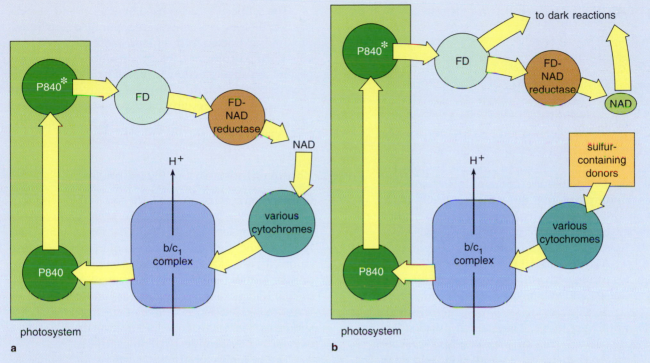

Figure 9-24 Cyclic (a) and noncyclic (b) electron flow in the aerobic green photosynthetic bacteria (see text). FD, ferredoxin; FD-NAD reductase, ferredoxin-NAD$^+$ oxidoreductase.

Electrons then flow to NAD$^+$ via a ferredoxin-NAD oxidoreductase complex, which may contain FAD as an internal carrier as in the equivalent complex of higher plants. From NADH, electrons are transferred via one or more cytochromes to the b/c_1 complex and then back to the P840 molecules at the reaction center of the photosystem. Transfer from the b/c_1 complex to the photosystem may be direct or may occur via additional cytochromes; the cytochromes on this side of the b/c_1 complex, like those delivering electrons to the complex from NADH, are generally types that do not occur in eukaryotic systems. Because the b/c_1 complex is present in the loop, H$^+$ is pumped across the membrane housing the system each time electrons cycle around the photosystem.

Noncyclic electron flow (Fig. 9-24b) uses electrons removed from inorganic sulfur compounds, including many of the substances used as donors by the purple sulfur bacteria. Electrons pass from the donors through one or more of a variety of cytochromes to reach the b/c_1 complex. At the end of the noncyclic pathway, after absorbing energy in the photosystem, electrons are delivered at high energy levels to ferredoxin as in eukaryotic photosynthesis. The electrons may remain with ferredoxin or be passed on to NAD. Either carrier may donate electrons to the dark reactions. Or, electrons may be passed from NAD to the respiratory electron transport system leading to oxygen as final electron acceptor.

As in purple bacteria, the same F$_o$F$_1$-ATPase active in oxidative phosphorylation uses the H$^+$ gradient established by photosynthetic electron transport as the energy source for ATP synthesis. All the components of the light reactions are associated with the plasma membrane in the green bacteria.

The Bacterial Dark Reactions The C$_3$ and C$_4$ pathways fixing CO$_2$ in purple and green bacteria proceed by the same mechanisms as in eukaryotes. Green bacteria can also fix CO$_2$ into organic molecules by reactions that reverse oxidative steps in pyruvate oxidation or the citric acid cycle. These reactions are catalyzed by unique enzymes that use ferredoxin, reduced in noncyclic photosynthesis, as a donor of high-energy electrons for CO$_2$ fixation.

Photosynthesis in Halobacteria One exceptional bacterial group, the *halobacteria*, possesses an incomplete and primitive photosynthetic mechanism that differs radically from those of the purple and green photosynthetic bacteria (see also p. 233). The halobacteria have no light-harvesting antennas, no photosystems, and no light-driven electron transport system. Instead, the photosynthetic membranes contain a light-absorbing molecule, *bacteriorhodopsin* (see Fig. 4-5), consisting of a polypeptide chain bearing the light-absorbing unit *retinal* (Fig. 9-25), a molecule almost identical to the visual pigment of animals.

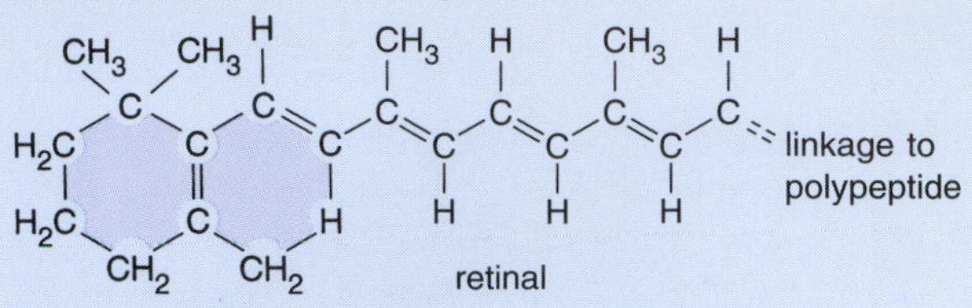

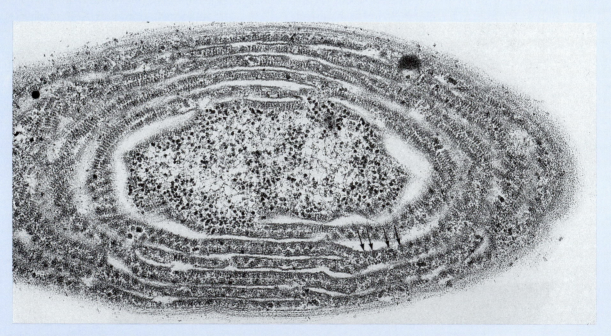

Figure 9-25 Retinal, the light-absorbing unit of bacteriorhodopsin.

Bacteriorhodopsin responds to light by pumping H^+ across the membrane containing the complex. During a pumping cycle the retinal unit alternately picks up and releases a hydrogen. The H^+ gradient established by light-induced pumping through bacteriorhodopsin drives ATP synthesis by a membrane-bound F_oF_1-ATPase as in other bacteria.

Studies conducted with isolated bacteriorhodopsin by E. Racker and W. Stoeckenius supplied some of the most compelling evidence supporting Mitchell's chemiosmotic hypothesis. They added purified bacteriorhodopsin to artificial phospholipid vesicles; while illuminated, the vesicles pumped H^+ and established a measurable H^+ gradient. Addition of an F_oF_1-ATPase to the vesicles allowed the system to synthesize ATP in response to the gradient, all in accordance with expectations of the Mitchell hypothesis.

The artificial vesicles used by Racker and Stoeckenius in their experiments were assembled from phospholipids extracted from a plant, and the F_oF_1-ATPase was purified from beef heart mitochondria. The experiments thus created a functional light-driven, ATP-synthesizing system with molecular components isolated from a bacterium (bacteriorhodopsin), a plant (the membrane phospholipids) and an animal (the F_oF_1-ATPase)!

Photosynthesis in the Cyanobacteria

Cyanobacteria carry out the light and dark reactions of photosynthesis by essentially the same pathways as eukaryotic algae and higher plants. Photosystems I and II resembling their eukaryotic counterparts, including chlorophyll *a* rather than bacteriochlorophylls, are present in these organisms. The cyanobacterial photosystem II includes an enzymatic assembly that allows water to be split as the source of electrons for photosynthesis. As a consequence, cyanobacteria release oxygen as a by-product of photosynthesis.

The cyanobacterial photosystems are connected into a Z pathway of the eukaryotic type. With the exception of a few cyanobacteria that substitute cytochrome c_2 for plastocyanin, the Z pathway in these prokaryotes includes the same electron carriers and

Figure 9-26 Photosynthetic vesicles in the cytoplasm of the cyanobacterium *Synechococcus*. Numerous phycobilisomes (arrows), the light-harvesting antennae of cyanobacteria, are attached to the vesicle membranes. × 51,000. (Courtesy of M. R. Edwards, New York State Department of Health, from *J. Cell Biol.* 50:896 [1971], by permission of the Rockefeller University Press.)

complexes as eukaryotic systems. The dark reactions in cyanobacteria also proceed by the same pathways used by the eukaryotic plants.

The genes encoding the F_oF_1-ATPase of cyanobacteria were recently isolated and sequenced by A. L. Cozens and J. E. Walker, S. E. Curtis, and others (see also p. 231). The sequences reveal that the cyanobacterial F_oF_1-ATPase is more closely related to the CF_oCF_1-ATPase of chloroplasts than to the F_oF_1-ATPase of bacteria.

Molecules carrying out the light reactions in cyanobacteria are bound to saclike membranes suspended in the cytoplasm (Fig. 9-26; see also Fig. 1-3). Thus the location of photosynthetic membranes in cyanobacteria is similar to the bacterial arrangement, even though the reactions of photosynthesis proceed according to the eukaryotic pattern.

One significant difference between cyanobacterial and most eukaryotic systems is in the form taken by the light-harvesting antennas. Chlorophyll *b* and the LHC I and II light-harvesting antennas typical of eukaryotes are absent in the cyanobacteria. Taking their place in these prokaryotes are protein-pigment combinations known as *phycobilisomes*, which can be seen as spherical particles attached to the cytoplasmic surfaces of photosynthetic membranes (arrows, Fig. 9-26). Phycobilisomes serve as light-harvesting antennas for photosystem II in cyanobacteria; two photosystem II complexes are associated with each phycobilisome. No accessory light-absorbing structures of any kind appear to be associated with photosystem I in cyanobacteria.

The primary pigments of phycobilisomes are the *phycobilins* (Fig. 9-27), hydrophilic, water-soluble pigments that contain a light-absorbing structure resembling the tetrapyrrole ring of hemoglobin. The pyrroles of the phycobilins, however, occur in linear rather than ring form.

Among eukaryotes, only one group, the red algae, uses phycobilisomes instead of LHC antennas. In red algae, phycobilisomes occur inside chloroplasts, on the surfaces of thylakoid membranes facing the stroma.

Cyanobacteria were the first organisms to release oxygen as a by-product of photosynthesis. As such,

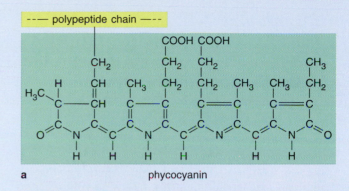

a   phycocyanin

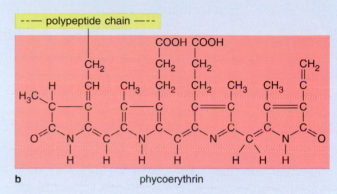

b   phycoerythrin

Figure 9-27 The cyanobacterial phycobilins phycocyanin **(a)** and phycoerythrin **(b)**. The pigments are combined in phycobilisomes with two related α- and β-polypeptides. Both phycocyanin and phycoerythrin occur in cyanobacteria; red algae may have either or both of these pigments.

they were critical to the first appearance of abundant oxygen in the earth's atmosphere and to the evolutionary development of aerobic organisms. Evidence from various sources indicates that cyanobacteria are the evolutionary progenitors of chloroplasts in the eukaryotic algae and green plants. The presence of phycobilisomes and phycobilins suggests that red algae may be the eukaryotic algae most closely related to the ancient cyanobacteria believed to have given rise to chloroplasts. (The probable evolutionary origins of chloroplasts are discussed in Chapter 26.)

Supplement 9-2

The Oxygenase Activity of RuBP Carboxylase and Photorespiration: An Evolutionary Fly in the Photosynthetic Ointment

The ability of RuBP carboxylase to act as an oxygenase impedes the progress of photosynthesis in many plant species. The phosphoglycolate product of the oxygenase activity, after conversion to glycolate by removal of a phosphate group, enters an oxidative pathway that can eventually yield CO_2 (see p. 264). Since the oxygenase activity of RuBP carboxylase uses O_2 and the remainder of the pathway evolves CO_2, the entire pathway is termed *photorespiration*. The pathway returns CO_2 from the fixed, organic form to the inorganic form and thus has the effect of reversing the dark reactions. Depending on the temperature and the concentrations of O_2 and CO_2 in the atmosphere, as much as 15 to 50% of the CO_2 fixed in photosynthesis may be lost again through photorespiration.

When RuBP carboxlyase acts as an oxygenase, the product of the first reaction of the C3 cycle is *phosphoglycolate* rather than 3PGA (Fig. 9-28a). This substance is subsequently dephosphorylated to glycolate, which enters microbodies (see Supplement 8-3) for oxidation. Oxidation occurs through the *glycolate pathway*, which occurs in microbodies of cells in leaves and other plant tissues (Fig. 9-28b). The enzyme oxidizing glycolate, *glycolate oxidase*, uses oxygen as final acceptor for the electrons removed in the oxidation and yields H_2O_2 along with glyoxylate as products. As in other microbody pathways, the H_2O_2 product is converted to H_2O and O_2 by catalase (see p. 243).

The glyoxylate produced by glycolate oxidation may enter another reaction series in microbodies, the *glyoxylate cycle* (see p. 245), in which it is converted to oxaloacetate or succinate. These acids may enter the citric acid cycle in mitochondria, in which oxaloacetate and succinate are oxidized to CO_2. By this means the CO_2 fixed in photosynthesis may wind up as inorganic CO_2 once again.

Plant microbodies also contain an aminotransferase that can add an amino group to glyoxylate, converting it into an amino acid, glycine. The glycine may enter mitochondria, where it is converted to serine, with the release of one carbon atom as CO_2:

$$2 \text{ glycine} \longrightarrow \text{serine} + NH_3 + CO_2 \qquad (9\text{-}6)$$

In effect, the release of CO_2 in this reaction also reverses carbon fixation in photosynthesis.

The oxygenase activity of RuBP carboxylase may reflect the fact that the enzyme evolved during primeval times, in an atmosphere rich in CO_2 and depleted in oxygen. Selection against the oxygenase activity may therefore be a relatively recent evolutionary development in the history of the enzyme, one that has perhaps not yet had time (in the evolutionary sense) to have much effect. It is also possible that the combined carboxylase-oxygenase activity is the only

Figure 9-28 Glycolate metabolism. **(a)** The reaction series producing glycolate, initiated when RuBP carboxylase uses oxygen instead of CO_2 as substrate. **(b)** The glycolate pathway, in which glycolate is oxidized to glyoxylate and H_2O_2.

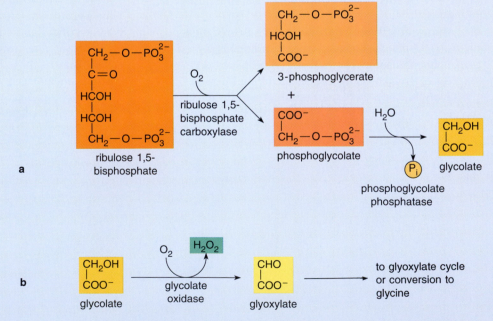

pattern by which the enzyme can work, and that significantly greater efficiency cannot be attained.

It is unclear why nature evolved the costly photorespiration pathway converting the product of RuBP oxygenase activity to CO_2. Although there is much controversy surrounding this point, photorespiration may simply be a means for ridding plant cells of toxic glycolate. That is, rather than evolving a means to eliminate the oxygenase activity of RuBP, plants with high photorespiration rates have instead developed a pathway to eliminate the glycolate product of the faulty oxygenase reactions. This adaptation allows plants to survive in the contemporary high O_2/low CO_2 atmosphere in spite of the oxygenase activity of RuBP carboxylase.

Plants with high photorespiration rates, especially at elevated temperatures, include many important crops—rice, barley, wheat, soybeans, tomatoes, potatoes, and tobacco, to name a few. When grown experimentally in hothouses at CO_2 concentrations high enough to eliminate photorespiration, some of these crops increase in growth by as much as five times by dry weight. This demonstrates the extent to which photorespiration reduces the efficiency of photosynthesis in these plants.

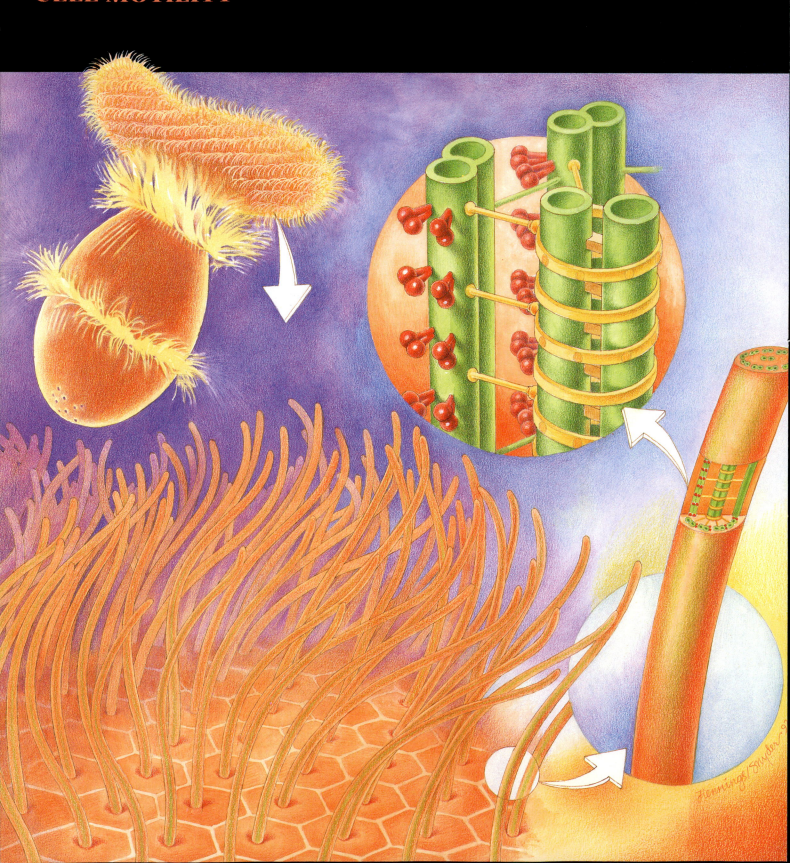

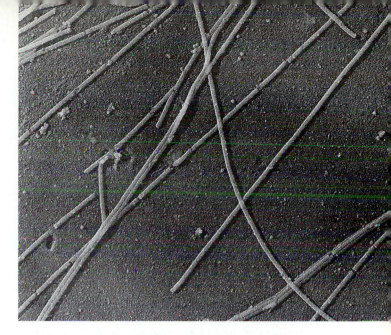

Figure 10-1 Microtubules isolated from erythrocytes of the salamander *Taricha* and prepared for electron microscopy by shadowing. × 42,000. (Photograph by the author.)

A ll cells are capable of movement. In the most simple form shared by all cells, cell motility involves slow streaming movements of the cytoplasm that move the fluid contents of the cell from place to place. All eukaryotic cells, in addition, are able to produce the more organized movements that divide the chromosomes during cell division. Further, many cell types are capable of highly systematized and rapid motions such as muscle contraction and the beating of flagella or cilia. In many-celled organisms, cellular movements are coordinated to produce the movements of body parts or the entire individual.

Two different structures, *microtubules* and *microfilaments*, are involved in most of these movements. Microtubules are unbranched cylinders about 25 nm in diameter with an open central channel (Figs. 10-1 and 10-2; see also Figs. 1-14 and 11-1). Microfilaments (see Fig. 11-1) are extremely fine, unbranched, solid fibers of much smaller diameter, about 5 to 7 nanometers—not much thicker than the wall of a microtubule. The two elements are assembled from different structural proteins—microtubules from *tubulins*, and microfilaments from *actin*.

Microtubules and microfilaments act both separately and in coordination to produce cellular movements. Flagellar and ciliary beating, for example, is based on microtubules. Microfilaments produce muscle contraction in animals and cytoplasmic streaming in both animals and plants. In animal cells, membranous organelles such as mitochondria and vesicles move through the cytoplasm primarily via microtubule-based systems; in plants and fungi, depending on the organelle or species, either system may be responsible for movement. In many plants, for example, microfilament-based systems move chloroplasts through the cytoplasm, and microtubule-based systems move mitochondria. The two motile elements coordinate their activities in animal cell division, in which microtubules divide and distribute the chromosomes and microfilaments divide the cytoplasm.

The molecular mechanisms responsible for movements are similar in microtubules and microfila-

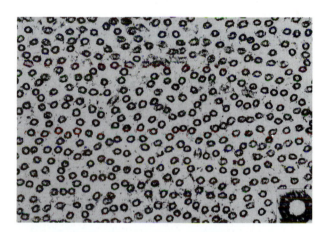

Figure 10-2 Microtubules in cross section. The inset at the lower right shows a single microtubule at higher magnification, in which the individual wall subunits are visible. The microtubules were polymerized from tubulin in the test tube. Main figure × 42,000; inset × 380,000. (Courtesy of W. L. Dentler, Jr.)

ments—both elements produce movement by an active sliding mechanism. Microtubules slide over microtubules, or microfilaments over microfilaments, and cell organelles slide over the surface of either motile element. These motions are produced by active, ATP-driven crossbridges that "walk" over the surfaces of microtubules or microfilaments by making attachments, swiveling actively and forcefully over a short distance, and then releasing (see Figs. 10-13 and

11-9). Several different motile proteins form microtubule crossbridges; best known of these are *dynein* and *kinesin*. Microfilament crossbridges consist of a single protein type, *myosin*.

Both microtubules and microfilaments also produce motion by controlled growth or disassembly. The resultant changes in length push or pull attached cellular structures. Both structures also act as cytoskeletal elements providing support to cell structures rather than motility.

This chapter describes the structure, function, and biochemistry of microtubules and the motile structures containing them, including flagella and cilia. The motile systems of bacterial cells are discussed in Supplement 10-1. The structure and biochemistry of microfilaments and their motile functions are covered in Chapter 11. Chapter 11 also gives examples of the relatively few eukaryotic motile systems based on elements other than microtubules or microfilaments. The structural roles of microtubules and microfilaments in the cytoskeleton are described in Chapter 12. Chapters 24 and 25 outline the activities of microtubules and microfilaments in cell division.

MICROTUBULE STRUCTURE AND ASSEMBLY

Microtubules are similar in structure in all eukaryotes, including animals, plants, fungi, and protists. The similarities extend to the molecular level: All microtubules are assembled from related proteins, which assemble and disassemble by essentially the same mechanisms.

Microtubule Structure

Highly magnified cross sections show that the wall of a microtubule consists in most cells of a circle of 13 spherical subunits, each about 4 to 5 nm in diameter (see inset to Fig. 10-2). Longitudinal views show that the subunits line up in parallel, lengthwise rows called *protofilaments* to form the wall (see Figs. 10-3 and 10-4b).

Each spherical wall particle is a unit of tubulin, the microtubule protein, which occurs in two related forms known as α- and β-*tubulin*. These proteins link into a dumbbell-shaped *heterodimer* consisting of one α- and one β-tubulin unit (Fig. 10-4). Each visible wall subunit is half of a dumbbell. The "dumbbells"

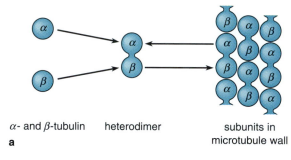

α- and β-tubulin   heterodimer   subunits in microtubule wall

a

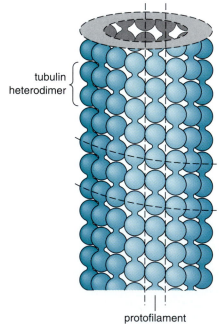

tubulin heterodimer

b   protofilament

Figure 10-4 Tubulin heterodimers and protofilaments. **(a)** The relationship among α- and β-tubulin, the tubulin heterodimer, and the visible wall subunits of microtubules. **(b)** The arrangement of heterodimers in microtubule walls.

Figure 10-3 Negatively stained microtubules isolated from sperm cells of the plant *Pteridium*. The arrangement of the wall subunits into longitudinal protofilaments is clearly visible in the region enclosed by brackets. × 170,000. (Courtesy of R. Barton and Academic Press, Inc., from *J. Ultrastr. Res.* 20:6 [1967].)

line up end to end in longitudinal rows to form protofilaments.

In some cellular structures, microtubules occur in a double or triple form consisting of one complete microtubule with one or two partial, *C*-shaped microtubules fused to it at one side (Fig. 10-5). Microtubules fused in double rows as *doublets* occur in flagella and cilia (see Fig. 10-11) and in triple rows as *triplets* in *centrioles* or *basal bodies*, structures that give rise to flagella and anchor them in the cytoplasm (see Fig. 10-18).

The Tubulins

The α- and β-tubulins both have molecular weights near 50,000. In most species the two tubulins share from 40% to 50% of their sequences in common, indicating that the genes encoding them evolved from a single ancestral gene. Because all known eukaryotes have the two tubulin types, the split of this ancient gene must have occurred very early in eukaryotic evolution.

Within a given tubulin type, conservation of structure is very high. All known α-tubulins, for example, are about 89% identical in amino acid sequence; all known β-tubulins are about 87% identical in sequence. In spite of the overall sequence conservation, most species have several variants, or *isotypes*, of the α- and β-tubulins (Table 10-1), each showing

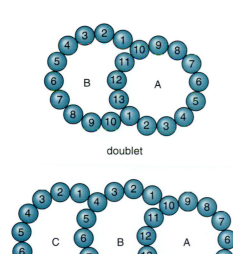

doublet

triplet

Figure 10-5 The arrangement of protofilaments in microtubule doublets and triplets. Doublets occur in flagella; triplets occur in centrioles (compare with Figs. 10-11 and 10-18). The doublet and triplet microtubules are labeled as the A, B, and C subtubules as shown, beginning with the complete microtubule of the set as the A subtubule.

small differences in amino acid sequence and each encoded in a separate gene. The significance of the different tubulin isotopes is unknown. Although the different isotypes are developmentally regulated, none seems to occur exclusively in any particular microtubule-based structure.

A third tubulin type, γ-*tubulin*, discovered by C. E. and B. R. Oakley, has been detected in a variety of species, including fungi, *Drosophila, Xenopus,* and humans, and is likely also to be generally distributed among eukaryotes. Rather than assembling into microtubules, γ-tubulin appears to be part of microtubule-organizing regions such as the *spindle pole body* in fungi and other lower eukaryotes and the *cell center* or *centrosome* in higher eukaryotes. These regions give rise to new microtubules containing α- and β-tubulin during cell division (see below and pages 702 and 704).

Polymerization of Tubulins into Microtubules

In living cells, microtubules assemble from a pool of heterodimers in the cytoplasm. The balance between assembled microtubules and the pool of subunits is finely tuned, so that microtubules assemble and disassemble continually at all stages of the cell cycle.

Biochemistry of Microtubule Assembly In the early 1950s, S. Inoué noted that certain treatments, such as exposure to temperatures near 0°C, elevated pressure, or the microtubule poison *colchicine* (see below), cause the spindle in dividing cells to disassemble and disappear. The spindle reassembles quickly after removal of the disturbance, sometimes within seconds or minutes. From these observations Inoué proposed that spindle fibers, later discovered to consist of bundles of microtubules, exist in equilibrium with a pool of unassembled subunits in the cytoplasm.

Table 10-1	Tubulin Isotypes	
Organism	Number of α-tubulin Isotypes	Number of β-tubulin Isotypes
Fungi		
Aspergillus	2	2
Physarum	4	3
Saccharomyces	1	2
Chlamydomonas (alga)	2	2
Drosophila	4	4
Chicken	5–6	7
Mouse	6	5
Human	5–6 (?)	6

The subunits in the cytoplasmic pool were later established to be tubulin heterodimers, allowing the equilibrium to be written as:

$$\text{tubulin heterodimers} \rightleftharpoons \text{microtubules} \quad (10\text{-}1)$$

A number of physical and chemical factors affect the equilibrium. The nucleoside triphosphate GTP must be available and Ca^{2+} must be kept at very low concentrations for the reaction to progress significantly toward microtubule assembly. The assembly proceeds most rapidly at a pH of about 6.8. Temperatures in physiological ranges promote microtubule assembly; reduced temperatures and elevated pressure push the equilibrium in the opposite direction.

GTP plays a critical role in microtubule assembly. Tubulin heterodimers must bind GTP before they can assemble into microtubules. One GTP is bound by the α-tubulin and one by the β-tubulin subunit of each heterodimer. At some time after microtubule assembly the GTP bound to the β-polypeptide is hydrolyzed to GDP and phosphate; the GTP bound to the α-subunit remains unchanged. Although the role of GTP is not completely understood, its binding to both tubulin subunits probably causes conformational changes in the heterodimer that make it bind strongly to the ends of existing microtubules, thereby pushing the equilibrium in the direction of polymerization.

R. C. Weisenberg and others found that heterodimers can self-assemble into microtubules in the test tube. Microtubules, in fact, were among the first cell structures to be assembled in a cell-free system (Figure 10-2 shows microtubules assembled in the test tube).

Microtubule Polarity At cellular tubulin concentrations, one end of a microtubule typically assembles or disassembles about two to four times faster than the other. This characteristic is termed *polarity*; the end with the faster rate of assembly or disassembly is the *plus end* of a microtubule; the end with the slower rate is the *minus end* (*plus* and *minus* in this sense refer to relative rates of assembly, not to charge). It is unknown whether the α- or β-end of tubulin heterodimers "points" toward the plus end of microtubules.

Microtubule polarity can be directly determined by several methods. One highly useful technique, worked out by S. R. Heidemann and J. R. McIntosh, exposes cells to tubulin after cell membranes have been broken with a detergent. Under the conditions used in the technique, the added tubulin forms partial, hook-shaped microtubules along the sides of existing microtubules (Fig. 10-6). The hooks extend in a clockwise or counterclockwise direction depending on microtubule orientation in the sections. If the hooks point in a clockwise direction, the plus end in the sec-

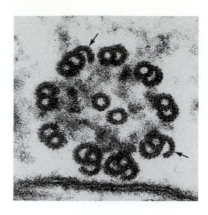

Figure 10-6 The hook technique for detecting microtubule polarity. Partial microtubules, which appear as hooks in cross sections, are grown on the side of existing microtubules. If the hooks point in a clockwise direction, as in this micrograph (arrows), the plus ends of the microtubules extend toward the observer. This preparation shows the microtubules of a flagellum from the protozoan *Tetrahymena*. × 190,000. (Courtesy of J. R. McIntosh and U. Euteneuer, from *Proc. Nat. Acad. Sci.* 78:372 [1981].)

tion is pointed toward the observer, as in Figure 10-6. If the hooks point in the counterclockwise direction, the minus end is directed toward the observer.

The hook technique revealed a definite correlation between microtubule polarity and motility (Fig. 10-7). If cell structures such as mitochondria are linked to microtubules by dynein crossbridges, the crossbridges "walk" the organelles toward the minus end of a microtubule (Fig. 10-7a). Dynein crossbridges between two microtubules are fixed to one microtubule and walk the fixed microtubule toward the minus end of the opposite microtubule (Fig. 10-7b and c). Kinesin crossbridges work in the opposite direction, sliding attached organelles toward the plus ends of microtubules (Fig. 10-7d). Cells evidently regulate the direction of movement in microtubule-based systems by controlling microtubule polarity and the type of crossbridge (see below).

Dynamic Instability of Microtubules at the Balanced State Careful observations by M. Kirschner and T. Mitchison revealed an unexpected characteristic in the behavior of microtubules at the balanced state. Under conditions in which microtubules are expected to remain approximately the same length, many individual microtubules can actually be seen under the darkfield microscope (see Appendix p. 784) to constantly lengthen or shorten. Typically lengthening proceeds relatively slowly, but shortening is so rapid that Kirschner and Mitchison characterized it as "catastrophic." They termed the overall pattern of lengthening and shortening *dynamic instability*. In spite of

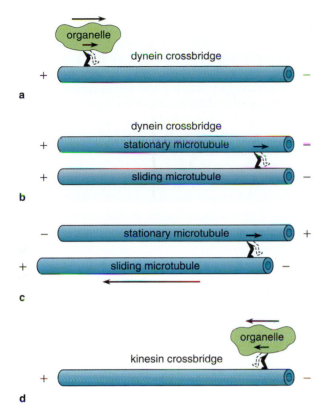

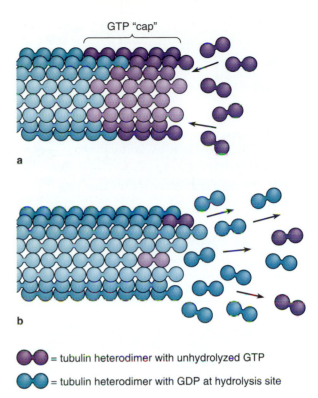

GTP "cap"

●●●● = tubulin heterodimer with unhydrolyzed GTP

●●●● = tubulin heterodimer with GDP at hydrolysis site

Figure 10-7 Movement of dynein and kinesin crossbridges with respect to the plus and minus ends of microtubules. **(a)** Dynein crossbridges walk cell structures such as vesicles or organelles toward the minus end of a microtubule. **(b)** Dynein crossbridges anchored to one microtubule walk toward the minus end of an adjacent microtubule. As a result, the adjacent microtubule is pushed in the direction of its plus end **(c)**. **(d)** Kinesin crossbridges walk vesicles toward the plus end of a microtubule.

Figure 10-8 Dynamic instability of microtubules (see text). **(a)** The slowly growing state, in which the microtubule is stabilized by a cap of heterodimers with unhydrolyzed GTP. **(b)** The catastrophic disassembly state, in which heterodimers with links to GDP are exposed at the microtubule tip. Microtubules are considered to switch randomly between the states in **(a)** and **(b)** at equilibrium.

the constant assembly and disassembly, the average length of microtubules and the amount of tubulin assembled into microtubules remain the same in a sample exhibiting dynamic instability.

Kirschner and Mitchison proposed that dynamic instability depends on the integrity of a "cap" of heterodimers with unhydrolyzed GTP at the plus end of a microtubule (Fig. 10-8). If GTP hydrolysis in the microtubule lags sufficiently, so that the end of a microtubule is completely covered by heterodimers linked to unhydrolyzed GTP, the microtubule is stable and additional GTP-containing heterodimers can add to the plus end from the heterodimer pool. The microtubule therefore grows slowly in length (Fig. 10-8a). If GTP hydrolysis within the microtubule proceeds rapidly enough to expose GDP-containing heterodimers at the plus end, the microtubule becomes unstable because GDP-containing heterodimers bind together relatively weakly (Fig. 10-8b). As a result, the microtubule disassembles catastrophically from

this end. If enough GTP-containing heterodimers add to the plus end to restore the cap, the microtubule is "rescued" and growth resumes. Individual microtubules are considered to switch randomly and repeatedly between the slowly growing and rapidly disassembling phases.

Dynamic instability has been observed with Nomarski (see Appendix pp. 784 and 785) and darkfield microscopy in living cells as well as in the test tube. It has great significance for the expected behavior and function of microtubules in cells, especially in systems such as the spindle in which microtubules are delicately balanced between assembly and disassembly.

Microtubule Treadmilling at the Balanced State Depending on tubulin concentration, stable microtubules can achieve a balance point at which heterodimers add to the plus end while the minus end is still disassembling. In this state it is possible that heterodimers may *treadmill* through the microtubules—that is, heterodimers may add at the plus end, be pushed

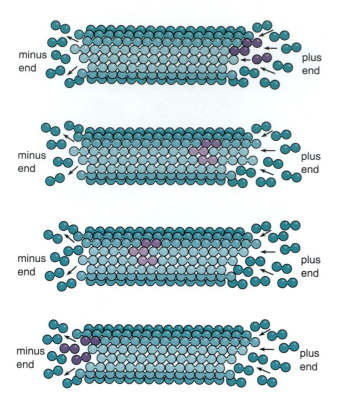

Figure 10-9 Movement of a group of tubulin heterodimers (in purple) through a microtubule by treadmilling as subunits are added continually at the plus end and released at the minus end. Whether microtubule treadmilling actually takes place in living cells remains controversial.

toward the minus end as more subunits add to the plus end, and eventually be released as they reach the minus end (Fig. 10-9).

It is still uncertain whether microtubule subunits actually treadmill in living cells—some experiments appear to detect treadmilling, and others find no evidence for the process. For example, P. Wadsworth and E. D. Salmon injected microtubule subunits, covalently linked to a light-sensitive dye, into dividing mammalian cells. After the cells assembled spindles incorporating the dye, they were exposed to a narrow beam of laser light. The laser bleached segments of the spindle microtubules that contained the dye, forming a spot that could be easily recognized in the light microscope. The spot remained stationary with respect to the microtubule ends as the chromosomes moved to the ends of the spindle, instead of moving as would be expected if treadmilling were taking place.

Most experiments testing treadmilling have produced the same result. However, in a recent experiment by Mitchison the bleached segment was observed to move slowly through stationary spindle microtubules, suggesting that treadmilling was actually taking place. Mitchison used a different marker in his experiments, a molecule that fluoresces when exposed to light at a certain wavelength. To explain these contradictory results, Mitchison proposed that the light-sensitive dyes used in the other experiments may form chemical crosslinks when exposed to light. The crosslinks, according to this idea, may have fixed the tubulin heterodimers in place in the microtubules and artificially stopped their treadmilling.

If treadmilling actually occurs in cells, it could provide another basis for microtubule-based motility in addition to sliding and growth. In this mechanism, proposed by R. L. Margolis and L. Wilson, an organelle would link to a heterodimer at the plus end of a treadmilling microtubule. As the heterodimer moves longitudinally through the microtubule, the organelle would be carried along, apparently moving over the microtubule surface. The movement would be polar, resembling the plus-to-minus-end movement produced by dynein.

Factors Regulating Microtubule Polymerization: Ca^{2+} and MAPs

Cells precisely regulate the time, place, and degree of microtubule assembly and the stability of assembled microtubules. This regulation depends primarily on two factors: (1) Ca^{2+} concentration and (2) the kind of *microtubule-associated proteins (MAPs)* combined with microtubules. These factors interact to provide a highly sensitive and balanced group of both short- and long-term controls. The short-term controls, which depend primarily on Ca^{2+} concentration, allow quick, localized adjustments in microtubule numbers and stability. The long-term controls, dependent primarily on the MAPs combined with microtubule proteins, regulate assembly or disassembly of microtubule-based structures that may persist through much of the cell cycle.

Ca^{2+} Control by changes in Ca^{2+} concentration depends on the exquisite sensitivity of the heterodimer-microtubule equilibrium to the concentration of this ion. At very low Ca^{2+} concentrations, equivalent to 1×10^{-8} to $1 \times 10^{-6}\,M$, microtubules assemble rapidly. As Ca^{2+} concentration increases to $1 \times 10^{-3}\,M$, polymerization drops to one-half its maximum level; at $2 \times 10^{-3}\,M$, Ca^{2+} inhibits polymerization completely. The almost instantaneous response of the equilibrium to alterations in Ca^{2+} concentration provides the primary mechanism for short-term regulation of microtubule assembly.

The low cytoplasmic Ca^{2+} concentrations at which microtubules assemble are characteristic of most eukaryotic cells. Calcium ions are maintained at this low level by active transport pumps in the plasma membrane and ER, which constantly move Ca^{2+} from the cytoplasm to the cell exterior and into ER vesicles (see p. 131). Localized disassembly is triggered by controlled Ca^{2+} release into the cytoplasm via gated calcium channels (see p. 128) in the ER or plasma membrane. One mechanism producing controlled Ca^{2+} release by ER channels is the $InsP_3/DAG$ regulatory pathway, in which Ca^{2+} is released as a second messenger in response to binding of hormones or growth factors at the cell surface (see p. 161).

MAPs   The MAPs, first discovered by R. D. Sloboda and his coworkers, regulate microtubule stability by combining directly with tubulin heterodimers as they assemble into microtubules. The final levels of MAPs in microtubules may be as high as 5% to 20% of the total tubulin content by weight. When integrated into microtubule structure, MAPs generally stabilize microtubules as long-term controls and reduce their tendency to separate into individual heterodimers. Some MAPs also probably serve as *nucleation sites* for microtubule assembly; that is, they provide sites on which the first subunits forming a microtubule are brought together. A very few known MAPs push the tubulin-microtubule equilibrium in the direction of microtubule assembly.

MAPs range from small proteins of 20,000 to 30,000 daltons to types in excess of 300,000 daltons. Mammalian brain cells, for example, contain a group of relatively low-molecular-weight MAPs collectively identified as the *tau* MAPs and several high-molecular-weight types called *MAP-1A, MAP-1B, MAP-2, MAP-3, MAP-4* and *MAP-5.*

Some brain-cell MAPs can also be detected in other cell types. One or more of the proteins in the MAP-1 and MAP-4 groups, for example, are found in microtubules in a variety of mammalian and avian cells. Other MAPs are restricted to a single cell type or to individual microtubular structures within cells. Most of our information regarding MAPs and their activities has come from the study of animal systems, particularly brain tissues in higher vertebrates. Little is known about MAPs in plants and fungi.

Different MAPs provide microtubules with varying degrees of stability. Flagellar microtubules, for example, are exceptionally stable and insensitive to many agents, such as elevated pressure, reduced temperature, or colchicine, that disrupt microtubules in other locations. In contrast, spindle microtubules are easily disassembled by these and other treatments.

Brain-cell microtubules provide an interesting example of the relationship of MAPs to microtubule stability. The majority of microtubules isolated from mammalian brain tissue are stable only at temperatures above 15°C; they disassemble rapidly below this temperature. However, a small fraction of mammalian brain microtubules resists disassembly even at temperatures approaching 0°C. The cold stability of this fraction depends on a MAP not present in the cold-sensitive brain microtubules. MAPs with similar properties also account for the general cold stability of brain, spindle, and cytoplasmic microtubules in animals such as arctic fishes, which normally live with body temperatures near 0°C (the MAPs with this function are sometimes called *STOPs*, for *Stable Tubule Only Proteins*).

In many cellular systems the ability of MAPs to stabilize microtubules is modified by the addition of phosphate groups. The phosphorylations, carried out by protein kinases specific for certain MAPs, generally reduce the ability of MAPs to stabilize microtubules and push the equilibrium in the direction of disassembly. In effect, the phosphorylations add a system that fine-tunes long-term regulation of microtubule assembly and stability by MAPs. The ability of STOPs to cold-stabilize mammalian brain microtubules, for example, is reduced through phosphorylation by a specific protein kinase.

MAPs and Ca^{2+} interact extensively in the regulation of microtubule assembly. MAPs may increase or decrease the sensitivity of microtubules to Ca^{2+} concentration or may make microtubules sensitive to Ca^{2+} only when the ion is combined with calmodulin, forming an active Ca^{2+}/calmodulin complex (see Information Box 6-2). In most or all of the systems in which Ca^{2+} works through calmodulin, the Ca^{2+}/calmodulin complex regulates microtubule assembly by modifying the activity of a MAP.

Cold-sensitive brain microtubules, for example, disassemble rapidly if exposed to Ca^{2+}. The effect of Ca^{2+} is evidently exerted directly on the tubulins or heterodimers of these microtubules. The cold-stable fraction of brain microtubules, in contrast, is insensitive to elevated Ca^{2+} concentrations as long as the MAP that provides cold stability is present. Calmodulin by itself has no effect on the cold-stable brain microtubules. However, addition of both calmodulin and Ca^{2+}, permitting formation of the active Ca^{2+}/calmodulin complex, causes immediate disassembly of the cold-stable microtubule fraction. Combination with the Ca^{2+}/calmodulin complex evidently causes a conformational change in the MAP that reduces its ability to stabilize microtubules.

Some of the protein kinases adding phosphate groups to MAPs are cAMP-dependent. The phosphate groups added by these kinases are removed by phosphatases dependent for their activity on the presence of Ca^{2+} or calmodulin. The activity of cAMP-dependent protein kinases and Ca^{2+}/calmodulin-dependent phosphatases also links microtubule regulation by the MAPs to the major cAMP and $InsP_3$/DAG regulatory pathways controlled by cell surface receptors.

Not all MAPs are regulatory in function. Some act as linkers between microtubules or between microtubules and other cell structures. Some of these MAPs, such as those interlinking microtubules, microfilaments, and intermediate filaments, form bonds assembling the cytoskeletal network supporting cell structures. The interlinking ability of many MAPs, including the tau group and MAP-2, depends on their structure as two-domain proteins. One domain recognizes and binds sites on the heterodimers of a microtubule; the second domain recognizes and links to other microtubules or cell structures.

Microtubule Organizing Centers (MTOCs)

Microtubules may appear at essentially any location in the nucleus or cytoplasm. In many places where they assemble, particularly when the numbers are small, microtubules polymerize from apparently structureless regions of the cytoplasm or nucleus. However, large-scale production of microtubules usually occurs in distinct structures. These specialized microtubule-generating sites, termed *microtubule organizing centers* (*MTOCs*), take one of four different forms:

1. *Centrioles* (also termed *basal bodies*; see Figs. 10-18 and 10-19), which generate the microtubules of flagella.

2. *Cell centers*, or *centrosomes*, collections of dense material located near the nucleus, which organize cytoplasmic microtubules in animal cells. The same centers also organize spindle microtubules during animal cell division, where they are known as *asters* (see Figs. 24-15 and 24-17).

3. *Kinetochores*, disc- or platelike structures on the surfaces of chromosomes, which attach and may also give rise to spindle microtubules (see Figs. 24-11, 24-12, and 24-14).

4. *Spindle pole bodies*, dense, platelike layers of material, which give rise to the spindle in many fungi (see Fig. 24-20).

Each of the MTOCs has been successfully isolated from cells. When supplied with tubulin under polymerizing conditions, all are capable of generating microtubules in the test tube. In all these structures, microtubules grow from MTOCs with their minus ends anchored in the MTOC, and their plus ends directed outward.[1] Because the plus end, at which further subunits are added, grows away from the MTOC, MTOCs are likely to act only as nucleation centers, providing the "seeds" from which microtubule assembly proceeds. Once assembly is under way, subunits add to the newly formed plus end, which grows outward from the MTOC.

MTOCs may provide nucleation centers by adjusting local conditions, such as Ca^{2+} concentration, to greatly favor heterodimer polymerization. Alternatively, they may provide organizing sites for MAPs that promote microtubule assembly. This possibility is supported by the fact that antibodies developed against some MAPs react positively with MTOCs. The recently discovered γ-tubulin, which is localized in microtubule-organizing regions of the cell center, may also promote microtubule assembly, possibly by acting as a template.

Tests of the molecular components of MTOCs reveal that proteins and, possibly, RNA are present. Although the proteins might function as MAPs, the possible role of RNA in MTOCs remains unknown.

Microtubule Poisons

The importance of microtubules to cellular activities makes them a sensitive target for biological defenses, including substances that act as microtubule "poisons" by interfering with the motile or cytoskeletal roles of microtubules. Several of these substances, most notably *colchicine* and *taxol* (Fig. 10-10), have proved to be highly useful in medicine and research.

Colchicine, an alkaloid extracted from plants in the genus *Colchicum* (including autumn crocus and meadow saffron), has been known since ancient times as an animal poison. (Alkaloids are cyclic organic compounds that react as bases.) During the 1700s, mild doses of colchicine were found to be effective in the treatment of gout, an application for which it is still used today. Investigators discovered in the 1930s that one of the primary cellular effects of colchicine is complete arrest of mitosis. Further research revealed that colchicine combines specifically with tubulin heterodimers and causes rapid disassembly of the more labile types of microtubules, including those of the spindle, cytoskeleton, and nerve cells. Existing flagellar microtubules, in contrast, are unaffected by colchicine, apparently because of the presence of MAPs that

---

[1] Inside living cells, spindle microtubules connect to kinetochores at their plus ends, so that their minus ends extend outward, away from the kinetochore. This arrangement is believed to result from the activity of kinetochores in attaching to previously assembled microtubules rather than acting as MTOCs (see p. 708 for details).

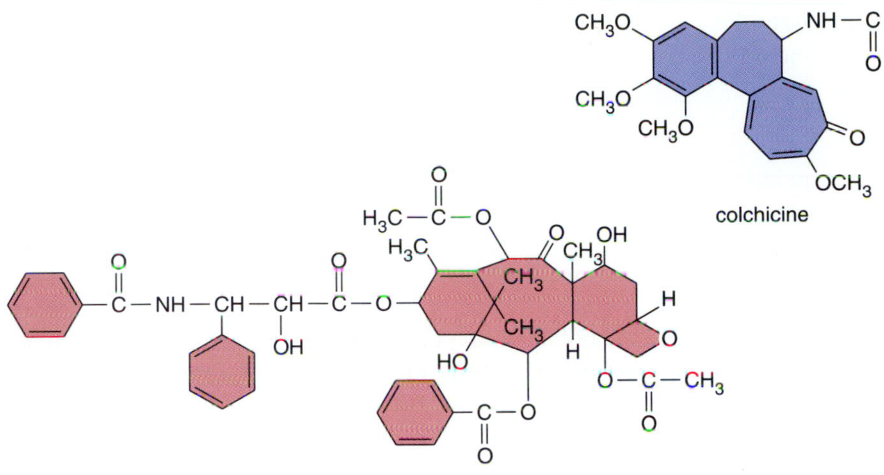

Figure 10-10 The microtubule poisons colchicine and taxol.

give extra stability to these microtubules. In general, microtubules in animals are most sensitive to colchicine; plants and many types of fungi, algae, and protozoa are more resistant to the drug's effects.

In plants, resistance to colchicine apparently depends on substitutions in the amino acid sequence of the α-tubulins, which may inhibit colchicine binding by altering the folding conformation of the site binding the poison. The reduced binding affinity for colchicine explains why plants such as *Colchicum* can survive with relatively high concentrations of colchicine in their tissues. There are also colchicine-resistant animal mutants, in which resistance also depends on one or more amino acid substitutions in α-tubulin.

The molecular basis for colchicine activity is not completely understood. Heterodimers linked to colchicine are still capable of binding to the plus ends of existing microtubules. However, when bound, the colchicine-heterodimer complexes evidently block addition of further heterodimers at the plus ends. Presumably the capped microtubules then disassemble from their minus ends.

Colchicine combines with tubulin specifically enough to make it an invaluable tubulin or microtubule "marker." The first effective methods for isolating and purifying tubulins depended on colchicine made radioactive by the attachment of tritium (^{3}H) atoms. Once marked by the radioactive colchicine, tubulin could be extracted and purified from cells simply by collecting the radioactive fraction separated by centrifugation.

The destruction of microtubules by colchicine makes the poison a highly useful probe for microtubule-based motility. Since it interacts with few or no other cellular molecules, inhibition of a particular cellular motion by colchicine provides a good first indication that the motion is based on microtubule

structure and function. The method is not foolproof because the microtubules of some species are insensitive to the poison, and some types of microtubule-based motion—flagellar motion, for example—are unaffected by colchicine.

Taxol, an alkaloid extracted from *Taxus brevifolia* (the western yew), has effects on microtubules opposite to those of colchicine. Taxol stabilizes microtubules by pushing the equilibrium in the direction of the polymer. It is so effective in this role that it causes most of the available tubulin heterodimers to assemble into highly stable microtubules. Use of the alkaloid has made it possible to extract intact microtubules from a variety of cell types, including many plants, in which microtubules are so sparsely distributed and unstable that isolation and purification of microtubule proteins had previously been difficult or impossible.

Colchicine and taxol have proved to be useful in medicine as treatments reducing the growth of both benign and malignant tumors. They are effective in this role primarily because, by interfering with spindle function, they slow or stop the uncontrolled cell division characteristic of tumors. Taxol has been so effective in cancer treatment that an intensive effort is now underway to find new sources for the drug (the supply of western yew is limited). This effort led recently to the first successful synthetic assembly of the drug.

THE MOTILE FUNCTIONS OF MICROTUBULES

Sliding, produced by the action of dynein, kinesin, and possibly other microtubule crossbridges capable of converting chemical to mechanical energy, is the predominant microtubule-based motile mechanism.

This mechanism underlies motions of the nucleus and cytoplasmic organelles and of the spindle and flagella.

Dynein and Microtubule Sliding

Of the two major microtubule crossbridges, dynein is the better known. The structure of this crossbridge and its activity in converting the chemical energy of ATP into the mechanical energy of movement are likely to be representative of kinesin and other microtubule motors. The crossbridging mechanism of dynein also has direct parallels to the activity of myosin in microfilament-based motile systems.

Dynein Structure Dynein has been studied most extensively in flagella, where it can be seen in sections as short, hooked or bobbed *arms* extending from one set of double microtubules to the next (Fig. 10-11). Freeze-fracture preparations (see p. 791) reveal some details of dynein substructure. In these preparations a dynein arm appears as two or three globular heads connected to a basal unit by short stalks, much like a bouquet of flowers (Fig. 10-12). Analysis of dynein isolated from flagella shows that a dynein arm contains as many as 12 polypeptides with a combined molecular weight ranging from 1.2 to 1.9 million.

At any level, two dynein arms, known as the *inner* and *outer* arms, are attached to flagellar microtubules (see Fig. 10-11). Separate analysis of the outer and inner arms has been made possible by the use of mutants lacking one of the two arm types or by chemical techniques allowing the arms to be removed separately. These methods reveal that the two arms have different groups of polypeptides and that the inner arms occur in at least three different forms, each with distinct polypeptide types. A recently characterized nonflagellar dynein, called *cytoplasmic dynein*, also contains a distinct group of polypeptides. The dyneins therefore make up a family in which the members have related but distinct properties. One or more of the dyneins are probably distributed universally among eukaryotic cells.

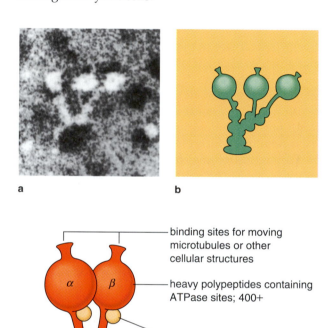

a   b

binding sites for moving microtubules or other cellular structures

α   β

heavy polypeptides containing ATPase sites; 400+

other polypeptides:
~55 to 125 (intermediate)
~20 (light)

binding site for stationary microtubule

c

Figure 10-12 Dynein structure. **(a)** A single, isolated dynein arm as seen in the scanning transmission electron microscope. × 425,000. (Courtesy of K. A. Johnson; reproduced from *J. Cell Biol.* 96:669 [1983] by copyright permission of the Rockefeller University Press.) **(b)** Tracing of the dynein arms in **(a)**. Although some variations are noted in different species, the polypeptides usually include two or three heavy components with molecular weights in excess of 400,000; these large polypeptides correspond to the globular head units. ATPase activity is associated with the head units. Also present are variable numbers of polypeptides of intermediate (55,000 to 125,000) and low (about 20,000) molecular weight; most of these occur in the stalks and base. **(c)** Tentative positions of the polypeptides in the dynein complex. Molecular weights are given in thousands.

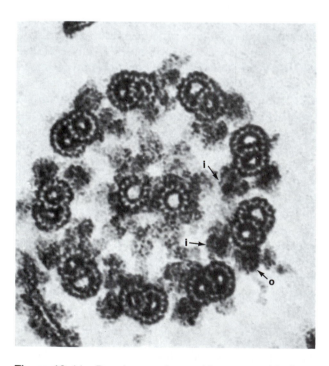

Figure 10-11 Dynein arms (arrows) in a sea urchin (*Lytechinus*) flagellum in cross section. Tannic acid has been added to the fixing solution to outline the microtubule subunits in a negative image. i, inner dynein arms; o, outer dynein arms. × 590,000. (Courtesy of K. Fujiwara, from *J. Cell Biol.* 59:267 [1963], by permission of the Rockefeller University Press.)

Of the various dynein polypeptides, the most significant for the crossbridging function are the *heavy chains*, which occur in these crossbridges in numbers equal to the number of globular head units. Flagellar dyneins contain two or three heavy chains, and cytoplasmic dyneins two. Each heavy chain, with molecular weight averaging from about 400,000 to 550,000, forms a head and stalk of a dynein microtubule motor. A rat cytoplasmic dynein heavy chain, for example, was recently deduced by Z. Zhang and his colleagues from its gene sequence to contain 4464 amino acids with a total molecular weight of 532,213. The heavy chain head contains binding sites for ATP and microtubules and carries out the steps converting chemical energy to the mechanical energy of microtubule sliding. The tail, with its associated smaller polypeptides, forms the "cargo" binding site—the site attaching structures moved by the microtubule motor. In general the cargo binding site is specialized to bind either a microtubule or one of a group of membrane-bound organelles, including the cell nucleus, mitochondria, or Golgi, secretion, or endocytotic vesicles. Binding specificity of the tail is probably provided by the smaller polypeptides.

Dynein Crossbridges and Microtubule Sliding Much remains to be learned about the molecular basis of microtubule crossbridging. However, observations of the biochemistry of ATP breakdown by the arms, and the effects of this breakdown on crossbridge binding and structure, have allowed researchers to fill in some details of the crossbridging cycle.

Several of these observations come from experiments with flagella or cilia in sea urchin sperm, the protozoan *Tetrahymena*, and the green alga *Chlamydomonas*, all favorite subjects for dynein studies. In all these cells, flagella or cilia become locked into nonmoving, rigid bends or waves when ATP supplies are exhausted. Under the electron microscope the dynein arms in this *rigor state*, as it is called, extend at approximately right angles to the direction of movement and are tightly attached at both ends to the surfaces of two adjacent microtubules. Addition of fresh ATP to flagella in rigor restores their flexibility and often restarts flagellar beating. As part of the restoration, ATP is hydrolyzed. Examination of ATP-restored flagella shows that many of the arms have released their attachment at one end; the unattached arms now extend from the microtubule surface at an angle of 45°.

These and other observations establish several important characteristics of the crossbridging mechanism:

1. Dynein arms undergo conformational changes that adjust their angle of attachment to the microtubule surface between approximately 90° and 45°.

2. ATP binding, but not hydrolysis, is required to release dynein arms attached in the 90° position.

3. ATP is hydrolyzed at some point after release of dynein arms; hydrolysis precedes or accompanies movement of the arms to the 45° position.

4. Release of the products of ATP hydrolysis (ADP and phosphate) does not occur until dynein arms reattach to an adjacent microtubule after ATP hydrolysis.

The hypothetical crossbridging cycle shown in Figure 10-13, advanced by P. Satir and others, combines these observations and shows how dynein arms may operate to produce a sliding force between microtubules. The crossbridging cycle shown in the figure is greatly simplified; the actual mechanism, which is not completely understood, is undoubtedly more complex. The cycle may be considered to begin with a dynein arm tightly linked to an adjacent microtubule at the 90° position (Fig. 10-13a). Binding an ATP molecule causes a conformational change at the tip of the dynein arm, altering the binding site so that it no longer fits the adjacent microtubule. Under these conditions the arm releases (Fig. 10-13b) and hydrolyzes its bound ATP. The products of hydrolysis remain bound to the arm (Fig. 10-13c). ATP hydrolysis causes a conformational change in the arm, shifting it to the 45° position and reactivating the microtubule binding site at its tip (Fig. 10-13d). In this condition the arm is like a compressed molecular spring storing much of the energy released by ATP hydrolysis. Reattachment to the adjacent microtubule releases the products of ATP hydrolysis (Fig. 10-13e) and triggers release of the arm from its 45° position. In response the arm swivels forcefully through an arc of 45°, sliding the attached microtubule by an equivalent distance (Fig. 10-13f). The arm is now ready to bind a second ATP and repeat the cycle. Each attach-swivel-release cycle slides the adjacent microtubules past each other by an increment of about 16 nm, or approximately the length of two heterodimers. The combined action of the hundreds or thousands of arms between the doublets produces a smooth and continuous sliding motion.

In systems in which dynein or other crossbridges move organelles along microtubules the base of the crossbridge is attached to the organelle. The opposite end of the crossbridge "walks" the organelle along a microtubule toward its minus and by attaching, swiveling, and releasing from the microtubule surface.

Microtubule Sliding Powered by Dynein in Flagella and Cilia

Much of our information about dynein structure and function was developed through the study of flagella

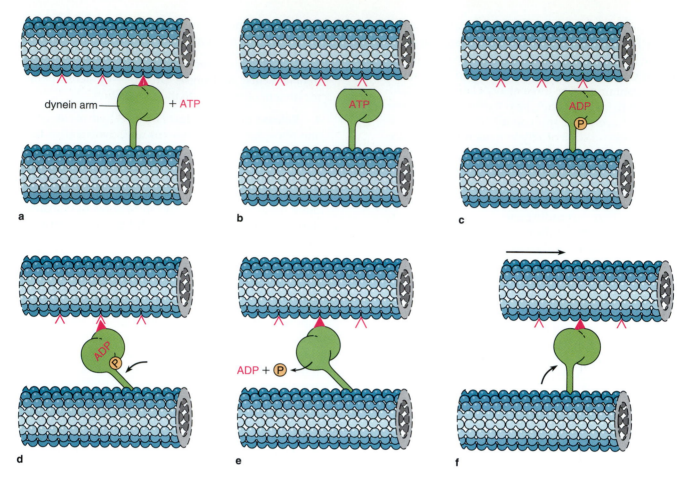

Figure 10-13 Steps in a provisional dynein crossbridging cycle, shown in simplified form (see text). Dynein crossbridges also move cell structures such as mitochondria along microtubules instead of sliding one microtubule over another. In this type of motility the structure would be substituted for the bottom microtubule in each part of this figure. The crossbridge base would be fixed to the structure, and the tip would move the structure along the microtubule at the top of each part by attaching, swiveling, and releasing from the microtubule surface.

and cilia. These motile structures, by which cells swim through a liquid medium or move fluids over their surfaces, occur in almost all eukaryotic organisms.

The 9 + 2 System of Microtubules in Flagella and Cilia   With very few exceptions, the arrangement of microtubules in flagella and cilia, called the *9 + 2 system* or *flagellar axoneme*, consists of a circle of nine fused, double microtubules, the *peripheral doublets*, arranged around two central microtubules, the *central singlets* (see Figs. 10-11 and 10-14). The circle of peripheral doublets has an outer diameter of about 200 nm, placing the 9 + 2 system just within the resolving power of the light microscope.

Each of the two central singlets is a complete microtubule. The peripheral doublets contain one complete microtubule, to which is joined a partial microtubule that appears *C*-shaped in cross section

(see Fig. 10-5). In the 9 + 2 system the complete microtubule of each peripheral doublet is the *A subtubule*; the partial microtubule is the *B subtubule* (see Fig. 10-14). Dynein arms connect in pairs to the A subtubule.

The 9 + 2 system is held together by a complex system of connecting elements. The central singlets are connected by a surrounding *sheath* and a *bridge* that runs directly between them. The A subtubule of each doublet is connected to the central sheath by a *spoke*. Another connecting element, the *nexin link* (from the Latin *nexere* = "to bind"), runs between the peripheral doublets, usually attaching the A subtubule of one doublet to the B subtubule of the next.

Longitudinal sections of flagella reveal that the arms, spokes, and other connecting elements repeat at regular intervals (Fig. 10-15). The distances between the elements are all multiples of the basic 8-nm length of a tubulin heterodimer, suggesting that a heterodi-

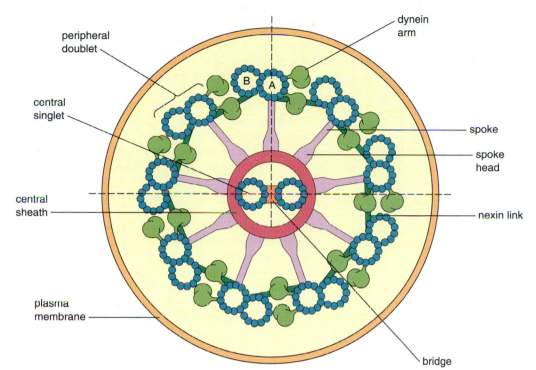

Figure 10-14 Microtubules and connecting elements in the 9 + 2 system of a flagellum. The complete A subtubules of a doublet usually have walls made up of a circle of 13 protofilaments; the B subtubules consist of a partial circle usually containing 10 to 11 protofilaments. The direction pointed by the dynein arms allows the point of view with respect to the tip or base of the flagellum to be determined. If the arms extend from the A subtubules in a clockwise direction, the flagellum is being viewed from base to tip. Compare with Figure 10-6.

mer "dumbbell" has a variety of binding sites that may link to dynein arms, nexin links, spokes, or central sheath elements and bridges. Depending on the position of a heterodimer in the walls of flagellar microtubules, one or more or none of these elements may attach.

The 9 + 2 system appears to be the same in cilia and flagella, which differ primarily in their number per cell, their relative length, and their beat pattern. Flagella occur in numbers of only one or a few per cell; cilia number in hundreds to thousands. Flagella are at least two to three times longer than the longest cilia, which usually range from 5 to 25 μm in length. In some sperm cells, particularly in insects, flagellar length may range up to several millimeters. (In *Drosophila*, the sperm flagellum is longer than the entire adult male animal!) Both flagella and cilia propel swimming cells through a liquid medium. Cilia also move fluids over the surfaces of cells that remain stationary within the body of an animal.

Figure 10-15 Arrangement of arms and linking elements in the 9 + 2 system. Depending on the species, the spokes are spaced evenly in groups of two or three along the A subtubules of the doublets. Entire spoke groups, containing two or three spokes per repeating unit, repeat at a 96-nm interval. The sheath elements contacted by the spoke heads are spaced at intervals of 16 nm. Dynein arms occur in pairs at intervals of 24 nm along the A subtubule; nexin links are spaced at intervals of 96 nm.

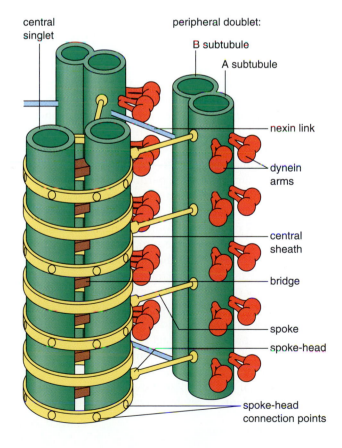

Sperm flagella typically beat in a series of waves that travel continuously in a flat plane from the base to the tip (Fig. 10-16a). In a few protozoan species, flagellar waves beat in a helical path rather than a flat plane. Cilia, in contrast, typically bend stiffly from the base in an oarlike "power" stroke, with the body of the cilium extended and almost straight (Fig. 10-16b). The bend then travels from the base to the tip to accomplish the recovery stroke. The beating in either case may range between 10 and 50 or more cycles per second.

Both beat patterns can be observed in the flagella of some species. In the single-celled alga *Chlamydomonas*, for example, the two flagella beat in the ciliary pattern to propel the cell forward and switch to the flagellar pattern to move it in reverse.

Almost all eukaryotic organisms possess at least some cell types that develop flagella or cilia at various stages of the life cycle. Among the major plant and animal groups, only the flowering plants have no flagellated cells of any type. Many protists have flagella or cilia; the sperm cells of most animals and the male gametes of some fungi and of plants from algae to primitive gymnosperms swim by means of flagella. Cells with flagella or cilia also occur regularly in the body tissues of most animals. In humans, ciliated cells line the respiratory tract, the ventricles of the brain, ducts in the male and female reproductive systems, and epithelia in other parts of the body. In these locations, cilia move fluids over cell surfaces. Cilia also occur in highly modified form in the rods and cones of the retina and hair cells of the organ of Corti in the inner ear; these structures have a sensory rather than motile function.

Exceptions to the 9 + 2 arrangement of microtubules are rare and limited almost exclusively to sperm flagella in a few animal groups. The exceptions include a 9 + 0 pattern, in which the central singlets are missing, as in the sperm tails of a mayfly and an annelid worm. Some other species, among earthworms for example, have sperm cells with a 9 + 1 pattern, in which there is a single central tubule. A few other variations, some that differ considerably from the 9 + 2 pattern, are found in the sperm cells of insect species. Generally these modifications are limited to sperm cells; cilia in body tissues of these animals have the usual 9 + 2 arrangement.

All the sperm cells with variations of the standard 9 + 2 system are motile. However, the flagellar beat is slower and less powerful and the beating differs from the standard pattern. The fact that sperm cells with unusual microtubule arrangements are still motile indicates that the 9 + 2 system, rather than being absolutely essential for flagellar movement, is probably the most efficient of several possible arrangements.

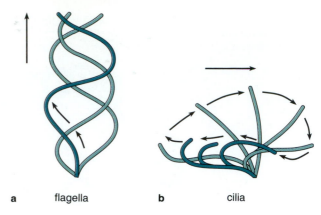

a flagella   **b** cilia

Figure 10-16 Flagellar and ciliary beating patterns. **(a)** The flagellar beating pattern, which consists of a series of smooth waves vibrating in a flat plane. The waves start at the base of the flagellum and proceed toward the tip. **(b)** The ciliary beating pattern, in which the shaft bends at the base in an oarlike power stroke. In the recovery stroke (dark green) the bend travels from base to tip.

Microtubule Sliding and the Flagellar Beat   As long ago as 1959, B. A. Afzelius suggested that flagellar beating is produced by active microtubule sliding, powered by the arms extending from the doublets. A series of major experiments have since provided clear and elegant support for this proposal.

The ATPase activity of dynein was first discovered in flagella. During the 1970s, I. R. Gibbons noticed that when flagella were stripped of their plasma membranes by exposure to detergents, the arms could be removed or added back by adjustments in salt concentration. When the arms were removed, the 9 + 2 systems lost their ATPase activity. The lost activity could be detected among the proteins released from the demembranated flagella. Adding the arms back to the 9 + 2 systems restored their ATPase activity. Gibbons coined the name *dynein* for the flagellar ATPase, meaning "force protein."

Several experiments later demonstrated that microtubules slide actively and forcefully in the production of flagellar movements. The most graphic was carried out by K. E. Summers and Gibbons, who isolated flagella and removed their membranes by treating them with a detergent. The isolated, demembranated flagella were then treated briefly with a protein-digesting enzyme, which left the dynein arms intact but partly digested the spokes and nexin links. Addition of ATP to these digested preparations caused the peripheral microtubule doublets, no longer held by their links, to slide actively and forcefully out of the 9 + 2 system (Fig. 10-17).

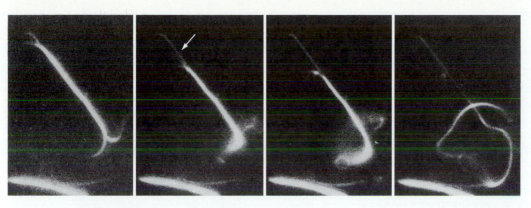

Figure 10-17 Successive darkfield light micrographs showing microtubule doublets (arrow) sliding actively from an isolated 9 + 2 system. ATP was added in the first frame. × 1,350. (Courtesy of I. R. Gibbons, from *Proc. Nat. Acad. Sci.* 68:3092 [1971].)

From their work, Summers and Gibbons proposed, in agreement with other investigators, that dynein arms produce the force sliding doublets in the 9 + 2 system. (The sliding involves movement of one doublet along another, not movement of A subtubules over B subtubules.) The spokes and nexin links, according to their model, act as elastic connectors that hold the doublets together and prevent them from sliding out of the 9 + 2 system during movement. The restriction imposed by the spokes and links forces the entire flagellum to bend to accommodate the internal displacement of the doublets due to sliding. Thus, according to Summers and Gibbons's model, the dynein arms produce the force for sliding and the spokes and nexin links convert the sliding into flagellar bends.

More recent experiments generally support these conclusions. Among the most definitive are experiments by R. B. Vallee, B. M. Paschal, and their colleagues, who isolated dynein arms from sea urchin flagella and attached them to the surface of a glass slide. Microtubules were then added to the slide in a solution containing ATP and other factors required for active dynein crossbridging. The dynein arms cycled actively between the glass and the microtubules, causing the microtubules to glide smoothly over the surfaces of the slides.

Other confirming evidence comes from the study of mutants, which gives insights into the possible functions of other flagellar structures. For example, in *Chlamydomonas* mutants lacking spokes and central singlets, flagella are paralyzed and incapable of movement even though dynein arms are present. Additional "bypass" mutations are able to cancel the effects of the missing spokes and central singlets, producing flagella that can move in altered waveforms and only in the flagellar pattern. These results indicate that spokes and central singlets contribute to the form taken by the flagellar bends but do not generate the force for bending. The results also suggest that spokes or central singlets may contribute

"switches" turning the flagellar beat on or off; when missing, the beat is locked in the off position unless a second mutation releases the doublets and dynein arms from the control. (Information Box 10-1 describes a human mutation in which both dynein arms are missing in flagella and the curious set of symptoms produced by this defect.)

Many flagellated cells, particularly among the protozoa and algae, are able to start and stop flagellar beating or switch between different beating patterns. *Paramecium*, for example, can quickly stop and reverse its ciliary beating if an obstacle is encountered; *Chlamydomonas* can switch almost instantly between a ciliary and flagellar beating pattern. Similarly, sperm cells of most animals are quiescent until their release, when a "switch" activates movement of the flagellum.

These controls appear to depend on changes in Ca^{2+} concentration and cAMP-dependent protein phosphorylations. For example, C. B. Lindemann, J. R. Tash and A. R. Means, and others found that initiation of flagellar movement in sperm and other flagellated cells is dependent on cAMP; a protein kinase activated by cAMP probably adds phosphate groups to one or more as yet unidentified flagellar proteins to start the movement. Flagellar beating is turned off by a Ca^{2+}/calmodulin-activated phosphatase that removes the phosphate groups.

Switching between flagellar and ciliary beats appears to be controlled by adjustments in Ca^{2+} concentration. In *Paramecium* and *Chlamydomonas* and many other cell types, low Ca^{2+} concentrations induce the ciliary pattern typical of normal swimming; elevated Ca^{2+} concentrations trigger the flagellar pattern typical of avoidance. The molecular steps linking alterations in Ca^{2+} concentration to beat switching remain unknown.

Centrioles, Basal Bodies, and Generation of Cilia and Flagella The microtubules of cilia and flagella arise directly from centrioles. These structures contain a

Dynein Arms and the Immotile Cilia Syndrome

In 1933 a German physician, M. Kartagener, noted an odd coincidence between three inherited human abnormalities: acute bronchitis, sinusitis, and reversal of the position of the heart from the left to the right side of the body. Later it was noted that patients with these conditions frequently also had chronic headaches and, if male, were sterile. The complete syndrome is inherited with a frequency of 1 in 15,000 persons.

The pieces of this medical puzzle were put together in 1975 by B. A. Afzelius and his colleagues at the Wenner-Gren Institute in Stockholm. Afzelius discovered that sterile males with Kartagener's syndrome had immotile flagella in sperm and other body cells. In the flagella the 9 + 2 system lacked dynein arms (see the figure in this box). In some cases the spoke heads or inner sheath and one or both central singlets were also missing.

Afzelius's discovery explained the connection between male sterility, bronchitis, sinusitis, and the headaches. The cilia in cells lining the respiratory system were paralyzed and unable to maintain the flow of mucus that normally removes irritating and infectious matter from the lungs and respiratory tract. This deficiency explains the

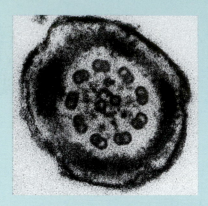

sinusitis and bronchitis. Presumably an insufficient flow of fluids in the ventricles of the brain, normally maintained by ciliated cells lining these cavities, produced the headaches. The connection between nonmotile cilia and the reversed position taken by the heart during embryonic development remains unexplained.

---

system of microtubules clearly related to the 9 + 2 system of flagella (Fig. 10-18). Nine sets of triplet microtubules make up the outer circle of the centriole. Each triplet contains one complete microtubule, the A subtubule, to which are attached a row of two partial microtubules, the B and C subtubules (see also Fig. 10-5b). Centriole triplets are thus analogous in structure, location, and arrangement to the doublets of the flagellar 9 + 2 system of microtubules. The triplets are embedded in deposits of dense material that vary in pattern from one species to another.

There are no central microtubules in centrioles equivalent to the central singlets of flagella. Centrioles appear almost structureless in the center except near one end, where the interior contains a "cartwheel" pattern of radiating spokes. The cartwheel has a central hub, with nine spokes that radiate from the center and connect with the A subtubule of each peripheral triplet (see Fig. 10-18).

During generation of a flagellum the A and B subtubules of each triplet lengthen at the end of the centriole opposite the cartwheel, giving rise to the A and B subtubules of the 9 + 2 system (Fig. 10-19a).

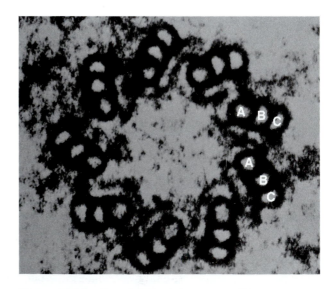

Figure 10-18 Centriole from a mouse cell in cross section. The cartwheel is faintly visible in the centriole lumen. The A, B, and C subtubules are marked for two of the triplets. × 360,000. (Courtesy of E. de Harven and Academic Press, Inc.; from *The Nucleus*, ed. A. J. Dalton and P. Haguenau, 1968.)

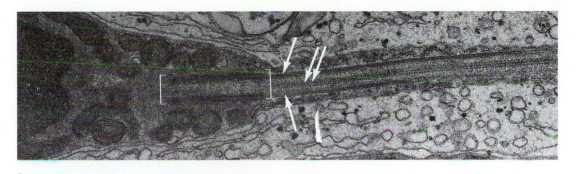

a

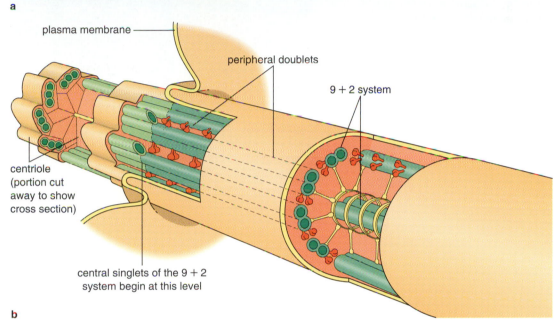

plasma membrane

peripheral doublets

9 + 2 system

centriole
(portion cut
away to show
cross section)

central singlets of the 9 + 2
system begin at this level

b

Figure 10-19 Centrioles, basal bodies, and the generation of flagella. **(a)** The centriole persisting as the basal body (bracket) in the flagellum of a frog sperm. The peripheral triplets of the basal body have given rise directly to the peripheral doublets of the 9 + 2 system (single arrows). The central singlets (double arrow) arise without any direct connection to centriole microtubules. × 19,500. (Courtesy of B. R. Zirkin.) **(b)** Relationship of the microtubules of the basal body to the 9 + 2 system of microtubules in a flagellum.

The central singlets of the 9 + 2 system arise at the border between the centriole and the growing axial complex without any connections to centriole microtubules. In most flagella no structures arise from the C subtubules. The elongating 9 + 2 system is covered by a sheathlike extension of the plasma membrane. When development is complete, the centriole remains attached to the microtubules of the 9+2 complex as a *basal body* (Fig. 10-19b).

Centrioles apparently have little or no motile function in cilia or flagella, because flagella broken from cell surfaces without their basal bodies can beat in a normal wave pattern when supplied with ATP. Instead, basal bodies may serve simply as cytoplasmic anchors for the flagellum. They may also contribute to the exceptional stability of flagellar microtubules

by serving as caps for their minus ends, preventing disassembly of tubulin heterodimers from these ends.

Dynein and Kinesin in Other Cytoplasmic Movements

Kinesin was first identified as a microtubule cross-bridging "motor," along with a cytoplasmic dynein, in studies of the movements transporting substances through the axons of nerve cells (see Fig. 5-16). Structures such as vesicles are transported simultaneously in both directions through axons, which may be as long as a meter or more in some animals. This vesicle movement was found to depend primarily on microtubules, which are packed in bundles that extend lengthwise in axons (Fig. 10-20). Tests for microtubule polarity show that the microtubules are aligned uniformly with their plus ends directed toward the axon terminals.

Movement of structures along microtubules can be observed in both intact axons and in cytoplasm that has been extruded by squeezing axons (Figure 10-21a shows a vesicle moving along the surface of an extruded axon microtubule). With darkfield or Nomarski light microscopes, vesicles can be seen to move in opposite directions on the same microtubule.

Kinesin was identified as the "motor" moving organelles toward the plus ends of microtubules—toward axon tips—in elegant experiments carried out in 1984 by R. D. Vale, T. S. Reese, and M. P. Sheetz. These investigators extracted and purified a protein from squid nerve axons that, when coated on the surfaces of plastic beads, could make the beads move over microtubules in the plus direction. When attached to glass microscope slides, the protein could make isolated microtubules glide actively over the surface of the glass. The movements proceed only if ATP is added to the experimental systems. The newly discovered motor was named *kinesin*, from the Greek *kinein* = to move. The Experimental Process essay by Sheetz on p. 300 describes events leading to the discovery of kinesin.

After its initial discovery, kinesin was found to be present as a cytoplasmic microtubule motor in many organisms, including the nematode *Caenorhabditis*, *Drosophila*, sea urchin, and humans. At least 27 different kinesin types have been discovered, all related but with distinct properties, indicating that the kinesins form an extended family with considerably more variation than the dyneins. All members of the kinesin family, however, share the same basic structure (Fig. 10-21b), including four polypeptides—two heavy chains of 110,000 to 130,000 daltons and two light chains of 60,000 to 70,000 daltons. Neither polypeptide type appears to be related to dynein. The N-terminal ends of the heavy chains form globular head units; the C-terminal ends extend into alpha helices that wind around each other in a coiled coil (see Information Box 2-4). The two globular head units contain the sites hydrolyzing ATP and binding microtubules. At its tip the tail broadens into a region that contains the light chains and the site binding vesicles and other organelles moved by kinesin. Almost all the known kinesins move their cargoes toward the plus end of microtubules.

The motor moving organelles in the opposite direction, toward the minus ends of the axonal microtubules, was identified by B. M. Paschal and R. B. Vallee and their colleagues as the form of dynein now known as cytoplasmic dynein. The two microtubule motors, cytoplasmic dynein and kinesin, probably ac-

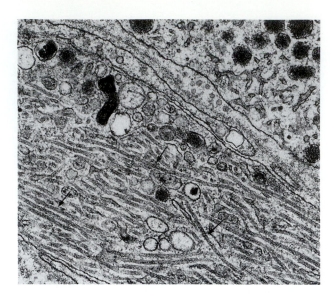

Figure 10-20 Microtubules (arrows) in the axon of a rat neuron. × 30,000. (Courtesy of P. Dustin and J. Flamend-Duran, from *Microtubules*, 2nd ed., New York: Springer-Verlag, 1984.)

count for two-directional transport of structures in axons. If kinesin is attached to a structure, it moves along axonal microtubules toward the axon tip; if cytoplasmic dynein is attached, movement is toward the cell body.

Members of the kinesin family and cytoplasmic dynein have been detected in a variety of cell types and species. The two motors produce a wide range of microtubule-based cytoplasmic movements, including the two-way vesicle traffic between the Golgi complex and the plasma membrane. This traffic, which forms part of the processes of exocytosis and endocytosis (see p. 591), moves along microtubule tracks in which the microtubules are typically arranged with their minus ends toward the cell interior and their plus ends toward the plasma membrane. The movement stops if cells are exposed to colchicine, indicating that microtubule-based mechanisms are directly involved. In this traffic, kinesin is expected to move vesicles toward the plasma membrane and dynein toward the cell interior. Both motors may also be involved in the movements dividing the chromosomes in mitosis and meiosis (see Chapters 24 and 25).

The amount of force generated by single kinesin motors was recently measured in elegant experiments by S. C. Kuo and M. P. Sheetz and by S. Block and K. Svoboda and their colleagues. The experiments, which used a newly developed technique called "optical tweezers" (see Information Box 10-2), followed the

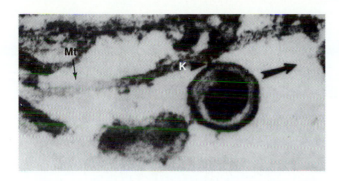

Figure 10-21 Vesicle movement by kinesin. **(a)** A vesicle moving along a microtubule in cytoplasm extruded from a squid axon. K, the kinesin crossbridge connecting the vesicle to the microtubule (MT). × 80,000. (Courtesy of R. J. Lasek; reproduced from *J. Cell Biol.* 101:2181 [1985], by copyright permission of the Rockefeller University Press.) **(b)** Kinesin structure (see text). **(c)** A kinesin molecule stepping along a microtubule.

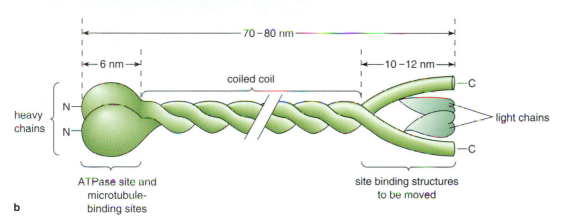

b

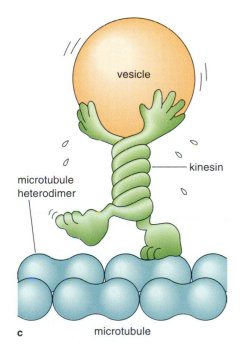

c

motions along microtubules of single kinesin motors attached to microscopic silica or latex beads. Using the optical tweezers, the investigators found that a force opposing movement by single kinesin motors could be increased to values of about 2 to 5 piconewtons before the motors stalled and stopped moving, indicating that a single kinesin molecule can exert forces in this range. The Block group found that as the

opposing force approached stalling values, the kinesin motors moved in jerks, with the distance traversed in each jerk equivalent to about 8 nm—the length of a tubulin heterodimer. This suggests that the kinesin motor moves along a microtubule by "stepping" from heterodimer to heterodimer, possibly by alternate attachment and release cycles by the two heavy chain heads (Fig. 10-21c).

Other microtubule-based mechanisms generating force for cellular movements may be based on treadmilling or microtubule growth. Evidence for these mechanisms is weak, compared to the impressive body of findings supporting sliding movements generated by dynein and other crossbridges, and limited primarily to observations of the spindle, in which some parts of the movement of chromosomes by the spindle may be produced by treadmilling or microtubule growth (see pp. 708–716 for details).

Microtubules assemble by polymerization of dumbbell-shaped heterodimers consisting of one α- and one β-tubulin polypeptide. The assembly, which proceeds to a finely balanced equilibrium between a heterodimer pool and assembled microtubules, is regulated in cells by alterations in Ca^{2+} concentration and combination of heterodimers with MAPs. Alterations in Ca^{2+} concentration provide short-term controls that may lead toward either assembly or disassembly of microtubules; combination with MAPs generally increases microtubule stability. Control by MAPs is

The Experimental Process

Identification of Microtubule Motors Involved in Axonal Transport

Michael P. Sheetz

MICHAEL P. SHEETZ received his B.A. from Albion College in 1968 and his Ph.D. from the California Institute of Technology in 1972. He was a research fellow for two years at the University of California, San Diego. Before joining Duke University Medical Center as Chairman of the Department of Cell Biology, Dr. Sheetz had been a professor at Washington University Medical School in St. Louis since 1985. Before that, he was in the Department of Physiology at the University of Connecticut Health Center at Farmington. Dr. Sheetz' major research contributions have been in the fields of cell motility and membrane structure.

By 1983 the question of how neurons maintain long axonal processes had evolved into the question of how membranous vesicles are transported back and forth between the synapse and the cell. In the 1970s studies of polypeptide assembly in neurons found no detectable protein synthesis in axons or synapses. Instead, newly synthesized proteins are carried from the cell body to the synapse in two major transport pools at two quite different rates. The slower pool (with a rate of 0.01–0.05 μm/sec) contains components of the cytoskeleton (actin, tubulin, and neurofilament proteins). The fast pool (with a rate of 1–5 μm/sec) is composed of vesicular proteins, which presumably move as small axonal vesicles. Over the next two years, two motor activities, one of which was later identified with the protein kinesin,[1] were identified in axons. These findings together with other morphological analyses of axons provided convincing evidence that membranous organelles are transported by ATP-dependent motors along microtubules.

As is often the case, it was the combination of such factors—and some serendipity—which resulted in the identification of the microtubule motor kinesin. Before 1985 no microtubule motors had been identified except for flagellar dynein. Although regions of axons that are rich in the vesicles transported by fast axonal transport are also rich in microtubules, other transport filaments such as actin filaments could have been aligned along microtubules. The identification of the microtubule motor proteins was made possible by the combination of a number of factors including a new technique in light microscopy called video-enhanced differential interference contrast (DIC) microscopy, *in vitro* motility systems, and the unique advantages of squid axoplasm—the cytoplasm extruded from squid giant axons [see p. 138].

The technology of video-enhanced DIC microscopy came from developments in the video and defense industries which allowed subtraction of video images in real time and also contrast enhancement. Images from the video camera were contrast enhanced well beyond what the eye could do, and this brought out noise in the microscope optics. Since much of the noise was the same in both out-of-focus and in-focus images, subtracting an out-of-focus from an in-focus image would give a clean image.[2] Single microtubules (0.025 μm in diameter) or vesicles of 0.050 μm in diameter could be detected routinely, which enabled the first visualization of fast axonal transport.[2] Video-enhanced DIC microscopy revealed small vesicles that moved rapidly along transport filaments at 1–2 μm/sec.

At the same time, an *in vitro* assay for myosin motility was developed using myosin-coated latex spheres. The behavior of the spheres was similar to that of the small vesicles in axoplasm. Namely, the myosin-coated spheres would show Brownian movement until they attached to actin cables from the alga *Nitella*. After binding to the actin, the myosin beads would move at a rate of about 2 μm/sec, much like the axonal transport vesicles.[3] This suggested that the transport vesicles were powered by myosin or by another motor that moves on the transport filaments. This also suggested that the motor protein interacts with the vesicle surface, moving the vesicles in the direction defined by the transport filament. The major difference between the actin/myosin-based motility and axoplasmic transport was that although actin-based movement is unidirectional, the transport filaments support bidirectional movements.

A minor digression is warranted here to describe the limitations of light microscopy, including video-enhanced DIC microscopy. From basic physical principles, the limit of resolution of light microscopy is about 0.2 μm; thus, objects separated by less than that distance appear as a single object. This means that video-enhanced DIC microscopy can detect a single microtubule—which is 0.025 μm in cross section—by inflating its diameter (utilizing the phenomenon of light diffraction), but it cannot resolve individual filaments within bundles of actin microfilaments or microtubules. Thus, it was necessary to analyze the same transport filament in both the light and electron microscopes to confirm that a single microtubule can support bidirectional transport (Fig. A).

Efforts to isolate the axoplasmic transport motor(s) were greatly aided by the serendipitous discovery that microtubules would move on glass coated with axoplasmic supernatant freed of vesicles. Because of the *in vitro* myosin motility studies, this suggested that axoplasmic supernatants contained a microtubule motor which would bind to glass in an active form. An assay was developed following the movement of added microtubules over axoplasm fractions bound to a glass slide.[4]

By adding various fractions derived from axoplasm and several brain sources, a protein termed *kinesin* was identified as a motor driving fast axonal transport.

The direction of movement driven by the kinesin motor remained to be identified. This movement was determined by using microtubules polymerized from a centrosome. Centrosomes seed the polymerization of microtubules with their plus or fast-growing ends outward. In this system, purified kinesin moved toward the plus end of the microtubules. In the axon it was found that the microtubules are arrayed in parallel with their plus ends pointed toward the periphery. Thus, kinesin must produce fast transport in the anterograde direction, that is, toward the axon tips. At the same time a factor was defined which was separable from kinesin and would move objects in the retrograde direction, away from the axon tips.

Using the same assay of microtubule movement on glass, it was subsequently found that MAP1C was the retrograde motor and that it shared many properties with dynein (hence the name *cytoplasmic dynein*, to distinguish it from axonemal dynein).[5] Many motors have been added to the list and it seems that a large number of different cellular processes involve the directed movement of objects on microtubules.

References

[1]Vale, R. D.; Reese, T. S.; and Sheetz, M. P. Identification of a novel force-generating protein, kinesin, involved in microtubule-based motility. *Cell* 42:39–50 (1985).

[2]Allen, R. D.; Metuzals, J.; Tasaki, I.; Brady, S. T.; and Gilbert, S. P. Fast axonal transport in squid giant axon. *Science* 218: 1127–28 (1982).

[3]Sheetz, M. P., and Spudich, J. A. Movement of myosin-coated fluorescent beads on actin cables *in vitro. Nature* 303(5912): 31–35 (1983).

[4]Kron, S. J., and Spudich, J. A. Fluorescent actin filaments move on myosin fixed to a glass surface. *Proc. Natl. Acad. Sci. USA* 83:6272–76 (1986).

[5]Paschal, B. M.; Shpetner, H. S.; and Vallee, R. B. MAPIC is a microtubule-activated ATPase which translocates microtubules in vitro and has dynein-like properties. *J. Cell Biol.* 105:1273–1282 (1987).

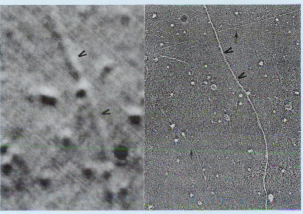

a

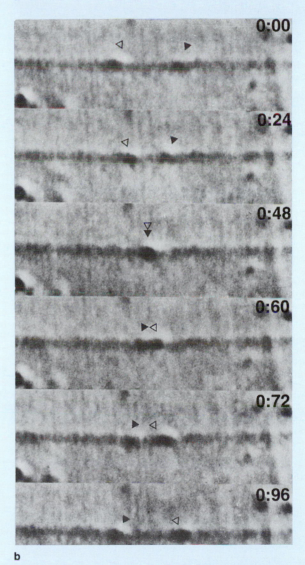

b

Figure A **(a)** Corresponding video-enhanced DIC image left) and electron micrograph (right) of a single microtubule (open arrowheads). **(b)** A series of video-enhanced DIC images showing bidirectional movement of two vesicles (triangles) on a microtubule.

Optical Tweezers for Microscopic Objects

A new optical technique developed by A. Ashkin and his colleagues makes it possible to hold and move objects as small as individual microtubules and microtubule motors and measure forces generated by the motors. The technique depends on the fact that as an object diffracts (bends) or reflects light, small mechanical forces are created on the object in opposition to the direction of diffraction or reflection. Although very small in magnitude—in the piconewton range—the forces are strong enough to move or trap microscopic objects (1 piconewton = 10^{-12} Newtons = 0.1 µdyne).

In order to move microscopic objects, a laser light beam is focused on the object through a high-resolution light microscope objective lens (see Appendix p. 783). The focus produces a spot of light that is brightest in the center and diminishes in a gradient of intensity in all directions from the center. If the refractive index of the object (see p. 784) is higher than that of the surrounding medium, bending of the light beam by the object produces a force that tends to move the object toward regions of brighter light. If the object is small enough, it will move toward brighter regions until it reaches the brightest center of the focused spot, where it is trapped and held. Once held by these "optical tweezers," the object can be moved about at will by traversing the microscope stage.

With high-resolution lenses the spot of light acting as optical tweezers can be focused on individual refracting structures inside cells such as vesicles or even microtubules. For the work with kinesin the microtubule motor was attached to a microscopic latex or silica bead that was used as the refracting object. When added to a preparation of microtubules and supplied with ATP, the beads could be observed moving along the microtubules. The beads were then gripped by the optical tweezers. Once held by the tweezers, the force required to oppose the movement and stop the kinesin motor could be calculated.

---

fine-tuned by addition and removal of phosphate groups, in control mechanisms linked in animal cells to the cAMP or $InsP_3$/DAG receptor-response pathways. Adjustment of the balance by these factors allows cells to precisely regulate microtubule assembly and disassembly, so that microtubules appear and disappear at controlled times and locations in the nucleus and cytoplasm.

Microtubules produce cellular motion primarily by the activity of crossbridging proteins of the dynein and kinesin families, which hydrolyze ATP and undergo conformational changes that convert chemical energy released by ATP hydrolysis into mechanical energy. Other possible motile mechanisms involving microtubules are controlled growth and treadmilling.

Microtubules are characteristic eukaryotic structures that have no direct parallels among prokaryotes. The universal occurrence of microtubules in eukaryotes indicates that they arose as part of the structural and functional adaptations leading to the first appearance of eukaryotic cells. (Bacterial flagella, which are based on proteins and a motile mechanism that are completely different from those of eukaryotic flagella, are described in Supplement 10-1.)

For Further Information

Suggestions for Further Reading

Allen, R. D. 1987. The microtubule as an intracellular engine. *Sci. Amer.* 256:42–49 (February).

Bayley, P. M. 1990. What makes microtubules dynamic? *J. Cell Sci.* 95:329–334.

Berg, H. C. 1975. How bacteria swim. *Sci. Amer.* 233: 36–44 (August).

Burns, R. G. 1991. α-, β-, and γ-tubulins: Sequence comparisons and structural constraints. *Cell Motil. Cytoskel.* 20:181–189.

Caplan, S. R., and Kara-Ivanov, M. 1993. The bacterial flagellar motor. *Internat. Rev. Cytol.* 147:97–164.

Chapin, S. J., and Bulinski, J. C. 1992. Microtubule stabilization by assembly-promoting MAPs: A repeat performance. *Cell Motil. Cytoskel.* 23:236–243.

Cleveland, D. W. 1990. Microtubule MAPping. *Cell* 60: 701–702.

Doetsch, R. N., and Sjoblad, R. D. 1980. Flagellar structure and function in eubacteria. *Ann. Rev. Microbiol.* 34: 69–108.

Dustin, P. 1980. Microtubules. *Sci. Amer.* 243:66–76 (August).

Dustin, P. 1984. *Microtubules*, 2nd ed. New York: Springer-Verlag.

Endow, S. A., and Titus, M. A. 1992. Genetic approaches to microtubule motors. *Ann. Rev. Cell Biol.* 8:29–66.

Fosket, D. E., and Morejohn, L. C. 1992. Structural and functional organization of tubulin. *Ann. Rev. Plant Physiol. Plant Molec. Biol.* 43:201–240.

Gelfand, V. J., and Bershadski, A. D. 1991. Microtubule dynamics: Mechanism, regulation, and function. *Ann. Rev. Cell Biol.* 7:93–116.

Glover, D. M., Gonzalez, C., and Raff, J. W. 1993. The centrosome. *Sci. Amer.* 268:62–68 (June).

Goldstein, L. S. B. 1993. With apologies to Sheherazade: Tails of 1001 kinesin motors. *Ann. Rev. Genet.* 27:319–351.

Jones, C. J., and Aizawa, S. 1991. The bacterial flagellum and flagellar motor: Structure, assembly, and function. *Adv. Microbiol. Physiol.* 32:109–172.

Joshi, H. C. 1993. γ-Tubulin: The hub of cellular microtubule assemblies. *Bioess.* 15:637–643.

Kimble, M., and Kuriyama, R. 1992. Functional components of MTOCs. *Internat. Rev. Cytol.* 136:1–50.

Kuo, S. C., and Sheetz, M. P. 1993. Force of single kinesin molecules measured with optical tweezers. *Science* 260: 232–234.

Lee, G. 1990. Tau proteins: An update on structure and function. *Cell Motil. Cytoskel.* 15:199–203.

Macnab, R. M. 1992. Genetics and biogenesis of bacterial flagella. *Ann. Rev. Genet.* 26:131–158.

Murray, J. M. 1991. Structure of flagellar microtubules. *Internat. Rev. Cytol.* 125:47–93.

Svoboda, K., Schmidt, C. F., Schnapp, B. J., and Block, S. M. 1993. Direct observation of kinesin stepping by optical trapping interferometry. *Nature* 365:721–727.

Tash, J. S. 1989. Protein phosphorylation: The second messenger signal transducer of flagellar motility. *Cell Motil. Cytoskel.* 14:332–339.

Tucker, J. 1992. The microtubule organizing center. *Bioess.* 14:861–867.

Vale, R. D. 1992. Microtubule motors: Many new models of the assembly line. *Trends Biochem. Sci.* 17:300–304.

Vale, R. D. 1993. Measuring single protein motors at work. *Science* 260:169–170.

Vallee, R. B. 1990. Molecular characteristics of high molecular weight microtubule-associated proteins: Some answers, many questions. *Cell Motil. Cytoskel.* 15:204–209.

Vallee, R. 1993. Molecular analysis of the microtubule motor dynein. *Proc. Nat. Acad. Sci.* 90:8769–8772.

Vallee, R. B., and Sheptner, H. S. 1990. Motor proteins of cytoplasmic microtubules. *Ann. Rev. Biochem.* 59:909–932.

Walker, R. A., and Sheetz, M. P. 1993. Cytoplasmic microtubule-associated motors. *Ann. Rev. Biochem.* 62:429–451.

Wiche, G., Oberkanins, C., and Himmler, A. 1991. Molecular structure and function of microtubule-associated proteins. *Internat. Rev. Cytol.* 124:217–273.

Williamson, R. E. 1993. Organelle movements. *Ann. Rev. Plant Physiol. Plant Molec. Biol.* 44:191–202.

Review Questions

1. Describe the structure of microtubules. What are protofilaments?

2. What are α- and β-tubulin? What relationships do these proteins have to the subunits visible in microtubule walls? What are heterodimers?

3. What are tubulin variants or isotypes?

4. How does GTP enter into microtubule assembly?

5. What are MAPs? What effects do MAPs have on microtubule assembly?

6. Outline the effects of Ca^{2+} on microtubule assembly. How do Ca^{2+}, calmodulin, and MAPs interact in microtubule assembly?

7. What is microtubule polarity? What is the relationship of polarity to cellular motility? What is treadmilling?

8. What are MTOCs? What major types of MTOCs occur in cells? How might MTOCs work in the initiation of microtubule assembly?

9. What is colchicine? Taxol? How do these substances affect microtubules? Why have they been useful in medicine and research?

10. Outline the 9 + 2 system of microtubules and connecting elements in flagella.

11. What experiments demonstrated that the dynein arms of the 9 + 2 system have ATPase activity?

12. Outline the structure of dynein. How is the dynein crossbridging cycle believed to work? Where and how does ATP enter the system? In what direction do dynein crossbridges "walk" with respect to microtubule polarity?

13. What experiments demonstrated that microtubules actually slide in flagella?

14. Compare the structure and function of cilia and flagella.

15. What mechanisms may regulate flagellar beating? What is the relationship of Ca^{2+} concentration to flagellar regulation? Of phosphorylation?

16. What is kinesin? What evidence links kinesin to axonal transport? Compare the activity of kinesin and dynein in microtubule-based motility.

17. Outline the structure of centrioles. How are centrioles related to the 9 + 2 system of flagella? What are the relationships among centrioles, flagella, and basal bodies?

Supplement 10-1

Bacterial Flagella

Many bacteria are propelled through a liquid medium by movements of protein fibers that extend from the cell surface (see Figs. 10-22 and 1-6). Most bacteria possessing the structures have a single flagellum; some species, however, have as many as 10 to 12, often extending in a tuft from one end of the cell. Although the fibers are called *flagella*, they have no structural or functional relationships to eukaryotic flagella. In fact, bacterial flagella generate motion by a mechanism that is completely without parallels in eukaryotes.

Structure of Bacterial Flagella

Bacterial flagella consist of a *hook* and *filament* extending from the cell surface, anchored in a *basal body* embedded in the plasma membrane and cell wall (Fig. 10-23). The filament is a long, slender protein fiber about 20 nm in diameter and from 10 to 20 to a hundred μm in length. The diameter of a filament is thus much smaller than a single microtubule in a eukaryotic flagellum.

Flagellar filaments consist of a hollow tubular polymer of a single protein called *flagellin*, which occurs in a distinct type in each species. Flagellin units assemble in 11 spiraled rows called *protofilaments* to form the wall of a filament (Fig. 10-24). The inner channel, or lumen, extending through the filament is about 6 nm in diameter. Protofilaments are able to slide over each other, allowing them to unwind and rewind between left-handed and right-handed helices without disturbing the tubular structure of the filament. The alternate helical patterns allow swimming bacterial cells to change direction (see below). Either helical arrangement twists the entire filament into a corkscrewlike path.

The hook contains a polymer of a single polypeptide type that is different from the filament protein. Like the lattice formed by the filament monomers, the hook polymer can alternate between several forms that change its overall shape. The hook structure is believed to form a sort of "universal joint" that connects the filament to the basal body, which consists of a central *shaft* bearing a series of *rings* that fit into the plasma membrane and cell wall (see Fig. 10-23; bacterial cell wall structure is outlined in Supplement 7-1).

The Flagellar Motor

The bacterial basal body forms a biological motor that rotates the flagellum in its socket in the wall, much like the propeller of a boat. The S and M rings on the shaft of the basal body (see Fig. 10-23) are believed to

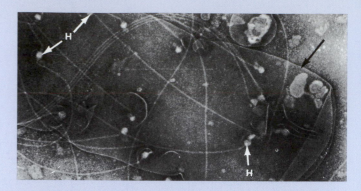

Figure 10-22 Flagella isolated with a portion of the cell wall (arrow) from the bacterium *Proteus mirabilis*. The basal structures of the flagella (H) appear as spherical particles in this preparation. × 50,000. (Courtesy of W. Van Iterson and North-Holland Publishing Company, Amsterdam.)

interact in some way with a pair of motor proteins, *MotA* and *MotB*, to produce the rotational force. The L and P rings are thought to form a bearing through which the shaft rotates in the cell wall.

Evidence that bacterial flagella rotate like propellers was developed in several ingenious experiments by M. R. Silverman and M. I. Simon. Their experiments were designed to get around the difficulty that bacterial flagella are too small to be seen directly to rotate in the light microscope. In one experiment they tethered bacterial cells to a glass slide by coating the slide with antibodies against the flagellar protein. Reaction with the antibodies fixed the flagellar filaments to the slide and prevented them from rotating. Under these conditions the bacterial cells could be seen to rotate instead. In a second experiment, Silverman and Simon attached polystyrene spheres, large enough to be visible in the light microscope, to the surface of an untethered bacterial flagellum by coating the spheres with antiflagellin antibody. Examination of living cells showed that the spheres revolved around the flagellar axis, exactly as expected if the flagella move by rotating in their wall sockets.

Several experiments demonstrated that rotation of the flagellar motor in most bacteria is driven by an H^+ gradient across the plasma membrane. Initially, S. H. Larsen and his colleagues found that agents that make the plasma membrane leaky to H^+, thereby destroying the H^+ gradient, stop flagellar motility even though ATP reserves remain high and available. Later, H. C. Berg and his colleagues showed that an experimentally imposed H^+ gradient can restart flagellar motility in cells that have exhausted their energy reserves and stopped moving. About 1000 protons pass through the flagellar motor for each complete revolution.

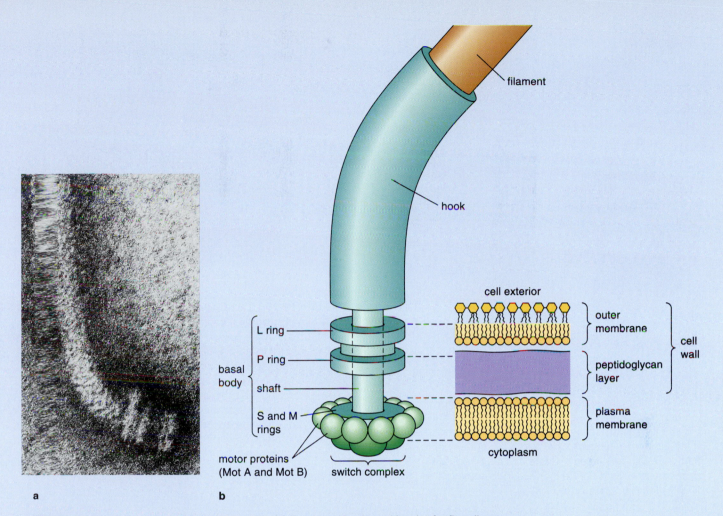

Figure 10-23 Structure of bacterial flagella. **(a)** The basal body and hook of a flagellum isolated from *E. coli*, a gram-negative bacterium. × 46,000. (Courtesy of M. L. DePamphilis and Julius Adler.) **(b)** Structures of the basal body in a gram-negative bacterium and their relationships to the plasma membrane, cell wall, and outer membrane. The entire basal body contains as many as 20 or more different polypeptides in gram-negative bacteria. The switch complex consists of three proteins that switch the flagellum between clockwise and counterclockwise rotation. The basal body of gram-positive bacteria is similar except that there are only two rings on the shaft, one fitting into the plasma membrane and one into the cell wall. The innermost ring is considered to have two functional parts, equivalent to the S and M rings of gram-negative basal bodies. The single outer ring, which fits into the relatively thick peptidoglycan sheath of gram-positive bacteria, is probably equivalent to the P ring of gram-negative cells.

The flagellar motor has some unusual characteristics that limit the possibilities for the rotating mechanism. The speed of rotation and the twisting force remain constant over a wide temperature range. Over this range the efficiency of energy conversion from the H^+ gradient to the mechanical energy of flagellar rotation is very high, approaching 100%. Moreover, the motor works with the same efficiency and temperature relationships in either the clockwise or counterclockwise direction. These observations collectively rule out possibilities such as enzymatically catalyzed formation and breakage of covalent or hydrogen bonds or conformational changes in proteins as the basis for motor function. These mechanisms are all temperature-dependent and would be expected to increase significantly in rate with increases in temperature, rather than remaining constant as the flagellar motor does.

In view of these restrictions, S. Khan and H. C. Berg proposed a model for the mechanism propelling the bacterial motor involving the M ring and the Mot proteins, found by genetic experiments to be necessary for flagellar function. Bacterial mutants lacking the Mot proteins have paralyzed flagella. The MotA protein appears to be the motor protein using the H^+

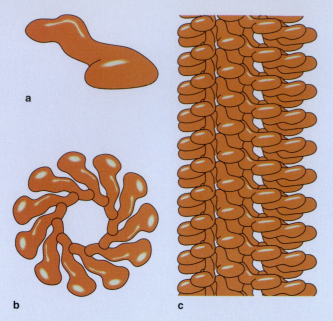

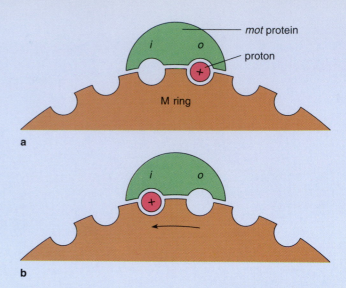

Figure 10-25 The Khan and Berg model for the bacterial flagellar motor in simplified form. In **(a)** a proton has just arrived at the point where the *o* channel meets the M ring. In **(b)** rotation of the M ring carries the proton from the *o* to the *i* channel, from which it can enter the cell. Another proton-binding site (these sites are spaced evenly around the edge of the M ring) now faces the *o* channel, and the process is ready to repeat. Reversal of the motor could be accomplished by a flip in position of the *mot* protein, so that the positions of the *o* and *i* channels are reversed.

Figure 10-24 Arrangement of flagellin polypeptides in the filament of a *Salmonella* flagellum. **(a)** A single flagellin polypeptide. **(b)** A filament in cross section, showing the central lumen. The lumen is wide enough to allow passage of single flagellin polypeptides, which flow through the filament to reach the tip during filament growth. **(c)** The helical packing of flagellin polypeptides in the filament. (Redrawn from originals courtesy of K. Namba, I. Yamashita, and F. Vonderviszt. Reprinted by permission from *Nature* 342:648 [89]; copyright © 1985 Macmillan Magazines Ltd.)

gradient to drive flagellar rotation; MotB is believed to form an elastic link tying MotA to the plasma membrane. In bacteria, oxidative electron transport pushes H^+ to the cell exterior, producing an H^+ gradient that is high outside and low inside (see p. 240). According to the Khan and Berg model (Fig. 10-25) the MotA protein forms a channel in the membrane that conducts hydrogen ions (protons) from outside the plasma membrane to the cell interior. The channel has two segments, one (*o* in Fig. 10-25), that conducts protons from outside the cell to the M ring, and another (*i* in Fig. 10-25) that conducts protons from the M ring to the cell interior. For a proton to pass from outside to inside the cell along the concentration gradient, it must shift in the region of the M ring from the *o* to the *i* channel in the MotA protein. The shift occurs by movement of the M ring, which must rotate for the proton to pass from the *o* to the *i* channel.

Flagellar Rotation and Bacterial Swimming Behavior

The Silverman and Simon experiments answered a long-standing question about bacterial swimming be-

havior. Bacterial cells typically swim smoothly in a more or less straight direction for one to several seconds, tumble or "twiddle" randomly for a much shorter period of about 0.1 second, and then take up smooth swimming again. The tethering experiments revealed that shifts between smooth swimming and tumbling reflect changes in the direction of flagellar rotation. During straight running the motor turns counterclockwise. To initiate the tumbling motion the motor switches to clockwise rotation. Tumbling ceases as the motor switches back to counterclockwise rotation.

The molecular changes underlying the switch between smooth swimming and tumbling depend on the helical winding of protofilaments in the flagellar filaments. During the counterclockwise rotation that produces smooth swimming the protofilaments of a flagellar filament are twisted into a left-handed helix. When the motor reverses direction, the left-handed helix unwinds and the filament rewinds into a right-handed helix. Of the two helices the left-handed one is more stable; if left undisturbed or rotated in a counterclockwise direction, flagellar filaments quickly wind into a helix in this direction.

Unwinding and rewinding of the protofilament helix begin at the base of the flagellar filament and proceed toward the tip. At the point where the helical change is taking place the smooth corkscrew traced by the flagellar filament assumes a random kink or bend. The bend destroys the directional force created by the flagellum and causes the cell to tumble randomly. If the clockwise rotation persists long enough, the entire flagellar filament rewinds into a right-handed helix, producing an unkinked corkscrew that allows a smooth swimming pattern to resume. Usually, however, the motor rotation returns to the counterclockwise direction before the entire filament has time to rewind.

The return to smooth swimming takes place through a momentary pause in rotation. During the pause the untwisting force is removed from the flagellar filaments, and they snap back into the stable left-handed helix. As counterclockwise rotation begins, swimming returns to the smooth mode.

The shift between smooth swimming and tumbling allows bacterial cells to swim toward favorable chemical stimuli and avoid noxious ones. When swimming cells encounter increasing concentrations of substances, such as glucose or galactose, that are registered by cell receptors as a favorable chemical environment, the tumbling mode is suppressed. As a consequence, the cells continue to swim toward the region of greatest concentration of the favorable stimulus. If a substance is encountered that registers as unfavorable, such as acetic acid, the straight running mode is suppressed and tumbling occurs much more frequently. The greater frequency of tumbling, caused by more frequent motor switching, changes the swimming direction more often and increases the probability that the swimming cell will escape the unfavorable stimulus. The responses to noxious chemicals depend on receptors that are linked to the flagellar motors by a series of intermediate proteins in the bacterial cytoplasm. (Supplement 6-1 describes the receptor-response mechanisms of bacteria.)

MICROFILAMENTS AND MICROFILAMENT-GENERATED CELL MOTILITY

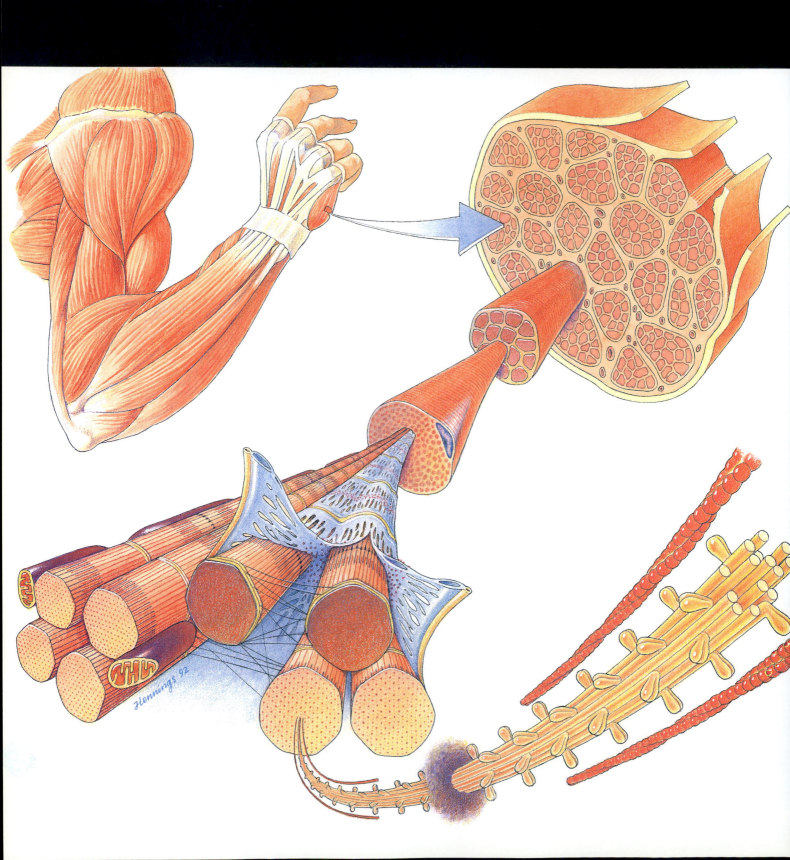

Microfilaments (also called *actin filaments*) produce many kinds of movements in eukaryotes. These movements, like those generated by microtubules, are based either on active microfilament sliding or on pushing movements developed by directional microfilament growth. Besides producing motion, microfilaments, along with microtubules and intermediate filaments, form supportive cytoskeletal networks in the cytoplasm.

Microfilament sliding, which accounts for most microfilament-generated motility, is powered by protein crossbridges that cycle in an attach-pull-release cycle to move one microfilament with respect to another or to move other cellular structures along microfilaments. The movements that result from microfilament sliding take two different major forms in eukaryotic cells, *cytoplasmic streaming* and *contraction*.

In cytoplasmic streaming, segments of the cytoplasm flow actively and directionally from one region to another. The elements moved by cytoplasmic streaming, which is relatively slow, include a wide variety of molecules, particles, and organelles within cells. More organized forms of cytoplasmic streaming push cell borders into extensions or, in *ameboid motion*, move entire cells.

In contraction, sliding microfilaments forcibly constrict or shorten cell segments or whole cells. In less organized forms the sliding generates cytoplasmic movements that narrow or constrict cells during embryonic development and form the cleavage furrow that separates the cytoplasm of animal cells during cell division. Contractile networks arranged in layers just under the plasma membrane produce capping movements that sweep receptors to one end of the cell surface and probably also take part in ameboid motion. At the other end of the spectrum, the most highly organized contractile microfilament systems underlie muscular movements in animals. These movements include the slow and powerful contractions of smooth muscle cells and, in the ultimate state of microfilament organization, the rapid and equally powerful contractions of skeletal and cardiac muscle.

In movement by directional assembly, microfilaments grow in length and push attached cell structures in the direction of growth. This mechanism powers the rapid and spectacular extension of a filament from the head of some invertebrate spermatozoa during fertilization, for example. Directional microfilament assembly may also be responsible for part of ameboid movement.

The microfilaments underlying these varied cellular movements assemble from subunits consisting of the protein *actin*. Another protein, *myosin*, is the ATP-driven crossbridge that generates the sliding forces responsible for cytoplasmic streaming and contraction. Both actin and myosin are families of related proteins rather than a single type.

This chapter discusses the structure and biochemistry of microfilaments and their functions in cell motility. The roles of microfilaments in the cytoskeleton are considered in the following chapter. Supplement 11-1 describes a number of motile systems that depend on neither microtubules nor microfilaments.

STRUCTURE AND BIOCHEMISTRY OF MICROFILAMENTS

Microfilament Structure

Microfilaments appear in sections as thin, dense fibers about 7 to 9 nm in diameter (Fig. 11-1; see also Fig. 1-15). When isolated and prepared for electron microscopy by negative staining (see p. 790), microfilaments can be seen to consist of two linear chains of roughly spherical subunits wound into a double helix that turns once in every 70+ nm (Fig. 11-2a and b). Each subunit in the twisted chains is an individual actin molecule (Fig. 11-2c).

Microfilaments vary considerably in length. They may contain from as few as 10 to 20 to more than a thousand actin subunits. In many cellular systems, microfilaments include another protein, *tropomyosin*, as a regular constituent (see Fig. 11-10). This protein serves as a fibrous reinforcement along actin chains and in some systems takes part in regulation of microfilament sliding (see below).

Microfilaments are often difficult to identify directly in cells viewed under the electron microscope because they resemble other cytoplasmic fiber types such as intermediate filaments. This problem is circumvented by a technique known as *myosin decoration*, developed in 1963 by H. E. Huxley. In the technique, microfilaments are reacted with a segment of the myosin molecule. Attachment of the myosin segments produces an "arrowhead" pattern that is unmistakable and characteristic only of microfilaments

(Fig. 11-3a). The arrowheads also indicate the *polarity* of microfilaments and the direction of microfilament-based motions (see below). The arrowheads reflect binding of the myosin segments, one for each actin subunit, in a double spiral that follows the actin double helix around a microfilament (Fig. 11-3b).

The Actins

The actins, which occur in all eukaryotes, are one of the most abundant groups of cellular proteins. They were first isolated and identified by F. B. Straub in 1942. X-ray and optical diffraction studies (see Appendix) show that an actin molecule folds into a structure consisting of two lobes of unequal size separated by a cleft or depression on one side (see Fig. 11-2c). Distributed over the structure are separate binding sites for myosin, other actin molecules, Mg^{2+}, and either ATP or ADP. Actin is distinguished biochemically by its unique ability to greatly stimulate myosin's activity as an ATPase (the name *actin* is derived from this property).

Most actins contain 374, 375, or 376 amino acids and have total molecular weights near 42,000. Although some unicellular eukaryotes possess only a single actin type, most fungi, plants, and animals have two or more distinct actins that vary to some extent in amino acid sequence. Reptiles, birds, and mammals have no less than six different actins, some of them restricted to certain cell types such as skeletal, cardiac, or smooth muscle. Higher plants may also have actins of six or more different types.

All these actin types are closely related in sequence. Among all eukaryotes the various actins differ in sequence by only about 15%; yeast and rabbit actins, for example, are 88% identical in sequence. These close sequence relationships suggest that all actins arose from a single ancestral type in very early evolutionary lines leading to eukaryotes.

Assembly of Actin Molecules into Microfilaments

Actin molecules assemble reversibly into microfilaments, reaching a balance similar to the tubulin-microtubule equilibrium (see p. 284):

$$\text{actin molecules} \rightleftharpoons \text{microfilaments} \quad (11\text{-}1)$$

To assemble into microfilaments, individual actin molecules must combine with Mg^{2+} and a molecule of ATP; both the ion and ATP bind in the cleft of an actin molecule (see Fig. 11-2c). The ATP is hydrolyzed as assembly proceeds; however, ATP breakdown lags significantly behind polymerization. As a result, the

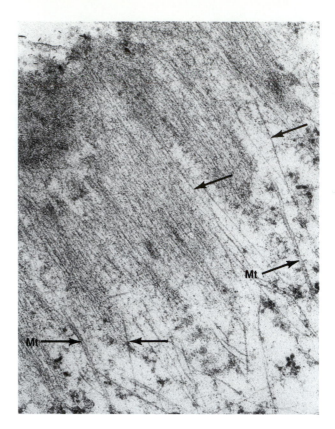

Figure 11-1 Microfilaments (arrows) in human capillary endothelial cells. Several microtubules (MT) are also visible. × 50,000. (Courtesy of K. G. Bensch, from *J. Ultrastr. Res.* 82:76 [1983].)

tips of polymerizing microfilaments contain actin subunits with ATP still attached. Just behind the tips, hydrolysis occurs so that all or almost all the subunits at points farther along the microfilaments are linked to ADP and phosphate instead of to ATP. The ATP-linked subunits at the tips may help stabilize the microfilaments, because ATP-actin subunits are about 10 times slower to disassemble from microfilaments than ADP-actin. These characteristics of the assembly reaction indicate that ATP binding, like the analogous binding of GTP during microtubule polymerization (see p. 284), induces a conformational change in actin that pushes the equilibrium in the direction of polymerization.

Microfilament Polarity and Treadmilling

The similarities between microfilament and microtubule assembly extend to the relative rates at which the two ends of a microfilament add or release actin subunits. If microfilaments previously decorated by myosin are placed in an actin solution under condi-

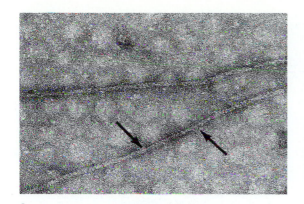

a

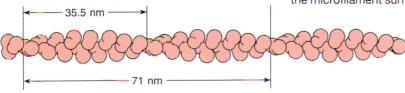

b

Figure 11-2 Microfilament and actin structure. **(a)** Negatively stained microfilaments isolated from a protozoan, *Acanthamoeba*. Individual actin subunits are visible at some locations (arrows). (Courtesy of T. D. Pollard.) **(b)** The double helix of actin subunits in a microfilament. The double helix makes a complete turn only once in every 71 nm and a half turn every 35.5 nm, so that the two chains appear to cross over each other at this interval when the helix is viewed from the side. Each actin subunit is a bilobed, or kidney-shaped, particle with a deep cleft separating the lobes **(c)**. Sites binding ATP and Mg^{2+} are located in the cleft; interactions between actin subunits in filament assembly probably occur along the top and bottom margins in the view shown in **(c)**. Recent evidence from X-ray and optical diffraction studies suggests that each actin subunit is aligned with its long axis almost perpendicular to the long axis of a microfilament, turned so that the two lobes face the microfilament surface.

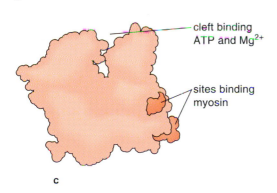

c

tions favoring polymerization, additional actin subunits, undecorated by myosin, clearly add more rapidly to one end than to the other (Fig. 11-4). Under physiological conditions the end assembling more rapidly (the *plus end*) adds subunits about 12 times faster than the other end (the *minus end*). For this reason, microfilaments are said to have *polarity*. (In this case, as with microtubules, the terms *plus*, *minus*, and *polarity* refer to relative rates of assembly and not to charge.) Structures sliding over a microfilament move toward the plus end.

Microfilament polarity can be directly determined by myosin decoration. The myosin arrowheads on the decorated segment point toward the more slowly polymerizing minus end; the plus end lies at the end marked by the tails of the arrowheads. Because the arrowheads point toward the minus end of a microfilament, this end is often called *pointed*; the opposite, plus end is termed *barbed*.

At certain actin concentrations, subunits may release from the minus end of microfilaments while subunits are still adding at the plus end. Under these

a

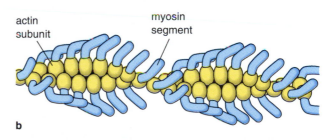

b

Figure 11-3 Myosin decoration, used to identify microfilaments in the electron microscope. **(a)** The "arrowhead" pattern produced when actin microfilaments are reacted with the HMM or S1 segments of myosin molecules. (Courtesy of J. A. Spudich, with permission from *J. Mol. Biol.* 72:619 [1972]. Copyright by Academic Press, Inc. [London] Ltd.) **(b)** The arrangement of myosin segments that produces the arrowhead pattern on an actin microfilament.

Figure 11-4 Correlation between myosin decoration and microfilament polarity. An original microfilament marked by myosin decoration, showing the arrowhead pattern, was placed in a solution favoring addition of subunits at both ends. The newly added actin chains appear smooth in the micrograph. Only a short length has added at the minus, or "pointed," (P) end compared to the long length added to the plus, or "barbed," (B) end. The arrowheads clearly point toward the minus end. (Courtesy of T. D. Pollard.)

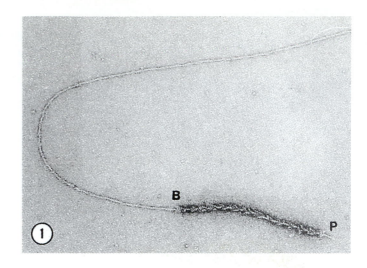

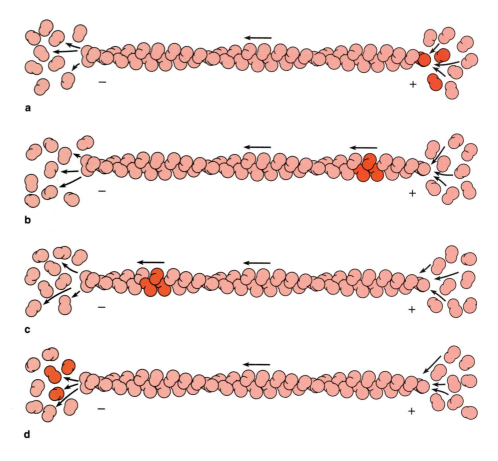

Figure 11-5 Movement of a group of actin subunits through a microfilament by treadmilling, in which microfilament subunits continuously add at the plus end and release at the minus end. A given group of subunits (in darker red) adds to the plus end in (a); as further subunits add at this end, the added group is pushed toward the minus end (b). The added group continues to treadmill through the microfilament (c) until its subunits release at the minus end (d).

conditions, actin molecules *treadmill* through microfilaments by adding at the plus end, moving along the microfilament as more subunits add at the plus end, and finally releasing at the minus end (Fig. 11-5). The same treadmilling process may also take place in microtubules under certain conditions (see p. 285).

Microfilament treadmilling, originally discovered by T. D. Pollard and A. Wegner and their colleagues, has as yet been demonstrated only in the test tube. Among the observations that make treadmilling seem

unlikely as a source of motility in living cells is the fact that the direction in which structures are observed to move along microfilaments, from the minus to the plus end, is opposite to the direction of treadmilling.

Factors Modifying Microfilament Polymerization

Under cellular conditions the actin-microfilament equilibrium tilts significantly in the direction of assembled microfilaments. The proportion of unassembled actin molecules in many cell types, however, is much larger than expected from the characteristics of the equilibrium reaction. This distribution depends on a wide variety of *actin-binding proteins* (*ABPs*) that link either to unassembled actin molecules or to assembled microfilaments and generally push the equilibrium toward the left. These ABPs thus have activities opposite to microtubule associated proteins (MAPs), which typically adjust the tubulin-microtubule equilibrium to the right and stabilize microtubules in the assembled form (see p. 287).

The Actin-binding Proteins   More than 60 known ABPs bind to actin and adjust the actin-microfilament equilibrium primarily in the direction of unassembled subunits (Table 11-1). Some of these proteins are widely distributed among eukaryotes, and some occur only in a single cell type of one species.

Although the various ABPs are numerous and diversified in structure, they adjust the actin-microfilament equilibrium primarily by one of three mechanisms. In one a protein links to unassembled actin molecules and blocks their polymerization into microfilaments. Primary among the ABPs acting in this way is *profilin*, which is widely distributed among eukaryotic cells.

The remaining mechanisms, *capping* and *severing*, affect assembled microfilaments rather than actin subunits. Capping proteins such as *brevin* bind to actin molecules at the ends of existing microfilaments, most of them at plus ends, and prevent further assembly and microfilament growth. Many of these ABPs cap very short microfilaments containing only a few subunits, effectively removing actin monomers from the pool available for polymerization.

Severing ABPs such as *fragmin* and *gelsolin* bind to individual actin subunits at points between the plus and minus ends of a microfilament. The binding alters the subunits so that they separate from their neighbors and introduce microfilament breaks. Severing proteins regulate microfilament length in the cytoplasm and also disassemble microfilament networks. Many proteins acting by this mechanism, including fragmin and gelsolin, remain attached to the severed fragments and prevent their reassembly. They therefore function as capping as well as severing proteins.

Two other groups of ABPs, the *crosslinkers* and *bundlers*, affect the form taken by microfilament networks rather than adjusting the actin-microfilament equilibrium. The crosslinkers, which are typically elongated, fibrous molecules, form crosslinks between existing microfilaments, or "crosswelds" where microfilaments cross over each other. The crosslinks, formed by ABPs such as *spectrin*, *fodrin* and *gelactin*, establish microfilament networks that make up parts of the cyto-

skeleton. In many cells these networks become extensive enough to convert the cytoplasmic regions containing them into a semisolid gel. A few crosslinking ABPs, such as *MARCKS* (*M*yristoylated *A*lanine-*R*ich *C* *K*inase *S*ubstrate), also link microfilaments to the plasma membrane. Some MAPs linked to microtubules, such as MAP-2 and several tau proteins (see p. 287), also bind microfilaments, adding interlinks that tie microtubules and microfilaments into cytoskeletal networks.

Bundling ABPs, including *villin* and *fimbrin*, link microfilaments in parallel, side-by-side fashion. The microfilament bundles created by these ABPs form cytoskeletal structures such as the internal cores supporting *microvilli*, the fingerlike extensions of cells in some locations such as the intestinal lining (see Fig. 12-9 and p. 353). Some of the bundling ABPs, including fimbrin, also work as crosslinkers, adding to their ability to form cytoplasmic networks.

The activities of ABPs are readily reversible, allowing cells to push the actin monomer-microfilament equilibrium quickly in either direction or to assemble or disassemble microfilament networks rapidly in local regions of the cytoplasm. Many ABPs, including fragmin, gelsolin, villin, and brevin, are inactive at the low Ca^{2+} concentrations typical of the cytoplasm and become active only when Ca^{2+} levels rise through release of this ion into the cytoplasm (see p. 131). Some ABPs vary in type of activity depending on Ca^{2+} concentration. Villin, for example, acts as a microfilament bundler at low Ca^{2+} concentrations and severs microfilaments at high Ca^{2+} levels. The effects of Ca^{2+} may be direct, through binding of the ion to the protein, or indirect, through the activity of the Ca^{2+}-dependent control protein calmodulin (see Information Box 6-2).

The direct or indirect calcium sensitivity of many ABPs adds the assembly of microfilaments and microfilament networks to the long list of cellular processes regulated by relatively small and practically instantaneous changes in Ca^{2+} concentration. The effects of Ca^{2+} also link microfilament regulation to the $InsP_3$/DAG pathway triggered by cell surface receptors, in which Ca^{2+} is released as a second messenger (see p. 161). The effects of Ca^{2+} and the ABPs provide living cells with a delicately balanced control of the time, place, and form of microfilament organization in the cytoplasm.

Many ABPs—at least 70 at the most recent count—are also regulated by addition or removal of phosphate groups. The phosphorylation, catalyzed by protein kinases of various kinds, activates some APBs and inhibits others. Some are regulated by both Ca^{2+} and phosphorylation; in some systems a protein kinase phosphorylating an ABP is activated by the Ca^{2+}/calmodulin complex. MARCKS, for example, crosslinks microfilaments and binds them to membranes when the ABP is dephosphorylated. (MARCKS

Table 11-1 Some Actin-binding Proteins

Protein	Molecular Weight	Calcium Sensitivity	Caps	Severs	Crosslinks	Bundles	Attaches to Plasma Membrane	Sources
Fragmin	42,000	+	+	+				Blood plasma
β-actinin	34,000–37,000	–	+					Muscle cells
Gelsolin	90,000–95,000	+	+	+				Macrophages, platelets, blood plasma, brain
Villin	95,000	+	+	+		+		Intestinal brush border
Brevin	90,000	+	+					Blood plasma
Severin	40,000	+	+	+				*Dictyostelium*
Filamin	250,000–270,000	–			+			Smooth muscle, macrophages
Spectrin	225,000–260,000	–			+		+	Many mammalian cell types
Fodrin	235,000–240,000	–			+			Widely distributed
α-actinin	100,000–105,000	+			+		+	Platelets, fibroblasts, muscle cells
Gelactin	23,000–38,000	–			+			Protozoans (soil amebae)
Fascin	58,000	–			+	+		Sea urchin oocytes
Vinculin	130,000	–			+	+	+	Muscle cells, fibroblasts
Talin	215,000	–			+		+	Smooth muscle
MARCKS	68,000–87,000	+			+		+	
Fimbrin	68,000	–			+	+		Intestinal brush border
MAP-2	280,000	–			+			Brain
					(To microtubules)			
Tau	55,000–62,000	–			+			Brain
					(To microtubules)			
Profilin	12,000–15,000	–			Binds actin subunits, prevents assembly into microfilaments			Widely distributed

has a "greasy foot" that anchors its N-terminal end to the cytoplasmic side of the plasma membrane.) Phosphorylation detaches MARCKS from the plasma membrane; both phosphorylation and combination of MARCKS with the Ca^{2+}/calmodulin complex inhibit its crosslinking ability. Because many of the protein kinases phosphorylating ABPs are activated by the cAMP or $InsP_3$/DAG pathways, these control mechanisms also link microfilament regulation to pathways triggered by cell surface receptors.

Microfilament Poisons The actin-microfilament equilibrium, like the tubulin-microtubule equilibrium, is the target of poisons produced by living organisms as biological defenses. Two of these poisons, the *cytochalasins* and *phalloidin* (Fig. 11-6), are made by fungi, and are both highly useful in research.

The closely related family of molecules making up the cytochalasins (from *cyto* = cell and *chalasis* = relaxation) is produced by *Helminthosporium* and other molds. The cytochalasins, like many microtubule poisons, are alkaloids—cyclic organic molecules that react as bases. They inhibit microfilament assembly and promote breakdown of existing microfilaments. Twenty different cytochalasins, each with minor substitutions in chemical groups, have been isolated from fungal molds.

The cytochalasins inhibit microfilament polymerization by capping the plus end. The capping is so effective that one molecule of cytochalasin can completely block further growth of a microfilament. Cytochalasins may also sever microfilaments into shorter lengths and disrupt microfilament networks, converting gelled regions of the cytoplasm into a more liquid form.

Phalloidin is produced by mushrooms, among them *Amanita phalloides* (the "death cap" mushroom). This microfilament poison is a cyclic peptide that binds to unassembled actin molecules and greatly increases their affinity for both the plus and minus ends of microfilaments. As a result, phalloidin has effects opposite to the cytochalasins and interferes with microfilament function by pushing the actin-microfilament equilibrium far in the direction of assembly.

Both phalloidin and at least one cytochalasin (*cytochalasin D*) are highly specific and combine only with actin or microfilaments. In doing so, both poisons are highly effective in arresting microfilament-based motility. These properties have made them indispensable as laboratory probes for microfilament-based motility inside living cells. If a motile activity is arrested or significantly inhibited by phalloidin or cytochalasin, it can be assumed as a first approximation that the motility is based on microfilaments. Because phalloidin promotes microfilament assembly and stabilizes microfilaments in the assembled form, it has also made it possible to identify and isolate microfilaments

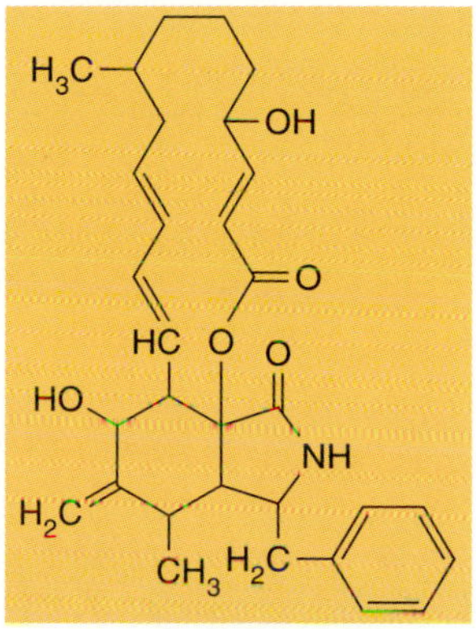

cytochalasin B

a

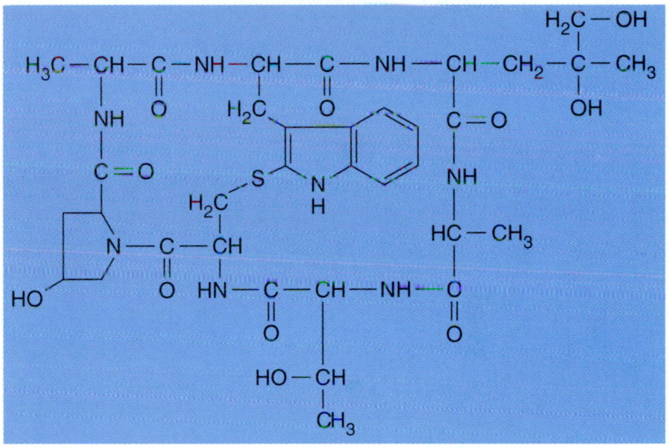

phalloidin

b

Figure 11-6 Microfilament poisons synthesized by fungi. (a) Cytochalasin *B*, which pushes the actin-microfilament equilibrium in the direction of monomers. The cytochalasins may also introduce breaks in microfilaments. (b) Phalloidin, which pushes the equilibrium in the direction of assembled microfilaments.

in groups such as plants, in which identification and analysis were previously difficult or impossible.

Parallels in the Biochemistry of Microfilaments and Microtubules

There are many parallels in the biochemistry of microfilaments and microtubules. Both structures are polymers assembled from protein monomers (microtubules from tubulin and microfilaments from actin) in an equilibrium reaction that can be readily tipped in either direction. The balance is regulated and controlled by a battery of regulatory proteins—MAPs in microtubules and a long list of ABPs in microfilaments. Many of the MAPs and ABPs are directly or indirectly regulated by Ca^{2+} and phosphorylation, providing a link between microtubule and microfilament activity and the fundamental cellular regulatory mechanisms

Table 11-2 Myosin Classes		
Myosin Class	Source	Function
I	Many animals, fungi, algae, higher plants	Movements of plasma membrane and membranous organelles; cytoplasmic streaming; ameboid movement
II		
muscle	Many animals	Movements of striated, smooth, and cardiac muscle
nonmuscle	Many organisms	Cytoplasmic division (cytokinesis); capping; ameboid movement
III	*Drosophila* photoreceptor cells	Unknown
IV	*Acanthamoeba*	Unknown
V	Chicken, yeast	Possibly involved in membrane motility
VI	*Drosophila*, pig	Unknown
VII	*Drosophila*, pig	Unknown

based on calcium ions and the cAMP and $InsP_3$/DAG second messengers. The two motile elements are also the targets of natural poisons, most notably colchicine and the cytochalasins, which disassemble microtubule and microfilament polymers respectively, and taxol and phalloidin, which push their respective microtubule and microfilament targets in the direction of stable polymers. As the following section shows, many parallels also exist between the crossbridging mechanisms that generate microfilament- and microtubule-based motility.

MYOSIN AND MICROFILAMENT-BASED MOTILITY

The power for microfilament sliding is supplied by myosin crossbridges, which generate motion by an attach-pull-release cycle. The activity of myosin crossbridges has a parallel in microtubule-based motility, in which dynein, kinesin, or other crossbridging proteins provide "motors" that slide structures along microtubules.

Myosin Structure

The myosins are a large family of molecules that occur in at least seven classes designated myosins I to VII (Table 11-2). The best known of these microfilament motors are myosins I and II. Myosin I is involved in several types of membrane-associated movement in nonmuscle cells, probably including ameboid motion and the movement of organelles in cytoplasmic streaming. In some cells, myosin I serves as a linker molecule tying microfilaments into cytoskeletal structures. Myosin II is responsible for the microfilament sliding that takes place in muscle cells, as well as several nonmuscle cell movements including cytoplas-

mic division, capping of substances bound to the plasma membrane, and possibly ameboid motion in coordination with myosin I in some organisms.

All myosin types are assembled from one or two large polypeptide subunits, the *myosin heavy chains*, and one or more smaller peptide subunits, the *myosin light chains*. Myosin I consists of a single, 110,000- to 190,000-dalton heavy chain in association with one or more light chains ranging from 14,000 to 27,000 daltons; myosin II contains two heavy chains of about 200,000 daltons, each in combination with two light chains of 16,000 to 20,000 daltons.

The heavy chains typically fold into a globular *head unit* and a *tail*. The light chains associate with the head units. The head units of both myosin types contain similar amino acid sequences, with one site that binds actin and another that binds and hydrolyzes ATP (Fig. 11-7). The hydrolytic activity of the myosins is unique among cellular ATPases because it is activated by actin and strongly inhibited by Mg^{2+}. In both myosin types the head units are responsible for the crossbridging action that produces microfilament sliding.

The tail of myosin I is very short and not conspicuously fibrous. In contrast, a myosin II molecule (Fig. 11-8) has a long tail consisting of two alpha-helical polypeptide chains, one extending from each head unit. The two alpha-helical chains twist around each other into a double helix or *coiled coil* (see Information Box 2-4). A myosin II molecule is thus a double-headed structure with a long, twisted tail. The tail contains binding sites that assemble myosin II molecules into superstructures called *thick filaments* (see below).

One of the two light chains in each head unit of a myosin II molecule, called the *essential light chain*, can be removed only by relatively drastic treatments, such as exposure to strong alkali, that denature myosin and destroy its function. The second light chain, the *regu-*

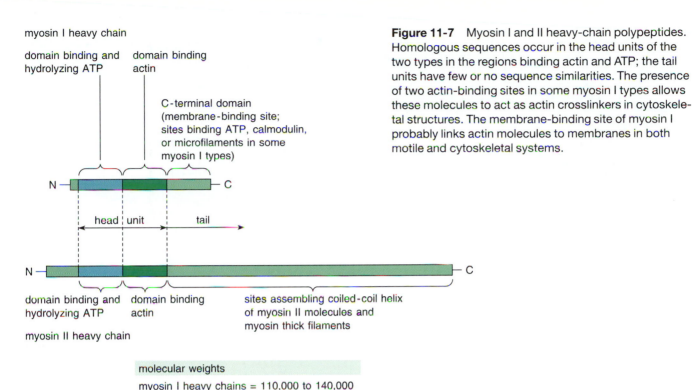

myosin I heavy chain

domain binding and hydrolyzing ATP

domain binding actin

C-terminal domain (membrane-binding site; sites binding ATP, calmodulin, or microfilaments in some myosin I types)

N — C

head : unit   tail

N — C

domain binding and hydrolyzing ATP

domain binding actin

sites assembling coiled-coil helix of myosin II molecules and myosin thick filaments

myosin II heavy chain

molecular weights

myosin I heavy chains = 110,000 to 140,000
myosin II heavy chains = 175,000 to 240,000

Figure 11-7 Myosin I and II heavy-chain polypeptides. Homologous sequences occur in the head units of the two types in the regions binding actin and ATP; the tail units have few or no sequence similarities. The presence of two actin-binding sites in some myosin I types allows these molecules to act as actin crosslinkers in cytoskeletal structures. The membrane-binding site of myosin I probably links actin molecules to membranes in both motile and cytoskeletal systems.

latory light chain, is easily removed from the remainder of the myosin molecule by mild treatments such as adjustments in salt concentration. The regulatory light chain is so called because it controls the crossbridging activity of myosin in some systems, as in smooth muscle (see below).

Myosin II molecules can be split into defined subparts by protein-digesting enzymes such as trypsin. Trypsin digestion splits the molecule about one-third of the way along the tail from the C-terminal end (see Fig. 11-8a). The fragment containing the heads, called *heavy meromyosin* (*HMM*), retains the ATPase activity of the myosin molecule. The remaining tail segment is called *light meromyosin* (*LMM*). The tail may break at the site between the HMM and LMM units because the alpha helices of the tail unwind in this region, producing a region in which the amino acid chain is more exposed to the enzyme. Some investigators have proposed that this site serves as a flexible "hinge" that may act in the myosin crossbridging cycle (see below).

More extensive digestion with trypsin or other proteinases splits the HMM fragment at the point where the heads attach to the tail, suggesting that another more exposed, flexible segment exists in this region. The break produces two fragments, *S1* and *S2*. The S1 fragments are the individual head units, and the S2 fragments are the portion of the tail between the heads and the breakpoint between HMM and LMM. The ATPase activity of myosin is associated with the S1 fragments and therefore the head structures; each S1 unit also contains sites binding actin

and the essential and regulatory light chains. Since there are two head units, each myosin II molecule has two ATPase and two actin-binding sites. The ability of S1 or HMM segments to bind actin forms the basis for the myosin decoration technique (see Fig. 11-3).

Myosin I, originally discovered by T. D. Pollard and E. D. Korn in *Acanthamoeba*, has since been detected in such diverse species as *Dictyostelium*, *Drosophila*, vertebrates, algae, and higher plants and is probably generally distributed among eukaryotes. In vertebrates it occurs in the microvilli of intestinal cells (see p. 353), where it may serve a primarily cytoskeletal role by linking microfilament bundles to the plasma membrane. Many species, including *Acanthamoeba*, *Dictyostelium*, and mammals, have several distinct genes encoding myosin I molecules. At least three myosin I genes encoding subtypes known as IA, IB, and IC have been identified in mammals.

Myosin II, the muscle-cell myosin of animals, was discovered as a major muscle protein by W. Kuhne in 1863. Different species and cell types have variants of this myosin with distinct amino acid sequences. Striated, smooth, and nonmuscle cells within the same species, for example, differ in the amino acid sequences of both the heavy and light chains, activity as an ATPase, and solubility. In higher vertebrates at least 11 different myosin II heavy chain variants occur in specific cell types such as fast and more slowly contracting skeletal muscle, the atrium and ventricles of the heart, smooth muscle, and nonmuscle cells. Myosin light chains also occur in distinct types.

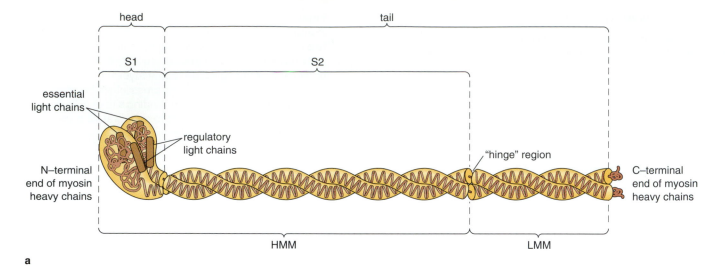

a

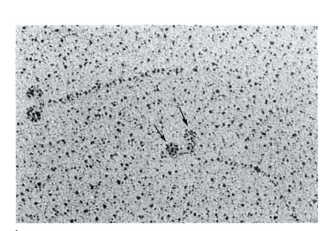

b

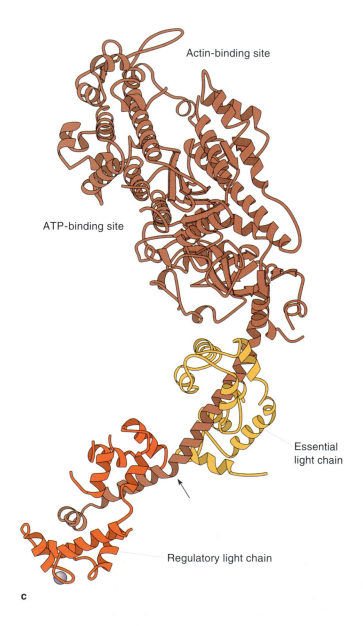

Actin-binding site

ATP-binding site

Essential light chain

Regulatory light chain

c

Figure 11-8 **(a)** Myosin II structure. Two myosin heavy chains wind into a coiled-coil structure to form the tail and into a less regular conformation to form the double head structure. One essential and one regulatory myosin light chain are also present in each subunit of the double head structure. HMM (heavy meromyosin), LMM (light meromyosin), S1, and S2 are fragments produced by protease digestion of myosin molecules. **(b)** Myosin molecules isolated from chicken skeletal muscle and prepared for electron microscopy by freeze-drying and shadowing. The tail and double head structure (arrows) are visible. (Courtesy of T. Wallimann, from *Eur. J. Cell Biol.* 30:177 [1983].) **(c)** Structure of the S1 head unit of myosin II as deduced from X-ray diffraction. One of the unusual features of the head unit is the long alpha helix (arrow) that extends through much of the structure. The essential light chain may stabilize this helix. (Courtesy of I. Rayment and the AAAS, from *Science* 261:50 [1993].)

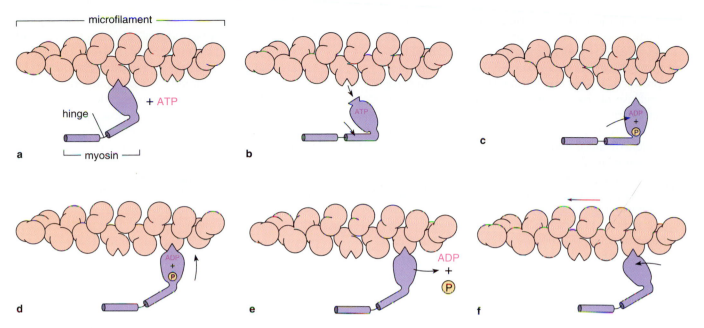

Figure 11-9 A simplified version of the crossbridging cycle that hydrolyzes ATP to power microfilament sliding (see text).

The Myosin Crossbridge Cycle

Much of the evidence that myosin crossbridges move through an attach-pull-release cycle to generate microfilament-based motion comes from X-ray diffraction and electron microscope studies of striated muscle by H. E. Huxley and J. Hanson and by A. F. Huxley and R. Niedergerke. In 1954 these investigators proposed that microfilament sliding is responsible for muscle contraction. Later, in 1969, H. E. Huxley suggested the basic principle of a myosin crossbridging cycle as the power source for the sliding. The Huxley model was fleshed out with biochemical details by R. W. Lymn and E. W. Taylor and others, who studied the interactions of actin and myosin both inside muscle cells and in the test tube.

The observations made by these and other investigators established several important characteristics that are critical for the crossbridging model:

1. Myosin crossbridges in striated muscle extend toward microfilaments at angles that vary between about 45° and 90°.

2. Myosin unattached to actin can bind and hydrolyze ATP. However, the products of the reaction, ADP and inorganic phosphate, cannot be released, nor the reaction fully completed, until myosin binds actin.

3. Binding between myosin and actin promotes release of the products of ATP breakdown and makes available a large increment of free energy.

4. Once the products of ATP hydrolysis are released,

myosin crossbridges must bind fresh ATP to release from actin.

The last observation was developed through experiments with muscles deprived of ATP, in which myosin crossbridges become locked to actin microfilaments in a *rigor state*. (This is the cause of *rigor mortis*, the stiffness of the limbs in animals after death.) Adding fresh ATP releases the crossbridges and restores flexibility to the muscles. Whether crossbridges in rigor are locked at 90° or 45°, or at various angles between these limits, is still being debated. We will assume for the purposes of this discussion that the locked, rigor position is at 45°.

Figure 11-9, based on the Huxley-Lymn-Taylor model, combines these observations and shows in a simplified way how myosin crossbridges may use ATP energy to produce active microfilament sliding. The cycle begins with the myosin crossbridge bent at a 45° angle and tightly linked to an actin subunit (Fig. 11-9a). The products of ATP breakdown have been released, and the ATP-binding site on the myosin crossbridge is active. Binding an ATP molecule (Fig. 11-9b) changes the conformation of the actin-binding site on the crossbridge, greatly reducing its affinity for the actin subunit. Under these conditions the crossbridge releases. As the crossbridge separates from actin, its bound ATP is hydrolyzed, but the products of the hydrolysis, ADP and P_i, remain bound to the crossbridge. ATP breakdown causes another conformational change in the crossbridge, shifting it to the 90° position and reactivating the actin-binding site at its tip (Fig. 11-9c).

In this condition the crossbridge may be considered as a compressed molecular spring storing much of the energy released by ATP breakdown. Attachment to an actin subunit in the microfilament (Fig. 11-9d) releases the products of ATP breakdown (Fig. 11-9e) and triggers release of the crossbridge from its 90° position. The crossbridge then swivels forcefully to the 45° position, pulling the attached microfilament through an equivalent distance, about 10 nm (Fig. 11-9f). The crossbridge is now ready to bind another ATP molecule and repeat the cycle. The direction of movement by a crossbridge is toward the barbed end of a microfilament.

Note in Figure 11-9e that as the myosin crossbridge makes its attachment to the adjacent microfilament during the crossbridge cycle, the head unit and part of the tail are shown as bending outward from a hinge point located partway along the tail. Whether a hinge actually exists at this point or whether the entire myosin tail bends through a smooth arc to accommodate movement of the head toward actin remains a subject of intensive debate. Other unanswered questions concern the magnitude of the crossbridge movement (the arc through which the head moves may be considerably less than 45°), which segments of the head undergo conformational changes, and whether one or both heads of a myosin molecule are involved in a single cycle.

Whatever the uncertainties about the crossbridging cycle, the ability of myosin to slide structures along microfilaments was graphically confirmed recently by the experiments of M. P. Sheetz and J. A. Spudich. These investigators isolated microfilaments by breaking open cells of the alga *Nitella*, in which microfilament bundles form bands just under the plasma membrane that move chloroplasts through the cytoplasm. The isolated bundles were attached to the surface of a microscope slide and washed free of cytoplasm. Microscopic plastic beads coated with the HMM fragments of rabbit skeletal myosin II molecules were then added to the slides; when supplied with ATP, the beads "walked" actively along the microfilament bundles at speeds comparable to the rate at which microfilaments slide in skeletal muscle. (The Experimental Process essay by Sheetz on p. 300 discusses his experiments with myosin-coated beads and describes equivalent experiments with microtubules.)

Spudich and Sheetz and another group, T. Yanagida and T. Harada and their coworkers, repeated the same approach using only S1 head units (see Harada and Yanagida's essay on page 322). The head units by themselves were able to slide beads along microfilaments. This finding indicates that the head units alone can accomplish all the essential reactions of the crossbridging cycle, including ATP binding and breakdown, binding and release between actin and myosin, and the conformational changes accomplishing movement. The finding also makes it seem unlikely that portions of the myosin tail, such as the hinge region, are essential for crossbridging. Because single head units can produce sliding motion, the results indicate further that the crossbridging cycle does not necessarily require a coordinated "walking" interaction between both heads of a myosin molecule.

Recently J. T. Finer, R. M. Simmons, and J. A. Spudich used the optical tweezers technique (see Information Box 10-2) to measure the force exerted by myosin crossbridges. In the experiment, microscopic latex beads were attached to both ends of short actin microfilaments and held by the optical tweezers. Myosin molecules were fixed to the surface of a glass microscope cover slip, and ATP was added to the preparation. The Spudich group found that forces averaging 3 to 7 piconewtons could be applied to the microfilaments before stalling movement, indicating that myosin head units develop forces in this range during the power stroke. This value is comparable to the 2 to 5 piconewton force produced by a single kinesin crossbridge acting on microtubules when measured by equivalent techniques (see p. 298) and is somewhat higher than the 1 piconewton force estimated in the Harada and Yanagida experiments (see their essay in this chapter). When the opposing force applied by the tweezers approached values stalling the crossbridging cycle, the microfilaments moved in jerks averaging 11 nm, indicating that this is the distance traversed by a single power stroke.

Regulation of Crossbridging

Many years of experimentation with vertebrate striated and smooth muscle have revealed that two major pathways regulate the myosin crossbridging cycle. One, *actin-linked regulation*, involves proteins that bind to actin and form regulatory units located on the surfaces of microfilaments. The second, *myosin-linked regulation*, is based on the regulatory myosin light chains. Both systems are controlled by changes in cytoplasmic Ca^{2+} concentrations.

Actin-linked Regulation in Striated Muscle Regulation linked to actin is best understood in striated muscle, including both skeletal and cardiac. This form of actin-linked regulation depends on two proteins, *tropomyosin* and *troponin*, that bind to microfilaments in striated muscle. The two proteins are believed to control the crossbridging cycle by undergoing Ca^{2+}-dependent conformational changes. The changes alternately block and expose the sites on actin bound by myosin during the crossbridging cycle.

Tropomyosin is a fibrous molecule built up from two alpha-helical chains wound into a coiled coil similar to the myosin tail (Fig. 11-10a). The two polypeptide chains of a tropomyosin molecule, depending on the cell type, may be identical or different; in striated muscle the two chains are different. Although its name suggests that tropomyosin is a form of myosin, the two molecules are actually unrelated. Distinct forms of tropomyosin occur in striated muscle, smooth muscle, and nonmuscle cells.

A tropomyosin molecule is slightly more than 40 nm in length, just long enough to stretch over a row of seven actin subunits in a microfilament. Tropomyosin molecules extend lengthwise along the microfilament in two end-to-end chains, one on either side near the "grooves" in the actin double helix (Fig. 11-10b).

Troponin is built up from three polypeptide subunits called *troponin C* (*TnC*), *troponin T* (*TnT*), and *troponin I* (*TnI*; see Fig. 11-10b). TnC and TnI are globular polypeptides; TnT is an elongated, fibrous molecule about one-third the length of tropomyosin. TnC, which closely resembles calmodulin in structure and function, provides the link between Ca^{2+} concentration and control of microfilament sliding by the actin-linked mechanism. (Calmodulin, in fact, can be substituted for TnC with little effect on troponin function.) Troponin occurs in equal numbers with tropomyosin molecules in striated muscle microfilaments.

Elegant work in the S. Ebashi and H. E. Huxley laboratories revealed how troponin and tropomyosin probably interact to regulate the myosin crossbridge cycle in muscle cells. The first clues to the role of these proteins in actin-linked regulation were discovered by Ebashi and his coworkers, who noted that Ca^{2+} is required for the interaction of actin and myosin isolated from striated muscle cells. At the very low Ca^{2+} concentrations typical of resting muscle cells the actin and myosin of muscle do not interact even if ATP is present. At higher Ca^{2+} concentrations the crossbridge cycle proceeds if ATP is available.

Ebashi and his coworkers later found that Ca^{2+} regulates the crossbridging cycle through troponin and tropomyosin. Muscle actin and myosin stripped of troponin and tropomyosin, they discovered, react together without dependence on Ca^{2+} concentration. Adding troponin and tropomyosin reestablishes the Ca^{2+} sensitivity. On this basis, Ebashi and his colleagues proposed that, when combined with Ca^{2+}, troponin and tropomyosin turn muscle contraction on; when free of Ca^{2+}, the control proteins switch the system off.

Huxley and his colleagues worked out the probable molecular basis for this regulatory mechanism. X-ray diffraction indicated that tropomyosin molecules lie in or near the surface grooves in the F-actin double helix (see Fig. 11-10b). During the initiation of microfilament sliding in muscle, when the mechanism is switched from "off" to "on," the Huxley group noted a change in the X-ray patterns indicating that tropomyosin moves to a position closer to the grooves.

From these observations, Huxley and Ebashi proposed that when the Ca^{2+} concentration is at the rest-

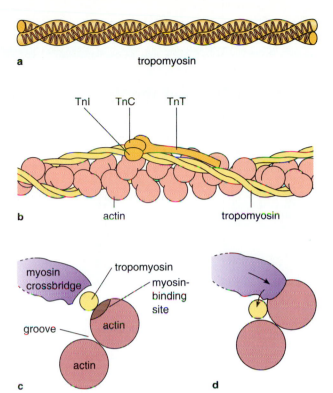

Figure 11-10 Actin-linked muscle regulation by the troponin-tropomyosin complex. **(a)** The coiled-coil structure of tropomyosin. **(b)** The arrangement of tropomyosin and the TnC, TnI, and TnT subunits of troponin on striated muscle microfilaments. **(c)** and **(d)** The actin-linked control mechanism as seen in a cross section of a microfilament (see text). In the blocking position **(c)**, tropomyosin covers the binding sites for myosin crossbridges on the actin subunits. Movement of tropomyosin toward the microfilament groove **(d)** exposes the myosin binding sites and triggers the crossbridging cycle.

ing level, troponin is tightly bound to tropomyosin. The binding holds the tropomyosin molecule slightly outside the microfilament grooves, in a position in which it covers the actin sites bound by myosin crossbridges (Fig. 11-10c). As Ca^{2+} concentration rises, troponin C binds Ca^{2+} and undergoes a conformational change that reduces its affinity for tropomyosin. Release of tropomyosin allows it to move into the groove. This shift exposes the myosin binding sites on the actin chains (Fig. 11-10d), permitting attachment of actin and myosin and initiating the crossbridging cycle.

Actin-linked regulation through the Ca^{2+}-sensitive interactions between troponin and tropomyosin controls microfilament sliding in all types of striated muscle, including both voluntary and cardiac muscle. Because troponin cannot be detected in smooth muscle or nonmuscle cells, actin-linked regulation involving troponin may be characteristic only of striated

Do the Two Heads of the Myosin Molecule Function Independently or Cooperatively?

Yoshie Harada and Toshio Yanagida

YOSHIE HARADA is a research associate of the Department of Bio-physical Engineering, Osaka University, Osaka, Japan. She grad-uated from the Faculty of Science of Ibaraki University in 1982 and received her D. Eng. in biophysics from the Faculty of Engi-neering Science of Osaka University in 1988. TOSHIO YANAGIDA is a professor in the Department of Biophysical Engineering at Osaka University. He graduated from the Faculty of Engineering Science of Osaka University in 1969 and received his D. Eng. in biophysics there in 1975. He was the recipient of the Osaka Sci-ence Award in 1989 and the Tsukahara Award in 1990.

The myosin molecule consists of two heads, each of which contains an enzymatic active site and an actin-binding site. The fundamental problem of whether the two heads function independently or cooperatively in the generation of sliding force has been studied by methods using suspensions of purified proteins,[1] actinomyosin threads,[2] precipitated molecules,[3] and chemical modifica-tion of muscle fibers.[4] No clear conclusion was reached about the action of myosin heads from these experiments.

We approached this question using a new *in vitro* motility assay. Several years ago, we demonstrated that single actin filaments labeled with a complex of fluores-cent dye and phalloidin, which stabilizes the filament structure of actin and is strongly fluorescent, can be re-solved and clearly observed continuously under a fluo-rescent microscope.[5] This allowed us to follow the sliding movement of single actin filaments interacting with myo-sin bound to a substrate *in vitro*.[6] This *in vitro* motility assay is very simple and reproducible, and it has since been widely used for studies on the mechanism of move-ment in muscle and nonmuscle cells. In the further appli-cation of this technique, we developed a method to hold and manipulate a single actin filament by glass micro-needles. The needles can also be used to measure sliding forces generated by single actin filaments.[7,8]

Using these techniques, we measured the sliding ve-locity of the actin filament on one-headed myosin mole-cules[9] and the force generated as the actin filament interacts with myosin subfragment-1 *in vitro*.[7] In these ex-periments, actin and myosin were obtained from rabbit

skeletal muscle and purified by conventional methods.[10] Actin filaments at a concentration equivalent to 2.5 μM in actin monomers were labeled with phalloidin-tetrame-thylrhodamine by incubating them overnight at 4°C in a solution containing 5 μM fluorescent phalloidin, 100 mM KCl, and 10 mM HEPES buffer at pH7.0.[5] The fluores-cently labeled actin filaments were observed under an in-verted microscope equipped with epifluorescence optics and illuminated by a mercury arc lamp. The fluorescent images obtained were videotaped with a high-sensitivity camera.[5] Using this equipment, the actin filaments could be clearly and continuously observed on a TV monitor. In order to minimize photo-bleaching and denaturation of proteins by the strong illumination, oxygen was removed from the assay solution by adding to it glucose-glucose-oxidase, catalase, and a reducing agent.[10]

Single-headed myosin was obtained by digestion of molecules with papain and purified.[2] The purity was checked; contamination by double-headed myosin mole-cules was found to be only 1%. Short myosin filaments, formed from the purified preparation of single-headed myosin dissolved in a high ionic solution, were coated onto a glass microscope slide. After washing off unbound myosin filaments, fluorescently labeled actin filaments at a concentration equivalent to about 10 nM in monomers were added to the slide in the presence of ATP. Smooth and fast movement over the glass slide was observed at a velocity of 6 $\mu m/s$ at 23°C, almost the same as that pro-duced by double-headed myosin (7 $\mu m/s$). Although the preparation contained less than 2% molar ratio of double-headed myosin, we were able to confirm that the contam-inating double-headed myosin was not responsible for the movement. This was done by observing the move-ment of actin filaments along hybrid myosin filaments that contained single- and double-headed myosin mixed in various molar ratios.

Another problem was the possibility that two adja-cent single myosin heads might interact cooperatively to act as two-headed myosin molecules rather than acting individually and singly, because the density of heads on the thick filaments was very high. This possibility was excluded by observing the rate at which actin filaments moved over myosin molecules from which most of the heads had been removed. This was done in order to re-duce the spatial concentration of single heads. The actin filaments moved just as fast along myosin filaments from which >90% of the heads had been thinned as before the thinning. Furthermore, the movement of thin filaments containing the tropomyosin-troponin complex over the single-headed myosin was found to be regulated by cal-cium ions, although the velocity curve was slightly different from that obtained when double-headed myosin was used. These results demonstrated collectively that cooperative interaction between two myosin heads is not essential to the development of the sliding movement.[9]

These tests did not tell us whether single-headed myosin can produce significant sliding force. For this purpose, we used a new technique for holding and manipulating single actin filaments. One end of the actin filament was caught and held by a very fine glass microneedle whose surface had been previously coated by N-ethylmaleimide-treated myosin to increase its affinity for actin.[7] The other end of the actin filament was brought into contact with the myosin-coated surface of a glass slide. In the presence of ATP, the actin filament moved and bent the needle. The force due to interaction between actin and myosin was determined by measuring the degree of bending of the needle. The stiffness of needles used was 5 to 20 pN/μm.

The method for producing myosin coats on glass slides had to be altered for this experiment. For previous observations of the movement of actin filaments at zero load, we had coated the surface with myosin molecules assembled into short filaments. But this preparation was not suitable for the force measurements because the density of myosin heads on the surface was not homogeneous and consequently the force varied. Therefore, we coated the glass surface, which had been treated with silicone, with myosin in monomeric form. Electron microscope observation showed that the surface was homogeneously coated with myosin heads.[10] It was found that monomeric myosin applied in this way could move actin filaments as fast as myosin filaments.[11]

Using the monomer-coated slides, the force generated by a single actin filament 1 μm long interacting with the myosin-coated surface was about 30 piconewtons (Harada et al. unpublished data). The number of myosin heads in the vicinity of an actin filament of this length was estimated to be about 100 from the density of myosin heads on the surface. Since the orientation of myosin heads was random and the heads bound to the surface would be able to interact with only one side of the actin filament, the number of heads that could participate in the force generation was probably one fourth of the 100 heads, i.e., about 25 heads per μm of actin filament. Thus, the force per head was estimated to be roughly 1 pN, which is comparable to the force exerted by double-headed myosin molecules in muscle.

To minimize damage to the myosin heads during the initial isolation and purification techniques, we digested myosin molecules with proteinase for only a brief period. This reduced the chance that the proteinase might introduce breaks within the heads. This was important to our results because myosin heads with breaks from digestion can move the actin filament as fast as normal ones, but can generate little force.

In conclusion, the results show that single-headed myosin subfragment-1 is sufficient to produce the movement of actin filaments at zero load.[11] The force measurements demonstrated that S-1 myosin fragments bound to the silicone-treated glass surface produced a force as large as intact, double-headed myosin. Therefore, cooperative interaction between the two heads of intact myosin is essential for inducing neither the sliding movement of actin filaments nor the force. But it is not yet known why myosin has double-headed structure.

References

[1] Tonomura, Y. *Muscle proteins, muscle contraction and cation transport.* Tokyo: University of Tokyo Press (1972).

[2] Cooke, R., and Franks, K. E. *J. Mol. Biol.* 120:36 (1978).

[3] Marggossian, S. S., and Lowey, S. *J. Mol. Biol.* 74:312 (1971).

[4] Chaen, S.; Shimada, M.; and Sugi, H. *J. Biol. Chem.* 261:13632 (1986).

[5] Yanagida, T.; Nakase, M.; Nishiyama, K.; and Oosaswa, F. *Nature* 307:58 (1984).

[6] Kron, S. J., and Spudich, J. A. *Proc. Nat. Acad. Sci.* 83:6272 (1986).

[7] Kishino, A., and Yanagida, T. *Nature* 334:74 (1988).

[8] Ishijima, A.; Doi, T.; Sakurada, K.; and Yanagida, T. *Nature* 352: 301 (1991).

[9] Harada, Y.; Noguchi, A.; Kishino, A.; and Yanagida, T. *Nature* 326:805 (1987).

[10] Harada, Y.; Sakurada, K.; Aoki, T.; Thomas, D. D.; and Yanagida, T. *J. Mol. Biol.* 216:49 (1990).

[11] Toyoshima, Y., et al. *Nature* 328:536 (1987).

muscle. Other actin-linked regulatory pathways, however, may operate in smooth muscle and nonmuscle cells (see below).

Myosin-linked Regulation   Control of the crossbridging cycle by elements linked to myosin was discovered primarily through research on smooth muscle by A. Szent-Gyorgyi and his coworkers. These control systems operate through a pathway in which elevated Ca^{2+} concentrations induce phosphorylation of the regulatory light chains. When phosphorylated, the regulatory light chains permit operation of crossbridging cycle; if the phosphate groups are removed, the cycle stops. The control depends on the regulatory light chains; if these polypeptides are removed from myosin, changes in Ca^{2+} concentration have no effect and crossbridge cycling and microfilament sliding proceed as long as ATP is available. In some smooth muscle cells, Ca^{2+} release is regulated by hormones, through surface receptors that trigger the $InsP_3$/DAG pathway and cause release of Ca^{2+} as a second messenger (see p. 161).

Phosphorylation of the light chains in myosin-linked control systems depends on the activity of an enzyme, *myosin light chain kinase* (*MLC kinase*). MLC kinase includes a calmodulin subunit that provides Ca^{2+}-sensitivity to the system. When Ca^{2+} is present at "resting" concentrations, calmodulin takes on a conformation that inhibits MLC kinase, and no phosphor-

ylation of myosin light chains takes place. As Ca^{2+} concentration rises above the resting level, calmodulin binds Ca^{2+} and undergoes a conformational change that activates the kinase. Under these conditions, phosphate groups are added to the light chains, and the crossbridging cycle is turned on.

Switching the mechanism off depends on *myosin light chain phosphorylase* (*MLC phosphorylase*), an enzyme that removes phosphate groups from the myosin light chains. MLC phosphorylase is continually active. As the Ca^{2+} concentration falls to the resting level, Ca^{2+} is released from calmodulin and MLC kinase is inhibited. Under these conditions, phosphate groups are quickly removed from the regulatory light chains by the constant MLC phosphorylase activity, and the crossbridging cycle stops. (Fig. 11-11 summarizes the steps in myosin-linked regulation.)

In many smooth muscle types, myosin regulation via the light chains is supplemented by an actin-linked mechanism based on another Ca^{2+}/calmodulin-activated protein, *caldesmon*. Caldesmon also occurs as an actin-linked regulatory protein in many nonmuscle cell systems. Caldesmon links directly to the microfilaments of smooth muscle cells, which also contain tropomyosin. When Ca^{2+} is present at resting levels, caldesmon takes a form in which it binds actin and blocks the sites bound by myosin. Elevated Ca^{2+} concentrations activate calmodulin, which in turn binds to caldesmon, inducing a conformational change

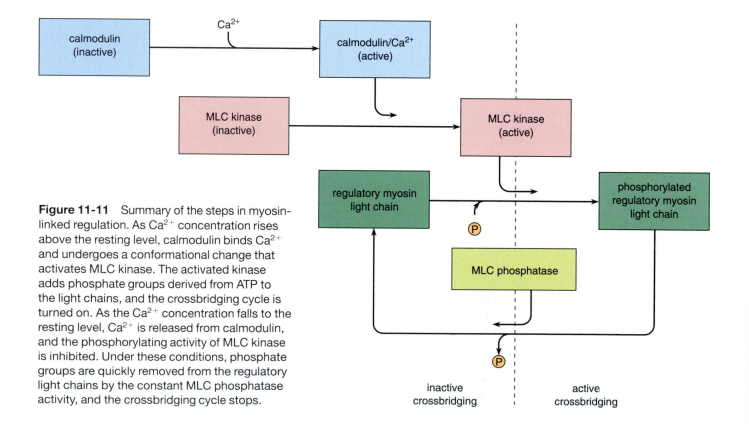

Figure 11-11   Summary of the steps in myosin-linked regulation. As Ca^{2+} concentration rises above the resting level, calmodulin binds Ca^{2+} and undergoes a conformational change that activates MLC kinase. The activated kinase adds phosphate groups derived from ATP to the light chains, and the crossbridging cycle is turned on. As the Ca^{2+} concentration falls to the resting level, Ca^{2+} is released from calmodulin, and the phosphorylating activity of MLC kinase is inhibited. Under these conditions, phosphate groups are quickly removed from the regulatory light chains by the constant MLC phosphatase activity, and the crossbridging cycle stops.

that greatly reduces the affinity of caldesmon for actin. Caldesmon then releases from actin; the release exposes the myosin-binding sites on microfilaments. Caldesmon may directly cover and uncover the myosin-binding sites, or it may work indirectly by binding and releasing tropomyosin, in a mechanism similar to the troponin-based pathway of striated muscle cells. Still another Ca^{2+}/calmodulin-dependent regulatory protein, *calponin*, regulates contraction in some smooth muscle cells in an actin-linked mechanism similar to that of caldesmon.

Although myosin-linked regulation was discovered primarily through research on smooth muscle cells, this regulatory pathway is apparently also widely distributed among nonmuscle cells in eukaryotes. In many nonmuscle systems the myosin-linked pathway is supplemented by regulation of actin by caldesmon and calponin, as it is in smooth muscle cells.

All these regulatory mechanisms involve myosin II molecules in various locations. Relatively little is known about myosin I regulation. In the soil protozoan *Acanthamoeba*, Korn and his coworkers found that myosin I is activated by phosphorylation of the heavy chain head unit, in a reaction catalyzed by a heavy chain kinase. At least some forms of myosin I are believed to be regulated by Ca^+, because calmodulin forms a light chain associated with the head unit in these myosins. However, changes in Ca^+ concentration have not as yet been definitely linked to myosin I activity in these types.

MOTILE SYSTEMS BASED ON MICROFILAMENTS

Studies of microfilament structure and function began with investigations of striated muscle. These studies led to discovery of actin and myosin, microfilament sliding, and its source of power in the crossbridging cycle.

Striated Muscle Cells

Striated Muscle Structure   Striated muscle was originally given its name because it shows a pattern of regularly spaced crossbands under the light microscope (Fig. 11-12a). A striated muscle, such as the biceps muscle of the upper arm, consists of a large number of greatly elongated, cylindrical or ribbonlike cells called *muscle fibers* (Fig. 11-12b and c). Individual muscle fibers may be as long as 40 millimeters. Muscle fibers are striated; their appearance is due to crossbanding in long, fibrous strands called *myofibrils* that are arranged parallel to the long axis of the cells

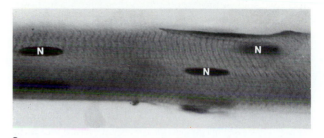

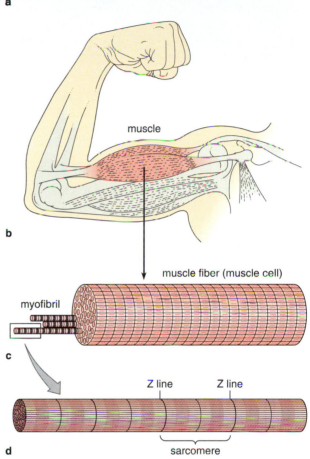

Figure 11-12   Striated muscle structure. **(a)** Striated chicken pectoralis muscle under the light microscope. The muscle fibers run horizontally in the micrograph. The cross-striations from which striated muscle takes its name run vertically. × 920. **(b)**, **(c)**, and **(d)** Relationships among muscles, muscle fibers, myofibrils, and sarcomeres. (Micrograph courtesy of N. N. Malouf. Copyright by Academic Press, Inc., from *Exp. Cell Res*. 122:233 [1979]. Diagrams redrawn from an original courtesy of D. W. Fawcett, from D. W. Fawcett and W. Bloom, *A Textbook of Histology*, 10th ed., Philadelphia: Saunders, 1975.)

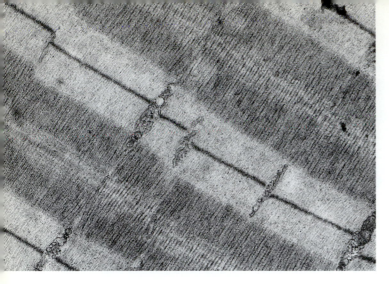

Figure 11-13 Myofibrils of fully contracted frog semi-tendinosus muscle in the electron microscope. × 34,000. (Courtesy of J. G. Tidball.)

(Fig. 11-12d). The myofibrils are surrounded by layers of cytoplasm containing multiple nuclei (visible in Fig. 11-12a) and large numbers of mitochondria. The cytoplasm surrounding myofibrils also contains a system of membranous tubules called the *sarcoplasmic reticulum (SR)*, related to the smooth endoplasmic reticulum of other cell types. The SR forms part of the system controlling muscle contraction (see below). The multiple nuclei in muscle fibers, which may number as many as 100 per cell, reflect the fact that in embryos these elongated cells develop through end-to-end fusion of many individual embryonic cells, whose nuclei persist individually in mature fibers.

The repeating units responsible for the striated appearance of myofibrils and muscle cells are clearly visible in the electron microscope (Figs. 11-13 and 11-14a). Each unit, about 2.0 to 2.5 μm long in resting muscle, is called a *sarcomere*. The boundaries of a sarcomere are marked by dark, slender transverse bands called *Z lines*. The broader light and dark bands between the Z lines are formed by overlapping *thin* and *thick filaments* that run lengthwise within the sarcomeres.

The thin filaments are single actin microfilaments in combination with troponin and tropomyosin. Myosin decoration shows that thin filaments connect at their plus, or barbed, ends to either side of a Z line and extend with their minus, or pointed, ends toward the center of a sarcomere. Z lines contain several proteins, including α-*actinin, filamin, synemin*, and *Z protein*, which probably crosslink microfilaments and hold them in register in this region. In the midregion of a sarcomere, thin filaments overlap with thick filaments, which are assemblies of myosin molecules. In vertebrates, each thick filament contains some 300 to 400 myosin molecules; an entire muscle fiber contains, in total, about 1 billion myosin molecules. Myosin head units in the thick filaments form crossbridges between the thick and thin filaments in the zones of overlap.

When thick filaments are isolated from striated muscle and prepared for electron microscopy by negative staining, crossbridges can be seen to extend from all regions of the thick filaments except for a "bare zone" in the middle (Fig. 11-15a). This appearance is believed to reflect an arrangement in which individual myosin molecules are held parallel to the long axis of the thick filament, running in opposite directions from the two ends. The heads are directed toward the filament ends, and the tails extend toward the midregion (Fig. 11-15b). Only tails are present near the center, arranged in an overlapping pattern that produces the bare zone. Cross sections suggest that at any point, vertebrate thick filaments are built up from a bundle of nine parallel myosin molecules (Fig. 11-15c).

Cross sections of sarcomeres show that thick and thin filaments overlap in a highly regular pattern. In vertebrates this pattern forms a *double hexagon*, with thick and thin filaments held in a ratio of 1:2 (Fig. 11-16a). Other patterns, both highly ordered and irregular, occur in invertebrates, giving a variety of thick-to-thin filament ratios and spacings in different invertebrate cell types and species. (Figure 11-16b to d shows some of these patterns.)

In relaxed muscle, thin filaments extend only about halfway along the thick filaments. This arrangement leaves open spaces that appear as a broad, somewhat lighter band, the *H band*, running transversely across the center of each sarcomere (see Fig. 11-14b). A narrow, dense transverse band, the *M band*, runs through the middle of the H band at the center of the sarcomere. As muscle contracts, the width of the H band and the space between the ends of the thick filaments and the Z lines (the I bands in Fig. 11-14b and c) both narrow by an equivalent distance. This observation, and the fact that neither the thick nor the thin filaments of sarcomeres become shorter during contraction, formed the basis for the first proposals that microfilaments slide to produce movement.

The identity of the thick and thin filaments with myosin and actin was established in 1953 by Huxley and Hanson, who took advantage of the fact that actin and myosin are released from striated muscle by different salt concentrations. At salt concentrations that release myosin, thick filaments disappeared or became fainter in muscle sections. At salt concentrations that release actin, thin filaments lightened or disappeared. Later investigations using labeled antibodies confirmed that the thick filaments contain myosin and the thin filaments are actin microfilaments.

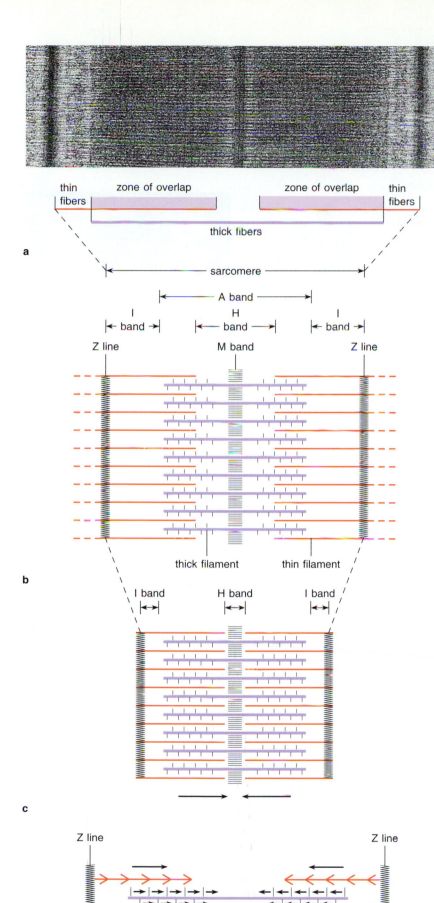

Figure 11-14 Microfilament sliding in sarcomere contraction. **(a)** A single sarcomere from frog semitendinosus muscle. × 44,000. (Courtesy of J. G. Tidball.) **(b)** Sarcomere structure in the relaxed state. **(c)** Sarcomere under contraction; the thin filaments are sliding inward along the thick filaments, reducing the width of the I and H bands by an equal distance and shortening the entire sarcomere. **(d)** Opposite polarity (arrows) of the thin filaments and the myosin crossbridges. Because of their opposite polarity, the attach-pull-release cycle of the myosin head units pulls the Z lines together and actively shortens the sarcomeres.

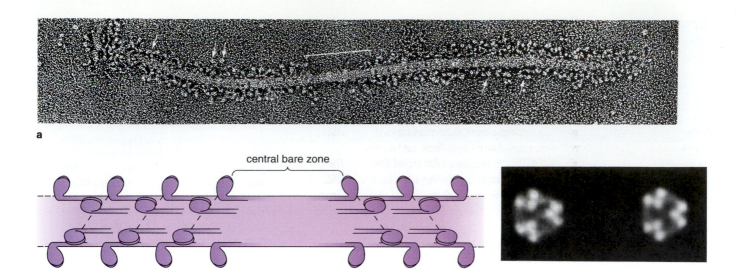

central bare zone

Figure 11-15 Thick filament structure. **(a)** A thick filament isolated from striated muscle, prepared for electron microscopy by negative staining. Myosin head units (arrows) are distributed all along the thick filament except for the central bare zone (bracket). × 116,000. (Courtesy of A. Elliott, from *J. Mol. Biol.* 131:133 [1979].) **(b)** The antiparallel or end-to-end arrangement of myosin molecules in thick filaments. **(c)** A composite, computer-averaged cross-sectional image of a thick filament from vertebrate striated muscle. The image shows the average density distribution of many superimposed thick filament cross sections. The density is reversed, so that regions of heaviest density appear lightest in the image. The image shows nine subunits, indicating that the tails of nine myosin molecules associate to form a thick filament. (Courtesy of F. A. Pepe.)

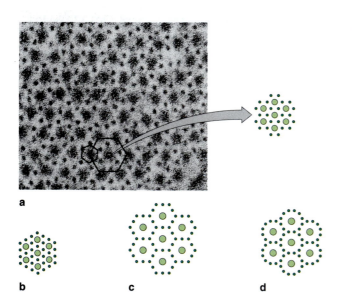

◀ **Figure 11-16** Arrangement of thick and thin filaments in sarcomeres. **(a)** Vertebrate sarcomere in a cross section made in the region of overlap between thick and thin filaments. The filaments form a double hexagon pattern (dotted lines). Each thick filament is surrounded by six thin filaments arranged hexagonally, and the thick filaments make up a larger hexagon. × 115,000. **(b)** The arrangement of thick and thin fibers in insect flight muscle and various other arthropod muscles (**c** and **d**). (Micrograph courtesy of H. E. Huxley, with permission from *J. Mol. Biol.* 37:507 [1968]. Copyright by Academic Press, Inc. [London] Ltd. Diagrams courtesy of F. A. Pepe; reproduced from *J. Cell Biol.* 37:445 [1968], by copyright permission of the Rockefeller University Press.)

Striated Muscle Motility and Its Regulation As a consequence of the arrangement of thick and thin filaments in sarcomeres, the myosin head units slide the thin filaments on either side of the sarcomere toward the M band. The sliding draws the Z lines closer together and actively shortens or contracts the entire sarcomere. The contraction of all the sarcomeres in a muscle fiber forcibly shortens the entire muscle and moves structures to which the muscle is attached.

The control of striated muscle contraction by motor nerves is linked to the troponin-tropomyosin system by a complex mechanism that regulates Ca^{2+} concentration in muscle cells. The sarcoplasmic reticulum (SR), the system of smooth ER sacs surrounding

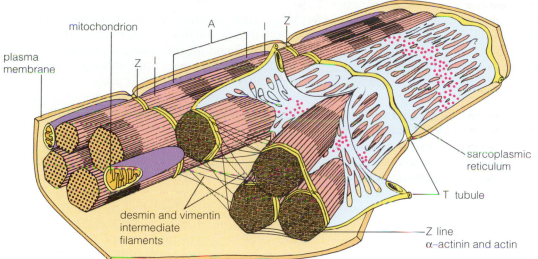

Figure 11-17 The arrangement of sarcomeres, myofibrils, sarcoplasmic reticulum, and T tubules in muscle fibers. The Z bands are held in place and reinforced by desmin and vimentin intermediate filaments. (Redrawn from an original courtesy of E. Lazarides, from *Nature* 283:249; copyright © 1980 Macmillan Magazines Ltd.)

the myofibrils (Fig. 11-17), directly regulates the Ca^{2+} concentration of the muscle cells. The SR membranes contain a Ca^{2+}-ATPase pump (see p. 131) that continuously removes Ca^{2+} from the cytoplasm surrounding the myofibrils and pushes it into the SR vesicles. Ca^{2+}-ATPase pumps in the plasma membrane also remove Ca^{2+} from the cytoplasm. This continuous pumping activity keeps the Ca^{2+} concentration of the muscle cell cytoplasm at the resting level, too low for activation of muscle contraction.

SR membranes are connected to the muscle cell surface by long, tubular invaginations of the plasma membrane called *T tubules* (T = transverse; see Fig. 11-17 and 11-18). The T tubules closely adjoin but do not fuse directly with the SR membranes.

At some point a motor neuron makes contact with the plasma membrane of a muscle cell via a synapse (see p. 141). An arriving impulse causes a voltage change in the muscle cell plasma membrane that is propagated over the entire muscle cell surface and downward into the T tubules. When the impulse arrives at the T tubule–SR junctions, gated Ca^{2+} channels in the SR membranes snap open in response to the arriving impulse and the SR membranes suddenly become "leaky" to their stored Ca^{2+} (see Fig. 11-18 and p. 131). At the elevated concentrations produced by this sudden release, troponin binds Ca^{2+}, causing tropomyosin to move away from the position in which it blocks the interaction between actin and myosin. This movement turns on the crossbridging cycle, and the muscle cell contracts.

When nerve impulses cease to arrive at the muscle cell surface, the gated Ca^{2+} channels close. Ca^{2+} flow into the muscle cell cytoplasm stops, and the Ca^{2+}-ATPase pumps quickly remove the remaining Ca^{2+}. As a consequence, Ca^{2+} concentrations drop to

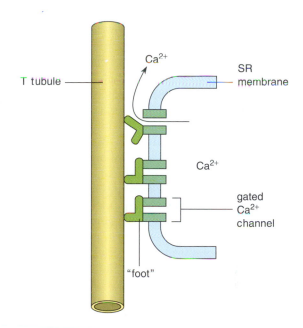

Figure 11-18 The relationship of T tubules and SR membranes in a mammalian muscle. Each gated Ca^{2+} channel has a "foot" connecting it to the SR; the gate of the channel at the top is open.

the resting level, and the troponin-tropomyosin system loses its bound Ca^{2+}. Under these conditions, tropomyosin is pulled from the microfilament grooves into the blocking position, the crossbridging cycle is switched off, and contraction stops.

Unwieldy and complex as it sounds, the control mechanism provides almost instantaneous regulation of skeletal muscle activity. The mechanism provides us and other animals with the ability to carry out voluntary movements that are characteristically rapid and powerful, yet delicately and precisely controlled.

Cardiac Muscle Cardiac muscle cells also contain striated myofibrils consisting of sarcomeres linked end to end and an internal system of T tubules and SR membranes that regulate contraction. There are several significant differences, however, in other features of the structure and function of cardiac muscle cells. Cardiac muscle cells are much shorter than skeletal muscle cells and usually contain only a single nucleus, reflecting the fact that the embryonic cells forming cardiac muscle remain single and do not fuse end to end. Cardiac muscle cells may connect to several neighboring cells, tying the cells into a tightly linked, integrated network. The network causes heart muscle to contract in all directions to produce a squeezing or pumping action rather than the lengthwise contraction characteristic of skeletal muscle.

No motor neurons connect directly to cardiac cells, and there are no synapses between neurons and muscle cells within the heart. The stimulus for contraction originates within the heart itself, in specialized groups of cardiac muscle cells called *pacemakers*, which generate impulses for contraction in a regular, periodic rhythm. An impulse triggered in a pacemaker travels rapidly throughout the heart by passing directly from the plasma membrane of one cardiac cell to the next. As in skeletal muscle cells, impulses passing over the plasma membranes are conducted into cardiac muscle cells by T tubules. Arrival of the impulse at the SR triggers Ca^{2+} release as in skeletal muscles, and contraction is switched on. As the signal originating from the pacemakers ceases, the flow of Ca^{2+} into the muscle cell cytoplasm stops and contraction is switched off. The wave of contraction triggered by the pacemakers travels through the heart from top to bottom, first pumping blood from the auricles into the ventricles, and then from the ventricles into the arteries leaving the heart.

The sarcomeres of skeletal and cardiac muscle are the most complex and organized assemblies of microfilaments known. Like the arrangement of microtubules in the 9 + 2 system of flagella, sarcomeres evidently provide the most efficient arrangement of microfilaments for the generation of motility. Measurements of the power generated by striated muscle, in fact, indicate an output about 10 times higher than that of flagella and place this microfilament-based system on a par with gasoline engines in power output per weight.

Smooth Muscle Cells

Smooth muscle occurs in the walls of the gut, air passages, blood vessels, and the urogenital tract of vertebrate animals. The cells of smooth muscle, so called because they show no regular striations in either the light or electron microscope, are slender, elongated, and pointed at their tips. Like cardiac cells, each smooth muscle cell has a single nucleus. However, no T tubules extend inward from the plasma membrane in smooth muscle cells and, although smooth ER membranes are present, they appear primarily as small, spherical vesicles scattered just under the plasma membrane and in groups in the cell interior.

Thin and Thick Filaments in Smooth Muscle Cells Actin thin filaments are distributed more or less uniformly in the cytoplasm of smooth muscle cells, aligned roughly with the long axis of the cell (Fig. 11-19). These microfilaments contain actin and tropomyosin in the same proportions as in skeletal and cardiac muscle cells, but in distinct forms. Caldesmon is present in one-fourth the amount of tropomyosin in the thin fibers of many smooth muscle cells. This control protein, rather than troponin, contributes to control of contraction in these cells (see below).

The myosin thick filaments of smooth muscle cells are somewhat flattened or ribbonlike. They lie parallel to the long axis of the smooth muscle cell, distributed among the thin filaments. No central bare zone is visible in isolated smooth muscle thick filaments. Instead, myosin head groups are distributed along the whole length of the fibers, probably in opposing directions on opposite sides of a thick filament. The proportion of thick filaments is much lower in smooth muscle cells—smooth muscle contains about 1 thick filament for every 12 to 18 thin filaments, as compared to 1:2 and similar thick:thin ratios in striated muscle.

Smooth muscle thin filaments, which typically occur in bundles, anchor in *dense bodies* within the cytoplasm or at the plasma membrane. The microfilaments anchor with their plus ends linked to the dense bodies, in an orientation equivalent to the attachment of thin filaments to Z lines in striated muscle. Like Z lines, dense bodies also contain α-actinin. The dense bodies therefore are probably the functional equivalents of the Z lines in striated muscle. The microfilaments between any two dense bodies, with their associated thick filaments, resemble a relatively unorganized sarcomere.

During contraction of smooth muscle cells, thin filaments slide over thick filaments, drawing the dense bodies closer together and, through the connections of dense bodies to each other and to the plasma membrane, forcibly shorten the entire cell. The loosely organized arrangement of thin and thick filaments allows smooth muscle cells to contract and relax over much greater lengths than skeletal and cardiac muscle cells. This allows organs containing smooth muscle in their walls, such as the stomach, intestine, urinary

a

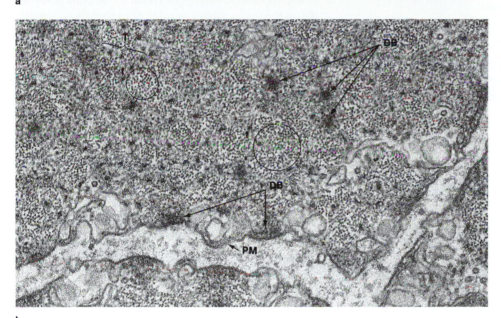

b

Figure 11-19 The arrangement of thick and thin fibers in smooth muscle. (a) Smooth muscle cell of rabbit vascular tissue. Myosin thick filaments (arrows) are surrounded by actin thin filaments in a pattern that is less highly ordered than the filament pattern of striated muscle. × 40,000. (Courtesy of A. V. Somlyo, with permission from *J. Mol. Biol.* 96:17 [1975]. Copyright by Academic Press, Inc. [London] Ltd.) (b) Smooth muscle cell of the digestive system of the tapeworm *Taenia coli* in cross section. Thin filaments are circled. T, thick filaments; DB, dense bodies; PM, plasma membrane. (Courtesy of P. Cooke.)

bladder, and uterus, to make major adjustments in size to accommodate, move, or expel their contents.

Control of Smooth Muscle Contraction   There are no close connections between neurons and smooth muscle cells equivalent to the nerve-muscle synapses of striated muscle cells. Instead, most smooth muscles are controlled by neurotransmitters (see p. 142) or hormones released into the body circulation. The response of smooth muscle to these substances depends on receptors in the muscle cell plasma membrane that are tailored to recognize and bind the controlling molecule.

Once bound to their activating substance, the receptors indirectly trigger the opening of gated Ca^{2+} channels in the ER or plasma membrane. In many smooth muscle cells the receptors triggering Ca^{2+} release do so through the $InsP_3$/DAG second messenger pathway. Opening the membrane channels allows Ca^{2+} to flow into the cytoplasm from the extracellular medi-

um or stores inside the ER cisternae. In other smooth muscle cells, receptor binding opens membrane channels admitting ions such as K^+, leading to a change in the voltage difference across the membrane (see p. 139). The potential change opens voltage-gated Ca^{2+} channels in the plasma membrane, allowing Ca^{2+} to flow from the extracellular medium into the muscle cell cytoplasm. Smooth muscle cells in some organs are interconnected by gap junctions (see p. 171) that allow ions to flow directly from one cell to the next. In this way an ion flow altering plasma membrane potential and stimulating internal Ca^{2+} release can be transmitted directly between cells, spreading a wave of contraction through the smooth muscle of an entire organ.

The Ca^{2+} released into the smooth muscle cytoplasm by these stimuli triggers contraction primarily through the myosin-linked pathway. Ca^{2+} activates calmodulin, which in turn activates MLC kinase, causing phosphorylation of the regulatory light chains and triggering the attach-pull-release cycle of myosin. This regulatory pathway is supplemented in many vertebrate smooth muscle cells by the actin-linked pathways involving caldesmon or calponin.

Relaxation of some smooth muscle types is induced by hormones bound by receptors at cell surfaces. In these cells the receptors activate cAMP-based pathways, leading to stimulation of cAMP-dependent protein kinases that add phosphate groups to MLC kinase. The phosphorylations reduce the activity of MLC kinase, thereby reducing phosphorylation of myosin light chains and inhibiting smooth muscle contraction.

Microfilament-Based Motility in Nonmuscle Cells

The distinctive biochemical properties of actin and myosin, as determined from studies of muscle cells, armed investigators with means to detect the two proteins and microfilament-based motility in nonmuscle cells. The myosin decoration technique has been used to detect actin microfilaments in protists, in the nonmuscle cells of a wide variety of animals, and in fungi, algae and higher plants. The amino acid sequences of nonmuscle actins are distinct but still closely related to those of striated and smooth muscle actins. Distinctive myosin I and II types have also been isolated and identified among the cellular proteins of all eukaryotes. These discoveries led to the concept that all eukaryotic cells contain motile systems based on actin and myosin.

Nonmuscle actin and myosin molecules are responsible for a variety of movements, including cytoplasmic streaming, ameboid motion, capping, the movements of cell layers during embryonic development, and the furrowing movements that divide the cytoplasm during cell division. In all these systems, myosin crossbridges are believed to generate the force for movement, in a process equivalent to the sliding filament mechanism of striated muscle cells.

Cytoplasmic Streaming   The active flowing movements of cytoplasm known as cytoplasmic streaming occur in all eukaryotic cells. This movement has been studied most extensively in algal cells. In one such study, B. A. Palevitz and P. K. Hepler used the myosin decoration technique to study cytoplasmic streaming in the alga *Nitella*, in which chloroplasts move along microfilament bundles underlying the cortical gel. Myosin decoration confirmed that the bundles consist of microfilaments; Palevitz and Hepler suggested that active streaming in *Nitella* is generated by myosin molecules attached to chloroplasts, which "walk" by their crossbridging cycles over microfilaments anchored in the cortical gel. This interpretation is supported by the experiments of Sheetz and Spudich, described earlier in this chapter (see p. 320), in which myosin-coated beads could be seen to move along the surfaces of microfilament bundles exposed by stripping the cortex from *Nitella* cells.

Microfilament-based cytoplasmic streaming has also been demonstrated in higher plants, particularly in cambium, vascular, and root tip cells. J. Heslop-Harrison and Y. Heslop-Harrison showed recently that antibodies developed against bovine myosin react strongly with the surfaces of plant organelles transported by cytoplasmic streaming in pollen tubes, indicating that myosin probably "walks" the organelles along tracks supplied by microfilaments. Little is known about the mechanisms regulating cytoplasmic streaming in the algae or higher plants.

Cytoplasmic streaming can also be observed in protozoan, fungal, and animal cells. In these cells the streaming movements are halted by drugs such as the cytochalasins and other treatments that interfere with microfilament structure or function. Although cytoplasmic streaming is therefore generated by microfilaments in these organisms, the organization and location of the microfilaments and myosin molecules producing the movement remain unknown.

Ameboid Motion   Ameboid motion, which is most often associated with protozoans such as *Amoeba proteus*, also occurs in fungi and in related forms in all animals. In animals, ameboid motion of individual cells takes place regularly during embryonic development. Certain cells, such as the white blood cells of vertebrates, retain the capacity for ameboid movement in adults. Other cell types, although nonmotile in their normal locations in adult tissues, may migrate by ameboid motion under certain conditions. Fibroblasts, for example, may break their normal attachments and migrate by ameboid motion into damaged tissue areas as part of the reconstruction taking place in response to wounding. Many other normally nonmotile body cells may become motile and travel by ameboid motion if they become transformed into cancer cells or if they are grown in tissue culture outside the body.

Ameboid motion in an organism such as *Amoeba* seems simple enough when observed in the light microscope. At some point along the cell margin a cytoplasmic lobe forms and swells outward. The cytoplasm of the cell flows into the lobe, eventually carrying the cytoplasmic organelles and the nucleus with it. Close examination of an ameboid cell shows that the cytoplasm just under the cell surface, except at the tip of an advancing lobe, is semisolid or gelled, forming a sort of cortical cytoplasmic tube through which the more liquid internal cytoplasm flows toward the tip.

The rate at which the liquid cytoplasm flows through the cortical tube is generally much faster than the rates noted for cytoplasmic streaming.

Actin and myosin were linked to ameboid motion by a series of important discoveries in protozoan and slime mold amebas. T. D. Pollard and others showed that extracted ameba cytoplasm contains microfilaments that form typical arrowheads when reacted with HMM. Actin isolated from ameboid cells was found to be almost identical to striated muscle actin in structure and chemical properties. Myosin was also identified in the isolated ameba cytoplasm, typically in smaller amounts with respect to actin than in striated muscle. As far as can be determined, troponin does not function in ameboid motile systems. The movements generated through the interaction of actin and myosin in ameboid cells require ATP and also start, stop, and undergo alterations in rate in response to changes in Ca^{2+} concentration. Exposure of the cells to cytochalasin disperses the cortical actin gel and slows or stops ameboid motion.

Thin sections of ameboid cells reveal that microfilaments form a network just under the plasma membrane. This layer corresponds to the cortical layer of gelled or semisolid cytoplasm in these cells. The layer is thinnest at the tip of an advancing pseudopod and becomes gradually thicker toward the tail

end of the cell. Recent work by E. D. Korn and his colleagues with fluorescent antibodies against myosin I or II demonstrated that myosin II occurs at highest concentrations in midregions and near the tail of ameboid cells; myosin I is concentrated in the cortical region near the leading edges of an advancing lobe.

These observations clearly indicate that actin and myosin generate the force for ameboid motion. Just how the actin-myosin system produces motion, however, remains a matter of controversy. D. L. Taylor and his colleagues proposed that contraction of the cortical gel in midregions and near the rear of an advancing ameba forces the liquid internal cytoplasm passively through the tube toward the tip, much like squeezing a toothpaste tube (Fig. 11-20). The fact that myosin II concentrations are highest in midregions and at the rear of the cortical tube supports this proposal and indicates that this myosin may be responsible for squeezing the tube. Other supporting evidence comes from the observation that extracted cortical gels are capable of contraction when supplied with ATP.

Cortical contraction by myosin II–microfilament interaction at the tail end of an advancing cell is not the whole story in ameboid motion, however. Some ameboid cell types can be punctured or lysed, thus eliminating any internal hydrostatic pressure developed by cortical contraction, without stopping local

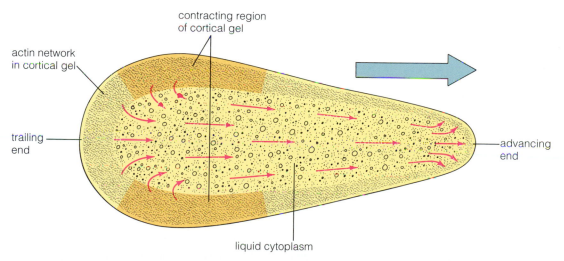

Figure 11-20 A possible mechanism producing part of the force for ameboid motion. At the trailing end the gelled cortical tube, consisting of a crosslinked network of microfilaments, continually disassembles by actin depolymerization or breakage of microfilament crosslinks. At the advancing end the tube continuously extends through microfilament assembly or formation of microfilament crosslinks. The liquid inner cytoplasm, containing the actin or microfilaments removed from the trailing end of the cortical gel, is forced passively through the tube by contraction of the cortical gel in the midregion and near the trailing end of the cell (in darker color). Myosin II molecules may be responsible for contracting the cortical gel; myosin I may power microfilament sliding advancing the leading edge of a pseudopod.

extensions of the pseudopod tip. In cells such as fibroblasts, which produce very thin, flattened pseudopods called *lamellipodia*, the pseudopods can carry out limited forward movements if sliced away from the cell body.

These observations suggest that pseudopods may advance through a combination of hydrostatic pressure induced by cortical contraction at the rear of the cell and local interactions of microfilaments within the crosslinked networks at the leading edge. The microfilament interactions at the leading edge may include active sliding powered by myosin I and directional polymerization. Some investigators have suggested that penetration of water into the crosslinked actin gel at the leading edge may produce swelling that also contributes to pseudopod movement.

Some cell types such as fibroblasts form transient attachments with the underlying substrate during ameboid motion. The attachments, called *focal contacts* (see also p. 184), form under advancing lamellipodia as well as regions under the trailing cell body. These focal contacts, particularly near the advancing cell margin, may form local sites against which sliding microfilaments push or pull during ameboid movement.

There is good evidence that, whatever the mechanism, ameboid motion is controlled by local changes in Ca^{2+} concentrations. In one experiment, Taylor injected *aequorin*, a protein that emits light when it binds with Ca^{2+}, into ameboid cells. The aequorin fluoresced brightly in the tail regions of advancing amebas and also in pulses at other sites of contraction, indicating that Ca^{2+} was released in these regions. The Ca^{2+} controlling ameboid movement is thought to be stored and released by the ER.

Other experiments using antibodies against calmodulin and MLC kinase showed that both these elements of Ca^{2+}-based regulation by the myosin-linked pathway are present in actively moving regions of ameboid cells. Depending on the cell type, the kinase adds phosphate groups to myosin light chains, heavy chains, or both polypeptide types to regulate motility. In general, phosphorylation of myosin light chains initiates or speeds movement, as it does in muscle cells. Heavy-chain phosphorylation, depending on the cell type, may either speed or slow movement.

During ameboid motion the cortical gel probably remains stationary as the underlying liquid cytoplasm moves toward the tip of the pseudopod. The gel is thought to disassemble continually in the thick region at the rear end of the cell to accommodate forward movement of the cell. The actin subunits and short microfilament lengths released by the disassembly are carried with the liquid cytoplasm toward the advancing tip, where they reassemble into a new network in the thin cortical layer. Formation and breakdown of the cortical gel at its front and rear ends may be regulated by Ca^{2+}, through its effects on Ca^{2+}-sensitive microfilament capping, crosslinking, and severing proteins. Many of these proteins, including gelsolin, fragmin, filamin, and gelactin, can be detected in ameboid cells of different types. In general, elevated Ca^{2+} concentrations would induce disassembly of the gel, and resting concentrations would favor cortical gel assembly.

Other Microfilament-Based Motile Systems Microfilaments have been established as the basis for several additional cellular movements. In all of these motile mechanisms, including capping, division of the cytoplasm in animal cells (furrowing), and contraction of cell layers during embryogenesis, microfilaments identified by myosin decoration are attached to or associated with the elements being moved. Myosin can also be detected in the moving regions. Treatment with cytochalasin or antibodies developed against actin or myosin generally stops the motion. In many cases, elements of the myosin-linked, but not the actin-linked, regulatory pathway have been detected in association with microfilament systems generating the motion.

Capping (see p. 108) involves movements that sweep substances bound to the cell surface into a localized mass, the "cap," at one end of the cell. The capped substances are later taken into the cytoplasm by endocytosis. The microfilaments responsible for capping are attached to the cytoplasmic portions of receptor molecules embedded in the plasma membrane. Sliding movements generated by these microfilaments sweep the receptors, and the substances bound to them on the outside surface of the plasma membrane, into the cap (see Fig. 4-17). Myosin and MLC kinase can also be detected in association with the microfilaments in the capping region.

The furrow that divides the cytoplasm during animal cell division is produced by a ring of microfilaments and myosin molecules lying just under the plasma membrane (see Fig. 24-31). The ring contracts by microfilament sliding, gradually cutting the cytoplasm in two (see p. 718 for details). Similar microfilament rings extending around the margins of embryonic cells are responsible for contractions that bend or fold tissue layers, such as the fold that produces the neural tube early in embryonic development.

Many other cell movements have been linked to microfilaments. These include changes in shape that take place during activation of blood platelets in response to wounding in invertebrates, contraction of blood clots, elevation of a conelike extension of the cytoplasm that engulfs the sperm nucleus at the surface of oocytes during fertilization in some species, and movements of tube feet in echinoderms such as starfish. The great variety of these and other micro-

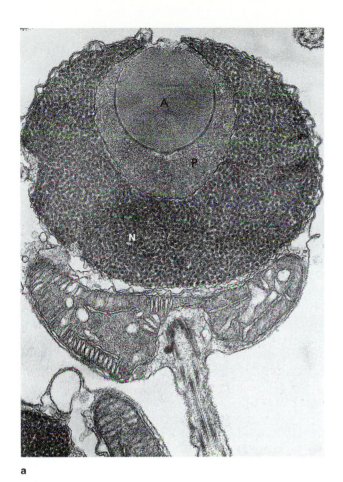

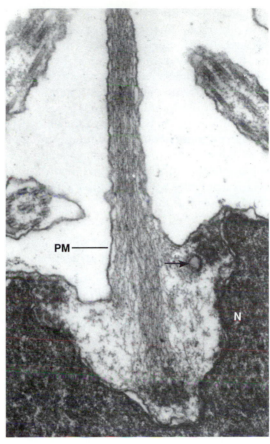

a

b

Figure 11-21 Extension of the acrosomal filament by microfilament growth in *Thyone* sperm cells. **(a)** An unreacted *Thyone* sperm cell, showing the acrosome (A), the cup-shaped region containing unpolymerized G actin (P), and the nucleus (N). × 54,000. **(b)** A *Thyone* sperm cell after extension of the acrosomal filament (arrow). Parallel microfilaments are visible within the filament. N, nucleus; PM, plasma membrane. × 64,000. (Courtesy of L. G. Tilney; reproduced from *J. Cell Biol.* 69:51 [1976] and *J. Cell Biol.* 77:536 [1978], by copyright permission of the Rockefeller University Press.)

filament-based motile systems indicates that the interaction of myosin crossbridges with microfilaments provides a highly versatile cytoplasmic motor that developed in many directions during the evolution of eukaryotes.

Movements Generated by Microfilament Growth

In systems in which microfilaments generate motion by directional growth, microfilaments extend forcefully by adding actin subunits. The prime example of motion produced by microfilament growth—the extension of a filament by sperm cells during fertilization—occurs in echinoderms and some other invertebrates. The extended filament firmly attaches the sperm cell to the surface of the egg.

L. G. Tilney studied filament extension in sperm cells of *Thyone* (a sea cucumber). In unreacted *Thyone* sperm, unassembled actin is stored in a cup-shaped deposit at the anterior end of the sperm head (Fig. 11-21a). As the sperm cell contacts the surface of a *Thyone* egg, the actin assembles rapidly into a parallel bundle of microfilaments (Fig. 11-21b). Tilney and his coworkers found that the actin molecules in unreacted sperm cells are linked to four different polypeptides that prevent their conversion into microfilaments. One of the polypeptides, in molecular weight and biochemical properties, closely resembles profilin, an actin-binding protein that stabilizes actin in unassembled form. According to Tilney, initiation of the acrosomal reaction in *Thyone* involves sudden release of the actin in the cup from its combination with the inhibiting proteins. No trace of myosin can be found in either the cup or the filament; the abrupt extension of the filament appears to depend on microfilament assembly.

A group of investigators including J. A. Theriot, T. J. Mitchison, L. G. Tilney, J. M. Sangerand, D. A. Portnoy, and their colleagues recently discovered that the bacterium *Listeria*, a mammalian pathogen, moves rapidly through the cytoplasm of its host cells by riding a wave of polymerizing microfilaments. In some way that is not as yet understood, one or more proteins secreted by the bacterium induce actin monomers to assemble into microfilaments at one end of the bacterial cell. The polymerization produces a "tail" that grows rapidly, pushing the bacterium through the cytoplasm as it grows.

Microfilaments produce a variety of movements in both muscle and nonmuscle cells. The motile function of microfilaments in almost all these systems is compatible with the sliding filament model derived from studies of striated muscle, in which motion is powered by myosin crossbridges walking along microfilaments by an attach-pull-release cycle. Energy for the movement is provided by ATP hydrolysis during the crossbridging cycle. Regulation of the crossbridging cycle may be through control proteins linked either to actin or myosin. In a few systems, actin microfilaments produce motion by assembling rapidly and directionally.

The activities of microfilaments in generating motility clearly parallel the functions of microtubules in eukaryotic cells (as outlined in Chapter 10). The existence of these parallel systems raises one of the fundamental questions of cell biology. Why have two separate but equivalent mechanisms, based on entirely different proteins, evolved and persisted in all eukaryotic cells to accomplish the same apparent ends of motility and cell support? The answer must lie in the distinct characteristics of the polymerization and sliding reactions of the two systems, which make one or the other more favorable for production of a given motion or supportive element at a certain time and place in the cell. Evolution has also explored other pathways for the generation of motion in eukaryotes in addition to microtubule- and microfilament-based systems. Some of these are described in Supplement 11-1.

For Further Information

Suggestions for Further Reading

Bretscher, M. S. 1987. How animal cells move. *Sci. Amer.* 257:72–90 (December).

Carlier, M.-F. 1991. Actin: Protein structure and filament dynamics. *J. Biolog. Chem.* 266:1–4.

Carraway, K. L. 1990. Membranes and microfilaments: Interactions and role in cellular dynamics. *Bioess.* 12: 90–92.

Cheney, R. E., Riley, M. A., and Mooseker, M. S. 1993. Phylogenetic analysis of the myosin superfamily. *Cell Motil. Cytoskel.* 24:215–223.

Condeelis, J. 1993. Life at the leading edge: The formation of cell protrusions. *Ann. Rev. Cell Biol.* 9:411–444.

Dubriel, R. R. 1991. Structured evolution of the actin cross-linking proteins. *Bioess.* 13:219–226.

Ebashi, S. 1991. Excitation-contraction coupling and the mechanism of muscle contraction. *Ann. Rev. Physiol.* 53: 1–16.

Endow, S. A., and Titus, M. A. 1992. Genetic approaches to molecular motors. *Ann. Rev. Cell Biol.* 8:29–66.

Febvre-Chevalier, C., and Febvre, J. 1986. Motile mechanisms in the actinopods (Protozoa): A review with particular attention to axopodal contraction/extension, and movement of non-actin filament systems. *Cell Motil. Cytoskel.* 6:198–208.

Finer, J. T., Simmons, R. M., and Spudich, J. A. 1994. Single myosin molecule mechanics: Piconewton forces and nanometer steps. *Nature* 368:113–119.

Fukui, Y. 1993. Toward a new concept of cell motility: Cytoskeletal dynamics in ameboid movement and cell division. *Internat. Rev. Cell Biol.* 144:85–127.

Goodson, H. V., and Spudich, J. A. 1993. Molecular evolution of the myosin family: Relationships derived from comparisons of amino acid sequences. *Proc. Nat. Acad. Sci.* 90:659–663.

Huxley, A. 1988. Muscular contraction. *Ann. Rev. Physiol.* 50:1–16.

Kuroda, K. 1990. Cytoplasmic streaming in plant cells. *Internat. Rev. Cytol.* 121:267–307.

Luna, E. J., and Hitt, A. L. 1992. Cytoskeletal–plasma membrane interactions. *Science* 258:955–964.

Milligan, R. A., Whittaker, M., and Safer, D. 1990. Molecular structure of F-actin and location of surface binding sites. *Nature* 348:217–221.

Pollard, T. D., and Cooper, J. A. 1987. Actin and actin-binding proteins. A critical evaluation of mechanisms and functions. *Ann. Rev. Biochem.* 55:987–1035.

Pollard, T. D., Doberstein, S. K., and Zot, H. G. 1991. Myosin-I. *Ann. Rev. Physiol.* 53:653-681.

Rayment, I., Rypniewski, W. R., Schmidt-Bäse, R., Smith, R., Tomchick, D. R., Benning, M. M., Winkelmann, D. A., Wesenberg, G., and Holden, H. M. 1993. Three-dimensional structure of myosin subfragment-1: A molecular motor. *Science* 261:50–53.

Sobue, K., and Sellers, J. R. 1991. Caldesmon, a novel regulatory protein in smooth muscle and nonmuscle actomyosin systems. *J. Biolog. Chem.* 266:12115–12118.

Stossel, T. P. 1993. On the crawling of animal cells. *Science* 260:1086–1094.

Tan, J. L., Ravid, S., and Spudich, J. A. 1992. Control of nonmuscle myosins by phosphorylation. *Ann. Rev. Biochem.* 61:721–759.

Theriot, J. A., Mitchison, T. J., Tilney, L. G., and Portnoy, D. A. 1992. The rate of actin-based motility of intracellular *Listeria monocytogenes* equals the rate of actin polymerization. *Nature* 357:257–260.

Trybus, K. M. 1991. Regulation of smooth muscle myosin. *Cell Motil. Cytoskel.* 18:81–85.

Walsh, M. P. 1992. Calcium-dependent mechanisms of regulation of smooth muscle contraction. *Biochem. Cell Biol.* 69:771–800.

Wertman, K. F., and Drubin, D. G. 1993. Actin constitution: Guaranteeing the right to assemble. *Science* 258: 759–760.

Williamson, R. E. 1993. Organelle movements. *Ann. Rev. Plant Physiol. Plant Molec. Biol.* 44:181–202.

Zof, A. S., and Potter, J. D. 1987. Structural aspects of troponin-tropomyosin regulation of skeletal muscle contraction. *Ann. Rev. Biophys. Biophys. Chem.* 16:535–559.

Review Questions

1. What types of cellular movements are based on microfilaments?

2. Outline the structure of microfilaments. How are actin subunits arranged in microfilaments?

3. Outline the structure of actin molecules. What binding sites occur on actin?

4. What conditions are necessary for actin to assemble into microfilaments?

5. What is microfilament polarity? How is polarity related to the assembly of microfilaments? What is microfilament treadmilling?

6. What kinds of proteins modify the assembly of microfilaments? What are capping, severing, crosslinking, and bundling proteins? What functions do these modifiers have inside cells?

7. What effects do the cytochalasins and phalloidin have on microfilaments? In what ways have these substances been useful in microfilament research?

8. In what ways are the structure and biochemistry of microfilaments and microtubules similar? Different?

9. Outline the structure of a myosin molecule. What are the myosin heavy chains? Light chains?

10. What is the basis for the myosin decoration technique? What is the relationship of myosin decoration to microfilament polarity?

11. Outline the crossbridging cycle that powers microfilament sliding.

12. What are troponin and tropomyosin? How do these molecules interact with actin and myosin in the regulation of the crossbridging cycle? What is the role of Ca^{2+} in this mechanism?

13. Outline the mechanism of myosin-linked regulation of the crossbridging cycle. What is the relationship of Ca^{2+} and phosphorylation to myosin-linked regulation?

14. How are actin and myosin arranged in striated muscle cells?

15. Define myofibril, muscle fiber, sarcomere, sarcoplasmic reticulum, T tubule, Z line and M band.

16. How were the thick and thin fibers of striated muscle identified as actin and myosin? What happens to the thick and thin fibers when muscle contracts?

17. Outline the role of the SR and T tubules in the control of voluntary muscle contraction.

18. What are the similarities and differences between the striated and cardiac forms of striated muscle? Compare the mechanisms regulating contraction of the two striated muscle types.

19. Outline the structure of smooth muscle cells. How are microfilaments and myosin thick filaments arranged in smooth muscle?

20. What mechanisms control the contraction of smooth muscle? What is caldesmon? Calponin? Compare the regulation of smooth and striated muscle.

21. What is cytoplasmic streaming? What is ameboid motion? How might the cortical gel, the plasma membrane, and the liquid inner cytoplasm interact in the production of ameboid motion?

22. What is capping? Furrowing? What roles do microfilaments play in these movements?

Supplement 11-1
Motility Without Microtubules or Microfilaments

Not all cellular movements depend on the activities of microtubules and microfilaments. Other motile mechanisms have appeared as evolutionary adaptations, including contractions produced by conformational changes in Ca^{2+}-sensitive proteins and changes in cell shape caused by alterations in osmotic pressure. Although generally not as conspicuous as microtubule- or microfilament-based systems, motile mechanisms based on Ca^{2+}-sensitive contractile proteins are highly efficient and specialized in some groups, especially among the protozoa. In plants, changes in cellular osmotic pressure are widely employed as a motile device.

Motile Systems Based on Ca^{2+}-Sensitive Contractile Proteins

Movements produced by Ca^{2+}-sensitive proteins depend on conformational changes induced by Ca^{2+} binding and release. As Ca^{2+} binds to these molecules, segments of the protein undergo conformational changes, leading to active contraction. The reverse alterations are produced as Ca^{2+} is released. Conformational changes occurring as Ca^{2+} is bound or released are not unusual; many proteins, including calmodulin, respond to calcium binding or release by altering their folding patterns.

Ca^{2+}-sensitive contractile proteins have been studied extensively in *vorticellids*, protozoans with a bell-shaped cell body attached to a substrate by a long, slender stalk (Fig. 11-22). Vorticellids respond to mechanical and chemical disturbances by contracting the stalk almost instantaneously into a tight coil. Ultrastructural studies by W. B. Amos showed that a fibrous bundle running the length of the stalk, the *spasmoneme*, contracts to convert the stalk from the extended to the helical form. Interspersed among the fibers of the stalk are tubules of the smooth ER. Amos found that, rather than actin, myosin, or tubulin, two major Ca^{2+}-binding proteins, collectively called *spasmin*, make up 80% of the mass of proteins isolated from the stalks. Among the remaining proteins is a Ca^{2+}-ATPase active transport pump.

Amos and his colleagues discovered that changes in Ca^{2+} concentration within ranges typical of the cytoplasm could induce repeated cycles of contraction and relaxation in spasmonemes in the total absence of ATP. From this information and the fact that membranous vesicles in the stalk had previously been shown to be capable of Ca^{2+} storage and release, Amos proposed that contraction of the fibrous elements within the stalk is directly powered by changes in Ca^{2+} con-

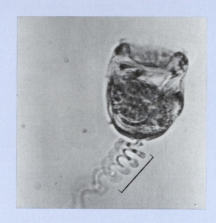

Figure 11-22   A vorticellid with stalk contracted into a tight coil (bracket). $\times$ 300.

centration. To initiate contraction, Ca^{2+} is released from the vesicles by a mechanism analogous to Ca^{2+} release by the SR in striated muscle. The Ca^{2+} binds to contractile proteins of the stalk, initiating conformational changes that directly shorten the fiber, possibly by altering each protein molecule in the fiber from an extended to a helical form. Relaxation of the stalk is induced by the Ca^{2+}-ATPase pump, which returns Ca^{2+} to the vesicles. The resulting drop in Ca^{2+} concentration causes release of the Ca^{2+} bound to the contractile fiber, which responds by returning to the extended form.

Proteins with similar Ca^{2+}-dependent contractile properties may also occur in *flagellar rootlets*, which anchor the basal bodies of flagella in many eukaryotic species (Fig. 11-23). The rootlets, through their connections to basal bodies, are believed to maintain flagella or cilia in a position that orients the plane of flagellar beating. J. L. Salisbury and his colleagues discovered that rootlets of the alga *Tetraselmis* change extensively in length depending on the presence of Ca^{2+} in the medium. The rootlets are fully extended when cells are fixed in calcium-free solutions (Fig. 11-23a). When exposed to Ca^{2+} during the fixation processes, the rootlets contract to less than half of their extended length (Fig. 11-23b). At the same time the two flagella of *Tetraselmis* undergo radical changes in position. Isolation and analysis of the *Tetraselmis* flagellar rootlets yield a group of calcium-sensitive polypeptides with no apparent similarities to microtubule or microfilament proteins. Addition and removal of Ca^{2+} convert the purified rootlet proteins reversibly between fibrous (low Ca^{2+}) and globular (high Ca^{2+}) forms, indicating that the rootlet polypeptides can undergo

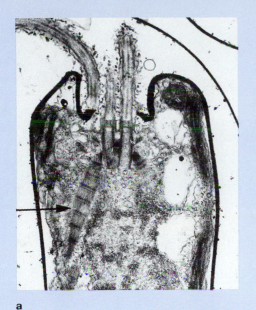

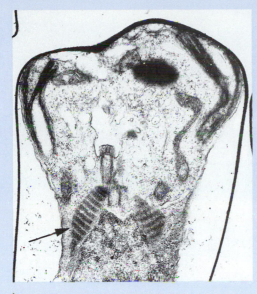

a

b

Figure 11-23 Flagellar rootlets (arrows) of the green alga *Tetraselmis* in extended **(a)** and contracted **(b)** form. The cell shown in **(a)** was fixed in a calcium-free medium; **(b)** was exposed to Ca^{2+} during fixation. $\times$ 15,000. (Courtesy of J. L. Salisbury; reproduced from *J. Cell Biol.* 99:962 [1984] by copyright permission of the Rockefeller University Press.)

Ca^{2+}-dependent conformational changes between extended and contracted states. Similar results have been obtained with flagellar rootlets in other algae and protozoa.

Significantly, Salisbury discovered that antibodies raised against flagellar rootlets and the contractile protein of spasmonemes also react with proteins in the cell centers (see p. 288) of mammalian cells. On this basis, Salisbury suggested that a "heart" of spasmin may beat at the center of every animal cell! Thus it may well be that Ca^{2+}-sensitive contractile proteins of the spasmin type, although reaching their highest evolutionary specialization and development among the protozoa, are widely distributed as a minor but still vital motile system among animal cells.

Movements Generated by Osmosis in Plants

During the day the leaves of many plants change position to face the sun as it moves across the sky. These movements have been shown to depend on changes in the internal osmotic pressure of cells located in a bulblike organ, the *pulvinus*, located at the base of the leaf. Cells on the side of the pulvinus opposite to the direction of movement swell, and cells on the side toward the direction of movement shrink to produce the motion. The alterations in osmotic pressure responsible for these changes result from movements of ions such as K^+ and Cl^- across the plasma membranes of the pulvinus cells.

In a few plants, such as the sensitive mimosa and Venus's fly trap, leaf movements due to changes in cellular pressure take place in response to touch or stimuli such as heat or electrical shock. Movement of the touch-sensitive leaves is rapid, compared to the 24-hour cycle of light-induced leaf movement. In mimosa the response is propagated from the point of stimulus at a rate of about 2 centimeters per second, so that leaves fold progressively along a stem or branch. Electrical measurements show that movements of K^+ and other ions are responsible for propagation of the response, in a change that resembles propagation of an action potential in neurons.

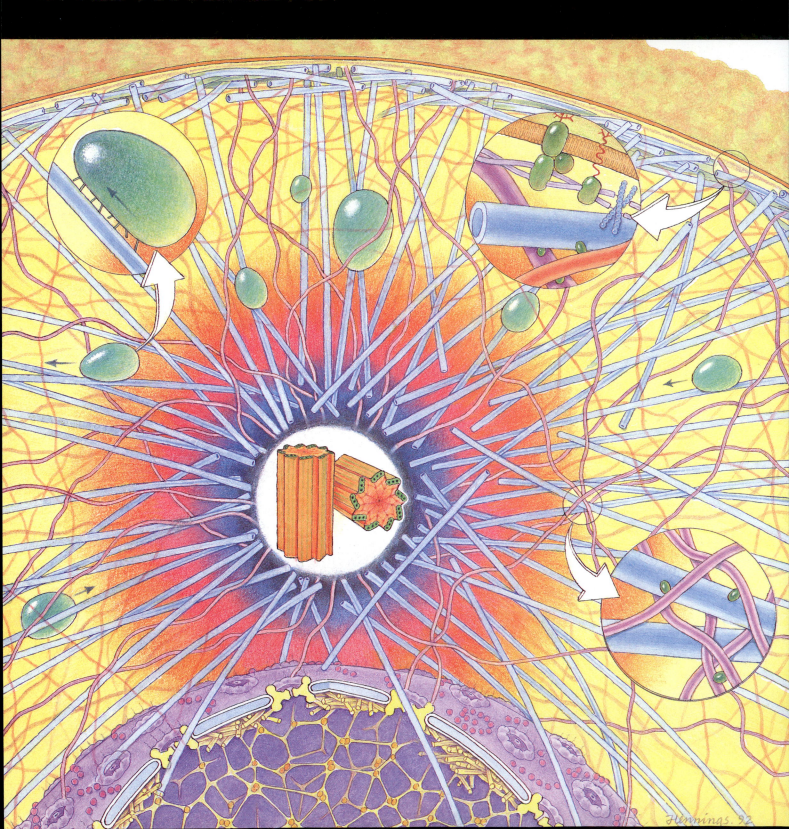

The material surrounding the organelles and structures inside cells was originally thought to be simply an unstructured solution of proteins. The nucleus and cytoplasmic organelles were considered to be freely suspended within this solution, with no fibrous network or other supportive structures in the ground substance to stabilize their positions.

We now know that, rather than being an unstructured, freely mobile suspension or solution, the cytoplasm is supported and stabilized by a complex cytoskeletal network. One or more of at least three structures—microtubules, microfilaments, and intermediate filaments—make up the major scaffolding of the cytoskeleton. A host of additional proteins crosslinks these cytoskeletal elements into an integrated system that organizes and supports the membranes, organelles, and background substance of the cytoplasm and, in many systems, helps to fix the cell in its environment.

Cytoskeletal systems extend throughout the cytoplasm from the plasma membrane to the nuclear envelope; a specialized matrix is suspected to support the nuclear envelope and the nuclear interior as well. The cytoskeleton also anchors the cell to extracellular structures via linking proteins that extend through the plasma membrane.

A major fraction of total cellular proteins, as much as one-third in some cells, is cytoskeletal material. Rather than being static and fixed in position in cells, this material has been found to change in makeup and structure as cells develop and differentiate, move, and alternate between cycles of growth and division. The importance of the cytoskeleton to cell function is underlined by the drastic effects of some of the known mutations in cytoskeletal elements, not only on individual cells but on the entire organism. Several human diseases, in fact, are caused directly by cytoskeletal faults.

The structure and biochemistry of two of the three major elements forming cytoskeletal supports—microtubules and microfilaments, have been described in Chapters 10 and 11. This chapter begins with a discussion of the remaining major cytoskeletal element, the intermediate filaments. The interactions of intermediate filaments with microtubules and microfilaments in the cytoskeleton are then presented. The primary emphasis of the chapter is on the cytoplasm; the systems known or suspected to support the nuclear envelope and nuclear interior are described in Chapter 13.

STRUCTURE AND BIOCHEMISTRY OF INTERMEDIATE FILAMENTS

Intermediate filaments (Figs. 12-1 and 12-2) are so called because their diameters, which average about 10 nm, range between those of microfilaments (about 5 to 7 nm) and microtubules (about 25 nm). Although all intermediate filaments are assembled from related proteins, they are much more varied in molecular structure than either microtubules or microfilaments. Five major classes of intermediate filaments have been discovered in the cytoskeletal structures of various animal cells—the *keratin, vimentin, desmin, neurofilament,* and *glial filament* classes (Table 12-1). These intermediate filament classes are assembled from related but distinct types of proteins, the *cytokeratins, vimentin, desmin, neurofilament,* and *glial fibrillary acid* proteins. Other filament types assembled from related polypeptides have been detected in cells, including a major class, the *lamins,* associated with the nuclear envelope and *nestin,* a recently discovered intermediate filament type occurring in embryonic nerve cells that give rise to the peripheral nervous system. (Chapter 13 details the lamins and their arrangement in the nuclear envelope.) In all, more than 40 different intermediate filament proteins have been described.

Intermediate filaments of all five major cytoplasmic classes occur throughout vertebrate animals. Intermediate filaments have also been observed in nervous, epithelial, and muscle tissue in most of the invertebrate phyla, and in slime molds and protozoa as well. No cytoskeletal elements related to intermediate filaments have as yet been definitely identified in plants. However, C. W. Lloyd and his coworkers have found fiber bundles in carrot and other plant cells that cross-react with antibodies developed against animal intermediate filament proteins; others have detected proteins related to the vimentin-desmin group and the lamins in plants (see also p. 387). S. J. McConnell and M. P. Yaffe found that the *mdml* gene of the yeast *Saccharomyces* encodes a protein with amino acid sequence related to intermediate filaments. Defects appear in the distribution of nuclei and mitochondria during cell division when the gene is in mutant form.

Intermediate filaments are most often identified in intact cells by the fluorescent antibody technique, in which antibodies made against intermediate filament proteins are linked chemically to a fluorescent

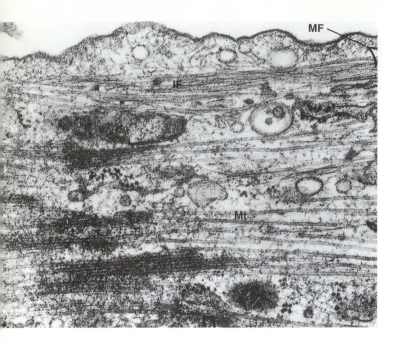

Figure 12-1 Intermediate filaments (IF), microtubules (Mt), and microfilaments (MF) in the region of the cleavage furrow of a dividing mammalian cell. × 52,000. (Courtesy of K. McDonald, from *J. Ultrastr. Res.* 86:107 [1984].)

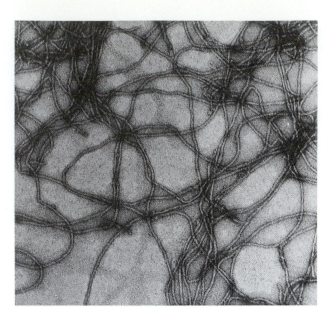

Figure 12-2 Vimentin intermediate filaments isolated from Chinese hamster cells and prepared for electron microscopy by negative staining. × 115,000. (Courtesy of P. M. Steinert, from *J. Biol. Chem.* 256:1428-1431 [1982].)

Table 12-1   Intermediate Filament Classes and Their Polypeptides

Class	Polypeptides	Number	Type	Molecular Weights	Occurrence
Keratin	Cytokeratins I and II	30	I and II	40,000–70,000	Epithelial cells
Vimentin	Vimentin	1	III	57,000	Cells of mesodermal origin
Desmin	Desmin	1	III	55,000	Skeletal, cardiac, and smooth muscle cells
Glial	Glial fibrillary acid protein	1	III	51,000	Glial cells
Neurofilament	Neurofilament	3	IV	70,000, 160,000, and 210,000	Neurons
Lamin	Lamins A, B, and C (in mammals)	3	V	72,000 (A), 66,800 (B), and 65,000 (C)	Possibly in all eukaryotic cells

marker (Fig. 12-3; see Appendix p. 786 for details of the technique). Preparations of this type have revealed the unexpectedly wide distribution of intermediate filaments in animal cells and have allowed the various classes of intermediate filaments to be identified with specific cell types.

Intermediate Filament Structure

Sectioned intermediate filaments appear in the electron microscope as solid, cylindrical fibers with diameters ranging between extremes of about 7 and 12 nm and averaging about 10 nm. They may extend for long distances in the cytoplasm, lying singly or linked into bundles or networks. Unlike individual microtubules and microfilaments, which usually lie in more or less straight courses in the cytoplasm, intermediate filaments are often bent into wavelike forms. This suggests that intermediate filaments are elastic structures that are more flexible than microtubules or microfilaments. Crosslinks can frequently be seen to extend between intermediate filaments and

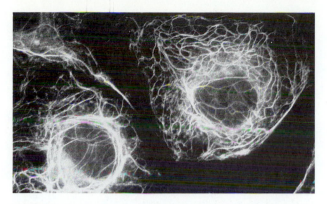

Figure 12-3 Keratin intermediate filaments stained by the fluorescent antibody technique in a cultured cell. (Courtesy of B. Geiger.)

other cell structures, such as organelles, the plasma membrane, microtubules, microfilaments, or groups of ribosomes.

Intermediate filaments can be unwound to some degree by exposing them to phosphate ions or other charged molecules at low concentrations. The unwound filaments reveal a twisted, ropelike structure (Fig. 12-4), suggesting that intermediate filaments contain several subfibers twisted in helical form.

Proteins of all major intermediate filament types have been at least partially sequenced, either directly or through the sequences of the genes encoding them. The sequences confirm that the protein subunits making up intermediate filaments are members of a single large family of related molecules. The sequences also reveal several structural features that are shared by all proteins of the family (Fig. 12-5). All contain a central region more than 300 amino acids in length that can wind into four relatively rigid, rodlike alpha helical segments separated by three less structured regions

(Fig. 12-5a). The four alpha-helical segments contain a repeating pattern of amino acid residues typical of alpha-helical chains that can wind by twos into a coiled coil (Fig. 12-5b; the coiled coil is discussed in detail in Information Box 2-4).

At either end of the fibrous central region the amino acid chain folds into globular end segments. These end segments are highly diverse in amino acid sequence and probably account for many of the distinctive properties of individual intermediate filament types. The globular segments, particularly those at the C-terminal end, are also primarily responsible for variations in molecular weight among the different intermediate filament proteins, which range from about 40,000 to as much as 210,000 (see Table 12-1).

The central rod segment, with its characteristic pattern of alpha-helical central segments, and the globular end caps distinguish intermediate filament proteins from all other proteins. These characteristics allowed identification of the lamins associated with the nuclear envelope, previously not known to be intermediate filaments, as members of this family. Undoubtedly more proteins belonging to the intermediate filament family remain to be discovered.

The different intermediate filament types fall into a number of related subgroups (see Table 12-1). Two of the subgroups are the *type I* and *type II* cytokeratins, which combine in equal numbers to form the keratin class of intermediate filaments. The *type III* subgroup includes closely related proteins forming three major intermediate filament classes, the desmin, vimentin, and glial filaments (desmin, vimentin, and glial fibrillary acidic protein, respectively). Another class III intermediate filament type designated as *peripherin* was recently detected in embryonic nerve cells. Each of the intermediate filaments in this subgroup is assembled from a single protein type. The *type IV* subgroup forms

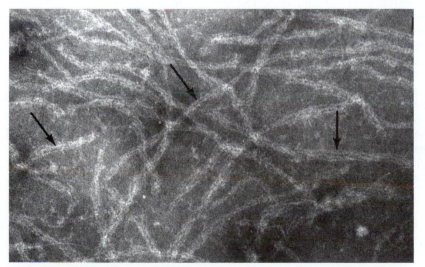

Figure 12-4 Human keratin intermediate filaments exposed to phosphate ions. The treatment unwinds the filaments slightly, showing subdivision into twisted subfibers called protofilaments (arrows). × 180,000. (Courtesy of U. Abei; reproduced by permission from *J. Cell Biol.* 97: 1131 [1983], by copyright permission of the Rockefeller University Press.)

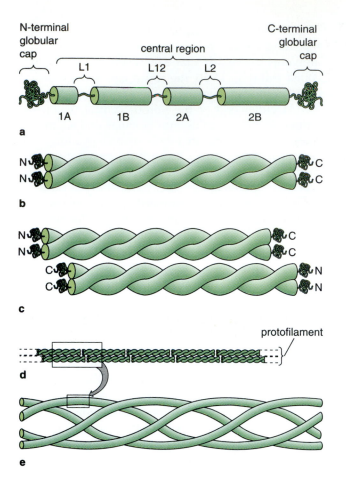

Individual Intermediate Filament Types

The Keratins   As many as 30 different cytokeratins (from the prefix *kerat* = horn) in different combinations form a wide variety of keratin intermediate filaments. They occur in the epithelial cells covering and lining body structures, including the outer skin, the lining of body cavities and ducts, and other tissues such as the corneal epithelium. In epithelial cells keratin filaments form networks under the plasma membrane, extend throughout the cytoplasm, and cover the surface of the nucleus (as in Fig. 12-3). Among the most prominent of these intermediate filaments are the *tonofilaments* that anchor desmosomes to the underlying cytoplasm (see p. 167 and Fig. 6-10). Keratin filaments of this type can be recognized in association with desmosomes in epithelial cells throughout the animal kingdom.

The bulk of the dead *stratum corneum* layer of skin in humans and other mammals is made up largely of cytokeratins. Keratin filaments also make up the primary content of external body structures produced by epithelial cells in mammals, including wool, hair, and horns. The so-called "hard" cytokeratins in animal wool, hair, and horns are stabilized by numerous disulfide (—S —S—) linkages in their globular end segments, which make these structures highly resistant to physical and chemical disturbance.[1] The Experimental Process essay by E. V. Fuchs on p. 345 describes her elegant experiments with keratin filament structure and its relationship to human disease.

Vimentin, Desmin, and Glial Filaments   The closely related vimentin, desmin, and glial intermediate filaments, whose proteins make up the type III subgroup, are widely distributed in animal cells. Most cells containing vimentin intermediate filaments (from *vimentus* = wavy), including those of connective tissues, blood cells, bone, and cartilage, share a common origin in embryonic mesoderm. Vimentin filaments occur with keratin filaments in some epithelial cells, such as those of the lens and iris of the eye, and with glial filaments in some glial cells. In most of these cell types, vimentin filaments are distributed in patterns

Figure 12-5   Intermediate filament structure. **(a)** An individual intermediate filament polypeptide. The amino acid chain in the rodlike central region winds into four alpha-helical segments (1A, 1B, 2A, and 2B), interrupted by three segments that are more flexible (L1, L12, and L2). Within the central region the alpha-helical segments contain many sequential repeats of a seven–amino acid sequence, in a pattern typical of polypeptides that can form a coiled coil. The central region is capped at either end by irregularly folded globular segments. Most of the sequence and size variation between different intermediate filament types is in the globular end caps. **(b)** A coiled-coil double pigtail formed by two intermediate filament polypeptides. **(c)** A tetramer consisting of two coiled-coil pigtails. Within the tetramer the pigtails are staggered and arranged in head-to-tail or antiparallel fashion, with the C-terminal ends of one pigtail next to the N-terminal ends of the other. The two pigtails may be twisted around each other in the tetramer to form another level of coiling. **(d)** The end-to-end linkage of tetramers to form a protofilament; each black bar is one of the two pigtails of a tetramer. Individual protofilaments are visible in Fig. 12-4. **(e)** A proposed structure for a fully assembled intermediate filament, in which four protofilaments wind around each other in a superhelical structure. (The structure would be tightly packed in native intermediate filaments, not loosely wound as in this diagram.) Each cylinder in **(e)** is a protofilament.

---

[1]Contending with these stabilizing disulfide linkages is a major human preoccupation. To create "permanent waves" in the hair or to remove unwanted curls, a series of chemical treatments is used first to break disulfide linkages to make the hair pliable and then to reform the linkages to set the hair in desired patterns.

Of Mice and Men: Genetic Skin Diseases Arising from Defects in Keratin Filaments

Elaine V. Fuchs

ELAINE FUCHS received her Ph.D. in Biochemistry from Princeton University in 1977, working under the supervision of Dr. Charles Gilvarg. From Princeton, Dr. Fuchs went to the Massachusetts Institute of Technology, where she was a Damon Runyon–Walter Winchell postdoctoral fellow, studying with Dr. Howard Green. In 1980, she became an Assistant Professor in Biochemistry at the University of Chicago; she was promoted to Associate Professor of Molecular Genetics and Cell Biology in 1985, and to Professor in 1989. She was also appointed to the Howard Hughes Medical Institute in 1988, where she is currently an investigator. She has received a number of academic honors and awards, including a Searle Scholar Award, NIH Career Development Award, Presidential Young Investigator Award, and R. R. Bensely Award from the American Association of Anatomists.

The epidermis provides the protective interface between various chemical and physical traumas of the environment and the rest of the bodily organs. It manifests its protective function by building an extensive cytoskeletal network of 10 nm intermediate filaments (IFs) composed of keratin proteins. Keratins (molecular mass 40–70 kilodaltons) are a family of proteins that can be subdivided into two distinct sequence groups, type I and type II, based on sequence homology. Type I and type II keratins are coexpressed as specific pairs, which first form obligatory heterodimers, then assemble into tetramers, and finally associate into higher ordered structures, leading to IFs with a 1:1 ratio of the two proteins.[1,2] In the early 1980s, Werner Franke's laboratory discovered that during epidermal development, embryonic basal cells are the first to express detectable levels of the type I keratin K14 (50 kd) and type II keratin K5 (58 kd). Subsequently, we found that the synthesis of this keratin pair increases greatly later in development, when cells begin to differentiate into epidermis. As basal epidermal cells differentiate and move outward toward the skin surface, they reduce K5/K14 expression and switch on a new pair of keratins, K1 (67 kd) and K10 (56.5 kd).[3] In the fully differentiated squamous cell, these keratins constitute approximately 85% of total cellular proteins. Thus, keratins are to an epidermal cell what globins are to a red blood cell.

Since embarking on keratin research nearly 15 years ago, a major question that we have asked is whether there might be defects in keratin genes giving rise to genetic skin diseases, as there are defects in globin genes giving rise to genetic blood diseases such as sickle cell anemia and thalassemias. One way to address this question is to genetically alter keratin sequences and evaluate the consequences of these defects on protein function. At the time we embarked on these studies, the function of keratin filaments was largely unknown. However, it was well established that these proteins could self-assemble into IFs, forming a distinct cytoskeletal network in keratinocytes. Therefore, after isolating and characterizing the genes encoding the human keratins K5 and K14, we began to engineer defects in the coding sequence of the K14 gene. In order to monitor the expression of the mutant keratin in cells that expressed the normal K14, we replaced sequences encoding the antigenic carboxy-terminal 5 residues of the K14 protein with sequences encoding the antigenic portion of another protein, the neuropeptide substance P. This enabled us to use an antibody to K14 to identify the wild-type K14 protein and an antibody to substance P to identify the mutant K14 protein.

We first introduced the tagged wild-type and mutant keratin genes into cultured epidermal cells. Using a technique known as gene transfection, a calcium phosphate-DNA precipitate was layered onto the surface of the keratinocytes. Approximately 10% of the cells engulfed this precipitate, enabling the mutant keratin genes to enter these cells. When these genes were expressed, the wild-type and mutant keratin proteins recognized the native keratin K5 in epidermal cells, formed heterodimers and then heterotetramers, and finally integrated into the keratin filament network.[4] The small substance-P tag did not interfere with the ability of the keratin to form keratin networks, as judged by double immunofluorescence staining with antibodies against substance P and K14. In addition, some mutations that we engineered had no effect, and the keratin filament network containing the mutant protein appeared normal as judged by this same technique. However, some mutant keratins caused gross distortions in the keratin filament network, producing withdrawal from the plasma membrane and disruption of the network (Figure A).

Did these mutant proteins interfere with IF formation, or did they distort some intracellular interaction(s) involving keratins? To evaluate this, we conducted *in vitro* filament assembly studies using purified human K5 and mutant K14 obtained from bacterial clones. A bacterial promoter was used to express the human K14 and K5 cDNAs in *E. coli*, which do not make keratins or other IF proteins. In these bacteria, the keratins accumulated in large cytoplasmic aggregates, called inclusion bodies.[2] These bodies could be isolated and solubilized in a 6.5 M urea buffer. In the presence of both K5 and K14 (or mutant K14), stable tetramers formed, which could be isolated by anion exchange chromatography. When dialyzed to remove the urea, these tetramers then assembled into IFs. In contrast to wild-type K14/K5 tetramers, which formed long, uniform IFs (Figure B, frame A), some mutant K14/K5 tetramers assembled into filaments that

Figure A Expression of keratin 14 mutants in human keratinocytes using gene transfection. The circled cell was transfected with a dominant negative mutant K14 gene, whose product integrated into and distorted the keratin network. Bar represents 25 μm. (Courtesy of Anthony Letai and Dr. Kathryn Albers; see also reference 4.)

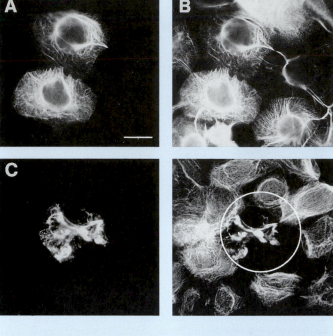

Figure B *In vitro* assembly of human K14 and K5 proteins from genetically engineered bacteria. Bacteria were genetically engineered to produce wild-type K14 and K5 human proteins, and mutant K14 proteins. Purified K5 and K14 proteins were solubilized in a 6.5 *M* urea buffer, and dialyzed against assembly buffer. (A) Filaments assembled from wild-type K5 and K14; (B) filaments assembled from wild-type K5 and a K14 mutant that caused filament aggregation; (C) filaments assembled from wild-type K5 and a K14 mutant that completely disrupted IF formation; (D) filaments assembled from wild-type K5 and the same K14 mutant shown in C, but this time in the presence of wild-type K14, at a 99% wild-type to 1% mutant K14 ratio. Note that this was a very potent dominant negative mutant: even 1% of the mutant protein was enough to interfere with filament elongation, resulting in shorter filaments than wild-type. Bar represents 100 nm. (Courtesy of Dr. Pierre Coulombe; see also reference 4.)

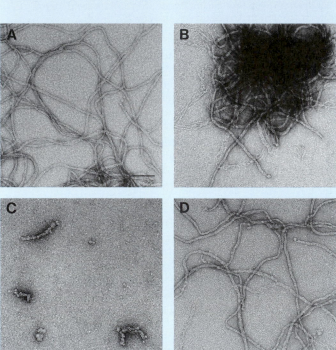

were aggregated and sometimes frayed at their ends (frame B), and some could not assemble into 10 nm structures at all (frame C).[4] Those mutants that were unable to assemble with K5 into IFs could nevertheless interfere with the assembly of wild-type K5 and K14: when as little as 1% of the mutant K14 protein shown in frame C was mixed with 99% wild-type K14 and 100% wild-type K5, the filaments formed were short and somewhat irregular (frame D). This is called a *dominant negative mutant*:

the mutant recognized the wild-type proteins and interfered with their function. Most dominant negative mutants of K14 were deletions or point mutations within the ends of a 310-amino-acid-residue domain predicted to be largely α-helical and involved in the intertwining of K14 and K5 polypeptides into a coiled-coiled heterodimer.

At this point in our studies, we wondered how our dominant negative mutant keratins might behave in the epidermis of skin: what happens when the native keratin filament network of a basal epidermal cell is disrupted? To answer this question, we used the human K14 promoter to drive the expression of a dominant negative K14 mutant in the basal epidermal cells of the skin of transgenic mice.[5] For this technique, we introduced the mutant keratin gene, called a *transgene*, into fertilized single-cell mouse embryos. The mouse embryos were then transferred to the oviduct of a foster mother mouse, who was mated the night before with a vasectomized male. Thus, the foster mother had the right hormones to nurture and carry the embryos to birth. When the mice were born, some of them contained the transgene in every cell of their body. These mice were from embryos that integrated at least one copy of the transgene into their own chromosomal DNA before their first cell division.

Our transgenic animals exhibited morphological and biochemical symptoms typical of humans who have an autosomal dominant skin disease known as *epidermolysis bullosa simplex (EBS)*.[5,6] Upon incidental trauma such as suckling, the mouse skin blistered. Microscopic examination of blistered skin sections revealed extensive cell breakdown in the basal epidermal layer where the mutant keratin transgene was expressed. The basal cells also showed disorganized keratin filament networks, often with shorter filaments. In severe cases, clumps of keratin protein were abundant in the cytoplasm of these cells.[5] Interestingly, the cells always lysed in a specific zone of the columnar basal epidermal cell, beneath the nucleus and above the plasma membrane. The pattern of cell breakage suggested to us that the function of the keratin filament networks in these basal epidermal cells is to impart mechanical integrity, without which the columnar cells become fragile and then lyse upon incidental trauma. In addition, these studies provided us with an important clue to the possible genetic basis for a sometimes life threatening and psychologically devastating human skin disease that affects about 1 person in 50,000.

At this point, our studies turned to the human skin disease EBS.[7] In conjunction with dermatologists, we obtained skin samples from EBS patients and cultured their epidermal cells. We examined the keratin networks in cultured EBS cells and discovered changes similar to those we had seen when genetically engineered mutant keratins were expressed by gene transfection in cultured human keratinocytes. In addition, the IFs assembled from the keratins isolated from EBS epidermal cells were shorter than normal. When the K5 and K14 mRNAs and

genes from two different EBS patients were isolated and characterized, we found two different point mutations in the K14 genes, both leading to an amino acid change at arginine 125. In one case, the mutation resulted in a cysteine 125 mutation, and in the other, a histidine 125 mutation.[7] Interestingly, when 51 different IF sequences sharing 25–99% amino acid identity were examined, the residue corresponding to this position was either arginine or lysine in 50 sequences. Moreover, this residue was within one of two small segments of the K14 protein that we had shown through our mutagenesis studies to be highly critical for IF assembly.

We do not yet know the extent to which EBS diseases have as their basis defects in keratin genes. EBS may have multiple causes, and defects in posttranslational modification of keratins or defects in as yet unidentified proteins that associate with, degrade, or sever keratin filaments in basal epidermal cells may also play a role in the disease. Finally, we would predict that some EBS cases should have defects in K5 as well as K14 genes. As more cases are examined, the etiology of EBS should become clearer.

In conclusion, our studies have shown that by starting with a protein and using a molecular genetic approach, we have been able to gain valuable insights into our understanding of the function of the protein, as well as developing an animal model for the study of a human skin disease. This approach should be broadly applicable to a number of fundamental questions in cell biology, where proteins have been extensively studied, but aspects of their function and relations to human diseases are unknown.

References

[1] Aebi, U.; Haner, M.; Troncoso, J.; Eichner, R.; and Engle, A. *Protoplasma* 145:73–81 (1988).

[2] Coulombe, P. A., and Fuchs, E. *J. Cell Biol.* 111:153–69 (1990).

[3] Fuchs, E. *J. Cell Biol.*, 111:2807–14 (1990).

[4] Coulombe, P. A.; Chan, Y.-M.; Albers, K.; and Fuchs, E. *J. Cell Biol.* 111:3049–64 (1990).

[5] Vassar, R.; Coulombe, P. A.; Degenstein, L.; Albers, K.; and Fuchs, E. *Cell* 64:365–80 (1991).

[6] Haneke, E., and Anton-Lamprecht, I. *J. Invest. Dermatol.* 78:219–23 (1982).

[7] Coulombe, P. A.; Hutton, M. E.; Letai, A.; Hebert, A.; Paller, A. S.; and Fuchs, E. *Cell* 66:1301–11 (1991).

similar to keratin filaments—in the tonofilaments of desmosomes and in networks that underlie the plasma membrane, extend through the cytoplasm, and surround the nucleus.

Desmin filaments (from *desmos* = link) appear as bracing elements in skeletal, cardiac, and smooth muscle cells. The primary location of desmin filaments in striated muscle is in association with the Z lines of sarcomeres (see Fig. 11-14), where they frequently occur along with vimentin filaments. In smooth muscle cells, desmin intermediate filaments are connected to dense bodies, which are the functional equivalents of Z lines in these cells (see Fig. 11-17 and p. 330). Glial filaments (from *glia* = glue) are found in *glial* cells, including astrocytes and some Schwann cells, which form a supportive tissue surrounding neurons in the central nervous system.

Neurofilaments Intermediate filaments of the neurofilament class are characteristic of neurons, where they occur in greatest numbers in axons, the long processes that extend from nerve cell bodies (see Figs. 12-6 and 5-15). In axons, neurofilaments occur in bundles arranged parallel to the long axis of these cell projections, in the same pattern as the microtubules found in axons (see p. 298). The proteins forming neurofilaments occur in multiple types in vertebrates; in birds and mammals, three related proteins assemble into neurofilaments. In spite of their prevalence in some types of neurons, the function of neurofilaments is unknown. Presumably they, along with the microtubules of axons, reinforce these structures in the longitudinal direction.

Intermediate Filament Assembly

Intermediate filament proteins assemble spontaneously into highly stable filaments in the test tube at neutral pH, with no requirements for particular ions, modifying proteins, or energy in the form of ATP or other nucleotides. Assembly appears to depend on associations set up by the central rod domain of intermediate filament polypeptides, particularly near its ends. In order to induce intermediate filaments to disassemble in the test tube, they must be exposed to salt at very low concentrations (much lower than the concentrations typical for living cells) or to strongly denaturing agents, such as urea, at very high concentrations.

Research by several groups, including those of E. Fuchs, Z. E. Zehner, N. Geisler, and S. A. Lewis and N. J. Cowan, indicates that individual protein subunits first associate by twos into the coiled-coil pigtails shown in Figure 12-5b during assembly into intermediate filaments. In these dimers the two polypeptides are in parallel alignment, with their N-terminal

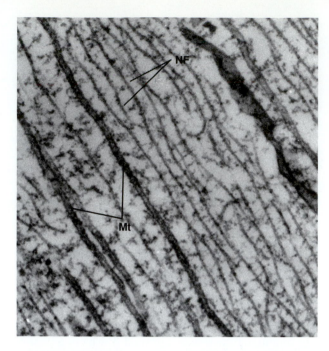

Figure 12-6 Neurofilaments (NF) and microtubules (Mt) in an axon of a human neuron. × 96,000. (Courtesy of P. Dustin and J. Flament-Durand, from *Microtubules*, 2nd ed., New York: Springer-Verlag.)

ends directed toward the same end of the structure. To form the next level of structure, two of the coiled-coil pigtails probably line up side by side to form a *tetramer* containing four intermediate filament proteins (Fig. 12-5c). Within the tetramer the two coiled-coil pigtails are slightly staggered and lined up in antiparallel fashion, with the C-terminal ends of one pigtail next to the N-terminal ends of the other. The tetramers then line up end to end during or after their assembly, forming a cablelike structure called a *protofilament* (Fig. 12-5d). Measurements of protofilament mass indicate that they probably contain two rows of tetramers. Thus a protofilament, which measures 4.5 nm in diameter, may be an *octamer* of intermediate filament polypeptides. Examination of slightly unraveled intermediate filaments indicates that four protofilaments intertwine to form the complete structure (Fig. 12-5e). If protofilaments are indeed octamers, a fully assembled intermediate filament would contain 32 individual polypeptides in cross section.

Although the extreme conditions required for disassembly in the test tube might indicate that intermediate filaments are permanent, static elements once assembled inside cells, they are actually dynamic structures that readily form and disassemble (see below). The mechanisms disassembling intermediate filaments in cells are not fully known, but they may include phosphorylation of intermediate filament

Table 12-2 Some Intermediate Filament–Associated Proteins (IFAPs)

IFAP	Molecular Weight	Intermediate Filament Associations	Cell Types	Tentative Function
Epinemin	44,500	Vimentin	Muscle and other mesodermally derived cells	Unknown
Filaggrin	~30,000	Keratin	Epidermal cells	Bundling
Paranemin	280,000	Desmin, vimentin	Embryonic muscle cells (myocytes)	Crosslinking
Plectin	300,000	Vimentin	Wide variety of cells of mesodermal origin	Crosslinking
Synemin	230,000	Desmin, vimentin	Muscle cells, some erythrocytes	Crosslinking

proteins. Phosphorylation at sites in the N-terminal segments has been observed to take place before the disassembly of vimentin and desmin filaments. The lamins, which regularly disassemble from the nuclear envelope as this membrane system breaks down during cell division (see p. 708), become phosphorylated just before disassembly and lose their phosphate groups as they reassemble at the close of division (for details, see p. 717).

Intermediate Filament–Associated Proteins (IFAPs)

A gradually lengthening list of proteins has been detected in association with intermediate filaments of the different classes. Generally these proteins, the *intermediate filament–associated proteins* (*IFAPs*; see Table 12-2), are retained with the filaments as they are purified and cleaned of cellular debris and contaminants. Antibodies developed against isolated IFAPs also react strongly with intermediate filaments inside cells, verifying a close association between IFAPs and intermediate filaments.

The work of assigning functions to IFAPs identified by these and other methods has only just begun. However, most IFAPs evidently work either as cappers that cover intermediate filament tips to arrest further assembly or as crosslinkers that tie intermediate filaments into networks by end-to-end or end-to-side linkages. One crosslinking IFAP, *filaggrin* (see Table 12-2) acts as a *bundler* by tying keratin-based intermediate filaments into parallel arrays. Some IFAPs act as *interlinkers* tying intermediate filaments to other elements of the cytoskeleton.

Intermediate Filament Regulation

Cells are capable of quickly modifying or disassembling their intermediate filament systems under certain conditions. During mitotic cell division, intermediate filament networks of many cell types are extensively rearranged and reduced or, in some cell types, disappear entirely. In damaged mammalian nerve axons, neurofilaments initially break down and then reorganize as the axon regenerates. Reorganization appears to take place by assembly from a cytoplasmic pool of existing subunits without synthesis of new intermediate filament proteins.

Other modifications occur as cells progress through development and differentiation, including changes in intermediate filament type. Both vimentin and desmin intermediate filaments coexist in some embryonic muscle cells. As many muscle cell types mature into adult form, vimentin disappears. Vimentin is also found in some embryonic nerve cells, but not in fully differentiated, adult neurons. The combinations of cytokeratin proteins in keratin filaments also change as some epidermal cells mature.

Because phosphorylation has been correlated with intermediate filament disassembly, at least some of the dynamic changes noted in intermediate filament structure and arrangement probably depend on reversible addition of phosphate groups. Changes in intermediate filament distribution and networking may also be produced by alterations in crosslinking and bundling IFAPs.

INTERMEDIATE FILAMENTS, MICROTUBULES, AND MICROFILAMENTS IN THE CYTOSKELETON

In animal cells, intermediate filaments occur with microtubules and microfilaments in cytoskeletal systems that extend throughout the cytoplasm. These systems support major cytoplasmic regions and maintain the position of the nucleus. In these supportive roles each major cytoskeletal element contributes characteristics that depend on its particular structural and biochemical properties, as altered by MAPs, actin-binding proteins, and IFAPs.

Cytoskeletal Functions

In their cytoskeletal role, microtubules, arranged singly, in networks, and in parallel bundles, evidently provide rigidity to cytoplasmic regions. When arranged in parallel bundles, as in spindle and nerve axons, microtubules also establish *polarity* by setting up definite ends to which elements being transported or separated may be delivered. They probably also provide elasticity to cellular projections and possibly contribute to the tensile strength of cellular parts and regions subjected to a stretching force. The cytoskeletal activities of microtubules are modified by a variety of MAPs (see p. 287), which regulate microtubule assembly and disassembly and set up crosslinks between microtubules and other elements in the cytoplasm. The activity of many MAPs is regulated by Ca^{2+}, working either directly or through calmodulin, and by cAMP-activated enzymes. These mechanisms tie microtubule assembly and crosslinking in the cytoskeleton to the major Ca^{2+} and cAMP-linked systems regulating cellular activities (see pp. 159 and 161).

Microfilaments also occur singly, in networks, and in parallel bundles in the cytoskeleton. The degree of crosslinking of cytoskeletal microfilaments, regulated by a seemingly endless variety of actin-binding proteins, gives cytoplasm containing the networks consistencies ranging from liquid to a rigid gel. In parallel bundles, microfilaments provide tensile strength to cytoplasmic regions. Other microfilament bundles support extensions of the plasma membrane such as microvilli (see Fig. 12-9). Many of the actin-binding proteins are regulated by Ca^{2+} and cAMP, also placing microfilament assembly and crosslinking under control of the major cellular regulatory systems.

Although there is little in the way of direct evidence, intermediate filaments are generally assumed also to be supportive structures in the cytoplasm. From their observed physical characteristics, intermediate filaments are expected to provide cytoplasmic regions with elasticity and resistance to breakage during stretching. This conclusion is supported by experiments such as those conducted by E. V. Fuchs, in which mutant epithelial cells with defective keratin intermediate filaments were subject to breakage when under physical stress (see Fuch's essay on p. 345).

Microtubules, microfilaments, and intermediate filaments are tied together in the cytoskeleton by a variety of interlinking proteins. Some of the interlinkers form ties between two of the cytoskeletal elements, such as microtubule-microfilament or microtubule-intermediate filament links, and some can interlink all three. The tau group of MAPs (see p. 287), for example, can form microtubule-microfilament interlinks; MAP-2 and another interlinker, *ankyrin* (see p. 314) have binding affinities for all three cytoskeletal elements. Some interlinkers can form connections with membrane proteins as well, tying the cytoskeletal elements to the plasma membrane and the boundary membranes of organelles in the cytoplasm. The known interlinkers are regulated by enzymes forming part of the InsP$_3$/DAG and cAMP-based pathways linked to receptors at the cell surface. (See Chapter 6 for details of these pathways.)

Interlinking proteins integrate the cytoskeleton to such an extent that none of the cytoskeletal elements is likely to act independently in the cytoplasm. As a consequence, modification or disturbance of one element is probably transmitted by interlinkers to all parts of the system. The cytoskeleton is thus a complex network, tied by the interlinking proteins into an integrated supportive structure.

Methods for Studying the Cytoskeleton

The cytoskeleton has been studied by an impressive list of methods and techniques. Antibodies against cytoskeletal elements have been particularly valuable in light and electron microscope studies. The light microscope techniques in which antibodies are tagged with fluorescent dyes have been used effectively in living as well as chemically fixed cells. These techniques allow intermediate filaments, microtubules, and microfilaments to be individually identified in the cytoskeleton (as in Fig. 12-3).

Cytoskeletal structures are also isolated from cells and analyzed biochemically. Here again, antibodies have proved highly useful as a means to detect individual cytoskeletal proteins. Other important methods involve protein crosslinkers (see p. 377), which are used to determine the "nearest neighbors" of cytoskeletal elements inside cells. These methods are frequently combined with microtubule and microfilament poisons, such as colchicine and the cytochalasins, and anti-intermediate filament antibodies to inhibit or destroy individual elements of the cytoskeleton. These methods have been used most extensively to investigate animal cells, particularly those of birds and mammals.

In molecular studies these approaches are combined with isolation and sequencing of genes encoding cytoskeletal elements. In some approaches, genes are altered and reintroduced into cells by genetic engineering techniques so that the effects of induced mutations on particular proteins can be assessed.

MAJOR CYTOSKELETAL STRUCTURES

The cytoskeleton is a fully integrated system that extends throughout the cytoplasm. However, it may be considered to have three major subdivisions supporting distinct cytoplasmic regions: (1) the plasma membrane and cell surface, (2) the cortical region underlying the plasma membrane, and (3) the inner cytoplasm between the cortex and the nucleus.

Cytoskeletal Systems Associated with the Plasma Membrane

The systems associated with the plasma membrane are the most varied cytoskeletal structures of animal cells. Depending on the cell type, these systems may carry out many functions in addition to simply reinforcing the plasma membrane. Some provide support and connections for motile mechanisms acting on or through the plasma membrane, such as ameboid movement and capping, or they may stabilize surface receptors and contacts between cells and the extracellular matrix. Other membrane-associated cytoskeletal systems reinforce the plasma membrane, anchor cell junctions, and stiffen surface projections such as microvilli (see below).

These membrane specializations are found primarily in animal cells. In plants the plasma membrane-associated cytoskeleton appears to be restricted in most cells to microtubules and limited numbers of microfilaments. The microtubules probably stabilize the system depositing cellulose microfibrils in the cell wall (see p. 194). The limited development of cytoskeletal systems associated with the plasma membrane in plants probably reflects the fact that cell walls are the primary structures supporting and reinforcing the cell perimeter in these organisms.

Membrane Support and Reinforcement in Mammalian Erythrocytes Much of our information about plasma membrane–associated cytoskeletal systems in animals comes from research with mammalian erythrocytes (red blood cells). These cells consist of little more than a plasma membrane enclosing ribosomes and a solution of hemoglobin; the nucleus and cytoplasmic organelles are eliminated as the cells mature. When placed in distilled water, mammalian erythrocytes swell and burst, spilling out the hemoglobin and leaving "ghosts" consisting of essentially pure plasma membranes with their underlying cytoskeleton. The absence of internal organelles greatly simplifies the task of identifying the plasma membrane–associated cytoskeletal proteins and working out their functions.

Figure 12-7 summarizes the results of work with erythrocyte ghosts. The membrane is reinforced by a combination of short actin microfilaments with fibers of *spectrin*, an actin-binding protein. Actin and spectrin are linked to each other and to the plasma membrane by proteins called *ankyrin* and *band 4.1* protein (Fig. 12-7a). Another protein, *adducin*, increases the number of spectrin molecules binding to actin. Myosin and tropomyosin are also present as part of

the microfilaments. Another protein of the membrane cytoskeleton, *band 4.9 protein* (also called *dematin*), may link actin microfilaments into bundles. (Band 4.1 and 4.9 proteins are named according to their positions when erythrocyte ghost proteins are isolated and run on electrophoretic gels—see Appendix p. 796.)

The short microfilaments of the erythrocyte membrane cytoskeleton contain only 10 to 20 actin subunits reinforced by tropomyosin. The primary molecule of the cytoskeleton is spectrin, which may number as many as 100,000 copies per erythrocyte ghost. Each spectrin molecule, about 100 nm long, is a four-part structure built up from two copies each of two closely related polypeptides, α- and β-*spectrin*. The polypeptides contain as many as 36 repeats of a variable 106-amino acid sequence. A number of proteins, including the actin-binding protein α-actinin, also contain the 106-amino acid repeat characteristic of spectrin. Each spectrin molecule has binding sites for actin, ankyrin, adducin, band 4.1 protein, and other spectrin molecules. The adducin- and spectrin-binding sites, which occur at the tips of the molecules, allow spectrin molecules to link together and to actin in an extended, meshlike network (as in Fig. 12-7b).

Ankyrin sets up links between spectrin and a major transport molecule of the erythrocyte plasma membrane, the *anion transporter*, which moves negatively charged ions across the membrane. The link between ankyrin and the anion transporter may be stabilized by another protein known as *band 4.2*. Band 4.1 protein also links spectrin to another integral protein of the plasma membrane, *glycophorin*. The entire network, including spectrin, ankyrin, adducin, actin, and the band 4.1, 4.2, and 4.9 proteins, makes up as much as 40% of the total protein content of mammalian erythrocyte ghosts.

A mammalian erythrocyte survives about four months in the circulatory system, during which it makes about 500,000 circuits around the body. The membrane cytoskeleton probably maintains the typically flattened, discoid shape of these cells and reinforces the plasma membrane against the stresses of passage through narrow blood vessels and turbulent regions of the circulatory system such as the heart valves.

Several common hereditary deficiencies in humans are related to mutations affecting the erythrocyte membrane cytoskeleton. Most of these mutations cause various forms of *hereditary hemolytic anemia*, in which a weakened membrane cytoskeleton makes red blood cells fragile and easily broken during circulation. Different mutations weaken the cytoskeleton by reducing the binding affinities of spectrin, ankyrin, band 4.1, 4.2, 4.9 proteins, glycophorin, or the anion transporter. Duchenne muscular dystrophy, characterized by degeneration of muscle tissue, appears to

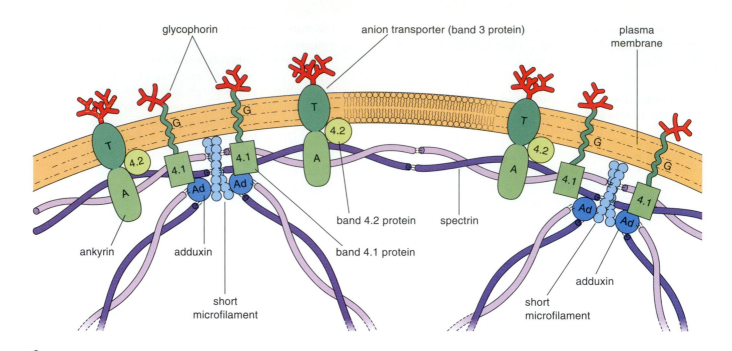

glycophorin

anion transporter (band 3 protein)

plasma membrane

band 4.2 protein   spectrin

ankyrin   adduxin   band 4.1 protein

adduxin

short microfilament

short microfilament

a

Figure 12-7 The cytoskeletal network underlying erythrocyte plasma membranes. **(a)** Interaction of microfilaments, spectrin, ankyrin (A), adducin (Ad), band 4.1 protein (4.1), band 4.2 protein (4.2), the anion transporter (T), and glycophorin (G) in the cytoskeletal system reinforcing the erythrocyte plasma membrane. Spectrin consists of four fibrous polypeptide subunits wound into a loose double pigtail; each pigtail consists of two different but related α- and β-spectrin polypeptides. **(b)** A cytoskeletal network isolated from a human erythrocyte membrane. The spectrin molecules (s) form the network itself; actin, band 4.1, and adducin proteins occur at the junctions of the network (arrows). Ankyrin molecules are visible on the spectrin fibers at the midpoints between the junctions (circles). × 70,000. (Micrograph courtesy of J. Palek and S.-C. Liu; reproduced from *J. Cell Biol.* 104:527 [1987], by copyright permission of the Rockefeller University Press.)

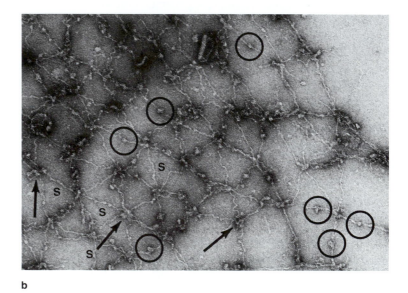

b

be caused by absence of *dystrophin*, a protein related to spectrin. In normal skeletal, cardiac, and smooth muscle cells, dystrophin links on the cytoplasmic side of the plasma membrane to a complex of integral membrane glycoproteins. On the extracellular side of the plasma membrane the complex binds to laminin molecules (see p. 186) in the extracellular matrix. When dystrophin is absent, the resulting lack of reinforcement and attachment to the extracellular matrix leads to tearing of the plasma membrane and muscle cell death.

Research using antibodies against spectrin, ankyrin, adducin, and band 4.1, 4.2, and 4.9 proteins revealed that the plasma membranes of many animal cell types are supported by an underlying network containing these or related proteins. These discoveries led to the concept that a cytoskeleton similar to that of mammalian erythrocytes may be associated with the plasma membranes of all animal cells. In many of these cells, and in the nucleated erythrocytes of vertebrates other than mammals, microtubules and intermediate filaments also occur in the cytoskeletal network underlying the plasma membrane. At least some of the same plasma membrane–associated elements have also been detected in protozoa and fungi. Fluorescent antibodies against spectrin have even been found to react in regions just under the plasma membrane in carrot, maize, potato, and bean, partic-

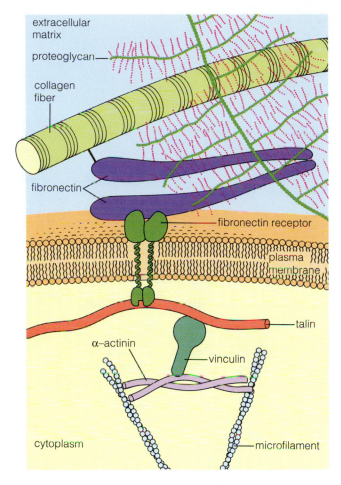

Figure 12-8 Linkage of the cytoskeleton to the extracellular matrix via membrane receptors. On the external side of the plasma membrane the receptor binds fibronectin, which links in turn to major elements of the extracellular matrix including collagen and the proteoglycans. On the cytoplasmic side of the membrane the receptor forms linkages through talin, vinculin, and α-actinin to microfilaments of the cytoskeleton.

ularly in structures containing thin-walled cells such as the tip of pollen tubes and embryonic root cells.

The plasma membrane–associated cytoskeletons of nonerythrocyte cells function along the same lines as those of erythrocytes—they reinforce the plasma membrane and maintain cell shape and the positions of integral membrane proteins. Some integral membrane proteins of these cells link at their outer ends to elements of the extracellular matrix. The fibronectin receptor, for example, which is a member of the integrin family of surface receptors, links to fibronectin (see p. 185) embedded in the extracellular matrix (Fig. 12-8; see also Fig. 7-9). On the cytoplasmic side of the plasma membrane the fibronectin receptor makes linkages via actin-binding proteins such as *talin, vinculin,* and α-actinin to microfilaments in the membrane cyto-

skeleton. These transmembrane attachments between the cytoskeleton and the extracellular matrix further stabilize the plasma membrane and anchor cells in position in tissues and organs.

Several observations have made it obvious that the cytoskeletal networks supporting the plasma membrane are flexible structures capable of short-term adjustments and dynamic change. During capping (see p. 108) the mechanism sweeping integral membrane proteins into the cap at one end of the cell also frequently sweeps and compresses the actin-spectrin network underlying the membrane into the same region. Much of the flexibility necessary for capping is provided by the long, elastic spectrin molecules and their linkage into a compressible network.

Other alterations that take place more slowly and persist for longer periods of time are noted during cell development and differentiation or when cells respond to stimulation by hormones or other agents at the cell surface. Many of these long-term alterations are regulated by addition of phosphate groups to proteins of the membrane cytoskeleton, including ankyrin, adducin, band 4.1, and band 4.9 proteins, through the activity of cAMP or Ca^{2+}/calmodulin-dependent protein kinases. Addition of phosphate groups reduces both the binding affinities of these components and the extent and rigidity of the cytoskeleton. The binding of band 4.1 protein to spectrin, for example, is regulated by phosphate groups added by a cAMP-dependent protein kinase. Addition of the phosphate groups weakens the binding of 4.1 protein to spectrin by about five times. The alterations in band 4.1-spectrin binding probably open the cytoskeletal network and loosen its attachment to the plasma membrane.

Specialized Plasma Membrane–Associated Cytoskeletons in Cell Junctions and Microvilli Specialized cytoskeletal structures occur in regions where animal cells are held together by junctions (see Chapter 6). In desmosomes, which tie epithelial cells into sheets, intermediate filaments of the keratin or vimentin type serve as supports anchoring the junctions in the underlying cytoplasm. The intermediate filaments make connections to dense, plaquelike structures lining the inner side of the plasma membrane in the junction (see Fig. 6-10). Another junction, the adherens junction, is anchored to an underlying network of microfilaments (see Fig. 6-12). Adherens junctions, with their associated microfilaments, also transmit contractile forces between cells in muscle tissue, particularly between individual cells in cardiac muscle.

Another membrane-associated cytoskeletal structure, and one of the most striking, is the system reinforcing microvilli, fingerlike projections that extend from the surfaces of cells lining the intestinal cavity and kidney tubules (Figs. 12-9 and 12-10). In this system, parallel bundles of microfilaments reinforce the microvilli in the longitudinal direction. These bundles extend as *rootlets* into the cytoplasm just under the microvilli, where they anchor in the *terminal web*, a network of microfilaments, spectrin, and cytokeratin-based intermediate filaments underlying the cell border. The intermediate filaments of the terminal web run in a network from one rootlet to the next and make connections at the sides of the cell to the desmosomes. Microfilaments extend in a similar pattern except that their connections at the side of the cell are made to adherens junctions. The junctions connect the cells to their neighbors; tight junctions in the same regions seal the spaces between the cells to the passage of ions and molecules. The internal structures of microvilli and the terminal web are stabilized and interconnected by various actin-binding proteins (see Fig. 12-10).

Microvilli may number a thousand or more per cell, tightly packed into an array known as the *brush border* of the intestinal and kidney cells bearing them. They increase the surface area of the plasma membrane, maximizing the surface area available for absorption in intestinal cells and for both absorption and secretion in kidney cells.

Cytoskeletal Structures of the Cortex and Inner Cytoplasm

Cytoskeleton of the Cell Cortex   In many cell types the cytoplasm near the plasma membrane is markedly "stiffer" than more internal regions. The gel-like nature of the cortical region depends in most cells on a network of microfilaments held together by crosslinking proteins such as filamin. These cortical networks are most highly developed in protozoan and animal cells that move by ameboid motion and in some animal oocytes and plant cells.

In ameboid cells the cortical network forms a peripheral "tube" through which the more liquid internal cytoplasm flows during movement (see Fig. 11-20). The cortical gel of ameboid cells probably has a motile as well as cytoskeletal function—contraction of the cortical tube, through an interaction between actin and myosin, may supply some of the force moving the liquid internal cytoplasm (see p. 332).

Both intermediate filaments and microfilaments occur in the cortical networks of animal cells. Although the intermediate filaments may reinforce the microfilament network of the cortex, they have not as yet been identified as contributing to any particular function such as ameboid motion.

The cortical region of cells that swim by means of flagella is frequently reinforced by bundles or networks of microtubules. The microtubules provide a

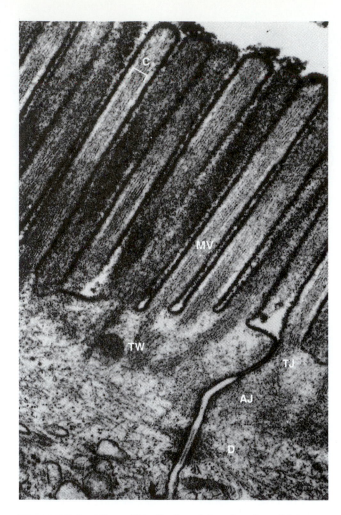

Figure 12-9   Microvilli in the brush border of a rat intestinal epithelial cell. MV, microvillus; C, core; TW, terminal web; TJ, tight junction; AJ, adherens junction; D, desmosome. × 600,000. (Courtesy of D. R. Burgess, from *Cold Spring Harbor Symp. Quant. Biol.* 46:845 [1982].)

cytoskeletal scaffold that anchors the basal bodies and rootlets (see pp. 295 and 338) of the flagella. Cortical reinforcements of this type occur in protozoa and in the swimming gametes of many animals, fungi, and plants. The most complex of these systems occur in the ciliate and flagellate protozoans, which have complicated microtubule networks that run in various directions and interconnect basal bodies and flagellar rootlets (an example is shown in Fig. 12-11). The cortical microtubules of these protozoans, along with unidentified fibers of various kinds, evidently reinforce the cell against the stresses of flagellar beating and maintain the basal bodies and shafts of the flagella in a fixed orientation at the surface of the organism.

The Interior Cytoskeleton   In animal cells the cytoplasm between the cortex and the nucleus is sup-

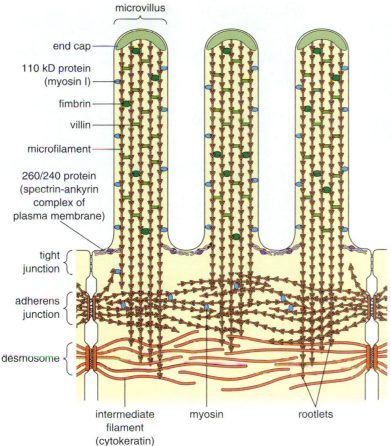

microvillus

end cap

110 kD protein (myosin I)

fimbrin

villin

microfilament

260/240 protein (spectrin-ankyrin complex of plasma membrane)

tight junction

adherens junction

desmosome

intermediate filament (cytokeratin)

myosin

rootlets

Figure 12-10   Molecular structure of microvilli and the terminal web. Each microvillus of a brush border is supported by a longitudinal core consisting of a bundle of 10 to 50 parallel microfilaments, aligned in a uniform polarity with their plus ends toward the tips of the microvilli. (The microfilaments are diagramed as they would appear if decorated by myosin—see p. 309). Two bundling proteins, fimbrin and villin, form crosslinks that tie microfilaments into the core bundle. The 110 kD protein forming the lateral links binding the core to the plasma membrane is a form of myosin I. The terminal web below the brush border contains myosin and the 260/240 protein, a form of spectrin that can bind both microfilaments and intermediate filaments. (The numbers refer to the molecular weights in thousands of the α- and β-polypeptides of this spectrin.) Also present as microfilament crosslinkers and modifiers in the terminal web are α-actinin, filamin, and vinculin. An as-yet-unknown protein forms the dense cap anchoring the core bundle to the plasma membrane at its outermost tip.

ported by systems that include all three cytoskeletal elements. These systems evidently hold the nucleus and many of the cytoplasmic organelles and structures in position in the cell.

Microtubules and intermediate filaments are the most prominent elements in the interior cytoskeleton of animal cells. Frequently both elements radiate from one or a few centers near the nucleus, surround the nucleus, and extend through the cytoplasm to the cell cortex (see Figs. 12-3 and 12-12). For microtubules the center of radiation is the *cell center* or *centrosome*, a region that acts as a microtubule organizing center (MTOC) from which cytoskeletal microtubules originate and extend (see pp. 288 and 701). The microtubules radiating from the cell center are generally arranged with their minus ends nearest the cell center and their plus ends directed toward the periphery.

The centers of intermediate filament organization are frequently located near cell centers so that microtubules and intermediate filaments appear to radiate from almost the same points. The radiating arrangement of intermediate filaments in these cells seems to be dependent on the microtubule system, because dispersal of microtubules by antimicrotubule agents such

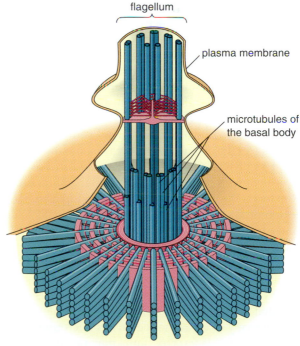

flagellum

plasma membrane

microtubules of the basal body

Figure 12-11   The system of microtubules and other, unidentified cytoskeletal structures supporting a basal body and flagellum in the flagellate protozoan *Codosiga botrytis*. The system is representative of the complex cortical cytoskeletal systems occurring in ciliate and flagellate protozoa. (Courtesy of D. J. Hibberd and The Company of Biologists Ltd., from *J. Cell Sci.* 17:191 [1975].)

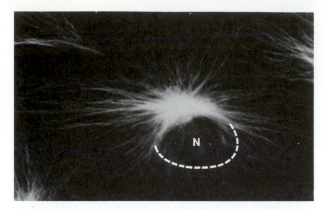

Figure 12-12 Microtubules, marked by fluorescent antibodies, radiating from the cell center of a cultured African green monkey cell. The dashed line marks the boundary of the nucleus (N). × 1,800. (Courtesy of J. Izant, from *Modern Cell Biology*, vol. 2 [J. R. McIntosh, ed.]. Copyright © 1983 by Wiley-Liss, a division of John Wiley and Sons, Inc.)

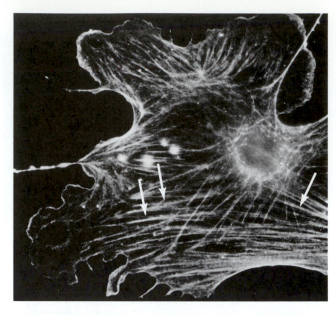

Figure 12-13 Stress fibers (arrows) in a cultured gerbil fibroblast cell. × 1,000. (Courtesy of J.-C. Lin and Plenum Publishing Co., from *Cell and Muscle Motility* [R. M. Dowben and J. W. Shay, eds.], vol. 2, 1982.)

as colchicine, low temperature, or high pressure causes the intermediate filament network to collapse. Injection of antibodies against intermediate filaments, which causes aggregation and collapse of the intermediate filament network, in contrast, has no detectable effect on the pattern of radiating microtubules.

Microtubules radiating from the cell center apparently support several major cytoplasmic systems, including segments of the ER, Golgi complex, secretory vesicles, lysosomes, and at least some mitochondria. Usually the Golgi complex is held in position between the cell center and the plasma membrane; the ER extends throughout much of the cytoplasm. Disruption of microtubules by agents such as colchicine causes Golgi membranes to disperse and ER membranes to cluster near the nucleus. Collapse of the intermediate filament network by injection of anti–intermediate filament antibodies usually has no effect on the positions of cytoplasmic organelles. This indicates that, even though intermediate filaments are located in many of the same regions as microtubules in the deeper cytoplasm, crosslinks may occur only between the organelles and microtubules.

Ameboid motility in cultured animal cells seems to be related to the position of the cell center and its radiating microtubules. In these cells, advancing lobes of cytoplasm move outward from the side of the cell containing the cell center. Ameboid movement, as a consequence, takes place in the direction "pointed" by the cell center. If the microtubules radiating from the

cell center are destroyed, the ameboid movements of cultured cells become random in direction.

Microfilaments in the Interior Cytoskeleton: Stress Fibers Microfilament networks, although also present in deeper regions of the cytoplasm, are usually less concentrated than in the cortex. One microfilament system, however, the *stress fiber* (Fig. 12-13; see also Fig. 1-16b), is prominent in deeper layers of a few animal cell types, including epithelial cells and fibroblasts. Stress fibers are thick bundles of microfilaments that extend in straight paths in the cytoplasm, usually making connections to the plasma membrane at points where the cell makes attachments to external structures. Besides microfilaments, stress fibers contain myosin, α-actinin, tropomyosin, and caldesmon, a protein regulating the motile interactions of actin and myosin (see p. 324). Vinculin and talin occur at points where stress fibers contact the plasma membrane.

Stress fibers contain all the elements required for active contraction, with bands of myosin, α-actinin, and caldesmon and other molecules in a repeating pattern reminiscent of striated muscle sarcomeres. However, stress fibers appear to be much more characteristic of stationary than motile cells. They are dismantled when cells containing them break loose from their extracellular connections and become motile. Microfilaments reorganize into stress fibers as these cells attach, flatten, and cease moving. Stress fibers also disassemble when the cells containing them enter cell division. At that time the cells lose their flattened shape, round up, and break most of their contacts with the substrate. For these reasons many investigators consider it likely that stress fibers are primarily cytoskeletal structures that reinforce the characteristically flattened or extended shape of the cells containing them. Tension developed in stress fibers by myosin crossbridging probably contributes to the flattening.

The sites where stress fibers contact the plasma membrane, called *focal contacts*, are specialized sites of interaction with the extracellular matrix. In the region of focal contacts, cell surface receptors of the integrin family (see p. 185) link the plasma membrane to elements of the extracellular matrix including collagen, laminin, and fibronectin. On the cytoplasmic side of the plasma membrane these receptors connect to the microfilaments of stress fibers through crosslinks set up by vinculin, α-actinin, and talin (as in Fig. 12-8). Most focal contacts are disassembled when stress fibers break down during cell movement or division. However, some that persist or form near the advancing cell margins may provide temporary attachment sites that serve as leverage points against which cells push during locomotion (see also p. 334).

Both stress fibers and focal contacts react positively with fluorescent antibodies against plectin. The presence of this IFAP indicates that stress fibers and focal contacts may link to intermediate filaments.

Cytoskeletal systems also support deeper regions of the cytoplasm in protozoan, fungal, and plant cells. Microtubules and microfilaments are the prominent supportive elements in the interior cytoskeletal systems of most cells in these groups. In at least some protozoa and fungi, however, limited networks of intermediate filaments occur along with microtubules and microfilaments.

In plants, although proteins related to intermediate filaments have been detected in some cells, microtubules and microfilaments seem to be the primary cytoskeletal elements supporting the inner cytoplasm. Microtubules occur singly in widely scattered arrays in the inner cytoplasm. Microfilaments extend in a sparse network throughout the cytoplasm and, in some cell types, surround the nucleus in a cagelike array. Microtubule poisons have no noticeable effect on the locations taken by the Golgi complex or ER in plant cells. W. Hensel and others found, however, that exposing plant cells to the microfilament poison cytochalasin disturbed the distribution of ER vesicles. Thus microfilaments instead of microtubules may maintain the position of the ER in plant cells.

There are indications that some enzymes and other factors associated with major biochemical pathways of the cytoplasm, such as glycolysis and protein synthesis, are tied to the interior cytoskeleton rather than simply suspended in free solution in the cytoplasm. Linkage to the cytoskeleton might organize the enzymes into arrangements that facilitate their interactions. As a consequence, the progress of biological reactions inside cells may differ to a greater or lesser extent from interactions observed in free solutions in the test tube.

DYNAMIC CHANGES IN THE CYTOSKELETON

The types and distribution of elements in the cytoskeleton change as cells develop, enter division, and move. Changes also occur in the cytoskeleton in disease, including cancer. These changes confirm that the cytoskeleton is a dynamic structure that is readily altered and adjusted as cells carry out different activities.

Cytoskeletal Alterations in Developing Cells

The cytoskeleton of many cell types changes extensively during development. Both the arrangement of existing cytoskeletal elements and their rate of synthesis may be altered. Striated muscle, one of the cell types in which cytoskeletal development has been studied most completely, illustrates both types of alteration.

Striated muscle fibers develop through fusion of individual embryonic cells called *myocytes*. Before fusion, myocytes have a cell center from which microtubules radiate throughout the cytoplasm. Intermediate filaments also radiate from one or more centers near the nucleus. The intermediate filaments of myocytes are primarily of the vimentin class, but some desmin filaments are also present. As striated muscle develops, myocytes fuse end to end and develop into muscle fibers, with extensive synthesis and alignment of microfilaments in sarcomeres. During these developmental changes, vimentin synthesis slows and intermediate filament assembly becomes concentrated almost entirely on the desmin type. The desmin filaments become associated with the Z-line margins and regions of the plasma membrane overlying the Z lines (see Fig. 11-17). In some muscle types, vimentin filaments are retained in mature cells in the same regions. As these changes occur, microtubules become aligned in roughly parallel bundles extending primarily in the long axis of the muscle fibers instead of radiating from cell centers.

Cytoskeletal Rearrangements During Cell Division

The cytoskeleton also changes in highly ordered patterns as cells cycle through interphase growth and cell

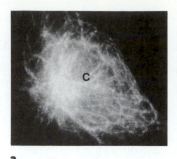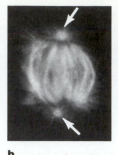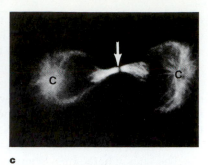

a   b   c

Figure 12-14 Microtubule reorganization in dividing mammalian cells (see text). **(a)** The single cell center at the interphase preceding division (C). **(b)** The fully formed spindle, with the cell center divided into two parts (arrows). **(c)** The close of division, with the cell centers (C) in separate daughter cells. The arrow in **(c)** marks the plane of cytoplasmic division; a remnant of the spindle persists in this region for a brief period. ([**a** and **c**] courtesy of S. H. Blose, from *Cell Motil.* 1:417 [1981]. Copyright © 1981 by Wiley-Liss, a division of John Wiley and Sons, Inc. **(b)** courtesy of B. R. Brinkley; reproduced from *J. Cell Biol.* 113:1091[1991] by copyright permission of the Rockefeller University Press.)

division. Of the three major cytoskeletal elements, microtubules undergo the most dramatic alterations. During late interphase in animal cells the cell center (Fig. 12-14a) splits into two parts, which move apart and take up positions on opposite sides of the nucleus. At the same time, almost the entire complement of cytoskeletal microtubules breaks down and reorganizes into the spindle, which stretches between the separated cell centers (Fig. 12-14b). After division of the replicated chromosomes by the spindle, cytoplasmic division cuts through the spindle midpoint and segregates the cell centers into separate daughter cells (Fig. 12-14c). As division is completed, the spindle microtubules disappear and the cytoskeletal arrangement of microtubules grows from the cell center, which takes up its characteristic position near the nucleus.

Microfilaments are also rearranged in dividing animal cells. In many cell types, interior microfilament networks break down, and microfilaments become concentrated in the spindle. Their role in spindle function, if any, is unknown. Experimental interference with microfilaments generally has no effect on division of the chromosomes (for details, see p. 712). In animal cells a band of microfilaments organizes around the cell periphery and contracts to divide the cytoplasm. Once division is complete, the microfilament networks typical of nondividing cells reappear. The loss of stress fibers and focal contacts in fibroblasts undergoing division has already been noted.

Many of the changes in organization of cytoskeletal microtubules and microfilaments are regulated by reversible phosphorylation of MAPs and actin-binding proteins. The protein kinases catalyzing these phosphorylations are activated by the cAMP- and

InsP$_3$/DAG-based mechanisms responding to growth hormones and other signals arriving at the cell surface. Combination with the Ca^{2+}/calmodulin complex, activated as part of the InsP$_3$/DAG pathway in many systems, also regulates microtubule and microfilament breakdown. For example, the actin-binding protein MARCKS (see p. 313) crosslinks the cortical actin network and ties it to the cytoplasmic side of the plasma membrane. MARCKS is regulated by protein kinase C and the Ca^{2+}/calmodulin complex. Phosphorylation by protein kinase C (activated as part of the InsP$_3$/DAG pathway) or combination with Ca^{2+}/calmodulin releases MARCKS from its attachment to the plasma membrane and inhibits its ability to crosslink microfilaments, leading to breakdown of the cortical actin network.

The degree of change in intermediate filament organization and distribution during division varies greatly in different animal cell types. In some cells no rearrangements occur and the existing intermediate filament network is simply separated into two parts as the cytoplasm divides. In other cells the intermediate network collapses into a layer over the nuclear envelope as cytoskeletal microtubules disassemble and reorganize into the spindle. In a few cell types, intermediate filaments disassemble and disappear almost completely during cell division. Generally intermediate filaments become more highly phosphorylated as cells make the transition from interphase to cell division.

Major changes in the organization of microtubules and microfilaments also accompany growth and division in higher plant cells. Research by R. W. Seagull, M. Osborn, K. Weber, B. E. S. Gunning, and others

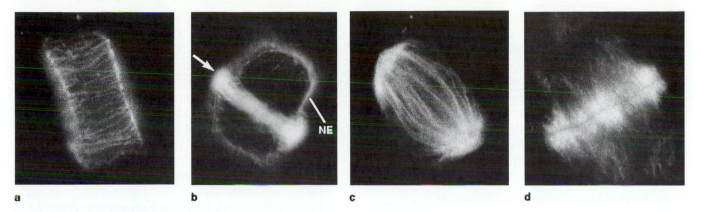

a   b   c   d

Figure 12-15   Microtubule reorganization in dividing onion root tip cells (see text). **(a)** The predominantly cortical distribution of microtubules at interphase. × 1,200. **(b)** Rearrangement of microtubules into the preprophase band (arrow). Microtubules also faintly outline the nuclear envelope (NE). × 1,500. **(c)** Concentration of microtubules into the spindle during cell division. × 1,500. **(d)** Organization of microtubules into the plane of cytoplasmic division. × 1,600. (Courtesy of S. M. Wick; reproduced from *J. Cell Biol.* 89:685 [1981] by copyright permission of the Rockefeller University Press.)

demonstrated that during interphase most microtubules in plant cells are concentrated in a layer under the plasma membrane (Fig. 12-15a). These are the microtubules that may stabilize enzyme complexes assembling cellulose microfibrils. As mitotic cell division nears, many of the microtubules break their linkages to the plasma membrane and assemble into a thick belt surrounding the cell (Fig. 12-15b). Microtubules also run from the belt to the nucleus. The peripheral belt of microtubules, the *preprophase band* (*PPB*), marks the level at which a new cell wall will form to divide the cytoplasm after nuclear division is complete. As cell division begins, the PPB disassembles and microtubules form in a layer around the nucleus. These microtubules grow and increase in number, eventually organizing into the spindle as the nuclear envelope breaks down early in mitosis (Fig. 12-15c). After division is complete, the spindle disassembles.

Spindle disassembly proceeds rapidly until all that remains is a residual layer of short microtubules at the former spindle midpoint. In the layer the short microtubules are oriented in the direction of the former spindle axis. The microtubule layer extends laterally until it contacts the cell walls (Fig. 12-15d). The new wall separating the daughter cells forms in this residual layer of microtubules. After cytoplasmic division is complete, the microtubule layer disassembles and the cell's complement of microtubules reorganizes into the cortical arrangement typical of interphase.

Investigations by C. W. Lloyd, A. M. Lambert, P. K. Hepler, and others showed that in at least some plant cells the distribution of microfilaments parallels the microtubule rearrangements during the cell cycle.

Microfilaments are distributed in a network in the cell cortex and in sparse collections in cytoplasmic interior of nondividing cells. As the PPB forms preceding division, many of the networks break down and microfilaments become concentrated in the PPB along with microtubules. Later, as the spindle forms, many of the microfilaments in the PPB become distributed in the spindle. As in animal spindles, the function of microfilaments in higher plant spindles remains unknown. Not all the microfilaments disappear from the region of the PPB. Some persist in the region of the band and possibly contribute later to the location of the new wall separating the cytoplasm at the close of division (see p. 719 for details).

The regular cycle of cytoskeletal alterations taking place during division of animal and plant cells indicates that the rearrangements are highly programmed through a master schedule, controlled ultimately through changes in gene activity in the cell nucleus (discussed further in Chapter 22).

Cytoskeletal Changes Related to Motility

Cultured cells show a variety of cytoskeletal alterations related to motility. As movement begins, the cell center or centrosome, with its radiating microtubules, migrates to the side of the nucleus facing the direction of movement. We have already noted disassembly of stress fibers and focal contacts that takes place as fibroblasts break connections to a substrate and become mobile. In some cell types, intermediate filaments also change from an extended network into a dense cap over one side of the nucleus as cells leave fixed positions and begin to move.

As movement stops, cells revert to the cytoskeletal systems characteristic of the nonmotile state. In some cases the types of intermediate filaments assembled are different in moving and nonmoving cells. For example, A. Ben-Ze'ev showed that bovine epithelial cells in the motile state synthesize large quantities of vimentin and limited amounts of the cytokeratins. When contact is made with other cells, which arrests the movement, cytokeratins are synthesized in large quantities, and vimentin synthesis is greatly reduced.

Cytoskeletal Abnormalities and Disease

A number of diseases are associated with disturbances of the cytoskeleton, particularly in cells undergoing transformation from the normal to the cancerous state. Typically microfilaments in transformed cells change from their organized arrangement, in which they are concentrated in specific cell regions, to a generalized, more evenly distributed network. The proportion of actin in disassembled form increases. Stress fibers, if present, generally break down and disappear. Linkages between actin and the plasma membrane, made via proteins such as talin and vinculin, are frequently broken. Often intermediate filaments revert to arrangements typical of embryonic rather than fully differentiated, adult cell types.

Contacts with neighboring cells are frequently lost as these changes take place. As part of these surface alterations, cell junctions such as desmosomes and tight junctions disappear. Partly as a result of the breakage of contact with other cells, the transformed cells divide more rapidly and, when not dividing, move constantly among neighboring cells. These characteristics—breakage of normal cell contacts, increased division rate, and invasive movements—are the hallmarks of cancer cells. They are also responsible for much of the disruptive effects of cancer cells on the body (see p. 651 for details).

Because many body cells have distinct types of intermediate filaments, the tissue of origin for tumor cells can frequently be determined by identifying their intermediate filaments. For example, the presence of intermediate filaments containing cytokeratins indicates that the tumor is a carcinoma originating from epithelial tissues. The particular cytokeratins in carcinomas often allow cancer subtypes to be identified—squamous cell carcinomas can be distinguished from adenocarcinomas, for example, by the combinations of cytokeratin types in the tumor cells. The determinations, which are important in determining the type of treatment to be employed, are routinely carried out by means of antibodies developed against each intermediate filament protein type.

Cytoskeletal rearrangements are also observed in many other diseases of humans and other animals, such as the previously noted alterations observed in persons with Duchenne muscular dystrophy and some forms of hereditary anemia. Other changes include intermediate filament clumping and disarray in alcoholic cirrhosis of the liver and disturbances in cytokeratin-based intermediate filaments in *hereditary epidermolysis bullosa simplex*, a disease characterized by separation and blistering of the skin (see also the Experimental Process essay by Fuchs on p. 345). All these cytoskeletal disorders point to a central role of the cytoskeletal system in the maintenance of normal cell activities.

The cytoskeletal system of each cell type provides the foundation for cellular motility, determines cell shape, and anchors the nucleus and major cytoplasmic organelles in the cytoplasm. Cytoskeletal frameworks also reinforce the plasma membrane and fix the positions of junctions, receptors, and connections to the extracellular matrix. The cytoskeleton is a dynamic system that changes in arrangement and composition as cells develop and differentiate, enter mobile and immobile stages, and pass through cycles of growth and division.

Some cytoskeletal structures undoubtedly remain to be identified and chemically characterized. Many questions also persist about the cellular controls regulating the assembly and crosslinking of individual cytoskeletal elements and the unknown cellular factors that integrate the cytoskeleton into a functional whole. And much still remains to be discovered concerning the role of the cytoskeleton in diseases such as cancer.

Suggestions for Further Reading

Aderem, A. 1992. Signal transduction and the actin cytoskeleton: The Roles of MARCKS and profilin. *Trends Biochem. Sci.* 17:438–443.

Albers, K., and Fuchs, E. 1992. The molecular biology of intermediate filament proteins. *Internat. Rev. Cytol.* 134: 243–279.

Bennett, V., and Gilligan, D. M. 1993. The spectrin-based membrane structure and micron-scale organization of the plasma membrane. *Ann. Rev. Cell Biol.* 9:27–66.

Bretscher, A. 1991. Microfilament structure and function in the cortical cytoskeleton. *Ann. Rev. Cell Biol.* 7:337–374.

Brown, S. C., and Lucy, J. A. 1993. Dystrophin as a mechanochemical transducer in skeletal muscle. *Bioess.* 15: 413–419.

Burridge, K., and Fath, K. 1989. Focal contacts: Transmembrane links between the extracellular matrix and the cytoskeleton. *Bioess.* 10:104–108.

Coulombe, P. A., and Fuchs, E. 1990. Elucidating the early stages of keratin filament assembly. *J. Cell Biol.* 111:153–169.

Davies, K. A., and Lux, S. E. 1989. Hereditary disorders of the red cell membrane skeleton. *Trends Genet.* 5:222–227.

Fuchs, E., and Coulombe, P. A. 1992. Of mice and men: Genetic skin diseases of keratin. *Cell* 69:899–902.

Fukui, Y. 1993. Toward a new concept of cell motility: Cytoskeletal dynamics in ameboid movement and cell division. *Internat. Rev. Cytol.* 144:85–127.

Hepler, P. K., Cleary, A. L., Gunning, B. E. S., Wadsworth, P., Wasterneys, G. O., and Zhang, D. H. 1993. Cytoskeletal dynamics in living plant cells. *Cell Biol. Internat. Rep.* 17:127–142.

Kelley, R. B. 1990. Microtubules, membrane traffic, and cell organization. *Cell* 61:5–7.

Kreis, T. E. 1990. Role of microtubules in the organization of the Golgi apparatus. *Cell Motil. Cytoskel.* 15:67–70.

Lloyd, C. W. 1987. The plant cytoskeleton: The impact of fluorescence microscopy. *Ann. Rev. Plant Physiol.* 38: 119–139.

Luna, E. J., and Hitt, A. L. 1992. Cytoskeleton–plasma membrane interactions. *Science* 258:955–964.

MacKae, T. H. 1992. Towards an understanding of microtubule function and cell organization: An overview. *Biochem. Cell Biol.* 70:835–841.

Marchesi, S. L. 1991. Mutant cytoskeletal proteins in hemolytic disease. *Curr. Top. Membr. Trans.* 38:155–174.

Nixon, R. A., and Shea, T. B. 1992. Dynamics of neuronal intermediate filaments: A developmental perspective. *Cell Motil. Cytoskel.* 22:81–91.

Proulx, P. 1991. Structure-function relationships in intestinal brush-border membranes. *Biochim. Biophys. Acta* 1071: 255–271.

Quinlan, R. A., and Stewart, M. 1991. Molecular interactions in intermediate filaments. *Bioess.* 13:597–600.

Shoeman, R. L., and Traub, P. 1993. Assembly of intermediate filaments. *Bioess.* 15:605–611.

Skalli, O., and Goldman, R. D. 1991. Recent insights into the assembly, dynamics, and function of intermediate filament networks. *Cell Motil. Cytoskel.* 19:67–79.

Steinert, P. M., and Bale, S. J. 1993. Genetic skin diseases caused by mutations in keratin intermediate filaments. *Trends Genet.* 9:280–287.

Terasaki, M. 1990. Recent progress on structural interactions of the endoplasmic reticulum. *Cell Motil. Cytoskel.* 15:71–75.

Vorobjev, T. A., and Nadezhdina, E. S. 1987. The centrosome and its role in the organization of microtubules. *Internat. Rev. Cytol.* 106:227–293.

Weber, K., and Osborn, M. 1985. The molecules of the cell matrix. *Sci. Amer.* 253:110–120 (October).

Williamson, R. E. 1991. Orientation of cortical microtubules in interphase plant cells. *Internat. Rev. Cytol.* 129: 135–206.

Review Questions

1. Compare the dimensions and structure of intermediate filaments, microtubules, and microfilaments.

2. Outline the classes of intermediate filaments and their protein subunits. In what cell types does each class of intermediate filament occur?

3. Compare the reactions assembling intermediate filaments, microtubules, and microfilaments. What factors might regulate intermediate filament disassembly inside cells? What information indicates that intermediate filaments can be disassembled inside cells?

4. Outline the structure of an intermediate filament protein. In what ways are the different intermediate filament proteins similar, and in what ways do they differ? How are intermediate filament proteins believed to assemble into intermediate filaments?

5. What are IFAPs? What functions do IFAPs carry out in the cytoskeleton?

6. Outline the roles and structures of microtubules, microfilaments, and intermediate filaments in the cytoskeleton. How is Ca^{2+} concentration related to the cytoskeleton?

7. What are cytoskeletal interlinkers? What roles do these molecules play in the cytoskeleton?

8. What are the major subdivisions of the cytoskeleton? What methods have been employed to study the arrangement of the cytoskeleton in these subdivisions?

9. Outline the cytoskeletal system supporting the plasma membrane in red blood cells. Why have red blood cells been used to study the cytoskeleton associated with the plasma membrane?

10. What is the relationship of the cytoskeleton to cell junctions?

11. What are microvilli? The terminal web? How are microtubules and intermediate filaments organized in these structures?

12. Describe the cytoskeleton of the cell cortex. How does the cytoskeleton in this region contribute to ameboid motion?

13. What is the cell center? What is the relationship of the cell center to the distribution of microtubules and intermediate filaments in the cytoskeleton? To cell motility?

14. What are stress fibers? Focal contacts? What molecular types occur in these structures? What evidence indicates that stress fibers are cytoskeletal rather than motile structures?

15. Compare the cytoskeletons supporting the inner cytoplasm in plant and animal cells.

16. What changes occur in the cytoskeleton in relation to motility? To cell development? To cell division? What is the relationship of these changes to the development of cancer?

17. Compare the changes taking place in the microtubule-based cytoskeletons of plant and animal cells in relation to the division cycle.

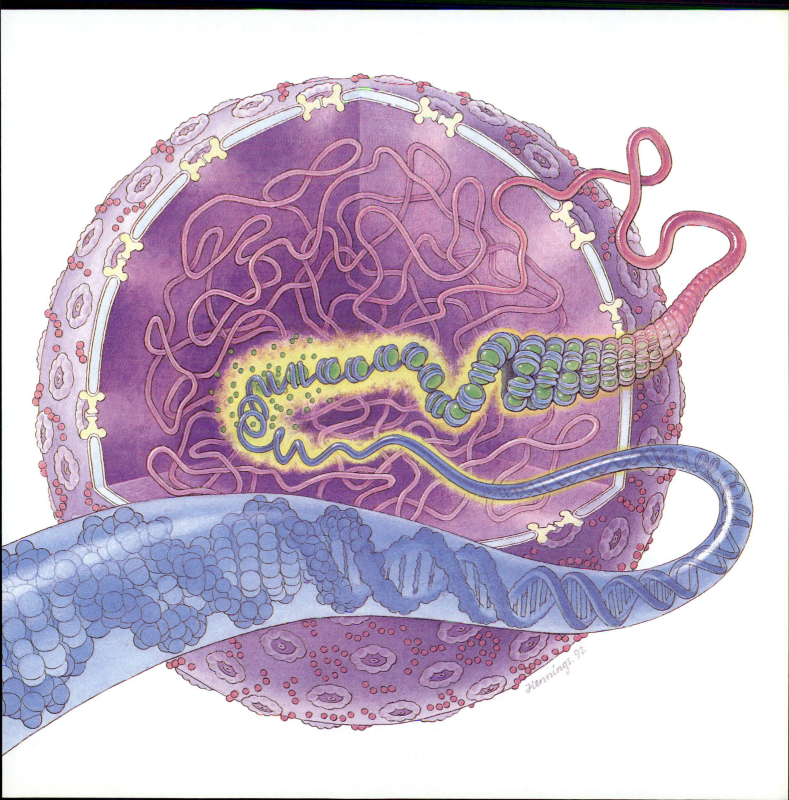

The discovery of DNA structure in 1953 by J. D. Watson and F. H. C. Crick set off the molecular revolution in biology. The discovery not only revealed the structure of DNA but also indicated how genetic information is encoded and utilized, replicated, and passed on in cell division. Later findings, aided by methods for sequencing DNA and RNA, worked out the detailed structure of genes and the RNA molecules copied from them and the amino acid sequences of the proteins encoded in the genes. The excitement continues today, with current emphasis on gene regulation, the total sequence structure and organization of the cell nucleus, the reactions processing RNA copies into finished form, and the mechanisms assembling and distributing proteins in the cell.

The major discoveries of this molecular revolution and the emphasis of current research are the topics of this and the following chapters. This chapter concentrates on the structure of DNA and its associated proteins and the organization of DNA and proteins into chromatin in the cell nucleus. The nuclear envelope, the membrane system that separates the nucleoplasm from the cytoplasm, is also described. The equivalent structures of prokaryotes are outlined in Supplement 13-1. Chapters 14 and 15 describe the genes coding for mRNA and other RNAs and the transcription and processing of their RNA products. Further chapters outline protein synthesis and the mechanisms regulating transcription and translation, the organization of the total complement of coding sequences in the nucleus, and the mechanisms sorting and distributing newly synthesized proteins inside and outside the cell. These chapters concentrate on the synthetic activities of cells during *interphase*, the period of growth between cycles of cell division. The mechanism of DNA replication and the molecular and structural events accomplishing cell division are the subjects of later chapters.

DNA STRUCTURE

The Watson–Crick Model

The model for DNA structure proposed by Watson and Crick was deduced primarily from X-ray diffraction evidence developed by M. H. F. Wilkins and his coworker, R. Franklin. (The X-ray diffraction technique is described in Appendix p. 802.) This evidence gave clues to the overall dimensions of the DNA molecule and indicated that the nucleotide chains forming its backbone twist into a helix. Other evidence, obtained by E. Chargaff from chemical analysis, indicated that the bases adenine and thymine occur in equal quantities in DNA; the quantities of guanine and cytosine are also equal. The structures of the individual nucleotides and the patterns by which they link into chains (see Chapter 2) had been known since the turn of the century.

From this information Watson and Crick proposed their model for DNA, in which two nucleotide chains wind into a right-handed double helix (Fig. 13-1; see also Fig. 2-29). The "backbones" of the two nucleotide chains, which consist of alternating sugar and phosphate groups, are separated from each other by a regular distance, approximately 1.1 nm, across the center of the helix. This space is exactly filled by base pairs consisting of one purine (adenine or guanine) and one pyrimidine (thymine or cytosine; see Fig. 13-2), lying in a flat plane roughly perpendicular to the long axis of the molecule. (The planes are viewed from their edges in Fig. 13-1a.) Each turn of the double helix was proposed to include 10 base pairs. The shapes of the DNA bases allow the base pairs to fit together like the pieces of a jigsaw puzzle. The A-T pairs are stabilized by hydrogen bonds (see p. 36), two between A-T pairs, and three between G-C pairs.

The packing of atoms in the DNA double helix produces two surface grooves that spiral around the surface of the molecule. The two grooves are of equivalent depth, but one, the *major* or *wide groove*, is wider than the other, the *minor* or *narrow groove* (see Fig. 13-1).

Watson and Crick proposed that the two nucleotide chains of the DNA double helix run in opposite directions or are *antiparallel*. This feature of DNA structure is most easily seen if the double helix is unwound and laid out flat, as in Figure 13-3. If the phosphate linkages holding together the chain on the left are traced from the bottom to the top of the figure, they extend from the 5'-carbon of the sugar below to the 3'-carbon of the sugar above each linkage. On the other chain the 5' → 3' linkages run from the top to the bottom of the figure. The antiparallel arrangement of the backbone chains is significant because it requires that a new RNA or DNA chain being copied from an existing chain *must run in the opposite direction from the template*.

Several stabilizing forces hold the DNA double helix together. One is hydrogen bonding between base pairs in the interior of the molecule. Although individually relatively weak, the collective effect of the many hydrogen bonds along a length of double helix sets up a strong and stabilizing attraction between the

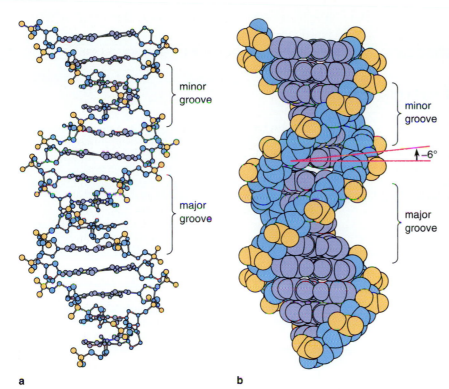

Figure 13-1 The DNA double helix in the B conformation. **(a)** Positions of atoms and bonds in the DNA double helix. Atoms are shown as circles and bonds as straight lines. The bases, which lie in flat planes, are seen on edge in this view. **(b)** Space-filling diagram, in which the spaces occupied by the individual atoms are shown as spheres. The number of base pairs per turn in the B conformation averages about 10.5; within the molecule the bases tilt slightly from the perpendicular, at an angle of $-6°$. The minus sign indicates that the tilt is counterclockwise. (Redrawn from originals courtesy of W. Saenger, from *Principles of Nucleic Acid Structure*. New York: Springer-Verlag.)

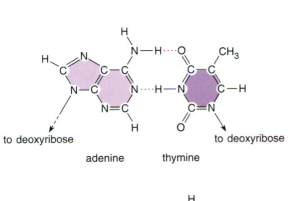

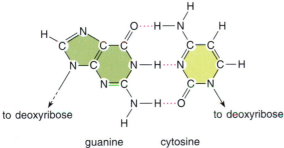

Figure 13-2 The adenine-thymine (A-T) and guanine-cytosine (G-C) base pairs of DNA. Adenine and guanine are purines, and thymine and cytosine are pyrimidines. Hydrogen bonds formed between the purine-pyrimidine base pairs are shown by dotted lines. As shown, A-T base pairs form two stabilizing hydrogen bonds, and G-C pairs form three.

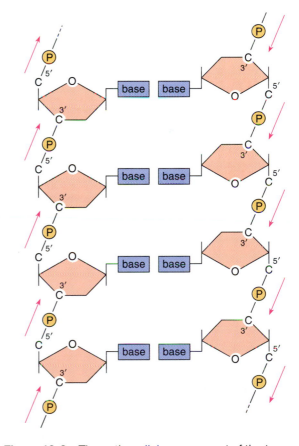

Figure 13-3 The antiparallel arrangement of the two nucleotide chains in DNA, in which the chains run in opposite directions. The arrows point in the 5' → 3' direction.

DNA Melting, Renaturation, and Hybridization

The fact that hydrogen bonds provide one of the primary forces holding the two nucleotide chains of a DNA double helix together makes DNA structure sensitive to temperature. As the temperature rises, kinetic motion of the two nucleotide chains and collisions with molecules in the surrounding solution increasingly disturb the hydrogen bonds holding double-helical structure together. Eventually, if the temperature rises sufficiently, so many of the hydrogen bonds are disturbed that the two chains come apart. As the chains separate, DNA loses its highly ordered structure and the molecule is said to *melt*.

Because G-C base pairs are held together by three hydrogen bonds and A-T base pairs by two, G-C base pairs are more resistant to separation at elevated temperatures. Therefore the greater the number of G-C as opposed to A-T base pairs in a DNA sample, the higher the temperature required to melt the DNA. The effect is approximately linear; each additional G-C base pair adds about 0.4°C to the temperature required to melt half of the DNA sample (T_m). Mammalian DNA, which contains on the average about 40% G-C and 60% A-T pairs, melts with a T_m of 87°C. Prokaryotic DNA, which typically has about 60% G-C base pairs, melts with a T_m of 95°C. These temperatures can be lowered significantly by adding agents, such as formamide, that reduce the strength of hydrogen bonding.

The process reverses if melted DNA is cooled slowly. During cooling, random collisions provide opportunities for complementary base pairing and rewinding of chains into intact double helices. The rewinding process is called *reannealing* or *renaturation*. Renaturation is facilitated if the DNA sample is broken into short lengths of a hundred nucleotides or so. The short lengths increase the number of random collisions between nucleotide chains in the cooling solution and thereby increase the probability that complementary chains will "find" each other and rewind.

DNA renaturation is much used as a method for determining how closely DNA samples from different sources are related in sequence. Because they share many sequences in common, DNA nucleotide chains from two closely related organisms can wind into hybrid double helices in which one nucleotide chain originates from one of the two species and the opposite chain from the other species. DNA from less closely related organisms share fewer sequences and form proportionately fewer hybrid helices when mixed and cooled after melting. *DNA hybridization* studies of this type have been of great value in several areas of research, particularly in evolution.

The accuracy and sensitivity of DNA hybridization are greatly improved if DNA nucleotide chains from the same species are prevented from winding into "native" double helices, in which both nucleotide chains originate from the same organism. To accomplish this, DNA from one of two species under study is melted and immobilized in single-chain form by adding it to nitrocellulose filter paper (nitrocellulose strongly binds single nucleotide chains). The filter paper is then washed free of unbound DNA and treated chemically so that any remaining binding sites for single nucleotide chains are covered. DNA from the other species, usually made radioactive by inclusion of labeled thymine, is then melted and added. Because no other binding sites on the nitrocellulose paper are available, the only DNA chains absorbed by the filter will be those that can form hybrid double helices with the nucleotide chains of the first sample fixed to the paper. The amount of the second sample attaching to the paper, measured by the radioactivity of the paper, provides a measure of the extent to which the two DNA samples share the same sequence information and is therefore a measure of the "relatedness" of the two organisms.

---

two nucleotide chains. (The effects of temperature on hydrogen bonding in the DNA double helix and the application of this effect to DNA studies are outlined in Information Box 13-1.) The double helix is also stabilized by hydrophobic associations (see p. 35) between the stacked bases in the interior of the molecule. In this region the bases, which are primarily nonpolar, pack tightly enough to exclude water and form a stable, nonpolar environment.

The A-T and C-G base pairing suggested to Watson and Crick a possible basis for DNA duplication during cell division. Because of the pairing restrictions a given sequence in one chain is compatible with only one sequence in the opposite chain (Fig. 13-4a). In the language of molecular biology, one chain is said to be *complementary* to the other. To account for DNA replication, Watson and Crick proposed that the two chains of a replicating molecule separate (Fig. 13-4b), and each serves as a *template* for making a complementary copy. The result is two new double helices with a sequence of base pairs identical to that of the original double helix (Fig. 13-4c). Much evidence, discussed in detail in Chapter 23, confirms that replication actually occurs in this pattern. Complementarity also underlies transcription of RNA copies from DNA (see Fig. 14-2).

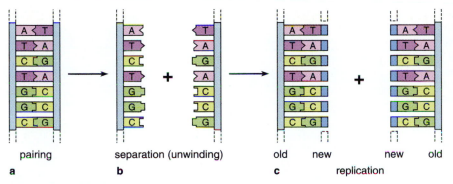

Figure 13-4 Complementarity and DNA replication. **(a)** The base pairing rules determine that the sequence A-T-C-T-G-G-C in one chain restricts the opposite chain to the complementary sequence T-A-G-A-C-C-G. Complementarity provides the basis for replication, in which each of the old chains unwinds **(b)** and serves as a template for the synthesis of a new complementary chain according to the base-pairing rules **(c)**. The result is two molecules, each consisting of one old and one new chain, that are exact duplicates of the original molecule.

a pairing **b** separation (unwinding) **c** old new replication new old

Table 13-1 DNA Conformations

Conformation	Base Pairs per Turn	Vertical Rise per Base Pair (A)	Angle Turned per Base Pair	Tilt of Base Pairs	Groove Width Major (A)	Groove Width Minor (A)	Groove Depth Major (A)	Groove Depth Minor (A)	Approximate Diameter (A)
A	11	2.56	32.7°	+20°	10.9	3.7	2.8	13.5	23
B	10.1–10.6	3.38	36.0°	−6°	5.7	11.7	7.5	8.5	20
C	9.3	3.32	38.6°	−8°	4.8	10.5	7.9	7.5	19
D	8	3.04	45.0°	−16°	1.3	8.9	6.7	5.8	
E	7.5	3.25	48.0°						
Z	12	3.70	−30.0°	−7°	8.8	2.0	3.7	13.8	18

All the detailed X-ray diffraction and other work carried out since Watson and Crick announced their model have supported the fundamental double-helical structure they proposed for DNA. The two investigators, with Wilkins, received the Nobel Prize in 1962 for the model, which has proved to be one of the most significant and far-reaching discoveries in the history of biology.

Refinements to the Watson-Crick Model: B-DNA and Other DNA Conformations

The X-ray data used by Watson and Crick were obtained from DNA preparations that had been oriented in parallel by drawing them into a fiber. Such preparations, which are at best only semicrystalline, produce somewhat blurred X-ray diffraction patterns. (The clearest diffraction patterns are produced by the highly ordered atoms of crystals; see Appendix p. 802.) The DNA form investigated by Watson and Crick, known as *B-DNA*, was one of three conformations, *A*, *B*, and *C*, that could be produced by drawing DNA fibers under differing conditions of hydration. In the years following Watson and Crick's original discoveries, methods were developed for casting true crystals of short DNA molecules of fixed sequence. X-ray diffraction of these crystals produced a greatly refined picture of DNA structure and allowed the positions of individual atoms to be precisely located.

This research revealed that DNA can take on several conformations, designated *D*, *E*, and *Z*, in addition to the A, B, and C forms (Table 13-1). Of the conformations, only the B and Z conformations are known to be common in cellular DNA. The Watson-Crick B conformation is the form taken by most DNA in both eukaryotic and prokaryotic cells. Although DNA can readily assume the A conformation (Fig. 13-5a), this form is actually the primary conformation taken by double-helical regions of RNA molecules. Of the remaining right-handed conformations, C-DNA seems only to be a laboratory curiosity, formed under extreme conditions that are not believed to occur inside living cells. The D and E conformations can be taken only by DNAs that contain certain sequences, such as regularly alternating A-T base pairs. D-DNA has been detected in nature only in *bacteriophage T2*, a virus infecting bacterial cells. The Z conformation (Fig. 13-5b), in which DNA winds to the left instead of the right, is believed to occur in a small proportion of cellular DNA.

All DNA molecules of any source, natural or artificial, and of any sequence can take on the A or B conformations. Double-helical regions in RNA molecules

Figure 13-5 DNA in the A and Z conformations. **(a)** Space-filling model of A-DNA. RNA double helices or hybrid double helices consisting of one DNA and one RNA nucleotide chain may also take the A conformation. The tilt of the base planes from the horizontal in A-DNA is +20°. The minor groove is deeper, and the major groove is more shallow than in the B form. **(b)** DNA wound into the left-handed Z conformation. Note the zigzag pathway (red lines) followed by the phosphate groups in Z-DNA (compare with Fig. 13-1b). Within Z-DNA the base pairs are flipped through 180° and are thus upside down with respect to their position in B-DNA. The edges of the flipped base pairs fill in most of the surface region analogous to the major groove of B-DNA, so that Z-DNA has a shallow major groove and a deep surface groove equivalent in position to the minor groove of B-DNA. (Redrawn from originals courtesy of W. Saenger, from *Principles of Nucleic Acid Structure*. New York: Springer-Verlag.)

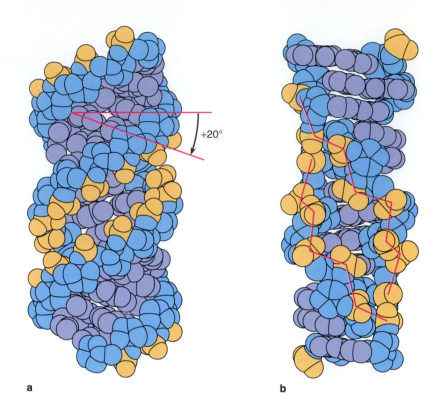

a   b

are limited to the A conformation because the —OH group at the 2'-carbon of the ribose sugar interferes with the degree of packing required to form the B conformation (an —H links at this position in deoxyribose; see Fig. 2-28). RNA double helices in the A conformation form regularly in regions of RNA molecules that fold back upon themselves in "hairpin" formations (see Information Box 13-2 and p. 397). *Hybrid* double helices, produced by one DNA and one RNA nucleotide chain winding together, may approximate the A conformation or take on a conformation with dimensions between the A and B forms. Hybrid helices occur as temporary intermediates during transcription, when an RNA copy is made from a DNA chain.

Z-DNA, discovered in 1979 by A. Rich and his coworkers, was found through X-ray diffraction of crystals of short DNA molecules with alternating G-C sequences, in the form

$$
\begin{array}{ccccccc}
-C & -G & -C & -G & -C & -G- \\
\vdots & \vdots & \vdots & \vdots & \vdots & \vdots \\
-G & -C & -G & -C & -G & -C-
\end{array}
$$

To their surprise the X-ray diffraction patterns indicated that the DNA in the crystals was twisted into a left-handed rather than a right-handed double helix. (The events leading to Rich's discovery are related in his Experimental Process essay on p. 370.) The helix was designated as Z-DNA because a line connecting the phosphate groups zigzags rather than following

the relatively smooth pathway observed in B-DNA (see Fig. 13-5b).

Z-DNA is less stable than B-DNA. In Z-DNA the negatively charged phosphate groups are brought closer together than in B-DNA, increasing their mutual repulsion and favoring a shift from the Z to the B form. The phosphate groups also extend farther from the surface of the double helix in B- than in Z-DNA, allowing water molecules and ions of opposite charge to form a neutralizing coat that also reduces their mutual repulsion, adding to the stability of the B form. However, certain factors or conditions increase the stability of the Z conformation and favor a shift from B to Z. One is DNA sequence; Z-DNA is favored by sequences that contain alternating purines and pyrimidines, as in the alternating G-C sequence used by Rich and his coworkers for their studies. Other factors stabilizing Z-DNA are addition of methyl (—CH$_3$) groups to cytosines in the DNA sequence (see Fig. 13-6), winding of the entire DNA double helix into left-handed supercoils or loops (known as *negative supercoils*; see below), and *Z-DNA–binding proteins*. Methylation of cytosines, which is common in nature, adds a strongly hydrophobic group that is exposed to water molecules in B-DNA but is protected from water in a hydrophobic pocket in Z-DNA. Negative supercoiling adds a left-handed twisting force that tends to wind DNA into the Z rather than B conformation. As it happens, DNA in both prokaryotic and eukaryotic DNA is subjected to negative supercoiling. The Z-DNA–binding proteins, discovered in the nuclei of *Drosophila*, human,

Inverted Sequences (Palindromes)

Inverted sequences are symmetrically repeated DNA sequences that are the same in both chains of a double helix when read in the 5' → 3' direction:

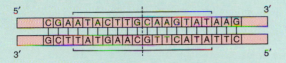

In this example the inverted repeat is enclosed by the brackets. The vertical dotted line marks the center of symmetry or *dyad axis* of the inverted sequence. An inverted sequence is usually not contiguous on either side of the dyad axis. Instead, a noninverted sequence usually separates the two halves of an inverted sequence (the noninverted sequence is boxed):

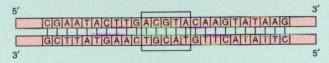

Inverted sequences, which occur commonly in natural DNA, can wind into a regular DNA helix in the B conformation. Because inverted sequences are symmetrical on both sides of the dyad axis, they can also wind into an *X*-shaped or cruciform structure. The noninverted sequence separating the inverted sequences extends as unpaired loops at the top and bottom of the cruciform:

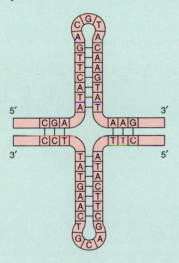

An RNA molecule copied from an inverted sequence may fold back on itself to form a double-helical segment in the region of the sequence. Because the RNA copy exists only as a single chain, the structure formed is a *hairpin* instead of a cruciform. For example, the RNA sequence copied from the top half of the DNA palindrome shown above:

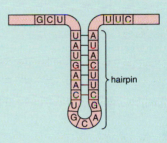

Hairpins copied from inverted sequences in DNA are a common and important feature of RNA structure, particularly in rRNA and tRNA molecules (as, for example, in the rRNA shown in Fig. 15-2). Probably many, if not most, of the inverted sequences in DNA have no function other than encoding the regions that, when transcribed, form hairpins in RNA copies. However, some inverted sequences forming cruciforms in DNA may serve as recognition sites for proteins regulating transcription; other inverted sequences mark the boundaries of foreign sequence elements inserted into a host DNA molecule (see p. 536).

wheat, bacterial, and other cells, can combine specifically with Z-DNA and "fix" the double helix in stable left-handed form.

The fact that proteins exist in eukaryotes that can bind and stabilize Z-DNA is one line of evidence indicating that this conformation is actually present inside cells. Another comes from anti–Z-DNA antibodies, which occur in persons with certain autoimmune diseases; production of these antibodies would not seem likely unless the Z form actually exists in the

The Discovery of Left-Handed Z-DNA

Alexander Rich

ALEXANDER RICH has an A.B. and M.D. from Harvard University. After postdoctoral work at Caltech he went to NIH from 1954 to 1958 as section head in Physical Chemistry. In 1955–56, he worked at the Cavendish Laboratory in Cambridge, England. At NIH and in England, he discovered the three-dimensional structure of collagen. During that period, he and his NIH coworkers discovered the formation of triple-stranded nucleic acid complexes. In 1958 he moved to the Biology Department at M.I.T. His work included discovery of polysomes in protein synthesis, discovery of DNA in chloroplasts, and determination of the three-dimensional structure of yeast phenylalanine transfer RNA. He was a member of the Viking mission team that searched for active biology on the surface of Mars.

One of the most important ways we have of learning about the three-dimensional structure of biological molecules, especially macromolecules, is through single-crystal x-ray diffraction analysis. In the early years of research on the nucleic acids, most of the x-ray diffraction work was carried out on DNA fibers. These studies are often useful for obtaining the general conformation of a polymer molecule but they are unable to produce a detailed picture. The first single-crystal x-ray diffraction analysis of the double helix came in 1973 when my colleagues and I solved the structure of two dinucleoside phosphates, ApU and GpC.[1] Both of these crystallized as double helical fragments held together by Watson–Crick hydrogen bonds. Furthermore, the crystals diffracted to 0.8 Å resolution, that is, atomic resolution. Every atom could be seen in the lattice, including water molecules and ions. If we could obtain this amount of detail from even larger nucleic acid fragments, we would learn a great deal more. By 1974 we were able to determine the three-dimensional structure of yeast phenylalanine transfer RNA.[2] However, this was carried out at about 2.5 Å resolution, which gave a reasonable view of the molecule, but the details still eluded us.

Shortly thereafter I met Jacques van Boom, a Dutch organic chemist who was highly skilled in synthesizing oligonucleotides. We resolved to collaborate in solving some nucleic acid structures. He would do the synthesis and we would try to crystallize and solve them. Little was known about the properties of small oligonucleotides, so we decided to concentrate on those containing guanine-cytosine base pairs, which are somewhat more stable than adenine-thymine base pairs. The first material he made for us was the alternating sequence d(CGCG), which is self-complementary. It crystallized, and we decided then to go a step further, to the self-complementary hexamer d(CGCGCG). My postdoctoral fellow Andrew Wang discovered that this would crystallize readily. Remarkably, the crystals diffracted x-rays to 0.9 Å resolution, again atomic resolution. This excited us very much because we thought that we could find out for the first time in some detail what the DNA double helix really looks like.

In x-ray diffraction studies, the intensity of the diffracted beams is registered on photographic film or with radiation counters. However, we lose the phases of the beams relative to each other. To reconstruct a three-dimensional electron density map and actually "see" the atoms, we have to solve the phases. When you believe you know the structure of the molecule, you can use a technique called a "translation-rotation search" to solve the diffraction pattern. Using a computer program, we took various models—A-DNA, B-DNA, and other variants—and searched through the diffraction data looking for a good fit. None of these models fit the diffraction data we obtained for the hexamer, so we decided to do the more laborious but more certain method of obtaining heavy atom derivatives. These would modify the diffraction pattern enough so that we could determine the relative phases of the diffracted beams and thus solve the structure. We found that a number of heavy metal ions could be diffused into the crystal lattice, and their position was determined by measuring the difference between their diffraction pattern and the diffraction pattern of the native molecule. Given enough of these, we eventually were able to determine the relative phases of the diffracting rays and from this reconstruct a three-dimensional electron density net.

In an x-ray diffraction experiment, solution of the structure consists of obtaining a three-dimensional electron density map. The electrons are clustered around the atoms and from this we can discern the structure of the molecule. In those days, the electron density was drawn on transparent plastic sheets in layers, which were then piled atop one another to view the entire three-dimensional distribution. When this was done with the first crystal of d(CGCGCG), we observed a very unusual molecule. We had expected to see ordinary B-DNA, but in fact it became apparent that the chains coiled about each other in a left-handed rather than a right-handed array. Nonetheless, the familiar guanine-cytosine base pairs were clearly visible in the electron density map, so the map couldn't be entirely wrong. I asked Andrew Wang whether he was certain that he had piled up the plastic sheets in the correct order. If, for example, they were piled in inverted order, the helix would have an inverted sense. However, after careful checking, he concluded that they had indeed been correctly assembled and that the molecule was, to our great surprise, a left-handed double helix![3]

It took us some time to understand the relationship of this left-handed helix with its zigzag backbone to right-handed B-DNA with its smooth backbone (see Fig. 13-5). It quickly became apparent that all the guanines in the structure had rotated about the glycosyl bond into the less common *syn* conformation, while the cytosines had the normal *anti* conformation. [*Syn* and *anti* refer to two energetically favored positions of the base with respect to the sugar in a nucleotide. In the more common *anti* conformation, the bulk of the base is positioned away from the sugar; in the *syn* conformation, the base rotates to a position closer to the sugar.] Further inspection revealed that in order to go from right-handed B-DNA to left-handed Z-DNA, the base pairs actually have to turn upside down. This act of turning upside down converts the sense of the helix from right to left and also is responsible for the change in the guanine residues to the *syn* conformation.

We began to read more widely about this sequence, and I rediscovered a paper of Pohl and Jovin,[4] written some seven years previously, in which they studied the circular dichroism of the polymer poly (dG-dC). They clearly showed that in a high salt solution, 5 *M* NaCl, the circular dichroism spectrum nearly inverted itself. Although the theory of circular dichroism is not well enough developed to come to a firm conclusion, it seemed likely to us that the crystal of left-handed Z-DNA (named for the zigzag backbone) might be the high salt form of poly (dG-dC).

Proving that required a different kind of analysis. For this we chose laser Raman spectroscopy.[5] The Raman spectrum is related to the manner in which the atoms in a molecule vibrate. These vibrations are more or less independent of whether the molecule is constrained in a lattice as in a crystal or freely tumbling in solution. We were able to show that the DNA in the crystal had a Raman spectrum identical to that found in 5 *M* NaCl and quite distinct from that of poly (dG-dC) in a 0.1 *M* NaCl solution. Thus it was clear that left-handed Z-DNA was the form of poly (dG-dC) stabilized in 5 *M* NaCl.

Although the first crystalline work was done with sequences with alternating cytosine and guanine residues, later work incorporated adenine-thymine base pairs. These can convert to the Z form, although not as readily as CG base pairs. The sequences that prefer Z-DNA formation are those that have alternating purines and pyrimidines with CG sequences being most important for this process. Z-DNA can form even in the absence of regularly alternating purines and pyrimidines, but somewhat more energy is needed for it to form.

Therefore, Z-DNA may be regarded as a higher energy state of the DNA double helix. The B-DNA conformation is the ground state, and energy has to be applied to the molecule to stabilize its conversion into the Z form.

The discovery of Z-DNA was startling to many research workers. It had been known for quite a while that DNA could undergo conformational changes. However, the magnitude of the conformational change from right-handed to left-handed DNA is much greater than other changes, and it stimulated a great deal of chemical and molecular biological investigation into its properties.

Several studies have been carried out that describe the formation of Z-DNA *in vivo*. The method for detecting the change often involves changes in the chemical reactivity of DNA when it goes from the B to the Z conformation. These have proven especially useful for studies of DNA in prokaryotic organisms.[6] In eukaryotic organisms, monoclonal antibodies have been used to study Z-DNA formation in physiologically active mammalian nuclei.[7] Nuclei have been prepared that are permeable to macromolecules but still maintain the ability to carry out DNA replication and transcription. The amount of Z-DNA formed in these nuclei is determined by the torsional strain of the DNA. Processes such as transcription that generate negative supercoiling increase the amount of Z-DNA in the nuclei, while the continued activity of the relaxing enzyme, topoisomerase I, results in a decrease of Z-DNA formation. Selective regions of individual genes have been identified that flip from the B to the Z form during transcriptional activity. However, further work will be needed to fully define the biological role of these conformational changes within cells.

References

[1]Day, R. O.; Seeman, N. C.; Rosenberg, J. M.; and Rich, A. *Proc. Nat. Acad. Sci. USA* 70:849–53 (1973); Rosenberg, J. M.; Seeman, N. C.; Kim, J. J. P.; Suddath, F. L.; Nicholas, H. B.; and Rich, A. *Nature* 243:150–54 (1973).

[2]Kim, S. H.; Suddath, F. L.; Quigley, G. J.; McPherson, A.; Kim, J. J.; Sussman, J. L.; Wang, A. H.-J.; Seeman, N. C.; and Rich, A. *Science* 185:435–39 (1974).

[3]Wang, A. H.-J.; Quigley, G. J.; Kolpak, F. J.; Crawford, J. L.; van Boom, J. H.; van der Marel, G.; and Rich, A. *Nature* 282:680–86 (1979).

[4]Pohl, F. M., and Jovin, T. M. *J. Mol. Biol.* 67:375–96 (1972).

[5]Thamann, T. J.; Lord, R. C.; Wang, A. H.-J.; and Rich, A. *Nuc. Acids Res.* 9:5443–57 (1981).

[6]Rahmouni, A. R., and Wells, R. D. *Science* 246:358–63 (1989).

[7]Wittig, B.; Dorbic, T.; and Rich, A. *J. Cell. Biol.* 108:755–64 (1989).

body. Anti–Z-DNA antibodies produced by injecting Z-DNA into a test animal also react positively with nuclei and chromosomes, indicating that Z-DNA is present.

Several functions have been proposed for Z-DNA. One possibility is that Z-DNA segments supply a reservoir of left-handed turns that can unwind the right-handed turns of B-DNA. This would allow long DNA molecules to unwind locally, greatly facilitating RNA transcription. A second proposal is that Z-DNA segments, through their ability to promote DNA unwinding, form part of the control mechanisms regulating RNA transcription. Addition or removal of methyl groups or Z-DNA–binding proteins, according to this proposal, would regulate the shift between the Z and B forms. A third possibility is that stretches of Z-DNA promote genetic recombination, in which two helices unwind, break, and exchange segments (see Chapter 25 for details of the recombination mechanism).

The various DNA conformations, rather than being rigidly fixed, are flexible structures that vary to a limited degree in essentially all of their dimensions. The primary source of variation is the nucleotide sequence in a DNA region; depending on the sequence, such characteristics as groove width, base packing, pitch, and numbers of base pairs per turn of the helix vary slightly but significantly. These local variations are critical to recognition of individual sequences by regulatory and other proteins combining with DNA (see p. 489). DNA conformations are also modified by proteins binding to the double helix, which may induce bends of various degrees or even unwind the helix locally.

DNA can also take on a triple-helical form in which a third, single DNA nucleotide chain fits along the wide groove of two chains wound into a double helix. The triple structure, originally discovered by A. Rich, D. Davies, and G. Felsenfeld, is stabilized by hydrogen bonds formed between bases of the added DNA chain and the edges of base pairs exposed in the wide groove. In the triple helix, for example, T in the third chain can recognize and bind specifically to the A of A-T base pairs in the double helix, and C can bind to the G of G-C base pairs. The binding patterns allow the third nucleotide chain to recognize and bind specific complementary sequences in an intact double helix. Although considered until recently to be only another laboratory curiosity, the DNA triple helix is attracting growing interest as a structure that may take part in cellular processes in which a single DNA chain interacts with a double helix, as in the initial steps of genetic recombination (see Chapter 25). It is also considered possible that a single chain extending from the end of the double helix may fold back at the ends of chromosomes, forming a triple-helical struc-

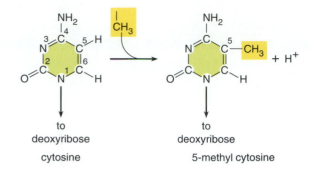

Figure 13-6 Addition of a methyl group to cytosine to form the modified base 5-methyl cytosine. The modification is common in natural DNAs.

ture that stabilizes the ends and protects them from enzymatic attack (see p. 699). There is also interest in the possibility of inducing stable triple helices as a means to block transcription or replication of infecting viral DNAs.

Chemical Modification of DNA

DNA is commonly modified inside cells by addition of a methyl group to cytosine, creating the modified base *5-methyl cytosine* (Fig. 13-6). The degree to which cytosines are methylated varies widely in different groups of organisms. In bacteria, less than 1% of the cytosines in DNA are methylated; the DNA molecules in protozoa, fungi, and most invertebrates similarly contain few or no methylated cytosines. Vertebrate DNA is methylated to about 2% to 8% of the total cytosine content. In higher plants, one-third or more of the cytosines may be methylated.

DNA methylation is of interest because an inverse correlation has been established in vertebrates between the amount of this modification and the level of mRNA transcription. The DNA of many mRNA-encoding genes is methylated at certain sites when the genes are inactive in transcription and demethylated at these sites when active. In many cell lines the patterns of methylation in these genes are reproduced and passed on during cell division. One mechanism by which methylation may regulate transcription, as noted, is by favoring the conversion from B- to Z-DNA. (The possible relationships between methylation and the regulation of mRNA transcription are discussed further in Chapter 17.)

DNA Supercoiling

In B-DNA the number of base pairs per turn (10.1 to 10.6; see Table 13-1) and the length occupied by one full turn can vary over only relatively narrow limits.

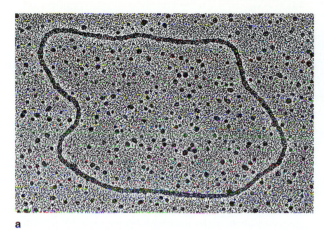

a

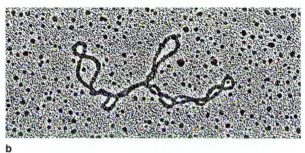

b

Figure 13-7 A circular DNA molecule in the relaxed (a) and supercoiled form (b). Each twist in the DNA circle is a supercoil. (Courtesy of L. W. Coggins.)

Turns cannot be compressed or expanded in B-DNA as in a rubber band; instead, if a strong twisting force is applied in either the positive or negative direction, DNA flips into supercoils to compensate for the strain induced by the twisting (Fig. 13-7). One supercoil forms for each full turn imposed on the DNA. The loops resemble the "knots" that appear when a coiled telephone cord is twisted. As noted, all natural B-DNAs are exposed to forces that tend to twist them in the direction opposite to the right-handed turns of the double helix, forming negative supercoils.

The origins of negative supercoils differ in prokaryotes and eukaryotes. In prokaryotes, supercoils are induced by enzymes called *topoisomerases*.[1] When introducing supercoils, these enzymes introduce a break in one or both nucleotide chains of a B-DNA double helix, rotate or twist the chains, and close the breaks so that both nucleotide chains are continuous again. (Figure 13-8 shows the operation of a topoisomerase I enzyme, which breaks one of the two nucleotide chains; topoisomerase II enzymes break both chains.) Energy for the supercoiling is supplied by ATP, which is hydrolyzed at a rate of at least one ATP for each supercoil wound into DNA. Topoisomerases can also work in reverse to remove supercoils from DNA. Removal of supercoils, called "relaxation" of the DNA, is powered by release of the supercoils and does not require an energy input.

In eukaryotes, topoisomerases evidently participate in the reactions unwinding DNA during replication and genetic recombination rather than maintaining the degree of negative supercoiling (see pp. 664 and 747 for details). Instead, eukaryotic DNA is held in negative supercoils by combination with histone proteins, which join with DNA to form a beadlike unit

[1]Topoisomerases are named for *topology*, the mathematical study of the possible ways in which objects of constant geometry, such as DNA, can be bent or twisted without disturbing their fundamental structure.

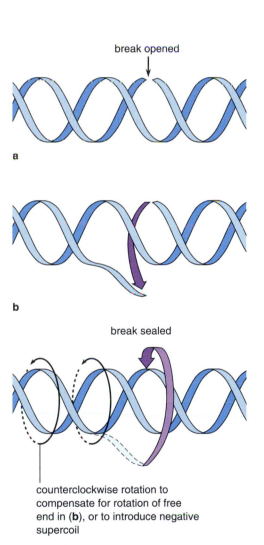

break opened

a

b

break sealed

counterclockwise rotation to compensate for rotation of free end in (**b**), or to introduce negative supercoil

Figure 13-8 Activity of topoisomerase I in introduction of supercoils in DNA. (a) The enzyme opens a single chain break in one of the two chains of a DNA double helix by opening a phosphate linkage. (b) The free end is rotated through one full turn and covalently rejoined (c). If the ends of the DNA molecule are not free to turn, one supercoil will be introduced for each complete rotation of the free end.

called a *nucleosome*, the fundamental structural unit of chromatin (see below and Fig. 13-10). The DNA turns in a left-handed direction around nucleosomes, much like a rope around a pulley. In effect the winding imposes one negative supercoil for each nucleosome.

Negative supercoiling has several important consequences in natural DNAs. Negative supercoils favor local unwinding of the DNA helix, exposing unpaired nucleotide chains for processes such as transcription, replication, and genetic recombination. Negative supercoiling also favors conversion of B- to Z-DNA and modifies the binding affinity of proteins for DNA, in some cases enhancing protein-DNA binding and in other cases inhibiting it. The degree of supercoiling may be a factor in gene regulation, particularly in prokaryotes.

PROTEINS ASSOCIATED WITH DNA IN CHROMATIN

The DNA in eukaryotic nuclei is closely associated with two major groups of proteins, the *histones* and *nonhistones*. Histones occur in approximately equal quantities by weight with DNA; nonhistones range from as little as 20% to approximately the same as the DNA by weight. The three elements together—DNA, histones, and nonhistones--collectively form the *chromatin* of eukaryotic nuclei. The total DNA complement of the nucleus, with associated histone and nonhistone proteins, is broken into individual lengths; these are the *chromosomes* of the nucleus.

Although the functions of the two classes of chromosomal proteins overlap to some extent, histones primarily form structural parts of the chromatin, maintaining DNA in specific three-dimensional coiling and folding arrangements. Nonhistones are primarily functional molecules that catalyze and regulate the activity of DNA in transcription and replication.

The Histones

Histones are distinctive molecules that differ from most other cellular proteins in their abundance, relatively small size, and strongly basic charge. These distinctive features contributed to their early discovery in 1884 by A. Kossel, not long after the discovery of DNA. (DNA was discovered by F. Miescher in 1871.)

Histone Structure   The histones are easily extracted by exposing chromatin to acids or salts at elevated concentrations. Removal in this way demonstrates that histones are held in chromatin by electrostatic attractions rather than covalent bonds. Five major histone types, *H1, H2A, H2B, H3,* and *H4*, are released

in sequence from chromatin as the salt concentration is gradually raised from cellular to higher levels (Table 13-2). With very few exceptions the treatment releases the same five histone types from all eukaryotes.

The strongly basic or positive charge of the histones results from their high content of lysine and arginine residues. The charged amino acids are concentrated in end segments that extend as "arms" from histone molecules (Fig. 13-9). The positive charges bind histones to DNA by interacting with the negatively charged phosphate groups on the surface of the DNA helix. Hydrophobic residues are concentrated in a globular domain from which the arms extend. The hydrophobic domain may set up histone-histone interactions in chromatin and may also interact with hydrophobic regions of the DNA double helix.

H2A, H2B, H3, and H4 are called *core histones* because they form a beadlike core structure around which DNA wraps to form nucleosomes. H1 is sometimes called the *linker histone* because it is believed to combine near a segment of DNA that extends as a link from one nucleosome to the next (see Fig. 13-10).

Comparisons among different species show that the core histones, particularly those of higher eukaryotes, are among the most highly conserved proteins known. Even where substitutions occur, many are conservative; that is, one amino acid is replaced by another of similar size and chemical properties. The degree to which the core histones are conserved indicates that the amino acid sequence in most segments is essential for their functions, and that mutations changing the sequence of these segments are likely to be lethal.

H3 and H4 are the most strongly conserved histones of higher eukaryotes. For example, only 2 amino acids out of a total of 102 differ in the H4 histones of the calf and garden pea. H2A and H2B, although less faithfully repeated among higher eukaryotes, are still among the most highly conserved proteins known.

The linker histone, H1, is less conserved. Among mammals alone, H1 histones show about 20% variation in sequence; as evolutionary relationships become more distant, the sequence diversity among H1 histones becomes more marked. The H1 histones of higher plants, for example, vary in about 75% of their sequences from animal H1s.

Except for H4, each histone type occurs in *variants* that differ slightly in sequence within species. Most vertebrate species have two to three variants of the H2A, H2B, and H3 histones that differ at one or two positions in their amino acid sequences. H1 histones are by far the most variable; most cell types have four to six H1 variants, some of which differ extensively in sequence.

The histone variants have several interesting characteristics. In vertebrates, particularly among mammals,

Table 13-2 Histones

Histone Type	Approximate Molecular Weight	Number of Amino Acids	Approximate Content of Basic Amino Acids
H1	17,000–28,000	200–265	27% lysine, 2% arginine
H2A	13,900	129–155	11% lysine, 9% arginine
H2B	13,800	121–148	16% lysine, 6% arginine
H3	15,300	135	10% lysine, 15% arginine
H4	11,300	102	11% lysine, 4% arginine

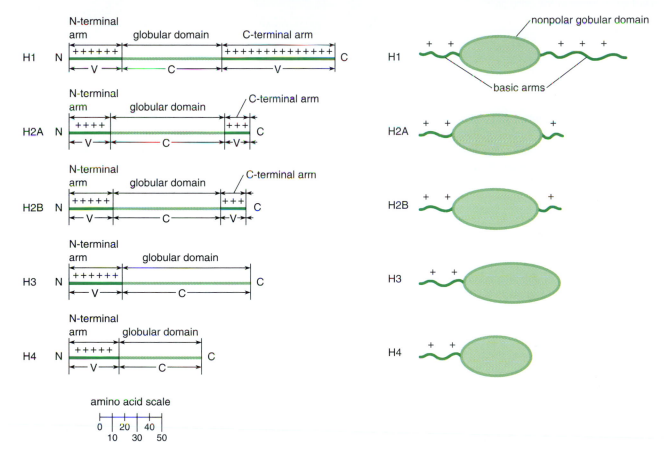

Figure 13-9   Sequence and domain structure of the histones. Each of the histones is divided into extended, charged domains and a globular hydrophobic domain. The extended and globular domains correspond to regions that are variable (V) and conserved (C) in amino acid sequence between the histones of different species. Although the interactions of the charged and nonpolar segments are still largely unknown, it appears that both regions coordinate in binding histone molecules to each other and to DNA.

the same variants are highly conserved and repeated over a wide range of species. The variants change in relative amounts and appear and disappear at fixed times in coordination with embryonic development and cycles of growth and division. Characteristic histone variants also appear in many species during meiosis, particularly during meiotic prophase, when chromosome pairing and recombination take place (for de-

tails see Chapter 25). These controlled changes indicate that the different combinations of histone variants are critical to the normal progress of cells through development and cycles of growth and division. The effects of the variants on DNA structure and function, however, are still unknown.

As part of the biochemical and structural alterations that produce motile male gametes in both plants

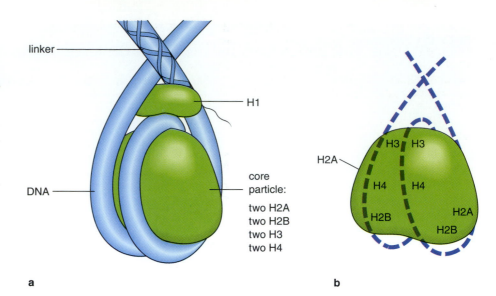

Figure 13-10 Nucleosome structure. **(a)** Approximately two turns of DNA wrap around a wedge-shaped core particle consisting of two molecules each of H2A, H2B, H3, and H4. The DNA is held with the midpoint of the two turns at the point of the wedge. H1 binds in the region where the DNA enters and exits the nucleosome, presumably stabilizing the DNA at this point. The entire structure is about 10 nanometers in diameter. **(b)** The possible locations of individual histones in the core particle, as deduced from X-ray diffraction.

In figure (a): linker — H1 — DNA — core particle: two H2A, two H2B, two H3, two H4

In figure (b): H2A — H3 — H3 — H4 — H4 — H2B — H2A — H2B

a

b

and animals the histones may be partly or completely replaced by a group of much smaller and even more basic proteins, the *protamines*. Replacement of histones by protamines is apparently related to the extreme compaction of DNA that takes place during the development of motile male gametes.

Chemical Modification of the Histones   The histones are chemically modified by addition of chemical groups to individual amino acid residues. Among the most significant of these modifications is addition of acetyl, methyl, or phosphate groups to amino acids in the charged arms of the histone chains. The modifications eliminate one positive charge for each addition and presumably reduce the strength of the electrostatic attraction of histones to DNA.

The various histone modifications are correlated with many important cellular mechanisms, including nucleosome assembly, gene regulation, DNA replication, cell division, and compaction of DNA. Histones are temporarily acetylated before being added to DNA during nucleosome assembly and then deacetylated when assembly is complete. This temporary acetylation may reduce the electrostatic attraction between the histones and DNA enough to permit precise control of nucleosome assembly. Histones associated with many but not all genes become more highly acetylated as the genes change from inactive to active form. In this case, acetylation may loosen chromatin structure, through reduction of the DNA-histone attraction, enough to allow access to DNA by enzymes catalyzing RNA transcription. (Further details of histone acetylation and gene regulation are presented in Chapter 17.)

Addition of methyl groups to lysines and histidines in H1, H2B, H3, and H4 is correlated with DNA replication in some cells. Phosphorylation, particularly of H1, occurs just before and during cell division and is reversed as cells finish dividing and enter the next phase of growth (for details, see Fig. 22-4 and Chapter 22). Increases in H1 phosphorylation are thought to promote packing of chromatin into assemblies such as clumps of heterochromatin in interphase nuclei (see Fig. 13-15) or the rodlike chromosomes into which chromatin packs during cell division (see Fig. 24-2).

The Nonhistone Proteins

Nonhistones are defined broadly as proteins, excluding the histones, that are associated with DNA in chromatin. Among this group are proteins that regulate individual gene activity in transcription. As such they provide the central controls of cell activity and differentiation. In contrast to the relatively uniform nature of histones, nonhistones are a varied group including proteins that are neutral, positively, or negatively charged and range in size from polypeptides smaller than histones to some of the largest proteins of the cell.

Types of Nonhistone Proteins   Proteins occurring in the nonhistone group fall into several clearly defined classes:

1.   Regulatory proteins controlling the activity of genes in RNA transcription

2.   Enzymes catalyzing reactions of transcription, replication, recombination, DNA repair, and modification of DNA and chromosomal proteins

3.   Proteins contributing to maintenance of chromatin structure

The best known and characterized nonhistones are the enzymes of class 2 above. These enzymes and their activities in RNA transcription and DNA replication and repair are the subjects of Chapters 14, 15, and 23. Chapter 25 describes the enzymes catalyzing genetic recombination. Intensive research into the nonhistones regulating RNA transcription is rapidly adding to the list of proteins in this group; these proteins and their functions in gene regulation are the subject of Chapter 17. Relatively little is known of nonhistones contributing to the maintenance of chromatin structure.

Nucleosomes

Histones combine with DNA to form nucleosomes, the fundamental structural units of chromatin. Nucleosomes pack DNA in a stable coiled form in eukaryotic nuclei and may also contribute to gene regulation by holding DNA in a state in which it is not readily available to enzymes catalyzing transcription.

Nucleosome Structure   In a nucleosome approximately two left-hand turns of DNA wind around a *core particle* consisting of two each of the four core histones— two H2A, two H2B, two H3, and two H4 (Fig. 13-10). H1 is located at the point where the DNA enters and leaves the core particle, where it probably ties DNA to the nucleosome core, much like a clasp on a string tie. A segment of DNA, the *linker*, connects one nucleosome to the next.

About 165 base pairs of DNA are included in the two turns of DNA wrapped around the core particle. This length includes a major segment held so tightly to the core particle that it is strongly protected against enzymatic attack (see below). This segment, 146 ± 1 base pairs in length, occupies about 1.8 turns around the core particle. The remaining DNA of the two turns, about 10 base pairs on either side of the central 146, is more susceptible to enzymatic attack, indicating that it is lifted slightly from the core particle surface. These more susceptible segments are probably the part of the DNA associated with H1. The 146–base-pair segment held tightly to the core particle is closely similar in length among all eukaryotes.

The linker segment tying one nucleosome to the next lies outside the two turns that loop around the nucleosome core particle. The linker length is variable, ranging over limits from a minimum of less than 10 to a maximum of about 90 base pairs. Generally linker length is shortest in lower eukaryotes, intermediate in plants, and longest in higher animals. In some species, linkers become longer or shorter as genes change between active and inactive form. Average linker length also varies as embryonic development proceeds. These variations may reflect differences in the types and degree of chemical modification of H1 variants associated with the linker region or changes induced by combination of regulatory nonhistone proteins with the DNA.

Nucleosomes can be assembled in the test tube from only the five histones and DNA; no other proteins are required. Essentially any B-DNA molecule can wind with the histone proteins into nucleosomes, including the DNA molecules of viruses and bacteria that are not complexed with histones in their natural locations, and artificial DNAs. The ability of nucleosomes to assemble from DNA of essentially any natural or artificial source indicates that DNA sequence is of little or no significance in the assembly process. However, nucleosomes do bind preferentially to some sequences, probably because these sequences permit DNA to bend more readily around the nucleosome core. RNA does not form nucleosomes in either the single- or double-helical form.

The Discovery of Nucleosomes   Early evidence that DNA is organized in nucleosomes came from observation of chromatin under the electron microscope by A. L. and D. E. Olins and by C. L. F. Woodcock. For their observations, chromatin was extracted from cells and treated with salt solutions at concentrations expected to reduce the electrostatic attraction of the histones for DNA, but not concentrated enough to remove the core histones. After this treatment the chromatin appeared in the electron microscope as strings of spherical beads (Fig. 13-11).

Other evidence came from enzymatic digestion of chromatin, pioneered by D. Hewish and L. Burgoyne, using a group of enzymes called *DNA endonucleases*. These enzymes break both sugar-phosphate backbone chains of a DNA molecule, causing the DNA to separate completely at the points of attack. (Information Box 13-3 presents details of the enzymes and the approach used in studying chromatin.) Hewish and Burgoyne found that brief digestion of extracted chromatin by an endonuclease cut DNA into pieces roughly 200 base pairs long or into larger pieces that were multiples of this length. More prolonged digestion shortened the DNA pieces to about 145 base pairs and released histone H1 from the chromatin. DNA of the regular 200– or 145–base-pair lengths was obtained only if the DNA digested by the endonucleases was associated with histones; DNA with no attached histones was cut into random lengths.

Further evidence was obtained by R. D. Kornberg, working with protein crosslinkers. (Protein crosslinkers are molecules that link together proteins occurring in close proximity in natural structures.) Kornberg found that prolonged treatment of chromatin with crosslinkers produced structures in which the H2A, H2B, H3, and H4 histones were locked by twos into

Figure 13-11 Nucleosomes (arrows) in chromatin isolated from an erythrocyte nucleus. (Courtesy of H. Zentgraf.)

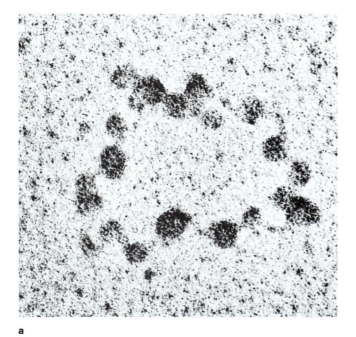

a

an *octamer*, indicating that an octamer of core histones is a fundamental structure in chromatin. H1 was not linked to the other histones by the crosslinkers, indicating that it is located at some distance from the remaining histones in nucleosomes.

Chemical analysis indicated that the histones occur in a 1H1:2H2A:2H2B:2H3:2H4 ratio in chromatin. In 1974, Kornberg combined all of this information into the current model for nucleosome structure, in which two turns of DNA wrap around an octamer of core histones. Kornberg reasoned that the 200 base-pair lengths released by brief digestion reflect a break made by the enzymes at a single exposed point on the linker between nucleosomes. More prolonged digestion slowly hydrolyzes the DNA associated with H1 in the nucleosome, releasing H1 and leaving a resistant piece about 145 base pairs long—the DNA protected from enzymatic attack by its tight association with the histone octamer. The core particle, consisting of two pairs of each of the core histones, and the location of a single H1 outside the core particle explained the crosslinking data and the 1:2:2:2:2 ratio observed for the histones in chromatin.

Several experiments immediately confirmed Kornberg's model and verified predictions developed from his hypothesis. One particularly interesting experiment by J. D. Griffith employed SV40, a mammalian virus in which the DNA circle of the virus is complexed with histones (Fig. 13-12a). Kornberg's model predicted that wrapping DNA through two turns around the nucleosome core particle compacts it to about 1/6.8 of its fully extended length (Fig. 13-12b). To test this prediction, Griffith first measured the circumference of the SV40 DNA circle with nucleosomes present and in the fully extended state with histones removed and nucleosomes absent. The two circumferences obtained, 1.48 for the extended state and 0.2 μm with nucleosomes present, are close to the 6.8:1 packing ratio predicted by Kornberg.

Other techniques, including X-ray diffraction and detailed electron microscopy, have confirmed that the

◀ **Figure 13-12** Evidence from SV40 minichromosomes supporting Kornberg's nucleosome model. **(a)** The circle of nucleosomes in the SV40 minichromosome. × 1,300,000. **(b)** The difference between DNA at its fully extended length (x), and DNA wrapped around nucleosomes (1/6.8x). (Micrograph courtesy of J.D. Griffith, from *The Molecular Biology of the Mammalian Genetic Apparatus* (P.O.P. Ts'o, ed.). Elsevier/North Holland, 1977.)

x

b

$\frac{1}{6.8} X$

c

DNA Endonucleases and Nucleosome Periodicity

The endonuclease digestion technique used by Hewish and Burgoyne to detect periodicity in chromatin is one of the basic methods used in chromatin research. The technique uses a group of enzymes that break sugar-phosphate bonds in both nucleotide chains of the DNA double helix. The breaks cause the DNA double helix to separate completely at points attacked by the enzymes.

Three different DNA endonucleases are commonly applied in this research: *DNase I, DNase II,* and *micrococcal nuclease* (also called *Staphylococcal nuclease*). In application of the technique, chromatin in intact form, with histones still attached, is exposed to one of the endonucleases. After digestion the enzyme is removed from the solution or inhibited, and the histones are separated from the DNA by exposure to high salt. Histone removal liberates the broken DNA pieces, which are run on an electrophoretic gel (see Appendix p. 796) to separate them by length. Typically the DNA pieces from such preparations separate into a series of regularly spaced bands called a "ladder," with the shortest DNA in the band at the bottom of the gel (see the figure in this box). The regular spacing of the bands in the ladder indicates that the DNA length of each heavier band is a regular multiple of the first band. If the first band contains DNA pieces 200 base pairs in length, for example, the next higher band contains pieces 400 base pairs in length, the next 600, and so on.

Micrograph courtesy of S. Spiker, reproduced with permission from *Ann. Rev. Plant. Physiol.* 36:235, © 1985 by Annual Reviews, Inc.

DNA is wound around the outside of the core particle in a two-turn coil and extends between nucleosomes as a linker. X-ray diffraction studies by A. Klug, P. J. G. Butler, and J. T. Finch and their colleagues, carried out with crystals cast from core particles retaining their tightly bound DNA, indicated the locations of individual histones within the core octamer (see Fig. 13-10b).

Changes in Nucleosomes During Transcription and Replication   One of the major unsolved problems of nucleosome structure concerns the relationship of nucleosomes to transcription and replication. What happens to nucleosomes as DNA unwinds for transcription and replication?

Changes in nucleosome structure undoubtedly occur as DNA enters these functions. One line of evidence supporting this conclusion comes from DNA endonuclease digestion. As genes go from the inactive to active state in transcription and replication, they become markedly more sensitive to digestion by the DNA endonucleases, indicating that nucleosomes open up in some way or are displaced.

Several mechanisms may combine their effects to alter nucleosomes in active chromatin. One is histone modification. The histones of active DNA, for example, are typically more highly acetylated than those of inactive genes. The acetylation, by reducing the electrostatic attraction of the histones for DNA, probably makes nucleosomes easier to unwind or displace from active sites. The tight attachment to DNA of non-histone proteins regulating genes probably also accounts for some of the nucleosome displacement noted in active genes. Another source of nucleosome unwinding or displacement may be through activity of enzyme complexes involved in transcription and replication. Once attached to the DNA and active in catalyzing transcription or replication, these complexes probably bind strongly enough to displace or unfold nucleosomes and open the DNA into more extended form. (Further details of nucleosome alteration during transcription and replication, including indications that nucleosomes are displaced rather than unfolding extensively during these processes, are taken up in the chapters on these topics—Chapters 14, 15, 16, and 23.)

ORGANIZATION OF CHROMATIN IN EUKARYOTIC NUCLEI

Nucleosomes in Chromatin Fibers

Nucleosomes wind or fold into thicker structures to form the chromatin fibers of eukaryotic nuclei. In nuclei that have been fixed and sectioned for electron microscopy (see Appendix p. 790), chromatin fibers appear as masses of irregularly folded fibers averaging about 10 nm in diameter, with no nucleosomes visible (Fig. 13-13a). If isolated at salt concentrations typical of the cell interior, chromatin fibers appear as more or less smooth fibers about 25 to 30 nm in diameter, containing tightly packed nucleosomes (Fig. 13-13b). Why isolated chromatin fibers appear two to three times thicker than those in fixed and sectioned nuclei, and which diameter is closest to the true diameter of chromatin fibers, are still unanswered questions.

Many attempts have been made to work out the patterns in which nucleosomes fold or wind into chromatin fibers. The most successful of these was the research of J. T. Finch and A. Klug and their colleagues, who used X-ray diffraction and other techniques to study isolated chromatin. From their results, Finch and Klug proposed that nucleosomes wind into a regular coil, the *solenoid*, to form chromatin fibers (Fig. 13-14). The proposed solenoid contains six to eight nucleosomes per turn, with a total outside diameter of 34 nm. A 10- to 11-nm periodicity detected in chromatin fibers by X-ray diffraction, according to Finch and Klug, reflects the length of one complete turn of the solenoid coil.

Maintenance of the solenoid is considered to depend on histone H1. If H1 is removed, isolated chromatin appears in the extended beads-on-a-string form or as irregular fibers with neither nucleosomes nor higher-

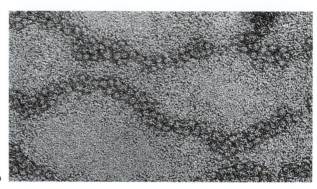

Figure 13-13 Chromatin fibers. **(a)** Chromatin fibers in a section, in which they appear about 10 nm in diameter. Typically no evidence of nucleosomes can be seen in sectioned chromatin. × 98,000. (Courtesy of J. S. Kaye; reproduced from *J. Cell Biol.* 31:159 [1966], by copyright permission of the Rockefeller University Press.) **(b)** Chromatin fibers isolated from mouse cells, in which nucleosomes appear to be wound or folded into superstructures 25 to 30 nm in diameter. (Courtesy of J. B. Rattner; reproduced from *J. Cell Biol.* 81:453 [1979], by copyright permission of the Rockefeller University Press.)

order structures visible. The H1 molecules must be complete, with their C- and N-terminal arms (see Fig. 13-9) intact, for higher-order structures to be present. For these reasons the arms of H1 molecules are believed to extend and set up linkages between nucleosomes that hold them in the solenoid.

Because the packing of nucleosomes in chromatin fibers probably prevents access to DNA by the enzymes of transcription and replication, the fibers probably unwind for these activities to take place. And, as has been noted, the increased sensitivity of active chromatin to DNA endonuclease digestion indicates that even nucleosomes probably unfold or disassemble

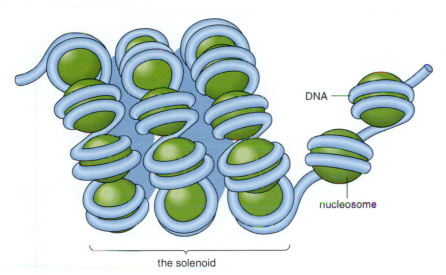

Figure 13-14 The nucleosome coil or solenoid proposed by J. T. Finch and A. Klug for the structure of chromatin fibers.

DNA

nucleosome

the solenoid

from DNA as it enters transcription or replication. All of this points to the conclusion that chromatin in intact nuclei is highly dynamic, with different folding conformations that reflect its activity.

The coiling of DNA around nucleosomes and the further winding of nucleosomes into chromatin fibers greatly compact the DNA of eukaryotic nuclei. A human nucleus, for example, contains in total about a meter of DNA. Winding the DNA into nucleosomes and chromatin fibers is estimated to shorten this length by at least a factor of 10,000. This total compaction makes accommodation of the DNA complement in the cramped volume of the cell nucleus feasible and, no doubt, protects the DNA from chemical and mechanical damage. DNA packing and protection from enzymatic attack may in fact be the primary functions of nucleosome and solenoid structures.

Chromatin in Eukaryotic Nuclei

In interphase nuclei, chromatin fibers are typically distributed between more open regions in which the fibers are loosely packed (known as *euchromatin,* from *eu* = typical) and regions of dense, tight packing (known as *heterochromatin,* from *hetero* = different; see Figs. 1-7 and 13-15). Frequently heterochromatin is located in a layer just inside the nuclear envelope and surrounding the nucleolus (as in Fig. 13-15). Transcription appears to be confined to euchromatin and the outer margins of heterochromatin; the DNA within heterochromatin masses appears to be completely inactive.

Packing of chromatin into heterochromatin has been correlated with inhibition of genetic activity by several methods. For example, in cells exposed to tritiated uridine, a radioactive substance used by cells to

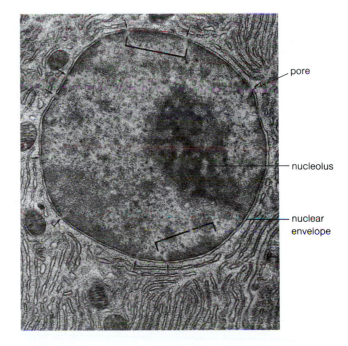

pore

nucleolus

nuclear envelope

Figure 13-15 Euchromatin and heterochromatin in eukaryotic nuclei. The nucleus is packed with chromatin fibers in both extended form as euchromatin and condensed form as heterochromatin (brackets). Typically masses of heterochromatin are concentrated in a layer just inside the nuclear envelope (NE), as they are in this nucleus. The heterochromatin layer is interrupted in the vicinity of pore complexes (arrows) in the nuclear envelope. × 10,000. (Courtesy of D. W. Fawcett and W. B. Saunders, Inc.)

synthesize RNA, label is incorporated only in euchromatin and the borders of the heterochromatin masses. No incorporation takes place within the masses of heterochromatin. Another line of evidence comes from an examination of the *Barr body*, a large block of heterochromatin that appears in the nuclei of mammalian females, including humans. The Barr body is formed from one of the two X chromosomes, which condenses into a tightly packed mass. Genetic tests show that most of the genes in the heterochromatic X chromosome are inactive.

Two types of heterochromatin are recognized. Chromatin that remains tightly packed throughout interphase and is never transcribed is called *constitutive* heterochromatin. DNA containing noncoding sequences repeated hundreds of thousands or even millions of times (see p. 532) forms the bulk of constitutive chromatin in eukaryotic cells. Packed chromatin containing coding sequences that may unfold and become active in transcription at some point in the cell cycle is termed *facultative heterochromatin*.

The *nucleolus* (see Figs. 13-15 and 15-10), suspended in the chromatin of eukaryotic nuclei, is a body formed by regions of one or more chromosomes that contain many repeats of genes encoding ribosomal RNAs (rRNAs). Within the nucleolus can be seen chromatin fibers and material known as *nucleolar fibrils* and *granules*. This material represents successive stages in the assembly of ribosomal subunits—the fibrils are early rRNA transcripts in association with processing proteins and some ribosomal proteins; the granules are late stages in which processed rRNA types are combined with ribosomal proteins in nearly complete ribosomal subunits. The shape and size of the nucleolus varies depending on the rate at which ribosomal subunits are being assembled. (Details of nucleolar structure, rRNA transcription, and ribosomal subunit assembly are presented in Chapter 15.)

Two major structures—the *nuclear envelope* and the *nuclear matrix*—may maintain chromatin in position inside the nucleus. Chromatin fibers frequently appear to be attached to the inner surface of the nuclear envelope, particularly at the margins of the pore complexes. These connections may keep the chromatin fibers of different chromosomes in separate regions of the nucleus. The connections probably involve linkages between chromatin fibers and the lamins, which form major cytoskeletal elements reinforcing the inner surface of the nuclear envelope (see below). Attachments to the nuclear envelope also appear at some division stages, as in the early stages of meiosis, when the chromosomes are attached to the nuclear envelope at their tips (see p. 730).

Chromatin may also be attached to the nuclear matrix, a supportive network believed by many investigators to extend throughout the nuclear interior. The degree to which the nuclear matrix organizes the chromatin remains controversial, because the fibers of the matrix become visible only in cells that have been exposed to relatively drastic treatments. Usually these include exposure to a detergent, which extracts lipids and soluble proteins, enzymes that digest DNA and RNA, and a concentrated salt solution. The combined treatment, originally developed by R. Berezney and D. S. Coffey, leaves a residual, insoluble network that outlines the nucleoplasm and nuclear envelope and, to some degree, the cytoplasm (Fig. 13-16). The portion that remains inside the nucleus, excluding the residual material that outlines the nuclear envelope, is the nuclear matrix.

One of the most interesting things about the nuclear matrix is a group of DNA sequences that resist the digestion and extraction procedure and remain associated with the matrix fibers. These sequences include a high proportion of genes that are active in RNA transcription. When replicating nuclei are used for the preparations, the residual matrix contains a high proportion of sequences active in replication. These findings suggest that the enzyme complexes replicating and transcribing DNA are preferentially associated with the nuclear matrix and that the matrix may organize the enzymes into assemblies that increase the efficiency of replication and transcription.

Attachments to the nuclear matrix are also thought by some investigators to organize the chromatin in looplike segments called *domains*. According to this idea, each long DNA molecule of a chromosome is attached to the matrix at intervals of about 50,000 to 150,000 base pairs, forming loops that extend from the matrix attachments. Within a domain loop the DNA is topologically restrained—that is, it is held so that supercoils cannot unwind if the DNA on either side of the loop is broken. The domains are believed to organize the chromatin into segments that act as units in DNA replication and perhaps transcription; during the folding and packing that convert chromatin into the short, rodlike chromosomes appearing during cell division the domains are believed to fold together through collection of the loop bases on a central chromosome axis or *scaffold* (see also p. 696) formed by elements of the matrix.

There is actually little evidence directly supporting the existence of domains. One line is taken from the appearance of chromosomes from dividing cells that have been treated to extract most of their proteins. Such chromosomes appear as a complex series of DNA loops extending from a central "scaffold" of residual protein (see Fig. 24-9). Another line comes from chromatin that has been digested very briefly by DNA endonucleases. In these preparations the DNA

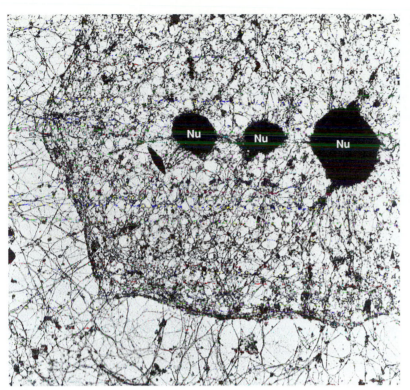

Figure 13-16 Nuclear matrix (NM) prepared by extraction of a cultured human cell by detergents, deoxyribonuclease, ribonuclease, and salts at high concentrations. The positions occupied by nuclear pore complexes before extraction are marked by ring-shaped deposits (arrows). CM, cytoplasmic matrix or cytoskeleton; Nu, nucleolus. × 11,000. (Courtesy of D. G. Capco, from *Cell* 29:847 [1982]. Copyright Cell Press.)

is initially released in lengths approximating those of the proposed domains. Often the ends of the lengths released contain sequences rich in A-T base pairs; the same types of sequences, called *MARs* (for *m*atrix *at*tachment *r*egions) are noted in DNA segments that remain attached to elements of the nuclear matrix.

It has proved difficult to determine how much of the nuclear matrix shown in Figure 13-16 is real and how much is an artifact produced by precipitation of proteins with other functions by the preparation techniques. Proteins that might be regular parts of the nuclear matrix have been difficult to identify; only topoisomerase II and *Rap1*, a protein binding the ends (called *telomeres*; see p. 699) of chromosomes, are widely accepted as parts of the matrix. As of yet, the proteins forming the supportive network of the matrix have not been determined.

In spite of these difficulties, it is likely that the nuclear interior is actually organized by a matrix or "nucleoskeleton" of some kind. One line of evidence comes from experiments with DNA hybridization. When radioactive DNA sequences of an individual chromosome are hybridized with the DNA of intact nuclei by the *in situ* technique (see p. 439), radioactivity is confined to a discrete region in the nucleus, with little evidence of overlap or intermingling of material from different chromosomes. Another line comes from the experiments of J. B. Lawrence and D. Spector and their colleagues, who found that in intact cells, processing reactions converting initial RNA transcripts to

finished form also occur in discrete, localized regions in the nuclear interior. Others found that DNA replication also occurs at localized sites within the nucleus, as if the DNA threads through these locations as it replicates. These localizations would be unexpected if the nuclear interior is a completely open, unstructured solution in which the chromatin is able to move freely.

THE BOUNDARY OF THE EUKARYOTIC NUCLEUS: THE NUCLEAR ENVELOPE

In eukaryotic cells the nucleus is surrounded and separated from the cytoplasm by the nuclear envelope. This membrane system also separates the major synthetic mechanisms of eukaryotic cells: RNA transcription and DNA replication is confined inside the nucleus, and protein synthesis in the cytoplasm.

Nuclear Envelope Structure

The nuclear envelope consists of three primary structural elements—the *outer membrane*, the *inner membrane*, and the *pore complexes* (see Figs. 13-15 and 13-17). The outer membrane, which faces the cytoplasm, is covered on its cytoplasmic side with ribosomes. At scattered points this membrane makes connections with the ER (see Fig. 20-3). The inner membrane, which faces the nucleoplasm, bears no ribosomes. In most

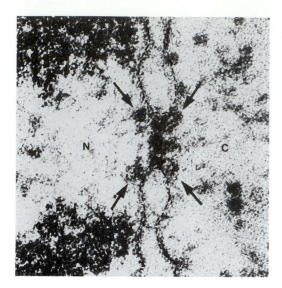

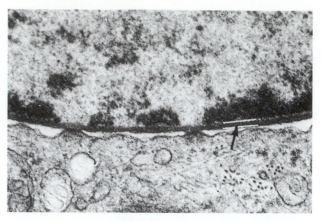

a

Figure 13-17 A segment of the nuclear envelope of an onion root tip cell including a pore complex in cross section. The outer and inner nuclear membranes are traced by the dashed lines. The annular material of the pore complex fills in the pore and extends into both the nucleoplasm (N) and the cytoplasm (C). The spherical subunits of the annular material are clearly visible (arrows). *N*, nucleoplasm; *C*, cytoplasm. × 234,000. (Courtesy of W. W. Franke, from *Phil. Trans. R. Soc. London B* 268:67 [1974].)

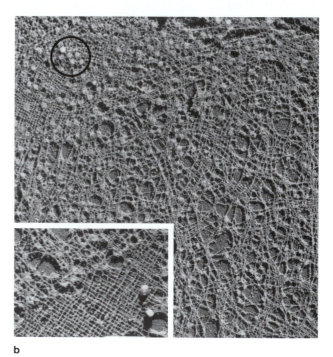

b

cells this membrane is lined on its inner surface with the lamins, intermediate filament proteins that may reinforce the nuclear envelope (Fig. 13-18a). The lamins appear in preparations isolated from *Xenopus* oocytes as a layer of crosshatched filaments on the surface of the inner nuclear membrane facing the nucleoplasm (Fig. 13-18b). In mammalian cells the lamin layer is a less regular meshwork about 10 nm in thickness.

The outer and inner nuclear membranes enclose a narrow space between them, the *perinuclear compartment*. The perinuclear compartment, which is about 40 nm wide, surrounds the nucleus and, through connections made by the outer nuclear membrane with the ER, is continuous with the space enclosed in the ER cisternae. Proteins synthesized by the ribosomes on the outer nuclear membrane can enter the perinuclear compartment and eventually make their way to the ER cisternae and other destinations in the cellular secretion pathway (for details, see Chapter 20).

At many points the nuclear envelope is perforated by *pores*, circular openings about 80 to 90 nm in diameter (see Fig. 13-17). At the margins of a pore the outer and inner membranes meet and become continuous. The pores are filled in by a dense, nonmembranous mass known as the *annulus* or *annular material*, which sets up a barrier to free diffusion of larger mol-

Figure 13-18 The nuclear lamina and the lamins.
(a) The nuclear lamina (arrow) lining the inner membrane of the nuclear envelope. The lamina is a fibrous layer of lamins, a group of proteins related to the intermediate fibers of the cytoskeleton. (Courtesy of D. W. Fawcett.)
(b) The network of lamin fibers underlying the nuclear envelope of a *Xenopus* oocyte, as seen in an isolated and shadowed preparation. The inset shows a segment of the network in which the lamin fibers are arranged in a regular mesh. Main figure × 10,000; inset × 18,000. (Courtesy of U. Aebi; reprinted by permission from *Nature* 323: 560, copyright 1986 Macmillan Magazines Ltd.)

ecules such as proteins. A pore is therefore not simply an open window between the nucleus and cytoplasm. The combination of a pore and annulus is the *pore complex*.

The annulus is roughly doughnut-shaped and extends beyond the membrane surfaces, curling around the pore margins on both the interior and exterior of the nuclear envelope like a grommet. When viewed face on, the annulus appears as a radially symmetrical, eight-sided structure about 120 nm in diameter, with eight spherical subunits spaced around its margin (Figs. 13-19 and 13-20). The spherical subunits appear to be connected to the center of the annulus by eight spokelike arms. Studies following the transport of ribonucleoproteins through pore complexes (see below) show material passing through the center of the pore complex, indicating that there is an orifice or duct at the pore center. Fibers extend from the margins of the pore complexes into both the nucleoplasm and cytoplasm. On the nuclear side the fibers may join into a basketlike structure. The fibers on either side are believed to provide "docking" sites to which molecules bind before transport through the pores.

The number of pore complexes varies with the activity of the nucleus in transcription. Most cells in plants and animals have about 15 to 20 pore complexes per square micrometer of nuclear envelope surface, totaling several thousand per nucleus. Highly active cells, such as developing oocytes, may have as many as 70 pore complexes per square micrometer and approach 50 million per nucleus. At the upper limit, pore complexes may take up as much as 30% of the total nuclear envelope surface.

Molecules of the Nuclear Envelope   Nuclear envelopes are difficult to isolate without contamination from other cellular structures. As a result, different methods yield widely varying contents of lipid, carbohydrate, and protein molecules. Some substances, however, are common to almost all preparations and are generally agreed upon as part of the nuclear envelope.

Essentially all preparations include the lamins as major envelope proteins. The lipid content of the nuclear envelope membranes clearly resembles that of the ER. Many enzymes, and other proteins and glycoproteins, also seem to be common to both the nuclear envelope and ER.

The molecular similarities between the nuclear envelope and the ER are not surprising, because of the direct connections frequently observed between the two systems. Further, the nuclear envelope is apparently reconstructed from elements of the ER when cell division is complete. During division in higher eukaryotes the nuclear envelope breaks down; the pore complexes disappear and the remaining membranes move into the cytoplasm and become indistinguishable from the ER. As division ends, nuclear envelopes reassemble from ER vesicles that surround the chromosomes and gradually fuse together. As membrane fusion progresses, pore complexes appear in the developing nuclear envelopes. (For further details of the nuclear envelope cycle during cell division see Chapter 24.)

Some enzymes, proteins, and glycoproteins appear to be unique to the nuclear envelope. Many of these molecules are probably parts of the pore complexes. Among these is an Mg^{2+}-dependent enzyme resembling nonmuscle myosins (see p. 316) that can break down both ATP and GTP. This enzyme, which is probably located in the pore complexes, has been linked to mechanisms transporting RNA from nucleus to cytoplasm (see below).

G. Blobel and L. Gerace and their coworkers and others have used fluorescent antibodies to identify several proteins of nuclear pore complexes. In this technique, antibodies are developed against proteins purified from isolated nuclear envelopes. The antibodies are attached to a fluorescent dye and reacted with cells to determine which bind to pore complexes. A group of 8 to 12 glycoproteins ranging in molecular weight from 45,000 to 210,000 has been provisionally identified as nuclear pore proteins by this method. Most of these glycoproteins, called *nucleoporins*, have an internal series of repeated amino acids and an unusual distribution of many single acetylglucosamine sugar units as their carbohydrate component. Many of the antibodies reacting with mammalian pore complexes also bind to nuclear envelopes in yeast cells, indicating that the same or similar nucleoporins are widely distributed among eukaryotes.

One of the nucleoporins found in mammals, *gp62*, is most abundant. Antibodies against gp62 block nuclear transport, indicating that this glycoprotein is central to pore complex function. Other proteins identified with nuclear pores include gp210, a glycoprotein with a transmembrane segment that may link pore complexes to the nuclear membranes. There are probably many nuclear pore proteins remaining to be identified: The total mass of the pore complex, estimated at about 120,000,000 daltons, or about 30 times the mass of a ribosome, has room for many more.

The Lamins   The lamins were identified as a form of intermediate filaments by sequencing studies and research with antibodies carried out by M. W. Kirschner, F. D. McKeon, G. Blobel, M. Osborn, and K. Weber. These tests revealed that lamins contain the internal arrangement of amino acids typical of intermediate filament proteins. Like other intermediate filament proteins, the lamins have globular end regions and a rodlike central portion (see p. 343 and Fig. 12-5).

Three different lamin proteins, *lamins A*, *B*, and *C*, are found in mammals (see Table 12-1). Lamin B appears early in development and occurs in all mammalian cells; lamins A and C, which are closely related (identical in their first 566 amino acids), are added as

Figure 13-19 Isolated pore complexes (circles) prepared for electron microscopy by negative staining. × 105,000. (Courtesy of P. N. T. Unwin; reproduced from *J. Cell Biol.* 93:63 [1982], by copyright permission of the Rockefeller University Press.)

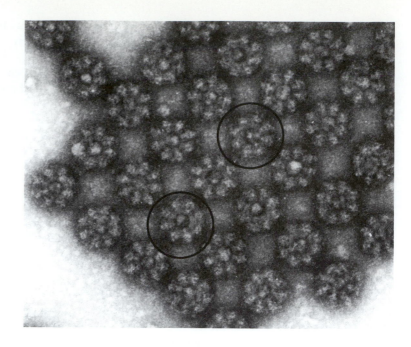

Figure 13-20 Possible structure of a nuclear pore complex. The eight-sided structure consists of an outer circle of spherical subunits that occur on both the outer and inner sides of the nuclear envelope. Fibers about 3 to 5 nm in diameter can sometimes be seen to extend outward from the subunits for some distance into both the nucleoplasm and the cytoplasm. On the nuclear side the fibers may join into a basketlike structure. The subunits are connected by radial spokes to a central structure containing an aperture that passively admits ions and molecules up to diameters of 9 to 10 nanometers. Larger complexes such as ribosomal subunits and other ribonucleoprotein complexes are moved through the aperture by an active mechanism that uses ATP or GTP hydrolysis as an energy source.

▼

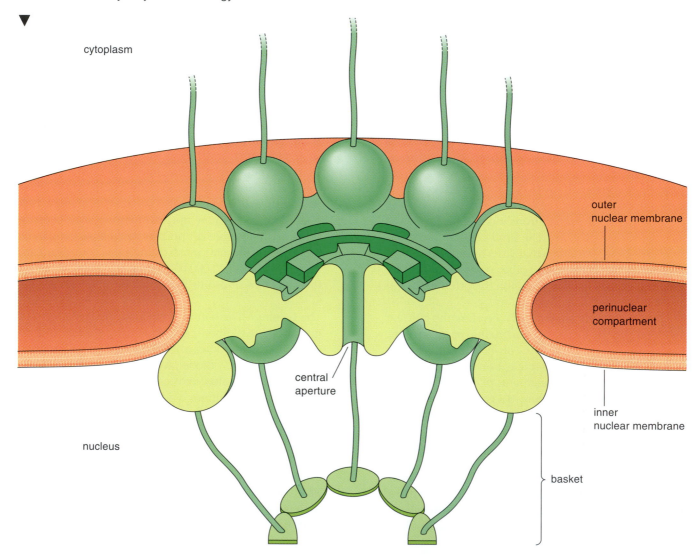

cells differentiate into mature forms. R. Forsner and L. Gerace found several integral membrane proteins of the inner nuclear membrane, which they term *LAPs* (for *l*amin *a*ssociated *p*roteins), that bind each of the three mammalian lamins.

Birds have two lamins equivalent to mammalian lamins A and B. Other vertebrates and invertebrates have one or two types of lamins. Although lamins also occur in plants, fungi, and protists, the number of lamins in these organisms and their relationships to animal lamins remain unknown.

Lamin fibers break down and reassemble on the same cycle as the nuclear envelope during cell division. The cycle of lamin breakdown and reassembly has been linked by Gerace and his coworkers to phosphorylation. Just before the lamins disassemble at the beginning of mitosis, their level of phosphorylation increases by about four times. They remain highly phosphorylated until the nuclear envelope reappears at the close of division, when the extra phosphate groups are removed and the lamins reassemble into fibers on the inner surface of the nuclear envelope. This phosphorylation-dependent pattern of assembly and disassembly has been successfully duplicated in the test tube.

The lamins are believed to provide a skeletal support that reinforces the nuclear envelope. They may also provide some of the linkages holding the nucleus in position in the cytoplasmic intermediate filament network—both vimentin- and desmin-based intermediate filaments can bind directly to lamin B. Linkages to cytoplasmic intermediate filaments may be limited to the nuclear pores, the only locations where lamins are known to be exposed directly to the cytoplasm.

Lamins probably also serve as linkers between the nuclear envelope and chromatin. The use of chemical crosslinkers as "nearest neighbor" probes typically produces crosslinks between DNA and the lamins. There is a molecular basis for this linkage; the central rod domain of both the A and C lamins has sites that can bind chromatin. These observations indicate that the lamins help maintain parts of the chromosomes in position in the nucleus.

Transport Through the Nuclear Envelope

Because fully functional, completely assembled ribosomes can be detected only in the cytoplasm (see p. 442), all cellular proteins must be synthesized in this location. All the proteins found in the nucleus, therefore, including histone and nonhistone proteins, and enzymes catalyzing functions such as transcription and replication, must therefore enter the nucleus from their place of synthesis in the cytoplasm. Some proteins, after combination with RNA into ribonucleoproteins of various kinds, pass again through pore complexes

in the opposite direction. Among these are the ribosomal proteins of the large and small ribosomal subunits. Many experiments show that this traffic in proteins and ribonucleoproteins between the nucleus and cytoplasm passes through and is regulated by the nuclear pore complexes.

Only certain large molecules normally pass through the pore complexes. Some proteins, such as albumin, penetrate into the nucleus only very slowly or not at all when injected into the cytoplasm. Others of equivalent or larger size, such as the lamins, enter rapidly. Significantly, the proteins penetrating the nucleus more rapidly are those that normally occur inside the nucleus in living cells. Large RNA-protein complexes, such as ribosomal subunits and mRNA-protein complexes, are selectively transported through the pores in the opposite direction, from nucleus to cytoplasm.

Research by Feldherr with *nucleoplasmin*, a molecule that facilitates nucleosome assembly, indicates that the ability of pore complexes to differentiate favored molecules depends on recognition of sites on the transported molecules. Nucleoplasmin is a large, 165,000-dalton protein consisting of five identical polypeptides, each with an extended tail region. Removal of the tails by proteinase digestion completely inhibits selective transport of nucleoplasmin into the nucleus, even though the total molecular weight of the molecule is reduced by more than 100,000 by the removal.

Research following the penetration of fragments of nuclear proteins into the nucleus shows that the sites recognized by pore complexes consist of amino acid sequences that act as "signals" for routing through the pores. The signals are evidently recognized and bound by receptors in the pore complex. After recognition and binding, the proteins are transported through the pores into the nucleus, in a mechanism that requires ATP or GTP breakdown by the pore complex ATP/GTPase. (Further information on the signals routing proteins to the nucleus and other destinations inside and outside the cell is presented in Chapter 20.)

Several experiments have demonstrated that transport of large molecules through the pore complexes requires energy. For example, D. E. Schumm and her coworkers discovered that isolated rat liver nuclei can export ribonucleoprotein particles containing mRNA only if the isolation medium contains a source of ATP or GTP. Replacement of ATP or GTP by substances that resemble ATP or GTP but cannot be broken down as an energy source completely inhibits ribonucleoprotein transport. Inhibition of the myosin-like enzyme breaking down ATP or GTP in nuclear pore complexes also stops export of ribonucleoprotein from the nucleus.

While passage of larger molecules is regulated by the pore complexes, these structures appear to be freely open to diffusion of most ions and small molecules. Several experiments confirm that most ions can pass freely through pore complexes. The electrical resistance measured across the nuclear envelope is very low, indicating that charged particles such as ions can pass without hindrance between nucleus and cytoplasm. Free diffusion of small molecules has also been shown by experiments such as those of E. Kohen, who injected pigment molecules of different sizes into the cytoplasm of living cells. The rates at which molecules passed through the nuclear envelope were recorded by noting the time required for colors due to the pigments to appear in the nucleus. Molecules up to the size of monosaccharides were able to penetrate the nucleus essentially as fast as free diffusion.

There is some evidence, however, that not all ions penetrate freely into the nucleus. For example, Ca^{2+} can be detected in cells by several dyes that fluoresce brightly when they combine with the ion. When the cytoplasm fluoresces brightly in cells, indicating Ca^{2+} release by control mechanisms such as the $InsP_3$/DAG pathway (see p. 161), the nucleus remains dark. Thus the nuclear pores may be able to selectively exclude this ion.

Other experiments indicate that the upper limit for free movement through the pore complexes is a molecular diameter of about 9–10 nm. C. M. Feldherr found that plastic-coated gold particles up to but not exceeding diameters of 9 to 10 nm can pass readily through the nuclear envelope. Similar results were obtained by P. L. Paine and his coworkers using dextran molecules of different sizes. Larger molecules can clearly be seen to pass through an opening in the center of pore complexes (Fig. 13-21).

In some cells, such as developing oocytes, early embryonic cells, and tumor cells, the cytoplasm accumulates membrane deposits that appear as stacks of nuclear envelopes complete with pore complexes (see Fig. 13-24). The structure and possible functions of these cytoplasmic deposits, called *annulate lamellae*, are described in Supplement 13-2.

The nucleus of eukaryotic cells contains DNA, histones, and nonhistone proteins as major molecular constituents. These three elements combine to form the chromatin fibers of the nucleus. Most of the DNA in the nucleus winds into the B conformation; limited segments may also wind in reverse into Z-DNA. Hybrid DNA-RNA double helices approximate the A conformation or assume a conformation between the A and B forms.

The five histones are complexed with DNA primarily as structural proteins. Four histones, H2A, H2B, H3, and H4, combine by twos to form the core particle of nucleosomes, the basic structural units of chromatin. The remaining histone, H1, ties DNA to the nucleosome and possibly also sets up linkages that wind nucleosome chains into chromatin fibers. The nonhistone proteins serve primarily functional roles in the nucleus as molecules regulating gene activity and catalyzing transcription, replication, DNA repair, and recombination.

Chromatin is distributed between euchromatin, a dispersed, open network active in RNA transcription, and heterochromatin, densely packed masses that are transcriptionally inactive. Part of the euchromatin containing rRNA genes forms the nucleolus. The chro-

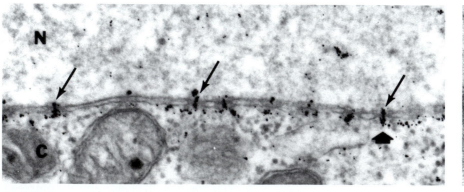
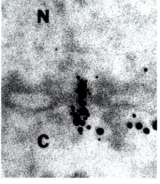

a   b

Figure 13-21 (a) Gold particles coated with a nuclear protein, nucleoplasmin, moving from the cytoplasm (C) to the nucleus (N) through pore complexes (arrows) of a *Xenopus* oocyte. × 60,000. (b) One of the pore complexes, with penetrating gold-nucleoplasmin particles, at higher magnification. × 200,000. (Courtesy of C. M. Feldherr, reproduced from *J. Cell Biol.* 99:2216 [1984], by copyright permission of the Rockefeller University Press.)

matin is probably organized and held in place by elements of the nuclear matrix, a supporting network of as yet unidentified proteins.

The nucleus is surrounded by the two concentric membranes of the nuclear envelope. The envelope is perforated by pore complexes, which freely admit most ions and small molecules but regulate transport of proteins and ribonucleoprotein complexes. The selective transport function allows some large molecules and molecular complexes to pass between the nucleus and cytoplasm and keeps others, such as the enzymes characteristic of the nucleus and cytoplasm,

restricted to their respective compartments. On an even larger scale the nuclear envelope obviously functions to separate the major organelles of the nucleus and the cytoplasm—or, as one investigator put it, "to keep the chromosomes in and the mitochondria out."

The nuclear envelope may also provide nucleoskeletal attachments that, with the nuclear matrix, organize the chromatin and enzyme complexes within the nucleus. Through the ribosomes attached to its outer membrane the nuclear envelope also acts as part of the cell's protein synthesis machinery. (The nuclear region of prokaryotes is described in Supplement 13-1.)

For Further Information

Chromatin, condensation during cell division, *Chs. 24 and 25*
DNA
  in mitochondria and chloroplasts, *Ch. 21*
  replication, *Ch. 23*
  sequence organization in the genome, *Ch. 18*
Heterochromatin and transcriptional regulation, *Ch. 17*
Histones and nonhistones in transcriptional regulation, *Ch. 17*
Nuclear envelope, cycle during cell division, *Ch. 24*
  in protein synthesis and secretion, *Ch. 20*
Nucleolus and rRNA synthesis, *Ch. 15*
Nucleosomes and replication, *Ch. 23*
  and transcription, *Chs. 14 and 15*
Nucleotides and nucleic acids, *Ch. 2*
Protein synthesis, *Ch. 16*
Regulation, transcriptional, *Ch. 17*
RNA transcription and processing, *Chs. 14 and 15*

Suggestions for Further Reading

Adams, R. L. P., Knowler, J. T., and Leader, D. P. 1986. *The Biochemistry of the Nucleic Acids*, 10th ed. New York: Chapman and Hall.

Bauer, W. R., Crick, F. H. C., and White, J. H. 1980. Supercoiled DNA. *Sci. Amer.* 243:118–133 (July).

Bradbury, E. M. 1992. Reversible histone modifications and the chromosome cell cycle. *Bioess.* 14:9–16.

Brennan, R. G., and Matthews, B. W. 1989. Structural basis of DNA-protein recognition. *Trends Biochem. Sci.* 14:286–290.

Dingwall, C., and Laskey, R. 1992. The nuclear membrane. *Science* 258:942–947.

Drilica, K., and Roviere-Yaniv, J. 1987. Histonelike proteins of bacteria. *Microbiol. Rev.* 51:301–319.

Felsenfeld, G. 1985. DNA. *Sci. Amer.* 253:118–133.

Fey, E. G., Bangs, P., Sparks, C., and Odgren, P. 1991. The nuclear matrix: Defining structure and functional roles. *Crit. Rev. Eukary. Gene Express.* 1:127–143.

Forbes, D. J. 1992. Structure and function of the nuclear pore. *Ann. Rev. Cell Biol.* 8:495–527.

Franke, W. W. 1987. Nuclear lamins and cytoplasmic intermediate filament proteins: A growing multigene family. *Cell* 48:3-4.

Freeman, L. A., and Garrard, W. T. 1992. DNA supercoiling in chromatin structure and gene expression. *Crit. Rev. Eukary. Gene Express.* 2:165–209.

Grunstein, M. 1992. Histones as regulators of genes. *Sci. Amer.* 267:68–74B (October).

Hanover, J. A. 1992. The nuclear pore: At the crossroads. *FASEB J.* 6:2288–2295.

Hernandez-Verdun, D. 1991. The nucleolus today. *J. Cell Sci.* 99:465–471.

Heslop-Harrison, J. S., and Bennett, M. D. 1990. Nuclear architecture in plants. *Trends Genet.* 6:401–405.

Hoffman, M. 1993. The cell's nucleus shapes up. *Science* 259:1257–1259.

van Holde, K. E. 1988. *Chromatin.* New York: Springer-Verlag.

Holliday, R. 1989. Untwisting Z-DNA. *Trends Genet.* 5:355–356.

Jackson, D. A. 1991. Structure-function relationships in eukaryotic nuclei. *Bioess.* 13:1–10.

Johnson, P. F., and McKnight, S. L. 1989. Eukaryotic transcriptional regulatory proteins. *Ann. Rev. Biochem.* 58:799–839.

Kessel, R. G. 1992. Annulate lamellae: A last frontier in cellular organelles. *Internat. Rev. Cytol.* 133:43–120.

Kornberg, R. D., and Klug, A. 1981. The nucleosome. *Sci. Amer.* 244:52–64 (February).

Panté, N., and Aebi, U. 1993. The nuclear pore complex. *J. Cell Biol.* 122:977–984.

Pettijohn, D. E. 1988. Histone-like proteins and bacterial chromosome structure. *J. Biolog. Chem.* 263:12793–12796.

Pienta, K. J., Getzenberg, R. H., and Coffey, D. S. 1991. Cell structure and DNA organization. *Crit. Rev. Eukary. Gene Express.* 1:355–385.

Rennie, J. 1993. DNA's new twists. *Sci. Amer.* 266:122-132 (March).

Saenger, W. 1984. *Principles of Nucleic Acid Structure.* New York: Springer-Verlag.

Schmid, M. B. 1988. Structure and function of the bacterial chromosome. *Trends Biochem. Sci.* 13:131–135.

Sigee, D. C. 1984. Structural DNA and genetically active DNA in dinoflagellate chromosomes. *Biosyst.* 16:203–210.

Silver, P. A. 1991. How proteins enter the nucleus. *Cell* 64: 489–497.

Spector, D. L. 1993. Macromolecular domains within the cell nucleus. *Ann. Rev. Cell Biol.* 9:265–315.

Turner, B. M. 1991. Histone acetylation and control of gene expression. *J. Cell Sci.* 99:13–20.

Verheijen, R., van Venrooij, W., and Ramaekers, F. 1988. The nuclear matrix: Structure and composition. *J. Cell Sci.* 90:11–36.

Watson, J. D. 1968. *The Double Helix.* New York: Atheneum.

Zlatanova, J. S., and van Holde, K. E. 1992. Chromatin loops and transcriptional regulation. *Crit. Rev. Eukary. Gene Express* 2:211–224.

Review Questions

1. What structures occur in eukaryotic nuclei? What is chromatin? Does chromatin occur in prokaryotes?

2. Outline the Watson-Crick model for DNA structure. What major sources of evidence did Watson and Crick use in development of their model? Why are A-T and G-C base pairs the only ones commonly expected in DNA?

3. What forces hold the DNA double helix together?

4. What is complementarity in DNA structure? What is the significance of complementarity for transcription and replication? How does the antiparallel arrangement of the two nucleotide chains of a DNA molecule affect transcription and replication?

5. Outline the structures of A- and Z-DNA. What is the biological significance of these conformations? What conditions promote Z-DNA stability? What evidence makes it seem likely that Z-DNA exists inside cells?

6. What is DNA supercoiling? What effects does supercoiling have on DNA structure and function?

7. What major kinds of histones occur in eukaryotic nuclei? What characteristics distinguish these proteins? What are histone variants?

8. What are nonhistone proteins? What functions are nonhistones believed to carry out in the nucleus?

9. Outline nucleosome structure. What major evidence was used by Kornberg in his development of the nucleosome model? What is the relationship between the DNA lengths obtained by DNA endonuclease digestion and nucleosome structure?

10. What evidence indicates that nucleosome structure is altered during transcription and replication?

11. How are nucleosomes believed to wind into chromatin fibers?

12. What is heterochromatin? Euchromatin? What is the relationship of these patterns of chromatin organization to transcription?

13. What is the nuclear matrix? What techniques are used to reveal the nuclear matrix? What are the possible functions of the nuclear matrix? What are chromatin domains? What is the possible relationship of these domains to the nuclear matrix?

14. Outline the structure of the nuclear envelope and pore complexes.

15. What functions are carried out by the nuclear envelope? What evidence indicates that the nuclear envelope is related to the ER?

16. What evidence indicates that most ions and small molecules can pass freely through the nuclear envelope? What evidence indicates that large molecules and complexes are transported selectively by the envelope?

17. What observations show that the selective transport of large molecules takes place through the pore complexes and requires energy?

18. What are lamins? How are lamins arranged in the nuclear envelope? What evidence indicates that the lamins are a type of intermediate filament protein?

19. What functions are the lamins believed to carry out in the nuclear envelope? What happens to the lamins during cell division?

Nuclear Structure in Prokaryotes

The nuclear material in prokaryotes, organized in the *nucleoid* (Fig. 13-22), is suspended directly in the cytoplasm, without separation from the cytoplasmic structures by a nuclear envelope or other membranes. No structural elements can be seen in the nucleoid except tightly packed and folded fibers. No nuclear background substance or matrix is identifiable, and no nucleolus is present. The term *prokaryote* reflects these differences and implies that the nuclear structure of bacteria and cyanobacteria is a primitive form that preceded the eukaryotic nucleus in evolution.

Bacterial Nucleoid Structure

Bacterial nucleoids appear as densely packed masses of fibers about 3 to 5 nm in diameter, not much larger than the 2-nm diameter of a DNA double helix. In some cells the fibers pack into regularly folded waves (as in Fig. 13-22a) of unknown significance. In many bacterial cells the edges of the nucleoid appear ragged or indefinite in outline, indicating that some fibers may extend from the margins into the surrounding cytoplasm. Genes active in transcription are probably located in these peripheral extensions of the nucleoid.

Because bacterial ribosomes attach to messenger RNAs and initiate protein synthesis while transcription is still in progress (see p. 487), some of the ribosome-filled regions outside the recognizable borders of the nucleoid probably include loops of nucleoid fibers that are active in simultaneous transcription and translation.

When isolated by methods designed to reduce breakage, the DNA of a bacterial nucleoid appears as a single, large molecule in the form of a closed circle. Other lines of evidence, including mapping by genetic crosses, also indicate that bacterial DNA exists as a closed circle. In *E. coli* the nucleoid circle contains 1360 μm of DNA, equivalent to about 4 million base pairs. Other bacteria have circles ranging from a minimum of about 250 μm (in mycoplasmas) to a maximum not far in excess of the *E. coli* circle, about 1500 μm. All this DNA, which occurs primarily or exclusively in the B conformation, is packed into cells only 1 to 2 μm long.

Many bacteria contain one or more smaller DNA circles called *plasmids* in addition to the nucleoid. These smaller circles, which in total contain about 0.1% to 5% as much DNA as the nucleoid, are dispersed throughout the bacterial cytoplasm. Plasmids

a

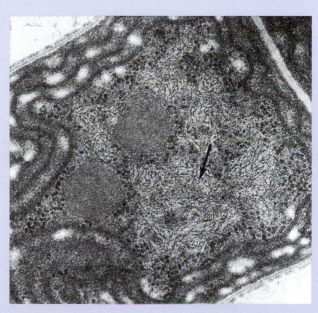

b

Figure 13-22 Prokaryotic nucleoids. **(a)** The nucleoid in a bacterium, *Bacillus megaterium*. The nuclear fibers are packed in a regularly folded pattern that produces a series of transverse bands or waves (arrows). × 130,000. (Courtesy of P. Giesbrecht, from *Inner Structures of Bacteria* [W. van Iterson, ed.]. New York: Van Nostrand Reinhold, copyright 1984.) **(b)** The nucleoid (arrow) in the cyanobacterium *Phormidium*. × 62,000. (Courtesy of M. R. Edwards.)

Table 13-3   Histonelike Proteins of *Escherichia coli*

Designation	Molecular Weight	Possible Function
H	28,000	May inhibit DNA replication; cross reacts with antibodies against eukaryotic histone H2A
HU	9,000	DNA folding; stimulation of transcription
IHF	10,000–11,000	Participates in DNA packing; stimulates recombination
FIS	11,200	May participate in DNA packing
H1 (or H-NS)	15,000	May participate in DNA packing
HLPI	17,000	Unknown

contain replication origins (see p. 680) and are duplicated and distributed to daughter cells along with the main DNA circle during division. Their inclusion as a separately replicating system has made plasmids highly useful as a vehicle for introducing recombinant DNA into bacterial clones. (See Supplement 14-1 for details of the cloning technique.) Many plasmids contain genes derived from the main DNA circle of the nucleoid or from viruses.

Bacterial Nucleoid Proteins

A number of different proteins can be detected in association with bacterial DNA. These include small, positively charged histonelike proteins, enzymes catalyzing transcription and replication, and regulatory proteins equivalent in function to some of the regulatory nonhistone proteins of eukaryotes.

Several types of histonelike proteins have been isolated from *E. coli* (Table 13-3). Two of these, *HU* and *IHF*, can wind bacterial DNA into compact, nucleosomelike structures in the test tube. HU contains basic and hydrophobic amino acids in about the same proportions as eukaryotic histones. However, no clustering into distinct basic and hydrophobic domains is discernible as in the histones, and there are no similarities in amino acid sequence between HU and histones. Whatever its structural similarities and differences to eukaryotic chromosomal proteins, HU appears to be generally distributed and highly conserved among different prokaryotes, including both bacteria and cyanobacteria. More distantly related forms of HU have also been detected in mitochondria and chloroplasts. Estimates of the total quantity of HU and other histonelike proteins in bacterial cells vary from enough to bind a minimum of 20% to as much as 100% of the bacterial DNA.

Although HU and IHF can wind bacterial DNA into nucleosomelike structures in the test tube, there is no evidence directly indicating that these proteins

have the same function inside living bacterial cells. Experiments with DNA endonucleases, for example, fail to reveal any regular periodicities in bacterial DNA similar to those reflecting nucleosome structure in eukaryotes. Despite the lack of direct evidence, HU in particular is widely believed to be a structural protein compacting DNA in bacteria and cyanobacteria.

Histonelike proteins with sequence similarities to eukaryotic histones have been discovered in a few bacteria. The sequence of the *AlgP* protein of *Psuedomonas* is 44% identical with sea urchin histone H1 over a stretch of 167 amino acids; the *Hcl* protein of *Chlamydia* is 35% identical with a 106–amino acid sequence of the sea urchin histone. It is unknown whether these histonelike proteins are products of an ancient ancestral gene shared by the prokaryotic and eukaryotic histones, introduction into the bacteria of genes derived from eukaryotes, or convergent evolution of unrelated proteins. Convergent evolution is a possibility in spite of the sequence similarities because much of the sequence identity between Algp, Hcl, and sea urchin H1 is in groups of arginine and lysine residues, which could have been assembled simply by selection for clusters of positively charged amino acids.

There is little doubt that the histonelike proteins or other elements increase the packing of bacterial DNA. Repulsion by the negatively charged phosphate groups spaced along the double-helical backbone limits the minimum volume into which naked DNA can be packed. The volume inside an entire *E. coli* cell is about 1000 times smaller than the minimum volume expected to be occupied by an uncomplexed *E. coli* DNA circle. This indicates that the histonelike proteins or other molecules with similar properties neutralize the phosphate groups and allow DNA to be packed into the nucleoid.

One interesting bacterial group, *Thermoplasma*, contains a histonelike protein, HT_a, that may be linked to the ability of these organisms to survive in hot springs with temperatures approaching 60°C. HT_a, a small, strongly basic protein containing 89 to 90 amino acids, evidently stabilizes the *Thermoplasma* DNA from heat denaturation at elevated temperatures. HT_a contains one segment with slight, but statistically significant sequence similarities to eukaryotic histones.

Nucleoid Structure in Cyanobacteria

Comparatively little is known about the structural organization of DNA in cyanobacteria. Different cyanobacteria contain DNA quantities ranging from about the same to as much as twice the amount per cell in *E. coli*. The fibers visible in the nucleoids of cyanobacteria, at about 5 to 7 nm (see Fig. 13-22b), are slightly larger in diameter than the fibers in bacterial nucleoids, suggesting that cyanobacterial DNA may have greater supercoiling or a more elaborate superstructure.

Although cyanobacterial DNA usually appears in linear form when isolated, T. M. Roberts and K. E. Koths detected a small number of DNA circles among the DNA isolated from the cyanobacterium *Agmenellum*. This finding suggests that cyanobacterial DNA may actually occur in the same circular form as bacterial DNA. The linear molecules usually isolated may reflect breakage of the longer cyanobacterial DNA circles during isolation.

Cyanobacteria contain at least two histonelike proteins. One, already noted, is closely related to the HU protein of bacteria. The second, a 16,000-dalton protein, may also interact with cyanobacterial DNA to form compact structures.

Dinoflagellates: Eukaryotes with Prokaryotic DNA Organization

One group of organisms has a nuclear system with both prokaryotic and eukaryotic features. These are the *dinoflagellates*, single-celled organisms that are variously classified as protists or algae or in a separate division of their own. They are primarily marine creatures that live by photosynthesis.

The cytoplasmic structures of dinoflagellates are typically eukaryotic and include mitochondria, chloroplasts, and other internal membrane systems characteristic of eukaryotes. In addition, a nuclear envelope separates the nuclear region from the cytoplasm. The organization of dinoflagellate DNA, however, is unique among eukaryotes. Suspended within the nucleus in different dinoflagellate species are from 4 to more than 200 small, rod-shaped bodies, the *chromosoids*, which each resemble a bacterial nucleoid (Fig. 13-23). Within the chromosoids are tightly packed masses of fibers 3–8 nanometers in diameter, folded into a pattern that resembles the ordered folds visible inside some bacterial nucleoids.

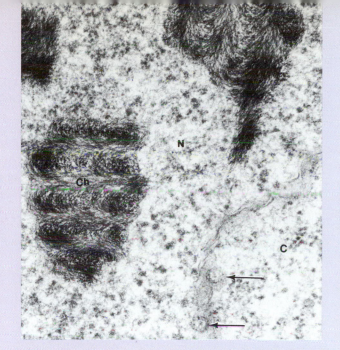

Figure 13-23 Nucleoidlike bodies, the chromosoids (C) in the nucleus of the dinoflagellate *Gyrodinium*. The nuclear envelope, complete with pore complexes (arrows) is also visible. N, nucleoplasm; C, cytoplasm. $\times$ 47,000. (Courtesy of D. F. Kubai and H. Ris, reproduced from *J. Cell Biol.* 40:508 [1969] by copyright permission of the Rockefeller University Press.)

No proteins equivalent to the histones of other eukaryotes can be detected in dinoflagellates. P. J. Rizzo and his coworkers found that dinoflagellate DNA is complexed with one or two types of histonelike proteins that differ in each species examined. No nucleosomelike bodies are produced by the interaction between these proteins and DNA.

Although the structural organization of dinoflagellate DNA appears to be prokaryotic, the DNA sequence organization shares at least one characteristic with eukaryotes. J. R. Allen and others discovered that a large proportion of dinoflagellate DNA consists of repeated sequences that apparently have no coding function. Sequences of this kind are limited to a few percent of the genome in prokaryotes but are common in eukaryotes.

Nuclear division in some dinoflagellates also exhibits some features that are prokaryotic and some that are eukaryotic. The part prokaryotic, part eukaryotic nature of the dinoflagellates makes these unusual organisms the subject of much evolutionary interest and speculation. Are the dinoflagellates a step in the main line of evolution of eukaryotes from prokaryotes, or are they an evolutionary offshoot that has persisted until today? Although no definite answer can be given, most evolutionists believe that present-day dinoflagellates are living fossils, survivors of a side branch of the evolutionary tree that diverged just before the appearance of fully eukaryotic cells.

Annulate Lamellae

The cytoplasm of some cell types contains masses of membranes of unknown function with many structural similarities to the nuclear envelope. These membranes, called *annulate lamellae (AL)*, consist of stacks of layered membranes, complete with pore complexes, that resemble segments of the nuclear envelope (Fig. 13-24). Often direct connections can be seen between AL and ER membranes (as in Fig. 13-24).

The pore complexes of annulate lamellae appear to be identical to those of the nuclear envelope. The pores even appear to be capable of transport—in Feldherr's experiments some substances, marked by heavy metals to make them visible in the electron microscope, could be seen to bind and enter the central aperture of the pore complexes of annulate lamellae as well as the nuclear envelope.

Cytochemical tests reveal that AL contain RNA in high concentrations. ATPase activity is also associated with the pore complexes of AL, as in the pores of the nuclear envelope. Antilamin antibodies fail to react with AL, indicating that these proteins are not present in the stacks.

AL may develop from the nuclear envelope or from the ER. The lamellar membranes first appear as flattened, closed sacs without pore complexes that extend from the nuclear envelope or ER. Pore complexes then differentiate rapidly in the sacs. Tests for the presence of RNA become positive as pore complexes appear.

AL are observed most frequently in animal oocytes, where they appear early in oocyte development and persist through fertilization and into the first few division cycles of the embryo. They also appear in the cytoplasm of developing sperm cells in some animals; however, they disappear as these cells reach maturity. AL have also been detected as transient structures in a wide variety of body cells in animals and in cultured plant cells, developing pollen cells, and some algae and protozoa. Significantly, many types of animal tumor cells develop AL. They may also be induced to appear in many cell types by exposure to such diverse stimuli as reduced temperatures, certain hormones or antibiotics, virus infections, antitubulin drugs such as colchicine, and cAMP. The variety of cell types in which they appear has led to the opinion that essentially any cell type may, under certain conditions, develop AL.

Many hypotheses have been proposed for the functions of AL, including transport or storage of ribonucleoprotein complexes originating in the nucleus. However, AL do not appear to be a vehicle for transferring ribonucleoprotein from the nucleus to the cytoplasm because RNA does not appear until after differentiation of pore complexes, which usually occurs in developing AL membranes located at some distance from the nucleus.

Storage seems a more likely possibility. According to this idea, ribonucleoprotein complexes, such as

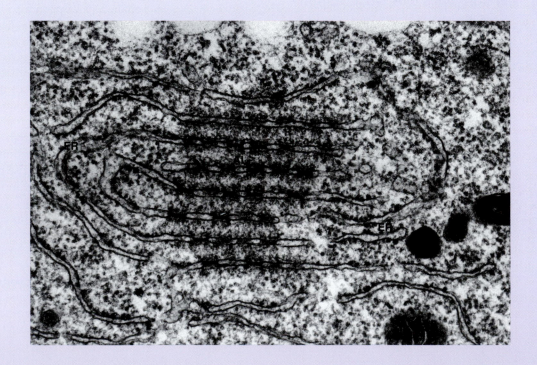

Figure 13-24 A stack of annulate lamellae (AL) in the cytoplasm of a *Xenopus* oocyte, showing direct connections to rough ER membranes (ER). Typical pore complexes are visible in the membranes (arrows). AL deposits may reach very large size in the cytoplasm. × 38,000. (Courtesy of J. P. Stafstrom, reproduced from *J. Cell Biol.* 98:699 [1984] by copyright permission of the Rockefeller University Press.)

particles containing mRNA in combination with proteins, are stored in the AL of oocytes and early embryonic cells by binding to the pores of these structures after they enter the cytoplasm. As AL disperse later in embryonic development, the mRNA is released to the cytoplasm in an active form. Their presence in tumor cells, according to this line of reasoning, is related to the fact that tumor cells typically undergo partial reversion to an embryonic state.

Another idea proposes that AL, as an addition or alternative to the storage function, may serve as reservoirs of membranes and pore complexes needed for rapid nuclear envelope assembly. This hypothesis is supported by the observation of an inverse relationship between AL and nuclear envelope membranes in some species. In these species, particularly in insects such as *Drosophila*, the cytoplasm of developing eggs contains large AL deposits. As rapid nuclear division takes place in these eggs, in which thousands of nuclei are produced within a few hours of fertilization, the deposits gradually disappear, possibly by conversion into nuclear envelopes.

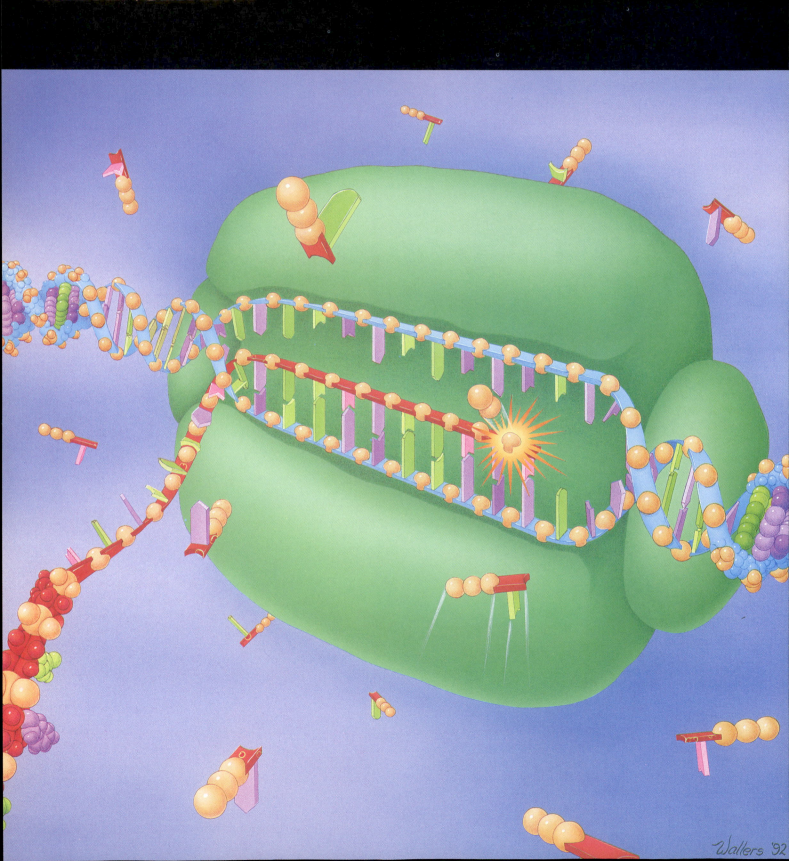

Transcription, the mechanism by which genes are copied into RNA, produces three major RNA types that interact in protein synthesis—messenger RNA (mRNA), ribosomal RNA (rRNA), and transfer RNA (tRNA). Messenger RNAs carry the encoded information required to make proteins of all types. Ribosomal RNAs form both structural and functional parts of ribosomes, the multienzyme complexes that use the information encoded in mRNAs to assemble amino acids into proteins. Transfer RNAs link to amino acids and match them with the coding sequences in mRNAs, working as a molecular dictionary that translates the nucleic acid code into the amino acid sequences of proteins.

Each major RNA type is transcribed in a preliminary form known as a *precursor*. The precursor, typically larger than the finished RNA product, is then *processed* by several biochemical steps. One step involves *clipping* and *splicing* reactions that remove surplus nucleotide sequences. Further processing may add nucleotides to one or both ends and chemically modify individual bases of the precursor. A group of small RNA molecules known as *small nuclear RNAs* (*snRNAs*) participates in reactions processing mRNAs and rRNAs into finished form. Proteins that form structural and functional parts of the finished product are added as processing takes place. The finished product, an RNA molecule linked to proteins in a ribonucleoprotein particle, moves through the nuclear envelope to the sites of protein synthesis in the cytoplasm.

This chapter describes eukaryotic mRNA transcription, the genes encoding the individual mRNA types, and their transcription and processing into finished RNA copies. Supplement 14-1 outlines techniques used for cloning and sequencing the DNA of genes. Transcription and processing of rRNA, tRNA, snRNAs, and another small RNA type, *small cytoplasmic RNAs* (*scRNAs*), are discussed in Chapter 15. Prokaryotic mRNA, tRNA, and rRNA genes and their transcription are described in Supplement 15-1. The regulation of RNA transcription is discussed in Chapter 17.

GENERAL FEATURES OF RNA STRUCTURE, TRANSCRIPTION, AND PROCESSING

RNA Structure

Most eukaryotic and prokaryotic RNAs are single nucleotide chains. However, many regions of RNA molecules wind into a variety of double-helical structures by folding and twisting back on themselves. Most common of these structures is a *hairpin* or *stem-loop* configuration in which a fold-back double helix is capped at its tip by an unpaired loop. (Figure 14-1 shows some of the primary types of fold-back structures.) Hairpins and other double-helical structures form in regions containing a symmetrically reversed sequence known as an *inverted repeat* (see Information Box 13-2). In addition to the usual A-U and G-C pairs, flexibility in the double helices and spatial rearrangements of the bases also allow G-U, G-A, A-C, A-A, G-G, and other nonstandard base pairs to form. The double-helical segments of RNA molecules wind into the A conformation (see p. 367 and Fig. 13-5).

The sequence of bases in an RNA molecule is its *primary structure*; the number and position of hairpins and other fold-back helices is its *secondary structure*. RNA molecules also fold into higher-level, three-dimensional *tertiary structures*. As yet, few tertiary structures have been deduced except for those of a few small RNA types such as 5S rRNAs and tRNAs (see Figs. 15-4 and 15-14).

Depending on the RNA type, secondary and tertiary structures may be more important to RNA functions than primary sequence. In many regions of ribosomal RNAs, for example, sequences can be altered extensively without affecting function as long as the pattern of hairpins and other secondary structures is preserved. The secondary and tertiary regions are particularly important as recognition sites for enzymatic, structural, and regulatory proteins active at all levels of RNA function from transcription, processing, and RNA transport through the mechanisms of protein synthesis.

RNA Transcription

RNA transcription is a polymerization reaction in which individual nucleotides link sequentially into a chain. The reaction, catalyzed by *RNA polymerases*, requires a DNA *template*, Mg^{2+} or Mn^{2+}, and the four RNA nucleoside triphosphates *adenosine triphosphate* (*ATP*), *guanosine triphosphate* (*GTP*), *cytidine triphosphate* (*CTP*), and *uridine triphosphate* (*UTP*). RNA nucleotides contain ribose rather than deoxyribose as a five-carbon sugar; the ATP used in RNA transcription is identical to the ATP synthesized in glycolysis and mitochondrial oxidations.

Figure 14-1 Some common secondary structures in RNA molecules. In the paired regions the RNA chain winds into a double helix in the A conformation.

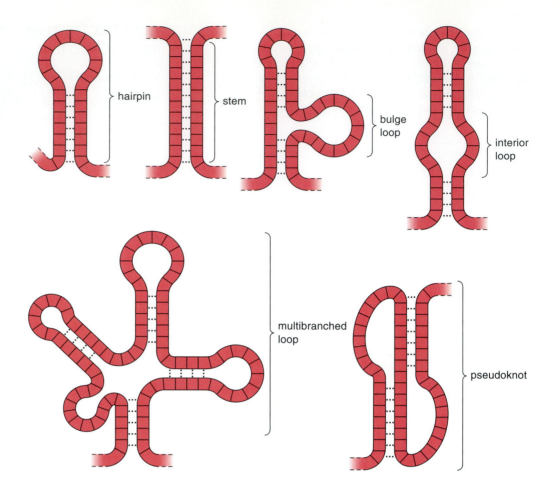

hairpin

stem

bulge loop

interior loop

multibranched loop

pseudoknot

A broad outline of transcription is shown in Figure 14-2. The mechanism starts as an RNA polymerase binds to the DNA template and recognizes the first base to be copied, which in Figure 14-2a is shown as guanine. According to base-pairing rules, the presence of guanine at this site causes RNA polymerase to bind CTP from among a pool of the four nucleoside triphosphates constantly colliding with the enzyme (Fig. 14-2b).

In response to binding CTP the enzyme undergoes a conformational change causing it to "read" the next base exposed on the DNA template, which in Figure 14-2 is shown as adenine. The presence of adenine at this site induces the enzyme to bind UTP from the nucleotides in the surrounding pool. Once bound to the enzyme, the UTP is held opposite the adenine on the DNA template, in a position favoring formation of the first linkage of the new RNA chain (Fig. 14-2c). RNA polymerase then catalyzes the linking reaction, in which the last two phosphates of the nucleotide most recently bound to the enzyme (UTP in Fig. 14-2c) are split off. The remaining phosphate binds to the 3'-carbon of the first nucleotide (Fig. 14-2d). The reaction creates a *phosphodiester linkage* between the 3'-carbon of the first sugar and the 5'-carbon of the second. Removing the terminal phosphates from UTP releases a large increment of free energy and so greatly favors formation of the linkage that the reaction is essentially irreversible.

In response to the formation of the first phosphodiester linkage the enzyme undergoes another conformational change, causing it to move to the next base exposed on the DNA template, shown as thymine in Figure 14-2. The enzyme then binds an ATP from the medium and catalyzes formation of the second phosphodiester linkage. This linkage, as before, is formed at the expense of two phosphate groups split off from the nucleotide most recently bound by the RNA polymerase. The polymerization continues, adding nucleotides one at a time into the growing chain until the enzyme reaches the end of the DNA template. At this point the newly synthesized RNA molecule and the enzyme are released from the DNA template, terminating transcription.

Figure 14-2 shows that the first nucleotide in the RNA chain retains all three of its phosphates linked to the 5' carbon of the sugar. These phosphates mark the beginning or *5' end* of a newly synthesized RNA molecule. The opposite, *3' end* is marked by the —OH group retained at the 3'-carbon of the last nucleotide bound into the RNA chain. Because a 5'-carbon marks the beginning and a 3'-OH marks the end of a newly synthesized RNA chain, transcription is said to proceed in *the 5' → 3' direction*.

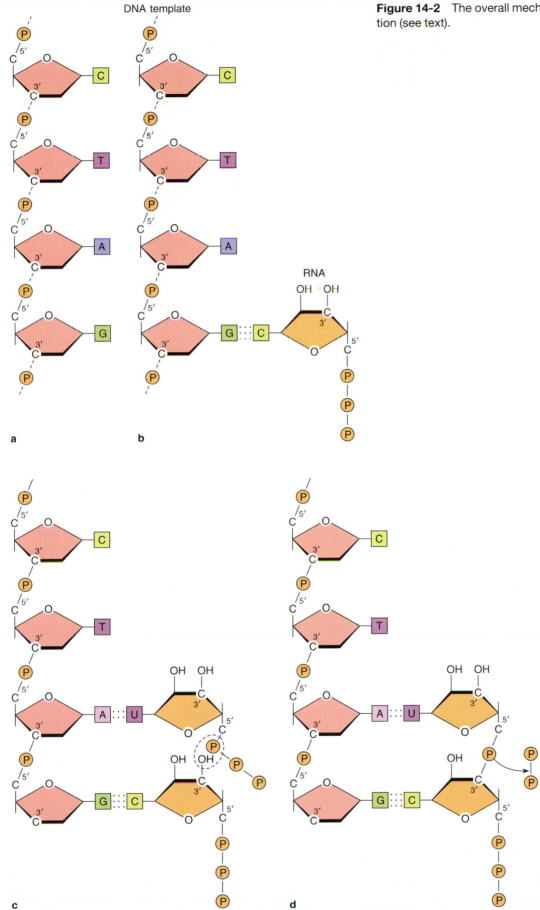

Figure 14-2 The overall mechanism of RNA transcription (see text).

DNA template

RNA

a

b

c

d

Transcription proceeds in three distinct phases: *initiation, elongation,* and *termination.* Initiation includes attachment of RNA polymerase to the DNA template and binding by the enzyme of the first nucleotide to be placed in the RNA molecule. Elongation involves the reactions in which RNA nucleotides are added sequentially according to the DNA template. In termination, RNA polymerization ends and the enzyme and completed RNA transcript are released from the DNA template.

A number of accessory proteins called *transcription factors,* although not directly part of RNA polymerase, are required for most efficient progress of transcription. In a sense, transcription factors act as accessory catalysts that speed individual steps of RNA synthesis.

Initiation   Random collisions between RNA polymerase molecules and DNA lead to initiation if they occur in a specific region of the DNA called the *promoter.* This region contains sequence elements that bind several transcription factors in combination with RNA polymerase and indicate the first base to be copied into an RNA transcript. The promoter also includes sequences involved in regulation of transcription.

Several transcription factors called *initiation factors* are necessary for RNA polymerases to recognize and bind tightly to the promoter. Without these factors an RNA polymerase may bind loosely to DNA sequences of essentially any type without initiating RNA transcription. Evidently the initiation factors bind first to the promoter, forming an active complex that fits one or more sites on the RNA polymerase enzyme. The enzyme then binds tightly to the initiation factors and the promoter, and the DNA unwinds in the promoter region. For mRNA genes in particular, a number of *regulatory proteins* participate in initiation by indicating which of the many protein-encoding genes are to be copied. These proteins recognize sequences located primarily outside, but near the promoter (details of these proteins and their activities are presented in Chapter 17).

Of the two DNA nucleotide chains fully exposed by the unwinding, only one contains the correct promoter sequences and acts as a template. The opposite, nontemplate chain is complementary to the promoter and template but does not contain encoded information. The location of the promoter sequences indicates which chain is the template by fixing the initiation complex to this side of the DNA double helix. The template is not always the same nucleotide chain of the DNA double helix; different genes may have their template chains on either side of the DNA helix. For a given gene, however, the nucleotide chain being copied remains the same; RNA polymerase does not switch back and forth between the two chains of the double helix within the boundaries of a gene.

Binding between RNA polymerase and the promoter is tight enough to protect the DNA in the region of contact from attack by enzymes or chemicals such as strong bases. This characteristic is used to identify and determine the length of the promoter through a technique called *DNA footprinting.* (Information Box 14-1 explains the technique.) The footprint made by RNA polymerase indicates that during initiation the enzyme binds tightly to a DNA segment about 75 base pairs long, extending from about 55 nucleotides in advance of and 20 nucleotides past the first nucleotide to be copied.

The final step in initiation takes place as RNA polymerase binds the first RNA nucleotide to be placed in the RNA transcript. This binding creates the 5' end of the new transcript.

Elongation and Termination   Once the first base is added, elongation begins and RNA nucleotides add sequentially until the polymerase reaches the end of the template. During each addition the enzyme: (1) binds a nucleoside triphosphate and matches it to the template, (2) splits two phosphate groups from the nucleoside triphosphate and forms the phosphodiester linkage, and (3) moves to the next template base to be copied. During the shift to elongation some of the initiation factors are released, and one or more *elongation factors* may add to the enzyme complex to speed steps in the sequential addition of bases.

Much interest centers on the almost perfect fidelity with which complementary bases are inserted into the growing RNA chain. Some of the accuracy depends on base pairing between the DNA template and the incoming RNA bases. However, research with *base analogs,* molecules that are similar to the nucleoside triphosphates, indicates that correct base pairing in transcription depends on conformational changes in the enzyme as well as base pairing. Some analogs closely resemble ATP, UTP, CTP, or GTP in structure but cannot form the usual Watson–Crick hydrogen bonds with DNA bases. Nevertheless, RNA polymerase substitutes the analogs correctly into the RNA transcript for the regular RNA bases. Evidently conformational changes occur in the active site in response to the template DNA base encountered by enzyme. The changes alter the active site so that binding by the correct complementary nucleotide is favored.

There are also indications that RNA polymerases may have the ability to *proofread*—that is, to go back and remove the most recently added bases if incorrect pairs are formed. The enzymatic activity necessary for proofreading (a 3' → 5' *exonuclease* activity that works in the direction opposite to transcription and removes the most recently added base or bases; see p. 668) was recently discovered to be associated with the RNA polymerase transcribing mRNA genes.

While elongation is in progress, the DNA template

Footprinting

Footprinting is widely used as a method to reveal sites on DNA bound by RNA polymerase enzymes and other DNA-binding proteins, including transcription factors and regulatory proteins. In the technique a DNA segment containing the region bound by a protein is produced in many copies by cloning or the polymerase chain reaction (see Supplement 14-1 for details of these techniques). The cloned DNA segments are usually labeled at their 5' ends. One sample of the DNA, unbound by protein, is treated with a DNA endonuclease or an alkali adjusted in concentration so that each DNA segment in the sample is broken at only one or a few sites. The breakage produces a number of fragments that are sorted out according to length by running them through an electrophoretic gel (see part **a** of the figure in this box). Autoradiography is used to detect the bands on the gel so that only fragments that start with the 5' end intact are detected (as in DNA sequencing techniques; see p. 423). The fragments are therefore sorted out by length from the beginning of the DNA sequence.

A second end-labeled sample of the DNA is then combined with the protein of interest and treated with the DNA endonuclease or alkali. The same collection of fragments is produced by the treatment except in regions where the DNA is protected from enzymatic or chemical attack by combination with the protein (see part **b** of the figure in this box). When run on the gel, the fragments sort out by length, but missing from the gel are fragments corresponding to the region protected by combination with the protein. The missing fragments correspond to a blank region in the gel in which no bands appear. This blank region is the "footprint" of the protein. Comparison of the gels made from the two samples allows the position of the missing bands and the region bound by the protein to be fixed. The footprint can be precisely located in a DNA sequence by comparing its sequence with sequencing gels (see p. 423) made from the same sample.

unwinds continuously just in front of the RNA polymerase complex and rewinds just behind it (Fig. 14-3). How the local DNA unwinding is accomplished remains unknown. The RNA polymerases may, as some investigators have suggested, include "unwindase" and "rewindase" capabilities as part of their active sites. Alternatively, DNA topoisomerases (see p. 373) may form part of the enzyme-factor complex carrying out RNA transcription. These topoisomerases unwind DNA by introducing breaks in one or both nucleotide chains and allowing the free ends to rotate around each other. Some evidence for this possibility is taken from the fact that DNA topoisomerases become concentrated in regions of chromosomes that are active in RNA transcription, and can be detected among the factors associated with RNA polymerases during transcription. In addition, S. J. Brill and his colleagues have shown that transcription stops in yeast mutants in which the topoisomerases thought to be involved in transcription are inactive.

As the new RNA molecule is transcribed, it winds temporarily with the template chain of the DNA into a short hybrid RNA-DNA helix of about 10–15 base pairs (see Fig. 14-3). Beyond this length the growing RNA chain unwinds from the DNA. The template DNA chain rewinds with the nontemplate chain into an intact DNA molecule as the RNA polymerase passes.

In prokaryotes, RNA transcription terminates in response to DNA sequences that signal the end of the region to be copied into RNA (see Supplement 15-1

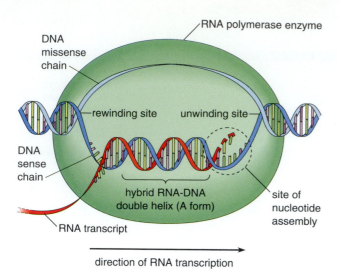

DNA
missense
chain

RNA polymerase enzyme

rewinding site   unwinding site

DNA
sense
chain

hybrid RNA-DNA
double helix (A form)

site of
nucleotide
assembly

RNA transcript

direction of RNA transcription

Figure 14-3 DNA unwinding and rewinding at the site of transcription. The transcript forms a hybrid RNA-DNA double helix with the DNA sense chain as part of the mechanism.

Table 14-1 Eukaryotic RNA Polymerases

RNA Polymerase Type	RNAs Transcribed	Ionic Requirement
I	28S, 18S, and 5.8S rRNAs	Mg^{2+}
II	mRNAs, most snRNAs	Mg^{2+}
III	tRNAs, 5S rRNAs, one snRNA, and scRNAs	Mn^{2+}

for details). Sequences signaling termination also appear in some eukaryotic genes, such as those encoding rRNAs. In eukaryotic mRNA genes, termination may be coupled to processing reactions rather than occurring in response to specific signal sequences in the DNA (see below). *Termination factors* contribute to separation of RNA polymerase and elongation factors from the template and release of the transcript in prokaryotes and possibly also in eukaryotes.

The RNA Polymerases RNA polymerases are among the largest and most complex proteins of living organisms. These enzymes are similar in overall structure in both prokaryotes and eukaryotes; each polymerase consists of 2 large polypeptide subunits and about 6 to 10 smaller ones. Typically the two largest polypeptides and a few of the smaller polypeptides form a *core enzyme* that is capable of transcribing essentially any DNA sequence, natural or artificial, into an RNA copy. The remaining smaller polypeptides restrict the transcribing capabilities of the enzyme to natural DNAs containing a promoter and, in eukaryotes, to certain gene types.

Only one type of RNA polymerase, which transcribes all RNA classes, occurs in prokaryotes (see Supplement 15-1). In contrast, eukaryotes have three types, designated *RNA polymerase I, II and III* (Table 14-1). The three eukaryotic polymerases share two major subunits of 220,000 and 140,000 daltons and three smaller subunits with molecular weights between about 4000 and 28,000. The remaining small subunits are evidently responsible for the differences in specificity between the three polymerases. The C-terminal end of the largest subunit extends as a tail domain containing as many as 50 repeats of a seven–

amino acid sequence. This tail figures importantly in interactions with proteins promoting the initiation of transcription. Mutations removing the tail from RNA polymerase II, for example, are lethal to yeast, *Drosophila*, and the mouse.

Eukaryotic RNA polymerase I, which transcribes all rRNAs except one type, 5S rRNA, is active in the nucleolus. RNA polymerase II transcribes mRNA genes and also most genes encoding the snRNAs that participate in RNA processing. Both RNA polymerase I and II initiate transcription from promoters that lie upstream, in 5'-flanking regions of their genes. RNA polymerase III transcribes 5S rRNAs, all tRNAs, and the remaining snRNAs and other RNAs not copied by RNA polymerase II. In contrast to the promoters recognized by RNA polymerases I and II, the promoters of most genes copied by RNA polymerase III lie inside the gene, within the sequence copied into the RNA transcript.

The three eukaryotic RNA polymerases are common to all eukaryotic organisms including animals, fungi, protozoa, and plants. Although some differences in total polypeptide number are detected in the RNA polymerases of the different major eukaryotic groups, the same division of labor between RNA polymerases I, II, and III is noted in all cases.

Transcription factors participate with all three eukaryotic polymerases in the reactions assembling RNAs. The factors have been best studied in the initiation phase of transcription, in which separate factors allow each polymerase type to recognize its promoters in DNA. For mRNA genes, initiation factors cooperate with regulatory proteins in indicating which genes are to be copied.

RNA Processing

All the major eukaryotic RNA types are transcribed in the form of precursors that are modified into mature form by processing reactions. The reactions—including clipping and splicing, addition of nucleotides to either end of a transcript, and chemical modification of bases—begin while transcription is still in progress and continue after assembly of the RNA precursor is complete.

Clipping is catalyzed by *RNA exonucleases* and *endonucleases* and, in some cases, by *ribozymes*—RNA molecules with the capacity to act as enzymes (see p. 86). RNA exonucleases nibble nucleotides one at a time from the 3' end of the precursors. RNA endonucleases make cuts at points within an RNA molecule. Ribozymes may catalyze either type of reaction. The positions at which nibbles or cuts are made are determined by sequence information or secondary structure, or a combination of both, within the RNA precursors. Once cuts are made to remove an internal segment, the free ends are rejoined into a continuous RNA molecule.

Addition of nucleotides, catalyzed by many specific enzymes, may occur at either the 5' or 3' ends of RNA precursors. Because the nucleotides are added after transcription, their sequence does not correspond to template sequences in the DNA.

Most chemical modifications change the RNA bases into other forms. (Figure 14-4 shows some of the more common modified bases.) Among the most common modifications is addition of methyl groups to U or C or A bases or to the ribose sugars. Other frequent modifications are removal of amino groups or addition of a hydroxymethyl (—CH_3OH) group. These and other less frequent chemical changes produce a wide variety of modified bases in finished RNA molecules. In some RNAs the ribose sugar is modified by addition of methyl or other chemical groups.

Base modifications are most frequent in tRNA and rRNA molecules. However, mRNAs also contain a few modified bases, particularly at the 5' and 3' ends. Base modifications have a significant effect on the three-dimensional conformations taken by RNA molecules. Undoubtedly they are also important as recognition signals in reactions between RNAs and other molecules, particularly proteins.

Proteins associate with the various RNA types as processing takes place. These proteins participate in the processing reactions, fold the RNAs into their finished three-dimensional form, and contribute to activity of the RNAs in processes such as protein synthesis. The proteins associated with an RNA type vary as the processing reactions proceed and the RNA takes up its functions in the cell.

TRANSCRIPTION AND PROCESSING OF MESSENGER RNA

mRNA transcription is the first step in the sequence of events leading to protein synthesis and ultimately to determination of cell structure and function. The precursors from which mRNA molecules are processed, called *pre-mRNAs* or *hnRNAs* (hn = *heterogeneous nuclear*), may be as much as 10 times longer than the finished product.

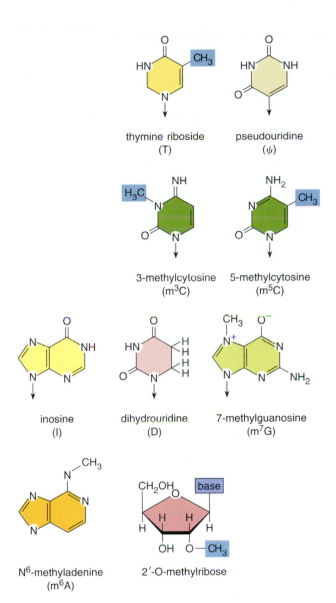

Figure 14-4 Some examples of the most common modified bases and a methylated sugar appearing in RNA molecules.

Studies of the structure of pre-mRNAs, mRNAs, and the genes coding for them have produced many unexpected results. Most surprising was the discovery that most mRNA genes are laid out in interrupted form, with intervening segments that are copied into pre-mRNAs but removed during processing. The intervening segments may be removed in different combinations, producing different finished mRNAs copied from the same gene but coding for related but distinct proteins. Also unexpected was the finding that finished mRNAs contain long spacer sequences at their leading and trailing ends. These sequences are copied from the gene but do not contain codes for amino acids and are not translated into proteins. Other segments added to the extreme 5' and 3' ends of mRNAs during processing do not appear in complementary form in the DNA coding for pre-mRNAs.

mRNA Structure

All mature eukaryotic mRNAs (Fig. 14-5) contain a continuous sequence of nucleotides coding for synthesis of a protein or polypeptide. The coding sequence lies internally in a eukaryotic mRNA molecule, separated from the ends by stretches of nucleotides that are copied from the template but not translated into amino acid sequences. The *5' untranslated region (5' UTR)* lying in advance of the coding sequence is enclosed by a structure at the 5' end of an mRNA known as the *5' cap*. The *3' untranslated region (3' UTR)* following the coding sequence is followed in most, but not all mRNAs by a series of A's at the 3' end known as the *poly(A) tail*. Starting from the 5' end a typical eukaryotic mRNA thus consists of a 5' cap → 5' UTR → coding sequence → 3' UTR → poly(A) tail.

The 5' Cap The cap structure at the 5' end of eukaryotic mRNAs consists of three nucleotides (Fig. 14-6). The first is a guanine nucleotide added during processing, reversed so that its 3'-OH faces the beginning end of the mRNA. The 5'-carbon of the added guanine is joined to the 5' carbon of the second nucleotide in the 5' cap by a chain of three phosphate groups that forms a 5' → 5' linkage at this point rather than the usual 3' → 5' linkage. The remaining two nucleotides in the cap are transcribed from the DNA template and are the first and second nucleotides of the unprocessed pre-mRNA. They remain in their original orientation, joined to each other and to the succeeding nucleotides of the 5' untranslated segment by the usual 3' → 5' linkages. The three nucleotides of the cap are modified by addition of methyl groups at various positions (see caption to Fig. 14-6 for details).

In different mRNAs the two nucleotides following the initial guanine nucleotide may have any of the four A, U, G, or C bases. These variations and differences in the sites of methylation create a family of related but distinct cap structures. In most cases a given mRNA type has only one of the several 5' cap structures.

All eukaryotic mRNAs have a 5' cap. 5' caps also occur on most but not all mRNAs of the viruses infecting eukaryotic cells. Among the viral mRNAs lacking a 5' cap are polio and picornavirus mRNAs. (Prokaryotic mRNAs are typically uncapped—see p. 454.) The 5' cap facilitates the initiation phase of eukaryotic protein synthesis (for details, see p. 461); removal of the cap slows initiation by about 85% to 90%. Addition of 5' caps to mRNAs that normally lack them, such as bacterial mRNAs, increases their rate of utilization in protein synthesis by eukaryotic ribosomes by an equivalent factor.

The 5' cap has other functions. It serves as a signal, recognized by pore complexes in the nuclear envelope, for transport from the nucleus to the cytoplasm. The cap may also protect mRNAs from degradation by ribonucleases.

The 5' Untranslated Region The 5' UTR following the 5' cap varies in length from as few as 10 to more than 200 nucleotides in different eukaryotic mRNAs. A variety of inverted sequences in this segment provides ample opportunities for secondary structures to form. Relatively few sequence features are conserved in the 5' UTR of different mRNAs, even among the mRNAs of closely related proteins such as those in globin, immunoglobulin, or actin families.

One sequence that is conserved in the 5' UTRs of different mRNAs lies just upstream of the AUG initiator codon for protein synthesis. Functions have been worked out for sequences of this type by genetic techniques in which individual bases are substituted or deleted. The effects of this approach, called *substitution-deletion analysis*, are then noted in living cells or in the test tube. Substitution-deletion analysis by M. Kozak and A. Shatkin showed that a purine, usually A, must be present at the position three nucleotides upstream for the initiator codon to be recognized by a ribosome. Other nucleotides, although not absolutely required, occur frequently enough in the region just upstream of the AUG to allow construction of the average or

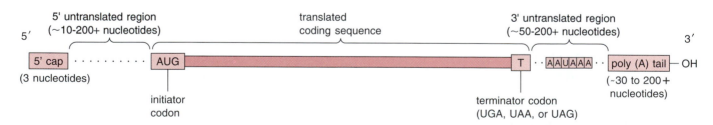

Figure 14-5 Structural features of a eukaryotic mRNA molecule. Some eukaryotic mRNAs have no poly(A) tail; the 5' cap is absent in some mRNAs of viruses infecting eukaryotic cells.

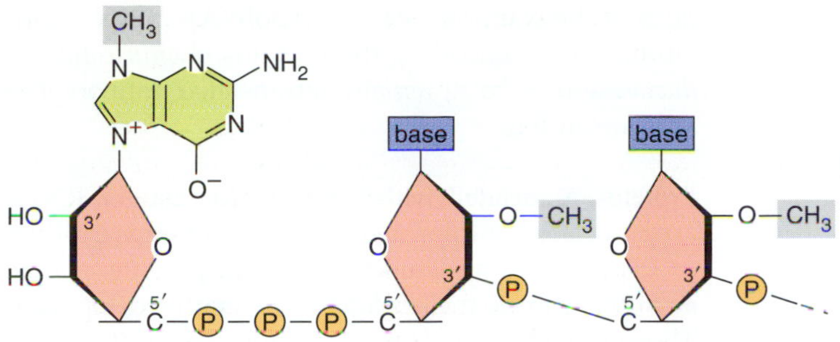

Figure 14-6 The 5' cap of a eukaryotic mRNA molecule. Methyl groups are shaded. The initial guanine, which is added to the 5' end of the transcript during processing, is reversed so that its 3'-OH faces the 5' end of the mRNA. The base of this guanine nucleotide is methylated at its 7 position. The second and often the third nucleotide are also methylated. The methyl group of the second and third nucleotides, however, is usually attached to the sugar rather than to the base. The base of the second nucleotide is also methylated in some mRNAs.

consensus sequence . . . GCCGCCPuCCAUGG . . . for this region (the initiator codon is underlined). In some mRNAs, additional AUGs lie upstream of the initiator codon in the 5' UTR. These lack the indicator sequence, however, and are ignored as starting sites by ribosomes.

In some mRNAs the 5' UTR contains sequence elements that promote entry of the mRNA into protein synthesis. For example, the 5' UTR of the AIDS virus mRNA contains a segment that increases translation of the mRNA by about a thousand times if proteins of the AIDS virus are present. Other mRNAs are increased in translation rate by an equivalent factor if supplied with the 5' UTR of the AIDS mRNA and the viral proteins. The 5' UTR of some mRNAs stored in oocytes has a sequence recognized and bound by "masking" proteins that prevent translation of the message until removal of the mask after fertilization.

The Coding Sequence The coding sequence begins with the AUG initiator codon and is read three nucleotides at a time until one of three *terminator codons UGA, UAA,* or *UAG* appears. The terminator codon, which does not code for an amino acid, acts as a signal telling the ribosome to stop protein synthesis.

The coding segment may include from less than a hundred to thousands of nucleotides, depending on the size of the protein or polypeptide encoded in the message. The coding segments of most eukaryotic mRNAs fall between limits of about 300 to 1500 nucleotides and code for proteins containing about 100 to 500 amino acids. Among the shortest coding segments are those of mRNAs coding for the protamines (see p. 376), which in some cases include less than 100 nucleotides. At the other extreme are very long coding segments such as that of the silk fibroin protein of silkworms, which has 14,000 nucleotides.

The 3' Untranslated Region The 3' UTR following the coding segment shows few similarities in sequence or length among different eukaryotic mRNAs. However, a variety of inverted sequences provide opportunities for secondary structures to form as in the 5' UTR. In addition, the 3' UTR of almost all mRNAs contains the short sequence *AAUAAA* located from 10 to 20 nucleotides upstream of the poly(A) tail (see Fig. 14-5). The AAUAAA sequence, which is highly conserved (AUUAAA, the only common variation, occurs in about 12% of mRNAs), is a processing signal that fixes the length of the 3' untranslated segment and indicates the site for attachment of the poly(A) tail.

The 3' UTRs of some mRNAs are as short as 50 to 60 nucleotides. Although the limit for most is about 200 nucleotides, a few are as long as 600 nucleotides or more. Some variation is even noted in the length of this segment in mRNAs encoding the same protein. These length variations have no apparent effect on the function of mRNAs in protein synthesis.

Sequences within the 3' UTR appear to regulate mRNA stability. Some mRNAs persist for a few hours or less in the cytoplasm; others are retained for as long as a hundred hours or more (see Table 17-2). In many cases the stability characteristics of a given mRNA type can be transferred to other mRNAs by transferring its 3' UTR. For some mRNAs stored in oocytes, sequence structures in the 3' UTR determine the location in the oocyte cytoplasm in which the mRNA is to be stored.

The Poly(A) Tail The unbroken stretch of adenine nucleotides forming the poly(A) tail varies in length from about 30 to more than 200 nucleotides. Most mRNAs have a poly(A) tail. However, some of the normally tailed mRNAs can also be detected in the cytoplasm in tail-less form. These alternate forms are called the *poly(A)⁺* (tailed) and *poly(A)⁻* (tail-less) forms. A relatively few mRNAs, including those for the protamines and most histones, occur regularly without poly(A) tails.

The poly(A) tail is complexed with a *poly(A) binding protein.* Mutations eliminating the protein in yeast are lethal, indicating that its functions in binding the poly(A) tail are vital. The tail and its attached mRNA are much more resistant to enzymatic breakdown when associated with the binding protein, which may account for the lethality of mutations eliminating the protein.

Although the poly(A) tail is one of the most striking and characteristic structures of mRNA molecules, its function remains uncertain. However, stability is considered a likely function because mRNAs that normally possess tails, if injected into cells with their tails removed, are quickly degraded. The same mRNAs in tailed form may persist for long periods of time. The tail may thus protect some mRNAs from breakdown by ribonucleases in the cytoplasm. Normally tail-less mRNAs, such as histone mRNAs, and the poly(A)⁻ forms of other mRNAs, seem to be as stable in the cytoplasm as those with poly(A) tails. Other sequence structures, such as specific fold-back structures at the 3' end or attached proteins, may protect these mRNAs from degradation.

Presence or absence of the poly(A) tail has varying effects on protein synthesis depending on the cell type and developmental stage. In most cells, both poly(A)⁺ and poly(A)⁻ mRNAs are able to direct protein synthesis; however, those with tails are translated more frequently. In some cell types, such as fertilized eggs of the clam *Spisula*, some poly(A)⁻ mRNAs, including those encoding the proteins *ribonucleotide reductase* and *cyclin A*, remain untranslated until poly(A) tails are added.

Proteins Associated with mRNAs Messenger RNAs are associated with proteins throughout transcription, processing, and protein synthesis. The proteins change as mRNAs leave the nucleus and enter the cytoplasm. Those initially associated with mRNAs are involved primarily with the processing reactions converting pre-mRNAs into finished form. Later proteins maintain mRNA structure, promote interactions between mRNAs and ribosomes in protein synthesis, and possibly protect mRNAs from enzymatic breakdown. The total number of associated proteins is significantly reduced as mRNAs become mature and link with ribosomes.

Some proteins associated with mRNAs are common to most or all mRNA types, and some vary according to the cell type. Presumably those that vary

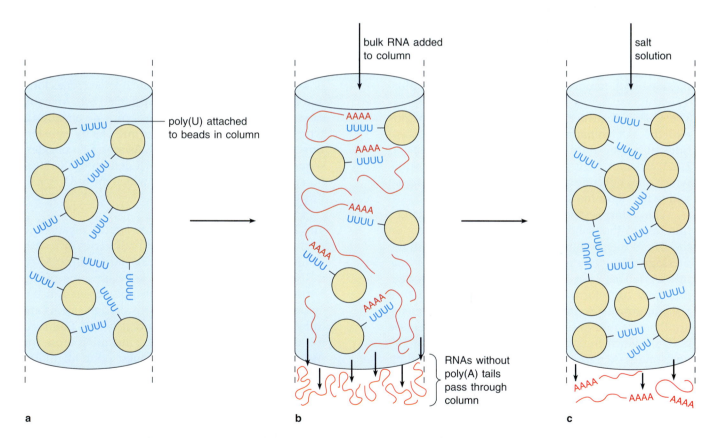

Figure 14-7 Use of affinity chromatography to separate mRNAs from a bulk RNA sample containing many types of cellular RNAs. **(a)** The chromatography column containing plastic beads to which segments of poly(U) RNA are attached. **(b)** The bulk RNA sample is poured through the column; mRNA molecules are trapped on the beads by base pairing between their poly(A) tails and the poly(U) segments. Other RNA types pass through the column. **(c)** The mRNA is released from the column as a purified sample by a salt solution.

with the cell type regulate the rate at which mRNAs join with ribosomes in protein synthesis or control the rate at which mRNAs are degraded. Among the proteins common to all mRNAs are at least four associated with the 5' cap, which take part in the initial reactions linking mRNAs to ribosomes, and the binding protein associated with poly(A) tails.

Isolating mRNAs for Study

mRNAs are isolated from cells by several methods. One takes advantage of the poly(A) tail, which binds readily to artificially synthesized poly(U) or poly(T) molecules by complementary base pairing. The poly(U) or poly(T) is linked to filter paper or to the plastic beads of a separatory column (see Appendix p. 800) and a solution of bulk cellular mRNAs is poured over the paper or column. The bound poly(U) or poly(T) in effect catches mRNAs by the tail and allows their quick separation from the mixture (Fig. 14-7). Specific mRNAs are isolated from these samples by a similar approach using complementary DNA segments as the "catching" agent bound to the filter paper.

Single mRNAs are also isolated by using antibodies made against the protein encoded in the mRNA. When added to cell extracts, such antibodies will often react with incompletely finished proteins emerging from ribosomes. Isolation of the ribosomes tagged by the antibody by cell fractionation techniques (see Appendix p. 794) isolates the mRNA encoding the protein. mRNAs can also be extracted in nearly pure form by isolating ribosomes from cells that concentrate protein synthesis on a single polypeptide, such as hemoglobin in maturing red blood cells.

Once isolated and purified, mRNA molecules can be sequenced in the form of a complementary DNA copy called a *cDNA*. The cDNA is made initially by a reaction catalyzed by *reverse transcriptase*, an enzyme that uses an RNA molecule as a template for making a DNA copy. The cDNA copy is then increased to quantities large enough for sequencing by cloning or the polymerase chain reaction (DNA cloning, the polymerase chain reaction, and methods used for DNA sequencing are described in Supplement 14-1).

Pre-mRNAs

Pre-mRNAs average from 5 to 10 times longer than their corresponding mature mRNA products. Early indications that pre-mRNAs are actually the precursors of mRNAs came from several sources, including *DNA-RNA hybridization* (see p. 366). Either pre-mRNA or mRNA, if mixed with denatured (single-stranded) whole-cell DNA from the same cell type, can wind together or hybridize with the same DNA segments. This means that both pre-mRNA and mRNA contain

sequences that are complementary to the same gene and are evidently copied from the same DNA. If mixed together before hybridization, pre-mRNA and mRNA *compete* for the same sequences in DNA during hybridization, providing further evidence that both are copied from the same DNA template, and that one is the precursor of the other.

Comparisons between mRNA genes, hnRNAs, and the mature mRNAs processed from them led to the discovery of intervening sequences interrupting the genes encoding mRNAs, first reported by R. J. Roberts and P. A. Sharp in 1957 (Roberts and Sharp received the Nobel Prize in 1993 for their pioneering research). Many of the first experiments used DNA-RNA hybridization to detect the intervening sequences. For example, P. Leder and his colleagues hybridized mouse β-globin pre-mRNA and mature mRNA to the DNA of the β-globin gene and examined the hybrids in the electron microscope. The β-globin pre-mRNA molecules hybridized with essentially the entire β-globin gene. The mature β-globin mRNA molecules covered a much smaller segment of the DNA of the gene; in addition, a large loop of unhybridized DNA extended from the middle of the gene (Fig. 14-8). The loop showed that a segment of the DNA that hybridized with the β-globin pre-mRNA had no counterpart in the processed mRNA. Thus a sequence within the transcribed region of the β-globin DNA was copied into the pre-mRNA but was later removed during processing. The β-globin gene therefore occurs in the DNA in

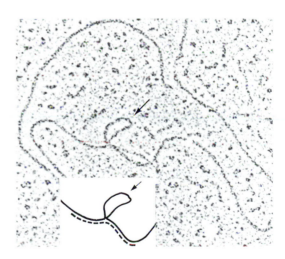

Figure 14-8 The DNA of a mouse β-globin gene (solid line in inset) hybridized with a fully processed mRNA molecule (dashed line in inset) encoded in the gene. A large, unpaired loop of DNA (arrow) extends outward from the middle of the hybridized region. The loop is an intron, a segment of the DNA that is copied into a pre-mRNA but spliced out in the processing reactions producing the finished mRNA. (Micrograph courtesy of S. M. Tilghman.)

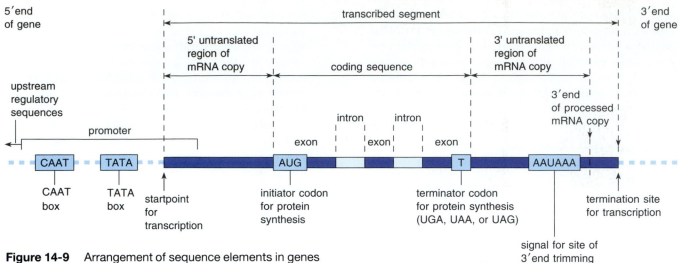

Figure 14-9 Arrangement of sequence elements in genes coding for mRNA molecules in eukaryotes (see text).

interrupted form. The intervening sequences interrupting the code are called *introns*; the segments retained in mature mRNAs are called *exons* (see Fig. 14-9).

Direct sequencing of genes, pre-mRNAs, and mature mRNAs provided the final and most definitive evidence that pre-mRNAs are the precursors of mRNAs. The sequencing studies also revealed that most but not all eukaryotic protein-encoding genes are interrupted by introns. Depending on the gene, from 1 to more than 60 introns may be present. The α2 collagen gene of chickens, for example, contains 51 introns.

Pre-mRNA molecules become associated with proteins while they are under transcription. The proteins are believed to hold pre-mRNA molecules in specific three-dimensional forms and to take part in the processing reactions. Also found in association with pre-mRNA particles are a group of snRNAs known as the *U snRNAs* (because uridine occurs in high proportions among their bases). Like the pre-mRNA-associated proteins, these snRNAs play both structural and functional roles during processing reactions (see below).

mRNA Gene Structure

The genes coding for mRNA molecules, like the pre-mRNA molecules copied from them, are much longer than their corresponding mRNAs (Fig. 14-9). The beginning or 5' end of an mRNA gene is so called because it corresponds to the 5' end of the pre-mRNA copied from it. In advance of the first nucleotide copied into the pre-mRNA lies the *5'-flanking region* of the gene, which contains the promoter and other upstream sequences regulating initiation of transcription. The promoter provides regulatory, recognition, and binding sites for the RNA polymerase II enzyme and indi-

cates the first DNA nucleotide to be transcribed, the *startpoint*.

The *startpoint* nucleotide is numbered +1. All nucleotides of the gene are numbered from this point, those upstream of the startpoint as −1, −2, −3, and so on, starting with the nucleotide immediately to the left of the startpoint. Those to the right or downstream of the startpoint are numbered as +2, +3, +4, and so on.

The gene segment transcribed into a pre-mRNA contains the sequences corresponding to the 5' UTR, the coding segment, and 3' UTR of the finished mRNA. At some point in the transcribed portion of the gene is the AUG initiator codon, which marks the beginning of the segment coding for a polypeptide. At a point farther downstream is a codon corresponding to one of the three terminator codons marking the end of the coding segment. One or more introns (shown in light blue in Fig. 14-9) may be inserted in the coding segment. The final sequences of the gene, downstream of the segment copied into pre-mRNA, are known collectively as the *3'-flanking sequences*.

Almost all eukaryotic mRNA genes contain the code for a single protein or polypeptide. However, because exons may be spliced together in different combinations during pre-mRNA processing (see below), several different mRNA types can be produced by a single gene.

Individual sequences within any segment of a protein-encoding gene are written in the 5' → 3' direction either as an RNA equivalent or *as the DNA sequence of the nontemplate chain*. The nontemplate chain is used because it runs in the 5' → 3' direction used for writing nucleotide sequences and because its sequence is identical to that of the RNA copied from the gene except for the substitution of T for U. Individual trip-

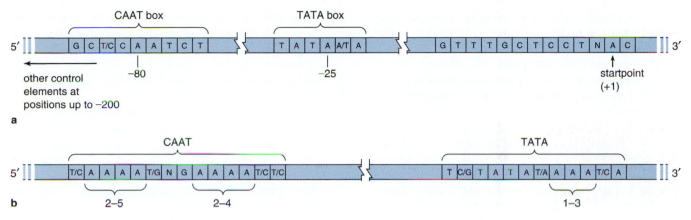

Figure 14-10 Sequence elements of the promoter in animals and lower eukaryotes (a) and plants (b). The CAAT and TATA boxes are shown as consensus sequences; N indicates that any of the four nucleotides may occur with approximately equal frequency at that site.

lets or the coded message can therefore be read directly from the nontemplate chain in the same direction as in the mRNA. For example, the initiator codon AUG is identified in the nontemplate DNA chain directly as AUG or as ATG. DNA sequences in nontranscribed regions of the gene, such as the promoter, are also written as the DNA sequence of the nontemplate chain.

The Promoter The promoter contains several important sequence elements that appear in more or less altered form in most eukaryotic protein-encoding genes. Similar sequences also occur in the promoters of the genes of viruses that infect eukaryotic cells. One such sequence is the *TATA box*, actually the consensus sequence TATA$\overset{A}{\underset{T}{}}$A in animals and lower eukaryotes, which appears at a site some 20 to 30 bases upstream of the startpoint (Fig. 14-10; the positions marked $\overset{A}{\underset{T}{}}$ indicate that either A or T may appear at this point). Substitution-deletion analysis has shown that the TATA box functions in initiation of transcription, including binding RNA polymerase and indicating the nucleotide to be used as the startpoint.

The fact that the TATA box indicates the startpoint has been demonstrated by several experiments. For example, S. Hirose and his coworkers found that substituting T for A at the second or fourth positions of the TATA box in the gene coding for the silk protein of silkworms greatly reduces the precision of startpoint location, yielding RNA copies that start randomly near the normal startpoint. Naturally occurring mutations involving substitutions in the TATA box, one of which is responsible for a form of the human disease *thalassemia* (see below), have the same effect. Not all mRNA genes have a TATA box; many of these *TATA-less* genes encode "housekeeping" proteins that

are synthesized continuously in cells. Evidently in these genes other sequence elements function in initiation and indicate the startpoint.

Other short sequence elements occur less frequently in mRNA gene promoters. In general these sequence elements bind regulatory proteins that modify the rate of initiation. One of the more common of these elements is the *CAAT box*, actually the consensus sequence GC$\overset{T}{\underset{C}{}}$CAATCT in animals and lower eukaryotes, centered roughly at about −80 in the promoter (see Fig. 14-10).

Other sequence elements that modify the rate of initiation have been identified at points from about 85 to as many as 200 or more bases upstream of the startpoint. Like the CAAT box, these upstream sequences are recognized and bound by regulatory proteins. *Enhancers*, short elements some 50 to 100 nucleotides long, have the curious ability to increase the rate of initiation when inserted experimentally at practically any location near a gene—upstream, downstream, or even within introns. The only requirements seem to be reasonable proximity, within a thousand base pairs or so of the gene, and location on the same DNA molecule as the gene being transcribed.

Enhancers have been found in association with many eukaryotic mRNA genes, including those coding for antibodies, myosins, hemoglobin, ovalbumin, and at least one histone protein, an H2A histone of chickens. Few sequence features appear to be common among enhancers except that many in higher eukaryotes contain an eight-nucleotide *octamer* sequence that is recognized and bound by regulatory proteins. Mutations or substitutions altering or eliminating enhancer sequences greatly reduce the rate of transcription of their target genes, in most cases by a factor of 100 or more.

How enhancers work is one of the more intriguing questions of molecular biology. According to the *loop model*, proposed by H. Echols, R. Tjian, and M. Ptashne and their colleagues, a regulatory protein binds the enhancer as one of the initial steps in initiation (Fig. 14-11a). The protein bound at the enhancer then interacts with other proteins bound at the promoter (Fig. 14-11b), throwing the length of DNA between the enhancer and the promoter into a loop. The combined regulatory proteins, according to the model, take on a conformation that binds RNA polymerase II more avidly than the promoter complex alone

(Fig. 14-11c) and thereby enhances the rate of transcription. Among the several lines of evidence supporting the loop model are electron micrographs directly showing loop formation by interaction of proteins at the enhancer and promoter sites (Fig. 14-11d and e).

Sequence Elements in the Transcribed Region: Exons and Introns Other than the initiator and terminator codons for protein synthesis, the sequences of primary interest in the coding segment of mRNA genes are those defining exons and introns. These sequences have been identified by several methods, including

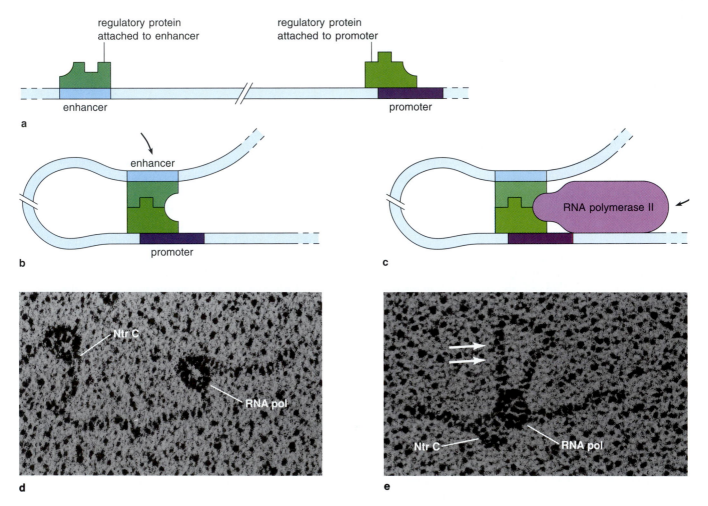

Figure 14-11 The loop model for the action of enhancers in increasing gene transcription (see text). **(a)** The regulatory protein bound to the enhancer. **(b)** Interaction between the regulatory protein and a protein bound to the promoter creates a loop in the DNA. **(c)** The combination sets up a conformation with high binding affinity for RNA polymerase II, which adds to the complex and initiates transcription. **(d)** The enhancer/promoter region of the *glnA* operon of *E. coli*, showing the regulatory protein NtrC bound to the enhancer and RNA polymerase (RNA pol) bound, presumably loosely, to the promoter. The two proteins can be distinguished because NtrC is more lightly coated by the tungsten metal used to shadow the preparation (see Appendix p. 790). **(e)** Formation of a loop between the regions bound by NtrC and RNA polymerase. Presumably the interaction leads to tight binding by the enzyme and initiation of transcription. ([**d** and **e**] courtesy of H. Echols, from *J. Biolog. Chem.* 265:14699 [1990].)

direct sequence comparisons of mature mRNAs with either unprocessed pre-mRNAs or the genes encoding them.

Analysis of the boundaries between exons and introns reveals that almost all introns interrupting mRNA genes, with very few exceptions, begin at their 5' ends with the two nucleotides *GU* and close at their 3' ends with the two nucleotides *AG*, a characteristic known as the *GU/AG rule* (or, when considered in DNA, as the *GT/AG* rule). Other sequences surrounding the bounding GU and AG and lying within introns are repeated frequently enough to allow construction of the consensus sequences shown in Figure 14-12. The GU/AG boundaries and the internal sequences are recognized as signposts by the splicing mechanisms removing introns (see below). In total, different introns average in length from about 200 to 400 nucleotides, but they may include from fewer than 50 to 30,000 or more nucleotides. Some of the longer introns may contain other genes, complete with their own regulatory sequences and introns.

The majority of known eukaryotic genes and the genes of viruses infecting eukaryotic cells contain at least one intron. Exceptions include histone genes that are active during the period in which DNA is repli-cated; in most species these histone genes occur entirely in uninterrupted form. A relatively few other genes, such as the zein seed storage protein gene of maize, interferon genes, and some globin genes of insects, are also intronless.

Why so much of the DNA of eukaryotic genes is occupied by introns remains uncertain. Why do eukaryotes expend considerable amounts of energy and raw materials to store and copy nucleotide sequences that are only to be removed and degraded from RNA transcripts? One advantage brought by introns is the capacity for coding mRNAs for different proteins in the same gene, processed out of a common pre-mRNA by including some exons or introns and retaining others (called *alternative splicing*; see below). In some genes, such as those coding for hemoglobins, immunoglobins, and insulin, exon-intron junctions fall at divisions between major structural or functional domains in the encoded proteins. This correspondence has given rise to the hypothesis that some proteins appeared in evolution through the exchange of exons between genes, in a process called *exon shuffling*. This mechanism might have allowed proteins with new but related functions to evolve by the assembly of new combinations of existing domains, already selected by

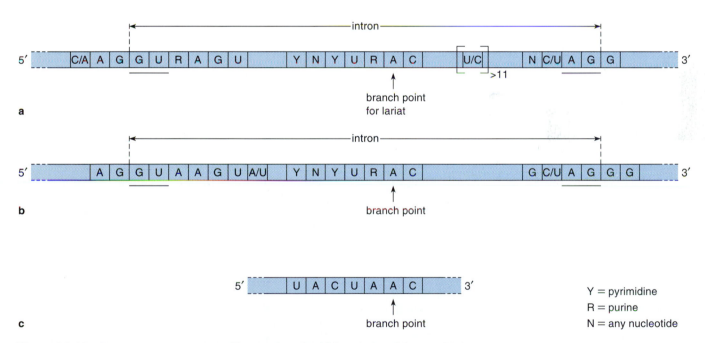

Figure 14-12 Consensus sequences of introns in animal (**a**) and plant (**b**) pre-mRNAs. Plant introns also contain a majority of G-C base pairs distributed throughout the intron. The underlined *GU* at the 5' end and the underlined *AG* at the 3' end are essentially invariant and occur in almost all intron/exon junctions. The branch point is the site at which the 5' end of the intron, once clipped free, joins to form a lariat during the processing reactions removing introns. (**c**) The invariant sequence around the branch point in yeast genes.

the evolutionary process. Evolution by this mechanism would be much more efficient than one-by-one changes in amino acids at random points. (For a further discussion of exon shuffling as a basis for evolutionary change, see Chapter 26.)

There are also genes, such as those encoding actins, in which no correlation between the positions of intron/exon boundaries and protein domains can be detected. F. H. C. Crick has proposed that such introns may represent random insertions of "selfish DNA"—functionless DNA segments that have in some way acquired the GU/AG boundary sequences making them an intron. (Two possible routes of random insertion are through viruses that regularly remove and insert DNA segments into genes, and *transposable elements*, DNA sequences that move from place to place within chromosomes; see p. 535.) Perfect removal of the intron from the mRNA copied from the gene, guaranteed by its processing signals, would ensure that the inserted DNA has no significant effect on the protein encoded by the gene and the organism containing it. Such selfish inserts would thus be protected from evolutionary selection and maintained permanently in the DNA from generation to generation.

Whether selfish or functional, introns are maintained in all eukaryotes in spite of the considerable phosphate-bond energy required for their replication and transcription. Similar introns also occur in genes coding for tRNA and rRNA in eukaryotes (see Chapter 15 for details).

Sequences at the 3' End of mRNA Genes In most mRNA genes the segment lying downstream of the coding sequence contains the AATAAA sequence (transcribed as AAUAAA in pre-mRNA), which marks the site at which the 3' end of the pre-mRNA is to be clipped off. Most single base substitutions in the AAUAAA sequence interfere with the clipping reaction, yielding mRNAs with greatly extended 3' ends, and reduce the efficiency of the reaction adding the poly(A) tail.

At least one human disease, a form of thalassemia, is caused by mutation of the AAUAAA signal to AAUAAG. This mutation causes faulty processing of an α-hemoglobin protein necessary for normal function of hemoglobin in erythrocytes. The result is anemia ranging from mild to severe, depending on whether the individual is homozygous or heterozygous for the mutation.

Transcription of Pre-mRNAs

Initiation of pre-mRNA transcription requires an interaction of regulatory proteins and transcription factors with the promoter, enhancer sequences if present, and RNA polymerase II. Investigations by R. G. Roeder and P. A. Sharp and their colleagues and others indicate that at least seven transcription factors—*TFIIA, TFIIB, TFIID, TFIIE, TFIIF, TFIIH,* and *TFIIJ*—interact with the promoter in the region of the TATA box and with RNA polymerase II to initiate tight enzyme binding, unwinding of the DNA, and initiation of transcription at the startpoint (Fig. 14-13). A large number of cofactors, many of them incompletely identified and characterized, join with the transcription factors in promoting the most efficient initiation of transcription. Many of the cofactors differ in different genes, cells, and species, indicating that they provide variations in the specificity of initiation. Of the various factors and cofactors, TFIID appears to be central to initiation; this factor binds first to the TATA box, in a reaction probably accelerated by TFIIA. TFIID is a large complex containing a *TATA-binding protein (TBP)* and several other factors. TFIIB then joins; the D-B complex stimulates binding of RNA polymerase in association with TFIIF. After the polymerase binds, the remaining three factors, TFIIE, TFIIH, and TFIIJ, add to the total complex in that order. As assembly of the complex becomes complete, the long C-terminal tail of the RNA polymerase is phosphorylated, possibly by a kinase activity associated with TFIIH. In the final step of initiation, ATP is hydrolyzed; at this point the DNA unwinds along the stretch occupied by the RNA polymerase and transcription begins. Nonhydrolyzable molecules resembling ATP bind to the complex, but transcription does not begin. The energy released by ATP hydrolysis probably powers unwinding of the DNA.

TBP, the subunit of TFIID binding the TATA box, is a saddle-shaped protein that induces a sharp bend in the DNA as it binds. This bending distorts the DNA region including the TATA box into a configuration that may stimulate binding of RNA polymerase and the remainder of the transcription factors. The distortion probably facilitates unwinding of the DNA for initiation of transcription. Recent evidence indicates that TBP is also required for transcription initiation by RNA polymerases I and III, even in genes without a TATA box (see below). Thus TBP works as a universal transcription factor in eukaryotes.

A variety of regulatory proteins modify the initiation process. Some of these proteins bind directly to TFIID and modify its affinity for the TATA box. Others bind to other initiation factors or to DNA sequences located elsewhere in the promoter or upstream sequences. Presumably these proteins, which serve as gene-specific regulatory factors, modify the conformation of the initiation complex or the DNA to enhance or inhibit initiation (see Chapter 17 for details).

Once initiation has been accomplished, elongation factors add to the complex. The enzyme then proceeds along the DNA template, adding bases one at a time until transcription terminates in regions well beyond the AAUAAA signal. Termination does not appear to require specific sequences. Instead, the enzyme appears simply to continue transcription after passing the AAUAAA signal until the pre-mRNA transcript is clipped at a point just beyond its AAUAAA sequence. Presumably the clipping releases a factor necessary for termination or induces a conformational change in the DNA or the segment of the transcript remaining with the enzyme that stops transcription. In support of this conclusion is the fact that an intact AAUAAA signal is required for termination. As transcription stops, the enzyme and the portion of the RNA transcript remaining with the enzyme are released from the DNA.

Processing Pre-mRNAs

The reactions processing pre-mRNA transcripts take place entirely in the nucleus. Addition of the 5' cap occurs while transcription is still in progress. Soon after transcription is complete, surplus nucleotides are cleaved from the 3' end, and the poly(A) tail is added. As these reactions progress, introns are removed.

Adding the Cap and Tail   The 5' cap is synthesized by a group of processing enzymes that are evidently loosely associated with the pre-mRNA ribonucleoprotein particle. The capping base is added directly to the first nucleotide copied from the startpoint, with no clipping reactions shortening the 5' end of the pre-mRNA. The capping nucleotide and one or two of the succeeding nucleotides are then methylated to complete the cap structure.

At the opposite end of the pre-mRNA the transcript is cleaved at a point 10 to 20 nucleotides past the AAUAAA signal. The enzyme *poly(A) polymerase* then assembles the poly(A) tail by adding A's one at a time to the 3' end of the pre-mRNA, using ATP as the source for the adenines.

Removing Introns   The splicing reactions removing introns break the sugar-phosphate bonds at the boundaries of introns and rejoin the free ends into a continuous mRNA molecule. Breakage occurs precisely at the intron-exon junctions, so that no nucleotides are added to or removed from the ends of the exons at either side of the introns. Otherwise, intron removal would produce a *frameshift*; that is, it would throw the coding triplets out of register, so that all triplets past the faulty point of removal or addition would be read incorrectly during protein synthesis.

Intron removal proceeds by a number of steps that involve participation of five U snRNAs, *U1, U2, U4, U5,* and *U6.* Each snRNA is associated with several proteins; the entire complex formed by the snRNAs

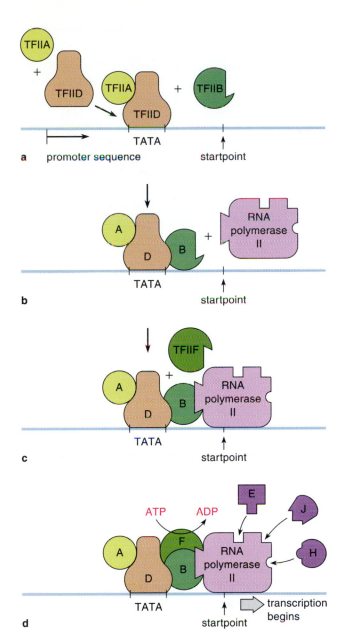

Figure 14-13   Interactions among initiation factors, RNA polymerase II, and the promoter during initiation of transcription (see text).

is known as a *spliceosome.* Research by J. A. Steitz and P. A. Sharp and their coworkers and others indicates that the snRNA-protein complexes bind to the intron and fold it into a conformation that triggers its removal. (Figure 14-14 shows the splicing reactions and the sequence in which the snRNAs are believed to interact.) As part of its removal the intron is joined into a *lariat,* a branched RNA structure. Intron removal, in all its complexity and precision, can proceed in the test tube as long as the ATP and the snRNAs in the form of a spliceosome are present.

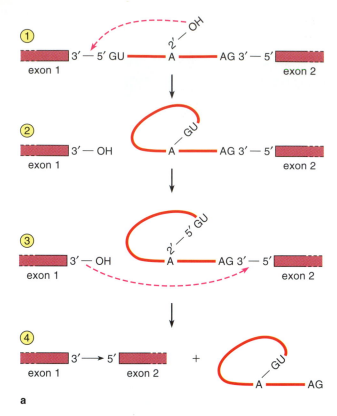

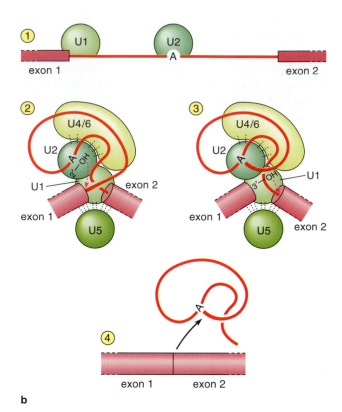

a

b

Figure 14-14 The splicing reactions removing introns, and the interactions of snRNAs in spliceosomes. **(a)** The overall sequence of reactions removing introns. The reaction depends on chemical reactivity of 2'- or 3'-OH groups of bases of the intron, held in positions allowing them to attack the 3' → 5' phosphodiester linkages at the intron boundaries. In step 1 a 2'-OH of the adenine located at the branch site (identified by an arrow in Fig. 14-12) attacks the 3' → 5' bond at the beginning or 5' end of the intron. The attack breaks the bond at the beginning of the intron and creates a 2' → 5' linkage between the base at the beginning of the intron and the central adenine (step 2). The bond makes a branch at the central adenine and transforms the intron into a looped structure, or *lariat*. The 3'-OH at the end of the first exon then attacks the 3' → 5' linkage at the end of the intron (step 3). The attack breaks the bond, releasing the intron, and creates a new 3' → 5' linkage that completes the splice by joining the first exon directly to the second exon (step 4). **(b)** The participation of U snRNAs in the splicing reactions, worked out by J. A. Steitz, C. F. Lesser, and C. Guthrie and their colleagues through sequence comparisons among the U snRNAs and the intron and by crosslinking techniques that identify nearest neighbors in the splicing complex. In

step 1 of these interactions, U1 binds at the 5' end of the intron, and U2 to the branch site containing the internal adenine. The binding occurs through base pairing between regions of the U snRNAs and sequences in the intron and its 5' boundary. Step 2 sets up the conformation allowing the intron and the U snRNAs to act as ribozymes splicing the intron. U4 and U6, linked together by mutual base pairing into a combined structure, pair with U2 and the intron. U5 pairs with sequences surrounding the 3' and 5' boundaries of the intron, and possibly with sequences in U1. The combined pairing brings the exons together and folds the intron into a conformation favoring breakage of the bond at the beginning of the intron by the internal adenine (red dashed arrow in step 2). The breakage creates the lariat (step 3) and allows the 3'-OH at the end of the first exon to attack the end of the intron (red dashed arrow in step 3). Breakage of this bond releases the intron and joins the exons (step 4). Proteins of the spliceosome maintain the snRNAs in positions favoring the reactions; each step of the process is also stimulated by other proteins acting as *splicing factors*. In all, there may be as many as 50 to 100 different proteins involved in splicing, either as parts of the spliceosome or as splicing factors.

Key evidence supporting the roles of snRNAs in intron removal was developed in the Steitz laboratory through an interesting relationship to the human autoimmune disease *lupus erythematosus*. Lupus patients develop antibodies that attack proteins in their own

bodies. One group of these autoantibodies strongly inhibits intron splicing in either living cells or the test tube. These antibodies were found to react specifically with proteins associated with the U snRNAs in spliceosomes. The U snRNAs taking part in intron removal have been detected in vertebrate and invertebrate animals, in plants, and in the yeast *Saccharomyces cerevisiae.*

The reactions removing introns from pre-mRNAs resemble a pattern of RNA self splicing occurring during RNA processing in lower eukaryotes and inside mitochondria and chloroplasts. The self-splicing reactions, which remove intervening sequences known as *group II* introns from RNA precursors (see pp. 435 and 619 and Fig. 15-7b), produce a lariat from the introns and depend on secondary structures that are closely similar to the pairing configurations of snRNAs with the introns of pre-mRNAs. The similarities are so close that it is considered likely that intron removal from pre-mRNAs by spliceosomes evolved from the group II mechanism and employs at least some self-catalyzed reactions in which RNA segments act as ribozymes.

Splicing reactions normally remove introns from within a single pre-mRNA transcript. In some organisms, however, splicing reactions join separately transcribed mRNA segments, copied from different regions of the genome, into a continuous mRNA. The process is known as *trans* splicing. (When considered in this regard the usual form of splicing, involving removal of introns from within a single mRNA transcript, is called *cis* splicing.) All known cases of *trans* splicing within the nucleus, observed in lower eukaryotes such as *Euglena*, trypanosomes, and the nematode worm *Caenorhabditis*, involve addition of a short, capped segment copied from one site in the DNA to the 5' end of a pre-mRNA transcribed from a different site to complete the 5' UTR. The separate segments have boundary sequences allowing them to be recognized by spliceosomes. In chloroplasts and plant mitochondria, a distinct pattern of *trans* splicing, based on

group II introns, joins separately transcribed exons into continuous mRNAs (see p. 620).

Alternative Splicing   The pre-mRNAs of many genes follow alternative splicing pathways that vary in the particular introns to be removed. Because of alternative splicing a DNA segment can be either an exon or an intron, depending on whether it is retained or removed during processing.

For example, the pre-mRNA of a gene active in the thyroid gland and hypothalamus of higher vertebrates has six potential exons (labeled A to F in Fig. 14-15a), separated by five introns. In the thyroid gland, exons A or B and C, D, and E are linked together, and all other introns and exons are removed by the splicing reaction, producing a finished mRNA that codes for the thyroid hormone *calcitonin* (Fig. 14-15b). In the hypothalamus, exons A or B, and C, D, and F are linked together, and other introns and exons removed, to produce an mRNA coding for a distinct polypeptide, the *calcitonin gene–related polypeptide (CGRP;* Fig. 14-15c), which may be involved in the function of taste receptors. In these alternative splicing reactions, either the A or B exon may be retained or eliminated in either protein; segment E is an exon in the calcitonin pathway and an intron in the CGRP pathway. Segment F is an intron for calcitonin and an exon for CGRP.

The list of pre-mRNAs processed differentially by alternative splicing lengthens constantly as more examples of this mechanism are discovered. Among others, the list includes pre-mRNAs for amylase, lamins, collagens, growth hormone, prolactin, fibrinogen, lens crystallin, myosin light and heavy chains, troponin T, tropomyosin, protein kinases C, fibronectin, vimentin, antibody polypeptides, and, in higher plants, a subunit of RuBP carboxylase. The distinct but

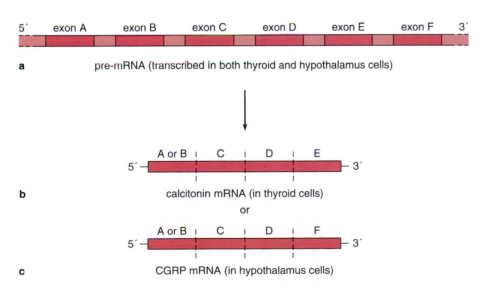

a   pre-mRNA (transcribed in both thyroid and hypothalamus cells)

b   calcitonin mRNA (in thyroid cells)

or

c   CGRP mRNA (in hypothalamus cells)

Figure 14-15   Alternative splicing of the same pre-mRNA to produce a calcitonin or a CGRP mRNA (see text). **(a)** The pre-mRNA, which contains six potential exons (A through F), separated by five introns. **(b)** Linkage of exons A or B and C, D, and E to produce calcitonin mRNA. **(c)** Linkage of exons A or B and C, D, and F to produce CGRP mRNA.

related proteins produced by alternative splicing have different functions or are routed to different final locations in the cell. The length of the list indicates that alternative splicing has been much used in evolution to increase the variety of proteins produced by the cells of an organism. In some cases alternative splicing routes determine whether or not a protein is functional.

Proteins regulating alternative splicing have been identified in a few organisms, including yeast, *Drosophila*, and *Xenopus*. In many cases it appears that, in the absence of regulatory proteins, alternative splicing follows a "default" pathway in which the introns removed have internal and surrounding sequences closest to the usual consensus sequences as shown in Figure 14-12. Their consensus sequences are recognized readily by spliceosomes and splicing factors, and the default introns are consistently removed. Some regulatory proteins recognize and bind sequences in the boundaries of default exons and block their removal. Other regulatory proteins bind sequences in the boundaries of introns with sequences that depart to a greater or lesser extent from the consensus and are therefore recognized less efficiently by spliceosomes. Binding by the regulatory proteins "recruits" spliceosomes and splicing factors to these alternative introns, leading to their removal.

For example, in *Drosophila* males an mRNA encoded in the *doublesex* (*dsx*) gene is spliced by a default pathway into a form that blocks female sexual differentiation. Research by M. Tian and T. Maniatis showed that in females, regulatory proteins encoded in two *transformer* genes bind to the *dsx* pre-mRNA and promote recognition by the splicing apparatus of an intron retained in the default pathway. Removal of this intron produces the female form of the *dsx* mRNA, which is ineffective in blocking female differentiation.

Other Processing Reactions Many pre-mRNAs are methylated at a few internal sites, usually at adenines in the 3' untranslated region, as a part of the processing reactions. At least some methylated adenines are retained in the finished mRNA. The functional significance of the methylated sites is unknown.

A few pre-mRNAs undergo an unusual processing reaction called *RNA editing*, in which sequences within the coding portion are altered after transcription so that they do not match the DNA from which they were copied. Almost all known examples of RNA editing, which were discovered through comparisons of gene sequences with those of mature mRNAs, involve (1) addition of one or more U's at specific sites in the coding sequence, (2) deletion of U's, or (3) conversion of C to U. The evolutionary origins and functional significance of RNA editing remain uncertain.

Although most examples of RNA editing involve mRNAs transcribed inside mitochondria, one example of nuclear RNA editing in mammals was recently found by S.-H. Chen and L. M. Powell and their coworkers. In the editing of the pre-mRNA encoding apolipoprotein-B, a C is converted to U in intestinal but not in liver cells. The change from C to U converts a codon specifying an amino acid into a terminator codon, producing a shortened version of the protein in intestinal cells. Surrounding sequences or secondary structures in the mRNA may indicate the C to be converted to U (for further information on RNA editing in mitochondria, see Chapter 21).

Research by J. B. Lawrence and D. Spector and their coworkers and others with fluorescent antibodies against enzymes and factors active in mRNA transcription and processing indicates that these reactions take place at discrete, localized sites in the nucleus rather than being generally distributed throughout the nucleoplasm. The sites appear as 20 to 50 bright "speckles" in nuclei examined by fluorescence microscopy (see p. 786) after exposure to the antibodies. The localization to specific sites may reflect organization of the enzymes and factors required for mRNA transcription and processing into large assemblies by elements of the nuclear matrix (see also p. 383).

ALTERATIONS IN CHROMATIN STRUCTURE DURING TRANSCRIPTION

Chromatin structure changes extensively in actively transcribing regions. Some of the changes probably reflect an opening of chromatin structure required for DNA unwinding and attachment of RNA polymerases. Other alterations, in regions within or near promoters, probably result from combination of regulatory proteins with the DNA.

The rate at which chromatin is digested by DNA endonucleases increases in characteristic patterns as most genes become active in transcription. (DNA endonucleases are the enzymes that gave some of the first clues leading to the discovery of nucleosomes; see p. 377 and Information Box 13-3.) Genes that are about to be transcribed are typically digested two to five times more rapidly by DNA endonucleases than genes in chromatin that is to remain inactive. This generally increased sensitivity to endonucleases, which extends over the entire length of a gene, including both its 5' and 3' flanking regions, probably reflects unwinding from the solenoid coil to a more extended chain of nucleosomes (see p. 377 for details of these structures).

At some point before transcription begins, scattered segments of the chromatin become highly sensitive to endonuclease digestion. These regions, called *hypersensitive sites* (see also p. 506), may be more than a hundred times more sensitive to one particular endonuclease, DNAse I, than the chromatin of totally inactive genes. The hypersensitive sites, which may include lengths of DNA ranging from about 30 to as many as 300 to 400 base pairs, are concentrated primarily in the 5'-flanking region of mRNA genes. These sites are believed to be DNA lengths from which nucleosomes have been dislodged by regulatory proteins.

As transcription begins, sensitivity of the entire gene to endonuclease digestion increases to some 5 to 10 times greater than completely inactive chromatin. This increased sensitivity extends from about the start-point through several hundred nucleotides into the 3'-flanking region. In addition to the increased sensitivity to endonucleases, some or all of the DNA lengths generated by endonuclease digestion of chromatin lose their regular periodicities (see p. 377) and become random. These changes are thought to reflect a change in nucleosome structure from the compact to a more open or unfolded state in the DNA being transcribed. The entire sequence of changes is typical of most mRNA genes: (1) moderately increased sensitivity, extending over relatively long stretches within and on both sides of the genes before transcription begins, (2) the appearance of hypersensitive sites just before transcription begins, and (3) more highly increased sensitivity in the transcribed and 3'-flanking region as transcription is initiated.

There is presently much controversy over the fate of nucleosomes at sites being transcribed. However, it is generally agreed that DNA wrapped around nucleosome core particles cannot unwind enough to allow transcription by the RNA polymerase complex. Therefore it seems likely that the association of DNA with nucleosomes must be altered in some way to accommodate passage of RNA polymerase. There are several possibilities for this alteration. The association between DNA and the nucleosome core particle surface might be loosened enough to allow the DNA double helix to unwind for transcription. Or the core particle might unfold, opening like a clamshell to permit DNA unwinding and passage of the enzyme complex. Still another possibility is that nucleosomes might be displaced by the enzyme—that, as the enzyme passes, a nucleosome might be dislodged briefly but completely and rewind with the DNA immediately after the enzyme passes.

While there is some experimental support for each of these possibilities, the weight of evidence now seems to fall on the side of a brief but complete displacement of nucleosomes by RNA polymerase as the enzyme passes. For example, E. M. Bradbury and his coworkers found that binding together the histones of core particles with chemical crosslinkers, expected to prevent nucleosomes from unfolding, had no adverse affects on the rate of RNA transcription.

Direct examination of transcribing genes also supports the idea that nucleosomes are displaced rather

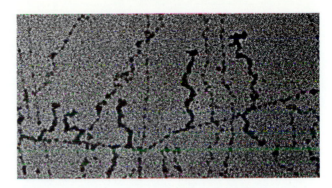

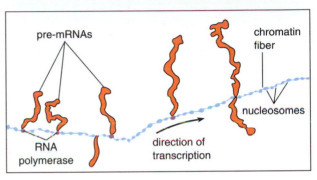

Figure 14-16 Transcribing chromatin isolated from the milkweed bug *Oncopeltus*. × 66,000. (Micrograph courtesy of C. D. Laird and V. E. Foe, from *Cell* 9:131 [1976]. Copyright Massachusetts Institute of Technology.)

than unfolding. Figure 14-16 shows an active mRNA gene that has been isolated and prepared for electron microscopy; pre-mRNAs are visible as thick fibers branching off from the DNA chain. RNA polymerase transcription complexes can be seen as dense particles at the bases of the transcripts. Nucleosomes generally seem to be displaced from the sites under transcription but are visible on the DNA stretches between sites occupied by the RNA polymerase complexes.

Therefore it seems likely that the general increase in sensitivity to endonuclease digestion in transcribed regions reflects an alteration of nucleosome structure loosening the winding of DNA around nucleosome core particles to some extent, permitting nucleosomes to be easily displaced for transcription. As a transcribing complex reaches a DNA length wound around a nucleosome, the nucleosome at that site is displaced rather than unfolding. After the enzyme passes, the nucleosome rewinds with the DNA.

Transcription of mRNA supplies the informational molecules required for direction of protein synthesis. mRNA transcription, carried out by the polymerase II enzyme, proceeds through stages of initiation, elongation, and termination, enhanced by a series of transcription factors. (The equivalent reactions transcribing mRNAs in prokaryotes are described in Supplement 15-1.) The initial pre-mRNA copied from mRNA genes is typically much longer than the final mRNA product because of the inclusion of introns, copied from intron sequences in the DNA. During the processing reactions converting pre-mRNAs to mature mRNAs a capping nucleotide is added to the 5' end of the transcript, introns are removed from coding segments, and the 3' end is clipped. Individual nucleotides are modified chemically to other forms at sites in mRNAs, particularly in the 5' cap. As a part of 3'-end processing, a poly(A) tail is added to many mRNAs. The spectrum of proteins added to mRNAs changes as processing takes place. Initial proteins, such as those associated with the U snRNAs taking part in clipping out introns, are primarily processing enzymes and factors. As mRNAs mature, processing proteins are replaced by others forming structural and functional parts of mRNAs.

As genes become active in transcription, chromatin structure changes, probably including unwinding of chromatin fibers and loosening of nucleosome structure. These changes are reflected in an increased sensitivity to digestion by DNA endonucleases.

For Further Information

Suggestions for Further Reading

von Ahsen, U., and Schroeder, R. 1993. RNA as a catalyst: Natural and designer ribozymes. *Bioess.* 15:299–307.

Archambault, J., and Friesen, J. D. 1993. Genetics of eukaryotic RNA polymerases I, II, and III. *Microbiol. Rev.* 57: 703–724.

Ausio, J. 1992. Structure and dynamics of transcriptionally active chromatin. *J. Cell Sci.* 102:1–5.

Barrell, B. 1991. DNA sequencing: Present limitations and prospects for the future. *FASEB J.* 5:40–45.

Cattaneo, R. 1991. Different types of RNA editing. *Ann. Rev. Genet.* 25:71–88.

Cech, T. R. 1987. The chemistry of self-splicing RNA and RNA enzymes. *Science* 236:1532–1539.

Chastain, M., and Tinoco, I., Jr. 1991. Structural elements in RNA. *Prog. Nucleic Acid Res. Molec. Biol.* 41:131–177.

Conaway, R. C., and Conaway, J. W. 1993. General initiation factors for RNA polymerase II. *Ann. Rev. Biochem.* 62: 161–190.

Dreyfuss, G., Maniatis, M. J., Piñol-Roma, S., and Burd, C. G. 1993. hnRNP proteins and the biogenesis of mRNA. *Ann. Rev. Biochem.* 62:289–321.

Erlich, H. A., Gelfor, D., and Sninsky, J. J. 1991. Recent advances in the polymerase chain reaction. *Science* 252: 1643–1651.

Gilbert, W. 1981. DNA sequencing and gene structure. *Science* 214:1305–1312.

Greenblatt, J. 1991. RNA polymerase-associated transcription factors. *Trends Biochem. Sci.* 16:408–411.

Gross, D. S., and Garrard, W. 1988. Nuclease hypersensitive sites in chromatin. *Ann. Rev. Biochem.* 57:159–197.

van Holde, K. E., Lohr, D. E., and Robert, C. 1992. What happens to nucleosomes during transcription? *J. Biolog. Chem.* 267:2837–2840.

Kerpolla, T. K., and Kane, C. M. 1991. RNA polymerase: Regulation of transcription elongation and termination. *FASEB J.* 5:2833–2842.

Kornberg, R. D., and Lorch, Y. 1992. Chromatin structure and transcription. *Ann. Rev. Cell Biol.* 8:563–587.

Kunkel, G. R. 1991. RNA pol III transcription of genes that lack internal control regions. *Biochem. Biophys. Acta* 1088:1–9.

Lamond, A. I. 1993. The spliceosome. *Bioess.* 15:595–603.

Lewin, B. 1990. *Genes IV*. New York: Oxford University Press.

Maniatis, T. 1991. Mechanisms of alternative pre-mRNA splicing. *Science* 251:33–34.

Martin, K. J. 1991. The interaction of transcription factors and their adapters, coactivators, and accessory proteins. *Bioess.* 13:499–503.

Mattaj, I. W., Tollervey, D., and Séraphin, B. 1993. Small nuclear RNAs in messenger RNA and ribosomal RNA processing. *FASEB J.* 7:47–63.

Maxam, A. M., and Gilbert, W. 1977. A new method of sequencing DNA. *Proc. Nat. Acad. Sci.* 74:560–564.

McKeown, M. 1992. Alternative mRNA splicing. *Ann. Rev. Cell Biol.* 8:133–155.

Mullis, K. B. 1990. The unusual origin of the polymerase chain reaction. *Sci. Amer.* 262:56–65 (April).

Nussinov, R. 1990. Signal sequences in eukaryotic upstream regions. *Crit. Rev. Biochem. Molec. Biol.* 25:185–224.

Parry, H. D., Scherly, D., and Mattaj, I. W. 1989. "Snurpogenesis": The transcription and assembly of U snRNA components. *Trends Biochem. Sci.* 14:15–19.

Patthy, L. 1991. Exons—original building blocks of proteins? *Bioess.* 13:187–192.

Reeder, R. H., Labhart, P., and McStay, B. 1987. Processing and termination of RNA polymerase I transcripts. *Bioess.* 6:108–112.

Rickwood, D., and Hames, B. D. 1982. *Gel Electrophoresis of Nucleic Acids.* London: IRL Press.

Sanger, F. 1981. Determination of nucleotide sequences in DNA. *Science* 214:1205–1210.

Southeimer, E. J., and Steitz, J. A. 1993. The U5 and U6 snRNAs as active components of the spliceosome. *Science* 262:1989.

Steitz, J. A. 1988. Snurps. *Sci. Amer.* 258:56–63 (June).

Steitz, J. A. 1992. Splicing takes a Holliday. *Science* 257: 888–889.

Symons, R. H. 1992. Small catalytic RNAs. *Ann. Rev. Biochem.* 61:641–671.

Wahle, E. 1992. The end of the message: 3'-end processing leading to polyadenylated messenger RNA. *Bioess.* 14: 113–118.

Watson, J. D., and Tooze, J. 1981. *The DNA Story: A Documentary Story of Gene Cloning.* New York: Freeman.

White, T. J., Arnheim, N., and Erlich, H. A. 1989. The polymerase chain reaction. *Trends Genet.* 5:185–188.

Review Questions

1. Outline the structure of RNA. What are inverted sequences? Hairpins? What is the conformation of RNA double helices? Give two examples of the importance of secondary structure to RNA functions.

2. Describe the overall steps in transcription. What is a phosphodiester linkage? What is the 5' end of an RNA transcript? The 3' end? What structural features mark the 5' and 3' ends of a newly synthesized RNA molecule?

3. Summarize the properties and functions of eukaryotic RNA polymerases I, II, and III. What major RNA types are transcribed by each of the eukaryotic RNA polymerases?

4. What types of promoters are recognized by each of the polymerase enzymes?

5. Outline the overall steps of initiation, elongation, and termination of transcription. What factors are important in these mechanisms?

6. What is the template chain of a DNA molecule with reference to RNA transcription? The nontemplate chain? In what direction are DNA and RNA sequences read? Write an RNA coding sequence containing four codons (include some U's) and indicate how the sequence reads in the template and nontemplate chains of the DNA.

7. What major steps occur in RNA processing?

8. Diagram an mRNA gene, the pre-mRNA copied from the gene, and a mature mRNA molecule processed from the pre-mRNA.

9. How were pre-mRNAs identified as the precursors of mRNAs? What are introns? Exons? How were introns discovered?

10. What is the promoter of an mRNA gene? What are the major functions of the CAAT and TATA boxes? What is a consensus sequence? What is the startpoint?

11. What is an enhancer? How do enhancers differ from promoters? How are enhancers believed to work? What is the meaning of "upstream" and "downstream" with reference to mRNA genes?

12. Diagram the major sequence features of introns. What mechanisms may account for the existence of introns? What is "selfish DNA"?

13. Outline the steps occurring in initiation, elongation, and termination of mRNA transcription. What transcription factors participate in the initiation of mRNA transcription?

14. What processing steps occur in intron removal? What are snRNAs? Spliceosomes? How are snRNAs thought to participate in intron removal? What is a lariat? What is *trans* splicing?

15. What is alternative splicing? What is the significance of alternative splicing to protein structure and function?

16. What changes typically occur in the sensitivity of DNA to endonuclease digestion as genes go from an inactive to an active state? What are hypersensitive sites? What is the possible significance of these changes to chromatin structure?

17. Outline the steps followed in cloning a DNA sequence. What is a restriction endonuclease? Why is *Eco*RI especially useful for cloning DNA? What is a DNA library? (See Supplement 14-1.)

18. Outline the steps in a polymerase chain reaction. What is a DNA primer? How are the primers supplied for a polymerase chain reaction? (See Supplement 14-1.)

19. Summarize the steps in DNA sequencing by the Maxam and Gilbert and Sanger techniques. What is a dideoxynucleotide? Why does replication stop at a site at which a dideoxynucleotide has been entered? (See Supplement 14-1.)

20. How are protein sequences deduced from nucleic acid sequences? What is the significance of the AUG start codon and the terminator codons UAG, UUA, and UGA to this deduction? (See Supplement 14-1.)

DNA Cloning, the Polymerase Chain Reaction, and DNA Sequencing

DNA sequencing is the most direct and informative method for studying the structure of the genes encoding mRNAs and other rRNA types. The various RNAs can also be sequenced indirectly by making a DNA molecule complementary to the RNA sequence of interest. Such complementary DNAs (cDNAs) can be made readily in the test tube by reverse transcriptase enzymes.

One of the basic requirements for DNA sequencing is that the DNA sample must be available in quantities not usually obtainable directly from cells. Instead, samples are increased in quantity for sequencing by one of two techniques, *DNA cloning* or *the polymerase chain reaction*.

DNA Cloning

In DNA cloning, developed in the laboratories of H. W. Boyer, P. Berg, and S. N. Cohen, a DNA sequence of interest is introduced into a bacterium or virus, which is then cultured under conditions that promote maximum growth. During the growth the introduced DNA sequence is replicated along with the native DNA of the bacterium or virus. When growth reaches desired levels, the DNA sequence can be extracted in numbers equivalent to the number of bacterial cells or viruses produced.

The cloning technique depends on a group of *restriction endonucleases* extracted from bacteria. These enzymes recognize short DNA sequences and cut the DNA at or near these sequences (Table 14-2). The most useful restriction endonucleases are those that attack *inverted sequences* in such a way that the fragments produced by the digestion have complementary, single-stranded ends. (Inverted sequences contain a sequence element that repeats in reverse order on opposite nucleotide chains of a DNA molecule.) For example, the much-used restriction endonuclease *Eco*RI recognizes and cuts the inverted sequence

$$\downarrow$$
$$———G—A—A—T—T—C———$$
$$———C—T—T—A—A—G———$$
$$\uparrow$$

at the arrows, producing DNA fragments with complementary single-chain ends

$$———G \qquad A—A—T—T—C———$$
$$———C—T—T—A—A \qquad G———$$

Since *Eco*RI attacks only this sequence, all the fragments produced by the cutting action of the enzyme have single-chain ends capable of complementary base pairing.

In most DNAs the sequences cut by *Eco*RI and other restriction endonucleases appear at random points spaced several thousand base pairs apart. The enzymes can therefore cut essentially any natural DNA molecule into a finite number of fragments. When the restriction endonuclease cuts inverted repeats in the pattern catalyzed by *Eco*RI, all the fragments produced are capable of pairing with any other fragment produced by the same enzyme.

The DNA fragments produced in this way are cloned in bacteria by taking advantage of *plasmids*—small, circular DNA molecules that exist in the cytoplasm of bacterial cells without connection to the major DNA circle of the bacterial nucleoid (see p. 391). Although unconnected to the nucleoid, plasmids are replicated and passed on during bacterial cell division.

To clone a sequence (Fig. 14-17), both the DNA sequence of interest and plasmids isolated from a bacterium are digested with a restriction endonuclease such as *Eco*RI, producing fragments with complementary ends (Fig. 14-17a and b). When the DNA is exposed to conditions that promote base pairing, DNA circles form that contain both the plasmid DNA and the DNA of interest (Fig. 14-17c). The single-chain breaks remaining in the circles are then covalently closed by *DNA ligase*, an enzyme with this activity (Fig. 14-17d; DNA ligase activity is described in detail on p. 671). The circles including the introduced DNA are termed *recombinant DNA* because the end product of the technique, with DNA segments from different sources joined covalently into a continuous DNA molecule, resembles the outcome of genetic recombination (see p. 739).

The recombinant DNA circles are next introduced into a bacterium such as *E. coli*. Exposing *E. coli* cells to an elevated temperature (42°C) for a few minutes in the presence of Ca^{2+} induces the living cells to take up the recombinant plasmids. The plasmids are added at concentrations low enough to make it likely that a single cell will take up only one recombinant circle. As long as the circle includes a site from the original plasmid that can initiate replication (called a *replication origin*; see p. 680), it is duplicated in the bacterial cytoplasm and passed on to daughter cells during division. After many rounds of replication and division a clone of cells grown from one of the original cells taking up a recombinant circle will contain many copies of the DNA sequence of interest.

Table 14-2   Bacterial Restriction Endonucleases

Enzyme	Source	Sequences Recognized and Sites Cleaved
EcoRI	E. coli	↓ G – A – A – T – T – C C – T – T – A – A – G ↑
HindIII	Haemophilus	↓ A – A – G – C – T – T T – T – C – G – A – A ↑
HpaII	Haemophilus	↓ C – C – G – G G – G – C – C ↑
HhaI	Haemophilus	↓ G – C – G – C C – G – C – G ↑
BamHI	Bacillus	↓ G – G – A – T – C – C C – C – T – A – G – G ↑

To harvest the bacteria the plasmids are extracted and purified by techniques such as centrifugation. The sequence fragment is then cut from the plasmids by digestion with the same restriction endonuclease used to generate the original fragments, followed by centrifugation, gel electrophoresis, or other techniques for separating and concentrating the sample. (Centrifugation and gel electrophoresis are described in Appendix pp. 795 and 796.)

Viruses that infect bacteria are used for cloning by a similar approach, by digesting the DNA of the virus and the sequence of interest with a restriction endonuclease. Bacteriophage *lambda*, which infects *E. coli*, is much used in this method of cloning. *Eco*RI digestion cuts a nonessential segment from the center of the lambda DNA that can be replaced with a fragment from a DNA sample digested with the same enzyme. Infecting an *E. coli* culture with the recombinant viruses leads to the production of millions to billions of viral particles containing the DNA sequence of interest. After cloning, the DNA, now also multiplied into millions or billions of copies, is extracted from the virus particles and separated from the lambda DNA by digestion with *Eco*RI.

When the entire DNA complement of an organism is used for cloning, the technique creates what is known as a *DNA library* containing cloned fragments of all the DNA sequences of the organism. Following extraction of the cloned DNA, individual sequences

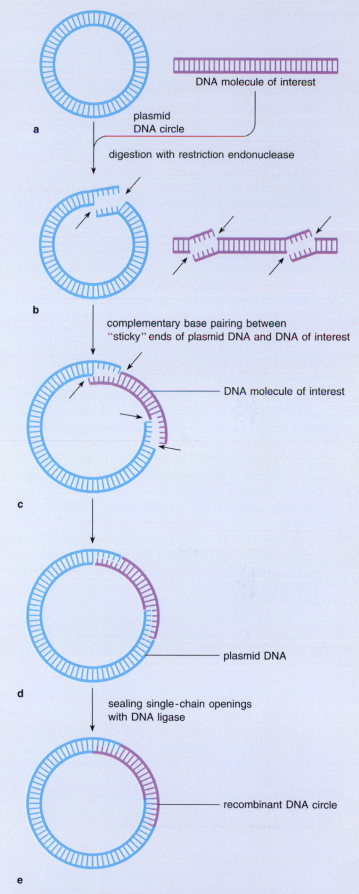

Figure 14-17   Production of recombinant DNA for cloning by restriction endonuclease digestion (see text).

DNA Cloning, the Polymerase Chain Reaction, and DNA Sequencing   **421** can be identified and isolated from the library by hybridization with DNA probes (see Appendix p. 798) complementary to the sequences of interest.

The Polymerase Chain Reaction

Producing multiple DNA copies by cloning recombinant molecules is a demanding procedure that requires elaborate biochemical techniques and considerable time. The polymerase chain reaction, developed in 1983 by K. B. Mullis and F. Faloona, has greatly simplified production of multiple DNA copies from a sample. (Mullis received the Nobel Prize in 1993 for his development of the polymerase chain reaction, which has found wide application in all branches of molecular biology.) The reaction takes advantage of several characteristics of *DNA polymerase*, the enzyme that catalyzes DNA replication, and the availability of laboratory equipment capable of automatically synthesizing short DNA molecules of any desired sequence.

In the replication reaction catalyzed by DNA polymerase the two nucleotide chains of the molecule being duplicated unwind (Fig. 14-18a). Each chain serves as template for synthesis of a complementary copy assembled by sequential base pairing (Fig. 14-18b and c). Wherever a thymine (T) is the base being copied in the template, a nucleotide containing adenine (A) is placed in the copy; wherever a cytosine (C) appears in the template, a nucleotide containing guanine (G) is placed in the copy. Similarly, an A in the template causes a T to be placed in the copy, and a G in the template causes a C to be placed in the copy. A template chain can be copied only in one direction. Because the two nucleotide chains of a DNA molecule are *antiparallel*, meaning that they run in opposite directions (see p. 364), replication proceeds in opposite directions in the two chains (indicated by vertical arrows in Fig. 14-18b).

DNA polymerase can add nucleotides only to the end of an existing nucleotide chain. Therefore, for replication to start and proceed, an existing chain called a *primer* must already be in place, paired with the end of the template chain at which replication is to begin.

These features of the replication process form the basis of the polymerase chain reaction. In a commonly used application of the technique (Fig. 14-19) a DNA double helix containing a sequence of interest is unwound into separate nucleotide chains by heating the sample to 90°C (Fig. 14-19a). To the mixture is added DNA polymerase, a supply of the four A, T, C, and G nucleoside triphosphates required for replication, and a large excess of short, artificially synthesized DNA chains that are complementary to the ends of the unwound template chains. The added DNA chains, about 20 to 30 nucleotides in length, serve as primers for the reaction. The mixture is cooled to 60°C, allowing the artificial primers to wind with the ends of the template chains (Fig. 14-19b). Replication then takes place, in which the DNA polymerase enzymes assemble complementary copies of the template chains starting from the artificial primers (Fig. 14-19c). This step completes the first cycle of the polymerase chain reaction. The reaction mixture now contains twice as many DNA molecules as the starting mixture.

The second cycle is initiated by again heating the reaction mixture, causing the newly synthesized double helices to unwind (Fig. 14-19d). The mixture is cooled, allowing additional copies of the artificial primer chains to rewind with the ends of the template chains, as in the first cycle (Fig.14-19e). The DNA polymerase enzymes make copies using the artificial chains, as before (Fig. 14-19f). At the completion of this cycle the number of DNA molecules has again doubled. Each time the heating and cooling cycle is repeated, the number of DNA molecules in the sample doubles. Cycling time is short enough for the sequential reactions to make hundreds of billions of DNA copies in a few hours.

The DNA polymerase often used in the polymerase chain reaction is one originally isolated from the bacterium *Thermus aquaticus*, a species that lives in hot springs. The enzyme, produced in quantity for the

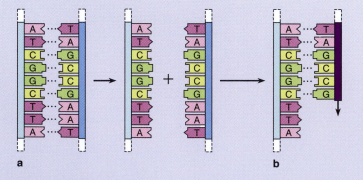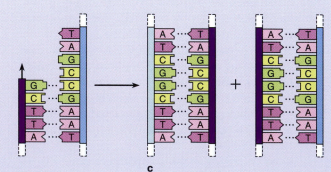

Figure 14-18 Replication by DNA polymerase (see text).

chain reaction by genetically engineered bacteria, is not denatured when the reaction mixture is heated to 90°C.

DNA cloning and the polymerase chain reaction are used to increase the quantity of desired DNA sequences for other purposes as well as DNA sequencing. These include the preparation of DNA to be introduced into cells for genetic engineering, studies of the interaction of regulatory proteins with genes, mapping chromosomes, and DNA fingerprinting, in which DNA samples are matched to determine the identity of blood or semen samples in paternity and criminal proceedings (see Information Box 18-2).

Sequencing DNA

Once obtained in large quantities by cloning or the polymerase chain reaction, the DNA sample is sequenced by one of two methods. One, developed by A. Maxam and W. Gilbert, uses chemical reagents that break DNA molecules at specific bases. The DNA fragments are run on electrophoretic gels, and the sequence is read directly from the series of bands sorted out by the gels. Since DNA copies can be made from RNA molecules by reverse transcriptase, the method can also be used indirectly to sequence RNA.

In the Maxam and Gilbert technique a quantity of the DNA molecule to be sequenced is first digested by a restriction endonuclease. The digestion breaks the DNA into groups of fragments. Within each group all the fragments are of uniform length and contain the same subpart of the DNA sequence. The fragment groups are separated by centrifugation or gel electrophoresis.

The DNA lengths of each fragment group are labeled at their 5' ends by enzymatic attachment of a radioactive phosphate group. A fragment group is then treated by reagents that modify and break the DNA chain at sites where specific bases appear. One sample from a fragment group is treated with the reagent *dimethyl sulfate*, which adds methyl groups to guanines and adenines (both purines) in the DNA fragments. The dimethyl sulfate treatment is carried out under conditions that methylate at random some but not all of the guanine and adenine bases. The DNA chain is unstable at sites containing the methyl groups and is easily broken at these points by treatment with an alkali at 90°C. The alkali treatment produces a family of DNA pieces from a fragment group, with each DNA piece broken at a different guanine or adenine in the sequence (Fig. 14-20a).

The pieces from a DNA fragment group are then run on an electrophoretic gel, which separates the family of pieces into individual bands according to length. A photographic emulsion is placed over the

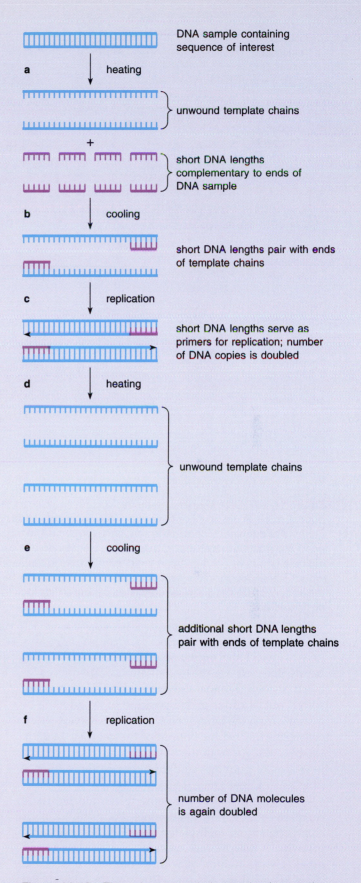

Figure 14-19 The polymerase chain reaction (see text).

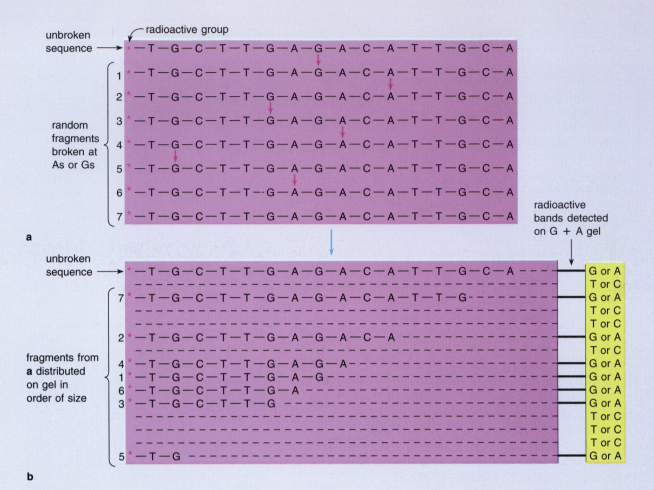

Figure 14-20 DNA sequencing by the Maxam and Gilbert technique (see text). To produce the G + A gel, a uniform class of DNA fragments **(a)** is broken randomly wherever guanine or adenine occurs in the sequence (arrows). The fragments are then sorted out according to length by running them on an electrophoretic gel **(b)**. The bands on the gel indicate points where a guanine or adenine occurs in the sequence; empty spaces show where T or C occurs in the sequence (see text). Fragments missing the beginning end of the sequence do not show up on the gel because they lack the radioactive marker and are not detected.

gels to detect the radioactive bands. Since the only bands detected are those containing DNA pieces labeled at their 5' ends, the radioactive fragments are sorted out in order of the position of an adenine or guanine from the labeled end, with successively shorter lengths toward the bottom of the gel (Fig. 14-20b). Reading the gel from the bottom to the top gives the positions of guanines and adenines in the sequence (the "G + A" gel in Fig. 14-21).

A modification of the same technique produces segments broken only at adenines. Points in the DNA chain containing methylated adenines are less stable in an acid than points containing methylated guanines. Gentle treatment with dilute acid breaks the chain at only these adenines. Running the end-labeled fragments sorts them out by length, with successively

shorter lengths toward the bottom of the gel. Reading this gel (the "A" gel in Fig. 14-21) from bottom to top and comparing it to the G + A gel distinguishes between G's and A's in the sequence.

A similar series of treatments using the reagent *hydrazine* produces DNA pieces ending in T or C. Hydrazine removes the T and C bases (both pyrimidines) from the DNA fragments, producing weak points that break when treated with alkali. Hydrazine is used under conditions in which the reaction is incomplete, so that only some of the T's or C's are removed at random points. The resulting fragments, ending at either T or C, produce the "T + C" gel shown in Figure 14-21.

A final step distinguishes T from C. By using the hydrazine reagent in the presence of 1 molar salt, the T's are protected and only C's are removed randomly

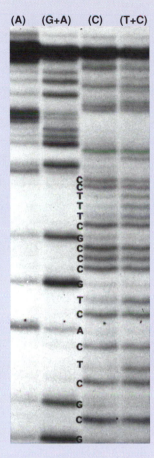

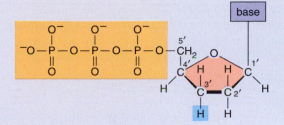

Figure 14-22 A dideoxynucleotide, in which an —H (boxed in blue) instead of an —OH is linked to the 3' carbon. Oxygens are therefore missing from both 2' and 3' carbons.

Figure 14-21 A, G + A, C, and T + C gels showing the sequence of a DNA fragment from a bacterial cell. The sequence is read directly from bottom to top. (Courtesy of A. M. Maxam.)

from the DNA fragment group. The fragments are then broken at the weak points by alkali, yielding a series of pieces ending with nucleotides that contained a C. This preparation produces the "C" gel shown in Figure 14-21.

The DNA sequence of the fragment group is read directly from the bottom to top of the aligned gels by comparing the G + A, A, T + C, and C gels at each level. Combining the sequences of all the fragment groups from the original molecule allows the entire sequence of the DNA molecule, or of a cDNA copy of an RNA molecule, to be reconstructed.

An alternate sequencing technique developed by F. Sanger and his colleagues is similar except that it uses DNA replication to provide the consecutive sequence lengths for gel electrophoresis. In the replication, modified forms of the four DNA nucleotides are used in which a single —H is bound to the 3' carbon of the deoxyribose sugar instead of an —OH (Fig. 14-22). A nucleotide in this form is known as a *dideoxynucleotide*. During DNA replication a new nucleotide is

normally added to the 3'-OH group of the most recently added nucleotide in the copy. Because the dideoxynucleotides have no 3'-OH available for addition of the next base, DNA replication stops whenever one of these nucleotides is present instead of an unmodified nucleotide.

In practice, groups of sequence fragments are obtained by digesting the DNA to be sequenced with a restriction endonuclease, as in the Maxam and Gilbert technique. Four different replications of a given fragment group are then run in the test tube, each with a different dideoxynucleotide added to the replication mixture along with all four of the unmodified A, T, G, and C nucleotides. In each run the dideoxynucleotide is added in low concentration, about 1% to 2% compared to the unmodified nucleotide with the same base. As a result, insertion of a dideoxynucleotide at any site is less probable than insertion of the unmodified nucleotide, and the points at which replication is stopped by an insertion are random. Use of a given dideoxynucleotide under these conditions therefore replicates the DNA of a fragment group to different lengths, each one stopping at a different nucleotide of the same base.

Suppose, for example, that the dideoxynucleotide used for one replication run contains the base adenine. This dideoxynucleotide is mixed in low proportions with the four unmodified nucleotides. At any point in the replicated copy calling for an adenine either the unmodified nucleotide or the dideoxynucleotide is inserted. Replication stops if an adenine dideoxynucleotide is inserted. Because the stopping points are random, replication for an extended time produces a group of sequence fragments in which each fragment starts at the same place but ends at a different A in the sequence (Fig. 14-23a). Running the fragments on an electrophoretic gel sorts them out by length, with the shortest sequence at the bottom of the gel (Fig. 14-23b). The adenine sequence is then read in order from the bottom of the gel to the top. Using the four dideoxynucleotides in turn and running the sequence fragments obtained gives A, T, C, and G gels that, when aligned side by side, allow the entire DNA sequence of a fragment group to be read in order from bottom to top.

Methods for sequencing DNA have been a primary source of the molecular revolution. In addition to revealing the structure of mRNA and other genes, DNA sequences provide insights into gene functions and the mechanisms by which genes are regulated. Sequence comparisons allow estimation of the degree to which genes of the same or different organisms are related, confirm evolutionary lineages determined by other more traditional methods, and establish lineages that were previously unsuspected. In some cases, comparisons of normal and mutant gene sequences have revealed the molecular basis of genetic deficiencies and hereditary disease.

Deducing Protein Sequences from DNA Sequences

Proteins can be sequenced indirectly from the DNA sequences of the genes encoding the proteins. The indirect sequencing involves starting at the three-nucleotide sequence (almost always AUG) that indicates

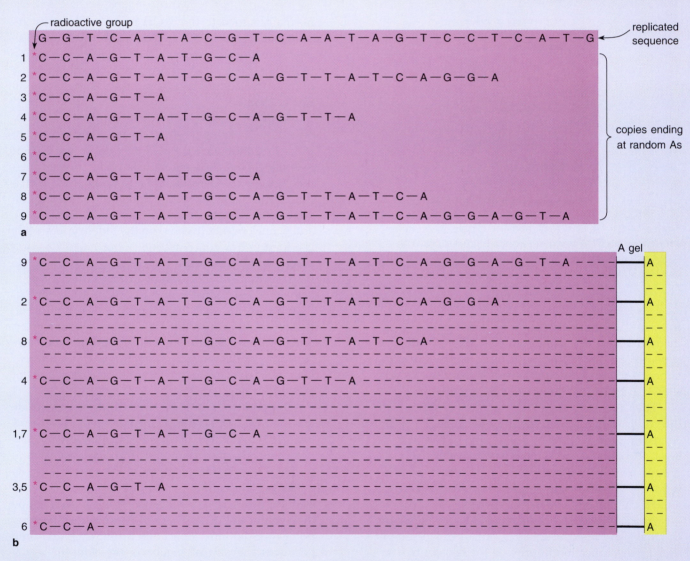

Figure 14-23 DNA sequencing by the Sanger technique (see text). To produce an adenine gel a DNA sequence is replicated in a cell-free system containing the four DNA nucleotides, with the adenine dideoxynucleotide added in low concentration. The presence of the adenine dideoxynucleotide causes replication to stop at random adenines in the sequence being copied **(a)**. Running the sequence segments on the electrophoretic gel **(b)** sorts them out by length; the sequence of A's can then be read directly from the bottom to the top of the gel.

the beginning of the coding portion of the gene. The amino acid sequence of the encoded protein is then worked out by reading the nucleotides three at a time until one of the three codons indicating the end of a coding sequence (UAG, UUA, or UGA) is reached. Because introns begin and end with the characteristic GU and AG combinations (see p. 411), they can be recognized and deleted from the nucleotides read to deduce the amino acid sequence.

Proteins are also sequenced indirectly from a cDNA copy of the mRNA transcribed from a gene. Either method is much faster than the laborious chemical methods used for direct protein sequencing, which can take months or even years to complete.

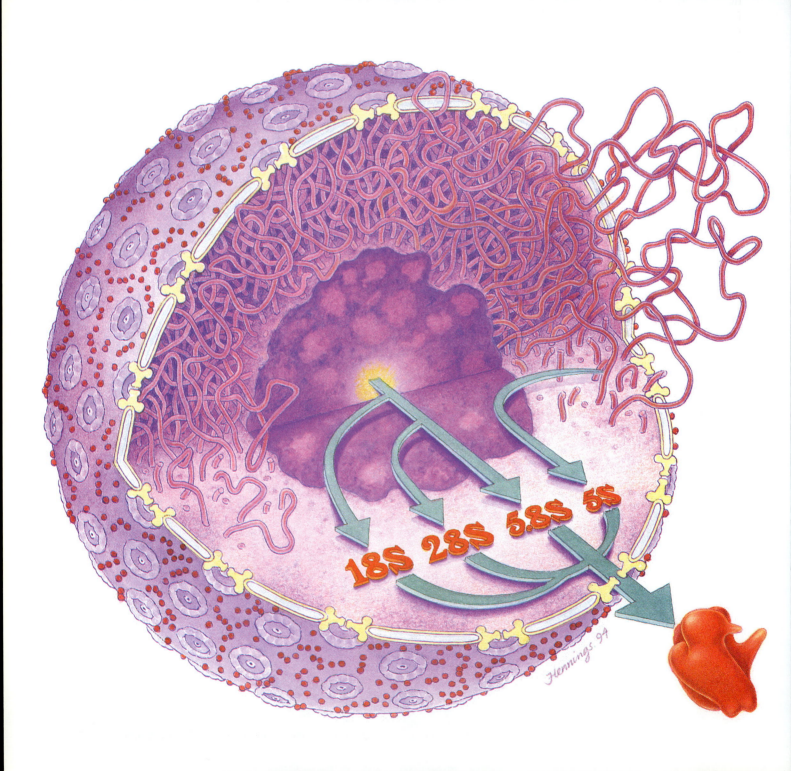

18S 28S 5.8S 5S

Hennings 94

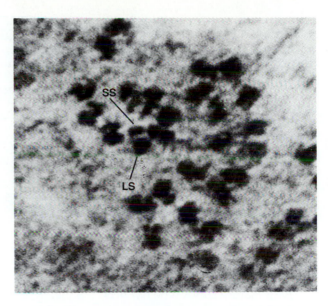

Figure 15-1 Ribosomes in a thin-sectioned rat liver cell. A cleft marking the division between the large and small subunit is visible in many of the ribosomes. LS, large subunit; SS; small subunit. (Courtesy of N. T. Florendo.)

R ibosomal RNA, transfer RNA, and small nuclear or cytoplasmic RNAs act in accessory roles in protein synthesis or as parts of the processing reactions converting RNA precursors into their finished forms. Each of these RNA types is transcribed in the form of a precursor that is subsequently processed to generate a mature rRNA, tRNA, or sn/scRNA molecule. Like those of mRNA processing, the reactions converting the precursors to mature forms may include clipping and splicing, removal of nucleotides from the ends or interior of the precursors, addition of nucleotides, chemical modification of individual bases, and addition of proteins to form ribonucleoprotein particles.

This chapter describes eukaryotic rRNAs, tRNAs, and sn/scRNAs, the structure of the genes encoding them, and the transcription and processing reactions assembling their precursors and processing them to finished form. The equivalent reactions of prokaryotes, including mRNA genes and their transcription, are described in Supplement 15-1.

TRANSCRIPTION AND PROCESSING OF RIBOSOMAL RNA

Ribosomal RNAs are transcribed in greater quantities than any other RNA type. This intensive rRNA synthesis supports the assembly of ribosomes, which are required literally in the millions per cell for protein synthesis to proceed at adequate levels.

The rRNAs

Ribosomal RNAs make up roughly half of the total mass of ribosomes in both prokaryotes and eukaryotes. Individual ribosomes are assembled from two subunits, one large and one small (Fig. 15-1). Each subunit consists of one or more rRNA molecules in combination with 30 to 40 different proteins, for a total of 70 to 80 proteins per eukaryotic ribosome (for details of ribosome structure, see p. 473).

Four distinct kinds of rRNA can be extracted from eukaryotic ribosomes, one from the small subunit and three from the large subunit. These rRNAs are usually identified in *Svedberg* or *S* units (see p. 795), which reflect the relative rates at which molecules descend in a centrifuge under standard conditions. The faster the molecules descend, the larger the S number, and, generally, the higher the molecular weight. In these units the single rRNA of the small eukaryotic ribosomal subunit is identified as *18S* rRNA, and the three rRNAs of the large subunit as *28S*, *5.8S*, and *5S* rRNA (Table 15-1). These numbers are consensus values; actual values vary depending on the species and methods used to prepare the rRNAs for centrifugation.

In eukaryotes, 28S, 18S, and 5.8S rRNA are transcribed by RNA polymerase I (see p. 402 for details of the eukaryotic RNA polymerases and their functions) in the form of a large pre-rRNA, which ranges from about 37S to 45S depending on the species. Each large pre-rRNA contains one 18S, one 5.8S, and one 28S sequence in that order, separated by transcribed spacers. Processing reactions release the individual rRNA sequences as separate molecules. Transcription and processing of large pre-rRNAs take place in the nucleolus.

The remaining rRNA type, 5S, is separately transcribed as a pre-5S rRNA by RNA polymerase III (see p. 402) on genes located outside the nucleolus. The pre-5S rRNA, which is slightly larger than the finished 5S molecules, subsequently enters the nucleolus for processing and assembly with 28S and 5.8S rRNA into large ribosomal subunits. 18S rRNA combines with a distinct group of ribosomal proteins to form small ribosomal subunits. In this chapter the precursor containing the 18S, 5.8S, and 28S rRNAs is termed the

Table 15-1 Ribosomes, Ribosomal Subunits, and rRNAs in Prokaryotes and Eukaryotes (with approximate molecular weights in parentheses)

Group	Intact Ribosomes	Ribosomal Subunits	rRNAs	Number of Nucleotides
Prokaryotes	70S (2,520,000)	30S (930,000)	16S (550,000)	~1480–1540
		50S (1,590,000)	23S (1,100,000)	~2900–2930
			5S (41,000)	116–120
Eukaryotes	80S (4,420,000)	40S (1,400,000)	18S (700,000)	~1790–1900
		60S (2,820,000)	28S (1,400,000)	~3400–4700
			5.8S (50,000)	158–163
			5S (41,000)	116–121

"large pre-rRNA" and the one containing 5S rRNA "pre-5S rRNA."

Each mature rRNA type has been completely sequenced in many eukaryotic species, with the sequences revealing many interesting and significant features. The sequence of a given rRNA type is very similar in closely related species but becomes increasingly divergent as it is compared in more distantly related organisms. The degree to which the primary structure of a given rRNA type is similar between two different species, as a result, reflects the evolutionary history and "relatedness" of the organisms. The dependence of rRNA sequences on evolutionary relationships is sufficient, in fact, to allow deduction of evolutionary trees based on rRNA sequence comparisons. The lineages constructed in this way correspond well with evolutionary trees based on other lines of evidence such as the fossil record. In some cases the rRNA-based trees have helped to resolve evolutionary relationships and origins of species and groups that were previously in doubt.

All the rRNAs contain inverted sequences that can form a wide assortment of hairpins and other secondary structures (see p. 397 and Fig. 15-2). Each rRNA type can fold into a secondary structure that with certain variations is common to all organisms and can be extended even to the equivalent rRNA types of prokaryotes, mitochondria, and chloroplasts. This is true in spite of the fact that the nucleotide sequences of a given rRNA type in distantly related organisms may be quite different. This shows that, with some exceptions, secondary rather than primary structure is the conserved feature of rRNA molecules. It follows from this conclusion that the secondary structure of rRNA molecules, rather than the nucleotide sequence, is the feature of greatest importance in ribosome assembly and function.

Among the exceptions to these conclusions are regions in which nucleotide sequence as well as secondary structure is conserved. In bacteria, for example, a sequence in the small-subunit rRNA (*16S rRNA*;

see Fig. 15-2a and Table 15-1) pairs with a complementary sequence in the 5' untranslated region of bacterial mRNAs during initiation of protein synthesis. The rRNA sequence, known as the *Shine–Dalgarno* sequence (see p. 486), and its complementary mate in bacterial mRNAs are highly conserved in bacteria.

Variations in the secondary structures of individual rRNA types among different species are limited primarily to certain regions of the molecules. Insertions and deletions of sequences in these regions, which are evidently less critical to function, also account for most of the differences in size between rRNAs of different species.

For example, bacterial 16S rRNA can fold into a pattern of hairpins and other paired structures equivalent to that of eukaryotic 18S rRNA (see Fig. 15-2). The patterns are equivalent even though bacterial 16S rRNA is significantly smaller and almost completely different in nucleotide sequence from eukaryotic 18S rRNA. Most of the variations in size and secondary structure between prokaryotic 16S and eukaryotic 18S rRNAs are confined to six variable regions.

All rRNAs undoubtedly fold further into tertiary, three-dimensional structures as they are processed and assembled into the ribosomal subunits. Investigations of these tertiary rRNA structures are just beginning to yield results. (Some details of this research are presented in the description of ribosome structure in Chapter 16.)

The functional significance of secondary and tertiary structures in rRNA remains largely unknown. However, the folding patterns probably serve as molecular "signposts" for rRNA processing, recognition of rRNA segments by proteins during assembly of the ribosomal subunits, and maintenance of ribosome structure. The folding patterns probably also provide recognition sites for associations among ribosomes, mRNAs, tRNAs, and the various factors linked to ribosomes during protein synthesis. Some secondary structures are likely to change between different forms as part of individual reactions of polypeptide assembly.

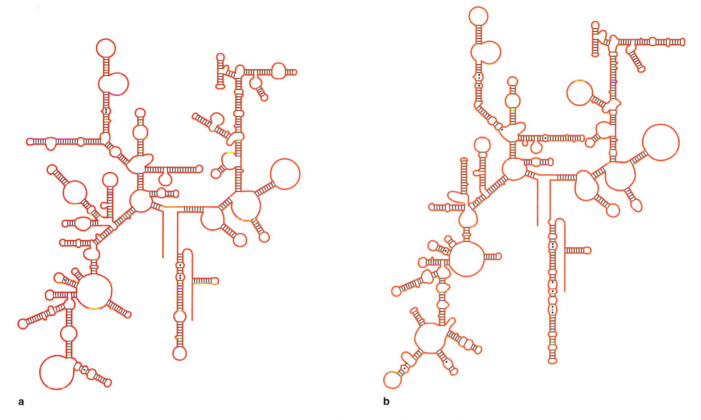

a

b

Figure 15-2 Secondary structures of a bacterial 16S rRNA from *E. coli* **(a)** and a eukaryotic 18S rRNA from a yeast cell (b). (Courtesy of H. F. Noller; reproduced, with permission, from *Ann. Rev. Biochem.* 53:119, © 1984 by Annual Reviews, Inc.)

The 28S and 5.8S rRNAs of eukaryotes show an interesting structural relationship to the major rRNA of the large prokaryotic ribosomal subunit, designated as 23S rRNA. Much of the 5.8S rRNA sequence is complementary to the 5' end of eukaryotic 28S rRNA. This allows a paired structure to form between the two rRNAs in this region (Fig. 15-3a). The 5' end of prokaryotic 23S rRNA, which is equivalent to eukaryotic 28S rRNA, can fold into essentially the same paired structure (Fig. 15-3b). Evidently during the evolution of eukaryotes from their prokaryotic ancestors a split occurred near the 5' end of the prokaryotic 23S rRNA, liberating a separate molecule that has persisted as eukaryotic 5.8S rRNA.

The smallest rRNA of the large ribosomal subunit, 5S rRNA, folds into a secondary structure containing four double-helical segments (Fig. 15-4a). These segments may fold further into a tertiary structure proposed by J. McDougall and R. N. Nazar (Fig. 15-4b). 5S rRNA is the most highly conserved of the various rRNA types. Of the approximately 120 nucleotides in this rRNA type, 23 are invariant or nearly so in all species, both eukaryotic and prokaryotic.

Despite its highly conserved nature, 5S rRNA variants with small differences in sequence occur within individual higher vertebrate species. In the amphibian *Xenopus*, in which the variants have been best studied, three different 5S rRNA types, differing in about 5% of their sequences, have been detected. Two of the variants make up the *oocyte-type* 5S rRNA, transcribed only in oocytes and early embryos. The remaining 5S rRNA variant, the *somatic-type*, is transcribed in all *Xenopus* cells. Multiple 5S variants have been detected in at least one other vertebrate species, the chicken. The functional significance of the 5S rRNA variants is unknown.

More 5S molecules have been sequenced in different species than any other RNA type. Information from these sequencing studies has provided the best and most complete of the RNA-based evolutionary trees.

rRNA Gene Structure, Transcription and Processing

The genes encoding large pre-rRNAs occur in multiple copies in all eukaryotes, concentrated in clusters at one or more locations in the chromosomes of each species. This arrangement is in distinct contrast to that of mRNA genes, which occur largely in single copies

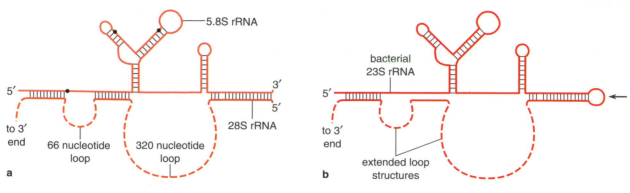

Figure 15-3 Relationship of eukaryotic 5.8S rRNA to the 5' end of prokaryotic 23S rRNA. **(a)** Pairing between 5.8S rRNA and the 5' end of 28S rRNA in eukaryotes. **(b)** The equivalent paired structure formed by fold-back hairpin at the 5' end of prokaryotic 23S rRNA. A single break at the arrow would convert this paired structure to the eukaryotic form, with a separate 5.8S rRNA molecule rather than the continuous structure shown. Dots mark the positions of modified bases.

scattered throughout the chromosome set. Each repeat in the large pre-rRNA gene clusters consists of a coding segment and a long *intergenic spacer* that contains sequences signaling initiation and termination of transcription (Fig. 15-5a).

Large Pre-rRNA Genes A large pre-rRNA gene (Fig. 15-5b) contains one copy each of the 18S, 5.8S, and 28S rRNA coding sequences, arranged in that order, with the 18S sequence closest to the 5' end of the gene. The three coding sequences are separated by *intragenic spacers* that are transcribed but are later cut out during large pre-rRNA processing.

Substitution-deletion experiments revealed that the promoters of most large pre-rRNA genes have two primary sequence elements that are recognized and bound by transcription factors during initiation. The *core promoter element* brackets the startpoint, in the region from about −30 to +15. The other sequence, the *upstream promoter element*, lies in the region from about −150 to −50. Individual sequences in the core promoter and upstream promoter element vary widely in different species. In general, no consensus sequences analogous to the TATA and CAAT boxes of mRNA genes are noted. This suggests that distinct regulatory proteins regulate initiation of large pre-rRNA transcription in different species. This conclusion is supported by experiments with cell-free systems. For example, among cell-free systems developed from mammalian species, large pre-rRNA genes isolated

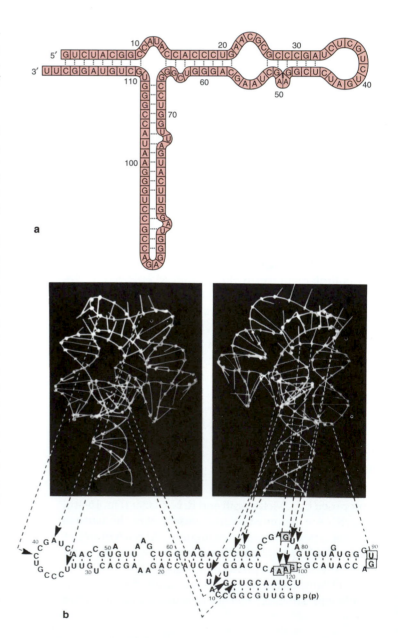

Figure 15-4 Structural models for 5S rRNA. **(a)** Fold-back double-helical structure in a human 5S rRNA. Bacterial 5S rRNAs can fold into essentially the same structure. **(b)** A tertiary structure proposed for 5S rRNA. ([b] courtesy of R. N. Nazar, from *J. Biolog. Chem.* 266: 4562 [1991].)

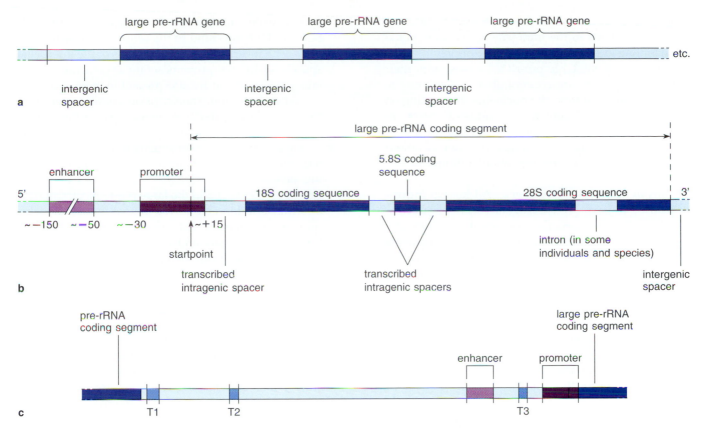

Figure 15-5 Large pre-rRNA genes. **(a)** The pattern of tandem repetition of large pre-rRNA genes and nontranscribed spacers in large pre-rRNA gene clusters. Tens, hundreds, or thousands of repeats occur in the clusters in different species. **(b)** The arrangement of coding sequences and intragenic spacers in large pre-rRNA genes. **(c)** The intergenic spacer of large pre-rRNA genes.

from the mouse can be successfully transcribed by RNA polymerase only in mouse cell extracts; attempts to transcribe mouse pre-rRNA genes with human cell extracts are unsuccessful.

The intergenic spacer separating large pre-rRNA genes (Fig. 15-5c) contains one or more termination signals for the gene repeat preceding it, with the last of the series often placed near the promoter for the following pre-rRNA gene. The multiple terminators appear to work primarily as a backup system, because transcription in most cases ends at the terminator lying farthest upstream. As a consequence, most of the intergenic spacer usually remains untranscribed. In many cases the terminators consist of inverted sequences preceding three or more A-T base pairs. The inverted sequences and A-T base pairs may act similarly in termination to one class of bacterial terminators (see p. 453 and below).

In many species the region in the intergenic spacer between the initial and final terminators contains repeated sequences in different combinations and types. Some of these sequence elements are recognized and bound by transcription factors and act as enhancers.

Additional promoters may also lie in this region. Although transcription rarely initiates from these upstream promoters, they greatly enhance initiation from the promoter bracketing the startpoint when they are present. The upstream promoters thus appear to act as additional enhancer elements, possibly by binding and "recruiting" transcription factors for the primary promoter. Even the final terminator sequence positioned just in advance of the primary promoter appears to have stimulatory effects on initiation. All of these sequence elements stimulating initiation of mRNA genes may be related to the intensity with which the genes are transcribed, which undoubtedly requires recruitment of transcription factors and RNA polymerase I molecules in great quantity.

With few exceptions the same arrangement of coding sequences and spacers occurs in the large pre-rRNA genes of all eukaryotes. The coding sequences show a greater or lesser degree of sequence conservation between different species depending, like the rRNAs they encode, on evolutionary relatedness. Few or no sequence homologies, however, appear in spacer regions within the gene, even between closely related species.

In some organisms, such as the amphibian *Xenopus*, sequence variations in the intragenic spacers are noted even among individuals of the same species.

Repeats of the large pre-rRNA genes, with their intergenic spacers, are concentrated in segments of the chromatin that form the nucleolus. Depending on the species, from dozens to thousands of copies are found in nucleolar chromatin. *Drosophila* has about 400 to 450 large pre-rRNA gene repeats in a haploid chromosome set; humans have about a thousand.

Introns in Large Pre-rRNA Genes All known introns appearing in large pre-rRNA genes occur about two-thirds of the way along the 28S coding sequence, in a region corresponding to one of the variable segments of the 28S secondary structure. The introns, which are spliced out during pre-rRNA processing, have been discovered in a number of eukaryotic organisms, including nuclear rRNA genes in the protozoans *Tetrahymena* and *Chlamydomonas* and the fungus *Physarum* and in mitochondrial and chloroplast rRNA genes of lower eukaryotes. Although no consensus sequences occur at the boundaries of these introns, all contain inverted sequences that can form similar patterns of secondary structures. The paired structures evidently arrange the introns into configurations that promote self splicing without intervention of enzymatic proteins (see Fig. 15-7).

Large Pre-rRNA Transcription and Processing Transcription of large pre-rRNA genes begins as the RNA polymerase I enzyme binds to the 5'-flanking promoter sequences. One or more transcription factors are required for recognition and binding; these factors, like their counterparts in mRNA transcription (see p. 412), bind to the promoters and set up a DNA-protein complex that is recognized and bound tightly by the polymerase. One of these factors, *SL1*, binds to the core promoter element. Although the core promoter contains no apparent sequence similarities to the TATA box of mRNA genes, the central polypeptide of SL1 is none other than *TPB*, the protein subunit of the TFIID initiation factor that binds the TATA box and plays a critical role in initiation of mRNA gene transcription (see p. 412). Thus TPB figures importantly in initiation by both RNA polymerase I and II (and also III; see below). SL1 has additional subunits that vary in properties between different species and seem to be the primary factors responsible for species specificity of the enzymatic mechanism transcribing large pre-rRNA genes in vertebrates. Another factor central to initiation of transcription of large pre-rRNA genes is *UBF*, which binds to both the core and upstream promoter elements and stimulates binding of both SL1 and RNA polymerase I to the promoter. UBF

appears to be common to many vertebrate species. Both SL1 and UBF also bind to the sequence elements acting as enhancers in upstream regions of the intergenic spacer. Additional proteins acting as cofactors stimulating transcription are also present in most species.

After initiation, transcription continues entirely through a large pre-rRNA gene, copying the coding sequences and spacers without interruption, and stops at one of the termination signals in the following intergenic spacer. Protein factors recognizing and binding the termination sequences may promote release of RNA polymerase I and the large pre-rRNA transcript.

Transcription is followed almost immediately by processing reactions that modify individual nucleotide bases or sugars and cut the 18S, 5.8S, and 28S rRNA molecules from the large pre-rRNA. Although the cutting sites and the sequence of cuts vary to some degree in different species, in most organisms initial cleavages by RNA endonucleases occur near or at the 5' and 3' ends of the 18S sequence, liberating this molecule as an almost completely processed entity (Fig. 15-6). These cuts leave a segment containing the 5.8S and 28S sequences, which are released by further cuts. While the clipping reactions are in progress, ribosomal proteins are added to the maturing rRNA molecules.

Although both RNA endonuclease and exonuclease activity can be detected in the nucleus, the specific enzymes cutting the 18S, 28S, and 5.8S molecules from the large pre-mRNA are still unknown. The sequence features recognized as signals by the enzymes to make the cuts are similarly unclear. No doubt some of the paired structures formed by inverted sequences within the large pre-rRNA and conformations set up by the combination of processing proteins with the pre-rRNA are important as recognition signals for the processing enzymes.

Several snRNAs, including U3, U8, U13, U14, and additional small RNAs known as E1, E2, E3, X, and Y, are present in nucleoli and probably participate in the reactions clipping mature rRNAs from the large pre-rRNA precursor. In nuclei in which U3 activity is inhibited, clipping at the 3' end of 18S rRNA stops; cleavage between the 18S and 5.8S segments appears also to be reduced to some extent. If U8 activity is inhibited, processing of both the 3' and 5' ends of 5.8S and 28S rRNA is faulty. Thus U3 and U8 appear to be involved somehow in these processing steps. The remaining small nucleolar RNAs have not as yet been linked to particular steps in large pre-rRNA processing.

The cleavage reactions releasing the 18S, 5.8S, and 28S rRNAs from the large pre-rRNA transcript were first traced in a classic series of experiments by R. P. Perry and others. In the experiments, cells were exposed to tritiated uridine, a radioactive precursor used

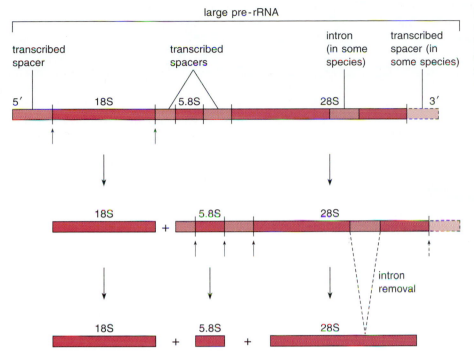

Figure 15-6 The probable sequence of cuts releasing rRNAs from the large rRNA precursor in eukaryotes (see text). The small vertical arrows indicate sites of enzymatic cleavage.

by cells to make RNA, followed after a brief interval by actinomycin D, a drug that halts RNA transcription. After blockage by actinomycin D, cells could process the rRNAs transcribed during exposure to the label but could not make any more rRNA copies. If cell extracts were made a few minutes after exposure to the label and actinomycin D block, the large pre-rRNA, which had an S value of 40S to 45S in the species used for the studies, was labeled. If cells were allowed to process rRNAs for about 10 to 15 minutes before extracts were made, the label was distributed between 18S rRNA and a larger 32S segment. By 20 to 30 minutes after exposure to label, the labeled 32S molecules disappeared and labeled 5.8S and 28S molecules could be found in the extracts. If the period between exposure and preparation of cell extracts was extended to several hours, radioactivity could be detected in cytoplasmic ribosomes. Besides revealing the sequence of cleavages producing the final 18S, 5.8S, and 28S rRNA molecules, the labeling experiments were the first to demonstrate that the large pre-rRNA transcript is actually the precursor of cytoplasmic 18S, 5.8S, and 28S rRNA. Sequencing studies confirmed later that this is indeed the case.

Splicing occurs as part of the processing reactions in species that contain introns in the large pre-rRNA, such as the protozoan *Tetrahymena*. T. R. Cech and his colleagues found that intron removal from *Tetrahymena* pre-rRNA can proceed in the test tube when the preparation is treated to remove, destroy, or denature proteins, using methods such as exposure to 8-M urea, SDS-phenol extraction, or digestion with proteinases.

Final proof that the reaction proceeds without the intervention of enzymatic proteins was obtained by cloning cDNA copies (see p. 420) of the *Tetrahymena* large pre-RNA gene in *E. coli*. RNA molecules transcribed from the cloned cDNA were able to carry out the splicing reaction in a cell-free medium, although no processing enzymes or other *Tetrahymena* proteins could possibly have been present. (*E. coli* cells do not possess any of the enzymes required for processing eukaryotic pre-rRNAs.) The interaction discovered by the Cech group was the first known example of a biological reaction catalyzed by RNA and not a protein. Cech received the Nobel prize in 1989 for his research leading to the discovery of self-catalyzed RNA interactions.

The pattern of secondary structures set up by the inverted sequences in the *Tetrahymena* intron, rather than specific sequences, is critical to the reactions removing the intron and splicing together the free ends generated by intron removal (Fig. 15-7a). Changes in specific nucleotide sequences of the intron have no effect as long as the pairing opportunities remain the same—as, for example, in the substitution of a G-C for an A-U base pair at a given point.

Introns in some mitochondrial and chloroplast genes are capable of forming similar patterns of secondary structure. These introns, if removed by the same reaction pathway as the *Tetrahymena* intron (see Fig. 15-7a), are classified together with the *Tetrahymena* intron as *group I introns*. Another group of introns that share a different pattern of secondary structures follow a self-splicing pathway in which a lariat is formed (Fig. 15-7b); these are known as *group II introns*. Group

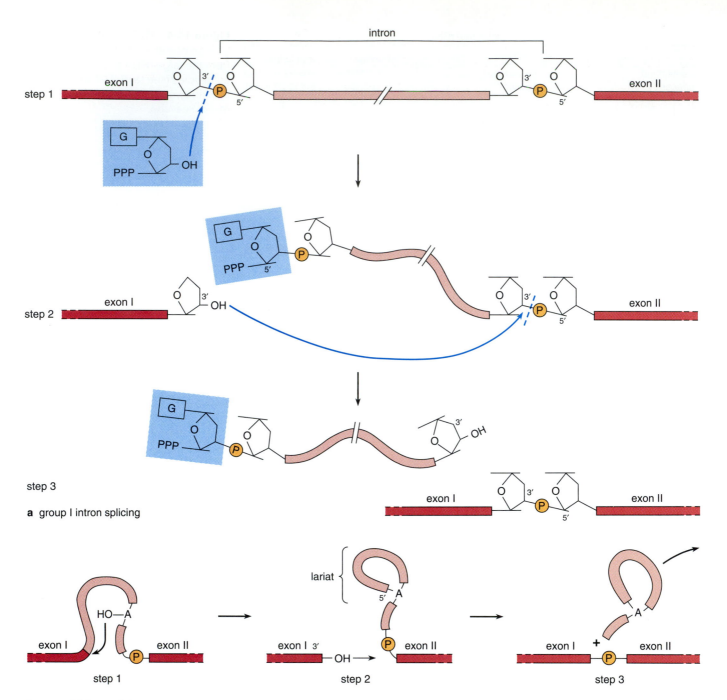

a group I intron splicing

b group II intron splicing

Figure 15-7 Self-catalyzed reactions steps removing group I and group II introns, as proposed by Cech and his colleagues. **(a)** Group I intron removal, in which guanosine acts as a cofactor. The first step in the sequence is initiated by guanosine, shown as GTP (in blue), which binds to the pre-rRNA transcript at the 5' end of the intron. Attachment of the guanosine and the folding pattern of the intron favor transfer of the phosphate group at the 5' boundary of the intron to guanosine (step 2). The reaction creates a break at precisely the 5' end of the intron. The role of guanosine is to provide a 3'-OH for attachment of the 5' end of the intron; hydrolysis of terminal phosphates is not required (GDP or GMP will also serve as 3'-OH donors for the reaction). In the third step the phosphate group at the 3' boundary of the intron is transferred to the —OH group at the 3' end of exon I, gener-

ated in the first reaction. This step breaks the 3' boundary of the intron and splices the free ends into a continuous and unbroken 28S sequence. **(b)** Reactions removing group II introns, in which a lariat is produced. In the first step of the sequence, an adenine within the intron instead of a guanosine cofactor donates an —OH group to start the reaction. As a result of this interaction, a lariat is created with the branch point at the A donating the —OH group (step 2). A phosphate group at the 3' boundary of the intron is then transferred to the —OH at the 3' boundary of exon I, as in step 3, to complete the splicing reaction. Because a lariat is formed as part of group II splicing, this pathway is considered to be a possible evolutionary precursor of the reactions in which spliceosomes remove introns from nuclear mRNA pre-mRNAs.

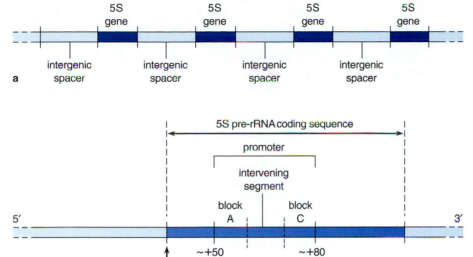

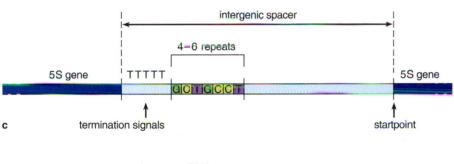

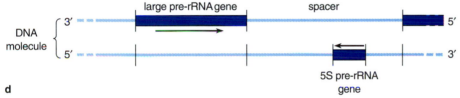

Figure 15-8 5S pre-rRNA genes. **(a)** The pattern of tandem repetition of 5S genes and intergenic spacers 5S gene clusters. **(b)** Structure of a 5S pre-rRNA gene. **(c)** Arrangement of sequences in the nontranscribed spacers of *Drosophila* 5S pre-rRNA genes. **(d)** Location of 5S pre-rRNA genes in yeast, in which the 5S and large pre-rRNA coding sequences occur on opposite DNA chains in the same DNA regions. The arrows indicate the direction of transcription of the yeast 5S and large pre-rRNA coding sequences.

II introns occur within organelle genes of some lower eukaryotes. (For further details of group I and II introns within organelle genes, see Chapter 21.)

5S rRNA Genes, Transcription, and Processing

5S Pre-rRNA Gene Structure   Genes encoding 5S pre-rRNAs also occur in clustered, multiple copies in eukaryotes (Fig. 15-8a). Each gene in a cluster, which contains a coding sequence about 120 base pairs long, is separated from the next by an intergenic spacer. Transcription begins at the 5' end of the coding sequence, so that there is no transcribed spacer at this end. In most eukaryotic species the coding sequence is followed by a short transcribed spacer about 15 to 50 base pairs in length; in some no transcribed spacer occurs at this end. No introns appear in the coding sequence of any known 5S pre-rRNA genes.

The intergenic spacer varies between 250 to 450 base pairs in length, with differences both between and within species. Although 5S pre-rRNA coding sequences are highly conserved among different organisms, few or no sequence similarities are discernible in the intergenic spacer.

Nucleotides can be substituted or deleted from any region within the spacer with little or no inhibition of 5S rRNA transcription in most species. However, transcription slows or stops if the transcribed segment of a 5S pre-rRNA gene is altered at points from about +50 to +80. This shows that the promoter lies entirely within the transcribed segment (see Fig. 15-8b). The internal promoter sequence of 5S pre-rRNA genes can successfully initiate transcription by RNA polymerase III of any DNA segment into which it is inserted as long as the required transcription factors are present.

Detailed studies of the internal promoter of 5S pre-rRNA genes show that it is divided into subregions with specialized functions (see Fig. 15-8b). The first segment, *block A*, contains a consensus sequence,

PuPuPyNNAPuPyGG, that is common to most of the genes transcribed by RNA polymerase III, including tRNA genes and most sn/scRNA genes (see below; in the consensus sequence Pu = a purine, Py = a pyrimidine, and N = any nucleotide). This segment fixes the startpoint for transcription and is recognized and bound by a transcription factor, *TFIIIA* (see below). The regions following block A in the 5S internal promoter, called the *intervening segment* and *block C*, are unique to 5S genes. These segments, particularly block C, are recognized by factors specific for 5S gene initiation and transcription (see below).

Termination of transcription is signaled by a cluster of A-T base pairs that lies just past the end of the transcribed segment. Termination usually occurs at the last of the A-T base pairs, which correspond to a series of U's in the 5S gene transcript. In some species, such as the sea urchin, an inverted sequence immediately precedes the series of A-T pairs, in a pattern that resembles the signals terminating transcription in some bacterial mRNA genes (see Supplement 15-1).

5S pre-rRNA genes are located in clusters on segments of the DNA outside the nucleolus. Yeast cells have 140 to 150 copies of 5S pre-rRNA genes per haploid chromosome set; *Drosophila* has about 160 copies; humans have about 500. Unusually large numbers of 5S genes are found in some species. *Xenopus*, for example, may have as many as 24,000 copies of the genes coding for 5S pre-rRNA per haploid genome! The *Xenopus* 5S rRNA genes, as noted, occur in somatic- and oocyte-types that are active at different times in the life cycle (see p. 431).

5S rRNA Transcription and Processing   In initiation of 5S rRNA transcription the internal promoter interacts with RNA polymerase III and three transcription factors, *TFIIIA*, *TFIIIB*, and *TFIIIC*. Two of these, TFIIIB and TFIIIC, are general factors that take part in the initiation of most or all genes transcribed by RNA polymerase III, including tRNA and snRNA/scRNA genes as well as 5S genes. Although there are no sequence elements resembling the TATA box of mRNA genes in the 5S promoter blocks, TBP, the segment of TFIID that binds the TATA box, also forms part of TFIIIB and plays a central role in 5S rRNA transcription initiation (see pp. 412 and 434). Thus TBP appears to be a universal transcription factor, necessary for transcription by all three RNA polymerases. The remaining 5S rRNA transcription factor, TFIIIA, binds only to 5S genes and evidently specifically identifies 5S genes for transcription by RNA polymerase III. In the interaction the 5S-specific TFIIIA factor binds first to the promoter, followed by the remaining factors and RNA polymerase III. Binding places the enzyme in a position to copy the nucleotide forming the startpoint for

transcription, which lies about 50 nucleotides upstream of the 5' end of the promoter.

Once transcription is initiated, the enzyme reads through the coding sequence and the short transcribed spacer at the 3' end. At this point the enzyme encounters the series of T's acting as a termination signal, and the transcript and enzyme release from the gene. The transcription factors may remain bound to the promoter region, maintaining the gene in an active state in which another RNA polymerase III molecule can bind and initiate transcription.

The 5S transcription factors can be interchanged between major taxonomic groups and still retain their activity in initiation. *Xenopus* 5S transcription factors, for example, can successfully initiate and maintain transcription in human 5S genes and vice versa. This interchangeability undoubtedly reflects the fact that the promoter sequence of 5S rRNA genes is reasonably well conserved between different species.

Reactions processing 5S pre-rRNA are limited essentially to removal of the surplus sequence that occurs at the 3' end of the transcript in some species. Cleavage or nibbling of these nucleotides by endo- or exonucleases generates a mature 5S rRNA molecule, ready for incorporation into large ribosomal subunits. In most species no bases or sugar units are modified chemically, and no nucleotides are added to either end during 5S pre-mRNA processing.

Alterations in Chromatin During rRNA Transcription

Active large pre-rRNA genes are digested more rapidly by DNA endonucleases than inactive genes, in patterns similar to active mRNA genes (see p. 416). The DNA of transcribed rRNA segments is also more susceptible than inactive segments to binding by chemical crosslinkers. Both of these lines of evidence support the conclusion that higher-order structures such as the solenoid (see p. 380) in the chromatin unfold in active large pre-rRNA genes. Sites hypersensitive to endonuclease digestion also appear in the promoter regions upstream of active large pre-rRNA genes in many species, in a pattern similar to the hypersensitive sites appearing in active mRNA genes (see p. 416).

In general the DNA lengths obtained after endonuclease digestion of active large pre-rRNA genes indicate that some of the DNA remains wound into nucleosomes and some is more or less unfolded or unwound. That is, some of the DNA released by the digestion appears in the regular 200 or 145 base-pair lengths characteristic of nucleosomes (see p. 377), and some appears in altered lengths. Although the sources of the regular and altered lengths are not known, the regular periodicities may represent DNA released from the intergenic spacer regions, in which direct electron microscopy shows that nucleosomes are retained. The altered lengths may be DNA segments released from the large pre-rRNA coding segments, in which the nucleosomes are presumably displaced by the RNA polymerase enzymes, which are closely packed in transcribed large pre-rRNA genes.

Active 5S rRNA genes are also more sensitive than inactive genes to endonuclease digestion, indicating that chromatin also unwinds and disassembles to a greater or lesser extent in these regions during RNA transcription. Hypersensitive sites also appear in active 5S rRNA genes, at points near the 5' end of the internal promoter and just upstream of the startpoint for transcription. 5S genes are so small that an entire gene and its nontranscribed spacer contain enough DNA to wrap around only two nucleosomes.

Actively transcribed large pre-rRNA genes have been observed directly in the electron microscope by O. L. Miller and D. L. Beatty and others. The genes in these preparations (Fig. 15-9) appear as long axial fibers coated at intervals with a brushlike matrix of shorter fibers. Work with enzymes and other techniques (see the Fig. 15-9 caption for details) demonstrated that the long axial fibers are DNA molecules or chromatin fibers of large pre-rRNA genes. The brushlike fibers of the matrices are large pre-rRNA molecules in the process of transcription. The granule at the base of each large pre-rRNA transcript is probably an RNA polymerase molecule transcribing the gene. The RNA polymerase enzymes are crowded along the gene, following each other closely as they transcribe the DNA. This appearance coincides with biochemical information, which indicates that large pre-rRNA sequences are among the most actively transcribed genes in the nucleus.

Each gene, represented by a length of axial fiber coated by large pre-rRNA transcripts, is separated from the next by a segment that carries no transcripts. These transcript-free regions are the intergenic spacers separating the large pre-rRNA genes. Visible nucleosomes appear to be limited to the intergenic spacers, in support of the conclusion that nucleosomes are displaced from sites under transcription by RNA polymerases (see p. 417).

rRNA and the Nucleolus

The nucleolus is an irregularly shaped body of variable size suspended within the nucleus of eukaryotic cells. Within the nucleolus are masses of granules intermixed with indistinct fibers somewhat smaller in diameter than chromatin fibers (Fig. 15-10; see also Fig. 13-15 and p. 382). The masses of fibers and granules are known as the *fibrillar* and *granular zones* of the nucleolus. Chromatin fibers can also often be discerned in and around the nucleolus (as in Fig. 15-10).

Several proteins and snRNAs can be detected in quantity in the nucleolus. RNA polymerase I, as might be expected, is present, along with initiation factors and a DNA topoisomerase (see p. 373), which may be involved in unwinding the ribosomal DNA for transcription. Ribosomal proteins are also present in large quantities. The snRNAs U3, U8, U13, U14, E1, E2, E3, X, and Y (see p. 447) are also detected in the nucleolus. At least some of these snRNAs, including U3 and U8, have been directly implicated in large pre-rRNA processing (see p. 434).

In many eukaryotes, all repeats of large pre-rRNA genes are clustered on one chromosome of a haploid set. In haploids, one chromosome therefore forms a single nucleolus. In diploids the chromosome carrying the large rRNA gene cluster occurs in two copies as a homologous pair. The nucleolar regions of the pair may remain separate, forming two distinct nucleoli, or may associate to form a single, large nucleolus. A few species, including humans, have exceptional distributions in which clusters of large pre-rRNA genes are distributed among several different chromosomes. In humans, five chromosome pairs contain large pre-mRNA genes. The clusters associate in different combinations to form a variable number of nucleoli in different human cells and tissues.

The clusters of large pre-rRNA genes are called *nucleolar organizers* (NORs) because of their capacity to form the nucleolus after cell division. During cell division, rRNA transcription stops in most eukaryotic cell types, and the nucleolus disassembles and disappears. After cell division is complete, rRNA synthesis resumes, and nucleoli reappear at the NOR locations (for details, see p. 716).

5S rRNA genes do not form bodies equivalent to the nucleolus. Although 5S genes are clustered and undoubtedly become covered with 5S pre-rRNAs during transcription, these regions cannot be distinguished from other nonnucleolar regions of the chromatin.

A classic experiment by M. L. Pardue and J. G. Gall provided some of the evidence showing that the DNA sequences coding for 18S and 28S rRNA are clustered in the nucleolus and that 5S rRNA genes are sited at other locations on the chromosomes. Pardue and Gall developed a technique called *in situ* hybridization, in which cells are placed on a microscope slide and the DNA is unwound in place in the chromosomes. The unwound, denatured DNA is then hybridized with radioactive rRNA added to the slide. After unhybridized, excess rRNA is washed away, a photographic emulsion is placed over the preparation to detect sites where the radioactive rRNA has hybridized. The Pardue and Gall preparations show that 18S

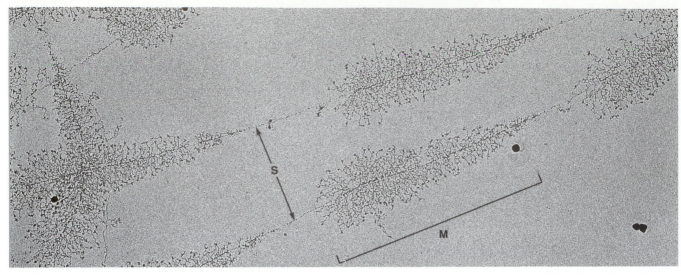

a

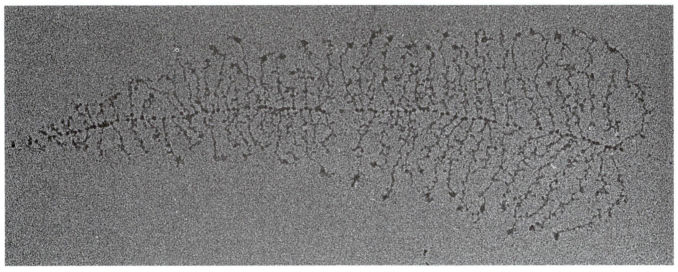

b

Figure 15-9 Transcribing large pre-rRNA genes isolated from the newt *Triturus*. **(a)** The long axial fibers (S) are the DNA molecules of the nucleolus. The matrix regions (M) are segments active in rRNA transcription. Large pre-rRNA transcripts extend at right angles from the DNA. The long axial fibers in such preparations are broken by DNAse, indicating that they are DNA molecules of rRNA genes; the fibers of the matrix regions are broken by RNAse, indicating that they are the rRNA product of the genes. × 26,000. **(b)** An active large pre-rRNA gene at higher magnification. The 5' end of the gene is located at the end where the matrix fibers are shortest. Longer, nearly complete large pre-rRNA molecules extend from the rDNA axis at the opposite, 3' end of the gene. The dark granules at the base of each ribonucleoprotein fiber are probably RNA polymerase molecules. × 46,000. (Courtesy of O. L. Miller, Jr. and Barbara R. Beatty, Oak Ridge National Laboratory.)

and 28S rRNA hybridize to chromosome regions forming the nucleolus, confirming that the sequences from which these rRNAs are transcribed are clustered in this location (Fig. 15-11a). 5S rRNA hybridizes at sites on many chromosomes outside the nucleolus (Fig. 15-11b), confirming the nonnucleolar location of 5S rRNA genes. In the example shown in Figure 15-11b, from *Xenopus*,

at least 15 of the 18 chromosomes carry 5S rRNA sites at their tips.

Further evidence that the nucleolus contains the genes coding for large pre-RNA was obtained by Brown and others by hybridizing large 28S, 18S, and 5.8S rRNAs to DNA isolated from the nucleolus and by direct sequencing of nucleolar DNA. The same

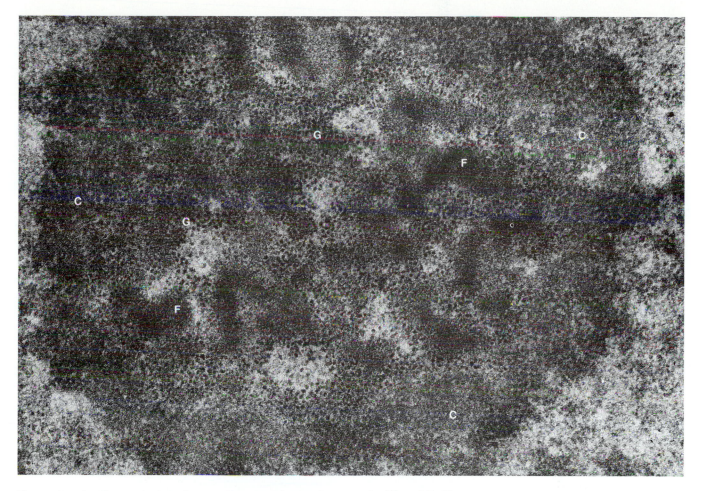

Figure 15-10 The nucleolus of a rat pancreas cell, showing granular (G) and fibrillar zones (F). Chromatin fibers (C) are visible at the margins of the nucleolus. × 50,000. (Courtesy of V. Marinozzi.)

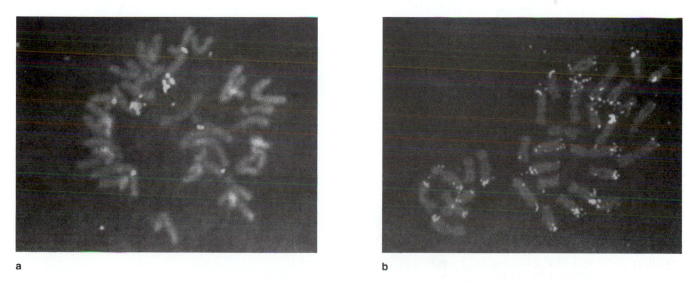

a

b

Figure 15-11 *In situ* hybridization of ribosomal RNAs to the metaphase chromosomes of *Xenopus*. **(a)** 18S and 28S rRNAs hybridize to sites on a single chromosome pair. These sites organize the nucleolus in interphase nuclei. **(b)** 5S rRNA hybridizes to the tips of many chromosomes at points outside the nucleolus. (Courtesy of M. L. Pardue.)

approaches have also confirmed that 5S rRNA genes in almost all species are located on chromosomes outside the nucleolus.

To date, only separate large and small ribosomal subunits have been detected in the nucleolus or nucleoplasm, and intact and fully functional ribosomes only in the cytoplasm. This and other evidence (see p. 461) indicates that protein synthesis is restricted to the cytoplasm in eukaryotic cells. (Figure 15-12 summarizes the events in rRNA synthesis.)

TRANSCRIPTION AND PROCESSING OF TRANSFER RNA

Transfer RNAs are small molecules containing from about 75 to 95 nucleotides. In addition to their distinctive structure, tRNAs differ from other small RNA types, such as 5S rRNA and the snRNAs, in their high content of modified bases. As many as 10% of the nucleotides in different tRNAs may be modified into other forms.

Cells contain many different kinds of tRNA molecules. Most correspond to codons specifying differ-ent amino acids in the nucleic acid code. There are 20 families of these tRNAs, one for each of the 20 amino acids. One set of tRNAs occurs in the cytoplasm of eukaryotic cells. Additional sets, containing distinct tRNA types, occur inside mitochondria and chloroplasts. Bacteria and cyanobacteria also have distinct tRNA families (see Supplement 15-1). A few tRNA molecules take part in reactions unrelated to protein synthesis, including cell wall synthesis in bacteria and serving as primers for assembly of cDNAs by reverse transcriptases.

tRNA Structure

The first nucleic acid molecule of any kind to be completely sequenced was a tRNA from yeast that binds to the amino acid alanine (Fig. 15-13a). This pioneering work was accomplished by R. Holley and his co-workers long before techniques for rapid nucleic acid sequencing were worked out. Holley's sequencing effort required seven years and used a full gram of alanine tRNA, extracted from 140 kilograms of commercial yeast. Holley received the Nobel Prize in 1968 for his pioneering work with tRNA structure and nu-

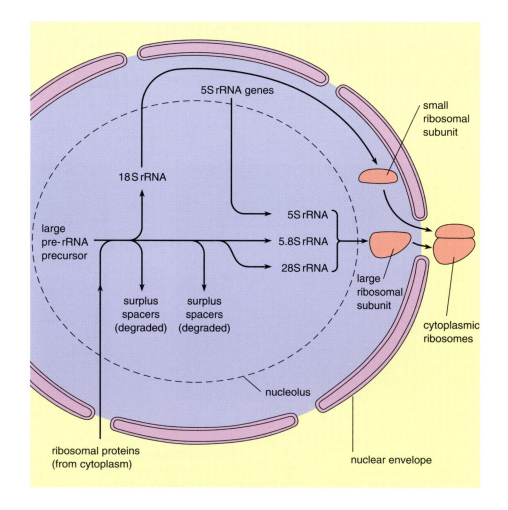

Figure 15-12 Summary of rRNA transcription and processing. As large pre-rRNA is transcribed within the nucleolus, it associates with processing proteins and ribosomal proteins entering the nucleus from their sites of synthesis in the cytoplasm. The initial processing cuts split large pre-rRNA into 18S rRNA and a segment containing 5.8S and 28S rRNA. The 18S rRNA segment, in association with the proteins of the small ribosomal subunit, is released from the nucleolus and enters the cytoplasm. The remaining segment from the original large pre-rRNA is split in the nucleolus to release the 5.8S and 28S rRNAs. These rRNAs associate with proteins of the large ribosomal subunit and 5S rRNA entering the nucleolar region from its sites of synthesis in the surrounding nucleoplasm to form large ribosomal subunits. Once their assembly is complete, the large subunits are also released from the nucleolus to enter the cytoplasm. The large and small subunits associate to form complete ribosomes during protein synthesis in the cytoplasm.

cleic acid sequencing. Since this first success, more than 400 different tRNA molecules have been fully sequenced.

As part of their original investigations, Holley and his colleagues noted that yeast alanine tRNA contains a series of inverted sequences that allows the molecule to fold into the "cloverleaf" structure shown in Figure 15-13. The cloverleaf was subsequently shown to be compatible with all tRNAs except for a few mitochondrial tRNAs that deviate to some degree from this pattern (see p. 612).

The tRNA cloverleaf contains four double-helical segments. At the top of the cloverleaf is the *acceptor stem*, set up by base pairing between the 5' and 3' ends of the molecule (Fig. 15-13b). A terminal CCA sequence at the 3' end of the molecule extends from this stem. The acceptor stem functions in binding an amino acid to the tRNA; the adenine at the end of the CCA sequence serves as the binding site. The next arm, moving clockwise around the molecule, is the *TψC arm*, so called because the sequence TψC occurs in its loop in almost all tRNAs (ψ is the symbol for the modified

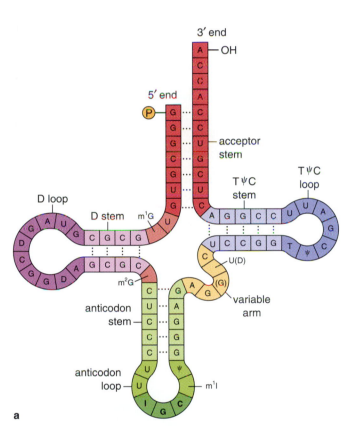

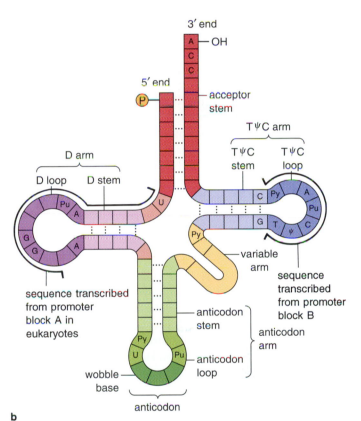

Figure 15-13 tRNA structure. **(a)** A eukaryotic alanine tRNA (from yeast), as originally determined by Holley. The nucleotides designated other than A, C, G, or U contain modified bases; the anticodon is indicated in boldface (in darkest green segment). **(b)** Summary of tRNA structure in both prokaryotes and eukaryotes (see text). The acceptor stem at the top of the diagram is set up by seven base pairs between regions at the 5' and 3' ends of the molecule. Moving clockwise around the molecule, the TψC arm is a hairpin containing five base pairs and a terminal loop with seven bases. Next in clockwise sequence is the variable arm, which may contain from 4 to 21 nucleotides. Below this arm is the anticodon arm, set up by a five-base pair stem and loop containing seven unpaired nucleotides. The final arm, the D arm, is built up from a stem containing either three or four base pairs and

a loop containing from 7 to 11 unpaired nucleotides. Spotted at fixed positions around the cloverleaf structure are the *invariant* bases (indicated by a letter). Most but not all tRNAs have the same bases at these positions. Other bases are *semiinvariant* and limited essentially to either a purine or pyrimidine at fixed points (indicated by Pu or Py). Like the equivalent structures in rRNA molecules, some of the base pairs holding the helices together in some tRNAs consist of unusual combinations such as G-A, G-U, G-ψ, and A-ψ pairs as well as the usual A-U and G-C pairs. Most of the invariant and semi-invariant bases occur in the loops of tRNA molecules, where they interact in forming tertiary folding structures and possibly also with other molecules during protein synthesis.

base *pseudouridine*; see Fig. 14-4). Next in clockwise sequence is the *variable arm*, which may contain from 4 to 21 nucleotides; most of the size differences among tRNAs depend on the number of bases in this arm. Below the variable arm is the *anticodon arm*. The three-nucleotide anticodon, which pairs with triplet codons on mRNA during protein synthesis (see p. 459), is carried at the tip of the loop of this hairpin. The final arm located clockwise from the anticodon arm is the *D arm*, so called because it contains the modified base *dihydrouridine (D)* in all tRNAs. Other common structural features of tRNAs are outlined in the caption to Figure 15-13.

Most modified bases in tRNA molecules appear at restricted sites in the sequence. The anticodon itself frequently contains modified bases, particularly in the position that pairs with the last base of mRNA codons. The base following the anticodon on its 3' side is almost always modified. The presence of modified bases in the anticodon means that unusual base pairs occur in the codon-anticodon association that forms the heart of RNA interactions in protein synthesis (see p. 478 for details). Experiments altering the modified bases within and adjacent to the anticodon have shown that the modified bases influence the fidelity of codon-anticodon pairing and help ensure that the reading frame of the mRNA's message is maintained—that is, that the bases in the mRNA are read three at a time.

During the early 1970s, A. Rich and his colleagues in America and J. Robertus, B. F. C. Clark, A. Klug, and their coworkers in England used X-ray diffraction (see Appendix p. 802) of crystals cast from purified tRNAs to work out the three-dimensional structure of tRNA molecules. Their investigations confirmed that tRNAs contain the cloverleaf arrangement proposed by Holley, folded further in three dimensions into an *L*-shaped structure (Fig. 15-14). The CCA acceptor group and the anticodon are exposed at the opposite ends of the *L*. Hydrogen bonding between atoms located at different points in the arms and hydrophobic interactions between nucleotide bases hold the molecule in its three-dimensional form.

tRNA Genes

In eukaryotes, tRNA genes occur in clustered, multiple copies. Each gene in a cluster consists of a short transcribed spacer, ranging from 3 to 10 nucleotides in length, followed by the coding sequence (Fig. 15-15). Another short transcribed spacer lies at the 3' end of the gene. As in the multiple copies of rRNA genes, each tRNA gene is separated from the next by an intergenic spacer.

Substitution-deletion experiments show that the promoter lies within the transcribed segment of tRNA genes as in 5S rRNA genes and most other sequences transcribed by RNA polymerase III. Additional sequences with critical effects on the initiation of tRNA transcription lie in the 5'-flanking region upstream of the transcribed segment.

The internal promoter of tRNA genes is split into two separate parts as in 5S rRNA. The first part, *block A*, is similar in function to the block A segment of 5S rRNA genes. The second half of the promoter, *block B*, falls at a variable distance from the A block. Nucleotides at individual positions between the two blocks may also have critical effects on promoter function in some tRNA genes. The two promoter blocks of tRNA genes serve both as sequences promoting initiation and as codes for parts of mature tRNA molecules (positions of the sequences corresponding to the two promoter blocks are indicated in Fig. 15-14b).

Sequences in the promoter blocks are highly conserved among eukaryotes. Sequence conservation in

Figure 15-14 Opposite side views showing the three-dimensional structure of tRNA molecules as deduced from X-ray diffraction. Most of the hydrogen bonds holding tRNAs in their three-dimensional form are set up between invariant and semiinvariant nucleotides in the loops at the tips of the D and TψC arms. (Courtesy of A. Rich.)

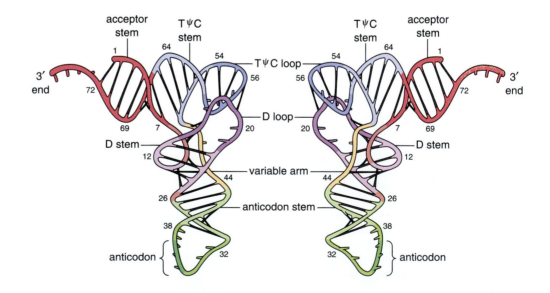

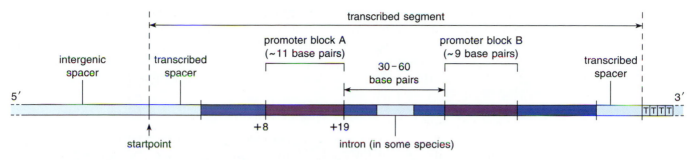

Figure 15-15 Eukaryotic tRNA gene structure (see text). The promoter sequences lie within the transcribed segment; some eukaryotic tRNA genes contain an intron at the position shown, which immediately follows the first nucleotide on the 3' side of the sequence corresponding to the anticodon in mature tRNAs.

these regions could reflect either requirements of transcription initiation or the structure and function of finished tRNA molecules. Of the two possibilities, promoter function is probably more significant for the block A segment, because the consensus sequence of this segment in tRNA genes, PuPuPyNNAPuPyGG, is shared with the block A of 5S rRNA genes.

Experiments by G. Ciliberto and his colleagues demonstrated the functional equivalence of the block A promoter regions of tRNA and 5S rRNA genes. These investigators constructed artificial genes by replacing the block A promoter of tRNA genes with the block A segment of a 5S rRNA promoter. The hybrid genes were fully functional in transcription when supplied with RNA polymerase III and the initiation factors necessary for tRNA gene transcription.

The 3' end of tRNA genes is marked by a series of four or more T's that serves as a termination signal. No sequence at the 3' end of eukaryotic tRNA genes corresponds to the terminal CCA sequence found universally at the 3' end of mature tRNA molecules, which is added during processing (see below).

Some eukaryotic tRNA genes contain an intron in their coding sequences. In all cases where introns have been detected in eukaryotic tRNA genes, which include yeast, mammals, and higher plants, the intron always occurs in exactly the same position—immediately following the first nucleotide on the 3' side of the sequence corresponding to the anticodon in mature tRNAs (see Fig. 15-16a).

No consensus sequences are noted in the introns of different tRNA gene families. Instead, the intron and the exons of pre-tRNAs form structures that provide the recognition signal for reactions removing the intron during processing (see below). In most tRNAs the intron contains a sequence that is complementary to the anticodon. Pairing between the anticodon and its complement in the intron probably contributes to the secondary structures triggering intron removal.

The total number of tRNA genes varies widely among different eukaryotic species. Yeast cells have about 400 tRNA genes per haploid genome; humans have about 1400. The champions in this regard, once again, are the amphibians; species such as *Xenopus* have some 8000 tRNA genes per haploid genome. In organisms such as yeast, with smaller totals, each tRNA type is encoded in some 18 to 20 copies; organisms at the other extreme, such as *Xenopus*, average nearly 400 copies of each tRNA gene.

tRNA Transcription and Processing

Transcription of tRNA genes requires the participation of at least two factors apparently identical to the TFIIIB and TFIIIC factors active in 5S gene transcription. These factors combine with the split promoter of tRNA genes before the RNA polymerase III binds to the promoter sequences. The transcription factors remain linked to the promoter as long as the gene is active, inducing successive rounds of enzyme binding and transcription. The TFIIIB and TFIIIC factors alone seem to be sufficient for initiation in the tRNA genes studied to date. Therefore, initiation of tRNA genes may not involve gene-specific factors equivalent to TFIIIA, which is specific and required only for 5S genes. TFIIIB, as previously noted, contains the universal transcription factor TBP.

The completed transcript of a tRNA gene is a pre-tRNA, longer than its finished tRNA counterpart at both its 5' and 3' ends and in some cases containing an intron that must be removed by splicing reactions. The surplus segments are removed from the ends of pre-tRNAs by RNA endonucleases. Little is known about the reactions processing the 3' end. However, cleavage of the surplus segment from the 5' end of pre-tRNAs is catalyzed in eukaryotes by *RNAse P*, an unusual enzyme that contains an snRNA, *M1 snRNA*, as part of its structure. Indications from studies in

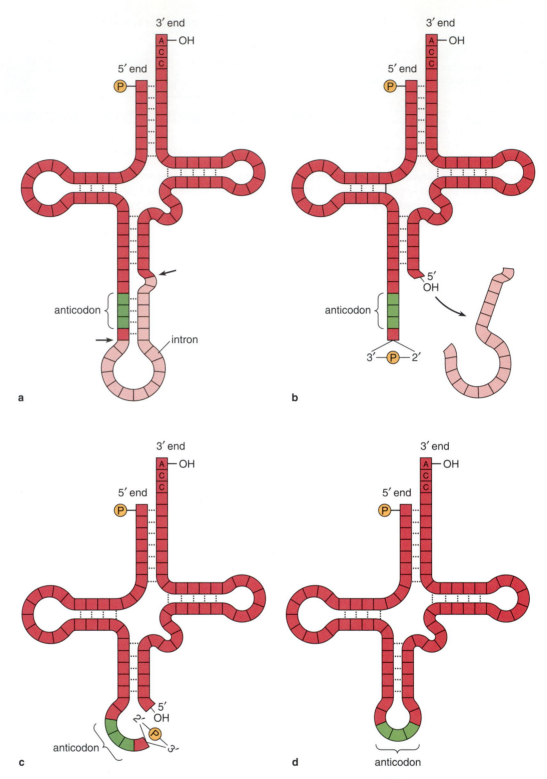

Figure 15-16 tRNA introns and their removal. **(a)** The introns of all tRNA genes occur in a position corresponding to the site between the second and third nucleotides on the 3' side of the anticodon of mature tRNA molecules (arrows). In intron removal an RNA endonuclease breaks the precursor at 5' and 3' ends of the intron (arrows in a). **(b)** The intron is released; note the unusual 2', 3' cyclic phosphate linkage and the 5'-OH group at the free exon ends generated by intron removal. **(c)** The rearrangement aligning the exon ends for the ligase reaction. **(d)** Covalent linkage of the exons into a continuous RNA chain by RNA ligase.

bacteria of the reaction cleaving the 5' end of pre-tRNAs, which is also catalyzed by RNAse P, are that the RNA rather than the protein of the endonuclease is active in catalysis (see Supplement 15-1). The 5' cleavage in tRNA processing is thus another reaction in which RNA molecules act as biological catalysts.

Following removal of surplus segments from the 5' and 3' ends of the maturing tRNA, a single processing enzyme, *tRNA nucleotidyl transferase*, adds the terminal CCA sequence, using CTP and ATP as donors. Base modifications take place while the other processing reactions are in progress. The modifications, carried out by a varied group of enzymes, may include methylations of both bases and sugar residues, acetylations, deaminations, addition of sulfhydryl and other groups, and rearrangements of the four original bases. Among the bases produced by these reactions is methylation of uracil to produce thymine, which occurs in tRNA molecules as well as in DNA. Few of the individual enzymes carrying out the modifying reactions have been identified in eukaryotes.

Introns are processed from pre-tRNAs in a two-step reaction that removes the intron and splices the free exon ends into a continuous molecule (Fig. 15-16). Intron removal is catalyzed by an RNA endonuclease that relies for its processing signals primarily on the exons of the precursor rather than on sequences within the intron itself. The exon regions recognized and bound by the RNA endonuclease place the enzyme in exactly the correct position to make the two breaks removing the intron. A base pair formed between the intron and the anticodon loop is also critical to the reaction. The free ends generated by intron removal are joined by an RNA ligase, in a reaction sequence requiring ATP and NAD. Yeast tRNA introns can be successfully spliced out by extracts made from *Xenopus* and wheat germ cells, indicating that the processing enzymes and reactions are probably essentially the same throughout the eukaryotes.

The reactions removing introns from tRNA molecules thus involve a third distinct splicing pathway. Intron removal from pre-mRNA molecules depends on recognition of consensus sequences at the intron boundaries, in reactions that include the participation of snRNAs. Removal of introns from pre-rRNAs depends on recognition of secondary structures in the intron. In rRNA intron removal, and probably also in the reactions splicing introns from mRNAs, segments of RNA molecules work as catalysts. In intron removal from pre-tRNAs, recognition of the intron and location of the splicing reactions depend primarily on segments that are retained in mature tRNAs—that is, on regions forming the exons. In contrast to mRNA and rRNA intron removal, tRNA splicing appears to depend entirely on protein-based enzymes.

SMALL NUCLEAR AND SMALL CYTOPLASMIC RNAs

A number of small RNA molecules transcribed in the nucleus carry out specialized functions that are of great importance to the mechanisms of transcription and protein synthesis. The small nuclear RNAs, or snRNAs, are a varied group that includes the U snRNAs (Fig. 15-17), which participate in pre-mRNA and pre-rRNA processing, and M1 RNA, which forms part of the RNAse P complex that takes part in tRNA processing. Other small RNAs that function primarily in the cytoplasm, the *small cytoplasmic RNAs (scRNAs)*, include *SRP-7S scRNA*, which forms part of the *signal recognition particle*. This ribonucleoprotein complex participates in the reaction attaching ribosomes to the endoplasmic reticulum during synthesis of proteins that become parts of membranes or are secreted to the cell exterior (see p. 579).

Transcription of these small RNAs is distributed between the RNA polymerase II and III enzymes. Except for one known type, U6 (and U3 in higher plants), the 14 known U snRNAs, so called because the base uridine appears frequently in their sequences, are transcribed by RNA polymerase II. The U snRNA products of this transcription are capped during processing by structures similar to the 5' caps of mRNA molecules. U6 and the remaining small RNAs are transcribed by RNA polymerase III. With the exception of U6 (and U3 in higher plants), none of the small RNAs transcribed by RNA polymerase III is capped.

All the U snRNAs, including U6, have upstream promoters. Within the promoters are *proximal* and *distal* sequence elements centered at -55 and -60; an enhancer sequence lies at a position centered around -175. Although U6 is transcribed by RNA polymerase III, its promoter also has a TATA box at -30. The proximal and distal sequence elements are recognized and bound by transcription factors that promote binding of RNA polymerase II or III. Among these factors is TFIID, with its TBP subunit, which binds strongly to the proximal element of U snRNA promoters, including those without recognizable TATA boxes. Curiously, experimentally induced mutations in the TATA box of U6 genes can switch their transcription from RNA polymerase III to II; alterations in the spacing between promoter sequence elements of U2 genes can switch their transcription from RNA polymerase II to III. It is obvious from these switches and from the universality of TBP as a central element of transcription initiation that the requirements of the three eukaryotic RNA overlap to a considerable extent. This is surprising, because until recently the promoter sequence requirements, transcription factors, and functions of the three polymerases were thought to be

Figure 15-17 The primary and secondary structures of human U1 snRNA. Each type of U snRNA has a combination of paired secondary structures and unpaired segments. One conspicuous secondary structure, a hairpin near the 3' end, is shared by all U snRNAs. In different species, sequence variations in a given U snRNA type are confined primarily to the paired structures; sequences in the unpaired segments are conserved. These relationships show that in some regions, secondary structure is more important to U snRNA function than specific sequences; in other regions the reverse is true. Presumably regions with conserved secondary structures interact primarily with proteins, and regions with conserved sequences interact primarily by pairing with other RNA molecules.

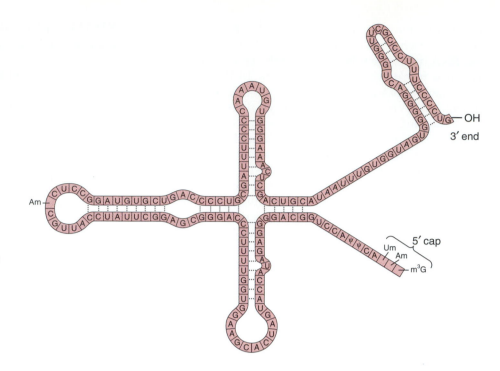

mutually exclusive. After the U snRNAs are processed to finished form, their 5' cap serves as a signal for transport through the nuclear envelope.

Each U snRNA links with several polypeptides to form ribonucleoproteins called *snRNPs* or *snurps*. Some of the polypeptides occur in many snurps—six core polypeptides, for example, are common to the U1, U2, and U5 snurps. The core polypeptides, known as the *Sm proteins*, are the antigens recognized by antibodies from lupus patients (see p. 414).

Anti-Sm antibodies, after extraction and purification, have been used as probes to locate and identify U snRNAs in humans and other species. Work with the antibodies and other techniques revealed that these molecules are widely distributed among organisms as diverse as dinoflagellate protozoans, garden peas, *Drosophila*, and humans. They are, in fact, probably universal among eukaryotes.

The DNA regions encoding a number of small nucleolar RNAs, including E1, E2, E3, X, and Y, which probably function in large pre-rRNA processing (see p. 434), have no promoter elements of any kind and are not separately transcribed. Instead they are included in the introns of various mRNA genes. The RNAs are released when the introns containing them are clipped from pre-mRNAs.

Only one small cytoplasmic RNA, the 7S scRNA occurring in the signal recognition particle (SRP), has been studied in detail. Analysis of the SRP by G. Blobel and his colleagues revealed that it contains the SRP 7S scRNA molecule in combination with several polypeptides. Sequencing shows that SRP 7S scRNA is related to the *Alu* family, a class of apparently non-

functional sequences that occur in many repeats in humans and other mammals. (Humans have 300,000 to 500,000 copies of the *Alu* sequence per haploid chromosome set.) Part of the *Alu* sequence is the same as the first 100 and last 40 nucleotides of an SRP 7S scRNA. This suggests that *Alu* sequences may be SRP 7S scRNA *pseudogenes*, that is, faulty, nonfunctional copies of the SRP 7S scRNA gene that have duplicated and persist in the genome. (For further details of pseudogenes in general, and the *Alu* pseudogene class in particular, see Chapter 18.)

Ribosomal genes encoding 18S, 5.8S, and 28S rRNAs are transcribed by RNA polymerase I in the nucleolus into a large rRNA precursor that contains one each of these sequences. Processing reactions release each of these rRNAs as individual molecules and chemically alter nucleotide bases at specific sites into modified forms. 5S pre-rRNA is transcribed outside the nucleus by RNA polymerase III and processed into mature 5S by enzymes that clip nucleotides from one end. 18S rRNA combines with ribosomal proteins to form the small ribosomal subunit; 5S rRNA enters the nucleus to combine with 28S and 5.8S rRNA and proteins in assembly of the large ribosomal subunit. The two subunits, after passage through the nuclear envelope, combine into complete ribosomes in the cytoplasm.

Transfer RNA genes are transcribed into pre-tRNA precursors by RNA polymerase III; processing of the precursor involves removal of surplus sequences at both its 5' and 3' ends and extensive alteration of nucleotide bases into modified form.

A variety of sn/scRNAs are transcribed by RNA polymerases II or III and processed into mature form. snRNAs, particularly those of the U series, act primarily as parts of the reactions processing mRNAs and rRNAs to mature form. The best-known scRNA, the 7S scRNA, forms part of the signal recognition particle, a ribonucleoprotein complex that promotes attachment of ribosomes assembling membrane or secretory proteins to membranes of the ER.

For Further Information

Suggestions for Further Reading

Abelson, J. 1992. Recognition of tRNA precursors: A role for the intron. *Science* 255:1390.

von Ahsen, U., and Schroeder, R. 1993. RNA as a catalyst: Natural and designed ribozymes. *Bioess.* 15:299–307.

Archambault, J., and Friesen, J. D. 1993. Genetics of eukaryotic RNA polymerases I, II, and III. *Microbiol. Rev.* 57:703–724.

Ausio, J. 1992. Structure and dynamics of transcriptionally active chromatin. *J. Cell Sci.* 102:1–5.

Cech, T. R. 1987. The chemistry of self-splicing RNA and RNA enzymes. *Science* 236:1532–1539.

Fournier, M. J., and Maxwell, E. S. 1993. The nucleolar snRNAs: Catching up with the spliceosomal snRNAs. *Trends Biochem. Sci.* 18:131–135.

Gabrielson, O. S., and Sentenac, A. 1991. RNA polymerase III(C) and its transcription factors. *Trends Biochem. Sci.* 16:412–416.

Helman, J. D., and Chamberlin, M. J. 1988. Structure and function of bacterial sigma factors. *Ann. Rev. Biochem.* 57:839–872.

Hernandez-Verdun, D. 1991. The nucleolus today. *J. Cell Sci.* 99:465–471.

Horowitz, M. S. Z. 1990. Structure-function relationships in *E. coli* promoter DNA. *Prog. Nucleic Acid Res. Molec. Biol.* 38:137–164.

Kunkel, G. R. 1991. RNA pol III transcription of genes that lack internal control regions. *Biochem. Biophys. Acta* 1088:1–9.

Kustu, S., North, A. K., and Weiss, D. S. 1991. Prokaryotic transcriptional enhancers and enhancer-binding proteins. *Trends Biochem. Sci.* 16:397–402.

Lewin, B. 1990. *Genes IV.* New York: Oxford University Press.

Luhrmann, R., Kastner, B., and Bach, M. 1990. Structure of spliceosomal snRNPs and their role in pre-mRNA processing. *Biochim. Biophys. Acta* 1087:265–292.

Mattaj, I. W., Tollervey, D., and Séraphin, B. 1993. Small nuclear RNAs in messenger and ribosomal RNA processing. *FASEB J.* 7:47–63.

Meréchal-Drouard, L., Weil, J. H., and Dietrich, A. 1993. tRNAs and tRNA genes in plants. *Ann. Rev. Plant Physiol. Plant Molec. Biol.* 44:13–32.

Noller, H. F. 1984. Structure of ribosomal RNA. *Ann. Rev. Biochem.* 53:119–162.

Parry, H. D., Scherly, D., and Mattaj, I. W. 1989. "Snurpogenesis": The transcription and assembly of U snRNA components. *Trends Biochem. Sci.* 14:15–19.

Phizicky, E. M., and Greer, C. L. 1993. Pre-tRNA splicing: Variation on a theme or exception to the rule? *Trends Biochem. Sci.* 18:31–34.

Reeder, R. H. 1990. rRNA synthesis in the nucleolus. *Trends Genet.* 6:390–395.

Richardson, J. P. 1990. RhO-dependent transcription termination. *Biochim. Biophys. Acta* 1048:127–138.

Scheer, U., and Benavente, R. 1990. Functional and dynamic aspects of the mammalian nucleolus. *Bioess.* 12:14–21.

Schlessinger, D., Bolla, R. I., Sirdeshmukh, R., and Thomas, J. R. 1985. Spacers and processing of large ribosomal RNAs in *Escherichia Coli* and mouse cells. *Bioess.* 3:14–18.

Sollner-Webb, B., and Mougey, E. B. 1991. News from the nucleolus: rRNA gene expression. *Trends Biochem. Sci.* 16:58–61.

Solymosy, F., and Pollak, T. 1993. Uridylate-rich snRNAs (U snRNAs), their genes and pseudogenes, and U snRNPs in plants: Structure and function. A comparative approach. *Crit. Rev. Plant Sci.* 12:275–369.

Srivastava, A. K., and Schlessinger, D. 1990. Mechanism and regulation of bacterial ribosomal RNA processing. *Ann. Rev. Microbiol.* 44:105–129.

Stragier, P. 1991. Dances with sigmas. *EMBO J.* 10:3559–3566.

Symons, R. H. 1992. Small catalytic RNAs. *Ann. Rev. Biochem.* 61:641–671.

1. What types of promoters are recognized by enzymes transcribing rRNAs, tRNAs, and sn/scRNAs?

2. What rRNA types are found in each ribosomal subunit in eukaryotes? In prokaryotes? What are the relationships in sequence and secondary structure among the rRNAs of different groups?

3. What RNA polymerase enzymes transcribe each rRNA type?

4. What evidence indicates that secondary structure is more important than primary sequence in many regions of rRNA molecules? What functional roles might secondary structures play in rRNA molecules?

5. Diagram the major sequence features of a large pre-rRNA gene, the large pre-rRNA copied from the gene, and the mature rRNA molecules.

6. Outline the reactions converting large pre-RNA precursors into mature form. What types of reactions remove introns from large pre-rRNA precursors?

7. Diagram the structure of a 5S pre-rRNA gene, the 5S pre-rRNA copied from the gene, and a mature 5S rRNA molecule.

8. What evidence indicates that the promoter of 5S rRNA genes is internal? How are the promoters of 5S rRNA genes structured? What transcription factors participate in 5S transcription?

9. How are large rRNA precursor and 5S pre-rRNA genes arranged in eukaryotes? What is the relationship between this arrangement and the nucleolus?

10. What is a nucleolar organizer? What structures are visible inside nucleoli?

11. Describe an experiment showing that large pre-rRNA genes are located in the nucleolus, and that 5S pre-rRNA genes are located on chromosomal sites outside the nucleolus.

12. Outline the structure of a tRNA molecule. What are invariant bases? Semiinvariant bases? Where does an amino acid attach to a tRNA molecule? Where is the anticodon located?

13. Diagram the structure of a tRNA gene. How are tRNA promoters structured? Compare the promoters of tRNA and 5S rRNA genes. What transcription factors participate in tRNA transcription?

14. Where do introns typically occur in tRNA genes?

15. Compare the mechanisms removing introns in mRNA, rRNA, and tRNA genes. In which of these mechanisms are RNA molecules believed to act as catalysts?

16. What are U snRNAs? Snurps? Which RNA polymerases transcribe genes encoding U snRNAs?

17. What is SRP 7S scRNA?

RNA Transcription and Processing in Prokaryotes

Although transcription proceeds by the same general mechanism in prokaryotes and eukaryotes, there are many differences between the two groups in the details of gene structure, transcription, and processing. All the major RNA classes—mRNA, rRNA, and tRNA—are transcribed by a single RNA polymerase in prokaryotes. Prokaryotic mRNA genes usually contain codes for more than one polypeptide instead of one as in eukaryotes. Introns are almost nonexistent in prokaryotic mRNA genes, and no clipping and splicing reactions are necessary to convert the transcripts of these genes to finished form. There is therefore almost no mRNA processing in most bacteria, except for a few internal methylations and the addition of a short poly(A) tail to some.

The essentially finished form of mRNA transcripts reflects the fact that transcription and translation occur simultaneously in prokaryotes—translation of the 5' end of mRNAs begins soon after initiation of transcription, well before transcription of the 3' end is complete. In addition, because there is no nuclear envelope in prokaryotes, the genes undergoing transcription and their mRNA transcripts are suspended directly in the cytoplasm, which contains ribosomes and all of the factors required for protein synthesis.

In contrast to bacterial mRNA genes, genes encoding rRNA and tRNA in bacteria include transcribed spacer segments that, like their counterparts in eukaryotes, extend the DNA segments occupied by the genes. As a result, newly transcribed rRNA and tRNA molecules contain extra nucleotides that are removed by processing reactions. The genes coding for rRNA and tRNA in bacteria also occur in multiple, repeated copies. Introns have been discovered in the tRNA genes of a few bacteria. Bacterial gene structure, transcription, and processing are generally considered to be representative of the cyanobacteria.

The Bacterial RNA Polymerase

The single bacterial RNA polymerase consists of a core enzyme containing several polypeptides. Three subunits, α, β, and β', make up the core; both β subunits are related in amino acid sequence to the two largest polypeptides of eukaryotic RNA polymerases (see p. 402). By itself the core enzyme is potentially capable of copying DNA from any source into an RNA transcript. However, in order to initiate transcription with high efficiency on bacterial promoters, the enzyme must combine with another polypeptide subunit called a *sigma factor*, which increases the selectivity of bacterial RNA polymerase for bacterial promoters by about a million times.

Distinct sigma factors enable RNA polymerases to distinguish between different bacterial gene groups and thus form part of the mechanism regulating transcription. At least five different sigma factors have been identified in *E. coli*. The primary sigma factor recognizing most promoter sequences in this bacterium has a molecular weight of 70,000 and is identified as σ^{70}. Other sigma factors are synthesized and combine with the core enzymes under certain conditions. For example, the sigma factor σ^{32} increases in concentration during heat shock and other forms of stress. This sigma factor promotes binding between the RNA polymerase and the promoters of genes encoding proteins that increase tolerance to stress. The promoters of bacterial viruses are typically recognized by sigma factors that are encoded in the viral DNA. The viral sigma factors, which cause RNA polymerase to bind most avidly to viral genes, enable viruses to dominate transcription in infected cells.

Bacterial Genes and Their Promoters

Bacterial genes are organized in transcriptional units called *operons*, which usually contain more than one coding sequence (Fig. 15-18). An individual coding sequence within an operon is termed a *cistron*. The coding sequences of bacterial viruses are organized in the same pattern. Most bacterial operons are controlled by a single promoter lying in the region in front of the transcribed segment. Some, however, have two or more promoters arranged tandemly in the 5'-flanking region. When two or more promoters are present, each is controlled by different regulatory factors, and each may independently initiate transcription of the operon.

Bacterial promoters, like their counterparts in eukaryotes, contain consensus sequences that provide recognition and binding sites for RNA polymerase, regulate the rate of initiation of transcription, and indicate the startpoint. All bacterial genes, including those encoding mRNAs, rRNAs, and tRNAs, are transcribed from upstream promoters with more-or-less similar consensus sequences. This similarity reflects the fact that no specialized RNA polymerases exist in eukaryotes; a single RNA polymerase makes RNA copies of all gene types. A few bacterial genes also have control sequences upstream of the promoter region that act as enhancers.

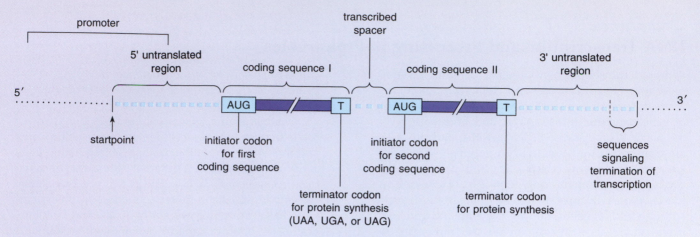

Figure 15-18 Structure of a prokaryotic mRNA gene or operon. Operons typically contain coding information for one to as many as eight or more polypeptides or proteins.

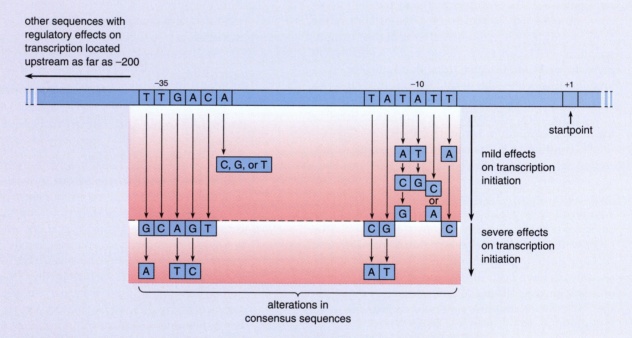

Figure 15-19 Sequence structure of bacterial promoters transcribed by the RNA polymerase enzyme linked to sigma factor σ^{70}. Promoters with "perfect" TTGACA and TATATT consensus sequences at the −35 and −10 positions bind RNA polymerase most strongly. Changes in the consensus sequences weaken RNA polymerase binding to the extent indicated by the lengths of the arrows extending downward from the sequences. (From data presented in P. H. von Hippel *et al.*, *Ann. Rev. Biochem.* 53:389 [1984].)

Two consensus sequences occur in *E. coli* promoters that are recognized by RNA polymerase in combination with sigma factor σ^{70} (Fig. 15-19). One, the short sequence TATATT, called the *−10 sequence* or *Pribnow box*, is usually centered about 10 nucleotides upstream of the startpoint. The second sequence, TTGACA, called the *−35 sequence*, usually begins about 35 nucleotides upstream of the startpoint. The startpoint for transcription usually lies about 7 to 8 nucleotides downstream of the −10 sequence. Other sequences with regulatory effects may lie upstream of the −35 sequence.

Bacterial promoters vary greatly in the strength with which they bind RNA polymerase, depending on how closely their −10 and −35 sequences fit the consensus sequences. Those with perfect copies of the consensus sequences bind RNA polymerase most strongly; those with base substitutions altering the consensus sequence bind RNA polymerase more weakly. (Figure 15-19 shows the effects of base substitutions at various points in the −10 and −35 sequences.)

The particular sequences present at the −10 and −35 positions, by binding the RNA polymerase more or less strongly, set a base-level rate for the initiation of transcription. The base level is adjusted by two overall types of regulatory proteins called *repressors* or *activators*—repressors reduce the level of transcription below the base level, and activators increase the rate above the base level. Promoters governed by repressors are by far the most common in the bacterial systems studied to date. (Further details of bacterial repressors and activators and their regulatory effects are presented in Chapter 17.)

Prokaryotic transcription proceeds through the same phases of initiation, elongation, and termination as in eukaryotes. In the initiation phase the core enzyme, in combination with a sigma factor, binds strongly to a promoter. Strong binding is very quickly followed by DNA unwinding and initiation of transcription at the startpoint. The sigma factor is released after transcription begins. The core enzyme continues transcribing the operon until it reaches termination signals at the 3' end of the gene.

Sequences signaling termination of bacterial operons occur in two types. One, the *type I*, or *rho⁻*, terminator, contains a short inverted sequence centered about 16 to 20 nucleotides upstream of the termination point. Also present in type I terminators is a series of four to eight A's, corresponding to a series of U's in the RNA transcript, lying just upstream of or including the termination point. The other bacterial terminator, *type II*, or *rho⁺*, lacks the series of A's and may or may not contain an inverted sequence. Type II ter-

minators, however, have sequences recognized by *Rho*, a factor promoting termination.

Termination takes place in type I terminators when the enzyme reaches the inverted sequence followed by a series of A's in the template chain. (Figure 15-20 outlines the type I termination mechanism.) In termination by type II terminators the Rho factor binds to its recognition sequences within the transcript of the type II terminator. Before binding, Rho binds ATP; binding of Rho to its recognition sequences triggers ATP hydrolysis and, according to current ideas, unwinds the RNA transcript from the DNA template.

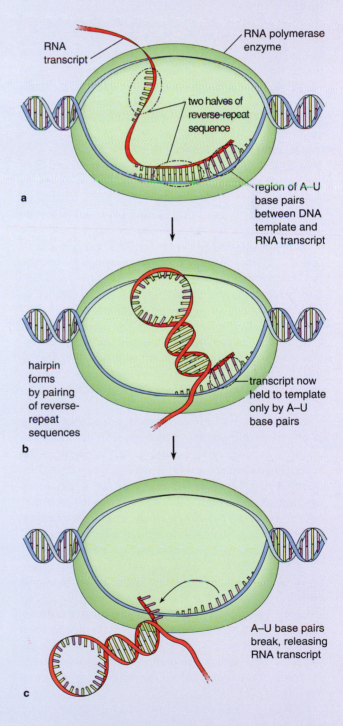

Figure 15-20 Termination by a type I terminator. Termination takes place when the enzyme reaches the inverted sequence followed by a series of A's in the template chain **(a)**. The inverted sequence, once transcribed, probably induces formation of a hairpin in the transcript, stabilized by the high proportion of G-C bonds within the hairpin structure (G-C base pairs are held together by three hydrogen bonds as compared to two for A-U base pairs—see Fig. 13-2). The hairpin in the transcript is believed to interrupt movement of the RNA polymerase just as it completes copying the series of A's in the template chain **(b)**. At this point the RNA transcript is held to the template chain only at its 3' end, in a region containing primarily A-U base pairs, which are relatively unstable. This instability, along with interference by the hairpin, is believed to cause release of the transcript and RNA polymerase at this point **(c)**. (Adapted from an original courtesy of T. Platt, from *Cell* 20:739[1980]. Copyright Cell Press.)

The unwinding stops transcription and releases the transcript and the polymerase enzyme from the DNA. Rho is a protein with 419 amino acid residues and distinct domains that interact with RNA and ATP. The molecule apparently combines into a hexamer containing eight Rho units arranged in a circle with a hole in the center. Presumably the RNA transcript wraps around the hexamer as part of termination.

mRNA Genes, Transcription, and Processing in Bacteria

Bacterial mRNA operons (see Fig. 15-18), like all genes in these organisms, are supplied with an upstream promoter. Within the promoter is the startpoint, the first DNA base to be copied into the mRNA transcript. Following the startpoint is a 5' untranslated region (UTR) that falls between the startpoint and the beginning of the first coding sequence or cistron. The cistron begins with the initiator codon for protein synthesis, usually AUG, and ends with one of three terminator codons, UGA, UAA, or UAG.

In most mRNA operons the second cistron is separated from the first by a short spacer sequence. The spacer, although copied into the mRNA transcript of the operon, contains no protein code and is not translated. Each successive cistron of an operon follows the same pattern: Each begins with the start codon for protein synthesis, ends with one of three terminator codons, and in most operons is separated from the next cistron by a spacer. Although most operons contain from three to four cistrons, as few as one to as many as eight or more may be included.

Following the last cistron of the operon is the 3' UTR that is transcribed but contains no coding triplets. At the end of the 3' UTR are sequences supplying the signals for termination of transcription. Operons are usually transcribed into a single mRNA by bacterial RNA polymerase, beginning at the startpoint, reading through all the coding sequences, and ending at the termination site. Some operons contain sequences capable of terminating transcription in the spacers between coding sequences. These internal sequences may be switched on or off as a regulatory mechanism to shorten the mRNA transcript and omit one or more of the downstream coding sequences. The switching mechanism, called antitermination, also occurs in mRNAs of bacterial viruses.

Processing is limited to methylation of a few bases in the transcript of some bacterial mRNAs and addition to some of a short poly(A) tail. No 5' cap structure (see p. 404) is added. Translation of a bacterial mRNA into polypeptides typically begins as soon as the 5' UTR and a few codons of the first coding sequence have been copied and continues while transcription of the remainder of the mRNA is in progress (see Fig. 16-21).

Bacterial rRNA Genes, Transcription and Processing

Bacterial rRNAs are significantly smaller than their eukaryotic counterparts and exist in fewer types (see Table 15-1). The small subunits of bacterial ribosomes contain 16S rRNA (see Fig. 15-2a), equivalent in secondary structure and function to the 18S rRNA of eukaryotes. Large ribosomal subunits in bacteria contain 23S rRNA equivalent to eukaryotic 28S, and a 5S rRNA. There is no separate rRNA corresponding to the 5.8S rRNA of eukaryotes; the sequence equivalent to eukaryotic 5.8S rRNA is included in the 5' end of prokaryotic 23S rRNA (see Fig. 15-3). The generally smaller S values for the major rRNAs of prokaryotes are reflected in the overall dimensions and structure of prokaryotic ribosomes, which are smaller and contain fewer proteins than eukaryotic ribosomes.

With the exception of bacterial 5S rRNA, no sequence homologies occur between bacterial and eukaryotic rRNAs. However, as noted on p. 430, each bacterial rRNA type can fold into secondary structures, including hairpins and other paired structures, that are similar to the eukaryotic forms. (Bacterial and eukaryotic 16S/18S rRNAs are compared in Fig. 15-2.) Bacterial 5S rRNA shares some sequence relationships as well as secondary structures with eukaryotic 5S— of the approximately 120 nucleotides of 5S rRNAs, 21 positions are the same in eukaryotes and prokaryotes.

Although no sequence homologies are noted between prokaryotic 16S and 23S rRNAs and their eukaryotic equivalents, many similarities exist between 16S or 23S rRNAs in different prokaryotes, especially among closely related species. The similarities extend to cyanobacterial, mitochondrial, and chloroplast rRNA types, which share sequence homologies with their bacterial 16S, 23S, and 5S rRNA counterparts (mitochondrial rRNAs are described in Chapter 21).

Different bacterial species contain from about 5 to 15 repeats of the rRNA operons. In *E. coli*, the best-studied species, seven copies of the rRNA operons occur in each cell, distributed at scattered points around the DNA circle. Individual rRNA operons in *E. coli* include one coding sequence for each of the three rRNA types, distributed in the gene in the order 16S → 23S → 5S (Fig. 15-21). Each coding sequence is separated from the next by a transcribed spacer. Transcribed spacers also occur at the 5' and 3' ends of the operon.

The rRNA operons of *E. coli* are *mixed operons* that contain tRNA codes as well as the 16S, 23S, and 5S coding sequences. One or two tRNA coding sequences are placed in the transcribed spacer separating the 16S and 23S coding sequences; some *E. coli* rRNA operons contain an additional one or two tRNA coding sequences following the 5S coding sequence.

Each rRNA gene of *E. coli* has two promoters,

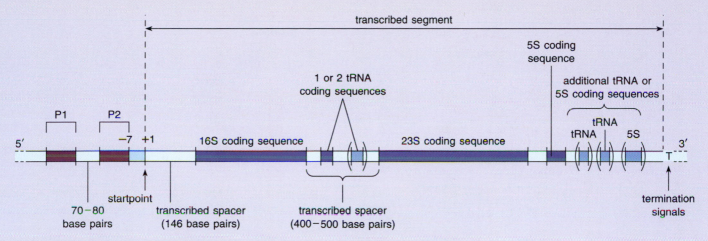

Figure 15-21 Arrangement of coding and other sequences in prokaryotic rRNA operons. The sequences coding for 16S, 23S, and 5S rRNAs and at least one tRNA sequence, shown in dark blue, are present in all prokaryotic rRNA genes in the indicated order. Additional tRNA or 5S rRNA sequences present in some genes are enclosed in parentheses. Each of the two promoters, P1 and P2, has equivalents of the −10 and −35 consensus sequences of bacterial mRNA promoters.

one following the other in the 5'-flanking region (see Fig. 15-21). The first promoter, *P1*, is centered about 300 nucleotides upstream of the startpoint; the second, *P2*, is centered about 110 nucleotides downstream of *P1*. Within each of the two promoters are sequences equivalent to the −10 and −35 consensus sequences of bacterial mRNA promoters. The two promoters apparently respond to different regulatory mechanisms, one to regulatory proteins active during periods of rapid growth, and the other to regulatory proteins active during periods of slower growth.

Following the startpoint is the transcribed spacer that precedes the 16S coding sequence in the rRNA operon. At the other end of the gene, at the 3' end of the transcribed spacer following the last coding sequence of the operon, lie sequence elements signaling termination of transcription, in patterns typical of type I terminators.

An entire rRNA operon is transcribed into a single pre-rRNA that includes the transcribed spacers and all the individual rRNA and tRNA sequences of the operon. The rRNAs and tRNAs are released from the precursor as separate molecules by processing reactions that begin before transcription is complete (Figure 15-22 outlines these reactions).

Bacterial tRNA Genes, Transcription, and Processing

Bacterial tRNA molecules closely resemble their eukaryotic counterparts in structure and function. Like eukaryotic tRNAs, all bacterial tRNAs can fold into a cloverleaf in two dimensions and an *L*-shaped structure in three dimensions.

E. coli has 77 tRNA sequences, coding for some 60 different tRNA types, distributed throughout the genome in pure operons containing only tRNA genes and in mixed operons. The tRNA coding sequences in both the pure and mixed operons are surrounded by transcribed spacer sequences (Fig. 15-23). In contrast to eukaryotic tRNA genes, in *E. coli* the CCA sequence forming the 3' end of mature tRNA molecules is encoded in the DNA in all the genes sequenced to date. In other bacteria, such as *Bacillus subtilis*, the CCA sequence element is included in some genes and absent in others. All bacteria apparently possess the enzymes necessary to add the CCA if it is incomplete or missing in the transcript.

Single promoters for bacterial tRNA genes occur upstream of the transcribed segments, in the 5'-flanking region of the genes. The tRNA promoters resemble those of mRNA operons in structure and contain representatives of the −10 and −35 consensus sequences. Sequences that enhance transcription also occur upstream of the promoter in tRNA operons, in the region from about −40 to −98.

Recently introns were discovered in several tRNA genes of purple photosynthetic bacteria, in archaebacteria (a bacterial group that also possesses other eukaryotic characteristics—see p. 778), and in cyanobacteria. All these introns have internal sequences typical of group I introns and are therefore probably capable of self splicing (see Fig. 15-7). The tRNA introns in archaebacteria were the first to be detected in bacterial genes of any type. The cyanobacterial intron, which occurs in a leucine tRNA gene, also appears in the same position in the equivalent leucine tRNA gene of chloroplasts.

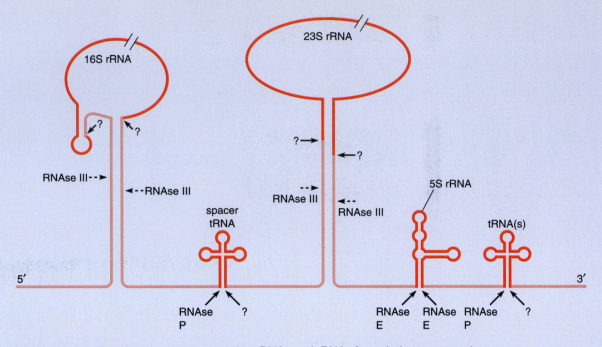

Figure 15-22 Processing reactions cutting rRNAs and tRNAs from their precursors in *E. coli*. Complementary sequences preceding and following the 16S and 23S rRNA sequences allow these segments to fold into two large hairpins, with the rRNA sequences held in the loops. The first cuts are made by an RNA endonuclease, RNAse III, in the stems of the hairpins (dashed arrows) to free the 16S and 23S rRNAs, still with surplus segments at their 5' and 3' ends. These initial cuts also release the tRNA in the spacer between the 16S and 23S segments and another segment containing the 5S rRNA and tRNA sequences at the 3' end of the pre-rRNA. The rRNA and tRNA sequences are then cut to their mature lengths (in dark red) by other enzymes (solid arrows). The enzymes removing surplus segments from 16S and 23S rRNA and the 3' ends of the tRNAs remain unknown. The 5' ends of the tRNAs are cut by RNAse P. The 5S rRNA is processed to its mature length by RNAse E, an endonuclease that cuts precisely at its 5' and 3' ends. The known enzymes were identified in mutants and through the activities of isolated and purified molecules in cell-free systems. (Redrawn from an original courtesy of D. Schlessinger; reproduced, with permission, from *Ann. Rev. Microbiol.* 44:105. © 1990 by Annual Reviews, Inc.)

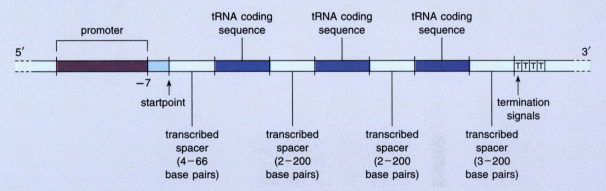

Figure 15-23 Arrangement of coding and other sequences in bacterial operons encoding only tRNA genes. The promoter resembles those of mRNA operons in structure and contain representatives of the −10 and −35 consensus sequences. Sequences with positive effects on transcription initiation also occur upstream of the promoter in the region from about −40 to −98. Terminators are type I, containing a reverse-repeat sequence followed by a series of T's.

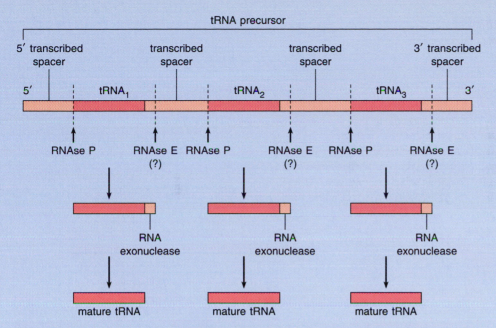

Figure 15-24 Processing of tRNA precursors transcribed from pure tRNA operons. The small vertical arrows indicate sites at which cuts are made. Initial breaks (not shown) may be made by RNAse III to release the tRNAs; the cuts leave the tRNAs with extra sequences still at their 5' and 3' ends. RNAse P processes the 5' tRNAs to mature length in a single reaction. The reactions processing the 3' ends take place in at least two steps: An endonuclease, possibly RNAse E, cuts the 3' end to a length including a few more nucleotides than the mature rRNA; the extra 3' nucleotides are then nibbled away one at a time by an as yet unknown exonuclease. The enzyme stops its 3' nibbling at the terminal CAA, if present; if no CCA is included in the precursor, the exonuclease trims to the nucleotide to which the 3'-terminal CCA will be attached.

All bacterial tRNAs are transcribed in the form of precursors with extra nucleotides at their 5' and 3' ends. The pre-tRNAs transcribed from tRNA genes containing multiple tRNAs include several coding sequences separated by transcribed spacers. The tRNAs are released from the precursors by a series of cuts and nibblings that involve several enzymes (Fig. 15-24). Attachment of the 3'-terminal CCA, if required, is carried out by the same series of reactions attaching the CCA to tRNAs in eukaryotes (see p. 413). While the cutting and trimming reactions are in progress, a battery of enzymes chemically modifies individual bases to other forms at precise locations in the tRNA sequence.

Of the various enzymes processing tRNA precursors, one, *RNAse P*, is of special interest because its catalytic activity depends on the RNA rather than the protein component of the molecule. Bacterial RNAse P, like its eukaryotic counterpart (see p. 445) contains a protein combined with M1 RNA. Experiments in the laboratory of S. Altman and S. Pace showed that 5' cleavage of tRNA precursors could be carried out by purified M1 RNA in protein-free solutions. Although catalysis proceeds most rapidly when both the protein and M1 RNA are associated together in an intact ribonucleoprotein particle, the protein component of RNAse P alone has no catalytic activity. This makes RNAse P another ribozyme, an RNA molecule with the ability to catalyze a biochemical reaction. Altman received the Nobel prize in 1989 with T. R. Cech for his work with ribozymes.

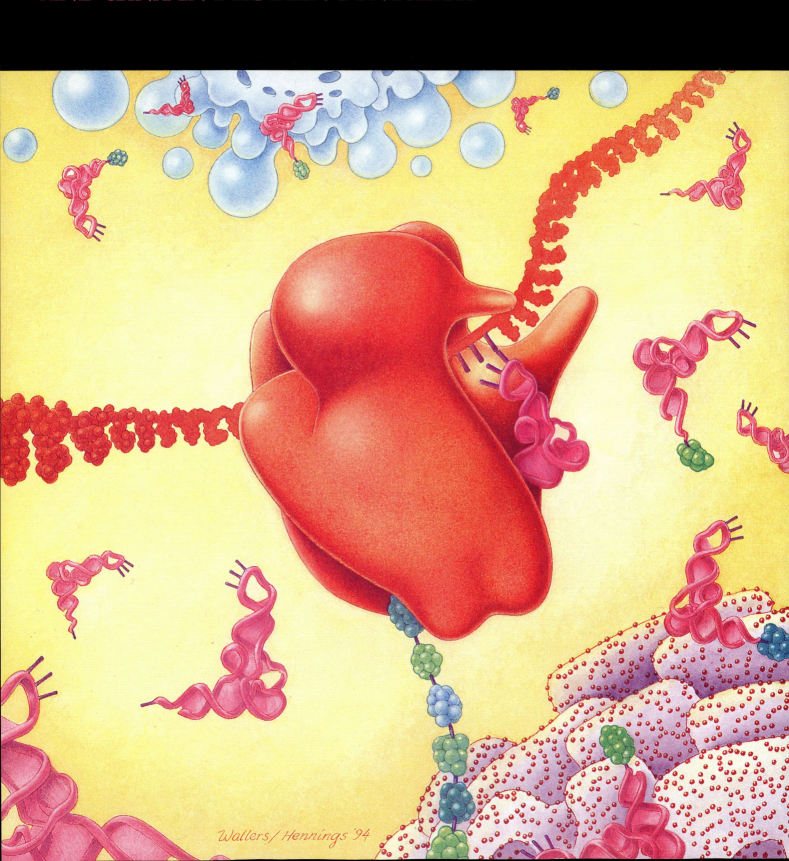

Wallers / Hennings '94

The mRNA, rRNA, and tRNA molecules transcribed in the nucleus interact in protein synthesis in the cytoplasm. Protein synthesis is called *translation* because the genetic code, represented by a sequence of nucleotides in an mRNA molecule, is translated into a sequence of amino acids during polypeptide assembly. In translation, ribosomes read along an mRNA molecule, gradually assembling an amino acid sequence corresponding to the coding sequence in the mRNA. The nucleotides of the mRNA are read three at a time as *codons*; each codon specifies addition of a particular amino acid at a corresponding position in the growing polypeptide chain.

Ribosomes are assemblies of rRNA and proteins that contain many binding sites and centers of catalytic activity. A ribosome contains one small and one large subunit, each containing rRNA in combination with proteins. In a sense, ribosomes can be considered as immense ribonucleoprotein enzymes with multiple active sites that recognize and bind mRNA and other required components and catalyze the reactions combining amino acids into a polypeptide chain.

Each amino acid enters protein synthesis attached to a tRNA molecule. The reactions combining amino acids and tRNAs, termed *amino acid activation*, ensure the accuracy of protein synthesis by linking an amino acid to a tRNA containing an anticodon for that amino acid. The reactions are called an *activation* because chemical energy released by ATP breakdown is conserved as amino acids attach to their respective tRNAs. This conserved energy, released as amino acids are transferred from tRNAs to a polypeptide chain, provides a major driving force for protein synthesis. The combination of an amino acid and its tRNA is an *aminoacyl-tRNA*.

The reactions, molecules, and structures of protein synthesis, including assembly of amino acids into polypeptides, ribosome structure, and attachment of amino acids to tRNA, are described in this chapter. The discovery and nature of the genetic code directing protein synthesis are also discussed. The chapter concentrates on protein synthesis in eukaryotes; the equivalent mechanisms of prokaryotes are described in Supplement 16-1.

POLYPEPTIDE ASSEMBLY ON RIBOSOMES

An Introduction to the Mechanism

As an introduction to protein synthesis, consider the synthesis of a hypothetical protein containing only two different amino acids, tryptophan and phenylalanine, in an alternating . . . Trp-Phe-Trp-Phe . . . sequence (Fig. 16-1). The information required for synthesis of this protein is encoded in an mRNA molecule in the sequence of adenine (A), uracil (U), guanine (G), and cytosine (C) nucleotides. These are taken by threes to form the nucleic acid code words, the codons. The mRNA codon for tryptophan is UGG; one of several codons for phenylalanine is UUC. In our example the mRNA contains an alternating sequence of the two codons UGG and UUC in the form . . . UGG/UUC/UGG/UUC . . . and so on.

During initiation of protein synthesis, mRNA attaches to a ribosome. Surrounding the ribosome is a pool of amino acids attached to their corresponding tRNAs, assembled through the reactions of amino acid activation. Among the tRNAs are those carrying tryptophan and phenylalanine. The anticodon of each tRNA can recognize and pair with the codon in the mRNA specifying the tRNA's amino acid (Fig. 16-1a). The anticodons in our hypothetical example contain the bases expected from the usual A-U, G-C base pairs. (In actuality the base pairs formed between the codon and anticodon frequently involve unusual base pairs—see below.) The tRNA carrying tryptophan is therefore considered to carry the anticodon ACC,[1] complementary to the mRNA codon UGG, and the tRNA carrying phenylalanine, to carry the anticodon AAG, complementary to the mRNA codon UUC.

At the ribosome the anticodons of tRNAs carrying either phenylalanine or tryptophan form complementary base pairs with the corresponding codons of the mRNA (Fig. 16-1b). If the sequence UUC is present at the ribosomal pairing site, a tRNA carrying the anticodon AAG attaches, carrying with it the amino acid phenylalanine. At the next triplet on the messenger the ribosome encounters the sequence UGG; a tRNA carrying tryptophan attaches at this point.

Phenylalanine and tryptophan are brought into close proximity by the pairing. Through energy originally derived from ATP during amino acid activation a peptide bond forms between the two amino acids. The enzymatic activity catalyzing formation of the

---

[1] To be technically correct, an anticodon, like all nucleic acid sequences, should be read in the standard 5' → 3' direction. The anticodon ACC, corresponding to the codon UGG, should therefore be read as CCA. However, for simplicity the anticodons in this example are given in the 3' → 5' direction.

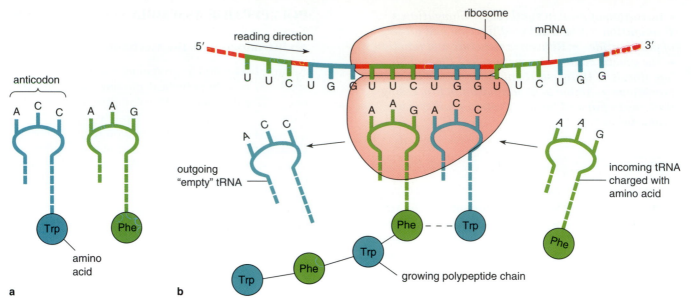

Figure 16-1 Overall reactions of polypeptide assembly. **(a)** The tRNAs for tryptophan (Trp) and phenylalanine (Phe) and the hypothetical anticodons ACC and AAG used in the text example. The tRNA diagrams show only a small segment of the tRNA cloverleaf. **(b)** The overall mechanism of polypeptide assembly, producing in this example a polypeptide chain containing tryptophan and phenylalanine in an alternating sequence (see text).

bond is part of the ribosome. The ribosome then moves to the next codon on the mRNA and the process repeats, this time attaching a phenylalanine to the growing peptide chain. As each successive amino acid is added to the polypeptide, its now-empty tRNA is released. Translation continues in this way until the ribosome reaches the end of the mRNA code, producing in this example a long polypeptide chain containing tryptophan and phenylalanine in an alternating sequence.

The reactions of polypeptide assembly fall into three overall phases. In the first phase, *initiation*, a large and small ribosomal subunit assembles with mRNA and a specialized aminoacyl-tRNA called an *initiator tRNA*. A series of *initiation factors* controls and speeds the process of initiation. Once the ribosomal subunits are assembled with mRNA and the initiator tRNA, the next phase, *elongation*, begins. In this phase, which is speeded by a group of *elongation factors*, aminoacyl-tRNA complexes bind to the ribosome in sequence according to the mRNA code. As the sequential pairing proceeds, amino acids are transferred from tRNAs into a gradually lengthening polypeptide chain. Elongation continues until the ribosome reaches the end of the coded message. At this point, protein synthesis enters its final phase, *termination*. During termination, mRNA and the completed polypeptide are released and the ribosomal subunits separate. *Termination factors* speed completion of protein synthesis.

All three phases require an energy input in the form of ATP or GTP hydrolysis; the phosphate bonds hydrolyzed during polypeptide assembly represent a significant energy input in addition to the ATP hydrolyzed as part of amino acid activation.

The reactions of polypeptide assembly were pieced together primarily in cell-free systems (see p. 795) developed by P. C. Zamecnik, W. V. Zucker, H. M. Schulman, D. Nathans, F. Lipmann, and their coworkers and by others. The systems, extracted from sources such as *E. coli*, rat liver cells, mammalian reticulocytes (immature blood cells), wheat germ, and yeast, include ribosomes, aminoacyl-tRNAs, and the initiation, elongation, and termination factors required for the most efficient progress of protein synthesis.

The initiation and other factors speeding steps in polypeptide assembly were discovered through research with the cell-free systems. Investigators noted almost immediately that the purest systems, those consisting of little more than ribosomes, mRNA, aminoacyl-tRNAs, and an energy source, assembled polypeptides very slowly or not at all. In order to proceed efficiently, the system had to retain much of the cytoplasmic solution surrounding ribosomes in the cell. The cytoplasmic substances in the solution necessary for polypeptide assembly were later identified as the initiation, elongation, and termination factors. The roles of these factors in protein synthesis were identified in experiments that involved purification of individual factors and observation of the effects of their addition, removal, or modification in cell-free systems.

Initiation

Eukaryotic initiation takes place in a series of steps worked out by W. C. Merrick, J. W. B. Hershey, W. F. Anderson, R. E. Thach, A. J. Shatkin, and their colleagues and by others. The steps involve interaction of the large and small ribosomal subunits, mRNA, an initiator aminoacyl-tRNA, GTP, and a large group of initiation factors. The reactions unite a small ribosomal subunit with an mRNA molecule, bind the initiator aminoacyl-tRNA, and add the large ribosomal subunit. This assembly is now ready to enter the elongation phase by adding the second aminoacyl-tRNA specified by the mRNA and forming the first peptide linkage.

The initiator tRNA is a specialized aminoacyl-tRNA that pairs only with the AUG appearing as the initiator codon at the beginning of both eukaryotic and prokaryotic protein codes (see pp. 405 and 454). A different aminoacyl-tRNA pairs with an AUG codon located internally in the message. Because AUG specifies the amino acid methionine, newly synthesized proteins begin with methionine in both eukaryotes and prokaryotes. The eukaryotic initiator tRNA, with its attached methionine, is identified as Met-$tRNA_i$, in which Met designates methionine, and the subscript i indicates that the tRNA is the specialized initiator type.

Steps in Eukaryotic Initiation In the first reaction of initiation (step 1 in Fig. 16-2) a ribosome separates into large and small subunits. Met-$tRNA_i$, prepared for its interaction by combination with GTP in a side reaction involving an initiation factor (step 2), then adds to the small subunit (step 3). The small subunit is now ready to add the mRNA, in a reaction (step 4) driven by ATP hydrolysis. Attachment takes place at the 5' cap of the mRNA; once attached, the small subunit moves or "scans" along the mRNA until it reaches the AUG initiator codon. The movement prepares the small subunit for the final step in initiation, addition of the large ribosomal subunit. This reaction (step 5), driven by hydrolysis of the GTP brought to the complex with the initiator tRNA, completes initiation. The completed ribosome, with mRNA and Met-$tRNA_i$ attached, is now ready to enter elongation.

Each reaction of initiation is speeded by initiation factors. Many of the factors add to the growing complex as initiation proceeds; with addition of the large subunit in the final step, all are released. (The Experimental Process essay by A. C. Lopo on p. 464 describes

a significant experiment identifying phosphorylation of an initiation factor as a critical regulatory control of initiation.)

Significance of the Initiator tRNA The specialized functions of the initiator tRNA allow protein synthesis to circumvent an operational problem with initiation. Ribosomes have two major binding sites for tRNAs. One, the *P site*, binds tRNAs linked to a polypeptide chain (termed *peptidyl-tRNAs*). The second site, the *A site*, binds aminoacyl-tRNAs, that is, tRNAs with a single linked amino acid. However, during initiation the initiator tRNA attaches to the P site, even though it carries only a single amino acid. The second aminoacyl-tRNA to bind to the ribosome, which contains a noninitiator tRNA, attaches to the A site. Juxtaposition of the two aminoacyl-tRNAs at the A and P sites sets up the conditions necessary for formation of the first peptide linkage.

Met-$tRNA_i$ has several unique structural features that cause the ribosome to recognize it as a tRNA bearing a peptide chain rather than a single amino acid (Fig. 16-3a). U. RajBhandary and his colleagues have shown that experimental transfer of one or more of these structural features to other tRNAs can convert them from their normal function in elongation to that of an initiator tRNA. One or more of the initiation factors may also contribute to the critical attachment of the initiator tRNA to the P site.

The 5' Cap and Initiation The cap structure at the 5' end of eukaryotic mRNAs (see p. 404) figures importantly in initiation. The cap is recognized by several initiation factors that attach to the mRNA and speed its binding to a ribosomal small subunit. If the cap is removed experimentally, the initiation phase essentially comes to a halt.

Only a very few mRNAs occur naturally without 5' caps. Among these are the capless mRNAs of several viruses infecting eukaryotic cells, including the poliovirus and the rhinoviruses responsible for the common cold. These viruses activate an enzyme that breaks down some of the initiation factors necessary for binding the capped mRNAs of host cells to ribosomes. As a result, translation of host cell mRNAs stops. The capless viral mRNAs are attached to ribosomes by a viral protein; this protein recognizes sequences in the 5' untranslated segment of the viral mRNAs and promotes attachment of the mRNAs to the host cell ribosomes. The mechanism effectively limits infected cells to synthesis of viral proteins.

The AUG Initiator Codon and Ribosome Scanning In 90% to 95% of eukaryotic mRNAs the initiator codon is the first AUG downstream of the 5' cap. This fact

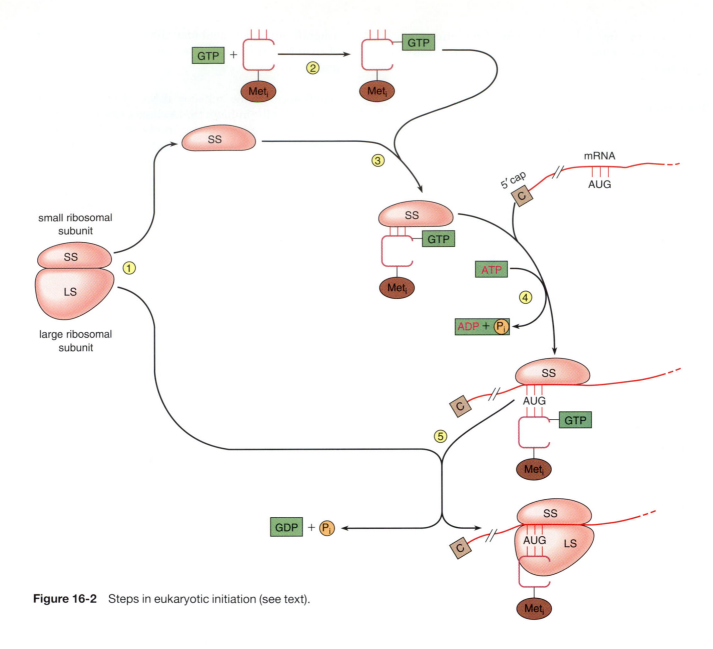

Figure 16-2 Steps in eukaryotic initiation (see text).

led to an initial hypothesis that the small subunit simply scans along the mRNA until it reaches the first AUG. However, additional research by M. Kozak and others revealed that a few eukaryotic mRNAs contain AUGs located upstream of the AUG initiator codon. In these mRNAs the small subunit ignores the upstream AUG sequences and uses instead only the AUG at the beginning of the coding sequence as the initiator for protein synthesis. The scanning hypothesis was modified to include the idea that sequences near the correct AUG initiator codon must contribute to its identification. Substitution-deletion studies by Kozak and A. Shatkin later identified the sequence indicating the correct AUG in higher eukaryotes as

GCCGCCPuCCAUGG (the AUG initiator codon is underlined in the sequence; Pu = a purine).

A few mRNAs have two initiator codons that may be used alternatively. The mRNA encoded by the *myc* gene, for example (*myc* is a gene implicated in the generation of cancer; see p. 647), has a GUG codon located some distance upstream of an AUG initiator codon. During initiation a ribosome may use the upstream GUG as the initiator codon or may read through this codon and use the downstream AUG as the initiator (GUG substitutes for AUG as an initiator codon in a few mRNAs). The alternate initiation points produce two proteins that differ in the presence or absence of a segment of amino acids at the N-terminal

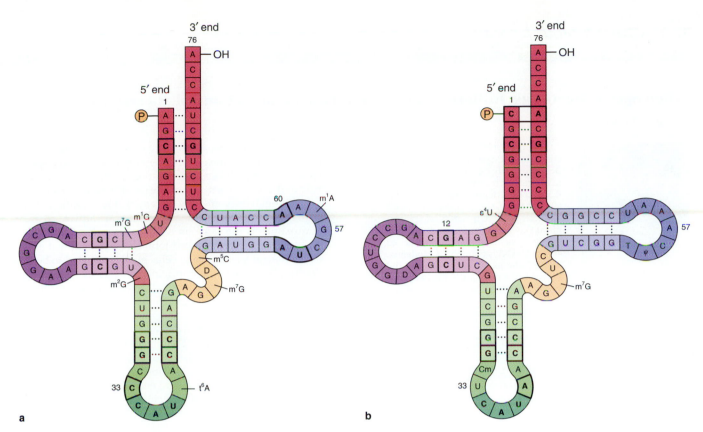

Figure 16-3 Structural features of initiator tRNAs in prokaryotes and eukaryotes. **(a)** The eukaryotic initiator tRNA has an AUC or AψC instead of the usual TψC sequence in the TψC arm, an A instead of the usual pyrimidine at position 60, and a C instead of the usual U at position 33 on the 5' side of the anticodon. The anticodon loop of eukaryotic initiator tRNAs is also unusually sensitive to digestion by S1 nuclease (an endonuclease that attacks single nucleotide chains), indicating that the anticodon loop may be more open and flexible in initiator than in elongator tRNAs. **(b)** In the bacterial initiator tRNA the last base pair at the tip of the acceptor stem (boxed), is "open" rather than linked by hydrogen bonds; this pair is hydrogen-bonded in all other bacterial tRNAs. An unmodified A occurs in the bacterial initiator tRNA on the 3' side of the anticodon; in all other bacterial tRNAs this A is always a modified base. Other base pairs unique to the bacterial initiator tRNA are in boldface. Both eukaryotic and prokaryotic initiatior tRNAs have a sequence of three successive G-C base pairs just before the anticodon loop; this sequence appears to be necessary for recognition and binding by the ribosomal P site.

end of the protein. The mechanism is one of several that allow two different proteins to be assembled from the same mRNA.

Elongation

During elongation, amino acids add one at a time to a growing polypeptide chain. The reactions of elongation involve mRNA-tRNA pairing at the two major active sites of the ribosome, the A and P sites. During this phase the A and P sites cycle through a three-step sequence that is repeated until the ribosome reaches the end of the mRNA coding segment: (1) aminoacyl-tRNA binding to the A site, (2) peptide bond forma-

tion, and (3) transfer of the peptidyl-tRNA formed at the A site by step 2 from the A to the P site. As part of step 3 the ribosome advances one codon farther along the mRNA. Elongation factors increase the rate of steps in the cycle; two GTPs are also hydrolyzed for each elongation cycle.

The Reactions of Elongation Figure 16-4 shows the reactions of a eukaryotic elongation cycle. The cycle shown is the first one following initiation; subsequent cycles of elongation follow the same overall pathway except that a peptidyl-tRNA instead of the initiator tRNA occupies the P site in steps 2 to 4. In a preliminary step, speeded by an elongation factor, an

The Experimental Process

When Sperm Meets Egg: Initiation Factors and Translational Control at Fertilization

Alina C. Lopo

ALINA C. LOPO's interests have focused on the cellular and molecular details of early development for most of her research career. After obtaining B.S. and M.S. degrees from the University of Miami, she completed doctoral research at the University of California at Davis and San Diego, then postdoctoral work at the U.C. San Francisco and U.C. Davis Schools of Medicine. In 1985 she joined the faculty of the Division of Biomedical Sciences at U.C. Riverside, where most of the work described here was carried out. She recently received an M.D. at the UCLA School of Medicine and is now in the residency training program at the UCLA Medical Center. In her spare time, she is a wife and mother.

A key consequence of fertilization is the dramatic activation of the metabolic machinery of the dormant egg. One of the many cellular processes that is rapidly and efficiently turned on at fertilization is protein synthesis. This translational activation is a fundamental step in the activation of development, since without the full activation of protein synthesis, development cannot be completed successfully. The mechanisms regulating translational activation at fertilization have been the subject of extensive investigation by cell, molecular, and developmental biologists since egg activation was first described by Tore Hultin in 1950. A puzzling aspect of this phenomenon is that the unfertilized egg contains all the necessary molecules to carry out protein synthesis. Although progress had been made in understanding how the ionic changes at fertilization influence translational activation, until relatively recently the macromolecular changes were poorly understood.

In the early 1980s work by several laboratories demonstrated that the regulation of protein synthesis at fertilization was, as in most other eukaryotic cells, primarily by regulation of the initiation phase. Working with John Hershey, we showed that possibly the earliest change in initiation was somehow related to the activity of an initiation factor, eIF4F,[1] which binds the 5' cap structure of mRNAs and greatly speeds attachment of mRNA to the small ribosomal subunit. In the preceding few years several groups, including Hershey's, had shown that, in rabbit reticulocyte lysate and in mammalian cells in culture, turning protein synthesis on or off correlated well with the phosphorylation or dephosphorylation of some initiation factors. This led us to ask whether similar changes took place in the egg following fertilization.

We focused our initial studies on the smallest subunit of eIF4F, the so-called cap-binding protein, or eIF4E. The reasoning behind this centered on the knowledge that the presence of the 5' or m⁷G cap on mRNA is an absolute requirement for efficient protein synthesis in eukaryotes. Previous studies suggested that the bulk of mRNAs stored in the egg during oogenesis were capped, so we thought the key to translational activation may reside in the level of activity of the cap-binding protein. If this protein were somehow activated at fertilization, then perhaps mRNA could enter the initiation pathway. At approximately the same time we were working on this problem, Roger Duncan and John Hershey obtained evidence that, at least in HeLa cells, eIF4E appeared to be dephosphorylated when protein synthesis was repressed following heat-shock treatment. This reinforced our belief that we were on the right track, that the change was biologically likely.

We purified the eIF4E protein from unfertilized eggs of the sea urchin using m⁷G affinity chromatography.[2] We used sea urchin eggs because it is easy and inexpensive to obtain very large quantities of pure unfertilized eggs. The large quantities were necessary because, based on evidence from mammalian cells in culture, eIF4E is of relatively low abundance in cells. The sea urchin egg eIF4E turned out to be quite similar in molecular weight and amino acid composition to the mammalian factor, which was being characterized by Bob Rhoads' group at the University of Kentucky and Nahum Sonenberg at McGill University. This was not surprising, because protein synthesis is a highly conserved cellular process.

We had improved our yields of pure sea urchin egg eIF4E approximately ten-fold, and we were soon ready to start making rabbit antibodies against the factor. Our plan was this: first, raise rabbit antibodies against sea urchin eIF4E. Then, prepare electrophoretic gels of homogenates from unfertilized eggs and from eggs fertilized and homogenized at 1-minute intervals up to 15 minutes after fertilization. Use each of these gels to prepare Western blots (see p. 799) and use our antibody to probe the blots and ask how many forms of the eIF4E were present. Since phosphate groups are charged, phosphorylation or dephosphorylation of a protein will change its charge. This would be observable as a change in position of a band in the gels.

Although this plan is quite straightforward conceptually, it took six months of very long hours and hard work to perfect the procedure and obtain results. Finally, we consistently obtained the following results: In the unfertilized egg, all of the eIF4E detectable by our method was in one band on the gel. However, no earlier than 4 and no later than 5 minutes after fertilization, a large fraction of the detectable eIF4E was converted to a more acidic form, as might be expected if it is phosphorylated.

While these results indicated we were on the right track, they did not conclusively prove that the protein was phosphorylated. There were other possible explanations. For example, acetylation of the protein would produce a change of similar magnitude and direction. There certainly is precedent for the covalent modification of proteins by acetylation, although it is less common than phosphorylation. Another even less likely possibility was that the protein was glycosylated. This, too, could have yielded a comparable change in mobility on the gel. Thus, we needed to establish the exact nature of the change before we could draw any meaningful conclusions from our data.

There were two ways we could go about demonstrating that the covalent modification was due to addition of a phosphate group. One way was to reverse the change *in vitro*, using an enzyme known to be specific for the removal of phosphate groups from proteins. Examples of such enzymes are alkaline phosphatase and acid phosphatase. While this works well in some systems, in our system we encountered enough problems early on to lead us to look for a more workable alternative. This was to give the unfertilized egg radioactive phosphate, then fertilize the eggs and show that, 4 to 5 minutes after fertilization, the eIF4E contained radioactivity. While relatively straightforward in concept, this experiment proved problematic in its execution for one reason: Unfertilized eggs are exceedingly impermeable to many substances, especially phosphate. However, it was essential that we get the radioactive phosphate into the egg prior to fertilization, since we had to show that the unfertilized egg eIF4E had no phosphate. To circumvent the problem, we preloaded the unfertilized eggs with a high level of radioactive phosphate by incubating a tiny amount of eggs for six hours with a very large amount of radioactive phosphate. By increasing the gradient in this way, we were able to load the phosphate pools in the eggs with radioactive material.

What we did next required mostly timing and speed. A labeled, unfertilized egg sample was homogenized and passed over the m^7G affinity column. The eIF4E would attach to the beads, while all other proteins would go through. We removed the eIF4E bound to the column by adding a large excess of m^7GTP to the column. An identical batch of radioactively preloaded unfertilized eggs was then fertilized, and homogenized at 5 minutes after fertilization. This fertilized sample was treated in a similar fashion, by passing over the m^7G affinity column. The two samples containing eIF4E were then analyzed by slab gel electrophoresis, and autoradiograms were prepared by exposing the gels to X-ray film. Much to our delight, when we examined the autoradiograms, we saw a radioactively labeled band right where eIF4E should be in the fertilized egg sample, but not in the unfertilized sample.

This showed that the mobility change we had observed was due to phosphate, but it raised another question: was eIF4E unlabeled in the fertilized egg because we were not getting radioactive phosphate into the ATP pools of the egg? The enzymes that catalyze the reaction of protein phosphorylation use the gamma-phosphate on ATP, not free phosphate. Since the unfertilized egg is metabolically quiescent, it was possible that we were getting hot phosphate into the egg but not into the ATP.

The question, an important one to answer, turned out to be easy to address. We simply ran a slab gel of the material that had not bound to the m^7G affinity column and prepared an autoradiogram from it. This material contained all the proteins in the unfertilized egg except eIF4E. The autoradiogram showed that many proteins were radioactively labeled in the unfertilized egg. This established that we were getting the radioactive phosphate into the ATP pools in the egg, and that protein kinases were capable of using this phosphate, even in the relatively quiescent egg, to phosphorylate proteins.

A final issue that concerned us was whether the eIF4E phosphorylation was unique to *Strongylocentrotus purpuratus*, the species we used in our work, or whether this was a more general phenomenon. We examined two other species of sea urchin that were evolutionarily distant from *S. purpuratus*, *Lytechinus pictus* and *Arbacia punctulata*. In all cases we were able to demonstrate phosphorylation of the eIF4E concomitant with the activation of protein synthesis.

Additional controls further strengthened our conclusions and provided more information on the possible triggers for phosphorylation of eIF4E. Looking back, it is interesting to note that about 75% of the time spent on this project was devoted to carrying out the experimental controls necessary to establish that we were not dealing with an artifact.

What did our results tell us? At the time we carried out this work, it was the first report of a specific biochemical change in the translational machinery that correlated with the activation of protein synthesis. However, it is important to keep in mind that like all scientific endeavor, our results raised as many, if not more questions than were answered. For example, what enzyme phosphorylates the eIF4E? How is it activated at fertilization? And, most important, does phosphorylation of eIF4E allow it to participate more efficiently in binding of the 5' cap, and if so, how?

References

[1]Lopo, A. C.; MacMillan, S.; and Hershey, J. W. B. *Biochemistry* 27:351 (1988).

[2]Lopo, A. C., and Hershey, J. W. B. *J. Cell Biol.* 101:221a (1985).

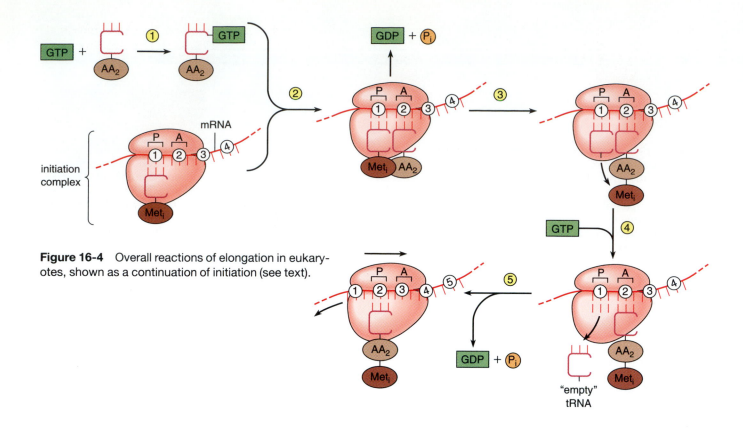

Figure 16-4 Overall reactions of elongation in eukaryotes, shown as a continuation of initiation (see text).

aminoacyl-tRNA links to GTP (step 1 in Fig. 16-4). The aminoacyl-tRNA then binds to the ribosome at the A site (step 2). The elongation factor and GTP bind to the ribosome with the aminoacyl-tRNA. The particular aminoacyl-tRNA binding at this step depends on formation of a correct codon-anticodon pair with the mRNA codon positioned at the A site. As the aminoacyl-tRNA binds, the GTP is hydrolyzed; the elongation factor and GDP are released from the ribosome along with inorganic phosphate. The energy released by GTP hydrolysis may contribute to aminoacyl-tRNA binding at the A site, or to conformational changes in the ribosome or tRNAs that are critical to the process (see below).

The ribosome, now with Met-tRNA$_i$ at the P site and an aminoacyl-tRNA at the A site, is ready for the next step in elongation, formation of the first peptide bond (step 3). In this step, which is catalyzed by a *petidyl transferase* activity of the large ribosomal subunit, the methionine carried by the initiator tRNA at the P site links to the amino acid carried by the tRNA at the A site (Fig. 16-5). Energy required for formation of the peptide linkage is derived primarily from hydrolysis of the bond linking methionine to the initiator tRNA. At the close of the reaction, both amino acids, linked into a dipeptide, are attached to the tRNA at the A site. By virtue of this attachment the tRNA at the A site is converted from an aminoacyl-tRNA into

a peptidyl-tRNA. An "empty" tRNA remains at the P site.

In the final reactions of elongation (steps 4 and 5 in Fig. 16-4) the empty tRNA releases from the ribosome, the ribosome advances by one codon along the mRNA, and the peptidyl-tRNA moves from the A to the P site. These reactions are speeded by another elongation factor and include breakdown of a second GTP. The sequence begins as the empty tRNA releases (step 4). The ribosome then moves along the mRNA through a distance equivalent to one codon (step 5). The movement, called *translocation*, carries the codon formerly at the A site, with its attached peptidyl-tRNA, to the P site and exposes a new codon at the A site. GTP is hydrolyzed as translocation takes place. There is now considerable evidence that the empty tRNA is not released directly from the P site. Instead it moves from the P site to a third location, the *E site* (E = exit) as the ribosome translocates. It is ejected from this location when the A site binds the next aminoacyl-tRNA entering the ribosome.

Three elongation factors speed steps in elongation. Fig. 16-6 outlines the roles of these factors in elongation and illustrates how factors in general promote various reactions in polypeptide assembly—they act much as enzymes do, binding to reacting molecules, speeding their interaction, and releasing unchanged at the close of the reaction.

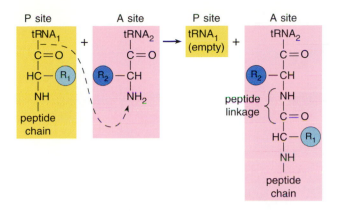

Figure 16-5 Reaction forming a peptide bond between the amino acids at the A and P sites of the ribosome. R' and R" indicate the side groups of the amino acids.

Completion of the initial elongation cycle readies the ribosome for another cycle of aminoacyl-tRNA binding, peptide bond formation, and movement of the ribosome and peptidyl tRNA. The only difference between the initial and subsequent elongation cycles is that the P site is occupied by a peptidyl-tRNA instead of an initiator tRNA at the beginning of each subsequent cycle (see Fig. 16-9). During each cycle the ribosome advances one codon and adds one amino acid to the growing polypeptide chain. As each cycle turns, two GTPs are hydrolyzed to GDP—one during aminoacyl-tRNA binding and one during translocation. The cycle, which turns up to six times per second in eukaryotes, repeats until the ribosome reaches a terminator codon at the end of the coding portion of the mRNA.

The growing polypeptide chain extends from the ribosome as elongation continues. If ribosomes are exposed to a proteinase during elongation, about 30 to 35 of the most recently added amino acid residues are protected from hydrolysis. This suggests that the growing polypeptide chain extends through an enclosed channel or groove in the ribosome that protects it from attack by the proteinase. Recent results indicate that there is actually a tunnel in the large subunit through which the growing polypeptide chain probably extends (see below).

At any given time, several ribosomes may be engaged in elongation on the same message, spaced along the mRNA like beads on a string. The entire structure, including the mRNA molecule and the attached ribosomes, is known as a *polysome* (Fig. 16-7). The number of ribosomes in a polysome depends on the length of the coding region of an mRNA molecule, ranging from a minimum of 2, observed, for example, in the very small mRNAs coding for protamines, to long complexes containing almost 100 ribosomes. In actively synthesizing cells about 70% of the ribosomal subunits available in the cytoplasm are linked to mRNAs in polysomes.

Accuracy of Elongation Ribosomes rarely make errors in the amino acids placed in polypeptide chains—on the average, only about 1 error for every 50,000 correct entries. Ribosomes also rarely make *frameshift errors*—that is, moving along the mRNA through a distance equivalent to either two or four nucleotides instead of three during an elongation cycle. Imperfect movement, if it occurs, shifts the reading frame so that the codons are read differently following the shifting point. Estimates of this error are even lower than codon-anticodon mispairs, as low as 1 in 100,000 elongation cycles.

The molecular basis for the accuracy of elongation is unknown. It is considered likely to involve more than codon-anticodon pairing, because some mutations in ribosomal proteins can increase or decrease the error rate. There may well be a *proofreading* mechanism in which the most recently added aminoacyl-tRNA is ejected from the ribosome if incorrect. Or, the ribosome may hydrolyze the most recently formed peptide bond if the wrong amino acid is attached.

Surprisingly, some mutations in ribosomal proteins *increase* the accuracy of protein synthesis. This unexpected effect may indicate that the error rate characteristic of unmutated, "wild-type" ribosomes is an evolutionary compromise between the maximum obtainable accuracy and other factors, such as the rate of protein synthesis or the amount of energy required, that might be adversely affected by the extra steps necessary for more perfect fidelity.

Some frameshifts are programmed rather than appearing as translation errors. Programmed frameshifts, usually moving the reading frame back or forward by one nucleotide (+1 or −1 frameshifts), occur in both prokaryotes and eukaryotes. In most the frameshift mechanism involves patterns of secondary structure in the mRNA that, on combination with a regulatory protein or other element, cause the ribosome to deviate from the usual three nucleotides traversed during a translocation. The controlled frameshifts, by establishing a new reading frame, produce alternative proteins from the same mRNA or shift production between active and inactive forms of a protein. For example, the eukaryotic protein *antizyme* inhibits an enzyme that catalyzes a critical step in the synthesis of polyamines, highly charged molecules of unknown function that are abundant in all living organisms. If the codons in the antizyme mRNA are read in perfect register, antizyme is made in incomplete, inactive form. If polyamines are made in excess of cellular requirements, some interact with the translation apparatus to

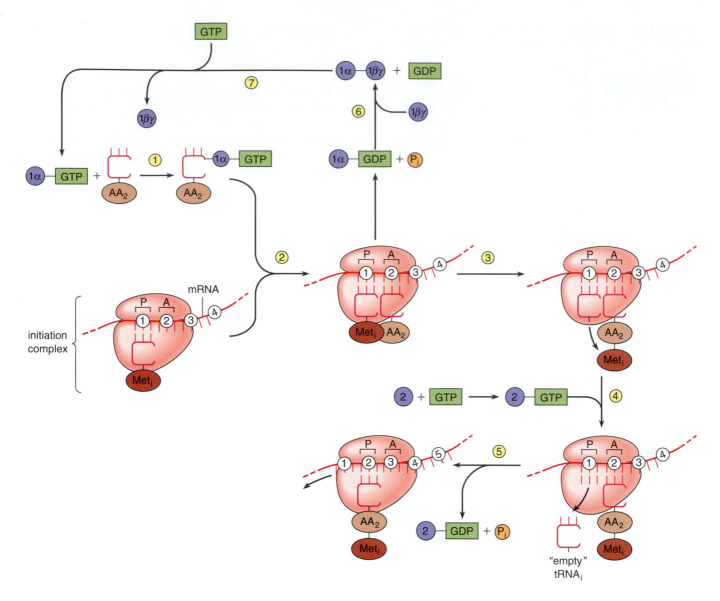

Figure 16-6 Roles of elongation factors in eukaryotic elongation. The cycle shown is the first one following initiation; later cycles follow the same pathway except that a peptidyl-tRNA instead of the initiator tRNA occupies the P site in steps 2 through 4. The reactions of eukaryotic elongation are facilitated by three elongation factors, *EF1α, EF2,* and *EF1βγ.* In the diagram the EF prefix has been eliminated for each elongation factor. *Met* indicates methionine; *AA₂* is the second amino acid specified by the code. In a preliminary reaction preparing an aminoacyl-tRNA for addition to the ribosome (step 1) an aminoacyl-tRNA reacts with the first elongation factor, EF1α, which is already combined with a molecule of GTP. The reaction forms the three-part or *ternary* complex EF1α·GTP·aminoacyl-tRNA, which is ready for addition to the ribosome. This complex binds at the A site (step 2). As the ternary complex binds, GTP is hydrolyzed; the resulting EF1α·GDP complex, relieved of its aminoacyl-tRNA, is released from the ribosome along with inorganic phosphate (P_i). The ribosome, now with Met-tRNA$_i$ at the P site and

an aminoacyl-tRNA at the A site, forms a peptide bond (step 3), which requires no elongation factors to proceed. As a preliminary to translocation of the ribosome along the mRNA, EF2 interacts with GTP to form an EF2·GTP complex. This complex binds to the ribosome, inducing a conformational change that favors release of the stripped tRNA from the P site (step 4). The ribosome then moves one codon along the mRNA, also moving the peptidyl-tRNA at the A site to the P site (step 5). As these events take place, GTP is hydrolyzed and EF2 is released from the ribosome as an EF2·GDP complex. A side cycle (steps 6 and 7) involving the third elongation factor, EF1βγ, regenerates the EF1α·GTP complex used in the first reaction of the elongation cycle. In this reaction sequence, EF1βγ first displaces the GDP residue attached to EF1α, forming an EF1α·EF1βγ complex (step 6). GTP then exchanges for EF1βγ in the complex (step 7). The regenerated EF1α·GTP complex is now ready to enter another elongation cycle.

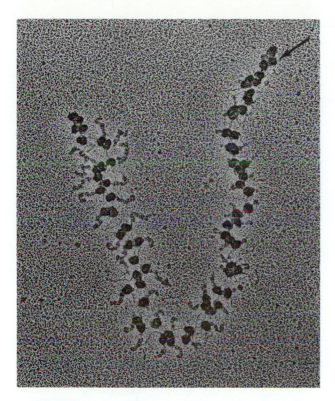

Figure 16-7 A polysome, consisting of a series of ribosomes reading the same mRNA, extracted from the gall midge *Chironomus*. Toward the beginning of the mRNA (arrow) the polypeptide chains extending from the ribosomes are shorter. X 68,000. (Courtesy of O. L. Miller, Jr., from *EMBO J.* 1:59(1982) by permission of IRL Press.)

induce a +1 frameshift, bypassing a terminator codon and producing the complete, active protein. Active antizyme then inhibits its target enzyme, leading to a reduction in polyamine synthesis.

Termination

Protein synthesis stops when the A site of a ribosome arrives at one of the three terminator codons UAA, UAG, or UGA on the mRNA. At this point the new polypeptide is released from the ribosome, the empty tRNA at the P site is ejected, and the ribosomal subunits separate from the mRNA. The reactions are catalyzed by a single termination factor, *RF* (for *Release Factor*) in eukaryotes.

In eukaryotes the appearance of one of the three terminator codons at the A site causes RF to bind to the A site along with GTP instead of an aminoacyl-tRNA (step 1 in Fig. 16-8). RF binding stimulates hydrolysis of the bond holding the polypeptide chain to the tRNA at the P site (step 2), catalyzed by the peptidyl transferase site of the large subunit. (The first step

in formation of a peptide linkage by the peptidyl transferase site is hydrolysis of the bond linking the peptide chain to the tRNA at the P site.) In termination, since no amino acid is located at the A site, the hydrolysis simply frees the polypeptide chain from the ribosome. As the polypeptide releases, the termination factor is ejected from the A site along with the empty tRNA at the P site (step 3). These reactions are accompanied by hydrolysis of the GTP bound with RF in step 1. As these steps run their course, the ribosomal subunits separate from the mRNA.

The overall steps in the initiation, elongation, and termination phases of protein synthesis are summarized in Figure 16-9. Formation of each peptide bond in a growing polypeptide chain involves a relatively large expenditure of cellular energy. Two phosphates are removed from an ATP molecule in amino acid activation; during elongation, two more phosphates are removed from GTP molecules: one during binding of an aminoacyl-tRNA to the A site and a second during each translocation. The four phosphate groups removed from GTP or ATP for each peptide bond formed release approximately $4 \times 7000 = 28,000$ cal/mol. Of this total energy release, about 5000 cal/mol are trapped in the formation of a peptide bond.

The reactions of protein synthesis are targets for several important regulatory controls, that speed or slow individual steps (for details, see Chapter 17; see also Lopo's Experimental Process essay). The reactions are also targets for an extensive array of antibiotics that kill or impair cells by interfering with individual steps in initiation, elongation, or termination. Many of the antibiotics affecting protein synthesis have also been vital to experiments isolating and identifying individual reactions of polypeptide assembly.

Chemical Modification and Folding of Newly Synthesized Polypeptides

The polypeptide chains released from ribosomes are incomplete structures that require two major finishing operations, chemical modification and folding. The two processes are interdependent because the final folding pattern of a protein depends to some extent on the types and degree of chemical modifications, and the sites recognized for modification often depend on the arrangements taken by a protein as it folds.

Chemical Modifications Reactions modifying polypeptide chains take three primary forms: (1) changes altering individual amino acid residues into other types, (2) covalent addition of organic units such as sugars or lipids to individual amino acids, and (3) enzymatic cleavage of one or more amino acids from

Figure 16-8 Termination of protein synthesis in eukaryotes (see text). *RF* indicates the release factor; *T* is one of the three terminator codons UAA, UAG, or UGA.

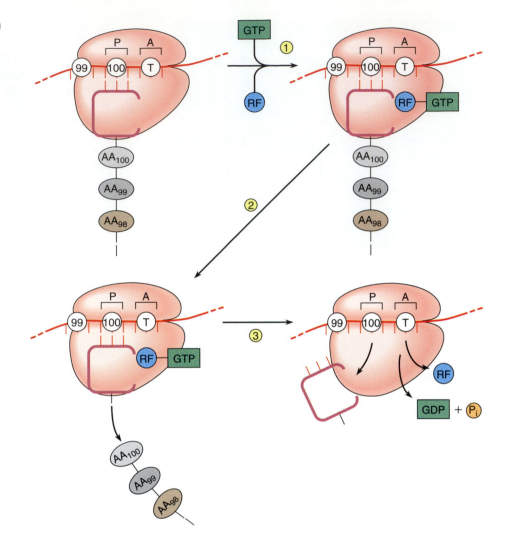

either end of the polypeptide chain. In rare instances, modifications include addition of extra amino acids to the N-terminal end of a polypeptide chain. The modifications, which are catalyzed by specific enzymes, take place at certain amino acid residues determined by the amino acid sequence and three-dimensional structure of the polypeptide under processing.

Chemical modifications include addition of methyl, acetyl, phosphate, hydroxyl, formyl, amide, amino, sulfate, and carboxyl groups. A number of the modifications, including acetylation and phosphorylation, are readily reversible. One of the reversible modifications, phosphorylation of serine, threonine, or tyrosine residues, is much used in regulation of the biological activity of enzymes and other proteins. In some cases, phosphorylation inhibits the activity of the target proteins, and in others it activates the protein receiving the phosphate groups (see Chapter 6 for details). A number of proteins are both activated and inhibited by phosphate groups added to different amino acid residues. In all, the various chemical alterations create more than 200 different types of modified amino ac-

ids. Thus completely processed proteins may include many more amino acid types than the original 20 entered during initial polypeptide assembly.

Among the most extensive of the reactions adding organic units to proteins are those linking sugar groups of various kinds, including glucose, mannose, fucose, galactose, amino sugars, and sialic acid (see Fig. 4-4). Addition of the sugars creates the glycoproteins, most of which become membrane or secreted proteins. The modifications adding sugars to membrane or secreted proteins take place in the ER and Golgi complex (see Chapter 20 for details). Another important class of proteins, the lipoproteins, is created by the addition of lipid units.

Enzymatic cleavage of newly synthesized polypeptides may remove a single amino acid or amino acid chains of various lengths. Many proteins are modified by removal of the initial methionine at the beginning of the amino acid chain. Larger segments of newly synthesized polypeptide chains are frequently removed as part of the changes converting proteins from inactive precursors to final, biologically active

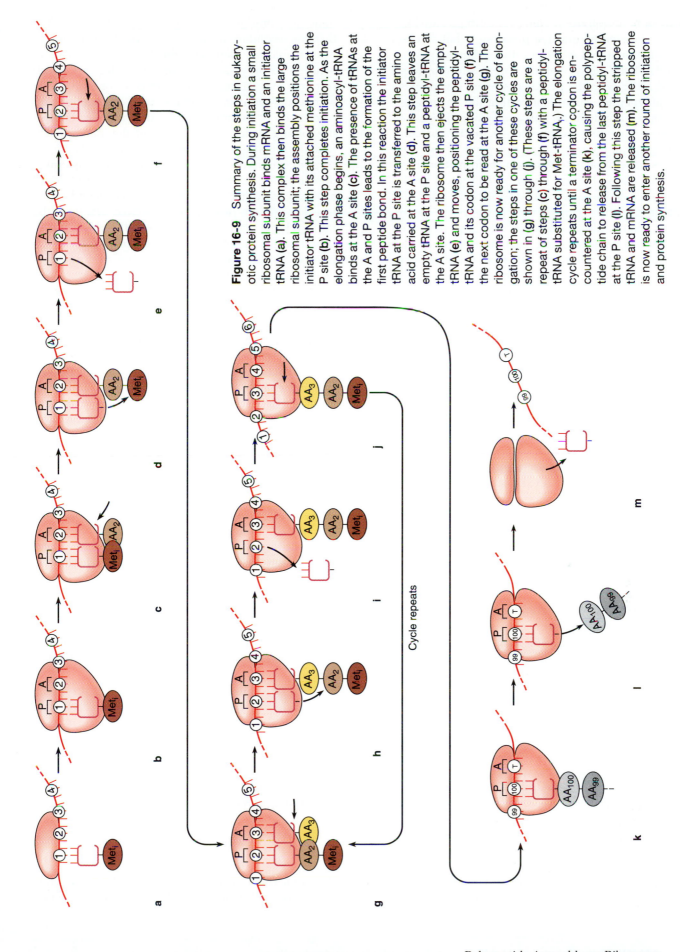

Figure 16-9 Summary of the steps in eukaryotic protein synthesis. During initiation a small ribosomal subunit binds mRNA and an initiator tRNA (a). This complex then binds the large ribosomal subunit; the assembly positions the initiator tRNA with its attached methionine at the P site (b). This step completes initiation. As the elongation phase begins, an aminoacyl-tRNA binds at the A site (c). The presence of tRNAs at the A and P sites leads to the formation of the first peptide bond. In this reaction the initiator tRNA at the P site is transferred to the amino acid carried at the A site (d). This step leaves an empty tRNA at the P site and a peptidyl-tRNA at the A site. The ribosome then ejects the empty tRNA (e) and moves, positioning the peptidyl-tRNA and its codon at the vacated P site (f) and the next codon to be read at the A site (g). The ribosome is now ready for another cycle of elongation; the steps in (g) through (j). (These steps are a repeat of steps (c) through (f) with a peptidyl-tRNA substituted for Met-tRNA_i.) The elongation cycle repeats until a terminator codon is encountered at the A site (k), causing the polypeptide chain to release from the last peptidyl-tRNA at the P site (l). Following this step the stripped tRNA and mRNA are released (m). The ribosome is now ready to enter another round of initiation and protein synthesis.

Cycle repeats

forms. Many digestive enzymes, for example, are initially synthesized as inactive precursors that can be transported or stored without cellular damage. Once released into the digestive cavity, processing enzymes secreted along with the precursors remove segments of the polypeptide chain, converting the enzyme to active form. Peptide hormones and viral proteins are also commonly processed from larger, inactive precursors. In general, removal of the polypeptide segment uncovers an active site or relieves a conformational restraint that allows the remainder of the protein to assume its final, active conformation. Most of the reactions removing protein segments take place in the Golgi complex and modify proteins destined to be secreted.

A few known processing reactions remove an intervening segment of amino acids from the interior of a protein, followed by joining of the free ends to produce a continuous amino acid chain. In effect, the splicing reaction removes an "intron" at the protein rather than pre-mRNA level (see p. 410). In one unusual variation of polypeptide splicing producing *concanavalin A* (a plant lectin, see p. 191) in the jack bean, the two polypeptide segments created by removal of the intervening segment are reversed in order, so that the former N- and C-terminal ends of the sequence link together in the interior of the chain. The splicing reaction appears to be self-catalyzed by the concanavalin A precursor protein.

Some peptide hormones are processed by cleavage pathways that release a variety of different polypeptides from the same precursor, depending on which of several alternate processing routes are used. The alternate pathways provide another mechanism increasing the number of proteins encoded within a single gene (others include alternative splicing of pre-mRNAs; see p. 415).

Polypeptide Folding Many completed proteins probably fold more or less spontaneously into their final, three-dimensional shapes. The final folding arrangement may be stabilized by disulfide (—S—S—) linkages between cysteine residues at different points in the polypeptide chain. The stabilizing disulfide linkages may form spontaneously or may be catalyzed by enzymes specialized for this activity. The conclusion that many proteins fold spontaneously is supported by experiments with a variety of proteins such as ribonuclease, cytochrome *c*, myoglobin, and lysozyme, which can refold into their fully active forms in the test tube and regenerate disulfide linkages after denaturation by heat and removal of disulfide linkages.

The processes folding newly synthesized polypeptides are believed to follow one or at most a few pathways determined by the amino acid sequence, rather than passing randomly through the many folding conformations possible for a given polypeptide.

Random sampling would require much more time than the seconds or milliseconds actually observed for protein folding. Generally the folding reactions are believed to progress from an initial stage in which secondary structures such as stretches of alpha helix and beta sheet form, combined with compaction and association of hydrophobic regions in the interior of the folding structure. In this form the incompletely folded protein is termed a *molten globule*. Hydrogen bonding and hydrophobic associations then assemble domains and subdomains, which proceed to "dock" against each other to achieve the final folded structure. For proteins destined to be secreted, many of the initial folding reactions take place as the polypeptides penetrate ER membranes during their synthesis on ribosomes attached to the ER. Final folding takes place in the ER lumen.

Proteins release energy as they fold into their final form, which represents a minimum energy state. As a result, reversing the process to unfold a protein is an energy-requiring process, and a protein tends to remain stable in its final conformation. Rather than being tightly fixed, the final folding conformation is flexible and may change in response to alterations in the medium or as a part of functional activity. Enzymatic and motile proteins, for example, regularly assume alternate folding conformations (see, for example, p. 85). Some proteins also unfold temporarily during transit through membranes.

For some proteins, folding is promoted by other proteins that direct the folding process along certain pathways. The folding factors, called *chaperones*, may work by combining with protein surfaces that are exposed during intermediate folding interactions, preventing aggregation and limiting further folding to the "correct" conformations. Chaperones also hold some proteins destined for insertion in membranes—particularly those of freely suspended organelles such as mitochondria—in intermediate, water-soluble conformations in which they can remain suspended in the aqueous cytoplasm until they collide with the target membrane.

In a sense the amino acid sequence of a protein, as modified by chemical alterations, is a second genetic code that spells out the final folding form to be taken by a protein. If enough of the folding code can be eventually understood, it may become possible to deduce not only the amino acid sequence of a protein from the sequence of the gene encoding it, but also its final folding conformation.

RIBOSOME STRUCTURE

Ribosomes have been investigated by electron microscopy, X-ray diffraction of crystals cast from ribosomes, and biochemical analysis to work out their three-dimensional structure, identify the proteins in the ribosomal subunits, and trace their arrangement with rRNAs. Many questions remain to be answered, particularly with respect to eukaryotic ribosomes, and research into both the physical and molecular structure of ribosomes continues at a rapid pace.

Major Structural Features of Ribosomes

The large and small ribosomal subunits have been examined in the electron microscope by J. A. Lake, G. Stöffler, and others. The subunits have been investigated primarily in isolated form, prepared by negative staining (see p. 790). This technique reveals several structural features that are remarkably similar in prokaryotic and eukaryotic ribosomes. Small ribosomal subunits (Figs. 16-10a and 16-11a) have a *head* and

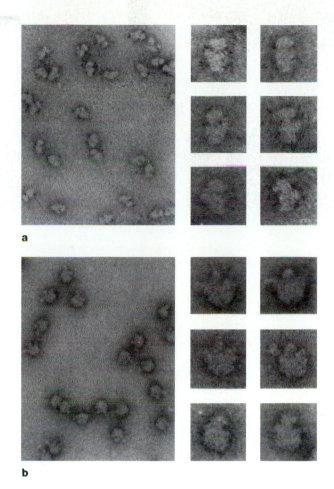

a

b

Figure 16-10 Isolated ribosomal subunits of the bacterium *E. coli*, prepared for electron microscopy by negative staining. **(a)** Small ribosomal subunits; **(b)** large ribosomal subunits. The large micrographs show a field of small or large ribosomal subunits; the insets to the right show enlarged subunits from different points of view. Main figures × 160,000; insets × 400,000. (Courtesy of G. Stöffler, reproduced with permission from *Ann. Rev. Biophysics Bioengineer.* 13:303. © 1984 by Annual Reviews, Inc.)

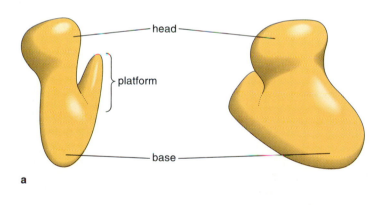

a

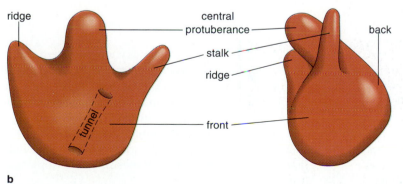

b

Figure 16-11 Structural features of the small **(a)** and large **(b)** ribosomal subunit of prokaryotes, primarily according to the models of J. A. Lake and his coworkers. The platform of the small subunit fits against the front of the large subunit in intact ribosomes.

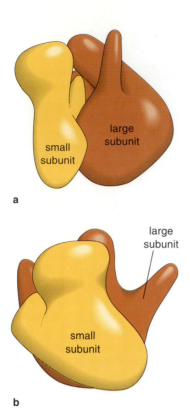

a

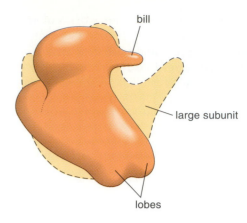

Figure 16-13   The bill and lobes, structural features characteristic of eukaryotic small ribosomal subunits.

large
subunit

small
subunit

b

Figure 16-12   The arrangement of the large and small subunit in an intact ribosome.

base, with an armlike *platform* extending from one side. Large subunits (Figs. 16-10b and 16-11b) have a prominent *central protuberance*, *stalk*, and *ridge* extending from one side. The platform surface of the small subunit fits into a depression on the large subunit in an assembled ribosome (Fig. 16-12).

Through images reconstructed from X-ray diffraction and other techniques, A. Yonath and H. G. Wittman and others found a tunnel about 10 nm long and 2.5 nm in diameter passing through the large subunit. The tunnel extends from the region containing the A and P sites to the part of the large subunit from which the newly assembled polypeptide chain exits the ribosome. This tunnel is probably the channel through which the newly assembled polypeptide chain travels on its way out of the ribosome.

These structures occur on both prokaryotic and eukaryotic ribosomal subunits. Eukaryotic small subunits also have several additional structural features (Fig. 16-13). One is a *bill* that extends from the head of the small subunit on the side opposite the cleft. Another feature peculiar to eukaryotes is a set of *lobes* at the end of the small subunit opposite the head. The lobes are believed to contain the additional sequences that make 18S rRNA larger than its 16S bacterial counterpart (see p. 430 and Supplement 15-1 for details).

Isolating and Identifying Ribosomal Proteins

Ribosomes are isolated by techniques (see p. 794) in which cells are broken open and the released contents are separated into fractions by centrifugation. Once separated and purified by centrifugation, ribosomes can be disassembled into individual rRNAs and proteins by adding salts such as CsCl, LiCl, or KCl in increasing concentrations. This shows that the individual proteins and rRNAs of ribosomal subunits are held together by noncovalent forces such as ionic attractions and hydrogen bonds.

The proteins released by salts are usually separated and purified by gel electrophoresis (see Appendix p. 796). The separated proteins are numbered in sequence as *S* and *L proteins* according to their subunit of origin and the rate at which they migrate through gels, with the largest proteins first in the sequence. For *E. coli* ribosomes, these techniques release 21 different proteins from the small subunit and 31 from the large subunit (see also Supplement 16-1). The proteins obtained from eukaryotic ribosomes vary according to the species studied and the isolation methods used, so that the number in each subunit is uncertain. S. Michel and his coworkers, for example, detected 32 proteins in the small subunit and 45 in the large subunit of yeast ribosomes, for a total of 77 distinct types. Other investigators, depending on the eukaryotic species and extraction techniques used, obtain numbers near these values.

Why eukaryotic ribosomes contain more proteins than prokaryotic ribosomes is unclear because the reaction sequences carried out by both types are basically similar. The additional proteins may bind the more numerous initiation factors characteristic of eukaryotes, take part in the more complex mechanisms regulating translation in this group, or promote linkage of ribosomes to membranes of the endoplasmic reticulum.

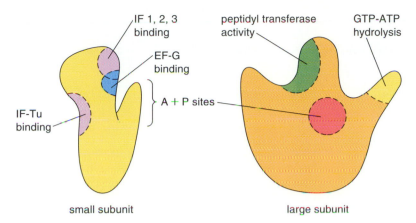

Figure 16-14 Tentative locations of functional sites on *E. coli* ribosomal subunits. The sites are believed to be distributed similarly in eukaryotic ribosomes. The platform of the small subunit is thought to be the site of codon-anticodon interaction. The platform is also believed to cooperate with nearby segments of the large subunit, particularly in the region at the base of the stalk, in formation of the A and P sites. The stalk of the large subunit contains sites binding elongation factors and hydrolyzing GTP and ATP. Messenger RNA is believed to thread through the gap or crevice between the small and large subunits. The eukaryotic bill of the small subunit (see Fig. 16-13) is involved in factor recognition in eukaryotes.

Locating Individual Proteins, rRNAs, and Functions in Ribosomes

Several methods have been employed in the laboratories of Lake, Stöffler, H. F. Noller, P. B. Moore, A. E. Dahlberg, R. Brimacombe, and others to locate individual protein and rRNA molecules within ribosome structure (see Information Box 16-1). These methods, used singly and in combination, have revealed the locations of many ribosomal proteins and the possible arrangement of rRNA molecules within the large and small subunits. By itself the work localizing rRNA and protein molecules in ribosomes would be little more than a sophisticated exercise in molecular morphology. Combining these results with other approaches, however, has allowed sites on the ribosomal surfaces, and the proteins and rRNA segments located in these sites, to be provisionally identified with ribosomal functions (see Fig. 16-14 and Information Box 16-1).

One of the most striking features revealed by these approaches is that ribosomal proteins interact cooperatively: No reaction can be identified with a single protein. In some reactions, as in binding of aminoacyl-tRNAs and elongation factors, proteins of both subunits evidently cooperate, making the A and P sites part of both subunits. The research also makes it obvious that rRNAs participate in many reactions of protein synthesis. Mutations in the bacterial small-subunit rRNA, for example, interfere with mRNA-ribosome association, tRNA pairing, maintenance of reading frame, and termination. Mutations in segments of the 23S rRNA of the bacterial large subunit inhibit peptide bond formation, activity of the A and P sites, and translocation. As a consequence, the rRNAs of ribosomes, once thought merely to provide a structural framework for assembly of the ribosomal proteins, are now considered likely to take an active part in many reactions of polypeptide assembly.

Some reactions of protein synthesis, including the central peptidyl transfer reaction linking amino acids into the growing polypeptide chain, may even be catalyzed by rRNAs acting as ribozymes (see p. 86). H. F. Noller and his coworkers found that residual structures produced after exposure of ribosomes to treatments designed to denature or remove proteins—including strong detergents and solvent extraction—could still catalyze peptidyl transfer. Enzymes breaking down RNA, however, quickly eliminated the peptidyl transferase activity. Further, T. R. Cech and his colleagues found that a modified ribozyme (one that in native form removes an intron) could catalyze a critical step of peptidyl transfer, hydrolysis of the bond linking a polypeptide chain to a tRNA.

Ribosome Assembly in the Test Tube

Disassembly techniques preserve individual ribosomal proteins and rRNAs of prokaryotes well enough to allow functional ribosomes to be reassembled in the test tube. The techniques for small subunit reassembly were worked out by M. Nomura and his colleagues, who put together functional small subunits from 16S rRNA and the small-subunit proteins of *E. coli*. Later, K. H. Nierhaus and F. Dohme reassembled functional large subunits from purified large-subunit proteins and rRNAs of *E. coli*.

Attempts to reassemble eukaryotic ribosomes have not been as successful, probably because the separation methods damage the ribosomal proteins. All that has been possible, in fact, is reassembly of ribosomal subunits from ribosome "cores"; that is, from particles that have not been disassembled past the point at which the most easily split proteins are removed.

Although the overall structures and functions of prokaryotic and eukaryotic ribosomes are similar, attempts to exchange prokaryotic and eukaryotic ribosomal proteins in reassembled ribosomes have been

Locating Ribosomal Proteins and rRNAs

Investigators have used many techniques in their attempts to unravel ribosome structure. Among the most productive has been the use of antibodies developed against individual ribosomal proteins. In this method, individual ribosomal proteins are injected into test animals to produce an antibody response. Antibodies developed against the proteins are isolated, purified, and reacted with intact ribosomal subunits. The antibodies, clearly visible in the electron microscope, mark sites occupied by the ribosomal proteins on the surfaces of the subunits.

Similar methods have been employed to locate segments of rRNA molecules in ribosomal subunits. Because nucleic acids are not usually recognized by antibodies, segments of the individual rRNA types are linked to chemical groups that do stimulate antibodies before being injected into a test animal. The antibodies developed in this way are then used to identify the locations of rRNAs linked to the same chemical groups in intact ribosomal subunits.

Locations of individual proteins in ribosomal subunits have also been determined by the use of chemical crosslinkers. The crosslinkers, which tie individual ribosomal proteins to their nearest neighbors, allow relative positions within the subunits to be assigned to each protein. Other approaches use enzymes, chemicals, or radioactive labels to mark segments of rRNA or protein molecules exposed at ribosomal surfaces. The results of different approaches such as experiments with antibodies and crosslinkers generally agree, indicating that the locations

of individual proteins determined by the different methods are probably correct.

Several elegant techniques have made it possible to link the proteins and rRNAs identified with ribosomal structures to individual reactions of protein synthesis. One is the *affinity label method*, in which a molecule reacting in protein synthesis, such as an initiation factor or aminoacyl-tRNA, is "activated" by linking it to a highly reactive chemical group. During interaction of the activated molecule with ribosomes in protein synthesis the reactive group is often transferred from the activated molecule to the ribosomal components reacting with it, thereby marking its site of interaction. In some cases, rather than being transferred, the reactive group forms a stable crosslink between the activated molecule and the ribosomal components it reacts with. Also useful in the research correlating ribosomal structure with function have been mutants, in which individual proteins and the functions they catalyze are faulty or missing from ribosomes. Partial assembly, in which individual proteins are left out when ribosomes are assembled experimentally from the individual rRNAs and proteins in the test tube, has provided evidence along the same lines.

Most of the work employing these methods has used *E. coli* ribosomes as experimental subjects because the number and identity of the proteins have been established, and the individual proteins have been separately isolated and sequenced (see Supplement 16-1 for details).

unsuccessful except for prokaryotic L7/L12 and eukaryotic L40/L41, which are interchangeable to a limited degree. Similarly, antibodies against prokaryotic ribosomal proteins show little or no cross reaction with eukaryotic ribosomal proteins. Relatively few antibody cross reactions are noted even among different eukaryotic groups—for example, only about 20% of the ribosomal proteins of rat and chicken ribosomes are found to cross react. These findings suggest that although overall functional steps are similar or even identical in ribosomes from different groups, the proteins carrying out these steps have diverged significantly during evolution. The variations noted in ribosomal proteins of different eukaryotic species may underlie some of the disagreement in results obtained by attempts to isolate the proteins—there may, in fact, be no "standard" number of ribosomal proteins shared by all eukaryotes.

Both the rRNA and protein components of eukaryotic ribosomes may also change as growth and de-

velopment progress. In *Xenopus*, for example, several distinct types of 5S rRNA are produced and are incorporated into ribosomes at different developmental stages (see p. 431). Equivalent changes have been detected in ribosomal proteins by S. Ramagopal, who found changes in the proteins extracted from ribosomes in different developmental stages of the slime mold *Dictyostelium*.

THE GENETIC CODE FOR PROTEIN SYNTHESIS

Solving the Code

The genetic code and its solution challenged investigators in scientific fields ranging from mathematics, chemistry, and physics to biology. Basically the problem was how codons for 20 amino acids could be spelled out from an alphabet consisting of only four different nucleotides. The codebreakers realized that the bases would have to be used in combinations of at least three to provide coding capacity for 20 amino acids. If used one at a time, only four different combinations could be written ($4^1 = 4$); if taken two at a time, only 16 could be spelled out ($4^2 = 16$). But 64 codons can be made if the four nucleotides are used in combinations of three ($4^3 = 64$), many more than needed to spell out codons for 20 amino acids.

In 1954, G. Gamov pointed out additional problems to be solved. One concerns separation of codons. In written English, spaces are used to separate words. But how are the codons separated in a DNA molecule, in which the bases are evenly spaced? One possibility was that the message is simply read by starting at some point and taking the following nucleotides three at a time. The first nucleotide would thus establish a *reading frame* for the coded message.

Gamov also considered the problem of *degeneracy* in the code. With 64 codons available in a triplet code, there are 44 more combinations than needed to specify 20 amino acids. In a nondegenerate code each amino acid would be specified by only one codon and the remaining 44 would be "nonsense" codes. In a degenerate code the same amino acid would be designated by two or more codons.

The number of bases in the codons and use of a reading frame in the genetic code were established experimentally by several different approaches. One experiment, carried out by S. Brenner and F. H. C. Crick in 1961, used mutants in which individual nucleotides were added or deleted in the DNA codes of *bacteriophage T4*, a virus infecting *E. coli*. Their experiment revealed that deletions or additions of one or two nucleotides within the coding sequence of a gene produced drastic changes in the proteins encoded in the gene. Additions or deletions of three bases, however, often had little effect.

Figure 16-15 explains these results. Figure 16-15a shows the normal form of the coding sequence of a gene. One or two additions or deletions shift the reading frame of the codons from the point of first mutation in the genes, completely disrupting the amino acid sequence encoded in the gene from the first mutation onward (Fig. 16-15b and c). Three consecutive additions or deletions, however, restore the reading frame after the third deletion (Fig. 16-15d). There is often little total disruption of the amino acid sequence if the three additions or deletions are closely spaced in the gene. These results are possible only if the code words are three nucleotides in length, and only if the code words are simply read three at a time by establishing an initial reading frame. Mutations that change the reading frame, as did the addition or deletion of one or two nucleotides in the Brenner and Crick experiments, are now known as *frameshift mutations*.

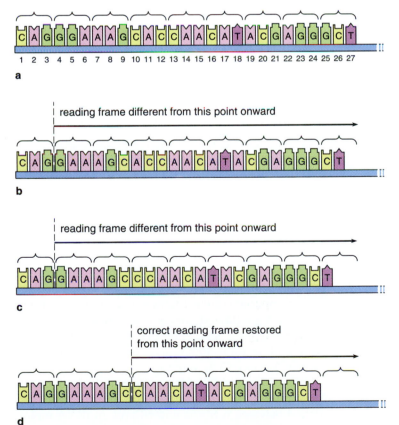

Figure 16-15 Effects of deletions or additions on the reading frame of the genetic code. **(a)** An unmutated sequence. **(b)** Effect of a single deletion at position 5 in the original sequence; the frame reads differently, with a total change in the codons, from the position of the deletion onward. **(c)** Effect of two deletions at positions 5 and 11 in the original sequence. The frame still reads differently from position 5 onward. **(d)** Effect of three deletions at positions 5, 11, and 12 in the original sequence; after the third deletion the original reading frame and codons are restored.

A second experiment, reported in 1966 by H. G. Khorana and his coworkers, confirmed that the codons are three nucleotides in length. They constructed artificial RNA molecules with different repeating sequences and used them as mRNAs in cell-free systems in the test tube. Repeating polymers of the form ABABABABABAB..., in which A and B are any two of the RNA nucleotides, coded only for the synthesis of a repeating *dipeptide* in cell-free systems. Assuming three-nucleotide codons, this is the expected result, because the code would be read as

$$\begin{array}{cccc} 1 & 2 & 1 & 2 \\ \end{array}$$
$$ABA/BAB/ABA/BAB...$$

That is, only two different codons would be contained in the message. Repeating *trinucleotides* of the form ABCABCABCABC... coded for polypeptides containing only one type of amino acid. This is also as expected, because in a triplet code this message would be read as

$$\begin{array}{cccc} 1 & 1 & 1 & 1 \\ \end{array}$$
$$ABC/ABC/ABC/ABC...$$

Repeating *tetranucleotide* polymers coded for polypeptides containing repeating sequences of four different amino acids. This is expected only if the message ABCDABCDABCDABCDABCDABCD... is read in triplets as

$$\begin{array}{cccccccc} 1 & 2 & 3 & 4 & 1 & 2 & 3 & 4 \\ \end{array}$$
$$ABC/DAB/CDA/BCD/ABC/DAB/CDA/BCD...$$

Khorana's results are therefore possible only if the codons are triplets.

The coding assignments of some individual codons were identified by noting the amino acids incorporated into proteins when artificial mRNAs were added to cell-free systems. The first experiments, carried out by M. W. Nirenberg and J. H. Matthei in 1961, tested artificial mRNAs assembled from a single, repeating nucleotide. An mRNA consisting simply of repeated uridines coded for a protein containing repeated phenylalanine residues and established that an RNA codon for phenylalanine is UUU. The same approach identified CCC as a codon for proline and AAA for lysine. The work with artificial mRNAs in Khorana's laboratory, as well as work by S. Ochoa and his colleagues, allowed the assignment of additional codons. About half of the codons were identified by these means.

The final experiments identifying the codons were carried out by Nirenberg and P. Leder, who set up a system testing individual codons. Nirenberg and Leder found that short artificial mRNAs containing only three nucleotides could bind to ribosomes and induce attachment of single aminoacyl-tRNAs. All 64 three-nucleotide codons were synthesized and added one at a time to ribosomes in cell-free systems. The particular aminoacyl-tRNA binding to the ribosomes in response was then identified. This approach allowed unambiguous assignment of all the codons (Fig. 16-16; by convention the codons are designated as they would appear in mRNA rather than DNA). Nirenberg and Khorana received the Nobel Prize in 1968 for their work in solving the genetic code.

Features of the Genetic Code

Inspection of the code reveals a number of interesting features. Three codons, UAA, UAG, and UGA, have no coding assignments. At first called nonsense codons, these codons were later found to specify termination of protein synthesis. A second feature is that there are two or more synonymous codons for all amino acids except two, methionine and tryptophan (Fig. 16-17). The code is therefore degenerate; with a few exceptions, each amino acid has either two or four synonymous codons. The exceptions are isoleucine with three codons and serine, arginine, and leucine with six each.

A third feature of the genetic code is its almost universal usage in living organisms. With some exceptions (see below), the same codons stand for the same amino acids in viruses and all organisms from bacteria to higher eukaryotes. This *universality* indicates that the code was established in its present form in the earliest living cells or even in life forms preceding the first fully evolved cells. (Evolution of the genetic code is discussed further in Chapter 26.)

Degeneracy and the Wobble Hypothesis

In 1966, Francis Crick advanced a hypothesis successfully explaining the pattern of degeneracy in the genetic code. He noted that all amino acids with the code XYA, where X and Y are any two letters of the code, also have another codon with the spelling XYG. Lysine, for example, is coded for by either of the two triplets AAA or AAG. Similarly, with one exception (isoleucine), all amino acids with the code XYU have another codon with the form XYC. Tyrosine, for example, is encoded in the two triplets UAU and UAC. If there are four codons for the same amino acid, both patterns are followed; that is, there are codons of the form XYA, XYG, XYU, and XYC for the same amino acid. Proline, for example, is encoded in the four triplets CCA, CCG, CCU, and CCC.

Crick reasoned that this pattern of degeneracy reflects an instability or "wobble" in some of the bases paired between mRNAs and tRNAs. In tRNA mole-

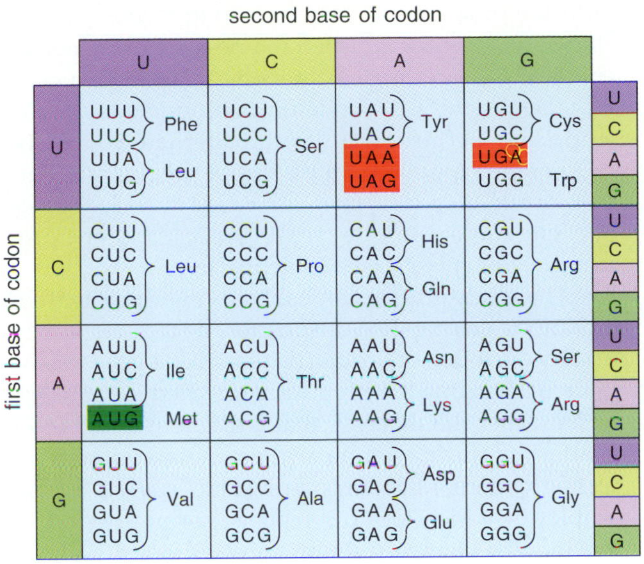

second base of codon

	U	C	A	G	
U	UUU ⎱ Phe UUC ⎰ UUA ⎱ Leu UUG ⎰	UCU ⎱ UCC ⎱ Ser UCA ⎰ UCG ⎰	UAU ⎱ Tyr UAC ⎰ UAA UAG	UGU ⎱ Cys UGC ⎰ UGA UGG Trp	U C A G
C	CUU ⎱ CUC ⎱ Leu CUA ⎰ CUG ⎰	CCU ⎱ CCC ⎱ Pro CCA ⎰ CCG ⎰	CAU ⎱ His CAC ⎰ CAA ⎱ Gln CAG ⎰	CGU ⎱ CGC ⎱ Arg CGA ⎰ CGG ⎰	U C A G
A	AUU ⎱ AUC ⎱ Ile AUA ⎰ AUG Met	ACU ⎱ ACC ⎱ Thr ACA ⎰ ACG ⎰	AAU ⎱ Asn AAC ⎰ AAA ⎱ Lys AAG ⎰	AGU ⎱ Ser AGC ⎰ AGA ⎱ Arg AGG ⎰	U C A G
G	GUU ⎱ GUC ⎱ Val GUA ⎰ GUG ⎰	GCU ⎱ GCC ⎱ Ala GCA ⎰ GCG ⎰	GAU ⎱ Asp GAC ⎰ GAA ⎱ Glu GAG ⎰	GGU ⎱ GGC ⎱ Gly GGA ⎰ GGG ⎰	U C A G

first base of codon (left margin)   *third base of codon* (right margin)

Figure 16-16 The genetic code, written by convention in the form in which the codons appear in mRNA. The three terminator codons, UAA, UAG, and UGA, are boxed in red; the AUG initiator codon is shown in green.

cules the three bases of the anticodon extend outward much like toes on a foot, with the bases at either end of the codon unsupported by close stacking against other bases (see Fig. 15-14). The lack of support allows the bases of the anticodon to take up a variety of positions in space, or to "wobble." Crick proposed, from the patterns noted in the degeneracy of the code, that the base of the anticodon most susceptible to wobble is the one that pairs with the third base of a codon in mRNA. The wobble permits this base greater freedom in hydrogen bonding, allowing a U in this position, for example, to pair with either A or G, or G to pair with either U or C.

The additional pairing opportunities due to base wobble, according to Crick's hypothesis, would lead to errors during polypeptide assembly except for the pattern of degeneracy in the genetic code. The pattern ensures that all the possible additional codons an anticodon might pair with because of wobble *still specify the same amino acid*. Degeneracy therefore compensates perfectly for any mRNA-tRNA mismatches that might occur as a result of wobble during protein synthesis.

Crick's *wobble hypothesis*, as it is known, and its predictions are confirmed by experiments in which natural messengers of known sequence are translated by cell-free systems containing less than the complete complement of aminoacyl-tRNAs. The result, in terms of protein synthesis inside cells, is that not all 61 aminoacyl-tRNAs are absolutely required for polypeptide assembly. The minimum number required if wobble occurs at the third base pair according to Crick's rules is 32.

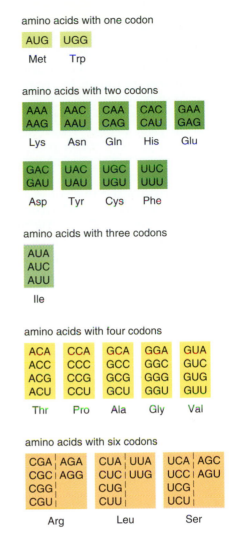

Figure 16-17 The genetic code arranged according to the pattern of degeneracy.

Wobble does not invariably occur at the third codon position. AUG, for example, specifies only methionine; the other codon expected for this amino acid by the wobble hypothesis, AUA, codes for isoleucine. The UGG codon for tryptophan similarly specifies only this amino acid; its expected synonym, UGA, is a terminator codon. The more restricted usage of the AUG and UGG codons indicates that for some reason, probably because of the structure of the tRNAs, wobble rarely occurs when these codons pair with their anticodons.

More recent work indicates that the third, wobble base of the anticodon is even more versatile in its pairing possibilities than anticipated by Crick's hypothesis. U. Lagerkvist found, for example, that in a cell-free system in which at least two valine tRNAs would be required to read the four valine codons GUU, GUA, GUC, and GUG according to Crick's wobble hypothesis, one tRNA was able to pair with all four. Other amino acids encoded by four different codons of the form XYN, in which the first two nucleotides are the same and only the third varies, were also found to be read successfully by single tRNAs. Lagerkvist proposed accordingly that the codons for many of these amino acids are read by a "two out of three" rule, in which only the first two bases of the codon are significant in codon-anticodon pairing. The third position may form essentially any type of base pair, including C-A, U-C, C-C, G-A, and G-G pairs unanticipated in Crick's wobble hypothesis.

Lagerkvist noted that the two-out-of-three rule applies primarily to amino acids in which one or both of the first two nucleotides is a C or G. The increased stability of the hydrogen bonding set up by the G-C base pairs, according to Lagerkvist, confers enough discrimination to the codon-anticodon association to make the third position unimportant. (Remember that G-C base pairs form three hydrogen bonds, while A-U pairs form only two; see Fig. 13-2.) The two-out-of-three rule for these amino acids means that the entire nucleic acid code can be read successfully by fewer than the 32 tRNAs predicted by Crick's wobble hypothesis. The actual number, probably in the lower 20s, is approached in mitochondria, in which the entire code is read successfully by only 22 different tRNAs (see Chapter 21).

The unusual base pairs permitted by wobble is only one unexpected feature of anticodons and their activity in protein synthesis. Anticodons also regularly contain modified bases, so that the usual A-U and G-C base pairs are not even possible. The modified base inosine (see Fig. 14-4) frequently appears in anticodons, for example; other modified bases also occur at lower frequencies in anticodons. These modified bases are found almost invariably at the wobble position in the anticodon, where they may modify or restrict the degree to which wobble mispairs take place.

Exceptions to Universality of the Genetic Code

One of the most surprising discoveries of recent years is that in some organisms or cellular organelles, codons have different meanings from the standard assignments (Table 16-1). One group of exceptions, the first to be discovered, occurs in mitochondria. In these organelles a number of coding substitutions were detected by experiments comparing the nucleotide sequence of mitochondrial genes with the amino acid sequence of the proteins encoded in the genes. The comparisons revealed that in all mitochondria examined to date, UGA codes for tryptophan in some proteins instead of for termination. This coding substitution is actually not too surprising, because UGA is the alternate codon expected for tryptophan according to the wobble hypothesis. Other substitutions noted in

mitochondria, among them human mitochondria, include AUA coding for methionine instead of isoleucine, and CUA coding for threonine instead of leucine. Once again, the AUA code for methionine is not too surprising, because AUA is the alternate codon for methionine expected from the wobble hypothesis. Further substitutions were later discovered in nuclear genes of ciliate protozoans, the yeast *Candida*, and one prokaryote, *Mycoplasma* (see Table 16-1).

These substitutions probably result primarily from mutations altering tRNA structure. The mutations may modify the codon-anticodon pairs formed by the tRNAs without changing the amino acid attached to the tRNA or may alter the tRNA so that a different amino acid is attached during amino acid activation without affecting codon-anticodon pairing. In *Candida*, for example, in which CUG is read as serine instead of leucine, the change rests on substitution of a single base in the anticodon loop. The substitution causes the tRNA to be charged with serine instead of leucine during amino acid activation.

A change in the genetic code of a somewhat different nature causes UGA, normally a terminator, to be read as the codon for an amino acid that is not one of the original 20 used in protein synthesis. The amino acid, *selenocysteine*, is carried by a tRNA with several distinct characteristics, including an unusually long variable loop and two additional base pairs in the D stem. The change has been found in two bacterial and three mammalian proteins; the mammalian plasma protein *selenoprotein*, for example, has 10 UGA sites at which selenocysteine is inserted. The insertions depend in part on the presence of a specialized sequence in 3' UTR of the selenoprotein mRNA. The unusual mechanism employing UGA for the insertion has prompted some investigators to claim selenocysteine as the "21st amino acid."

AMINO ACID ACTIVATION

The reactions of amino acid activation, which attach amino acids to their corresponding tRNAs, satisfy two requirements for protein synthesis. One is provision of a major part of the energy, conserved in the aminoacyl-tRNA complexes, required for formation of peptide bonds. The second is correct matching of amino acids and tRNAs, which provides the ultimate basis for the accuracy of translation.

The Reactions of Amino Acid Activation

Amino acid activation takes place in two steps, both catalyzed by the same enzyme, an *aminoacyl-tRNA synthetase*. In the first step an amino acid (AA) interacts with ATP. In the reaction, catalyzed by an ami-

Table 16-1 Some Exceptions to the Universality of the Genetic Code

Codon	Universal Assignment	Exceptional Assignment	Location
AGA	Arg	Ser	*Drosophila* mitochondria
AGA, AGG	Arg	Terminator	Mammalian mitochondria
AUA	Ile	Met	Yeast (*Saccharomyces*) and mammalian mitochondria
CGG	Arg	Trp	Plant mitochondria
CUG	Leu	Ser	Yeast (*Candida*) nucleus
CUN*	Leu	Thr	Yeast (*Saccharomyces*) mitochondria
UAA, UAG	Terminator	Gln	Ciliate nuclei
UGA	Terminator	Trp	Yeast (*Saccharomyces*), *Paramecium*, mammalian mitochondria; ciliate and *Mycoplasma* nucleus

* N = any of the four bases

noacyl-tRNA synthetase corresponding to the amino acid, the two terminal phosphates of ATP are split off and the remaining AMP is linked to the carboxyl group of the amino acid. The product of the reaction is an *aminoacyl-adenylate* complex (AA-AMP):

$$\text{AA} + \text{ATP} \xrightarrow{\text{aminoacyl-tRNA synthetase}} \text{AA-AMP} + \text{PP} \qquad (16\text{-}1)$$

The AA-AMP complex remains attached to the synthetase. The same enzyme molecule next binds a tRNA corresponding to the amino acid already linked to the enzyme. The second step in amino acid activation, in which the amino acid is transferred from AMP to the tRNA, forming an aminoacyl-tRNA (AA-tRNA) then takes place:

$$\text{AA-AMP} + \text{tRNA} \xrightarrow{\text{aminoacyl-tRNA synthetase}}$$
$$\text{AA-tRNA} + \text{AMP} \qquad (16\text{-}2)$$

In the reaction the amino acid is linked to the tRNA molecule by a covalent bond to the terminal adenine of the CCA sequence at the 3' end of the acceptor stem (see Figs. 16-18 and 15-13). The aminoacyl-tRNA end product is released from the enzyme along with AMP.

Conversion of ATP to AMP during amino acid activation releases a large increment of free energy, greatly favoring progress of the reaction in the direction of the aminoacyl-tRNA product. Much of the energy released by ATP hydrolysis is conserved in the aminoacyl-tRNA complex, which may be considered a "high-energy" molecule. The amount actually con-

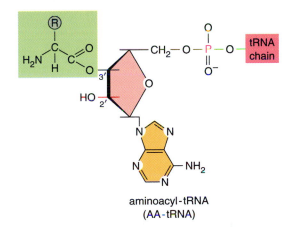

aminoacyl-tRNA
(AA-tRNA)

Figure 16-18 The final aminoacyl-tRNA product of amino acid activation, in which the amino acid is linked to the A of the terminal CCA sequence that lies at the tip of the acceptor stem in all tRNAs (see Fig. 15-13).

served depends on the particular amino acid. For most, hydrolysis of the bond between the amino acid and tRNA yields about 5000 to 7000 cal/mol at standard conditions. This conserved energy, as noted, provides much of the energy required to form peptide linkages during polypeptide assembly.

The Enzymes Catalyzing Amino Acid Activation

The aminoacyl-tRNA synthetases occur in 20 different types, one for each of the 20 amino acids used in polypeptide assembly. The enzymes, which vary widely in structure, contain one or two subunits known as the α- and β-polypeptides. In different synthetase enzymes, one or two copies of each subunit may be present, producing a group of proteins with one to four subunits and molecular weights ranging from about 37,000 to nearly 400,000. All possess sites that recognize an amino acid and a tRNA, bind ATP, and catalyze formation of the AA-AMP and AA-tRNA linkages.

Structural comparisons among the enzymes in both prokaryotes and eukaryotes reveal that they occur in two classes, *I* and *II*, that have enough distinct structural features to suggest two independent evolutionary origins for the synthetases. The primary difference between the two classes, which each contain 10 enzymes, lies in the domain binding ATP. Class I enzymes have a folded region with five parallel beta strands in this domain, with two sequence motifs, MSK and HIGH, in the one-letter abbreviations (see p. 54), common to all 10 members. Class II enzymes have a different arrangement of amino acids in this domain, including nine beta strands in an antiparallel array. Several functional differences are noted between the two classes, including attachment of amino acids to the 2'-OH of the terminal adenine of tRNAs by class I enzymes and to the 3'-OH by class II enzymes in most cases.

The Accuracy of Amino Acid Activation The aminoacyl-tRNA synthetases are remarkable for the accuracy with which they match tRNAs and amino acids—mistakes occur only about once or twice in every 10,000 amino acids. The accuracy is believed to depend on a proofreading mechanism that removes incorrectly attached amino acids rather than on near perfection in the initial linkage of amino acids. Many synthetases can bind initially to one or two or even several different amino acids as long as they are comparable in size and shape to the amino acid normally activated by the enzyme. The enzyme may even catalyze the reaction hydrolyzing ATP and forming the initial aminoacyl-adenylate (AA-AMP) complex. However, the reaction continues into the second step transferring the amino acid to a tRNA only if the amino acid is the correct one.

Evidently, once an AA-AMP complex forms, subsequent binding of a tRNA induces conformational changes in the enzyme. These changes position the amino acid for attachment to the tRNA only if the amino acid is correct. If an incorrect AA-AMP complex is attached to the enzyme, the conformational changes lead to hydrolysis of the complex and release of the amino acid, AMP, and tRNA. The process then begins again and runs to completion only if the correct amino acid binds to the enzyme.

The overall accuracy of protein synthesis is highly dependent on the fidelity of amino acid activation. Ribosomes evidently recognize only the anticodon of an aminoacyl-tRNA; if an incorrect amino acid is attached to the tRNA, a ribosome will place it in the polypeptide sequence as readily as the correct one. This has been demonstrated by experiments designed to "trick" ribosomes by linking an incorrect amino acid to a tRNA or by chemically modifying the amino acid carried by a tRNA to another form. If an incorrect amino acid is experimentally linked to a tRNA in this way, the ribosome enters the incorrect amino acid into a growing polypeptide according to the codon carried by the tRNA, with no recognition of the amino acid–tRNA mismatch.

Recognition of tRNAs by Synthetases P. Schimmel, J. Normanly, T. F. Donahue, W. H. McClain, T. A. Steitz, and O. C. Uhlenbeck and their coworkers and others have used a variety of techniques to work out the mechanisms by which the synthetases distinguish individual tRNAs. The methods include sequence comparisons among the several tRNAs linking to the same amino acid (called *isoaccepting* or *cognate* tRNAs); deletions, substitutions, or chemical modification of one or more nucleotides in tRNAs; X-ray diffraction of crystals cast from synthetase enzymes with their attached tRNAs; and chemical crosslinking. The crosslinking technique reveals segments of tRNA molecules held close enough to the synthetases to be attached by the molecule used as a crosslinker.

These experimental approaches have revealed several features of tRNA structure important in identification of tRNAs by synthetases (Fig. 16-19). In general the studies demonstrate that recognition depends on a relatively few nucleotides concentrated primarily in the anticodon, acceptor stem, and D stem. Most of these locations lie on the surfaces of tRNA molecules folded into their *L*-shaped three-dimensional form (see Fig. 15-14), particularly at the anticodon and along the inside corner of the *L*.

The anticodon serves as a primary recognition site for about half of the tRNAs. In a few the anticodon seems to be solely responsible for recognition by a synthetase, as in the tRNAs linking methionine, isoleucine, and valine in *E. coli* (Fig. 16-19a). Changes in the anticodon of the *E. coli* methionine tRNA from CAU to CUA, for example, with the remainder of the tRNA left intact, are enough to switch its recognition from methionine to glutamine synthetase. In others, while the anticodon remains primary in importance, changes in one or more nucleotides in other regions reduce the accuracy of tRNA recognition to some extent (Fig. 16-19b).

In a number of tRNAs, nucleotides outside the anticodon are primarily or exclusively responsible for tRNA recognition. (Figure 16-19c and d show positions important in two of these tRNAs.) Recognition of alanine tRNA, for example, seems to depend

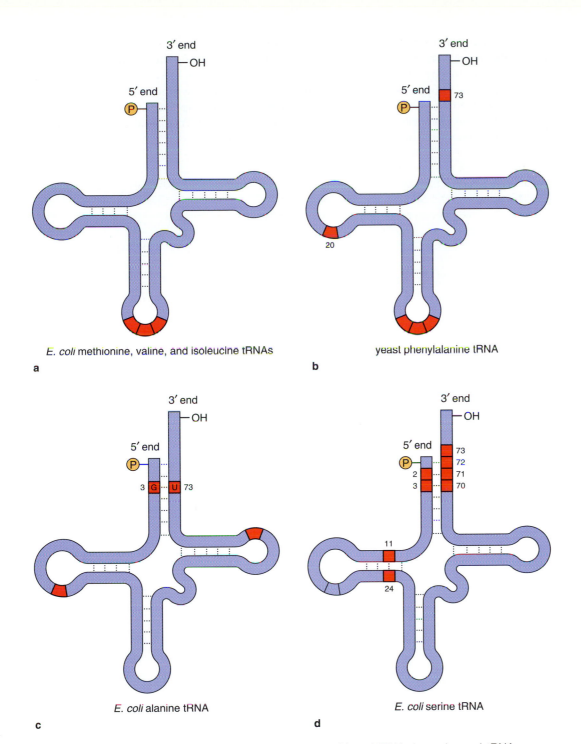

Figure 16-19 Sequence features important in the recognition of tRNAs by aminoacyl-tRNA synthetases. **(a)** tRNAs in which the anticodon is the sole recognition feature. *E. coli* methionine, valine, and isoleucine are examples of this type. **(b)** tRNAs in which the anticodon is primary, but nucleotides at other positions contribute to recognition. Yeast phenylalanine tRNA is an example of this type. **(c)** and **(d)** tRNAs in which sequence features other than the anticodon determine recognition. *E. coli* alanine and serine tRNAs are examples of this type. Positions important in recognition are shown in red.

almost exclusively on a single G-U base pair in the acceptor stem (see Fig. 16-19c). Substitutions in this base pair change the tRNA from recognition by the synthetase attaching alanine to the one attaching cysteine. The alanine tRNA is even recognized successfully by the synthetase if all elements other than the acceptor stem are eliminated. The differences in the sites recognized on tRNAs probably reflect the major structural differences among the synthetase enzymes. Few consistent differences are noted in tRNA recognition by class I and II synthetases except that the anticodon end generally appears more important for recognition by class I enzymes, and the acceptor end for class II enzymes.

For Further Information

Addition of sugar groups to proteins (glycosylation), *Ch. 20*
Amino acids, *Ch. 2*
Endoplasmic reticulum in protein synthesis, *Ch. 20*
Evolution of the genetic code, *Ch. 26*
Golgi complex in protein modification and secretion, *Ch. 20*
mRNA structure and transcription, *Ch. 14*
Phosphorylation and regulation of protein activity, *Ch. 6*
Posttranscriptional regulation, *Ch. 17*
Protein structure, *Ch. 2*
Protein synthesis in mitochondria and chloroplasts, *Ch. 21*
rRNA transcription and structure, *Ch. 15*
Secretion of proteins, *Ch. 20*
Transcriptional regulation, *Ch. 17*
tRNA transcription and structure, *Ch. 15*

Suggestions for Further Reading

Carter, C. W. 1993. Cognition, mechanism, and evolutionary relationships in aminoacyl-tRNA synthetases. *Ann. Rev. Biochem.* 62:715–748.

Cavarelli, J., and Moras, D. 1993. Recognition of tRNAs by aminoacyl-tRNA synthetases. *FASEB J.* 7:79–86.

Cooper, A. A., and Stevens, T. H. 1993. Protein splicing: Excision of intervening sequences at the protein level. *Bioess.* 15:667–674.

Fox, T. D. 1987. Natural variation in the genetic code. *Ann. Rev. Genet.* 21:67–91.

Georgopoulos, C., and Welch, W. J. 1993. Role of the major heat shock proteins as molecular chaperones. *Ann. Rev. Cell Biol.* 9:601–634.

Gethway, H.-J., and Sambrook, J. 1992. Protein folding in the cell. *Nature* 355:33–45.

Hendrick, J. P., and Harte, F.-U. 1993. Molecular chaperone functions of heat shock proteins. *Ann. Rev. Biochem.* 62:349–384.

Hershey, J. W. B. 1991. Translational control in mammalian cells. *Ann. Rev. Biochem.* 60:717–755.

Jakubowski, H., and Goldman, E. 1992. Editing of errors in selection of amino acids for protein synthesis. *Microbiol. Rev.* 56:412–429.

Kozak, M. 1991. Structural features in eukaryotic mRNAs that modulate the initiation of translation. *J. Biolog. Chem.* 266:19867–19870.

Lender, P., and Prat, A. 1990. Baker's yeast, the new workhorse in protein synthesis studies: Analyzing eukaryotic translation initiation. *Bioess.* 12:519–526.

Liljas, A. 1991. Comparative biochemistry and biophysics of ribosomal proteins. *Internat. Rev. Cytol.* 124:103–136.

Matthews, C. R. 1993. Pathways of protein folding. *Ann. Rev. Biochem.* 62:653–683.

McClain, W. H. 1993. tRNA identity. *FASEB J.* 7:72–78.

Merrick, W. C. 1992. Mechanism and regulation of eukaryotic protein synthesis. *Microbiol. Rev.* 56:291–315.

Moras, D. 1992. Structure and function relationships between aminoacyl-tRNA synthetases. *Trends Biochem. Sci.* 17:159–164.

Noller, H. F. 1991. Ribosomal RNA and translation. *Ann. Rev. Biochem.* 60:191–227.

Nomura, M. 1990. Early days of ribosome research. *Trends Biochem. Sci.* 15:244–247.

Parker, J. 1989. Errors and alternatives in reading the universal genetic code. *Microbiol. Rev.* 53:273–298.

Richards, F. M. 1991. The protein folding problem. *Sci. Amer.* 264:54–63 (January).

Riis, B., Rattan, S. I. S., Clark, B. F. C., and Merrick, W. C. 1990. Eukaryotic elongation factors. *Trends Biochem. Sci.* 15:420–424.

Rowling, P., and Freedman, R. B. 1993. Folding, assembly, and posttranslational modification of proteins within the lumen of the endoplasmic reticulum. *Subcellular Biochem.* 21:41–80.

Schulman, L. H. 1991. Recognition of tRNAs by aminoacyl-tRNA synthetases. *Prog. Nucleic Acids Res. Molec. Biol.* 41:23–87.

Seckler, R., and Jaenicke, R. 1992. Protein folding and protein refolding. *FASEB J.* 6:2545–2552.

Review Questions

1. Define the initiation, elongation, and termination stages of polypeptide assembly.

2. Outline the major steps in initiation of protein synthesis. What are initiation factors? How do initiator tRNAs differ from tRNAs active in elongation?

3. What are the A and P sites of ribosomes? How are these sites believed to operate in polypeptide assembly? Diagram the reaction taking place during formation of a peptide linkage.

4. Outline the steps taking place in termination of polypeptide assembly.

5. List the reactions in which phosphate bond energy is utilized in protein synthesis. How many phosphate groups would be hydrolyzed in synthesis of a protein 100 amino acids long?

6. What major modifications take place during processing of newly synthesized proteins? What kinds of organic and inorganic groups are added to proteins during processing?

7. Outline the structure of ribosomes. What structural features are common to prokaryotic and eukaryotic ribosomes? What features are unique to eukaryotic ribosomes? What rRNAs occur in each ribosomal subunit?

8. What information established that the codons of the genetic code are nucleotide triplets? What findings indicated that the genetic code is read by establishing a reading frame?

9. What experiments identified the coding assignments of the codons?

10. What does degeneracy mean with respect to the genetic code? What patterns are noted in degeneracy of the code? What is the wobble hypothesis? How does the mechanism proposed in the wobble hypothesis prevent mistakes during polypeptide assembly?

11. What does universality mean with respect to the genetic code? What is the possible relationship between tRNA function and the exceptions noted to the universality of the code?

12. Outline the steps in amino acid activation. What two major outcomes result from the reactions?

13. What features of tRNA structure are recognized by aminoacyl-tRNA synthetases? How does the accuracy of protein synthesis depend on the functions of tRNAs? Of the synthetases?

14. On what steps in amino acid activation and polypeptide assembly does the accuracy of protein synthesis depend? Which are more important to the accuracy by which amino acids are entered into proteins, nucleic acids or enzymatic proteins?

Supplement 16-1
Protein Synthesis in Prokaryotes

Polypeptide assembly proceeds in bacteria through the same phases of initiation, elongation, and termination as in eukaryotes. Although there are differences in detail, particularly in factors taking part in the reactions, the three stages proceed by similar overall mechanisms. Amino acid activation also proceeds by essentially the same reaction sequence in prokaryotes and eukaryotes.

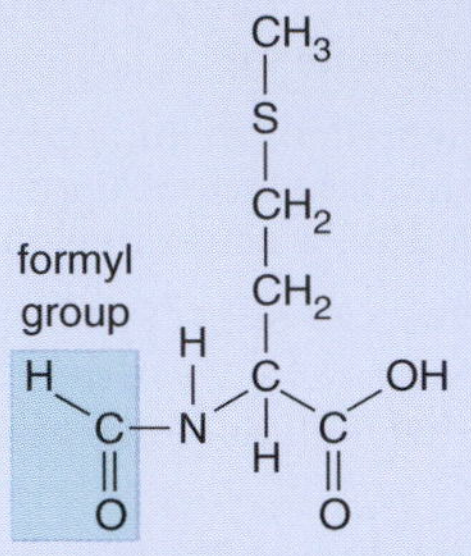

Polypeptide Assembly in Bacteria

Initiation of protein synthesis in bacteria takes place in a series of steps involving interaction of mRNA, large and small ribosomal subunits, an initiator tRNA, GTP, and several initiation factors. The first amino acid placed in a prokaryotic polypeptide chain is methionine, as it is in eukaryotes. In prokaryotes, however, the methionine carried by the initiator tRNA is modified into *formylmethionine* by the addition of a formyl (—C—H)

$\underset{\displaystyle O}{\overset{\displaystyle \|}{}}$

group that blocks the amino end of the amino acid (Fig. 16-20). The initiator tRNA of bacteria, with its attached formylmethionine, is identified as *fMet-tRNA$_i$*. Like its eukaryotic counterpart, the bacterial initiator tRNA differs in structure from the tRNAs participating in elongation (see Fig. 16-3b). These structural differences contribute to the ability of the bacterial initiator tRNA to combine with the P site of ribosomes during initiation.

In bacteria, binding of mRNA to the small ribosomal subunit and positioning of the AUG codon at the P site evidently depend on complementary base pairing between the mRNA and the 16S rRNA molecule of the small ribosomal subunit. The mRNA sequences that associate with the small subunit were determined from footprinting studies by J. A. Steitz (see Information Box 14-1). Steitz found that the portion of the mRNA protected by the association, some 30 to 40 nucleotides long, was approximately centered on the initiator codon. Subsequently, J. Shine and L. Dalgarno noted that a sequence near the 3' end of 16S rRNA is complementary to a segment of the mRNA footprint. Shine and Dalgarno proposed accordingly that binding and positioning of mRNA on the small ribosomal subunit depend on complementary base pairing between the mRNA and the 3' end of 16S rRNA. Their hypothesis later received direct support from experiments by Steitz and K. Jakes, who were able to isolate fragments of the 3' end of 16S rRNA paired in a complex with mRNA, held together by

Figure 16-20 The formylmethionine (fMet) placed as the first amino acid in prokaryotic protein synthesis. The formyl group is shaded in blue.

hydrogen bonds as predicted by the Shine–Dalgarno hypothesis. The mRNA sequence pairing with the 16S rRNA during bacterial initiation, the consensus sequence *AGGAGGU*, is now known as the *Shine–Dalgarno sequence*. The mechanism pairing mRNA with 16S rRNA in the small subunit seems to be peculiar to prokaryotes. In eukaryotes the ribosome binds to the 5' end of the mRNA and scans along the 5' untranslated segment until it encounters the AUG start codon.

The reactions of elongation and termination follow the same overall pathway as in eukaryotes. Three elongation factors, *EF-Tu*, *EF-G*, and *EF-Ts*, have the same functions in elongation as their respective eukaryotic counterparts EF1α, EF2, and EF1βγ (see Fig. 16-6). The complete cycle of elongation turns more rapidly in bacteria—about 15 to 18 times per second, as compared to the maximum of about 6 times per second in eukaryotes.

Processing of newly synthesized polypeptides in bacteria primarily involves removal of one or more amino acids from either end of the newly synthesized polypeptide chain and modification of individual amino acid residues to other forms. The entire formylmethionine is removed from the N-terminal end of some bacterial proteins during processing; in others, only the formyl group is removed.

The Proteins of Bacterial Ribosomes

The small ribosomal subunits of bacteria, after dissociation and disassembly, prove to contain a single rRNA type, 16S rRNA, in combination with 21 proteins. The prokaryotic large ribosomal subunit contains two rRNA types, 5S and 23S, in combination with 31 different proteins.

Using the numbering system based on the rate at which the isolated proteins migrate on electrophoretic gels (see p. 796), the small-subunit proteins are identified as S1 to S21, roughly in order of molecular weight, with the largest protein first. The 31 large-subunit proteins are numbered from L1 to L34, also with the largest protein first. Two numbers in the large-subunit sequence, L8 and L26, are unused because of misassignments early in the research identifying the proteins; another large-subunit protein, L7/12, originally thought to be two distinct proteins, later proved to be the same polypeptide except for the presence of a modifying acetyl group in L7 that is absent in L12. All the large- and small-subunit proteins occur in single copies in prokaryotic ribosomes except for L7/L12, which is present in four copies in the large subunit. The entire protein complement of *E. coli* ribosomes, as well as the *E. coli* rRNAs, has been completely sequenced.

Both bacterial transcription and translation can be seen in progress in Figure 16-21, a remarkable electron micrograph made by O. L. Miller and his colleagues. The thin strand running from upper left to lower right in the figure is a part of the DNA of a bacterial cell. Attached to the DNA at intervals are side branches; the branches are individual mRNA molecules, already attached to ribosomes and engaged in translation before their transcription is complete. The small, dense particles where the side branches join the DNA are RNA polymerase molecules transcribing the gene. The mRNA-ribosome complexes become longer toward the right in the micrograph, meaning that the direction of transcription runs from left to right in the picture. Simultaneous transcription and translation, in which translation of the 5' end of the mRNA mole-

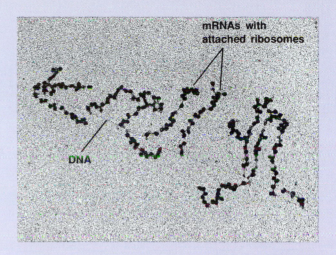

Figure 16-21 Simultaneous transcription and translation in *E. coli* (see text). × 57,000. (Courtesy of O. L. Miller, Jr., Barbara A. Hamkalo, and C. A. Thomas, Jr.)

cules is initiated before transcription is complete, occurs regularly in bacteria.

Actively growing *E. coli* cells are estimated to contain as many as 20,000 ribosomes, which densely pack the cytoplasm. Most of these ribosomes are engaged in protein synthesis in assemblies resembling Figure 16-21. Because the cell interior is packed with these assemblies, which include DNA and mRNA as well as ribosomes, much of the region recognized as cytoplasm probably contains extensions of the nucleoid. The nucleoid therefore probably extends throughout most of an active bacterial cell; the region recognized as the nucleoid likely contains only the tightly packed, inactive protein of the bacterial DNA.

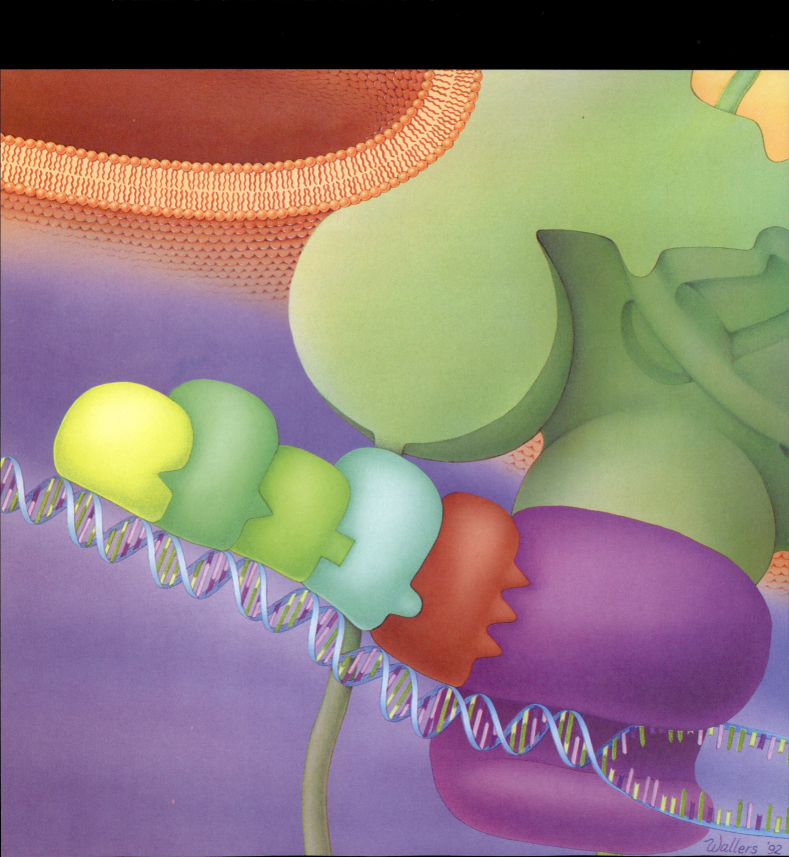

- *Transcriptional regulation* ▪ *Regulatory proteins*
▪ *Steroid hormone receptors* ▪ *Homeotic genes* ▪
Heat shock genes ▪ *DNA methylation* ▪ *Histone
modification* ▪ *Posttranscriptional regulation* ▪
Translational regulation ▪ *Posttranslational
regulation* ▪ *Transcriptional and translational
regulation in prokaryotes*

The genes encoding mRNAs, rRNAs, and tRNAs in eukaryotic cells may number as many as 30,000 or more. However, at any given time only a fraction of these genes, estimated to be about 5000 in most cells, is active. This is true even though most eukaryotic cells retain all the genes typical of the species and have the capacity to synthesize any of the RNAs and proteins encoded in the genes.

The difference between the number of potentially active genes and the RNAs and proteins actually assembled reflects controls at two primary levels. *Transcriptional regulation* determines which of the many genes in the nucleus are copied into RNAs and which genes are maintained in an inactive state. *Translational regulation* controls the rate at which mRNAs copied from active genes are used by ribosomes in protein synthesis.

These primary mechanisms are supplemented by controls at two additional levels. *Posttranscriptional regulation* controls events between transcription and translation, primarily the rates at which mRNAs are processed into finished form. *Posttranslational controls* adjust the levels of cellular proteins by regulating the rate at which newly assembled polypeptides are processed into finished form and the rate at which finished proteins are degraded.

Enzymes are among the proteins regulated in kinds and quantity at all these levels. By regulating enzyme synthesis, the mechanisms controlling transcription and translation extend indirectly to the reactions catalyzed by the enzymes, including those assembling all the remaining cellular molecules—lipids, carbohydrates, and nucleic acids.

The entire spectrum of controls provides cells with an exquisitely sensitive mechanism for regulating the kinds and numbers of all cellular molecules produced and the times and locations of their synthesis. The total mechanism underlies the development and maintenance of complex organisms, in which individual cells differentiate into mature forms and carry out specific, coordinated tasks. The mechanism also allows cells to respond and adjust to environmental changes and stimuli.

Although the molecular basis of transcriptional regulation is incompletely understood, it has become clear that cells are controlled at the most fundamental level by *regulatory proteins* with the capacity to recognize individual genes and adjust their activity. The transcriptional controls imposed by these proteins are the primary mechanisms underlying cell differentiation and development. Because regulatory proteins and their activities are so important to cell growth, division, and development, the research identifying and tracing the functions of these proteins is among the most exciting and fundamentally significant areas of cell and molecular biology.

This chapter describes the mechanisms controlling transcription and translation in eukaryotes and their coordination into integrated pathways of regulation. Comparisons with the regulatory mechanisms of prokaryotes are made in Supplement 17-1.

TRANSCRIPTIONAL REGULATION IN EUKARYOTES

The regulatory proteins controlling transcription in eukaryotes are part of a larger group of chromosomal proteins known as the *nonhistones* (see p. 376). The first nonhistone proteins to be definitely identified as gene regulators were those responding to stimulation of cells by steroid hormones, discovered in the 1970s. Because these regulatory proteins directly bind steroid hormones, labeling the hormones with radioactivity provided a convenient and effective way to tag the proteins for isolation and purification.

During the 1980s, technical breakthroughs enabled genes to be used as "probes" to detect and trap their regulatory proteins. In these techniques the DNA of a gene or its 5'-flanking control region is increased in quantity by cloning or the polymerase chain reaction (see Supplement 14-1). The DNA is immobilized by binding to nitrocellulose filter paper or to gel beads in a separatory column (Fig. 17-1a). A bulk preparation of cellular proteins is then poured over the paper or through the column (Fig. 17-1b); those able to recognize and bind the control sequences of the gene are trapped in the paper or column as they bind the DNA (Fig. 17-1c). After nonbinding proteins are washed away, the regulatory proteins recognizing the gene can be released in purified form by adding a salt solution (Fig. 17-1d).

More than a hundred types of eukaryotic regulatory proteins have been identified by these approaches, and the list grows constantly. Among them are examples from fungi and protozoa, insects, vertebrates, higher plants, and viruses infecting these forms. (Table 17-1 lists some of these proteins and the genes they regulate.)

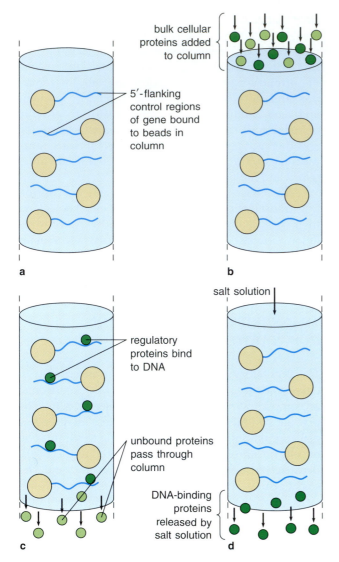

bulk cellular proteins added to column

5'-flanking control regions of gene bound to beads in column

a

b

salt solution

regulatory proteins bind to DNA

unbound proteins pass through column

DNA-binding proteins released by salt solution

c

d

Figure 17-1 A method using the 5'-flanking region of a gene to trap a regulatory protein recognizing sequences in the DNA. **(a)** The DNA of the gene is immobilized by binding to gel beads in a column. **(b)** A bulk preparation of cellular proteins is poured through the column. **(c)** Proteins with the ability to recognize and bind the control sequences of the gene are trapped in the column as they bind the DNA. **(d)** After nonbinding proteins are washed away, the regulatory proteins recognizing the gene are released in purified form when a dilute salt solution is poured through the column.

The isolated regulatory proteins are studied by a combination of biochemical, molecular, and genetic techniques, including genetic methods that permit individual amino acids to be changed or deleted, and grafting techniques that split off segments or domains of a regulatory protein and attach them to other proteins. *Drosophila* and the yeast *Saccharomyces cerevisiae* have been particularly useful in this research because many mutations previously identified through genetic crosses proved to alter regulatory proteins.

These studies have allowed regions within the regulatory proteins to be identified with such activities as recognition of DNA sequences and promotion of transcription by RNA polymerase. As it turns out, the regions with these activities have structures that are shared by many types of regulatory proteins (see below). The Experimental Process essay by G. C. Prendergast on p. 492 describes his elegant experiments applying these and other approaches to the structure and activity of the Myc regulatory protein.

One of the most striking features of the proteins isolated to date is that they control transcription by mechanisms that are universally shared among eukaryotic organisms. Many regulatory proteins can be interchanged among organisms as distantly related as yeast and humans and still operate successfully. This indicates that the regulatory mechanisms probably evolved with the earliest eukaryotes.

For genes encoding mRNAs, footprinting studies (see Information Box 14-1) have shown that regulatory proteins bind primarily to short control sequences in 5'-flanking regions of the genes. The control sequences, which vary between about 8 and 20 nucleotides in length, form parts of promoters and enhancers (see p. 409) and also occupy sites between these elements. Because they are parts of the same DNA molecule as the gene, the control sequences are called the *cis* elements of the regulatory system (*cis* = on the same side; see Fig. 17-2a). Some *cis* elements recognized by specific regulatory proteins are listed in Table 17-1.

The *cis* control elements have been studied by substitution/deletion experiments (see p. 404), in which individual bases are changed or eliminated and the effects on transcription noted. The control sequences have also been studied by recombinant DNA techniques in which a *cis* element, either derived from an organism or artificially synthesized, is placed in the 5'-flanking region of a gene encoding an enzyme that can be easily identified. The effects of the sequence on regulation of the gene are noted after the gene is introduced into a living cell. (Figure 17-3 outlines a typical technique used in this type of experiment.)

The regulatory proteins binding to *cis* elements, called the *trans* elements of the regulatory system (*trans* = on the opposite side; see Fig. 17-2a), interact with DNA and with other proteins to promote or inhibit binding of an RNA polymerase and initiation of transcription. Each protein binding to a *cis* element of a gene usually recognizes a distinct sequence in the 5'-flanking region.

Stimulation of transcription by proteins acting as *trans* elements is called *positive regulation*, and inhibition is termed *negative regulation*. Although most *trans* elements work as either positive or negative regulators, a few vary between positive and negative effects

Table 17-1 Some Proteins Regulating Specific Genes

Regulatory Protein (*trans* element)	Sequence Recognized (*cis* element)	Source	Motif*	Genes Regulated
ACF		Mammals		Serum albumin
Antennapedia		*Drosophila*	HTH	Homeotic gene
AP-2	CCCCAGGC	Mammals		MHC class I, proenkephalin, metallo-thionein, others
AP-3		Mammals		SV40 viral enhancer
AP-4		Mammals		Proenkephalin, others
APF		Mammals		Serum albumin
CDF		Sea urchin		Histone H2B
C/EBP	TGTGGAAAG	Mammals	LZ	Serum albumin, α-globin
CREB (ATF)	TGACGTCA	Mammals	LZ	Somatostatin, proenkephalin
Estrogen receptor	GGTCANNNTGACC	Birds, mammals	ZnF	Vitellogenin, ovalbumin, others
Fos/Jun (AP-1)	TGACTCA	Mammals	LZ	Growth regulation
GAL4		Yeast	ZnF	Enzymes of galactose metabolism
GCN4		Yeast	LZ	Genes encoding enzymes synthesizing amino acids
Glucocorticoid receptor	GAACANNNTCTTC	Mammals	ZnF	Prolactin, others
GT-1		Plants		RuBP carboxylase, chlorophyll-binding protein
HSF		Mammals, others		Heat shock proteins
ITF		Mammals		β-interferon
MAT proteins		Yeast	HTH	Mating type genes
NF-1 (CTF)	GCCAAT	Mammals		α-globin, β-globin, serum albumin, heat shock proteins
OCT-1 (OTF1)	ATTTGCAT	Mammals	HTH	snRNA, histone genes, others
OCT-2 (OTF2)	ATTTGCAT	Mammals	HTH	Antibody genes
Pit-1		Mammals	HTH	Growth hormone
SP1	GGGCGG	Vertebrates	ZnF	Wide variety
TUF		Yeast		Ribosomal proteins
Ultrabithorax		*Drosophila*	HTH	Homeotic gene

Adapted from data presented in P. F. Johnson and S. L. McKnight, *Ann. Rev. Biochem.* 58:799 (1989); C. Murre et al., *Cell* 56:777 (1989); and P. J. Mitchell and R. Tjian, *Science* 245:371 (1989).
* HTH = helix-turn-helix; LZ = leucine zipper; ZnF = zinc finger.

depending on the positions and functions of *cis* elements in the DNA and the effects of other regulatory proteins binding the control regions of the same gene.

Almost all regulatory proteins are modified in activity by addition of phosphate groups. The phosphate groups, added by protein kinases forming parts of regulatory cascades in the cell (some of them triggered by surface receptors; see. p. 154), may adjust activity of the regulatory proteins upward or downward. In some cases the same protein may be adjusted

upward or downward in activity by phosphate groups added in different segments. For example, the ability of the Jun regulatory protein to stimulate specific gene transcription is increased by addition of phosphate groups to serine residues at two positions in the N-terminal end of the protein and is decreased by addition of phosphate groups to a cluster of sites in the C-terminal end.

Regulatory proteins typically control genes cooperatively. This cooperative activity has several highly

Sequential Clues to the DNA-Binding Specificity of Myc

George C. Prendergast

GEORGE C. PRENDERGAST is senior research biochemist in Cancer Research at the Merck Sharp and Dohme Research Laboratories in West Point, Pennsylvania. He received his Ph.D. in molecular biology from Princeton University in 1989, providing the first evidence for gene regulation by the Myc protein. As an American Cancer Society Postdoctoral Fellow at the Howard Hughes Medical Institute of the New York University Medical Center, Prendergast identified the DNA binding specificity of Myc and isolated the murine form of Max, a DNA-binding partner protein for Myc. He is currently studying the biochemistry of the Myc and Ras proteins.

The chief characteristic of a cancer cell is its ability to grow uncontrollably. One of the most important reasons that cancer cells escape normal growth controls is because they have sustained extensive damage to the signaling machinery that regulates cell division. A number of the proteins that comprise this machinery are encoded by the cellular oncogenes, which are mutant in tumor cells. One of the first oncogenes to be identified, called *ras*, was cloned from tumor cell DNA that when introduced into normal cells could transfer the abnormal growth characteristics of the tumor cells. Since this first experiment in the late 1970s, many other cellular oncogenes have been identified and cloned. But the rapid progress in discovery of new oncogenes has outpaced work to determine how they normally work and why their function is altered in tumor cells.

The *myc* oncogene was identified a short time after *ras* and has since been found to be involved in many human cancers. Initial studies performed in Robert Weinberg's laboratory demonstrated that when introduced with *ras* into normal rat fibroblasts by DNA transfection, plasmids that continuously expressed *myc* from a strong promoter could force uncontrolled cell growth.[1] Later, it was shown that the Myc protein was localized in the cell nucleus and could nonspecifically adhere to DNA. These observations suggested the hypothesis that Myc may recognize specific DNA sequences and perhaps regulate transcription, DNA replication, or other nuclear processes important for cell growth.

Recent tests of this hypothesis greatly benefited from several key clues, all of which came from the comparison of amino acid sequences of Myc with other DNA-binding proteins. The first clue was the identification by Steven McKnight's group of an amino acid sequence pattern, or motif, called the "leucine zipper" (LZ) in the carboxy-terminus of Myc.[2] The LZ motif, named for the characteristic heptad repeat of leucines it contained, was originally discovered in the transcription factor C/EBP where it

was shown to be necessary for dimerization and DNA binding. Because C/EBP could bind DNA specifically, this suggested that Myc might too, since it also contained an LZ motif.

A second clue to Myc DNA binding specificity came from a comparison I made while in Edward Ziff's group between the amino acid sequences of Myc and another LZ oncogene protein called Fos. Many LZ proteins contain a section rich in basic amino acids located immediately upstream of the LZ motif. When dimerized by the LZ, this so-called basic motif could be demonstrated to form a sequence-specific DNA-binding domain in Fos, C/EBP, and other proteins. I noticed that Myc had a basic motif similar in sequence to Fos's but located in a different position approximately 40 amino acids upstream of the Myc LZ.[3]

At this time I resolved to test the notion that the Myc basic motif could mediate sequence-specific DNA binding, even though it appeared to be in a region of the protein different from other LZ proteins. But other clues clarifying the problem continued to come in quick succession. A third clue to Myc's DNA-binding capabilities came from David Baltimore's group, who discovered in Myc another amino acid sequence pattern, called the helix-loop-helix (HLH) motif.[4] Like the LZ, the HLH was originally identified in another protein, called E12, that was isolated through its ability to bind to a DNA element implicated in transcription of immunoglobulin genes. The E12 HLH motif was shown to be important for the ability of E12 to dimerize and specifically recognize DNA. The Myc HLH identified by Baltimore's group fit exactly into the 40-amino-acid space between the LZ and the Fos-like basic motif I had noticed in Myc. The proximity of this new dimerization motif to what I thought might be a DNA-binding region in Myc made me feel that I was on the right track.

First, I had to find a way to functionally dimerize the Myc basic motif. Studies of LZ proteins showed that a basic motif had to be dimerized to work, but analyses of the Myc caboxy-terminal region containing the HLH and LZ motifs suggested that it dimerized poorly. Therefore, to force dimerization of the Myc basic motif, I generated by recombinant DNA techniques a chimeric plasmid that would fuse the Myc basic motif to the E12 HLH (which had been shown in Baltimore's group to dimerize efficiently). The chimeric E12/Myc protein encoded on the plasmid could be produced by *in vitro* transcription and then by translation of the resultant RNA in a rabbit reticulocyte lysate. I could show that E12/Myc was capable of dimerizing through the E12 HLH by cotranslating *in vitro*-transcribed E12/Myc RNA along with that for normal E12, and then immunoprecipitating the protein mixture with anti-Myc antibodies. The presence of E12 protein in the immunoprecipitate indicated that it must have been bound to E12/Myc in the mixture, since E12 lacked the region reacting with the Myc antibody.

About this time, two more clues from the comparison of Myc amino acid sequences with newly cloned proteins suggested what DNA elements to screen for binding to the E12/Myc chimera. First, by comparing the DNA-binding sites of several HLH proteins whose specificity was known, a DNA sequence consensus pattern for recognition, CANNTG, was identified (called an E box, for enhancer box). Second, the basic regions of Myc and a novel HLH/LZ transcription factor called TFE-3, cloned by Holger Beckman in Tom Kadesch's group, were observed to be very similar.[5] TFE-3 could bind two DNA sequences, $E_{\mu E3}$ from the immunoglobulin heavy chain gene (containing the E box consensus CATGTG), and E_{USF} from the adenovirus major late promoter (containing the E box consensus CACGTG). Because of the similarity of the TFE-3 and Myc basic regions, I tested E12/Myc for specific binding to a number of DNA sequences that contained these E box consensus sequences.

To do this, I used a technique called the gel mobility shift assay. In different trials, the E12/Myc chimeric protein was incubated with various radiolabeled DNA oligonucleotides containing a particular E box sequence. Also included was a >1000-fold molar excess of nonspecific unlabeled DNA to prevent nonspecific binding by E12/Myc to the radiolabeled oligonucleotide probe. The mixtures were then separated on nondenaturing polyacrylamide gels and processed for autoradiography. A positive result would be indicated by the presence of a radioactive band toward the top of the gel, where the oligonucleotide would be bound to the slowly migrating protein, rather than at the bottom of the gel, where the unbound DNA would run. To verify the assay was working, positive controls were performed for specific DNA binding of TFE-3 to E_{USF} and $E_{\mu E3}$. Following tests with several oligonucleotides, I found that E12/Myc could specifically bind E_{USF}, suggesting that the Myc basic motif recognized the E box sequence CACGTG.[6]

To be certain of this result, two important control experiments had to be performed. First, it was important to show that the Myc basic motif was responsible for the activity, and not an E12-derived part of the chimera. Second, I had to verify that E12/Myc was recognizing the E box sequence and not another part of the oligonucleotide. For the first control, I constructed and expressed by recombinant DNA methods three mutant E12/Myc proteins with different missense mutations in the basic motif. When tested as before, all three mutants lost DNA-binding activity compared to normal E12/Myc. This argued that E12/Myc was recognizing DNA through the Myc basic motif. For the second control, I made mutant oligonucleotides containing mutations in and around the CACGTG E box sequence and tested all for binding to E12/Myc. Only those with mutations in or adjacent to the E box sequence affected DNA binding. This demonstrated that, as predicted, E12/Myc was recognizing the E_{USF} element in the oligonucleotide.

Did the DNA binding specificity of E12/Myc reflect that of Myc *in vivo*? If it did, one would predict that E12/Myc would interfere with the action of Myc when they were co-expressed in cells, through competition for specific DNA-binding sites. When it was introduced along with Myc and Ras into normal rat fibroblasts by DNA transfection, I found that E12/Myc could suppress the uncontrolled cell growth induced in its absence. In contrast, an E12/Myc mutant that lost the ability to specifically bind DNA because it had a mutation in the Myc basic motif also lost the ability to suppress Myc/Ras-induced cell growth. This control experiment demonstrated that suppression by E12/Myc required the DNA-binding activity of the Myc basic motif. Taken together, the data argued that the specificity of the E12/Myc chimera for DNA was the same as Myc itself.

What was the conclusion of my work? I gathered the first evidence for a specific Myc DNA-binding activity *in vitro* that could be argued to be relevant to Myc *in vivo*, thereby identifying the first known target for Myc activity in the cell. The set of experiments that led to this conclusion were prompted in part from numerous clues turned up by sequence comparisons. With further work, it should be possible to test Myc's potential function as a regulator of transcription, DNA replication, or other nuclear processes important for normal and tumorigenic cell growth.

References

[1]Land, H.; Parada, L. F.; and Weinberg, R. A. Tumorigenic conversion of primary embryo fibroblasts requires at least two cooperating oncogenes. *Nature* 304:596–602 (1983).

[2]Landschultz, W. H.; Johnson, P. F.; and McKnight, S. L. The leucine zipper: a hypothetical structure common to a new class of DNA binding proteins. *Science* 240:1759–64 (1987).

[3]Prendergast, G. C., and Ziff, E. B. DNA binding motif. *Nature* 341:392 (1989).

[4]Murre, C.; McCaw, P. S.; and Baltimore, D. A new DNA-binding and dimerization motif in immunoglobulin enhancer binding, daughterless, MyoD, and Myc proteins. *Cell* 56:777–83 (1989).

[5]Beckmann, H.; Su, L.-K.; and Kadesch, T. TFE3: A helix-loop-helix protein that activates transcription through the immunoglobulin enhancer μE3 motif. *Genes Dev.* 4:167–79 (1990).

[6]Prendergast, G. C., and Ziff, E. B. Methylation-sensitive sequence-specific DNA binding by the c-Myc basic region. *Science* 251:186–89 (1991).

Figure 17-2 Transcriptional control of genes by regulatory proteins. **(a)** The *cis* and *trans* elements of the control system. The *cis* elements are the sequences in DNA, usually upstream of a gene, recognized and bound by regulatory proteins, which are termed the *trans* elements of the system. **(b)** and **(c)** Control of genes in groups and networks by *cis* and *trans* elements. In **(a)**, genes A to D are controlled as a group by a common *cis* element recognized by a single control protein. In **(b)** a network in which one of the genes A to D controlled as a group by an initial regulatory protein encodes a second regulatory protein controlling another group of genes, E to H.

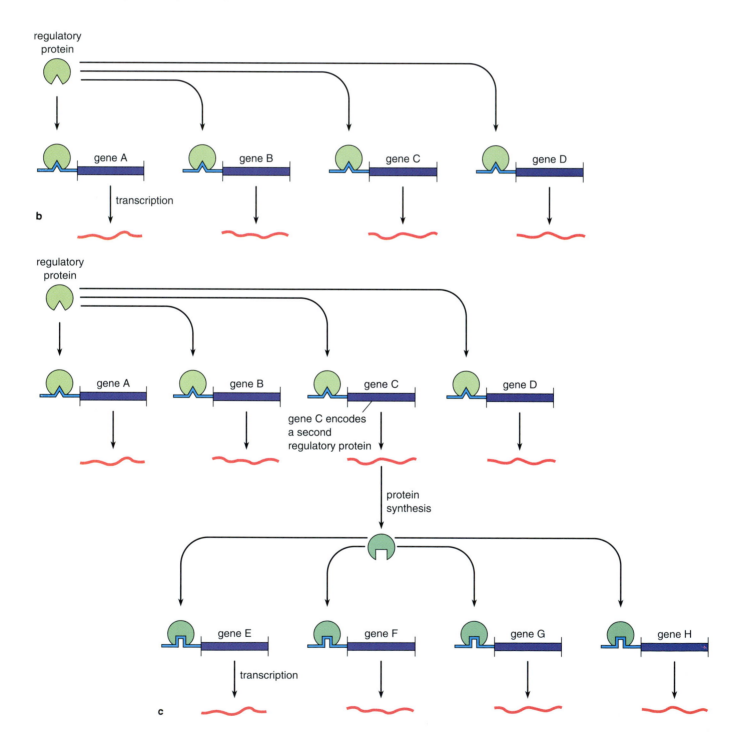

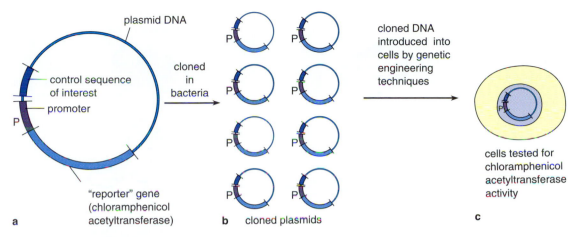

Figure 17-3 A technique for testing a control sequence. **(a)** The sequence is introduced into a bacterial plasmid by recombinant DNA techniques (see p. 420). The control sequence is placed into the 5'-flanking region of a "reporter" gene encoding an enzyme that is easily detected by its activity (bacterial chloramphenicol acetyltransferase is often used for this purpose). After cloning **(b)**, the copies are placed in recipient cells by genetic engineering techniques **(c)**. After an incubation period of about one hour, the cells are tested for relative amounts of enzyme activity. To evaluate sites within the control sequence, individual bases may be changed one at a time in consecutive experiments, and the amounts of enzyme activity compared.

significant consequences for gene control. By acting in different combinations, a relatively small number of proteins can specifically regulate a large number of genes. This circumvents a basic problem in gene regulation—if each gene were regulated by a single, distinct protein, the number of genes encoding regulatory proteins would have to be equal to the number of genes to be regulated. Regulating the regulators would require another set of genes of equal number, and so on until the coding capacity of any chromosome set, no matter how large, would be exhausted.

Cooperative interactions also allow gene activity to be adjusted upward or downward by degrees rather than simply being turned entirely on or off. In addition, regulation in combinations sets up control groups and *networks* (see Fig. 17-2a) in which multiple genes recognized by one or more regulatory proteins are linked in their regulatory responses. The networks become particularly extensive if genes encoding other regulatory proteins are among the group recognized by common factors (Fig. 17-2b).

Structures Common to Regulatory Proteins

Studies of the amino acid sequences of regulatory proteins, along with X-ray diffraction and other biophysical techniques, have revealed structural features or *motifs* common to several regulatory subgroups, including the *helix-turn-helix*, *zinc finger*, and *leucine zipper*. The helix-turn-helix motif (Fig. 17-4), discovered by T. A. Steitz and B. W. Matthews and their cowork-

ers, involves two alpha-helical segments in the regulatory protein, linked by a tight bend in the amino acid chain. One of the helices, called the *recognition helix*, is held in the protein in a position that fits exactly into the major or minor DNA groove. Amino acid side chains facing the groove recognize and bind a specific sequence in the DNA. The second helix backs up and stabilizes the recognition helix by interacting with portions of the DNA molecule. Generally regulatory proteins containing the helix-turn-helix motif combine by twos as *dimers*, in which two of the recognition helices are held in positions so that they fit precisely into two consecutive major or minor grooves in the DNA (as in Fig. 17-4b). Most dimers with the helix-turn-helix motif fit into the major groove.

The helix-turn-helix motif has been found in a number of regulatory proteins, including the *repressor* protein (see p. 514) controlling the tryptophan operon in bacteria, and many other bacterial repressors; a protein regulating genes for enzymes metabolizing galactose sugars in yeast; and more than 80 proteins able to recognize *homeodomain* sequences in animal DNA (homeodomain sequences are regulatory regions coordinating groups of genes that control the development of major embryonic structures; see below).

The zinc-finger motif (Fig. 17-5), worked out by R. S. Brown, D. Sander, and S. Argos, A. Klug, J. M. Berg, and their colleagues and others, consists of an approximate 30-residue sequence containing two cysteines and two histidines, or in a few cases, four cysteines. The cysteine and histidine residues are held

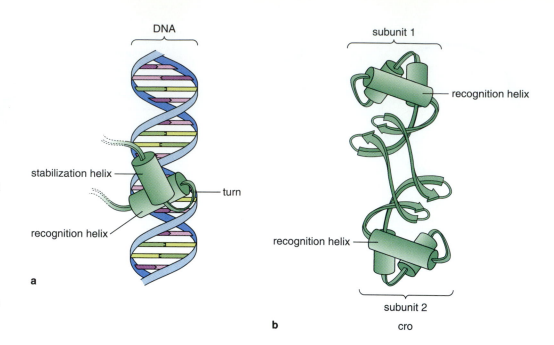

Figure 17-4 The helix-turn-helix motif, in which two alpha-helical segments are connected by a short turn. **(a)** Structure of the motif, showing how one of the two helices, the recognition helix, fits into the major groove of DNA. **(b)** Typical arrangement of two polypeptide subunits, each with a helix-turn-helix motif, as a dimer in which the recognition helices of the subunits fit into two consecutive turns of the DNA. ([b] redrawn from an original courtesy of T. A. Steitz, from *Proc. Nat. Acad. Sci.* 79:3097 [1982].)

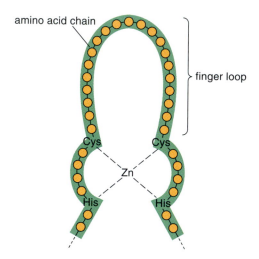

Figure 17-5 The zinc-finger motif of a loop of amino acids stabilized by interactions between a zinc atom and the side chains of cysteine and histidine residues. The tip of the finger recognizes and binds the edges of nucleic acid bases exposed in the DNA grooves.

in a four-part complex by a stabilizing zinc atom, forming the amino acid chain into a fingerlike loop that extends into the DNA and interacts with the edges of nucleic acid bases exposed in the major groove. Proteins with this motif usually have multiple zinc fingers, ranging from 2 to 10, extending into and interacting with nucleotide sequences in consecutive DNA regions. Several regulatory proteins in yeast, regulators controlling genes in coordination with steroid hormones (see p. 499), and *TFIIIA*, a protein that activates 5S rRNA genes (see p. 438), are among the regulatory proteins with zinc fingers.

The zinc-finger motif, which is the most widely distributed DNA-binding structure, has been found in more than 200 regulatory proteins. Zinc also occurs as a stabilizing atom in two additional DNA-binding structures. One of these is a *zinc-fold* containing about 70 amino acids, occurring in yeast regulatory proteins such as GAL4 (see below). The zinc-fold is stabilized by two zinc atoms interacting jointly with six cysteines. The second is a domain occurring in steroid hormone receptors (see below), consisting of about 70 amino acid residues stabilized by two zinc atoms, each interacting with four cysteines. The domain contains two amino acid alpha helices. One binds as a recognition helix in the DNA major groove; the second forms a stabilizing structure backing up the recognition helix.

The helix-turn-helix and zinc-finger motifs occur in several families of regulatory nonhistone proteins. Although the proteins within each family are not necessarily related in sequence, the DNA sequences recognized by the proteins of a family are similar. Within the families the particular DNA *cis* sequences recognized and bound depend on variations in the sequence of amino acids in the alpha-helical segments fitting into the wide or narrow groove of the DNA.

The leucine-zipper motif (Fig. 17-6), first proposed by S. L. McKnight and his colleagues, sets up a connecting region that holds two identical or different polypeptides into a two-part regulatory unit. The zipper consists of a coil of two alpha helices, one from each of the two proteins, held together by the "zipper" (Fig. 17-6a). The "teeth" of the zipper are a series of leucine residues, which occur at every seventh position in the alpha helices forming the coil. In this po-

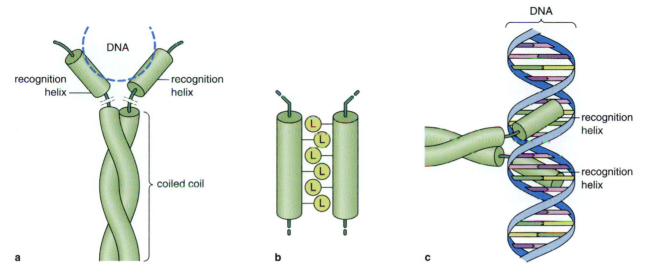

Figure 17-6 The leucine-zipper motif. **(a)** A coiled coil of alpha helices from two separate polypeptide subunits, stabilized by a leucine zipper. The arrangement holds two recognition helices in the DNA major groove. **(b)** The coiled coil unwound, showing the leucine residues that form the "teeth" of the zipper. **(c)** The fit of the recognition helices into the major DNA groove.

sition they form hydrophobic associations that hold the coil together (Fig. 17-6b). The zipper places two alpha helices, one from each polypeptide, in positions where they fit precisely into the major groove of the DNA (Fig. 17-6c).

The leucine-zipper motif occurs in regulatory proteins of yeast, higher plants, and humans and other vertebrates. Regulatory proteins containing a leucine zipper carry out vital cell functions in normal cells; in cancer cells the genes producing some of these proteins (*jun*, *myc*, and *fos*, for example; see p. 644) are faulty, so that the proteins promote transformation of cells to the cancerous state (for details, see Chapter 22). Leucine zippers also hold polypeptide subunits together in nonregulatory proteins.

Leucine zippers may hold together dimers of regulatory proteins with the helix-turn-helix or zinc-finger motifs. For example, the Myc regulatory protein, which contains a helix-turn-helix motif, is held in dimers with other regulatory proteins by a leucine zipper (see also Prendergast's essay on p. 492).

Additional DNA binding motifs have been discovered, including a *helix-loop-helix* motif, in which the turn linking the stabilizing and recognition helices is longer and more extended. Undoubtedly the list will grow as investigation continues in this area.

Regulatory Systems Controlling Specific Genes

Many control systems involving regulation of specific genes have been investigated. The following sections describe several of these systems that illustrate fundamental regulatory mechanisms. The first is a relatively simple system that illustrates basic principles of the interaction between regulatory proteins and DNA. A second example shows how multiple proteins join cooperatively in regulation. The final three examples show how genes are controlled in groups and networks by regulatory proteins that recognize *cis* elements in two or more genes.

A Relatively Simple System Involving Two Interacting Proteins: The GAL4/80 System of Saccharomyces

A pair of regulatory proteins of yeast cells, *GAL4* and *GAL80*, cooperate in a clear-cut system that serves as an excellent introduction to regulatory proteins and their activity. Although relatively simple, the *GAL4/80* system has provided some of the most significant and useful information about transcriptional regulation.

The genes regulated by the proteins GAL4 and GAL80 encode enzymes that carry out steps in galactose metabolism. Research by M. Ptashne and S. A. Johnston and their coworkers revealed that GAL4 recognizes and binds a DNA sequence upstream of several yeast genes; GAL4 binding increases transcription of these genes by about a thousand times and activates galactose metabolism.

The GAL4 protein, which binds to DNA as a dimer, has two major domains. One contains the region recognizing and binding *cis* elements in the DNA.

This domain contains a zinc-fold structure that interacts with the DNA. The other includes a region capable of directly or indirectly activating transcription by RNA polymerase II. The activating region may bind either RNA polymerase II or one of the transcription factors (see p. 412) required for transcriptional initiation by the polymerase enzyme. Division into two structural and functional regions—one recognizing and binding DNA and the other activating transcription—is typical of many eukaryotic regulatory proteins.

The way GAL4 activates transcription is still unknown. Ptashne proposed that GAL4, after binding to the upstream site in the DNA (Fig. 17-7a), interacts with a protein that binds to the promoter region. The protein interacting with GAL4 may be either TFIID, a transcription factor that binds to the TATA box (see p. 409), or RNA polymerase II. (TFIID is shown in Fig. 17-7 as the protein interacting with GAL4.) Interaction between GAL4 and the protein brings the promoter and enhancerlike regions together, forming a *loop* in the DNA (Fig. 17-7b). The combination promotes tight binding of RNA polymerase II to the promoter (as shown in Fig. 17-7c), and transcription begins.

GAL80 acts as a negative regulatory protein controlling the activity of GAL4. When no galactose is present in the cytoplasm of the yeast cell, GAL80 binds to GAL4 (Fig. 17-7d). Although GAL4 can still recognize and bind the enhancerlike region upstream of the galactose genes when GAL80 is attached, its ability to activate transcription is inhibited. In the presence of galactose or some of breakdown products, GAL80 undergoes a conformational change that releases it from GAL4 (Fig. 17-7e). With GAL4 no longer inhibited, transcription can proceed.

Ptashne's group found that GAL4 can activate transcription in organisms as diverse as insects, mammals, and higher plants if the *cis* element it recognizes is inserted into 5'-flanking regions of their genes. The yeast factor is also able to cooperate with mammalian regulatory proteins in gene activation. These observations indicate that the protein and its control mechanism have been highly conserved in evolution.

Cooperative Regulation of the Serum Albumin Gene

Cooperative regulation of genes by multiple regulatory proteins is the primary mechanism allowing cells to regulate large numbers of genes by a relatively few different proteins. Cooperative interactions also provide greater sensitivity to the regulatory mechanism and may contribute to linkage of genes into groups and networks.

The *serum albumin gene*, which is transcribed in liver and spleen cells in higher vertebrates, is a primary example of regulation by a combination of proteins. The product of the gene occurs in large

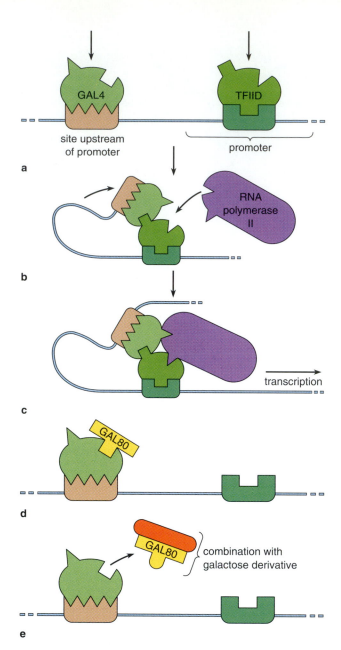

Figure 17-7 Regulation by GAL4/GAL80. After recognizing and binding a *cis* element upstream of the promoter **(a)**, GAL4 interacts with a protein that binds to the promoter **(b)**, here shown as TFIID, a transcription factor that binds to the TATA box. The interaction brings the promoter and enhancerlike regions together, forming a loop in the DNA. **(c)** This conformation induces tight binding of RNA polymerase II to the promoter, and transcription begins. **(d)** When no galactose is present in the cytoplasm, GAL80, a negative regulatory protein, binds to GAL4. Although GAL4 can still recognize and bind the enhancerlike region upstream of galactose genes when GAL80 is attached, its ability to activate transcription is eliminated. **(e)** In the presence of galactose or some of its metabolic derivatives, GAL80 undergoes a conformational change that releases it from GAL4. The release frees GAL4 from inhibition and allows initiation of transcription to proceed.

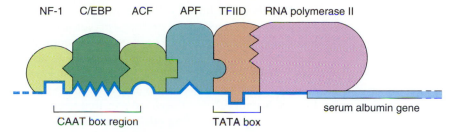

NF-1   C/EBP   ACF   APF   TFIID   RNA polymerase II

serum albumin gene

CAAT box region

TATA box

Figure 17-8 Control of the serum albumin gene by multiple regulatory proteins acting in coordination. Three regulatory proteins, NF-1, C/EBP, and ACF, bind in or near the CAAT box. C/EBP is a dimer held together by a leucine zipper. Substitutions or deletions that eliminate binding of any one of these factors significantly reduce but do not totally inhibit transcription of the serum albumin gene. The fourth regulatory protein, APF, binds to a sequence element between the CAAT and TATA boxes. APF appears to be highly specific for the serum albumin gene, and interference with its binding greatly reduces transcription of the gene. The combination of regulatory proteins increases the binding affinity of the RNA polymerase II enzyme to the promoter of the serum albumin gene, either by interacting directly with the enzyme or through an interaction with a transcription factor such as TFIID (an interaction with TFIID is shown).

quantities in the blood plasma of vertebrates. Three regulatory proteins, *C/EBP*, *NF-1*, and *ACF*, bind in or near the CAAT box (see p. 409) in the promoter of the serum albumin gene of liver cells (Fig. 17-8). C/EBP is a dimer whose two polypeptide subunits are held together by a leucine zipper. Substitutions or deletions in the sequences in or near the CAAT box that eliminate binding of any one of these factors significantly reduce but do not eliminate transcription of the serum albumin gene. A fourth regulatory protein, *APF*, binds to a *cis* element between the CAAT and TATA boxes. Sequence alterations that eliminate binding of this protein, which appears to be highly specific for the serum albumin gene, greatly reduce transcription of the gene.

Together these four regulatory proteins stimulate transcription of the serum albumin gene in liver cells to levels about 1000 times greater than other body cells. The primary protein giving specificity to the system appears to be APF; C/EBP, NF-1, and ACF are more general factors that regulate other genes as well. The molecular mechanism by which the combination of proteins regulating the serum albumin gene increases the rate of transcription is unknown. In some way the combination increases the binding affinity of RNA polymerase II to the promoter of the serum albumin gene, by interacting either directly with the enzyme or with one of the initiation factors such as TFIID (TFIID is shown in Fig. 17-8).

Genes Controlled in Groups and Networks by Regulatory Proteins

Many eukaryotic genes are controlled in groups or networks by common regulatory proteins. The essen-

tial feature linking regulation in groups or networks is the presence in two or more genes of a *cis* element that is recognized and bound by the same regulatory protein. The common sequence allows a single regulatory protein to increase or decrease transcription of all the genes containing the sequence as a unit. The GAL regulatory system of yeast provides a relatively simple example of this control pattern—several genes are activated by GAL4 through a common GAL4-binding *cis* element.

Several major networks controlled by common regulatory proteins have been discovered, including some that are of fundamental importance to embryonic development and the response of cells to environmental change. Among these are genes regulated by steroid hormone receptors in vertebrates, *homeotic genes*, which control major developmental pathways, and the so-called *heat shock genes*, activated under conditions of stress.

Regulation by Steroid Hormone Receptors Steroid hormones are small lipid molecules that coordinate cell function in higher animals. For example, the steroid hormones determining sexual development and activity in animals—progesterone, estrogen, and testosterone—control development of the reproductive system and secondary sex characteristics and regulate reproductive activities and behavior. These hormones exert their effects by entering cells and combining with *steroid hormone receptors*. The combination activates the receptors in their role as proteins regulating gene transcription. Cells may contain as many as 10,000 receptor proteins for a single steroid hormone.

Steroid hormones enter cells by penetrating directly through the plasma membrane (Fig. 17-9).

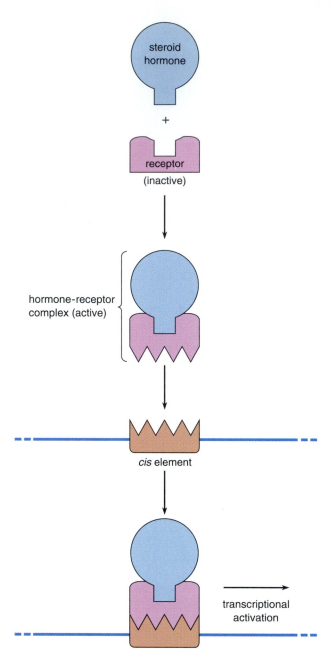

Figure 17-9 Regulation by steroid hormones. A steroid hormone penetrates into the cell, where it combines with its receptor. Combination with the hormone activates a binding site on the receptor for a *cis* element in the 5'-flanking control region of one or more genes. Binding of the activated receptor to the *cis* element activates transcription.

Inside they combine with a steroid receptor, usually in the nucleus. The combination induces a conformational change that activates the receptor, which then binds as a dimer to *cis* elements in 5'-flanking regions of specific genes, usually in segments upstream of the promoter. Linkage of the activated receptor to the *cis* elements activates mRNA transcription or inhibits transcription of previously active genes.

Whether and how a cell responds to a given steroid hormone depend on the presence of a receptor for the hormone, the number of genes containing *cis* elements recognized and bound by the activated receptor, and the effects of other proteins that modify the activity of the receptor. In chickens, for example, liver cells contain an *estrogen receptor* that specifically activates a gene encoding *vitellogenin*, a precursor of yolk proteins. Exposure to estrogens increases transcription of the vitellogenin gene from less than one transcript per day to about five transcripts per minute. Cells of the oviduct contain a distinct estrogen receptor that specifically activates a different gene, one encoding *ovalbumin*, a major protein of egg white. When stimulated by estrogens, transcription of the ovalbumin gene increases to levels so high that ovalbumin mRNA makes up 90% of the cellular total. The ovalbumin gene is not stimulated in liver cells, nor the vitellogenin gene in oviduct cells, in response to the hormone. Other body cells, even though exposed to estrogen in essentially the same concentrations, synthesize neither protein. Some of these cells contain no estrogen receptors of any kind, and some contain estrogen receptors that stimulate other genes.

As many as 50 to 100 genes within a cell may be controlled as a group by an activated steroid receptor. Each gene in the group shares one or more DNA sequences that are recognized and bound by an activated receptor. Some genes also contain sequences recognized by more than one type of activated receptor, placing these genes under the control of more than one steroid hormone. This arrangement, along with the fact that some genes controlled by steroid hormone receptors encode proteins regulating other genes, organizes the responding genes into networks that control major cell activities.

The identification of steroid hormone receptors and their functions, carried out in the laboratories of K. R. Yamamoto, R. M. Evans, B. W. O'Malley, P. Chambon, T. C. Spelsberg, G. S. McKnight, E. R. Barrock, and others, revealed that the regulatory proteins of this group share several structural features. All have at least three binding regions. One binds the hormone recognized by the receptor. Binding the hormone activates a second region containing two zinc ions holding together a structure containing an alpha-helical segment that binds the edges of bases exposed in the DNA major grooves. This region recognizes control sequences in the DNA of a gene regulated by the hormone. The third region directly or indirectly promotes initiation of transcription by RNA polymerase II or, in some cases, inhibits transcription.

The amino acid sequences of the DNA-binding sites and, to a lesser extent, the hormone-binding sites, are highly conserved among different steroid hormone receptors. This sequence conservation indicates that the regulatory proteins of this class belong to a gene family that probably descended from a single ancestral gene. The DNA sequences recognized and bound by different steroid hormone receptors, which are typically inverted repeats, are also similar. G. Klock and U. Strahle showed, for example, that changing only two nucleotides can alter a sequence recognized by the estrogen receptor to one binding an adrenal steroid receptor.

Experiments with isolated steroid receptors have shown that the regions binding DNA and the hormone occur in separate domains that are easily separated from each other by treatments such as mild proteinase digestion. In separated form the DNA-binding domain retains its ability to recognize and bind its DNA target sequence and to activate or inhibit transcription. The DNA-binding domain is continuously active when separated from the hormone-binding domain, indicating that the function of the hormone-binding domain is to inhibit the DNA-binding domain when no hormone is present.

Using recombinant DNA techniques (see p. 420), hybrid genes have been produced in which the code for a hormone-binding domain is attached to a sequence encoding a regulatory protein unrelated to steroid hormone receptors. In the protein encoded by the grafted gene the hormone-binding domain frequently carries out its inhibitory function when no hormone is present and activates the unrelated regulatory protein when its hormone is bound. This suggests that the hormone-binding domain folds into a conformation that covers and blocks active sites in other domains when no hormone is present and folds out of the way when bound to the hormone (Fig. 17-10).

More than thirty regulatory proteins have been discovered with structures placing them in the steroid hormone receptor family. Although most bind steroid hormones, some are activated by other substances, including thyroid hormones, retinoic acids, and vitamin D_3. A number of "orphan" receptors—as many as 25—have also been detected by searches for proteins related in structure to steroid hormone receptors. As yet the substances activating these receptors are unknown.

Homeotic Genes Another major control network is set up by regulatory proteins encoded in *homeotic genes*. These genes were first discovered in animals in the 1930s by E. B. Lewis, who noted that some single-gene mutations in *Drosophila* were able to transform one entire body part into a different one. One of these mutations, for example, causes a leg to form in a position normally occupied by an antenna. ("Homeotic" is derived from *homeosis*, a pattern in which one body part develops as a likeness of another.)

Although the functions of homeotic genes in *Drosophila* were unknown at the time of their discovery,

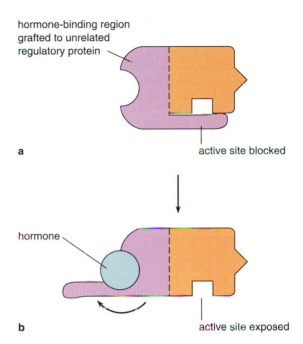

Figure 17-10 Possible activity of the hormone-binding region of a steroid hormone receptor when grafted to an unrelated regulatory protein. **(a)** When no hormone is bound, the hormone-binding region folds into a conformation that covers or blocks active sites on the grafted regulatory protein. **(b)** When the hormone is bound, the hormone-binding region undergoes a conformational change that exposes active sites on the grafted regulatory protein.

they were obviously "master" genes with the ability to determine the direction taken by major developmental pathways that involve activity of scores or even hundreds of individual genes. Eventually more than 20 homeotic genes were detected through mutations in *Drosophila*; they control such major embryonic processes as dorsal-ventral and anterior-posterior differentiation, segmentation, and eye, leg, and wing development.

During the early 1980s, several homeotic genes were isolated from *Drosophila* by D. S. Hogness and W. Bender, M. P. Scott and T. C. Kaufman, and W. J. Gehring and his coworkers. Gehring and W. J. McGinnis noted a sequence element common to the coding regions of all homeotic genes, the *homeobox sequence* (Fig. 17-11a). The sequence, about 180 nucleotides in length, encodes a protein segment with about 60 amino acid residues (Fig. 17-11b). This segment folds into a region of the protein known as the *homeodomain*. The homeodomain recognizes and binds strongly to a *cis* element in the 5'-flanking regions of all genes controlled as a unit by a homeotic gene. The homeodomain folds into four alpha-helical segments. The third helix, held with helix 2 in a helix-turn-helix motif, recognizes

Figure 17-11 Homeotic genes and homeodomain proteins. A sequence element common to all of the homeotic genes, the homeobox (a), encodes the homeodomain (b). The homeodomain, which contains four alpha-helical segments including a helix-turn-helix motif, binds to a recognition sequence occurring in the 5'-flanking regions of all genes controlled as a unit by a homeotic gene.

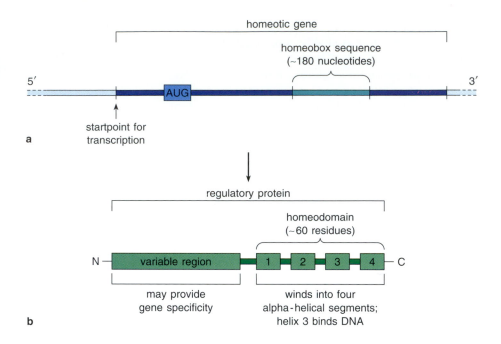

and binds an eight-base pair, or *octamer* sequence, in the 5'-flanking region of the genes controlled as a group by a homeotic gene. The third helix fits into the large groove of the DNA molecule and recognizes nucleotide sequences exposed in the groove.

The differences between the homeodomains of different regulatory proteins are usually small. For many the amino acid residue at position 9 in the third helix is critical to the ability of the homeodomain to recognize and bind specific DNA sequences. For example, J. Treisman and her coworkers found that the homeodomain of the *bicoid* regulatory protein of *Drosophila* has a lysine at this position; the homeodomain of another regulatory protein encoded by the *paired* gene has a serine at this position. If the serine of the paired protein is changed experimentally to lysine, this protein switches to binding the DNA sequence recognized by the bicoid protein.

Some of the regulatory proteins in this group, although controlling different genes, have identical homeodomains. In this case the ability of the proteins to recognize and bind different DNA sequences evidently depends on differences in amino acid sequences outside the homeodomains, as in the variable domain near the homeodomain (see Fig. 17-11b) or on interactions with other regulatory proteins.

Tests in other organisms using homeobox or homeodomain sequences as probes revealed that homeotic genes occur in all major animal phyla except sponges and coelenterates. The homeobox sequences in these genes are highly conserved and encode homeodomains that differ only in conservative amino acid substitutions; that is, the replacement of one amino acid by another of similar chemical properties. Only 5 con-

servatively substituted amino acids out of 60, for example, are different in the homeodomains of two homeotic regulatory proteins of *Drosophila* and *Xenopus*, an amphibian. More than 20 different homeodomain-containing proteins have been detected in humans.

Recently S. Hake and her colleagues discovered a gene controlling leaf development in maize that encodes a protein with a homeodomain. In maize plants with a mutant form of the gene called *knotted-1*, leaf veins are twisted into protrusions rather than lying in flat planes as in normal plants. Using the gene as a probe for DNA of similar sequence, these investigators found two other genes in maize encoding proteins with homeodomains. Others have found a gene in rice encoding a homeodomain protein with activity similar to maize *knotted-1*; homeodomain-encoding genes have also been detected in the plant *Arabidopsis*. Thus homeotic genes may be common in plants as well as animals and may have appeared in evolution before the split producing the major eukaryotic kingdoms.

Although the functions of most homeotic genes in vertebrates are unknown, at least some appear to control major developmental pathways as in *Drosophila*, particularly in the nervous and neuroendocrine systems. For example, E. M. DeRobertis and his coworkers tested the developmental effects of one homeodomain-containing protein by injecting antibodies against the protein into *Xenopus* embryos. In the injected embryos, regions of the central nervous system that would normally form parts of the spinal cord near the brain developed into more anterior structures of the hindbrain. The tested homeodomain protein therefore appears to direct development of major parts of the central nervous system along the anterior-posterior axis. Rather than controlling major developmental pathways, some homeotic genes of vertebrates appear to regulate more general mechanisms that affect both embryos and adults as a whole. For example, one regulatory protein encoded in a homeotic gene of mammals, *OCT-1*, activates a wide variety of genes that are expressed in almost all body cells, among them snRNA and histone genes.

The homeotic regulatory proteins of *Drosophila* are capable of activating genes when injected into mammalian cells. This, as well as the general conservation of homeotic genes among animals and higher plants, confirms that this system, like many mechanisms regulating genetic activity in eukaryotes, has ancient evolutionary origins and has been maintained largely intact for many millions of years. The genes controlled by the conserved systems, however, often carry out distinct and unrelated functions in different major taxonomic groups.

Heat Shock Genes   The stress-induced group collectively known as heat shock genes has been detected in highly conserved form in all organisms from bacteria to humans and higher plants. In addition to heat, the genes of this group are highly activated in animals by adverse conditions such as fever, inflammation, infection, wounding, cancer, and chemotherapy; in higher plants by conditions such as nutrient deprivation, water loss, and wounding. Bacterial heat shock genes are activated in response to adverse environmental conditions such as elevated temperatures and oxygen deprivation or by exposure to damaging substances such as heavy metals and ethyl alcohol. Many of the proteins encoded in heat shock genes are *chaperones* (see p. 472), molecules that direct protein folding along correct pathways. These proteins may aid refolding and reactivation of proteins damaged by stressful conditions. Other heat shock proteins inactivate proteins not required for responses to stress or promote degradation of damaged proteins. Most heat shock proteins are also produced in constant but reduced amounts in unstressed cells, where they function primarily in promoting folding of newly synthesized proteins and insertion of proteins in cellular membranes.

Heat shock proteins occur in several highly conserved families with molecular weights averaging 60,000 (the *hsp60* proteins), 70,000 (*hsp70* proteins), and 90,000 (*hsp90* proteins). Most of the hsp60 and hsp70 proteins act as chaperones. Some of the hsp70 proteins are more than 95% similar in amino acid sequence in organisms as distantly related as humans and yeast; hsp70 proteins are even 50% similar in sequence between humans and *E. coli*. The hsp90 proteins are chaperones that bind and stabilize many proteins important to cellular responses, including steroid receptors, protein kinases, calmodulin, actin, and tubulin. In some cases, association with hsp90 stabilizes cellular proteins in an inactive state. Other proteins are produced in response to stress in many species; many of these additional proteins are less highly conserved or unique to a species or larger taxonomic group. The combined effects of heat shock proteins alter cellular activities as diverse as DNA replication, protein synthesis and degradation, movement of proteins across ER and mitochondrial membranes, and endocytosis.

Genes encoding heat shock proteins are controlled as a group by one or more copies of a common control sequence in their 5'-flanking regions called the *heat shock element* (*HSE*). This sequence element, which is spaced from 80 to 150 nucleotides upstream of the startpoint for transcription, is recognized and bound by a family of regulatory proteins known as *heat shock factors* (*HSFs*). In unstressed cells, HSFs are present but inactive. A stress such as elevated temperature leads in some as yet undiscovered way to phosphorylation and activation of HSFs. The activated HSFs bind as trimers to HSE sequences, leading to activation of the heat shock genes as a group. Heat shock genes also contain sequence elements recognized and bound by other regulatory proteins.

The preceding examples clearly illustrate several principles that are common to many systems regulating specific genes in animals, plants, and lower eukaryotes:

1.   Regulation of specific genes depends on an interaction between control sequences in 5'-flanking regions of genes (*cis* elements), and regulatory proteins (*trans* elements). The control sequences may form parts of promoters or enhancers (see p. 409) or may lie outside these regions; the regulatory proteins, which may have positive or negative effects, may combine directly with the DNA, other regulatory proteins, transcription factors, or RNA polymerase II.

2.   Most regulatory proteins combining directly with DNA have at least two functional regions or domains—one recognizing and binding *cis* elements in the DNA and the other directly or indirectly activating transcription.

3.   The DNA-binding regions of many regulatory proteins contain common structural elements such as the helix-turn-helix or zinc-finger motif.

4.   Several to many regulatory proteins cooperate in the control of most genes.

5.   Regulatory proteins and the DNA sequences they recognize are generally highly conserved among eukaryotes. However, the genes controlled by the conserved systems may carry out distinct and unrelated functions in different taxonomic groups.

6. Genes are controlled in groups and networks by the presence of common *cis* elements.

7. Multiple regulatory factors acting in different combinations greatly expand the number of genes that can be specifically recognized and controlled by a relatively small number of different proteins.

The mechanisms by which regulatory proteins activate or deactivate transcription in their target genes remain uncertain. Positive regulatory proteins may directly promote binding of RNA polymerase II and initiation of transcription or may enhance binding of other proteins, such as transcription factors, that are necessary for initiation. Limited evidence points to TFIID (see p. 412) as a major target of regulatory proteins; this factor, which is central to initiation by RNA polymerase II, appears to have regions binding and interacting with regulatory proteins. Positive regulators may also unwind DNA in the promoter, remove blocking nonhistones, or open or unwind chromatin structures such as nucleosomes or solenoids (see p. 377 and below). Negative regulatory proteins may inhibit activity of positive regulators or block binding of positive regulators, transcription factors, or RNA polymerase.

Whatever the mechanisms employed, regulatory proteins allow genes to be specifically turned on or off or adjusted precisely in activity to match the needs of the cell for both long-term differentiation and development and short-term responses to changes in the environment.

MECHANISMS MODIFYING TRANSCRIPTIONAL CONTROL BY REGULATORY PROTEINS

Several control mechanisms have general effects on transcription rather than regulating particular genes. These mechanisms appear either to place individual genes or blocks of genes in a "ready" state, in which they can be activated by specific regulatory proteins, or to remove them from this state, so that they cannot be transcriptionally activated even if specific regulatory proteins are present.

Modifications in both DNA and the histones have been linked to the shift of genes between unavailable and ready states. The primary DNA modification correlated with genetic regulation is methylation; for the histones, chemical modification by acetylation and phosphorylation and substitution of histone types and variants have been implicated as general transcriptional controls.

DNA Methylation and Transcriptional Regulation

DNA methylation by addition of methyl ($-CH_3$) groups to cytosines is the best documented of the general controls adjusting gene activity. (The methylation produces the modified base 5-methyl cytosine; see Fig. 13-6.) Many inverse correlations between DNA methylation and transcriptional activity have been noted in vertebrates. Cytosines in many, but not all genes in vertebrates are more highly methylated when the genes are inactive in transcription. Conversion to active transcription is correlated in these genes with removal of methyl groups from some cytosines, particularly those in 5'-flanking regions of the gene.

In most of the methylations related to gene regulation the methylated cytosine is part of the short sequence CpG, called a *CpG doublet* (the *p* represents the phosphate group connecting the C and G nucleotides). Often, groups of CpG doublets are clustered into *CpG islands* in the 5'-flanking regions of genes. Although only as little as 3% to 4% of the total cytosine content of DNA is methylated in higher animals, as much as 80% to 100% of the CpG doublets may carry a methyl group, particularly in inactive genes.

A doublet, which has the form $\cdots \frac{GC \cdots}{\cdots CG}$ in double-stranded DNA, has two cytosines available for methylation. One or both of the cytosines may be methylated.

The enzyme responsible for DNA methylation in mammals, *(cytosine-5)-methyltransferase*, is a *hemimethylase*. The enzyme catalyzes addition of a second methyl group to hemimethylated CpG doublets—that is, to the remaining C of a doublet in which one of the two cytosines is already methylated. This activity may allow the pattern of methylation in cell lines to be preserved and passed on during DNA replication (Fig. 17-12).

The correlation between methylation and gene regulation has been followed in both bulk DNA and in individual genes. Bulk studies, for example, have followed methylation in such systems as the X chromosomes of mammalian females, including those of human females. In body cells, one of the two X chromosomes becomes almost completely inactive in transcription and folds into a block of heterochromatin. DNA methylation is significantly higher in the inactive X.

Methylation patterns in the globin genes of chickens and other vertebrates, which encode the polypeptide subunits of hemoglobin molecules, are among the best studied of individual genes. In mature sperm cells of the chicken, all the globin genes are highly methylated and inactive. Soon after fertilization, cytosines in the 5'-flanking region of the embryonic hemoglobin genes of red blood cells become demethylated. Shortly thereafter, transcription of embryonic globin genes begins in these cells. The genes encoding the globin proteins of adults remain highly methylated and inactive in the embryonic blood cells. Later in development, as hatching approaches, transcription switches from the embryonic to the adult globin genes in developing red blood cells. This change is correlated with addition of methyl groups to some, but not all cytosines in the 5'-flanking regions of the embryonic globin genes and removal of methyl groups from 5'-flanking cytosines in the adult globin genes. In cells such as fibroblasts, in which globin genes remain inactive throughout embryonic development and adult life, the 5'-flanking control regions remain fully methylated. Unmethylated globin genes injected into fibroblasts are rapidly transcribed, however.

Typically in such experiments, genes are observed that become demethylated and still remain inactive. Some globin genes in mammals, for example, become demethylated in placental tissues but remain inactive in globin mRNA synthesis. This observation shows that, although demethylation appears to be a necessary step toward transcriptional activation in genes controlled by this mechanism, demethylation by itself is not sufficient to initiate transcription. Other control steps, such as binding or release of specific regulatory proteins, are necessary for transcription to take place.

Many viruses infecting vertebrate cells also show a clear correlation between methylation and transcriptional activity. Although there are exceptions, the DNA of most of these viruses is unmethylated when the virus is in free form. On infection the viral DNA immediately becomes fully methylated. Subsequent activation of the viral genes is correlated with demethylation at specific sites within their 5'-flanking control regions.

Methylation as a general regulatory mechanism appears to be limited to vertebrates and to viruses infecting vertebrate cells. In many invertebrates, including *Drosophila*, and in fungi, methylation of cytosines cannot be detected in either active or inactive genes. In higher plants, cytosines in DNA sequences appear to be highly methylated at all times, whether genes are active or not. A mutation greatly reducing the amount of methylation also has no apparent effect on development in the higher plant *Arabidopsis*. The use of methylation as a modifying gene control only among the vertebrates suggests that this mechanism appeared relatively late in the evolution of the animal kingdom.

It is not clear how the added methyl groups control transcription. However, a methyl group added to the 5'-carbon of cytosine projects into the major groove of DNA, in the region where the recognition segments of regulatory proteins are believed to interact with the edges of the DNA bases. The projecting methyl group may block insertion of regulatory pro-

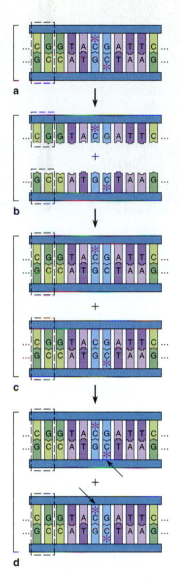

Figure 17-12 A mechanism by which addition of methyl groups to DNA by (cytosine-5)-methyltransferase might pass on methylation patterns during DNA replication. The dashed box encloses an unmethylated doublet; a fully methylated doublet is shown in blue (red asterisks indicate methylated C's). (a and b) The DNA unwinds for replication. (c) Replication produces two DNA molecules that are hemimethylated in the blue CpG doublet. (d) This doublet is then fully methylated (arrows) by the enzyme. The result duplicates the methylation pattern of the parent DNA molecule.

teins into the major groove of the DNA. There is some evidence supporting this idea; G. C. Prendergast and E. B. Ziff found that addition of a methyl group to a CpG doublet in the *cis* element . . . GGCCACGACC . . . blocked binding of the Myc regulatory protein to this sequence. It is also possible that general transcriptional regulation by methylation depends on regulatory proteins that recognize and bind methylated sequences and block their transcription. J. Boyes and A. Bird have identified several proteins able to bind methylated DNA, including one that binds methylated CpG doublets. Methylated genes were transcribed at unexpectedly high levels in mutants with low levels of this protein.

Histone Modifications and Genetic Regulation

The histone proteins organize DNA at several levels (see Chapter 13 for details). At the most fundamental level, histones wind DNA into nucleosomes, the primary structural units of chromatin. Interactions between H1 molecules of different nucleosomes are believed to wind nucleosome chains into a *solenoid*, the coil of nucleosomes considered to be equivalent to individual chromatin fibers (see Fig. 13-14). H1 interactions are also believed to fold chromatin into heterochromatin, in which chromatin fibers are tightly packed and inactive. Chemical modification of histones has been linked to changes in the organization of chromatin at each of these levels and to transcriptional regulation.

Chemical modification of histones by reversible addition of acetyl (CH_3-C-) groups has been most

$$\overset{\|}{O}$$

clearly correlated with gene regulation. Acetyl groups are added to the core histones (H2A, H2B, H3, and H4), which wind DNA into nucleosome core particles. Each core histone has from two to four lysine side chains available for addition of acetyl groups; in total, an individual nucleosome has 26 sites that can be modified by reversible addition of an acetyl group.

Evidence showing a positive correlation between addition of acetyl groups to core histones and transcriptional activity, developed through the investigations of V. G. Allfrey, E. M. Bradbury, M. A. Gorovsky, and others, has accumulated over many years. For example, in developing embryos, acetylation increases with transcriptional activity. In viruses in which the viral DNA is wound into nucleosomes by host-cell histones, such as the SV40 virus (see p. 378), the histones of active viral genes are about four times more highly acetylated than those of inactive genes.

The molecular effects of acetylation of core histones on chromatin structure are undetermined. The acetylations are concentrated in end segments of the core histones that extend as arms from the main mass of the proteins (see p. 374 and Fig. 13-9). These arms may project outward from the nucleosome core to interact with the DNA coiled around the nucleosome. Addition of acetyl groups may reduce the attraction of the arms for DNA, loosening the arms and allowing access by regulatory proteins and transcription factors. There is some evidence that this may actually be

the case—A. P. Wolffe and his coworkers found that acetylation of the arms, or removing them with a proteinase, increased binding of transcription factors to DNA containing nucleosomes. Reduction in the attraction of core histones for DNA could also alter chromatin structures at higher levels, including uncoiling solenoids or unfolding heterochromatin.

Addition of phosphate groups to H1, and to a lesser extent to one or more core histones, has been correlated with conversion of chromatin to an inactive state as heterochromatin. H1 phosphorylation has been proposed to induce packing of chromatin fibers into heterochromatin by increasing the attraction between H1 molecules in different chromatin regions.

There is considerable evidence indicating that, whatever the alterations to individual histone types, chromatin structure does change as genes shift between active and inactive states. The DNA of active chromatin is typically more susceptible to digestion by endonucleases (see p. 377). This greater susceptibility is considered to be caused by a more open chromatin structure that permits access to the DNA by the endonucleases. In the 5'-flanking regions of many active genes, short lengths of DNA become essentially as susceptible to attack by endonucleases as naked DNA; these *hypersensitive sites* are believed to reflect displacement of nucleosomes by regulatory proteins (see p. 506).

There is also evidence that nucleosome structure is altered by binding of regulatory proteins. For example, B. Pina and his coworkers found that in the promoter of a tumor virus associated with mouse mammary cancer a nucleosome is positioned at a precise location. The promoter is recognized and bound by a steroid hormone, which activates transcription of the viral DNA. Binding of the steroid receptor in this case does not displace the nucleosome but makes the DNA wound around the nucleosome more susceptible to endonuclease digestion, as if the DNA of the nucleosome is unwound to some extent by receptor binding.

Any unfolding of chromatin, by addition of acetyl groups or removal of phosphate groups, should make genes generally more available to RNA polymerases or regulatory proteins, thereby providing a nonspecific mechanism that enhances transcription. At the same time, removal of acetyl groups or addition of phosphate groups, expected to increase folding into nucleosomes, solenoids, or heterochromatin, should make genes in these structures less available.

Specific gene regulation depends on regulatory proteins (the *trans* elements), which can recognize and bind control sequences (the *cis* elements), usually located in the 5'-flanking regions of genes. Other, more generalized controls probably convert blocks of genes

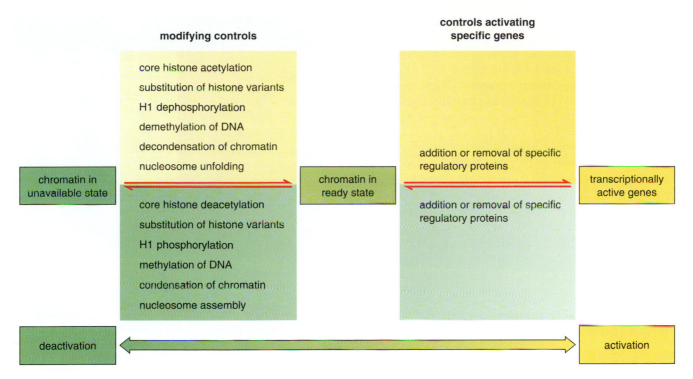

Figure 17-13 Summary of some of the possible levels of transcriptional regulation in eukaryotes. A potentially large number of modifying controls may convert individual genes or blocks of genes between an unavailable state in which they cannot be activated to a ready state in which they can be activated. Conversion between the ready and active states is moderated by specific regulatory proteins that can recognize the promoter, enhancer, and other control sequences of individual genes.

between an "unavailable" state, in which they cannot be activated for transcription, and a "ready" state, in which they may be selected for transcription by the specific regulatory proteins. Among these general controls are DNA methylation and modifications in chromatin structure resulting primarily from histone alterations, including acetylation and phosphorylation of core histones. These changes may convert chromatin structure over a range between an open state at one extreme, characteristic of genes that are available for activation of transcription by regulatory proteins, and a fully condensed state at the other extreme, in which genes cannot be activated even if regulatory proteins are present (Figure 17-13).

The entire combination of control mechanisms, which turn genes on and off at the proper times and places and in the correct sequences and combinations, sets up the regulatory sequences controlling embryonic development, adult growth, and maintenance of adult activities.

POSTTRANSCRIPTIONAL REGULATION

Before mRNAs copied from active genes can participate in protein synthesis, they must be processed,

transported to the cytoplasm, and made available to ribosomes. Regulatory mechanisms may be imposed during each of these steps to adjust the numbers and kinds of mature and available mRNAs that reach the cytoplasm. These controls, acting at steps between transcription and the initiation of translation, accomplish posttranscriptional regulation.

Posttranscriptional Regulation During mRNA Processing

Two processing steps have been established as posttranscriptional controls in eukaryotes—the reactions splicing introns from pre-mRNAs and the addition of poly(A) tails to 3' ends (see p. 413). In thyroid cells, for example, the pre-mRNA transcript of the calcitonin gene is processed by splicing together some exons but not others to produce *calcitonin*, a peptide hormone (see Fig. 14-15). In cells of the hypothalamus the pre-mRNA product of the same gene is processed differently, by splicing together another combination of exons to produce a distinct mRNA coding for a different polypeptide, *CGRP* (for *calcitonin gene-related polypeptide*). These observations indicate that the alternative splicing reactions are regulated in different patterns in thyroid and hypothalamus cells as a mechanism of posttranscriptional control.

Variations in the rate of poly(A) tail addition are employed as a regulatory mechanism by a virus infecting eukaryotic cells, the *adenovirus*. In this virus several genes, *L1*, *L2*, and *L3*, are transcribed at the same rates late in the infective process. (These genes code for coat proteins of the virus.) Initially poly(A) tails are added to the L1 pre-mRNA at two or three times the rate of the L2 and L3 pre-mRNAs. Later, when viral coat protein production is nearly complete, poly(A) addition to L1 pre-mRNAs decreases, and addition of poly(A) to L2 pre-mRNA increases, so that the rate of L2 mRNA processing exceeds L1 processing by several times. These differences in poly(A) tail addition regulate the rate at which the mRNA products of the L-genes complete processing and, ultimately, the rate at which the finished mRNAs reach the cytoplasm.

Posttranscriptional Regulation of mRNA Stability

Posttranscriptional control of mRNA stability occurs widely in eukaryotes and may in fact be a universal regulatory mechanism. Typically mRNAs have a finite life expectancy in the cytoplasm, and differences are frequently noted in the relative stability of individual mRNA types (Table 17-2). Most significantly, the life expectancy of individual mRNAs increases or decreases in coordination with cell activities or differentiation and development. When chicken liver cells are exposed to estrogens, for example, the vitellogenin mRNA has a half-life of nearly 500 hours. If estrogens are withdrawn, the half-life of vitellogenin mRNA drops to 16 hours—about 30 times shorter.

In many cases experimental transfer of the 3' untranslated region (UTR) from one mRNA to another transfers the stability characteristics of the donor mRNA, indicating that sequences or secondary structures in this region regulate mRNA breakdown. For example, the stability of most histone mRNAs is coordinated with DNA replication. During replication, histone mRNAs are relatively stable in the cytoplasm, with half-lives of about 60 minutes. At completion of DNA replication, or if replication is experimentally inhibited, histone mRNA half-lives drop to a few minutes.

Experiments by W. F. Marzluf showed that the last 20 nucleotides of histone mRNAs, including an inverted sequence that can form a hairpin (see p. 397), affect the stability of these mRNAs; histone mRNAs, with few exceptions, lack poly(A) tails. Removal of this 3'-end structure makes histone mRNAs unstable and leads to their rapid breakdown. Adding the last 20 to 30 nucleotides of histone mRNAs to other, nonhistone mRNAs gives the nonhistone types stability characteristics typical of histone mRNAs. These hybrid non-histone mRNAs are stable during DNA replication and are degraded rapidly when replication stops. Apparently the 3'-end hairpin of histone mRNAs protects them from degradation by the usual cytoplasmic enzymes hydrolyzing other mRNA types; after DNA synthesis is complete, a specialized ribonuclease recognizing the 3' hairpin appears and degrades the histone mRNAs.

The effects of transcriptional and posttranscriptional regulation combine to control both the kinds and numbers of mature and available mRNA molecules reaching the cytoplasm. Regulation of the kinds of mRNA reaching the cytoplasm primarily results from transcriptional controls. The distinction is not complete because some posttranscriptional regulatory mechanisms, such as alternative splicing, also regulate the kinds of mRNA produced. The numbers of mRNAs reaching the cytoplasm are controlled transcriptionally through differences in the rate of initiation of transcription, and posttranscriptionally through changes in the rates of mRNA processing and breakdown in the cytoplasm.

TRANSLATIONAL REGULATION

Translational regulation controls the rate at which mature mRNAs located in the cytoplasm are used in protein synthesis. This pattern of regulation is particularly important in animal eggs and early embryos. It is also important in cells in which rapid, short-term adjustments in levels of protein synthesis are critical

Table 17-2 Half-Lives of Some Mammalian mRNAs

Protein Encoded	Half-Life in Hours
β-actin (nonmuscle)	60
Cytochrome P_{450}	36
Dihydrofolate reductase	97
Fatty acid synthetase	48–96
Glucocorticoid receptor	2.3
Glucose 6-phosphate dehydrogenase	15
Glyceraldehyde 3-phosphate dehydrogenase	75–130
Heat shock protein 70 (hsp70)	2.0
Insulin receptor	9
Ornithine aminotransferase	19
Ornithine decarboxylase	0.5
Pyruvate kinase	30
Thymidine kinase	2.6
Tubulin	4–12
Tyrosine aminotransferase	2.0

Adapted from J. L. Hargrove and F. H. Schmidt, *FASEB J.* 3:2360 (1989).

to survival, because changes at this level in some cases can take effect within seconds or minutes after a stimulus.

Translational regulation may adjust the numbers of polypeptides synthesized upward or downward or may completely turn on or off the synthesis of a polypeptide or group of polypeptides. In either case, translational regulation usually proceeds without changing the number of mRNA molecules in the cytoplasm.

The processes of translation, from initiation through termination, include many steps potentially subject to regulatory controls. However, most known eukaryotic translational controls operate during initiation of protein synthesis. Regulation of protein synthesis at initiation rather than later stages represents an economy in the use of ribosomes, since controls inhibiting elongation or termination would trap ribosomes in inactive complexes.

Translational Controls Regulating Steps in Initiation

Several methods have been used to detect translational regulation by controls of individual initiation factors. One frequently used technique looks for modifications to initiation factors, such as phosphorylation or dephosphorylation, that are correlated with inhibition or activation of initiation. The effects of experimentally adding or removing phosphate groups to the factors are then noted.

Another method follows the effects of adding or removing individual initiation factors in either active or inactive form. It has been possible, for example, to identify factors inhibiting initiation by adding each factor in unmodified, active form and noting the effects on cells in which protein synthesis is downregulated.

Among the various eukaryotic initiation factors, one identified as *eIF2* (for *e*ukaryotic *I*nitiation *F*actor) is a frequent target of regulatory controls. This factor promotes the binding of an initiator tRNA to the small ribosomal subunit during initiation (see Fig. 16-4). The regulatory controls depend on addition of a phosphate group to eIF2, which converts it from active to inactive form.

The best-documented control system based on eIF2 regulates hemoglobin synthesis in developing red blood cells (Fig. 17-14). In these cells, translation of hemoglobin proteins depends on the supply of *hemin*, an iron-containing ring structure incorporated into the hemoglobin molecule. When hemin is in adequate supply, it combines with and inactivates a regulatory protein, *hemin-controlled inhibitor* (*HCI*; Fig. 17-14a). Under these conditions, protein synthesis proceeds at high levels in developing red blood cells. If hemin becomes unavailable, HCI is released from combination with hemin and becomes active. In active form, HCI acts as a protein kinase that specifically adds phosphate groups to eIF2 (Fig. 17-14b). The phosphorylation inhibits the factor's activity in initiation and halts protein synthesis. If hemin again becomes available, HCI is complexed, inhibiting its phosphorylating activity and releasing the eIF2 blockage.

As a result of the mechanism, translation runs in the presence of hemin and is halted in its absence. The regulatory control is effective in developing erythrocytes because protein synthesis becomes concentrated almost entirely in the assembly of hemoglobin molecules in these cells. Interruption of the assembly of other proteins when hemin supplies are low is not critical. The mechanism works rapidly enough to shut down protein synthesis within five minutes after hemin becomes unavailable.

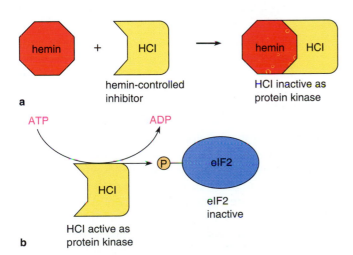

a

hemin + HCI → hemin HCI

hemin-controlled inhibitor

HCI inactive as protein kinase

eIF2 remains active

b

ATP → ADP

HCI active as protein kinase

P eIF2

eIF2 inactive

Figure 17-14 Translational regulation by hemin and HCI, the hemin-controlled inhibitor. Hemin is the iron-containing polyporphyrin ring incorporated into the hemoglobin molecule to complete its synthesis. **(a)** When hemin is in adequate supply, it combines with and inactivates HCI. Under these conditions, protein synthesis proceeds at high levels. **(b)** If hemin becomes unavailable, HCI is released from its complex with hemin and becomes activated. In the active form, HCI acts as a protein kinase that specifically adds phosphate groups to eIF2. Phosphorylation converts eIF2 to an inactive form and halts protein synthesis.

The system inhibiting hemoglobin mRNA translation works in cell-free preparations as well as in intact cells. Addition of unphosphorylated eIF2 to inhibited cell-free systems extracted from red blood cells releases the blockage, indicating that the initiation factor is actually the target of the regulatory mechanism.

Similar regulatory mechanisms based on eIF2 phosphorylation have been discovered in a variety of cell types. These mechanisms, which employ distinct protein kinases, are triggered by various unfavorable conditions including elevated temperature, ATP depletion, virus infection, general starvation, or deprivation of glucose or amino acids. All these conditions, through their effect in activating enzymes phosphorylating eIF2, effectively slow or shut down protein synthesis.

The eIF2 phosphorylation triggered by virus infection provides a natural defense against viruses. The defense is effective because the assembly of viral proteins as well as normal host proteins is inhibited.

Initiation factors other than eIF2 are used as control points in some systems. A. C. Lopo and her colleagues found that in sea urchin eggs, for example, eIF4E, an initiation factor promoting association of the mRNA 5' cap with the small ribosomal subunit, was present in an inactive, unphosphorylated form in unfertilized eggs. At fertilization a protein kinase is activated that rapidly phosphorylates the factor and releases protein synthesis from inhibition. (Lopo's Experimental Process essay on p. 464 describes her experiments identifying phosphorylation of eIF4E as a translational control in sea urchin eggs.)

One group of viruses, including the poliovirus and the rhinoviruses responsible for the common cold, employs a mechanism that favors translation of their own mRNAs by destroying the eukaryotic initiation factors recognizing and binding the 5' cap of mRNAs during initiation (see p. 461). The factors are degraded by proteinases activated by the viruses. Destruction of the initiation factors prevents infected cells from translating their own mRNAs, but the viral mRNAs, which lack 5' caps, are unaffected by loss of the initiation factors. A protein encoded by the viruses specifically recognizes and binds the viral mRNAs and promotes their attachment to ribosomes and entry into polypeptide assembly.

The virus causing AIDS, the *human immunodeficiency virus* (*HIV*), also employs a translational control that greatly favors initiation of its own mRNAs. The HIV DNA encodes a protein that combines with the 5' UTR of the viral mRNA, greatly speeding its translation. Addition of the HIV 5' UTR to any other mRNA, viral or not, speeds translation of that mRNA by a factor of some 1000 to 2000 times if the protein is pres-

ent. (For details of the AIDS virus and its mechanisms of infection, see Chapter 19.)

Translational Controls Regulating mRNA Availability

Some translational controls operate by removing mRNAs from the pool available for translation. These controls are particularly important in many animal oocytes, in which large quantities of mRNAs are stored in completely processed but inactive form. These mRNAs remain inactive until the egg has been fertilized and embryonic development is under way. The availability of many of these mRNAs depends on sequences in the 5' or 3' UTRs of the mRNAs and the combination of regulatory proteins with these sequences.

For example, the mRNA for a ribosomal protein of the small subunit *S19* (see p. 474) is synthesized in amphibian oocytes and stored in the cytoplasm in inactive form. The mRNA does not become active until the tailbud stage of the embryo. P. Mariottini and F. Amaldi found that if the 5' UTR of the S19 mRNA was transferred experimentally to an unrelated mRNA, one for chloramphenicol acetyltransferase (CAT), and introduced into early embryos, the CAT mRNA also remains untranslated until the tailbud stage. With its own 5' UTR the CAT mRNA enters protein synthesis immediately after introduction into early embryos. Presumably the S19 mRNA is regulated through a control protein that binds to a sequence element in the 5' UTR and blocks entry of the mRNA into protein synthesis. At the tailbud stage other elements, possibly other binding proteins, inhibit activity of the blocking protein and allow the S19 mRNA to initiate polypeptide assembly.

POSTTRANSLATIONAL REGULATION

Once their synthesis is complete, the numbers and kinds of proteins in eukaryotic cells are controlled by variations in the rate of protein breakdown. Some proteins persist for months or even years; others have half-lives on the order of hours. In some cases the rate of breakdown is altered depending on environmental conditions or the developmental stage of the cell.

Proteins in the lens of the eye in higher vertebrates normally persist without breakdown for the life of the individual. For humans these proteins may last 90 years or more without degradation. Another relatively long-lived protein is hemoglobin, which persists as long as red blood cells last—about three months in humans, for example. The relatively stable proteins with long half-lives are generally degraded in lysosomes when the organelles or structures containing them are taken in and digested (see p. 597).

Proteins with relatively short half-lives include steroid hormone receptors; homeotic regulatory proteins; and heat shock proteins, which are synthesized in response to elevated temperatures or other environmental stress. M. Rechsteiner and his coworkers noted that these and many other proteins with short half-lives contain four amino acids in relatively large proportions in their sequences—proline, glutamic acid, serine, and threonine. The Rechsteiner laboratory calls these proteins the *PEST* group, after the single-letter abbreviations for the amino acids.

Rechsteiner and others, including A. Hershko and A. Varshavsky, demonstrated that relatively short-lived proteins, rather than being degraded in lysosomes, are hydrolyzed by enzymes in the cytoplasmic solution. The proteins destined for cytoplasmic hydrolysis are identified for breakdown by the attachment of *ubiquitin*, a 76-amino acid protein whose name reflects the fact that it is ubiquitously present, in highly conserved form, in all eukaryotes. As many as 10 or more ubiquitin units may be added to lysine residues within the marked proteins, some of them in chains. During protein breakdown the ubiquitin units are split off intact and recycled.

Addition of ubiquitin depends on alterations at the N-terminal end of the proteins destined for degradation. Many proteins are modified posttranslationally (see p. 469) by addition of an acetyl group to their N-terminal end. This acetyl group is removed from proteins scheduled to be ubiquinated. In addition, an acidic residue, if present at the N-terminal end of the protein, is replaced by a basic amino acid, arginine, before ubiquitins are attached. The arginine is transferred directly from an arginine-tRNA complex. These N-terminal alterations are evidently required for recognition of the protein destined for breakdown by the enzymes attaching ubiquitin. The relationship of the PEST amino acids to this recognition, if any, is unknown.

Ubiquitin is among the most highly conserved proteins of eukaryotes. The sequence of this protein is identical in animals from insects to mammals and differs in only four amino acids between organisms as distantly related as yeast, animals, and higher plants. Conservation at this level indicates that the function of the protein is highly critical and that almost any alteration in its amino acid sequence is likely to be lethal.

Ubiquitin has other functions besides marking proteins for degradation. It is added as a posttranslational modification to some proteins that remain fully functional, such as a portion of histone H2A and H2B molecules in nucleosomes, and some cell-surface receptors. The structural and functional significance of ubiquitination in this case is unknown. Ubiquitin also apparently facilitates the folding of some newly synthesized polypeptides into their final three-dimensional form.

The total spectrum of transcriptional, posttranscriptional, translational, and posttranslational controls determines which proteins are made in cells and in what quantities. Which of the many controls operating at these levels is most significant in determining the numbers and kinds of proteins depends on the cell type and the protein. For most proteins, transcriptional controls are most important in determining numbers and kinds. For some, however, regulatory controls acting after transcription is complete are paramount. Many of the proteins working as "housekeepers" in the cell—proteins required for standard, nonspecialized maintenance of cell reactions—fall into this category. Many of these proteins are transcribed in constant, steady numbers; their final concentrations in the cytoplasm are adjusted primarily or exclusively by posttranscriptional regulation of the number of mRNAs made available for protein synthesis, or regulation of their rates of translation.

For Further Information

Suggestions for Further Reading

Atwater, J. A., Wisdom, R., and Verma, I. M. 1990. Regulated mRNA stability. *Ann. Rev. Genet.* 24:519–541.

Bird, A. 1992. The essentials of DNA methylation. *Cell* 70: 5–8.

Brandeis, M., Ariel, M., and Cedar, H. 1993. Dynamics of DNA methylation during development. *Bioess.* 15:709–713.

Churchill, J., and Campuzano, S. 1991. The helix-loop-helix domain: A common motif for bristles, muscles, and sex. *Bioess.* 13:493–498.

DeRobertis, E. M., Oliver, G., and Wright, C. V. E. 1990. Homeobox genes and the vertebrate body plan. *Sci. Amer.* 263:46–52 (July).

Felsenfeld, G. 1992. Chromatin as an essential part of the transcription mechanism. *Nature* 355:219–224.

Freedman, L. P., and Luisi, B. F. 1993. On the mechanism of DNA binding by nuclear hormone receptors: A structural and functional perspective. *J. Cellular Bioch.* 51:140–150.

Gehring, W. J. 1992. The homeobox in perspective. *Trends Biochem. Sci.* 17:277–280.

Gronemeyer, H. 1992. Control of transcriptional activation by steroid hormone receptors. *FASEB J.* 6:2524–2529.

Grunstein, M. 1990. Histone function in transcription. *Ann. Rev. Cell Biol.* 6:643–678.

Grunstein, M. 1992. Histones as regulators of genes. *Sci. Amer.* 267:68–74B (October).

Harrison, S. G. 1991. A structural taxonomy of DNA-binding proteins. *Nature* 353:715–719.

Hentze, M. W. 1991. Determinants and regulation of cytoplasmic mRNA stability in eukaryotic cells. *Biochim. Biophys. Acta* 1090:281–292.

Herschbach, B. M., and Johnson, A. D. 1993. Transcriptional repression in eukaryotes. *Ann. Rev. Cell Biol.* 9:479–509.

Hershey, J. W. B. 1991. Translational controls in mammalian cells. *Ann. Rev. Biochem.* 60:717–755.

Kolb, A., Busby, S., Buc, H., Garges, S., and Adhya, S. 1993. Transcriptional regulation by cAMP and its receptor protein. *Ann. Rev. Biochem.* 62:749–795.

Kornberg, R. D., and Lorch, Y. 1992. Chromatin structure and transcription. *Ann. Rev. Cell Biol.* 8:563–587.

Kozak, M. 1992. Regulation of translation in eukaryotic systems. *Ann Rev. Cell Biol.* 8:197–225.

Kuhlemeier, C. 1992. Transcriptional and post-transcriptional regulation of gene expression in plants. *Plant Molec. Biol.* 19:1–14.

Lyon, M. F. 1992. Some milestones in the history of X-chromosome inactivation. *Ann. Rev. Genet.* 26:17–28.

McCarthy, J. E. G., and Gualerzi, C. 1990. Translational control of prokaryotic gene expression. *Trends Genet.* 6:78–85.

McGinnis, W., and Kuziora, M. 1994. The molecular architecture of body design. *Sci. Amer.* 270:58–66 (February).

McKnight, S. L. 1991. Molecular zippers in gene regulation. *Sci. Amer.* 264:54–64 (April).

Morimoto, R. I., Sarge, K. D., and Abravaya, K. 1992. Transcriptional regulation of heat shock genes. A paradigm for inducible genomic responses. *J. Biolog. Chem.* 267:21987–21990.

Pabo, C. O., and Sauer, R. T. 1992. Transcription factors: Structural families and principles of DNA recognition. *Ann. Rev. Biochem.* 61:1053–1095.

Parsell, D. A., and Lindquist, S. 1993. The function of heat shock proteins in stress tolerance: Degradation and reactivation of damaged proteins. *Ann. Rev. Genet.* 27:437–496.

Proud, C. G. 1992. Protein phosphorylation in translational control. *Curr. Top. Cellular Regulat.* 32:243–369.

Ptashne, M. 1989. How gene activators work. *Sci. Amer.* 260:40–47 (January).

Rhodes, D., and Klug, A. 1993. Zinc fingers. *Sci. Amer.* 268:56–65 (February).

Richter, J. D. 1991. Translational control during early development. *Bioess.* 13:179–183.

Ross, J. 1989. The turnover of mRNA. *Sci. Amer.* 260:48–55 (April).

Sachs, A. B. 1993. mRNA degradation in eukaryotes. *Cell* 74:413–421.

Stragier, P. 1991. Dances with sigmas. *EMBO J.* 10:3559–3566.

Turner, B. M. 1991. Histone acetylation and control of gene expression. *J. Cell Sci.* 99:13–20.

Vollbrecht, E., Veit, B., Sinha, N., and Hake, S. 1991. The developmental gene *Knotted-1* is a member of a maize homeobox gene family. *Science* 350:241–243.

Wahli, W., and Martinez, E. 1991. Superfamily of steroid nuclear receptors: Positive and negative regulators of gene expression. *FASEB J.* 5:2243–2249.

Welch, W. J. 1993. How cells respond to stress. *Sci. Amer.* 268:56–64.

Yura, T., Nagai, H., and Mori, H. 1993. Regulation of the heat shock response in bacteria. *Ann. Rev. Microbiol.* 47:321–350.

Review Questions

1. At what levels is protein synthesis regulated in eukaryotic cells? What effect does this regulation have on synthesis of cellular molecules?

2. What are *cis* and *trans* elements, and what are their overall functions in transcriptional regulation? What is positive and negative regulation?

3. What structural and functional domains are shared by most regulatory proteins? How are these domains believed to interact in transcriptional regulation? How are enhancer and promoter sequences believed to interact?

4. Outline methods used to isolate regulatory proteins. What are the helix-turn-helix, zinc-finger, and leucine zipper motifs? How do these motifs function in transcriptional regulation?

5. What mechanisms expand the ability of a relatively few proteins to control multiple genes in eukaryotes? How are genes controlled in groups and in networks? What is the significance of cooperative gene regulation by multiple proteins?

6. Outline the mechanism by which steroid hormones regulate genes. What functional sites and domains are present on steroid hormone receptors? What structural motif is common to steroid hormone receptors?

7. What are homeotic genes? Homeoboxes? Homeodomains? What structural motif is common to homeotic regulatory proteins? What sequence elements are recognized in genes controlled by homeotic regulatory proteins?

8. Contrast the functions of homeotic genes in *Drosophila* and higher vertebrates.

9. How do controls modifying transcription, such as DNA methylation, differ in effect from regulatory proteins?

10. What base is primarily methylated in DNA? What effect does methylation have on transcription? What evidence links methylation to transcriptional regulation?

11. At what levels of chromatin organization might changes occur that affect regulation? What kinds of evidence link alterations in chromatin organization to transcriptional regulation?

12. What histone modifications are implicated in transcriptional regulation?

13. What is posttranscriptional regulation? At what levels might posttranscriptional regulation take place? What evidence indicates that regulation of this type actually occurs?

14. Contrast the mechanisms controlling the stability of histone and nonhistone mRNAs. What is the relationship between the stability of some mRNAs and protein synthesis?

15. What is translational regulation? At what steps of polypeptide assembly does translational regulation commonly occur? How is this pattern of regulation related to fertilization? To viral infections?

16. What is posttranslational regulation? Ubiquitin? What is the relationship of this protein to posttranslational regulation? What other functions does ubiquitin carry out in eukaryotic cells? What are PEST proteins?

17. Compare and contrast the overall mechanisms of transcriptional regulation in prokaryotes and eukaryotes (see Supplement 17-1).

18. What are operons? Operators? Repressors? Activators? Inducers? Contrast transcriptional regulation by repressors synthesized in active and inactive forms. What is autoregulation? (See Suppl. 17-1.)

19. How are sigma factors involved in regulation in bacteria? In cell differentiation? (See Suppl. 17-1.)

20. What forms of posttranscriptional regulation occur in bacteria? What is antisense RNA? How might antisense RNA be used to treat viral or bacterial infections? (See Suppl. 17-1.)

Supplement 17-1
Regulation in Prokaryotes

Although bacteria regulate both transcription and translation, their regulatory mechanisms differ in many details from those of eukaryotes. The opportunities that bacteria offer for genetic and biochemical analysis have revealed many of these details and have made it obvious that bacteria, although relatively simple genetically, have transcriptional and translational pathways that are highly versatile and varied.

Because bacteria normally live in constantly changing environments, most of their regulatory controls are easily reversible systems that allow rapid, short-term switches from one biochemical pathway to another. The systems are typically highly automated, tying gene expression in a direct, mechanical way to changes in the environment. Very few bacterial regulatory pathways lead to permanent cellular changes resembling the development and differentiation of eukaryotic cells.

Transcriptional Regulation in Bacteria

Bacterial genes are organized in *operons*—transcriptional units that usually contain more than one coding sequence (see Supplement 15-1 and Fig. 15-18). The coding sequences of bacterial viruses (bacteriophages) are organized in the same pattern. Each operon is controlled as a unit by one or more promoters and, in a few genes, by enhancerlike sequence elements located upstream of the promoters.

The *cis* control elements in 5'-flanking regions of bacterial operons are recognized and bound by regulatory proteins called *repressors* and *activators*. Repressors adjust the rate of initiation by their promoters downward from a base level that is usually relatively high. Thus repressor-regulated operons are characterized by a relatively high rate of transcription when free of their repressors and by a reduced rate when their promoters are complexed with a repressor. Activators work in the opposite way, by adjusting their promoters upward from a base level that is usually low. Activator-regulated operons are therefore characterized by a very low or essentially nonexistent transcription rate when free of their activators and a higher rate when their promoters are complexed with an activator. In the relatively few bacterial operons with enhancerlike sequences, these elements are recognized and bound by regulatory proteins with activities similar to the enhancer-binding proteins of eukaryotes.

Operons are also regulated by *sigma factors*, proteins that bind to RNA polymerase and enable it to recognize promoters with specific sequence elements (see p. 451). Sigma factors have essentially the same overall effect as activators—presence of the factors allows initiation and transcription of their specific operons; when the factors are absent, transcription of the operons is shut off.

Many operons are controlled by more than one regulatory mechanism, and many of the regulatory elements, such as repressors, activators, or sigma factors, can repress or activate more than one operon. The result is a complex network of superimposed controls that provides total regulation of transcription and allows almost instantaneous alterations when environmental conditions change.

Regulation by Repressors   Bacterial repressors were the first genetic regulatory proteins of any kind to be identified. They were discovered in the 1950s by F. Jacob and J. Monod of the Pasteur Institute in Paris, who were interested in the inheritance of genes controlling sugar metabolism in *E. coli*. These investigators found that three enzymes involved in lactose metabolism are encoded in the same transcriptional unit in *E. coli*. They termed the unit the *lac operon* (Fig. 17-15; Jacob and Monod were the first to call bacterial transcriptional units operons, and they introduced many terms and concepts still widely employed in studies of bacterial gene structure, transcription, and regulation.)

The *lac* operon was found to be controlled by a regulatory protein which was termed the *lac repressor*. The *lac* repressor, encoded in a gene separate from the *lac* operon, is synthesized in active form. In this form it binds to a site within the promoter and inhibits transcription of the *lac* operon (Fig. 17-16a; the sequence segment binding the repressor was termed the *operator*). RNA polymerase can still bind when the *lac* repressor is linked to the operator, even though the operator and the region bound by RNA polymerase overlap to some degree. However, the presence of the repressor prevents the tight binding of RNA polymerase necessary for initiation of transcription. This is the situation when no lactose is present in the medium.

The *lac* repressor also has a site that can recognize and bind a chemical derivative of lactose. If lactose becomes available, some of the derivative binds the repressor (Fig. 17-16b). The binding induces a conformational change in the repressor that greatly reduces its affinity for the *lac* promoter, and the repressor is released. RNA polymerase can now bind tightly to the promoter, and the operon is transcribed at its base level, which for the *lac* operon is high.

The repressor mechanism automatically adjusts transcription of the *lac* operator to the availability of

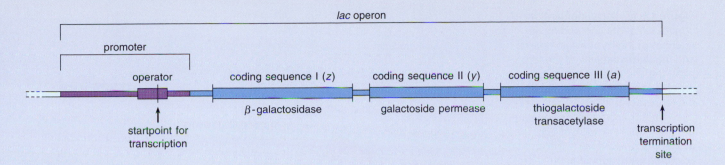

Figure 17-15 Arrangement of control elements and coding sequences in the *lac* operon of *E. coli*. The operator is the DNA site recognized and bound by the repressor. The first coding sequence (*z*) encodes β-*galactosidase*; the second (*y*) encodes *galactoside permease*, and the third (*a*) encodes *thiogalactoside transacetylase* (see text).

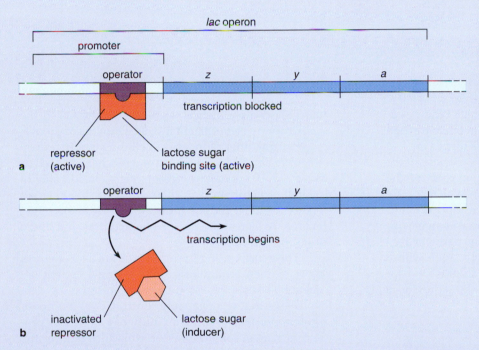

Figure 17-16 Regulation by the lactose repressor. **(a)** The repressor is active and binds the operator when no derivatives of lactose metabolism are available, blocking transcription. **(b)** If lactose becomes available, a derivative of the sugar binds the repressor and converts it to an inactive form in which it no longer binds the operator. Conversion to the inactive form allows tight binding of RNA polymerase and initiation of transcription. *z*, *y*, and *a*, coding sequences of the *lac* operon.

lactose. When lactose is present, the operon is turned on and the enzymes necessary to run the pathway are synthesized; when no lactose is available, the operon is turned off and few or none of the enzymes are synthesized. The mechanism ensures that the enzymes of the pathway are not made unless they are required.

The repressor-based mechanism can also reduce synthesis of the enzymes of a pathway when a product of the pathway appears in the medium. The *trp* operon, which encodes enzymes of the biochemical pathway synthesizing the amino acid tryptophan, works in this way (Fig. 17-17). The *trp* repressor has two binding sites, one for the promoter of the *trp* operon and one for tryptophan. However, in contrast to the *lac* repressor, the *trp* repressor is synthesized in

inactive form. In this form the repressor has no affinity for the *trp* promoter, and transcription of the operon takes place at the basal rate, which is relatively high. This is the situation when the supply of tryptophan is low in the cell or its surrounding medium (Fig. 17-17a).

If tryptophan becomes available or if cellular levels of the amino acid are high because of synthesis by the tryptophan pathway, excess tryptophan combines with the tryptophan-binding site on the repressor (Fig. 17-17b). The binding induces a conformational change that activates the promoter-binding site of the repressor. As a result, the repressor binds to the *trp* promoter, blocking tight binding by RNA polymerase and stopping transcription. The mechanism thus works to adjust synthesis of the enzymes of the tryptophan

Figure 17-17 Arrangement of control elements and coding sequences in the tryptophan operon and regulatory activity of the tryptophan repressor. **(a)** When tryptophan is unavailable in the medium, the repressor is inactive in binding the operator and transcription can proceed. **(b)** If tryptophan becomes available in the medium, the amino acid binds and activates the repressor. The activated repressor binds the operator and blocks transcription. *trp* A, B, C, D, and E, coding sequences of the *trp* operon.

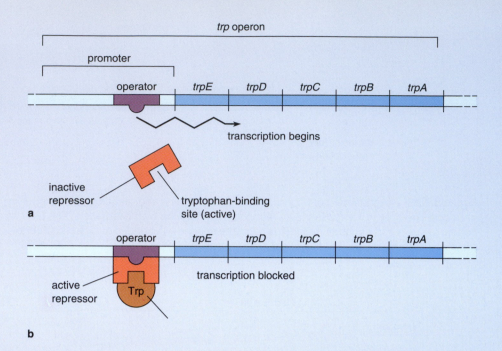

pathway to the availability of tryptophan. When tryptophan is available, the pathway is shut off, and when tryptophan is required, the pathway is turned on.

Figure 17-18 summarizes the two variations of the repressor mechanism. Note that in the form of the pathway regulating the *lac* operon, in which combination with a substance inactivates the repressor and activates the operon, the substance combining with the repressor is called the *inducer*. In the pathway regulating the *trp* operon, in which combination with a substance activates the repressor and shuts off the operon, the substance combining with the repressor is called the *corepressor*.

lac form: active repressor + inducer →
                  inactive repressor;
                  operon transcribed
                  at basal rate
trp form: inactive repressor + corepressor →
                  active repressor;
                  operon transcribed at
                  levels below basal rate

Jacob and Monod received the Nobel Prize in 1965 for their discovery and explanation of the operon and its regulation.

The *lac* repressor was successfully isolated and purified in 1967 by W. Gilbert and B. Müller-Hill. Since then it has been fully sequenced and its probable three-dimensional structure worked out. The protein has 360 amino acids; the fully active repressor consists of four repressor proteins forming a tetramer with a total molecular weight of 152,000. The tetramer con-

tains short alpha-helical segments positioned in the helix-turn-helix motif to fit into the DNA major groove.

Several other repressors, including the *trp* repressor, have also been sequenced and biochemically characterized. In general, repressors vary considerably in overall structure and in their effects on their target operons—some reduce transcription by their target operons to a greater or lesser extent; others shut down transcription entirely.

The genes coding for repressors are also operons, most of them containing only a single coding sequence. The operons encoding most repressors are *autoregulated*—that is, negatively controlled by their own protein products. As the quantity of a repressor protein increases, some repressor molecules bind to the promoter of their own operon and reduce its transcriptional activity. Autoregulation adjusts repressor synthesis so that there are always some repressor molecules of a given type in the cell.

The promoter sequence recognized and bound by an active repressor may lie downstream of the -10 sequence, as it does in the *lac* and *trp* operons, may overlap this sequence, or may overlap or lie upstream of the -35 sequence. (See p. 452 for an explanation of the -10 and -35 sequences and their activity in bacterial promoters.) Many operons are supplied with two or more sequences recognized and bound by the same or different repressors. The *lac* operon, for example, has two other sites that can be bound by the *lac* repressor in addition to the primary operator sequence. Additional repressor molecules bound to these sites increase the degree of repression. Multiple sequences bound by different repressors place an op-

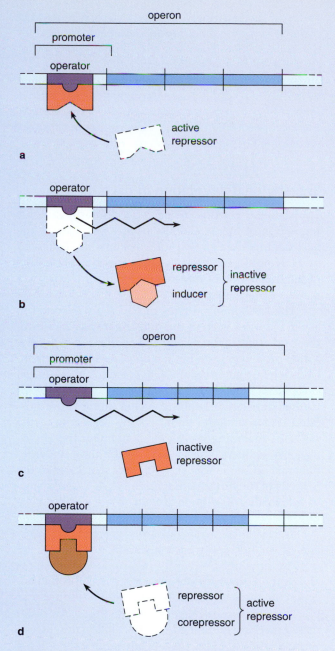

Figure 17-18 Summary of the two forms of the operon mechanism. (a and b) *lac* form, in which the repressor is synthesized in active form and is inactivated by the inducer; (c and d) *trp* form, in which the repressor is synthesized in inactive form and is activated by the corepressor. O, operator region of the promoter.

eron under the control of distinct regulatory pathways. This multiple control increases the extent of the automated networks regulating bacterial transcription.

Regulation by Activators Activators were originally discovered through investigation of mutations that affect several operons simultaneously. One of the first activators detected in these mutants was the *cAMP receptor protein* (*CRP*; also called the *catabolite activator*

protein, or *CAP*). It is a positive regulator of the *lac* and two other operons, the *gal* and *ara* operons, which encode enzymes metabolizing galactose and arabinose sugars.

CRP is activated by *cyclic AMP* (*cAMP*; see p. 158) through a pathway linked to the concentration of glucose in the cell or the surrounding medium. CRP is synthesized in inactive form, in which it has no affinity for the promoters of the *lac*, *gal*, and *ara* operons. When glucose concentrations are low, the enzyme converting ATP to cAMP is active, raising cAMP levels in the cell. Combination with cAMP induces a conformational change that converts CRP to active form. In this form, CRP binds to its target operons and increases their transcription rate above the low base level. Production of the enzymes encoded in the operons positively controlled by CRP allows sugars other than glucose to be metabolized as an energy source.

If glucose becomes available, a sensory pathway that is not as yet understood leads to inhibition of the enzyme converting ATP to cAMP. As a result, cAMP concentration falls, and CRP is converted to inactive form. This shuts down the operons synthesizing enzymes that metabolize other sugars and allows the cell to concentrate on glucose, the preferred metabolite.

CRP is a dimer consisting of two identical polypeptides. Like the *lac* repressor, CRP folds into a helix-turn-helix motif that holds two alpha-helical segments in the correct position to bind in the large groove of the DNA. Binding of CRP induces a sharp bend in the DNA that in some way facilitates initiation of transcription. The bend presumably induces tight binding by RNA polymerase, possibly by unwinding the DNA.

A gradually lengthening list of activators has been discovered in bacteria. Typically activators control groups of operons encoding enzymes that carry out major functions, such as transport, sulfate assimilation, anaerobic electron transport, responses to stress or DNA damage, and phosphate metabolism. Groups of operons concerned with major functions of bacterial viruses are also controlled by activators. Like the genes encoding repressors, many of those encoding activators are autoregulated—the regulator proteins act as repressors of their own operons.

In each system examined so far, activity of positive regulators depends on their combination with nucleotide-derived substances such as cAMP. Note that the role of cAMP in activating CRP is quite different from its function as a second messenger in eukaryotic cells (see Chapter 6 for details). There is no known instance in eukaryotes in which cAMP directly activates a transcriptional regulator as it does in bacteria.

Sigma Factors as Positive Regulators The bacterial RNA polymerase does not bind promoter sequences tightly enough to initiate transcription unless it is bound to a sigma factor (see p. 451). In *E. coli*, most operons are transcribed by RNA polymerase in combination with the "standard" sigma factor, a 70,000-dalton protein identified as σ^{70}. Under conditions of stress, such as elevated temperatures, oxygen deprivation, or exposure to damaging substances such as ethyl alcohol, another sigma factor, σ^{32}, is synthesized in quantity. This sigma factor, in combination with RNA polymerase, specifically recognizes promoters of operons encoding heat shock proteins (see p. 503) in *E. coli*. Another sigma factor, σ^{54}, appears in *E. coli* and promotes synthesis of a specialized group of operons under conditions in which nitrogen supplies are limited.

Positive regulation by sigma factors accounts for one of the rare examples of cell differentiation in bacteria. Under certain conditions the bacterium *Bacillus subtilis* undergoes *sporulation*, in which it differentiates into a quiescent, highly resistant form that can survive extended desiccation and lack of nutrients. During the process an initial cell, the *sporangium*, divides to produce two unlike products. One, the *forespore*, develops into the spore. The second, the *mother cell*, synthesizes a protective protein coat around the developing spore and then breaks down.

P. Stanier and his colleagues discovered that the entire mechanism is controlled by a series of five sigma factors that are synthesized only during sporulation. The first sporulation sigma, σ^{H}, is induced when *B. subtilis* encounters nutrient starvation. Transcription by RNA polymerase associated with σ^{H} activates a further series of genes encoding additional sigmas, leading to the fourth sigma in the cascade, σ^{G}, which appears in the forespore and activates operons encoding proteins involved in the production of the spore. Among the proteins synthesized in response to σ^{G} activity is the final sigma in the cascade, σ^{K}, which appears in the spore mother cell and activates operons concerned with formation of the protein coat. Under normal growth conditions the sporulation sigmas are not produced and the operons concerned with sporulation remain inactive.

The mechanisms controlling bacterial operons tie regulatory controls into networks that provide a balanced, precise, and sensitive adjustment of the entire cell to suit environmental conditions and allow maximum efficiency in use of available nutrients. The entire mechanism is aided by the fact that bacterial mRNAs are very short-lived, about 3 minutes on the average. This permits the cytoplasm to be cleared quickly of the mRNAs produced by one set of operons as they are replaced in activity by another set. In rapidly dividing bacteria, which may pass from one generation to the next in as little as 20 minutes, enzymes and other proteins are also quickly diluted in off-spring if their synthesis is halted. This is in distinct contrast to the regulatory processes of eukaryotic cells, particularly those imposing development and differentiation in higher eukaryotes, which are relatively long-term, preprogrammed, and usually irreversible under normal conditions. Although relatively little is known about regulatory pathways in cyanobacteria, the mechanisms controlling transcription in these organisms are believed to resemble those of bacteria.

Translational Regulation in Bacteria

Translational controls in bacteria primarily involve regulation of the availability of mRNAs. One of the best-documented prokaryotic systems, investigated by M. Nomura and his colleagues, involves regulation of the translation of ribosomal proteins in *E. coli* by a feedback mechanism.

The ribosomal proteins of *E. coli* are encoded in seven operons. Each of the mRNAs transcribed from these operons can be recognized and bound by a ribosomal protein encoded in the operon. As any of these ribosomal proteins accumulate in excess in the *E. coli* cytoplasm, they bind to their own mRNAs and block further translation. Binding occurs in the 5' untranslated region and the initial coding segments of the mRNAs, including the region containing the Shine–Dalgarno sequence (see p. 486). The binding evidently depends on sequences or secondary structures in the 5' UTRs of the mRNAs that resemble rRNA segments to which the ribosomal proteins bind in ribosomes (Fig. 17-19).

Translational regulation of this type, through blockage of an mRNA by a translation product of the mRNA, appears frequently in prokaryotic systems and in the viruses infecting bacteria. Each protein encoded in these mRNAs is capable of binding and blocking its own mRNA, providing a typically automated mechanism that adjusts translation downward when quantities of the protein appear in excess in the cell. As yet no feedback mechanisms regulating the availability of mRNAs in this manner have been reported in eukaryotes or the viruses infecting eukaryotic cells.

The availability of mRNAs in prokaryotes is also regulated by *antisense RNAs*—RNAs, transcribed from codes in the DNA, that have sequences complementary to segments in the 5' UTR of their target mRNAs. Pairing of the antisense RNAs with the 5' UTR, usually in the Shine–Dalgarno region, blocks initiation of protein synthesis and effectively reduces availability of the mRNAs to ribosomes.

Regulation of transcription or translation by antisense RNAs appears to be limited to prokaryotes. However, the existence of such RNAs offers a poten-

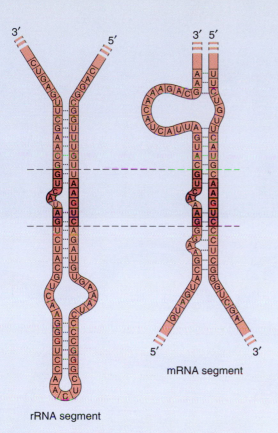

rRNA segment

mRNA segment

Figure 17-19 Similar sequences and secondary structure in a segment of bacterial 16S rRNA (left) bound by ribosomal protein S8 and in the mRNA encoding S8 (right). When present in excess, S8 binds the mRNA segment and blocks further translation of the mRNA. (From data presented in D. P. Cerretti et al., *J. Molec. Biol.* 204: 309 [1988].)

tially powerful method for controlling transcription or translation of specific mRNAs in eukaryotic cells, particularly those of infecting viruses or bacteria. The method has been used successfully in experiments to block mRNAs of the herpes simplex virus and several cellular genes, including those encoding β-actin, a heat shock protein, and several proteins implicated in the conversion of normal cells into tumor cells. Through this means it may eventually be possible to defeat infections or some forms of cancer by synthesizing antisense RNAs that can combine with and inactivate the mRNAs of the pathogen or block the growth of tumor cells.

ORGANIZATION OF THE GENOME AND GENETIC REARRANGEMENTS

The *genome* is the entire collection of genes and all other functional and nonfunctional DNA sequences in the nucleus of an organism. Among the genes are those encoding mRNAs, rRNAs, tRNAs, and sn/scRNAs; other functional sequences occur as regulatory elements or as sites where replication begins. Surprisingly, eukaryotic genomes also contain great numbers of apparently nonfunctional sequences, so many that they take up more of the available DNA than functional sequences in many species. Much of this nonfunctional DNA consists of *repetitive sequences*—sequence elements repeated thousands or even millions of times. Repetitive sequences inflate the genomes of many eukaryotes well beyond the amount of DNA needed for coding, regulation, and replication.

Another unexpected characteristic of genomes is that the arrangement of functional and nonfunctional sequences is not necessarily fixed. Existing sequences have been observed to move from one location to another or to be internally rearranged. Insertion of DNA sequences from outside the genome is also frequently observed. Many of these changes, known collectively as *genetic rearrangements*, are rare events that take place on the evolutionary time scale. However, some are relatively frequent and can be observed within the lifetime of single individuals.

The organization of genes and other sequence elements in eukaryotic genomes and genetic rearrangements are the subjects of this chapter. The structure of prokaryotic genomes and some of the major processes responsible for genetic rearrangements in prokaryotes are described in Supplement 18-1. The programmed genetic rearrangements producing active antibody genes in vertebrates are described in Chapter 19.

HOW EUKARYOTIC GENOMES ARE ORGANIZED: AN OVERVIEW

The genome of a eukaryotic cell is divided among several to many linear, individual DNA molecules. Each DNA molecule, held in association with histone and nonhistone proteins, is a *chromosome* of the organism (Fig. 18-1). The ends of a chromosome are the *telomeres*; telomeres function in replication and make the chromosome tips inert to chemical interactions and enzymatic attack. Internally a chromosome is divided into two *arms* by the *centromere*, the position at which spindle microtubules attach to the chromosome during cell division. Depending on the location of the centromere, the arms may be of equal or unequal length. Chromosomes also have *replication origins*—sites at which enzymes bind to initiate replication. Artificial or natural DNA molecules with these three features—telomeres, a centromere, and replication origins—can function effectively during replication and cell division and be maintained as permanent parts of the genome.

Some of the best evidence that each chromosome consists of a single DNA molecule comes from the yeast *Saccharomyces cerevisiae*, in which the number of DNA molecules separated by gel electrophoresis is equal to the number of chromosomes and to the number of linkage groups determined by genetic crosses.

A *haploid* or *monoploid* genome contains one copy of each chromosome. Usually each chromosome of a haploid genome contains a distinct group of genes and noncoding sequences. A *diploid* genome contains

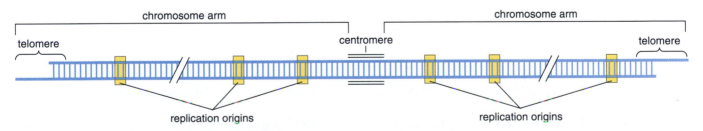

Figure 18-1 Essential features of a eukaryotic chromosome (see text). Each chromosome is a single, linear DNA molecule. At its tips are sequences forming the telomeres; internally the chromosome is divided into two arms by the centromere, the position at which spindle microtubules attach during cell division. Chromosome duplication begins from replication origins spaced along the chromosome.

two copies—a pair—of most or all of the chromosomes. The two copies of a chromosome pair have the same genes but may have different versions of the genes with distinct DNA sequences. The different versions are called *alleles* of a gene. Genomes with three (*triploids*), four (*tetraploids*), or more copies of each chromosome also occur with some frequency in nature.

The genes coding for mRNAs and other RNA types probably number from as few as 5000 or less in lower eukaryotes to perhaps as many as 100,000 or more in animals. In most eukaryotes, mRNA genes occur primarily in one copy per haploid chromosome set. Some mRNA genes, however, such as those encoding histones, tubulins, and actins, are regularly repeated. Repeats of mRNA genes, when present, may be identical or show individual variation and may be clustered or scattered in the genome. When variations occur in coding sequences among a group of repeated mRNA genes, the group is known as a *gene family*. The multiple genes encoding the tubulin, actin, histone, and hemoglobin proteins, among others, make up families of this type. Some gene families contain so many different members with varying degrees of relatedness that they are known as *superfamilies*. The large group encoding immunoglobulins and related proteins is the most extensive gene superfamily in vertebrates. In addition to fully functional sequences, many gene families contain *pseudogenes*—faulty copies that either remain untranscribed or, if transcribed and translated, produce nonfunctional proteins.

In contrast to mRNA genes, all rRNA and tRNA genes are repeated in eukaryotic genomes. In most eukaryotes the rRNA gene repeats, which typically number in the hundreds to thousands, are usually identical or vary in less than 1% of the copies. The genes encoding most of the rRNA types are clustered in a localized region on one or a few chromosomes called the *nucleolar organizer* (*NOR*). tRNA genes, repeated from tens to hundreds of times for each tRNA type, are usually scattered in small groups throughout the genome. sn/scRNA genes, in single or repeated copies, are also distributed throughout the genome. Pseudogenes also occur among the rRNA, tRNA, and sn/scRNA gene copies.

Repetitive sequences are classified as *moderately* repetitive if they occur in a few to a hundred thousand copies and *highly* repetitive if they occur in hundreds of thousands to millions of copies. Moderately repetitive sequences, which include both functional and nonfunctional elements, are distributed in clusters and also dispersed singly or in small groups throughout the genome. Highly repetitive sequences, which are primarily or exclusively nonfunctional, are usually clustered, often in groups near the tips of the chromosomes or surrounding the centromeres.

ORGANIZATION OF CODING SEQUENCES

mRNA Genes

Of the DNA containing codes for RNAs of various kinds in eukaryotes, most, 80% or more in most species, encodes mRNAs. Each mRNA gene has the sequence elements diagramed in Figure 14-9. The transcribed portion of an mRNA gene contains segments corresponding to the 5' untranslated region, coding segment, and 3' untranslated region. Introns may interrupt the coding segment of the gene. Upstream of the gene, in the 5'-flanking region, are promoter, enhancer, and other regulatory sequences.

Single-Copy mRNA Genes mRNA genes occurring in single copies are positioned one after another by the thousands in the genome. Between the 3' end of one gene and the 5'-flanking control elements of the next gene are apparently nonfunctional spacers usually containing moderately repetitive sequences.

Exceptions are noted in some species to the tandem arrangement of mRNA-encoding genes. In some cases, one gene, complete with its own 5'-flanking elements and internal introns, is located within an intron of another gene. More rarely, two genes overlap in the DNA. For example, J. P. Adelman and his colleagues found two rat genes, one encoding gonadotropin-releasing hormone and the other an unidentified protein, that occupy opposite nucleotide chains of the same DNA segment over a distance of some 500 nucleotides. Similar examples of gene overlap have been noted in *Drosophila*.

mRNA Gene Families Typically the multiple mRNA copies in gene families code for the most abundant proteins of the cell, such as myosins, tubulins, histones, collagens, hemoglobins, and immunoglobulins. The individual mRNA genes of these families may be clustered or dispersed, identical or nearly so in sequence, or more diversified.

For example, members of the actin and tubulin gene families show limited sequence variation and are dispersed throughout the genome without clustering in most species. Members of the histone family, which for individual histone types range from almost invariant (H4) to moderately different (H1), are clustered in some species and dispersed in others.

mRNA Pseudogenes Many mRNA gene families include one or more pseudogenes. These faulty copies evidently result from the operation of two independent mechanisms that produce recognizably different classes called *nonprocessed* and *processed pseudogenes*. Nonprocessed pseudogenes appear through mutations in gene copies that originated as duplicates of a single ancestral gene. The mutations, including nucleotide substitutions, deletions, and rearrangements, may occur in both transcribed segments and flanking regions. Some pseudogenes of this class are so altered in their 5'-flanking regions that they cannot form an initiation complex with the polymerase II enzyme and remain untranscribed. Others retain a functional promoter and can be transcribed but have disabling deletions or substitutions within the transcribed segments. The disabling alterations may interfere with translation, substitute critical amino acids, produce codon frameshifts (see p. 467), or change a codon for an amino acid to a terminator codon. Frameshifts or internal terminators result in proteins that contain stretches of "nonsense" amino acids or are significantly shortened. In any case, polypeptides translated from mRNAs of pseudogenes are nonfunctional. Nonprocessed pseudogenes have been detected for example, in the hemoglobin, immunoglobulin, interferon, and major histocompatibility complex (MHC) gene families of mammals.

Processed pseudogenes resemble a processed mRNA rather than an intact gene. In particular, 5'- and 3'-flanking sequences and introns are missing. Processed pseudogenes usually even contain a series of A-T base pairs corresponding to the poly(A) tail of a processed mRNA. Because none of the normal 5'-flanking sequences of the promoter are present, processed pseudogenes of mRNA families are usually not transcribed. Pseudogenes of this class have been detected in many mammalian gene families, including the tubulin, actin, and immunoglobulin families, and also as faulty copies of single-copy mRNA genes such as the tropomyosin, cytochrome *c*, and β-lipoprotein genes.

Processed pseudogenes are believed to originate from DNA copies of fully processed mRNAs that are subsequently inserted in the genome. The copies are probably made through the activity of *reverse transcriptase*, an enzyme that makes complementary DNA (*cDNA*) copies from an RNA template (Fig. 18-2). Reverse transcriptase is not known to occur normally in eukaryotic cells. However, many viruses and transposable elements common in eukaryotic cells (see below) encode reverse transcriptases that could copy mRNAs as well as their usual templates.

The characteristic that makes it seem likely that processed pseudogenes arise by insertion is the fact that they are typically flanked at either end by a short, directly repeated sequence. Flanking direct sequence repeats of this type are typical of DNA segments that have been inserted in the genome, often through the activity of enzymes encoded in infecting viruses (see below).

Although nonprocessed pseudogenes are widely distributed among eukaryotes, processed pseudogenes are restricted almost exclusively to mammals, in

which they may make up as much as 20% of the genome. Their prevalence in mammals may reflect the activity of *retroviruses* infecting mammalian cells. During their cycle of infection these viruses make a cDNA copy of their genetic information, which is stored in free viral particles as an RNA molecule. Besides the reverse transcriptase necessary to make this copy, retroviruses encode proteins capable of inserting a cDNA copy into the host cell DNA. Retrovirus-infected cells therefore possess all the enzymatic machinery necessary to produce and insert processed pseudogenes in the genome.

The chromosomal locations of nonprocessed and processed pseudogenes reflect their probable origins. Nonprocessed pseudogenes usually occur near the functional copies they were duplicated from, often within a gene family cluster as one or more of the repeats. Processed pseudogenes occur randomly in the genome, usually with no tendency to be placed near their normal counterparts.

DNA changes leading to formation of nonprocessed pseudogenes extend over evolutionary time; that is, over millions of years. The time involved can be estimated by comparing the numbers of accumulated mutations in nonprocessed pseudogenes with known or estimated mutation rates. Such comparisons show that it takes from one to two million years for a duplicate copy in a gene family to accumulate enough mutations to become a nonfunctional, nonprocessed pseudogene. Therefore most of the known nonprocessed pseudogenes appeared millions of years ago and are generally present in all members of a contemporary species. Often they are present in all members of major taxonomic groups, meaning that they date at least to the common evolutionary ancestor of the groups sharing them.

In contrast, processed pseudogenes can appear instantly when a DNA copy of an mRNA molecule is inserted into the genome. As with the nonprocessed type, the age of processed pseudogenes can be estimated by noting the number of mutations altering their internal sequences. Processed pseudogenes of the metallothionein and cytochrome *c* genes in humans, for example, are perfect copies of their respective mRNAs. Their insertions into the genome are therefore evolutionarily relatively recent. Their origins are not in the immediate past, however, because the same pseudogenes are also present in rats and mice, indicating that they arose at or somewhat before the evolution of the common ancestor of these species. Other processed pseudogenes, such as three β-tubulin pseudogenes of higher vertebrates, are highly mutated, indicating that they arose much earlier in evolution. (Figure 18-3 shows the distribution of genes, pseudogenes, and other sequence elements in the human α-globin gene cluster.)

Figure 18-2 Formation of a processed pseudogene through the activity of reverse transcriptase. **(a)** An mRNA-encoding gene is transcribed by RNA polymerase, producing a pre-mRNA transcript. After removal of introns and addition of the poly(A) tail during processing **(b)**, the finished mRNA is copied into complementary DNA (cDNA) by reverse transcriptase **(c)**. The cDNA is then used as template for replication of a complementary nucleotide chain, forming a double-stranded cDNA copy **(d)**. This reaction is also catalyzed by reverse transcriptase. The cDNA copy is then inserted by unknown mechanisms in the DNA of the genome **(e)**.

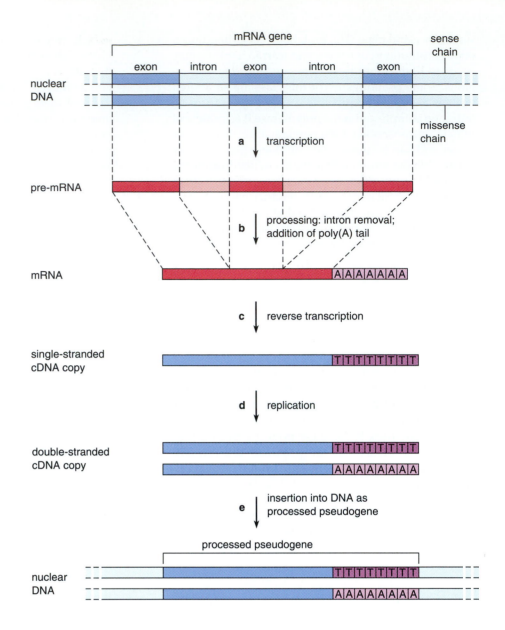

rRNA, tRNA, and sn/scRNA Genes

Genes encoding rRNAs, tRNAs, and sn/scRNAs are the most highly repeated coding sequences in eukaryotes. Pseudogenes of these RNAs tend to be as repetitive as their normal counterparts, in some cases outnumbering the functional copies by many times. This is particularly evident among the genes transcribed by RNA polymerase III (see p. 402), including 5S rRNA and sn/scRNA genes.

rRNA Genes Hundreds or thousands of large pre-rRNA gene repeats (these genes include 18S, 5.8S, and 28S rRNAs; see p. 432) are clustered in nucleolar organizer regions (NOR) at sites on one or several chromosomes of eukaryotic genomes (see Table 18-1). The NOR is so called because the nucleolus forms around it when rRNA transcription and the assembly of ribo-

somal subunits are in progress (see pp. 439 and 716). Although large pre-rRNA genes are repeated many times in NORs, the total DNA length occupied by these genes represents a relatively small fraction of the total coding DNA, only about 1–2% of the genome in most species. Pseudogenes of the large pre-rRNA genes are present in some species, as in some strains of *Drosophila melanogaster*.

Within NORs, individual genes are organized as shown in Figure 15-5b, with coding sequences for 18S, 5.8S, and 28S rRNAs separated by intragenic spacers. A long intergenic spacer separates one large pre-rRNA gene from the next in the clusters. Little or no variation occurs in the coding sequences of these genes from one repeat to the next within a species. Although the intergenic spacers vary somewhat, these sequence elements are also relatively uniform within

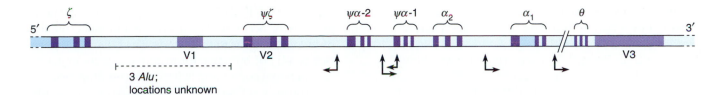

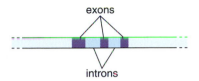

Figure 18-3 The human α-globin gene cluster. Closest to the 5' end of the cluster is the gene for an embryonic globin, ζ, followed by a nonprocessed ζ pseudogene, ψζ. Two nonprocessed pseudogenes of the adult α-hemoglobin genes, ψα-1 and ψα-2, follow. Two functional adult α-hemoglobin genes, α₂ and α₁, and another embryonic gene, θ, lie toward the 3' end of the cluster. *Alu* sequences are indicated by arrows; the vertical part of each arrow shows the location and the horizontal part shows the direction of the sequence. The *Alu* sequences show two patterns of accumulated mutations, indicating that they inserted into this cluster at two separate times in evolutionary history. Three variable regions, V1, V2, and V3, consist of variable numbers of repeated sequences. V1, for example, contains a 36 base-pair sequence that is repeated 32 times in some human individuals, and 58 times in others. Note that one variable region lies within the first intron of the ψζ pseudogene.

Table 18-1   Organization of rRNA and tRNA Sequences in Eukaryotic Genomes

Organism	Large pre-rRNA Gene Repeats	Large pre-rRNA Gene Organization	5S rRNA Gene Repeats	5S rRNA Gene Organization	tRNA Gene Repeats
Tetrahymena	1–2 copies in micronucleus; thousands of copies in macronucleus	One chromosome site in micronucleus; macronuclear copies extrachromosomal	325	Unknown	800
Yeast *(Saccharomyces)*	100–120	Clustered on single chromosome	140	Intermingled with large pre-rRNA genes in same cluster	320–400
Higher plants	1000–30,000	Clusters on one or several chromosomes of haploid set	Unknown	Unknown	Unknown
Drosophila melanogaster	130–250	Single clusters on two chromosomes (X and Y)	500	Single cluster	750
Sea urchin	~175	One cluster	~175	Unknown	Unknown
Xenopus laevis	400–600	One cluster	24,000	Clusters on most or all chromosomes	7800
Human	300	Single clusters on five chromosomes of haploid set	160	One major cluster	1300

a species. Much of the variation that does occur is limited to differences in the number of repeats of short sequence elements. In most species the repeated sequences of the intergenic spacer are found only in this location in the genome. In mammals, however, some repeats of the intergenic spacers are also distributed in other locations.

While sequence variation is limited in the inter- and intragenic spacers of large pre-rRNA genes within species, few similarities are noted in these sequence elements between different species, frequently even between closely related ones. The spacer sequences are thus generally highly conserved within but not between species.

With very few exceptions, 5S pre-rRNA genes are clustered in separate locations from NORs. Each repeated unit in a 5S gene cluster includes a coding sequence and an intergenic spacer. (Figure 15-8 shows the internal structure of a typical 5S pre-rRNA gene.) Repeats in the clusters vary in different species from hundreds to thousands of copies (see Table 18-1). Like the intergenic spacer of large pre-rRNA genes, 5S intergenic spacers contain combinations of repeated sequence elements that show limited variations in number and sequence within the clusters. Sequence differences in the coding regions of 5S rRNA genes, producing the 5S rRNA variants (see p. 431), are also noted in some species. 5S pseudogenes have been detected in both *Xenopus* and humans.

tRNA Genes   In most species, tRNA genes are dispersed on several or most chromosomes, in clusters containing either single or multiple tRNA types. (Table 18-1 lists tRNA repeat numbers for several eukaryotes; the internal structure of tRNA genes is shown in Fig. 15-15.) tRNA pseudogenes are common, particularly among higher eukaryotes; in mammals, some are repeated enough—as many as 100,000 copies per haploid genome—to make them major components of the nonfunctional repetitive sequences. The *C family* of tRNA pseudogenes of cows and goats, for example, contains a sequence that can be folded into a secondary structure equivalent to the stems and loops of a tRNA molecule, complete except for alterations in the acceptor stem. The anticodon sequence of the C family corresponds to that of a cysteine tRNA.

Many of the tRNA-derived pseudogenes have nearly perfect internal promoters (see p. 444 and Fig. 15-15), which may contribute to their wide dispersion and prevalence in mammalian genomes. The promoters are functional, allowing these pseudogenes to be transcribed by RNA polymerase III enzymes into nonfunctional tRNA transcripts. The nonfunctional tRNA copies increase the chance that some of the transcripts will be converted into cDNA by reverse transcriptase and reinserted at other locations in the genome. Because newly inserted processed tRNA pseudogenes formed in this way retain the internal RNA polymerase III promoter, each round of transcription and reinsertion increases the chance that more cDNA copies will be made and inserted.

sn/scRNA Gene Families   The genes encoding sn/scRNAs (see p. 447) occur in both clusters and as single genes in different eukaryotic species. snRNA genes occur in clusters of tandem repeats in sea urchins, for example; in yeast, snRNA genes occur in single copies. Repeats of functional sn/scRNA genes are relatively low in number, on the order of about 10 to 30

copies for each type. However, sn/scRNA pseudogenes may occur in much greater numbers, on the order of hundreds to thousands. The U1 snRNA gene (see p. 447) of humans, for example, occurs in about 30 functional copies and 500 to 1000 pseudogenes.

The only scRNA gene characterized in detail, the SRP 7S scRNA gene (see p. 448), encodes a small cytoplasmic RNA forming a part of the signal recognition particle (SRP). The SRP participates in the reactions attaching ribosomes to the ER during the synthesis of membrane and secreted proteins (see p. 579 for details). Although functional copies of the SRP 7S scRNA gene probably number fewer than 10 in most eukaryotic species, nonfunctional copies may occur in hundreds to many thousands. The *Alu* family of repeated sequences in humans and other primates, for example, which contains hundreds of thousands of repeats, is apparently a collection of SRP 7S scRNA pseudogenes (see Information Box 18-1).

Most sn/scRNA pseudogenes are flanked by direct sequence repeats at their 5' and 3' ends, suggesting that they were inserted in genomes as cDNA copies of processed RNA transcripts. Some snRNA pseudogenes, however, contain the 5'- and 3'-flanking sequences typical of functional genes, indicating that at least some are nonprocessed. Mixed snRNA pseudogenes also occur, containing both nonprocessed and processed elements; evidently this class originated through insertions of one pseudogene type into another.

Coding sequences occur in single and multiple copies, distributed individually or in clusters in the chromosomes of eukaryotic genomes. Genes encoding mRNAs are present singly or in multiple copies as families. The families include members that vary to a greater or lesser extent in sequence. The members of mRNA gene families may be clustered in one or a few locations or may be distributed singly throughout the genome. Genes encoding rRNAs, tRNAs, and sn/scRNAs typically occur in multiple numbers. Large pre-rRNA genes are clustered in one or a few locations in the genomes of most eukaryotes; in these locations the clusters constitute the nucleolar organizer regions around which the nucleolus forms. The genes for 5S rRNAs, tRNAs, and sn/scRNAs are distributed in small clusters throughout the genomes of most eukaryotes. All the major gene types also appear in the form of pseudogenes that, in some cases, outnumber functional copies. Depending on the gene, pseudogenes may be of the processed or nonprocessed type, or both.

REPETITIVE NONCODING SEQUENCES

The existence of repetitive noncoding sequences in eukaryotic genomes first came to light during the 1960s when two investigators, R. J. Britten and D. E. Kohne,

ALU Sequences

The *Alu* family of nonfunctional repeated sequences in humans and other primates (see the figure in this box) is named for the restriction endonuclease *Alu*I, which recognizes and cuts a short sequence element that occurs within most *Alu* repeats. A complete primate *Alu* element is about 300 base pairs in length and contains two similar but not identical repeats of a sequence closely resembling portions of the coding region of an SRP 7S scRNA. The two repeats are separated by an A-rich insert and followed by a short poly(A) segment. Both deletions and insertions occur in the two repeats of the SRP 7S sequence, including a 31-base pair insertion in the second repeat. The short poly(A) segment at the 3' end of an *Alu* element indicates that *Alu* sequences probably originated as processed pseudogenes of SRP 7S scRNA genes. In humans there is considerable variation among *Alu* sequences produced by mutations, deletions, or rearrangements. Many incomplete copies missing major portions are also present.

A short direct sequence repeat flanks both ends of the *Alu* unit in about 80% of the members of this family in humans. Such flanking direct repeats are hallmarks of mobile elements that can insert in new locations in the genome.

Alu sequences occur singly, in pairs, and in small clusters in the human genome. The α- and β-globin gene clusters of humans, for example, contain *Alu* sequences singly and in inverted pairs in the spacers between individual genes. *Alu* elements also occur as inserts in the introns of as many as 25% of other mRNA genes. In this location they are transcribed and appear as RNA copies in the pre-mRNA transcripts of these genes. The *Alu* copies are cut from the pre-mRNA transcripts as the introns are removed during processing, so that they have no effect on the proteins translated from the mRNAs.

The more complete *Alu* sequences contain an RNA polymerase III promoter, making them potentially active in transcription. Relatively few *Alu* sequences of humans are actually transcribed, however, possibly because necessary flanking sequences are missing. Those that are transcribed may be used for production of cDNA copies, leading to further rounds of insertion in the genome.

In different primate species, *Alu* repeats occur in 100,000 to 500,000 copies and make up from 3–8% of the total DNA of the genome. They occur in 300,000 to 500,000 copies in the human genome, where they constitute 6–8% of the total DNA. At this level there is one *Alu* sequence for every 5000 to 9000 DNA base pairs in the genome.

Alu sequences occur in all primates investigated to date, and also in rodents such as the rat and mouse. Indications are that *Alu*'s are continually inserted into the genomes of humans and other mammals at a rate of about one insertion every 100 years. Differences between individual *Alu* sequences because of mutations, DNA rearrangements, and deletions provide a measure of the time since their insertion. The *Alu*'s within human globin gene clusters, for example, differ in sequence from a "standard" *Alu* by about 14%, indicating that they were inserted some time ago, probably before the evolutionary split between apes and humans. Other *Alu* sequences have been inserted more recently.

Recently two human patients were discovered to suffer from genetic disabilities resulting from insertion of an *Alu* sequence into the coding regions of mRNA genes. One of these insertions, discovered by M. R. Wallace and her coworkers, was responsible for development of neurofibromatosis, characterized by tumors of the nervous system and other defects. The *Alu* insertion responsible for the gene mutation in this patient did not appear in his parents, indicating that it occurred in his lifetime.

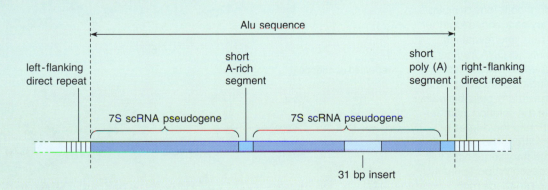

developed a method, now known as *reassociation kinetics*, for detecting them in bulk DNA samples. In this technique the DNA of an organism is isolated and purified, broken into pieces about 1000 base pairs in length, and heated to unwind the DNA into separate chains. The preparation is then cooled slowly, and the rate at which the chains rewind into double helices is noted.

Nonrepeated sequences rewind at a basal rate that is relatively slow because each nonrepeated DNA segment has only one pairing partner in the solution, which it must encounter by random collisions in order to rewind (Fig. 18-4a and b). Because they have more possible pairing partners, repeated sequences reassociate more rapidly as DNA preparations are cooled (Fig. 18-4c and d).

The rapidity with which repeated sequences rewind gives a measure of their degree of repetition. For the group rewinding most rapidly, defined as highly repetitive sequences, the time required for half of the sequences to rewind indicates repetition from hundreds of thousands to millions of times in the genome. Other sequences, defined as moderately repetitive, rewind at rates indicating repetition from hundreds to one hundred thousand times (Table 18-2).

Although rRNA and tRNA sequences were known at the time to occur in hundreds or thousands of copies, Britten and Kohne's results indicating that some sequence elements occur in millions of copies in eukaryotes were at first discounted. However, their results were later confirmed by several unrelated experimental approaches.

The most direct confirmation came from DNA sequencing. One series of experiments sequenced *satellite bands* obtained when the genomes of some species are broken into short fragments and centrifuged (see Appendix p. 795). When centrifuged, the DNA fragments of these species separate into a *main band*, which contains the bulk of the genome, and one or more separate, small satellite bands. The satellite bands separate from the main band because they contain higher proportions of A-T or G-C pairs than the bulk DNA, giving them a markedly different density. Sequencing of the satellite bands showed that each consists of a single sequence element, in many cases less than 10 base pairs in length. One of the satellites from *Drosophila melanogaster*, for example, contains nothing more than tandem repeats of the sequence AAGAG. Although the sequence is very short, hybridization studies showed that it pairs with a surprisingly large fraction of the DNA in the genome, so large that the AAGAG element must occur in more than a million copies.

Repetitive sequences have also been detected by means of restriction endonucleases, bacterial enzymes that recognize short sequence elements and cut DNA at sites containing the sequences (see p. 420 and Table 14-2). Many restriction endonucleases have been identified and purified, each specific in its recognition and cutting activity for a different DNA sequence. Use

Figure 18-4 Rewinding of single-chain DNA fragments from unique or repeated sequences. The a_1 and a_2 nucleotide chains are complementary. **(a)** DNA fragments containing unique, nonrepeated sequences. **(b)** After unwinding by heating, each nucleotide chain has only one pairing partner in the solution. On cooling, random collisions bringing complementary sequences together are relatively rare, and rewinding proceeds slowly. **(c)** DNA fragments containing repeated sequences. **(d)** After unwinding by heating, each single chain has many possible pairing partners in the solution, numbering in proportion to the degree of sequence repetition. The probability that complementary sequences will collide is increased proportionately, and rewinding proceeds more rapidly than in **(b)**.

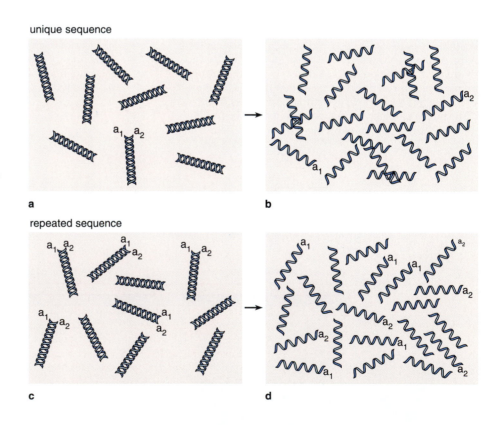

unique sequence

a

b

repeated sequence

c

d

of the enzymes allows repeated sequences to be detected and cut from the genome as fragments of uniform length. Once separated from the remainder of the DNA by techniques such as gel electrophoresis (see Appendix p. 796), the uniform fragments can readily be sequenced. (Figure 18-5 explains the technique in further detail.) The technique also confirmed that the genomes of all eukaryotes investigated contain both moderately and highly repetitive sequences and allowed the sequence structure of many repeated elements to be determined. An adaptation of the restriction endonuclease methodology led to "DNA fingerprinting," a technique that allows persons to be recognized through individual differences in repeated sequence elements (see Information Box 18-2).

Moderately Repeated Sequences

Functional Sequences   Among the functional sequences in the moderately repetitive group are genes encoding rRNA and tRNA and repeated mRNA genes such as those encoding the histones. The transcribed and nontranscribed spacers of these genes also contain functional elements, such as promoters and enhancers, that are repeated to some degree among different genes.

Moderately repeated functional sequences also occur at the telomeres (tips) of eukaryotic chromosomes. These sequences are thought to stabilize the tips, because if they are removed, chromosomes become highly susceptible to exonuclease digestion and

Table 18-2	Distribution of Unique and Repeated Sequences in Different Eukaryotes			
Organism	Unique (%)	Moderately Repetitive (%)	Highly Repetitive (%)	Reference
Dinoflagellate	40		60*	*Cell* 6:161 (1975)
Polytoma (an alga)	70	10	20	*Chromosoma* 49:19 (1974)
Wheat	25	50–68	10	*Heredity* 37:231 (1976)
Rana clamitans (green frog)	22	67	9	*Molec. Genet.* 5:3 (1976)
Chick	70	24	6	*Eur. J. Biochem.* 45:25 (1974)
Rat	65	19	9	*Biochemistry* 13:841 (1974)
Human	64	25	10	*J. Mol. Biol.* 63:323 (1972)

* Moderately and highly repetitive combined.

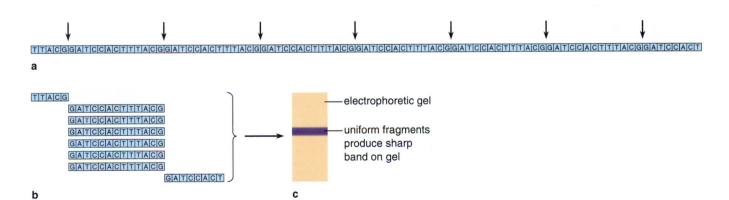

Figure 18-5   Detection of repeated sequences by restriction endonucleases, which cut DNA at specific short sequences. **(a)** A repeated sequence containing the element . .GGATCC. . . recognized by the restriction endonuclease *Bam*HI. Only one of the two nucleotide chains is shown in the diagram. Arrows indicate the points attacked by the enzyme. **(b)** The fragments produced by *Bam*HI digestion. The digestion produces a set of uniform fragments that, when run on an electrophoretic gel **(c)**, produce a sharp, single band. Digestion of a DNA sample containing several different classes of repeated sequences will produce multiple bands, depending on the presence and locations of the sequence recognized by the endonuclease.

DNA Fingerprinting

A forensic technique detecting small, individual differences in DNA sequence elements, developed in 1985 by A. J. Jeffreys, V. Wilson, and S. V. Thien, is gradually gaining acceptance as a method for identifying persons responsible for crimes such as murder and rape. The technique, *DNA fingerprinting*, has also been employed to determine paternity. The method depends on groups of repeated sequences called *hypervariable elements*. Although different in each individual, hypervariable elements have *core* sequences that are similar enough from one person to the next to be represented by an average or consensus sequence. The repeated DNA sequences are frequently preserved well enough in semen or bloodstains, even in stains that have been dried for some time, to allow their extraction and comparison with those of a suspect or victim.

In the technique (see part **a** of the figure in this box) a DNA sample obtained from fresh or dried semen or blood, after being increased in quantity by the polymerase chain reaction (see Supplement 14-1), is digested with a restriction endonuclease, yielding fragment classes of different sizes. The fragments are then run on an electrophoretic gel and extracted by Southern blotting (see Appendix pp. 796 and 798). The bands containing hypervariable sequences of interest are identified by hybridizing them with a radioactive probe consisting of DNA with the consensus sequence for the hypervariable elements.

These samples are compared with samples from the victim or suspect. In a rape case, for example, a DNA fingerprint made from the semen or blood of the suspect could be compared with a fingerprint made from a semen sample obtained from the vagina or clothing of the victim. In a murder case a DNA fingerprint made from bloodstains found on the suspect's person, clothing, or other articles could be compared with the DNA fingerprint of the victim. If comparisons show the DNA fingerprints obtained from these sources to be the same or closely similar, the results can be taken as valid evidence for or against conviction.

DNA fingerprints can also be used as an indicator of paternity because the DNA fingerprints of a parent and offspring are much more similar than those of unrelated persons. Similar DNA fingerprints, when taken along with other tests such as blood typing, can provide a good indication of whether a child has been fathered or mothered by a given person (part **b** of the figure in this box shows a fingerprint made to determine paternity).

To be dependable, DNA fingerprints must be made and compared with great care. Degradation of DNA in samples or contamination by bacteria can produce extra bands that invite errors in identification. The polymerase chain reaction can also produce extra bands (as in the figure in this box). Because of these difficulties, many scientists have proposed that laboratories and individual technicians preparing and comparing DNA fingerprints should be tested regularly and licensed only if competence is demonstrated.

Courts have experienced some operational problems with admission of evidence from DNA fingerprinting. Some experts estimate that the chance of an innocent person producing an incriminating DNA fingerprint is as small as one in hundreds of millions; others maintain that the chance may be as great as 1 in 50. Opinions this diverse have provided defense attorneys with ammunition for casting doubt on the validity of the technique. Some civil libertarians have also expressed concern that DNA fingerprinting constitutes an invasion of privacy. Further experience with the method, agreement on standards to be used for comparison of fingerprints, establishment of rules governing submission to the test, and licensing of technicians and laboratories promise to make the technique more useful as a source of definitive legal evidence.

readily link covalently to various cellular elements including other chromosomes. Readdition of a telomere from the same or a different chromosome restores stability. The telomere repeats also participate in replication as part of a mechanism that fills gaps left at the ends of DNA molecules (for details, see p. 674).

Telomere repeats are surprisingly uniform among eukaryotes. Humans and other vertebrates, for example, all have repeats of the sequence

5'-CCCTAA-3'
3'-GGGATT-5'

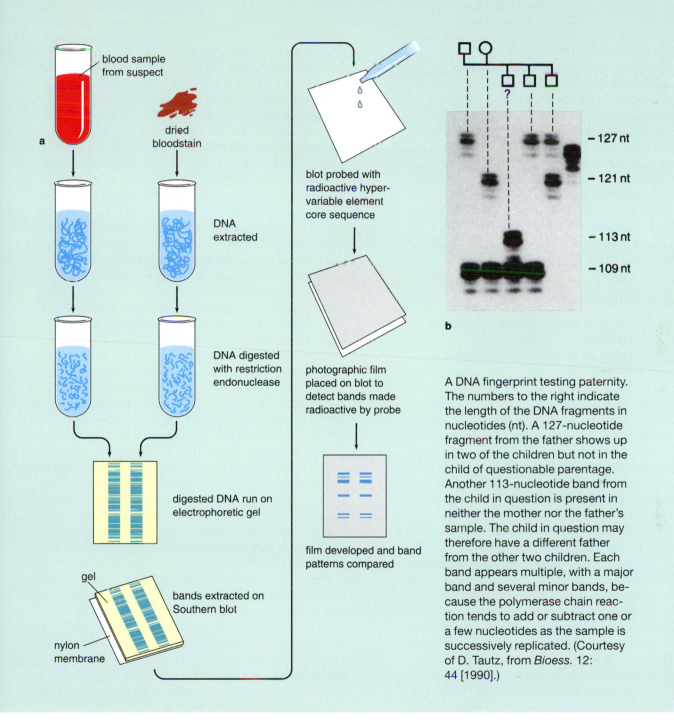

a

blood sample from suspect

dried bloodstain

DNA extracted

DNA digested with restriction endonuclease

digested DNA run on electrophoretic gel

gel

bands extracted on Southern blot

nylon membrane

blot probed with radioactive hypervariable element core sequence

photographic film placed on blot to detect bands made radioactive by probe

film developed and band patterns compared

b

– 127 nt

– 121 nt

– 113 nt

– 109 nt

A DNA fingerprint testing paternity. The numbers to the right indicate the length of the DNA fragments in nucleotides (nt). A 127-nucleotide fragment from the father shows up in two of the children but not in the child of questionable parentage. Another 113-nucleotide band from the child in question is present in neither the mother nor the father's sample. The child in question may therefore have a different father from the other two children. Each band appears multiple, with a major band and several minor bands, because the polymerase chain reaction tends to add or subtract one or a few nucleotides as the sample is successively replicated. (Courtesy of D. Tautz, from *Bioess.* 12: 44 [1990].)

at the telomeres of each chromosome. Fungi, plants, and protists have similar sequences. In humans the repeats of this sequence extend over 10,000 to 15,000 base pairs at each telomere. Telomere repeats are so similar among various eukaryotes that those of one species can generally be substituted for another, no matter how distantly related, without interfering with chromosome functions in replication and cell division.

Nonfunctional Moderately Repeated Sequences The moderately repeated class also contains nonfunctional sequences in quantity. Some of these sequences are dispersed throughout the genome in single copies or in repeats located between coding sequences. Other moderately repeated elements are clustered at one or more sites in the chromosomes.

Pseudogenes commonly form part of moderately repetitive nonfunctional sequences, especially in mammals. Some of these pseudogenes may be transcribed, especially those, such as tRNA and sn/scRNA pseudogenes, that contain internal RNA polymerase III promoters as part of their sequence structure. As far as can be determined, transcripts of these moderately repeated pseudogene sequences are nonfunctional.

Individual moderately repeated sequence elements have been studied in some detail in several mammals, including humans. The best-characterized of the moderately repeated nonfunctional sequences occurring in humans and other primates is the *Alu* family, first identified in 1979 by C. W. Schmid and his coworkers (see Information Box 18-1). tRNA pseudogenes are also a common element of the moderately repeated nonfunctional sequences of mammals.

Highly Repeated Sequences

The generally nonfunctional repeated sequences occurring in hundreds of thousands to millions of copies are typically clustered in one or a few locations in eukaryotic genomes. The clusters may contain tandem repeats of a single sequence or regularly alternating combinations of several sequences. In some clusters the tandem repeats are "pure," with no other sequence types inserted. Other clusters are interrupted at points by one or more repeats of a pseudogene or moderately repeated elements. In general the chromatin containing highly repeated sequences is tightly condensed as constitutive heterochromatin (see p. 382) in interphase cells and is totally inactive in RNA transcription.

The genome of *Drosophila melanogaster*, for example, has four different highly repeated sequence elements 5, 7, 10 and 12 base pairs long. These sequences are repeated millions of times and clustered at a few locations in the genome. The highly repetitive sequences of mammals generally are repeats of an element about 100 base pairs long that is characteristic of the particular mammalian group. Variants of the sequence occur in clusters in different, regularly alternating combinations. Different chromosomes within a species usually contain clusters with distinct repeat patterns. These clusters are often located at either side of the centromere and, in some species, near the telomeres.

One of the most striking characteristics of highly repetitive sequences is the near-perfect fidelity with which they are conserved within a species. Separate species, however, no matter how closely related, have different highly repeated sequences arranged in distinct numbers, patterns, and combinations. The degree of preservation within a species suggests that a correction mechanism must somehow recognize new mutations as they appear anywhere in the thousands or millions of repeats and must remove or replace nucleotides to return the mutated sequences to match a "standard" type. Curiously, preservation of highly repeated elements, although extensive, is usually not perfect; often a small percentage varies from the standard. This indicates that the unknown mechanism preserving these sequences is "leaky" and allows a small number of incorrect copies to persist.

Highly repeated sequences account for the unexpectedly large DNA content of genomes in some eukaryotic groups. Many relatively simple organisms, such as snails, have more DNA in their genomes than genetically more complex animals such as mammals. The differences are most extreme among amphibians—some salamanders have as much as 25 times more DNA per nucleus than humans. Similar variations are noted among plants, where wide differences in DNA content are noted even between closely related species. For example, some members of the genus *Vicia* differ in total DNA amount per nucleus by as much as five times. Humans, incidentally, fall about midway between the potato and the pea in DNA content per nucleus. Most of these differences in DNA content result from varying proportions of highly repetitive sequences rather than from differences in the number of coding and other functional sequences.

Many hypotheses have been advanced to account for the abundance of highly repetitive sequences in eukaryotes. One proposal maintains that these sequences are "evolutionary junk," the products of continued mutation and duplication of once-functional sequences. Another possibility is the opposite idea, that these sequences are generated by duplications and maintained as evolutionary raw material, as blocks of DNA that may eventually evolve into functional genes. A third idea is that highly repetitive clusters, which appear at localized sites on almost all the chromosomes of the set in many eukaryotes, may serve as recognition sites for such processes as chromosome pairing and crossing over during meiotic cell division. Related to this idea is still another proposal, that blocks of highly repetitive sequences act as passive spacers, positioning regions of the chromosomes where they can function most efficiently in pairing and recombination. Although all these proposals stand as more or less remote possibilities, there is no experimental evidence directly supporting any of them. The only information even suggesting that the sequences might be functional is the fidelity with which they are preserved within a species. The biological bases for the origin, existence, and preservation of so much apparently nonfunctional DNA in eukaryotes remains unknown.

Moderately repeated sequences may occur in the spacer regions between mRNA genes and in introns within the genes. Some of these moderately repeated elements are functional sequences, such as those of the promoter and other elements regulating transcription; others are nonfunctional sequences such as pseudogenes. Nonfunctional moderately repeated sequences may be dispersed throughout the chromosomes, placed singly or in small repeated groups between functional sequences. Or, they may be clustered in groups at localized sites on the chromosomes. Highly repetitive sequences are generally clustered rather than dispersed. Frequently the clusters of highly repeated sequences are located near the centromeres and telomeres. Table 18-3 summarizes the characteristics of moderately and highly repeated sequences; the hypothetical chromosome shown in Figure 18-6 reviews the arrangement of the different sequence elements of eukaryotic genomes. Information Box 18-3 describes efforts under way to sequence the human and other genomes completely.

GENETIC REARRANGEMENTS

Genomes were originally thought to contain sequences that, with rare exceptions, are constant in arrangement. However, the techniques of molecular biology have revealed that some sequence elements move about in the genome or are added or deleted with much greater frequency than previously expected.

Many of these changes, known collectively as *genetic rearrangements*, actually are very rare events that appear and spread slowly among populations as part of the evolutionary developments giving rise to new species. However, some occur rapidly enough to be observed within the lifetimes of individuals. As a consequence, the genome, once thought to be a relatively unchanging, genetic "Rock of Gibraltar" for each species, is now known to be in a dynamic state of flux.

Several mechanisms provide opportunities for rapid genetic rearrangements. One source of relatively rapid change is movable sequences known as *transposable elements* (TEs). These sequences can leave one location in the genome and insert in another, or generate copies that insert in new locations while leaving the original copy intact. Another source of rapid change is certain viruses that carry sequence elements relatively rapidly to new locations in the same genome and between the genomes of different individuals. At times, DNA fragments originating from other individuals are directly taken up by cells and integrated into nuclear genomes. DNA transfer from individual to individual may sometimes take place among totally unrelated organisms such as bacteria, higher plants, and animals. Transfer also takes place between the nucleus and the two major DNA-containing cytoplasmic organelles, mitochondria and chloroplasts (described in Chapter 21).

Genetic rearrangements also arise from breakage and random rejoining of chromosome segments. The rearrangements produced by this process, called *translocations*, have figured significantly in the evolution of some eukaryotic groups, particularly insects and higher plants. Tracing the translocation patterns in these groups has in some cases allowed evolutionary lineages to be reconstructed or clarified. Translocations have also been linked to generation of disabilities and diseases such as cancer.

Genetic rearrangements are detected by several methods, including direct sequencing, changes in the DNA fragment classes produced by restriction endonucleases, the appearance of spontaneous mutations

Table 18-3 Characteristics of Moderately Repeated and Highly Repeated Sequences		
Characteristic	Moderately Repeated	Highly Repeated
Number	$\sim 10^3 - 10^5$	$10^6 - 10^7$
Transcription	Some sequences transcribed	Rarely or never transcribed
Sequence structure	Complex	Usually simple
Arrangement in genome	Dispersed and clustered	Clustered, located primarily at centromeres and telomeres
Distribution in chromatin	Both heterochromatin and euchromatin	Condensed heterochromatin
Time of replication	With coding sequences	Late
Evolutionary origin	Many as pseudogenes or transposable elements	Unknown
Function	Some functional as mRNA or accessory RNA genes; others as regulatory, transcription, or processing signals	Most or all no known function

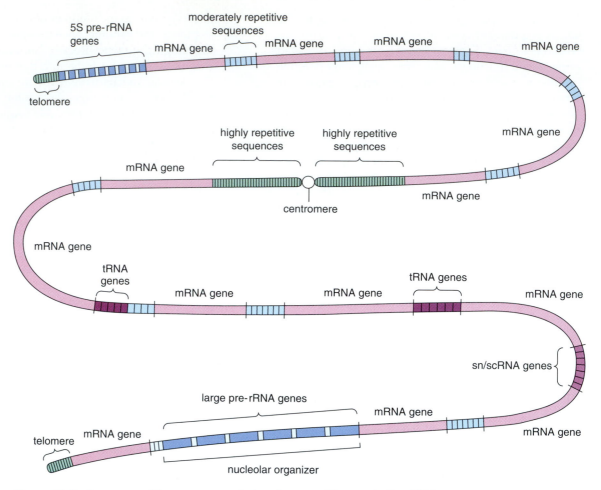

Figure 18-6 Summary of the arrangement of coding, repetitive, and other DNA sequences in eukaryotes. Genes encoding mRNAs occur in tandem over long stretches in the chromosome, mostly in single copies. Some, such as the genes coding for the histones, are repeated. Genes encoding rRNAs and tRNAs are repeated and occur in clusters. A cluster of large pre-rRNA genes forms the NOR, around which a nucleolus forms. The genes for 5S rRNA are usually clustered at some distance from the NOR, on the same or a different chromosome. The sn/scRNA genes may be located singly or clustered. Moderately repeated sequences may be dispersed singly throughout the chromosomes, placed in small repeated groups between functional sequences, or clustered in larger groups at localized sites. Highly repetitive sequences are usually clustered around the centromeres and, in some species, near the telomeres.

in genetic lines, and direct observation of alterations in chromosomes in the light or electron microscopes. Some genetic rearrangements, including movement of TEs and viral infections, take place commonly in prokaryotes as well as eukaryotes (see Supplement 18-1 for details).

The genetic rearrangements resulting from these mechanisms are of more than academic interest. The movement of TEs and chromosome translocations has been implicated in the development of some forms of cancer. The harnessing of rearrangement mechanisms, particularly those of transposable elements, shows promise as a means to change the genome of organisms at will. Purposeful rearrangements of this type,

called *genetic engineering*, may improve domestic plants and animals and correct human genetic disabilities.

The first indications that sequences can move from place to place in either prokaryotic or eukaryotic genomes dates back to research with maize conducted in the 1940s by Barbara McClintock. McClintock noted that some mutations affecting kernel and leaf color in maize appeared and disappeared spontaneously and rapidly under certain conditions and that "controlling elements" affecting these mutations appeared to move from place to place in the genome. Some of the mutations and movements of controlling elements occurred so rapidly that changes in their effects could be noticed at different developmental times in the same in-

Information Box 18-3

Sequencing Human and Other Genomes

As of yet, no eukaryotic genome has been entirely sequenced. The longest genome attempted, that of the cytomegalovirus, barely exceeds 200,000 base pairs. Recently completed in the laboratory of B. Barrell, this sequence represents only a tiny fraction of the DNA in eukaryotic genomes. However, an effort is now under way to sequence the human genome, which includes some 3.2×10^9 base pairs. Establishing the sequence will be very expensive—current estimates range from $1.00 to $10 per base pair, or about 3 billion to more than 30 billion dollars. If ever obtained, the complete sequence would fill the equivalent of 200 Manhattan telephone directories!

Scientists are presently divided over the advisability of sequencing the human genome. Those in favor of the project promise that obtaining the sequence would inevitably speed the search for mutations responsible for hereditary diseases and provide means for their diagnosis and treatment. The complete sequence is also expected to answer fundamental questions about genome organization and the mechanisms controlling genes in development and cell differentiation. Those opposed claim that the money and effort required to sequence the human genome would be better spent in research designed to target specific problems, restricted to segments of the genome that are functional. These opponents point out

that, since only 3–5% of the human genome is functional (the remaining 95–98% consists of introns, repeated elements, and other nonfunctional sequences), the complete sequence would represent little more than small islands of useful information "floating in a vast sea of genetic gibberish." Whatever the pros and cons, the project is now going ahead.

A more limited project, developed by J. C. Venter and S. Brenner, is designed to avoid sequencing nonfunctional DNA in the human genome. In this technique, mRNAs are isolated from active cells and copied into cDNAs. Only these cDNAs are sequenced, so that the sequences obtained are restricted to functional ones. More than 2500 previously unknown genes have been identified, and hundreds sequenced in Venter's laboratory by this approach, using brain tissue as the source of mRNAs. (Brain tissue is ideal because as many as 60% of human mRNA-encoding genes are active in brain cells.)

Other projects of lesser magnitude, including an effort to obtain the complete sequence of the bacteria *E. coli* (4,700,000 base pairs) and *Mycoplasma* (742,000 base pairs), the yeast *Saccharomyces cerevisiae* (20 million base pairs, the nematode worm *Caenorhabditis elegans* (100 million base pairs), and the plant *Arabidopsis thaliana* (common wall cress; 70,000,000 base pairs) are also under way.

---

dividual or in a single developing kernel (Fig. 18-7). McClintock reported her findings in the late 1940s and early 1950s, but her conclusions were not widely accepted because other investigators were not prepared to accept the idea that sequences could move from place to place in the DNA.

These attitudes began to change in the 1960s, when TEs were detected by molecular biologists working with bacteria. In the 1970s, examples were discovered in yeast and higher eukaryotes including mammals. McClintock was belatedly awarded the Nobel Prize in 1983 for her pioneering work with TEs, after these discoveries confirmed her early findings and showed that movable genetic elements are probably universally distributed among prokaryotes and eukaryotes.

Transposable Elements

Classes of Eukaryotic Transposable Elements The TEs of eukaryotes, although highly variable in internal

Figure 18-7 Maize kernels showing differences in color patterns due to mutations arising from movement of transposable elements.

structure, fall roughly into two classes with characteristics that depend on the mechanism by which they move from place to place in the genome. Members of the first class, called *transposons*, move by a mechanism that cuts the TE from its original location in the genome and inserts it in a different site. Most transposons are bounded by a short sequence that is repeated in inverted form at either end of the element. A central nonrepeated sequence bounded by the terminal repeats usually includes the code for at least one protein, a *transposase* involved in the reaction cutting the transposon from one location in the genome and inserting it in a new one. Most transposases have the ability to insert the transposon at essentially any location in the genome. Some, however, are restricted to insertion at certain sequence elements.

The second class, known as *retrotransposons* or *retroposons*, duplicates and moves to new locations via an RNA intermediate. The retroposon is transcribed into an RNA copy that is subsequently used to produce a cDNA copy by reverse transcriptase, along the lines shown in Figure 18-2. The cDNA copy is then inserted into the genome at a new location, leaving the original copy undisturbed and in place. Each round of movement increases the number of retroposons in the genome. Insertions that occur in body cells during the lifetime of an individual are lost when the individual dies. However, insertions that occur in reproductive cells may be passed on to become permanent features of the genome.

Insertion of TEs Most transposons and retroposons insert into genomes by creating a *staggered break* in the host DNA (Fig. 18-8). The host DNA is opened in such a way that a short, single-chain segment extends from both free ends (Fig. 18-8a). During insertion of the transposable element the gaps opposite the single chains are filled in by DNA replication (Fig. 18-8b and c). As a result, the base pairs originally included within the staggered break are duplicated as a short direct repeat at both ends of the inserted element. This mechanism accounts for the fact that TEs are often bracketed by short direct repeats. (The inverted repeats forming the boundaries of the TE itself lie just inside the host-derived direct repeats produced by the staggered break.) Because most TEs can insert at essentially any site in the DNA, the host-derived direct repeats differ in sequence at each insertion site. TEs that are clipped from one site and moved to another usually leave the direct repeat as a "scar" or "footprint" marking their former location (Fig. 18-8d and e).

Both the transposon and retroposon forms of TEs are detected primarily through DNA sequencing or through their effects on genes at or near their sites of insertion. When inserted within the coding segment of a gene, a TE usually destroys the functional activity of the protein encoded by the gene. Insertion within the 5'-flanking regions of a gene frequently destroys or at least inhibits transcription of the gene. At times, however, insertion of a TE has the opposite effect—transcriptional activity of the gene is increased, particularly if the transposable element blocks inhibitory sequences or happens to have picked up the control sequences of a highly active gene during one of its previous insertions. At times, excision of a transposon leaves unsealed openings in the DNA, producing chromosome breaks that may lead to rearrangements of chromosome segments. Even when a transposon is excised without leaving open breaks, the short direct repeat left as a footprint may interfere with genetic activity by interrupting 5'-flanking control sequences or introducing frameshift mutations (see p. 477) in the coding segment of genes.

Entire genes of the host cell DNA may become trapped within a TE and may move with the element to a new location during transposition. Such a trapped host gene may become continuously active through the effects of promoter sequences in the element. Host cell genes may also become abnormally active if moved in a TE to the vicinity of an enhancer or promoter of an intensely transcribed gene.

Once inserted in the genome, TEs become more or less permanent residents, duplicated and passed on during cell division along with the rest of the DNA. TEs that insert into reproductive cells may be inherited, thereby becoming a permanent part of the species.

Long-standing TEs are subject to mutation along with other sequences in the DNA. Such mutations may accumulate in the element, gradually altering it into a nonmobile, residual sequence in the DNA. The genomes of many eukaryotes are littered with the evolutionary junk of nonfunctional TEs created in this way.

TEs are currently being exploited as a means to introduce desired genes into organisms or to correct genetic defects. Some successes have already been obtained with this technique, and there are indications that TEs will provide the key to the genetic engineering of many species.

The Primary TEs of Mammals: Retroviruses A group of infective viruses, the *retroviruses*, are the primary source of TEs in mammals and other vertebrates. The TEs derived from these viruses range from harmless residents of the genome to highly dangerous pathogens.

A retrovirus, known as a *provirus* when inserted in the DNA of a host cell, consists of a set of terminal direct repeats enclosing a central nonrepeated sequence (Fig. 18-9). The terminal direct repeats include sequences capable of acting as enhancer, promoter, and termination signals for transcription. The central sequence, which includes from about 5000 to 10,000 base pairs in different retroviruses, usually contains

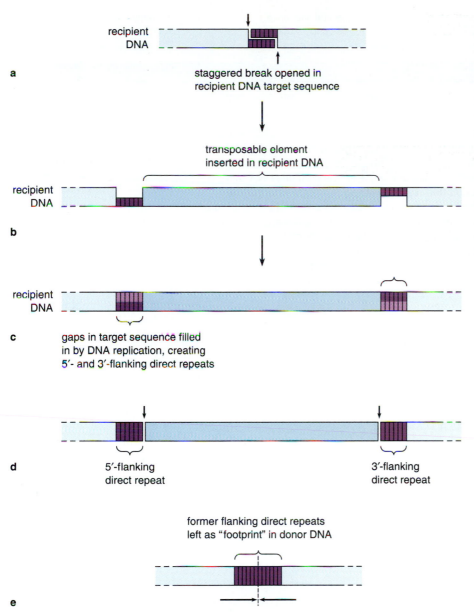

recipient DNA

staggered break opened in
recipient DNA target sequence

a

transposable element
inserted in recipient DNA

recipient DNA

b

recipient DNA

c

gaps in target sequence filled
in by DNA replication, creating
5'- and 3'-flanking direct repeats

5'-flanking
direct repeat

3'-flanking
direct repeat

d

former flanking direct repeats
left as "footprint" in donor DNA

e

Figure 18-8 Insertion and removal of transposable elements. **(a)** A staggered break is opened in a recipient DNA molecule. **(b)** The TE inserts at the staggered break, and **(c)** the single-stranded gaps remaining in the receiving DNA are filled in. The filling in creates 5'- and 3'-flanking direct repeats at the boundaries of the TE in its new location. **(d)** Removal of a TE occurs by cuts made by a transposase at either end of the element (arrows). **(e)** Removal leaves a "footprint" consisting of the former 5' and 3' direct repeats at the site formerly occupied by the TE.

codes for several proteins, including a reverse transcriptase and an *integrase* catalyzing viral duplication and insertion in the genome. Other encoded proteins form parts of free viral particles: coat proteins and a group of proteins associated with RNA in the core of the viral particle. As in other TEs, a retrovirus is flanked by short direct repeats derived from the host DNA when inserted into the genome.

A free retroviral particle contains two copies of the viral genome in the form of two single-stranded RNA molecules. The two RNA copies contain the same genes in the same order but may carry different alleles of the genes. On infection the RNA molecules of the retrovirus are released into the host cell, along with reverse transcriptase molecules that are included in the viral coat. The reverse transcriptase makes a double-stranded cDNA copy of either of the two RNA molecules of the virus, which is subsequently inserted in the genome by the integrase.

Essentially all humans and most other mammals contain from 1 to as many as 100 or more integrated retroviruses. In mice, where the most detailed investigations have been made, new insertions of retroviral elements appear in germ-line DNA on average about once in every 30 generations. Most of these originate from new cycles of infection.

The insertions may be harmless or may be pathogens that cause disease or genetic disturbances including spontaneous mutations, deletions, chromosome breaks, or translocations. Most notable among those with detrimental effects is the HIV (for *Human Immunodeficiency Virus*) retrovirus responsible for AIDS,

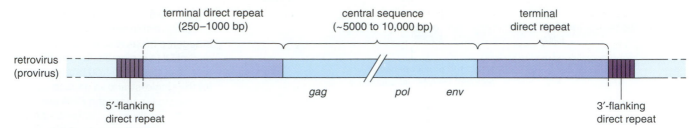

Figure 18-9 A mammalian retrovirus in the proviral form in which it inserts in the DNA. The vertical dashed lines mark the outer boundaries of the provirus; 5'- and 3'-flanking direct repeats outside the vertical dashed lines are derived from the host cell DNA. A retroviral genome has three coding regions. Located closest to the 5' end of the central sequence is the *gag* region, which encodes four proteins that associate with RNA in the core of a retrovirus in its free, infective form outside the host cell. Occupying the approximate center of the central sequence is the *pol* region, which contains the codes for a reverse transcriptase and an integrase catalyzing insertion of the retrovirus in new locations. At the 3' end of the central sequence is the *env* region, which contains codes for glycoproteins forming the coat of free retroviral particles.

and several that are associated with the development of cancer. (HIV infection and AIDS are described in Supplement 19-1.)

Some inserted retroviruses that induce cancer contain a host cell gene within their central sequences that triggers entry of cells into uncontrolled DNA replication and cell division. In its normal location the gene is transcribed only when its activity is required for progress of cells through the division cycle. Inclusion in a retrovirus, however, places the gene under the influence of the highly active retroviral promoter, which frequently makes the gene continually active. Retroviruses also activate genes related to cell division by moving them to the vicinity of an active host cell promoter or enhancer or by delivering an activated control element to the vicinity of a host cell gene. Any of these alterations may produce uncontrolled, rapid cell division. Normal cell genes converted to cancer-related forms by the activity of retroviral transposons or other mechanisms are known as *oncogenes*. (Oncogenes and the induction of cancer by retroviruses and other means are discussed in Chapter 22.)

Many long-standing retroviral insertions of human and other genomes have been so altered by additions, deletions, and rearrangements of sequences that they are no longer active as either infective or mobile elements. These elements add to the considerable burden of evolutionary debris that is replicated and passed from generation to generation.

Retroviruses are not the only viruses that insert their genomes into the host cell DNA. Copies of the SV40, adenovirus, and herpesvirus genomes, for example, also insert and become more or less permanent parts of the host genome. However, none of the DNA segments inserted by these viruses act as TEs—none, once inserted, move from place to place in the genome.

As previously noted, many processed pseudogenes, such as *Alu* elements, are bounded by short direct repeats, which indicate that these pseudogenes are inserted at a staggered break in the host DNA sequence by a mechanism similar or identical to that used by transposable elements. In keeping with this probable route of origin, the flanking repeats of most pseudogenes are unique to each site of insertion.

Many processed pseudogenes move to new locations in the genome via the RNA/cDNA route followed by retroviruses. This common route of insertion suggests that processed pseudogenes may have evolved by taking advantage of mechanisms moving retroviruses and other retrotransposons. Whatever their evolutionary origins, TEs and pseudogenes together make up a large proportion of the genome in most eukaryotes—about 20% in mammals.

Genetic Rearrangements by Translocations Chromosome rearrangements often occur as a result of breaks appearing in one or more chromosomes, most often induced by radiation such as X rays or the effects of chemical carcinogens. Breaks may also appear through excision of TEs, and through malfunctions in normal cell processes such as DNA replication, recombination, or production of antibodies in higher vertebrates. The free DNA ends exposed by the breaks are highly reactive and tend to reseal, through the formation of covalent linkages, if they come in contact with other free ends. Resealing may rejoin a broken chromosome segment in its normal position or may attach it at a new location in the same or a different chromosome. A translocation occurs if a broken chromosome segment becomes attached in a new location.

Many of the breaks responsible for translocations interrupt genes or their control regions, so that only part of a gene lying at the boundary of a translocation is moved to a new location. Such chromosome exchanges are often lethal to the cell containing them. Some translocations, however, arise from breaks in intergenic regions that allow transfer of chromosome segments with boundaries containing completely intact genes or control regions. Cells with chromosome rearrangements of this type often grow and divide. In this case the chromosome rearrangements are duplicated and passed on, founding a cell line in which all descendants share the translocations. Viable cells may at times also arise from rearrangements originating from breaks within the introns of mRNA-encoding genes. Rearrangements of this type may be another source of exon shuffling (see p. 411).

Cell lines with translocations may form a clone within an organism that has little or no effect on survival. Or, a clone may be sufficiently altered in its activities to interfere with the vital functions of the organism, as in the translocations responsible for some forms of leukemia and other cancers. Most cancer-causing translocations place an intact gene related to cell division within the control of a "strong" promoter or enhancer, in such a way that the gene is more or less continuously activated, triggering uncontrolled growth in a cell line (see p. 649 for details). Rarely translocations may have beneficial effects. If passed through the germ lines leading to the production of gametes, such changes may become established as the primary type of a species or may lead to the appearance of a new species. Beneficial translocations, although rare, appear to have been particularly important to the evolution of new species in some taxonomic groups, including insects and plants.

For Further Information

Antibody genes, *Ch. 19*
Cancer, *Ch. 22*
Chloroplast genes, *Ch. 21*
Cloning, *Supplement 14-1*
Introns, *Chs. 14 and 15*
Mitochondrial genes, *Ch. 21*
mRNA genes, eukaryotic, *Ch. 14*
  bacterial, *Supplement 15-1*
Recombination, genetic, *Ch. 25*
rRNA genes, eukaryotic, *Ch. 15*
  prokaryotic, *Supplement 15-1*
sn/scRNA genes, *Ch. 15*
Telomeres, in DNA replication, *Ch. 23*
tRNA genes, eukaryotic, *Ch. 15*
  prokaryotic, *Supplement 15-1*
Viruses and viral reproduction, *Ch. 2*

Suggestions for Further Reading

Barrell, B. 1991. DNA sequencing: Present limitations and prospects for the future. *FASEB J.* 5:40–45.

Bishop, J. E., and Waldholz, M. 1990. *Genome.* New York: Simon and Schuster.

Blackburn, E. H. 1990. Telomeres and their synthesis. *Science* 249:489–490.

Cohen, M., and Larsson, E. 1988. Human endogenous retroviruses. *Bioess.* 9:191–196.

Collins, F., and Galas, D. 1993. A new 5-year plan for the U.S. human genome project. *Science* 262:43–46.

Finnegan, D. J. 1989. Eukaryotic transposable elements and genome evolution. *Trends Genet.* 5:103–107.

Jeffreys, A. J., and Harris, S. 1984. Pseudogenes. *Bioess.* 1: 253–258.

Kao, F.-T. 1988. Human genome structure. *Internat. Rev. Cytol.* 96:52–88.

Lewin, B. 1990. *Genes IV.* New York: Oxford University Press.

Mandal, R. K. 1984. The organization and transcription of eukaryotic ribosomal RNA genes. *Prog. Nucleic Acid Res. Mol. Biol.* 31:115–180.

McElfresh, K. C., Vining-Forde, D., and Balazs, I. 1993. DNA-based identity testing in forensic science. *Biosci.* 43: 149–157.

Morgan, R. A., and Anderson, W. F. 1993. Human gene therapy. *Ann. Rev. Biochem.* 62:191–217.

Mulligan, R. C. 1993. The basic science of gene therapy. *Science* 260:926–932.

Murray, T. H. 1991. Ethical issues in human genome research. *FASEB J.* 5:55–60.

Notani, N. K. 1991. Plant transposable elements. *Subcellular Biochem.* 17:73–79.

O'Brien, S. J., Senanez, H. N., and Womack, J. E. 1988. Mammalian genome organization. An evolutionary view. *Ann. Rev. Genet.* 22:323–351.

Olson, M. V. 1993. The human genome project. *Proc. Nat. Acad. Sci.* 90:4338–4344.

Rechsteiner, M. C. 1991. The human genome project: Misguided science policy. *Trends Biochem. Sci.* 16:455–459.

Schmid, M. B. 1988. Structure and function of the bacterial chromosome. *Trends Biochem. Sci.* 13:131–135.

Shapiro, J. A., ed. 1983. *Mobile Genetic Elements.* New York: Academic Press.

Sharp, S. J., Schaack, J., Cooley, L., Burke, D. J., and Soll, D. 1985. Structure and transcription of eukaryotic tRNA genes. *Crit. Rev. Biochem.* 19:107–144.

Strange, C. 1989. Budding plant genome projects. *Biosci.* 39:760–762.

Trends Biotechnology, vol. 10, number 2, 1992. Entire issue on genome sequencing and mapping.

Varmus, H. 1987. Reverse transcriptase. *Sci. Amer.* 257: 56–64 (September).

Watson, J. D., and Cook-Deegan, R. M. 1991. Origins of the human genome project. *FASEB J.* 5:8–11.

Weill, C. R., and Wessler, S. R. 1990. The effects of plant transposable element insertion on transcription initiation and RNA processing. *Ann. Rev. Plant Physiol. Mol. Biol.* 41:527–552.

Wilde, C. D. 1986. Pseudogenes. *Crit. Rev. Biochem.* 19: 323–352.

Willard, H. F. 1990. Centromeres of mammalian chromosomes. *Trends Genet.* 6:410–416.

Review Questions

1. What is a genome? A chromosome? Define haploid or monoploid, diploid, triploid, and tetraploid.

2. Why do you suppose that mRNA genes evolved in primarily single copies and rRNA and tRNA genes in multiple copies in eukaryotic genomes? What are gene families?

3. Diagram the sequence structure of a typical eukaryotic mRNA gene, and identify the function of each sequence element entered in your diagram.

4. What are nonprocessed pseudogenes? Processed pseudogenes? What mechanisms are believed to be responsible for the appearance of each pseudogene type in eukaryotic genomes?

5. Are pseudogenes functional? Are they ever transcribed? What structural features indicate that some pseudogenes may be able to move from place to place in the genome?

6. Diagram the structure of a large pre-rRNA and a 5S rRNA gene and identify the functions of the sequence elements in your diagram. Where are these genes located in eukaryotic genomes?

7. What is a nucleolar organizer (NOR)? What is the structural relationship of the NOR to the nucleolus?

8. Diagram a tRNA gene and identify the functions of the sequence elements in your diagram. Where are tRNA sequences located in eukaryotic genomes? How are tRNA pseudogenes identified? Are tRNA pseudogenes likely to be transcribed? Why?

9. What evidence indicates that eukaryotic genomes contain large numbers of repeated sequences? What types of repeated sequences occur in eukaryotes? What functions do repeated sequences have in eukaryotic genomes?

10. What does the term *satellite* mean in reference to repeated sequences?

11. Where are repeated sequences located in eukaryotic genomes? What are centromeres? Telomeres? What is the relationship of repeated sequences to these structures?

12. Do you think that sequencing the human genome is worth the time and expense? Why?

13. What mechanisms account for genetic rearrangements in eukaryotes? What are transposons? How are transposable elements believed to appear in new locations in eukaryotic genomes?

14. What genetic effects may result from the movement of transposable elements to new locations?

15. What is a retrovirus? What is the relationship of retroviruses to cancer? How are retroviruses used in genetic engineering?

16. What are chromosome translocations? What is the relationship of this mechanism to cancer?

Supplement 18-1
Genome Structure in Prokaryotes

Prokaryotic genomes are organized in patterns that are fundamentally different from eukaryotic genomes. A prokaryotic nucleoid contains only a fraction of the DNA typical of eukaryotic cells—*E. coli* nucleoids, for example, contain only about 1500 μm of DNA, as compared to the centimeters or meters of DNA in a typical eukaryotic nucleus. Although cyanobacteria contain about 2 to 2.5 times more DNA per cell than bacteria, cyanobacterial DNA content is still only a fraction of that of eukaryotic cells.

Bacterial nucleoids also differ from their eukaryotic counterparts in containing a single DNA molecule, arranged as a covalently closed, continuous circle. As a consequence, all the genes of a bacterial genome are genetically linked. In eukaryotes, in contrast, the genome is divided among several to many DNA molecules, each one a separate linkage group. Because prokaryotic DNA is circular, there is no equivalent of telomere sequences in these organisms.

Another fundamental difference is the general economy of sequence usage in prokaryotes. Almost all the DNA directly makes up parts of genes or forms the 5' or 3' control elements regulating transcription. Several to many genes are grouped together in operons controlled by single promoters. Very few nonfunctional repetitive sequences can be detected in bacteria. Introns, too, are much rarer among prokaryotes—they have been detected in only a limited number of genes in a few bacterial groups).

This economy in sequence usage differs from the arrangement of sequences in most eukaryotes, in which nonfunctional sequences make up a significant part of the genome. The absence of superfluous sequences in prokaryotes may reflect the small size of bacterial and cyanobacterial cells, which have limited space for DNA. The time available for DNA replication is also limited in the rapid division cycles and short generation times characteristic of these organisms.

Another distinguishing characteristic of prokaryotic systems is the presence of *plasmids*, additional DNA circles suspended in the cytoplasm outside the nucleoid. Plasmids range from very small circles containing only a few hundred nucleotides to larger ones including many thousands of base pairs. Plasmid circles contain replication origins, so they are duplicated and passed on during cell division. Genes originating from the same or a different prokaryotic species or from viruses infecting bacteria may be included in plasmids. Bacterial plasmids frequently also contain transposons. In addition to their natural functions, plasmids are widely used to introduce DNA sequences into bacterial cells for cloning (see p. 420).

The Arrangement of Coding Sequences in Prokaryotes

Genes encoding mRNAs are distributed throughout the bacterial DNA circle. The arrangement of prokaryotic mRNA genes in operons including the codes for one to several polypeptides, controlled by the same promoter and transcribed as a unit into a single mRNA transcript (see Fig. 15-18), is fundamentally different from the disposition of mRNA genes in eukaryotic genomes.

The mRNA genes of prokaryotes are packed into the genome with little or no space separating the flanking control regions of one operon from the next or between individual coding segments within operons. Some coding segments within operons even overlap, so that nucleotides do double duty as parts of two adjacent genes.

Ribosomal RNA operons occur in multiple copies in bacteria, ranging from 2 or 3 up to a maximum of about 15 repeats. The rRNA genes may be arranged singly or distributed among several small clusters. The seven rRNA operons of *E. coli* are distributed singly at scattered points around the genome. Each rRNA operon includes the code for 5S rRNA as well as the 16S and 23S coding sequences. The rRNA operons of bacteria also contain from one to three tRNA sequences, some placed between the 16S and 23S codes and in some rRNA operons at the 3' end of the gene (see Fig. 15-21a).

tRNA genes are scattered around the genome. In *E. coli*, 46 different tRNA types are encoded in 77 coding sequences located singly, in clusters, and in mixed operons containing both tRNA and protein-encoding sequences. Some tRNA genes of *E. coli* occur in multiple copies ranging from two to a maximum of seven, arranged in the same or different clusters.

Repeated Sequences in Prokaryotes

Although repetitive sequences are limited in prokaryotes, some sequences in this category have been detected in a few species. *E. coli* and *Salmonella* contain a number of different moderately repeated sequence elements dispersed around the genome, ranging from about 4 to 40 base pairs in length. Among these are functional sequences serving as signals for genetic recombination and integration or excision of DNA segments. For example, an 8-base pair element known as the *Chi sequence* is located at more than 900 sites around the *E. coli* DNA circle. This sequence is recognized by enzymes catalyzing recombination. A repeated element of unknown function, an inverted

Figure 18-10 Summary of the arrangement of coding sequences in prokaryotes. Located at points around the circle are rRNA operons containing coding sequences for 18S, 23S, and 5S rRNAs and some tRNAs, spotted singly or concentrated in several clusters. tRNA operons are spotted at other locations around the circle in pure form and with protein-encoding sequences in mixed operons. Other segments are filled in with mRNA operons. The mRNA operons may be continuous, with no intervening spacers, or separated by small amounts of spacer DNA. Repetition of sequences is confined to rRNA and tRNA genes and a relatively small amount of repetitive DNA.

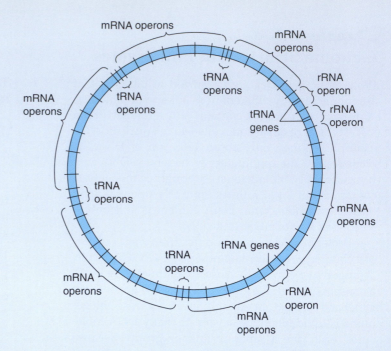

sequence about 25 to 30 base pairs long, occurs at 100 to 200 sites in the *E. coli* genome. These elements have been suggested as binding sites for proteins that tie the *E. coli* DNA circle into looplike domains.

It is not certain whether cyanobacteria share the same patterns of sequence distribution as bacteria, or even whether cyanobacterial genomes are circular. Few mRNA genes have been mapped in cyanobacteria, and the numbers and disposition of rRNA and tRNA genes in these organisms are mostly unknown. Plasmids have been detected in cyanobacteria, however, and studies tracing the rate at which DNA rewinds after dissociation indicate that there are few repeated sequences. These observations and the generally prokaryotic characteristics of the cyanobacteria suggest that the arrangement of genes and other sequence elements in these organisms is probably similar to those of bacteria. Figure 18-10 summarizes the arrangement of coding sequences in the circular DNA molecules of bacterial and, presumably, cyanobacterial genomes.

Genetic Rearrangements in Bacteria

Prokaryotic genomes are even more subject to genetic rearrangements than those of eukaryotes. Sequences are exchanged by recombination among the nucleoid, plasmids, and infecting viruses. Transposable elements also insert at a low but constant frequency in both the nucleoid and plasmids. In addition, bacteria at times take up and integrate DNA segments from the surrounding medium that originate from the breakdown of other cells. The more or less constant movement of sequence elements to and from the nucleoid

and plasmids provides an important source of variability that is vital to the survival of prokaryotic species in their constantly changing environments.

Transposable elements were first detected in bacteria by H. Saedler, P. Starlinger, J. A. Shapiro, and others in the late 1960s, about 20 years after McClintock discovered movable sequences in maize. Most bacterial TEs (Fig. 18-11) are bounded by a repeated sequence that is inverted at the two ends of the element. The terminal repeats enclose a central nonrepeated sequence usually encoding one or more proteins. At least some of the encoded proteins are enzymes that catalyze movement of the TEs. Like their eukaryotic counterparts, bacterial TEs are bracketed by short direct repeats, indicating that insertion takes place at a staggered break in the bacterial host DNA.

Bacterial TEs fall into the transposon rather than retroposon class (see p. 536). However, rather than being excised from one site and inserted in another, bacterial transposons move primarily by a process that involves replication of the original element and insertion of the replicated copy in the new location by a mechanism resembling genetic recombination. The mechanism, which is not completely understood, leaves the original element in place as well as inserting a copy in a new location. Excisions also occur, but evidently by a mechanism not directly connected with movement to a new location.

Bacterial transposons appear at a new location about once in 10^5 to 10^7 individuals, at rates comparable to the mutation rate in bacteria. Excisions occur more slowly, about once in 10^6 to 10^{10} individuals. These rates would suggest that transposons are con-

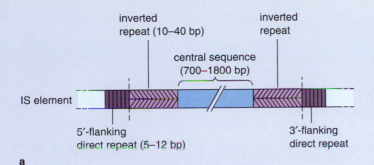

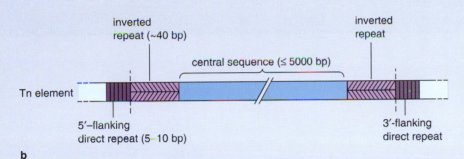

Figure 18-11 Typical bacterial transposable elements. **(a)** An IS element (IS = *Insertion Sequence*). The central sequence encodes from one to three polypeptides concerned with insertion of the transposon in new locations. **(b)** A Tn element (Tn = *Transposition*). The large central sequence of Tn elements encodes polypeptides catalyzing movement of the transposon and often includes one or more genes derived from the host cell.

stantly on the increase in bacteria and that eventually most of the DNA of bacterial genomes would be occupied by TEs. However, the maximum number of transposons is limited to about 1% of the genome in most bacterial species. In at least some bacterial transposons the limit is imposed by a repressor protein encoded in the transposon, which inhibits transcription of the genes encoding enzymes necessary for further transpositions.

Bacterial transposons may insert in plasmids as well as the main genome of a bacterial cell. Their insertion in plasmids facilitates transfer of transposons to other cells and species, since plasmids are frequently passed from cell to cell among prokaryotes and sometimes even from prokaryotes to eukaryotes.

Bacterial TEs, particularly a group known as *Tn elements* (see Fig. 18-11b), frequently contain one or more host cell genes. Many drugs such as penicillin, erythromycin, tetracycline, ampicillin, and streptomycin, once highly effective against bacterial pathogens, have been greatly reduced in their effects by the spread among bacteria of resistance genes carried in TEs. Most often the resistance-conferring genes encode enzymes that recognize and catalyze breakdown or neutralization of the antibiotics. The genes have made diseases caused by more than 50 different genera of bacteria either untreatable or highly resistant to standard antibiotics.

Bacterial TEs cause the same spectrum of genetic disturbances as eukaryotic elements—mutations, deletions, substitutions, inversions of DNA sequences, and modification of the transcriptional activity of nearby genes— when they appear in new locations in the genome (see p. 536). Although many of the genetic disturbances are deleterious, TE movement and insertion is sometimes an important and beneficial source of variability.

Evidence that the balance between deleterious and beneficial effects of TEs to bacteria leans toward benefit comes from several sources. One is development of drug resistance brought about by genes carried in TEs. Another line of evidence comes from experiments testing survival of bacterial strains with and without certain TEs. In these experiments, bacteria possessing the TEs consistently become the dominant cell type when placed together with cells lacking the elements. The advantage conferred by TEs in these experiments presumably results from the variability generated by mutations, gene rearrangements, or modifications of gene activity when the elements move to new locations. The benefits of the mechanism are especially marked in prokaryotes, in which offspring are produced in large numbers in generations that quickly come and go.

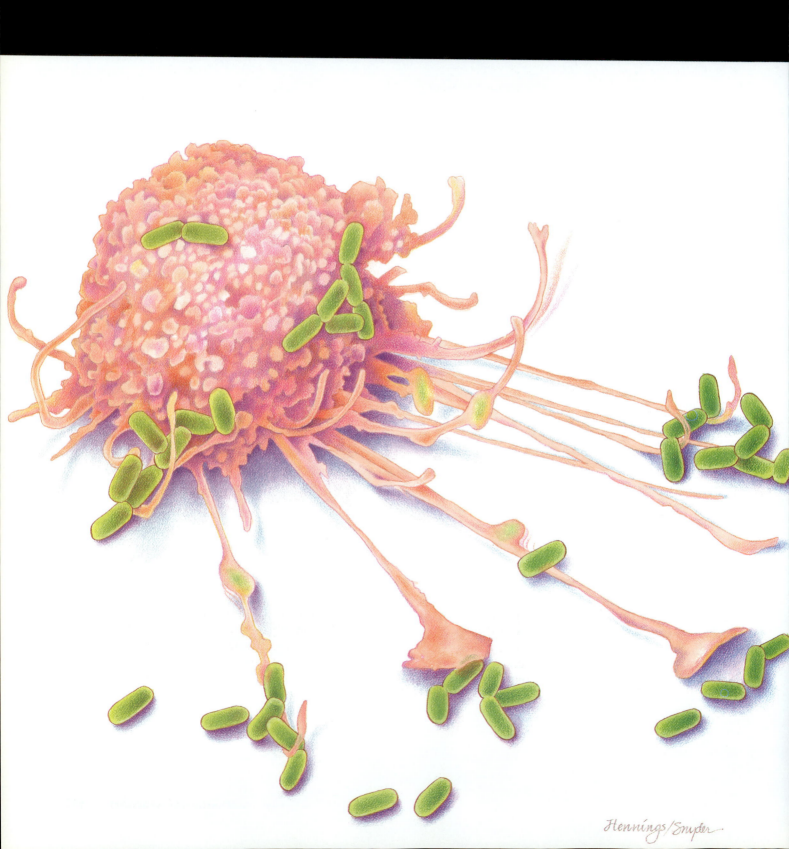

Hennings/Snyder

Antibodies provide the body's first line of defense against invading pathogens including bacteria, viruses, toxins, fungi, and protozoa. By combining with molecules from an invading pathogen, antibodies trigger a series of responses that, if successful, eliminate the pathogen from the body.

Molecules that stimulate antibody production are called *antigens*. Most natural antigens are proteins or polysaccharides. However, certain nucleic acid types such as Z-DNA (see p. 367) and other molecules can also stimulate antibody production. Substances acting as antigens may be freely suspended in solution in body fluids or may be a surface or internal molecule of a pathogen such as an infecting bacterium or virus.

Vertebrates are able to produce antibodies that specifically recognize and react with millions or even billions of different antigens, including man-made substances never encountered before by any living organism. For a time it was thought that each different antibody type must be encoded in a distinct gene, and therefore that much of the genome must be occupied by antibody genes. However, further research revealed that antibody diversity results from programmed genetic rearrangements that assemble a relatively small number of gene subunits into many different combinations to produce a very large number of distinct genes.

Antibodies are produced in *lymphocytes*, a type of leukocyte or white blood cell. The antibody-producing lymphocytes are called *B cells* because they originate and undergo much of their maturation in bone marrow. It is in these cells that the genetic rearrangements producing antibodies take place. Completed antibody molecules appear on the surfaces of B cells and are also secreted by the cells into the bloodstream and other body fluids.

Antibody production by B cells is coordinated with the activities of *T cells*, another type of lymphocyte. T cells are so called because they mature in the thymus. Another nonleucocyte cell type, the *macrophages*, directly engulf invading bacteria and foreign cells and stimulate T-cell development. The coordination of B cells, T cells, macrophages, and other leucocyte types in the immune response produces a total defense that effectively protects vertebrate animals, including humans, against most invading disease agents. In some cases the immune system is also effective in killing cancer cells.

This chapter describes antibody structure, the structure of the gene groups encoding antibodies, and the genetic rearrangements and other processes giving rise to functional antibody genes. Supplement 19-1 discusses AIDS (*Acquired Immune Deficiency Syndrome*), caused by a virus that is able to circumvent the immune system.

PRODUCTION OF ANTIBODIES

Antibodies bind tightly to specific antigens located on the surfaces of invading viruses or organisms, on the surfaces of infected cells, or suspended as free particles in the circulation. Binding by an antibody has one or more effects. By linking to the surfaces of invading pathogens, antibodies block reactive groups that would otherwise bind the pathogen to surface receptors on body cells and promote its entry into the cell interior. The presence of antibodies on a pathogen surface also marks the pathogen for engulfment and destruction by macrophages. For antigens suspended in the body circulation, combination with antibodies causes aggregation or precipitation of the antigen and also marks the antigen for engulfment and elimination.

Antibody Structure

Antibody molecules, also called *immunoglobulins*, are large, complex proteins built up of two pairs of polypeptide subunits, one larger and one smaller (Fig. 19-1a and b). The two smaller polypeptide subunits, each about 220 amino acid residues in length, are identical in structure; these are the *light chains* of an antibody molecule. The two larger chains, identical to each other but different from the light chains, each contain about 440 amino acid residues; these are the *heavy chains* of an antibody molecule. The light and heavy chains are built up from domains called *variable* and *constant* regions. As the names suggest, the variations in amino acid sequence that allow antibodies to recognize and bind different antigens are confined almost entirely to the variable regions. The constant regions are highly conserved between different antibodies.

An antibody light chain is structured from two domains, one variable and one constant. Each heavy chain has a single variable domain and a long constant region containing three to four domains.

The four-part complex formed by the polypeptides of an antibody approximates the Y-shaped structure in Figure 19-1a and b, in which the variable regions of the light and heavy chains combine to form the arms of the Y. The tips of the two arms of a given

Figure 19-1 Antibody structure. **(a)** The tetramer structure of antibodies, including two light and two heavy polypeptide chains held together by disulfide (—S—S—) linkages. Each light and heavy polypeptide chain consists of constant regions (C_L and C_H) and variable domains (V_L and V_H). The variable domains are divided internally between hypervariable (HV) regions, in which most of the variability in amino acid sequence between different antibodies is concentrated, and framework regions (FR), which are more conserved in structure between different antibodies. **(b)** A space-filling model of an antibody molecule, as deduced by X-ray diffraction. (Redrawn from an original courtesy of D. R. Davies, from *Proc. Nat. Acad. Sci.* 74:5140 [1977].)

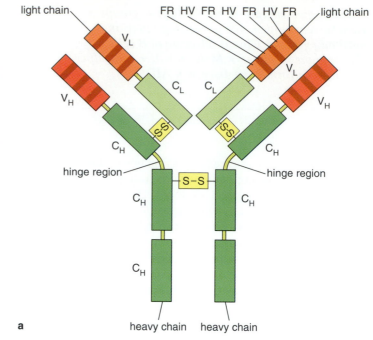

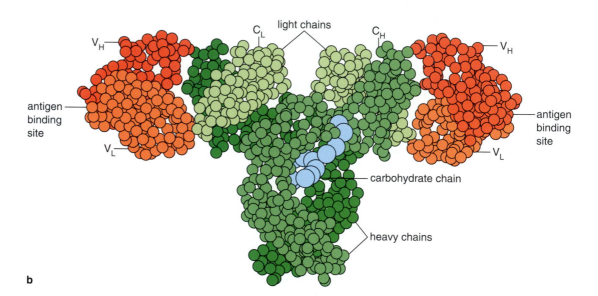

antibody have identical binding sites that recognize and bind a specific antigen. The arms are connected to the tail of the *Y* by flexible "hinges" that allow the arms to flex or rotate to accommodate antigen binding. The tail of the *Y* determines overall features of antibody function such as location in different regions of the body and whether the antibody is secreted or remains attached to cell surfaces. The four polypeptides are held in the *Y*-shaped structure by disulfide (—S—S—) linkages between the light and heavy chains (shown in Fig. 19-1a).

An antigen-binding site takes the form of a depression or cleft about 2 nm long and 3 nm wide at the tip of a *Y*-arm (Fig. 19-2). With these dimensions

an antibody can recognize and bind only a relatively small portion of most antigens. Recognition and binding depend on a close fit between the antibody site and the surface of the antigen and on opportunities for the formation of hydrophobic and ionic attractions. For protein antigens the portion recognized may include from as few as 3 or 4 to a maximum of about 16 amino acid residues. The presence of two binding sites allows antibodies to crosslink one antigen molecule to another, forming the aggregates or precipitates that are recognized for engulfment by macrophages.

Although the variable and constant regions of the light and heavy chains may differ extensively in amino

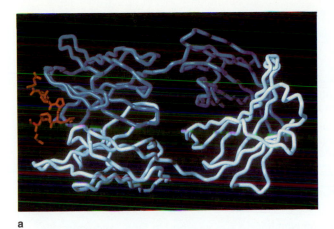

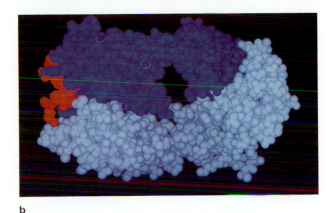

a

b

Figure 19-2   An antigen (in red) in combination with the binding site at the tip of one Y arm. The antigen is a synthetic 19-residue polypeptide homologous to a portion of the myohemo-erythrin molecule. **(a)** Tracing of the backbone of the amino acid chain in one Y arm. **(b)** Space-filling model of the region depicted in (a). (Courtesy of I. A. Wilson, from *Science* 244:712. Copyright 1990 by the AAAS.)

acid sequence, the domains forming these regions are remarkably similar in number of amino acid residues (about 110). The three-dimensional folding pattern of the amino acid chain within any of the domains is also similar. These similarities suggest that the constant and variable domains of both the light and heavy chains evolved by duplication of a single ancestral domain.

Within the variable region of either a light or heavy chain, three short segments, the *hypervariable* (*HV*) *segments*, display most of the variation in amino acid sequence between different antibodies. Relatively little variation occurs in the *framework* (*FR*) *segments*, the other segments of the variable region (see Fig. 19-1a). The folding patterns taken by the four framework segments hold the hypervariable segments in the antigen-binding clefts at the tips of the Y arms.

Antibody structure and function have been most thoroughly studied in mice and humans. Two families of light chains, λ (*lambda*) and κ (*kappa*), and one family of heavy chains occur in these species. Each family is encoded in a separate group of gene subunits: the *lambda light chain* (L_λ) gene group, the *kappa light chain* (L_κ) gene group, and the *heavy chain* (*H*) gene group.

A completed antibody molecule has either a pair of lambda light chains or a pair of kappa light chains, in combination with a pair of heavy chains. Antibodies therefore have the structure $L_{\lambda2}H_2$ or $L_{\kappa2}H_2$. Although antibodies reacting with distinct antigens have different variable regions, all the antibodies of the same type, formed in a single B cell to combine with a single, specific antigen, are identical in their variable regions. However, the antibodies of a single type, with identical variable regions, fall into several *classes* de-

pending on the combination of heavy chain domains used in their assembly.

Five different combinations of heavy chain domains produce the α-, γ-, δ-, ε-, or μ- heavy-chain classes of each antibody type. An antibody assembled with heavy chains of the α-class is termed an *IgA* antibody (where *Ig* = immunoglobulin); an antibody assembled with heavy chains of the γ-class is known as an *IgG* antibody. Similarly, antibodies with the δ-, ε-, or μ-heavy chain classes are known respectively as *IgD*, *IgE*, and *IgM* antibodies.

The five classes of a single antibody type all react with the same antigen. However, each has a different function and is found primarily in different locations in the body. IgA antibodies, for example, are the primary type occurring in body secretions such as tears, milk, and the fluids coating the surfaces of mucous membranes. IgD antibodies primarily anchor to the surfaces of B cells, where they serve as receptors. IgG antibodies are secreted into the blood, where they form the primary class of circulating antibody. (Table 19-1 summarizes the antibody classes and some of their functions.)

In short, the variable regions of the light and heavy chains determine the type of antigen bound by the antibody. The constant region of the heavy chain, which forms the tail of the Y, determines the class, overall functions, and primary location of the antibody molecule in the body.

Antibody molecules also contain carbohydrate groups forming from about 2% to as much as 14% of their structures by weight, depending on the antibody class (see Table 19-1). Antibodies are therefore glyco-proteins rather than ''pure'' protein molecules. The

Table 19-1 Antibody Classes and Their Functions

Class	Heavy Chain Type	Carbohydrate Content (%)	Functions and Locations
IgA	α	7–11	Major antibody in body secretions such as saliva, tears, intestinal secretions, respiratory tract secretions, and milk
IgG	γ	2–3	Major antibody of blood and lymphatic circulation; neutralizes toxins, viruses, and bacteria in body fluids; activates complement system and macrophages when bound to antigens
IgD	δ	10–14	Receptors on lymphocyte surfaces
IgE	ε	12	Cell surfaces; may be important in control of parasitic infections; elevated in persons undergoing allergic reactions
IgM	μ	9–12	Receptors on lymphocyte surfaces and in blood and lymphatic circulation; effective in binding large antigens with multiple binding sites such as viruses and bacteria; activates complement system and macrophages when bound to antigen

functions of the carbohydrate groups, which link covalently to the constant regions of the heavy chains, are unknown.

Antibody Gene Groups

In both mice and humans the three gene groups coding for the lambda and kappa light chains and heavy chains are located on separate chromosomes. In embryonic lymphocytes, before any of the genetic rearrangements associated with antibody production take place, the sequence elements forming the three gene groups are spread over considerable lengths of DNA. The heavy-chain gene group, for example, occupies some 200,000 base pairs of DNA in embryonic lymphocytes. Within the long stretches of DNA making up the gene groups the codes for subparts of the light and heavy chains are split into separate segments. Some segments occur only once; others are repeated in slightly different versions from a few to as many as hundreds or thousands of times.

The Kappa Light-Chain Group In the mouse kappa light-chain gene group of embryonic lymphocytes, two sets of segments contain sequences that contribute to the variable region (Fig. 19-3). One set, called the V_κ elements (V = variable), consists of 300 or more copies of different but related sequences, all coding for the 5' untranslated region and the first 95 amino acids of the variable region in the mRNA. Each copy has its own 5'-flanking control region, including a promoter. The portion of the variable region encoded in the V_κ elements includes the three hypervariable regions and the framework segments separating them. Following the cluster of V_κ elements, after a long spacer that may include as many as 25,000 base pairs,

lies another cluster of elements contributing to the variable region, the J_κ *elements* (J = joining). There are five J_κ elements in the mouse kappa light-chain gene group of the mouse, each 39 nucleotides in length and each separated from the next J_κ element by a spacer of 300 to 350 base pairs. The J_κ elements are all slightly different versions of the sequence coding for the last 13 amino acids of the variable region, which make up the last framework segment. During the rearrangements forming an active kappa light-chain gene, any one of the several hundred V_κ elements, each containing a slightly different version of the initial coding sequence, combines with any one of the J_κ elements to make up the first part of a mature kappa light-chain gene. This part consists of the 5'-flanking control region, the 5' untranslated region of the mRNA, and a single exon containing the entire code for the variable region (see below and Fig. 19-3).

The J_κ elements of the kappa light-chain gene are followed by a long intron of some 2500 to 4000 base pairs. At the 3' end of this intron lies a single exon encoding the C_κ element, the constant region of the kappa light chain. In the rearrangements forming a functional kappa light-chain gene the composite gene segment formed by random selection and joining of one V_κ and one J_κ element is linked to the long intron preceding the C_κ element, forming a complete V_κ-J_κ-C_κ kappa light-chain gene (see Fig. 19-3b).

Human kappa light-chain genes are similarly arranged. At least some of the multiple copies of the V_κ and J_κ sequence elements in mice and humans are probably nonfunctional "pseudoelements," as is the third J_κ element in mice, for example.

The Lambda Light-Chain Group The lambda light-chain gene group is more simply organized in the

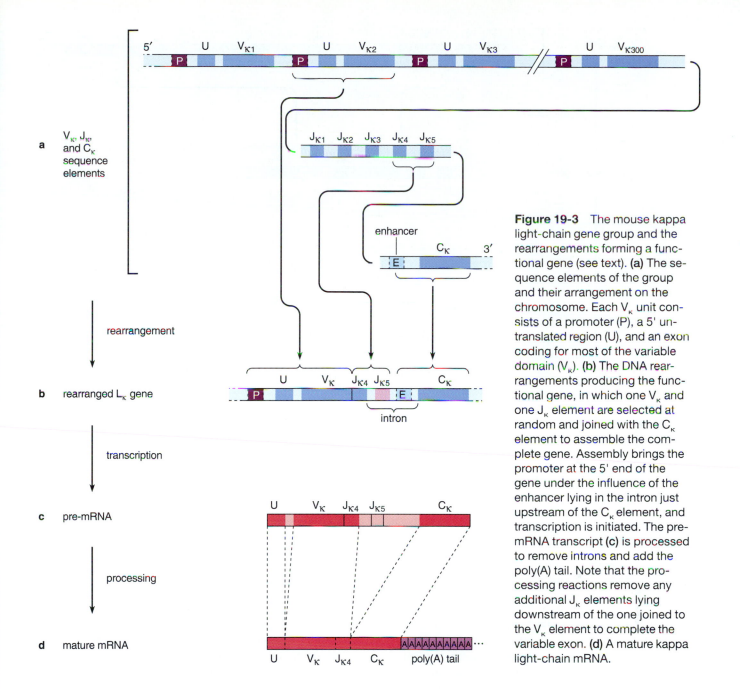

Figure 19-3 The mouse kappa light-chain gene group and the rearrangements forming a functional gene (see text). **(a)** The sequence elements of the group and their arrangement on the chromosome. Each V_κ unit consists of a promoter (P), a 5' untranslated region (U), and an exon coding for most of the variable domain (V_κ). **(b)** The DNA rearrangements producing the functional gene, in which one V_κ and one J_κ element are selected at random and joined with the C_κ element to assemble the complete gene. Assembly brings the promoter at the 5' end of the gene under the influence of the enhancer lying in the intron just upstream of the C_κ element, and transcription is initiated. The pre-mRNA transcript **(c)** is processed to remove introns and add the poly(A) tail. Note that the processing reactions remove any additional J_κ elements lying downstream of the one joined to the V_κ element to complete the variable exon. **(d)** A mature kappa light-chain mRNA.

mouse (Fig. 19-4). There are only two sets of sequence elements, each arranged with one V_λ element separated by a long spacer from two J_λ elements. Each J_λ element is followed by its own C_λ element. An intron of about 1250 base pairs lies between each J_λ and C_λ element. In the rearrangements forming complete lambda light-chain genes the V_λ element combines with either of the J_λ elements and its accompanying C_λ element. Since there are so few V_λ and J_λ elements in lambda light-chain genes, relatively little variability in antibodies is generated by the different possible combinations of these elements. Perhaps for this reason, mouse antibodies contain kappa light chains

much more frequently than lambda light chains. In humans the lambda light-chain gene group is more complex and contains multiple V and J elements in an arrangement resembling the kappa light-chain gene group in the mouse.

The Heavy-Chain Group The heavy-chain gene group of mice and humans contains a greater variety of sequence elements (Fig. 19-5). The heavy-chain gene group begins in mice with a cluster of different V_H elements estimated to number between 100 and 1000. Like the V elements of light-chain genes, each V_H element includes 5'-flanking control sequences, a 5'

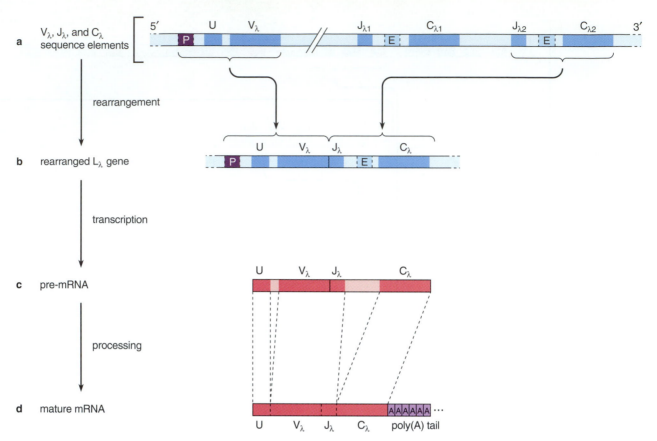

Figure 19-4 The mouse lambda light-chain gene group and the rearrangements forming a functional gene (see text). **(a)** The sequence elements of the group. Only one of the two combinations of V_λ, J_λ, and C_λ elements occurring in mice is shown; the mouse lambda gene family contains another combination of elements similar to the one shown, located on the same chromosome. **(b)** The DNA rearrangements producing a functional gene, in which one V_λ element joins with one of the two J_λ-C_λ elements. The combination brings the promoter at the 5' end of the assembled gene under the influence of the enhancer lying just upstream of the C_λ element, initiating transcription. The pre-mRNA transcript **(c)** is processed to remove introns and add the poly(A) tail; **(d)** shows the mature mRNA.

untranslated region, and coding sequences, in this case including the code for the first 94 amino acids of the heavy-chain variable region. These amino acids form the first two hypervariable segments of the variable region and their intervening framework segments. The V_H elements are followed by a long spacer that separates them from the next cluster of coding elements, a group of about 14 D_H *elements* (D = diversity). Each D_H element contains a different version of the sequence coding for the third hypervariable segment of the heavy-chain variable region. Each D_H element, about 13 nucleotides in length, is separated from its neighbor by a spacer. Another long spacer separates the D_H elements from a cluster of four J_H elements, each coding for the last framework segment of the heavy-chain variable region. In the rearrangements forming a complete heavy-chain gene, one each of the V_H, D_H, and J_H elements combine at random to form a V-D-J exon coding for the variable region of the heavy chain.

The cluster of J_H elements is separated from the C_H elements coding for the constant region by a long intron containing some 8000 base pairs. Eight C_H elements, each containing either three or four exons separated by introns of about 300 base pairs, occur in the mouse C_H cluster. These C_H elements encode the constant regions that specify the IgA, IgG, IgD, IgE, and IgM antibody classes. In the rearrangements forming completed antibody genes the variable exon formed by the combination of one V_H, D_H, and J_H element selected at random is joined to a constant region containing one or more of the five different C_H elements or their variations (see Fig. 19-5b). The heavy-chain gene group of humans is arranged in similar fashion, with at least 200 V_H, more than 20 D_H, 6 J_H, and 8 C_H elements.

Other Sequence Elements The light- and heavy-chain gene groups of both mice and humans have other important sequence elements. One is an enhancer (see

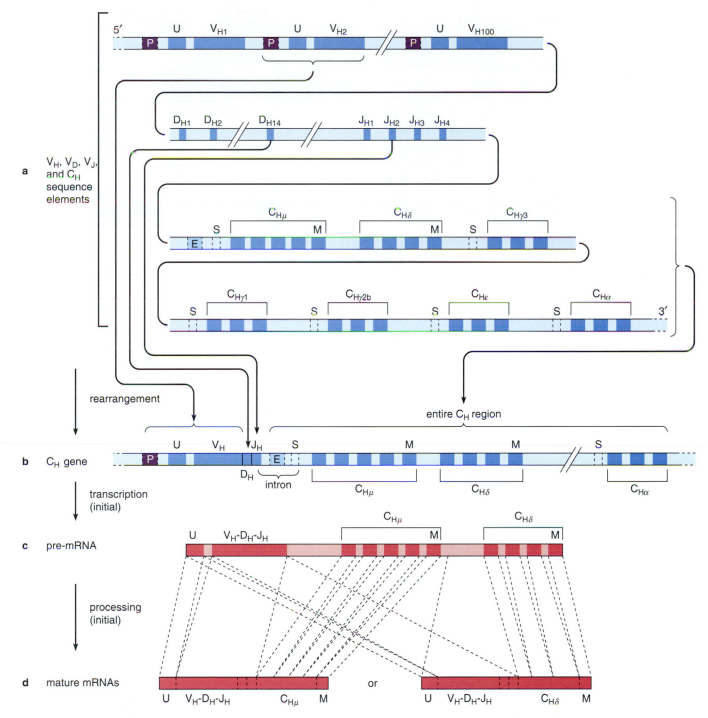

Figure 19-5 The mouse heavy-chain gene group and the rearrangements assembling a functional gene (see text). **(a)** The sequence elements and their arrangement on the chromosome. Four clusters of elements occur in the heavy chain gene group, the V_H, D_H, J_H, and C_H clusters. Each V_H element is supplied with a promoter (P) and a 5' untranslated region (U) as in the other antibody gene groups. The C_H elements consist of either four or five exons that each code for a heavy-chain constant domain; two of the C_H elements, $C_{H\mu}$ and $C_{H\delta}$, have in addition an extra exon (M) that encodes a hydrophobic amino acid "tail." The M exon, when retained in the mature mRNA and translated, forms an anchor that binds finished IgM and IgD antibodies to the lymphocyte plasma membrane. The S sites (S) lying just upstream of each C_H element (except the $C_{H\gamma}$ element) are clusters of repeated sequences that take part in class switching. **(b)** The DNA rearrangements producing a functional gene. Note that the entire collection of C_H elements is included as a unit in the final gene. **(c)** The pre-mRNA transcribed from the gene. Initially the first pre-mRNAs transcribed include both the $C_{H\mu}$ and $C_{H\delta}$ exons. One of these C_H elements is removed from the pre-mRNA during processing, producing a finished mRNA that encodes either a $C_{H\mu}$ or $C_{H\delta}$ heavy chain **(d)**. Initial processing includes the M exon with the mature mRNA, so that the antibodies assembled from the initial $C_{H\mu}$ or $C_{H\delta}$ heavy chains are linked to the plasma membrane. At later stages of lymphocyte development the M exon is eliminated during processing, so that the antibodies assembled from the heavy chains are secreted. Still later in lymphocyte development, one or more C_H elements may be deleted from the gene by further DNA rearrangements, accomplishing class switching.

p. 409), located in the intron separating the J and C elements (see Figs. 19-3 to 19-5). Repeats of a set of sequence elements, the *R sequences* (R = recombination), follow each V element, lie on both sides of the D elements of heavy-chain genes, and precede each J element (Fig. 19-6). These sequences, which occur in inverted forms in their various locations, take part in the programmed interactions rearranging the V, J, and C sequence elements into functional antibody genes (see below). Another important sequence element, the *S site* (S = switch), occurs in the intron just preceding every C_H element in the heavy-chain gene cluster except the $C_{H\delta}$ element (see Fig. 19-5). Each C_H element has its own version of the S site, which contains a 20 to 80 base-pair sequence unit repeated hundreds of times. The S sites take part in *heavy-chain switching*, a process by which a given antibody type changes from one class to another (see below). An *M exon*, if retained in the mRNA encoding the heavy chain, attaches the finished antibody to the surface of B cells. If the M exon is eliminated, the antibody is secreted into the body fluids (see p. 415).

The antibody gene families are members of a large and diverse *immunoglobulin* (*Ig*) superfamily encoding cell surface molecules of various kinds. Many of the receptors and other cell surface molecules taking part in the immune response, such as the T-cell receptors and the MHC surface molecules of lymphocytes and body cells (see below), are members of the Ig superfamily. The family also includes cell surface molecules that have no apparent functional relationship to immunity, such as the cell adhesion molecules N-CAM and I-CAM (see p. 166). The many members of the Ig superfamily share domains that resemble the variable and constant domains of antibodies. For this reason the entire Ig superfamily is thought to have evolved, through many branches, from a single ancestral gene encoding the original Ig-type domain.

The Genetic Rearrangements Assembling Antibody Genes

The rearrangements bringing together the separate sequence elements of antibody genes take place in embryonic B cells. In assembly of the variable regions of antibody molecules, these rearrangements delete or make nonfunctional the unused V, D, and J sequences, so that the final gene contains only one of each of these elements as functional units.

As an example of these events we shall examine the genetic rearrangements producing kappa light-chain and heavy-chain genes in the mouse, in which the process has been best studied. Similar events rearrange lambda chains, although possibilities for variation are much reduced because there are fewer genetic elements.

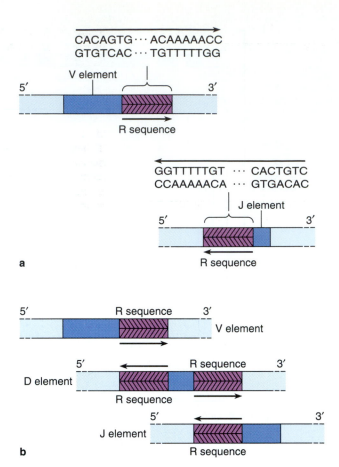

Figure 19-6 The R sequences involved in pairing V, D, and J elements in antibody genes. (a) The arrangement of R sequences in kappa and lambda light-chain genes. (b) The arrangement of R sequences in heavy-chain genes (see text).

Kappa Light-Chain Gene Rearrangements The first rearrangement of a mouse kappa light-chain gene brings a V_κ element next to a J_κ element. The intervening DNA is clipped out, and the V_κ and J_κ elements join into one continuous DNA sequence. Any unused J_κ elements to the right of the J_κ used in gene assembly ($J_{\kappa 5}$ in Fig. 19-3b) become part of the intron separating the variable exon from the constant exons. Although transcribed into the pre-mRNA copy of the gene (see Fig. 19-3c), the extra J's are clipped out with the intron during pre-mRNA processing and do not contribute to the mature mRNA (see Fig. 19-3d) or to the antibody protein it encodes. The extra V_κ elements lying upstream of the one making the V-J joint also remain nonfunctional (see below).

The R sequences flanking each V and J element contribute to the reactions bringing V and J elements together (Fig. 19-7). The R sequences take complementary forms that allow pairing between the elements. Following each V element, the R sequence takes the consensus form CACAGTG...ACAAAAACC. (The

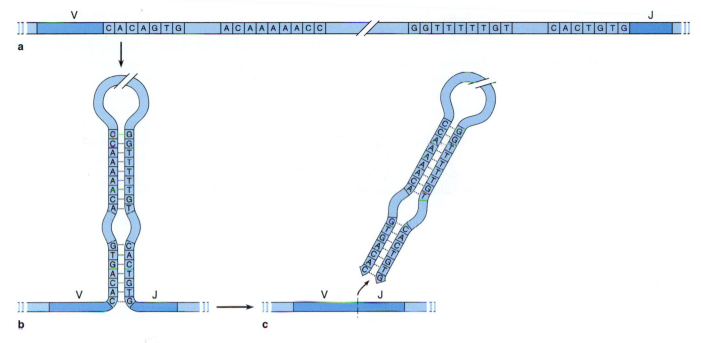

Figure 19-7 R sequence pairing that brings V and J elements together during antibody gene rearrangement. D elements, when present, are joined to V and J elements by a similar mechanism. **(a)** The R sequences occurring in inverted form on the 3' side of V elements and the 5' side of J elements (only one of the two DNA chains is shown). **(b)** Formation of a fold-back double helix between the inverted R sequences. The hairpin loop contains any intervening V and J elements. **(c)** Elimination of the hairpin and joining of the V and J elements into a continuous exon.

dots indicate a spacer that is either 11 or 22 base pairs in length.) Preceding each J element the consensus sequence appears in inverted, complementary form as GGTTTTTGT. . .CACTGTG (see Figs. 19-6a and 19-7a). The inverted arrangement of the sequences allows the 3' end of any V element to pair with the 5' end of any J element. The hairpin formed by the pairing holds the V-J pair in register, with the 3' end of the V abutting the 5' end of the J (Fig. 19-7b). Any V and J elements lying between the two that are brought together are trapped in the loop of the hairpin. The hairpin is clipped from the structure, and the free ends are joined into a continuous DNA molecule (Fig. 19-7c).

The V-J rearrangement places the variable exon next to the long intron originally separating the J_κ cluster from the C_κ element. This completes rearrangement of the kappa light-chain gene, which now consists of the 5'-flanking control region and a series of exons including those coding for the variable and constant regions (see Fig. 19-3b). The rearrangement also places the promoter in the 5' flanking region under the control of the enhancer in the long intron between the variable and constant exons. This activates the promoter, and transcription of the completed antibody gene begins. The promoters of any unused V_κ elements upstream of the 5'-flanking control region

are too far from the enhancer to be activated. These extra promoters therefore remain inactive, and transcription is initiated only at the promoter lying just upstream of the completed antibody gene.

Little is known about the enzymes and factors promoting V-J joining (or V-D-J joining; see below), except for two recently discovered *recombination activator proteins*, *RAG-1* and *RAG-2*, which are required for the rearranging mechanisms to proceed. It is presently unknown whether the RAGs are subunits of a putative *recombinase* enzyme catalyzing the joining reaction or are factors stimulating activity of the recombinase.

Heavy-Chain Gene Rearrangements Heavy-chain gene rearrangements (see Fig. 19-5) are more complex because of the additional D_H and constant region elements of this gene family. The first rearrangement brings one of the several D_H elements next to a J_H element through pairing of R sequences. In heavy-chain genes an R sequence lies on either side of each D element (see Fig. 19-6b). Since the R sequence on the 3' side of a D element is complementary to the R sequences on the 5' side of a J element, any D element can pair with any J element. After clipping to eliminate the DNA of the loop formed by the pairing, the paired D_H and J_H elements are linked into a continuous D_H-J_H sequence that codes for the third hypervariable segment and the framework segment that follows it.

The same mechanism involving R sequences adds a V_H element to complete the variable region of the gene. Each D element is immediately preceded by an R sequence that is complementary to an R sequence at the 3' end of each V element. Pairing between these R sequences brings a V_H element with its promoter, selected at random from the V_H cluster, next to the already joined D_H-J_H sequence. As before, the hairpin formed by the pairing is eliminated, and the V_H, D_H, and J_H elements are sealed into a complete variable region exon. Any J_H elements downstream of the one forming the variable exon become nonfunctional as part of the long intron separating the variable and constant exons (see Fig. 19-5b).

The initial rearrangement producing the heavy-chain gene leaves intact the entire chain of C_H elements, which lie downstream of the variable exon, separated from it by the long exon containing the enhancer (see Fig. 19-5b). Transcription of the heavy-chain gene, initiated through placement of the promoter within range of the enhancer, copies the first two C_H elements, producing a pre-mRNA that has sequences for two complete constant regions at its 3' end (see Fig. 19-5c). Evidently the first signal terminating transcription lies in the intron separating the δ constant element from downstream elements. The first of the heavy-chain elements codes for a constant region of the μ-class and the second for a constant region of the δ-class. During pre-mRNA processing, one or the other of these is eliminated, producing a finished mRNA that encodes a heavy chain of either the μ- or δ-type (see Fig. 19-5d). When combined with kappa and lambda light chains, these heavy chains produce finished antibodies of either the IgM or IgD class. As a consequence, the first antibodies produced by a maturing B cell are of these two classes.

Class Switching   The IgG, IgA, or IgE antibody classes are produced later in the maturation of B cells by class switching (Fig. 19-8). During early stages of B-cell differentiation, after the IgM and IgD classes have been made for a time, the cell switches to the next class, often IgG. The switch occurs by a further rearrangement of the heavy-chain gene involving only constant elements. In this rearrangement the S site following the enhancer pairs with an S site on the 5' side of one of the remaining C_H elements (Fig. 19-8a and b). The C_H elements between the interacting S sites are then deleted, and the free ends are rejoined (Fig. 19-8c). Transcription of the gene now produces a heavy-chain mRNA of the type specified by the C_H element placed at the end of the long intron by the rearrangement. In Figure 19-8c this is shown as the γ-constant region,

producing a heavy chain of the γ-class. When combined with light chains, the γ-heavy chains will produce an IgG antibody. Because each of the remaining C_H elements contains a signal terminating transcription downstream of its last exon, only the first element following the intron is transcribed.

Further class switches may take place by the same mechanism, changing the B cell from IgG to IgA or IgE production. Once a switch is made, a B cell cannot return to production of a previous antibody class because the constant elements coding for the earlier classes have been eliminated from the heavy-chain gene.

Additional Variability from Somatic Mutations   After the genetic rearrangements producing the finished light- and heavy-chain genes are complete, still another mechanism adds to the potential variability of antibodies. This additional variability results from activity of an as yet unexplained mechanism that introduces single base changes in the sequences coding for the three hypervariable regions. These *somatic mutations*, which occur at a rate as much as a million times higher than mutations in other genes, gradually alter the variable regions of the antibodies produced by the B cell. The mechanisms stimulating B cells for antibody production in effect select the variants produced by somatic mutation that most strongly bind an antigen (see below). Thus somatic mutation, along with the selection process, fine-tunes antibodies toward greater selectivity and binding strength for a stimulating antigen.

Estimates of the number of different antibodies that the entire B-cell population of an individual is capable of making range from many millions to 10 billion or more. The total variability explains why almost any foreign substance, natural or artificial, may by chance encounter antibodies capable of recognizing and binding it.

In maturing B cells a functional antibody gene is assembled in only one of the two chromosomes in the pairs carrying the antibody gene groups. The gene groups in the other chromosome of the pairs may undergo some rearrangement but do not produce functional genes that are transcribed into antibody mRNAs. The basis of the mechanism limiting effective gene rearrangement to only one of the two chromosomes in a pair, called *allelic exclusion*, remains unknown. However, there are indications that some of the proteins produced through transcription of the functional genes may inhibit rearrangement of the excluded ones. Some support for this hypothesis is taken from the results of experiments in which a fully rearranged antibody gene is introduced into embryonic B cells. In the cells, rearrangement of antibody genes of the same type (lambda, for example) as the introduced gene is inhibited.

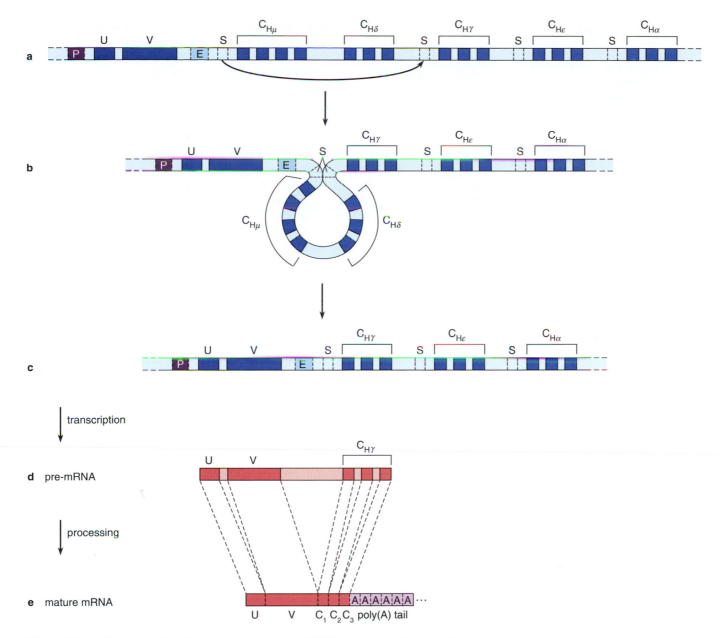

Figure 19-8 Class switching in heavy-chain genes. **(a)** The rearranged gene before class switching; only one $C_{H\gamma}$ element is shown to simplify the diagram. **(b)** Interaction between the S site following the enhancer and the one preceding the $C_{H\gamma}$ element brings these sites into close proximity. Elimination of the loop created by S-site interaction produces the switched gene **(c)**, in which the $C_{H\mu}$ and $C_{H\delta}$ elements have been eliminated. Transcription now reads from the promoter through the $C_{H\gamma}$ element, producing a pre-mRNA containing the exons of a γ-class heavy chain **(d)**. Processing yields a mature mRNA coding for a heavy chain of this class **(e)**. Further rearrangements between the S site following the enhancer and another S site farther downstream of the $C_{H\gamma}$ element will produce additional class switching.

Stimulation of B Cells Producing an Antibody Type: Clonal Selection

B cells continue to mature through the lifetime of the individual, yielding individual cells capable of synthesizing a specific antibody type. In early B cells the initial IgM or IgD antibodies, rather than being re-leased to the medium, are inserted as recognition molecules at the cell surface.

Unless stimulated by contact with an antigen that can combine with their exposed IgM or IgD antibodies, the early B cells remain inactive in cell division. If the B cell encounters an antigen capable of combining

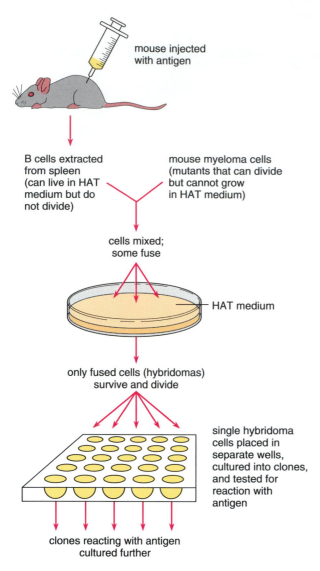

mouse injected
with antigen

B cells extracted
from spleen
(can live in HAT
medium but do
not divide)

mouse myeloma cells
(mutants that can divide
but cannot grow
in HAT medium)

cells mixed;
some fuse

HAT medium

only fused cells (hybridomas)
survive and divide

single hybridoma
cells placed in
separate wells,
cultured into clones,
and tested for
reaction with
antigen

clones reacting with antigen
cultured further

Figure 19-9 A technique used to produce mono-clonal antibodies. *HAT* medium (HAT = *h*ypoxanthine + *a*minopterin + *t*hymidine) contains aminopterin, which blocks direct synthesis of purine and pyrimidines, but contains thymidine and hypoxanthine. Normal cells can use thymidine and hypoxanthine to make purines and pyrimidines. The B cells removed from the mouse are normal cells that can live in HAT medium but do not survive in cultures for extended periods because they do not divide in the absence of T-cell stimulation and other interactions of the immune system. The myeloma cells are mutants that can divide and grow indefinitely in cultures but lack the enzymes necessary to make purines and pyrimidines from thymidine and hypoxanthine in HAT medium. The fused cells—hybridomas—contain genetic information from both the B and myeloma cells and can grow, divide, and survive indefinitely in cultures.

with the variable region of the antibody exposed at its surface (and interacts with a T cell; see below), cell division is triggered, leading to production of a rapidly dividing clone of B cells. Because they are descendants of the original stimulated cell, all the cells of the clone make antibodies that bind the same antigen.

Production of a clone through stimulation by an antigen is termed *clonal selection.* The process was proposed by Macfarlane Burnet some 30 years ago, long before the cellular and molecular mechanisms underlying the process were understood. Burnet, who based his hypothesis partly on ideas advanced by D. W. Talmage and N. K. Jerne, proposed clonal selection as a form of natural selection operating in miniature: Antigens select the cells recognizing them to reproduce and become dominant in the B-cell population. (Burnet received the Nobel Prize in 1960; Jerne was awarded the Prize in 1984.) The Experimental Process essay by G. Ada on p. 558 describes his experiments supporting clonal selection, which were critical to acceptance of the theory.

As division of a B-cell clone proceeds, somatic mutations, which increase greatly in frequency after the cell line is stimulated by antigen, lead to small differences in the variable regions of the antibodies. During further rounds of interaction with T cells the B cells most likely to be stimulated are those with variations of the antibody binding most strongly to the antigen. As a consequence, the antibody type produced by the clone, which at first may bind the antigen relatively weakly, becomes progressively stronger and more specific in binding capacity.

Class switching within the clone as individual B cells mature confronts the antigen with antibodies of the five different classes. After stimulation, most of the IgM and IgD antibodies, along with the remaining classes, are secreted from the maturing B cells rather than inserted into the cell surface. The change from membrane-bound to secreted forms of IgM and IgD antibodies results from a variation in pre-mRNA processing. The variation eliminates the final exon of the transcript (the M exon in Fig. 19-5d), which codes for the segment anchoring the IgM or IgD antibody molecules to the plasma membrane. Secretion of the various antibody classes leads to a full-blown immune response against the antigen, which, if successful, eliminates the antigen from the body.

Elimination of the antigen interrupts the immune response, and division of the clone is reduced to the low levels characteristic of unstimulated B cells. Most or all of the fully mature B cells of the clone gradually die and are eliminated from the bloodstream and other body fluids. (Life expectancy of mature B cells is a matter of days.) However, immature B cells of the same clone remain in the bloodstream in elevated numbers. These quiescent cells, with IgM and IgD antibodies exposed as cell surface receptors, remain ready as *memory B cells* that can provide a quick response if the antigen is again encountered by the individual.

The ability of a clone of B cells to produce antibodies against a single antigen is widely used as a means to produce *monoclonal antibodies* for scientific research and medicine (Fig. 19-9). In the technique, first developed by G. Kohler and C. Milstein, an animal such as a mouse or rat is injected with an antigen of interest. Antibody-producing B cells are then extracted from the spleen of the animal; because the B cells do not divide readily by themselves, they are placed in a culture vessel with *myeloma* cells, a cancerous lymphocyte type. The B cells and myeloma cells are cultured under conditions that cause them to fuse into large, composite cells called *hybridomas*. The hybridomas combine the characteristics of the two cell types: the antibody production of the B cells and the rapid, continuous division of the myeloma cells. The B cells are said to be "immortalized" by fusion into hybridomas with the myeloma cells. Nutritional conditions set up in the culture prevent myeloma cells from growing unless fused with B cells (see Fig. 19-9).

Single hybridoma cells are removed from the fusion culture and used to start clones. The clones can be grown to immense numbers, producing large quantities of the desired antibody. Because a clone is derived from a single hybridoma cell, the antibody produced is a single, highly specific type.

The monoclonal antibody technique circumvents a problem inherent in methods for antibody production in which the antigen is simply injected into an animal such as a horse or rabbit. The injected animal typically produces a wide spectrum of antibodies that react to different parts of the antigen with varying degrees of specificity. Frequently some of the antibodies also cross react with other molecules sharing structural characteristics with the antigen. The monoclonal technique produces a single antibody type that can be selected to be highly specific for the antigen of interest.

Monoclonal antibodies have some disadvantages. Cell fusion is a relatively rare event, so that many of the B cells are left in an unfused, nondividing state. As a result, much of the potential for specific antibody production remains untapped. Moreover, at present, monoclonal antibodies can be produced readily only with B cells derived from the mouse; human B cells, for unknown reasons, fuse with myeloma cells much less frequently than mouse B cells. Using mouse-derived monoclonals as treatments for human diseases frequently induces an immune response against the mouse antibodies in the patient, greatly reducing the effectiveness and persistence of the monoclonal antibodies.

INTERACTION OF B AND T CELLS IN THE IMMUNE RESPONSE

The T cells interacting with B cells in the immune response occur in two major types, *helper T cells* and *killer T cells*. Helper T cells stimulate the activity of B cells in division and antibody secretion. Without helper T cells, little or no B-cell response occurs. Killer T cells recognize and kill body cells containing infective viruses. Killer T cells also kill some kinds of tumor cells and foreign cells originating from outside the body.

T Cells and the MHC Genes

B and T cells recognize each other through cell surface molecules encoded in a gene group known as the *major histocompatibility complex* (MHC; see also p. 151). Research by M. L. Gefter, H. M. Grey, J. Smith, and their coworkers and others has shown that as part of this recognition, MHC molecules bind and display fragments of an antigen.

MHC molecules also play a role in recognition of foreign cells. Cells from different individuals of the same or different species have different MHC molecules on their surfaces, enabling lymphocytes to recognize a cell introduced from a different individual as foreign and to destroy it. As such, MHC molecules are one of the primary barriers to organ and tissue transplantation among humans. For this reason, MHC molecules are sometimes called *transplantation antigens*.

The MHC gene family, which occupies more than 25,000 base pairs of DNA, comprises major gene groups that are central to the immune response. Two of these groups encode the *class I* and *class II MHC molecules*, which display antigens at cell surfaces. Class I MHC molecules occur on all body cells of an individual except those of the immune system. Class II MHC molecules occur on B and T cells and other white blood cells. Class I and II MHC show enough similarities to antibodies to indicate that they are members of the immunoglobulin superfamily.

Both class I and class II MHC consist of two polypeptides (Fig. 19-10a). Only the larger of two polypeptides making up a class I MHC is encoded in the MHC gene group (the green segment in Fig. 19-10b). An analysis by D. C. Wiley and his coworkers showed that this polypeptide has three domains, which extend outward from the surface of the plasma membrane. The domain nearest the plasma membrane (domain III in Fig. 19-10b) includes a transmembrane segment anchoring the MHC in the plasma membrane. This domain demonstrates clear sequence similarities to the constant domains of antibody molecules. The other two domains (domains I and II in Fig. 19-10b)

The Experimental Process

Instruction or Selection: The Role of Antigen in Antibody Production

Gordon Ada

GORDON ADA worked first with Macfarlane Burnet and then with Gus Nossal at the Walter and Eliza Hall Institute in Melbourne, Australia (1948–68), and received his D.Sc. at the University of Sydney in 1959. He was Professor of Microbiology at the John Curtin School of Medical Research in Canberra until his retirement in 1987. He left Australia in 1988 to work at the Johns Hopkins School of Public Health in Baltimore, Maryland, becoming Director of their Center for AIDS Research. He has recently returned to Canberra.

Antibodies (antitoxins) were discovered, isolated, and shown to be proteins in the late 19th century. They occur in response to an infection, and extensive studies showed them to have a remarkable specificity in their interaction with antigens, approaching that of enzyme-substrate reactions. Until the 1960s, there was no understanding of the mechanisms of protein synthesis within cells.

The first attempt to explain antibody production by cells was by a remarkable German chemist, Paul Ehrlich. He proposed in 1901 that white blood cells possessed receptors at their surface. These receptors had "sidechains" which bound foreign substances (antigens) derived from infectious agents. This binding event stimulated the cell, which responded by producing these receptors in excess, and many receptors—the antibodies—were shed into the blood. This concept implied that these receptors occurred *naturally* in a sufficient range of specificities to bind components from many infectious agents.

It was soon found, however, that proteins from many other sources, such as milk, could induce formation of antibodies and bind to them. Karl Landsteiner's finding in the 1930s that antibodies could be formed to and would bind with exquisite specificity to *completely synthetic* compounds seemed to discredit Ehrlich's concept. How could the body recognize substances that had not existed previously? This led to the notion that antigen instructed a cell to make antibody of a complementary specificity. In one popular form of the Instructive Theory of antibody formation, the Template Theory, it was proposed that antigen entered the lymphocyte and acted as a template inside the cell; flexible precursor antibody molecules moulded themselves on this antigen template to assume a conformation with the correct antigen-binding specificity.

In the 1940s, Macfarlane Burnet pointed out that the Template Theory did not explain several facts about antibody formation, such as a second (memory) response to an antigen, in which antibody production is greater than in the primary response; and immunological tolerance, in which antigen administration failed to induce an antibody response. Building upon ideas recently reformu-

lated first by Niels Jerne and particularly by David Talmage which were reminiscent of Ehrlich's concept that antibodies existed normally, Burnet in 1957 proposed the Clonal Selection Theory of antibody formation. The crux of this theory was that in the animal host there existed populations of lymphocytes, each possessing immunoglobulin receptors of a single specificity. Antigen *selected* those cells with receptors that recognized polypeptide segments called epitopes on that antigen, causing the selected cells to differentiate and proliferate to become clones of antibody-secreting cells (ASCs). Allowing for increases in specificity due to somatic mutation, the secreted antibody from the progeny of a single precursor cell would have a specificity similar to the precursor's Ig receptors. Tolerance occurred due to the deletion of "forbidden clones," i.e., cells reacting against self components.

The Template and Clonal Selection theories were so different as to be irreconcilable. In the first, the cell produced a molecule whose specificity could be modified by a foreign agent, the antigen; in the second, antigen selected cells that were already genetically programmed to produce a molecule of a particular specificity. Which theory was correct? This essay described a series of experiments that established the essential correctness of the Clonal Selection Theory.

A colleague, Gus Nossal, working with Joshua Lederberg, argued that individual lymphocytes exposed to two different antigens would make antibodies either of both specificities (the Template Theory) or of only one specificity (Clonal Selection). They immunized rats with two chemically related but immunologically distinct proteins, the flagellar proteins of two different *Salmonella* bacteria. Individual ASCs isolated some time later from each rat were exposed in microdroplets first to one and then to the other antigen in the form of motile bacteria; when viewed in a microscope, each of 62 separate ASCs examined were found to immobilize only those *Salmonella* bacteria with flagella of a single specificity. This result certainly favored the Clonal Selection Theory. There was the possibility, however, that a precursor cell might have the capability of making antibody of many specificities but that the first contact with an antigen channeled the cell's response into a more restricted if not a single specificity.

Nossal and I then decided to try a different tack. The Template Theory predicted that an individual ASC must contain many thousands of molecules of intact antigen in order for the latter to act as a template in a cell for the entire time antibody was being synthesized and secreted. In fact, in earlier experiments when large amounts (milligrams) of a particulate antigen (visible by electron microscopy) had been administered, one group reported that ASCs seen in thin sections seemed to contain particles of antigen. We changed the methodology in two ways. As the immunogen, we used flagella because they are highly immunogenic, nanogram quantities inducing a strong antibody response; and we labeled it with the recently available carrier-free preparations of the isotope iodide-125 (^{125}I), in order that tiny amounts of the antigen could be traced in the body. Individual specific ASCs isolated from the draining lymph nodes of rats immunized with the labeled antigen were fixed on glass slides, dipped in photographic emulsion, exposed in the dark for 60 days, and then developed. We had calculated that a single grain (above background levels) over an ASC would correspond to about 10 molecules of antigen. In fact, the great majority of more than 200 cells examined were found to have no associated grains. A group in London led by John Humphrey later obtained similar results. Again, this result argued against the template theory. (As we now know, 25 years later, antigen is in fact taken into lymphocytes but it is degraded and does not persist to become a template.)

However, we now needed an approach that allowed direct examination of the interaction of antigen with precursor immunocompetent lymphocytes before they became ASCs. Two Israelis, Dov Sulitzeanu and David Naor, pointed the way.

Sulitzeanu and Naor showed that if a suspension of cells including lymphocytes from lymphoid tissues of unimmunized animals was exposed at 0°C to a labeled antigen (bovine serum albumin) for a short time, the unbound antigen washed away, and the cells autoradiographed, distinct labeling patterns were seen. Rather than all or most lymphocyte-like cells being labeled (a Template Theory prediction), only a very small minority (<1%) were labeled, and of these, some cells were labeled more than others. Pauline Byrt and I repeated and extended these findings with our antigens. Cells from the mouse thoracic duct, a source of almost pure lymphocytes, showed the clearest labeling pattern of all (Ref. 2, Fig. 9.2). In addition, we were able to show that if the cells were exposed to anti-Ig sera before exposure to labeled antigen, this prevented the subsequent binding of the antigen. We were seeking evidence that the antigen was binding to the cells via a specific Ig receptor, but it might be that the labeled cells were coated nonspecifically with cytophilic antibody. What was needed was a *functional* test to show this binding pattern was more than a coincidence.

Lymphocytes were known to be very sensitive to ionizing radiation, and I recalled that at a recent Cold Spring Harbor Symposium (1967), Richard Dutton had described how pulsing with tritiated thymidine had selectively killed dividing lymphocytes. It was not feasible for us to use tritiated antigen as the level of radioactivity of the labeled antigen would be far too low. However, we knew that the path length of emitted β rays from ^{125}I was about 10 μm, about the mean diameter of resting lymphocytes. Would the radiation from ^{125}I-labeled flagella adsorbed to a lymphocyte surface selectively damage that cell so that it would not proliferate and mature to become an ASC when later exposed to the same (but unlabeled) antigen? The experiment was worth a try!

Splenocytes from normal, unimmunized mice were exposed to ^{125}I-labeled flagella of one of two different specificities. After standing for 1 hour at 4°C, unadsorbed antigen was washed away, and the cells immediately adoptively transferred to mice with the same gene complement whose own immune lymphoid system had been destroyed by previous x-irradiation. Both groups of mice were then immunized with unlabeled flagella of both specificities, and the quantities of antibody reacting with both antigens measured some days later. It was hoped that mice which had received cells exposed to hot antigen would respond poorly if at all after immunization with antigen of the same immunological specificity, but give a normal response to flagella of the other specificity. Such a result would have demonstrated that cells to which the labeled antigen bound were necessary for the production of antibody of that specificity.

In the first experiment, however, no such differences were found. What was the reason for this result? Were the cells that had bound the antigen unimportant in an immune response? Or was the design of our experiment inadequate in some way? One possibility was that the short exposure of the labeled cell to radiation was insufficient to cause enough cellular damage.

To test this possibility, the experiment was repeated but after exposure to the antigen for 1 hr, followed by removal of unbound antigen. The cells were let stand in the test tube at 4°C for 16–20 hr before adoptive transfer and challenge. This time the expected results were obtained; exposure to labeled antigen of one specificity significantly reduced the ability of the lymphocyte population to respond to that antigen, but not to the immunologically distinct antigen.

This result was difficult to reconcile with any theory other than one requiring individual cells to have a very restricted potential for antigenic stimulation, such as the Clonal Selection Theory. The results were completely contrary to the prediction of the Template Theory. Together with amino acid sequence data on different immunoglobulins, which also were becoming available at that time, these findings sounded the death knell for any form of an Instruction Theory of Antibody Formation.

References

Burnet, F. M. A modification of Jerne's theory of antibody production using the concept of clonal selection. *Aust. J. Sci.* 20: 67–69 (1957).

Nossal, G. J. V., and Ada, G. L. In *Antigens, Lymphoid Cells, and the Immune Response.* New York: Academic Press (1971).

Ada, G. L., and Nossal, G. J. V. The clonal selection theory. *Scientific American* 257:62–69 (1987).

Silverstein, A. M. The history of immunology. In *Fundamental Immunology.* Ed. Paul, W. E. New York: Raven Press, 21–38 (1989).

Figure 19-10 Class I and II MHC.
(a) Overall domain structure of class I and II MHC. **(b)** A class I MHC showing alpha helices as coils and beta strands as arrows. The polypeptide subunit encoded in the MHC gene group is shown in green; the β-microglobulin polypeptide is yellow. The red arrow marks the site binding and displaying a processed antigen. (Redrawn from an original courtesy of D. C. Wiley. Reprinted by permission from *Nature* 319:506; copyright 1987 Macmillan Magazines Ltd.)

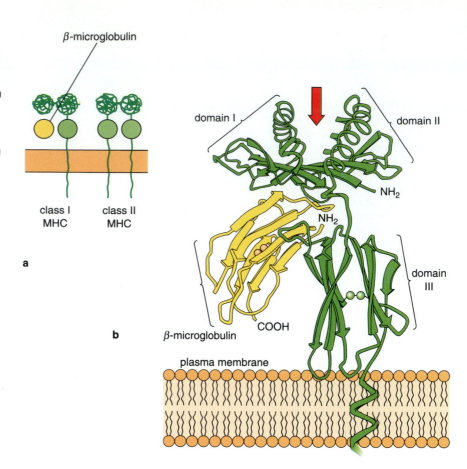

form a cleft that binds an antigen fragment (red arrow in the figure). The small polypeptide subunit of a class I MHC is a relatively small protein, β-*microglobulin*. Although encoded in a separate gene on a different chromosome, the β-microglobulin polypeptide contains a domain with clear sequence similarities to the constant regions of antibody genes, indicating that its gene is yet another member of the Ig superfamily.

Both polypeptides of a class II MHC are encoded in the MHC gene group. Each polypeptide has two domains (see Fig. 19-10a). The domain closest to the plasma membrane resembles domain III of a class I MHC polypeptide; the outer domain is similar to either domain I or II of a class I MHC polypeptide. The outer domains of the two class II MHC polypeptides combine to form a cleft displaying an antigen fragment.

Both class I and class II MHC have variable and constant domains. However, unlike antibody genes, each version of an MHC is encoded in its own gene. Thus differences between the MHC molecules of different individuals depend on different genes and alleles rather than genetic rearrangements. The number of genes and alleles is so extensive that no two individuals except identical twins are likely to have the same class I and II MHC.

The antigen fragments displayed by MHC molecules arise from cellular *processing* reactions. The peptide fragments arise from different sources for class I and class II MHC. For class II MHC, the best under-

stood of the two types, the fragments originate from foreign proteins taken into leukocytes by endocytosis (Fig. 19-11). Once inside the cytoplasm in an endocytic vesicle (see p. 591), the antigens are split by protein-digesting enzymes into fragments of about 8 to 20 amino acid residues (Fig. 19-11a and b). Subsequently, newly synthesized class II MHC are introduced into the vesicle membranes by fusion with vesicles from the ER and Golgi complex (Fig. 19-11c). Of the several to many antigen fragments produced by the processing reactions, some fit the binding sites of one or more of the class II MHC (Fig. 19-11d). The selected antigen fragments, bound to the MHC, are brought to the cell surface for display by fusion of the vesicle with the plasma membrane (Fig. 19-11e and f).

The peptide fragments displayed by class I MHC follow a similar pathway except for their intracellular origin. The proteins from which these fragments are derived originate from infecting viruses or other parasites or from the cell itself. The proteins are digested in the cytoplasm and then transported through ER membranes. In the ER the fragments join with Class I MHC undergoing assembly. From this point the fragments are delivered to the cell surface by the usual ER→Golgi complex→secretory vesicle route. For either class I or class II MHC the display of antigen fragments at the cell surface is termed *antigen presentation*. The two MHC pathways are not entirely exclusive; for example, class II MHC have been demonstrated

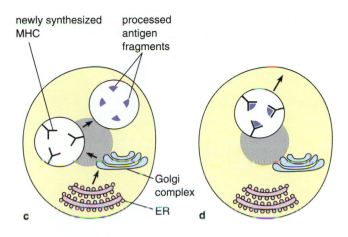

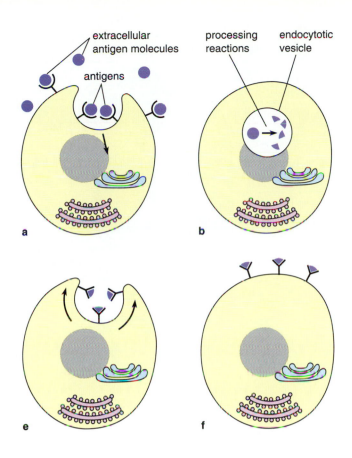

Figure 19-11 Endocytosis, processing, and display of antigen fragments at the cell surface by class II MHC (see text). N, nucleus.

to display polypeptide fragments of cellular origin in some cases.

MHC Binding by T Cells

T cells possess surface receptors that, like the antibodies synthesized by B cells, are tailored to recognize and bind specific antigens. Unlike an antibody, however, a T-cell receptor can bind an antigen only when it is displayed at a cell surface as a fragment linked to a class I or II MHC. The binding site of an MHC folds around the antigen fragment, forming a composite region that includes segments of both the antigen and MHC. This region is recognized and bound by the T-cell receptor (see Fig. 19-12b). Killer T cells can bind an antigen fragment only when it is displayed by a class I MHC, the type occurring on body cells. Helper T cells can bind an antigen fragment only when it is presented by a class II MHC, the type occurring on B cells and other leucocytes of the immune system.

The differences in MHC recognition by helper and killer T cells depend primarily on another surface receptor type, the *CD4/CD8 receptors* (see Fig. 19-12b). These receptors, which are also members of the immunoglobulin superfamily, recognize and bind constant regions on MHC molecules, CD4 binding to the constant region of class II MHC and CD8 to the constant region of class I MHC. Binding by CD4 or CD8 stabilizes the association between the T-cell receptor and

an MHC molecule displaying an antigen fragment. Either of the two receptors may be present on a T cell. If CD4 is present, the T cell interacts stably only with cells bearing a class II MHC and acts as a helper T. If CD8 is present, the T cell interacts stably only with cells bearing a class I MHC and acts as a killer T.

Although each T-cell clone can recognize only a single antigen, the entire population of T cells in an individual, like the antibodies of B cells, is collectively able to recognize and bind essentially any antigen, whether natural or artificial.

A T-cell receptor consists of two dissimilar polypeptides, each with a variable and a constant region (Fig. 19-12a). The variable region of both polypeptides extends away from the surface of T cells; the constant domain includes a transmembrane segment that anchors the receptor to the plasma membrane. The variable regions of the two polypeptides combine to form a binding site that recognizes an antigen fragment presented by an MHC (Fig. 19-12b).

A T-cell receptor is encoded in genes related to but distinct from antibody genes. The gene groups encoding T-cell receptor polypeptides, like the antibody heavy-chain genes, consist of clusters of V, D, J, and C elements separated by extensive spacers. One each of the V, D, J, and C elements is selected at random and linked together to make a continuous gene encoding a T-cell receptor polypeptide. The random assembly of V, D, and J elements produces an essentially

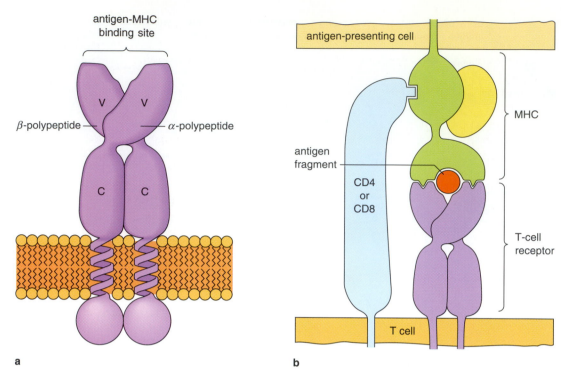

Figure 19-12 The T-cell receptor. **(a)** The receptor consists of two polypeptides, each containing a constant (C) and variable (V) domain. **(b)** A T-cell receptor binding to an MHC in combination with a polypeptide fragment. The combination is stabilized by the CD4/CD8 receptors also carried on T cells, which recognize and bind the constant regions of MHC. The diagram shows a class I MHC, which would be recognized by CD8.

endless variety of T-cell receptors, one for each T cell and its descendants.

Interaction of B Cells and Helper T Cells in the Immune Response

The combination recognized by helper T-cell receptors—a processed antigen fragment bound to a class II MHC—appears primarily on B cells. In B cells the antigen is first encountered and bound by IgM or IgA antibodies extending from the B-cell surface. The antibodies with the bound antigen are taken into the cell by endocytosis, where the antigen is released from the antibody, processed, and attached to a class II MHC. The MHC with the processed antigen is subsequently introduced into the plasma membrane for presentation (as in Fig. 19-11e and f).

When a helper T cell binds an antigen displayed by a class II MHC on a B cell, the helper T secretes several growth factors known as *interleukins* (*IL*). The interleukins stimulate the B cell to enter rapid division, forming a clone of cells making antibodies against the same antigen. The interleukins also stimulate the original B cell and its descendants in the clone to switch to secreted forms of the antibody.

Antibodies are secreted into body cavities and the body exterior by a mechanism involving a *poly Ig re-*

ceptor on the surfaces of epithelial cells. The receptor, another member of the Ig superfamily, can recognize and bind the constant regions of both IgA and IgM antibodies. The receptor binds antibodies circulating in the bloodstream; after entering the epithelial cells by endocytosis, vesicles containing the receptor-antibody complexes move through the cytoplasm to the opposite side of the cell. This side forms the surfaces of mucous membranes or lines a body cavity communicating with the cell exterior, such as the digestive tract, vagina, or the ducts of salivary, mammary, and tear glands. On this side of the epithelial cell, the vesicle fuses with the plasma membrane, placing the poly Ig receptors and their attached antibodies on the cell surface. At this point the poly Ig receptor is cleaved at its base, releasing the antibodies with a segment of the receptor still attached.

Secreted antibodies of the IgA class bind antigens in the body cavities or in secretions such as saliva, tears, and milk. On mucous membranes, such as those of the vagina in females, the released IgA antibodies form a sort of "antiseptic paint" that helps eliminate antigens and infective agents. A similar mechanism, involving a receptor carried on placental cells, transports IgG antibodies from the maternal circulation into the bloodstream of the developing fetus. (A blood barrier formed by cells of the placenta prevents the

B cells from the mother from directly mixing with the bloodstream of the fetus.)

The interactions between B and T cells takes place in organs of the lymphatic system, including lymph nodes, the spleen, and tonsils, where T cells are present in high concentrations. The lymphatic tissue filters and concentrates the antigens, where they are encountered by B cells migrating into the organs. Cells of the lymphatic organs also contribute to B- and T-cell activation.

Several drugs used to suppress activity of the immune system have helper T cells as their targets. *Cyclosporin A*, for example, is used routinely after organ transplants to reduce the chance that the immune system will reject the transplant. The drug, a small 11-amino acid polypeptide, evidently binds and blocks one of a series of molecules taking part in an internal response cascade triggered by the T-cell receptor. The blockage inhibits transcription of several key genes encoding interleukins and other proteins required for T-cell activation. A major problem with cyclosporin A and other immunosuppressive drugs is that, while they lower the chance of transplant rejection, they leave the treated individual significantly more susceptible to bacterial, fungal, and viral infections. Helper T cells are also a primary target of the virus causing AIDS (see Supplement 19-1).

Killer T Cells in the Immune Response

The combination bound by a killer T cell—a processed antigen presented by a class I MHC—usually appears on the surfaces of body cells infected by a microorganism or virus. Binding the class I MHC with its displayed antigen activates the killer T cell to destroy the infected cell by releasing a protein called *perforin*. Perforins have hydrophobic segments that penetrate into the plasma membrane of the infected cell. As they penetrate, the perforins assemble into tubelike structures that open wide pores in the plasma membrane (Fig. 19-13a). The pores permit free diffusion of ions and small molecules, leading to lysis and death of the infected cell (Fig. 19-13b). How killer T cells avoid killing themselves at the same time remains a mystery.

Killer T cells respond similarly when they encounter a class I MHC on a cell from a different individual. Evidently the foreign class I MHC is recognized as a class I MHC of the same individual

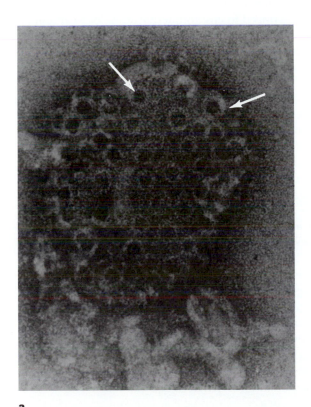

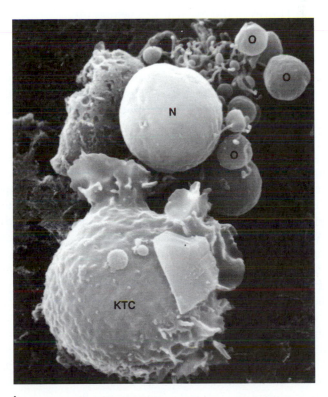

a b

Figure 19-13 Action of killer T cells in killing body cells displaying foreign antigens.
(a) Tubular pores (arrows) assembled by polymerization of perforin molecules released by a killer T cell. (Courtesy of E. R. Podack, from *Proc. Nat. Acad. Sci.* 82:8629 [1985].) **(b)** Lysis of a target cell by a killer cell. The lysed cell is toward the top of the figure. N, nucleus of the lysed cell; O, cytoplasmic organelles; *KTC*, the killer T cell. (Courtesy of G. Kaplan, from S. Joag *et al.*, *J. Cellular Biochem.* 39:239 [1989].)

would be if it carried a foreign antigen fragment. Tumor cells displaying altered surface molecules may also be recognized as foreign and lysed by killer Ts.

Malfunctions of the Immune Response

Although the complex mechanisms of the immune response defend vertebrates remarkably well against invading cells and molecules, the system is not foolproof. Some malfunctions of the immune system cause reactions against proteins or cells of the body itself, producing autoimmune diseases. In addition, some viruses and other pathogens have evolved means to escape surveillance and destruction by the immune system. A number of these pathogens, including the virus causing AIDS, even use elements of the immune response to promote infection.

Autoimmune Disease   One of the most remarkable features of the normal immune system is that foreign molecules are recognized and eliminated, but not molecules forming parts of the body. A major part of this *self-tolerance* depends on elimination by the thymus of T cells capable of recognizing body proteins as antigens. The absence of these T cells inhibits activation of B cells capable of making antibodies against body proteins. However, the mechanisms setting up self-tolerance are slightly "leaky," and almost all individuals probably produce antibodies against their own cells or proteins at some time. In most cases the effects of such anti-self antibodies are not serious enough to produce recognizable disease. However, in some individuals, about 5% to 10% of the population, the attack on self becomes a serious problem.

Several autoimmune diseases occur relatively frequently. *Insulin-dependent juvenile onset diabetes* results from an autoimmune attack on a specific protein on the surfaces of the β-cells producing insulin in the pancreas. The attack gradually eliminates the β-cells over a period of a few years until the individual is incapable of producing insulin. In *acquired hemolytic anemia* an autoimmune response destroys circulating red blood cells. *Lupus erythematosus* is caused by production of a wide variety of autoantibodies against blood cells, blood clotting factors, and internal cell molecules and structures such as mitochondria and the polypeptides of snRNAs (see p. 448). The disease is characterized by anemia, uncontrolled bleeding, and circulatory and renal difficulties produced by the deposition of antibody-antigen complexes in capillaries and kidney glomeruli. *Rheumatoid arthritis* results from an autoimmune attack on connective tissue, particularly in the joints, causing pain and inflammation. *Myasthenia gravis* is associated with production of a variety of autoantibodies, frequently including one against the receptor for acetylcholine, one of the primary elements in transmission of nerve impulses across synapses (see p. 141). The anti-acetylcholine receptor antibodies may account for the extreme muscular weakness of patients with this disease. In *multiple sclerosis*, antibodies against *myelin basic protein*, a component of the myelin sheaths (see p. 140) insulating the surfaces of nerve cells, seriously disrupt the nervous system.

Almost all autoimmune diseases are related to the genetic alleles determining the structure of MHC molecules, particularly class II MHC. Individuals possessing certain alleles are much more susceptible to development of autoimmunity. For example, J. A. Todd and H. O. McDevitt discovered that susceptibility of individuals to insulin-dependent juvenile onset diabetes is affected by differences in the amino acid at position 57 in the amino acid sequence of the β polypeptide of class II MHC; the side group of this amino acid extends into the cleft presenting an antigen. Individuals with MHC alleles placing alanine, valine, or serine in this position are much more susceptible to insulin-dependent juvenile onset diabetes; the presence of aspartic acid at this position provides resistance to the disease.

The molecular events initiating these and other autoimmune diseases are unknown in most cases. In some instances the onset of an autoimmune disease may be traced to environmental stress or to an injury that exposes body cells not normally associated with lymphocytes to cells of the immune system. In other cases, autoimmune disease is triggered by a massive infection in which the body is exposed to an antigen that stimulates production of antibodies that by chance can cross react with some body proteins. For example, antibodies developed against infecting streptococcus bacteria sometimes also react with human heart tissue, producing a form of rheumatic fever; antibodies against the Epstein-Barr and hepatitis B viruses are also active against myelin basic protein, the body protein attacked in multiple sclerosis. Antibodies produced through injection of a vaccine sometimes also cross react with body cells or molecules.

Many of the more serious autoimmune diseases appear early in life, during adolescent or early adult years. Another peak in onset occurs during the 40s or 50s. A general failure of immune tolerance, along with a generalized reduction in effectiveness of the immune system, often occurs in old age. For unknown reasons, most autoimmune diseases are more common—some as much as three or four times—in human females than in males.

Infective Agents That Defeat or Sidestep the Immune Response   A number of pathogenic viruses or organisms have evolved mechanisms that either directly defeat the immune system by inhibiting or destroying part of the response, or that circumvent it by one means or another. Several pathogenic bacteria, including those responsible for rheumatic fever, gonorrhea, meningococcal meningitis, and relapsing fever, regularly change their surface groups to avoid destruction by the immune system. A pathogenic protozoan, the trypanosome causing African sleeping sickness, also avoids detection by constantly changing its surface groups. In many viruses the "spikes" that bind the viral coat to cell surfaces during infection (see p. 20) have the segments bound by cell receptors located in clefts or pits that are too narrow for recognition and binding by antibody proteins. Other segments of the coat proteins exposed to antibodies are altered so rapidly by constant mutations that the virus avoids or reduces the effectiveness of an immune response. These systems are used by many viruses causing human disabilities, including those responsible for polio, influenza, the common cold, and AIDS (caused by the *human immunodeficiency virus*, or *HIV*).

Many viruses take advantage of features of the immune system to get a free ride to the cell interior. For example, AIDS has a surface molecule that is recognized and bound by a receptor, *CD4*, located on helper T cells (see Fig. 19-12). Binding to the receptor facilitates entry of the virus into T cells, their primary site of infection.

HIV causes catastrophic failure of the immune response, primarily through gradual destruction of helper T cells. Elimination of these vital links in B-cell activation eventually disables the immune system. Although incapacitation of the immune system does not directly kill a person infected with HIV, a host of secondary infections such as fungal diseases and pneumonia, unopposed by an immune response, usually does. (Further details of HIV and the characteristics of AIDS are presented in Supplement 19-1.)

USE OF ANTIBODIES IN RESEARCH

The ability of higher vertebrates to produce antibodies specifically recognizing and binding antigens of almost unlimited variety has been used in many lines of scientific research, particularly in identification and isolation of cellular molecules. By adding markers such as heavy metals or fluorescent dyes to antibodies, the cellular locations of molecules of various kinds can be highlighted for electron or light microscopy. Antibodies attached to gels in separatory columns (see Appendix p. 801) have been used to extract and purify single proteins from mixtures obtained from whole cells. Antibodies have also been widely used to interfere with the activity of specific molecules inside cells, thereby providing clues to their normal cellular functions. In addition, antibodies are being developed as a possible means for delivering toxic substances to specific cell types, such as those of tumors. For all these purposes, monoclonal antibodies provide exceptional and exclusive specificity for the molecules or systems under study.

Antibodies have also provided critical insights into the mechanisms by which enzymes operate. By using molecules with structural similarities to the transition state for a chemical reaction (see p. 78) as antigens, investigators have induced test animals to develop antibodies that function as catalysts. The resulting antibody-enzymes, or *abzymes*, have demonstrated the importance of the transition state in enzymatic catalysis and have also provided a means to produce "designer enzymes"—enzymes capable of catalyzing a desired reaction. (Details of abzymes and their importance to studies of enzymatic catalysis are presented in Chapter 3.)

For Further Information

Antibodies as artificial enzymes (abzymes), *Ch. 3*
Antibody gene rearrangements and cancer, *Ch. 22*
Cell surface receptors, *Ch. 6*
Endocytosis and exocytosis, *Ch. 20*
Enhancers, *Ch. 14*
Genetic rearrangements, *Ch. 18*
Promoters, *Ch. 14*
Recombination, *Ch. 25*
Transposable elements, *Ch. 18*
Viruses, *Ch. 1*

Suggestions for Further Reading

Ada, G. L., and Nossal, G. 1987. The clonal selection theory. *Sci. Amer.* 257:62–69 (August).

Anderson, R. M., and May, R. M. 1992. Understanding the AIDS pandemic. *Sci. Amer.* 266:58–66 (May).

Barber, L. D., and Parham, P. 1993. Peptide binding to major histocompatibility complex molecules. *Ann. Rev. Cell Biol.* 9:163–206.

von Boehmer, H., and Kisielow, P. 1990. Self-nonself discrimination by T cells. *Science* 248:1369–1373.

von Boehmer, H., and Kisielow, P. 1991. How the immune system learns about self. *Sci. Amer.* 265:74–81 (October).

Burton, D. R. 1990. Antibodies: The flexible adaptor molecule. *Trends Biochem. Sci.* 15:64–69.

Clark, E. A., and Ledbetter, J. A. 1994. How B and T cells talk to each other. *Nature* 367:425–428.

Davis, M. M. 1990. T cell receptor gene diversity and selection. *Ann. Rev. Biochem.* 59:475–496.

DeFranco, A. L. 1993. Structure and function of the B cell antigen receptor. *Ann. Rev. Cell Biol.* 9:377–410.

Gellert, M. 1992. Molecular analysis of V(D)J recombination. *Ann. Rev. Genet.* 26:425–446.

Goldberg, A. L., and Rock, K. L. 1992. Proteolysis, proteasomes, and antigen presentation. *Nature* 357:375–379.

Greene, W. C. 1993. AIDS and the immune system. *Sci. Amer.* 269:98–105 (September).

Grey, H. M., Sette, A., and Buus, S. 1989. How T cells see antigen. *Sci. Amer.* 261:56–64 (November).

Haseltine, W. A. 1991. The molecular biology of the human immunodeficiency virus type 1. *FASEB J.* 5:2349–2360.

Johnston, M. I., and Hoth, D. F. 1993. Present status and future prospects for HIV therapies. *Science* 260:1286–1293.

Levine, T. P., and Chain, B. M. 1991. The cell biology of antigen presentation. *Crit. Rev. Biochem. Molec. Biol.* 26:439–473.

Levy, J. A. 1993. Pathogenesis of HIV infection. *Microbiol. Rev.* 57:183–289.

Lieber, M. R. 1991. Site-specific recombination in the immune system. *FASEB J.* 5:2934–2944.

Mitsuya, H., Yarchoan, R., and Broder, S. 1990. Molecular targets for AIDS therapy. *Science* 249:1533–1544.

Mizel, S. B. 1989. The interleukins. *FASEB J.* 3:2379–2388.

Nelsen, B., and Sen, R. 1992. Regulation of immunoglobulin gene transcription. *Internat. Rev. Cytol.* 33:121–149.

Noelle, R. J., and Snow, E. C. 1991. T helper cell dependent B cell activation. *FASEB J.* 5:2770-2776.

Ramsdell, F., and Fowlkes, B. J. 1990. Clonal deletion versus clonal anergy: The role of the thymus in inducing self tolerance. *Science* 248:1342–1348.

Schwartz, R. H. 1993. T cell anergy. *Sci. Amer.* 269:62–71 (August).

Scientific American, vol. 269, September 1993. Entire issue on the immune system, its development, functions, and diseases.

Sinha, A. A., Lopez, M. T., and McDevitt, H. O. 1990. Autoimmune diseases: The failure of self tolerance. *Science* 248:1380–1388.

Smith, K. A. 1990. Interleukin-2. *Sci. Amer.* 262:50–57 (March).

Weiss, A. 1991. Molecular and genetic insights into T cell antigen receptor structure and function. *Ann. Rev. Genet.* 25:487–510.

Weiss, R. A. 1993. How does HIV cause AIDS? *Science* 260:1273–1279.

Young, J. D.-E., and Cohn, Z. A. 1988. How killer T cells kill. *Sci. Amer.* 258:38–44 (January).

Review Questions

1. What is an antigen? What types of substances act as antigens?

2. What are lymphocytes? B cells? T cells? What do the *B* and *T* signify in these names?

3. Outline the structure of an antibody molecule. What are the heavy chains of an antibody molecule? The kappa and lambda light chains? Variable and constant regions? Hypervariable segments? Framework segments?

4. What is the relationship between heavy chains and antibody classes? What are the major functions of each antibody class?

5. Outline the structure of the kappa light-chain gene group of the mouse. What are V, J, and C elements? Where are promoters located in the gene group? Enhancers?

6. Outline the structure of the heavy-chain gene group of the mouse. What are D elements? How many C elements occur in the mouse heavy-chain gene group? Diagram the sequence structure of an individual C element.

7. What are R sequences and S sites? How do these sequence elements function in the genetic rearrangements producing functional antibody genes?

8. Diagram the structure of a rearranged kappa light-chain gene and heavy-chain gene in the mouse.

9. Outline the genetic rearrangements taking place in production of a functional heavy-chain gene. From what sequence elements do the introns of rearranged antibody genes originate?

10. What is somatic mutation? What is the significance of this process to the specificity of antibodies? What is allelic exclusion?

11. What structural and functional alterations change antibody production from the membrane-bound to secreted forms of the IgM and IgD antibodies?

12. What structural alterations accomplish switching from IgM and IgD to IgG antibodies?

13. What is a B-cell clone? What events stimulate a B cell to found a clone?

14. What is the relationship between B-cell clones and the monoclonal antibody technique? What is a hybridoma? Why are monoclonal antibodies superior as a research tool to antibodies generated by injecting an animal with an antigen?

15. What is the difference between helper and killer T cells? What major functions are carried out by these cells during an immune response? What is a perforin? An interleukin?

16. What is a class I MHC? A class II MHC? What major functions are carried out by these surface molecules in the immune system?

17. What are the relationships among class I and II MHC and body cells, killer T cells, helper T cells, and B cells?

18. Compare the mechanisms producing variability in MHC and antibodies.

19. What is a T-cell receptor? What elements are recognized and bound by a T-cell receptor? What mechanisms generate variability in T-cell receptors?

20. What is a processed antigen? Trace the events leading from the initial encounter with an antigen to display of a processed antigen in a B cell. What cell types other than B cells display processed antigens?

AIDS (Acquired Immune Deficiency Syndrome)

AIDS appears to be a relatively recent human affliction. The first cases were reported in 1981, and the disease was probably established in the human population not too long before this time, perhaps some 30 to 40 years. The earliest known blood samples containing antibodies against the AIDS virus (the *human immunodeficiency virus* or *HIV*) were taken from African patients in Zaire in 1959. HIV-1, the virus responsible for the most common form of AIDS in the industrial world, was isolated and identified in 1983 in the laboratories of R. C. Gallo in the United States and L. Montagnier in France.

As of mid-1993 the AIDS virus was estimated to have infected 14 million people worldwide and to infect one new person each 15 seconds. By the year 2000 the total may reach 40 million. In North America more than 1 million people have been infected with the virus. The disease is most prevalent in sub-Saharan Africa, where 8 million people have been infected. In some cities of this region, one out of every three adults carries the virus. Over 2 million individuals worldwide have developed full-blown AIDS as a result of HIV infection; most of these individuals are dead.

HIV is a retrovirus (see p. 536). Like all retroviruses, its genome is encoded in a pair of RNA molecules when the virus is in the free, infective form (Fig. 19-14). In the free virus the nucleic acid core is surrounded by a protein coat, covered on its exterior surface by a lipid bilayer derived from a host cell. Extending from the lipid bilayer are *spike proteins* consisting of a basal glycoprotein, *gp41*, which anchors the spike in the membrane, and a surface glycoprotein, *gp120*, which extends from the basal unit.

The gp120 glycoprotein has the central role in HIV infection. This glycoprotein is recognized and bound by the CD4 receptor on the surfaces of helper T cells (see Fig. 19-12b). Some other cell types without the CD4 receptor, including macrophages, regulatory cells of the intestinal lining (chromaffin cells), glial cells of the brain (see p. 348), and cells of the duodenum, colon, and rectum, also apparently have receptors able to bind gp120. Once bound to a receptor by gp120, the viral particles enter the cell, probably by fusion of the viral membrane coat with the plasma membrane. Fusion introduces the naked viral particle, free of its boundary membrane, directly into the cytoplasm of the cell. At this point the virus loses its protein coat and its RNA core is copied into DNA by a reverse transcriptase (see p. 523) included among the proteins of the infecting viral particle.

One or more of the DNA copies of the viral genome then insert into the host cell genome. In this form the virus is protected from attack by the immune system, even though the infected individual may mount an impressive immune reaction against proteins of the virus. The integrated DNA contains the *gag, env,* and *pol* genes typical of retroviruses, which encode coat proteins and enzymes required for duplication, excision, and insertion of the virus. The viral genome also includes several genes that encode regulatory proteins. The regulatory proteins, unusually extensive for a retrovirus, allow the integrated retroviral DNA to exist in three primary states: (1) inactivity; (2) slow activity in which only a few active, infective particles are produced; or (3) rapid, intensive production of infective viral particles. Rapid viral production is often accompanied by lysis and death of the host cell.

The primary detrimental effect of HIV infection is a gradual but steady destruction of helper T cells. Elimination of the helper Ts disables the responses of macrophages and B cells in the immune response. As the infection progresses, the immune system becomes less and less effective, until at final stages the infected person is left essentially defenseless against infective microorganisms and viruses. It is still uncertain how HIV infection leads to T-cell destruction. The virus may directly kill T cells or may stimulate an immune response that kills the cells (or both).

Symptoms and Progress of AIDS

Many of the pathogens infecting individuals with AIDS are common organisms of the environment that rarely cause problems for persons with healthy immune systems. For example, people with advanced AIDS become especially susceptible to infection by a fungus, *Pneumocystis carinii*, that causes an otherwise rare form of pneumonia. Many AIDS patients also develop an otherwise rare skin cancer known as *Kaposi's sarcoma*, in which reddish-purple lesions appear on the skin.

AIDS progresses through a number of stages that are identified by the presence of antibodies to HIV, various symptoms, and the number of surviving helper T cells in the blood. Soon after the initial infection, usually within two to five weeks, symptoms resembling flu or mononucleosis may appear, including fatigue, fever, a rash, and headaches. These symptoms ordinarily last for only a few days to two weeks. During this initial stage, antibodies against HIV usually

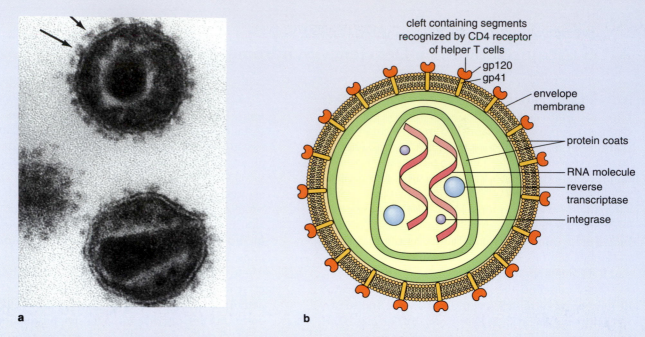

Figure 19-14 HIV-1, the primary virus responsible for AIDS. **(a)** Viral particles sectioned for electron microscopy. The gp120 spikes can be seen extending from the surfaces of the particles (arrows). (Courtesy of H. Gelderblum.) **(b)** Major structural features of HIV-1.

Labels in figure b:

cleft containing segments recognized by CD4 receptor of helper T cells
gp120
gp41
envelope membrane
protein coats
RNA molecule
reverse transcriptase
integrase

appear in the bloodstream, and the individual becomes infective to others.

After the initial stage, overt symptoms disappear. However, during this second, apparently symptomless stage the virus continues to proliferate, primarily in organs of the lymphatic system such as lymph nodes, and the individual remains infective. Although T cells begin a steady decline during the symptomless stage, their numbers remain high enough to provide an effective immune response.

After the symptomless period, which may last from 2 or 3 to 10 years or more (average 7 years), the first signs of problems with the immune system appear. As the disease progresses, T cells become reduced to 200 to 400 per ml of blood, as compared to normal levels of about 1000/ml. Among the many relatively mild afflictions appearing at this point may be persistent fungal or viral infections such as athlete's foot, shingles, thrush (white spots or ulcers in the mouth and throat caused by fungal infection), herpes outbreaks in the mouth and genital area, skin infections, and, in women, chronic fungal infections of the vagina. Chronic fever, diarrhea, night sweats, and weight loss are also common. The combination of symptoms at this stage is called the *AIDS-related complex*. During this period, which may last for several years, the symptoms may vary in severity and even disappear and reappear.

The advanced stages of the disease appear as T cells fall significantly below 200/ml in the blood. The individual now becomes highly susceptible to in-

fection by a variety of microorganisms, many of them common in the environment. Pneumonia caused by *Pneumocystis carinii*, tuberculosis, infection of the brain by *Toxoplasma*, and development of cancers such as Kaposi's sarcoma occur in many patients. Severe diarrhea and weight loss and often symptoms of mental impairment including memory loss and imprecision in perception, thought, and speech accompany these infections. The appearance of these conditions is usually taken as evidence of full-blown AIDS. As the T-cell count falls below 100/ml in the blood, additional pathogens may infect the body, including *cytomegalovirus*, which invades tissues in the retina and digestive systems, and *Mycobacterium avium*, a ubiquitous bacterium in the environment. *M. avium*, which is closely related to the bacterium causing tuberculosis (*M. tuberculosis*), may grow to massive numbers in the blood, lungs, liver, bone marrow, spleen, and digestive tract. Death usually follows within a short time after the T cell count falls below 100/ml.

Because of the variety of symptoms that may appear, there has been considerable controversy over defining when infected persons may be considered to have progressed from being a carrier to having AIDS. The definition is significant, because individuals who are diagnosed as having AIDS are eligible for treatment by state and local health departments and other agencies. In January 1991 the Federal Centers for Disease Control defined AIDS as a T-cell count of less than 200/ml of blood and one or more of the associated infections typical of AIDS. The new definition added many thousands of AIDS patients to the case-load who previously would have been diagnosed only as unaffected carriers.

Most AIDS patients in Western Europe and the Americas are infected with HIV-1, the virus first isolated by Gallo and Montagnier. Only a small percentage of AIDS patients in these countries are infected by a related virus, *HIV-2*. In some countries in Africa, however, infections with HIV-2 are common.

Transmission and Prevention of AIDS

Persons with HIV infections are potentially infective to others during any stage of the disease, including the early symptomless stage. HIV is transmitted primarily through direct passage of blood cells or fluids from the bloodstream of one individual to another or through transmission of infected leukocytes from the semen of one individual to the bloodstream of another during heterosexual or homosexual intercourse. The transfer may take place through breaks in the skin or mucous membranes exposed to blood or semen, through blood transfusions, or by contaminated needles. The virus can also be transmitted from mother to child during pregnancy, at birth, or during nursing.

Although infective viral particles appear in the vaginal secretions of infected women, the concentration in this location is evidently too low to be likely to cause infection as long as the vaginal mucosa is undamaged. Thus sexual intercourse must be traumatic—that is, must cause injury to the vaginal lining—for HIV to be readily transmitted from an infected female to a sexual partner by this route.

The virus also appears in saliva, tears, and urine, but again the concentration of viral particles in these body fluids is evidently too low to be readily infective. It is therefore considered difficult or impossible to catch AIDS through food, water, sneezing, coughing, or even sharing toothbrushes. Among the many families with AIDS patients living at home, there are no known cases of transmission of the disease to family members who are not sexual partners. Similarly, no one in the health professions has contracted the disease by nonsexual contact with AIDS patients. Some health workers, however, have contracted AIDS by accidental needle punctures when handling contaminated blood or through spills of blood on skin or mucous membranes in which breaks from rashes, cuts, or other sources are present.

AIDS prevention for most persons is straightforward, although not necessarily easy to achieve: avoidance of sexual intercourse with persons whose AIDS status is in question; and, for intravenous drug users, the use of previously unused, sterile needles. Condoms, if intact, are considered to give reasonable protection against infection by the virus during vaginal, oral, or anal intercourse. Contraction of AIDS through blood transfusions has essentially disappeared because all blood given by donors is now routinely screened for HIV antibodies. There is no risk of contracting AIDS by giving blood by the sterile techniques routinely used in hospitals and clinics.

Possible Origins of the Disease   The origins of the AIDS virus are uncertain. However, the virus was probably picked up by humans from primate populations, possibly in rural Africa. HIV-1 and HIV-2 are closely similar to the *simian immunodeficiency virus* (*SIV*), which infects other primates. From nucleic acid sequence comparisons and the locales occupied by the nonhuman primates carrying SIV, it appears possible that chimpanzees may have been the source of HIV-1—the SIV of chimpanzees is more closely related to human HIV-1 in nucleic acid sequence than any other primate SIV—and sooty mangabeys and rhesus macaques the source of HIV-2.

It is common for viruses to be relatively benign in their normal host species but highly virulent in other species (see also p. 652). In many nonhuman primates, SIV infections cause no significant symptoms unless the virus infects a primate other than its usual host. The SIV infecting African green monkeys, for example, causes no evident AIDS-like symptoms. The same virus causes AIDS, however, if transferred to Asian macaques. Viruses related to HIV and SIV also infect many other mammalian groups.

Research Toward Prevention or Cure   At the present time, human AIDS is incurable. The situation is not unusual for viral diseases, especially those caused by retroviruses. Unless a cure is found, as many as 50% to 100% of the people now infected with HIV will die within the next 10 years. The burden of caring for those with the disease is already daunting and certain to severely challenge the medical and economic resources of the entire world. Two drugs, *AZT* (*3'-azido-2',3'-dideoxythymidine*) and *DDI* (*dideoxyinosine*), slow progress of the disease. Both drugs, which are *dideoxynucleotides*, work by interfering with the reaction catalyzed by reverse transcriptase, in which a DNA copy is made from the RNA molecules carried in infecting HIV particles (dideoxynucleotides have no 3'-OH group available for formation of the 3'→5' linkages joining nucleotides together in DNA; as a result, DNA synthesis stops when a dideoxynucleotide is inserted—see p. 425). Although the drugs prolong the lives of AIDS patients, they are not cures. AZT also has side effects so severe that many persons are unable to tolerate it.

Efforts are under way to develop vaccines or antibodies against HIV. Basically the vaccines use killed whole virus or subparts of the virus as an antigen to be injected into an infected individual or person at risk. An immune reaction generated against the antigen used as a vaccine will then, it is hoped, provide the individual with antibodies that will also be effective against the live virus.

HIV is a formidable opponent to development of an effective vaccine because of its integration into the host cell genome. Its protection against antibodies in this location may make even successful vaccines work only as a prevention, not a cure—they may be limited to immunization against infection but have limited benefit to persons already infected. Constant mutation of coat proteins, particularly gp120, also provides protection to the virus. An antibody developed against one form of the gp120 spike protein, for example, may fail to recognize and combine with a mutated gp120.

Hopefully, this research will eventually lead to a vaccine that is effective in preventing, curing, or reducing HIV infections to a benign form. Some hope also lies in behavior modification, which has already proven effective in reducing the rate of new infections among male homosexuals. It is also possible that the human population will gradually develop resistance to HIV or that the virus will slowly mutate to a less virulent form. These changes have evidently occurred in some nonhuman primates infected by SIV in which the infections produce no serious symptoms. In the meantime, humans are well advised to pick their sexual partners with extreme care, to practice so-called safe sex, and to avoid contaminated needles.

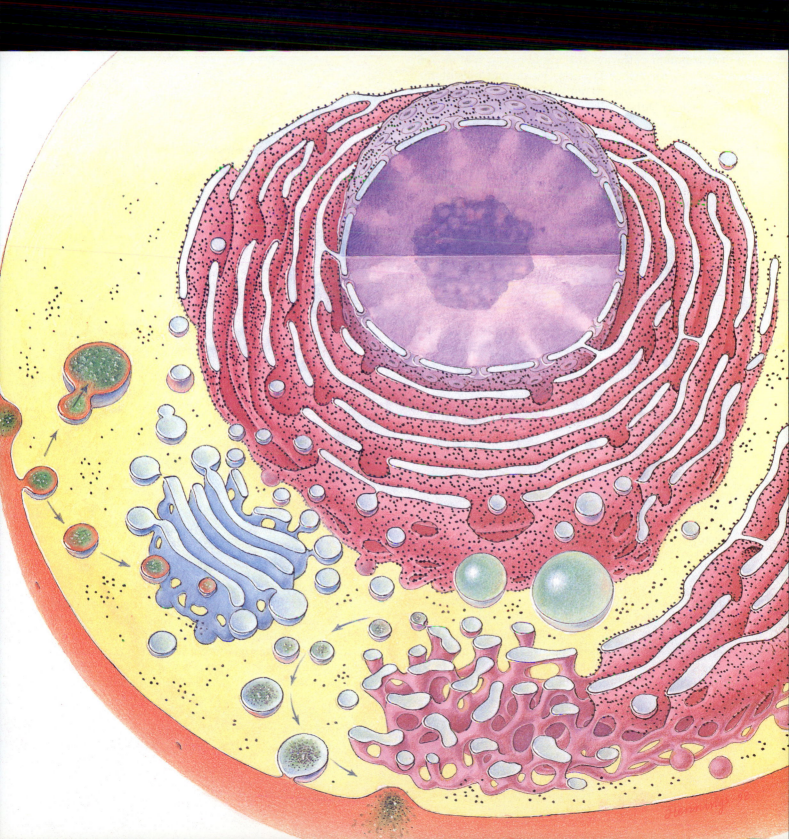

*• Endoplasmic reticulum • Golgi complex •
The signal directing proteins to the ER • Further
distribution within the ER-Golgi system • Signals
directing proteins to the nucleus and cytoplasmic
organelles • Exocytosis • Endocytosis • Clathrin
and coated pits • Lysosomes • Glycosylation of
glycoproteins*

One of the revelations of contemporary cell biology is that cells contain elaborate sorting and distribution pathways that route newly synthesized proteins to various locations inside and outside the cell. Those destined to remain inside the cell may become part of the soluble background substance of the cytoplasm or may be routed to the nucleus, various organelles, or the plasma membrane. Secreted proteins and those destined to become parts of extracellular structures such as the cell wall or extracellular matrix are routed to the outside.

Proteins are sorted to these locations by a remarkable group of cellular mechanisms that encompass every membrane and compartment of the cell. With the exception of proteins destined to enter the soluble cytoplasmic substance, which are simply released into the cytoplasmic solution after assembly, all proteins include segments that serve as *signals* directing them to a final cellular destination. The signals, which are molecular equivalents of addresses and zip codes, are recognized by receptors in sorting destinations including the endoplasmic reticulum (ER), nucleus, mitochondria, chloroplasts, microbodies, and the large central vacuoles of plant cells. For many proteins the initial target organelle is an entry point for complex routes of further distribution. For example, after distribution to the ER as an initial organelle, proteins may be routed to the Golgi complex, secretory vesicles, lysosomes, the plasma membrane, or cell exterior.

The distribution pathways routing proteins to the cell exterior form the basis of secretion or *exocytosis*. Typically, newly synthesized proteins following this pathway proceed from the ER, through the Golgi complex and secretory vesicles, to the plasma membrane and cell exterior. Proteins and other substances can enter cells by *endocytosis*, which essentially reverses the pathways of exocytosis between the plasma membrane and the Golgi complex. The fact that exocytosis and endocytosis use similar routes means that some way stations simultaneously sort and distribute proteins traveling in opposite directions.

As they pass through the ER and Golgi complex, many proteins and other molecules are chemically modified by processing enzymes. Among the most significant of the processing reactions are those adding sugar groups to proteins and lipids, thereby converting them to glycoproteins and glycolipids. Other enzymes add chemical groups such as phosphates or sulfates or remove segments of the molecules as they pass through the organelles. The ER is also the primary site at which membrane lipids are synthesized and started on their way to final cellular destinations.

This chapter outlines the major cellular mechanisms that sort, modify, and distribute proteins in eukaryotic cells. Exocytosis and endocytosis are also discussed. Supplement 20-1 compares these systems with mechanisms distributing proteins in bacteria.

THE ER AND GOLGI COMPLEX IN PROTEIN DISTRIBUTION

Much of cellular sorting and distribution is based on the ER and Golgi complex. Many proteins start their journeys by entering the membranes or internal compartments of the ER. From there they travel to the Golgi complex, where the majority are modified chemically before being sorted into vesicles carrying them to their final destinations.

Structure of the ER and Golgi Complex

The ER The ER is a collection of membranous tubules, vesicles, and flattened sacs that extends through more or less of the cytoplasm. ER membranes form continuous sacs that enclose a *lumen* or channel that is separated by the ER membranes from the surrounding cytoplasm. The ER occurs in two forms. One, the *rough ER*, is so called because it has ribosomes attached to membrane surfaces facing the surrounding cytoplasm (Figs. 20-1 and 20-2). Rough ER commonly extends into large, flattened sacs called *cisternae* (singular = *cisterna*; see Figs. 20-1b and 20-2). The other form, the *smooth ER*, has no ribosomes (see Fig. 20-1a); these membranes are primarily tubular in form and generally of smaller dimensions than rough ER cisternae. At some points the smooth and rough ER membranes may connect, forming a continuous, inner channel enclosed by the two systems.

Although most cells have both rough and smooth ER, their relative proportions vary considerably. In some cells, as in the pancreas, most ER membranes are rough. In others, such as epithelial cells, most are smooth. The total quantity of ER membranes also varies widely. The cytoplasm of some cells, such as pancreatic cells, is tightly packed with ER membranes; in others, as in many higher plant cells, ER membranes are sparsely distributed and occupy only a minor fraction of the cytoplasm (as in Fig. 1-18). All possible gradations between these extremes are found in nature.

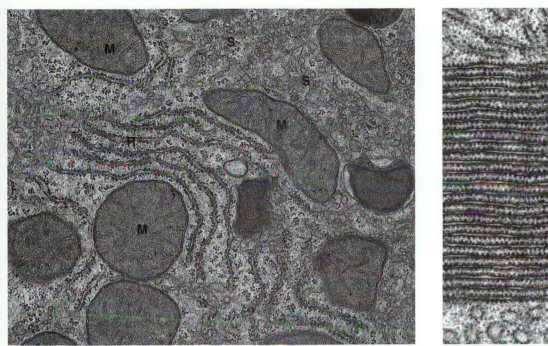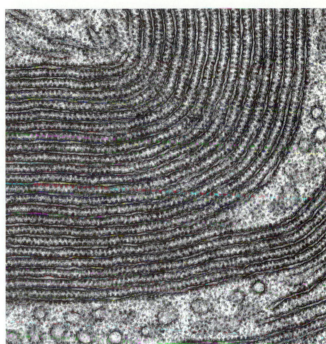

Figure 20-1 The endoplasmic reticulum (ER). **(a)** Rough (R) and smooth (S) ER in a hamster liver cell. M, mitochondrion. × 24,000. **(b)** Closely spaced cisternae of the rough ER in a pancreatic cell of the bat. × 31,000. ([a and b] from *The Cell* by D. W. Fawcett, 1966. Courtesy of D. W. Fawcett and W. B. Saunders Company.)

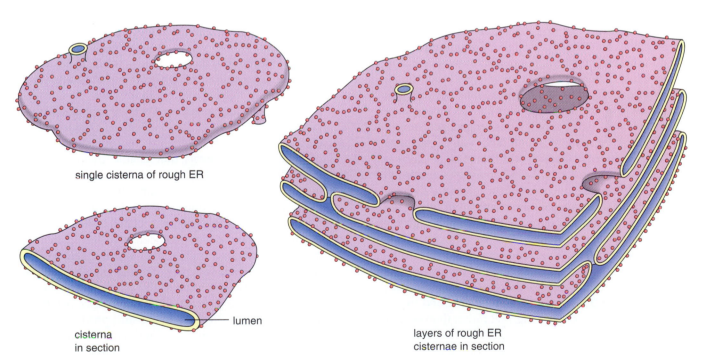

single cisterna of rough ER

cisterna
in section

lumen

layers of rough ER
cisternae in section

Figure 20-2 The arrangement of membranes in the rough ER.

The ER and Golgi Complex in Protein Distribution   **573**

The total quantity and relative proportions of rough and smooth ER membranes change readily depending on cell activity. Cells also vary considerably in the proportions of ribosomes suspended freely in the cytoplasm and attached to the rough ER.

In most cells, ribosomes also attach to the outermost membrane of the nuclear envelope, on the side facing the surrounding cytoplasm (Fig. 20-3). Direct connections can sometimes be seen between the outer nuclear membrane and ER membranes (arrow in Fig. 20-3), making the compartment between the two membranes of the nuclear envelope (the *perinuclear compartment*) continuous with the ER channels. The outer membrane shares many molecular constituents with ER membranes in addition to ribosomes. Thus in many ways the nuclear envelope can be considered as an extension of the ER that surrounds the nucleus. The pattern in which the nuclear envelope disassembles and reassembles during cell division also supports a close relationship with the ER—during breakdown at the beginning of the division sequence the membranes of the nuclear envelope separate into vesicles that become indistinguishable from the ER. At the close of division the process is reversed—the nuclear envelope apparently reforms from ER vesicles that surround the decondensing chromosomes (see Chapter 24 for details).

ER membranes can be isolated from cells and separated into rough and smooth fractions by sucrose gradient centrifugation (see p. 795; rough ER, because of its content of ribosomes, is significantly higher in density than smooth ER). Rough ER membranes isolated in this way prove to contain an extensive group of proteins. Several of these are integral membrane proteins associated with the sorting mechanisms that route newly synthesized proteins into or through the ER membranes (see below). Another group of rough ER proteins consists of membrane-bound enzymes that carry out initial steps in addition of sugar groups to glycoproteins and glycolipids. (Further sugar units are added in the Golgi complex; Supplement 20-2 describes the reactions assembling carbohydrates in the ER and Golgi complex in more detail.) The rough ER also contains enzymes and other factors concerned with assembly of polypeptide subunits into multisubunit proteins and the formation of disulfide linkages stabilizing the complexes. Some of the molecules associated with these activities are suspended in solution in the lumen of the rough ER rather than bound to ER membranes.

Enzymes and factors carrying out varied reactions not directly associated with protein synthesis and distribution also occur in both the rough and smooth ER. Lipids are synthesized by enzymes on the cytoplasmic side of the ER; as their synthesis progresses, they are inserted in the bilayer half of the ER

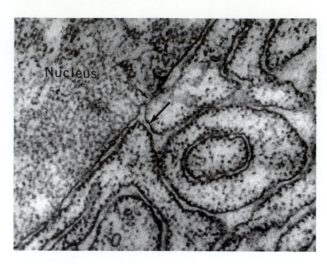

Figure 20-3 A connection (large arrow) between the outer membrane of the nuclear envelope and the rough ER in a mouse salivary gland nucleus. The outer membrane of the nuclear envelope is covered with ribosomes. The small arrow points to a pore complex of the nuclear envelope. × 62,000. (From *The Cell* by D.W. Fawcett, 1966. Courtesy of D. W. Fawcett, H. Parks, and W. B. Saunders Company.)

facing the cytoplasm. From this position, specialized proteins flip some lipids to the opposite bilayer half (see p. 113). The membrane lipids then travel to parts of the membrane systems linked directly or indirectly to the ER: the nuclear envelope, the Golgi complex, lysosomes, secretory and storage vesicles, and the plasma membrane (see p. 579). The traffic in phospholipids is so great that in highly active cells as much as 50% of the phospholipid content of the ER is transported to the Golgi complex every 10 minutes. Lipids synthesized in the ER also make their way to organelles such as mitochondria and chloroplasts by means of *lipid transfer proteins* that enable them to travel through the aqueous cytoplasm (see p. 114).

In addition, rough and smooth ER membranes carry out initial reactions in the oxidation of fats (see Supplement 8-1). Enzymes associated with other metabolic pathways have also been detected in the ER, including glycogen metabolism and the synthesis of steroid hormones. Enzymes of the ER membranes also detoxify a variety of potentially injurious substances by oxidations, hydroxylations, and other reactions. These substances include toxins and products of putrefaction in foods, and substances inhaled in smoke.

Central to the detoxification reactions is the P_{450} family of cytochromes, a group of closely related cytochromes that are anchored to ER membranes by a hydrophobic portion of their amino acid chains. The cytochromes of this family detoxify a wide range of substances, including acetone, alcohols, organic solvents, pesticides, and various drugs by a series of reactions leading to addition of an —OH group. The hydroxylation converts the toxic substances to soluble derivatives that are more easily excreted from the body.

The enzymes and factors carrying out ER functions not directly related to protein synthesis occur in both rough and smooth ER. However, these activities are likely to be more concentrated in smooth ER membranes. The relative proportions of rough and smooth ER membranes reflect the amount of ER activity unrelated to protein synthesis. In cells carrying out extensive lipid synthesis or detoxification and relatively little ER-based protein synthesis, for example, the majority of ER membranes are smooth.

In higher plants, tubular extensions of smooth ER extend through plasmodesmata, the openings that form cytoplasmic channels connecting adjacent cells (see p. 194). The tubular extensions evidently connect the ER compartments of adjacent cells. Although the functions of the ER connections are uncertain, they may participate in a communication system that integrates cell activities in plants.

The Golgi Complex   A Golgi complex[1] consists of a localized stack of flattened, saclike membranes (Fig. 20-4). No ribosomes occur on Golgi membranes, and ribosomes are also characteristically absent from the spaces between and immediately surrounding the sacs. An entire Golgi complex often appears to be cup-shaped in cross section, giving the structure convex and concave faces.

The individual sacs of a Golgi complex, called *cisternae* as in the ER, are typically dilated or swollen at

[1]The Golgi complex, named for Camillo Golgi, the scientist who first described it in the late 1800s, is also called the Golgi *apparatus*. Some plant cell biologists use the term *dictyosome* (from *dicty* = net) for a single stack of Golgi membranes and reserve the term Golgi complex (or apparatus) for the entire collection of dictyosomes in a cell.

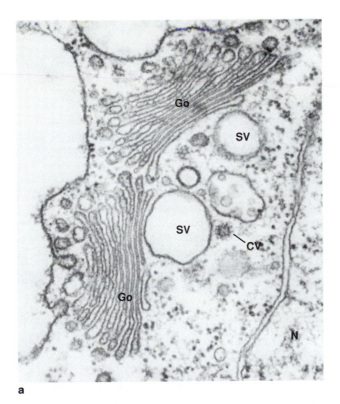

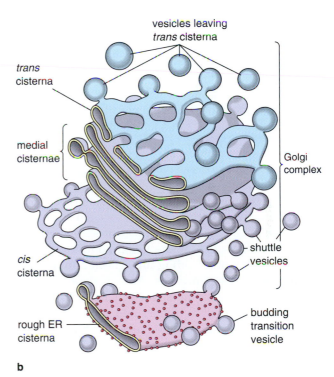

Figure 20-4   The Golgi complex. **(a)** Golgi complexes (Go) in a cell of the green alga *Chlamydomonas reinhardi*. Large secretion vesicles (SV) are present in the cytoplasm near the complexes. CV, coated vesicle; *N*, nucleus. × 80,000. (Courtesy of M. G. Farquhar; reproduced from *J. Cell Biol.* 91:77s [1981], by copyright permission of the Rockefeller University Press.) **(b)** The arrangement of *cis*, medial and *trans* cisternae and vesicular traffic in the Golgi complex. A cisterna of the rough ER is shown below the Golgi complex; the overall direction of progress of newly synthesized proteins through the vesicular traffic is from bottom to top of the diagram. The vesicles budding from the *trans* cisterna of the Golgi complex carry proteins destined primarily for secretory vesicles or lysosomes.

their margins. Dispersed around the cisternae are numerous vesicles of various sizes; some of these bud from or fuse with the edges of the cisternae. Both the swollen margins of the sacs and the vesicles surrounding the Golgi complex usually contain granular material consisting of proteins or glycoproteins moving between the Golgi cisternae or en route to final destinations in the cell. Much variation is noted in the size, shape, and degree of dilation of Golgi sacs and vesicles depending on cell type and activity.

Golgi complexes are usually closely associated with the rough ER, separated from it by a layer of small, protein-filled vesicles (see Fig. 20-4b). Some of these vesicles are *transition vesicles* that transport proteins from the ER to the Golgi complex. The side of the Golgi complex facing the ER is known as the *cis* face. At the opposite *trans* face, vesicles bud off and fuse to form the larger vesicles found on this side of the Golgi complex. Sacs between the *cis* and *trans* faces form the *medial* segment of the Golgi complex. *Shuttle vesicles* are believed to move proteins and glycoproteins between the Golgi sacs.

Many experiments have demonstrated that newly synthesized proteins move from the ER through the transition vesicles to reach the *cis* face of the Golgi complex. (Some details of these experiments are given later in this chapter.) After processing in the cisternae of the Golgi complex, the finished proteins are sorted in the *trans* cisternae and released in vesicles that eventually fuse and coalesce into secretory vesicles, lysosomes, or storage vesicles. Proteins are also evidently sorted to some extent as vesicles fuse in the traffic leaving the Golgi complex. Many proteins brought into the cell by endocytosis also enter the Golgi complex, where they are sorted and sent away again, frequently after processing, in vesicles of different types. (Fig. 20-5 summarizes vesicle traffic associated with the Golgi complex.)

The diverse activities of the Golgi complex in chemical processing and routing are reflected in an impressive assembly of enzymes. Among these are enzymes adding or removing sugar groups to complete the processing of glycoproteins and glycolipids. Other Golgi enzymes add sulfate, acetyl, or phosphate groups to glycoproteins and glycolipids, add fatty acids to proteins, or remove segments from the interior or ends of the amino acid chains of proteins.

The Golgi enzymes processing proteins by removing amino acid segments convert many proteins from precursor to active form. For example, many peptide hormones, such as insulin and glucagon, are synthesized in the ER in the form of inactive precursors that contain extra amino acid segments. The extra segments are removed by proteolytic enzymes in the Golgi complex. The proteolytic reactions also include alternate processing pathways (see p. 472) that release

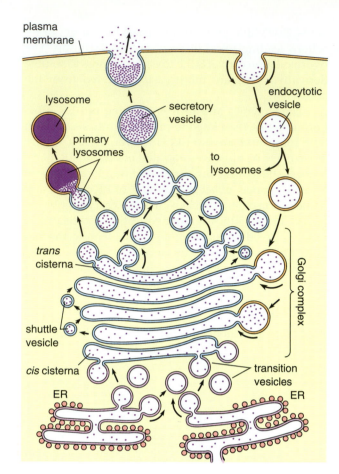

Figure 20-5 Vesicle traffic to and from the Golgi complex. Transition vesicles carry newly synthesized proteins from the ER to the *cis* cisterna of the Golgi complex. Proteins then move through the Golgi complex from the *cis* to the *trans* face in shuttle vesicles that bud from one cisterna and fuse with the next one in line (from the bottom to the top of the Golgi complex in the diagram). At the *trans* face of the Golgi complex, vesicles bud off and fuse to form lysosomes and secretory vesicles. Some material also enters the Golgi complex from outside the cell via endocytotic vesicles.

different peptide hormones from the same precursor. Some proteolytic reactions, particularly those converting hydrolytic enzymes from a precursor to active form, occur after placement of the proteins in vesicles leaving the Golgi complex or after secretion from the cell.

Also present are enzymes or complexes of more generalized function, such as an H^+-ATPase pump (see p. 131) that lowers the internal pH of cisternae at the *trans* face of the Golgi complex. Golgi membranes probably also contain receptors that sort and route proteins through the complex.

Some Golgi enzymes are restricted to animals or plants. Enzymes that add sugar groups to proteoglycans, the large molecular assemblies that form parts

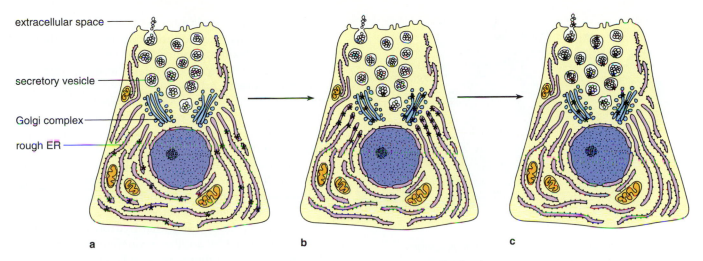

extracellular space

secretory vesicle

Golgi complex

rough ER

a   b   c

Figure 20-6   Results of a typical experiment tracking the progress of newly synthesized proteins through the ER and Golgi complex by autoradiography. The proteins have been made radioactive by exposing cells of the guinea pig briefly to labelled amino acids. **(a)** Three minutes after exposure; only the rough ER is labeled. **(b)** Three-minute exposure to the label followed by a 17-minute "chase" by exposure to unlabeled medium. Label is distributed over the rough ER, Golgi complex, and secretion vesicles. **(c)** Three-minute exposure followed by 117-minute chase. Label is concentrated over secretion vesicles. (Redrawn from an original courtesy of P. Favard. From *Handbook of Molecular Cytology*, North-Holland Publishing Company, Amsterdam, 1969.)

of the extracellular matrix (see p. 182), occur only in animal Golgi complexes. Plant Golgi membranes contain a complex collection of enzymes that assemble sugars into precursors of cell wall molecules.

The number of Golgi complexes per cell varies greatly in different tissues or species. Although the average number is about 20, some cell types may have many more. For example, corn root tip cells contain several hundred complexes; more than 25,000 have been counted in cells of *Chara*, a green alga. In plants, Golgi complexes are typically more or less evenly scattered through the cytoplasm. In animal cells, Golgi complexes are frequently concentrated near the ER or just outside the nucleus in a region close to the centrioles or cell center (see p. 355).

Microtubules serve as cytoskeletal supports holding both the ER and Golgi complexes in position in animal cells. Destroying the microtubule network with agents such as colchicine causes both the ER and Golgi complex to contract and lose their typical orientation in the cytoplasm. Elements of the ER and Golgi complex, as well as vesicle traffic moving between the Golgi complex and plasma membrane, may also travel along microtubules toward either the cell center or periphery. These movements are powered by "motors" such as dynein and kinesin (see p. 289) that use energy released by ATP hydrolysis to push the ER and Golgi cisternae and vesicles along the microtubules.

Evidence That Proteins Move Sequentially Through the ER and Golgi Complex

Movement of proteins through the ER and Golgi complex has been followed by a variety of techniques. The first evidence was developed by L. G. Caro and G. E. Palade, who gave pancreatic cells a pulse of radioactive amino acids and followed movement of label through the cells by autoradiography (see Appendix p. 791; Palade received the Nobel Prize in 1974 for his pioneering work tracing cellular distribution pathways). In this and many confirming experiments by later investigators, label first appeared in rough ER membranes and then, within several minutes, in rough ER vesicles and cisternae (Fig. 20-6a). Soon after this, within 10 to 30 minutes, label appeared in transition vesicles and *cis* cisternae of the Golgi complex (Fig. 20-6b). Subsequently label appeared in medial and *trans* cisternae of the Golgi complex and in secretory vesicles between the Golgi complex and the plasma membrane (Fig. 20-6c). Finally, within one to four hours after initial exposure, radioactivity appeared in the extracellular space, showing that the secretory vesicles eventually discharge their contents to the cell exterior. Similar experiments using labeled amino acids, sugar units, or antibodies developed against individual proteins have traced the routes followed by proteins destined to enter lysosomes or storage vesicles or to become components of the plasma membrane. Cell fractionation and centrifugation (see Appendix p. 794) have

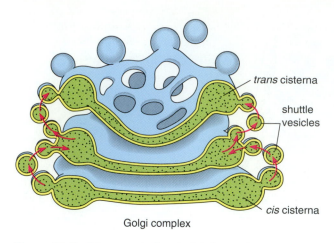

trans cisterna

shuttle vesicles

cis cisterna

Golgi complex

Figure 20-7 Movement of materials from *cis* to *trans* cisternae of the Golgi complex via shuttle vesicles.

also been widely employed to trace protein movement and distribution through the ER, Golgi complex, and further destinations.

In addition, movement of proteins and other molecules through the ER and Golgi complex has been followed by tagging them with markers making them visible in the light or electron microscope. Heavy metals are frequently used as markers for electron microscopy; for light microscopy, fluorescent dyes are commonly used. The marking approach is readily combined with antibody studies by attaching dyes or heavy metals to antibodies developed against proteins under study (for details of these techniques, see Appendix p. 791). Experiments using these and other techniques also traced the routes of proteins traveling in the opposite direction, from the plasma membrane or exterior to the cell interior, by endocytosis.

Among other significant conclusions, the results of these experiments show that the nuclear envelope functions similarly to the ER in protein synthesis—proteins synthesized by ribosomes attached to the outer membrane of the nuclear envelope enter the perinuclear compartment and then move to the ER and other elements of the distribution system. The experiments also indicate that individual Golgi cisternae remain more or less fixed in position, and that proteins and glycoproteins move from one cisterna to the next via shuttle vesicles (Fig. 20-7). In this mechanism, first proposed by M. G. Farquhar, the shuttle vesicles are thought to bud from one cisterna, travel the intervening distance, and fuse with the next cisterna in line. Eventually proteins and glycoproteins reach the cisterna at the *trans* face. At this point, budding forms secretory vesicles, lysosomes, or storage vesicles.

J. E. Rothman and his colleagues have pioneered biochemical investigation of the protein factors in-

volved in shuttle vesicle traffic. Using isolated mammalian systems, the Rothman group found that as shuttle vesicles bud from a Golgi cisterna, they become covered by a coat of proteins termed *COPs* (for *CO*ating *P*rotein). Addition of the coat appears to depend on another protein, *ARF*, which acts as an on-off switch for the coating process. ARF may stimulate coat formation when combined with GTP, and also act in the reverse manner, disassembling the coat, when its bound GTP is hydrolyzed to GDP. Addition of the coat is believed to aid vesicle formation by forcing the budding membrane into a spherical shape.

The coat is removed before the vesicles fuse with the next cisterna, in a step that hydrolyzes GTP. Fusion is promoted by another protein, termed *NSF* by Rothman, in a step that requires ATP to proceed. Several additional proteins called *SNAPs* (for *SO*luble *NSF* *A*ttachment *P*roteins) by the Rothman group bind NSF to a receptor on the vesicle surface. Rothman proposes that SNAPs and NSF work to promote vesicle fusion by binding the vesicle snap receptor (which Rothman terms *v-SNARE*, for *v*esicle *SNAP* *RE*ceptor) to a receptor on the target membrane to which the vesicle fuses (termed the *t-SNARE*). Different v- and t-SNARES may determine the precise locations to which vesicles are targeted within the pathway.

The proteins discovered by the Rothman group have counterparts in yeast, indicating that the mechanism forming and fusing vesicles is widely spread among eukaryotes. The protein encoded in the yeast *sec18* gene, which is necessary in normal form for vesicle traffic at all steps of the secretory pathway from the ER to the plasma membrane, is related to mammalian NSF. Another yeast gene affecting vesicle traffic, *YPT1*, encodes a GTP-binding protein.

MECHANISMS SORTING AND DISTRIBUTING PROTEINS

Signals Directing Proteins to Initial Destinations

The signals directing proteins to various cellular locations are segments of the polypeptide chain including from a few to nearly 100 amino acids, located at one or several positions. Almost all proteins targeted to the ER, mitochondria, or chloroplasts include a signal at the N-terminal end that directs the protein to the entry membrane. With few exceptions, the N-terminal signal is removed after the protein has entered the entry membrane. Further signal segments indicate whether the protein is to be inserted into the entry membrane and determine its orientation in the membrane. If the protein is to pass through the entry membrane, additional signals may direct it to membrane systems enclosed within the compartment bounded by the entry membrane or further locations along the distribution pathway. The signals are recognized by specific receptor molecules involved in sorting proteins and directing them to their final locations.

The N-terminal signals directing proteins to the ER, mitochondria, or chloroplasts consist of characteristic combinations of charged, polar, and nonpolar amino acids rather than specific amino acid sequences. The combinations fold or wind into secondary structures and patterns of charge and polarity to form the signal. Any of several amino acids can usually be substituted at a given position without altering an N-terminal signal, provided the substituted amino acid has similar chemical properties and, in some cases, a side group of similar size.

Proteins targeted to the nuclear interior have signals located within the protein chain. The nuclear signal also depends on patterns of amino acids with certain chemical properties rather than a specific sequence.

The identification and characterization of sorting and distribution signals have been greatly accelerated by the powerful techniques of molecular biology and genetic engineering. It has been possible to create hybrid proteins in which segments suspected to serve as signals are linked to segments of other proteins normally lacking signals of any kind. The protein to which the signal is attached, known as the *marker* or *reporter protein*, is usually a soluble cytoplasmic protein with readily identifiable enzymatic or other activity. The effect of an added signal in directing the protein to particular cellular compartments can then be traced by following the activity of the protein.

This approach is usually initiated at the DNA level. A hybrid gene is constructed by splicing a DNA sequence coding for a suspected signal to part or all of a gene coding for a marker protein. The hybrid gene is then introduced into a cell by transposable genetic elements, plasmids, direct injection into the nucleus, or other means. Once inside the host cell, the hybrid gene directs synthesis of the marker protein with the added signal. The same techniques have also been used to construct genes with nucleotide substitutions or deletions in regions encoding suspected signals. The effects on sorting and distribution of changes in individual amino acids at specific locations in the signal are then noted.

One of the most productive uses of these techniques links bacterial signals to eukaryotic proteins, or vice versa. This technique has established, among other findings, that many bacterial distribution signals can be read and interpreted correctly by eukaryotic cells, or eukaryotic signals by bacterial cells. This interchangeability indicates that the signals and distribution mechanisms are very ancient in origin and have been highly conserved in evolution.

Mechanisms Targeting Proteins to Pathways Beginning at the ER

The Signal Hypothesis   The idea that cells use signals for routing proteins was developed primarily from experiments with cell-free systems isolated from the dog, conducted by G. Blobel, B. Dobberstein, P. Walter, and their associates. These investigators noted that if ribosomes in a cell-free system (see p. 795) with no ER membranes are supplied with mRNAs for secretory proteins, the proteins synthesized were longer than those secreted in living cells by about 15 to 30 amino acids. The extra segment, which is usually located at the N-terminal end of the protein, has a high proportion of hydrophobic amino acids. If vesicles derived from ER membranes were added to the cell-free system, ribosomes translating the mRNAs attached to the vesicles, and the newly synthesized proteins became concentrated inside them. Analysis of the proteins inside the vesicles showed that the N-terminal sequence had been removed. An enzyme removing the segment could be isolated from the rough ER membranes. When extracted, it could remove the sequence segment from newly assembled secretory proteins.

In 1975, Blobel and Dobberstein combined these results into a *signal hypothesis* outlining how the N-terminal amino acid segment might direct proteins to the rough ER and insert them into this membrane system. Subsequent research by Blobel, his colleagues, and others added many details, including the activity of a *signal recognition particle* (*SRP*) that participates in the mechanism.

Figure 20-8 shows how the ER-directing signal mechanism is now considered to operate. The mRNAs for proteins to be synthesized on the rough ER contain a series of amino acids at the beginning of the coding sequence that spells out an ER-directing signal. These mRNAs initiate protein synthesis by the usual route, by binding to ribosomal subunits suspended in the soluble cytoplasm (see p. 461). As a ribosome assembles and begins to translate the encoded message, the first segment of the new polypeptide to emerge is the ER-directing signal (Fig. 20-8a and b). As the signal appears, it is bound by the SRP along with GTP (Fig. 20-8c and d). The ribosome-SRP complex is recognized and bound by a receptor in the ER membrane, the *SRP receptor* (Fig. 20-8e). Binding by the receptor attaches the ribosome firmly to the membrane. At this point the signal is transferred from the SRP to the ER membrane and, by virtue of its hydrophobic character, penetrates into the membrane. The growing polypeptide chain extends from

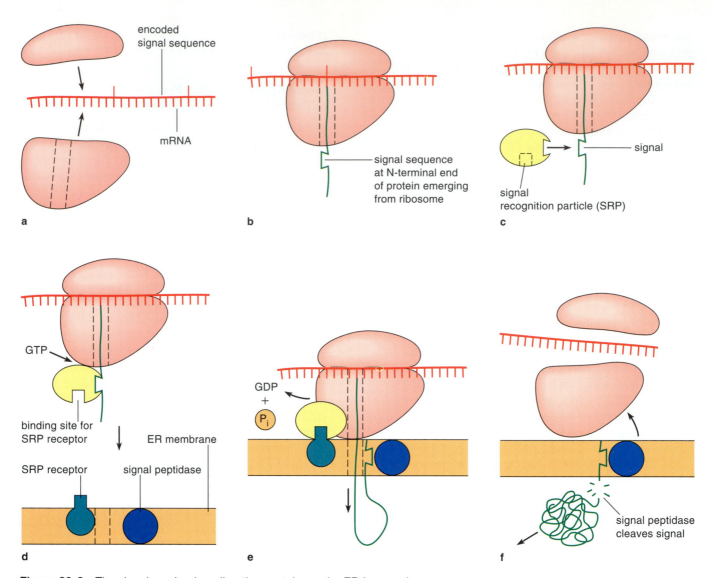

encoded signal sequence

mRNA

a

signal sequence at N-terminal end of protein emerging from ribosome

b

signal

signal recognition particle (SRP)

c

GTP

binding site for SRP receptor

ER membrane

SRP receptor

signal peptidase

d

GDP + P$_i$

e

signal peptidase cleaves signal

f

Figure 20-8 The signal mechanism directing proteins to the ER (see text).

the ribosome through the membrane as a loop following the signal. GTP hydrolysis releases the SRP from the SRP receptor and the ER.

Binding of ribosomes to the ER appears to open a channel or pore in the ER membrane. The channel may be formed by a group of integral membrane proteins that assemble as a ribosome attaches to the ER. Movement of the growing polypeptide chain may be stimulated by other proteins located in both the ER membrane and ER compartment. Among the latter are probably chaperones (see p. 472) that promote folding of the penetrating chain; the folding may help draw the chain through the membrane.

If the protein is to be secreted, the entire polypeptide chain beyond the signal eventually passes through the ER membrane and into the ER compartment. If the protein is to remain embedded in the membrane, one or more *stop-transfer signals* within the protein ar-

rest its movement through the membrane. Soon after extension of the growing polypeptide into or through the ER membrane an enzyme of the membrane, the *signal peptidase*, clips the N-terminal signal from the protein (Fig. 20-8f). The ribosomal subunits dissociate from the mRNA and release from the ER membrane in the usual pattern as they complete assembly of the polypeptide. The signal element remaining in the membrane is degraded.

Evidence Supporting the Signal Hypothesis The basic tenets of the signal hypothesis have been supported by an extensive series of experiments carried out by the Blobel group and others. One of the most definitive was accomplished by V. R. Lingappa and his co-workers using hybrid proteins. The marker protein used in the experiment was a chimpanzee α-globin, a protein that is normally synthesized on freely suspended ribosomes and remains in the cytoplasmic solution. Onto this gene the Lingappa group grafted a DNA sequence coding for the N-terminal end of a secreted *E. coli* protein, β-lactamase. The hybrid gene was added to a cell-free system derived from wheat containing the enzymes required for transcription and translation. To this were added ER membranes isolated from a dog. In the composite system the gene was transcribed, the mRNA translated, and the marker protein inserted into the ER cisternae. As insertion took place, the *E. coli* sequence was removed from the front of the α-globin protein. The experiment clearly showed that the added sequence is a signal directing a formerly cytoplasmic protein to the ER membranes and inserting it inside the ER cisternae. The experiment also confirmed that the bacterial signal could be recognized and read correctly by eukaryotic receptors and cleaved from the protein at exactly the correct site by the signal peptidase in the ER cisternae. (Lingappa's Experimental Process essay on p. 583 further describes his elegant experiment and its controls.)

The SRP and SRP receptor were discovered by Blobel and his coworkers. In 1981, Blobel and Walter and their coworkers found that, if ribosomes in a cell-free system were washed with a dilute salt solution, they were unable to attach to ER membranes. In the washing solution was found a large complex containing six different polypeptides and, surprisingly, an RNA molecule about 300 nucleotides in length. This complex, the SRP, allowed ribosomes to attach to the ER when added back to the cell-free preparations. Further investigation revealed that the SRP recognizes and binds the N-terminal signal as it emerges from a ribosome and binds the ribosome as well. Recognition of the signal depends on a 54,000-dalton protein of the SRP; the same protein binds GTP. The SRP was subsequently found to be recognized by a protein in ER membranes; this protein is the SRP receptor (also called the *docking protein*), which forms a part of ER membranes. The SRP receptor binds strongly to an SRP linked to a signal sequence.

The RNA molecule forming part of the signal recognition particle is now known as *SRP 7S scRNA*. Investigations by Walter and others showed that if this RNA is removed or broken by enzymatic attack, the SRP disassembles. Thus the scRNA probably acts as a backbone holding the SRP together. This conclusion is supported by the ability of SRP particles to self-assemble in fully functional form if the separate polypeptides are added to intact SRP 7S scRNA molecules.

SRP 7S scRNA molecules have been detected in yeast cells, amphibians, insects, and higher plants. When assembled with the mammalian SRP proteins, amphibian and insect scRNAs can form functional signal recognition particles. Even bacteria have RNAs with sequence similarities to SRP 7S scRNA and a pro-

tein similar to the 54,000-dalton protein of the SRP. (The plasma membrane acts as the counterpart of the ER in prokaryotes; see Supplement 20-1.) Thus the signal mechanism is probably universally distributed in a similar form throughout living organisms.

The stop-transfer signals that arrest progress of proteins through ER membranes were identified and evaluated in hybrid proteins. For example, C. S. Yost, J. Hedgpeth, and V. R. Lingappa constructed a hybrid protein with no stop-transfer sequences by linking 182 amino acids from the N-terminal end of bacterial β-lactamase to 142 amino acids from the C-terminal end of chimpanzee α-globin. The hybrid protein, when translated in a cell-free system complete with dog ER cisternae, passed completely through the ER membranes and wound up in the ER compartment. Adding a segment that contained the stop-transfer sequence of an antibody to the middle of the hybrid, between the β-lactamase and α-globin segments, produced a protein that remained anchored in the ER membranes. In the anchored hybrid the α-globin segment extended from the cytoplasmic side of the ER cisternae and the β-lactamase segment extended into the ER compartment.

Characteristics of ER-Directing Signals Compilation of hundreds of ER-directing signals has revealed that most contain three segments (Fig. 20-9a). At the beginning of the signal, from one to seven amino acids form a polar segment that usually includes from one to three positively charged residues such as lysine. This positively charged segment may promote initial attachment of the signal to the negatively charged ER membrane surface. (Many phospholipid molecules are negatively charged at their polar ends; see p. 95.)

Following the positively charged segment is a hydrophobic *core* that includes from at least 6 to 12 or more amino acids. The amino acids in this region are capable of forming an alpha helix or beta strand (see p. 56) long enough to span the hydrophobic interior of a membrane. The ability of the core to assume a hydrophobic membrane-spanning structure probably promotes insertion of the signal in the membrane interior.

Following the core is the third region of the signal, including the point at which the signal is cleaved after insertion of a protein into the membrane. In most signals this region contains amino acids with small, uncharged side chains at the first and third positions upstream of the cleavage site and often an amino acid with a bulky, polar or charged side group at the second position upstream of the cleavage site (Fig. 20-9b). These structural features are presumably chemical signposts recognized by the signal peptidase removing the signal. The signal peptidase is universal in its activity and can recognize and correctly cleave N-terminal signals from any source, either eukaryotic or prokaryotic.

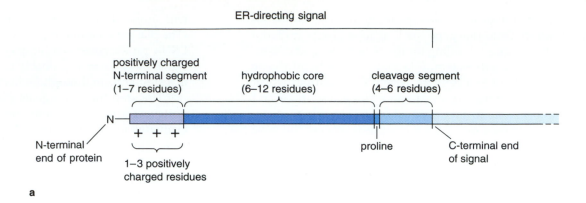

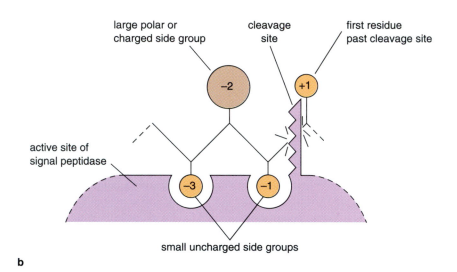

Figure 20-9 ER-directing signals. **(a)** Functional segments common to most ER-directing signals. **(b)** Arrangement of amino acid types at the cleavage site. This arrangement probably fits into the active site of the signal peptidase, precisely locating the cleavage site at the C-terminal end of the signal.

Most identified stop-transfer signals consist of a hydrophobic segment 8 to 20 amino acids in length, flanked at either end by charged or hydrophilic regions. In general the number of stop-transfer signals determines the number of membrane-spanning segments in a protein and the number of times the amino acid chain loops back and forth across the membrane.

Signals Directing Proteins to Further Destinations along the ER Route

Some proteins entering the ER stay there; others become integral elements of further stations along the pathway, including the Golgi complex, lysosomes, the central vacuole of plant and yeast cells, and the plasma membrane. Direction to any of these locations depends on the presence or absence of *postinsertion signals* that target proteins to their final destinations.

Secretion is evidently the fate of proteins that have only an ER-directing signal at the N-terminal end and no further routing information in the form of stop-transfer signals or postinsertion signals. For example, lysosomal enzymes as well as ER proteins are secreted if their routing signals are eliminated or when segments of the receptor mechanism distributing these proteins go awry. Secretion is thus the "default" pathway for ER-directed proteins. If one or more stop-transfer sequences follow the ER-directing signal, so that no postinsertion signals are present, placement in the plasma membrane is evidently the default pathway for the system.

In most cases the sequences or structures forming postinsertion signals for ER-directed proteins have proved difficult to work out because surrounding sequences greatly modify the activity of the internal signal. As a result, most postinsertion signals have been identified only as a region of unknown location and sequence within a longer stretch of amino acids in a

Redirecting Protein Traffic from Cytoplasm to the Secretory Pathway: A Test of the Signal Hypothesis

Vishwanath R. Lingappa

VISHWANATH R. LINGAPPA received the B.A. degree from Swarthmore College in 1975, after which he pursued graduate research with Gunter Blobel at The Rockefeller University. He received the Ph.D. degree in 1979 and the M.D. degree from Cornell University Medical College in 1980. After residency in internal medicine at the University of California Hospitals in San Francisco, he joined the UCSF faculty where he is Professor of Physiology and Medicine. A Board Certified internist, Dr. Lingappa is engaged in medical student teaching, research in molecular mechanisms of protein trafficking, and general internal medicine practice in San Francisco.

For several years after it was first proposed, the signal hypothesis[1] (see p. 579) was met with skepticism from some quarters. Some critics doubted that a small, discrete sequence could be the sole determinant of chain targeting to, and translocation across, the ER membrane, especially since considerable variation in amino acid sequence occurs between signal sequences even within a single species. Rather, they favored notions such as (1) that global features of protein folding distinguished secretory from cytosolic proteins, or (2) that other specific regions of secretory proteins in addition to the signal sequence might be involved in translocation, or (3) that a specific sequence in cytosolic proteins prevented their translocation. Others wondered if cell-free systems, from which most of the data in favor of the signal hypothesis had been derived, were an adequate reflection of events occurring *in vivo*.

We thought of a simple experiment that would provide a dramatic test of the predictions of the signal hypothesis, and thereby resolve some of the controversy. Our idea was to see if a signal sequence was sufficient to direct a normally cytosolic protein across the ER membrane, both in cell-free systems and in living cells. To do such an experiment we would have to take the DNA encoding the signal sequence from one protein and attach it, using molecular genetic techniques, in frame at the 5'-end of the DNA for the coding region of a normally cytoplasmic protein. The resultant hybrid DNA molecule would encode a protein composed of the signal sequence from the secretory protein followed by the "passenger" coding sequence of the cytosolic polypeptide. If mRNA encoding such a *chimeric protein* was transcribed from this DNA, and then translated into protein either in a cell-free system or in a living cell, the location of the protein would provide a powerful test of the signal hypothesis.

In planning experiments it is often useful to consider the possible outcomes in a preliminary "thought experiment" before actually investing a major amount of time and effort. We considered the following possibilities:

1. The signal hypothesis is incorrect or overly simplistic. Hence a signal sequence is either unnecessary or, if necessary, may not be sufficient for translocation of a normally cytosolic protein across the ER membrane. In either case, the chimeric protein would remain in the cytoplasm.

2. The signal hypothesis is correct. Hence a signal sequence, being both necessary and sufficient for translocation across ER membranes, would redirect an otherwise cytosolic protein into the secretory pathway.

3. Finally, regardless of whether the hypothesis is correct or incorrect, the experimental approach might be flawed, for either systematic or fortuitous reasons. For example, juxtaposition of the specific sequences from the two proteins composing the chimera might cause the chain to fold into an insoluble aggregate and hence be scored as nontranslocating. A different chimera might have given a different result, and the one we chose to study may have been an (unfortunate) exception to the general rule.

Experimental science is an inductive rather than a deductive process. Thus, a negative result (e.g., lack of translocation of a particular chimeric protein) is relatively uninformative because it does not distinguish between possibilities (1) and (3). On the other hand, a positive result (e.g., translocation of the chimera) suggests that the signal hypothesis is correct for at least a subset of proteins. It does not, however, rule out the existence of other proteins that are exceptions to these rules. A crucial feature of well-planned experiments is the inclusion of controls that allow the results to be properly interpreted. For example, it is necessary to confirm that the chosen cytoplasmic protein *without* the signal sequence is expressed exclusively in the cytoplasm.

Globin, the protein component of hemoglobin, was chosen as the cytoplasmic protein for these experiments. This is a small, highly conserved protein whose tertiary structure is known. A colleague provided a full-length cDNA clone of chimpanzee alpha globin for these studies. This full-length cDNA had been engineered into a site within the *E. coli* beta lactamase gene. Beta lactamase, being a secretory protein of *E. coli*, has a signal sequence. Moreover, a convenient restriction endonuclease site (see

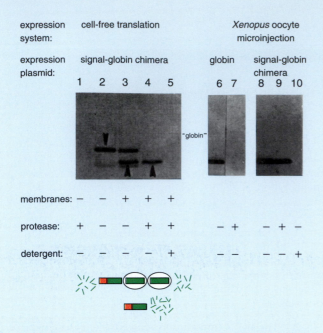

expression system: cell-free translation

Xenopus oocyte microinjection

expression plasmid: signal-globin chimera

1 2 3 4 5

globin

6 7

signal-globin chimera

8 9 10

~globin~

membranes: − − + + +

protease: + − − + +   − +   − + −

detergent: − − − +   − −   − +

Figure A Signal globin expression and translocation in a cell-free translation system (with cartoon interpretation) and in *Xenopus* oocytes.

p. 420) at the start of the globin coding region made deletions of the upstream lactamase codons beyond the signal sequence technically quite easy. Thus, for the initial experiments we chose to use the beta lactamase signal sequence. These DNA molecules were constructed behind an active promoter (see p. 409). This allowed us to make mRNA from the cloned DNA in a test tube and then translate this mRNA in a cell-free translation system.

In the absence of added ER membranes, the full chimera was synthesized, which was larger than authentic globin by the size of the signal sequence (Fig. A, lane 2, downward pointing arrowhead). When proteases were added to this translation product it was totally digested, as espected (lane 1). However if ER-derived membrane vesicles were present *during* translation, a different pattern was observed. A new product the size of authentic globin was seen (lane 3, upward pointing arrowhead), in addition to some residual full-length chains. When these products were digested by added proteases, the authentic globin-sized product was protected from digestion while the residual full-length chains were degraded (lane 4). Protection of authentic globin-sized chains from digestion by added proteases was lost if the vesicle bilayer was solubilized with a mild detergent, strongly suggesting that protection from proteases was due to translocation into the vesicle lumen (analogous to translocation into the lumen of the ER *in vivo*) and not due to intrinsic protease resistance of the protein itself. Moreover, if membrane vesicles were added after completion of protein synthesis, no authentic globin-sized products were observed and the protein remained completely protease sensitive.[2] This demonstrated that translocation of globin occurred only while the signal sequence-bearing chain was being synthesized. As a necessary control, glo-

bin without a signal sequence was shown not to be translocated even when membrane vesicles were present during its synthesis.[2]

To test whether these results applied to events occurring *in vivo*, we used microinjection to introduce the mRNA for authentic globin or the signal-globin chimera into the cytoplasm of *Xenopus* oocytes, large living cells with an active secretory pathway. mRNA so introduced is translated by ribosomes in the oocyte cytoplasm. Radiolabeled amino acids included in the medium which bathes the oocytes are taken up and incorporated into the newly synthesized proteins. Homogenization of the oocytes converts the ER into vesicles, and analysis of localization can be carried out either by fractionation of the homogenate into cytosol and vesicles or by protease digestion, as previously described. Authentic-sized globin was synthesized from both mRNAs (lanes 6 and 8). However, globin synthesized from the chimeric mRNA was protected from added protease (lane 9) unless detergent was added (lane 10), while that synthesized from globin mRNA not encoding a signal sequence was not protected from added proteases (lane 7). Moreover, fractionation demonstrated that authentic globin synthesized from globin mRNA was entirely in the cytosol while authentic-sized globin synthesized from the chimeric mRNA was entirely in the vesicles[3] (not shown). These data suggested that, in the case of the chimera, authentic globin was generated as a result of cleavage of the signal sequence from the growing chain as it translocated to the ER lumen. Thus, both *in vitro* and *in vivo*, the beta lactamase signal sequence was sufficient to direct chimpanzee alpha globin out of the cytoplasm and into the secretory pathway.[2,3]

An ironic twist to this story was that, some time later, an insect globin gene was cloned and found to have a remarkable structure. Globin is found in the cytoplasm of red blood cells in most animal species. However insects lack red blood cells. Instead, they secrete their globin into a fluid, called hemolymph, which bathes the cells. Thus insect globin, whose mature sequence is quite conserved with other animal globins, is naturally a secreted rather than a cytoplasmic protein. When the mRNA for insect globin was examined, it was found to encode a signal sequence at its amino terminus. Unbeknownst to us, we had been "scooped" by evolution 800 million years ago (when insects branched off from the lineage leading to vertebrates) on the neat idea of redirecting cytoplasmic globin into the secretory pathway!

Subsequent work over the years has shown that these findings are true for a wide range of signal and passenger sequences, although passenger sequences are not always passive in the translocation process. Presumably due to features of chain folding, the choice of the passenger can significantly influence the efficiency of every known step in the process of translocation. Furthermore, at least some passenger domains fold in such a way that they cannot be translocated by a given signal sequence. Nevertheless, the simple experiments described above established that, in principle, a signal sequence would redirect a cytoplasmic protein into the secretory pathway.

Although experiments are usually carried out to answer a specific question, they often are most valuable for the subsequent issues they raise and the new ways of looking at a problem that they foster. The simple experiment described above not only provided a test that many critics of the signal hypothesis found compelling, but also contributed to the realization that protein chimeras were a powerful approach to the study of specific sequences involved in protein trafficking. Subsequent work on trafficking of protein chimeras has allowed the study of protein sequences that target proteins to other intracellular organelles besides those of the secretory pathway (e.g., chloroplasts and mitochondria). Similarly, sequences that cause proteins to stop on their way across particular membranes (a crucial step in the assembly of integral membrane proteins) have been identified through the use of chimeric proteins. Finally, protein chimeras have shown that a signal sequence can be located far from the amino terminus and that translocation across the ER can be uncoupled from translation (see, for example, Ref. 4). These sorts of experimental manipulations have contributed to deciphering the intricate mechanisms of protein target-

ing, translocation, and trafficking and remain a fruitful line of investigation today.[5]

Acknowledgments

I thank R. Hegde for critical reading of the manuscript.

References

[1]Blobel, G., and Dobberstein, B. *J. Cell Biol.* 67:852 (1975).

[2]Lingappa, V. R.; Chaidez, J.; Yost, C. S.; and Hedgpeth, J. *Proc. Nat'l Acad. Sci. USA* 81:456 (1984).

[3]Simon, K.; Perara, E.; and Lingappa, V. R. *J. Cell Biol.* 104:1165 (1987).

[4]Perara, E.; Rothman, R. E.; and Lingappa, V. R. *Science* 232: 348–52 (1986).

[5]Chuck, S. L., and Lingappa, V. R. *Cell* 68:1–13 (1992).

Figure 20-10 The second-generation or anti-ideotype technique used to generate antibodies against signal receptors. For the KDEL receptor, a peptide with the KDEL signal is used as an antigen to develop the first generation of antibodies. One or more of these anti-KDEL antibodies might have an active site resembling that of the KDEL receptor. The anti-KDEL antibodies are used as antigens to develop a second generation of antibodies that can recognize the KDEL binding site of the first antibody. The second-generation antibodies are used as a probe for the KDEL receptor in ER or nearby membranes.

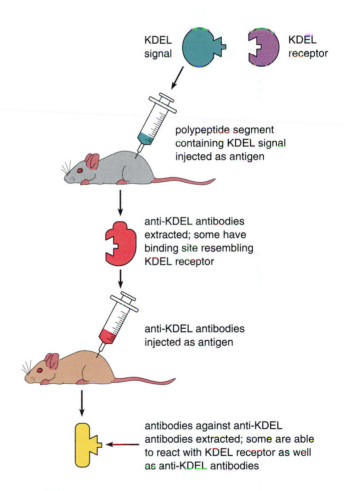

KDEL signal

KDEL receptor

polypeptide segment containing KDEL signal injected as antigen

anti-KDEL antibodies extracted; some have binding site resembling KDEL receptor

anti-KDEL antibodies injected as antigen

antibodies against anti-KDEL antibodies extracted; some are able to react with KDEL receptor as well as anti-KDEL antibodies

protein. However, it has been possible to identify several of these signals, including the KDEL signal retaining proteins in the ER and the mannose 6-phosphate signal directing proteins to lysosomes.

The KDEL Signal Retaining Proteins in the ER   The shortest route followed by ER-directed proteins retains them in the ER. These proteins were found by S. Munro and H. R. B. Pelham to share a sequence of four amino acids at their C-terminal ends, most commonly Lys-Asp-Glu-Leu, or *KDEL* in the one-letter code. Monro and Pelham found that removal of the ER retention signal causes the proteins to be secreted rather than retained in the ER. Conversely, addition of the KDEL signal to the C-terminal end of a non-ER protein causes retention in the ER.

D. Vaux and his colleagues used "second-generation" or *anti-ideotype* antibodies to identify a receptor for the KDEL signal and the probable mechanism by which the receptor operates (Fig. 20-10). Antibodies were first developed against the KDEL signal. If a membrane receptor recognizes and binds the KDEL

signal as part of the ER retention mechanism, it was considered possible that one or more of the anti-KDEL antibodies might have active sites resembling that of the KDEL receptor. Next the anti-KDEL antibodies were used as antigens to develop second-generation antibodies that could recognize the KDEL-binding site. These antibodies were then used as a probe for the KDEL receptor in ER or nearby membranes. The antibodies were found to bind to a membrane protein located primarily in vesicles adjacent to the ER or in the *cis* face of the Golgi complex.

These findings clearly support a hypothesis for salvage and retention of resident ER proteins advanced earlier by Munro and Pelham. According to their idea (Fig. 20-11), resident ER proteins are not fixed rigidly in the ER. Instead, these proteins are carried away in the transition vesicles budding from the ER and fusing with the Golgi complex, along with proteins addressed to sites in the ER-directed pathway. As the ER proteins enter the transition vesicles, they are bound and trapped by the KDEL receptor located in the vesicle membranes (Fig. 20-11a and b). These vesicles then fuse with the *cis* face of the Golgi complex (Fig. 20-11c). After reaching the *cis* cisterna, the KDEL receptors with their bound ER proteins are sorted and placed in vesicles that bud from the Golgi complex and return to the ER (Fig. 20-11d and e). As the vesicles fuse with the ER membranes (Fig. 20-11f),

the KDEL receptors release the ER proteins, and the cycle of release and salvage of ER proteins is ready to turn again.

Thus retention of proteins in the ER depends on a C-terminal KDEL signal, a membrane-bound receptor recognizing and binding the KDEL signal, and a *salvage compartment* consisting of vesicles in which ER proteins bound by the KDEL receptor are sorted out and returned to the ER. If no KDEL signal is present, the proteins leaving the ER continue on their pathway through the system. The same mechanism retaining proteins in the ER compartment also operates in yeast and plants. In these organisms, HDEL (H = histidine) serves as an ER-retaining signal as well as KDEL.

The Mannose 6-Phosphate Signal Directing Enzymes to Lysosomes Lysosomes are membrane-bound sacs containing a combination of enzymes capable of breaking down most biological substances (see p. 597). Many enzymes destined to enter lysosomes were found by S. Kornfeld and others to contain a postinsertion signal consisting of a mannose sugar with a phosphate

Figure 20-11 Mechanism proposed for salvage of resident ER proteins (see text).

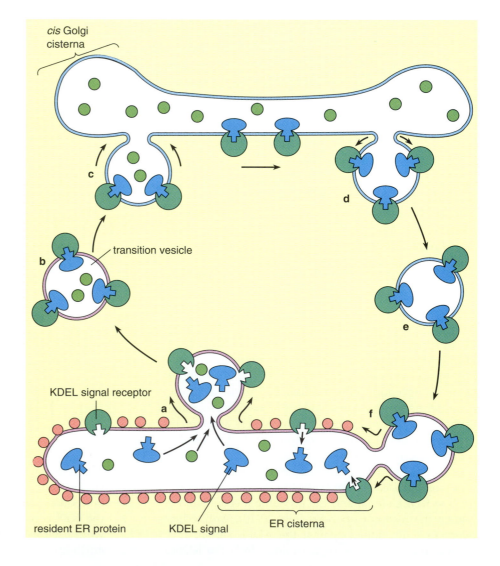

group added to its 6-carbon (the *mannose 6-phosphate signal*; Fig. 20-12a). The signal is part of a carbohydrate unit added in the ER and Golgi complex to enzymes destined for lysosomes.

Tests by W. J. Brown and Farquhar detected a protein in vesicles near the Golgi complex and in lysosomal membranes that acts as the receptor for the lysosome signal. The receptor, which is present in *trans* cisternae of the Golgi complex, is most active in binding the mannose 6-phosphate signal near pH 7. Binding weakens as pH falls below 6; at about pH 5, the pH characteristic of the lysosomal interior, binding affinity is completely lost.

These characteristics suggested that the mechanism sorting lysosomal proteins operates as shown in Figure 20-12b. The internal pH of the Golgi complex is only slightly acid in *trans* cisternae. As a consequence,

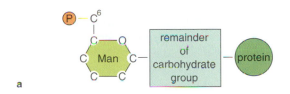

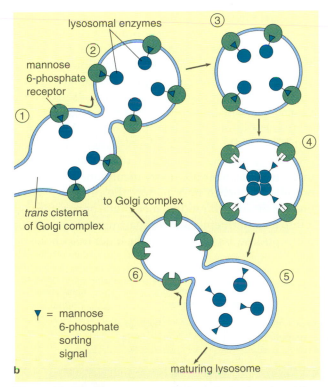

Figure 20-12 The mannose 6-phosphate signal and the mechanism directing proteins to lysosomes. **(a)** The mannose 6-phosphate signal, which forms part of a carbohydrate group added to lysosomal proteins in the ER and Golgi complex. **(b)** The mechanism sorting lysosomal enzymes by means of receptors and the mannose 6-phosphate signal (see text).

enzymes with the mannose 6-phosphate signal are bound tightly to the inner surfaces of the membranes containing the receptor (step 1 in Fig. 20-12b). These membranes pinch off as vesicles destined to form lysosomes (steps 2 and 3). As the internal pH of these vesicles falls because of the activity of H^+-ATPase pumps in the vesicle membranes, the receptors lose their affinity for the lysosomal enzymes and release them into the interior of the vesicles (step 4). After the enzymes are released, the receptors return to the Golgi complex, presumably by means of shuttle vesicles that pinch off from the maturing lysosomes and fuse with the Golgi (steps 5 and 6).

Brown and Farquhar found that treating cells with ammonium (NH_4^+) ions interfered with the lysosomal sorting mechanism. The ammonium ions became concentrated in lysosomes, raising their internal pH. Under these conditions the mannose 6-phosphate receptors failed to release the hydrolytic enzymes and remained linked to them in lysosomes. The linkage jammed the mechanism so that the receptors could not recycle to the Golgi complex. As a result, the supply of mannose 6-phosphate receptors in the Golgi complex became exhausted, and the lysosomal enzymes passing through the Golgi complex were secreted to the exterior of the cell instead of entering lysosomes. Freeing the receptors by removing NH_4^+ from the cells restored the sorting mechanism, switching the enzymes with the mannose 6-phosphate signal from secretion to their normal destination in lysosomes.

The carbohydrate-based signal directing proteins to lysosomes appears to be exceptional. If cells are treated with *tunicamycin*, an antibiotic that inhibits formation of the carbohydrate groups of glycoproteins (see Supplement 20-2), most of the glycoproteins, except for lysosomal enzymes, still reach their ultimate destinations in the cell. Thus for these nonlysosomal glycoproteins, which constitute the majority of sorted and distributed glycoproteins in the cell, carbohydrate groups are evidently not critical to the sorting process and probably do not contain essential routing signals.

Distribution to Locations Outside the ER-Golgi Pathway

Among the important cellular destinations lying outside the ER→Golgi complex→plasma membrane route are mitochondria, chloroplasts, and the nuclear interior. Each of these locations is served by a specialized signal that directs proteins to it. Rather than being assembled on the ER, the proteins addressed to these locations are made on ribosomes that remain freely suspended in the cytoplasm or, in some cases, attach to the surfaces of the organelle.

Distribution to Mitochondria and Chloroplasts Both mitochondria and chloroplasts contain DNA, ribosomes, and all the components necessary for synthesizing proteins (for details, see Chapter 21). However, the proteins encoded and assembled inside the two organelles are only a small part of their total complement. Most of their proteins are encoded in the cell nucleus and synthesized on cytoplasmic ribosomes.

Recent research has shown that the information sorting and distributing proteins to the organelles is spelled out by a signal mechanism with overall properties similar to ER-directed pathways. Among this research are experiments using hybrid proteins. For example, G. van den Broeck and his coworkers used this technique to study the signal directing the small subunit of RuBP carboxylase (see p. 264) to chloroplasts. They produced a hybrid protein by grafting the RuBP carboxylase signal onto the N-terminal end of a bacterial marker enzyme, *neomycin phosphatase*, which normally remains in solution in the bacterial cytoplasm. When supplied with the RuBP carboxylase signal and introduced into tobacco plants, the bacterial enzyme was routed correctly into tobacco chloroplasts. Similar experiments have shown that bacterial cytoplasmic proteins with grafted mitochondria-directing signals are directed successfully to mitochondria.

Mitochondrial Sorting and Distribution Signals Proteins are directed to the different mitochondrial membranes and compartments (see p. 210 and Fig. 8-2) by a variety of signal mechanisms. Most frequent of these is a "standard" pathway that initially directs proteins entirely through both boundary membranes to the mitochondrial matrix. After arrival in the matrix, proteins may stay in this location or be secondarily directed to the inner boundary membrane or intermembrane compartment. Other signal mechanisms direct proteins to the outer boundary membrane or provide additional pathways for certain proteins to the intermembrane compartment or inner boundary membrane.

The "standard" signals directing proteins to the mitochondrial matrix, which may contain as many as 70 amino acids, are typically longer than ER-directing signals. Different mitochondrial signals of this group almost all contain a long region near the N-terminal end with positively charged residues (Fig. 20-13). In a central region following the initial charged segment, many mitochondrial matrix-directing signals contain a stretch of about 20 to 25 polar and nonpolar amino acid residues, long enough to span a phospholipid bilayer. The amino acids of this segment are spaced so that, when arranged into an alpha helix or beta strand, the structure is polar on one side and nonpolar on the other. Structures of this type, termed *amphipathic*, are water soluble but can fold into arrangements allowing them to penetrate spontaneously into membranes (see Fig. 4-7).

Following the amphipathic sequence is a cleavage segment necessary for removal of the signal once insertion of the protein is complete. In many proteins directed to mitochondria a second signal region immediately follows the cleavage indicator for the first signal. The second signal consists of a hydrophobic stretch of amino acids bounded by charged residues and followed by its own cleavage signal; this signal, when present, determines routing within the mitochondrial interior.

Experiments with hybrid proteins show that the first signal directs proteins entirely through the outer and inner boundary membranes and into the mitochondrial matrix. A signal peptidase removes the signal as the protein reaches this location. If no second signal is present, the protein remains in solution in the matrix. If a second signal follows, removal of the first signal places the second at the N-terminal end of the protein. The second signal then directs the protein to insert into the inner boundary membrane. Depending on other sequence information, the protein may remain in the inner boundary membrane or pass through it into the intermembrane compartment. A second signal peptidase associated with the inner mitochondrial

Figure 20-13 Structure of N-terminal signals directing proteins to mitochondria. Chloroplast-directing signals are similar, except that the core of the first signal is neutral or hydrophobic.

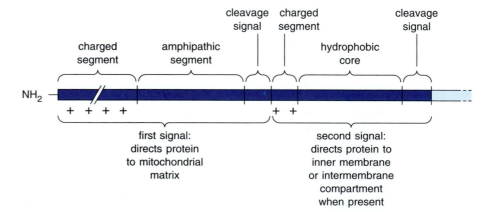

membrane or intermembrane compartment removes the second signal as these movements are completed.

The experiments with hybrid proteins also indicate that the signals directing proteins to mitochondria are universally recognized among eukaryotes. For example, the signal of a plant mitochondrial protein, a malate dehydrogenase enzyme of the watermelon, successfully inserts hybrid proteins into oocyte mitochondria of *Xenopus*, an amphibian. The universal effectiveness of these signals, and probably also of the receptors binding them, indicates that the mitochondrial sorting and distribution mechanism has very ancient origins among eukaryotic organisms and has changed relatively little in the course of evolution.

A number of receptors and other proteins have been identified with the standard system. Two proteins of the outer mitochondrial membrane, *MOM19* and *MOM72*, work as receptors for the standard mitochondria-directing signal. These transfer proteins to a multi-subunit *general insertion protein* that forms a channel directing proteins across the outer membrane. Additional proteins, *MP11* and *Mos6*, form a protein-conducting system in the inner boundary membrane. Chaperones in both the cytoplasm outside mitochondria and in the matrix fold proteins into conformations aiding their travel through the system.

The standard mitochondrial distribution system has interesting relationships to bacterial protein secretion. Mitochondria are believed to have evolved from symbiotic, bacterialike ancestors (see Chapter 26 for details). Their route of entry is believed to have been through enclosure in an endocytotic vesicle (Fig. 20-14). The inner mitochondrial membrane, according to this hypothesis, is the evolutionary remainder of the plasma membrane of the original bacterial symbionts; the outer membrane is a eukaryotic addition derived from the membrane of the endocytotic vesicle. As mitochondria evolved from the bacterial symbionts, most genes encoding the original bacterial proteins were transferred to the cell nucleus.

The two parts of standard mitochondrial signals are believed to reflect these origins. The first signal directing proteins to the mitochondrial matrix, which contains the amphipathic segment, is a eukaryotic evolutionary development necessary to get proteins inside the organelle. These proteins are encoded in genes that were transferred from the organelle to the nucleus during evolutionary development of mitochondria. Once inside the organelle, the proteins remain in the matrix if no second signal is present. These proteins supposedly had no membrane-directing signal in the bacterial ancestors of mitochondria and remained in solution in the prokaryotic cytoplasm.

The second signal, if present, is a remnant from the prokaryotic ancestor. In the ancestor this signal determined whether the protein would become part

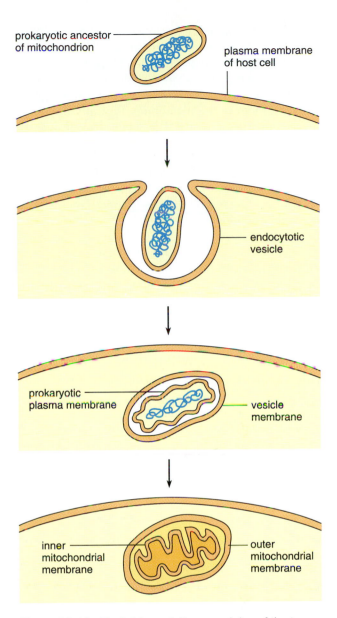

Figure 20-14 Probable evolutionary origins of the two mitochondrial boundary membranes (see text).

of the plasma membrane or be released to the outside of the membrane. In mitochondria the signal directs the protein to the inner membrane (the equivalent of the ancestral plasma membrane) or the intermembrane compartment (equivalent to the space outside the ancestral plasma membrane). In keeping with this proposed relationship, the second signal has structural features resembling the present-day bacterial signal directing insertion into the plasma membrane or release to the outside.

Superimposed on the standard pathway are other mitochondrial delivery systems that appear to be entirely eukaryotic in origin. Proteins such as the mitochondrial porin (see p. 236) destined to form part of the outer boundary membrane have an internal signal or domain that permits entry into this membrane, either directly by spontaneous insertion or via the MOM19 and MOM72 receptors and general insertion protein. For the outer membrane proteins the routing signal or other sequence elements act as stop-transfer signals trapping them in the membrane. Typically the outer membrane-directing signals are not cleaved and remain part of the mature protein. Other proteins, such as cytochrome *c*, have signals or domains that permit them to cross the outer boundary membrane and enter the intermembrane compartment, without first passing into the matrix as in the standard pathway.

Chloroplast-Directing Signals Chloroplasts also have a protein sorting and distribution system equivalent to the standard mitochondrial pathway. The N-terminal signals directing proteins to chloroplasts in this pathway range from about 30 to nearly 100 amino acids in length; most contain more than 50 amino acids. Most of the relatively few chloroplast signals that have been sequenced resemble the standard mitochondrial signal in overall structure. An initial signal is present that directs proteins to the stroma, equivalent to the mitochondrial signal directing proteins to the matrix. The chloroplast initial signal has few evident structural features except an enrichment in serine and threonine and few or no acidic residues. The core of the first signal is neutral or hydrophobic. The initial signal ends in a region that indicates a cleavage site.

In some chloroplast proteins a second signal with properties similar to prokaryotic signals follows the first, stroma-directing signal. The second signal begins with a few charged residues followed by a hydrophobic core sequence and ends with a cleavage signal closely resembling that of prokaryotic signals. The second signal, exposed at the N-terminal end when the initial, chloroplast-directing signal is removed, targets the protein to a thylakoid membrane. Depending on further signal information, the protein may remain embedded in the thylakoid membrane or pass into the thylakoid compartment. A signal peptidase associated with the thylakoid membrane removes the second signal as these movements take place. The second signal is thus the equivalent of the second signal in mitochondria directing proteins to the inner membrane and intermembrane compartment.

Chloroplasts are considered to have evolved from a prokaryotic ancestor similar to present-day cyanobacteria. The pathway directing proteins to chloroplasts probably followed a series of developments equivalent to those proposed for mitochondria, in which the initial chloroplast-directing signal is a eukaryotic development necessary to get proteins inside the organelle. The second signal is a prokaryotic remnant that originally directed proteins to photosynthetic membranes suspended in the cytoplasm; in chloroplasts this signal now directs proteins to the thylakoids. In addition to the standard pathway, chloroplasts also have other signal mechanisms directing proteins to destinations such as the outer boundary membrane.

The signals directing proteins to mitochondria and chloroplasts in plant cells differ sufficiently in structure to prevent misdirection of proteins from one organelle to the other. Hybrid proteins made by grafting a chloroplast signal onto a marker protein, for example, regularly enter chloroplasts; the same protein with a mitochondrial signal enters mitochondria. The differentiation undoubtedly involves specific signal recognition by receptors in the outer membranes of the organelles.

The work accomplished to date with chloroplast-directed proteins indicates that the signal mechanisms of this organelle, like the mitochondrial pathway, are ancient and highly conserved. Chloroplast-directed proteins of algae such as *Chlamydomonas*, for example, can be inserted and correctly cleaved by chloroplasts of a higher plant such as tobacco or vice versa.

Signals Directing Proteins to the Nuclear Interior Signals directing proteins to the nuclear interior differ fundamentally from ER- and organelle-directing signals. Nuclear signals can occur at any site within a protein; some consist of two segments occurring at separate sites in the amino acid chain. The only restriction appears to be that the signal must be exposed at the surface of the protein in its folded state. Unlike the ER- and many organelle-directing signals, the nuclear signal is generally retained when a protein reaches its destination in the nucleus. Nuclear signals are relatively short, strongly basic sequences with some interspersed neutral or hydrophobic residues (Table 20-1). These signals are evidently recognized and bound by receptors in nuclear pore complexes (see p. 384). After binding, an ATP-dependent mechanism pushes the protein through the pore complex into the nuclear interior.

Retention of the nuclear signal after proteins arrive in the nucleus may reflect the fact that in many cell types, particularly in higher eukaryotes, the nucleus breaks down during division and releases its contents to the cytoplasm. Retention of the nucleus-directing signal ensures that nuclear proteins return to the reforming nucleus as division becomes complete. The process can be imitated by injecting bulk nuclear proteins into the cytoplasm of essentially any cell; the injected proteins quickly enter and remain inside the nucleus. Several experiments with hybrid proteins by D. Kalderon and his associates and others demonstrated that addition of the nuclear signals can send normally cytoplasmic proteins to the nucleus.

Table 20-1 Some Nucleus-Directing Signals	
Protein	Signal
Adenovirus E1A	K R P R P
Histone H2B	C P P G K K R S K A
Human lamin A	S V T K K R K L E
Influenza virus nuclear protein	P K K A R E P
myb gene regulatory protein	P L L K K I K Q
Nucleoplasmin*	R P A A T K K A G E A K K K K L D K E D E
Polyoma virus large T antigen	V S R K R P R P A
Rabbit progesterone receptor	R K F K K F N K
SV40 T antigen	P K K K R K V
SV40 UP1	A P T K R K G S
Yeast ribosomal protein L3	P R K R
Yeast ribosomal protein L29	K T R K H R G and K H R K H P G

* The nucleoplasmin sequence contains two consecutive nuclear signals.

Signals Directing Proteins to Other Locations: Microbodies and Plant Vacuoles Microbodies (see p. 243) are small, cytoplasmic structures in which a single boundary membrane encloses an interior matrix. The organelles, which include peroxisomes, glyoxisomes, and glycosomes, contain batteries of enzymes catalyzing a variety of primarily oxidative pathways.

Research with more than 40 proteins directed to microbodies reveals that a short three-amino acid sequence at or near the C-terminal end serves as a routing signal. For most microbody proteins the signal takes the form Ser-Lys-Leu, or SKL. The signal remains part of the microbody proteins after entry into the organelle. Addition of the signal is enough to send proteins normally located in the soluble cytoplasm to microbodies.

The SKL signal appears to be universally employed among eukaryotes. A gene encoding *luciferase*, a microbody enzyme of fireflies, is expressed correctly, with insertion of luciferase in peroxisomes, in yeast and plant cells. Similarly, the C-terminal signal of a yeast peroxisomal protein, PMP20, can direct proteins to peroxisomes in mammalian cells.

Plant vacuoles contain a variety of proteins, including hydrolytic enzymes. For most proteins destined to enter these organelles, a short segment of 8 to 24 amino acids at or near the C-terminal end serves as a routing signal. Few similarities are noted among the routing signals of most vacuolar proteins except that the C-terminal domain is rich in hydrophobic amino acids. Many hydrolases, such as chitinase, occur in plant cells in both the cytoplasmic solution and in vacuoles. The vacuolar versions of these enzymes differ in the presence of the C-terminal, vacuole-directing signal.

For both microbodies and plant vacuoles, routing signals have been discovered only for proteins inside the organelles; little is known about signals directing proteins to their boundary membranes. However, the signal sorting proteins to the boundary membrane of plant vacuoles, known as the *tonoplast*, appears to be located in a transmembrane segment that acts as a stop-transfer sequence.

Eukaryotic cells employ elaborate systems of signals and receptors to deliver proteins to internal and external locations. These mechanisms evidently arose very early in the evolution of eukaryotes as a necessary part of the development of separate internal compartments, bounded and set apart by membranes, as specialized regions of the cell. These ancient mechanisms were highly conserved and passed on essentially intact as the various kingdoms of eukaryotic organisms evolved.

EXOCYTOSIS AND ENDOCYTOSIS

ER-directed proteins lacking further distribution signals are secreted to the cell exterior via exocytosis. Some of the final steps of exocytosis are reversed in endocytosis to provide a route of entry for selected materials from outside the cell. Endocytosis is more than a simple reversal of exocytosis, however; it employs receptors and other elements that do not operate in exocytosis.

Substances entering cells by endocytosis are enclosed in pockets that invaginate inward from the plasma membrane and pinch off as *endocytotic vesicles*. Once inside, the vesicles have various fates depending on the types of proteins or other substances enclosed within them. In some pathways the vesicles fuse with lysosomes, which exposes the proteins in the vesicles to breakdown by lysosomal enzymes. Other pathways deliver endocytotic vesicles to the Golgi complex, where the enclosed proteins are sorted and delivered to further destinations (see Figs. 20-5 and 20-16).

Segments of the plasma membrane constantly cycle between the cell interior and the cell surface through the combined workings of exocytosis and endocytosis. As exocytosis proceeds, vesicle membranes are introduced into the plasma membrane. This input of membrane material is counterbalanced by the removal of membrane segments from the plasma membrane by endocytosis. The balance of the two mechanisms maintains the surface area of the plasma membrane at controlled levels.

The vesicle movements of both exocytosis and endocytosis are apparently powered by microtubule-based mechanisms. Microtubules lie in tracks extending in the directions followed by vesicles in their travels between the Golgi complex and the plasma membrane, with the plus ends of the microtubules (see p. 284) directed toward the plasma membrane. J. Tooze and his coworkers and others observed vesicles sliding in both directions over the microtubules, as if their movements are powered by "motors" working on the microtubule surfaces. A series of experiments has shown that microtubule motors are indeed involved in this movement, with the kinesin motor moving vesicles toward the plasma membrane and another motor, a cytoplasmic dynein, moving them toward the cell interior (see pp. 290 and 297 for a description of these motors). As expected if microtubules are involved in this motion, the vesicle movements cease in cells exposed to colchicine, a drug that promotes microtubule disassembly.

Exocytosis

The experiments by G. Palade and others revealing the pathways of protein secretion led to the conclusion that all eukaryotic cells are capable of releasing proteins to the cell exterior by exocytosis. Different animal cells secrete a variety of proteins by exocytosis. Glandular cells secrete peptide hormones or milk proteins; epithelial cells lining the digestive tract secrete mucus and digestive enzymes; pancreatic cells secrete insulin and other peptide hormones and the precursors of several digestive enzymes including amylase and trypsin. Plant cells secrete structural proteins of the cell wall and cell wall enzymes such as peroxidase, acid phosphatase, and α-amylase.

Many cell types, such as the mucus-secreting cells of the intestinal lining, secrete proteins more or less continuously. Exocytosis of this type is termed *constitutive secretion*. Some cell types, such as neurons, pituitary cells, and mast cells, secrete certain proteins only in response to a stimulus. Exocytosis of this type is called *regulated secretion*. Mast cells, for example, store about 1000 secretory vesicles in their cytoplasm. On receiving an activating signal from outside, as much as 80% of these vesicles are released by exocytosis in a matter of minutes. Most cells secreting proteins regulatively also secrete others constitutively. Cells of the pituitary gland, for example, secrete the peptide hormone ACTH regulatively and laminin (a protein of the extracellular matrix), constitutively.

In cells secreting proteins in both patterns, proteins secreted constitutively are sorted into vesicles that move directly to the plasma membrane. Regulated proteins are placed in storage vesicles for varying lengths of time before the vesicles fuse with the plasma membrane. Proteins probably have a signal that directs them to one of the two pathways. Research by R. B. Kelley, who studied the effects of adding genes for foreign proteins to the pituitary cells, clearly supports this conclusion. Foreign proteins such as insulin and growth hormone, secreted regulatively in other cell types, were also sorted into storage vesicles and secreted regulatively in pituitary cells. Foreign proteins secreted constitutively in their normal locations were correctly sorted into vesicles destined for immediate secretion in the pituitary cells.

Constitutive secretion is probably the "default" pathway for the two mechanisms. Proteins to be secreted constitutively probably contain no diverting signal and are simply routed into vesicles for immediate secretion. Proteins secreted regulatively probably have a signal diverting them from the constitutive pathway into storage vesicles.

Regulated secretion appears to take place primarily or exclusively in response to stimuli from outside the cell, received by receptors forming parts of intracellular signaling pathways such as those employing InsP$_3$/DAG, Ca^{2+}, or cAMP as second messengers (see pp. 159 and 161). For example, Ca^{2+} concentration increases in pancreatic cells immediately before the regulated release of secretory vesicles containing insulin. Increases in Ca^{2+} concentration are also noted immediately before regulated secretion in neurons, mast cells, and in plant cells such as those of growing pollen tubes. In many cell types in which Ca^{2+} forms part of the pathway regulating secretion, experimentally added Ca^{2+} will also induce secretion. In other cells, increases in cAMP or GTP rather than Ca^{2+} concentration appear to be involved in regulated secretion.

A large number of proteins has been implicated in the mechanism triggering fusion of vesicles with the plasma membrane in regulated secretion. Much has been learned about this mechanism through the study of neurotransmitter release by the fusion of synaptic vesicles with the plasma membrane in nerve synapses (see p. 141). Although identified in synapses, the proteins in this system are believed to have counterparts in all regulated systems.

Some proteins, like *synaptobrevin*, are located in vesicle membranes; others, like *syntaxin*, form part of the plasma membrane. Other proteins in the cytoplasmic solution include NSF and SNAPs, originally discovered by the Rothman group in their study of shuttle vesicle fusion in the Golgi complex (see p. 578). The cytoplasmic solution also includes several GTP-binding proteins and other regulatory proteins activated by combination with Ca^{2+} or phosphorylation.

Supposedly Ca^{2+} release triggers a series of events leading to binding of synaptobrevin on vesicles with syntaxin in the plasma membrane, in an interaction promoted by NSF, SNAPs, and other cytoplasmic factors. Thus, in Rothman's terminology, synaptobrevin works as a v-SNARE and syntaxin as a t-SNARE. Linkage of synaptobrevin to syntaxin via NSF and SNAPs triggers fusion of vesicles with the plasma membrane and release of secretory proteins. Proteins related to many elements of the synaptic vesicle system have been detected in yeast, indicating that the mechanism has counterparts in all secretory cells.

Several potent neurotoxins target proteins working in synaptic vesicle fusion. The botulinum toxin released by *Clostridium* bacteria, for example, breaks down syntaxin and synaptobrevin; the tetanus toxin also attacks synaptobrevin. Destruction of these vital proteins impairs neurotransmitter release, with drastic effects on the nervous system.

Endocytosis

Most eukaryotic cell types carry out endocytosis by one or more of three distinct but closely related pathways. In one mechanism, *receptor-mediated endocytosis*, the substances taken in are bound to the cell surface by receptors. The receptors, which are glycoproteins anchored in the plasma membrane, recognize and bind only certain molecules from the medium.

A second pathway, *bulk-phase endocytosis*, simply takes in bulk fluid from the surrounding medium. No binding by cell surface receptors occurs, and any molecules that happen to be in solution in the extracellular fluid are taken in. Bulk-phase endocytosis apparently proceeds at a fairly constant rate in all eukaryotic cells. The endocytosis carried out by higher plant cells is primarily or exclusively of the bulk-phase type and evidently functions to remove segments of the plasma membrane introduced by exocytosis.

The third endocytotic pathway, *phagocytosis* (meaning "cell eating"), differs from both receptor-mediated and bulk-phase endocytosis in that the materials entering the cell are large aggregates of molecules, cell parts, or even whole cells. The process resembles receptor-mediated endocytosis to the extent that the materials taken in are bound by specific cell surface receptors. Both receptor-mediated and bulk-phase en-

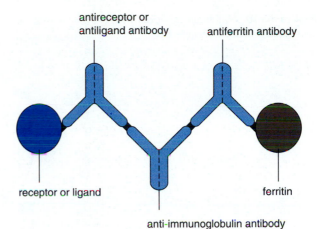

Figure 20-15 The three-way complex linking ferritin, an iron-containing marker that is visible in the electron microscope, to ligands or their receptors (see text).

docytosis are widely distributed among eukaryotic cells, but phagocytosis is more limited. In mammals, for example, only a few cell types, such as macrophages, can take in materials by phagocytosis.

Substances entering cells by endocytosis are followed in their travels by one or more of several experimental techniques. For light microscopy, molecules entering cells are frequently linked to a fluorescent dye such as fluorescein. For electron microscopy the substances are tagged by linking them to heavy metals. Antibodies are also widely used in endocytosis research. Both a receptor and the molecule it recognizes and binds (called the *ligand*) can be specifically tagged by linking them to antibodies. In one of the most frequently employed variations of this approach, different antibodies are developed against the receptor or its ligand, against the antibody proteins themselves, and against *ferritin*, an iron-containing protein that is visible in the electron microscope. Reacting the antibodies sets up a three-way complex that links ferritin specifically to a receptor or its ligand (Fig. 20-15).

Receptor-Mediated Endocytosis Of the three endocytotic mechanisms, receptor-mediated endocytosis has been studied experimentally in greatest detail, particularly in mammalian cells. The substances mammalian cells take in by receptor-mediated endocytosis include growth factors and peptide hormones; blood serum proteins, including molecules that carry metabolites such as cholesterol and other lipids, iron, and vitamins; lysosomal enzymes; antibodies; and defective serum proteins and glycoproteins. Some of these substances, such as serum proteins carrying iron and vitamins, are recognized and taken in by essentially all mammalian cell types; others, such as defective serum proteins and glycoproteins, are taken in by

Figure 20-16 Pathways in receptor-mediated endocytosis (see text).

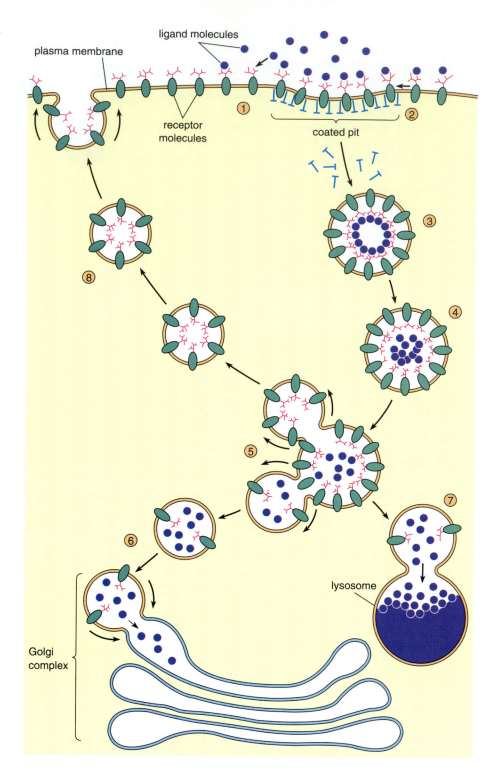

ligand molecules

plasma membrane

receptor molecules

coated pit

lysosome

Golgi complex

only a relatively few cells such as those of the liver. Some pathogens or toxins, including viruses and the diphtheria and cholera toxins, use receptor-mediated endocytosis as a pathway to enter cells.

Receptor-mediated endocytosis proceeds in several steps (Fig. 20-16). After binding by a receptor (step 1 in Fig. 20-16), substances entering cells are con-

centrated in depressions in the plasma membrane (step 2). The depressed membrane segments, called *coated pits*, are characterized by a thick layer of dense material that covers their cytoplasmic surface (see Fig. 20-17). The pits then invaginate and pinch free from the plasma membrane as endocytotic vesicles (step 3). As the vesicles form, they lose their surface

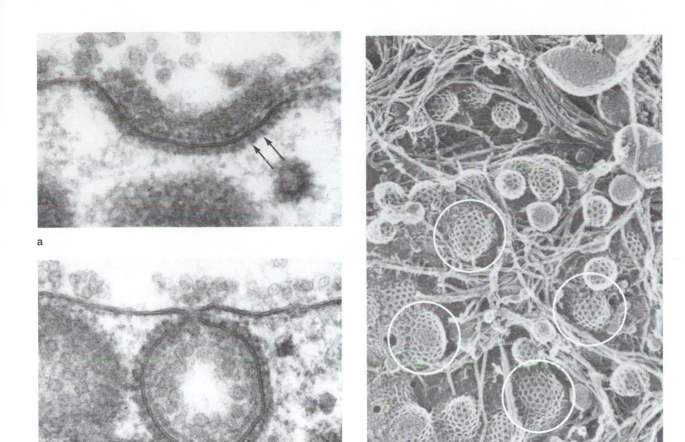

a

b

c

Figure 20-17 Formation of a coated pit **(a)** and vesicle **(b)** in a developing chick oocyte. In sections the coat appears as a covering of short, knobbed bristles on the cytoplasmic surface of the pit or vesicle membrane (arrows in **a**). × 135,000. (Courtesy of M. M. Perry.) **(c)** Coated pits (circled) viewed face on from the inside surface of a plasma membrane isolated from a mouse liver cell. The fibers (arrow) are elements of the cytoskeleton. × 80,000. (Courtesy of J. Heuser, from *Cell* 30:395 [1982]. Copyright Cell Press.)

coat. At this stage the vesicles may be spherical or extended into elongated and variously branched tubules. As they travel into the underlying cytoplasm along microtubules, the vesicles may fuse into larger structures. Multiple GTP-binding proteins participate in coated pit formation, invagination, and the release of endocytotic vesicles into the cytoplasm.

During vesicle movement and fusion, substances in the vesicles are sorted and in many cases released from their receptors (step 4). The sorting evidently involves an interaction between signals on the endocytosed ligand-receptor complexes and receptors on the endocytotic vesicles. For example, the receptor for transferrin, a blood protein that carries iron atoms, has the signal sequence YTRF in its N-terminal region. This signal, which is shared with several other surface receptors, is the primary indicator for endocytosis and sorting within endocytotic vesicles. The low-density lipoprotein (LDL, a carrier for cholesterol) receptor carries the signal NPVY; other typical sorting signals for endocytosed receptors are YRGV and YQTI. Induced mutations in these signals often leave the receptor stranded at the cell surface, unable to enter endocytosis. Other sequences surrounding these signals contribute to precise targeting to destinations within the cell. The sorted molecules may proceed to the Golgi complex or lysosomes within their original endocytotic vesicle, or in smaller vesicles that pinch off from the endocytotic vesicle (step 5).

The substances sorted into vesicles that travel to the Golgi complex (step 6) may remain there or be distributed further to the ER or perinuclear space.

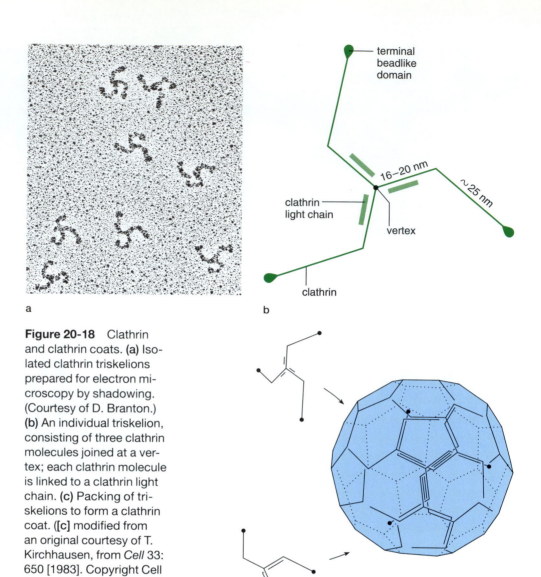

Figure 20-18 Clathrin and clathrin coats. **(a)** Isolated clathrin triskelions prepared for electron microscopy by shadowing. (Courtesy of D. Branton.) **(b)** An individual triskelion, consisting of three clathrin molecules joined at a vertex; each clathrin molecule is linked to a clathrin light chain. **(c)** Packing of triskelions to form a clathrin coat. ([c] modified from an original courtesy of T. Kirchhausen, from *Cell* 33: 650 [1983]. Copyright Cell Press.)

terminal beadlike domain

16–20 nm

~25 nm

clathrin light chain

vertex

clathrin

a

b

c

They may also be resorted and secondarily routed in vesicles to the plasma membrane or to lysosomes. Fusion of vesicles with lysosomes (step 7) exposes their contents to degradation by the lysosomal enzymes. The material in vesicles returning to the plasma membrane (step 8), primarily unbound receptors, is reintroduced into the plasma membrane by exocytosis. This accomplishes recycling of the receptors. Not all receptors are recycled; some are degraded in the vesicles that fuse with lysosomes.

Receptors for substances entering cells by endocytosis are usually present in the plasma membrane in many thousands of copies. Liver cells, for example, contain as many as 150,000 transferrin receptors; the LDL receptor occurs in about 20,000 copies in the plasma membrane of a typical mammalian cell such as a fibroblast.

The Proteins Forming Coated Pits The coats of coated pits appear as a basketlike network that covers the cytoplasmic side of the plasma membrane (Fig. 20-17). The basketlike appearance results from an interlocking network of polypeptides that trace out pentagons, hexagons, and other regular geometric patterns on the membrane surface.

B. M. F. Pearse and her colleagues found that isolated coated pits contain a 180,000-dalton polypeptide as a major constituent. Termed *clathrin* by Pearse (from *clathratus* = lattice), the polypeptide is capable of assembling reversibly into baskets in the test tube under various conditions. In the electron microscope an isolated clathrin subunit appears as a three-legged structure called a *triskelion* (Fig. 20-18). The molecular weight of intact triskelions lies near 600,000, indicating that the three-legged structure probably contains three clathrin polypeptides, linked together at the center of the triskelion. Each triskelion also contains three *clathrin light chains*, one associated with each clathrin molecule of the triskelion.

Clathrin does not link directly to the plasma membrane. Instead, it is bound to the membrane by a group of *adaptor* proteins that can bind both clathrin and the membrane. The adaptor proteins also bind the cytoplasmic extensions of receptors trapped in coated pits.

Clathrin and other coated-pit polypeptides have been detected in animals, plants, fungi, algae, and protozoa. The amino acid sequence of the clathrin polypeptide in yeast, deduced from the gene sequence by S. K. Lemmon and her coworkers, is 50% identical to mammalian clathrin.

Assembly and Disassembly of the Clathrin Coat Clathrin coats assemble with adaptor proteins from a pool of subunits in solution in the cytoplasm. Once an endocytotic vesicle pinches off from the plasma membrane, the coat disassembles from its surface and joins the cytoplasmic pool of disassembled subunits. At any time, about half the total clathrin complement of eukaryotic cells is estimated to be assembled into coats.

Assembly of the clathrin coat evidently proceeds in cells without an energy input. However, disassembly of the coats from fully formed vesicles is an energy-requiring process that involves an *uncoating protein* with ATPase activity. The activity of the uncoating protein depends on the presence of assembled baskets containing both clathrin and clathrin light chains. When activated, the ATPase hydrolyzes ATP, on the order of about three molecules of ATP per triskelion, and promotes rapid disassembly of clathrin baskets.

Not all plasma membrane segments involved in endocytosis form coated pits. Membrane coats are rare in cells taking in material by phagocytosis; the pockets forming in the plasma membrane in this process usually have smooth walls. Significantly, membrane invagination in phagocytosis is frequently observed to be ATP-dependent, in contrast to coated-pit formation, which is ATP-independent. This observation suggests that an energy-requiring motile system, possibly based on microtubules or microfilaments, may form membrane pockets in phagocytosis.

Clathrin coats also occur on some secretory vesicles budding from *trans* cisternae of the Golgi complex, particularly on those containing proteins released by regulated secretion. Secretory vesicles entering the constitutive pathway are coated instead by COPs.

Endocytotic Vesicles, pH, and Receptor-Ligand Sorting The vesicles that pinch off from coated pits during receptor-mediated endocytosis initially contain a mixture of different receptors, with each still linked to its ligand. Very shortly after a newly formed endocytotic vesicle separates from the plasma membrane, many of the receptors release their ligand into the vesicle interior.

Ligand release appears to depend on both pH changes in endocytotic vesicles and sensitivity of the receptor-ligand complexes to the changes. The pH of the medium immediately surrounding the cell, which is trapped and carried into endocytotic vesicles as they form, is initially near the neutral value of pH 7. Immediately after an endocytotic vesicle releases from the plasma membrane, its internal pH drops rapidly to 5 to 5.5. The pH drop can be measured experimentally by attaching pH-sensitive dyes such as fluorescein to ligands. (The dyes absorb light differently at different pH's.) The reduction in pH results from activity of an H^+-ATPase pump in the vesicle membranes. At the lower pH characteristic of the endocytotic vesicle some receptors, such as the LDL receptor, lose their affinity for their ligand and release it into the vesicle interior. Other receptors, such as the receptor binding EGF, are insensitive to the pH change and remain tightly bound to their ligand.

In many cases the sensitivity of receptors to pH changes in endocytotic vesicles determines whether receptors are degraded in lysosomes or recycled to the plasma membrane. Most receptors remaining bound to their ligands as endocytotic vesicle pH falls travel to lysosomes and are degraded with their ligand. Those releasing their ligand are sorted into membrane segments that pinch off from endocytotic vesicles and eventually return to the plasma membrane. As expected from this mechanism, agents that raise the pH of endocytotic vesicles, such as NH_4Cl, block ligand release by pH-sensitive receptors. At the artificially elevated pH the receptors are trapped with their ligands in the lysosomal pathway. As a consequence, they are degraded and fail to reappear at the cell surface.

Activities of Lysosomes in Endocytosis Lysosomes (Fig. 20-19), which occur in greatest abundance in animal cells, contain some 40 to 50 or more different hydrolytic enzymes including proteinases, lipases, phospholipases, phosphatases, nucleases, glycosidases, and sulfatases. The enzymes, in aggregate, are capable of breaking down all major classes of biological macromolecules into small, soluble subunits that can be transported across the lysosomal membrane into the surrounding cytoplasm. Hydrolytic enzymes equivalent to some of the lysosomal enzymes of

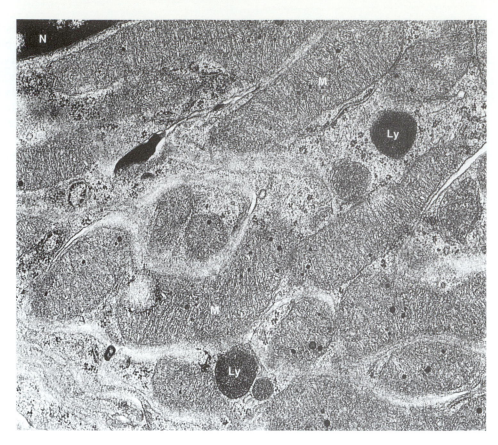

Figure 20-19 Lysosomes (Ly) in cells of mouse kidney. *N*, nucleus; *M*, mitochondrion. × 22,000.

animals are present in the central vacuoles of plant and yeast cells.

In general the hydrolytic reactions catalyzed by the lysosomal enzymes take place at about pH 4.5 to 5, which is characteristic of the lysosomal interior. The lysosomal enzymes, almost without exception, are glycoproteins containing complex carbohydrate groups. One of the many unanswered questions about lysosomes is how, given the impressive array of hydrolytic enzymes contained within these structures, the lysosomal boundary membrane remains intact.

The lysosomal boundary membrane contains characteristic proteins, among them an H^+-ATPase active transport pump that continually moves H^+ from the surrounding cytoplasm into the lysosomal interior, thereby maintaining the acidic pH characteristic of these organelles. Other proteins in the lysosomal boundary membrane transport small molecules resulting from digestion in the lysosomal interior to the surrounding cytoplasm.

Lysosomes participate in the degradation of cellular molecules and structures through a variety of pathways in addition to endocytosis. Entire cell organelles may be digested within lysosomes; the sig-

nificance of this digestion, known as *autophagy*, is obscure, except that it probably plays an important role in breakdown of organelles in cells undergoing differentiation or physiological stress. In another process the lysosome membrane ruptures and hydrolytic enzymes are released into the surrounding cytoplasm. If large numbers of lysosomes are involved, this process either results in or accompanies cell death. Both of these self-digestion mechanisms are part of normal development in animals and plants (as, for example, in the breakdown of cells and resorption of the tail in developing frogs; see also p. 632).

Lysosomal enzymes are also regularly secreted to the cell exterior. In this process, lysosomes act as secretory vesicles and release their hydrolytic enzymes to the exterior by fusing with the plasma membrane. Certain types of leucocytes, including eosinophils, for example, secrete lysosomal enzymes on encountering antigen-antibody complexes or when invading sites of infection. Lysosomal enzymes are also secreted to the cell exterior as part of normal cell function during processes such as bone resorption and reduction of uterine tissues after pregnancy.

For Further Information

Suggestions for Further Reading

Abeijon, C., and Hirschberg, C. B. 1992. Topography of glycosylation reactions in the ER. *Trends Biochem. Sci.* 17: 32–36.

Almers, W. 1990. Exocytosis. *Ann. Rev. Physiol.* 52:607–624.

Boulikas, T. 1993. Nuclear localization signals. *Crit. Rev. Eukary. Gene Expression* 3:193–227.

Caplan, M. J. 1991. Biogenesis and sorting of plasma membrane proteins. *Curr. Top. Membr. Transport* 39:37–86.

Chock, S. P., and Schmauder-Chock, E. A. 1992. The secretory granule and the mechanism of stimulus-secretion coupling. *Curr. Top. Cellular Regulat.* 32:183–208.

Chrispeels, M. J., and Tague, B. W. 1991. Protein sorting in the secretory system of plant cells. *Internat. Rev. Cytol.* 125:1–45.

Creutz, C. E. 1992. The annexins and exocytosis. *Science* 258:924–931.

Dahms, N. M., Lobel, P., and Kornfeld, S. 1989. Mannose 6-P receptors and lysosomal enzyme targeting. *J. Biolog. Chem.* 264:12115–12118.

DeBoer, A. D., and Weisbeek, P. J. 1991. Chloroplast protein topogenesis: Import, sorting, and assembly. *Biochim. Biophys. Acta* 988:1–45.

Driouich, A., Faye, L., and Staehelin, A. L. 1993. The plant Golgi apparatus: A factory for complex polysaccharides and glycoproteins. *Trends Biochem. Sci.* 18:210–214.

Ferro-Novick, S., and Novick, P. 1993. The role of GTP-binding proteins in transport along the exocytic pathway. *Ann. Rev. Cell Biol.* 9:575–599.

Glick, B. S., Beasley, E. M., and Schatz, G. 1992. Protein sorting into mitochondria. *Trends Biochem. Sci.* 17:453–459.

Holtzman, E. 1989. *Lysosomes.* New York: Academic Press.

Holtzman, E. 1992. Intracellular targeting and sorting. *Biosci.* 42:608–620.

Hong, W., and Tang, B. L. 1993. Protein trafficking along the exocytotic pathway. *Bioess.* 15:231–238.

Keen, J. H. 1990. Clathrin assembly proteins and the organization of the coated membrane. *Adv. Cell Biol.* 3: 153–176.

Kelley, R. B. 1990. Microtubules, membrane traffic, and cell organization. *Cell* 61:5–7.

Kiebler, M., Becker, K., Pfanner, N., and Neupert, W. 1993. Mitochondrial protein import: Specific recognition and membrane translocation of preproteins. *J. Membr. Biol.* 135:191–207.

Luzio, J. P., and Banting, G. 1993. Eukaryotic membrane traffic: Retrieval and retention mechanisms to achieve organelle residence. *Trends Biochem. Sci.* 18:395–398.

van Meer, G. 1989. Lipid traffic in animal cells. *Ann. Rev. Cell. Biol.* 5:247–275.

Mellman, I. M., and Simons, K. 1992. The Golgi complex: In vitro veritas. *Cell* 68:829–840.

Moore, P. J., Swords, K. M. M., Lynch, M. A., and Staehelin, L. A. 1991. Spatial organization of the assembly pathways of glycoproteins and complex polysaccharides in the Golgi apparatus of plants. *J. Cell Biol.* 112:589–602.

Osumi, T., and Fugiki, Y. 1990. Topogenesis of peroxisomal proteins. *Bioess.* 12:217–222.

Paulson, J. C., and Colley, K. J. 1989. Glycosyltransferases. *J. Biolog. Chem.* 264:17615–17618.

Pelham, H. R.B. 1990. The retention signal for soluble proteins of the endoplasmic reticulum. *Trends Biochem. Sci.* 15:483–486.

Pryer, N. K., Wuestehube, L. J., and Schekman, R. 1992. Vesicle-mediated protein sorting. *Ann. Rev. Biochem.* 61: 471–516.

Pugsley, A. P. 1993. The complete general secretory pathway in gram-negative bacteria. *Microbiol. Rev.* 57:50–108.

Rapoport, T. A. 1992. Transport of proteins across the endoplasmic reticulum membrane. *Science* 258:930–936.

Rothman, J. E., and Orci, L. 1992. Molecular dissection of the secretory pathway. *Nature* 355:409–415.

Salmond, G. P. C., and Reeves, P. J. 1993. Membrane traffic wardens and protein secretion in gram negative bacteria. *Trends Biochem. Sci.* 18:7–12.

Sanders, S. L., and Schekman, R. 1992. Polypeptide translocation across the endoplasmic reticulum membrane. *J. Biolog. Chem.* 267:13791–13794.

Schmid, S. L. 1992. The mechanism of receptor-mediated endocytosis: more questions than answers. *Bioess.* 14: 589–596.

Subrimani, S. 1993. Protein import into peroxisomes and biogenesis of the organelle. *Ann. Rev. Cell Biol.* 9:445–478.

Tatham, P. E. R., and Gomperts, B. D. 1991. Late events in regulated exocytosis. *Bioess.* 13:397–401.

Trowbridge, I. S., Collawn, J. F., and Hopkins, C. R. 1993. Signal-dependent trafficking in the endocytotic pathway. *Ann. Rev. Cell Biol.* 9:129–161.

Vitale, A., and Chrispeels, M. J. 1992. Sorting of proteins to the vacuoles of plant cells. *Bioess.* 14:151–160.

Wandersman, C. 1992. Secretion across the bacterial outer membrane. *Trends Genet.* 8:317–321.

Watts, C., and Marsh, M. 1992. Endocytosis: What goes in and how? *J. Cell Sci.* 103:1–8.

Weinhaues, U., and Neupert, W. 1992. Protein translocation across mitochondrial membranes. *Bioess.* 14:17–23.

Wirtz, K. W. A. 1991. Phospholipid transfer proteins. *Ann. Rev. Biochem.* 60:73–99.

Review Questions

1. Outline the structure of the ER and Golgi complex. In what ways are the two membrane systems similar? Different? In what ways is the nuclear envelope related to the ER?

2. What are transition vesicles? Shuttle vesicles? Secretory vesicles? Storage vesicles? Endocytotic vesicles?

3. Outline the primary functions of the ER and Golgi complex. What major groups of proteins occur in the ER and Golgi complex?

4. Outline the major steps in assembly of the carbohydrate groups of glycoproteins in the ER and Golgi complex.

5. What major lines of evidence indicate that proteins move from the ER, through the Golgi complex, and to the cell exterior?

6. What experimental results led to formulation of the signal hypothesis? Outline the events taking place when a ribosome initiates protein synthesis on an mRNA for a secretory protein.

7. What is the signal recognition particle? How does it function in ER-directed distribution mechanisms? What other elements and factors are believed to operate in direction and insertion of proteins into the ER, and how are they believed to interact?

8. What structural characteristics are common to most ER-directing signals? How are these elements believed to act in directing proteins to the ER?

9. What are stop-transfer signals? Postinsertion signals? How are signals in these catagories believed to operate in the insertion, sorting, and distribution of ER-directed proteins?

10. What is the salvage compartment, and how is it believed to interact with the KDEL signal to return resident proteins to the ER?

11. What signal determines distribution of proteins to lysosomes? What is the role of pH in the distribution of proteins to lysosomes?

12. Outline the sources of membrane proteins for the ER, Golgi complex, plasma membrane, nuclear envelope, mitochondria, and chloroplasts.

13. What signals route proteins to mitochondria and chloroplasts? What separate membranes and compartments may proteins be assigned to in the two organelles? Compare the signals of ER- and organelle-directed proteins.

14. Outline the mechanism directing proteins to the mitochondrial matrix and chloroplast stroma. What mechanisms distribute proteins further from these locations? In what ways are these distribution mechanisms related to prokaryotic secretion?

15. What signals send proteins to the nucleus? What is the role of nuclear pore complexes in the distribution of proteins to the nucleus? Compare the probable routes followed by a nuclear envelope protein and a chromosomal protein. How are proteins directed to microbodies?

16. What is the difference between constitutive and regulated secretion? What evidence suggests that constitutive secretion is the "default" pathway?

17. Trace the possible routes followed by a protein taken in by endocytosis. Compare receptor-mediated endocytosis, bulk phase endocytosis, and phagocytosis.

18. What cellular pathways may be followed by the receptors active in receptor-mediated endocytosis? What is the relation of pH changes to these pathways?

19. What are coated pits? How do coated pits function in endocytosis? What major protein groups occur in coated pits? What overall functions are associated with each protein group? What is a triskelion? Adaptor proteins?

20. What types of enzymes occur in lysosomes? How do lysosomes function in endocytosis?

21. Compare protein secretion in bacteria and eukaryotes. What are the characteristics of membrane-directing signals in bacteria? (See Supplement 20-1.)

Protein Sorting, Distribution, and Secretion in Prokaryotes

Although prokaryotes are more simply organized than eukaryotic cells, proteins are also routed to various locations in these organisms. In gram-positive bacteria (see p. 199), possible destinations are the cell interior, the plasma membrane, or the cell exterior including the cell wall. In the more complex gram-negative bacteria and cyanobacteria, different destinations include the cell interior, the plasma membrane, the periplasmic compartment between the plasma membrane and outer membrane, the outer membrane itself, and the cell exterior (see Fig. 7-21). Among the proteins secreted to the exterior are hydrolytic enzymes and toxins.

Proteins destined to remain inside bacterial cells are assembled on ribosomes suspended in the cytoplasmic solution. Proteins addressed to the plasma membrane, cell wall, or exterior may be assembled on ribosomes that are freely suspended or attached to the cytoplasmic surface of the plasma membrane.

With some exceptions, distribution of proteins to the plasma membrane in bacteria depends on an N-terminal signal similar to the ER-directing signal of eukaryotes. The bacterial signal includes an initial positively charged segment, a hydrophobic middle segment capable of spanning a membrane as an alpha helix or beta strand, and a terminal region specifying cleavage of the signal. These segments have properties similar to the ER-directing signals of eukaryotes except that the initial segment is frequently more positively charged, and the core often less hydrophobic. Once its function in directing insertion of proteins in the plasma membrane is complete, the N-terminal signal is removed from most prokaryotic proteins by a signal peptidase.

Insertion of newly synthesized proteins in the plasma membrane of *E. coli* requires a membrane potential and ATP. Several proteins have been implicated in the process through the effects of mutations that slow or stop movement of proteins through the plasma membrane. Two of these, *SecB* and *GroEL*, are chaperones in the cytoplasmic solution that hold proteins in conformations promoting their entry into the membrane. Another protein, *SecA*, is an ATPase attached to the cytoplasmic surface of the membrane. SecA probably acts in concert with *SecY*, an integral membrane protein that may be primarily involved in movement of proteins across the plasma membrane. ATP hydrolysis by SecA is considered likely to be associated with transfer of proteins from SecB or GroEL to SecY. Several other integral membrane proteins— *SecD*, *SecE*, and *SecF*—may join with SecY as parts of a protein-translocating complex in the *E. coli* plasma membrane.

Recently A. Dobberstein and P. Walter and their associates discovered a particle in *E. coli* that has some of the properties of the eukaryotic signal recognition particle. The possible prokaryotic counterpart to the SRP contains a 4.5S RNA that can be substituted for SRP 7S scRNA in eukaryotic SRP and a 48,000-dalton polypeptide that can take the place of the critical 54,000-dalton polypeptide in eukaryotic SRP. The prokaryotic 4.5S RNA and 48,000-dalton polypeptide also have sequence similarities to their eukaryotic counterparts. These findings indicate that *E. coli* and other bacteria have an SRP-like complex that recognizes N-terminal signals and assists binding of proteins to the bacterial plasma membrane as part of membrane insertion or secretion.

If proteins are to remain part of the plasma membrane, stop-transfer sequences or conformations hold the protein in the membrane. Proteins that are to be placed in the cell wall or secreted in gram positive bacteria, or released into the periplasm in gram-negative bacteria, pass entirely through the plasma membrane.

Relatively little is known about the signals and mechanisms directing proteins to the outer membrane of gram-negative bacteria or indicating that they are to be secreted by passing entirely through the outer membrane. Many of these proteins travel to the periplasmic space by means of an N-terminal signal that directs their insertion and passage through the plasma membrane. Once the proteins are in the periplasm, further information, probably involving folding conformations promoted by their amino acid sequences, directs their insertion in the outer membrane. Passage entirely through the outer membrane, which is highly impermeable to most substances, may involve receptor and translocator proteins in the outer membrane that are unique to each secreted protein.

In gram-positive bacteria a C-terminal signal indicates proteins that are to be retained in the cell wall after passage through the plasma membrane. The signal, about 35 amino acids in length, contains a sequence segment, LPXTG (in which X indicates any amino acid), which is highly conserved in many gram-positive wall proteins. Other segments of the wall-directing signal include a hydrophobic domain and a charged tail.

Although the bacterial pathways distributing newly synthesized proteins are simpler than eukaryotic routes, the interchangeability of signals demonstrates the close relationship of the two systems. As noted in the main part of this chapter (see p. 579), eukaryotic secretory and membrane proteins, when introduced by recombinant DNA techniques into bacterial cells, are correctly routed to the plasma membrane or secreted to the cell exterior. Similarly, bacterial secreted proteins such as β-lactamase are correctly sorted into secretory vesicles and released to the exterior when their genes are introduced into yeast cells or *Xenopus* oocytes. The similarities indicate that the evolutionary lineage of the ER-directing signal mechanism of eukaryotes probably extends to beginnings in ancient prokaryotic organisms or to still earlier ancestors of both present-day prokaryotes and eukaryotes.

Supplement 20-2

Reactions Adding Sugar Units to Proteins in the ER and Golgi Complex

The sugar units added in protein glycosylation include primarily glucose, galactose, mannose, fucose, acetylglucosamine and acetylgalactosamine (see Fig. 4-4 and p. 99). Before addition the sugar units are activated in the soluble cytoplasm surrounding the ER and Golgi complex by reaction with one of the nucleoside triphosphates UTP, CTP, or GTP. When mannose is activated, for example, UTP reacts with the sugar to produce a UDP-mannose derivative:

$$UTP + mannose \rightarrow UDP\text{-}mannose + PP_i \quad (20\text{-}1)$$

Some activated sugars are transported across ER and Golgi membranes in this form by proteins that exchange a nucleotide-sugar complex from outside for an unbound, or "empty," nucleotide on the inside. The sugar is then transferred directly from the nucleotide residue to the growing carbohydrate group by enzymes called *glycosyl transferases*, which are active within the organelle compartment. Energy for the additions is derived from hydrolysis of the nucleoside diphosphate-sugar complex, which retains much of the energy released by removal of the phosphates in reaction 20-1. The glycosyl transferase enzymes, themselves glycoproteins, are integral membrane proteins of the ER or Golgi complex.

Some sugars are transported into the ER and added to glycoproteins via a *carrier lipid* that acts as an intermediate in the reactions (Figure 20-20 shows the carrier lipid *dolichol*). In this pathway, one or more activated sugars are transferred from their nucleotide carrier to the carrier lipid on the cytoplasmic side of the ER membranes. Several sugars may be linked together on the carrier lipid, producing a partly assembled carbohydrate group on the cytoplasmic side of the ER. The sugar or carbohydrate is then transported across the ER membranes while linked to the carrier lipid and is transferred to the glycoprotein being modified.

The reactions linking sugars to glycoproteins take two primary forms depending on the amino acid side chains to which the carbohydrate groups link. *N-linked* carbohydrate groups are bound to an amino ($-NH_2$) group in the side chain of an asparagine residue in the protein. *O-linked* carbohydrate groups are bound to a hydroxyl ($-OH$) group on the side chain of serine or threonine (or of the modified amino acids hydroxyproline or hydroxylysine). Both *N-* and *O-linked* carbohydrate groups may be added to the same protein.

Figure 20-21 shows a series of reactions adding an *N-linked* carbohydrate group that is common to many organisms. The process begins on the cytoplasmic side of the ER, with transfer of two acetylglucosamine units and five mannose units from their nucleotide carriers to a carrier lipid (Fig. 20-21a to c). Once assembled to this extent, the complex is transported across the ER membrane by the carrier lipid. In the ER lumen, four more mannose units are added (Fig. 20-21d). This core structure is now transferred from the carrier lipid to an asparagine residue in the protein being glycosylated (Fig. 20-21e). In some organisms, this step completes the core structure; in others, three glucose residues are transferred from their nucleotide carriers on the lumen side of the ER to form the core

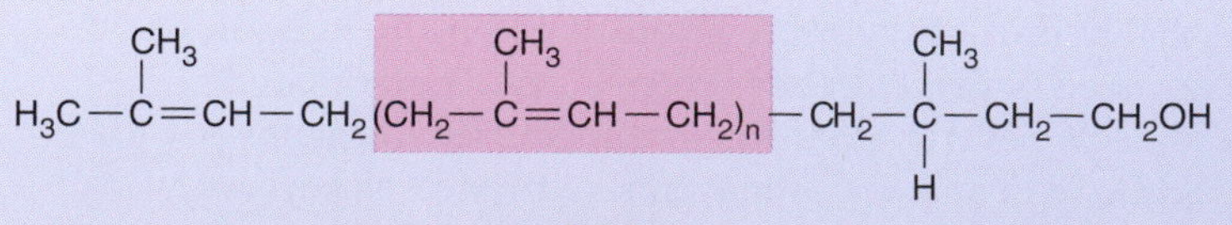

Figure 20-20 Dolichol, a membrane carrier lipid that transfers sugar groups from the cytoplasm into the ER lumen. The number of isoprenoid units (in red) varies from 15 to 18.

dolichol

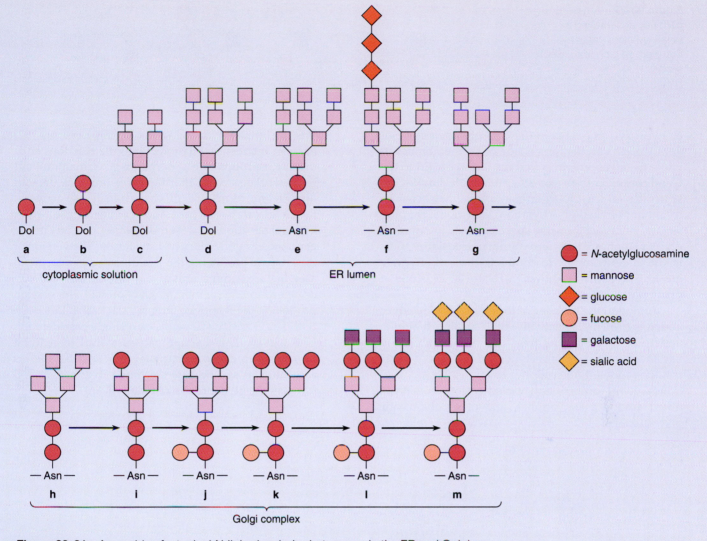

Figure 20-21 Assembly of a typical *N*-linked carbohydrate group in the ER and Golgi complex (see text). *Dol* = dolichol; *Asn* = asparagine.

shown in Figure 20-21f. The same core structure, with or without the glucose residues, is assembled by a closely similar pathway in the ER of plants, mammals, and yeast.

Later steps vary to some degree depending on the glycoprotein or organism. In most cases the three glucose residues and one mannose unit are removed while the glycoprotein is still in the ER (Fig. 20-21g). After this initial trimming the glycoprotein is transported in transition vesicles to the Golgi complex, where three more mannose residues are usually removed (Fig. 20-21h). Why the transient glucose and mannose residues are first added and then usually removed almost immediately is unknown; perhaps they serve as processing signals for later steps in the pathway or are necessary for movement of the glycoproteins to the Golgi complex.

In some plant glycoproteins the carbohydrate structure undergoes no further change and remains as the carbohydrate group of the mature glycoprotein. In other plant and in all mammalian glycoproteins, further glycosylations add capping sugars to assemble a more complex structure. The steps diagramed in Figure 20-21i to m outline a typical mammalian pathway removing mannose residues and adding acetylglucosamine, fucose, galactose, and sialic acid to complete the carbohydrate group.

Less is known about the reaction series adding *O*-linked carbohydrate groups to proteins. In the most simple *O*-linked core structures (Fig. 20-22), either a single mannose or acetylgalactosamine unit is transferred from a nucleotide carrier directly to the side-chain —OH group of a serine or threonine residue, or, more rarely, to an —OH group of hydroxyproline

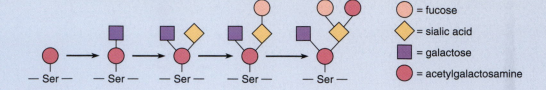

Figure 20-22 A typical pathway assembling a small *O*-linked carbohydrate group (see text). Ser, serine.

— Ser — — Ser — — Ser — — Ser — — Ser —

◯ = fucose
◇ = sialic acid
▪ = galactose
● = acetylgalactosamine

or hydroxylysine. This initial core glycosylation evidently occurs on the lumen side of ER membranes or while the glycoprotein is in transit from the ER to the Golgi complex. To this or other relatively simple core structures (at least four different cores have been detected in *O*-linked glycosylation), additional capping sugars, including the same spectrum used in N-linked glycosylation, are transferred from nucleoside carriers in the Golgi complex to complete the carbohydrate group. The pathway shown in Figure 20-22 is one of many possibilities.

Although the number of different sugar residues added to proteins is relatively small, an almost infinite variety of carbohydrate trees can be assembled by variations in the sequence, type, and number of sugar residues added in the capping reactions. These variations account for the highly varied glycoprotein types noted among plant, animal, and other eukaryotic glycoproteins.

The human A, B, and O blood antigens (see p. 152) have an interesting relationship to the reactions adding sugar groups to glycoproteins. F. Yamamoto and his coworkers found that the three antigens depend on different forms of a glycosyl transferase active in the Golgi complex. The A and B antigens result from base substitutions at seven different positions in the gene encoding the glycosyl transferase. The version of the enzyme producing the A antigen adds a terminal acetylgalactosamine unit at one site in the carbohydrate group of a membrane glycoprotein; the version producing the B antigen adds a galactose at this position. The O antigen depends on a single base deletion in the gene. The deletion causes a shift in reading frame that produces a completely inactive enzyme. As a result, the position glycosylated in the A and B forms of the glycoprotein remains vacant.

CYTOPLASMIC GENETIC SYSTEMS

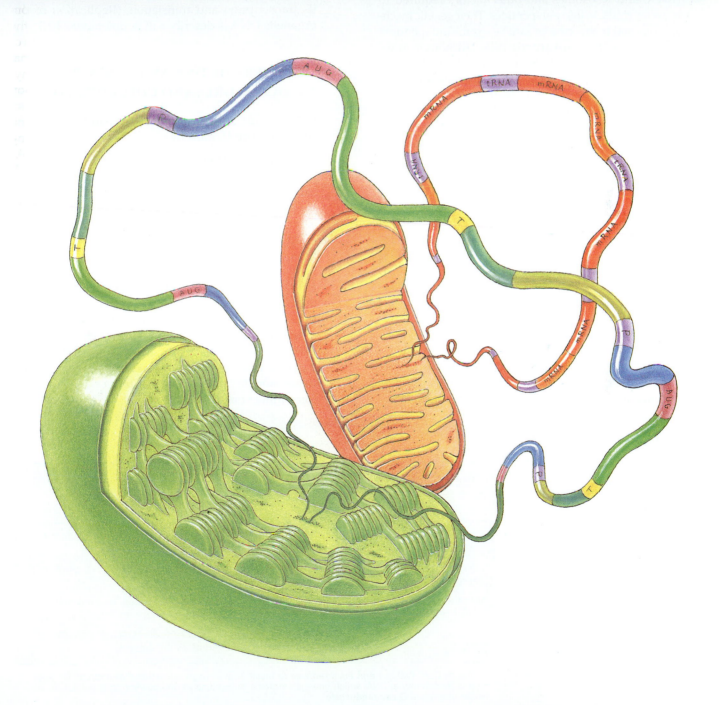

■ *Mitochondrial and chloroplast DNAs* ■
Organelle genes ■ *Genome arrangement* ■
Transcription and processing ■ *Translation* ■
Organelle genetic codes ■ *Origins of organelle*
genomes

Mitochondria and chloroplasts contain their own genetic systems, complete with information encoded in DNA molecules, enzymatic mechanisms for transcribing the information into mRNA, tRNA, and rRNA, and ribosomes and other factors required to translate mRNAs into polypeptides. Their genetic mechanisms make the two organelles semi-independent structures within eukaryotic cells. Although mitochondria and chloroplasts are residents in the cytoplasm of eukaryotic cells, their genomes and their transcription and translation mechanisms are primarily prokaryotic in nature.

The mitochondrial or chloroplast DNA molecules of several species have been completely sequenced. The sequences reveal that the total number of polypeptides encoded in the organelle DNAs includes only a fraction of the protein complement of either organelle. Most organelle polypeptides are encoded in the cell nucleus and synthesized in the cytoplasm outside mitochondria and chloroplasts. These polypeptides make their way into the organelles by the distribution mechanisms discussed in Chapter 20.

The study of mitochondrial and chloroplast genomes and the transcription and translation mechanisms of the two organelles has proved to be a fascinating exercise in its own right. Though obviously prokaryotic in origins, the genetic systems of the organelles have some features that are eukaryotic, and some that are unique and without parallels in other systems. In addition, the organelles have been an in-

valuable source of new information about the workings of genes in general and of important exceptions tempering the conclusions of molecular biology. Several of the most fundamental dogmas of the field have been revised as a result of discoveries made in mitochondria and chloroplasts, including one so basic as the supposed universality of the genetic code. In a very real sense the unusual and versatile genetic systems of the organelles illustrate what is possible within the boundaries of molecular rules.

This chapter discusses the genetic systems of mitochondria and chloroplasts, including the DNA of each organelle, its organization into the genome, and its transcription and translation. (Replication of the organelle DNA is described in Supplement 23-1.)

DISCOVERY OF DNA AND GENES IN MITOCHONDRIA AND CHLOROPLASTS

Cytoplasmic Inheritance of Mutations Affecting the Organelles

The first indications that mitochondria and chloroplasts have their own genetic systems appeared quite early in the history of genetics, from crosses carried out not long after the turn of the century by C. Correns and others. These investigators noted that some traits affecting chloroplast color were inherited primarily through the maternal line, with no evident linkage to nuclear genes. Inheritance through the maternal line in the plants studied paralleled inheritance of the cytoplasm, which comes primarily or completely from the egg cell, not the pollen. For this reason, Correns proposed that the unusual patterns of chloroplast inheritance depend on genes carried in the maternal cytoplasm, perhaps in chloroplasts themselves. By the 1950s, similar patterns of inheritance were noted for some mitochondrial traits in fungi.

Table 21-1 Some Human Diseases Caused by Mutations in Mitochondrial Genes	
Disease	Symptoms
Kearns–Sayre syndrome	May include muscle weakness, mental deficiencies, abnormal heartbeat, short stature.
Lebers hereditary optic neuropathy*	Vision loss from degeneration of the optic nerve, abnormal heartbeat.
Mitochondrial myopathy and encephalomyopathy	May include seizures, strokelike episodes, hearing loss, progressive dementia, abnormal heartbeat, short stature.
Myoclonic epilepsy	Vision and hearing loss, uncoordinated movement, jerking of limbs, progressive dementia, heart defects.

* Found by D. C. Wallace and his coworkers to result from a single substitution of guanine for adenine in mitochondrial DNA, which changes arginine to histidine at one position in subunit 4 of NADH-coenzyme Q oxidoreductase.

More recent genetic work revealed additional patterns of cytoplasmic inheritance in higher plants and animals. In humans, several diseases have been identified with mutations in mitochondrial genes (Table 21-1). The diseases are characterized by alterations in mitochondrial structure and enzyme activity and by damage to organ systems most dependent on mitochondrial reactions for energy: the central nervous system, skeletal and cardiac muscle, the liver, and the kidneys. Within families the defects are transmitted through the mother in nonmendelian patterns. (In mammals mitochondria of the offspring are usually derived entirely from the egg cytoplasm; mitochondria carried in the sperm cell are degraded soon after fertilization.)

The Search for DNA in the Organelles

DNA was first detected in mitochondria and chloroplasts in the 1920s. In 1924, E. Bresslau and L. Scremin found that mitochondria stained positively with the *Feulgen stain* for light microscopy, a procedure that specifically adds a chemical group giving DNA a red-magenta color. In the 1960s, electron microscopy by H. Ris and others revealed deposits of fibers in mitochondria and chloroplasts with the dimensions expected for DNA, located in the same sites that react positively with the Feulgen stain. Typically the DNA fibers visible in the organelles are much thinner than nuclear chromatin fibers, approaching the dimensions of the fibers in bacterial nucleoids (Figs. 21-1 and 21-2).

The light and electron microscope observations touched off a search for DNA in the organelles that was rewarded with the first isolation of mitochondrial DNA (*mtDNA*) by G. Schatz and M. M. K. Nass and their colleagues; soon after, R. Sager, A. Rich, and J. T. O. Kirk isolated DNA from chloroplasts (*ctDNA*). Analyses revealed that although mtDNA and ctDNA are primarily prokaryotic in structure and sequence organization, they also have some features resembling those of eukaryotes and some features that are unique—unlike the DNA of either prokaryotes or eukaryotes.

THE ORGANELLE DNAs

Isolation of the DNA

DNA is isolated from mitochondria and chloroplasts by several techniques. In one of the most-used procedures, cells are broken and the mitochondria or chloroplasts are separated by centrifugation. As part of the isolation procedure the organelles are exposed to deoxyribonuclease while their boundary membranes are still intact. This step destroys any DNA of nuclear origin that might contaminate the suspending solution. After successive centrifugations and washes to remove deoxyribonuclease and degraded DNA from the suspending medium, the boundary membranes of the organelles are disrupted, usually by exposure to detergents. Breakage of the membranes releases the organelle DNA, which is purified and concentrated by centrifugation.

The DNA isolated from mitochondria and chloroplasts, with very few exceptions, takes the form of closed circles (Figs. 21-3 and 21-4). The DNA is largely "naked"—that is, it has no associated structural proteins except for small quantities of a histonelike protein similar to the HU proteins of *E. coli* and other bacteria (see p. 392).

The DNA Circles of Mitochondria

mtDNAs vary considerably in size among major taxonomic groups. Among animals, mtDNA circles are

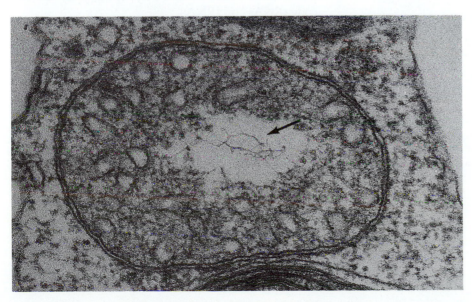

Figure 21-1 DNA fibers (arrow) in a mitochondrion of the brown alga *Egregia*. × 79,000. (Courtesy of T. Bisalputra and A. A. Bisalputra, reproduced from *J. Cell Biol.* 33:511 [1967], by copyright permission of the Rockefeller University Press.)

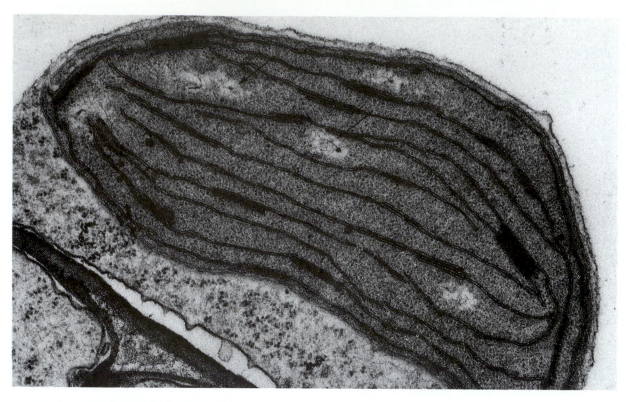

Figure 21-2 DNA fibers (arrows) in a maize chloroplast. × 37,000. (Courtesy of L. K. Shumway and T. E. Weier.)

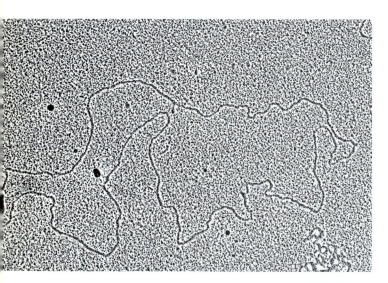

Figure 21-3 Mitochondrial DNA circle isolated from a mouse fibroblast cell. × 20,000. (Courtesy of M. M. K. Nass, from *Science* 165:25 [1969]. Copyright 1969 by the American Association for the Advancement of Science.)

relatively small and uniform in size, ranging from 4.5 to 5.5 μm of included DNA (about 16,000 to 20,000 base pairs) in most species. The human mtDNA circle, for example, contains 16,596 base pairs. The mtDNA of several animals, including *Drosophila, Xenopus, Ascaris*, the sea urchin, and humans, has been completely sequenced. The sequences reveal a remarkable similarity in gene complement among different animal mtDNAs—less than 40 genes, of which not much more than a dozen encode polypeptides; the rest encode rRNAs and tRNAs.

The mtDNA circles of plants are relatively large and vary over extreme limits from a minimum of about 35 μm to more than 700 μm of included DNA. Wide variations in size are even observed among mtDNA circles of closely related plant species. Fungal and protozoan mtDNA circles, with approximately 10 to 25 μm of included DNA, lie between the animal and plant extremes.

Despite their varied lengths, the mtDNAs of plants, fungi, and protozoa contain approximately the same number of genes as the relatively small animal mtDNAs. The DNA segments accounting for most of the differences in genome size among the mitochondria of different taxonomic groups contain sequences that are largely nonfunctional. In some, as in higher plants, much of the nonfunctional DNA consists of repeated sequences.

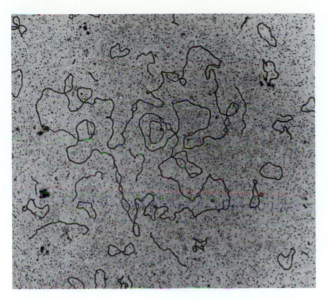

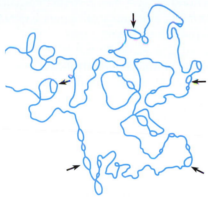

Figure 21-4 DNA circle from a pea chloroplast. The "eyes" (arrows) are regions where the nucleotide chains of the DNA have unwound. The small circles in the background are φX viral DNA, included for size comparison. × 15,000. (Courtesy of R. Kolodner and K. K. Tewari.)

In most cells the total complement of mtDNA circles is less than 1% of total cellular DNA. Higher percentages occur in a few cell types. In rapidly dividing yeast cells, mtDNA may make up as much as 18% of the cellular total; in large animal eggs, such as those of amphibians, mtDNA may represent as much as 99% of total cellular content. The large proportion of mtDNA in these eggs reflects the fact that much of their cytoplasmic volume is occupied by mitochondria. Dividing the mtDNA total by the number of mitochondria gives enough DNA in most cells for about 5 to 10 circles per mitochondrion.

Chloroplast DNA

The DNA of chloroplasts is also circular in all plant species except for one known exception, the green alga *Acetabularia*, which has linear ctDNA. Considerable differences in the size of ctDNA circles are noted among different plants, especially in green algae. In these algae the circles range from a minimum of 27 to as much as 200 μm of included DNA, or about 80,000 to 600,000 base pairs. Among most land plants, including bryophytes, ferns, and seed plants, variations in ctDNA circles are more limited and range from about 40 to 50 μm of included DNA, or about 120,000 to 160,000 base pairs. As in mitochondria, these size differences result primarily from variations in the content of apparently nonfunctional sequences; the gene complement of ctDNA in the chloroplasts of different species is remarkably similar.

The DNA circles of chloroplasts contain several times as many genes as those of mitochondria. The three ctDNA circles sequenced to date, from *Marchantia* (a liverwort), the angiosperms *Nicotiana* (tobacco), and rice, contain a similar complement of about 140 genes. As many as 100 of these genes encode polypeptides; the remainder encode rRNAs and tRNAs.

ctDNA may make up as much as 6% of the total DNA of a plant cell. At these levels there is enough DNA for about 50 to 100 circles per chloroplast. Chloroplasts thus typically contain more DNA circles than mitochondria.

GENES AND GENOME STRUCTURE IN THE ORGANELLES

Mitochondrial Genes

The research identifying mitochondrial genes was initially carried out by genetic crosses and biochemical analysis. Genes for several mitochondrial proteins, such as cytochrome *b*, were located inside the organelle in this way. Later, RNA-DNA hybridization (see Information Box 13-1, p. 366) was used as a test. In this technique the RNAs isolated from mitochondria are added to mtDNA; the regions in the organelle DNA forming hybrid pairs with the added RNAs can be assumed to code for the RNA types. This work established that mRNAs, rRNAs, and tRNAs are encoded inside the organelles. Direct sequencing, carried out by G. Attardi and his coworkers and others, confirmed the presence of the genes identified in mtDNA by the other techniques. The sequences also revealed the presence of several additional, previously unknown genes that have since been identified by comparing their encoded amino acid sequences with those of mitochondrial proteins.

These experimental approaches established that the mtDNA circles of all protozoan, fungal, plant, and animal species encode a relatively small number of proteins (Table 21-2). In higher vertebrates, in which

Table 21-2 Genes of Yeast, Mammalian, and Plant Mitochondria*

Gene	Yeast	Mammalian	Plant
Cytochrome oxidase			
subunit I	+	+	+
subunit II	+	+	+
subunit III	+	+	+
Cytochrome *b*	+	+	+
ATPase complex			
F_1 subunit α	−	−	+
F_o subunit 6	+	+	+
F_o subunit 8	+	+	+
F_o subunit 9	+	−	+
NADH-coenzyme Q oxidoreductase subunits	0 (other fungi 6)	7	6?
Ribosomal proteins	1	−	+
16S rRNA equivalent	+	+	+
23S rRNA equivalent	+	+	+
5S rRNA equivalent	−	−	+
tRNAs	24	22	~30
RNA component of RNAse P	+	−	?

* Genes for *maturase* or *transposase* polypeptides are also present within the introns of some fungal mitochondria.

the most complete studies have been carried out, the list includes genes for 13 polypeptides. The peptides are subunits of the major complexes transporting electrons and the ATPase using the mitochondrial H^+ gradient as an energy source for ATP synthesis. Only one of the electron transport complexes, the one delivering electrons from succinic acid to the quinone pool (complex II; see p. 228) does not contain at least one mitochondrially encoded subunit. The mtDNA circles of other organisms contain a similar gene complement, except that some fungi and plant mtDNAs encode one or more ribosomal proteins in addition to electron transport and ATPase subunits.

All mtDNA circles also contain rRNA and tRNA genes. The rRNA genes of animal, fungal, and protozoan mitochondria are limited in almost all species to one copy each of sequences coding for molecules equivalent to the 16S and 23S rRNAs of bacteria. Although one gene for an rRNA equivalent to prokaryotic and eukaryotic 5S rRNA is also present in plant mtDNA, no 5S equivalent is encoded in protozoan, fungal, or animal mitochondria. No known mitochondria of any species contain genes for a separate rRNA equivalent to eukaryotic 5.8S rRNA. A sequence equivalent to eukaryotic 5.8S, however, is included in the mitochondrial 23S rRNA equivalent at its 3' end, as it is in bacteria (see p. 431 and Fig. 15-3).

The tRNA genes of mitochondria make up an abbreviated list. Mammalian mitochondria, for example, usually contain only 22 tRNA genes, fungal mitochondria only 24, and higher plant mitochondria from as few as 15 to about 30. A number of organisms have greatly reduced numbers of tRNAs encoded in mitochondria—only 2 in a sea anemone and 3 in the alga *Chlamydomonas* and some *Paramecium* species. In higher plant mitochondria with reduced numbers of tRNA genes, those encoded inside the organelle are supplemented by tRNAs encoded in the nucleus. The limited tRNAs encoded in the sea anemone, *Chlamydomonas*, and *Paramecium* mitochondria are probably also supplemented by tRNAs originating from the cell nucleus. Most organelle tRNAs are unusual molecules that depart to a greater or lesser extent from the "standard" tRNAs encoded in the cell nucleus or in prokaryotes (see Fig. 21-6). With tRNA genes the grand total for mitochondrial genomes in higher vertebrates, including mammals, amounts to 37 genes: 2 rRNA, 22 tRNA, and 13 protein-encoding genes.

Sequence Structure of Mitochondrial mRNA Genes The sequence structure of animal mitochondrial mRNA genes reflects an unusual economy in the use of coding space (Fig. 21-5a). The genes lack individual promoters and other upstream control elements, and most have no segments equivalent to the 5' untranslated region (UTR) of nuclear-encoded mRNAs; they simply begin directly with the initiator codon for protein synthesis, which may be AUG, AUA, AUU, or AUC.

The 3' end of mRNA genes in animal mitochondria is similarly shortened. No 3' UTR is present, and even the terminator codon for protein synthesis is usually shortened to only a T or TA (as represented in the nontemplate chain; see p. 409). This abbreviated codon, appearing in the mRNA transcript as a U or UA, is extended into a complete UAA terminator when a short poly(A) tail is added to the mRNA during processing.

G. Attardi, D. P. Chang, D. A. Clayton, and D. F. Bogenhagen and their colleagues discovered that rather than initiating at each gene, transcription in animal mitochondria begins at one promoter for each nucleotide chain of the DNA circle and transcribes the entire chain as a unit (see below). In this sense, animal mitochondrial genomes consist of only two operons, one for each of the two nucleotide chains of the DNA circle.

The mRNA genes of fungi and plants use space more generously. Both 5' and 3' UTRs are present, and terminator codons are complete. In the fungi, 5' UTRs may be extremely long, up to more than 900 base pairs. Promoters are present for single genes or for groups of genes organized into operons. In yeast mitochondria 13 promoters serve as initiation sites for the various mRNA, tRNA, and rRNA genes of the

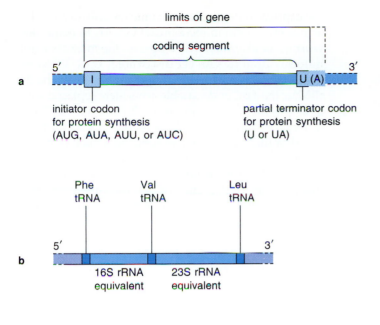

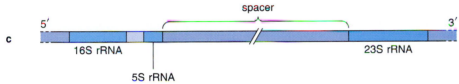

Figure 21-5 Mitochondrial gene structure. **(a)** mRNA gene structure in animal mitochondria. Individual mRNA genes have no upstream promoters or other control sequences and no code for 5' or 3' untranslated regions. The nucleotide sequence of the gene begins at the initiator codon for protein synthesis and in most cases ends with an incomplete terminator codon. The initiator and terminator codons are given in their mRNA equivalents. **(b)** rRNA gene structure in mammalian mitochondria. **(c)** rRNA gene structure in maize mitochondria. *Phe*, phenylalanine, *Val*, valine, *Leu*, leucine.

DNA circle; plant mitochondrial genomes may have as many as 25 promoters.

Introns occur in mitochondrial mRNA genes of *Saccharomyces* and other fungi and higher plants. In *Saccharomyces* the genes encoding two cytochrome oxidase subunits and cytochrome *b* contain introns; some individuals also possess an intron in the gene coding for the 23S rRNA equivalent. Plant mitochondrial genes encoding cytochrome oxidase, NADH dehydrogenase, and ribosomal proteins contain introns. The fungal and plant mRNA introns lack the GT and AG boundaries and internal sequences typical of eukaryotic nuclear mRNA introns. Instead, the introns of these mitochondria share combinations of inverted sequences that evidently guide the mechanisms clipping them from pre-mRNA transcripts of the genes. The introns are capable of self splicing in the test tube—that is, the mechanism removing them from the pre-mRNAs is catalyzed by the RNA of the intron itself (see p. 435 and below).

Mitochondrial rRNAs and Their Genes The ribosomal RNAs encoded in mtDNAs vary widely in size. Those of mammals, trypanosomes, and *Drosophila* are among the smallest rRNAs as yet described. The 16S and 23S rRNA equivalents of mammalian mitochondria, for example, centrifuge at values of only 12S and 16S; those of trypanosomes and *Drosophila* are even smaller. At the other extreme the rRNAs of plant mitochon-

dria, at 18S and 26S, approach the dimensions of cytoplasmic, rather than bacterial, ribosomal RNAs.

In sequence and secondary structure the mitochondrial rRNAs generally resemble prokaryotic rRNAs more than the rRNAs encoded in eukaryotic nuclei. The ribosomal rRNAs of plant mitochondria, which also include a 5S rRNA type, are in fact identical to bacterial rRNAs in secondary structure and closely similar in nucleotide sequence. Fungal mitochondrial rRNAs, though less similar in sequence to prokaryotic rRNA types than those of plants, still resemble the bacterial 16S and 23S rRNA equivalents more than their eukaryotic counterparts and are capable of forming secondary structures characteristic of bacterial rRNAs.

The rRNA sequences of animal mitochondria show fewer similarities to bacterial rRNAs than those of fungal mitochondria. However, even animal mitochondrial rRNAs can still form secondary structures similar to those of bacterial rRNAs. The similarities in secondary structure in spite of sequence divergence are most remarkable in *Drosophila*, in which some of the secondary structures, though still similar to their bacterial equivalents, contain pure or almost pure A-U base pairs.

Considerable variation is noted in the arrangement of rRNA genes. In yeast the genes for the small and large subunit rRNAs are located at widely separated sites on the DNA circle. In animals the two rRNA sequences are located together, with the small subunit at the 5' end; no spacer sequences separate the small and large subunit rRNA genes (Fig. 21-5b). The boundaries of the rRNA genes are marked by tRNA genes at their 5' and 3' ends. In maize mitochondria, rRNA genes are arranged as a unit with the 16S rRNA gene at the 5' end, followed closely by the 5S rRNA gene. A long spacer separates the 5S gene from the 23S gene, which is located at the 3' end of the unit (Fig. 21-5c). Thus, although maize mitochondria contain a 5S coding sequence, its placement differs from that of bacterial rRNA genes, where the 5S gene follows the 23S gene (see p. 454).

Introns occur in the mitochondrial 23S rRNA genes of some fungal strains and in green algae. In strains of *Saccharomyces* that have an intron in the 23S rRNA equivalent the intron contains a reading frame coding for a protein whose amino acid sequence resembles transposases in bacteria. (Transposases are enzymes involved in the reactions that excise transposons and insert them in new locations in the DNA; see p. 536.) The possible inclusion of a transposase suggests that the intron may carry information necessary for its removal and insertion as a movable genetic element. This may account for the fact that the element occurs in some yeast strains but not others and at times may appear in offspring of genetic crosses not expected to contain the intron.

Mitochondrial tRNA Genes and Their Products   The tRNA genes encoded in animal and fungal mtDNAs, although reduced in number, include at least one tRNA for each of the 20 amino acids. In mammalian mitochondria, with only 22 encoded tRNAs, only two amino acids (serine and leucine) are supplied with two tRNA genes; the remaining 18 are represented by only one tRNA gene each. Although many plant mitochondria, such as those of the liverwort, tobacco, and rice, contain at least one tRNA for each amino acid, some lack encoded tRNAs for one or more amino acids. In these mitochondria, as noted, the list made inside the organelles is supplemented by tRNAs imported from the surrounding cytoplasm.

The mitochondrial tRNAs generally resemble prokaryotic more than eukaryotic tRNAs; none includes an encoded terminal CCA sequence. However, some features of mitochondrial tRNAs are neither prokaryotic nor eukaryotic (Fig. 21-6). Mitochondrial tRNAs in animals are distinctly smaller and contain fewer modified bases than those of either the prokaryotic or eukaryotic cytoplasm. In the serine tRNA of mammalian and some other animal mitochondria the D arm is entirely missing; only five unpaired bases lie in the segment of the sequence normally occupied by this arm (see Fig. 21-6c; some other differences between cytoplasmic and mitochondrial tRNAs are noted in

the caption to the figure). In some organisms the TψC arm is missing in mitochondrial tRNAs. In nematode worms, for example, which possess the most aberrant mitochondrial tRNAs yet discovered, D. R. Wolstenholme and his coworkers found that in 21 of the 22 tRNAs, the TψC and variable arms were replaced with a simple loop containing 6 to 12 nucleotides. These differences may be related to some of the substitutions noted in the genetic code in mitochondria (see p. 480 and below). In spite of the sequence differences, animal mitochondrial tRNAs, with the exception of mammalian serine tRNA, can fold in three dimensions into a form similar to the *L*-shaped structure of other tRNAs (see Fig. 15-4).

Organization of Mitochondrial Genomes   The arrangement of genes in mtDNAs reflects two divergent pathways of evolutionary development—the line including protozoa, fungi, and higher plants, in which space utilization is relatively liberal, and the animal line, in which extreme economy is evident. Space utilization reaches its peak in mammals, in which it seems as if the evolutionary process has cut every dispensable nucleotide from mtDNA.

Fungi are the best studied of the species with liberal use of DNA sequences. In fungi such as *Saccharomyces* (outer circle in Fig. 21-7), coding sequences occupy only about half of the DNA circle. The remainder is taken up primarily by long, apparently nonfunctional sequences consisting of as much as 90% A-T base pairs. mRNA genes are scattered at intervals around the DNA circle; even the two rRNA genes are so widely separated that they lie almost at opposite sides of the genome. The only clustered genes are those for the tRNAs—coding sequences for 16 of the tRNA genes of yeast are concentrated in one region between the two rRNA sequences. All the mRNA, rRNA, and tRNA genes of yeast mitochondria, with the exception of a single tRNA gene, are located on the same nucleotide chain of the DNA double helix. Introns, which occur in two mRNA genes and in the 23S rRNA gene of some yeast strains, are so extensive that their total length alone approximates the entire DNA content of animal mitochondrial genomes! Several different sites around the yeast mtDNA circle serve as origins of replication.

With the exception of rRNA genes, which are located together in a single unit, most known genes of plant mtDNA circles are sited at widely spaced intervals, separated by noncoding sequences of varying length. Although essentially the same gene complement is present, there is great diversity in the arrangement of mitochondrial genes in plants.

Animal mitochondrial genomes (inner circle in Fig. 21-7) generally occupy only about one-fifth as much DNA as those of fungi and plants. In verte-

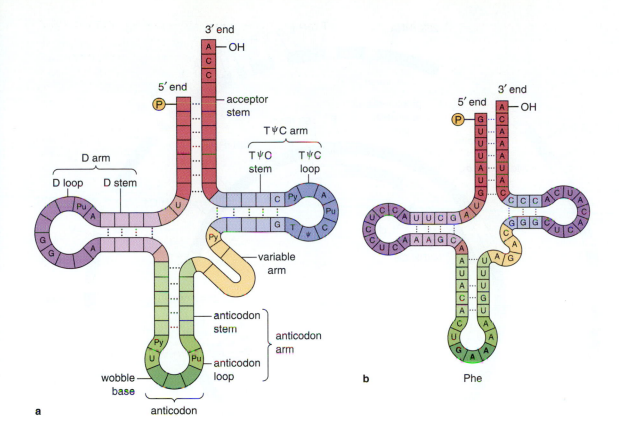

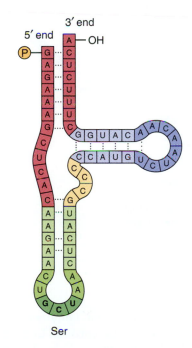

Figure 21-6 Mitochondrial and cytoplasmic tRNAs compared. **(a)** The typical cloverleaf pattern of cytoplasmic tRNAs. Invariant bases are named; semi-invariant bases are indicated as a Py (pyrimidine) or Pu (purine). **(b)** Human mitochondrial phenylalanine tRNA. **(c)** Human mitochondrial serine tRNA. The sequences were reconstructed from the genes for these tRNAs; the positions of modified bases in the mature forms is unknown. However, mitochondrial tRNAs typically contain fewer modified bases than their cytoplasmic equivalents. Note that some of the invariant bases of cytoplasmic tRNAs are different in the mitochondrial tRNAs.

brates most genes abut, with no separating nucleotides. A number of genes, including several encoding mRNAs and tRNAs, overlap to a greater or lesser extent. In humans, for example, the genes encoding ATPase subunits 6 and 8 overlap by 46 nucleotides; the tyrosine and cysteine tRNA genes overlap by 1 nucleotide. Most mRNA genes are separated by tRNA genes; this arrangement probably forms the basis for processing transcripts in animal mitochondria (see below).

Only one segment, about 5% to 7% of the DNA, is not transcribed in vertebrate mitochondria. This segment contains promoters for each nucleotide chain of the DNA and the replication origin of the DNA circle.

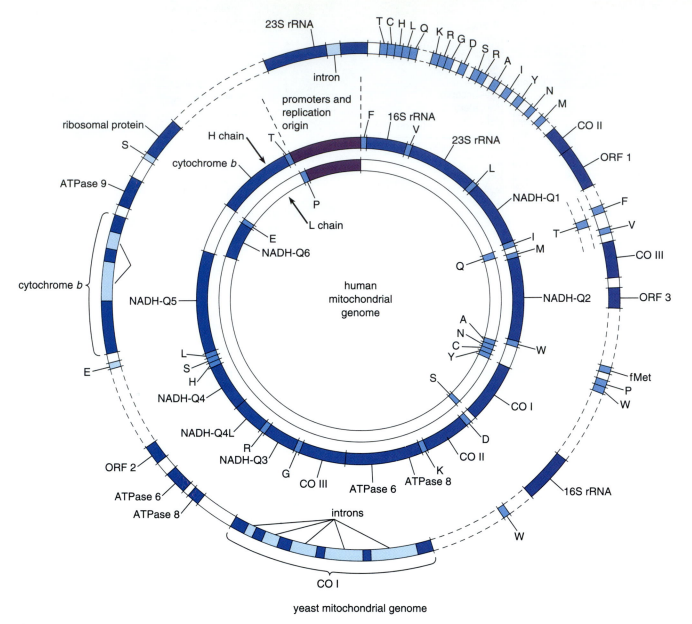

Figure 21-7 The yeast (outer circle) and human (inner double circle) mitochondrial genomes (not drawn to scale). In the yeast genome, as yet unsequenced regions are shown with dashed lines. CO I–III, cytochrome c oxidase subunit genes; NADH-Q 1–6 and 4L, NADH-coenzyme Q oxidoreductase genes. The tRNA genes are identified by the single-letter code for their respective amino acids (see Fig. 2-16).

(The function of the origins in mtDNA replication is described in Supplement 23-1.) Variations in the length of the nontranscribed segment account for most of the size differences noted in vertebrate mtDNA circles.

Within the transcribed region, genes are distributed on both nucleotide chains of the DNA circle. Most animal mitochondrial genes are located on one nucleotide chain of the DNA double helix identified as the *heavy* or *H chain*, which contains most of the G's of G-C base pairs and is significantly denser than the opposite *light* or *L chain*. Only one mRNA gene and eight tRNA genes are encoded in the L chain. In human mitochondria the L chain also encodes a small RNA type that may participate in DNA replication.

With few exceptions, mRNA, rRNA, and tRNA genes occur in the same order and arrangement in the mtDNA circles of all vertebrate species studied to date. However, the only invertebrate mtDNA circles studied in any detail, from *Drosophila*, *Ascaris*, and the sea urchin, show some differences in gene position

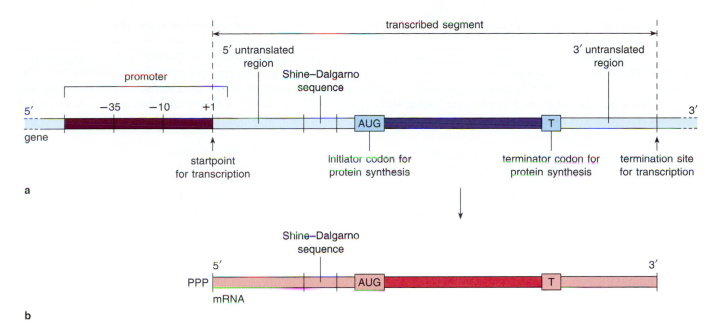

Figure 21-8 Chloroplast mRNA genes and their products. **(a)** A chloroplast mRNA gene. A promoter of the bacterial type is present, and the gene includes segments encoding 5' and 3' untranslated regions as well as a coding segment with complete initiator and terminator codons. **(b)** The mRNA copied from the gene. No poly(A) tail is added to chloroplast mRNAs during processing. The Shine–Dalgarno sequence promotes association of mRNAs with ribosomes during initiation of protein synthesis (see p. 486).

from each other and from the vertebrate pattern. These variations indicate that different DNA rearrangements appeared in the course of mitochondrial evolution in vertebrates and invertebrates. At the moment it is impossible to determine which of the patterns observed in vertebrates and invertebrates is closest to a common animal ancestral form.

Chloroplast Genes and Genomes

Chloroplast genomes typically contain many more genes than those of mitochondria. The ctDNA circle of *Marchantia*, sequenced by K. Ohyama and his colleagues, encodes about 100 polypeptides, two sets of rRNAs, and 37 tRNAs (among which 31 different types are represented). Of the protein-encoding genes, more than 70 have been identified; these include 20 or more ribosomal proteins, the large subunit of RuBP carboxylase, and subunits of photosystem II, the cytochrome b_6/f complex, ATPase, NADH dehydrogenase, and RNA polymerase. There are good indications that many of the remaining unknown sequences encode ribosomal proteins. The chloroplast genomes of tobacco, sequenced by K. Shinozaki and his colleagues, and rice, sequenced by J. Hiratsuka and coworkers, have an almost identical complement of genes. A number of chloroplast genomes contain genes that

have been altered to nonfunctional form (as *pseudogenes*; see p. 522) by deletions of coding nucleotides. For example, a gene encoding a large ribosomal subunit polypeptide (L23; see p. 474) has been altered to a pseudogene by a deletion in spinach, rice, and wheat chloroplasts. Sequencing, genetic crosses, and other methods indicate that with some exceptions the chloroplasts of other plants and algae probably contain a gene complement similar to those of *Marchantia*, tobacco, and rice.

Chloroplast mRNA Genes and Their Products Some chloroplast mRNA genes are organized as individual transcription units, each with its own upstream promoter, and others occur in operons in which several coding sequences are controlled by a single promoter. Each chloroplast promoter possesses consensus sequences equivalent to the −10 and −35 elements of bacterial promoters (see p. 452). Sequence structure within chloroplast mRNA genes also resembles the prokaryotic model (Fig. 21-8). Both 5' and 3' UTRs are present. The coding segment of chloroplast mRNA genes ends in a complete UAA, UAG, or UGA terminator codon. Termination signals lying downstream, at the 3' end of chloroplast mRNA genes, also clearly resemble their bacterial counterparts. The promoter and termination sequences are similar enough to those of bacteria to allow successful transcription of chloroplast genes by *E. coli* RNA polymerase.

Some chloroplast mRNA genes are interrupted by introns. The RuBP carboxylase gene, for example, is split into 10 exons and 9 introns in *Euglena*. The boundary sequences of these introns occur as highly conserved consensus sequences, suggesting that they may be spliced out by reactions similar to those in the cell nucleus. The chloroplast boundary sequences, however, differ from their nuclear counterparts.

Mature chloroplast mRNAs generally resemble the mRNAs of bacteria (see Fig. 21-8b). Each consists of a 5' UTR, a coding sequence initiated by an AUG codon and closed by a terminator, and a 3' UTR. Like bacterial mRNAs, the mature mRNAs of chloroplasts are capless. It does not appear, however, that poly(A) tails are added to the 3' end of chloroplast mRNAs as they are in bacteria.

Chloroplast rRNA and tRNA Genes and Their Products
The genes for chloroplast rRNAs (Fig. 21-9) include codes for 16S, 23S, and 5S rRNAs similar to bacterial rRNAs. However, unlike bacterial rRNA operons, the

rRNA genes of chloroplasts are split into two transcription units, with 16S and 23S genes in one unit and the 5S gene in the other. The 16S/23S unit begins with the 16S rRNA sequence, which is separated from the 23S coding sequence by a transcribed spacer. Like the rRNA operons of *E. coli*, this spacer contains codes for two tRNA genes. The two tRNA genes separating the rRNA sequences are split by introns, as are some other chloroplast tRNA genes (see below).

An intron occurs within the 23S rRNA coding segment in the chloroplasts of some green algal species such as *Chlamydomonas*. Like the 23S rRNA intron of yeast mitochondria, the *Chlamydomonas* intron contains a sequence coding for a polypeptide. The encoded polypeptide may increase the efficiency of the splicing reaction removing the intron containing it or may perform an unidentified, unrelated function.

Mature chloroplast rRNAs form secondary structures that are almost identical to the 16S, 23S, and 5S rRNAs of prokaryotes. The 5S rRNAs of chloroplasts show particularly marked similarities to cyanobacterial 5S rRNAs—one set of unique deletions occurs at the same sites only in the 5S rRNAs of cyanobacteria

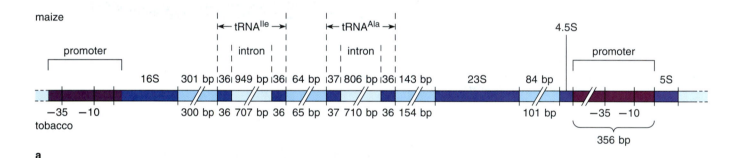

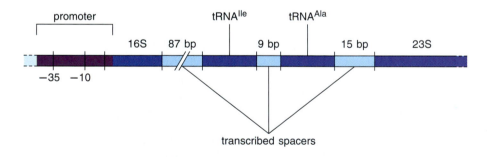

Figure 21-9 Chloroplast rRNA genes. **(a)** Arrangement of coding sequences in the rRNA genes of higher plant chloroplasts. Sequences that appear in mature rRNA molecules are darkest blue, transcribed spacers are medium blue, and introns are lightest blue. The numbers above and below the gene indicate the base pairs in the spacers, introns, and tRNA coding segments in maize and tobacco, respectively. **(b)** Arrangement of sequences in the region between the 16S and 23S coding sequences of *Euglena*, a green alga. Note that there are no introns in the tRNA genes, and that the transcribed spacers are much shorter. (From data in P. R. Whitfeld and W. Bottomley, *Ann. Rev. Plant Physiol.* 34:279 [1983].)

and chloroplasts. At many other sites, chloroplast 5S rRNAs share nucleotides in common with those of both bacteria and cyanobacteria.

Chloroplast tRNA genes occur singly and in operons containing several tRNA coding sequences. The tRNA operons are each supplied with a prokaryote-like 5' promoter with typical −10 and −35 consensus sequences. Some of the single tRNA genes, however, appear to have internal promoters resembling those of eukaryotic tRNAs (see p. 444). Most lack codes for a complete 3' terminal CCA or contain only a code for the first C; these nucleotides are added during processing as they are on nuclear-encoded tRNAs.

Introns are present in some tRNA genes of higher plant and green algal chloroplasts. The chloroplast tRNA introns differ fundamentally from those of nucleus-encoded tRNAs. In nuclear tRNA genes, introns fall only in one position, following the first nucleotide on the 3' side of the anticodon (see Fig. 15-16). In contrast, the introns splitting chloroplast tRNA genes, which include from 325 to more than 2500 base pairs, may occur at several different positions. Further, no single, uniform sequence pattern occurs within chloroplast tRNA introns. A few of the introns contain unidentified coding sequences. These variations in tRNA introns suggest that they may have arisen through several independent evolutionary events.

An intron was recently discovered in a gene encoding a leucine tRNA in cyanobacteria by D. A. Schub and J. D. Palmer and their coworkers. The intron, which closely resembles that of a chloroplast leucine tRNA, adds additional support to the hypothesis that chloroplasts evolved from ancient cyanobacteria.

Organization of Chloroplast Genomes   In general the genes and gene arrangement of chloroplasts are very similar to prokaryotic genomes, particularly those of cyanobacteria. These and other close similarities (see pp. 276 and 393) indicate that chloroplasts evolved from their prokaryotic ancestor more recently than mitochondria.

Chloroplast genomes show indications of development along two evolutionary pathways. In land plants, including bryophytes, ferns, and seed plants, chloroplast genomes are remarkably uniform in sequence arrangement, as in the *Marchantia* and tobacco genomes. In green algae, in contrast, chloroplast genomes show major differences in gene location even within closely related species. Thus one evolutionary pathway is marked by remarkable conservation in sequence organization, and the other by organizational diversity. The ctDNAs of *Euglena* and *Chlamydomonas*, in fact, show more differences in gene arrangement than the known DNA circles of all land plant chloroplasts.

Chloroplast genomes, like the protozoan-fungal-plant group of mitochondria, generally reflect liberal use of DNA sequences. Introns are present, spacers appear between genes, and there are considerable tracts of apparently nonfunctional DNA. In plants, much of the nonfunctional DNA is concentrated in an inverted sequence (see p. 369) that varies considerably in length among different species and taxonomic groups (Fig. 21-10). The inverted sequence also contains two copies of the rRNA genes; in many species, mRNA and tRNA genes also occur in the inverted region. These additional genes also occur in two copies, one in each half of the inverted repeat. Legumes such as the garden pea and broad bean have an exceptional arrangement in which one-half of the inverted sequence has been lost secondarily through a major deletion.

Both mRNA and tRNA genes are scattered throughout the nonrepeated region of chloroplast genomes, distributed approximately equally between the two nucleotide chains of the DNA double helix. Although several genes overlap by one or more nucleotides,

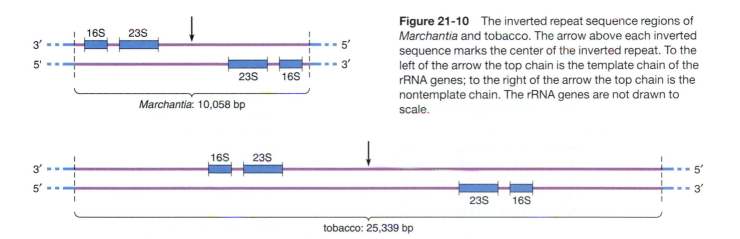

Figure 21-10   The inverted repeat sequence regions of *Marchantia* and tobacco. The arrow above each inverted sequence marks the center of the inverted repeat. To the left of the arrow the top chain is the template chain of the rRNA genes; to the right of the arrow the top chain is the nontemplate chain. The rRNA genes are not drawn to scale.

most are separated by fairly long intergenic spacers. The spacers contain the promoters of the genes or operons following them and, in some cases, additional, apparently nonfunctional sequences. Among these sequences in higher plant chloroplasts are several families of repeated sequences with elements ranging from about 100 to 1000 nucleotides in length.

Replication of ctDNA circles proceeds from two origins, in a pattern resembling bacterial DNA replication. In *Chlamydomonas* and possibly other chloroplasts the origins lie in the vicinity of the rRNA genes. (ctDNA replication is described in more detail in Supplement 23-1.)

Considering the long evolutionary history of land plants, it is remarkable that their chloroplast genes are so similar in arrangement. Most known variations can be accounted for by a relatively few sequence inversions that appeared at some time during the evolution of seed plants. The relatively minor differences in gene arrangement between *Marchantia* and tobacco, for example, can be explained by the inversion of a single DNA segment. Among the few exceptions to the general conservation of chloroplast gene arrangement among the land plants are the legumes. In addition to the deletion eliminating one-half of the inverted sequence in these plants, various sequence rearrangements have taken place in other parts of the genome.

TRANSCRIPTION AND TRANSLATION IN MITOCHONDRIA AND CHLOROPLASTS

Active transcription of mitochondrial and chloroplast genes was first detected by the use of radioactive uridine. When isolated and maintained in a suitable medium, both organelles were able to incorporate labeled uridine into transcribed RNA. Analysis of the labeled products of transcription confirmed that all three RNA classes are transcribed. Later experiments identified the RNA polymerase enzyme copying the DNA and made it possible to follow transcription and RNA processing in the test tube.

Transcription and Processing in Mitochondria

The H and L nucleotide chains of animal mitochondria are transcribed separately as complete units. Transcription of each chain begins at its promoter (in the purple region in Fig. 21-7). Because the chains are antiparallel, as in all DNA molecules, transcription of the H and L chains proceeds in opposite directions around the DNA circle.

The mitochondrial RNA polymerase is a relatively simple enzyme including a core subunit that carries out the polymerization reaction. Another subunit acts as a specificity factor with properties and structure resembling those of bacterial sigma factors (see p. 451). The subunit gives the enzyme the ability to recognize and bind the mitochondrial promoters. Curiously, the structure of the mitochondrial RNA polymerase core subunit resembles neither prokaryotic nor eukaryotic RNA polymerases. Instead, it is most like the RNA polymerases of two bacterial viruses, T3 and T7. This similarity suggests that the original prokaryotic symbiont from which mitochondria are derived was infected with a bacterial virus encoding the T3/T7-like RNA polymerase. For reasons that are not understood, the viral RNA polymerase was more suitable for mitochondrial transcription than the prokaryotic one and was selected. As a further curiosity, the viral-type RNA polymerase is encoded in a gene now located in the cell nucleus.

In addition to the two promoter sequences, mammalian mitochondria have a sequence upstream of each promoter that is recognized and bound by a transcription factor, termed *mtTFA* by its discoverer, D. A. Clayton. Binding by the transcription factor greatly stimulates transcription of mitochondrial DNA.

Once transcription of the H and L chains of animal mitochondria is complete, processing reactions convert the two transcripts into finished mRNA, rRNA, and tRNA molecules. These reactions include: (1) clipping, which cuts the individual RNA sequences from the transcripts; (2) addition of nucleotides to the 3' ends of mRNAs and tRNAs; and (3) chemical modification of bases in the maturing rRNAs and tRNAs.

Most of the clipping reactions are catalyzed by RNA endonucleases that cut out the tRNA molecules separating other coding sequences (Fig. 21-11). Clipping these tRNAs releases the rRNAs and most of the mRNAs from the transcripts. An enzyme with properties similar to bacterial RNAase P, recently identified in human mitochondria by Attardi and his colleagues, cuts the 5' ends of the tRNA molecules. (RNAase P recognizes and cuts tRNA precursors at this end in bacteria—see p. 457.) Presumably another as yet undiscovered RNA endonuclease cuts the 3' ends.

The second step of animal mitochondrial processing, addition of nucleotides to the released transcripts, places the terminal CCA sequence at the 3' ends of the tRNAs and a short poly(A) tail at the 3' end of mRNAs. Addition of the poly(A) tail completes the UAA terminator codon, which is encoded only as the initial U or UA when the mRNAs are first released.

The final step of RNA processing inside animal mitochondria, base modification, converts individual nucleotides in rRNA and tRNA molecules to other

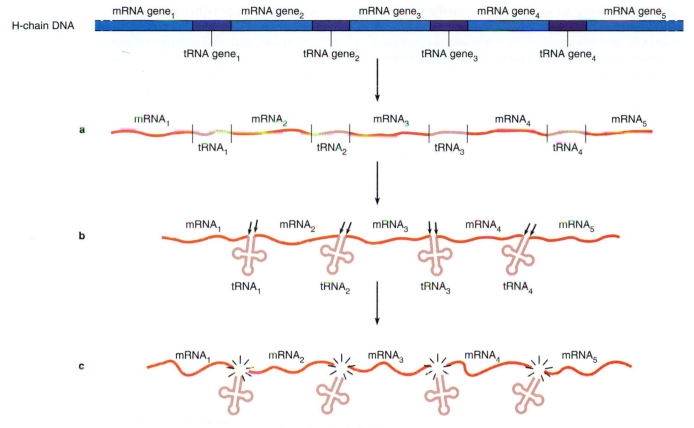

Figure 21-11 Processing reactions releasing mRNAs by clipping out punctuating tRNAs from the H-chain transcript. **(a)** A region of the H chain transcript with tRNA sequences marking the boundaries between mRNA sequences. After folding into their characteristic secondary structures **(b)**, the tRNA sequences are recognized by RNA endonucleases that clip the tRNAs at their 5' and 3' ends (arrows). **(c)** The clipping reactions release the mRNAs as well as the tRNAs.

forms. One of the more interesting results of these reactions is the probable creation of two different tRNAs from a single precursor. Only one methionine tRNA is encoded in animal mitochondria. However, polypeptide assembly in mitochondria and elsewhere requires two methionine tRNAs. One is an initiator tRNA that begins the process of polypeptide assembly by pairing with the initiator codon for protein synthesis (see p. 461); the other recognizes the AUG codon when it appears internally in the message. Evidently different modification pathways produce the two different methionine tRNAs from the same pre-tRNA in animal mitochondria.

In Protozoa, Fungi, and Plants Transcription and processing in the mitochondria of protozoa, fungi, and plants proceed along lines more closely resembling the equivalent reactions in bacteria. In yeast, transcription begins at multiple promoters and produces as many distinct transcripts containing different com-

binations of mRNA, rRNA, or tRNA precursors. Transcription also proceeds from multiple promoters in plant mitochondria.

Relatively little is known about the steps converting initial RNA transcripts into finished mRNA, rRNA, and tRNA molecules in most nonanimal mitochondria. However, the reactions identified in the most completely investigated nonanimal mitochondria, such as those of the yeast *Saccharomyces*, are similar to the overall reactions carried out in animal mitochondria: clipping or trimming, chemical modification of bases in rRNA and tRNA, and addition of nucleotides to the 3' ends of at least tRNA molecules. Poly(A) tails do not appear to be added to mRNAs in nonanimal mitochondria.

The processing steps of nonanimal mitochondria also include clipping and splicing reactions removing introns from both rRNA and mRNA genes. Secondary structures formed by fold-back double helices presumably indicate the sites to be broken and align the free ends generated by intron removal. These reactions in plant mitochondria include *trans* splicing (see p. 415), in which segments of pre-mRNAs transcribed as separate units from different DNA sites are joined to form a complete, continuous mRNA.

RNA Editing   An unusual pattern of mitochondrial RNA processing was recently discovered in a protozoan group, the trypanosomes, by R. Benne and J. E. Feagin and their coworkers. The reactions, called *RNA editing*, were detected when the investigators compared the sequences of mitochondrial mRNAs with their genes in the trypanosome mtDNA. Surprisingly the mRNAs contained U's not coded in the DNA or were missing U's that are encoded. The added or deleted U's correct the mRNA copy, eliminating frameshifts and other deficiencies that would otherwise make the protein translated from the mRNA nonfunctional. Thus the genes carry an incorrect code for the proteins, which is corrected by editing during mRNA processing. Examples of RNA editing have been discovered in other species, including fungi, plants, and higher animals. In fact, almost all mRNAs are edited in plant mitochondria. Most of these occur in mitochondria and involve the addition or deletion of one or more U's at specific sites or chemical conversion of C to U in mRNA coding sequences. Editing of sequences in tRNAs has also been discovered in mitochondria of *Acanthamoeba castelanii*, an ameboid protozoan; rRNA editing has been found along with mRNA editing in mitochondria of the slime mold *Physarum.*

The reactions of RNA editing provide one of several examples of a basic molecular dogma circumvented in an organelle genetic system, in this case the fundamental proposition that the sequence of amino acids in a protein is determined by the sequence of nucleotides in DNA. The genes encoding edited mRNAs code essentially for nonsense; sense, in terms of a series of amino acids forming a functional protein, is made only after the mRNAs encoded in the genes are edited.

Transcription and Processing in Chloroplasts

The close affinities of chloroplasts to prokaryotes extend to transcription and processing. The RNA polymerase enzymes of bacteria and chloroplasts are interchangeable and capable of transcribing genes from either source. In either case, however, the enzymes translate promoters of their own system more readily than those of the foreign system. Some chloroplast RNA polymerases, such as the one from *Chlamydomonas*, are sensitive to rifampicin, an antibiotic that inhibits bacterial RNA polymerase but not eukaryotic RNA polymerases. Conversely, an inhibitor of eukaryotic RNA polymerase, α-amanitin, has no effect on either chloroplast or bacterial RNA polymerases.

Transcription of chloroplast genes produces RNA copies that are subsequently processed into mature form. Although relatively little is known about the processing reactions of chloroplasts, these mechanisms probably include the same overall pathways as in nonanimal mitochondria. The processing reactions of chloroplasts, like those of fungal mitochondria, also include the removal of introns—from mRNA transcripts in green algal chloroplasts and from both mRNA and tRNA transcripts in the chloroplasts of higher plants. Both *trans* splicing and RNA editing also occur in chloroplasts.

Addition of nucleotides during processing may be limited to the terminal CCA of tRNA precursors, because poly(A) tails have not been detected on chloroplast mRNAs. Although almost nothing is known of the reactions removing introns, these mechanisms probably include the full spectrum of reactions taking place in mitochondria. And, because chloroplast tRNAs have unique intron structures, the mechanisms removing them may include additional clipping and splicing processes that remain to be discovered. Moreover, the introns of mRNA precursors in green algae, which have conserved boundary sequences equivalent to those of nuclear mRNAs, may be removed by pathways resembling those splicing introns from pre-mRNAs in the cell nucleus.

Polypeptide Assembly in Mitochondria and Chloroplasts

Experiments with isolated mitochondria and chloroplasts have shown that their ribosomes are fully active in protein synthesis. The isolated organelles, when supplied with amino acids and an energy source (an oxidizable substrate for mitochondria, or light for chloroplasts), can incorporate amino acids into proteins.

A full set of aminoacyl-tRNA synthetases catalyzing amino acid activation (see p. 480) is present in mitochondria and chloroplasts. In most cases the organelle synthetases differ in structure and properties from their counterparts in the surrounding cytoplasm, so much so that the enzymes of one system cannot successfully interact with the tRNAs of the other. However, in plant mitochondria in which some tRNAs are imported from the cytoplasm the tRNAs entering from outside can be activated inside the organelle.

Some of the departures from universality of the genetic code observed in mitochondria (see below and p. 480) probably depend on unusual interactions between aminoacyl-tRNA synthetases and the unique tRNA types found in mitochondria. For example, in yeast mitochondria a tRNA recognizing the codon CUA is recognized by the synthetase attaching threonine to tRNAs. In systems following the "universal" code, tRNAs pairing with CUA are recognized by a leucine aminoacyl-tRNA synthetase and are attached to that amino acid.

Polypeptide Assembly Inside the Organelles The mechanisms assembling polypeptides in mitochondria and chloroplasts, like amino acid activation, show clear affinities to prokaryotic systems. Both organelle and prokaryotic polypeptide assembly employ a methionine modified by a formyl group (fMet; see p. 486) as the first amino acid placed in a protein. Cytoplasmic protein synthesis in eukaryotes instead uses an unmodified methionine as the first amino acid. The mechanisms assembling polypeptides in mitochondria and chloroplasts are speeded by initiation and elongation factors similar to bacterial, rather than eukaryotic cytoplasmic counterparts.

Initial pairing between ribosomes and mRNAs in chloroplasts and fungal mitochondria also follows a pathway similar to bacterial initiation. In bacteria, initial binding of mRNAs to ribosomes depends on recognition and pairing between the Shine–Dalgarno sequence in the mRNA 5' UTR and a complementary sequence near the 3' end of 16S rRNA (see p. 486). Similar complementary sequences between the mRNAs and the small-subunit rRNA can also be detected in

chloroplasts and fungal mitochondria. The bacterial mechanism differs fundamentally from eukaryotic initiation, which depends on an interaction between ribosomes, initiation factors, and the 5' cap of eukaryotic mRNAs (see p. 461).

Although a prokaryotelike mechanism therefore probably binds mRNAs to ribosomes in chloroplasts and fungal mitochondria, a different and unique reaction with no affinities to either prokaryotic or eukaryotic systems must carry out this function in animal mitochondria. Neither the 5' cap required for eukaryotic-type initiation nor the 5' UTR necessary for prokaryotic-type Shine–Dalgarno pairing is present in mitochondrial mRNAs. The mechanism by which these unusual mRNAs associate with ribosomes during the initiation of protein synthesis remains unknown.

The Ribosomes of Mitochondria and Chloroplasts Ribosomes occur in the matrix of mitochondria and in the stroma of chloroplasts (Fig. 21-12). They may be freely suspended in the matrix or stroma or attached to membranes—mitochondrial ribosomes to the matrix side of cristae membranes and chloroplast ribosomes to the stroma side of thylakoid membranes.

The organelle ribosomes, like the reactions they carry out, show many similarities to prokaryotic ribosomes. The parallels are most apparent for chloroplast ribosomes, which are similar to prokaryotic ribosomes in size and protein composition. Functional

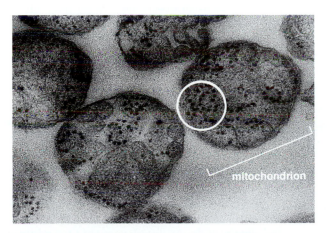
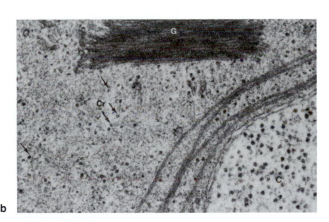

a   b

Figure 21-12 Organelle ribosomes. **(a)** Ribosomes (circled) inside mitochondria of a yeast cell (*Candida utilia*), in a section made parallel to the surfaces of the mitochondrial cristae. The ribosomes are probably attached to the surfaces of the cristae, in a pattern and function analogous to the attachment of ribosomes to the ER. × 46,000. (Courtesy of J. André, from *FEBS Lett.* 3:177 [1969].) **(b)** Ribosomes (Cr) inside a chloroplast of *Chlamydomonas*. The chloroplast ribosomes are distinctly smaller than the ribosomes in the surrounding cytoplasm (C). G, granum. × 56,000. (Courtesy of U. W. Goodenough and R. P. Levine, from *J. Cell Biol.* 44:547 [1970], by copyright permission of the Rockefeller University Press.)

ribosomes have even been assembled from chloroplast and bacterial large and small ribosomal subunits in either combination. Within these similarities, however, some proteins of chloroplast ribosomes clearly differ from their prokaryotic counterparts. Differences are even noted in the numbers and kinds of ribosomal proteins between the chloroplasts of some plant species.

Mitochondrial ribosomes show much greater variability. In different eukaryotic groups, mitochondrial ribosomes vary in size from particles smaller than bacterial ribosomes to structures approximating the weight and dimensions of the ribosomes in eukaryotic cytoplasm. The smallest mitochondrial ribosomes are found in animals. These ribosomes centrifuge down at 55-60S, a value significantly smaller than chloroplast ribosomes, which centrifuge down at 70S. The mitochondrial ribosomes of fungi, most protozoa, and plants are about the same size as prokaryotic ribosomes.

The variation in size among mitochondrial ribosomes is reflected in differences in ribosomal proteins. The patterns and numbers of bands produced in electrophoretic gels by mitochondrial ribosomal proteins of fungi, protozoa, plants, and animals are completely different. Further, antibodies against ribosomal proteins of mitochondria in one group do not cross react with other groups. Considerable diversity is also noted within major eukaryotic groups, even among closely related species. For example, D. E. Leister and I. B. Dawid showed that in *Xenopus laevis* and *Xenopus mulleri*, two species that are related closely enough to form viable interspecies hybrids, *X. laevis* has three mitochondrial ribosomal proteins not found in *X. mulleri*, and *X. mulleri* has four not found in *X. laevis*.

The dissimilarities noted among mitochondrial ribosomes also extend to the ribosomes of bacteria. No structural relationships can be detected among the proteins of bacterial and mitochondrial ribosomes in either distribution in electrophoretic gels or response to antibodies. Attempts to substitute components between mitochondrial or bacterial ribosomes, using either whole subunits or dissociated proteins, have also failed. These differences indicate that the evolution of ribosomal proteins in mitochondria has been unusually rapid and divergent as compared to chloroplast ribosomes.

Despite the differences, organelle ribosomes, with some exceptions, generally react to inhibitors much like prokaryotic ribosomes. Most notably, *cycloheximide*, which interferes with protein synthesis on cytoplasmic but not on bacterial ribosomes, has no effect on polypeptide assembly in either chloroplasts or mitochondria. Conversely, *chloramphenicol*, which inhibits prokaryotic but not eukaryotic protein synthesis, stops polypeptide assembly on both mitochondrial and chloroplast ribosomes.

These reactions provide a powerful approach for determining which polypeptides are synthesized inside mitochondria and which are assembled on cytoplasmic ribosomes. Exposing eukaryotic cells to cycloheximide limits cellular protein synthesis to polypeptides made inside mitochondria or chloroplasts; chloramphenicol limits synthesis to polypeptides made on cytoplasmic ribosomes. This approach has confirmed that the polypeptides synthesized inside the organelles are limited to the list encoded in the organelle DNA. The remaining proteins, which include the great majority of polypeptides making up the membranes and compartments of the organelles, are synthesized in the cytoplasm outside mitochondria and chloroplasts. The polypeptides assembled outside the organelles get inside through organelle-directing signals that form parts of the polypeptide chains (see p. 588 for details).

The Genetic Code in Mitochondria and Chloroplasts One of the most surprising and significant results from experiments with organelle protein synthesis was the discovery of departures from universality of the genetic code in animal and fungal mitochondria (see p. 480 for details). The mitochondrial exceptions were found when the amino acid sequences of several mitochondrial proteins were compared with the nucleic acid sequences of the DNA segments coding for them. The cytochrome oxidase subunit I polypeptide in yeast, bovine, and human mitochondria, for example, was found to contain tryptophan at five sites corresponding to UGA codons in the gene encoding the subunit. (UGA normally serves as a terminator codon.) Other exceptions substituting the code for one amino acid for another also occur (see Table 16-1).

Some of the coding changes in animal and fungal mitochondria may be related to the reduced numbers of tRNA types in the organelles of these groups. Reading all 61 codons with only 22 to 24 different tRNAs was evidently accomplished in mitochondria through structural changes in tRNAs expanding base wobble in codon-anticodon pairing (see p. 478), allowing some tRNAs to read more mRNA codons than their cytoplasmic counterparts. As an evolutionary by-product the expanded wobble may also have provided some tRNAs with the ability to read codons outside the universal pairing assignments, leading to substitutions in the genetic code. As noted, some of the coding substitutions probably also result from interactions between aminoacyl-tRNA synthetases and the unusual tRNAs of mitochondria, leading to charging of tRNAs with different amino acids.

ORIGINS OF THE ORGANELLE GENOMES

There are many unanswered questions about the origins of organelle genetic systems and their relationships to nuclear genomes. One of the primary questions involves the mechanism of gene transfer. If mitochondria and chloroplasts originated from ancient, symbiotic prokaryotes as most investigators agree, how were genes transferred from the developing organelles to the nucleus?

Observations of *promiscuous DNA* leave little doubt that genes can move between organelles and the cell nucleus. "Promiscuous DNA" refers to DNA sequences that occur in more than one of the three DNA-containing membrane-bound compartments of cells—the nucleus, mitochondria, and chloroplasts. For example, fragments of mitochondrial ribosomal RNA and cytochrome *b* genes, along with parts of the replication origin of yeast mitochondria, can be detected in the nuclei of yeast cells. The degree of sequence divergence between the copies in the mitochondria and nucleus indicates that the transfer took place about 25 million years ago. Mitochondrial sequences have been discovered in the nuclei of other species, including sea urchins, locusts, and humans. Transfer of chloroplast genes to the nucleus has also been detected. In spinach, sequence fragments sufficient to make up about six complete chloroplast genomes occur at various locations in the cell nucleus.

Chloroplast genes also appear in quantity in plant mitochondria, including those of maize, mung beans, spinach, and peas. In maize a large 12,000 base-pair segment originating from chloroplasts, containing sequences for 16S rRNA and several tRNAs, can be detected in mitochondria. Curiously, no sequences originating from either mitochondria or the nucleus have as yet been detected in chloroplasts in any plant species.

The mechanisms by which sequences might change places between the organelles and the nucleus are unknown. However, mitochondria and chloroplasts have sometimes been observed to fuse or at least form temporary connections, providing a possible means of DNA transfer. The organelles also sometimes break open, releasing their DNA circles into the cytoplasm. The liberated DNA circles or fragments could then enter the nucleus, either directly or through the activity of viruses.

While the mechanisms are unknown, experiments by P. E. Thorsness and T. D. Fox confirm that DNA sequences can move from an organelle to the nucleus. These investigators introduced genes into yeast mitochondria and detected gene transfer from the mitochondria to the nucleus at the rate of two migrations for every 100,000 cells per cell division (see the Experimental Process essay by Thorsness on p. 624).

Other questions surround the complement of genes still existing inside the organelles. Why are any genes still inside the organelles, and why are the groups remaining inside so similar in different taxonomic groups?

The 13 or so proteins encoded inside mitochondria probably make up less than 5% of the total in the organelle, which may easily amount to 300 or more different polypeptides. Although chloroplasts encode many more proteins than mitochondria, the polypeptides encoded inside probably still represent only about 15% to 20% of the total. With very few exceptions the polypeptides encoded and synthesized in either mitochondria or chloroplasts are subunits of more complex proteins including polypeptides encoded in the cell nucleus. Maintenance of the relatively few genes remaining inside the organelles requires more than 100 polypeptides, including enzymes, transcription and translation factors, and ribosomal proteins, almost all encoded in the cell nucleus and synthesized in the cytoplasm outside mitochondria and chloroplasts.

It is possible that the genes remaining inside mitochondria and chloroplasts encode proteins that, for reasons such as hydrophobicity, cannot be successfully synthesized in the cytoplasm and transported into the organelles. Another possibility is that transfer is still in progress, and that the genes remaining inside the organelles represent a transitory step on the way to eventual transfer of all organelle genes to the nucleus. Supporting this conclusion is the fact that, against the background of general similarity in gene complement, some differences are noted in the locations of a few genes in the organelles in different organisms. The gene for mitochondrial ATPase subunit 9 is located in the nucleus in animals and in at least one fungus, *Aspergillus*; in yeast this gene is in mitochondria. In *Neurospora*, copies of the gene are present in both the nucleus and mitochondria, but only the nuclear copy is active. Similarly, genes for both subunits of RuBP carboxylase are present in red algal chloroplasts; in green algae and higher plants, only the gene for the larger of the two subunits remains in the organelle.

If gene transfer is still in progress, it is likely that movement from the organelles is probably much less probable today than during the very ancient period when the organelles first took up residence in the cytoplasm of cells destined to form eukaryotes. Initially the developing organelles and the nuclei of their host cells presumably had similar, prokaryotelike genetic systems. During this period, transfer of genes in functional form would have been relatively easy. However, as eukaryotic features appeared in cell nuclei, including distinct promoter and regulatory sequences, the chance that an organelle gene could be transferred in functional form was greatly reduced. The present collection of organelle genes may therefore represent a sort of evolutionary cul-de-sac from which functional organelle genes may now move to the nucleus only at vanishingly small rates.

The Experimental Process

Escape of DNA from Mitochondria to the Nucleus During Evolution and in Real Time

Peter E. Thorsness

PETER THORSNESS attended The Colorado College and received a B.A. in chemistry in 1982. He pursued graduate studies in biochemistry at the University of California, Berkeley. He investigated principles of metabolic regulation and regulation of enzymes via covalent modification in the laboratory of Daniel E. Koshland, Jr., receiving his Ph.D. in 1987. Dr. Thorsness then accepted a postdoctoral fellowship from the American Cancer Society and joined Thomas D. Fox's research group at Cornell University. In 1991, he joined the faculty of the Department of Molecular Biology at the University of Wyoming.

The concepts and principles of evolution have been the subject of much controversy. Perhaps the most troubling aspect of evolution for a molecular biologist is the inability to experimentally determine the true path of evolution for a given protein or gene—scientists are often left to "best guess" the process of evolution. On the cellular level, it has long been proposed that eukaryotic cells arose by an endosymbiotic process in which one organism (the pro-eukaryote) was colonized by a second (the promitochondrion in the case of the mitochondrial compartment).[1]

Present day mitochondria contain only about a half-dozen protein coding genes plus the tRNAs and rRNAs necessary for translation; the nucleus of the cell contains hundreds of additional genes that are necessary for functional mitochondria. Since pro-mitochondria were presumed to have had a full complement of genetic material, mitochondrial evolution is proposed to have followed a path of loss or transfer of genetic material to the nucleus. In addition to this circumstantial case for gene transfer, there is evidence for the presence of pseudogenes of mitochondrial origin in the nuclei of a variety of organisms. To test for the presence of a pathway for transfer of genetic information between intracellular compartments, a researcher would ideally be able to place a DNA of specific sequence in the mitochondrial compartment and have a method to detect its presence in the nucleus if it ever made the move. However, for a number of years it was impossible to do this test as it was not feasible to genetically transform mitochondria with standard techniques.

A major technical advance was made in 1988 when a method was developed to introduce DNA into mitochondria of the budding yeast Saccharomyces cerevisiae. The method, biolistic transformation, was amazingly simple in design, but the obvious is often overlooked. Soon after this technique was developed, I joined the laboratory of Thomas D. Fox at Cornell University in Ithaca, New York, who had developed a particularly useful version of the biolistic transformation process.[2] Dr. Fox's scheme took advantage of several useful genetic tricks that are largely available only when working with S. cerevisiae. This yeast can grow well with no mitochondrial DNA whatsoever as long as it has a fermentable carbon source. Yeast that lack mitochondrial DNA make a particularly good target for mitochondrial transformation since plasmid DNAs introduced into their mitochondria are not subject to the rapid recombination processes that can occur between different mitochondrial DNAs. To transform the yeast we used a plasmid DNA that contained both the nuclear gene URA3, to mark the presence of the DNA in the nucleus by complementation of the recessive mutation ura3-52, and a mitochondrial gene, COX2, to mark the presence of the DNA in mitochondria.[3]

The transformation procedure involved precipitation of the plasmid DNA onto 0.5 μm tungsten particles. An explosive charge accelerated the DNA-coated particles at a lawn of yeast cells on an agar plate in an evacuated chamber. Yeast that were pierced by a particle, survived, and received plasmid DNA were identified by complementation of the nuclear mutation ura3-52. A small subset of those cells also had DNA delivered to the mitochondrial compartment. These mitochondrial transformants were identified by a marker rescue technique. Nuclear transformants from the biolistic transformation were replica-plated to a lawn of haploid yeast cells of opposite mating type whose mitochondrial genome lacked the COX2 gene. Neither the transformants nor the tester strain were able to grow on nonfermentable carbon sources (such as glycerol or ethanol) due to their incomplete mitochondrial genomes. However, a diploid formed from one haploid that contained the mitochondrial gene COX2 on the plasmid and a second haploid that contained the rest of the mitochondrial genome would be able to utilize nonfermentable carbon sources and grow. Using this screen, we were able to isolate the 1 transformant in 2000 that had DNA delivered to both the nucleus and mitochondria.

At this point in our experiment we had a yeast that was a uracil prototroph (cell that can make uracil) because it carried the plasmid in the nucleus that complemented the ura3-52 mutation, and also had the COX2 gene in mitochondria. As our goal was to detect the movement of DNA from one compartment in the cell to another, it was necessary to isolate versions of this transformed yeast that contained the plasmid only in the nucleus or only in mitochondria. Subsequently, migration of DNA from mitochondria to nucleus would be revealed by the appearance of uracil prototrophs within a pool of what was formerly entirely uracil auxotrophs (cells that cannot make uracil). Likewise, the movement of DNA from the nucleus to mitochondria could be followed by the appearance of COX2 DNA in mitochondria that formerly were devoid of any mitochondrial DNA. The DNA in mitochondria would be scored by rescuing a *COX2* mutation in a tester strain, as previously described. It is important to note that we assumed the *URA3* gene, when present in the cell only in mitochondria, would not be expressed because of the altered genetic code of mitochondria and the lack of protein synthesis in these mitochondria. Likewise, the *COX2* gene is not functional when located in the nucleus because of differences in genetic codes. It was possible to isolate both a strain that contained the plasmid DNA only in mitochondria and a strain that contained the plasmid DNA only in the nucleus by taking advantage of the mitotic instability inherent with plasmids propagated in these compartments.

The next question was straightforward: Do cells that contain the plasmid DNA only in their mitochondria ever move it to their nucleus and consequently become uracil prototrophs? Not knowing what to expect, we grew up the uracil auxotroph strain and plated about 10^8 cells on an agar plate lacking uracil. After several days of incubation the plate had given rise to hundreds of colonies, indicating complementation of the *ura3-52* mutation and thus the presence of DNA with this information in the nucleus. When we measured the rate at which these uracil prototrophs arose, we found it exceeded the normal mutation rate for a nuclear gene in yeast. The rate of DNA escape and migration from mitochondria to the nucleus was on the order of 2×10^{-5} events/cell/cell division (Fig. A). Frankly, we were somewhat astonished that, at least for yeast and for an evolutionary time scale, DNA was leaking out of mitochondria all the time. As for any other experiment directed at questions of evolution, our results only provided evidence for a pathway for transfer of genetic information to the nucleus from mitochondria, and did not prove that evolution actually happened in that fashion.

Of course we conducted several control experiments to convince ourselves that DNA had been transferred from mitochondria to the nucleus. In the first, we showed that the nuclear mutation *ura3-52* never reverts; yeast cells lacking the plasmid containing the wild-type copy of the *URA3* gene never became uracil prototrophs. A related but even more important observation was that yeast which previously had plasmid DNA in mitochondria or in the nucleus but had lost it during nonselective outgrowth never gave rise to uracil prototrophs. In fact, only those yeast containing the plasmid DNA in mitochondria were capable of generating uracil prototrophs from a clonal population of cells that were originally exclusively uracil auxotrophs.

What about the migration of DNA in the opposite direction—from nucleus to mitochondria? We ran essentially the same trials as for detecting the migration of DNA from mitochondria to nucleus: growth of a large number of yeast cells containing the plasmid DNA only in the nucleus, followed by genetic analysis of what DNA was in mitochondria. Despite examining over 10^{11} yeast, we never detected DNA in mitochondria that had formerly resided in the nucleus. That is not to say that DNA cannot migrate in that direction, but we estimated that the rate of migration from nucleus to mitochondria cannot be any more frequent than 1×10^{-10} events/cell/cell division.

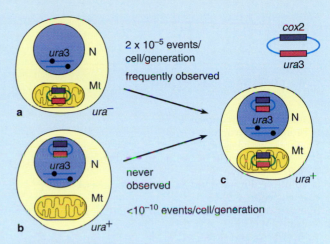

Figure A Experimental scheme. A plasmid carrying the nuclear gene *URA3* and the mitochondrial gene *COX2* could be maintained in either the nucleus (N), the mitochondrion (Mt), or both compartments. **(a)** Cells containing the plasmid in their mitochondria frequently gave rise to cells containing the plasmid in both the nucleus and mitochondria **(c)**. **(b)** Cells containing the plasmid in the nucleus failed to give rise to cells with the plasmid in both compartments at detectable frequency. All strains carried a nonreverting chromosomal *ura3* mutation.

The concepts of evolution stimulated our research approach, but they were not the only factors. We also were interested in the prospects for using this novel genetic assay to analyze the mitochondrial compartment. We found that the escape of DNA from mitochondria was increased by changing the temperature at which the yeast were grown; yeast grown at 37°C or 16°C rather than the optimal 30°C had a two-fold higher rate of escape. Other environmental stresses such as freezing cells in 15% glycerol at −70°C or incubation of cells with $1M$ sorbitol also induced the escape and migration of DNA from mitochondria to the nucleus without cell growth. Likewise, a different nuclear genetic background also influenced the rate of DNA escape from mitochondria. Different strains of yeast had rates of DNA escape that varied five-fold. To further extend our genetic analysis of the mitochondrial compartment, we have isolated mutant strains of yeast with an increased rate of DNA escape from mitochondria. We hope that the isolation and analysis of nuclear genes that complement mutations affecting the rate of DNA escape from mitochondria will generate new insights into the biogenesis and metabolism of mitochondria.

References

[1]Margulis, L. *Symbiosis in cell evolution: Life and its environment on the early earth*, San Francisco: Freeman (1981).

[2]Fox, T. D.; Sanford, J. C.; and McMullin, T. W. *Proc. Nat'l. Acad. Sci. USA* 85:7288 (1988).

[3]Thorsness, P. E., and Fox, T. D. *Nature* 346:376 (1990).

For Further Information

Suggestions for Further Reading

Bonen, L. 1993. *Trans*-splicing of pre-mRNA in plants, animals, and protists. *FASEB J.* 7:40–46.

Clayton, D. A. 1992. Transcription and replication of animal mitochondrial DNAs. *Internat. Rev. Cytol.* 141:217–232.

Dietrich, A., Weil, J. H., and Marechal-Drouard, L. 1992. Nuclear-encoded tRNAs in plant mitochondria. *Ann. Rev. Cell Biol.* 8:115–131.

Gray, M. W., Hanic-Joyce, P. J., and Covello, P. S. 1992. Transcription, processing, and editing in plant mitochondria. *Ann. Rev. Plant Physiol. Plant Molec. Biol.* 43:145–175.

Gray, M. W., and Covello, P. S. 1993. RNA editing in plant mitochondria and chloroplasts. *FASEB J.* 7:64–71.

Grivell, L. A. 1983. Mitochondrial DNA. *Sci. Amer.* 248:78–89 (March).

Gruissem, W. 1989. Chloroplast RNA: Transcription and processing. In A. Marcus, ed., *The Biochemistry of Plants: A Comprehensive Treatise*, vol. 15. New York: Academic Press, pp. 151–191.

Hadjuk, S. L., Harris, M. E., and Pollard, V. W. 1993. RNA editing in kinetoplastid mitochondria. *FASEB J.* 7:54–63.

Hanson, M. R., and Folkerts, O. 1992. Structure and function of the higher plant mitochondrial genome. *Internat. Rev. Cytol.* 141:129–172.

Kuroiwa, T. 1991. The replication, differentiation, and inheritance of plastids with emphasis on the concept of organelle nuclei. *Internat. Rev. Cytol.* 128:1–62.

Lonergan, K.M., and Gray, M. W. 1993. Editing of transfer RNAs in *Acanthamoeba castellani* mitochondria. *Science* 259:812–816.

Nagley, P., and Devenish, R. J. 1989. Leading organellar proteins along new pathways: The relocation of mitochondrial and chloroplast genes to the nucleus. *Trends Biochem. Sci.* 14:31–35.

Osawa, S., Watanabe, K., and Muto, A. 1992. Recent evidence for evolution of the genetic code. *Microbiol. Rev.* 56:229–264.

Palmer, J. D. 1990. Contrasting modes and tempos of genome evolution in land plants. *Trends Genet.* 6:115–120.

Poulton, J. 1992. Mitochondrial DNA and genetic disease. *Bioess.* 14:763–768.

Shadel, G. S., and Clayton, D. A. 1993. Mitochondrial transcription initiation. Variation and conservation. *J. Biolog. Chem.* 268:16083–16086.

Sugiura, M. 1992. The chloroplast genome. *Plant Molec. Biol.* 19:149–168.

Thorsness, P. E., and Fox, T. D. 1990. Escape of DNA from mitochondria to the nucleus in *Saccharomyces cerevisiae*. *Nature* 346:376–379.

Wallace, D. C. 1992. Diseases of the mitochondrial DNA. *Ann. Rev. Biochem.* 61:1175–1212.

Wolstenholme, D. R. 1992. Animal mitochondrial DNA: Structure and evolution. *Internat. Rev. Cytol.* 141:173–216.

Review Questions

1. What patterns of inheritance distinguish genes carried in mitochondria and chloroplasts from those carried in the cell nucleus?

2. Outline two procedures used to isolate organelle DNA.

3. Compare the physical structure of mtDNA and ctDNA with the DNA of prokaryotes and eukaryotes.

4. List the genes typically occurring in mitochondria. What mitochondrial systems contain polypeptide subunits encoded in the organelles?

5. What major differences are noted in the genes included in animal, fungal, and plant mitochondria?

6. Diagram the sequence structure of mRNA genes in animal and nonanimal mitochondria. What sequence features are shared with prokaryotic mRNA genes? With eukaryotic mRNA genes?

7. List the major chloroplast reaction systems containing polypeptides encoded inside the organelles.

8. What sequence features are shared between chloroplast and prokaryotic mRNA genes? Between chloroplast and eukaryotic mRNA genes?

9. Compare the locations of promoters in animal, nonanimal, chloroplast, and prokaryotic genomes.

10. What are the H and L chains of animal mitochondrial genomes? How are these chains transcribed?

11. What structural features serve as processing signals in animal mitochondrial RNA transcripts? What processing steps convert mRNAs from precursor to finished form in animal mitochondria? What processing step completes the terminator codons in animal mitochondrial mRNAs?

12. What is RNA editing? What is the significance of RNA editing to the molecular dogma stating that information flows from DNA to RNA to proteins?

13. Compare the RNA polymerases of mitochondria, prokaryotes, and eukaryotes.

14. What major features of protein synthesis are similar in mitochondria, chloroplasts, and bacteria?

15. Compare initiation of protein synthesis in animal mitochondria, chloroplasts, bacteria, and eukaryotic ribosomes.

16. What features of codon-anticodon pairing in protein synthesis may be responsible for the ability of mitochondria and chloroplasts to carry out protein synthesis with a reduced number of tRNAs?

17. What features of the mitochondrial genetic system may be responsible for persistence of substitutions in the genetic code in this organelle?

18. What major lines of evidence indicate that genes have moved from mitochondria and chloroplasts to the cell nucleus?

In most eukaryotic cells, cytoplasmic growth, the subject of previous chapters in this book, leads to DNA replication and cell division. Universal among eukaryotes is the division pathway in which DNA replication is followed by *mitosis*. Mitotic cell division is responsible for growth and production of body mass in many-celled eukaryotes and maintenance of populations of single-celled organisms such as protozoa. Replication and mitosis are distinguished by the generally perfect fidelity with which the genetic information of the parent cell is duplicated and passed on to the products of division.

A second pattern of cell division, replication followed by *meiosis*, is limited to eukaryotes that reproduce sexually. Meiosis reduces the chromosome number by one-half and leads directly or indirectly to production of male and female gametes. The cell products of meiosis differ from their parent cells in both DNA sequence information and quantity per nucleus.

The mitotic pattern of cell division occurs in a repeating series of stages called the *cell cycle*. Cytoplasmic growth takes place during *interphase*, the stage of the cell cycle preceding a mitotic division. Near the end of interphase, DNA replication produces two identical copies of the genetic information. During the following mitotic division the replicated DNA molecules, complete with histone and nonhistone proteins, are divided into two genetically equivalent nuclei. Division of the cytoplasm encloses the nuclei in separate daughter cells. The two cell products then enter interphase of the next cell cycle.

The cell cycle is regulated by a variety of genetic and biochemical factors, many of which remain poorly understood. One of the most intensive investigative efforts of cell and molecular biology is devoted to discovering the fundamental molecular switches that convert cells from cytoplasmic growth to cell division. The motivation for this research is more than academic, because faulty cell cycle regulation leads to the uncontrolled growth characteristic of cancer.

The results to date have made it apparent that, although many factors may modify progress of cells through the division cycle, the ultimate controls of cell division are genetic. One experimental approach leading to this conclusion is identification of genetic mutants in which the cell cycle is altered or arrested at various points. Identification of the proteins encoded in the mutant genes and their functions has brought investigators close to the ultimate switches controlling the cell cycle. Among the most important of these proteins are the *cyclins* and *cyclin-dependent protein kinases*, which combine into regulatory complexes providing central controls of the cell cycle.

Critical information linking genetic activity to cell cycle control has also come from investigation of tumors. This research has established that many tumors are caused directly or indirectly by the activity of aberrant genes that induce cells to enter rapid and uncontrolled cycles of division. Study of these cancer-causing genes, called *oncogenes* (from the Greek *oncos* = tumor), has revealed that almost all have a normal cellular counterpart called a *proto-oncogene*. Identification of proto-oncogenes and their encoded proteins, made possible through research into oncogenes, has led to fundamentally important insights into genetic controls of the normal cell cycle.

This chapter describes the cell cycle and its regulation, and how it goes awry in cancer. Later chapters take up further events in the growth and division of both mitotic and meiotic cells, including DNA replication (Chapter 23), the mechanisms of mitosis (Chapter 24), and the highly specialized division sequences of meiosis (Chapter 25). The bacterial cell cycle and cell division are discussed in Supplement 24-1.

OVERVIEW OF THE MITOTIC CELL CYCLE

The mitotic cell cycle alternates regularly between interphase and mitosis. Each of these major stages is further divided into substages defined by major biochemical and structural alterations of the cell.

Interphase

Interphase begins as a cell from a previous division enters the *G1* stage of interphase, a period of growth in which proteins and other cellular molecules are synthesized, but the DNA remains unreplicated (the *G* refers to the "gap" or break in DNA synthesis at this time). At some point, if the cell is going to divide, DNA replication begins. The initiation of DNA replication ends G1 and begins the *S* period of interphase (from *S* = DNA synthesis). During S the entire nuclear complement of DNA is precisely duplicated. As replication is completed, S ends and the cell enters the final stage of interphase, *G2*. Most cell types remain in

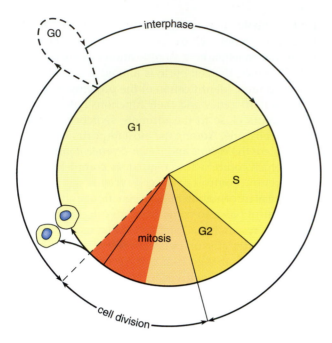

Figure 22-1 The cell cycle. The G1 period of interphase may be of variable length, but for a given cell type the timing of S, G2, and mitosis is usually relatively uniform. G1 may be prolonged by entry into a state of arrest, the G0 stage (dashed lines). Cytoplasmic division or cytokinesis (red segment) usually begins while mitosis is in progress and reaches completion as mitosis terminates.

G2 only briefly; at the end of G2, which marks the end of interphase, mitosis begins. After mitotic division is complete, the two cell products of the division enter G1 of the next interphase.

Each stage of interphase is characterized by a distinct pattern of cellular synthesis. During G1, most of the proteins, carbohydrates, and lipids typical of a given cell type are synthesized, but no DNA replication takes place. Although assembly of the major classes of cellular molecules usually continues during S and G2, S is uniquely marked by DNA replication. The histone and nonhistone chromosomal proteins associated with DNA are also duplicated during S. G2 is characterized by synthesis of a group of proteins necessary for progress through mitosis.

Within a species almost all variations in the duration of interphase involve alterations in G1. S and G2 are relatively uniform within a species and generally proceed without delays once S begins. Thus the molecular switches initiating S, in effect, trigger the remainder of the cell cycle. In most mammalian cells, passage from the beginning of S to the end of G2 requires about 10 to 14 hours. A few exceptions to the usually brief duration of G2 have been noted; in the protozoan *Amoeba*, for example, most of the cell cycle and most cell growth take place during G2 rather than G1.

Figure 22-2 A fully replicated set of human chromosomes, packed into the rodlets visible during mitosis. Each chromosome is double as a result of DNA replication and duplication of chromosomal proteins during the previous interphase. × 1,500. (Courtesy of S. Brecher.)

Mitosis

During mitosis (see Figs. 24-1 and 24-2) the chromosomes, each consisting of a fully replicated DNA molecule complexed with histone and nonhistone proteins, fold or *condense* into thick, rodlike structures. The condensed chromosomes are longitudinally double as a result of DNA replication and duplication of the chromosomal proteins during the previous interphase (Fig. 22-2). The two parts of each chromosome, which are normally exact duplicates of each other and contain the same genetic information, are called *chromatids.*

As part of the early stages of mitosis, RNA transcription gradually slows and stops and protein synthesis drops to minimum levels. The halt in rRNA synthesis is reflected structurally in gradual disappearance of the nucleolus; as protein synthesis slows, vesicle traffic between the ER, Golgi complex, and plasma membrane diminishes and exocytosis and endocytosis stop.

Also during early stages of mitosis the nuclear envelope breaks down and the *spindle* (see Fig. 24-18) forms by assembly of large numbers of microtubules. In the spindle, microtubules are arranged in a roughly parallel array, establishing two ends or *poles* in the dividing cell. The spindle moves into the position formerly occupied by the nucleus, continuing to add microtubules until it fills most of the cytoplasm. Each chromosome then attaches to spindle microtubules and moves to the spindle midpoint. The attachments are made such that the two chromatids of each chromosome link to microtubules leading to *opposite* spindle poles. Forces developed by the spindle microtubules and their associated "motor" proteins (see pp. 290 and 297) separate the two chromatids of each chromosome and deliver them to the opposite poles.

After reaching the spindle poles, the chromatids unfold and return to the extended interphase condition, and the nuclear envelope reforms. At this point the chromatids, now dispersed into a state resembling interphase chromatin, are the chromosomes of the next cell cycle. As the daughter nuclei take shape, RNA transcription resumes and the nucleolus reappears. Protein synthesis increases in rate, and the vesicle movement associated with exocytosis and endocytosis commences. By this time, cytoplasmic division has cut off the daughter nuclei in separate cells, which are now in G1 of the next interphase. The result of DNA replication during interphase and the subsequent mitosis and cytoplasmic division is production of two cells, each with a nucleus containing genetic information equivalent to that of the parent cell entering division.

The time required for one complete turn of the cell cycle, including both interphase and mitosis, varies widely in different cell types and species. Cells in tissues such as the intestinal epithelium in mammals and other vertebrates move through the cell cycle rapidly, requiring only 10 hours or so to pass through interphase and mitosis. Cancer cells may proceed through the cell cycle even more rapidly. The most rapid cycles occur in the early divisions of some animal embryos. Other cells proceed through the cycle very slowly, on the order of days, months, or even years, or do not divide further. For the cells of a species, most of these variations are due to differences in the length of G1. In the most rapidly dividing embryonic cells, G1 is almost nonexistent.

Alterations in the Cell Cycle

Cell Cycle Arrests During Interphase   Some cells, such as those of the central nervous system or muscle tissue in mammals, stop dividing entirely once their development and differentiation are complete. In plants, most of the bulk tissues of stems, roots, and leaves are completely arrested in division except for *meristems* and *cambiums*, regions of persisting division from which differentiating cells originate. In most arrested cell types in both plants and animals the cell cycle stops in G1.

Cultured cells can be induced to enter cell cycle arrest by a number of treatments, including nutrient deprivation and exposure to inhibitors of RNA or protein synthesis. Division arrest also occurs spontaneously when cultured cells growing on a surface come into contact with one another. This *density-dependent* or *contact inhibition of division*, as it is called, is believed to be a normal cell response that helps main-

tain fully developed animal tissues in a nondividing state. Tumor cells frequently fail to display density-dependent inhibition and continue to divide even when in contact.

L. Lajtha and a number of other investigators proposed that during arrest, cells enter a specialized state, termed *G0*, that constitutes a temporary or permanent exit from the cell cycle (see Fig. 22-1). The idea that a specialized G0 state exists depends mainly on observations that cultured cells arrested in division by treatments such as nutrient deprivation do not begin to divide immediately if the arresting treatment is terminated. Instead, an interval of one to several hours (10 hours or more for mammalian cells in culture) passes before the cell cycle restarts, as if a number of specialized blocks to division, representing the G0 state, must be removed before switches restarting the cell cycle can be turned on.

There are exceptions in which arrest occurs at stages of the cell cycle other than G1. A small percentage of adult plant and animal cells arrests in G2, with DNA fully replicated but with no tendency to enter mitosis. G2 arrest, for example, occurs in some kidney cells of higher vertebrates.

Variations in the Coordination of Interphase, Mitosis, and Cytoplasmic Division   In most dividing cells the cell cycle proceeds through interphase, mitosis, and cytoplasmic division in a coordinated succession. However, replication, mitosis, and cytokinesis are potentially separable and in some organisms proceed independently.

In a few cell types of both animals and plants, DNA replication occurs one or more times without subsequent mitosis or cytoplasmic division. This mechanism, for example, produces the *polytene* (from *poly* = many and *tene* = thread) nuclei of *Drosophila* and some other insects. In these nuclei, the chromosomes form thick, cablelike structures because of the repeated replication; each chromosome may contain a thousand or more replicated DNA molecules.

More frequent, but still relatively uncommon, are cells that develop multiple nuclei by replicating their DNA and undergoing mitosis without cytoplasmic division. For example, multinucleate cells arise in mammalian liver by this process.

Another variation occurring in some developmental systems eventually produces cells containing single nuclei with normal chromosome and DNA quantities, but in a pattern in which mitosis and cytoplasmic division are widely separated in time. This occurs regularly, for example, in developing insect eggs. In these eggs, replication and mitosis proceed initially without cytoplasmic division and produce an early embryo with hundreds or thousands of nuclei suspended in a common cytoplasm. Eventually the cytoplasm divides, enclosing the nuclei in separate cells. A similar process occurs in initial development of endosperm tissue in plants such as barley.

A Controlled End to the Cell Cycle: Programmed Cell Death (Apoptosis)

It may seem strange to think of cell death as a controlled and regulated consequence of the cell cycle. However, in many-celled organisms, particularly in animals and plants, programmed cell death is an important part of normal development and body maintenance. In animals, for example, programmed cell death takes place in many developmental pathways, including loss of the tail in frog embryos and disappearance of webbing between the fingers and toes in human development. Controlled cell death is especially important in development of the immune system, particularly for mechanisms that eliminate T cells capable of reacting against self (see p. 564). In adult mammals, programmed cell death underlies formation of fur or hair and the cornified surface layers of the skin. In higher plants, controlled cell death is a regular part of development of many adult structures, including the xylem elements of vascular tissue and the woody parts of stems. Programmed cell death is also called *apoptosis*, from a Greek word that describes the shedding of leaves.

Like progress through the cell cycle, programmed cell death is ultimately under genetic control. In many animal systems the genes controlling cell death are activated by signals in the form of hormones and other factors, which arrive at the cell surface as a sort of metabolic death notice. Failure of the internal cellular mechanisms triggering programmed cell death, so that groups of cells survive that would normally die, is one of several important causes of cancer (see below). Conversely, investigations into the pathways controlling cell death may lead to methods for directing the elimination of tumors and other pathogenic tissues.

Programmed cell death follows several pathways and takes varied forms in different tissues and species. In mammals, many cell types undergoing programmed cell death display a characteristic series of biochemical and structural changes, including shrinkage and compaction of the cytoplasm, condensation and fragmentation of DNA and chromatin in the nucleus, and blebbing and breakage of cellular membranes. Release of hydrolytic enzymes from lysosomes may accelerate breakdown of cellular materials. Often cells undergoing programmed cell death, or fragments of the dying cells, are engulfed by macrophages.

CELL CYCLE REGULATION: FACTORS MODIFYING THE CELL CYCLE

Progress through the eukaryotic cell cycle is modified by a variety of factors. In many cases these factors were initially discovered through observations of their fluctuation in coordination with the cell cycle. Later each was found to be capable of influencing the progress of cells through division if experimentally manipulated. More recent research has begun to reveal the genetic basis of many of the modifying factors. The important modifying factors will be discussed before the genes involved in cell cycle regulation are considered because in most cases the genes were discovered through the effects of mutations on the various factors modifying the cell cycle.

Among the most significant of the cell cycle modifiers are those operating through receptors in the plasma membrane. Controls of this type are vital to multicellular organisms, in which they provide the primary means by which cell division is coordinated in development and maintenance of various organ systems. In these controls, combination of a surface receptor with an external protein such as a growth factor or hormone transmits a signal into the cytoplasm that starts a cascade of reactions leading directly or indirectly to cell division. Many oncogenes encode faulty cell receptors or proteins forming parts of the internal reaction cascades triggered by receptors.

Steroid hormones also act as external factors regulating cell division. In systems controlled by steroid hormones the receptor is located inside the cell rather than at the cell surface (see p. 499 for details). Once bound to the steroid hormone, which penetrates into the cell, the activated receptor acts as a regulatory protein that directly activates or inhibits gene transcription. In some cases the genes activated by steroid hormones directly or indirectly regulate cell division.

Although not a steroid hormone, the thyroid hormone also penetrates directly into cells and activates a receptor that directly binds as a regulatory protein to the control regions of genes. One oncogene, *erbA*, encodes an altered form of the thyroid hormone receptor (see Table 22-2 and below).

Other factors that may regulate or influence the cell cycle include alterations in transport of ions across the plasma membrane and the attainment of a critical cytoplasmic size or volume. Growth to a certain cytoplasmic volume, usually about twice the volume of a cell just completing division, is the primary factor regulating cell division in most single-celled organisms. In most cases the means by which these factors modify cell cycle regulation remain unknown. However, it is assumed that they directly or indirectly activate genes controlling the cell cycle or stimulate or inhibit the products of these genes.

Regulation of Growth and Division Through Surface Receptors

Surface receptors are glycoprotein molecules that span and project from both the exterior and cytoplasmic surfaces of the plasma membrane (see p. 154). The receptor segment extending from the outside surface of the plasma membrane is tailored to recognize and bind proteins that act as growth factors or hormones. Usually peptide hormones or growth factors are secreted into the extracellular medium by one cell type and are bound by receptors on other cell types of the same organism. For example, the much studied *platelet-derived growth factor* (*PDGF*) is secreted by platelets (nonnucleated, membrane-bound derivatives of white blood cells), phagocytes, vascular endothelial cells, and smooth muscle cells. Among the target cells with PDGF receptors are fibroblasts of the skin and tendons, vascular smooth muscle cells, and glial cells of the central nervous system.

The reaction cascades triggered when a surface receptor binds a peptide hormone or growth factor include intermediate or final steps in which protein kinases—enzymes that add phosphate groups to target proteins—are activated. Addition of phosphate groups by the activated kinases promotes or inhibits activity of the target proteins, leading directly or indirectly to regulation of cell division.

The Three Major Pathways Activated by Surface Receptors   Cell surface receptors exert their effects on cell division by one of three major pathways (Fig. 22-3; for details of the pathways, see Chapter 6). In the simplest pathway (see Figs. 22-3a and 6-2) the cytoplasmic extension of the surface receptor itself acts as a protein kinase. Binding a hormone or growth factor induces a conformational change in the receptor that activates its protein kinase segment. In the active form the protein kinase adds phosphate groups to tyrosine residues in target proteins, among them some that directly or indirectly regulate cell division. The PDGF receptor, for example, works directly as a protein kinase in this way.

The surface receptors with built-in protein kinase activity also work indirectly, by triggering a series of reactions that leads to phosphorylation by other protein kinases not directly linked to the receptor (see Fig. 6-2c). In these pathways the protein kinase activity of the receptor phosphorylates the cytoplasmic extension of the receptor itself. This self-phosphorylated state is recognized as a signal by an *adaptor* protein;

binding of the adaptor to the receptor triggers a series of reactions leading to activation of other protein kinases, among them one particularly important to the cell cycle, MAPK (for *m*itogen-*a*ctivated *p*rotein *k*inase). Several regulatory proteins activating genes controlling cell division are among the proteins phosphorylated and activated by MAPK. Central to the series of reactions leading from the receptor to MAPK is *Ras*, a GTP-binding protein (G protein; see p. 157) that acts as a major on-off switch for the pathway. When Ras is bound to GTP, it is in the "on" state and activates the remainder of the pathway. When the bound GTP is hydrolyzed to GDP, through a GTPase activity that forms part of the protein, Ras is in the "off" state and the pathway is shut down (see p. 155 for details). In many cancer types the Ras protein is present in a mutated form in which its GTPase activity is faulty, leaving Ras and its pathway permanently switched on and causing continued activation of genes triggering the cell cycle.

The two remaining major control systems triggered by cell surface receptors activate protein kinases only indirectly, through reaction cascades that involve generation of *second messengers* in the cytoplasm. One of these pathways uses *cyclic AMP* (*cAMP*) as a second messenger (Fig. 22-3b; see also Fig. 6-8). Binding peptide hormones or growth factors by surface receptors leads to activation of the *adenylate cyclase* enzyme converting ATP to cAMP. cAMP then activates protein kinases that add phosphate groups to serine and threonine residues in their target proteins.

The other major indirect pathway uses *inositol triphosphate* (*InsP$_3$*) and *diacylglycerol* (*DAG*) as second messengers (see Fig. 22-3c; see also Fig. 6-9). Binding a hormone or growth factor by the receptor indirectly activates *phospholipase C*, an enzyme that breaks down the membrane phospholipid *phosphatidylinositol* to yield InsP$_3$ and DAG. The primary effect of InsP$_3$ is to release Ca^{2+} from stores in the ER, which acts as another second messenger for the pathway. Ca^{2+}, along with the DAG also liberated as a second messenger, activates another set of protein kinases that also add phosphate groups to serine and threonine residues in their target proteins. Ca^{2+} also acts as a second messenger regulating other systems, either directly or through combination with calmodulin (see p. 164 and Information Box 6-2). The cAMP and InsP$_3$/DAG pathways frequently interact through steps in which enzymes and other factors of one pathway are stimulated or inhibited by those of the other. Many oncogenes producing uncontrolled cell division encode growth factors, surface receptors, protein kinases, or other elements forming part of the cAMP or InsP$_3$/DAG pathways (see below).

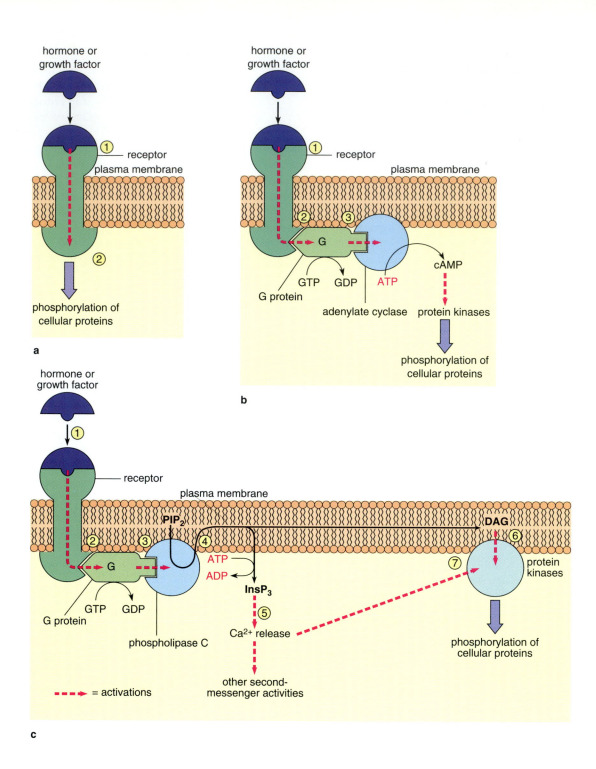

a

b

c

----- = activations

The protein types phosphorylated in each cell type differ depending on the surface receptors, the interactions of the pathways, and the particular group of protein kinases activated by the pathways. The ultimate functions of many of these phosphorylated proteins in cell cycle regulation remain unknown. However, in some cell types, proteins phosphorylated by the pathways include some that regulate transcription of genes important to cell division control, among them several identified through the study of onco-genes (*myc*, *jun*, and *fos*; see Table 22-2 and below). Proteins phosphorylated by the pathways also directly or indirectly modify activity of the cyclin/ cyclin-dependent protein kinase complexes working as central controls of the cell cycle (see below). Other proteins phosphorylated by the pathways include enzymes carrying out oxidative reactions, enzymes of the pathways themselves, cytoskeletal proteins, transport complexes of the plasma membrane, and unidentified proteins known only by molecular weight.

Figure 22-3 Three major pathways by which cell surface receptors for hormones and growth factor stimulate cell activity. **(a)** The most simple pathway, in which combination of the receptor with the hormone or growth factor (step 1) induces a conformational change that activates a protein kinase site (step 2) that forms part of the receptor itself. Addition of phosphate groups to tyrosine residues in cellular proteins by the protein kinase may stimulate or inhibit activity of the proteins. **(b)** The pathway generating cAMP as second messenger. Combination of the receptor with the hormone or growth factor (step 1) activates a G protein on the cytoplasmic side of the plasma membrane (step 2). The G protein, which is linked to GTP in active form, stimulates adenylate cyclase (step 3). The activated adenylate cyclase catalyzes conversion of ATP to cyclic AMP (cAMP; step 4). The cAMP activates protein kinases (step 5) that add phosphate groups to serine and threonine residues in target proteins. The phosphorylations may stimulate or inhibit the target proteins. **(c)** The pathway generating inositol triphosphate (InsP$_3$) and diacylglycerol (DAG) as second messengers. Combination of the receptor with the hormone or growth factor (step 1) leads to activation of a G protein on the cytoplasmic side of the plasma membrane (step 2), which, in turn, activates phospholipase C (step 3). This enzyme catalyzes breakdown of phosphatidyl inositol (PIP$_2$) into the two second messengers InsP$_3$ and DAG (step 4). InsP$_3$ is released into the cytoplasm, where it triggers release of Ca^{2+} from stores in the ER. The Ca^{2+}, along with the DAG released in step 4, activates protein kinases (steps 6 and 7) that add phosphate groups to serine and threonine residues to target proteins.

Surface Receptors and Density-Dependent Inhibition of Cell Division   Cell surface receptors may also be involved in density-dependent inhibition of the cell cycle. This phenomenon is easily observed in cultured cells grown on a glass or plastic surface. In such cultures, division usually stops as the cells form a continuous, unbroken layer one cell in thickness. The inhibition can be released by scraping cells from a portion of the surface. Cells at the free edges of the "wound" created in this way are released from inhibition and enter successive rounds of division until the layer is again continuous and all cells are in contact with their neighbors on all sides.

Although little is understood about the molecular basis for density-dependent inhibition, the initial signals that cells have come in contact evidently depend on glycoproteins in the plasma membrane. Many experiments have shown that if contact-inhibited cells are exposed to dilute solutions of proteinases or enzymes attacking carbohydrate groups, the inhibition of cell division is broken. The membrane glycoproteins implicated by these experiments may be receptors for proteins located on the surfaces of other cells.

In one cell type, mouse fibroblasts, cells inhibited by contact secrete a protein that can inhibit division in other fibroblasts. The protein joins a gradually lengthening list of secreted signals known as *negative growth factors*, which can turn off division when bound by cell surface receptors. Among the proteins included in this group are *interferons*, proteins that can inhibit growth of a variety of cell types, including cells infected by viruses. Negative growth factors apparently work by binding to the surfaces of cells that contain receptors able to recognize and attach them. Binding in some unknown way interrupts or inactivates the cytoplasmic reaction pathways that would normally turn on cell division when positive growth factors such as epidermal growth factor (EGF) or PDGF bind to the cell surface.

Alterations in Ion Transport

Changes in the cellular concentrations of Ca^{2+}, Na$^+$, and K$^+$ are regularly observed in coordination with the cell cycle. A long series of experiments demonstrated that blockage of these changes can interrupt the cell cycle, indicating that the ions may have important modifying effects on the progress of cells through division.

The cytoplasmic concentrations of Na$^+$ and K$^+$ in nondividing cells are controlled primarily by the Na$^+$/K$^+$-ATPase pump located in the plasma membrane (see p. 131 for details). This active transport pump moves Na$^+$ out of the cell and K$^+$ inward at the expense of ATP. The activity of the pump establishes Na$^+$ at low concentrations inside the cell and high concentrations outside, and K$^+$ in an opposite high inside/low outside gradient.

A stimulus that triggers cell division, such as binding of a growth factor by a cell surface receptor, causes an almost immediate change in the flow of these ions. Within seconds after a division stimulus, Na$^+$ flows inward and K$^+$ outward, evidently in response to a signal that opens channels for these ions in the plasma membrane. Blockage of the early Na$^+$ flow by drugs that prevent the Na$^+$ channels from opening or by removal of Na$^+$ from the extracellular medium can completely inhibit cell division in some cell types.

The initial flow of Na$^+$ and K$^+$ is followed after some minutes in many cell types by operation of another transport complex that exchanges Na$^+$ outside for H$^+$ inside. This exchange adds to movement of Na$^+$ into the cell. In addition, removal of H$^+$ causes an increase in cytoplasmic pH that is characteristic of many cells leaving division arrest. In some cases the transport protein exchanging Na^+ for H^+ is activated by protein kinases forming part of the $InsP_3$/DAG pathway.

Equivalent changes are noted in the flow of Ca^{2+}. In nondividing cells, Ca^{2+} is kept at low concentration in the cytoplasm by Ca^{2+}-ATPase pumps that remove the ion from the cytoplasmic solution (see p. 131). These pumps, which operate in the plasma membrane, mitochondria, and the ER, produce a concentration difference in which Ca^{2+} is low in the soluble cytoplasm and high outside the cell and in the ER and mitochondria. A division stimulus, such as binding of a growth factor at the cell surface, causes a rapid rise in cytoplasmic Ca^{2+} concentration. Depending on the cell type, some or all of this Ca^{2+} is released from the ER by the $InsP_3$ second messenger, which opens Ca^{2+} channels in the ER membranes. Soon after Ca^{2+} release the ER channels close and the Ca^{2+}-ATPase pumps return cytoplasmic concentration of the ion to low levels again. The Ca^{2+} "spike" lasts long enough to modify cell cycle regulation. At least part of this modification is exerted through a variety of Ca^{2+}-dependent protein kinases that operate at important steps of several regulatory cascades.

The importance of Ca^{2+} as a modifying control has been demonstrated in experiments extending over many years of cell cycle research. Generally removal of Ca^{2+} from the extracellular medium, or experimental changes that prevent Ca^{2+} from entering the cytoplasm, block entry into cell division. Conversely, treatments introducing Ca^{2+} into the cytoplasm, such as addition of a calcium ionophore (a molecule that makes the plasma membrane leaky to Ca^{2+}), can trigger entry of some cells into S and mitosis. These experiments demonstrate the critical role of Ca^{2+} in the G1–S transition. (Changes in Ca^{2+} concentration are also vital to regulation of events in mitosis, such as the spindle mechanism separating chromatids; see Chapter 24 for details.)

Histone and Nonhistone Modifications

Changes in the histone and nonhistone chromosomal proteins, particularly involving addition or removal of phosphate groups, are also keyed to the cell cycle. A number of the alterations have been linked to cell cycle regulation.

Some nonhistones are phosphorylated during release of the cell from division arrest or during the G1–S transition. Some of these early phosphorylations may involve nonhistone proteins that control the activity of genes involved in cell cycle regulation. Other nonhistone phosphorylations rise during S and G2 and peak during mitosis. These phosphorylations may be related to chromosome condensation or, possibly, to the general transcriptional arrest taking place during mitosis. Some nonhistone phosphorylations, as has been noted, are carried out by protein kinases activated in the regulatory cascades linked to the cell surface.

The histones are also phosphorylated in patterns keyed to the cell cycle. L. R. Gurley and his colleagues followed changes in the levels of histone phosphorylation as cultured Chinese hamster cells progress through the cell cycle (Fig. 22-4). H1 phosphorylation begins and rises as the cells approach and enter S. Phosphorylation of this histone increases further during G2, reaching an overall peak during metaphase of the following mitosis. After metaphase, H1 phosphorylation drops sharply. By the time the hamster cells reach the following G1, H1 phosphorylation is at its lowest levels. H3 also undergoes limited phosphorylation in coordination with the cell cycle in the Chinese hamster cells, beginning late in G2 and progressing until metaphase, when levels again drop to a minimum.

Detailed work in the Gurley laboratory revealed that the regular variation in H1 phosphorylation is the sum of three distinct events involving additions of phosphate groups at separate sites. One site, which Gurley termed the *G1 site*, is phosphorylated as the cells break their G1 arrest and prepare to enter S. Phosphorylation of the second site, the *S site*, begins as the cells initiate DNA replication. The final phosphorylations involve as many as four separate *M sites*. Phosphorylation of these sites begins just before prophase and extends through metaphase of mitosis (see Fig. 22-4). Removal of all these phosphate groups, which are attached to specific serine and threonine residues in H1 molecules, begins after metaphase and is completed by the following G1.

The initial G1 phosphorylation in this series may be part of the sequence of changes breaking arrest and preparing the chromatin for replication. The S phosphorylations may be directly related to replication. The addition of phosphate groups to H1 during prophase and metaphase of mitosis may be part of the molecular changes leading to chromosome condensation. The phosphorylations may activate H1-H1 linkages that fold chromatin fibers from a more extended state to the compact form characteristic of mitotic metaphase (see also p. 694).

Similar results were obtained by E. M. Bradbury and his coworkers in experiments with the slime mold *Physarum*. In *Physarum*, H1 phosphorylation remains constant during S, rises during G2, and peaks at metaphase of mitosis. H1 phosphorylation falls rapidly after metaphase, bottoming out at about one-half the metaphase level before the next S begins. (G1 is nonexistent, or nearly so, in *Physarum*.) H3 phosphorylation is also noted in *Physarum* at metaphase, as in the Chinese hamster cells.

The rise in H1 phosphorylation in *Physarum* coincides with the beginning of chromosome condensation. A role for H1 phosphorylation in condensation is also indicated by observations in yeast cells in which H1 phosphorylation is experimentally increased; the chromosomes in these cells condense early. All this is supported by recent work with the cyclin/cyclin-dependent protein kinase complexes. One of the major proteins phosphorylated by a cyclin-dependent protein kinase is histone H1.

Cytoplasmic Volume

In most cell types, DNA replication and mitotic division are delayed until the cytoplasm reaches a critical size or volume that is more or less fixed for each cell type. Without this relationship the average size of cells would vary widely, or increase or decrease as division proceeds in a cell line. Cytoplasmic volume is the most critical factor regulating the cell cycle in populations of single-celled organisms; it also figures importantly in control of cell division in many-celled forms.

The relationship between cytoplasmic volume and cell division is one of the oldest observations of cell cycle controls. In the late 1920s, M. Hartmann discovered that amputating part of the cytoplasm of an *Amoeba* to keep the cytoplasmic volume below a minimum size prevented the protozoan from dividing. The amputated cells reentered division cycles if allowed to grow to normal size. Equivalent observations were made with other lower eukaryotes and more recently with cultured mammalian cells. Generally each eukaryotic cell type is characterized by a narrow range in size distribution; growth of the cytoplasm to a volume about twice that at the close of the previous division is usually necessary for the next division to proceed. If cells are made abnormally small by some means, G1 is usually extended until the critical volume is reached.

During early embryonic development, cells in many animal and plant species naturally bypass the requirement for growth between divisions. In many embryos there is no growth between divisions for some period just after fertilization, producing *cleavage* divisions in which average cytoplasmic volume is smaller in each successive cell generation. By the time the cleavage divisions run their course, cells of the embryo may be many thousands of times smaller than the fertilized egg. In *Xenopus*, for example, there are 12 sequential cleavage divisions following fertilization, producing an early embryo of about 4000 cells.

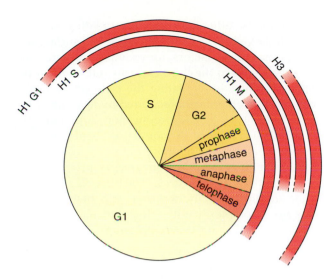

Figure 22-4 Phosphorylation of H3 and the G, G1, S, and M sites of histone H1 during the cell cycle in Chinese hamster cells, according to the experimental results of L. R. Gurley and his coworkers. The bars outside the cell cycle indicate periods of the cycle during which the H1 and H3 sites are phosphorylated.

There is no cell growth between these divisions, so that the 4000 cells occupy approximately the same total volume as the original fertilized egg.

The variety of modifying cell cycle controls indicates that there are many possible mechanisms by which the progress of cells through growth and division may be modified. Many cell types have evolved distinct combinations of these regulatory mechanisms. The multiplicity of the pathways undoubtedly reflects the many interrelated steps at which increases or decreases in activity of genes or their products can advance, delay, or stop the cell cycle.

GENES REGULATING THE CELL CYCLE

Recent research has begun to reveal the genes and gene products active and modified in cell cycle regulation. This research has progressed primarily through the two avenues of investigation outlined in the introduction to this chapter: identification of mutations affecting the cell cycle, and the study of oncogenes.

The genes identified by this research fall roughly into three categories. In the first are genes acting as fundamental controls that directly regulate passage through the cell cycle. Particularly important among these genes are those controlling passage through the G1–S boundary and the transition from G2 to mitosis. Genes in the second category encode proteins that modify the activity of genes in the first category, or their products, rather than directly regulating passage through the cell cycle. Genes encoding cell surface receptors for growth factors, for example, fall into this category. Genes in the third category, rather than directly or indirectly regulating the cell cycle, encode enzymes and other factors required to maintain critical activities such as DNA replication. Although the genes in this category do not act as regulatory elements, mutations or alterations in these genes, which deprive the cell of required enzymes or proteins, can alter or stop the division cycle. The three categories of genes are not mutually exclusive, and their interactions are so complex that it is often difficult to assign a gene to one of the categories with complete confidence.

Cell Cycle Genes Discovered Through Study of Mutants

Mutants affecting the cell cycle have been identified by several genetic approaches. Greatest progress has been made through study of *temperature-sensitive mutants*. Organisms with mutations of this type grow and divide normally within a limited temperature range (called the *permissive temperature*) but are inhibited at a temperature above or below this range (called the *restrictive temperature*). Temperature-sensitive mutants can be grown at permissive temperatures in quantities large enough for extensive genetic and biochemical analysis. Raising or lowering the temperature to restrictive levels then activates the mutation and allows its characteristics and the gene carrying it to be studied.

Cell Cycle Genes in Yeast   Research detecting cell cycle–related genes through temperature-sensitive mutants has progressed to the greatest extent in the yeasts *Saccharomyces cerevisiae* and *Schizosaccharomyces pombe*, in which some 150 mutations affecting more than 30 different genes have been described by P. Nurse and his colleagues and others. (Table 22-1 lists some of these genes in *S. cerevisiae*.) The group identified in these species, which includes genes provisionally assigned to all three categories—fundamental controls, modifiers, and maintenance—confirms the complexity of cell cycle control even among relatively simple eukaryotes.

The central genes controlling the cell cycle, those encoding cyclins, cyclin-dependent protein kinases, and their modifiers, were first discovered in yeast. In the system investigated by Nurse and his coworkers in *S. pombe*, cyclin is encoded in the *cdc13* gene; the cyclin-dependent protein kinase is encoded in the *cdc2* gene. Both genes are required in normal form for passage through the cell cycle; each has counterparts in all eukaryotes. The proteins encoded in the genes combine into the *cyclin/Cdc2* complex; tight associa-

Table 22-1	Some Cell Cycle Genes in *Saccharomyces cerevisiae*
Gene	Function of Encoded Protein
cdc2	DNA polymerase III (equivalent to DNA polymerase δ in higher eukaryotes)
cdc6	Initiation of DNA replication
cdc7	Initiation of DNA replication
cdc8	Thymidylate kinase (catalyzes synthesis of precursors for DNA replication)
cdc9	DNA ligase
cdc17	DNA polymerase I (equivalent to DNA polymerase α in higher eukaryotes)
cdc21	Thymidylate synthase (catalyzes synthesis of precursors for DNA replication)
cdc25	Activates *cdc28* gene product
*cdc28**	Cyclin-dependent protein kinase
cdc34	Adds ubiquitin groups (see p. 511) to histones H2A, H2B
cdc35	Adenylate cyclase
cdc36	May inhibit *cdc28* gene product
cdc39	May inhibit *cdc28* gene product

*Homologue of *cdc2* in *S. pombe* and higher eukaryotes.

tion with cyclin is necessary for Cdc2 to be active as a protein kinase. When activated, the cyclin/Cdc2 complex carries out key phosphorylations on serine and threonine residues in target proteins that trigger entry into mitosis. (Cdc2 is also called $p34^{cdc2}$ in both yeast and higher eukaryotes in reference to its molecular weight of 34,000.)

The activity of Cdc2 as a serine/threonine protein kinase within the cyclin/Cdc2 complex is directly modified by the products of several other genes including *wee1* and *cdc25*, which also have counterparts in all eukaryotes, and the product of another gene, *mik1* (Fig. 22-5). Wee1 adds an inhibiting phosphate to a tyrosine located at position 15 in the Cdc2 protein. The phosphorylated tyrosine occupies a position in the region of the protein binding ATP as part of its activity as a protein kinase; addition of the phosphate evidently blocks this active site of the protein. When tyrosine 15 is phosphorylated, Cdc2 and the cyclin/Cdc2 complex are inactive. Cdc25 is a protein phosphatase that specifically removes the inhibiting phosphate from tyrosine 15. The Mik1 gene product is a protein kinase that adds an inhibiting phosphate group to Cdc2. Whether the cyclin/Cdc2 complex is active or inactive at critical points in the cell cycle depends on the balance of activity of Wee1, Mik1, and Cdc25.

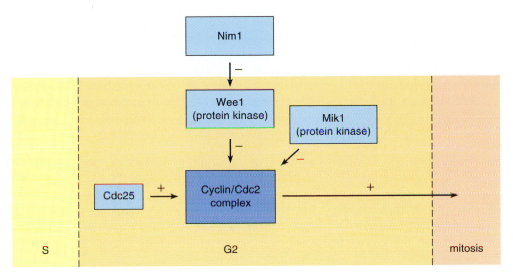

Figure 22-5 Gene products active in the transition from G2 to mitosis in *S. pombe*. The product of the *cdc2* gene, the Cdc2 protein kinase, combines with the product of the *cdc13* gene, cyclin, to form an active mitosis-inducing complex equivalent to MPF in higher eukaryotes. The *cdc25* gene encodes a protein that promotes activity of the Cdc2 protein kinase; the products of *wee1* and *mik1*, both protein kinases, phosphorylate Cdc2 and inhibit its activity. Another protein kinase, *Nim1*, adds an inhibiting phosphate to Wee1. The activity of the Cdc2/cyclin complex depends primarily on the balance of activity of the *cdc25* protein kinase against the *wee1* and *mik1* gene products.

The products of other genes modify the activity of these direct regulators of the cyclin/Cdc2 complex; the *nim1* gene product, for example, is a protein kinase that adds an inhibiting phosphate group to Wee1.

Research with *S. cerevisiae* has revealed similar involvement of the cyclin/Cdc2 complex in cell cycle regulation, except that a series of at least nine cyclins joins in different combinations with the Cdc2 protein kinase.[1] The different combinations modify the activity and targets of the associated Cdc2 protein kinase.

The multiple cyclins of *S. cerevisiae* are varied in active concentration at different stages of the cell cycle through a combination of transcriptional controls, posttranscriptional modifications, and selective breakdown by proteinases. The sequence of cyclin activations produces a series of distinct cyclin/Cdc2 complexes that control successive stages of the cell cycle. At any stage the activity of a cyclin/Cdc2 complex in triggering the next stage depends on the bal-

ance of modifying factors such as Wee1 and Cdc25. The distinct complexes fall roughly into three groups controlling passage through critical points in the cell cycle: one active during late G1, controlling the transition from G1 to S and start of the cell cycle; a second active during S and controlling genes encoding enzymes and factors necessary for DNA replication; and the third controlling the transition from G2 to mitosis. By metaphase of mitosis all cyclins have been degraded and do not reappear until the sequence for the next cell cycle is initiated during the following G1. The controlling mechanism in higher eukaryotes is similar, except that higher eukaryotes have multiple cyclin-dependent protein kinases as well as cyclins (see below).

One of the most significant findings in the *S. cerevisiae* system is identification by C. Koch, K. Nasmyth, and others of gene groups controlled directly or indirectly by phosphorylations carried out by activated cyclin/Cdc2 complexes. One of these gene groups, activated in G1 in preparation for entry into S, shares an upstream control element with the consensus sequence CACGAAA, known as the *SCB box*. A major regulatory protein recognizes the SCB box and turns on genes required for movement toward the G1–S boundary. The regulatory protein is itself activated

[1]The Cdc2 protein kinase is a product of a gene originally named *cdc28* in *S. cerevisiae*; although given a different name, its product is essentially the same cyclin-dependent protein kinase, with the same activities, as the Cdc2 protein of *S. pombe*. For the sake of clarity it will be called the Cdc2 protein kinase in this chapter.

directly or indirectly through phosphorylations carried out by one of the earliest cyclin/Cdc2 complexes to appear in active form in the cell cycle. In some manner, activation of this early cyclin/Cdc2 complex in actively growing yeast populations is triggered by growth of cells to a certain size, dependent on the supply of nutrients in the environment.

Among the proteins made through transcription of genes controlled by the SCB box are two additional cyclins that form complexes with Cdc2. These later cyclin/Cdc2 complexes activate a different major regulatory protein, one that recognizes an upstream control element with the consensus sequence ACGCGTNA (in which N = any nucleotide). The group of genes sharing this control sequence, known as the *MCB box*, includes at least 17 encoding enzymes and factors required for DNA replication and 2 further cyclins that, in combination with the Cdc2 protein kinase, direct entry into S. After S is complete, the last cyclins to appear form complexes with the Cdc2 protein kinase that carry out phosphorylations moving cells from G2 to mitosis.

Cyclins and Cyclin-Dependent Protein Kinases in Higher Eukaryotes Complexes between cyclins and cyclin-dependent protein kinases regulate the cell cycle by mechanisms closely similar to those of the yeasts *S. cerevisiae* and *S. pombe*. A series of cyclin/cyclin-dependent protein kinase complexes occurs in a sequence that triggers successive stages in the cell cycle. The targets of the cyclin-dependent protein kinases are determined by the particular cyclin and protein kinase present in a given complex. Whether a complex is active at any stage, particularly during the transitions from G1 to S and G2 to mitosis, depends on the balance of modifying factors such as Wee1 and Cdc25, which inhibit or promote activity of the complex. For cells of higher animals the primary modifying signals tipping the balance of modifying factors toward active cell division come through reaction cascades linked to cell surface receptors. There are also inhibitors of the cyclin/cyclin-dependent protein kinase complexes that are keyed to important checkpoints in the cell cycle. For example, specific inhibitors prevent entry into S if DNA damage during G1 by agents such as radiation or chemical carcinogens has not been repaired; entry into mitosis is blocked if the DNA has not fully replicated.

The cyclin-dependent protein kinases encoded in the *cdc2* genes of humans are about 60% identical to the yeast *cdc2* gene products. P. C. L. John and his colleagues found that antibodies against *cdc2*-encoded protein kinases react positively with algal cells and higher plants such as wheat, carrot, and oats, indicating that an equivalent gene is probably universally distributed among eukaryotes. Similar results indicating universal distribution have also been obtained with antibodies against cyclins.

Cyclins and cyclin-dependent protein kinases were actually discovered in eukaryotes before they were identified in yeast, as proteins active in promoting cell division in early *Xenopus* and surf clam embryos. Cells removed from *Xenopus* embryos during cleavage divisions continue to divide for some time in synchrony with those still in the embryo, as if an internal molecular clock keeps the divisions on schedule. The first insight into the nature of the clock was found by M. W. Kirschner, who found a protein, termed *MPF* (for *m*aturation *p*romoting *f*actor, later changed to *m*itosis promoting factor), whose activity cycled in coordination with the divisions.

M. J. Lohka and his coworkers extracted and purified MPF and found that it consists of two polypeptides with molecular weights of 34,000 and 45,000. The smaller polypeptide of the MPF complex is a serine/threonine protein kinase that was subsequently identified as the product of the *cdc2* gene. The larger polypeptide turned out to be a cyclin, first discovered by T. Hunt and J. Ruderman as a protein that alternated in concentration in coordination with the cell cycle in cleaving *Spisula* (surf clam) eggs. Later found to be equivalent to the *cdc13* gene product of *S. pombe*, cyclin was directly implicated in cell cycle control by an extended series of experiments in sea urchins and other organisms. (The Experimental Process essay by Hunt on p. 642 tells the story of his discovery of cyclin and the exciting developments stemming from the research.)

Further research by Kirschner, Hunt, and others including A. W. Murray, D. Beach, P. Nurse, W. D. Dunphy, H. Richardson, and J. Newport and their colleagues revealed the series of cyclin/cyclin-dependent protein kinase complexes controlling the cell cycle in eukaryotes (Fig. 22-6). MPF is now known to be one complex of the series, one that controls the transition from G2 to mitosis.

At least six major cyclin types, identified as *cyclin A, B, C, D, E,* and *F,* have been identified in higher vertebrates. Some of these major cyclins, such as cyclin D, occur in two or more subtypes. The cyclins combine with one of at least five closely related cyclin-dependent protein kinases encoded by genes in the *cdc2* family. In eukaryotes these cyclin-dependent protein kinases are identified as *CDKs* numbered CDK1 to CDK5. Of these CDKs, CDK2 is active in cyclin/CDK combinations during all major stages of the cell cycle. Figure 22-6 shows some of the known combinations of these CDKs with cyclins, as well as the periods of the cell cycle in which the cyclin/CDK complexes are active.

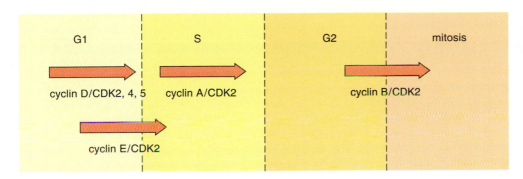

Figure 22-6 Successive complexes between cyclin and cyclin-dependent protein kinases in the eukaryotic cell cycle (see text).

Each cyclin/CDK complex has a different set of target proteins to which the CDKs add phosphate groups; the phosphorylations, depending on the proteins and the positions at which the phosphate groups are added, promote or inhibit activity of the target proteins. In general, cyclins C, D, and F, in combination with CDKs 2, 3, 4, and 5, are active during late G1 and control progress toward the G1/S boundary; passage through the boundary and entry into S appear to be controlled by a cyclin E/CDK2 complex. During S a cyclin A/CDK2 complex is the controlling factor; passage from G2 to mitosis depends on cyclin B in combination with CDK2. Various modifying factors adjust the balance of activity of the successive cyclin/CDK complexes. Among the proteins directly or indirectly activated by the complexes are regulatory proteins controlling transcription of genes required for cell division and proteins forming parts of the mechanisms of mitosis.

Of the various cyclins, the cyclin B/CDK2 complex controlling passage from G2 to mitosis and the factors modifying and balancing its activity have been best studied in eukaryotes (Fig. 22-7). The mechanism resembles the equivalent cyclin/Cdc2 complex and modifying factors operating during the G2 to mitosis transition in *S. pombe*. CDK2 is assembled more or less continuously and remains in approximately constant concentrations in the cytoplasm throughout the cell cycle. Cyclin B concentration is at a minimum following a previous mitosis. Following entry into S, cyclin B synthesis begins and the cyclin assembles with CDK2 in the cyclin B/CDK2 complex. The complex remains inactive at this point because, as it is assembled, CDK2 is phosphorylated at the inhibiting tyrosine 15 by a vertebrate equivalent of the Wee1 protein kinase. Just before the G2–mitosis transition, cyclin B/CDK2 becomes fully active when CDK2 is dephosphorylated by a vertebrate equivalent of the Cdc25 phosphatase.

In active form, cyclin B/CDK2 catalyzes a series of phosphorylations that initiate mitotic division. The addition of phosphate groups to histone H1, for ex-

ample, which is thought to trigger chromatin condensation, is catalyzed by the complex at early mitotic stages; other phosphorylations by the cyclin B/CDK2 complex cause nuclear envelope breakdown through phosphorylation of the lamins (see pp. 707 and 717) and promote spindle assembly and activity. Phosphorylation of components of the transcription apparatus by the activated cyclin B/CDK2 complex inhibits RNA transcription during mitosis.

Full activation of cyclin B/CDK2 leads to degradation of the cyclin B subunit and inhibition of the complex just before anaphase begins. Degradation occurs when, near the end of its period of activity, the cyclin B/CDK2 complex activates a series of enzymes that catalyze breakdown of cyclin B. The breakdown is necessary for cells to exit mitosis and enter the subsequent G2; if cyclin B breakdown is inhibited, or if intact cyclin B is added during mitosis, cells arrest their progress and fail to complete division. As cyclin B breaks down in the normal course of events, CDK2 is rephosphorylated by the Wee1 equivalent and remains in the cytoplasm in inactive form until combination with cyclins during the following G1.

The cyclins and cyclin-dependent protein kinases and their encoding genes are thus strongly implicated as central controls triggering the transition from G2 to mitosis. The presence of similar cyclins in similar form in a wide range of eukaryotes, including algae, fungi, higher plants, and animals, indicates that cell cycle controls based on these polypeptides were established very early in evolution and have remained as critical regulatory elements through a billion years of evolutionary history. (Cell cycle control in the bacterium *E. coli* is described in Information Box 22-1.)

Cell Cycle Genes Identified Through Study of Oncogenes

As noted, most oncogenes are altered forms of genes that also occur in unaltered form as proto-oncogenes in normal body cells. As such, they are specialized

Finding Cyclins and After

Tim Hunt

TIM HUNT did his Ph.D. on hemoglobin synthesis with the late Asher Korner at Cambridge University. From 1968 through 1970 he worked on the control of globin synthesis with Irving London at the Albert Einstein College of Medicine in New York. Between 1971 and 1990 he shared a laboratory with Richard Jackson at Cambridge, studying rabbit reticulocyte lysates and teaching biochemistry. Since then, he has been at ICRF Clare Hall Laboratories, doing single-minded research on cyclins, cdc2s, and the control of the cell cycle.

The discovery of cyclin was a lucky accident. It came about through teaching summer courses at the Marine Biological Laboratory, Woods Hole. I first went there in 1977 to teach protein synthesis in the embryology course; the *quid pro quo*, apart from the beautiful surroundings, was the supply of sea urchin eggs in the same place as well-equipped laboratories. For the first two summers we studied the increase in the rate of protein synthesis which occurs in these eggs after fertilization. We never did find out, really, what turned on protein synthesis in sea urchin eggs, and it slowly dawned that what looked like a big stimulation was really only a small relief from overwhelming inhibition. Going from 99.9% inhibition to 98% inhibition is equivalent to going from 0.1% activity to 2% activity, a 20-fold increase in rate.

One of the instructors in the 1979 embryology course was Joan Ruderman. Her graduate student, Eric Rosenthal, had recently started to investigate a different kind of change in protein synthesis in fertilized eggs of the clam, *Spisula*. His elegant experiments showed that the patterns of protein synthesis changed soon after fertilization in these cells; new mRNAs were recruited onto ribosomes, and some of the previously translated mRNAs stopped being used.[1] This made sense, because changing an egg into an embryo presumably needs new proteins. But we had no idea what the new proteins were, and called the main ones *A, B,* and *C* in descending order of molecular weight. Sequencing of cDNA by Nancy Standart identified protein *C* as the small subunit of ribonucleotide reductase.

It was much more difficult to understand the situation in sea urchin eggs, because of a series of careful experiments performed by Bruce Brandhorst, another Woods Hole habitué. Unlike what happened in clams, Brandhorst could find no change in the pattern of protein synthesis after fertilization of sea urchin eggs.[2] His data were very convincing.

Thus the paradox. It had long been known that if protein synthesis was inhibited after fertilization of sea urchin eggs, DNA replication was permitted but entry into mitosis was blocked. How could this be, if the eggs

(1) made so little protein after fertilization, compared to what they already possessed, and (2) there was no difference in the kind of proteins synthesized?

By 1982, our studies of protein synthesis and its control in sea urchin eggs had explored many alleys without great success. On July 22, in a quiet moment, I did a simple experiment, purely out of curiosity, to compare the patterns and rates of protein synthesis in fertilized and parthenogenetically activated sea urchin eggs. I added ^{35}S-methionine to the eggs, and took samples at regular intervals for running on an SDS-polyacrylamide gel. The result was intriguing and completely unexpected. The most strongly labeled band at early times faded away after about an hour, whereas most of the other bands got stronger. Repetition of the experiment with carefully controlled batches of fertilized eggs put the matter beyond doubt, for in the next experiment, the protein appearing as the strongly labeled band—which we termed *cyclin*— came and went twice. Its disappearance preceded cell division by about 10 minutes. When division was inhibited by colchicine, or by inhibition of DNA synthesis, the destruction of cyclin was delayed or failed to occur. We could see cyclin continuing to cycle several hours after fertilization by labeling older embryos for 30 minutes, and then adding emetine to block further synthesis. The cyclin band went away, while all the others stayed.

These results were written up in a state of high euphoria.[3] People had been looking for so-called "periodic proteins" for several years, for it was widely suspected that if cells engaged in different activities at different stages of their cycle, they would contain different sets of enzymes. I do not recall having seen anyone point out what seems plain in retrospect, namely that for such a mechanism of restricting certain activities to particular phases of the cycle to work, the proteins in question would have either to be very unstable, or to be eliminated by some special mechanism when they were no longer required. Such appeared to be the case with cyclin.

Two years later, Joan Ruderman and I made very detailed studies of the stability of cyclins during the cell cycle, and found that they were stable proteins during interphase, and very rapidly destroyed just before the metaphase → anaphase transition. Clams contained two cyclins, corresponding to the two translationally regulated proteins *A* and *B*.[4,5]

John Gerhart had given a seminar on MPF in 1979, which had introduced me to this fascinating and important entity, and he turned up on the very day that cyclin was discovered. He told me about experiments that he, Marc Kirschner, and Mike Wu had been doing to measure MPF levels during meiosis in *Xenopus* embryos.[6] I was particularly excited by the news that after its initial, autocatalytic appearance, new protein synthesis was required for MPF to reappear after meiosis II. Could cyclin be a component of MPF? Or was cyclin perhaps the enzyme that catalysed the activation of MPF? If the latter, did it act directly and enzymatically, or by titrating out a hypothetical "anti-MPF"?

The descriptive phase of cyclin studies came to an end in 1986, when Katherine Swenson discovered that clam cyclin *A* mRNA promoted maturation in frog oocytes. This was a very important development, because it clearly demonstrated that cyclins could promote entry into M-phase. The question then arose whether frogs (or indeed, any organism apart from marine invertebrates) had cyclins. One day in Berkeley, Mike Wu offered to show me a real MPF assay. I thought it would be fun to try in parallel if maternal mRNA would score in this assay, as Swenson's experiments implied. Fortunately, Eric Rosenthal was in Berkeley working on oogenesis in *Urechis caupo*, an invertebrate worm that lives in the mudflats of Bodega Bay. He gave me some oocyte and embryo RNA, and Wu injected it into oocytes. The ones injected with two-cell embryo mRNA duly matured, whereas the oocyte mRNA-injected cells just sat there. We decided to see if *Xenopus* mRNA could do the same, and were thrilled to find that mRNA from mature or fertilized eggs was active. All the controls looked good, and we guessed this meant that *Xenopus* has cyclins.

Back in England, my graduate students, John Pines and Jeremy Minshull, were making good progress cloning and sequencing cyclin mRNA from *Arbacia*. Minshull decided to make a cDNA library from *Xenopus* and screen it with an oligonucleotide based on a consensus sequence developed from clam cyclin *A* and sea urchin cyclin *B*. Since they were, evolutionarily speaking, so far apart, it was likely this sequence would find any cyclin (indeed, flies, cows, and soybeans all eventually yielded up cyclins to oligonucleotide 40). In early 1987, we saw our first vertebrate *B*-type cyclin.

By now, cell-cycle studies had taken an enormous boost from the work of Paul Nurse and Melanie Lee at the ICRF in London, who found the human version of the yeast cell cycle control gene, cdc2.[7] In 1988, Joan Ruderman and David Beach showed that clam cyclin immunoprecipitates possessed protein kinase activity, and contained a polypeptide that probably corresponded to clam cdc2.[8] That same summer, various consortia of frog and yeast workers showed that MPF contained cdc2,[9,10] and when antibodies to frog cyclins became available, the large subunit of MPF proved to be a *B*-type cyclin.[11]

Meanwhile, how could we directly show that cyclin was really essential for driving cell cycles? Two approaches put the matter beyond doubt, both relying on the cell-free system from frog eggs that Fred Lohka had originally developed in Yoshio Masui's laboratory.[12] This preparation could replicate the DNA of added nuclei and then enter mitosis, which, like mitosis in intact sea urchin eggs, required protein synthesis. Minshull, Blow, and I used complementary oligonucleotides capable of pairing with cyclin mRNA to eliminate this mRNA. We found that without cyclin mRNA, DNA replication was OK, but mitosis failed.[13]

At about this time, Andrew Murray did a much more daring experiment in Kirschner's laboratory. He re-

moved all the endogenous mRNA from the frog egg extract using ribonuclease, and then added back Pines's *Arbacia* cyclin B mRNA and a ribonuclease inhibitor. The only protein made by the extract was cyclin, and amazingly, the nuclei replicated and entered mitosis.[14] In fact, they entered mitosis not once, but two or three times, and at each mitosis the cyclin was destroyed. A simple construct lacking the first 90 codons of the cyclin mRNA gave rise to a stable protein, because the "destruction box" had been removed. When this mRNA was added to the extracts, they entered mitosis but got stuck there![15] This convinced even the most hardened sceptics that cyclin synthesis was needed to get into mitosis, and its destruction was required for the cell to return to interphase.

1988 was also the year that a nest of new cyclins turned up in yeast. It was especially exciting that some of them, discovered independently by Fred Cross, Bruce Futcher, and Steve Reed, appeared to be involved in control of the G1-S transition. Thus it emerged that the periodic protein of sea urchin eggs was a representative of a family of cell cycle regulators, whose function now emerges as the control, in time and in space, of the cdc2 family of protein kinases. The size and variety of these families, their richness for the optimist or complexity for the pessimist, presents a considerable challenge. The pressing question now, it seems to me, is to identify the key substrates whose phosphorylation can bring about cell cycle transitions. Life was simpler in 1982.

References

[1] Rosenthal, E. T.; Hunt, T.; and Ruderman, J. V. *Cell* 20:487–94 (1980).

[2] Brandhorst, B. P. *Dev. Biol.* 52:310–17 (1976).

[3] Evans, T.; Rosenthal, E. T.; Youngblom, J.; Distel, D.; and Hunt, T. *Cell* 33:389–96 (1983).

[4] Westendorf, J. M.; Swenson, K. I.; and Ruderman, J. V. *J. Cell Biol.* 108:1431–44 (1989).

[5] Hunt, T.; Luca, F. C.; and Ruderman, J. V. *J. Cell Biol.* 116:707–24 (1991).

[6] Gerhart, J.; Wu, M.; and Kirschner, M. *J. Cell Biol.* 98:1247–55 (1984).

[7] Lee, M. G., and Nurse, P. *Nature* 327:31–35 (1987).

[8] Draetta, G., et al. *Cell* 56:829–38 (1989).

[9] Dunphy, W. G.; Brizuela, L.; Beach, D.; and Newport, J. *Cell* 54:423–31 (1988).

[10] Gautier, J.; Norbury, C.; Lohka, M.; Nurse, P.; and Maller, J. *Cell* 54:433–39 (1988).

[11] Gautier, J., et al. *Cell* 60:487–94 (1990).

[12] Lohka, M. J., and Masui, Y. *Science* 220:719–21 (1983).

[13] Minshull, J.; Blow, J. J.; and Hunt, T. *Cell* 56:947–56 (1989).

[14] Murray, A. W., and Kirschner, M. W. *Nature* 339:275–80 (1989).

[15] Murray, A. W.; Solomon, M. J.; and Kirschner, M. W. *Nature* 339:280–86 (1989).

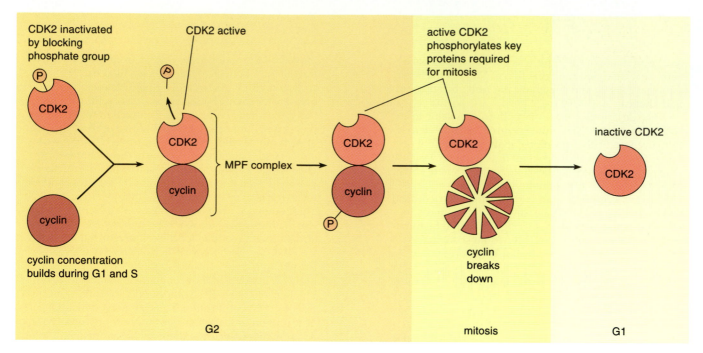

Figure 22-7 Interactions of cyclin and the CDK2 protein kinase during the transition from G2 to mitosis. During G2, cyclin, which gradually increases in concentration during S and G2, combines with CDK2 to form the MPF complex. At this point the complex is inactivated by a phosphate group added by Wee1 to tyrosine 15 of CDK2, in the site binding ATP. Late in G2, Cdc25 removes the blocking phosphate from CDK2 and activates the kinase. Among the targets of CDK2 are histone H1, the lamins of the nuclear envelope, and microtubule proteins; phosphorylation of these target proteins initiates mitosis. Soon after mitosis begins, cyclin breaks down through enzymatic activity induced by the cyclin B/CDK2 complex itself. Cyclin breakdown inactivates the complex by the end of metaphase.

examples of mutants, in this case associated with development of cancer.

Identifying Oncogenes Many oncogenes were first identified through the study of cancer induced by viral infections, particularly by a group of viruses infecting mammals, the *retroviruses* (see p. 536). Although the majority of retroviral infections appear to be harmless or to cause other disabilities, some have been linked to tumor formation in humans and other mammals. In many of these cases the retroviruses carry oncogenes or cause disruptions in the genome by altering a proto-oncogene to an oncogene. (The mechanisms by which retroviruses carry or alter host genes are outlined in Chapter 18.) In either case, identification of the oncogene has frequently led to discovery of the corresponding proto-oncogene and its normal role in the cell cycle.

Among the most productive of the investigators active in this effort were J. M. Bishop and H. E. Varmus, who received the Nobel Prize in 1989 for their research establishing the relationship between oncogenes and their normal proto-oncogene counterparts.

Among other important discoveries, Bishop and Varmus and their colleagues showed that the DNA sequence of the *src* oncogene was also present in a closely related proto-oncogene form in normal cells.

Oncogenes have also been identified by *transfection*, a genetic engineering technique in which the DNA of human cancer cells is introduced into cultured mouse cells. Frequently the mouse cells incorporate the introduced DNA and are transformed into types that grow much more rapidly than normal cells. The human oncogene responsible for the transformation is then isolated from the mouse cells by cloning techniques. (Figure 22-8 outlines a typical method used in this approach.) Among the pioneers using transfection to isolate human oncogenes were J. Cooper and R. A. Weinberg.

Not all cellular functions affected by oncogenes are directly or indirectly related to regulation of cell division. Besides growing through uncontrolled cell division, cancer cells *metastasize*: that is, they break loose readily from their tissues of origin to lodge in other body locations where they grow and divide to form additional tumors (see below). A few oncogenes

Cell Cycle Control in *E. Coli*

DNA replication and cytoplasmic division are initiated in *E. coli* when the cell doubles in length and volume after a previous division. At this time, DNA replication and cytoplasmic division are triggered by transcription of a single gene, *DnaA*. About 20 to 40 molecules of the protein encoded in *DnaA* recognize and bind to a region of the DNA circle that serves as the *replication origin*, the site at which DNA replication begins. This region, *OriC*, contains four sequence boxes bound by the DnaA proteins. Binding by DnaA forces the DNA to unwind in A-T–rich sequence elements forming part of *OriC* (A-T base pairs, held together by two base pairs instead of three as in G-C pairs, unwind more easily). The unwound region of *OriC* provides access to the DNA by the enzymes of replication, and replication proceeds from the origin (see p. 680 for further details of *OriC* and the enzyme complex replicating *E. coli* DNA). The sensory mechanism linking cell size to transcription of the *DnaA* gene remains unknown.

Two controls modify entry into division in *E. coli*. One is linked to the stress response in *E. coli* and involves a gene transcribed if the DNA is damaged by chemical alterations or breakage or when cells are exposed to unfavorable environmental conditions such as elevated temperature (see p. 518). The gene, *SulA*, encodes a protein that blocks DNA replication. The other control depends on methylation of the A's in A-T base pairs in GATC elements in and near *OriC*. After a previous round of replication, methylation of A's in the newly synthesized nucleotide chain proceeds rapidly in all locations of the *E. coli* DNA circle except in the GATC elements in and near *OriC*. Methylation of these A's is delayed for about 30% to 40% of the cell cycle; as long as they remain unmethylated, DNA replication is blocked. The delayed methylation is the primary control of cell cycle length in *E. coli* growing under optimal conditions.

The remaining cell cycle genes of *E. coli* consist of two operons encoding proteins concerned with division of the cytoplasm and its regulation. Through the activity of these proteins the cell wall grows inward at a region halfway between the ends of the cell and cuts the cytoplasm in two. (Details of the genes and proteins accomplishing cytoplasmic division in *E. coli* are presented in Supplement 24-1.)

---

promote metastasis rather than, or in addition to, being cell cycle controls.

Oncogene Types About 100 oncogenes have been discovered through the study of retroviruses, transfection experiments, and other techniques. (Table 22-2 lists some oncogenes commonly found in tumors of humans and other animals.) The proteins encoded in most of these oncogenes prove to encompass a variety of types including (1) protein kinases, (2) polypeptide growth factors, (3) surface receptors for hormones or growth factors, (4) G proteins active in reaction pathways activated by surface receptors, (5) steroid or thyroid hormone receptors, or (6) regulatory proteins controlling gene activity.

In addition to the oncogenes in these classes, several recently discovered genes known as *tumor suppressor genes* (also termed *recessive oncogenes* or *anti-oncogenes*) have the unusual effect of inhibiting tumor growth promoted by oncogenes. The genes have this beneficial effect if they are present in normal form in at least one of the two chromosomes of the set. However, their conversion to a mutated, inactive form may be associated with development of cancer. Tumor suppressor genes have generated considerable interest because of their potential use as a means to control cancer. Other oncogenes include some that in normal form encode proteins promoting programmed cell death; even a cyclin has been implicated in cancer—a cyclin D subtype, cyclin D1, has been implicated in changes inducing uncontrolled cell division when synthesized in abnormally high amounts.

Oncogenes Encoding Protein Kinases Among the oncogenes, those encoding protein kinases are among those most frequently identified in tumors. The majority, including *src*, *abl*, *fps*, *yes*, *fgr*, *met*, and *ros* (see Table 22-2), encode protein kinases that add phosphate groups to tyrosine residues in their target proteins. Other protein kinases, products of the *mos*, *raf*, and *mil* oncogenes, phosphorylate serine or threonine residues in their target proteins.

Some of the target proteins phosphorylated by these protein kinases undoubtedly take part in pathways that directly or indirectly regulate the cell cycle or affect other functions that go awry in cancer cells. For example, the kinases encoded in two of the genes in this group, *src* and *ros*, add phosphate groups to membrane phospholipids taking part in the $InsP_3$/DAG pathway of cellular activation (see p. 161). The preponderance of protein kinases among the direct or

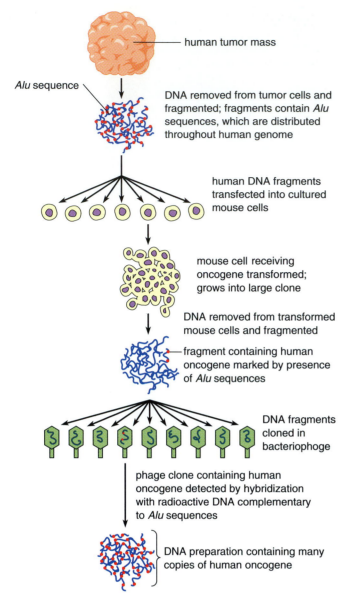

human tumor mass

Alu sequence

DNA removed from tumor cells and fragmented; fragments contain *Alu* sequences, which are distributed throughout human genome

human DNA fragments transfected into cultured mouse cells

mouse cell receiving oncogene transformed; grows into large clone

DNA removed from transformed mouse cells and fragmented

fragment containing human oncogene marked by presence of *Alu* sequences

DNA fragments cloned in bacteriophoge

phage clone containing human oncogene detected by hybridization with radioactive DNA complementary to *Alu* sequences

DNA preparation containing many copies of human oncogene

Figure 22-8 A technique used for identifying human oncogenes by transfection of cultured mouse cells. (In transfection, cells take up DNA fragments from the surrounding medium and integrate them into their own DNA.) The human DNA can be identified among the mouse DNA sequences by the presence of *Alu* repeats (see p. 527), which are distributed widely in the human genome.

indirect oncogene products reflects the fact that addition of phosphate groups to cellular proteins is a primary mechanism of cell cycle regulation in eukaryotes. In general the oncogene forms of the protein kinases are produced in a state of activity that promotes cell division, in situations in which their normal counterparts would be either less active or inactive.

At least some of the oncogenes encoding protein kinases promote malignancy by increasing the ten-

dency of cancer cells to metastasize. Among the target proteins phosphorylated by *src*, for example, is the surface receptor for fibronectin, one of the major proteins of the extracellular matrix (see p. 185). The phosphorylation, which takes place on the cytoplasmic extension of the fibronectin receptor, is believed to inhibit binding between the receptor and fibronectin, thereby loosening attachment of the cell to the extracellular matrix and promoting its tendency to break loose and migrate.

Oncogenes Encoding Growth Factors or Surface Receptors Only one known oncogene, *sis*, encodes a polypeptide growth factor. The protein encoded in *sis*, a segment of PDGF, is able to induce division in cells with a PDGF receptor. Fibroblasts, smooth muscle cells, and the glial cells of the central nervous system are among the cell types with PDGF receptors. As such, they are susceptible to tumor formation by exposure to the product of the *sis* oncogene.

Several oncogenes, including *erbB* and *fms*, encode growth factor receptors. The *erbB* oncogene encodes a receptor for *epidermal growth factor* (*EGF*), a growth factor stimulating division of many body cells; *fms* encodes a receptor for *colony stimulating factor* (*CSF*), a growth factor that normally triggers growth in macrophages, the white blood cells that scavenge foreign cells and substances from the body. Another oncogene, *ros*, probably also encodes a growth factor receptor.

Oncogenes Encoding G Proteins The G proteins encoded in *ras* oncogenes play a central role in some pathways of the cytoplasmic reactions signaling that a surface receptor has bound to its target protein (see Fig. 6-2c). Three *ras* genes, H-*ras*, N-*ras*, and K-*ras*, have been identified in mammals; each has a mutant form implicated in the generation of cancer.

The faulty forms of the G proteins encoded in *ras* oncogenes generally have reduced ability to break down GTP. This limits the self-deactivation function of these proteins (see p. 158). In this form the G proteins remain active in stimulating the regulatory pathways in which they take part, which include activation of genes promoting cell division. The result, if other conditions are met, is uncontrolled cell division. The *ras* oncogenes are found in a very large percentage of tumors of certain kinds, as in 97% of pancreatic carcinomas and 20% to 40% of cancers of the colon and rectum. They are also common in cancers of epithelial tissues and the blood.

An Oncogene Encoding a Steroid or Thyroid Hormone Receptor Only a single known oncogene, *erbA*, encodes a receptor for steroid or thyroid hormones. The DNA sequence of this oncogene closely resembles that

Table 22-2   Some Oncogenes and Their Original Sources

Oncogene	Original Cancer Type	Original Source	Activity or Product	Cellular Location
abl	Abelson leukemia	Mouse	Tyr protein kinase	Cytoplasm
erbA	Erythroblastosis	Chicken	Thyroid hormone receptor	Nucleus
erbB	Erythroblastosis	Chicken	EGF receptor	Plasma membrane
ets*	Myeloblastosis	Chicken	Regulatory protein	Nucleus
fes*	Feline sarcoma	Cat	Tyr protein kinase	Cytoplasm
fgr	Feline sarcoma	Cat	Tyr protein kinase	Plasma membrane
fms	Feline sarcoma	Cat	CSF receptor	Plasma membrane
fos	Osteosarcoma	Mouse	Regulatory protein	Nucleus
fps*	Sarcoma	Chicken	Tyr protein kinase	Cytoplasm
jun	Sarcoma	Chicken	Regulatory protein	Nucleus
met	Osteosarcoma	Mouse	Tyr protein kinase	Cytoplasm
mil†	Sarcoma	Chicken	Ser/Thr protein kinase	Cytoplasm
mos	Sarcoma	Mouse	Ser/Thr protein kinase	Cytoplasm
myb	Myeloblastosis	Chicken	Regulatory protein	Nucleus
myc	Myelocystosis	Chicken	Regulatory protein	Nucleus
neu	Neuroblastoma	Rat	EGF receptor-like protein	Plasma membrane
p53	Many cancers		Tumor suppressor gene	Nucleus
raf†	Sarcoma	Mouse	Ser/Thr protein kinase	Cytoplasm
ras	Sarcoma	Rat	G protein	Plasma membrane
rb	Retinoblastoma	Human	Tumor suppressor gene	Nucleus
rel	Reticuloendotheliosis	Turkey	Regulatory protein	Nucleus
ros	Sarcoma	Chicken	Tyr protein kinase; probable GF receptor	Plasma membrane
sis	Sarcoma	Monkey	PDGF	Secreted
ski	Carcinoma	Chicken	Regulatory protein	Nucleus
src	Sarcoma	Chicken	Tyr protein kinase	Cytoplasm
trk	Carcinoma	Human	Tyr protein kinase	Cytoplasm
yes	Sarcoma	Chicken	Tyr protein kinase	Cytoplasm

*Same oncogene initially identified in different organisms.
†Same oncogene initially identified in different organisms.
  EGF, epidermal growth factor; CSF, colony stimulating factor; GF, growth factor; PDGF, platelet-derived growth factor.

of genes encoding normal receptors for thyroxin and a particular class of steroid hormones, the glucocorticoids. These and other normal forms of the steroid or thyroid hormone receptors are activated by binding the hormone; in active form the receptor recognizes and binds the control regions of specific genes and regulates their transcription. The *erbA* oncogene encodes a faulty receptor that is continuously active in gene regulation, whether or not the steroid or thyroid hormone is bound.

Oncogenes Encoding Nuclear Regulatory Proteins   The proteins encoded in *ets, fos, jun, myb, myc, rel,* and *ski* are known or suspected to regulate genes fundamentally important to cell cycle regulation. The normal forms of these oncogenes are activated early during the G1–S transition, as expected if they are fundamental controls of the cell cycle. Among the many genes activated by the regulatory proteins are some known to promote cell division, such as other oncogenes and genes encoding growth factors. The *myc* product, for example, can bind directly to DNA and probably regulates genes controlling the G1–S transition. (The

Experimental Process essay by G. C. Prendergast on p. 492 describes his research identifying the DNA-binding activity of Myc, the *myc* gene product.)

Tumor Suppressor Genes   The normal counterparts of tumor suppressor genes encode products that inhibit cell division. A mutation eliminating the activity of one of the two copies of a gene in this group usually has no noticeable effect, because the remaining copy is still active in suppressing cell division. A mutation eliminating the activity of the second copy of the gene, however, may lead to uncontrolled cell growth. The genes are thus "recessive" in the sense that both copies must be in mutant form to produce an effect. The recessive nature of tumor suppressor genes is in stark contrast to "dominant" oncogenes such as *ras* or *myc*, in which a single copy of the oncogene can promote uncontrolled growth.

In some human families, for example, individuals inherit one normal and one mutated copy of the *rb* gene, located at a site on chromosome 13 of the set. The gene encodes the *RB* protein, which has been shown by P. D. Robbins and his coworkers to bind

DNA and repress transcription of *fos* and other regulatory proteins activating genes related to cell division. As long as one normal copy of the gene is present, individuals are not unusually susceptible to developing cancer. However, the chromosome region containing the normal copy is highly subject to deletions; deletions impairing the remaining normal copy of the *rb* gene in these individuals lead to development of a form of *retinoblastoma*, a cancer of the eye most commonly found in children. The development of two other cancers of children, *Wilms' tumor* (a cancer of kidney cells) and *neuroblastoma*, follows a similar pattern. (Neuroblastoma, a cancer of sympathetic neurons, is the most common solid tumor of children.) Loss of normal *rb* genes in both chromosomes has been demonstrated in adult bone, lung, and prostate cancers in addition to the tumors occurring in children. RB in normal form stops the cell cycle by several activities, including binding and blocking regulatory proteins stimulating transcription of genes required for cell division as in retinoblastoma. Normal RB also binds the cyclin A/CDK2 complex and blocks its role in promoting the transition from G1 to S.

Another tumor suppressor gene encoding a protein known as *p53* is implicated in a wide range of human cancers when inactivated, including bladder, brain, breast, cervical, colon, esophageal, laryngeal, liver, pancreatic, prostate, skin, and stomach cancer and some leukemias. In fact, mutant forms of p53 are found in more human cancers than any of the oncogenes—70% of colon cancers, 50% of lung cancers, and 40% of breast cancers, for example, exhibit impaired forms of the p53 protein. In normal form, p53 works as a regulatory protein that stimulates transcription of some genes and inhibits others. Among the genes stimulated by p53 is one encoding another protein, *p21*, which is a powerful inhibitor of CDK kinases. p53 also combines with and inactivates a variety of proteins promoting cell division, including some required for DNA replication, and activates pathways leading to programmed cell death. A long list of point mutations in the p53 gene impair its encoded protein by changing single amino acids at different positions; some cancer-causing substances, such as cigarette smoke, produce characteristic disabling alterations in the p53 gene. Several proteins of viruses causing cancer (see below) work by binding and inactivating either the p53 or RB proteins. More than 10 genes with a suspected tumor suppressor function have been detected in addition to *rb* and *p53*.

Introduction of normal copies of the *rb* and *p53* genes has suppressed growth of some cultured tumor cells; presumably these are cells in which these genes are doubly mutated into oncogenic form. The successes with cultured cells give hope that introduction of normal tumor suppressor genes may provide a means to control at least some forms of cancer.

Alterations Changing Proto-Oncogenes into Oncogenes

Sequencing the known oncogenes has revealed that many are altered in some way as compared to their normal cellular counterparts. The characteristics of these alterations have provided clues to the genetic factors giving rise to cancer. In addition, they have afforded valuable insights into the patterns by which their normal counterparts, the proto-oncogenes, may work to regulate cell division.

The alterations distinguishing oncogenes from proto-oncogenes fall into several clearly defined classes involving: (1) changes in the DNA sequence of the gene or its control regions, (2) movement or *translocation* of part or all of a gene to a different location in the chromosomes, or (3) *amplification* of the gene to produce extra copies. Often more than one of these changes is noted in the same oncogene, producing a copy that is extensively altered from its normal form. In most cases the alterations activate the gene or its products, leading directly or indirectly to changes associated with the transformation of normal cells to cancer cells.

Most of the gene alterations involved in development of cancer occur in body cells, not in reproductive cells. Therefore, although the primary causes of these tumors are genetic, they are not hereditary. They die with the individual and are not passed on to offspring. A relatively few alterations producing oncogenes, however, occur in reproductive cells and are passed on, giving the persons inheriting them an increased likelihood of developing cancer.

Sequence Changes in Coding or Control Regions Mutations altering the DNA sequence of coding segments, promoters, or enhancers are among the most common alterations converting proto-oncogenes to oncogenes. The oncogenic forms of the *ras* gene detected in cancer cells, for example, usually differ from their normal counterparts by single base substitutions in one of three positions in the coding sequence. The alterations cause amino acid substitutions in corresponding positions in the encoded G protein. In the *ras* oncogene associated with human bladder carcinoma, for example, a mutation leads to substitution of valine for glycine at position 12 in the amino acid sequence of the G protein encoded in the gene. In one form of human lung cancer the *ras* oncogene contains a mutation substituting leucine for glutamine at position 61 in the encoded G protein.

Changes in DNA sequence leading to formation of oncogenes may also involve deletion of one or more nucleotides from a DNA sequence. The *myc* gene in its normal, proto-oncogene form, for example, consists of three exons separated by two introns (Fig. 22-9a). In many of its oncogenic forms the first exon and intron and the control sequences in advance of the gene are missing (Fig. 22-9b). Although the effects of the deletions are unknown, removal of the first exon, which is normally copied into the *myc* mRNA but not translated into a protein sequence, may alter RNA processing to increase the amounts of *myc* mRNA available to ribosomes in the cytoplasm. Removal of the promoter region from the *myc* gene does not prevent transcription of the gene, because additional signals capable of acting as promoters form parts of the second exon.

Translocations and Oncogene Formation Translocations involve gross exchanges in which whole chromosome segments separate and attach to sites in other chromosomes. In some translocations, breaks occur in two chromosomes simultaneously, setting up a reciprocal exchange of segments between the chromosomes involved (see also p. 538).

Translocations carry genes included in the relocated segments to new positions in the chromosomes. The new location often places the gene in a control environment that is distinctly different from its previous position. Controls that limited activity of the gene in its previous location may be absent in the new environment, causing the gene to be more active. In some cases only part of the gene is translocated, so that important control sequences are left in the original location. The gene in its new location may come under the influence of a highly active promoter or enhancer belonging to another gene, also making the gene more active. If the gene is related to cell cycle control, the translocation may result in rapid and uncontrolled cell division.

Chromosome translocations have been observed in association with many forms of human cancer, including several types of leukemias and *Burkitt's lymphoma*. In one line of Burkitt's lymphoma a reciprocal exchange has taken place between segments of chromosomes 8 and 14 of the human set (Fig. 22-10a). The original break in chromosome 8 in this line occurred at the *myc* proto-oncogene, separating *myc* between its first and second exons. In chromosome 14 the break occurred within the segment coding for the immunoglobulin heavy-chain constant region (see p. 549). The broken segments then exchanged places, moving the shortened *myc* gene next to the heavy-chain coding region of chromosome 14 (Fig. 22-10b). The translocation increases the concentration of the Myc protein, either through an increase in the rate of *myc* transcription or, if the first exon is lost, through an increase in processing of its mRNA. One possibility for the increase in transcription is that the *myc* gene is brought

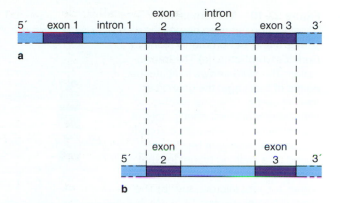

Figure 22-9 The *myc* gene in normal and oncogene forms. **(a)** The normal form, which has three exons and two introns. **(b)** A common oncogenic form of *myc*, in which the first exon and intron are missing. Promoterlike sequences forming part of the second exon allow transcription of the oncogene.

under the influence of enhancer elements in the immunoglobulin gene. Another is that movement to the new location separates *myc* from as yet unidentified sequence elements or other genes that regulate or suppress its activity in its normal location on chromosome 8.

In a translocation responsible for *chronic myelogenous leukemia* the *abl* gene moves from its normal location as a proto-oncogene at one end of human chromosome 9 to a site on chromosome 22 containing the normal form of the gene *bcl*. Joining the two genes produces an *abl-bcl* combination that is highly active as an oncogene. The translocation produces a characteristically shortened chromosome 22 known as the *Philadelphia chromosome*, which is a hallmark of chronic myelogenous leukemia.

Leukocytes of the B-cell and T-cell types (see p. 545) are frequently involved in leukemias developing as a result of chromosome translocations. Their involvement probably results from a failure of the breakage and rearrangement mechanisms that normally assemble genes of the immunoglobulin family into antibody genes in B cells and receptors in T cells (see p. 552). A breakage occurring as part of these rearrangements evidently remains open, providing an opportunity for segments broken from chromosomes at other sites to join. For unknown reasons the site on chromosome 8 containing the *myc* gene frequently shares in reciprocal translocations with the immunoglobulin genes to create a *myc* oncogene. Perhaps an arrangement of chromosomal proteins in the *myc* region also creates an exposed site that is easily broken.

Amplification in Oncogene Production Gene amplification results from random overreplication of restricted

Figure 22-10 A chromosome translocation between chromosomes 8 and 14 commonly found in Burkitt's lymphoma. **(a)** The translocation: The solid arrows indicate the points of breakage; the dashed arrows show how the broken chromosome segments exchange places in the translocation. The light and dark crossbands represent bands produced by a staining technique that identifies individual human chromosomes for light microscopy. **(b)** The region of chromosome 14 at which the fragment from chromosome 8 joins. The translocation places the second and third exons of the *myc* gene next to exons encoding heavy-chain segments in an antibody gene.

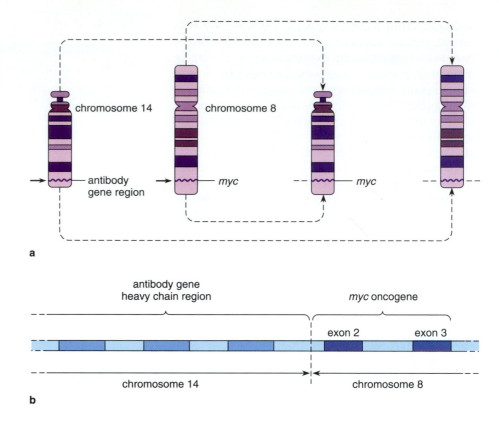

segments of the DNA (see p. 683). The long chromosomes of eukaryotic cells contain multiple sites from which DNA replication proceeds. Normally each site is activated once during an S period, so that the entire DNA molecule of the chromosome is duplicated only one time. Rarely one or more replication sites is activated more than once during an S phase, causing a segment of the DNA molecule to be overreplicated, or amplified. DNA amplification produces extra copies of genes or other DNA sequences, increasing the "dosage" of these elements in the cell. If an amplified segment happens to contain a gene related to cell division, the extra copies of the gene may lead to overproduction of the gene's encoded protein and stimulation of division.

Gene amplification is frequently observed in association with human and other animal cancers. Amplified copies of the *myc* gene, for example, appear in several human tumors, including neuroblastomas and some breast and colon cancers. In cells of mouse adrenocortical tumors the *ras* gene is amplified from 30 to 60 times. Amplification of the oncogenes in each of these examples results in increased transcription of the gene, contributing to conversion of the affected cells to malignant types.

Amplification also frequently accompanies other forms of chromosome alterations. For example, in leukemias involving translocations between the *myc* and immunoglobulin gene regions, the *myc* oncogene is often amplified in its new location.

Oncogene Formation Without Alterations in Gene Sequences In some instances, normal genes are converted to oncogenes without changes in the coding or control sequences of the gene itself. This takes place, for example, in the process termed *insertional activation*, in which a retrovirus inserts into the DNA near a gene related to cell division. Insertion of the retroviral DNA brings the proto-oncogene under the influence of "strong" viral promoters or enhancers. Stimulation by the nearby viral elements activates the gene and promotes cell division. For reasons that are not understood, a hotspot for retroviral insertion lies near the *myc* gene, close enough for the gene to become activated to oncogenic form when the viral DNA inserts. Insertional activation of the *myc* gene in this pattern takes place, for example, in a cancer of birds known as *avian leukosis*.

Conversion of a single proto-oncogene to an oncogene is usually not sufficient in itself to produce cancer. Instead, cancer cells are commonly characterized by several active oncogenes, sometimes as many as three or more. In one study of human malignancies by D. J. Slamon and his coworkers, for example, tumors from 54 patients with 20 different kinds of cancer were examined. More than one oncogene was identified in each of the 54 tumors; three oncogenes, *myc*, *fos*, and *ras*, were active in most. These observations support the conclusion that the genes controlling cell division form parts of complex networks in which the balance of activity among many genes, rather than control by single genes, is responsible for activation or deactivation of the division process. They also add support to the generally held conclusion that cancer develops through successive changes involving the additive effects of several mutations or other changes (see below).

THE CHARACTERISTICS OF CANCER

All forms of cancer are characterized by uncontrolled cell division and metastasis, in which tumor cells break loose and spread from their sites of origin to lodge and grow in other locations in the body. Carcinomas—including tumors of the skin, colon and rectum, lung, breast, and prostate—are by far the most common human cancers.

Effects of Cancer Cells

The detrimental effects of cancerous growths result from interference with the activity of normal cells, tissues, and organs or from loss of vital activities due to conversion of essential cells from functional to nonfunctional forms. Solid tumors destroy surrounding normal tissues by compressing them and interfering with blood supply and nerve function. Tumors may also break through barriers such as the outer skin, internal membranes and epithelia, or the gut wall. The breakthroughs cause bleeding and infection and destroy the separation of body compartments necessary for normal function. Both compression and breakthroughs can cause pain that may become extreme in advanced cases.

Tumors of blood cells convert dividing cell lines from functional types to nonfunctional forms that crowd the bloodstream but are unable to carry out required activities such as the immune response. Some tumors of glandular tissue upset bodily functions through excessive production and secretion of hormones. When the total mass of tumor tissue becomes large, the demands of the actively growing and dividing cancer cells for nutrients may deprive normal cells of their needed supplies, leading to generally impaired body functions, muscular weakness, fatigue, and weight loss.

Tumors with these destructive characteristics are described as *malignant*. Not all tumors are malignant; some types of unprogrammed tissue growth, such as common skin warts, usually cause no serious problems to their hosts and are therefore classified as benign. Benign tumors, in contrast to the malignant types, grow relatively slowly and do not invade surrounding tissues or metastasize. Often benign tumors are surrounded by a fibrous capsule of connective tissue that prevents or retards expansion or breakup.

Some initially benign tumors, including even common warts, however, may change with time to take on malignant characteristics.

Much has been learned about the biochemical and structural characteristics of cancer cells by the study of cells in culture. Normal body cells placed in culture eventually become quiescent and stop dividing. In contrast, tumor cells, or cells that have undergone some of the genetic and other alterations leading to development of the cancerous condition, may continue to divide indefinitely. Such cells are said to be *transformed*. Among the changes noted in transformed cultured cells, in addition to continued division, are increases in metabolic rates and transport of substances across the plasma membrane, loss of cytoskeletal structures and reduced attachment to other cells, partial or complete loss of specialized structures, and loss of contact inhibition of movement. Some transformed cells become self-sustaining by secreting their own growth factors. Transformed cells may show from a few to all of these characteristics.

Viruses and Cancer

Viral infections contribute to many forms of cancer by introducing oncogenes or other genes promoting division or by inhibiting tumor suppressor genes or their encoded proteins. Two viral groups—retroviruses and *DNA tumor viruses*—have been linked to generation of cancer. The tumors caused or promoted by viruses are common enough to account for about 15% of cancer worldwide.

Retroviruses   The retroviral genome, which is encoded in RNA molecules in free viral particles, is integrated into host cell genomes in the form of a DNA copy (see p. 537). All mammalian individuals, including humans, are likely to contain several to many infecting retroviruses integrated into their DNA. Some retroviruses carry genes capable of inducing uncontrolled cell division and many other characteristics of malignant cells. The division-inducing genes provide an evolutionary advantage to the retroviruses carrying them because the cell lines they infect grow in large numbers and thereby increase the capacity of the virus for reproduction and further infections. The included genes are particularly advantageous to a retrovirus if the growth they induce is slow enough to permit extended survival of the host as an agent causing further infections.

DNA Tumor Viruses   The DNA tumor viruses contain DNA as their hereditary material. Although these viruses may or may not integrate into the host cell genome as a part of their normal cycles of infection, integration usually occurs when they are associated with cancer. Unlike the retroviruses, none of the DNA tumor viruses contains oncogenes that are recognizable as abnormal forms of host cell genes. Instead, the genes causing uncontrolled growth in these viruses appear to be of purely viral origin.

Many of the groups to which the DNA tumor viruses belong are among the most common viruses infecting humans and other mammalian species (Table 22-3). Some of the more than 50 known *papillomaviruses* are relatively harmless and cause benign tumors such as skin and venereal warts in humans. Others, however, have been implicated in cervical carcinomas, which are a leading worldwide cause of cancer deaths among women. About 90% of cervical carcinoma cells are found to be infected by one or both of the papillomaviruses *HPV-16* and *HPV-18*. Some of the normally benign tumors induced by papillomaviruses can also become malignant if exposed to factors promoting the development of cancer, such as ultraviolet light or X rays.

Several members of another group, the *herpesviruses*, are suspected but not as yet proved to be responsible for human cancers. Two of these viruses, *herpes simplex I and II*, are almost universally spread among human populations. In most persons, herpes simplex I causes lesions (cold sores) in skin or mucous membranes in many parts of the body; herpes simplex II causes genital lesions. Usually the herpes lesions are painful but otherwise not dangerous. However, her-

pes simplex II is frequently present in cervical tumor cells. It is not presently known whether the virus contributes to cancer of the cervix or is present in most cervical tumors as a coincidence, simply because it infects so much of the human population.

Links to cancer are stronger for another herpesvirus, the Epstein–Barr virus. Most people become infected with this virus very early in life and show no symptoms. However, some individuals infected with the virus develop cancer. The Epstein–Barr virus has been implicated as a cause of one form of Burkitt's lymphoma that is common in parts of Africa; a nasopharyngeal carcinoma linked to infections by the virus is most common in China. Cancer is apparently more likely to be associated with Epstein–Barr viral infections if the immune system of the affected individual is suppressed due to factors such as extreme stress or other infections.

There are indications that, among other DNA tumor viruses, the *hepadnaviruses*, which are responsible for human diseases such as hepatitis B, can produce human cancers. Persons with lifelong, residual infections of the hepatitis B virus are several hundred times more likely to contract liver cancer than uninfected individuals, particularly if they develop cirrhosis of the liver from alcoholism or other causes. The appearance of cancer related to the infection in these people may be delayed by as much as 20 to 30 years after initial exposure to the virus.

Some additional DNA tumor viruses induce malignant tumors primarily or exclusively when they infect organisms other than their normal hosts. The *adenoviruses* fall into this category. Most humans experience several adenovirus infections during their lifetimes. Typically the infections cause respiratory and intestinal upsets but not cancer. The human adenovirus, however, can induce malignant tumors when injected into newborn hamsters. The *papovaviruses*, another group in this category, includes the SV40 virus of monkeys. The SV40 virus is not associated with cancer in its natural monkey hosts but is capable of transforming cultured mouse and human cells to malignant types, especially if other oncogenes such as *ras* are also present.

The growth-inducing genes carried by DNA tumor viruses, as noted, have no obvious counterparts among the gene complement of their host cells. SV40 viral DNA, for example, encodes *T antigens*, which promote DNA replication and cell division. DNA encoding the T antigens is capable of inducing DNA replication when included in DNA transfected into a suitable cell type. At least some of the effects of the SV40 T antigens are also exerted through their ability to combine with and inactivate the products of tumor suppressor genes, including RB and p53. Similarly the adenovirus *E1A* gene is capable of inducing cell division in transfection experiments. The protein encoded

Table 22-3	Viral Groups in Which DNA Tumor Viruses Occur
Viral Group	**Tumor Type**
Adenoviruses	Tumors usually induced only in species other than natural host
Hepadnaviruses animal hepatitis (woodchuck, duck, ground squirrel)	Liver cancer (?)
human hepatitis B	Liver cancer (?)
Herpesviruses cytomegalovirus Epstein–Barr virus	Burkitt's lymphoma, nasopharyngeal carcinoma
herpes simplex I herpes simplex II	Cervical cancer (?)
Papillomaviruses	Skin warts, venereal warts, cervical carcinoma in humans, alimentary carcinoma in cattle
Papovaviruses BK JC polyoma SV40	Various cancers in host and other species

in this gene combines with and inactivates the *rb* gene product; the product of another adenovirus gene, *E1B*, combines with p53. The E1A and E1B products together are able to transform cultured cells into fully malignant types. Papillomaviruses also encode two proteins, *E6* and *E7*, that combine with and inactivate P53 and RB respectively.

Why many of the DNA tumor viruses cause cancer only in species outside their normal hosts is unknown. It is possible that in their normal hosts the effects of viral genes are balanced and kept in check by host cell genes that have evolved this function. In nonhost species the viral genes may enter a cellular environment in which the controlling genes are missing or have activities that differ sufficiently from the normal host types to release the viral genes from control. In this situation the resulting uncontrolled cell division provides a significant step leading to the development of a malignant tumor. Immune suppression is also often present as a contributing factor in cancers induced by DNA tumor viruses.

Multistep Progression in Cancer

For most cancers, fully malignant cancer cells do not usually develop from a single cause. Multiple alterations, sometimes requiring many years to appear, are necessary for full establishment of malignancy. This feature is known as the *multistep progression* of cancer. In most cases the complete sequence of steps leading from an initiating alteration to full malignancy is unknown.

An initiating genetic alteration leading eventually to development of cancer may be induced by a long list of factors, including exposure to radiation or certain chemicals, insertion of a retrovirus, generation of random mutations during replication, random DNA amplification, or random loss of a DNA segment.

The initial changes may also be inherited. A small percentage of human cancers, about 5%, show a strong genetic predisposition. Depending on the gene and cancer type, as many as 100% of the persons receiving the oncogenes or faulty tumor suppressor genes involved are likely to develop cancer. Some forms of cancer of the large intestine or colon, for example, are strongly predisposed cancers. Some cancers, including breast and ovarian cancer, show limited disposition in some family lines—members of these families show a somewhat greater tendency to develop the cancer than individuals in other families.

Steps from the initiating change to a fully malignant state usually include conversion of additional genes to oncogenic form. Also important during intermediate stages are further alterations to the initial and succeeding genes that increase their activation. The initial conversion of a proto-oncogene to an oncogene by a translocation, for example, may be compounded at successive steps by sequence changes and amplification. Progression in some cases is also advanced by alterations that reduce the activity of tumor suppressor genes. These changes may be as significant to the full development of cancer as conversion of proto-oncogenes to oncogenes.

At any stage along the way a change required to advance progression toward the fully malignant state may happen soon after a previous change or only after a considerable passage of time. Or, it may not occur at all, leaving the transformation at an intermediate stage without development of malignancy within the lifetime of the individual.

The last stage in progression to malignancy is often metastasis. The separation and movement of cells from a primary tumor to secondary locations may occur through development of motility by the cancer cells or may result from penetration into the circulatory or lymphatic systems, which carry the malignant cells throughout the body. Movement to secondary locations may be followed by growth of the displaced cells into additional tumors at these sites.

Metastasis is promoted by changes during transformation that alter cell surface molecules of tumor cells and break their connections to other cells or to molecules of the extracellular matrix such as collagen, fibronectin, laminin, and the glycosaminoglycans (see Chapter 7). Often changes in cytoskeletal elements linked to the plasma membrane, such as microfilaments and microtubules, are associated with metastasis. For example, phosphorylations catalyzed by the *src* protein kinase reduce the affinity of the fibronectin receptor for this extracellular matrix protein; cells with *src* and some other oncogenes also secrete proteinases that break connections to the extracellular matrix and aid their movement into and through surrounding tissues.

Relatively few of the cells breaking loose from a tumor survive their passage through the body to lodge and grow in new locations. Most metastasizing cells are destroyed by various factors, including deformation by passage through narrow capillaries and destruction by blood turbulence around structures such as the heart valves and vessel junctions. Tumor cells often develop changes in their MHC (see p. 557) that permit their detection and elimination by the immune system, particularly in the bloodstream. Unfortunately, the rigors of travel through the body may work to select for survival the cells that are most malignant—that is, those most able to grow uncontrollably and spread by metastasis. Methods used to treat cancer, such as chemotherapy, may have the same effect—cells most resistant to the treatment may survive, greatly reducing or eliminating the effectiveness of the treatment.

Some genetic alterations taking place in the multistep progression of specific cancers have been traced. For example, B. Vogelstein and his coworkers reconstructed a sequence of alterations leading to malignancy in the development of colorectal cancer (Fig. 22-11). The proposed sequence, one of several possible pathways for progression of this cancer, begins with a deletion leading to loss of a gene on chromosome 5 of the human set. This gene is missing on one of the two chromosomes in pair 5 as a hereditary deficiency in some persons who have a higher incidence of colorectal cancer. Loss of the gene leads to limited proliferation of intestinal epithelial cells, producing a benign tumor known as an *adenoma class I* that grows as a *polyp* from the intestinal epithelium. At some point the progression is furthered by mutation of a *ras* proto-oncogene to oncogene form, increasing proliferation of the tumor to a still benign but faster-growing form known as an *adenoma class II*. Progression continues through deletion of a gene on chromosome 18 possibly encoding a tumor suppressor gene; this change advances the adenoma to *class III*, which is still benign. Finally, an alteration appears on chromosome 17 mutating or deleting the gene encoding the tumor suppressor protein p53. Loss of p53 converts the class III adenoma into a malignant carcinoma; eight to 10 further genetic changes increase the capability of the carcinoma cells to break loose and migrate. Other progressions leading to colorectal cancer involve the same alterations in a different order or distinct gene changes. The alteration in the gene carried on chromosome 5, however, is apparently the first step leading to colorectal cancer in most progressions.

The Role of Chemicals and Radiation in Multistep Progression of Cancer

Many natural and artificial agents initiate or promote development of cancer. Most of these agents, collectively called *carcinogens*, are chemicals or forms of radiation capable of inducing chemical changes in DNA. However, carcinogens are not limited solely to agents altering DNA; in some cases, carcinogens may act by modifying RNAs or proteins or by increasing the rate of DNA replication and cell division.

Chemical Carcinogens   The most directly potent chemical carcinogens (Fig. 22-12) are *alkylating agents*—substances capable of substituting a monovalent organic group such as a methyl or ethyl group for a hydrogen atom. Other molecules, such as many of the *polycyclic hydrocarbons*, are not directly carcinogenic in themselves. Instead, chemical derivatives of these substances, produced by natural biochemical processes in the body, convert them to cancer-inducing molecules. Ironically, one of the primary routes for conversion of substances such as the polycyclic hydrocarbons into cancer-causing derivatives is detoxification, carried

out primarily in the liver (see p. 574). Detoxification normally converts poisons into nontoxic, soluble substances that can be readily cleared from cells and excreted. In the case of the polycyclic hydrocarbons, detoxification converts these usually insoluble molecules into soluble derivatives that are highly reactive inside cells. Polycyclic hydrocarbons, incidentally, are the primary carcinogens of tobacco and other smokes produced by the incomplete combustion of organic substances or fossil fuels. Tobacco smoke, the most frequent cause of cancer, is estimated to be responsible for one-third of cancer cases in the United States.

The *phorbol esters* provide an interesting example of a group of chemicals that promotes the growth of cancer cells rather than inducing mutations. These molecules resemble the DAG second messenger generated by the InsP$_3$/DAG pathway, which allows them to stimulate the segment of the pathway activating protein kinase C. The phosphorylations catalyzed by protein kinase C, in turn, modify the cell cycle in the direction of more rapid division. The phorbol esters are especially effective in activating protein kinase C because they are metabolized very slowly inside cells. Thus their effects last much longer than those of DAG, which is typically degraded very quickly after its formation. Although phorbol esters usually promote the growth of cancer cells, they sometimes have the opposite effect and partially restore normal function, particularly in cells transformed by the *ras* and *sis* oncogenes.

Radiation   Primary among the radiation sources implicated in cancer are X rays, ultraviolet light, radon gas, and the radiation emitted by certain isotopes such as uranium-235. These sources of radiation have been linked to development of several types of cancer, including skin, bone marrow, breast, and lung tumors. Ultraviolet light is absorbed directly by DNA; the absorbed energy converts chemical groups in the DNA bases into more reactive forms in which they interact with each other or with surrounding molecules. The other forms of carcinogenic radiation are *ionizing*: They promote formation of chemically active groups of various kinds when they are absorbed by cellular molecules surrounding the DNA. The reactive groups can also cause breaks or mutations in DNA or chemically alter many other molecules important in biological systems, including RNA and proteins.

Some of the alterations induced by radiation may promote transcription of genes related to cell division or may modify controls regulating the cell cycle. Other changes may promote loss of mature cell structures or metastasis. One of the more peculiar characteristics of radiation-induced cancer is the potentially long delay between initial exposure and development of full malignancy, as long as 30 to 40 years in some instances.

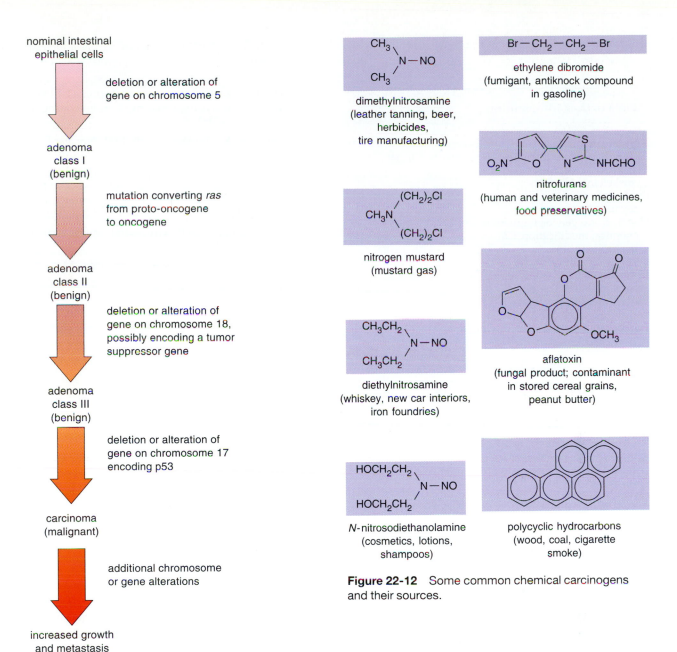

Figure 22-11 One of several possible sequences in the multistep progression of colorectal cancer (see text).

nominal intestinal
epithelial cells

deletion or alteration of
gene on chromosome 5

adenoma
class I
(benign)

mutation converting *ras*
from proto-oncogene
to oncogene

adenoma
class II
(benign)

deletion or alteration of
gene on chromosome 18,
possibly encoding a tumor
suppressor gene

adenoma
class III
(benign)

deletion or alteration of
gene on chromosome 17
encoding p53

carcinoma
(malignant)

additional chromosome
or gene alterations

increased growth
and metastasis

dimethylnitrosamine
(leather tanning, beer,
herbicides,
tire manufacturing)

ethylene dibromide
(fumigant, antiknock compound
in gasoline)

nitrofurans
(human and veterinary medicines,
food preservatives)

nitrogen mustard
(mustard gas)

diethylnitrosamine
(whiskey, new car interiors,
iron foundries)

aflatoxin
(fungal product; contaminant
in stored cereal grains,
peanut butter)

N-nitrosodiethanolamine
(cosmetics, lotions,
shampoos)

polycyclic hydrocarbons
(wood, coal, cigarette
smoke)

Figure 22-12 Some common chemical carcinogens and their sources.

The most common natural source of carcinogenic radiation is the ultraviolet component of sunlight. Prolonged exposure to sunlight can lead to development of skin cancers, including some, such as melanoma, which are highly malignant. Another natural source of radiation in many localities is radon gas released from the earth. Although the evidence is somewhat controversial, inhalation of radon gas has been linked to development of leukemia and lung, kidney, and prostate cancers. The highest estimates implicate radon as the primary cause of 15,000 new cases of cancer each year in the United States. In many areas, radon gas seeps through foundations into houses,

particularly in homes with cellars or built on concrete slabs. The greatest exposure to ionizing radiation from artificial sources in the United States is from medical and dental X rays. Although the conclusion is controversial, diagnostic X rays are considered by some investigators to be an important cause of cancer, particularly leukemias.

The best defense against both radiation and chemical carcinogens is to avoid them. The elimination of cigarette smoking alone would probably reduce cancer deaths by as much as one-third in the American and Western European populations—not to mention an additional reduction in deaths and disabilities due to smoking-induced heart disease. Avoidance of overexposure to sunlight and reduction of radon concentration in houses, which is relatively inexpensive, would probably also significantly reduce cancer deaths.

For Further Information

Amplification, DNA, *Ch. 23*
Antibody genes and their rearrangements, *Ch. 19*
Cell surface receptors, *Ch. 6*
Cytoskeleton, *Ch. 12*
DNA replication, *Ch. 23*
    repair, *Ch. 23*
Histones and genetic regulation, *Ch. 17*
    chemical modification, *Ch. 13*
    in chromatin structure, *Ch. 13*
Immune response, *Ch. 19*
Nonhistones and genetic regulation, *Ch. 17*
    chemical modification, *Ch. 13*
    in chromatin structure, *Ch. 13*
Retroviruses, *Ch. 18*
Translocations, chromosome, *Ch. 18*

Suggestions for Further Reading

Aaronson, S. A. 1991. Growth factors and cancer. *Science* 254:1146–1153.

Andrews, B. J., and Herskowitz, I. 1990. Regulation of cell cycle–dependent gene expression in yeast. *J. Biolog. Chem.* 265:14057–14060.

Andrews, B. J., and Mason, S. W. 1993. Gene expression and the cell cycle: A family affair. *Science* 261:1543–1544.

Baserga, R., and Rubin, R. 1993. Cell cycle and growth control. *Crit. Rev. Eukary. Gene Expression* 3:47–61.

Bishop, J. M. 1990. Retroviruses and oncogenes II. *Biosci. Rep.* 10:473–491.

Bogurski, M. S., and McCormick, F. 1993. Proteins regulating Ras and its relatives. *Nature* 366:643–654.

Bradbury, E. M. 1992. Reversible histone modifications and the chromosome cell cycle. *Bioess.* 14:9–16.

Cohen, S. M., and Ellwein, L. B. 1990. Cell proliferation in carcinogenesis. *Science* 249:1007–1011.

Donachie, W. D. 1993. The cell cycle of *Escherichia coli*. *Ann. Rev. Microbiol.* 47:199–230.

Dumont, J. E. 1989. The cAMP-mediated stimulation of cell proliferation. *Trends Biochem. Sci.* 14:67–71.

Evans, C. W. 1992. Cell adhesion and metastasis. *Cell Biol. Internat. Rep.* 16:1–10.

Feldman, M., and Eisenbach, L. 1988. What makes a tumor cell metastatic? *Sci. Amer.* 259:60–85 (November).

Forsberg, S. L., and Nurse, P. 1991. Cell cycle regulation in the yeasts *Saccharomyces cerevisiae* and *Schizosaccharomyces pombe*. *Ann. Rev. Cell Biol.* 7:227–256.

George, A. M., and Cramp, W. A. 1987. The effects of ionizing radiation on structure and function of DNA. *Prog. Biophys. Molec. Biol.* 50:121–169.

Gupta, S. K., Gallego, C., and Johnson, G. L. 1992. Mitogenic pathways regulated by G protein oncogenes. *Molec. Biol. Cell* 3:123–128.

Gutman, A., and Wasylyk, B. 1991. Nuclear targets for transcriptional regulation by oncogenes. *Trends Genet.* 7:49–54.

Haluska, F. G., Tsujimoto, Y., and Croce, C. M. 1987. Oncogene activation by chromosome translocation in human malignancy. *Ann. Rev. Genet.* 21:321–345.

zur Hausen, H. 1991. Viruses in human cancer. *Science* 254:1166–1173.

Henderson, B. E., Ross, R. K., and Pike, M. C. 1991. Toward the primary prevention of cancer. *Science* 254:1131–1138.

Johnson, G. L., and Vaillancort, R. R. 1994. Sequential protein kinase reactions controlling cell growth and differentiation. *Curr. Opinion Cell Biol.* 6:230–238.

Kirschner, M. 1992. The cell cycle then and now. *Trends Biochem. Sci.* 17:281–-285.

Knudson, A. G. 1993. Antioncogenes and human cancer. *Proc. Nat. Acad. Sci.* 90:10914–10921.

Koch, C., and Nasmyth, K. 1994. Cell cycle regulated transcription in yeast. *Curr. Opinion Cell Biol.* 6:451–459.

Levine, A. J. 1993. The tumor suppressor genes. *Ann. Rev. Biochem.* 62:623–651.

Liotta, J. 1992. Cancer cell invasion and metastasis. *Sci. Amer.* 266:54–63 (February).

Lowy, D. R., and Willumsen, B. M. 1993. Function and regulation of RAS. *Ann. Rev. Biochem.* 62:851–891.

McKinney, J. O., and Heintz, N. 1991. Transcriptional regulation in the eukaryotic cell cycle. *Trends Biochem. Sci.* 16:430–435.

Murray, A. W. 1992. Creative blocks: Cell cycle checkpoints and feedback controls. *Nature* 359:599–604.

Murray, A. W., and Kirschner, M. W. 1991. What controls the cell cycle. *Sci. Amer.* 264:56–63 (March).

Norbury, C., and Nurse, P. 1992. Animal cell cycles and their control. *Ann. Rev. Biochem.* 61:441–470.

Pitot, H. C., and Dragan, Y. P. 1991. Facts and theories concerning the mechanisms of carcinogenesis. *FASEB J.* 5:2280–2286.

Pouyssegur, J., and Seuwen, K. 1992. Transmembrane receptors and intracellular pathways that control cell proliferation. *Ann. Rev. Physiol.* 54:195–210.

Reed, S. I. 1992. The role of p34 kinases in the G1 to S-phase transition. *Ann. Rev. Cell Biol.* 8:529–561.

Sherr, C. J. 1993. Mammalian G1 cyclins. *Cell* 73:1059–1065.

Solomon, E., Borrow, J., and Godward, A. D. 1991. Chromosome aberrations and cancer. *Science* 254:1153–1160.

Stetler-Stevenson, W. G., Aznavoorian, S., and Liotta, L. A. 1993. Tumor cell interactions with the extracellular matrix during invasion and metastasis. *Ann. Rev. Cell Biol.* 9:541–573.

Travali, S., Koniecki, J., Petralia, S., and Baserga, R. 1990. Oncogenes in growth and development. *FASEB J.* 4:3209–3214.

Varmus, E. 1990. Retroviruses and oncogenes I. *Biosci. Rep.* 10:413–430.

Williams, G. T., and Smith, C. A. 1993. Molecular regulation of apoptosis: Genetic controls on cell death. *Cell* 74:777–779.

Review Questions

1. Outline the stages of the cell cycle. What major events take place in G1? In S? G2? What stage is most variable in duration? Compare the total DNA quantity per cell in G1 and G2 with that of a gamete in a diploid organism.

2. What features of mitosis ensure that the products of division receive exactly the same complement of chromosomes?

3. What are chromosomes and chromatids? At what stages do chromatids exist in the cell cycle? When are chromatids converted to chromosomes?

4. What is the G0 stage of the cell cycle? At what other stages may the cell cycle arrest? What evidence indicates that replication, mitosis, and cytoplasmic division are potentially separable?

5. What major systems regulate cell division through receptors on the cell surface? Outline the operation of the systems. What are G proteins? How do they operate in the reaction cascades triggered by cell surface receptors?

6. What are second messengers? How are second messengers generated in the reaction systems triggered by cell surface receptors?

7. What are protein kinases? What is their role in the pathways linked to surface receptors?

8. What is density-dependent inhibition of cell division? How is this mechanism related to receptors on the cell surface?

9. Outline the changes noted in ion transport during the cell cycle. How might these changes be related to cell surface receptors?

10. What alterations are noted in histone and nonhistone chromosomal proteins during the cell cycle? How might these changes be related to cell surface receptors? To cyclin/CDK complexes?

11. What evidence relates cell volume to cell cycle regulation? What natural systems bypass regulation by cell volume?

12. How are temperature-sensitive mutants used in the study of cell cycle genes? What major functions are carried out by genes affecting the cell cycle? List one yeast gene operating in each of the functional categories.

13. What is cyclin? The cyclin-dependent protein kinases (Cdc's or CDKs)? MPF? How do the cyclins and CDKs interact in regulation of the cell cycle? What factors modify activity of the cyclin/CDK complexes? What role do phosphorylations play in the interactions between cyclins, CDKs, and their modifiers? Trace the interactions of cyclin B and CDK2 during the transition from G2 to mitosis.

14. What primary methods have been used to identify oncogenes? What types of cellular activities are carried out by oncogenes? Give examples of oncogenes carrying out these activities.

15. What are proto-oncogenes? What is the relationship between oncogenes and proto-oncogenes? What alterations convert proto-oncogenes into oncogenes?

16. What are recessive oncogenes or tumor suppressor genes? What characteristics distinguish the operation of recessive and other oncogenes in cancer cells?

17. How are chromosome translocations related to the development of cancer? Why are translocations significant in the development of blood cancers?

18. What major characteristics of malignant tumors are responsible for their destructive effects?

19. What is multistep progression? What factors are likely to initiate progression of cells from a normal to malignant state?

20. What is the relationship of tumor suppressor genes to the development of cancer?

21. What is metastasis? What factors promote metastasis?

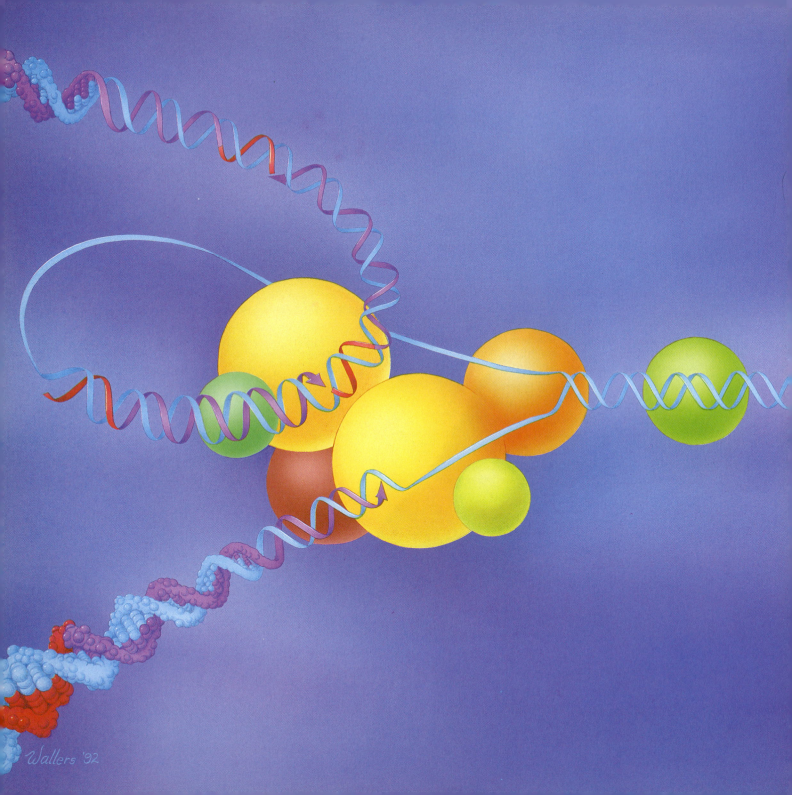

During late interphase, cells replicate their DNA in preparation for mitotic or meiotic division. The replication, in which the parental DNA molecule is duplicated into two exact, sequence-by-sequence copies, is remarkable for its fidelity. Once replication is complete, many of the few mistakes that do occur are corrected by repair mechanisms that scan the DNA to detect and repair base mismatches and other irregularities. The result is nearly perfect duplication of the genetic information of the parental cell, ready to be divided and placed in daughter cells by the division mechanisms.

While extremely rare, replication errors do slip by the correction mechanisms in very small numbers. These mistakes persist as *mutations*, which constitute any difference in sequence from the parental template that appears and persists in the replicated copies. The few mistakes remaining after DNA replication and repair are highly important to the evolutionary process because they are the primary source of the variability acted upon by natural selection.

During the S phase of interphase, histone and nonhistone chromosomal proteins are duplicated as well as the parental DNA. The newly synthesized chromosomal proteins combine with the replicated DNA to reproduce the arrangement of these proteins

in the G1 chromosomal parent. Synthesis of chromosomal proteins is almost as significant to cell reproduction as DNA replication, because duplication of the histones and nonhistones preserves the patterns of genetic regulation and cell differentiation characteristic of a cell type.

This chapter describes the enzymatic mechanisms replicating DNA and repairing errors that arise during replication, along with the characteristics of the few errors that persist as mutations. Duplication of chromosomal proteins is also considered. DNA replication in mitochondria and chloroplasts is outlined in Supplement 23-1.

SEMICONSERVATIVE DNA REPLICATION

When J. D. Watson and F. H. C. Crick discovered the molecular structure of DNA, they pointed out that a possible replication mechanism is inherent in its structure (Fig. 23-1; see Chapter 13 for details of Watson and Crick's discovery and the molecular structure of DNA). Because the two nucleotide chains of a DNA double helix are complementary, each can serve as a template for the synthesis of the missing half when the chains separate by unwinding. The mechanism, producing replicated molecules that each consist of one "old" nucleotide chain used as the template and one "new" nucleotide chain assembled on the template, is termed *semiconservative replication* (Fig. 23-2a).

Soon after Watson and Crick made their discovery, it became apparent that DNA replication might also be *conservative* (Fig. 23-2b). In conservative replication the two nucleotide chains of the original molecule, after unwinding and serving as templates, would rewind into an all-"old" molecule. The two new nucleotide chains would separate from their templates and wind together into an all-"new" molecule.

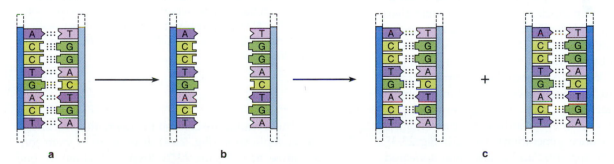

Figure 23-1 DNA replication. The two nucleotide chains of a DNA double helix (a) unwind (b); each half of the original molecule serves as a template for a complementary chain assembled according to the A-T and G-C base-pairing rules. The result is two DNA molecules that are exact duplicates of the original molecule entering replication (c).

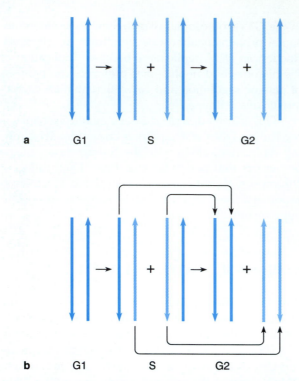

a G1   S   G2

b G1   S   G2

Figure 23-2 Two hypothetical pathways for DNA replication. Template chains are dark blue and copies are light blue. **(a)** Semiconservative replication, in which each copy chain remains wound with its template after replication, producing two daughter molecules that are one-half old and one-half new. **(b)** Conservative replication, in which parental chains rewind into an all-"old" helix after replication; copy chains wind together into an all-"new" DNA chain. All living organisms replicate DNA by the semiconservative pathway.

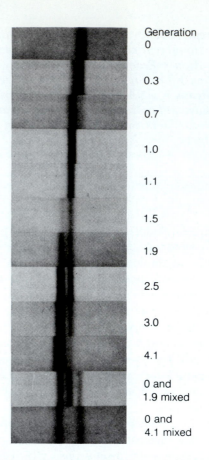

Generation
0
0.3
0.7
1.0
1.1
1.5
1.9
2.5
3.0
4.1
0 and
1.9 mixed
0 and
4.1 mixed

Figure 23-3 Meselson and Stahl's results demonstrating semiconservative replication in *E. coli*. Actively dividing cells were transferred from a growth medium containing a ^{15}N source to one containing a ^{14}N source; after transfer, DNA was extracted at intervals and centrifuged. The photograph shows DNA bands separated by density in the centrifuge; the density increases toward the right. The three bands in the next-to-lowest frame correspond, reading from left to right, to ^{14}N-^{14}N DNA, ^{14}N-^{15}N hybrid DNA, and ^{15}N-^{15}N DNA. After one round of replication (generation time 1.0), all of the DNA is of intermediate density, consistent with molecules containing one ^{14}N and one ^{15}N chain. This result ruled out conservative replication, which would have produced two bands, one at the ^{14}N level and one at the ^{15}N level. At two generations (generation times 1.9 to 2.5) there are two distinct bands, one hybrid ^{14}N-^{15}N helix and one all-new, ^{14}N-^{14}N helix. This result is consistent only with semiconservative replication. (Courtesy of W. Meselson and F. W. Stahl.)

The definitive experiment demonstrating that DNA replication is semiconservative was conducted by W. Meselson and F. W. Stahl in 1958 with the bacterium *E. coli*. Meselson and Stahl grew *E. coli* in a medium containing a heavy isotope of nitrogen (^{15}N) for several generations, long enough for the DNA to become completely labeled. The bacteria were then transferred to a ^{14}N medium. At the time of transfer, and at intervals afterward, DNA extracts were made and analyzed by buoyant density centrifugation (see Appendix p. 796). DNA extracted at the time of transfer from ^{15}N to ^{14}N medium centrifuged into a single band characteristic of ^{15}N DNA (generation 0 in Fig. 23-3). After an interval sufficient for most of the cells to replicate once (generation 0.3 to 1.0 in Fig. 23-3), the ^{15}N band disappeared and a new band appeared that was intermediate in density between ^{15}N and ^{14}N DNA. After longer periods of growth (generations 1.1 to 4.1 in Fig. 23-3) a new, less dense band characteristic of pure ^{14}N DNA appeared. The intermediate ^{15}N-^{14}N band did not disappear, however.

These results are consistent only with semiconservative replication (Fig. 23-4). If replication were conservative, as in Figure 23-2b, two distinct bands, one characteristic of ^{14}N and one of ^{15}N DNA, would appear in the centrifuge after one cell generation. In this case the two ^{15}N nucleotide chains of the replicating DNA would separate and serve as templates for copies made with ^{14}N DNA. After replication the original,

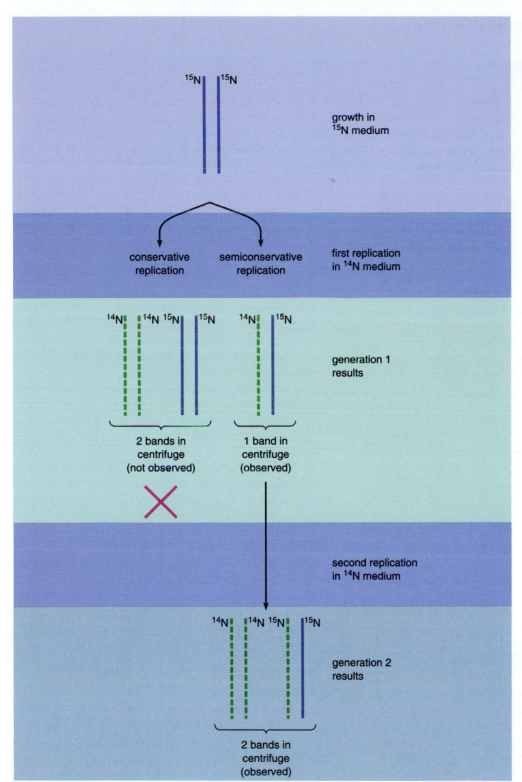

Figure 23-4 The combinations of ^{14}N and ^{15}N DNA and the number of bands expected in the centrifuge if replication is conservative or semiconservative.

^{15}N ^{15}N

growth in
^{15}N medium

conservative
replication

semiconservative
replication

first replication
in ^{14}N medium

^{14}N ^{14}N ^{15}N ^{15}N ^{14}N ^{15}N

generation 1
results

2 bands in
centrifuge
(not observed)

1 band in
centrifuge
(observed)

second replication
in ^{14}N medium

^{14}N ^{14}N ^{15}N ^{15}N

generation 2
results

2 bands in
centrifuge
(observed)

old ^{15}N nucleotide chains would reassociate, as would the new ^{14}N chains. Therefore after conservative replication the two DNA products would be identifiable after one generation as separate ^{15}N and ^{14}N bands. The production of a single band of intermediate density after one generation, actually observed by Meselson and Stahl, showed instead that all of the DNA molecules were hybrids containing one ^{15}N and one ^{14}N chain. This result eliminated conservative replication as a possibility and was the expected result if replication is semiconservative as in Figure 23-2a. At about the same time, another investigator, J. H. Taylor, showed by an equivalent experiment that DNA also replicates semiconservatively in eukaryotes.

THE REACTIONS OF DNA REPLICATION

An Overview of the Mechanism

Replication of a template DNA chain resembles RNA transcription with several important exceptions and additions. Assembly of individual DNA nucleotides into a chain is catalyzed by a group of enzymes known as *DNA polymerases*. The enzymes use as raw materials the four DNA nucleotides in the form of nucleoside triphosphates: *deoxyadenosine triphosphate (dATP), deoxyguanosine triphosphate (dGTP), deoxycytidine triphosphate (dCTP),* and *thymidine triphosphate (TTP)*. The three nucleotides dATP, dGTP, and dCTP differ from their counterparts in RNA synthesis by the presence of deoxyribose rather than ribose as the five-carbon sugar (indicated by the *d* in front of triphosphate; see Fig. 2-26). In addition, thymidine triphosphate is used instead of uridine triphosphate in the assembly of DNA nucleotide chains.

Another fundamental difference is that DNA polymerases require a *primer*—they can initiate replication only by adding nucleotides to the end of an existing nucleotide chain that acts as a primer for the reaction. RNA polymerases, in contrast, can put the first nucleotide of a new chain in place with no primer requirement. Although either DNA or RNA chains can act as primers for DNA replication, RNA is generally used in nature as the primer. At some point after DNA synthesis is initiated, the RNA primer is removed and replaced by DNA.

Polymerization of a new DNA chain proceeds from the primer as shown in Figure 23-5. DNA polymerase binds to the exposed 3'-OH group of the primer (Fig. 23-5a) and recognizes the first base to be copied from the template, shown as a guanine in the figure. According to base-pairing rules, presence of a guanine at this site causes the enzyme to bind dCTP from the nucleoside triphosphates in the surrounding medium (Fig. 23-5b). All four nucleoside triphosphates collide and bind weakly to the DNA polymerase enzyme, but normally only the nucleotide forming a correct base pair with the template base, in this case dCTP, proceeds from initial weak binding to tight binding.

Tight binding holds dCTP opposite the guanine of the template, in a position favoring a covalent linkage between the 3'-OH group at the end of the primer and the innermost phosphate group bound to the 5'-carbon of the dCTP (Fig. 23-5c). The last two phosphates of dCTP are split off and the remaining phosphate is bound to the oxygen of the 3'-OH, creating a *3' → 5' phosphodiester linkage* (see also p. 364) between the primer and the added nucleotide.

In response to formation of the first phosphodiester linkage, DNA polymerase moves to the next base on the DNA template, shown as a thymine in the figure. The enzyme then binds a dATP nucleotide from the medium and catalyzes formation of the second phosphodiester linkage. As before, the linking reaction splits off two phosphate groups from the nucleoside triphosphate most recently bound by the polymerase. The process then repeats, adding complementary nucleotides in succession into the growing DNA chain.

Each nucleotide added to an exposed 3'-OH group provides the 3'-OH group for the next assembly reaction. As a result, a 3'-OH group is always present at the newest end of the growing chain, and synthesis is said to proceed in a 5' → 3' direction. All known DNA polymerases add nucleotides only in this direction.

Because the two chains of a DNA double helix are *antiparallel* (see p. 364), the newly synthesized DNA chain in Figure 23-5 runs in a direction opposite to the template chain. In the figure the 5' end of the newly synthesized chain is at the bottom and its 3' end is at the top. The template chain extends in the opposite (antiparallel) direction, with its 5' end at the top of the figure and its 3' end at the bottom.

While the basic polymerization reaction is thus similar to RNA transcription (except for the primer requirement), the entire reaction of DNA replication differs fundamentally from RNA transcription because replication is semiconservative: the two nucleotide chains of the template DNA double helix must unwind and separate completely for replication to be semiconservative. Another fundamental difference between transcription and replication is related to unwinding. DNA unwinding and replication proceed at a fork that apparently moves *unidirectionally* along the template DNA (Fig. 23-6). To produce a unidirectional fork, the two nucleotide chains of the template DNA must unwind and replicate simultaneously in the same overall direction. However, because the two chains of a DNA double helix are antiparallel, only one is presented to the replication mechanism in the required 3' → 5' direction. How is the other, "wrong-way" chain replicated in the same overall direction followed by the moving fork? This problem is solved by a mechanism that replicates the wrong-way chain in short bursts that actually run in the direction opposite to fork movement (Fig. 23-7). The short lengths produced by this *discontinuous replication*, as it is called, are then covalently linked into a continuous nucleotide chain.

The enzymes and factors meeting the special requirements of DNA replication—priming, unwinding, and unidirectional replication—have been pieced together through research with a variety of prokaryotic and eukaryotic systems and with viruses infecting both prokaryotic and eukaryotic cells. Longest and most successful of the research efforts in DNA repli-

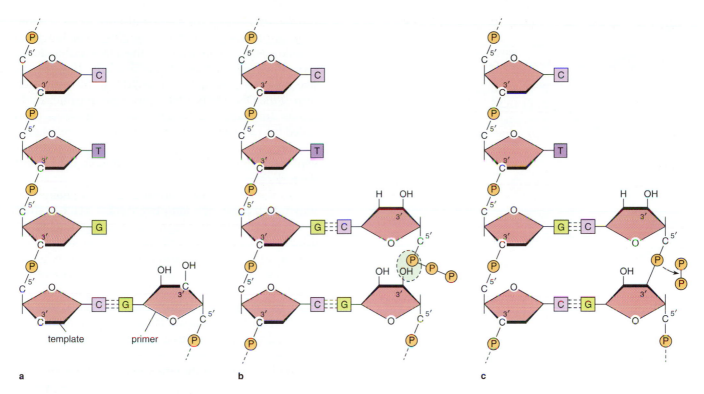

Figure 23-5 The polymerization reaction linking DNA nucleotides into a copy chain during replication (see text).

cation have been those using *E. coli*, particularly in the laboratory of A. Kornberg. (Kornberg received the Nobel Prize in 1959 for his work in DNA replication.) Much of Kornberg's research was conducted with temperature-sensitive mutants (see p. 638), in which cells could be grown in large quantities at permissive temperatures and then analyzed for the effects of the mutations at restrictive temperatures. By this means a particular mutant could be identified with a missing function in DNA replication at the restrictive temperature, allowing the protein encoded in the gene in many cases to be identified with a single reaction in the replication process. Also successful has been research with several viral systems, primarily viruses infecting *E. coli* and a few other bacteria.

In more recent years, research in DNA replication has been extended to eukaryotic systems. The yeast *Saccharomyces* and several viruses infecting eukaryotic cells, including the *adenovirus* and a virus of monkeys, *SV40*, have been particularly useful in this research. The SV40 system, developed by T. J. Kelly, B. W. Stillman, and C. R. Wobbe and their coworkers, has been especially valuable because the virus encodes only one protein, a *T antigen*, which acts in the initiation of replication. The remaining enzymes and factors of replication are supplied by the eukaryotic host cell and can be identified as such. Recently T. Tsurimoto,

Figure 23-6 Pattern of grains traced in a photographic emulsion by a radioactively labeled, replicating *E. coli* DNA molecule. The DNA has partially replicated. The two forks (arrows) are the advancing sites of replication. × 260. (From "The Bacterial Chromosome" by J. Cairns, *Scientific American* 214 [1966]. Copyright 1966 by Scientific American, Inc. All rights reserved.)

T. Melendy, and Stillman were able to construct the first completely defined eukaryotic cell-free replicating system, including SV40 DNA, the T antigen, and eight purified eukaryotic proteins. (The functions of the proteins are outlined below.)

Mechanisms Solving the Special Problems of DNA Replication

The research with prokaryotic and eukaryotic systems has allowed a mechanism to be pieced together that, with some differences in detail, applies to both groups

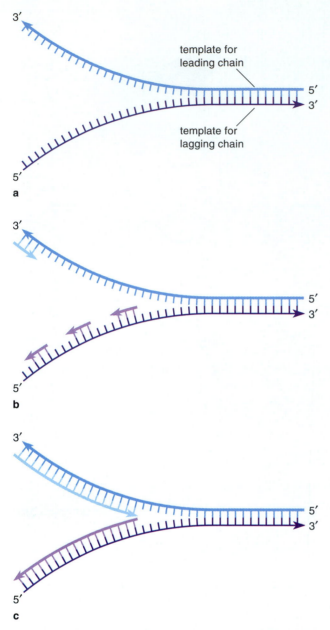

Figure 23-7 How antiparallel template chains are replicated unidirectionally at a fork. The template chain presented to the replication complex in the "wrong" 5' → 3' direction—the chain on the bottom in **(a)**—is copied in short bursts that run counter to the direction of fork movement **(b)**. The short lengths are then linked into a continuous chain **(c)**. The overall effect is unidirectional synthesis in the direction of fork movement.

of organisms and also to many of the viruses infecting prokaryotes and eukaryotes. Each of the three mechanisms peculiar to DNA replication—unwinding, priming, and unidirectional fork movement—involve the coordinated activity of several enzymes and factors (Fig. 23-8).

Unwinding   Unwinding depends on an *unwinding enzyme*, or *helicase*, that moves along the template double helix just in front of the DNA polymerase, separating the template chains as it goes (Fig. 23-8a). The helicase uses at least one ATP to unwind each turn of the DNA helix. Once unwound, the nucleotide chains are stabilized by *DNA binding proteins* (also called *helix destabilizing proteins* or *single-strand binding proteins*), which bind avidly to single chains and prevent them from rewinding in the region just behind the fork.

As the helicase unwinds the template chains, the unwinding twists the template double helix in front of the fork in the same right-handed direction as the double helix turns. The long helix in front of the fork cannot rotate readily to compensate for this twisting force; nor can turns be compressed in advance of the fork because the number of turns per unit length is fixed in DNA at an almost constant value (see Table 13-1). Instead, the double helix in front of the fork is forced into one positive supercoil (see p. 372) for each full turn unwound.

Rather than accumulating in front of the fork, the supercoils are relieved by a *DNA topoisomerase*, which introduces breaks in the DNA in front of the fork, allows the DNA to rotate around the breaks, and reseals the DNA. A single cycle of breakage, rotation, and resealing by a DNA topoisomerase eliminates one supercoil. Either or both of two types of topoisomerases may operate to relieve supercoils in different organisms: *DNA topoisomerase I*, which introduces a break in one of the two DNA chains to remove a supercoil, and *DNA topoisomerase II*, which introduces breaks in both chains (see below).

Priming and Polymerization   As the DNA unwinds, primers are laid down by a specialized RNA polymerase called a *primase*. The primase attaches to a template chain and assembles an RNA primer about 5 to 10 nucleotides long (see Fig. 23-8a). RNA primers are laid down in the 5' → 3' direction on both sides of the fork, in the direction of unwinding on one chain and in the opposite direction on the other. The assembly displaces the binding proteins from the regions where primers are laid down, freeing them to cycle ahead for reuse at the advancing unwinding site.

The short RNA primers are used as starting points for replication by DNA polymerases (Fig. 23-8b). As in primer synthesis, polymerization proceeds in the 5' → 3' direction on both sides of the fork, so that nucleotides are added in the unwinding direction on one chain and in the opposite direction on the other.

The DNA polymerases are highly accurate and usually enter only the correct complementary base opposite a base on the template chain. However, base mismatches do occur at a regular but low frequency. Most mismatches arising during replication are corrected, either by a *proofreading* mechanism carried out by the DNA polymerases themselves or by reactions that repair mismatched bases after replication is complete (see below).

Once the primers have served their functions, they are removed, either by a DNA polymerase or an RNAase specialized for this function (Fig. 23-8c). The gaps created by primer removal are filled in by a DNA polymerase, which adds nucleotides until the last remaining bases in the gaps are paired and filled in (Fig. 23-8d). Since the DNA polymerase cannot link the last nucleotide added to the 5' end of the next segment, a single-chain "nick" is left open at the 3' end of the filled gap. The nicks are closed by the final major enzyme acting in DNA replication, *DNA ligase*. This enzyme, which uses ATP or NAD as an energy source, forms the final phosphate linkages required to seal the copies into continuous DNA nucleotide chains (Fig. 23-8e; Table 23-1 summarizes the activities of the major enzyme classes and factors in DNA replication).

Unidirectional Fork Movement   Replication in the same direction on both sides of a replication fork, as shown in Figure 23-7, is more apparent than real. On the template chain exposed in the "correct" 3' → 5' direction, only a single primer is usually laid down in the direction of unwinding. The DNA copy assembled on this template chain is called the *leading chain*. On the template chain exposed in the "incorrect" 5' → 3' direction, primers are laid down at regular intervals in the direction opposite to unwinding. The DNA copy assembled on this template chain is the *lagging chain*. The primers are laid down in the lagging chain about every 1000 to 2000 nucleotides in prokaryotes and every 100 to 200 nucleotides in eukaryotes. The lengths of DNA assembled in the lagging chain, after primer removal and gap filling, are sealed into a continuous nucleotide chain by DNA ligases (see Fig. 23-7c).

Evidence that primers are actually laid down at short intervals in synthesis of the lagging chain was first obtained in a classic experiment carried out in 1968 by R. Okazaki and his coworkers. In their experiment, replicating bacteria were exposed to a "pulse" of tritiated thymidine lasting a few seconds. If DNA was isolated from the bacteria immediately after the pulse, radioactivity was found to be associated almost entirely with short pieces of DNA 1000 to 2000 nucleotides in length. Short RNA primers could be detected at the 5' ends of these fragments. If longer periods were allowed between the pulse of label and DNA extraction, the radioactivity was associated with DNA of high molecular weight. These results showed that initial, short DNA pieces were later covalently linked into longer continuous strands, as expected if replica-

tion of the lagging chain is discontinuous. Initial synthesis of short, discontinuous DNA lengths, now called *Okazaki fragments*, has since been detected in a wide variety of prokaryotes, eukaryotes, and viruses.

Replication advances at a rate of about 500 to 1000 nucleotides per second in prokaryotes and about one-tenth this rate in eukaryotes. The entire process is so rapid that the RNA primers and gaps left by discontinuous synthesis persist for only seconds or parts of a second. As a consequence, the enzymes and factors of replication operate only in the vicinity of the fork. A few micrometers behind the fork the new DNA chains are fully continuous and wound with their template chains into complete DNA double helices.

The Enzymes and Factors of DNA Replication

Although the overall pattern of DNA replication is basically the same in prokaryotes and eukaryotes, details of the reaction sequence vary significantly in the two groups of organisms. The variations depend primarily on differences in the properties and types of the enzymes and factors carrying out replication. Among the enzymes and factors the DNA polymerases are most varied and distinct (Table 23-2). The central reaction of DNA replication, addition of nucleotides to the end of an RNA primer, is catalyzed by one DNA polymerase in prokaryotes and at least two different types in eukaryotes. Other distinct DNA polymerases catalyze subsidiary reactions such as gap filling after primer removal and repair of replication errors. The enzymes and factors catalyzing other reactions of DNA replication—unwinding, priming, primer removal, and DNA ligation—also differ to a greater or lesser extent in properties and organization between prokaryotes and eukaryotes.

Table 23-1	Major Enzymes and Factors of DNA Replication
Enzyme or Factor	**Activity**
Helicase	Unwinds DNA helix
Binding proteins	Stabilize DNA in single-chain form
DNA topoisomerases	Relieve supercoils in front of replication fork
Primase	Synthesizes RNA primers
DNA polymerases	Assemble DNA chains on primers; fill gaps left after primer removal
DNA ligase	Seals nicks left after DNA primases fill gaps left by primer removal

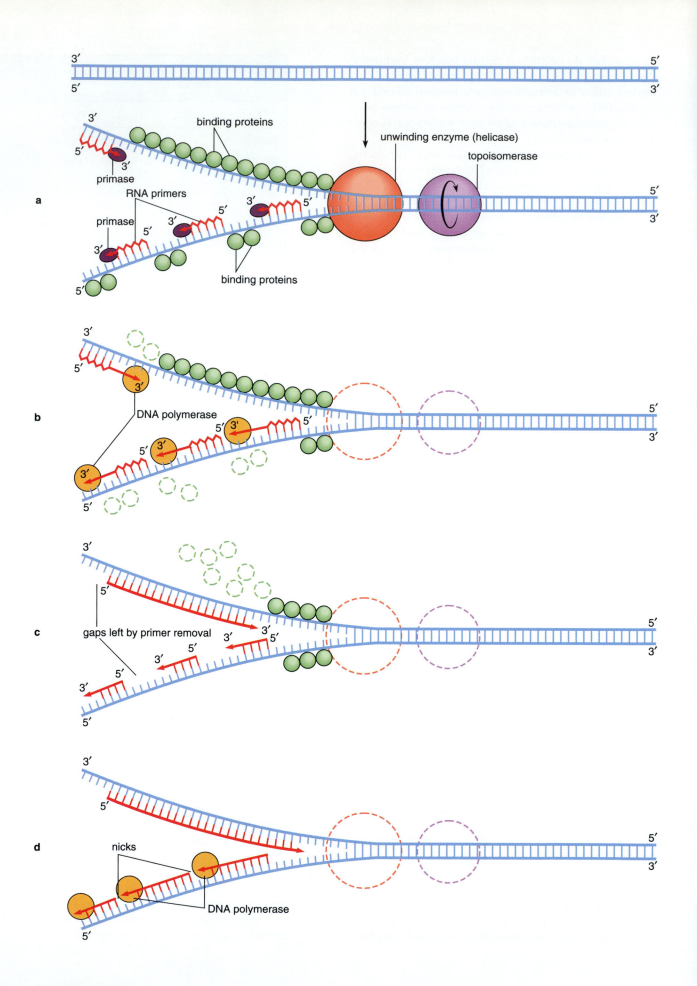

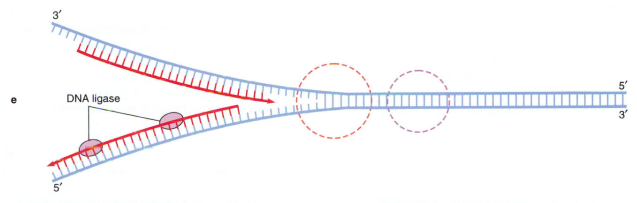

e DNA ligase

Figure 23-8 The enzymes and factors catalyzing DNA replication in diagrammatic form. In living cells the enzymes are organized into complexes that coordinate in several substeps. **(a)** Unwinding and primer synthesis. The template chain is unwound by a helicase that hydrolyzes ATP as an energy source for unwinding. Just in front of the unwinding site a topoisomerase relieves supercoils generated by the unwinding. As the chains unwind, DNA-binding proteins stabilize them in single-chain form and prevent them from rewinding. These proteins are displaced as short RNA chains are laid down as primers by a primase. **(b)** After primers are laid down, they are extended as new DNA chains by a DNA polymerase. **(c)** The primers are then removed, leaving gaps at their former sites. **(d)** The gaps are filled in by another DNA polymerase. Gap filling leaves a single-chain nick because the DNA polymerase cannot join the 3' end of the chain it has just completed to the 5' end of the next chain in line. **(e)** The single-chain nicks are closed by DNA ligase, making the nucleotide chains continuous. Primer synthesis, removal, gap filling, and nick sealing occur primarily in the lagging chain.

Table 23-2 Prokaryotic and Eukaryotic DNA Polymerases

Enzyme	Direction of Synthesis	Exonuclease Activity	Probable Functions
Prokaryotic			
polymerase I	$5' \rightarrow 3'$	$5' \rightarrow 3'$ $3' \rightarrow 5'$	Gap filling after primer removal; DNA repair
polymerase II	$5' \rightarrow 3'$	$3' \rightarrow 5'$	Gap filling after primer removal; DNA repair
polymerase III	$5' \rightarrow 3'$	$5' \rightarrow 3'$; $3' \rightarrow 5'$	Primary replication enzyme
Eukaryotic			
polymerase α	$5' \rightarrow 3'$	None*	Primary replication enzyme (with polymerase δ); DNA repair
polymerase β	$5' \rightarrow 3'$	None*	DNA repair
polymerase γ	$5' \rightarrow 3'$	$3' \rightarrow 5'$	Primary replication enzyme of mitochondria and chloroplasts
polymerase δ	$5' \rightarrow 3'$	$3' \rightarrow 5'$	Primary replication enzyme (with polymerase α)
polymerase ϵ	$5' \rightarrow 3'$	$3' \rightarrow 5'$	DNA repair; may cooperate with polymerases α and δ in primary replication

*May be associated with other proteins that have $3' \rightarrow 5'$ exonuclease activity.

Prokaryotic DNA Polymerases Three DNA polymerases, designated *I*, *II*, and *III*, have been identified in *E. coli* through research in the laboratories of Kornberg, Cairns, M. L. Gefter, and others (see Table 23-2). Each of these enzymes can work in the forward 5' → 3' direction as a DNA polymerase assembling nucleotides into a chain, or in reverse as a *3' → 5' DNA exonuclease* removing nucleotides one at a time from the end of a nucleotide chain. Two, DNA polymerases I and III, can also remove nucleotides in the forward direction as *5' → 3' exonucleases.*

The roles of the three prokaryotic DNA polymerases in replication were unscrambled primarily through studies of temperature-sensitive mutants in *E. coli.* Mutants deficient in polymerase III activity at restricted temperatures were completely unable to polymerize nucleotides into DNA, indicating that this enzyme is the primary replication enzyme of *E. coli.* Mutants lacking polymerase I activity at restricted temperatures assembled DNA normally except for short segments left open in newly assembled nucleotide chains. This indicates that polymerase I fills the gaps left by primer removal.

Mutants lacking polymerase II activity at restricted temperatures were able to replicate their DNA with no apparent problems, leaving the role of this enzyme in doubt. However, mutants lacking both polymerase I and polymerase II activity showed greater deficiencies in gap filling at restricted temperatures than mutants lacking polymerase I alone. This suggests that polymerase II may act in an auxiliary role, facilitating gap filling by DNA polymerase I. Taken together, these findings indicate that polymerization of the new DNA chains on templates, starting from primers, is catalyzed by polymerase III. DNA polymerase I fills the gaps left after primer removal, with DNA polymerase II working in coordination or possibly functioning as a backup for DNA polymerase I.

E. coli DNA polymerase III is a complex enzyme that contains at least 10 different polypeptide subunits. Some occur in multiple numbers, giving a total of about 20 for the enzyme. Polypeptide α, a major subunit present in two copies in the enzyme, is a "core" subunit that carries out the fundamental polymerization of nucleotides into a chain. The other *E. coli* enzyme directly implicated in replication, DNA polymerase I, is a relatively simple molecule consisting of a single polypeptide subunit.

The ability of the prokaryotic DNA polymerases to work backward as a 3' → 5' exonuclease and remove nucleotides one at a time from a DNA chain underlies a correction mechanism known as *DNA proofreading.* The proofreading mechanism, first proposed by D. Brutlag and Kornberg, provides DNA polymerase III with the ability to correct most of its own base-pairing errors during DNA replication. Ev-

idently DNA polymerase III adds nucleotides to a growing chain only if the most recently added base is correctly paired (Fig. 23-9a). If the most recently added nucleotide is incorrectly paired (Fig. 23-9b) and is unstable because of the mispairing, the reverse, 3' → 5' exonuclease activity of polymerase III is favored. As a result, the enzyme goes backward and removes the mispaired base or bases until a correct pair, forming a stable association resistant to disruption, is reached (Fig. 23-9c). At this point, forward activity of the enzyme again becomes favored, and it returns to the 5' → 3' polymerization reaction (Fig. 23-9d).

Several lines of evidence support proofreading by DNA polymerase III. For example, the error rate of bacterial DNA polymerase III with functional exonuclease activity is about one mispair for every million nucleotides assembled in a cell-free system. If the enzyme's exonuclease activity is experimentally inhibited, the error rate increases to about one mistake for every 1000 to 10,000 nucleotides assembled.

Another factor reducing the error rate is a DNA repair mechanism that detects and corrects most of the mismatched bases that escape proofreading (see below). The combination of the two processes in living cells achieves a final error rate as low as one mistake for every billion to 10 billion nucleotides assembled.

The characteristics of the proofreading mechanism may explain why the replicating enzymes must proceed from a primer and why RNA rather than DNA primers are used in DNA replication. If a DNA polymerase was able to start synthesis of a new DNA chain with no existing nucleotides in place, the first few nucleotides would likely be mispaired. This is because the bases in very short double helices have greater freedom of movement, allowing them readily to take up alternate configurations that permit "nonstandard" base pairs to form. As a consequence, the enzyme would reverse and remove the mispaired bases. The initial situation would likely be repeated, causing the enzyme to cycle back and forth, adding and removing bases without being able to proceed to stable elongation of the copy chain. An RNA primer, in contrast, is laid down by primases with no proofreading ability. The primases ignore mispairs and assemble an RNA chain that is long enough to restrict base pairs primarily to the correct ones toward its 3'-OH end, from which the DNA polymerase begins polymerization. Mispairs in the primers are unimportant to the final outcome because they are removed and replaced later in replication.

Eukaryotic DNA Polymerases Work with eukaryotes has established that higher organisms possess at least five DNA polymerases, designated α, β, γ, δ, and ε (see Table 23-2). All five eukaryotic enzymes, like their

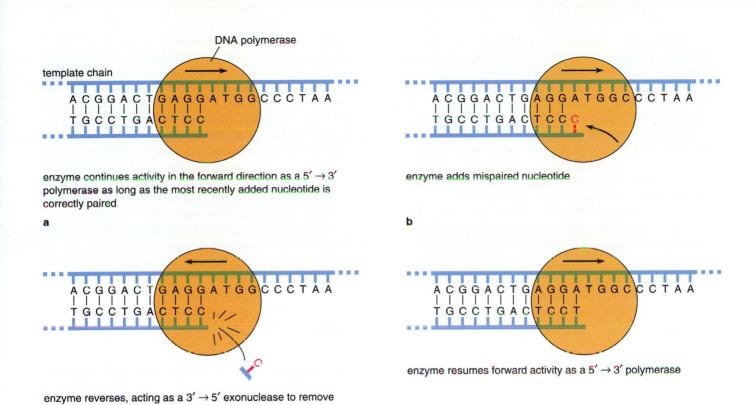

enzyme continues activity in the forward direction as a 5' → 3' polymerase as long as the most recently added nucleotide is correctly paired

a

enzyme adds mispaired nucleotide

b

enzyme reverses, acting as a 3' → 5' exonuclease to remove mispaired nucleotide

c

enzyme resumes forward activity as a 5' → 3' polymerase

d

Figure 23-9 Proofreading by bacterial DNA polymerase III (see text).

bacterial counterparts, polymerize DNA chains only in the 5' → 3' direction.

Several lines of evidence initially implicated DNA polymerase α as the primary replication enzyme of eukaryotes. Among this evidence were observations by H. J. Edenberg, S. Anderson, M. L. DePamphilis, and others that eukaryotic replication slows significantly or stops when the system is exposed to inhibitors or antibodies against DNA polymerase α. Temperature-sensitive mutations also implicated polymerase α as a primary replication enzyme. However, later work by J. J. Byrnes and others revealed that agents inhibiting DNA polymerase δ also greatly slow or inhibit eukaryotic DNA replication. In addition, a factor called *PCNA* (for *proliferating cell nuclear antigen*), previously found necessary for DNA replication to progress in eukaryotes, was discovered to work only in association with DNA polymerase δ.

These findings led to the proposition that DNA polymerases α and δ cooperate in DNA replication. Other characteristics of the enzymes supported this conclusion. Although lacking the 3' → 5' exonuclease activity necessary for proofreading, DNA polymerase α has two subunits that can work as a primase. DNA polymerase δ has no priming ability, but it has the 3' → 5' exonuclease activity required for proofreading and can also act as a helicase.

These properties suggest that the two DNA polymerases cooperate through a mechanism in which polymerase δ catalyzes synthesis of the leading chain and polymerase α assembles the lagging chain. The primase activity of polymerase α is consistent with its involvement in lagging-chain synthesis, in which frequent primers must be laid down. In support of this hypothesis is the finding that in the completely purified SV40 cell-free system only the lagging chain is assembled if polymerase δ is omitted from the reaction mixture.

Although the combination of enzymes and factors that accomplish proofreading remains unclear, there is good evidence from experiments by T. A. Kunkel and others that proofreading actually takes place in eukaryotes. Measurements of mispairing when exonuclease activity is inhibited show that DNA polymerases α and δ place the wrong base in DNA copies once in several thousand times. The proofreading function of polymerase δ, when active, improves the fidelity by about 100 times. Polymerase δ is therefore believed to supply the proofreading function for the coordinated synthesis. The combination of proofreading and mismatch repair (see below) sets the accuracy of eukaryotic replication at very high levels, equivalent to one mispairing for every billion to trillion nucleotides placed in a growing DNA chain.

Because DNA polymerase γ is found primarily in association with mitochondria and chloroplasts in eukaryotic cells, this enzyme is thought to be specialized for DNA replication in these organelles (see Supplement 23-1). Polymerase ε, which has structure and properties similar to polymerase δ, has been implicated in DNA repair and possibly in coordination with polymerases α and δ in the primary replication reaction. DNA polymerase β is apparently absent from protozoa, fungi, and at least some higher plants, indicating that its functions are probably not central to the replication mechanism. Other research indicates that DNA polymerase β may work in coordination with polymerases α and ε in DNA repair (see below).

Eukaryotic DNA polymerases have proved difficult to isolate intact. For this reason, the subunit composition of the enzymes is still uncertain. However, DNA polymerase α appears to consist of at least four different polypeptide subunits. The largest of these subunits is the core unit catalyzing polymerization of a nucleotide chain, and the two smallest have primase activity. DNA polymerase δ has at least two subunits plus the PCNA factor, which is closely associated with the enzyme. PCNA evidently works as a sort of sliding clamp holding DNA polymerase δ tightly to the DNA, greatly improving its *processivity*—its ability to continue replication for long stretches without releasing from the DNA. Another factor, *RFC*, which contains five polypeptide subunits and associates with DNA polymerase δ, may facilitate primer recognition and binding by polymerase δ. Most of the various polypeptides of the primary replication complex, including those of polymerases α and δ and the PCNA factor, are highly conserved among eukaryotes from yeasts to humans.

The Priming Enzymes   The primases working in DNA replication have the capacity to incorporate DNA as well as RNA nucleotides during their polymerization reactions. However, the primers synthesized for DNA replication usually consist only of RNA because of the greater affinity of the primases for RNA than for DNA nucleotides as reactants.

The primase of prokaryotes is closely associated with the helicase unwinding DNA for replication, forming a so-called *primosome* that catalyzes unwinding and priming as a coordinated reaction. The primase activity of eukaryotes, as noted, is associated with DNA polymerase α, forming a complex that carries out priming and DNA polymerization in a coordinated sequence.

Enzymes Removing Primers   Primer removal in prokaryotes appears to depend on the ability of DNA polymerase I to work simultaneously as a 5' → 3' polymerase and a 5' → 3' exonuclease (Fig. 23-10). In its role as a 5' → 3' exonuclease the enzyme nibbles RNA bases one at a time from the primer, starting at the single-chain nick left between a newly synthesized DNA segment and the RNA primer (Fig. 23-10a). As it removes the RNA bases, it works simultaneously as a 5' → 3' polymerase, adding a new DNA nucleotide as each RNA nucleotide is removed (Fig. 23-10b). This *displacement synthesis*, as it is called, continues until the enzyme reaches the end of the primer. The reaction leaves a single-chain nick at the 3' end of the segment formerly occupied by the primer (Fig. 23-10c), which is sealed by DNA ligase.

Some experiments indicate that another enzyme, *ribonuclease H*, may also be involved in primer removal in prokaryotes, possibly in a secondary or backup role. The enzyme has the capacity to recognize and remove the RNA chain of hybrid RNA-DNA helices. Primer removal, although not completely inhibited, is delayed in *E. coli* mutants with defective forms of ribonuclease H. The same enzyme participates in primer removal in eukaryotes.

The Enzymes and Factors Unwinding DNA   The activity of DNA-binding proteins in stabilizing DNA in the single-stranded form is well documented in *E. coli* and several prokaryotic and eukaryotic viruses. In *E. coli*, temperature-sensitive mutants with defects in the DNA-binding protein *SSB* stop DNA replication immediately at restrictive temperatures. C. R. Wobbe and his coworkers and others found that *RPA*, a eukaryotic DNA-binding protein, carries out a similar function in the SV40 system.

The helicase unwinding DNA in prokaryotes, like so many of the enzymes and factors of replication, was first implicated by studies of temperature-sensitive *E. coli* mutants in Kornberg's laboratory. These studies implicated a protein called *DnaB* as the *E. coli* helicase. As noted, the bacterial helicase is complexed with the primase in the primosome, in a combination that is associated with the lagging chain. The helicase activity associated with polymerase δ carries out this function in eukaryotes.

The DNA-binding proteins and helicases gain initial access to the DNA in prokaryotes through the activity of the protein *DnaA*, which recognizes and binds the *OriC* origin of replication at one point in the bacterial DNA circle (see p. 680 and below). Binding by DnaA opens the DNA in the region of *OriC*, forming an unwound site that is extended by the helicase and stabilized in single-stranded form by the SSB DNA-binding protein. In the SV40 system the T antigen provides an initial unwinding function equivalent to bacterial DnaA. Whether nonviral eukaryotic systems typically contain initial unwinding factors equivalent in function to DnaA or the T antigen remains unknown.

The DNA topoisomerases relieving supercoils in front of the replication fork have been identified in both prokaryotes and eukaryotes. The enzyme unwinding supercoils in *E. coli* is a DNA topoisomerase II also known as *DNA gyrase*. Agents that inhibit DNA gyrase, such as the drug novobiocin, strongly inhibit DNA replication in *E. coli*. In yeast or cell-free systems replicating the DNA of the eukaryotic virus SV40, both DNA topoisomerases I and II seem to be active in DNA replication. Of the two, however, only DNA topoisomerase I can be eliminated without stopping the process. DNA topoisomerase II may therefore be the primary enzyme that unwinds positive supercoils in eukaryotes as well as prokaryotes.

The DNA topoisomerases also carry out steps in other vital cell processes such as DNA repair, genetic recombination, and release of tangles created during replication and recombination. Without release of the tangles, chromosomes would be unable to separate during mitosis and meiosis (for details, see p. 709).

DNA Ligase DNA ligase, which closes the single-chain nicks remaining in newly synthesized chains after the gaps left by primer removal are filled, catalyzes the last major step in DNA replication. The activity of DNA ligase was first identified in bacteria and viruses with mutant forms of the enzyme. If newly replicated DNA was unwound by heating (see p. 366) in the mutants, the "new" nucleotide chains separated into short pieces, indicating that single-chain nicks were left open during replication. The activity of DNA ligase in sealing nicks was subsequently characterized by I. R. Lehman and his coworkers. (Figure 23-11 outlines the mechanism.) Since its initial discovery in prokaryotes and viruses, DNA ligase has been isolated from a variety of eukaryotes as well.

Organization of the Enzyme Systems Replicating DNA The distinct properties of prokaryotic and eukaryotic enzymes are reflected in differences in their organization at the replication fork (Fig. 23-12). A dimer formed by two bacterial DNA polymerase III molecules is considered to sit astride the replication fork, with one half of the dimer catalyzing assembly of the leading chain and the other half assembling the lagging chain (Fig. 23-12a). The primosome, consisting of the combined helicase-primase activities, is thought to work with the DNA-binding proteins on the lagging chain just in advance of the polymerase III complex. Immediately in front of the fork is DNA gyrase, the topoisomerase II unwinding positive supercoils in the bacterial system.

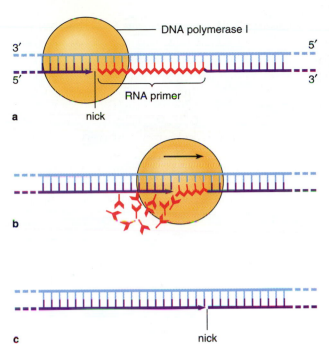

Figure 23-10 Displacement synthesis by bacterial DNA polymerase I. **(a)** The enzyme starts at a single-chain opening or nick between a segment of newly synthesized DNA and the next primer in line. **(b)** The enzyme then displaces the primer one nucleotide at a time, using its $5' \rightarrow 3'$ exonuclease activity; new DNA nucleotides are added at the same time in the $5' \rightarrow 3'$ direction by the polymerase activity of the enzyme. **(c)** After the primer is replaced by displacement synthesis, a single-chain nick is left that is sealed by DNA ligase.

Eukaryotic systems are believed to be organized similarly, with several exceptions (Fig. 23-12b). DNA polymerases α and δ have been proposed to coordinate their activities in replication, with polymerase δ catalyzing assembly of the leading chain and polymerase α the lagging chain. The primase activity associated with polymerase α assembles the short primers required for discontinuous synthesis of the lagging chain. Note from Figure 23-12b that the helicase unwinding the fork is part of the polymerase δ complex, and that supercoils can be released by either DNA topoisomerase I or II.

Stillman's group has found indications that DNA polymerase δ may take part in synthesis of both the leading and lagging chains. According to their proposal, the eukaryotic replication complex contains two molecules of DNA polymerase δ; one of these assembles the leading chain, and the other takes over synthesis of the lagging chain after the DNA polymerase α/primase complex has primed an Okazaki fragment and laid down a few DNA bases. This mechanism would permit DNA polymerase δ to proofread

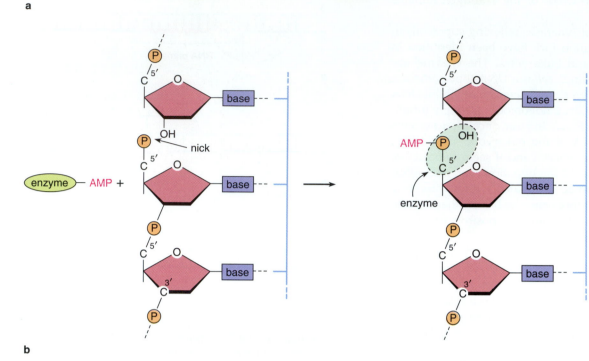

$$\text{enzyme} + \text{NAD} \longrightarrow \text{enzyme} - \text{AMP} + \text{NMN} \quad \text{or} \quad \text{enzyme} + \text{ATP} \longrightarrow \text{enzyme} - \text{AMP} + \text{PP}$$

a

b

Figure 23-11 Activity of DNA ligase in sealing single-chain nicks left in new DNA nucleotide chains by DNA polymerases. In the first step in bacteria **(a)**, DNA ligase hydrolyzes NAD into AMP, which remains attached to the enzyme by a covalent bond, and *NMN* (*nicotinamide mononucleotide*), which is released. In eukaryotes the ligase splits the two terminal phosphates from ATP in the first step and retains the AMP residue through a covalent linkage. In either case the ligase-AMP product retains much of the free energy released by the hydrolysis. In the second step **(b)** the AMP group is transferred from the enzyme to the 5'-phosphate exposed at a nick in a newly replicated DNA chain. The ligase then catalyzes reaction of this activated group with the 3'-OH exposed at the other side of the nick **(c)**. In this step the 5'-phosphate group is split from AMP and attached to the adjacent 3'-oxygen, closing the nick by a phosphodiester linkage. The AMP split off by the reaction is released to the medium.

c

the lagging as well as the leading chain (DNA polymerase α does not possess the 3'→5' exonuclease activity required for proofreading).

In Figure 23-12a and b the lagging chain is shown looped around the enzyme. The loop, first proposed by B. M. Alberts and his colleagues, allows the two halves of bacterial polymerase III, or the combined polymerase α/polymerase δ complex of eukaryotes, to carry out assembly of both leading and lagging chains in the same overall direction and progress as a unit in the direction of fork movement.

R. Berezny and R. T. O'Keefe and their colleagues and others found that, rather than being distributed evenly throughout the nucleus, replication appears to be concentrated in some 50 to 250 localized sites in eukaryotic nuclei. The research was accomplished by attaching a fluorescent dye to antibodies specific for 5-bromodeoxyuridine (BrdU), a base analog that can

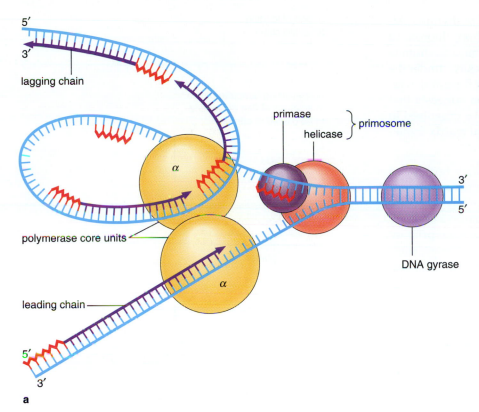

5′
3′
lagging chain

primase
helicase } primosome

α

polymerase core units

α

leading chain

5′
3′

a

3′
5′

DNA gyrase

Figure 23-12 Proposed organization of the enzymes carrying out
unwinding, priming, and DNA polymerization in prokaryotes (a) and
eukaryotes (b). In the prokaryotic
complex the two α-subunits are the
core units of the DNA polymerase III
complex. The loop shown in the
lagging chain in both (a) and (b) allows the DNA polymerases to move
in the same direction on the lagging
and leading chains. PCNA and RF-C
are additional polypeptides forming
part of the eukaryotic complex;
PCNA is necessary for polymerase
δ activity and increases the number
of bases the polymerase can copy
without releasing; RF-C may facilitate primer recognition and binding
by polymerase δ.

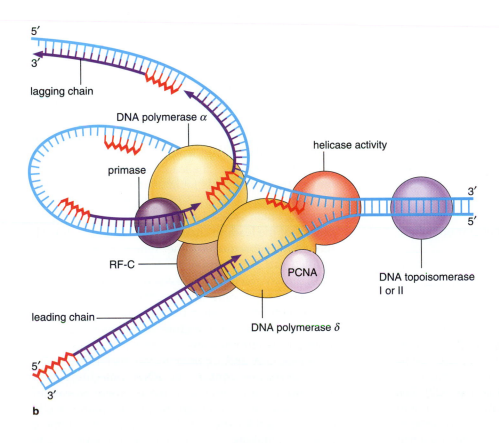

5′
3′
lagging chain

DNA polymerase α

helicase activity

primase

RF-C

PCNA

DNA topoisomerase
I or II

3′
5′

leading chain

DNA polymerase δ

5′
3′

b

be used by DNA polymerases instead of thymine. After exposure to BrdU during replication, fluorescent anti-BrdU antibodies were found to bind to discrete spots in nuclei, producing bright "speckles" under the fluorescent microscope rather than a diffuse and overall background glow. The localization suggests that replication enzymes and factors may be organized in units by attachment to a nucleoskeletal matrix framework (see p. 382) within eukaryotic nuclei.

The complexes carrying out the primary replication reactions in prokaryotes and eukaryotes are similar except for several details. Prokaryotes have a single primary polymerase, III; eukaryotes have two, α and δ, acting in coordination. In prokaryotes the primase is associated with the helicase in the primosome complex; in eukaryotes the primase forms part of polymerase α. In eukaryotes, helicase activity may be associated with polymerase δ.

Telomere Replication in Eukaryotes: Replacing the Primer of the Leading Chain

The linear DNA molecules of eukaryotes present special problems to the replication mechanism at the end of the leading chain. Removal of the primer laid down to begin replication of the leading chain leaves an unfilled gap opposite the 3' end of the template; because no 3'-OH group is available to act as primer, none of the DNA polymerases can initiate synthesis to fill the gap (Fig. 23-13).

If left unfilled, the gap would make the leading chain shorter than its template by a length equivalent to the primer. Because the same problem would appear at successive replications, the DNA would become progressively shorter. Eventually the DNA molecule would become too short to replicate and would disappear from the genome. Obviously this does not happen, at least not in reproductive cells; in these cells the total length of replicating eukaryotic chromosomes is maintained through successive rounds of divisions and preserved from one generation to the next.

Elegant work by E. W. Blackburn and her colleagues showed that replacement of the primer of the leading chain depends on a group of repeated sequences that caps the ends, or *telomeres* (from *telo* = end, and *mere* = segment), of all eukaryotic chromosomes. These terminal repeats typically consist of hundreds of copies of short sequences that are rich in G and T bases in the nucleotide chain that has its 3' end at the telomere (see Fig. 23-14 and p. 529). For example, the protozoan *Tetrahymena* has repeats of the sequence TTGGGG at the telomere in this nucleotide chain; *Saccharomyces* has variations of the sequence TGTGGG in this position; humans and other vertebrates have variations of the sequence TTAGGG, and higher plants have the sequence TTTAGGG.

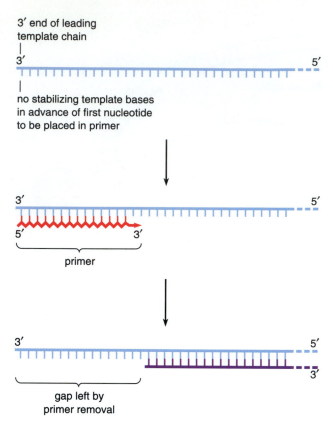

Figure 23-13   The gap left by primer removal at the beginning or 5' end of the leading chain. There is no available 3'-OH group for a DNA polymerase to use as a starting point for filling the gap.

Research by Blackburn, C. W. Greider, and J. W. Szostak revealed that a specialized enzyme, *telomerase*, maintains the telomere sequences and, thereby, the length of the leading chain. Telomerase does not maintain leading-chain length by directly filling the gap left at the end of the leading chain by primer removal. Instead, it adds telomere repeats *to the end of the template chain* (Fig. 23-14a and b). The extended template chain is used for primer synthesis and assembly of the leading chain as usual (Fig. 23-14c). Although the subsequent primer removal still leaves an unfilled gap at the end of the leading chain (Fig. 23-14d), the extended template chain lengthens the leading chain correspondingly and thus compensates for the inability of the system to fill the gap left by primer removal. The system therefore maintains the average length of the chromosome.

Blackburn and Greider found that the *Tetrahymena* telomerase contains an RNA molecule with a sequence complementary to the telomere repeats of the species. They proposed that, rather than directly copying a DNA template, the enzyme adds telomere repeats by making copies from its own RNA template (Fig. 23-15).

Their proposal is supported by experiments in which the telomerase of one species is used to fill in

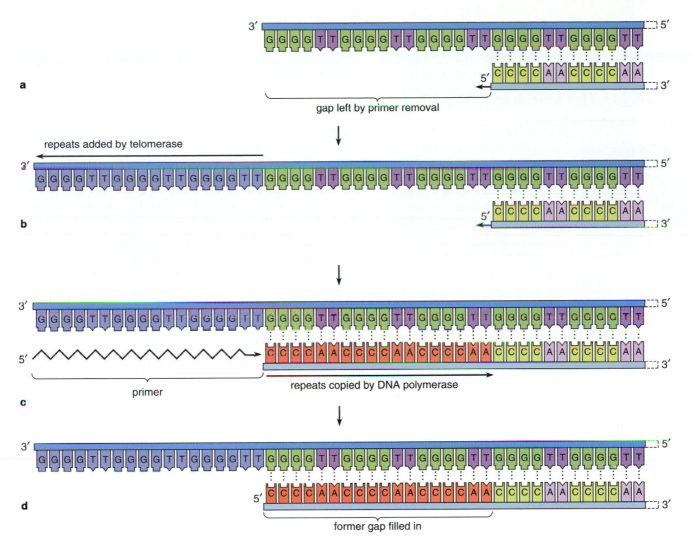

Figure 23-14 The indirect mechanism by which telomerases are believed to maintain the lengths of eukaryotic chromosomes. **(a)** The gap left in the leading chain by primer removal. **(b)** The telomerase extends the template chain by adding telomere repeats. **(c)** The extended chain is used as template for DNA synthesis by insertion of a primer and copying by DNA polymerase. **(d)** After primer removal the entire mechanism has filled in the original gap left by primer removal. The mechanism is shown as it would occur in the protozoan *Tetrahymena*, which has the telomere repeat sequence TTGGGG when read in the conventional 5' → 3' direction. (Adapted from an original courtesy of C. W. Greider; reprinted by permission from *Nature* 337:331. Copyright © 1989 by Macmillan Magazines Ltd.)

telomere sequences in another. The telomerase of *Saccharomyces*, for example, can add telomere sequences at the ends of *Tetrahymena* DNA molecules. The yeast telomerases, however, add the typical yeast telomere repeats to the ends of *Tetrahymena* telomeres. In another experiment, Blackburn and her colleagues carried out site-directed mutagenesis to alter the RNA molecule associated with the telomerase. When the RNA molecule was altered in sequence, the sequence inserted by the telomerase was also altered to remain complementary to the telomerase RNA.

In general the telomerases of all eukaryotes seem to be completely interchangeable and able to recognize and add sequences to the telomeres of any species. The mechanism is therefore evidently highly conserved and very ancient among eukaryotes. (Greider's Experimental Process essay on p. 676 describes the experiments leading to the discovery of telomerase and its critical activity in telomere replication.)

Although telomere sequences are maintained by telomerases in mammalian reproductive cells, the telomere repeats of somatic cells gradually decrease

Telomerase: The Search for a Hypothetical Enzyme

Carol W. Greider

CAROL GREIDER was raised and educated in Davis, California. She received her bachelor's degree from the University of California at Santa Barbara in 1983. She did her graduate work with Dr. Elizabeth Blackburn at U.C. Berkeley where she initially discovered the telomerase enzyme. After receiving her Ph.D., in 1988 she took a position as a Cold Spring Harbor Fellow at Cold Spring Harbor Laboratory, on Long Island, New York. Dr. Greider now has an independent laboratory working on telomerase at Cold Spring Harbor where she is a Senior Staff Investigator.

Replication of the molecular end of a chromosome presents a unique problem. It was first pointed out by Watson[1] that DNA polymerases would not be able to completely replicate the ends because they polymerize DNA only in the 5' to 3' direction and they require a primer. One would imagine that at each round of division a region at the end of the chromosome would not be fully replicated (see Fig. 23-13), and after many rounds of division chromosomes would be significantly shorter. A number of theoretical mechanisms were proposed for how chromosome ends might replicate. However, until the structure of telomeres was characterized, it was not possible to test the numerous models that had been proposed.

Telomeres were first cloned and characterized in ciliated protozoa; they consist of tandem repeats of very simple sequences, e.g., (TTGGGG)n in *Tetrahymena*. The number of repeats on each chromosome end is variable; thus telomeres characteristically appear as 'fuzzy' bands on Southern blots. In 1983 Szostak and Blackburn showed that *Tetrahymena* telomeres would function as telomeres in the yeast *Saccharomyces cerevisiae*. This cross-kingdom conservation of telomere function not only allowed yeast telomeres to be cloned, but it also provided evidence for how telomeres might be replicated.[2] When the *Tetrahymena* (TTGGGG)n telomeres were cloned back out of yeast, they had yeast telomeric $G_{1-3}T$ sequences added onto them.[3] This result—along with the fact that telomeres normally have a variable number of (TTGGGG)n repeats, and evidence that under some circumstances trypanosome and *Tetrahymena* telomeres increase in length—suggested that there must be a mechanism, maybe an enzyme, that adds telomeric sequences *de novo* onto chromosome ends. We imagined that this mechanism could overcome the end-replication problem indirectly: sequence loss due to incomplete replication might be restored by *de novo* addition of telomere repeats.[3]

In Liz Blackburn's laboratory at U.C. Berkeley in April 1984, I set out to look for such a hypothetical enzyme activity that would add telomeric sequences onto chromosome ends. The lesson I learned from looking for an enzyme activity that had never been reported before was this: The assay you choose is most important; try as many different assays as possible. At the time, we did not have detailed information about the molecular structure of the chromosome end. Liz Blackburn and I developed an assay in which we hoped to get specific addition of DNA nucleotides with labeled phosphate (^{32}P-dNTPs) onto a restriction fragment having telomeric sequences at the end to compare with ends having nontelomeric sequences. Initially, I tried using a large DNA fragment that had (TTGGGG)n repeats at only one end. This fragment was incubated in a buffer with *Tetrahymena* cell extracts and ^{32}P-dNTPs. After incubation, the fragment was digested at two internal sites with a restriction enzyme to generate three pieces, one with the (TTGGGG)n end, one with the nontelomeric sequence end, and one internal piece as a control. We then compared the label incorporated into the different fragments to determine if there was some specificity for label incorporation into telomeric ends. Because we did not know what the molecular end structure was, we first treated the fragment with either of two different single-strand exonucleases to generate either 5' or 3' overhangs, or left the fragment blunt-ended. We tried many different extract preparations and buffer conditions as well as different combinations of ^{32}P- and unlabeled nucleotide. In all the experiments both end fragments incorporated labeled nucleotide and, depending on the extract used, the internal band often was also labeled. The fragments with a recessed 3'-end labeled best, as they could be filled in by endogenous DNA polymerases. Having a large number of hot bands in an agarose gel was not very informative.

To look more closely at the labeled products, I began running the fragments out on sequencing gels after incubation with the *Tetrahymena* extracts and ^{32}P-dNTPs. In retrospect this was a very good move; it brought us one small step closer to the 'right' assay. We hoped not only to see labeling but to visualize any increase in size, which would be indicative of *de novo* addition. Such an increase of a few nucleotides would not be seen on an agarose gel but could be resolved on a high-resolution gel of the kind used for DNA sequencing. A 'fill in' reaction by DNA polymerase should stop at the length of the original input DNA, while *de novo* addition might generate products that are larger than the input DNA fragments. After a number of these experiments, I got a result suggesting (if you were being optimistic) that a fragment with a (TTGGGG)$_n$ 3' overhang was getting longer. Liz suggested in the next experiment that I try using as a substrate $(TTGGGG)_4$ oligonucleotide, which a postdoc in the lab, Eric Henderson, was characterizing. In the very first experiment where I used this oligonucleotide, an elongation pattern with a six-base periodicity was apparent. That was Christmas 1984, over nine months after we started looking for activity. Subsequently, the six-base periodicity seen in that first successful experiment turned out to be the result of the novel enzyme *telomere terminal transferase* or *telomerase*, which we had set out to look for.

The next year was spent trying to determine whether this activity was in fact a novel enzyme and not some side reaction of DNA polymerase that was fooling us.[4] However, the initial breakthrough, in my mind, was having an activity to work on and to characterize in the first place. Using synthetic DNA oligonucleotides and sequencing gels in the assay made the difference. Using oligonucleotides, we could obtain a 100-fold molar excess of primer over what we had been adding when we were using DNA restriction fragments. In retrospect, we approached the 'right' assay slowly but continuously; we just kept making new and reasonable changes to the assay.

Having identified the activity and shown that $(TTGGGG)_n$ repeat synthesis was independent of added template, the next big question was: How does an enzyme know how to synthesize

$$TTGGGGTTGGGGTTGGGG \ldots$$

over and over without a template? A plausible and testable (!) model came to mind: the enzyme carries its own template with it. Half the answer came quickly. I treated active telomerase fractions with micrococcal nuclease or RNase A and found both would destroy enzyme activity. This suggested that RNA was somehow required for telomerase activity. The other half of the answer, that the essential RNA component does provide the template for $(TTGGGG)n$ repeat synthesis, took several more years to prove.

First I had to purify the enzyme biochemically and identify the RNA species that copurified (the lability of

telomerase activity made this a large task).[5] I then sequenced the 159-nucleotide copurifying RNA directly and used the sequence information to clone the gene. The RNA contained the sequence CAACCCCAA, exactly what you would expect for a template. But having an RNase-sensitive enzyme and an RNA that copurified with a CCCCAA sequence did not prove that this RNA had anything to do with telomerase. It could have been a cruel joke (on me) that an unrelated RNA copurified with telomerase activity and had that sequence. To prove that the 159-RNA species was required for telomerase I used a trick that was popular in the field of RNA splicing, where most factors (snRNPs) have essential RNA components. DNA oligonucleotides complementary to the RNA were incubated with telomerase and RNase H. RNase H will cleave the RNA of a DNA/RNA duplex. With specific oligonucleotides, RNase H cleaved the 159-RNA and inactivated telomerase. Thus that particular RNA was essential for telomerase activity.[6]

The final proof for the template mechanism came a year later when workers in Liz Blackburn's lab showed that mutations in the CAACCCCAA sequence of the telomerase RNA gene would template the synthesis of mutant telomeres *in vivo* in *Tetrahymena*.[7] Today telomerase activity has been identified in several ciliates as well as human cells. A telomerase-mediated dynamic length equilibrium is widely expected to be the mechanism used by most eukaryotes to overcome the end-replication problem.

References

[1]Watson, J. D. *Nature New Biology* 239:197–201 (1972).

[2]Szostak, J. W., and Blackburn, E. H. *Cell* 29:245–55 (1982).

[3]Shampay, J.; Szostak, J. W.; and Blackburn, E. H. *Nature* 310: 154–57 (1984).

[4]Greider, C. W., and Blackburn, E. H. *Cell* 43:405–13 (1985).

[5]Greider, C. W., and Blackburn, E. H. *Cell* 51:887–98 (1987).

[6]Greider, C. W., and Blackburn, E. H. *Nature* 337:331–37 (1989).

[7]Yu, G.-L.; Bradley, J. D.; Attardi, L. D.; and Blackburn, E. H. *Nature* 344:126–32 (1990).

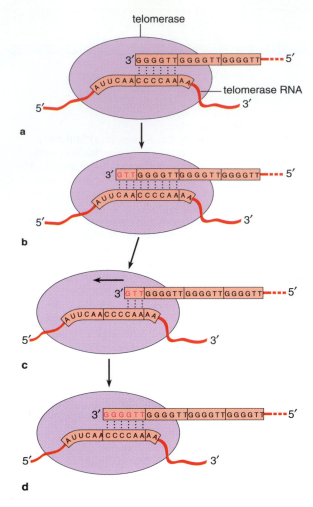

telomerase

telomerase RNA

Figure 23-15 The mechanism proposed by Blackburn and Greider by which telomerase adds telomere repeats to the template chain, using the *Tetrahymena* system as model. **(a)** The RNA of the telomerase pairs with the terminal telomere repeat. **(b)** The telomere is elongated by addition of the first three nucleotides of the repeat; synthesis occurs in the usual 5' → 3' direction using the RNA of the telomerase as template. **(c)** The enzyme moves so that the 3' end of the RNA template now pairs with the three nucleotides just added to the telomere. **(d)** The remainder of the repeat is filled in, using the RNA of the telomerase as template. This completes the telomere repeat; more repeats may be added by additional cycles from **(a)** through **(d)**. (Modified from an original courtesy of C. Greider, from *Bioess.* 12:363 [1990].)

in number as cells undergo repeated divisions and as individuals age. For this reason, some investigators have suggested that telomerase activity is high only in reproductive cells but reduced or even absent in body cells. The reduced activity in body cells may lead to eventual loss of telomeres and development of chromosome instability, contributing to the debilitating effects of old age in humans and other mammals.

Replication of the circular DNA molecules of prokaryotes, mitochondria, chloroplasts, and viral genomes with circular DNA molecules presents no equivalent problems to the mechanism because there are no nucleotide chains with free ends. As a result, a nick opened at any point around the circle provides a 3'-OH group from which primer insertion or gap filling may proceed.

Repair of Replication Errors

Proofreading produces replicated DNA copies that are highly faithful to their template molecules. However, even with the low error rates of the replication mechanism, small numbers of incorrectly paired bases persist in the replicated copies. The total numbers are magnified by the relatively large amounts of DNA replicated in prokaryotic and especially eukaryotic cells and by the frequency with which replication takes place.

Repair of incorrectly paired bases, called *mismatch repair*, forms part of the biochemical arsenal of all prokaryotes and eukaryotes. The final level of fidelity reached by proofreading and mismatch repair makes the error rate for replication equivalent to the characteristically low rates at which mutations occur in prokaryotes and eukaryotes. (The characteristics of mutations and their effects are discussed in Information Box 23-1.)

The "correct" A-T and G-C base pairs fit together in the DNA double helix like pieces of a jigsaw puzzle, with overall dimensions that maintain separation of the sugar-phosphate backbone chains at a regular 1.1-nm distance (see p. 364). Mispaired bases are too large or too small to fit perfectly within the interior of an undisturbed DNA helix and cannot form the patterns of hydrogen bonds characteristic of the normal base pairs. As a result, base mismatches cause local distortions in the structure of the DNA double helix, which provide recognition points for the enzymes catalyzing mismatch repair.

In prokaryotes an enzyme complex catalyzing mismatch repair evidently scans along newly replicated DNA molecules. A protein encoded in the *mutS* gene has a primary role in recognizing and binding mispaired bases. Once the MutS protein binds, another protein closely associated with MutS, *MutH*, opens a nick in one chain of the DNA in the vicinity of the mismatch (Fig. 23-16a). A helicase then unwinds the DNA and endonucleases nibble away the unwound chain including the mismatched base, leaving a gap in the DNA that may approach 1000 nucleotides in length. (The mismatch repair system is often called *long-patch* repair in reference to the relatively long length of DNA removed.) The gap left by the

Mutations and Their Effects

Even with the safeguards built into DNA replication by proofreading and mismatch repair, a small but significant number of errors escape detection and persist in the replicated copies. These persistent errors are the primary source of mutations.

Though the final error rates are very low, the large amounts of DNA replicated and the frequency of replication magnify the numbers that persist. In a bacterium such as *E. coli*, for example, with about 4 million base pairs, an error rate of 1×10^{-10} means that an incorrectly matched base pair is likely to leak past the proofreading and repair mechanisms about once every two to three replications. Although the total per cell is small, the rapidity of bacterial replication and cell division and the large number of cells in bacterial populations can produce millions of persisting mistakes per generation. In eukaryotes such as humans, with about 3×10^{10} base pairs per nucleus, incorrectly paired nucleotides are likely to appear and persist somewhere in the genome several times in the replication of each cell. With literally billions of replications taking place each day in the human body (the divisions maintaining red blood cells alone proceed at the rate of 2 million per second!), the total number of mistakes generated by mispaired bases reaches numbers on the order of at least millions or more per day per individual.

Mutations also arise from other sources, such as the action of chemical carcinogens and radiation on DNA (see p. 654). Another frequent source is methylated cytosines, which tend to break down into thymines, producing a G-T mismatch in a site formerly occupied by a G paired with 5-methylcytosine.

The Characteristics of Mutations

Any change in a DNA molecule that alters its sequence, so that it no longer represents an exact copy of its parental DNA molecule, constitutes a mutation. Mutations in

functional sequences of the genome have varying effects depending on their positions. Mutations in the promoter or other control regions of a gene may alter transcriptional activity but have no effect on the amino acid sequence of the protein encoded in the gene. Similarly, mutations in DNA segments specifying the 5' or 3' untranslated regions of mRNAs may alter the activity of the mRNAs in protein synthesis but still not affect the amino acid sequence of the encoded protein.

Mutations within the coding sequence of a gene, however, may change a codon for one amino acid to a codon for another. For example, a single substitution of C for A at the third position of the codon GAA changes the amino acid specified at this position from glutamine to asparagine (parts **a** and **b** of the figure in this box). The effects of a change in a single amino acid may range from negligible to severe, depending on the position of the change in a polypeptide chain and the degree of difference in size and chemical properties between the original and substituted amino acid.

Base substitutions may also change a codon for an amino acid to a terminator codon. This type of change is likely to have drastic effects on the activity of a protein because the entire amino acid sequence after the position of the new terminator codon is lost.

Not all mutations in coding regions cause amino acid substitutions or termination. For example, mutation of the GAA codon for glutamine to GAG causes no change in the amino acid sequence of the encoded protein because both GAA and GAG code for the same amino acid (parts **a** and **c** of the figure). Alterations that cause no change in the amino acid sequence of the encoded protein are called *silent mutations*.

Most mutations in the coding regions of mRNA genes are detrimental; a very few are beneficial. They affect only the individual if they occur in somatic cells. However, mutations persisting in reproductive cells can be passed on to offspring. Whether detrimental or beneficial, they are the ultimate source of variability for the evolutionary process.

a

| C C C | A G C | C G A | G A A | G G G | G C C | A A G | gene |
| proline | serine | arginine | glutamine | glycine | alanine | lysine | protein encoded in gene |

b

| C C C | A G C | C G A | G A C | G G G | G C C | A A G | gene |
| proline | serine | arginine | aspara gine | glycine | alanine | lysine | protein encoded in gene |

c

| C C C | A G C | C G A | G A G | G G G | G C C | A A G | gene |
| proline | serine | arginine | glutamine | glycine | alanine | lysine | protein encoded in gene |

Effects of mutations on proteins encoded in a gene. **(a)** The gene sequence and the corresponding sequence of amino acids in a protein. A mutation changing adenine to cytosine at the position marked by the arrow would cause substitution of asparagine for glutamine in the encoded protein **(b)**. A change from adenine to glycine at the same position **(c)** would lead to no change in the amino acid sequence because the codons GAA and GAG both stand for glutamine.

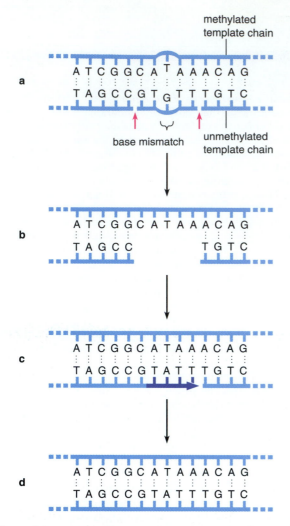

methylated template chain

a

A T C G G C A T A A A C A G
T A G C C G T G T T T G T C

base mismatch   **unmethylated template chain**

b

A T C G G C A T A A A C A G
T A G C C   T G T C

c

A T C G G C A T A A A C A G
T A G C C G T A T T T G T C

d

A T C G G C A T A A A C A G
T A G C C G T A T T T G T C

Figure 23-16 Excision repair of mismatched bases in *E. coli*. **(a)** Repair enzymes recognize a mispaired region, break one chain of the DNA, and remove a series of bases including the mismatched base. **(b)** The excision leaves a gap in the DNA, which is filled in **(c)** by a DNA polymerase using the intact template chain as a guide. **(d)** The single-chain nick left after gap filling is sealed by DNA ligase to complete the repair.

removal (Fig. 23-16b) is then filled by DNA polymerase III using the undamaged template chain as a guide (Fig. 23-16c). The repair is completed by DNA ligase, which seals the single-chain nick left after gap filling (Fig. 23-16d). Some 99% of the mispaired bases left uncorrected by proofreading in prokaryotes are estimated to be corrected by mismatch repair.

How do the enzymes carrying out mismatch repair "know" which is the template base and which is the mispaired copy in a base mismatch? In prokaryotes, at least, one answer evidently lies in a preference of the repair enzymes for unmethylated DNA chains, combined with a brief delay in methylation of the copy chain during replication. Bacterial DNA chains

are highly methylated at all times in the cell cycle except for a brief period just after their initial assembly during replication. Evidently excision repair takes place during this brief period, when the template chain is highly methylated and the copy chain still unmethylated. Since the enzymes prefer unmethylated DNA, excisions removing mispaired nucleotides are made in the copy rather than the template chain. There is some evidence that the repair enzymes also recognize DNA chains with single-chain nicks, as in the newly assembled lagging chain before closure by DNA ligase.

Equivalent mechanisms repairing mismatched bases have also been demonstrated in eukaryotes. Proteins related to bacterial MutS, for example, have been detected in yeast, frogs, mice, and humans. At the moment the basis for chain selection in eukaryotic mismatch repair remains unknown. However, although relatively few bases are methylated in eukaryotic DNA chains as compared to bacterial DNA, the degree of methylation might also provide one signpost distinguishing between template and copy chains, as it does in bacteria. Detection of single-chain nicks may also contribute to recognition of newly synthesized chains in eukaryotes.

REPLICATION ORIGINS AND REPLICONS

Replication Origins

DNA synthesis begins at localized sequences that act as *replication origins*. In many respects, replication origins function similarly to promoters of transcription. The origins are recognized by proteins that bind and destabilize the DNA. The destabilization opens or "melts" the DNA in the region of the origin and arranges it into a conformation that is recognized and bound tightly by the replication enzymes.

The *E. coli* DNA circle has a single replication origin, *oriC*, a 245-base pair segment that contains several repeated sequence elements (Fig. 23-17). Some of the sequence elements are rich in A-T base pairs, which probably make the DNA easier to unwind in the origin (an A-T base pair is stabilized by only two hydrogen bonds, in contrast to the three bonds stabilizing G-C base pairs; see Fig. 13-2). In *E. coli* the DnaA protein binds to the origin and promotes chain separation; the helicase unwinding DNA then binds, followed by the primosome and DNA polymerase. Initiation leads to formation of two replication forks, which proceed in opposite directions from the origin (Fig. 23-18a). The two forks meet and complete replication at a point approximately 180° from the origin.

Specific DNA sequences serving as replication origins have as yet been identified in only one eukary-

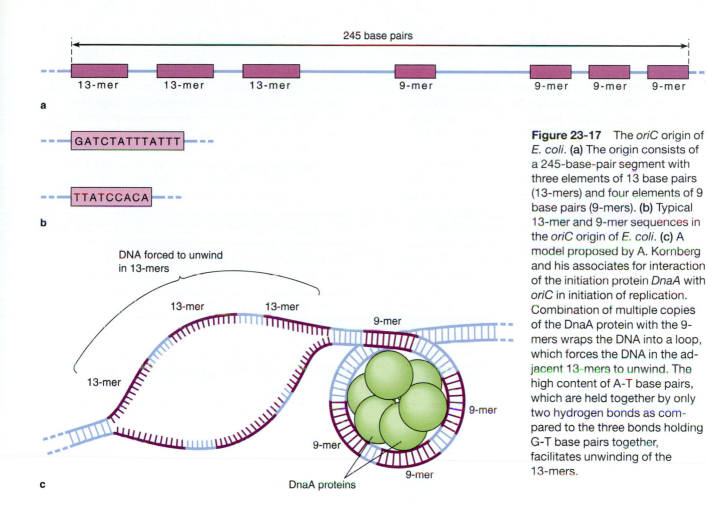

Figure 23-17 The *oriC* origin of *E. coli*. **(a)** The origin consists of a 245-base-pair segment with three elements of 13 base pairs (13-mers) and four elements of 9 base pairs (9-mers). **(b)** Typical 13-mer and 9-mer sequences in the *oriC* origin of *E. coli*. **(c)** A model proposed by A. Kornberg and his associates for interaction of the initiation protein *DnaA* with *oriC* in initiation of replication. Combination of multiple copies of the DnaA protein with the 9-mers wraps the DNA into a loop, which forces the DNA in the adjacent 13-mers to unwind. The high content of A-T base pairs, which are held together by only two hydrogen bonds as compared to the three bonds holding G-T base pairs together, facilitates unwinding of the 13-mers.

otic species, the yeast *Saccharomyces cerevisiae*. The origins were identified by adding various *Saccharomyces* DNA sequences to DNA circles and testing which could promote initiation of DNA replication in living cells. The origins, termed *ARS* in yeast (for *autonomously replicating sequences*), all contain at least one copy of the consensus sequence ^T_ATTTA$^{TA}_{CG}$TTT^T_A and several additional elements rich in A-T base pairs. B. Stillman and his coworkers have isolated an *origin recognition complex (ORC)*, containing six protein subunits, that recognizes and binds the ARS consensus sequence.

About 400 *ARS* are distributed through the yeast genome, some with variations in internal sequences. W. L. Fangman and his coworkers and others found that *ARS* with different sequences are activated at distinct times in S, possibly through different ORCs that specifically recognize the sequence subtypes. The controlled activation of *ARS* subgroups replicates subparts of the DNA in a definite order, with some regions replicating early in S and others later. Replication of the genome in a definite order appears to be typical of all eukaryotes.

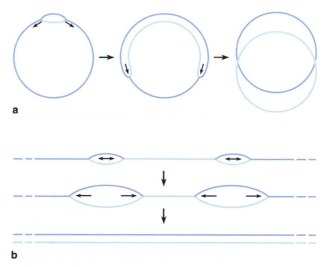

Figure 23-18 Replication from a single origin in the circular DNA molecule of a prokaryote **(a)** and from multiple origins in the linear molecules of eukaryotes **(b)**.

Only a very few replication origins have been identified in higher eukaryotes. Two have been identified by H. Cedar and his coworkers in humans, one just upstream of a β-globin gene, and the second downstream of a gene encoding the dihydrofolate reductase enzyme. Rather than a specific sequence at which replication begins, the two origins appear simply to be zones including thousands of base pairs in which replication is more likely to begin than in other regions. The preference for these zones as origins depends on chromatin structure, that is, association of the DNA with histone and nonhistone proteins. If proteins are removed from the DNA, the two zones are no more likely than other sequences to initiate replication. This seems generally to be the case with higher eukaryotic DNA cleaned of proteins—if exposed to replication enzymes in cell-free systems, or if injected into nuclei capable of replication, initiation can occur at essentially any site. Therefore it seems likely that in higher eukaryotes, chromatin structure, rather than specific sequences, sets up zones in which the DNA is more accessible to the factors and enzymes initiating replication.

Replicons

Higher eukaryotes, in which genomes are considerably larger than in *Saccharomyces*, typically have many more replication zones or origins. Estimates of the total have been made by a technique originally worked out in 1968 by J. A. Huberman and A. D. Riggs. These investigators labeled replicating DNA by exposing actively dividing Chinese hamster cells briefly to radioactive thymidine. After labeling, the replicated DNA was extracted and prepared for light microscope autoradiography. In the preparations, radioactivity could be detected in segments spaced at 15- to 60-μm intervals along the DNA molecules, as expected if replication proceeds simultaneously at multiple sites. The spacing indicates that each hamster chromosome, which contains several centimeters of DNA on the average, probably initiates replication from thousands of sites. (Figure 23-19 shows the results of a similar experiment with *Triturus* DNA.)

Later work established that replication in all eukaryotic chromosomes proceeds from multiple sites. A length of DNA whose replication is initiated from one of these sites is termed a *replicon*. A replication origin lies near the center of each replicon. Activation of an origin produces two replication forks that move in opposite directions (see Fig. 23-18b and c) until they meet forks from other replicons. Replicons vary in length from as short as 4 μm, or 13,000 base pairs, to as long as 280 μm, or 900,000 base pairs in different species. In all there may be from 20,000 to as many as 100,000 or more replicons per haploid genome in higher eukaryotes.

Rather than activating all replicons simultaneously, higher eukaryotic cells, like yeast cells, initiate replication in specific subgroups or clusters of replicons at certain times during S. The patterns in which replicons are activated have been traced by exposing cells in S to a pulse of 5-bromodeoxyuridine (BrdU), followed by a fluorescent staining to reveal segments of the chromosomes incorporating BrdU during the pulse. The segments taking up BrdU are repeated from cell to cell at early, middle, or late S within a species, showing that subregions of the chromosomes replicate at more or less fixed times within S.

Typically genes in active segments of the chromosomes replicate early, and genes in inactive or heterochromatin segments replicate later in S. In some species, entire chromosomes may enter replication early or late. In mammalian females, for example, the X chromosome that is active in transcription replicates early in S and the inactive X replicates later in body cells.

The pattern of replicon activation in eukaryotic cells changes with development. H. G. Callan found that in amphibian (*Triturus*) embryos the S phase runs its course in only one hour at the blastula stage. Later, at the neurula stage, S lasts four to six hours. In adult *Triturus* body cells, S lasts 48 hours, and in cells about to enter meiosis, it extends to 200 hours. By preparing autoradiographs after cells were exposed to radioactive precursors at these stages, Callan observed that the variations are correlated with differences in the distances that separate replicating segments in the DNA. During the neurula stage, for example, the segments taking up label are spaced about 40 μm apart (see Fig. 23-17a). In premeiotic cells, spacing between the replicating segments extends to 100 μm or more (see Fig. 23-17b). Therefore differences in the overall rate of replication in *Triturus* appear to be correlated with the number of replicons activated per unit length of chromosome. Similar results have been obtained with a variety of eukaryotes. Not all organisms regulate the duration of S by variations in the number of replicons activated; some, such as mouse cells in the interphase before meiosis, increase the duration of S by slowing the rate of DNA unwinding and replication.

Normally each replicon is activated only once during a given S period so that no segments of the genome are overreplicated. Although the basis for this control is unknown, it is believed to depend on one or more proteins that combine with and block the origin of each replicon once its replication is complete. Support for this interpretation comes from experiments in which cells in the S stage of interphase are fused with cells in G1 or G2 (see p. 110). In S + G1 fusions, factors carried by the S cells can induce G1 nuclei to enter replication. An S + G2 fusion, however, induces no further replication in the G2 nucleus, as if the origins in a G2 cell are indeed blocked and unable to initiate

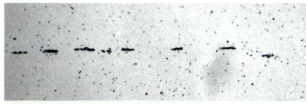

a

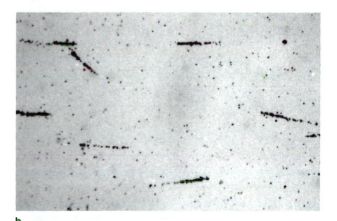

b

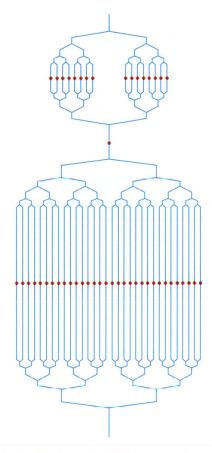

Figure 23-19 Replication origins detected by autoradiography in **(a)** premitotic cells of an embryo and **(b)** premeiotic cells of the newt *Triturus*. The origins are more widely spaced in the premeiotic cells, indicating that the longer S phase typical of the interphase before meiosis in this species depends on activation of fewer origins per unit length. × 450. (Courtesy of H. G. Callan, from *Biol. Zbl.* 95:531 [1976].)

Figure 23-20 The branched DNA segments produced by multiple activations of replicons (the red dots indicate replication origins). At some points the number of activations of replicons is reduced, so that the degree of amplification is lower. (Modified from an original courtesy of C. D. Laird and V. E. Foe, from *Cold Spring Harbor Symp. Quant. Biol.* 38:311 [1973].)

replication. Nuclei in mitosis also fail to enter replication when fused with an S cell, indicating that the block to replication persists until the next G1. Presumably the blocking proteins are removed during the mitosis–G1 transition.

Although replicons in most eukaryotic cell types are normally activated only once during S, in certain cells some replicons are activated repeatedly, so that portions of the genome are copied more than once (Fig. 23-20). Local increases in the number of DNA sequences by multiple activations of replicons is one source of *DNA amplification*, a process by which extra copies of genes are generated.

DNA amplification may be programmed as a normal part of development or may be a random, unscheduled event leading to alterations in cell function. Programmed DNA amplification takes place in cells in which rRNAs or proteins are required in greater quantities than can be produced through maximum activity of the normal gene complement. In developing *Drosophila* oocytes, for example, genes encoding proteins forming the outer egg coat are regularly amplified, thereby increasing the oocyte's capacity to make the coat proteins. Unscheduled DNA amplification underlies the development of drug re-

sistance in some cell types and is implicated in the development of some types of cancer (see p. 649).

DUPLICATION OF CHROMOSOMAL PROTEINS

Although DNA replication is the most critical event of the S phase for distribution of hereditary information to daughter cells, duplication of the chromosomes in preparation for cell division also requires doubling of structural, regulatory, and other proteins forming regular parts of the chromatin. The five histone proteins occur in chromatin in precise ratios with DNA that must be maintained to preserve chromatin structure as the DNA quantity doubles during S. Nonhistone regulatory proteins must be duplicated and combined with both daughter DNA molecules to preserve the pattern of gene regulation typical of the cell type. The doubling of histone and nonhistone proteins ensures

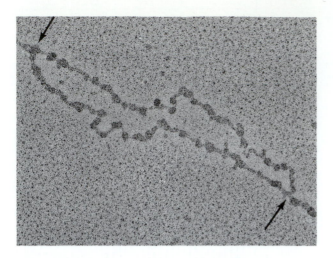

Figure 23-21 Replicating chromatin isolated from a *Drosophila* embryo, showing two replication forks (arrows). Nucleosomes are visible in close proximity to the replication forks in both replicated and prereplicative segments, indicating that DNA is free of nucleosomes only in the immediate vicinity of the replication fork. × 96,000. (Courtesy of S. L. McKnight, from *Cell* 12:795 [1977]. Copyright Massachusetts Institute of Technology.)

that the products of chromosome duplication are structurally and functionally the same as the chromosomes entering the process and are fully prepared for mitotic division.

Synthesis of Chromosomal Proteins

The five histones associated with DNA in chromatin—H1, H2A, H2B, H3, and H4—are assembled in close coordination with DNA replication. Transcription of the genes encoding the five histones is initiated in late G1, just prior to S. There is some evidence that regulatory proteins stimulating transcription of the histone genes are produced or become activated as part of the events switching cells from G1 to the division pathway. For example, G. S. Stein and G. L. Stein and their coworkers found that *HiNFD*, a regulatory protein promoting histone H4 transcription, is barely detectable at G1 but becomes highly active during S.

Histone mRNAs are also regulated posttranscriptionally at the level of pre-mRNA processing. The 3' ends of histone mRNAs are clipped by processing reactions that include activity of an snRNA, U7, which forms complementary base pairs with the 3' end of histone mRNAs. M. L. Birnstiel and his coworkers discovered that the 3' segment pairing with U7 snRNA is masked in nondividing cells. In cells at S the segment is exposed, allowing U7 to pair with histone pre-mRNAs and participate in the processing reactions.

Fully processed histone mRNAs appear in the cytoplasm as soon as S begins. These mRNAs are immediately used by ribosomes to assemble the histone proteins. The newly assembled histones are subsequently routed into the nucleus for assembly with DNA into nucleosomes (see p. 377). Because nucleosome assembly follows closely behind replication forks, the time during which replicated DNA remains in the naked, uncomplexed form is very limited (Fig. 23-21).

Histone gene transcription peaks early in S but continues at elevated levels as long as the DNA is replicating. The accumulated histone mRNAs, which have a half-life during S in higher eukaryotes of about 45 to 60 minutes, drive synthesis of histone proteins at high levels throughout S. As DNA replication becomes complete, transcription of the histone genes shuts down, and newly synthesized and processed histone mRNAs cease to enter the cytoplasm. Cessation of histone gene transcription is coupled with a rapid drop in the half-life of histone mRNAs to about 10 to 15 minutes. This halt in histone gene transcription and greatly increased mRNA breakdown rapidly clear the cytoplasm of histone mRNAs. As a result, further synthesis of histone proteins essentially stops.

Thus histone synthesis starts through transcriptional regulation of the histone genes and posttranscriptional control of histone pre-mRNA processing as S begins, and stops through a combination of transcriptional and posttranscriptional controls as S ends. The signals turning off histone transcription probably involve destruction or deactivation of regulatory proteins that recognize and bind control sequences in the promoters of histone genes. N. B. Pardey and W. F. Marzluff showed that breakdown of histone mRNAs at the end of S is regulated by an interaction between hydrolytic enzymes and a folded hairpin structure unique to histone mRNAs that occurs near the end of the 3' untranslated segment. Addition of the histone hairpin structure gives unrelated RNAs the stability characteristics of histone mRNAs. A globin mRNA to which the histone mRNA 3' structure has been added, for example, is degraded rapidly at the close of S along with the histone mRNAs (see also p. 508).

Not all histone synthesis is confined to the S phase. In many cell types, histones are synthesized during G1 and G2 at about 5% to 8% of their level during S. This steady histone assembly is probably a "housekeeping" function that compensates for histones lost from the chromosomes as a result of random histone turnover or breakdown during G1 and G2. The histone mRNAs synthesized at steady levels lack the 3' hairpin structure characteristic of S-phase histones and, unlike the S-phase histones, have poly(A) tails.

In contrast to the marked increases in histone synthesis during DNA replication, the synthesis of bulk nonhistone proteins does not change appreciably during S. This is not too surprising, since many non-histone proteins are enzymes associated with the more or less constant chromosomal functions of transcription or proteins combining with the RNA products of transcription (see p. 376). It is likely, however, that against this background of relatively constant non-histone protein synthesis, at least some regulatory proteins and any nonhistones forming part of the structural framework of chromosomes are duplicated during S in patterns similar to the histones.

Nucleosomes and DNA Replication

Histone proteins combine with DNA to form nucleosomes, the structural subunits of chromatin. In nucleosomes, two molecules each of the core histones—H2A, H2B, H3, and H4—form an *octamer* around which approximately two turns of DNA are wrapped. The DNA is stabilized by a single H1 molecule that lies in a position outside the core structure (see Fig. 13-10 and pp. 377–380).

For replication to proceed, nucleosomes probably unfold or disassemble temporarily from the DNA during the brief period when a replication fork passes. This conclusion is supported by electron micrographs of replicating chromatin, which show that the forks proper are free of nucleosomes. Measurements of the nucleosome-free region around replication forks reveal that only some 200 to 300 base pairs of DNA are exposed, roughly one to one and a half times the length of DNA wrapped around a nucleosome. This suggests that the short DNA lengths exposed at the fork reassociate with the histones or that nucleosomes refold as soon as enough DNA is replicated in either copy to wrap completely around a nucleosome core. It is still uncertain whether nucleosomes dissociate completely from the DNA as a replication fork passes or merely unfold or open out in some way without detaching to accommodate passage of the replication enzymes.

Electron micrographs also show that nucleosomes are present in their usual numbers and separation in both DNA chains behind the fork. This means that new nucleosomes must be added behind the fork to bring the total up to twice the G1 amount. In nucleosome assembly, H3 and H4 add to DNA first, followed by H2A and H2B to complete the core particle. It is still undetermined whether old nucleosomes are partially or completely conserved or whether new nucleosomes are assembled entirely from new histones.

It is probably no coincidence that Okazaki fragments in higher eukaryotes are about 100 to 200 base pairs in length. The average distance between nucleosomes is about 200 base pairs in eukaryotic DNA, suggesting that a primer is laid down on the lagging chain as each nucleosome unwinds and releases its 200–base-pair unit of DNA.

Essentially nothing is known about the assembly of nonhistone proteins on newly replicated DNA. However, there is little doubt that patterns of nonhistone distribution on DNA template molecules are exactly duplicated and passed on during replication. In cultured cells, for example, the locations of sites hypersensitive to DNA endonucleases, which reflect specific associations between nonhistone regulatory proteins and DNA (see p. 416), are faithfully reproduced and passed on for many generations. Although there is as yet no evidence to support the opinion, most investigators assume that nonhistone regulatory proteins add to the newly assembled DNA molecules as they replicate, in the brief period before they wind into nucleosomes. Supposedly DNA control sequences are fully exposed and available for binding by the regulatory proteins only during this period.

In addition to chromosomal proteins, the pattern of DNA methylation associated with gene regulation is duplicated during S in higher vertebrates. Cytosines in short CG doublet sequences are often methylated in inactive genes, and unmethylated in active genes (see p. 504). During S a hemimethylase enzyme adds methyl groups to C's in newly synthesized nucleotide chains to duplicate the pattern of methylation in the parental DNA molecules entering replication (p. 504 and Fig. 17-12 explain the process).

For Further Information

Suggestions for Further Reading

Baker, T. A., and Wickner, S. H. 1992. Genetics and enzymology of DNA replication in *E. coli*. *Ann. Rev. Genet.* 26: 447–477.

Bambara, R. A., and Jessee, C. B. 1991. Properties of DNA polymerases δ and ε, and their roles in eukaryotic DNA replication. *Biochim. Biophys. Acta* 1088:11–24.

Blackburn, E. H. 1991. Structure and function of telomeres. *Nature* 350:569–573.

Blackburn, E. H. 1992. Telomerases. *Ann. Rev. Biochem.* 61: 113–129.

Burhans, W. C., and Huberman, J. A. 1994. DNA replication origins in animal cells: A question of context? *Science* 263:639–640.

Clayton, D. A. 1992. Transcription and replication of animal mitochondrial DNA. *Internat. Rev. Cytol.* 141:217–232.

DePamphilis, M. L. 1993. Eukaryotic DNA replication: Anatomy of an origin. *Ann. Rev. Biochem.* 62:29–63.

Diffley, J. F. X. 1994. Eukaryotic DNA replication. *Curr. Opinion Cell Biol.* 6:368–372.

Drake, J. W. 1991. Spontaneous mutation. *Ann. Rev. Genet.* 25:125–146.

Fairman, M. P. 1990. Nucleosome segregation—divided opinions? *Bioess.* 12:237–239.

Fangman, W. L., and Brewer, B. J. 1991. Activation of replication origins in yeast chromosomes. *Ann. Rev. Cell Biol.* 7:375–402.

Greider, C. W. 1990. Telomeres, telomerase, and senescence. *Bioess.* 12:363–369.

Gruss, C., and Sogo, J. M. 1992. Chromatin replication. *Bioess.* 14:1–8.

Hamlin, J. L., Leu, T.-H., Vaughn, J. P., Ma, C., and Dijkwel, P. A. 1991. Amplification of DNA sequences in mammalian cells. *Prog. Nucleic Acids Res. Molec. Biol.* 41:203–239.

Heinhorst, S., and Cannon, G. C. 1993. DNA replication in chloroplasts. *J. Cell Sci.* 104:1–9.

Heintz, N. 1991. The regulation of histone gene expression during the cell cycle. *Biochim. Biophys. Acta* 1088:327–339.

Heywood, L. A., and Burke, J. F. 1990. Mismatch repair in mammalian cells. *Bioess.* 12:473–477.

Kunkel, T. A. 1992. DNA replication fidelity. *J. Biolog. Chem.* 267:18251–18254.

Laskey, R. A., Farman, M. P., and Blow, J. J. 1989. S phase of the cell cycle. *Science* 246:609–614.

Lindahl, T., and Barnes, D. E. 1992. Mammalian DNA ligases. *Ann. Rev. Biochem.* 61:251–281.

Marians, K. J. 1992. Prokaryotic DNA replication. *Ann. Rev. Biochem.* 61:673–719.

Matson, S. W., and Kaiser-Rogers, K. A. 1990. DNA helicases. *Ann. Rev. Biochem.* 59:289–329.

McHenry, C. 1991. DNA polymerase III holoenzyme: Components, structure, and mechanism of a true replicative complex. *J. Biolog. Chem.* 266:19127–19130.

Meyer, R. R., and Lane, P. S. 1990. The single-strand DNA-binding protein of *E. coli*. *Microbiol. Rev.* 54:342–380.

Modrich, P. 1991. Mechanisms and biological effects of mismatch repair. *Ann. Rev. Genet.* 25:229–253.

Moysis, R. K. 1991. The human telomere. *Sci. Amer.* 265: 48–55 (August).

Osley, M. A. 1991. The regulation of histone synthesis in the cell cycle. *Ann. Rev. Biochem.* 60:827–861.

Palladino, F., and Gasser, S. M. 1994. Telomere maintenance and gene repression: A common end? *Curr. Opinion Cell Biol.* 6:373–379.

Radman, M., and Wagner, R. 1988. The high fidelity of DNA replication. *Sci. Amer.* 259:40–46 (August).

So, A. G., and Downey, K. M. 1992. Eukaryotic DNA replication. *Crit. Rev. Biochem. Molec. Biol.* 27:129–155.

Waga, S., and Stillman, B. 1994. Anatomy of a DNA replication fork revealed by reconstitution of SV40 DNA replication *in vitro*. *Nature* 207:207–212.

Wang, T. S.-F. 1991. Eukaryotic DNA polymerases. *Ann. Rev. Biochem.* 60:513–552.

Wintersberger, E. 1994. DNA amplification: New insights into its mechanism. *Chromosoma* 103:73–81.

Review Questions

1. Define conservative and semiconservative replication. Outline Meselson and Stahl's experiment showing that replication is semiconservative.

2. Describe the steps taking place in assembly of a DNA chain on a DNA template. How do these steps differ from the assembly of an RNA chain on a DNA template?

3. What is a primer? Why are primers required in replication?

4. How are the problems of template unwinding and unidirectional replication solved?

5. Define the roles of DNA binding proteins, helicases, unwinding enzymes, DNA polymerases, DNA topoisomerases, and DNA ligase in replication. What is the difference in activity between DNA topoisomerases I and II?

6. What are the leading and lagging chains in replication? What are Okazaki fragments, and how do they function in replication?

7. Compare the DNA polymerases of prokaryotes and eukaryotes. What evidence indicates that polymerase III of prokaryotes and polymerases α and δ of eukaryotes are the primary replication enzymes?

8. What is proofreading? What evidence indicates that proofreading actually takes place?

9. What combinations of enzymes or activities are believed to accomplish proofreading in prokaryotes and eukaryotes?

10. What are replication origins? Replicons? What evidence indicates that replicons are activated in a definite sequence in eukaryotes?

11. How are replicons related to differences in duration of the S stage? To DNA amplification?

12. What is mismatch repair? How is it believed to take place?

13. What mechanism appears to ensure that mismatches are repaired in the newly synthesized DNA copy instead of the template chain in bacteria?

14. What regulatory processes control synthesis of histones during the cell cycle?

15. Compare the initiation and progress of replication in *E. coli*, a eukaryotic chromosome, a mitochondrion, and a chloroplast. (See Supplement 23-1.)

Supplement 23-1
DNA Replication in Mitochondria and Chloroplasts

Mitochondria and chloroplasts contain all enzymes and factors necessary to replicate their DNA. This has been amply demonstrated in isolated mitochondria and chloroplasts, which are able to incorporate the four nucleoside triphosphates into newly synthesized DNA.

Both organelles contain DNA polymerases with properties distinct from those active in the cell nucleus. The DNA polymerase of mitochondria, as demonstrated by W. J. Adams and G. Kalf and by A. Weissbach and his coworkers, is DNA polymerase γ (see Table 23-2), which can act in the forward direction as a 5' → 3' polymerase and has the 3' → 5' exonuclease activity required for proofreading. Chloroplasts contain a similar enzyme. A DNA topoisomerase with similarities to bacterial DNA topoisomerase II has also been detected in mitochondria, along with a primase and DNA-binding proteins related to bacterial SSBs. The DNA polymerases of mitochondria and chloroplasts, and evidently all the other enzymes and factors concerned with DNA replication, are encoded in the cell nucleus rather than in the organelles.

The pattern of DNA replication inside both mitochondria and chloroplasts differs significantly from replication pathways observed in the nucleus. Much of this difference rests on the fact that the DNA molecules of the two organelles, like the DNA of bacteria, are circular rather than linear. In mitochondria, for example, replication begins from an origin at one point on the circle, as it does in bacteria. In contrast to bacterial replication, however, only one replication fork, instead of two, progresses around the circle from the origin.

D. D. Chang and Clayton showed that mitochondrial replication in animals begins in the nontranscribed region, at a point just downstream of the transcription promoter for the L chain (Fig. 23-22; see p. 618). Replication is primed by an RNA transcript that is initiated at the L-chain promoter. This primer is made by the same RNA polymerase that transcribes the H and L chains in mRNA, rRNA, and tRNA synthesis. For replication, Clayton and his coworkers found that the RNA transcript is clipped off soon after its initiation, leaving a short RNA segment still attached to the DNA. This segment serves as the primer for DNA replication of a new H chain. DNA replication begins at the 3' end of the primer and proceeds around the template L chain. Because the template L chain is presented to the enzyme in the correct 3' → 5' orientation, replication of the copy proceeds continuously in the required 5' → 3' direction. The opposite H chain, rather than acting as a template for discontinuous synthesis of the lagging chain, is initially pushed out as a single-stranded *displacement loop* (Figs. 23-22b and c and 23-23).

In mammalian mitochondria, displacement of the H chain continues until about two-thirds of the L chain has been replicated. At this point, correspond-

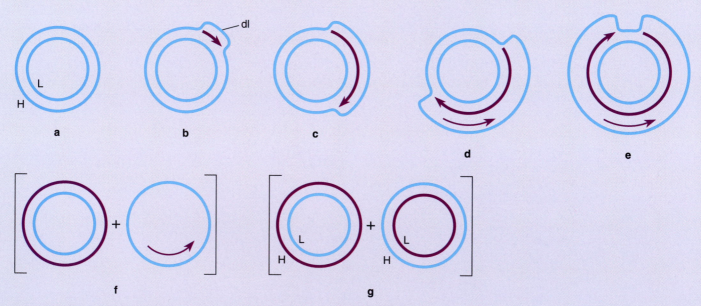

Figure 23-22 DNA replication in mammalian mitochondria (see text). The light blue lines indicate template chains; the dark blue lines, newly synthesized copies. (Modified from an original courtesy of D. L. Robberson.)

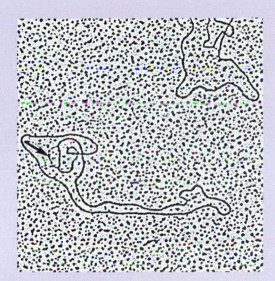

Figure 23-23 A single-stranded displacement loop (arrow) in a replicating mitochondrial DNA circle from a mouse cell. The loop is the H chain, displaced by replication of the L chain. (Courtesy of H. Kasamatsu.)

ing to a site between genes for cysteine and asparagine tRNAs on the L chain (see Fig. 21-7), replication begins in the opposite direction on the template H chain (Fig. 23-22d). Replication of the H chain, since it proceeds in a direction opposite to L-chain replication, also occurs continuously in the 5' → 3' direction. As a consequence, both template chains act as leading chains for continuous replication, and no discontinuous Okazaki fragments are generated. As replication of the L chain is completed, the H chain is released as a free circle that finishes its replication separately (Fig. 23-22e to g).

Chloroplast replication initially resembles the mitochondrial pattern except that replication proceeds simultaneously from two origins separated by about 7000 base pairs, forming two displacement sites that open toward each other (Fig. 23-24a to c). As the loops meet, they form two regular replication forks that proceed in opposite directions around the circle as in bacteria (Fig. 23-24d and e). The forks advance until replication of the circle is complete (Fig. 23-24f).

Replication of mitochondrial DNA is not confined to the S phase. Instead, mitochondria appear to replicate their DNA almost continuously during interphase. Individual mitochondrial DNA molecules may replicate several times or not at all. Chloroplasts, although more restricted in their periods of DNA synthesis during interphase, also replicate their DNA independently of S.

At some time during interphase, usually in advance of mitosis and cytokinesis, mitochondria and chloroplasts divide by a mechanism that parcels out the replicated DNA circles and other organelle structures. The organelle division mechanism, which resembles prokaryotic cell division, is described in Chapter 24.

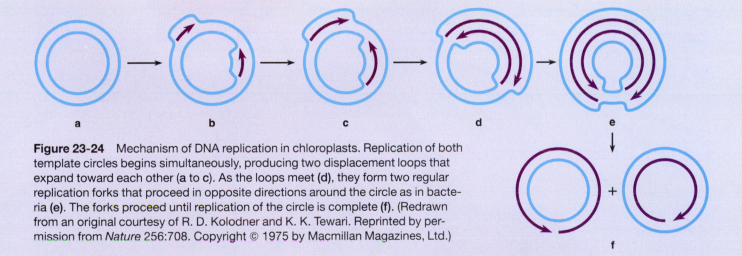

a   b   c   d   e   f

Figure 23-24 Mechanism of DNA replication in chloroplasts. Replication of both template circles begins simultaneously, producing two displacement loops that expand toward each other (a to c). As the loops meet (d), they form two regular replication forks that proceed in opposite directions around the circle as in bacteria (e). The forks proceed until replication of the circle is complete (f). (Redrawn from an original courtesy of R. D. Kolodner and K. K. Tewari. Reprinted by permission from *Nature* 256:708. Copyright © 1975 by Macmillan Magazines, Ltd.)

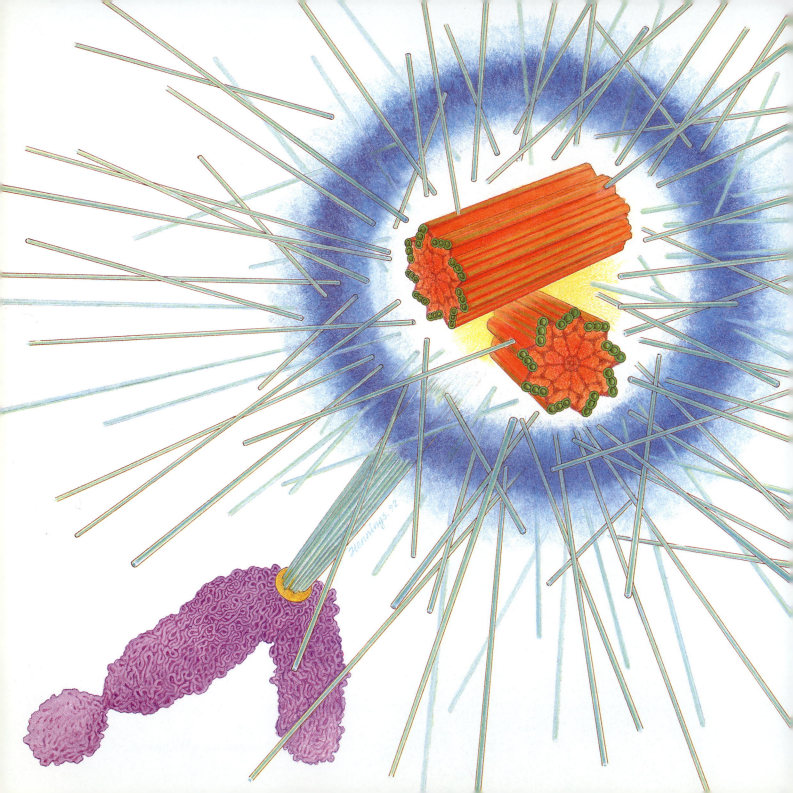

The mechanisms of mitosis divide the replicated chromosomes, and cytoplasmic division (*cytokinesis*) divides the cytoplasm. The major steps and outcome of these processes have been understood in overall terms since the late 1800s. However, mitosis and cytokinesis include many details that have not been explained completely at the molecular level, including mechanisms as basic to cell division as the movement of chromosomes by the spindle.

The realization that the cell cycle is fundamentally important to development and differentiation has renewed interest in nuclear and cytoplasmic division. Interest has also been spurred by research into the genetics of the cell cycle, which has revealed that many mutations capable of interrupting the cycle affect steps during mitotic and cytoplasmic division. Attention has further come from a renewed appreciation of the many fundamental cellular systems that come into play during mitosis and cytokinesis, including the microtubule- and microfilament-based motile systems that figure so prominently in division of the nucleus and cytoplasm.

Current interest in mitosis and cytokinesis centers in several areas, which are the subjects of this chapter. These include the factors folding chromatin into chromosomes for division, the structure of chromosomes during mitosis, the role of centrioles and the cell center in mitosis, the mechanisms moving the chromosomes during anaphase, breakdown and reformation of the nuclear envelope and the nucleolus, the relationships of microtubules and microfilaments to cytoplasmic division, and the division of mitochondria and chloroplasts. In some of these areas, recent research has provided answers to questions posed for nearly a century. In others the present state of knowledge amounts to not much more than the conclusions reached by classical cytologists more than a hundred years ago.

AN OVERVIEW OF MITOSIS AND CYTOKINESIS

During mitosis the fully replicated DNA molecules of the nucleus, with their associated histone and nonhis-tone proteins, condense into thick, rodlike structures, the *chromosomes*. The chromosomes appear longitudinally double as they condense, as a result of DNA replication and duplication of chromosomal proteins during the previous interphase (see Chapter 23 for details). The two duplicate parts of each chromosome, called *sister chromatids*, are normally exact copies of each other and contain exactly the same genetic information. Mitosis separates sister chromatids and delivers them to opposite ends of the dividing cell, where they become enclosed in separate daughter nuclei.

The Stages of Mitosis

Although an essentially continuous process, mitosis is traditionally separated into four stages for convenience in study. The first stage, *prophase* (from *pro-* = before; Figs. 24-1a and 24-2a), begins as the long chromatin fibers of the chromosomes start to fold and pack into thicker structures. By the end of prophase the folding process, called *condensation*, has packed the chromosomes into short, relatively thick double rods (Figs. 24-1b and 24-2b).

While chromosome condensation is in progress, the nucleolus becomes smaller and eventually disappears completely in most species. Disappearance of the nucleolus reflects a gradual reduction in RNA transcription of all kinds, which terminates completely as prophase draws to a close. Protein synthesis and vesicle movement between the ER, Golgi complex, and plasma membrane also drop to minimum levels in most cells by the end of prophase.

During this period the *spindle* (see Figs. 24-18 and 24-21) assembles in the cytoplasm. In the assembly, spindle microtubules align in roughly parallel arrays extending from one end of the cell to the other. This orientation sets up two ends or *poles* in the cell. The microtubules of the spindle, with their associated motile proteins or "motors" (see pp. 290 and 297), develop forces later in mitosis that separate sister chromatids and deliver them to opposite poles.

In most higher eukaryotes, prophase ends and the next stage, *metaphase*, begins (from *meta-* = between; Figs. 24-1b and c and 24-2b and c) when the nuclear envelope breaks down. Separations appear in the nuclear envelope that convert the previously continuous membranes into vesicles. These vesicles drift into the surrounding cytoplasm, where they lose their pore complexes and become morphologically indistinguishable from cisternae of the rough ER. The spindle then moves into the region formerly occupied by the nucleus, where each chromosome attaches to spindle microtubules and moves to the spindle midpoint.

Much of the precision of mitosis depends on the pattern by which the chromosomes attach to spindle microtubules. Each of the two chromatids of a chromosome makes attachments to microtubules leading

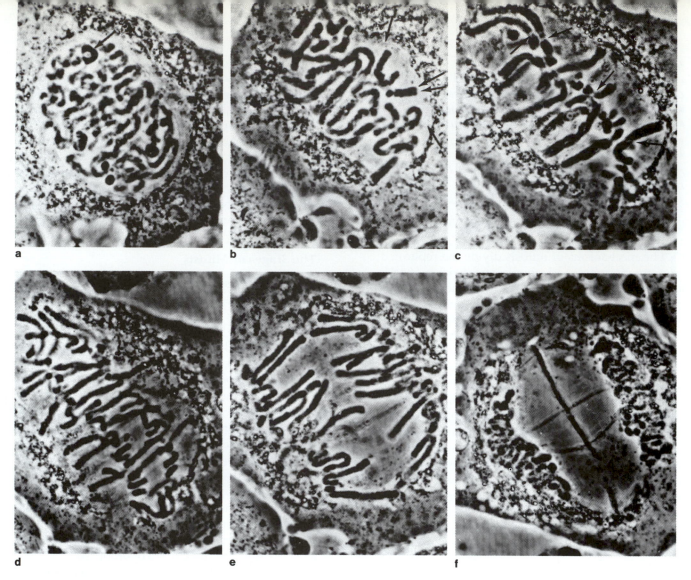

Figure 24-1 Mitosis in the blood lily *Haemanthus*, photographed in living tissue by phase contrast light microscopy. **(a)** Prophase. The chromosomes have condensed into visible structures. The nucleolus (arrow) is still present. **(b)** Early metaphase. The nuclear envelope has broken down, but fragments (arrows) are still present at some locations. Division of the chromosomes into chromatids can be distinguished at some points (double arrow). **(c)** Late metaphase. The kinetochores of each chromosome have attached to the spindle and lie in a plane at the spindle midpoint (the arrows indicate the kinetochore regions of several chromosomes). The chromosome arms extend from this plane in random directions. **(d)** Early anaphase. The two chromatids of each chromosome have separated and moved part of the distance to opposite spindle poles. **(e)** Late anaphase. The kinetochore regions of each chromatid have reached the poles. The arms of many chromosomes trail into the midregion of the spindle. **(f)** Telophase. The chromosomes at the poles have decondensed and are no longer individually visible. × 700. (From *Cinematography in Cell Biology*, 1963. Courtesy of M. Bajer and A. S. Bajer and Academic Press, Inc.)

to *opposite* spindle poles. This opposite attachment ensures that sister chromatids are pulled to opposite poles when the spindle develops its motile forces at the next stage of mitosis.

The third stage, *anaphase* (from *ana-* = apart or back; Figs. 24-1d and e and 24-2d and e) begins as the chromatids start their movement to opposite spindle poles. The anaphase movement, which is not com-

pletely understood, involves an interplay of microtubules, microtubule motors of the dynein and kinesin families (see pp. 290 and 297), and adjustments in the equilibrium between tubulin subunits and assembled microtubules.

As the chromatids reach the poles, anaphase ends and the fourth and final stage of mitosis, *telophase*, begins (from *telo-* = end; Figs. 24-1f and 24-2f). During

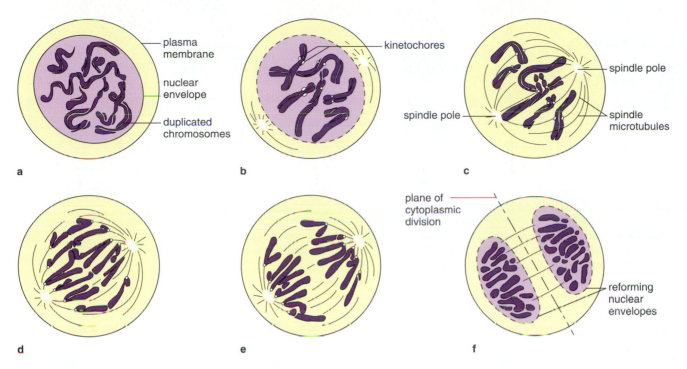

Figure 24-2 The stages in Figure 24-1 in diagrammatic form.

telophase the chromatids decondense and return to the extended interphase form. As decondensation proceeds, vesicles apparently derived from the ER surround the chromatid masses. These vesicles gradually extend and join until the chromatids, by now dispersed into a state resembling interphase chromatin, are surrounded by a continuous nuclear envelope. By this time the chromatids are considered to be the chromosomes of the next cell cycle.[1] As the telophase nucleus takes shape, pore complexes differentiate in the nuclear envelope, RNA transcription resumes, and the nucleolus reappears. Protein synthesis increases in rate as telophase nears completion, and vesicle movement between the ER, Golgi complex, and plasma membrane commences. Each daughter nucleus is now in the G1 stage of the next interphase.

Most of the spindle disappears during telophase except for a layer of short microtubules that persists at the former spindle midpoint. This microtubule layer takes part in cytoplasmic division (see below).

The result of DNA replication during interphase and the subsequent mitosis is production of two daughter nuclei, each with genetic information equivalent to the parent nucleus. Because the chromosomes attach individually to the spindle, the process works equally well with haploid, diploid, or polyploid cells. The total time for completion of mitosis varies in dif-

ferent cell types and species; in most organisms, passage through the four stages requires from one to four hours.

The pattern of mitosis described in this overview applies to most higher plants and animals. Many variations occur in protists, fungi, and lower plants. However, few of these variations result in significant differences in the outcome of the process.

Cytokinesis

In most cells, division of the cytoplasm begins during anaphase. In animal cells, protists, and many fungi and algae the first indication of cytokinesis is the appearance of a *furrow* in the plasma membrane. The furrow, which extends completely around the dividing cell, gradually deepens along a plane through the former spindle midpoint (Fig. 24-3 and 24-30), marked by the short lengths of microtubules persisting in this region. The furrow continues to deepen until the daughter nuclei are cut off in separate cells, each with its own cytoplasm and a complete plasma membrane. The force cutting the cytoplasm in two is generated by contraction of a band of microfilaments encircling the cell under the advancing furrow (see Fig. 24-31).

Cytokinesis in higher plants also follows a plane marked by the layer of microtubules persisting at the former spindle midpoint (see Figs. 24-4, 24-32, and 24-33). Rather than being cut by a gradually deepening

[1]Some authorities consider that chromatids are converted to chromosomes at the instant sister chromatids are separated at anaphase.

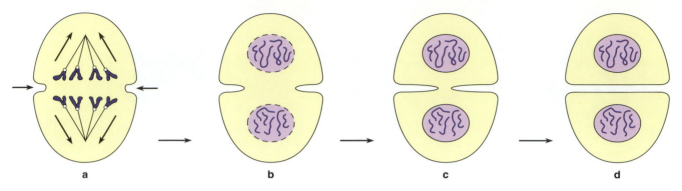

Figure 24-3 The furrowing mechanism that divides the cytoplasm in animal cells. The furrow begins as an indentation running completely around the cell in the plane of the former spindle midpoint (arrows in **a**). It deepens progressively (**b** and **c**) until the daughter nuclei are enclosed in separate cells (**d**). The force for the division is generated by a band of microfilaments that extends around the cell at the apex of the furrow.

Figure 24-4 Cytoplasmic division in higher plants, which begins as a layer of vesicles collects in the plant of the former spindle midpoint (**a**). The vesicles, which contain precursors of cell wall molecules, fuse (**b**) to produce a new cell wall and plasma membranes separating the daughter cells.

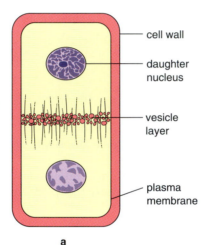

cell wall

daughter nucleus

vesicle layer

plasma membrane

a

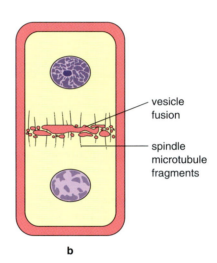

vesicle fusion

spindle microtubule fragments

b

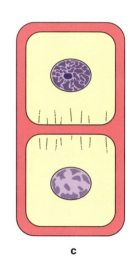

c

furrow, the microtubule layer is invaded by membrane-bound vesicles originating from the Golgi complex. The vesicles, which contain cell wall material, gradually fuse until a continuous wall, lined on either side by plasma membranes, encloses the daughter nuclei in separate cells.

THE EVENTS OF PROPHASE

Chromosome Condensation

Condensation packs the duplicated chromosomes from their interphase state, in which they may be as much as several centimeters in length, into compact rods a few microns long. Chromosomes continue condensing throughout prophase and reach their most compact state by the end of metaphase, when they are short enough to be divided by the spindle without tangling or breaking.

Condensation appears to be initiated when a cyclin B/CDK2 complex (also called *MPF*; see p. 641 and below) is fully activated at the G2–prophase transition. The activated CDK2 protein kinase phosphorylates a number of proteins critical to entry into mitosis, including histone H1. This phosphorylation coincides with chromosome condensation and may be a step that promotes H1-H1 interactions that lead to folding and packing of chromatin fibers.

Condensation apparently proceeds without major alterations in the molecular structure of individual chromatin fibers. Many observations of condensed chromosomes have demonstrated that nucleosomes are retained. Further, DNA is released in the usual 200 and 140 base-pair lengths characteristic of nucleosomes (see p. 377) if condensed chromosomes are digested with DNA endonucleases. Although the folding of individual chromatin fibers within the condensed chromosomes appears irregular, it is probably ordered and precisely determined by the condensa-

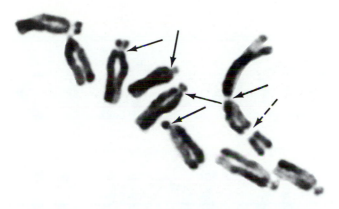

Figure 24-5 Fully condensed chromosomes of the plant *Vicia*. Solid arrows indicate primary constrictions, the site at which the chromosomes connect to spindle microtubules; the dashed arrow indicates the secondary constriction containing the nuclear organizer. × 1700. (Courtesy of S. Wolff.)

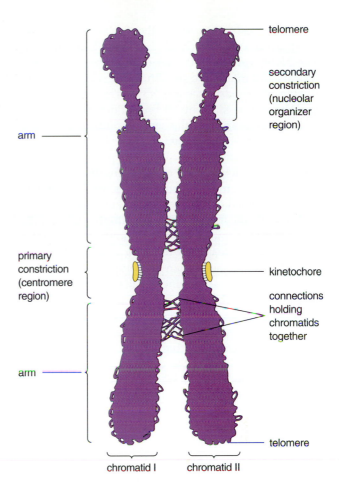

Figure 24-6 Major structures and regions of a fully condensed chromosome. The chromosome is divided into two identical units, the sister chromatids, which are the result of DNA replication and the duplication of chromosomal proteins during the previous interphase. Each chromatid consists of a single, linear DNA molecule with its associated histone and nonhistone proteins. A narrowed region on each chromatid, the primary constriction or centromere, carries the kinetochore, the structure that links to spindle microtubules during mitosis. On either side of the primary constriction are the arms of the chromosome; each arm terminates at its tip in a telomere. In at least one chromosome or chromosome pair of the set a secondary constriction marks the nuclear organizer site (NOR), a region containing many repeats of rRNA genes. The nucleolus reforms at this site at the close of mitosis.

tion mechanism. Otherwise, condensation would not be expected to produce the characteristic sizes and shapes observed for individual chromosomes, which are repeated from cell to cell within a species (see below).

One question that intrigued cell biologists for many years is how the very long chromatin fibers of chromosomes are able to condense from their extended interphase state without tangling. It now appears that *DNA topoisomerase II*, the enzyme that can pass one DNA helix entirely through another (see p. 373), relieves any interchromosome tangles occurring during condensation. T. Uemara and his colleagues found in the yeast *Schizosaccharomyces pombe* that DNA topoisomerase II activity is absolutely required for condensation to proceed normally; tangles do indeed occur and persist, leading to failure of chromatids to separate normally at metaphase, if this topoisomerase is inactive.

Structure of Fully Condensed Chromosomes

Overall Chromosome Structure Fully condensed chromosomes show several characteristic features in most eukaryotes (Figs. 24-5 and 24-6). At some point along a condensed chromosome, both chromatids show a differentiated region collectively called the *centromere* or *primary constriction*. This region, which is usually smaller in diameter than other parts of the chromosome, contains the *kinetochores*, sites at which spindle microtubules attach to sister chromatids during metaphase. A kinetochore appears in most species as a disc- or ball-like structure on the surface of a chromatid (see Figs. 24-11 and 24-12). Each chromatid has its own kinetochore, so that there are two kinetochores per chromosome in the centromere region.

The position of the centromere, which is fixed for a given chromosome of a set, may be in the approximate middle of the chromosome or at a point closer to either end. The chromatin on either side of the centromere forms the *arms* of the chromosome; at the tip of each arm is the *telomere*.

Other structures are frequently visible on one or more of the fully condensed chromosomes of a species. Usually at least one chromosome or chromosome pair of a set has a narrowed region in addition to the primary constriction called the *secondary constriction* or *nucleolar organizer* (*NOR*; see Figs. 24-5 and 24-6). The NOR contains a cluster of large pre-rRNA gene repeats and marks the site where the nucleolus forms during interphase (see pp. 439–442). The nucleolus persists at this site until it breaks down at late prophase, reappearing at the same location during telophase and the subsequent G1.

The patterns and locations of primary and secondary constrictions, the relative lengths of the arms, and the structure of the telomeres are constant for each condensed chromosome in a eukaryotic species. Collectively the morphology and number of the entire set of chromosomes form what is known as the *karyotype* of the species (shown for *Vicia faba* in Fig. 24-5; Fig. 22-2 shows the human karyotype). In many cases the karyotype is so distinctive that a species can be identified from this characteristic alone.

Although karyotypes are often highly distinctive, some of the chromosomes within a karyotype are so similar in some species that they cannot be separately distinguished. This is the case in the human karyotype shown in Figure 22-2. This problem is circumvented in mammals and other higher vertebrates by staining techniques that produce a unique pattern of crossbands in each chromosome (Figs. 24-7 and 24-8). The crossbands, which reflect underlying differences in molecular structure among the chromosomes, have made it possible to identify unambiguously each condensed chromosome of humans and many other animal species. In turn, this unambiguous identification has allowed mutant genes to be identified with individual chromosomes of the set, and the chromosome rearrangements taking place in association with disorders such as cancer to be traced (see p. 649).

A number of investigators have proposed that ordered chromatin folding during condensation depends on an arrangement of chromatin fibers into loops radiating from a central core or *scaffold* that runs the length of each chromosome. The primary evidence for this hypothesis comes from studies of condensed chromosomes by U. K. Laemmli and his coworkers. These investigators treated chromosomes with heparin and dextran sulfate, negatively charged molecules that extract most of the proteins from chromatin. After the extraction the position of each chromosome is marked by a residual scaffold (Fig. 24-9). Surrounding the scaffold are whorls of DNA fibers, which can be traced as loops extending from the scaffold. The individual loops contain from 50,000 to 150,000 base pairs of DNA. Among the proteins identified in scaffolds by Laemmli and his coworkers is DNA topoisomerase II, the enzyme implicated in removing tangles in condensed chromosomes.

Scaffolds remain somewhat controversial because, with one exception, they have been seen only in chro-

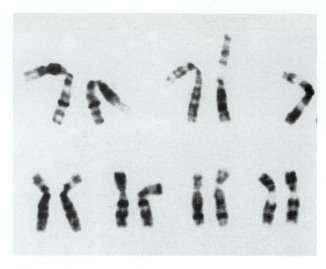

Figure 24-7 The crossbands produced in condensed human chromosomes by G banding (G = giemsa, the stain used in the technique). In the technique, chromosomes are exposed to a warm solution of salts (0.15 *M* NaCl and 0.015 *M* sodium citrate) or the proteinase trypsin, followed by giemsa staining. The giemsa stain contains a mixture of acidic and basic dyes, including azure A, azure B, and eosin. Although the molecular basis for the G-banding technique is not understood, the staining technique evidently denatures proteins in regions containing repetitive sequences (see p. 532), causing these regions to aggregate into bands that take up the giemsa stain. (Courtesy of W. Schnedl. © Academic Press, Inc., from *Int. Rev. Cytol.*, Suppl. 4:237 [1974].)

mosomes prepared for examination by the Laemmli technique. The exception is a preparation made by T. Hirano and T. J. Mitchison, in which metaphase chromosomes were reacted with an antibody that combines specifically with a group of proteins present during mitosis. The antibody, tagged with a fluorescent dye, bound along the central axis of metaphase chromosomes, revealing a central scaffoldlike structure similar to those in Laemmli's preparations. The scaffolds seen in these preparations may reflect an underlying organization of the chromatin fibers of condensed chromosomes as loops extending from a central axis (Fig. 24-10). The loops may represent *domains* of the genome, constituting segments of genetic material that operate as a unit in replication and other functions, often proposed as part of eukaryotic chromosome structure (see also p. 382). During condensation a protein spaced at sites between the domains may act as a "folder" by setting up linkages that pull the domains together as loops (see Fig. 24-10). These proteins, arranged along the central axis of fully condensed chromosomes, may be among the proteins precipitated into scaffolds by the Laemmli technique.

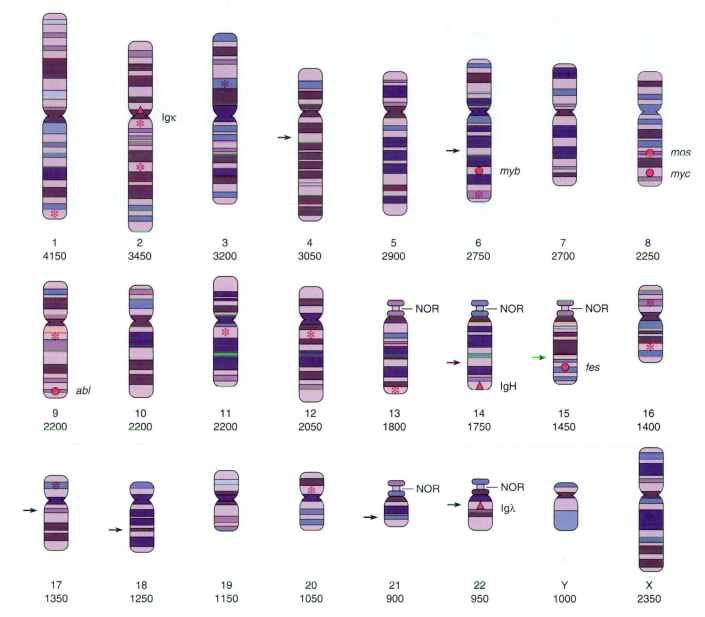

Figure 24-8 The pattern of G bands in human chromosomes, showing sites occupied by antibody genes (triangles), several oncogenes (dots), and sites prone to breakage. *abl*, *fes*, *mos*, *myb*, and *myc*, oncogenes; I$_g$H, antibody heavy-chain genes; Igκ, kappa light-chain genes; Igλ, lambda light-chain genes. Sites where translocations frequently occur are marked with an arrow. The numbers give the gene content estimated for each chromosome. (Modified from an original courtesy of J. J. Lunis, from *Science* 221:227 [1983]. Copyright 1983 by the American Association for the Advancement of Science.)

Kinetochores and Centromeres Kinetochores appear disclike in many groups, including mammals, some insects and protozoa, slime molds, and green algae (Fig. 24-11). A disclike kinetochore is usually structured from three more layers—an outermost dense layer; a middle, less dense and faintly fibrous zone; and an innermost, coarsely granular layer closely applied to the underlying chromatin. In some species the outermost surface of the kinetochore is coated with fibers that extend outward from its surface as the *corona*. Kinetochores appear as ball-like, uniformly dense masses in higher plants (Fig. 24-12).

Relatively little is known about the molecular constituents of kinetochores. The best evidence showing that proteins form part of kinetochore structure in higher eukaryotes comes from the use of antibodies

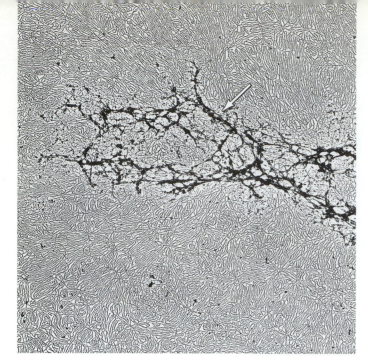

Figure 24-9 The residual "scaffold" (arrow) left after treatment of a human metaphase chromosome to remove chromosomal proteins. The DNA, which extends in loops from the scaffold framework, can be seen as the whorls of lines in the background. × 14,000. (Courtesy of U. K. Laemmli, from *Cell* 12:817 [1977]. Copyright Massachusetts Institute of Technology.)

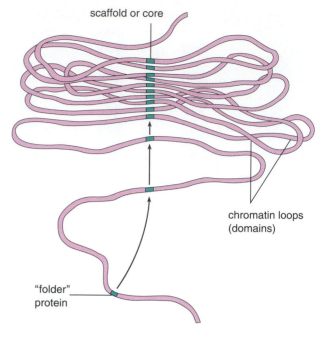

scaffold or core

chromatin loops
(domains)

"folder"
protein

Figure 24-10 Possible arrangement of chromatin fibers in condensed chromosomes, in which the fibers are held as loops radiating from a central core. The core is formed by the association of "folder" molecules along the central axis of a chromatid.

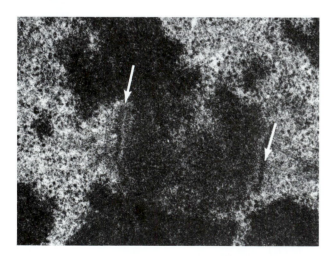

Figure 24-11 The two kinetochores (arrows) at the primary constriction of a Chinese hamster metaphase chromosome. Three layers are clearly visible in the kinetochore on the left. Because of the thinness of the section, only short segments of the chromosome arms on either side of the primary constriction are included in the micrograph. Kinetochores of this type range from 0.4 to 0.9 μm in diameter in different species (about 2.2 μm in mammals) and from 50 nanometers to as much as 0.1 μm in total thickness. × 49,000. (Courtesy of L. I. Journey.)

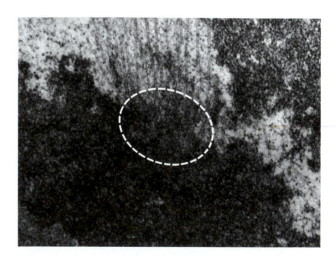

Figure 24-12 A ball-like kinetochore (outlined) on a chromatid of the plant *Haemanthus*. × 29,000. (Courtesy of A. Bajer and Springer-Verlag.)

produced in patients with *scleroderma*, an autoimmune disease in which antibodies are developed against several cell structures including centromeres and kinetochores. The antikinetochore antibodies react with a group of proteins that seems to be common to the kinetochores of many higher organisms, including those of both animals and plants (Table 24-1).

Kinetochores also react positively with stains for RNA. The RNA stain is reduced or eliminated by techniques removing RNA, such as ribonuclease digestion or cold perchloric acid extraction. The positive RNA stain suggests that some of the proteins associated with kinetochores may be ribonucleoproteins. If this is indeed the case, the function of the RNA component is unknown.

During the initial steps of metaphase, kinetochores attach to the spindle by "capturing" the ends of existing spindle microtubules. The microtubules attached by a kinetochore collectively form a *spindle fiber*, which in most higher eukaryotes contains from about 15 to 35 microtubules. Recent research into the anaphase movement of chromosomes indicates that kinetochores, besides serving as attachment sites for spindle microtubules, take an active part in generation of the force for chromosome movement (see below).

The centromere region bearing the kinetochores appears to contain a specialized group of DNA sequences. These sequences may serve as recognition sites for the proteins forming the kinetochores. In the yeast *Saccharomyces cerevisiae*, L. Clarke and J. A. Carbon and others isolated centromere DNA by identifying the minimum sequences capable of functioning as centromeres—that is, as sequences allowing DNA molecules containing them to attach and be separated correctly by the spindle (see below). Sequencing revealed that centromeres of this species contain from about 120 to 140 base pairs of DNA and include three characteristic sequence elements (Fig. 24-13a). Mitchison and his coworkers identified a complex of three proteins, the *CBF3* complex, that recognizes and binds the *S. cerevisiae* centromere sequences. The complex can link both to DNA and to microtubules and includes a microtubule motor with dyneinlike motile properties—that is, a motor that can produce movement toward the minus ends of microtubules.

The degree to which the particular group of centromere sequences of *S. cerevisiae* are related to those of other organisms is uncertain. Even the related yeast *Schizosaccharomyces pombe* has radically different centromere DNA (Fig. 24-13b), and the *S. cerevisiae* centromere sequences are not recognized by the human kinetochore proteins identified by scleroderma antibodies. Further, the *S. cerevisiae* sequences have no ability to induce microtubule attachment when introduced into the cells of higher eukaryotes or even into *S. pombe*. These differences suggest that each species

Table 24-1   Proteins Identified in Centromeres and Kinetochores

Protein	Molecular Weight	Tentative Location
CENP-A	18,000	In chromatin under kinetochore; probably a centromere-specific histone H3 variant
CENP-B	80,000	In centromere chromatin; binds to DNA sequences in centromere
CENP-C	140,000	In kinetochore; GTP-binding protein
CENP-D	50,000	In kinetochore; kinesinlike

Adapted from H. F. Willard, *Trends Genet.* 6:410 (1990).

or higher taxonomic group may have distinctive centromere DNA sequences.

In most higher eukaryotes, long tracts of repetitive sequences occur on either side of the centromere region (see p. 532 for details). These flanking repetitive sequences range from simple to complex in different species and vary in content between different chromosomes of a set and, in some cases, even between different individuals of the same species. In some species the flanking repeated sequences occur in other regions of the chromosomes as well, frequently in regions just inside the telomeres (see below). Although generally known as *centromere DNA*, or *C-DNA*, these flanking sequences probably have little or no relationship to the function of the centromere in binding kinetochore proteins. They may simply be functionless "filler" DNA that flanks centromeres, or they may contribute to the elements holding sister chromatids together until anaphase (see below).

Telomeres   Specialized short, repeated sequences occur in telomeres in all eukaryotic species (see p. 529). E. W. Blackburn, C. W. Greider, and J. W. Szostak showed that these sequences are recognized by *telomerases*, a group of RNA-containing enzymes that solve a problem inherent in replicating the end of the leading template chain (see p. 674). Besides taking part in DNA replication, telomeres are essential for the functional and biochemical stability of chromosomes. Chromosomes lacking telomeres are highly sensitive to enzymatic breakdown and readily interact chemically with other cellular structures including other chromosomes. Frequently, as a result of such interactions, chromosomes fuse into multiple structures that cannot be distributed correctly by the spindle. The stabilizing effects of telomeres appear to depend on "capping" of DNA molecules by the telomere repeats.

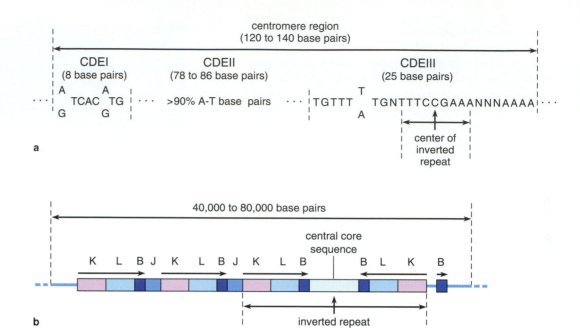

Figure 24-13 Centromere DNA sequence structure in *Saccharomyces cerevisiae* and *Schizosaccharomyces pombe*. **(a)** *S. cerevisiae* centromere DNA consists of three sequences, CDEI, CDEII, and CDEIII. CDEI is an eight-nucleotide sequence (Pu indicates a purine, either A or G). Although conserved in *S. cerevisiae* centromere DNA, elimination of most or all of this sequence has little effect on attachment to the spindle but disturbs separation of the chromatids at anaphase. The CDEII sequence is nonconserved except for a preponderance of A-T base pairs, which make up more than 90% of this element. Experimental changes in CDEII have little effect as long as the A-T content is not reduced significantly below 90%. However, changes in CDEIII amounting to as little as one base substitution, especially in the CCG element at the center of the inverted repeat, can completely eliminate the ability of chromosomes to attach to the spindle. **(b)** Sequence structure of the centromere region in *S. pombe*. The elements B, J, K, and L are different types of repeated sequences. Essentially the entire segment shown is necessary for full centromere activity. **(a)** and **(b)** are not drawn to relative scale.

Centromeres, Telomeres, Replication Origins, and Artificial Chromosomes The basic elements necessary for DNA molecules to function stably as chromosomes in cell division have been nicely worked out by a series of recombinant DNA experiments with the plasmids (see p. 420) of yeast cells. Plasmids are independent DNA circles that do not normally connect to the spindle. Instead, they are simply distributed randomly to daughter cells during cytoplasmic division. Szostak, Blackburn, and A. W. Murray found that inclusion of an *S. cerevisiae* centromere sequence in plasmids allowed them to connect to the spindle and be distributed with the chromosomes during mitosis. If the plasmid circles were opened, creating linear DNA molecules unprotected by telomere sequences at their tips, the circles became unstable and were lost. However, addition of yeast telomere sequences to the opened tips made the linear plasmids stable and able to maintain their capacity for normal distribution during mitosis.

Murray and Szostak found that addition of a yeast replication origin (an *ARS* element; see p. 681) to a linear plasmid sequence containing a centromere and telomeres completed their cell cycle functions. These *artificial chromosomes* replicated during interphase and were distributed correctly by the spindle to daughter cells during mitosis in *S. cerevisiae*. These experiments defined the minimum elements necessary for replication and distribution of linear genetic information to daughter cells via mitosis: a DNA molecule with a replication origin, a centromere, and telomeres at both ends. Besides demonstrating the minimum structural elements necessary for a linear DNA molecule to act as a chromosome, the artificial chromosomes have proved to be highly useful as a means of cloning DNA in large quantities in yeast

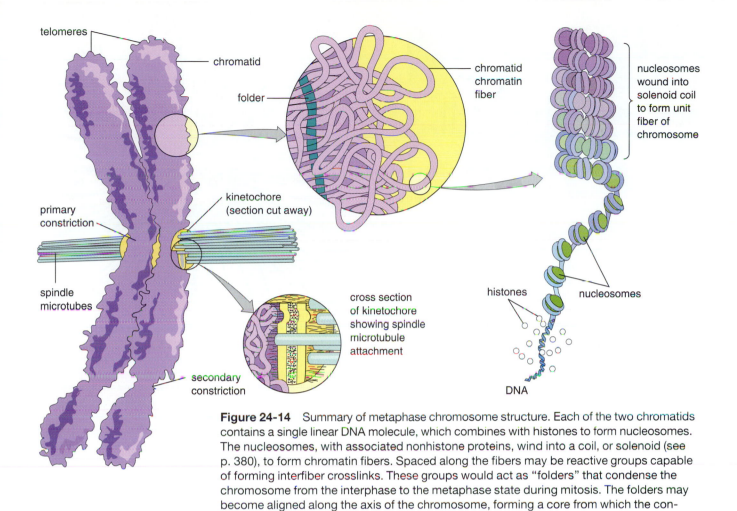

Figure 24-14 Summary of metaphase chromosome structure. Each of the two chromatids contains a single linear DNA molecule, which combines with histones to form nucleosomes. The nucleosomes, with associated nonhistone proteins, wind into a coil, or solenoid (see p. 380), to form chromatin fibers. Spaced along the fibers may be reactive groups capable of forming interfiber crosslinks. These groups would act as "folders" that condense the chromosome from the interphase to the metaphase state during mitosis. The folders may become aligned along the axis of the chromosome, forming a core from which the condensed chromatin fibers extend as looped domains. The final folding pattern produces the features recognizable in fully condensed chromosomes: a primary constriction (centromere), arms of characteristic length and thickness, and telomeres. Secondary constrictions associated with the nucleolar organizer region may be located on one or more chromosomes or pairs of the set. Two kinetochores, one for each chromatid of the chromosome, form the attachment sites for spindle microtubules within the primary constriction.

cells for sequencing or as a vehicle to introduce genes for genetic engineering.

The condensed chromosomes produced by the folding of chromatin fibers are characterized by several essential structural and functional features. Each is divided lengthwise into two sister chromatids, which are the products of DNA replication and the duplication of chromosomal proteins during the S phase preceding division. At some point the chromosome contains a centromere region that carries the kinetochores, one for each sister chromatid. The kinetochores attach the chromosomes to the spindle. The tips of the chromosomes are defined by the telomeres, structures that figure in replication and stabilize the chromosomes against chemical interactions. A centromere, telomeres, and at least one replication origin are the minimum requirements for a DNA molecule to function as a chromosome (Figure 24-14 summarizes the structure of fully condensed chromosomes).

Spindle Formation

In higher eukaryotes the spindle assembles according to one of two basic patterns depending on whether an *aster* is present at each spindle pole during division. Asters develop from the *cell center* (see p. 355), a system of microtubules radiating from a region located

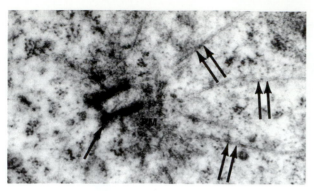

a

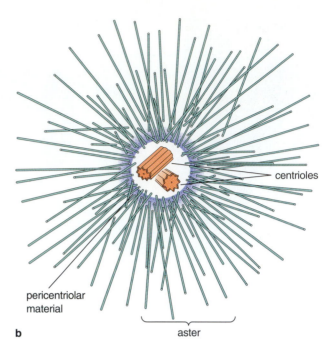

b

Figure 24-15 Centrioles and asters. **(a)** One of a pair of centrioles (single arrow) at the center of an aster from chick spleen. Microtubules (double arrows) of the cell center, or aster, are anchored in dense pericentriolar material (PM) near the centriole but do not make direct contact with the centriole microtubules. × 41,000. (Courtesy of J. André.) **(b)** The arrangement of microtubules, centrioles, and pericentriolar material (in blue) in a cell center, or aster.

In figure 24-15b the labels read: centrioles, pericentriolar material, aster.

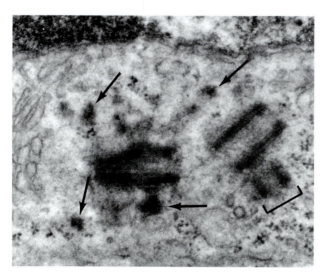

a

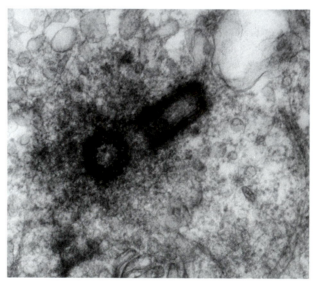

b

Figure 24-16 Centriole duplication. **(a)** Centrioles duplicating in a rat cell. The plane of section has caught one of the procentrioles (bracket) arising as a bud from one end of a parent centriole. Several deposits of pericentriolar material (arrows) are also present. × 60,000. (Courtesy of R. G. Murray, A. S. Murray, and A. Pizzo, from *J. Cell Biol.* 26:601 [1965], by copyright permission of the Rockefeller University Press.) **(b)** The duplicated centrioles and surrounding pericentriolar material in a cultured mammalian cell. × 55,000. (Courtesy of C. L. Reider. Reprinted from *Electron Micr. Rev.* 3:269, copyright 1990, with kind permission from Elsevier Science Ltd., The Boulevard, Langford Lane, Kidlington OX5 IGB, UK.)

near one side of the nucleus (Fig. 24-15). Inside the cell center is a pair of *centrioles* usually arranged at right angles to each other (as in Figs. 24-15b and 24-16b; centrioles are barrel-shaped structures containing nine sets of microtubule triplets arranged in a circle—see p. 295 and Fig. 10-18).

Asters figure in spindle formation in most animal cells and in the cells of some lower plants. At the beginning of S the centrioles within the cell center separate slightly and begin to duplicate. The duplication, which continues throughout S, G2, and into mitotic

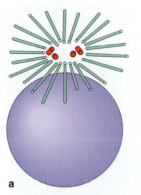

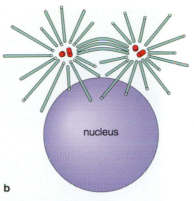

nucleus

a   b

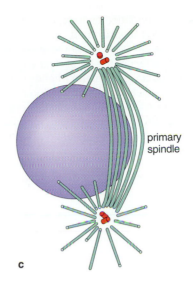

primary spindle

c

Figure 24-17   Formation of astral spindles. **(a)** The original pair of centrioles duplicates, producing two centriole pairs. **(b)** While the centrioles are duplicating, the aster divides into two parts, with each part containing a parent and daughter centriole. Microtubules lengthen between the asters as they separate. **(c)** Separation of the asters continues until they lie at opposite sides of the nucleus. The mass of microtubules extending between them forms the primary spindle. After the nucleus breaks down at the close of prophase, the primary spindle moves into the location formerly occupied by the nucleus. The microtubules and centrioles are not drawn to scale.

prophase in some species, produces two pairs of centrioles. Centrioles duplicate in an interesting pattern in which each produces a small, budlike extension, the *procentriole* (Fig. 24-16a). A procentriole grows at right angles from one end of the parent centriole and gradually increases in length during S, G2, and prophase. By the end of prophase, duplication is usually complete and the centriole copies are indistinguishable from the parents (Fig. 24-16b).

As prophase begins, the cell center separates into two parts, each half containing one centriole of the original pair and one of the new copies (Fig. 24-17a). The two parts, now the asters of the developing spindle, continue to separate until they reach opposite ends of the nucleus (Figs. 24-17b and c). As the asters move apart, microtubules lengthen in the direction of movement and fill the space between them. By late prophase, when the asters are fully separated, the microtubules extending between them form a large mass around one side of the nucleus. This mass is the *primary spindle* (Fig. 24-17c). When the nuclear envelope breaks down at the close of prophase, the spindle moves into the region formerly occupied by the nucleus. This type of spindle is identified as an *astral spindle* because of the asters at its poles (Fig. 24-18a).

In cells without a cell center, no changes associated with spindle formation take place until early prophase. At this time, light microscopy shows a clear space developing in a narrow zone around the nucleus (Fig. 24-19a). The clear zone grows until it takes on a spindle shape and completely surrounds the nucleus (Fig. 24-19b). Electron microscopy reveals that

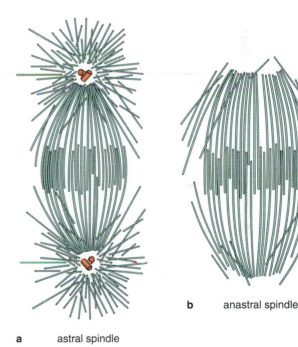

a   astral spindle

b   anastral spindle

Figure 24-18   Astral and anastral spindles. **(a)** An astral spindle, with centrioles and asters at its tips, is typically more pointed at the poles. **(b)** An anastral spindle, in which microtubules are less convergent at the poles, producing a structure that is broader at the tips. The presence or absence of centrioles and asters has no apparent effect on spindle function in division of the chromatids. Note that microtubules originating from the poles overlap at the spindle midpoint in both spindle types.

Figure 24-19 Formation of an anastral spindle in a living cell of the plant *Haemanthus*, as observed under polarized light. The developing spindle appears bright because of the effect of the parallel arrangement of microtubules on polarized light. **(a)** At early stages the spindle appears as a clear zone surrounding the nucleus. **(b)** A later stage in which the spindle is more fully developed and extends toward the poles. N, nucleus. × 700. (Courtesy of S. Inoué and A. Bajer and Springer-Verlag.)

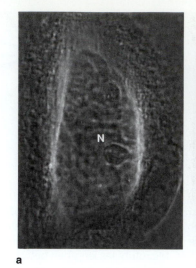

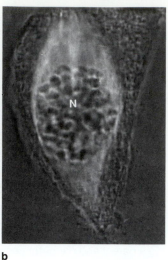

a   b

the clear zone is filled initially with masses of randomly oriented microtubules. As development proceeds, the microtubules orient into more or less parallel arrays that converge at opposite sides of the nucleus and form the poles of the spindle. This *anastral spindle* (see Fig. 24-18b) fills the region formerly occupied by the nucleus when the nuclear envelope breaks down. Spindle formation follows the anastral pathway in higher plants lacking cell centers and centrioles, including all angiosperms and most gymnosperms. Some animal cells, such as certain protozoa and the developing eggs of many higher animals, also lack asters or centrioles. The presence or absence of asters and centrioles has no apparent effect on spindle function in mitosis.

A different pattern of spindle formation occurs in lower eukaryotes including some algal and fungal species. In these organisms the spindle forms between *spindle pole bodies* (*SPBs*; Fig. 24-20a), which function similarly to the asters of higher organisms. SPBs consist of a series of layers of dense plates or plaques separated by less dense layers, usually embedded partially or completely in the membranes of the nuclear envelope. Microtubules extend from the outer dense plaque into the surrounding cytoplasm during interphase (Fig. 24-20b).

During interphase a single SPB is located at one side of the nucleus (24-20c). Late in interphase the SPB divides into two parts, and microtubules elongate from the inner surfaces of the SPBs into the nucleoplasm (Fig. 24-20d). As the SPBs separate, the microtubules extending from the inner plaques elongate and interdigitate (Fig. 24-20e). The two SPBs continue to migrate until they take up final positions at opposite sides of the nucleus. The microtubules extending between the SPBs now form a spindle that stretches across the nuclear interior (Fig. 24-20f). In most species with SPBs the nuclear envelope remains intact

during these developments, so that the spindle is located inside the nucleus. Later in division the nuclear envelope may remain intact, develop perforations in the regions of the poles, or break down completely. In species in which the nuclear envelope remains partially or completely intact, a furrow separates the nucleus into two parts after the chromosomes have been divided by the spindle.

Microtubule motors have been implicated in the movements separating spindle pole bodies and asters. In both *S. cerevisiae* and *S. pombe*, spindle pole bodies duplicate but fail to separate in cells with mutations in several genes that encode kinesinlike proteins; similar effects are noted in higher organisms with mutations in genes encoding kinesinlike proteins. Antibodies against dyneinlike proteins also interfere with separation of the asters in some higher organisms. Thus in different species either kinesinlike or dyneinlike motor proteins, or both types acting together, may separate the poles during spindle formation.

MTOCs and Spindle Formation The microtubules giving rise to astral spindles grow from a region surrounding but not directly connected to the centrioles. In the electron microscope the origins of microtubule growth appear as deposits of dense material from which microtubules radiate. In astral spindles these dense deposits are known as the *pericentriolar material* or *satellite bodies* (see Figs. 24-15 and 24-16).

Several experiments have shown that the pericentriolar material, and not the centrioles themselves, is the *microtubule organizing center* (*MTOC*; see p. 288) from which spindle microtubules grow. R. R. Gould and G. G. Borisy, for example, found that isolated pericentriolar material can act as a center for microtubule assembly if supplied with tubulin under polymerizing conditions. In most isolated cell centers, no microtubules formed directly from the centrioles. When

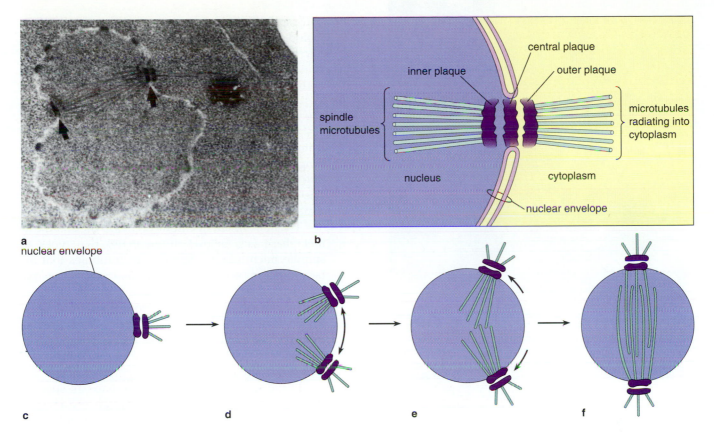

Figure 24-20 Spindle pole bodies (SPBs). **(a)** SPBs (arrows) at opposite sides of a *Saccharomyces cerevisiae* nucleus. Spindle microtubules extend from the SPBs into the nucleus on the nuclear side of the nuclear envelope and from the SPBs into the cytoplasm on the opposite side. × 42,000. (Courtesy of A. E. M. Adams and J. R. Pringle. Reproduced from *J. Cell Biol.* 107:1409 [1988] by copyright permission of the Rockefeller University Press.) **(b)** Major structural elements of an *S. cerevisiae* SPB. **(c through f)** show function of SPBs in generating the spindle in an organism such as *S. cerevisiae*. **(c)** An SPB at interphase. Late in interphase the SPB divides into two parts **(d)**. As the two SPBs move apart, microtubules grow from their inner surfaces into the nucleoplasm. These microtubules elongate and interdigitate as the SPBs continue to separate **(e)**. By the time the SPBs have reached opposite sides of the nucleus, the spindle has fully formed between them **(f)**.

they did form, these microtubules extended outward from the centriole triplets, in a pattern resembling flagellar rather than spindle microtubule formation.

These observations indicate that centrioles are simply passengers during astral spindle formation, and that rather than giving rise to the spindle, as often proposed, they are passively separated and moved to the spindle poles by the dividing asters. In these locations they are incorporated into separate daughter cells at the following mitosis and cytoplasmic division. Thus the developing spindle microtubules probably divide the centriole pairs and place them at opposite ends of the cell, just as the spindle divides and moves the chromosomes later in mitosis. This distribution ensures that daughter cells receive centrioles

and are capable of generating the 9 + 2 system of microtubules of flagella, which seems to be the primary or exclusive function of centrioles (see p. 295).

L. Clayton, C. M. Black, and C. W. Lloyd recently used antibodies to test plant cells for the presence of MTOCs with properties similar to the pericentriolar material of astral spindles. Antibodies were obtained from scleroderma patients, who make autoantibodies against pericentriolar material as well as kinetochores. The antibodies, when linked to a fluorescent dye and reacted with onion root tip cells, stained the zone around the nucleus in which spindle microtubules first appear at early prophase. As mitosis progressed to anaphase, the stain became more concentrated at the poles. Thus microtubule organizing structures

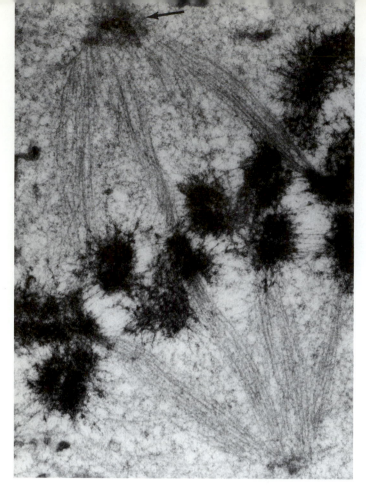

Figure 24-21 The fully developed spindle of a mammalian cell. Only kinetochore microtubules have been caught in the plane of section. One of the centrioles is visible in cross section in the aster at the top of the micrograph (arrow). × 14,000. (Courtesy of C. L. Rieder. Reprinted from *Electron Micr. Rev.* 3:269, copyright 1990, with kind permission from Elsevier Science Ltd., The Boulevard, Langford Lane, Kidlington 0X5 IGB, UK.)

similar to the pericentriolar material of animal cells probably also generate the anastral spindles of higher plants.

The identity of most components of the pericentriolar material remains unknown. However, M. Kirschner and Y. Zheng and their coworkers discovered that a newly described tubulin type, γ-tubulin (see p. 283), is localized in the pericentriolar material. These investigators found that antibodies against γ-tubulin react positively with human, mouse, *Drosophila*, and *Xenopus* pericentriolar material and with the spindle pole bodies of the fungi *Aspergillus* and *Schizosaccharomyces*. Because the antibody reaction is most intense during prophase through metaphase, when spindle microtubule assembly is most active, γ-tubulin is thought to take part in the microtubule-organizing function of the pericentriolar material, possibly by forming initial templates from which microtubules grow. B. A. Palevitz and his coworkers have also detected γ-tubulin in the regions of the nucleus organ-

izing spindles in higher plants, and in other MTOCs including the region at the spindle midpoint in which microtubules persist and proliferate after chromosome division in the plane of cytoplasmic division in both animal and plant cells.

The Fully Organized Spindle Depending on the species, a completed spindle (Fig. 24-21) may contain from tens to many thousands of microtubules. The fungus *Phycomyces*, for example, has only 10 microtubules in its spindle; the rat kangaroo has about 1500. The spindle of *Haemanthus*, a flowering plant, is estimated to contain 10,000 microtubules.

The attachment of chromosomes to the spindle at metaphase (see below) differentiates two types of spindle microtubules. Spindle microtubules that attach to chromosomes are known as *kinetochore microtubules* (*KMTs*). These microtubules extend without breaks or overlaps from a kinetochore to either pole of the spindle. Other microtubules, which extend from the poles but do not make connections to chromosomes or kinetochores, are known as *interpolar microtubules* (*IMTs*).

In the spindles of most and perhaps all species, IMTs originating from one pole overlap and interdigitate in the region of the spindle midpoint with those from the opposite pole (Fig. 24-22a; see also Fig. 22-18). The overlapping arrangement creates two *half spindles*, in which the microtubules, including both KMTs and IMTs, are oriented with their plus ends toward the spindle midpoint. (The plus ends are the ends on which tubulin subunits assemble most rapidly; see p. 284.) Because of this orientation the microtubules in the zone of overlap run in opposite directions, or are *antipolar* (Fig. 24-22b). Both the overlapping arrangement of IMTs and the antipolar arrangement of the two half spindles are significant for the mechanisms generating the anaphase movement (see below).

Fully organized spindles are in a finely balanced state of equilibrium with a pool of unassembled tubulin molecules. Microtubules continually assemble and disassemble within the structure, so that there is a constant turnover of microtubule subunits. Many treatments, including exposure to colchicine, cold, hydrostatic pressure, and ultraviolet radiation, can shift the equilibrium in the direction of unpolymerized subunits and cause almost instantaneous disassembly of the spindle. If the dispersing agent is removed, the spindle reassembles rapidly, often within seconds or minutes. This ready conversion between assembled and disassembled microtubules also figures importantly in the anaphase movement of chromosomes.

Several molecular components can be detected in spindles. J. R. McIntosh and M. P. Sheetz and their colleagues found that antibodies against dynein react

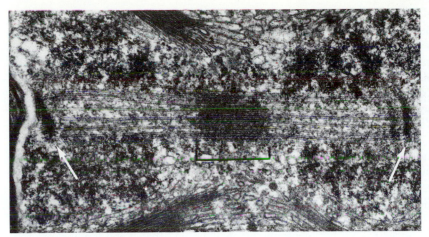

a

Figure 24-22 Overlap of interpolar microtubules at the spindle midpoint. **(a)** Spindle of the diatom *Fragilaria* showing a zone of overlap (bracket) between interpolar microtubules at the spindle midpoint. Spindle pole bodies (arrows) are visible at both poles. (Courtesy of D. H. Tippit, from *J. Cell Biol.* 79:737 [1978], by copyright permission of the Rockefeller University Press.) **(b)** Arrangement of interpolar microtubules in the zone of overlap. Microtubules originating from either pole extend past the spindle midpoint, forming two overlapping half spindles. Note that the IMTs are arranged uniformly with their minus ends at the spindle poles and their plus ends directed toward the midpoint.

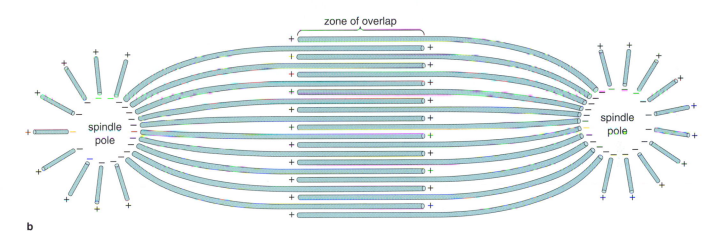

b

in the zone of overlap of IMTs at the spindle midpoint, at the poles, and in or near kinetochores. Kinesin has also been detected in the spindle midpoint and the kinetochores. Creatine phosphate, a substance capable of taking up and releasing energy in the form of phosphate groups, and calmodulin, the control protein activated by Ca^{2+} (see p. 165), are also present. Other molecular components include a Ca^{2+}-dependent ATPase associated with pumping calcium across membranes; actin and myosin; and a variety of microtubule-associated proteins (MAPs; see p. 287). B. M. Alberts and his coworkers, for example, developed antibodies to more than 50 *Drosophila* proteins that bind to microtubules as MAPs. When linked to a fluorescent dye and added to cells, some of the antibodies bound primarily to the asters, others to pericentriolar material or kinetochores, and some to the entire spindle.

Also present in the spindle are many small, ER-derived vesicles, which become most concentrated at the poles. These vesicles are probably involved in regulating Ca^{2+} concentration in the spindle and, by this means, regulating spindle assembly and function (see below). By the end of prophase the spindle is fully

formed, primed and set for the central events of mitosis—the attachment of chromosomes and the anaphase movement.

Breakdown of the Nucleolus and Nuclear Envelope

Nucleolar Breakdown In most organisms the nucleolus becomes progressively smaller during prophase and finally disappears toward the end of this stage. As the nucleolus disappears, rRNA synthesis halts.

Some of the molecules of the disaggregated nucleolus probably diffuse into the nucleoplasm and later, after the nuclear envelope breaks down at the close of prophase, intermix freely with the cytoplasmic solution. Several experiments indicate, however, that a significant portion of the disaggregated nucleolar material collects on the surfaces of the condensing chromosomes and is carried with the chromosomes to daughter nuclei during anaphase. This material, which includes partially processed pre-rRNAs and ribosomal proteins, evidently contributes to reformation of the nucleolus during telophase (see below).

Breakdown of the Nuclear Envelope Disassembly of the nuclear envelope at the close of prophase begins as openings appear in the envelope membranes at scattered sites. The inner and outer membranes fuse at the openings, producing separate flattened, closed vesicles that diffuse into the cytoplasm. For a time, pore complexes persist in the vesicles, but soon these disappear and the remaining sacs become morphologically indistinguishable from the ER.

Disassembly of the nuclear envelope and its reformation during telophase are paralleled by the behavior of the lamins, the cytoskeletal proteins that line the inner membrane of the nuclear envelope (see p. 385). Work with fluorescent antibodies by L. Gerace and his associates showed that just before the nuclear envelope begins to break down at late prophase in mammalian cells, the lamins disassemble. One type, *lamin B* (see p. 385), remains bound to the membranous vesicles released from the nuclear envelope to the cytoplasm; the other two mammalian lamins, *A* and *C*, either enter the cytoplasmic solution or bind to the surfaces of the metaphase chromosomes. The lamins reassemble on the inner surface of the nuclear envelope as it reforms during telophase (see below).

Lamin disassembly appears to be controlled by phosphorylation (see also p. 387). Just before the lamins are released from the disintegrating nuclear envelope in late prophase, their phosphorylation increases by at least four times. The lamins remain in a highly phosphorylated state until telophase, when they are dephosphorylated as the nuclear envelope reassembles.

The protein kinase of the activated cyclin B/CDK2 complex has been shown by M. Peter and his coworkers to be capable of adding phosphate groups to the same sites on the lamins that are phosphorylated during nuclear envelope breakdown in mitotic cells. G. Desev and his colleagues found that active cyclin B/CDK2 can also directly induce the lamins to disassemble when added to isolated nuclei. These findings indicate that activated cyclin B/CDK2 triggers nuclear envelope breakdown by adding phosphate groups to the lamins.

EVENTS DURING METAPHASE

Breakdown of the nuclear envelope leaves the chromosomes scattered throughout the region formerly occupied by the nucleus. During metaphase the chromosomes move actively from these scattered locations to the spindle midpoint. This active movement, called *congression*, depends on connections made by the chromosomes to spindle microtubules. By the time the chromosomes reach the spindle midpoint, the connections are sorted out so that the kinetochores of sister chromatids are correctly attached to microtubules leading to opposite spindle poles.

R. B. Nicklas established that congression depends on spindle microtubules. Nicklas detached chromosomes from the spindle of living cells with a microneedle and moved them to distant locations in the cytoplasm. The detached chromosomes remained motionless for a time and then returned rapidly to the spindle midpoint. Electron micrographs made from cells at various stages in this sequence showed that the return movement did not begin until the chromosomes reattached to spindle microtubules. Nicklas's experiments also showed that if the kinetochore connections are scrambled by disturbing the chromosomes with a microneedle, sister kinetochores reattach correctly to microtubules leading to opposite spindle poles. Both dynein and kinesin, which have been detected by fluorescent antibodies in kinetochores, may coordinate activities moving the chromosomes to the spindle midpoint and maintaining them in this position during metaphase.

THE EVENTS OF ANAPHASE

Separation and movement of sister chromatids to opposite spindle poles depend on several distinct processes. Chromatid separation is an autonomous process that can take place independently of the spindle. The subsequent movement of chromatids to the poles, the anaphase movement proper, is completely dependent on spindle microtubules and their motor proteins. The anaphase movement results from a combination of separate but coordinated movements associated with interpolar and kinetochore microtubules.

Separation of Sister Chromatids

If the spindle is disassembled experimentally by exposing metaphase cells to microtubule "poisons" such as colchicine (see p. 288), chromosomes are left without microtubule attachments at the former spindle midpoint. Under these conditions, sister chromatids still separate more or less in unison at the time at which anaphase would normally begin. The chromatids of chromosomes cut from the spindle by a microneedle also separate at the same time as those of chromosomes remaining attached to the spindle. These observations make it clear that separation of sister chromatids depends on a mechanism that is independent of the spindle or spindle microtubules.

It was once thought that sister chromatids do not separate until replication of centromere DNA, which

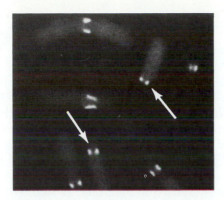

Figure 24-23 Kinetochores in prophase chromosomes reacted with fluorescent antibodies against kinetochore proteins. Each chromosome clearly has two kinetochores (arrows). (Courtesy of B. R. Brinkley and R. P. Zinkowsi. Reproduced from *J. Cell Biol.* 113:1091 [1991], by copyright permission of the Rockefeller University Press.)

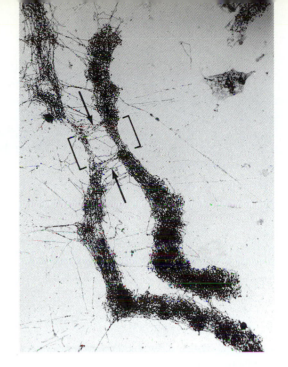

Figure 24-24 Fibrous connections (arrows) that appear to anchor the chromatids together on either side of the centromere region (brackets) of an isolated human chromosome. × 12,000. (Courtesy of J. G. Abuelo and D. E. Moore, from *J. Cell Biol.* 41:73 [1969], by copyright permission of the Rockefeller University Press.)

was supposed to take place just before anaphase begins. The replication was thought to free sister chromatids from each other so that they could move to the poles. However, more recent investigations measuring uptake of DNA precursors such as tritiated thymidine show that no significant fraction of the DNA remains unreplicated after the S phase preceding mitosis. Further, no replication can be detected as the anaphase movement begins. R. M. McCarroll and W. L. Fangman showed directly that in yeast, centromere sequences replicate during S along with the rest of the DNA. Work with fluorescent antibodies (see p. 791) by B. R. Brinkley and his colleagues also shows that the kinetochores are duplicated and clearly double by the G2 stage preceding mitosis. (Figure 24-23 shows double kinetochores on prophase chromosomes.) Thus, delayed centromere replication cannot be responsible for sister chromatid separation.

A number of observations suggest that the connections holding chromatids together before anaphase, rather than involving centromeres, lie just outside the centromere region. At metaphase, early tension developed by spindle microtubules frequently pulls the centromere region apart. The chromatids are still held together, however, at sites above and below the centromere. In the electron microscope the connections can be seen as fibers that extend between the chromatids (Fig. 24-24). These connecting fibers may be tangled chromatin loops, which are digested by DNA topoisomerase II to separate the chromatids at the beginning of anaphase. This possibility is supported by T. Uemara's experiments, in which DNA topoisomerase II was found to be required in *S. pombe* for separation of the chromatids at anaphase as well as for chromosome condensation.

Components of the Anaphase Movement

Once sister chromatids separate, their movement to the poles is clearly dependent on the spindle. The movement stops immediately if the spindle is disrupted by antimicrotubule agents such as colchicine. If the disrupting agent is removed, anaphase movement does not resume until the spindle reassembles. Connections between chromatids and spindle microtubules are also necessary for the movement to proceed. If the connections are experimentally broken, the chromatids remain stationary until they reattach to spindle microtubules.

The total distance moved by chromatids during anaphase has two components. One results from a reduction in the distance between the kinetochores and the poles; the second is produced by an increase in the total length of the spindle, so that the poles become farther apart (Fig. 24-25a). The component of the total anaphase movement caused by a reduction of the kinetochore-to-pole distance is termed *anaphase A*; that caused by lengthening of the entire spindle is *anaphase B* (Fig. 24-25b).

The rate at which both components of the anaphase movement take place, about 1 to 2 μm per minute, is very slow compared to other cellular motions such as cytoplasmic streaming, flagellar beating, or muscle contraction. The anaphase movement, in fact, occurs at approximately the rate at which microtubules lengthen or shorten through the addition or release of microtubule subunits. This observation suggests that, although part or all the anaphase movement may be powered by microtubule motors such as dynein or kinesin, microtubule assembly/disassembly is the rate-limiting process.

Mechanisms of the Anaphase Movement

Many hypotheses have been advanced to explain the anaphase movement. For many years there was little definitive evidence favoring one hypothesis over another. Recent investigations, however, have narrowed the possibilities and clearly support distinct, microtubule-based mechanisms for the anaphase A and B movements.

Anaphase A   Much has been learned about anaphase A from experiments by T. Mitchison, M. W. Kirschner, D. E. Koshland, and G. J. Gorbsky in which kinetochore microtubules were given reference marks that allowed their movement to be traced during anaphase. In one group of these experiments, spindle microtubules were combined with a dye molecule that bleaches when it is exposed to light (Fig. 24-26a). The marked spindles were exposed to a narrow beam of light that bleached a stripe across the kinetochore microtubules perpendicular to the spindle axis (Fig. 24-26b). The bleached segment of the kinetochore microtubules remained in the same position, at the same distance from the pole, as anaphase A progressed (Fig. 24-26c). This finding indicates that kinetochore microtubules

Figure 24-25 The two components of the anaphase movement. **(a)** Measurements by H. Ris showing relative contributions of kinetochore and interpolar microtubules to the anaphase movements in one cell type. 1, distance separating chromosomes from the poles; 2, distance between sister chromatids; 3, total distance between the poles. At 15 minutes, anaphase separation begins (curve 2). During the period from 15 to 30 minutes, as the anaphase movement takes place, the distance between the kinetochores and the poles becomes shorter and the separation between the poles increases. The total distance traveled by the chromatids is the sum of the two components. (Redrawn from "How Cells Divide" by D. Mazia. *Scientific American* 205:100 [1961]. Copyright © 1961 by Scientific American, Inc. All rights reserved.) **(b)** The two components in diagrammatic form. Anaphase A (curve 1 in **a**) is the component of the total movement contributed by the decrease in distance between the kinetochores and the poles, in which the kinetochore microtubules grow shorter. Anaphase B (curve 3 in **a**) is the component of the total movement contributed by lengthening of the entire spindle.

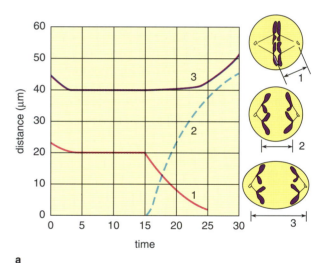

a

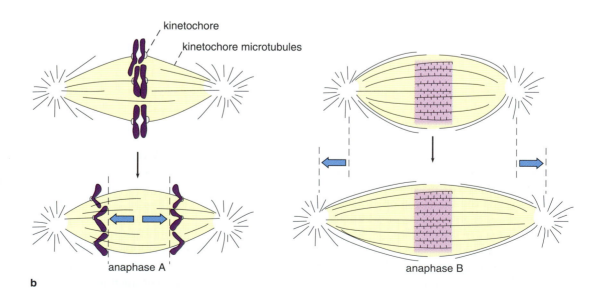

b

remain stationary during anaphase A and that the kinetochores move by sliding along or over these microtubules.

Because kinetochore microtubules become shorter as anaphase A progresses, they must disassemble at the end nearest the chromosome as the kinetochore passes over them. Evidence directly supporting this conclusion was obtained in labeling experiments by Mitchison and Kirschner and their colleagues. During metaphase, labeled tubulin was incorporated at the kinetochore end of the KMTs, indicating that new subunits are continually added at this end before anaphase begins. At anaphase the label disappeared from the KMTs, as expected if microtubules disassemble at the chromosome end as the kinetochores move over them.

The movement of kinetochores is therefore evidently much like a locomotive moving along a railroad track, except that the track is disassembled as the locomotive passes over it. The mechanism generating the motile force is presently undetermined. Both kinesinlike and dyneinlike proteins have been detected in kinetochores, and either or both motors may move chromosomes along kinetochore microtubules. Because the movement is toward the minus ends of kinetochore microtubules, a dyneinlike motor would seem the most likely candidate for this motion (see p. 291; kinetochore microtubules are arranged uniformly with their minus ends toward the poles). However, S. A. Endow and her coworkers found kinesinlike proteins capable of producing movement toward microtubule minus ends, admitting the possibility that these motors, as well as dynein, may be involved in anaphase A. (Figure 24-27 shows one possibility for the action of a microtubule motor in producing kinetochore movement.)

Anaphase B   The accumulated evidence of more than two decades of investigation clearly indicates that anaphase B occurs by active microtubule sliding. The sliding, probably powered by a microtubule motor, occurs between IMTs in the zone of overlap at the spindle midpoint.

The first indications that IMTs slide at the spindle midpoint, a mechanism originally proposed by J. R. McIntosh, D. H. Tippet, and J. D. Pickett-Heaps and their colleagues, were obtained by the laborious method of sectioning entirely through spindles from one pole to the other and tracing the path of each microtubule. The sectioning confirmed that IMTs originating from opposite poles overlap in the spindle midpoint and showed that as anaphase B proceeds, the degree of overlap decreases by an amount equivalent to the increase in distance between the poles.

Further evidence that microtubule sliding is responsible for anaphase B was obtained by H. Masuda,

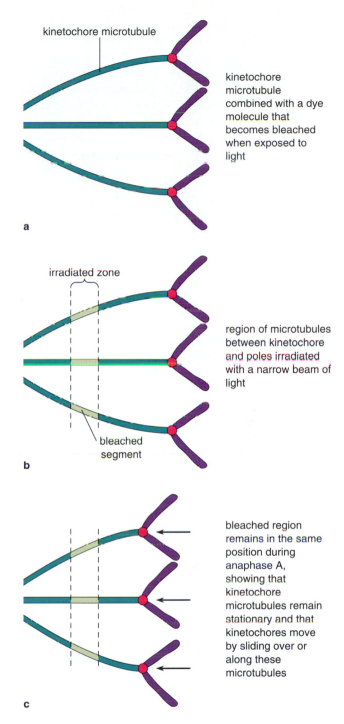

Figure 24-26   An experiment by G. J. Gorbsky and his colleagues demonstrating that kinetochore microtubules remain stationary during anaphase A movement and that kinetochores must move by sliding over or along kinetochore microtubules (see text).

K. L. McDonald, and W. Z. Cande using labeling techniques. These investigators labeled tubulin subunits with biotin, a marker molecule that can be detected in the light microscope by anti-biotin antibodies combined with a fluorescent dye. The labeled subunits

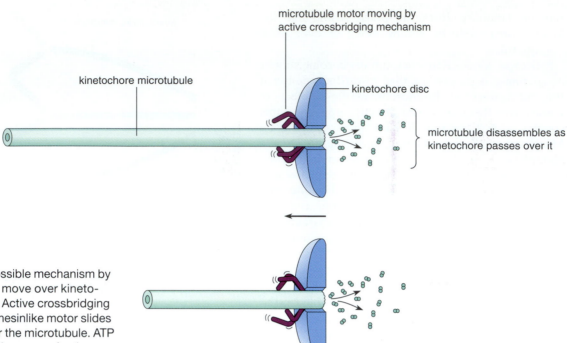

kinetochore microtubule

microtubule motor moving by
active crossbridging mechanism

kinetochore disc

microtubule disassembles as
kinetochore passes over it

Figure 24-27 A possible mechanism by which kinetochores move over kinetochore microtubules. Active crossbridging by a dyneinlike or kinesinlike motor slides the kinetochore over the microtubule. ATP hydrolysis supplies the energy for the crossbridging cycle. As the kinetochore passes, the microtubule disassembles.

were added to isolated spindles under conditions that favored microtubule polymerization. The labeled tubulin subunits were added to the plus ends of microtubules overlapping at the spindle midpoint (Fig. 24-28a and b). The labeled tubulin was then removed from the medium and further polymerization was carried out with unlabeled tubulin (Fig. 24-28c). Although the zone of overlap lengthened, the distance between the poles did not increase as either the labeled or unlabeled tubulin was added, indicating that the force for anaphase B is unlikely to be produced by microtubule assembly.

After incorporation of labeled and unlabeled tubulin, ATP was added to the isolated spindles. The addition caused the poles to move apart; at the same time, the segments labeled by biotin moved over each other in a pattern indicating that progressive sliding occurred in the zone of overlap between the IMTs originating from opposite poles (Figs. 24-28d and e; the Experimental Process essay by Cande on p. 714 gives details of the experiment).

It remains unclear whether a dyneinlike or kinesinlike microtubule motor, or both motors working in coordination, power anaphase B. Vanadate, an inhibitor of flagellar dynein, and antidynein antibodies both stop anaphase B. Anti-dynein antibodies also react with the spindle, especially in the zone of overlap. However, kinesin can also be detected in the spindle; Cande and his coworkers have found that an inhibitor of kinesin, AMPPNP, also stops anaphase B move-

ment. In general, Cande and his colleagues find that the anaphase B motor seems more kinesinlike than dyneinlike in isolated spindles under conditions that support spindle elongation (see Cande's essay).

One problem with dynein as an anaphase B motor is that this motor typically produces movement toward the minus ends of microtubules, opposite to the direction of anaphase B sliding. However, a few examples of dyneinlike motors producing movement toward microtubule plus ends have been observed, and it is possible that one of these unusual dyneins is involved in anaphase B.

From time to time the idea has been advanced that microfilaments participate in the anaphase movement. This hypothesis is based primarily on the observation that actin microfilaments and myosin can be identified in the spindle, with the microfilaments near kinetochores in orientations that make it possible that they participate in anaphase. Spindles in some species also stain strongly with fluorescent antibodies developed against either actin or myosin, particularly near the kinetochores. However, addition of cytochalasin B, which interferes with microfilament-based motility (see p. 315), has no effect on chromosome movement. Similarly, antibodies against actin or myosin fail to interrupt either anaphase A or B. One of the most telling experiments of this type was carried out by P. Kiehart, Inoué, and I. Mabuchi, who injected antiactin and antimyosin antibodies into dividing cells. The antibodies completely stopped movements known to be

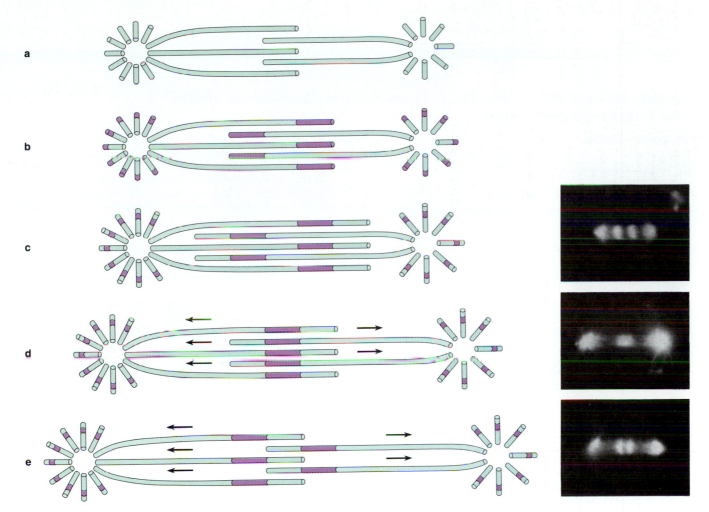

Figure 24-28 The experiment conducted by H. Masuda, K. L. McDonald, and W. Z. Cande demonstrating microtubule sliding in the zone of overlap at the midpoint in biotin-labeled spindles (see text). The labeled zones are shown in purple in the diagram and appear as regions of bright fluorescence in the micrographs. Micrographs × 2,400. (Courtesy of W. Z. Cande. Reproduced from *J. Cell Biol.* 107:623 [1988], by copyright permission of the Rockefeller University Press.)

based on microfilaments, such as the furrowing responsible for cytoplasmic division, but had no effect on the anaphase movement.

Anaphase A probably occurs through activity of kinetochores moving actively along KMTs. Although the mechanism producing the force for this movement is still undetermined, it is probably produced by microtubule motors driving the kinetochores along KMTs. Microtubule disassembly is also involved directly or indirectly in anaphase A, because KMTs disassemble as the kinetochores pass over them. Anaphase B appears to take place through forceful sliding between microtubules originating from opposite poles. The sliding, also driven by microtubule motors, decreases the zone of overlap between half spindles at the spindle midpoint. The rate of microtubule assembly or disassembly probably limits the rate

of both anaphase A and anaphase B. Both dynein- and kinesinlike motors have been implicated in generation of force for both anaphase A and B. The motors acting in anaphase may be different members of the dynein or kinesin families from those powering other stages of mitosis, such as separation of the poles, formation of the spindle, and congression of the chromosomes to the spindle midpoint. Some mutant genes encoding kinesinlike motors affect separation of the poles, for example, but not the anaphase movement.

Regulation of the Anaphase Movement

Although there are some conflicting results and differences in detail, several lines of evidence have linked

Spindle Elongation *in vitro:* The Relative Roles of Microtubule Sliding and Tubulin Polymerization

W. Zacheus Cande

W. Z. CANDE studies the mechanism of mitosis and meiosis in diatoms, yeast, and maize. He is a Professor of Cell and Developmental Biology in the Department of Molecular and Cell Biology at the University of California, Berkeley, and has been a member of the Berkeley faculty since 1976.

Although substantial progress has been made in describing spindle ultrastructure, relatively little is known about how the spindle operates at a biochemical level. *In vitro* model systems have been important tools for analyzing complex motility events such as muscle contraction and flagellar beat. Muscle cell models have provided the bridge that allowed us to understand the roles of individual proteins, such as myosin, during muscle contraction. Until recently, the mitosis field lacked this type of system for understanding how the spindle generates force for moving chromosomes. In part this is due to the number of mechanochemical events involved. Anaphase chromosome movement consists of at least three distinct events, and the conditions required for the reactivation of one phase of movement may not be compatible with maintenance of the other phases of chromosome separation *in vitro*. In most spindles, chromosomes move poleward (anaphase A) as the kinetochore microtubules shorten. Then, or at the same time, spindles elongate (anaphase B) and the polar or nonkinetochore-attached microtubules [IMTs in this book] grow and are rearranged as the spindle increases in length. Finally, in some cells, there may be interactions between elements of the surrounding cytoplasm and the poles, mediated by astral microtubules, which help move the reforming nuclei further apart. This last phase may not be present in all cells; however, it certainly plays a major role in fungi where spindles are intranuclear and cytoplasmic microtubule arrays are attached to the nuclear envelope and persist throughout mitosis.

The major challenge in isolating functional spindles from dividing cells is developing a medium that will stabilize the spindle microtubules against the rigors of isolation yet still preserve spindle function. This is a difficult task since the distinction between unnatural fixation and physiologically useful stabilization may only be a matter of degree. As was first demonstrated by Shinya Inoué in the late 1950s using polarization optics, the spindle is a very dynamic structure and the spindle fibers, i.e., the microtubules, are in a dynamic equilibrium with a subunit pool.[1]

With respect to organization of cytoskeletal components, the diatom spindle stands in comparison to other spindles much as striated muscle does to smooth muscle and other nonmuscle motility systems; it is a representative, yet much more highly ordered system.[2] In the diatom spindle, the machinery responsible for anaphase A and B are spatially separated, thus allowing the possibility of studying one process independently of the other. Second, the central spindle, the structure responsible for spindle elongation, is constructed of two interdigitating sets of microtubules that originate from platelike pole complexes. These are arranged in paracrystalline arrays and antiparallel microtubules display specific nearneighbor interactions in the zone of microtubule overlap. Since the zone of microtubule overlap is visible even by light microscopy, it is possible to monitor changes in overlap zone organization during spindle elongation. This degree of order is also found in the mammalian spindle, but only after midbody formation during telophase. My colleague, Kent McDonald, and I reasoned that the unique structural attributes of the diatom spindle that make it so useful for morphological studies would also be advantageous for developing an *in vitro* system for studying spindle elongation. For example, we thought that the highly ordered and crossbridged microtubules of the diatom central spindle would more easily withstand the rigors of isolation than the microtubules of the more loosely organized mammalian spindle.

McDonald and I used the large spindles (300 microtubules per half spindle, spindle length 12 μm) isolated from the diatom *Stephanopyxis turris* to describe the behavior of microtubules in the zone of microtubule overlap *in vitro*.[3] After screening numerous species from several collections, we selected this diatom because of its natural mitotic synchrony in culture and the presence of a spindle that was large enough for light microscopy yet not so large as to prohibit quantitative analysis in the electron microscope. The homogenization medium we used was designed to stabilize microtubules and the isolation procedure was kept as simple as possible. Populations of mitotic cells were collected by filtration, cell walls disrupted by mechanical shear, and spindles and nuclei were then collected on coverslips by low-speed centrifugation. Each coverslip contained hundreds of spindles, embedded in chromatin. After ATP addition, the spindles increased in length and a prominent gap developed between the two half-spindles, which was bridged by only a few microtubules. As visualized online with video microscopy and polarization optics or with populations of spindles using indirect immunofluorescence, the two half-spindles slid completely apart with a concurrent decrease in the extent of the zone of microtubule overlap.[3]

These results can best be interpreted by imagining that the microtubules of one half-spindle push off against the microtubules of the other half-spindle generating sliding forces between them. Our observations are consistent with models that postulate mechanochemical interactions in the zone of microtubule overlap that generate the forces necessary for anaphase B. Movement apart of the two half-spindles is not likely to be due to the autonomous swimming of each half-spindle, since it is limited by the extent of microtubule overlap.

The changes in distribution of newly incorporated tubulin in the spindle midzone give further support to our interpretation of the mechanism of force generation.[4] When spindles are incubated in biotin-labeled tubulin before ATP addition, labeled tubulin is incorporated as new microtubules nucleated at the poles and into the spindle midzone as two bands that flank the original zone of microtubule overlap [see Fig. 24–28 b and c]. Ultrastructural studies demonstrate that the new tubulin adds onto the ends of the original half-spindle microtubules in the midzone and does not form new microtubules next to the old ones. Thus, this new tubulin is a marker for the free ends of the original half-spindle microtubules. After ATP addition the spindle elongates, and the two bands in the midzone become one [see Fig. 24–28d]. This change can be explained if the microtubules containing the newly incorporated tubulin in each half-spindle slide over each other toward the center of the overlap zone. We have also used this method of labeling the ends of microtubules in the overlap zone to demonstrate that microtubule sliding occurs during spindle elongation in the small spindles of the diatom *Cylindrotheca fusiformis* and in the fission yeast *Schizosaccharomyces pombe*, where it is difficult to measure changes in spindle length.

Our observations eliminate several possible mechanisms of force generation. First, the poles are not pulled apart *in vitro*. There are no cytoplasmic structures attached to the isolated spindles that could generate this force. Second, microtubule polymerization cannot be responsible for generating the forces required to move the spindle poles apart because spindle elongation *in vitro* can occur in the absence of tubulin polymerization. Finally, it has been suggested that tension generated during spindle formation, stored in chromatin and kinetochore microtubules, and released during chromosome-to-pole movement may be responsible for spindle elongation. However, spindles elongate in the absence of chromatin and with kinetics inconsistent with this model.

When tubulin derived from brain microtubules is included in the reactivation medium, the extent of spindle elongation is no longer limited to the size of the original overlap zone, and some spindles elongate by the equivalent of 3 to 5 times their length. However, it is possible to uncouple tubulin polymerization from spindle elongation.[4] In the absence of ATP, tubulin is incorporated into the spindle and the zone of microtubule overlap increases in extent. After removal of tubulin and subsequent addition of ATP, the spindle elongates to an extent determined by the size of the new overlap zone, not the original overlap zone. Spindle elongation under these conditions is sensitive to the same inhibitors in the presence or absence of added tubulin, suggesting that microtubule polymerization does not contribute to the generation of forces required for anaphase B. Thus, the process of tubulin polymerization is distinct from that of force generation and its role is to define the extent of polymer that can slide through the zone of microtubule overlap. Depending on the relative rates of microtubule assembly and sliding as defined experimentally by the relative tubulin and ATP concentrations, it is possible *in vitro* for the overlap zone to increase, decrease, or remain constant as the spindle elongates.

The pharmacology of spindle elongation is similar in fission yeasts, diatoms, and mammalian cells, suggesting that a similar force-generating enzyme is used in all eukaryotic cells. It has several novel features: like kinesin it is inhibited strongly by the nonhydrolyzable ATP analog AMPPNP; however, unlike kinesin it is sensitive to low concentrations of N-ethyl maleimide and vanadate. Recently, in collaboration with Takashi Shimizu, we have described the substrate specificity of the anaphase B motor using 20 different ATP analogues and other nucleotides, and have shown that the motor is more kinesinlike than dyneinlike in the range of nucleotides that will support spindle elongation. We are now using biochemical and molecular approaches to directly identify and characterize the anaphase B motor in diatoms.

References

[1] Inoué, S., and Sato, H. Cell motility by labile association of molecules: The nature of mitotic spindle fibers and their role in chromosome movement. *J. Gen. Physiol.* 50:259–92 (1967).

[2] Pickett-Heaps, J. D., and Tippit, D. H. The diatom spindle in perspective. *Cell* 14:455–67 (1978).

[3] Cande, W. Z., and McDonald, K. L. *In vitro* reactivation of anaphase spindle elongation using isolated diatom spindles. *Nature* (Lond.) 316:168–70 (1985).

[4] Masuda, H.; McDonald, K. L.; and Cande, W. Z. The mechanism of anaphase spindle elongation: Uncoupling of tubulin incorporation and microtubule sliding during *in vitro* spindle reactivation. *J. Cell Biol.* 107:623–33 (1988).

controlled changes in Ca^{2+} concentration to regulation of the anaphase movement. Among the most significant are the results of experiments by M. Poenie, C. H. Keith, P. Hepler, and M. L. Shelanski and their colleagues and others using indicators such as *aequorin* and *quin-2*, which fluoresce in the presence of Ca^{2+}. Injection of these molecules during prophase or until late metaphase produces only very low levels of fluorescence, indicating that Ca^{2+} is present at low concentrations in the spindle up to this time. However, just before anaphase begins, one or more bursts of fluorescence lasting some 20 seconds are emitted by the indicators. The bursts indicate that Ca^{2+} is released into the cytoplasmic solution in the vicinity of the spindle microtubules as the chromosomes begin their movement to the poles.

Injecting Ca^{2+} into mitotic cells in metaphase also speeds entry into anaphase; if injected at anaphase, Ca^{2+} increases the rate of anaphase movement. Conversely, reducing Ca^{2+} concentration at metaphase by exposing cells to a calcium chelator such as EGTA, which binds Ca^{2+} and removes it from solution, or a Ca^{2+} buffer, which also removes excess Ca^{2+} from solutions, slows entry of cells into anaphase.

Through the use of antimonate ions, which form an insoluble, electron-dense precipitate when complexed with calcium, S. M. Wick and Hepler showed that the vesicles concentrated in the spindle contain stored calcium ions. The spindle is stained by fluorescent antibodies against the Ca^{2+}-ATPase pump of muscle cell ER membranes; injection of the same antibody into living cells blocks mitosis at metaphase. It has also been possible to detect the polypeptides of the calcium pump among the proteins isolated from spindles.

These observations suggest that the Ca^{2+} concentration of resting spindles is maintained at low concentrations by Ca^{2+}-ATPase pumps, which continually transport Ca^{2+} from the cytoplasmic solution into the spindle vesicles. At the onset of anaphase, gated Ca^{2+} channels open briefly in the vesicle membranes, releasing the ion into the cytoplasm surrounding the spindle microtubules. This increase in Ca^{2+} concentration is part of the pathway triggering anaphase and maintaining the anaphase movement.

Ca^{2+} may have its most significant effects in promoting microtubule disassembly, particularly at the kinetochores as dyneinlike or kinesinlike motors push the chromosomes toward the poles along kinetochore microtubules. As noted, this disassembly appears to be the rate-limiting reaction of anaphase A. The ion may work directly on calcium-binding proteins or through control pathways dependent on calmodulin activated by combination with Ca^{2+} (see p. 165); both protein types have been detected in the spindle. There are indications that Ca^{2+} is released during mitosis through activity of the $InsP_3$/DAG cellular response pathway in higher animals (see p. 161).

THE EVENTS OF TELOPHASE

During telophase the spindle disassembles except for short microtubules that persist at the former spindle midpoint and take part in cytoplasmic division. The chromosomes gradually decondense and reach the extended interphase state. During decondensation the nucleolus reappears at NOR sites on the chromosomes and the nuclear envelope reforms. Of these alterations, greatest progress has been made in working out the molecular steps underlying reorganization of the nucleolus and reformation of the nuclear envelope.

Reorganization of the Nucleolus

During telophase the nucleolus reappears on one or more chromosomes as a small, spherical body that gradually increases in size until the typical interphase structure is established. B. McClintock was the first to demonstrate that a specific chromosomal site is required for reappearance of the nucleolus at telophase. Working with maize, McClintock succeeded in obtaining mutant individuals in which the nucleolus failed to reappear at telophase. Instead, numerous small, unattached nucleoluslike bodies were distributed among the chromosomes. McClintock proposed the term *nucleolar organizer* (*NOR*) for the site necessary for normal reappearance of the nucleolus at telophase. We now know that the NOR contains many repeats of large pre-rRNA genes (see p. 439).

Several experiments indicate that some of the material reorganized into the nucleolus is probably derived from rRNA and protein molecules released earlier, when the nucleolus breaks down during prophase. For example, N. Paweletz and M. C. Risueno used a silver staining technique that specifically tags nucleolar proteins with dense deposits, making them visible in the electron microscope. During interphase the silver technique stains only the nucleolus. As the nucleolus breaks down during mitotic prophase, the densely staining material becomes distributed over the condensing chromosomes. At telophase the dense material once again becomes associated with the reforming nucleolus at the NORs. Other evidence shows that rRNA can be detected in association with isolated metaphase chromosomes. Analysis shows that this rRNA includes complete and partially processed large pre-RNAs.

These observations suggest that the cycle of nucleolar disappearance and reappearance during mitosis reflects the following molecular changes. At the beginning of prophase the nucleolus contains pre-rRNA molecules in association with ribosomal proteins, all at various stages in processing into ribosomal subunits. As prophase progresses, rRNA transcription and processing stops, and the partially processed ribosomal precursors are released when the nucleolus breaks down. These collect on the surfaces of the condensing chromosomes and may also disperse into the cytoplasm. At telophase the partially processed ribosomal subunits return to the NOR and assemble at this site. Processing of the preexisting material then resumes. Transcription of new rRNA also begins at this time, so that, as telophase proceeds into interphase, a progressively greater part of the nucleolus is newly synthesized material.

Presumably dispersal of partially processed pre-rRNAs and ribosomal proteins during metaphase and anaphase is an adaptation that reduces the mass of the chromosomes bearing NORs. Dispersal of the nucleolus during anaphase would also reduce the chance of tangling and nonseparation of the chromatids containing NORs.

Reformation of the Nuclear Envelope

When the nuclear envelope begins to reappear during telophase, flattened membranous vesicles apparently identical to ER cisternae become closely applied to the surfaces of the decondensing chromatids (Fig. 24-29). These vesicles increase in number and gradually fuse until the chromatids are enclosed by an envelope consisting of two concentric membranes separated by a narrow space. Pore complexes develop as the membranes fuse into a continuous layer. By the end of telophase the daughter nuclei are completely separated from the cytoplasm by a fully differentiated nuclear envelope. The envelope develops in such close association with the surfaces of the chromatids that the cytoplasm appears to be largely excluded from the newly formed nuclei.

The lamins, which are dephosphorylated as the nuclear envelope reforms, reassemble into a fibrous framework on the surface of the inner nuclear membrane facing the nucleoplasm. Recent work by Gerace and his colleagues indicates that during telophase in mammals, lamins A and C, if not already bound, link tightly to the surfaces of the chromosomes. Both of these lamins have sites that can bind chromatin; both are dephosphorylated as they associate with telophase chromosomes during reformation of the nuclear envelope. The lamin B molecules carried on the ER vesicles then contact the lamin A and C proteins. This contact brings the vesicles into close proximity to the chromosome surfaces, and the nuclear envelope, with its supporting lamin network, reassembles.

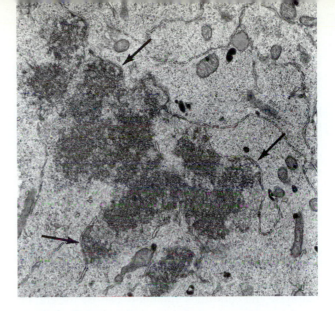

Figure 24-29 Segments of the nuclear envelope (arrows) reappearing at the margins of decondensing chromatids in a human cell at telophase. The membrane segments extend and fuse until the chromosomes are surrounded by a complete and continuous nuclear envelope. × 44,000. (Courtesy of G. G. Maul and Academic Press, Inc., from *J. Ultrastr. Res.* 31:375 [1970].)

Several integral proteins of the vesicles assembling into the reforming nuclear envelope recognize and bind the lamins. Lamin B is bound by a *lamin B receptor* that specifically recognizes this lamin. The receptor appears to bind lamin B to the vesicles formed when nuclear envelopes break down as well as to intact and reforming nuclear envelopes. Gerace and his colleagues recently discovered another group of integral nuclear envelope proteins, which they term *lamin-associated proteins* (*LAPs*), that in different combinations bind lamins A, B, and C and also chromatin. One of these proteins, *LAP2*, attaches nuclear envelope vesicles to chromatin even before most lamins bind, indicating that it may be important in the earliest steps of nuclear envelope reformation. Like the lamins, the LAPs are phosphorylated during mitosis; phosphorylation of LAP2 inhibits its binding to both lamins and chromatin. Fusion of vesicles during reformation of the nuclear envelope appears to involve many of the same GTP-binding proteins and other factors that take part in vesicle fusion in the ER/Golgi complex secretory pathway (see p. 578). Some of the inhibitors that block vesicle fusion in the secretory pathway, for example, also block vesicle fusion in reforming nuclear envelopes.

The proteins forming parts of the nuclear matrix and solution surrounding the chromatin probably return from the cytoplasm to the nucleus by two major routes. Some probably attach to the surfaces of the chromosomes as they condense during prophase, remaining attached in this position throughout mitosis until the newly formed nuclear envelope separates the nucleus from the cytoplasm at telophase. Others return by virtue of nucleus-directing signals forming parts of their amino acid sequence (see p. 590). These proteins are left in the cytoplasm as the nuclear envelope reorganizes, but they return to the nucleus via cellular sorting and distribution mechanisms that direct them through the nuclear pore complexes during or soon after telophase.

CYTOKINESIS

Cytoplasmic division begins in most cells during late anaphase and continues into telophase. By the end of telophase, when the daughter nuclei are fully formed and surrounded by complete nuclear envelopes, cytokinesis is usually complete. The two primary mechanisms by which cytokinesis takes place, furrowing in animals and cell plate formation in plants, share one major feature in common. In both systems a band of overlapping microtubules persisting at the former spindle midpoint is intimately involved in the early stages of cytoplasmic division.

Furrowing

As anaphase and telophase progress in animal cells, the spindle breaks down rapidly except for microtubules that persist at the midpoint. The persisting microtubules are the remnants of the zone of overlap at the former spindle midpoint and retain the interdigitated, antipolar arrangement characteristic of this region. Other structures, such as vesicles, may also move to the equatorial layer. The fully developed layer, with associated microtubules and vesicles, is the *midbody* (Fig. 24-30).

As the midbody develops, a furrow appears in the plasma membrane, extending around the cell at the level of the midbody. The furrow gradually deepens and compresses the midbody until it is only a narrow, fibrous connection between the daughter cells. The midbody may persist in this form for a time, but in most cells it breaks down and disappears soon after furrowing is complete. In some cells the residual midbody is engulfed by one of the daughter cells. The furrow, once induced, may form in as little as 20 seconds; cytoplasmic division by furrowing in some cells is complete in as little as 10 minutes.

A series of experiments showed that development and inward extension of the furrow depend on actin microfilaments. Fine filaments about 5 to 7 nm in diameter are visible at the apex of the furrow (Fig. 24-31), running in a complete band around the dividing cell. The HMM decoration technique (see p. 309) produces

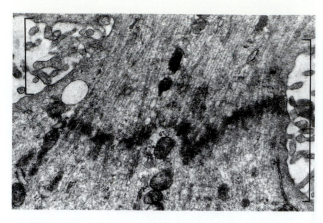

Figure 24-30 The midbody in a human cell. Dense material surrounding microtubule remnants at the former spindle midpoint forms a continuous layer across the cell (see text). The deepening furrow is visible on either side of the cell opposite the dense layer (brackets). × 35,000. (Courtesy of A. Krishan and R. C. Buck, from *J. Cell Biol.* 24:443 [1965], by copyright permission of the Rockefeller University Press.)

"arrowheads" on these fibers, indicating that they are actin microfilaments. Fluorescent staining using both antiactin and antimyosin antibodies produces bright fluorescence just under the furrow, demonstrating that both components of the actin-myosin contractile system are present in the band. Further evidence that furrow formation and contraction depend on microfilament activity comes from observations that antiactin and antimyosin antibodies, as well as microfilament poisons such as cytochalasin B, stop furrowing in most animal cells.

Changes in Ca^{2+} concentration have been implicated in regulation of furrowing. T. G. Hollinger and A. W. Schuetz, for example, induced formation of cleavage furrows in frog eggs by injecting Ca^{2+} just below the cell surface. Injection of fluorescent Ca^{2+} indicators also shows that a brief rise in cytoplasmic Ca^{2+} concentration occurs in coordination with the onset of furrowing.

In most animal cells the spindle is located symmetrically in the center of the cell. As a result, cytoplasmic division cuts the dividing cell into two more or less equal parts, following a plane that extends through the former spindle midpoint. In a few animal cell types, such as the cells of some early animal embryos, the spindle is located asymmetrically, leading to unequal division of the cytoplasm.

In general, experiments displacing the spindle before anaphase also displace the future plane of cytoplasmic division. Spindle displacement after anaphase usually has no effect on the plane of division. Therefore molecular interactions during prophase

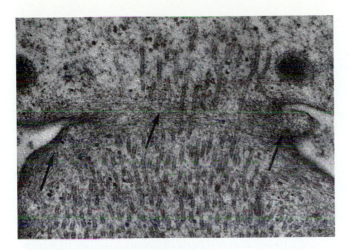

Figure 24-31 Microfilaments (arrows) under the plasma membrane in the region of an advancing furrow in a rat cell. × 31,000. (Courtesy of D. Szollosi, from *J. Cell Biol.* 44:192 [1970], by copyright permission of the Rockefeller University Press.)

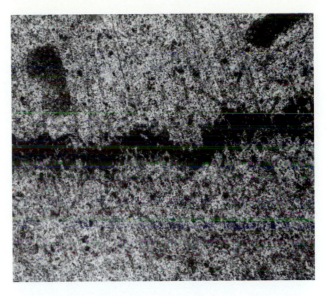

Figure 24-32 The phragmoplast in *Haemanthus*, consisting of a layer of dense material enclosed in vesicles, collected around microtubule remnants at the former spindle midpoint. × 18,000. (Courtesy of A. Bajer and Springer-Verlag.)

and metaphase, probably between microtubules extending from the spindle and asters and the cell cortex, fix the position at which the furrow will form. Once the position is fixed, displacing the spindle has no effect.

Cell Plate Formation

In plants the spindle also disassembles after anaphase except for short lengths of overlapping, antipolar microtubules that persist at the former spindle midpoint. Vesicles originating from the ER and Golgi complex move into the developing layer at the former spindle equator until a continuous layer forms (Fig. 24-32). At first this layer, termed the *phragmoplast*, is confined to the area of the cell formerly occupied by the spindle midpoint. The layer then grows laterally by the accumulation of additional microtubules, microfilaments, and vesicles, until the phragmoplast spreads entirely across the dividing cell. Experiments using radioactive tracers have shown that the vesicles contain precursors of cell wall materials such as cellulose and pectins.

As the phragmoplast expands, vesicle fusion begins in its older regions (Fig. 24-33a and b; see also Fig. 24-4). The fusion, which amounts to a specialized form of exocytosis, dumps cell wall material into a gradually expanding layer between the daughter cells. As in exocytosis in other cell types (see p. 592), a rise in Ca^{2+} concentration can be detected in the regions of vesicle fusion. Fusion continues until the entire phragmoplast is replaced by a continuous cell

wall (Fig. 24-33c). The new cell wall is lined on either side by plasma membranes derived from the vesicle membranes. When the plasma membranes and new cell wall are fully formed, the partition at the midregion is termed the *cell plate*.

The cell plate does not entirely cut off cytoplasmic communication between daughter cells. Numerous narrow openings, lined by extensions of the plasma membrane, persist in the cell plate and remain as direct cytoplasmic connections between the daughter cells (arrows in Fig. 24-33c; see also Fig. 7-20). These connections, called *plasmodesmata* (see p. 194), evidently form where tubular extensions of the ER are caught in the developing cell plate.

Cytoplasmic division by phragmoplast and cell plate formation occurs in almost all plant cells. Exceptions among plants are noted in only a few cell types, such as the microspores giving rise to pollen. In these cells, cytoplasmic division occurs by a process resembling furrowing.

The plane of cytoplasmic division in many plant cells appears to be determined well before mitosis begins. In these cells the nucleus takes up a position, usually in the center of the cell, some 8 to 20 hours prior to mitosis. The nucleus becomes surrounded by a band of radiating microtubules and microfilaments that extends outward as a disclike layer and contacts the cell wall. In the region of contact a belt of microtubules and microfilaments called the *preprophase band* (*PPB*) forms around the cell periphery (see Fig. 12-15). The PPB persists for a time but usually disappears before the onset of mitosis. The position taken by the

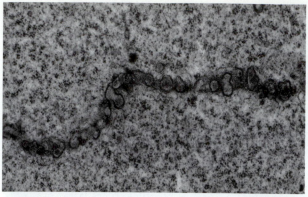

a

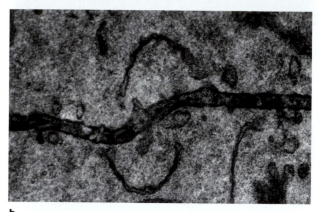

b

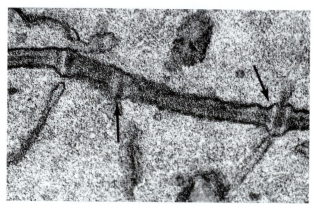

c

Figure 24-33 Fusion of phragmoplast vesicles to form the cell plate. **(a)** An early stage in *Haemanthus*, in which vesicle fusion is under way. × 25,000. (Courtesy of W. G. Whaley, M. Dauwalder, J. E. Kephart, and Academic Press, Inc., from *J. Ultrastr. Res.* 15:169 [1966]. **(b)** A later stage in the grass *Phalaris*, in which vesicle fusion is nearly complete. × 59,000. **(c)** A fully formed cell plate in *Phalaris*; several plasmodesmata (arrows) remain as direct connections between the cytoplasm of daughter cells. × 70,000. ([b and c] courtesy of A. Frey-Wyssing, J. R. Lopez-Saez, K. Muhlethaler, and Academic Press, Inc. from *J. Ultrastr. Res.* 10:422 [1964].)

PPB marks the site at which the cell plate will contact the old cell wall during cytokinesis.

Once the PPB has marked the future site of cytoplasmic division, displacement of the nucleus or spindle usually has no effect on the location at which the cell plate attaches to the side walls of the dividing cell. If the spindle and developing phragmoplast are displaced experimentally, the developing cell plate curves at its edges so that it still meets the original wall at the site previously marked by the PPB. Some investigators have proposed that a layer of microfilaments persisting in the region of the former PPB directs the margins of the developing cell plate to the cell wall.

PPBs are characteristic of ordered tissues in which the position of cells is critical for development of plant form. Less organized tissues such as callus and liquid endosperm enter division without forming PPBs. In most ordered tissues the position taken by the nucleus and the subsequent formation of the PPB are symmetrical, so that the plane of division cuts the dividing cell into two equal parts. However, in some cells, such as those giving rise to the guard cells of stomata, the nucleus and PPB take up an asymmetric position, so that the future plane of cytoplasmic division produces daughter cells of unequal size.

Formation of the PPB, spindle, and phragmoplast involves an essentially complete reorganization of the cytoskeleton. Microtubules and microfilaments, which are located primarily in the cortex of interphase plant cells, become concentrated in the PPB as this structure forms. As dividing cells enter prophase, the PPB disassembles, and microtubules reassemble into the spindle. During anaphase the spindle disassembles, and microtubules reorganize into the phragmoplast. After mitosis is complete, the arrangement typical of interphase cells reappears (see also p. 359).

The midbody in animals and both the PPB and phragmoplast in plants have been shown to contain γ-tubulin during their development. Therefore the new microtubules appearing around the periphery of the midbody and phragmoplast and in PPBs are probably added through activity of these regions as MTOCs.

Distribution of Mitochondria and Chloroplasts

The distribution of mitochondria and chloroplasts during mitosis and cytokinesis is complicated because both organelles contain their own DNA and are partly autonomous (see Chapter 21 for details). Both organelles replicate their DNA during interphase (see Supplement 23-2). At some point following replication, both organelles divide by a process that resembles furrowing (Figures 24-34 and 24-35 show the division process in mitochondria). As part of the division

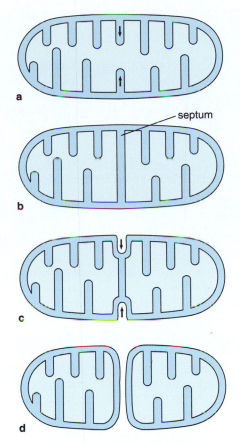

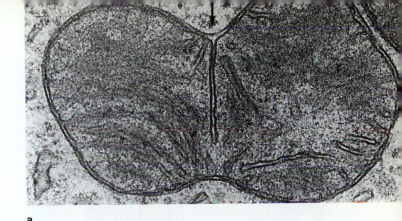

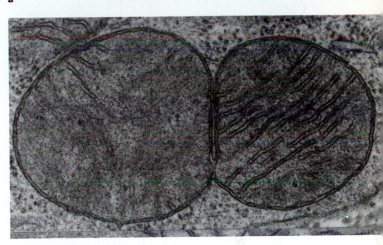

Figure 24-34 Steps in division of a mitochondrion; chloroplast division follows a similar pattern. **(a)** The inner boundary membrane grows inward; **(b)** growth continues until the inner membrane cuts the matrix into two parts as a septum extending completely across the organelle. **(c)** The outer membrane then invaginates into the septum; **(d)** invagination continues until the organelle separates into two parts. As a part of the division, internal structures, including soluble and membrane-bound enzymatic systems, DNA circles, and ribosomes, are distributed in approximately equal portions between the two division products.

Figure 24-35 Dividing animal mitochondria, in stages comparable to Figure 24-34b **(a)** and 24-34c **(b)**. (Micrographs by T. Kaneseki, from D. W. Fawcett, *The Cell*, W. B. Saunders Co., 1987.)

the internal structures of both organelles, including soluble and membrane-bound enzymatic systems, DNA circles, and ribosomes, are distributed in approximately equal portions between the two division products.

In most cells, organelle DNA replication and division proceed throughout much or all of the cell cycle. From the initiation of G1 onward, division increases the number of the organelles until a maximum number characteristic of the cell type is reached. In rapidly dividing cells the maximum organelle count is reached late in G2, just before the next cytoplasmic division takes place. Organelle numbers are smallest just after cytoplasmic division is complete.

The relative quantities of mitochondria, chloroplasts, and other cell organelles distributed to daughter cells depend in most systems simply on the volume of cytoplasm received during cytokinesis. If the plane of cytoplasmic division cuts through the approximate center of the cell, the daughter cells receive roughly equal number of cytoplasmic organelles. If the plane of cytoplasmic division is asymmetric, the number of cytoplasmic organelles delivered to daughter cells varies proportionately. This relative distribution reflects the fact that there are no processes equivalent to the spindle mechanisms that precisely organize and divide cytoplasmic organelles. Only the fact that daughter cells both receive cytoplasm, usually in equivalent amounts, ensures that the cytoplasmic organelles are distributed approximately equally.

Cytoplasmic division in animal and plant cells completes the process of cell division. The new plasma membranes cut the cytoplasm of the parent cell into two portions and distribute the various cytoplasmic organelles, including mitochondria, chloroplasts if present, and vesicles of the ER and Golgi complex, between daughter cells.

Table 24-2 Some Mutations Affecting Mitosis or Cytokinesis

Mutation	Encoded Protein	Effect of Mutation
Saccharomyces cerevisiae		
cdc31	Similar to Ca^{2+}-binding proteins	SPBs do not duplicate; spindle does not form.
cin8	Similar to kinesin	SPBs duplicate but do not separate.
esp1	Unknown	Extra SPBs form.
kar1	53kD	Defects in SPB duplication and chromatid separation.
ndc1	Unknown	Chromatids do not separate; go to same pole.
spa1	56kD	Extra SPBs form.
top2	DNA topoisomerase II	Chromatids do not separate.
Schizosaccharomyces pombe		
cdc2	cdc2 protein kinase	Failure to enter mitosis; no chromosome condensation or nuclear envelope breakdown.
cut7	Similar to kinesin	Spindle formation faulty.
dis1, 2, 3	Unknown	Chromatids do not separate.
nda2	α1-tubulin	Spindle does not assemble.
nda3	β-tubulin	Spindle does not assemble.
nuc2	Unknown	Spindle does not elongate.
Drosophila		
asp	Unknown	Abnormal spindle; mitosis arrests at metaphase.
gnu	Unknown	Asters divide, but spindle does not form.
mrg	Unknown	Asters do not duplicate; forms spindle with single pole.
polo	Unknown	Multiple asters form.
Mammals		
MS1-1	Unknown	Cleavage furrow does not form.
RCC1	45kD, binds DNA	Faulty chromosome condensation.
Tax18	Unknown	Spindle assembly blocked; taxol can reverse effects.
ts2	Unknown	Chromosomes do not attach to spindle.
ts111	Unknown	Cleavage furrow does not form.
ts546	Unknown	Similar to *ts2*.
ts654, 655	Unknown	Faulty chromosome condensation.
ts687	Unknown	No anaphase movement.
ts745	Unknown	Asters do not divide; forms spindle with single pole; no chromosome division.

GENETIC REGULATION OF MITOSIS AND CYTOPLASMIC DIVISION

Division of the chromosomes and cytoplasm is an incredibly complex process with many subparts and events that must be closely coordinated for a successful outcome. As a consequence, there undoubtedly are many control points that regulate the process and ensure that all the events occur in the correct manner and sequence. These control points have been investigated by some of the same methods described in Chapter 22 for study of the entire cell cycle: identification of genetic mutants and detection of factors that modify portions of mitosis or cytoplasmic division.

A number of genes have been identified in both yeasts and higher organisms encoding factors that control, modify, or are necessary for parts of mitosis or cytoplasmic division to proceed (Table 24-2). Among the most significant are the cyclin and *cdc2* genes (see p. 638), which have been detected in organisms from yeast to mammals. These genes encode the cyclin and cyclin-dependent protein kinases, particularly cyclin B and CDK2, which when fully activated trigger the primary events of mitotic prophase: condensation of the chromosomes, formation of the spindle, and nuclear envelope breakdown. The active cyclin B/CDK2 complex has been directly implicated in the H1 phosphorylation associated with condensation of the chromosomes and the lamin phosphorylation accompanying nuclear envelope breakdown, and may carry out phosphorylations regulating spindle assembly and function. The cyclin B/CDK2 complex must break down or become inactive for cells to progress from metaphase to anaphase. In most cells, cyclin B is hydrolyzed by a specific proteinase activated at the metaphase–anaphase transition, and a phosphate group is removed from CDK2 to accomplish inactivation of the complex.

Many additional genes affecting segments of mitosis and cytoplasmic division have been detected, particularly in *S. cerevisiae*, *S. pombe*, and *Drosophila* (see Table 24-2). In *S. cerevisiae*, for example, several genes have been identified, including *cdc31*, *spa1*, *esp1*, and *kar1*, that in mutant form interfere with normal formation or function of spindle pole bodies. In the mutants the SPB does not duplicate; nor does it form the chromosome attachments necessary for separation of chromatids. Other mutations alter or block spindle function. Either the spindle does not assemble, or it assembles incompletely; or, if assembled, there is no anaphase movement. In one unusual set of mutants, certain chromosomes of the complement fail to be included in daughter nuclei. Other mutants affect chromatid separation generally, so that chromatids remain locked together.

For many of these genes, little is known of their encoded products or how they exert their effects. However, several of the mutations affecting spindle form or function involve alterations in tubulin. Presumably these genes merely provide the tubulin necessary for spindle assembly rather than acting as regulatory elements. Others encode kinesinlike proteins. One of the mutants blocking chromatid separation, *top2*, encodes topoisomerase II. The failure of chromatids to separate in *top2* mutants indicates that the enzyme normally eliminates intertwined loops produced during replication or tangles formed as chromosomes condense.

The mitosis and cytoplasmic division mutants detected in higher eukaryotes in addition to the cyclin and *cdc2* genes have similar effects. Several of the mutants identified in mammals inhibit spindle formation, either partially or completely; others form spindles that cannot produce the anaphase movement.

One general characteristic made clear by these studies is the potential independence of many mitotic processes. Although the spindle may not form or chromatids may not separate, chromosomes eventually decondense and the nuclear envelope still forms at approximately the usual times. This indicates that some segments of the complex mitotic pathway are controlled by separate banks of genes that do not depend for their operation on completion of earlier steps.

Although it has not as yet been possible to identify the proteins encoded in many of these genes or their mode of action, genetic studies, at the very least, confirm that individual steps of mitosis and cytoplasmic division are under genetic control and that the operation of a series of genes, working in defined and regulated cascades, is responsible for the complex events of cell division. This order and precision depend ultimately on specific genes, activated at the right time and place before and during cell division.

The entire cycle of replication, mitosis, and cytoplasmic division, in all of its elegant complexity, occurs countless billions of times in the growth, development, and maintenance of structure in many celled eukaryotes. In humans, mitotic divisions following fertilization of the egg produce an adult with an estimated total of 65 trillion cells. To maintain the supply of red blood cells alone in humans, more than 2 million mitotic divisions occur per second in each individual. These repeated division cycles of mitosis occur almost without error throughout the lifetime of the organism.

Suggestions for Further Reading

Balczon, R. D., and Brinkley, B. R. 1990. The kinetochore and its roles during cell division. In K. W. Adolph, ed., *Chromosomes, Eukaryotic, Prokaryotic, and Viral*, pp. 167–189. Boca Raton, Fl.: CRC Press.

Baskin, T. I., and Cande, W. Z. 1990. The structure and function of the mitotic spindle in flowering plants. *Ann. Rev. Plant Physiol. Plant Molec. Biol.* 41:277–315.

Bensh, D., Stillman, B., and Watson, J. D., eds. 1991. *The Cell Cycle. Cold Spring Harbor Symp. Quant. Biol.*, vol. 56. Cold Spring Harbor, N.Y.: Cold Spring Harbor Laboratory Press.

Bickmore, W. A., and Sumner, A. T. 1989. Mammalian chromosome banding—an expression of genome organization. *Trends Genet.* 5:144–148.

Blackburn, E. H. 1990. Telomeres: Structure and synthesis. *J. Biolog. Chem.* 265:5919–5921.

de Boer, P. A. J., Cook, W. R., and Rothfield, L. I. 1990. Bacterial cell division. *Ann. Rev. Genet.* 24:249–274.

Brinkley, B. R. 1991. Chromosomes, kinetochores, and the microtubule connection. *Bioess.* 13:675–681.

Clarke, L. 1990. Centromeres of budding and fission yeasts. *Trends Genet.* 6:150–154.

Dingwall, C., and Laskey, R. 1992. The nuclear membrane. *Science* 258:942–947.

Donachie, W. D. 1993. The cell cycle of *Escherichia coli*. *Ann. Rev. Microbiol.* 47:199–230.

Endow, S. A. 1993. Chromosome distribution, molecular motors, and the claret protein. *Trends Genet.* 9:52–55.

Georgatos, S. D., Meier, J., and Simos, G. 1994. Lamins and lamin-associated proteins. *Curr. Opinion Cell Biol.* 6:347–353.

Glover, D. M., Gonzalez, C., and Raff, J. W. 1993. The centrosome. *Sci. Amer.* 268:62–68 (June).

Gorbsky, G. J. 1992. Chromosome motion in mitosis. *Bioess.* 14:73–80.

Hayles, J., and Nurse, P. 1992. Genetics of the fission yeast *Schizosaccharomyces pombe*. *Ann. Rev. Genet.* 26:373–402.

Hepler, P. K. 1992. Calcium and mitosis. *Internat. Rev. Cytol.* 138:239–268.

Hiraga, S. 1992. Chromosome and plasmid partition in *E. coli*. *Ann. Rev. Biochem.* 61:283–306.

Hogan, C. J., and Cande, W. Z. 1990. Antiparallel microtubule interactions: Spindle formation and anaphase B. *Cell Motil. Cytoskel.* 16:99–103.

Joshi, H. C. 1993. γ-tubulin: The hub of cellular microtubule assembly. *Bioess.* 15:637–643.

Kellogg, D. R., Moritz, M., and Alberts, B. M. 1994. The centrosome and cellular organization. *Ann. Rev. Biochem.* 63:639–674.

Kuriama, R., and Nislow, C. 1992. Molecular components of the mitotic spindle. *Bioess.* 14:81–88.

Luthenhaus, J. 1990. Regulation of cell division in *E. coli*. *Trends Genet.* 6:22–25.

McIntosh, J. R., and Hering, G. E. 1991. Spindle fiber action and chromosome movement. *Ann. Rev. Cell Biol.* 7:403–426.

McIntosh, J. R., and McDonald, K. L. 1989. The mitotic spindle. *Sci. Amer.* 261:48–56 (October).

Murphy, M., and Hayes-Fitzgerald, M. 1990. *cis*- and *trans*-acting factors involved in centromere function in *Saccharomyces cerevisiae*. *Molec. Microbiol.* 4:329–336.

Murray, A. W., and Szostak, J. W. 1987. Artificial chromosomes. *Sci. Amer.* 257:62–68 (November).

Page, B. D., and Snyder, M. 1993. Chromosome segregation in yeast. *Ann. Rev. Microbiol.* 47:231–261.

Rattner, J. B. 1991. The structure of the mammalian centromere. *Bioess.* 13:51–56.

Rieder, C. L. 1990. Formation of the astral mitotic spindle: Ultrastructural basis for the centrosome-kinetochore interaction. *Electron Micr. Rev.* 3:269–300.

Sawin, K. E., and Endow, S. A. 1993. Meiosis, mitosis, and microtubule motors. *Bioess.* 15:399–407.

Scheer, U., and Benavente, R. 1990. Functional and dynamic aspects of the mammalian nucleolus. *Bioess.* 12:14–21.

Schlessinger, D. 1990. Yeast artificial chromosomes: Tools for mapping and analysis of complex genomes. *Trends Genet.* 6:248–258.

Schulman, L., and Bloom, K. S. 1991. Centromeres: An integrated protein/DNA complex required for chromosome movement. *Ann. Rev. Cell Biol.* 7:311–336.

Stearns, T., Evans, L., and Kirschner, M. 1991. γ-tubulin is a highly conserved component of the centrosome. *Cell* 65:825–836.

Sumner, A. T. 1990. *Chromosome Banding*. London: Unwin Hyman.

Walker, R. A., and Sheetz, M. P. 1993. Cytoplasmic microtubule-associated motors. *Ann. Rev. Biochem.* 62:429–451.

Warn, R. M. 1990. Cytokinesis genetics takes wing. *Trends Genet.* 6:309–310.

Review Questions

1. What is condensation? When does it begin? What characteristics of the condensation mechanism suggest that it may be built into the chromosomes themselves?

2. What is the relationship of phosphorylation to condensation? Of DNA topoisomerase II activity to condensation?

3. What is the significance of condensation to mitotic division?

4. Define chromosome, chromatid, centromere, kinetochore, chromosome arm, primary and secondary constriction, and telomere. What functions are telomeres known or believed to carry out?

5. What is a karyotype?

6. What is the chromosome scaffold or core?

7. What is an "artificial chromosome"? What are the minimum components of an artificial chromosome?

8. Outline the two major patterns of spindle formation. What are centrioles and asters? What are IMTs? KMTs? Spindle pole bodies?

9. What characteristics indicate that spindles are in equilibrium with a pool of microtubule subunits?

10. List the major components of spindles other than microtubules.

11. What experimental results indicate that separation of the chromatids is a distinct process from the anaphase movement? What results rule out the possibility that late centromere replication or kinetochore duplication is responsible for separation of the chromatids?

12. What is the possible role of DNA topoisomerase II in chromatid separation?

13. What is anaphase A? Outline one experiment indicating that during anaphase A, kinetochores move along KMTs.

14. What is anaphase B? Outline two experiments indicating that anaphase B occurs by microtubule sliding at the spindle midpoint. What evidence indicates that a dynein- or kinesinlike ATPase is active in anaphase B?

15. What evidence implicates Ca^{2+} as a major factor in regulation of the anaphase movement? What effects might increases in Ca^{2+} concentration have on the spindle?

16. Outline the structure of centrioles. What is the probable relationship of centrioles to spindle formation and function?

17. What pattern is followed in centriole duplication? What is the major cellular function of centrioles?

18. What is pericentriolar material? What is the possible relationship of this material to spindle formation?

19. Outline the nucleolar cycle during mitosis. What is the nucleolar organizer region? What is the relationship of rRNA genes to this region?

20. Outline the cycle of nuclear envelope breakdown and reformation during mitosis in higher organisms. What is the relationship of the nuclear envelope to the ER? How do nuclear proteins return to the nucleus at the close of mitosis?

21. What are the lamins? What is the relationship of the lamins to nuclear envelope breakdown and reformation? What role does phosphorylation play in the breakdown and reassembly of the lamins during mitosis? What happens to the lamins after nuclear envelope breakdown?

22. Outline the events taking place in cytoplasmic division by furrowing. What is the midbody? What evidence indicates that microfilaments provide the force for furrowing?

23. Outline the events taking place in cytoplasmic division in plants. What is the phragmoplast? Cell plate? Preprophase band?

24. What events apparently fix the plane of cytoplasmic division in plants? What is the relationship of cytoplasmic division in plants to exocytosis?

25. In what ways is cytoplasmic division in plants similar to the process in animals? In what ways is it different?

26. What evidence indicates that mitosis and cytoplasmic division are under genetic control? What is the possible role of the cyclin B/CDK2 complex in the regulation of mitosis?

Supplement 24-1

Cell Division in Bacteria

Although there is actually little in the way of evidence to support the proposal, division of the nuclear material in bacteria has long been believed to involve either the plasma membrane or the cell wall. According to a model first proposed by F. Jacob, the DNA circle of bacterial cells is considered to be attached to the plasma membrane or, by means of connections made through the plasma membrane, to the cell wall (Fig. 24-36a). The connection is proposed to be in the region of the DNA circle containing the replication origin. The attachment site is replicated along with the DNA, producing two complete DNA circles that are both attached to the plasma membrane (Fig. 24-36b and c). After replication is complete, the membrane or cell wall grows between the attachment points, separating the DNA circles and placing them in opposite ends of the cell (Fig. 24-36d and e). Cytoplasmic division (Fig. 24-36e and f) encloses the DNA molecules in separate

daughter cells (Fig. 24-36g). The replicated DNA circles of bacterial cells are tied together, like two links in a chain, as a result of the replication process. A topoisomerase (topoisomerase IV in *E. coli*) breaks and recloses the chains to relieve the tangles.

The Jacob model is supported by observations that bacterial DNA, when isolated, is frequently attached to fragments of the plasma membrane. Various experiments tracing the insertion of new wall material, however, indicate that new molecules are added throughout the cell wall rather than in a growth region restricted to a site between the separating DNA circles. Thus, if membrane attachments do indeed separate the replicated DNA circles of bacteria, the localized growth powering the separation must take place in the membrane itself rather than in the cell wall.

S. Hiroga and his colleagues have discovered a

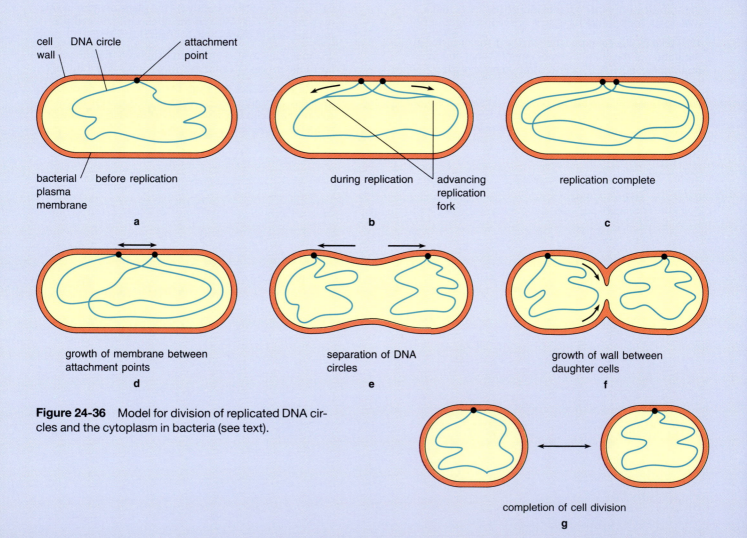

Figure 24-36 Model for division of replicated DNA circles and the cytoplasm in bacteria (see text).

gene, *mukB*, that in mutant form prevents migration of replicated DNA circles to opposite ends of the cell in *E. coli*. When the gene is in mutant form, cells produced by division frequently lack DNA circles (the designation of the gene comes from the Japanese word *mukaku*, meaning "anucleate"). The gene encodes the largest single protein yet discovered in *E. coli*, a 176,826-dalton protein containing 1543 amino acids that has a head and coiled-coil tail and resembles myosin in structure. The structure of MukB and the production of anucleate cells when it is in mutant form suggest that the protein may be a motor powering chromosome separation in *E. coli*.

Cytoplasmic division in bacteria takes place through a process that superficially resembles furrowing in animal cells. A *septum* of wall material gradually extends into the cytoplasm from the inner surface of the cell wall until the cell is cut in two (Fig. 24-37). The septum is lined by an inward extension of the plasma membrane, so that the cell products each retain a complete and unbroken surface membrane when division is complete.

A number of temperature-sensitive mutations have been found to affect septum formation in *E. coli*. Three of these mutant genes, *minC*, *minD*, and *minE*, appear to affect the site at which septum formation takes place. The protein encoded in *minD* activates the *minC* protein; in active form, MinC can induce septum formation at essentially any site in the *E. coli* cell, including polar regions that do not normally form septa. In some way the protein encoded in *minE* limits activity of MinC, and septum formation, to the cellular midpoint.

A series of proteins, including *FtsA, FtsI, FtsQ, FtsW*, and *FtsZ*, all encoded in the same operon, is essential for septum growth. Of these proteins, FtsZ is of special interest because, before septum formation begins, it becomes organized in a ring around the cell at the position in which the septum forms. FtsZ is a GTPase that becomes active in hydrolyzing the nucleotide during septum formation. FtsI is an enzyme with transglycolase and transpeptidase activity. Its

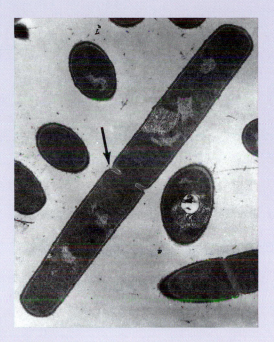

Figure 24-37 Growth of the septum (arrow) dividing the cytoplasm in the bacterium *Bacillus cereus*. × 13,000. (Courtesy of L. Santo.)

normal activity is essential for the reactions that insert new molecules into the peptidoglycan layer. FtsI binds penicillin and is one of the targets of this antibiotic (for this reason, FtsI is often identified as *PBP3*, for *p*enicillin-*b*inding *p*rotein). Penicillin kills bacterial cells by blocking closure of bonds opened to allow wall expansion during cell growth. The wall, weakened by the open bonds, bursts because of the high osmotic pressure developed inside the cell (see p. 119). The entire process of septum formation is inhibited by the protein encoded in the *su1A* gene, which inhibits FtsZ. The gene is activated when DNA is damaged in *E. coli* (the *su1A* product blocks DNA replication as well as septum formation), or when cells are exposed to environmental stress such as elevated temperature.

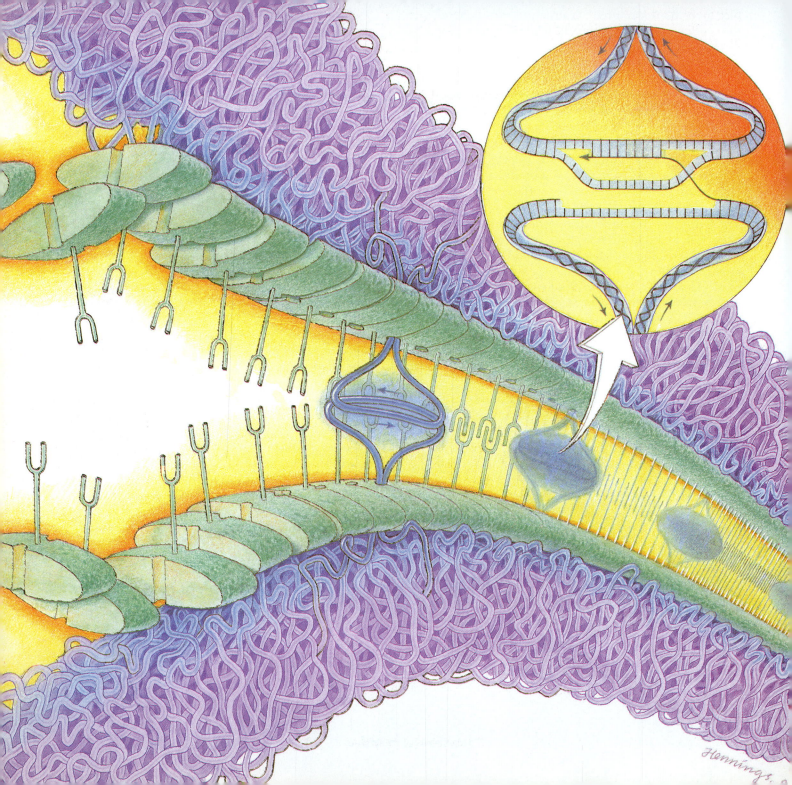

Meiosis takes place only in eukaryotes that reproduce sexually. The meiotic division sequence, which precedes formation of male and female gametes, ensures that the gametes receive half the number of chromosomes present in the body or *somatic* cells of the species. The union of gametes in fertilization returns the chromosome number to the somatic value. Without the reduction in chromosome number that occurs during meiosis, gamete union would double the number of chromosomes in each generation. (The derivation of the term *meiosis*, from the Greek *meioun* = to diminish, reflects this reduction.) Because meiosis reduces the chromosome number by half, the division mechanism is restricted to organisms that have at least the diploid number of chromosomes.

The reduction mechanism is simple in broad outline. In meiosis a single replication is followed by two sequential divisions of the nucleus, which halves the chromosome number. This pattern is in contrast to mitosis, in which replication and division alternate equally.

Besides reducing the chromosome number, meiosis includes two processes that have great significance for the development and survival of sexually reproducing organisms. One is *recombination*, a mechanism by which genetic information is mixed into new combinations. Recombination generates an almost infinite variety of genetic types in each generation. The second process is synthesis of the RNAs and proteins required for gamete production and early stages of embryonic development. This synthesis reaches its highest levels in the development of animal oocytes, which may grow to very large size during meiosis.

This chapter describes the major events of meiotic cell division, including the processes that reduce the chromosome number and mix genetic information into new combinations, and the highly specialized forms of RNA and protein synthesis that take place during meiosis.

OVERVIEW OF MEIOSIS

Meiosis is preceded by a premeiotic interphase during which the DNA replicates, chromosomal proteins are duplicated, and many of the proteins and other molecules required for cell division are assembled. Premeiotic interphase is followed by the two sequential meiotic divisions. During the first of these divisions, which is highly specialized and differs significantly from a mitotic division, the chromosome number is reduced, recombination takes place, and, in animals in particular, RNAs and proteins are synthesized in quantity. The second meiotic division proceeds much like a mitotic division in the same species.

Premeiotic Interphase

The interphase preceding meiosis has G1, S, and G2 stages that resemble their counterparts in a premitotic interphase (see p. 629). However, there are some differences. Typically DNA replication is more extended, so that the S phase takes longer than premitotic S in the same species. Depending on the organism, the greater duration of S may reflect activation of fewer replicons (see p. 680) per unit time or a reduction in the rate at which the DNA unwinds and replicates.

In addition to prolongation of S, a small part of the DNA, about 0.1% to 0.2%, remains unreplicated until meiotic prophase. The unreplicated fraction, first detected by H. Stern and Y. Hotta in premeiotic cells of lily, consists in this plant of about 5000 to 10,000 DNA segments ranging from 1000 to 5000 base pairs in length, distributed throughout the genome. According to Stern and Hotta, a protein, which they term the *L protein*, binds to these sequence segments during premeiotic interphase and blocks their replication. A similar DNA fraction is delayed in replication in the mouse and probably in other sexually reproducing eukaryotes as well. The extended S phase of premeiotic interphase and the unreplicated DNA fraction are probably related in some way to chromosome pairing or genetic recombination (see below).

Meiosis

Each of the two meiotic divisions is separated for convenience in study into the same four stages as mitosis: *prophase*, *metaphase*, *anaphase*, and *telophase* (see p. 691). The stages are identified as part of the first or second meiotic division by a I or II, as in *prophase I* or *prophase II*. An abortive interphase of greater or shorter duration, called *interkinesis*, may separate the two meiotic divisions. No S phase nor any additional DNA replication takes place during interkinesis.

There are many variations in the details of meiosis in different eukaryotes. In general, the spindle forms and functions as in a mitotic division of the same species. Depending on the organism and sex, spindle formation may involve centrioles or spindle pole bodies or may proceed without the presence of obvious polar structures. This chapter concentrates on meiotic division as it takes place in most higher plants and animals.

Meiotic Prophase I

Although continuous, prophase I of meiosis is divided into five sequential substages according to major structural characteristics and activities of the chromosomes: (1) *leptotene*, in which initial chromosome condensation takes place; (2) *zygotene*, in which the chromosomes pair; (3) *pachytene*, in which recombination occurs; (4) *diplotene*, during which extensive RNA transcription and protein synthesis take place in some species; and (5) *diakinesis*, in which condensation becomes complete and final alterations prepare the chromosomes for division. The relative complexity of prophase I is reflected in the time meiotic cells spend in the stage, which may extend for weeks, months, or even years depending on the species.

Leptotene   The onset of leptotene (from *leptos* = fine or thin; Figs. 25-1a and 25-2a) is marked by the initiation of condensation. As in mitosis, condensation involves folding and packing of chromatin fibers from the extended interphase state into thicker, threadlike structures (see p. 694).

Although leptotene is superficially similar to the beginning of mitotic prophase, several differences can be seen in the structure and arrangement of the chromosomes. One is that sister chromatids are so closely associated that they cannot be separately distinguished. As a result, chromosomes appear as single rather than double structures. A second difference, which also becomes obvious as soon as the chromosomes begin to condense, is that the telomeres, or tips of the chromosomes, are attached to the nuclear envelope (diagramed in Figs. 25-2a through 25-2c). The chromosome arms extend from the connections in loops or folds into the nucleoplasm. The telomere attachments may contribute to the mechanisms aligning and pairing the chromosomes (see below).

Zygotene   Leptotene ends and zygotene (from *zygon* = yoke or Y) begins as the chromosomes begin to pair (Figs. 25-1b and 25-2b). In diploid organisms, each chromosome, except for one type of sex chromosome (see Information Box 25-1), is present in two copies as a *homologous pair*. One member, or *homolog*, of each pair is derived from the male parent of the organism and the other from the female parent. The two homologs contain the same genes aligned in the same order, but differing forms of the genes with distinct DNA sequences, called *alleles*, may be present in either homolog. Chromosome pairing, called *synapsis*, proceeds until the two homologs of each homologous pair are aligned along their entire length. The name given to this stage refers to the *Y*- or fork-shaped appearance of the chromosomes at the sites where pairing is in progress.

Initially the homologs move from their original, often widely separated locations in the nucleus and begin pairing at one or more scattered points. Initial pairing frequently, but not always, begins where homologs are attached by their telomeres to the nuclear envelope—the attachment sites move together, and pairing of the chromosomes proceeds inward from the telomeres. The molecular mechanisms accomplishing this first stage of pairing, in which homologous chromosomes may approach from distances amounting to many micrometers, is unknown. Pairing extends from the initial sites in zipperlike fashion until the homologs are completely paired.

Frequently chromosomes of different pairs become entangled during synapsis (Fig. 25-3). The tangles reflect the extended and intertwined arrangement of chromatin fibers during interphase. By the time recombination begins, essentially all interlocking tangles are resolved and eliminated. The tangles are evidently released by DNA topoisomerase II, which can break both nucleotide chains of one double helix, pass a second DNA helix entirely through the first, and correctly reseal the breaks (see p. 373).

The experiments of Stern and Hotta indicate that one part of the DNA left unreplicated in premeiotic interphase in the lily, consisting of segments 3000 to 5000 base pairs in length, completes its replication as the chromosomes pair. Stern and Hotta term these segments *zygDNA* because they replicate during zygotene.

Some of the zygDNA sequences are transcribed during zygotene. The transcription produces a *zygRNA*, which makes up a large fraction of the total RNA of lily meiotic cells at this stage. The zygRNA persists only until recombination begins. Pairing does not take place in the lily if zygDNA replication is inhibited during zygotene, and neither unreplicated zygDNA nor its zygRNA product can be detected in mitotic cells. These observations suggest that zygDNA and its zygRNA product are involved somehow in chromosome pairing during meiotic prophase. A similar group of zygDNA sequences and zygRNA can also be detected in the mouse, suggesting that zygDNA may be generally distributed and active in synapsis.

Although they are closely paired, the synapsed homologs remain separated by a regular space about 1.5 to 2 μm wide. Electron microscopy shows that this space is filled by the *synaptonemal complex* (Fig. 25-4), a ribbonlike structure that assembles as homologs come together. The complex is probably a supportive protein framework that is intimately involved in pairing and recombination (see below). In synapsis it may

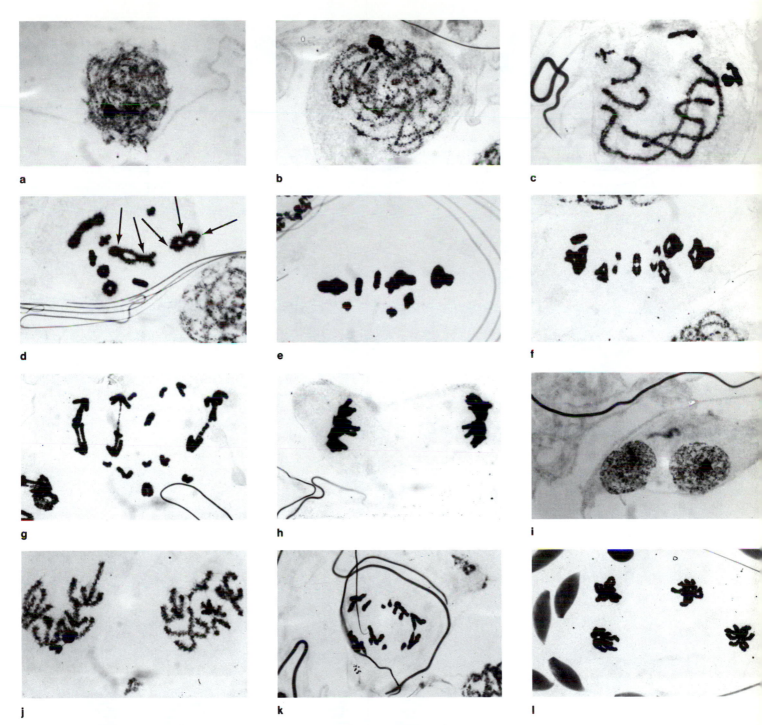

Figure 25-1 Stages of meiosis in the grasshopper *Chorthippus*. **(a)** Leptotene; **(b)** zygotene; **(c)** pachytene; **(d)** diplotene, showing several crossover points (arrows); **(e)** diakinesis; **(f)** metaphase I; **(g)** anaphase I; **(h)** telophase I; **(i)** interkinesis; **(j)** prophase II; **(k)** anaphase II, showing only one of the two spindles; **(l)** telophase II, showing both spindles. × 2,200. (Courtesy of J. L. Walters.)

function to bring the homologs together or to reinforce and stabilize the chromosomes after they have paired.

Pachytene Pachytene (from *pachus* = thick; Figs. 25-1c and 25-2c) begins as pairing becomes complete. As the name suggests, condensation proceeds during pachytene until the chromosome arms are much thicker than in earlier stages. The fully paired homologs are called either *tetrads*, referring to the fact that each consists of four chromatids, or *bivalents*, referring to the fact that each consists of two chromosomes. Visually

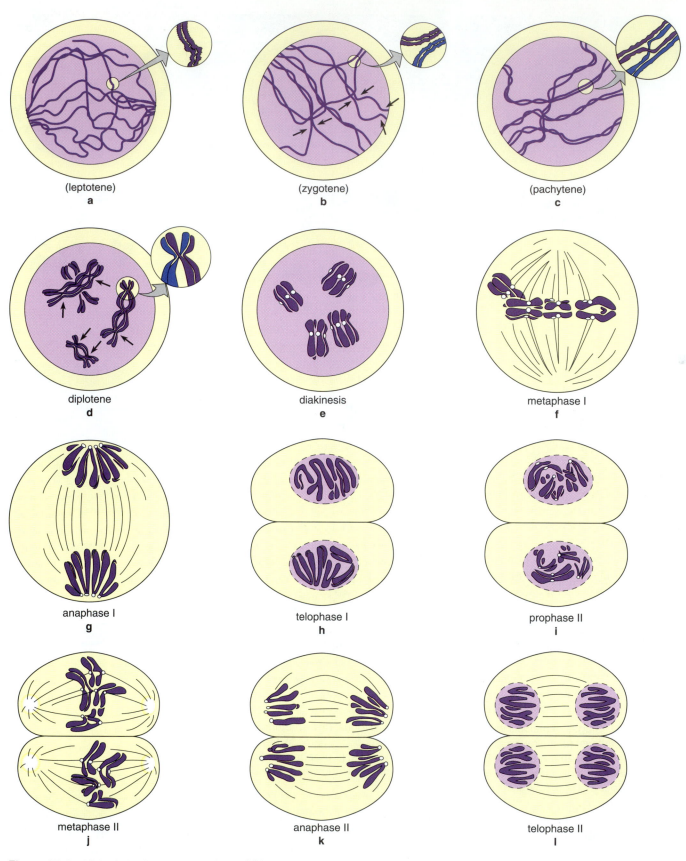

Figure 25-2 Meiosis in diagrammatic form. (a) Leptotene. Although both chromatids are shown for diagrammatic purposes in the magnified circles for parts a, b,
and c of this figure, the split between the two chromatids
of a chromosome is not actually visible during these
stages. (b) Zygotene. Pairing is in progress at several
points (arrows). (c) Pachytene. (d) Diplotene. The chromosomes are held in pairs only by crossovers (arrows)
remaining from recombination during pachytene. All four
chromatids usually become visible in the tetrads at this
time. (e) Diakinesis. (f) Metaphase I. (g) Anaphase I. (h) Telophase I. (i) Prophase II. (j) Metaphase II. (k) Anaphase II.
(l) Telophase II.

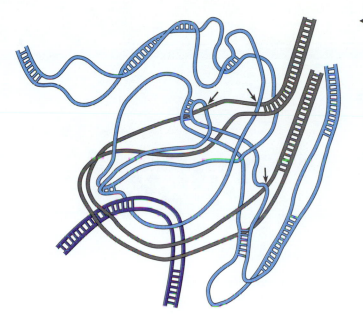

◄ **Figure 25-3** Diagram of tangles (arrows) created between chromosomes of different tetrads during leptotene pairing in a meiotic cell of rye (*Secale*). By pachytene the tetrads are untangled and all interlocks are released, presumably by breakage and rejoining of DNA helices by topoisomerase II. The two chromosomes of one homologous pair are shown in light blue, the second in dark blue, and the third in gray. Paired segments are shown as regions with crosslines between homologs. (Redrawn from an original courtesy of M. Abirached-Darmency and Springer-Verlag, from *Chromosoma* 88:299 [1983].)

Figure 25-4 The synaptonemal complex, assembled at pachytene between a pair of chromosomes in a grasshopper. The inset shows the relationship of the synaptonemal complex to the chromatin fibers of the paired chromosomes. × 74,000. (Micrograph courtesy of P. B. Moens, from *J. Cell Biol.* 40:542 [1969], by copyright permission of the Rockefeller University Press.)

▼

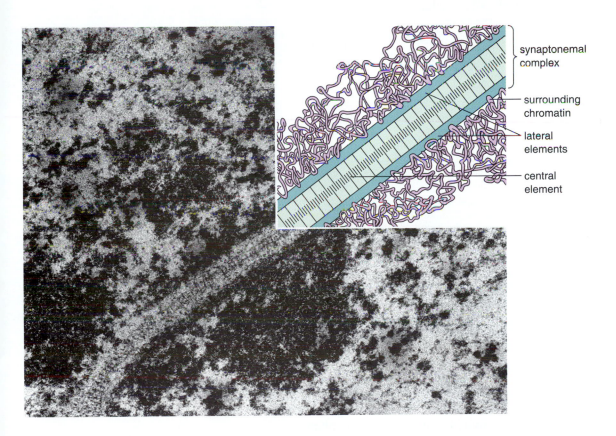

synaptonemal complex

surrounding chromatin

lateral elements

central element

the chromosomes appear as bivalents because only chromosomes, but not individual chromatids, can be discerned at this stage. Pachytene may last for days or even weeks. During this time, homologous chromosomes remain closely paired and retain their attachments to the nuclear envelope.

The primary event of pachytene is genetic recombination. In recombination the chromatids of opposite homologs exchange segments and generate new combinations of alleles. The second, remaining fraction of the DNA left unreplicated during premeiotic S in the lily, which Stern and Hotta term *P-DNA*, replicates during pachytene. The P-DNA, made up of segments ranging from 1000 to 3000 base pairs, includes moderately repetitive sequences and genes encoding enzymes active in DNA nicking and repair. Some of these genes are transcribed during pachytene. The delayed replication of the P-DNA segments, as Stern and

Division of the Sex Chromosomes in Meiosis

In most animals and a few plants, one or more chromosomes are different in structure or number in opposite sexes. The chromosomes that are regularly different in opposite sexes are called *sex chromosomes* (see the figure in this box). Chromosomes that are the same in both sexes are *autosomes*. In most species with sex chromosomes, one sex contains a homologous pair of sex chromosomes and the other sex contains only one chromosome of this pair. By convention the sex chromosomes present in a homologous pair in one sex are called *X chromosomes*. In the opposite sex the single X may exist alone or in association with another sex chromosome, the *Y chromosome*, which does not occur in the other sex. In most animals, females are XX and males are XY or XO (the O indicates that no Y is present). This distribution is reversed in a few animal groups; in butterflies and moths, copepods, and birds, males are XX and females are XY.

In the sex with two X chromosomes the two X's pair normally and pass through the meiotic divisions in the same pattern as the autosomes. In XY individuals the X and Y chromosomes may or may not contain homologous segments capable of pairing during prophase I. (In humans a short region at the tip of one arm of the X and Y is capable of pairing.) Whether pairing occurs or not, the X and Y line up on the spindle along with the rest of the chromosomes at metaphase I of meiosis. Both the X and Y contain two chromatids at this stage, as do the rest of the chromosomes. In most species, particularly where the X and Y undergo partial pairing and recombination, the kinetochores of both chromatids of the X make connections to one pole, and the kinetochores of the Y make connections to the opposite pole, in a pattern similar to the autosomes. In this case the X and Y separate and move to opposite poles of the spindle at anaphase I. Anaphase II separates the chromatids of these chromosomes, so that the haploid products of meiosis receive either an X or Y chromatid. Less frequently the two kinetochores of each sex chromosome make opposite spindle connections at metaphase I, along the lines of a mitotic division, with

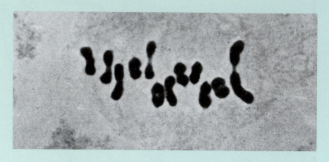

Sex chromosomes (X and Y) in the plant *Silene*. (Courtesy of H. E. Warmke, from *Amer. J. Bot.* 33:648 [1946].)

the result that both poles receive an X and a Y chromatid at anaphase I. The X and Y make opposite spindle connections at metaphase II, so that the final haploid products still receive either an X or a Y.

In XO individuals (common in insects), both chromatids of the single X usually go to the same pole of the spindle at anaphase I. The other pole receives no X. At anaphase II the nucleus receiving the X chromosome divides again and distributes one X chromatid to each of the two division products. The other nucleus, which received no X at anaphase I, yields two O nuclei without X chromosomes. The reverse pattern is also noted, in which the two chromatids of the X make opposite spindle connections and separate at anaphase I. In the second meiotic division the single X chromatid moves to one of the two poles, leaving the other with no X. The result of the two patterns is the same: Of the four nuclei resulting from meiosis, two contain an X chromosome and two have no X (O nuclei). At fertilization, gametes containing an X or Y (or no Y) fuse to form the diploid zygote, restoring the XX or XY (or XO) complement of sex chromosomes.

Hotta suggest, may be connected with the recombination mechanism.

In the developing oocytes of many animals, another pattern of replication makes extra copies of DNA segments containing large pre-rRNA genes during pachytene. The extra copies produced by this pattern of replication, called *rDNA amplification* (see also p. 683), are released to the nucleoplasm. During the diplotene stage the amplified rDNA copies form extra nucleoli that actively synthesize rRNAs at this time. The rDNA amplification is evidently a mechanism that increases the cell's capacity to synthesize rRNAs

and assemble the large numbers of ribosomes appearing in the oocyte cytoplasm during the next stage of prophase I.

Diplotene The close of pachytene and the beginning of diplotene (from *diplos* = double; Figs. 25-1d and 25-2d) is marked by pronounced changes in chromosome structure and activity. At this stage, sister chromatids become individually visible, making each chromosome appear clearly double (the name given to the stage reflects this double appearance). As a consequence, all four chromatids of a tetrad can be in-

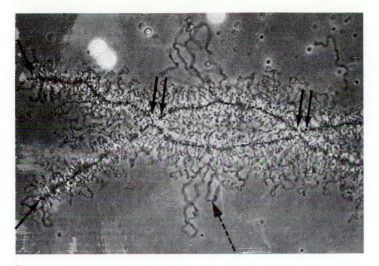

Figure 25-5 Lampbrush chromosomes. **(a)** Part of a lampbrush chromosome isolated from the newt *Triturus*, as seen in the phase contrast light microscope. The single arrows indicate the two chromosomes of the tetrad, held together at two points by crossovers (double arrows). Multiple paired loops extend from each chromosome; one loop of each pair (dashed arrow) extends from one chromatid of the homolog, and the second loop from the other chromatid. × 400. (Courtesy of J. G. Gall.) **(b)** The relationships among loops, chromatids, and chromosomes in a lampbrush chromosome.

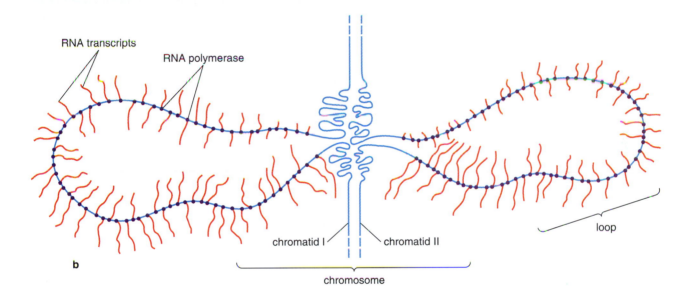

dividually distinguished. The separation between homologs also widens during diplotene. At the same time, the synaptonemal complex disappears from most sites between homologs, and the chromosomes lose their attachments to the nuclear envelope.

The points at which homologous pairs remain attached to each other during diplotene (arrows in Fig. 25-1d) are sites at which two of the four chromatids cross over between the homologs (shown diagrammatically in Fig. 25-2d). These crossing places, called *crossovers* or *chiasmata* (singular *chiasma* = "crosspiece"), reflect chromatid exchanges taking place during recombination in the previous pachytene stage. The number and position of the crossovers vary in different cells undergoing meiosis in the same species. However, there is usually at least one crossover per chromosome arm. Electron microscopy reveals that small segments of the synaptonemal complex persist at the crossovers,

possibly as reinforcements holding the homologs together.

In many animals the chromosomes decondense to a greater or lesser extent and become active in RNA transcription during diplotene. The degree of decondensation varies from barely discernible (as in Fig. 25-1d) to almost complete reversion to an interphaselike state. In some organisms, particularly female amphibians, birds, and reptiles, decondensation at this stage produces a *lampbrush* configuration in which much of the chromatin forms loops that extend outward from the chromatids (Fig. 25-5). The loops are sites of intensive RNA transcription throughout diplotene.

Decondensation and the formation of lampbrush chromosomes reflect intensive activity in transcription of mRNA, rRNA, and tRNA during the diplotene stage. Structurally the rRNA synthesis is marked by extensive growth of nucleolar material. If extra rDNA sequences have been produced by DNA amplification, these also become active in transcription and develop into extra nucleoli called *accessory nucleoli*. The accessory nucleoli take part in intensive synthesis of rRNAs and assembly of ribosomal subunits during diplotene.

The cytoplasm of animal oocytes grows extensively during diplotene through production of ribosomes, multiplication of cytoplasmic organelles, and synthesis of large quantities of proteins, fats, carbohydrates, and other cytoplasmic components. In birds and reptiles, oocytes become truly massive during diplotene growth—the yolk of a hen's egg, for example, is a single, large cell produced by diplotene growth. Most of this growth is limited to the cytoplasm; the oocyte nucleus remains microscopic or in some species becomes just large enough to be visible to the naked eye.

Diplotene may be greatly extended. In developing amphibian oocytes this part of meiotic prophase may last up to nearly one year. In humans, oocytes reach diplotene in unborn females at about the fifth month of fetal life and remain arrested in this stage during the remainder of prenatal life, through birth and childhood, and until the individual reaches sexual maturity. Then, just before ovulation, one oocyte each month breaks diplotene arrest and continues the meiotic sequence. The time between the onset and completion of diplotene in human females may thus amount to as much as 50 years or more. Intensive RNA and protein synthesis, however, occurs only during the part of the stage preceding arrest in the developing fetus.

Diakinesis   At the close of diplotene the chromosomes recondense and pack further into short rodlets. The loops of lampbrush chromosomes, if present, are withdrawn, all RNA transcription ceases, and nucleoli disappear. These changes mark the transition to the final stage of meiotic prophase I, diakinesis (from *dia* = across and *kinesis* = movement; Figs. 25-1e and 25-2e). The name refers to the fact that chromosomes become so compact at this stage that the chiasmata are forced to move laterally, toward the tips of the chromosomes. The chromosomes are now ready to enter metaphase of the first meiotic division.

Meiotic prophase I produces four results of significance to the outcome of meiosis: (1) chromosome condensation, (2) synapsis, (3) recombination, and (4) synthesis of most or all of the RNA, protein, lipid, and carbohydrate molecules required for gamete differentiation and the early stages of embryonic development. The remaining events of meiosis are concerned primarily with the divisions that reduce the chromosome number.

The Meiotic Divisions

Near the end of prophase I the spindle for the first meiotic division forms. Centrioles, if present, complete their replication, and the two resultant centriole pairs migrate within their asters to opposite sides of the nucleus and form the spindle poles as they do in mitosis. Centrioles persist throughout meiosis in male animals, so that sperm cells receive centrioles. In the oocytes of many female animals the centrioles disappear at some point during the meiotic divisions, so that the mature eggs of these species contain no centrioles. These are replaced during fertilization by centrioles introduced by the sperm cell.

Metaphase I   As in mitosis, breakdown of the nuclear envelope provides a convenient reference point for the transition from prophase I to metaphase I (Figs. 25-1f and 25-2f). The developing spindle invades the position formerly occupied by the nucleus, and the tetrads, left scattered by breakdown of the nuclear envelope, move to the spindle midpoint. Except for the pairing of homologous chromosomes into tetrads, meiosis at this point resembles the transition from prophase to metaphase in mitosis.

The first major divergence from the mitotic pattern—and, in fact, the distinctive event of the first meiotic division—occurs as the tetrads attach to spindle microtubules. In each tetrad the two sister chromatids of one homolog attach to microtubules leading to one spindle pole, and the sister chromatids of the other homolog attach to microtubules leading to the opposite pole (Fig. 25-6a). This pattern differs fundamentally from mitosis, in which sister chromatids make connections to opposite spindle poles (Fig. 25-6b).

A series of observations by R. B. Nicklas demonstrated clearly that the metaphase I attachments are functions of the chromosomes, and not of factors in the surrounding cytoplasm. If a metaphase I tetrad was pushed into a metaphase II spindle, connections were still made as they would be in a metaphase I spindle, with both chromatids of a chromosome connected to the same pole. Conversely, the two chromatids of a metaphase II chromosome, if pushed into a metaphase I spindle, made connections leading to opposite spindle poles as they normally would at metaphase II. This behavior indicates that the metaphase I kinetochore connections and the opposite connections made in metaphase II depend on whether the chromosomes are paired as tetrads or are single. Pairing as tetrads during metaphase I, in turn, depends on chiasmata. An extended series of observations made over many years has shown that if chiasmata are disturbed or are absent from a homologous pair, the homologs usually come apart and make random, metaphase II–like attachments to the spindle at metaphase I.

Although the connections made by homologs are normally opposite, mistakes do occur at a low but constant rate. These mistakes cause both chromosomes of a homologous pair to move to the same pole at anaphase I. As a result of this *nondisjunction* of homologs, two of the nuclei resulting from meiosis receive an extra copy of one chromosome, and two are deprived entirely of a copy. The complete loss of a chromosome is usually fatal to a zygote formed by a deprived gamete. However, zygotes receiving an extra copy, which gives them three copies of one chromosome instead of two, are often viable but develop into individuals that have varying degrees of abnormality. In humans, for example, *Down's syndrome* results from nondisjunction of chromosome 21 during formation of one of the two gametes uniting to form the individual, giving the individual three copies of this chromosome.

Anaphase I Anaphase I begins as the tetrads are pulled apart and homologs start their movement toward opposite spindle poles (Figs. 25-1g and 25-2g). As a result of the movement, homologous pairs are separated and the poles receive the haploid number of chromosomes. Anaphase I therefore accomplishes the reduction in chromosome number that is one of the primary outcomes of meiosis. However, each chromosome still contains two chromatids at this stage.

Different homologous pairs separate independently from each other at anaphase I, so that any combination of chromosomes of maternal and paternal origin may be delivered to a spindle pole. The random combinations delivered to the poles contribute to the total genetic variability of the products of meiosis. (The random combinations account for the independent assortment of alleles first noted in genetic crosses by Mendel.) The 23 chromosome pairs of humans, for example, allow 2^{23} different combinations of maternal and paternal chromosomes to be delivered to the poles, producing a potential 8,400,000 different types of gametes from this source of variability alone. Even without recombination, therefore, the possibility that two children of the same parents could receive the same combination of maternal and paternal chromosomes is 1 chance out of $(2^{23})^2$, or one in 70 trillion! The further variability introduced by recombination, which mixes chromatid segments randomly between maternal and paternal chromatids, makes it practically impossible for meiosis to produce two genetically identical gametes in humans and most other sexually reproducing organisms.

Telophase I and the Interphase Between Divisions I and II The degree to which meiotic cells proceed through

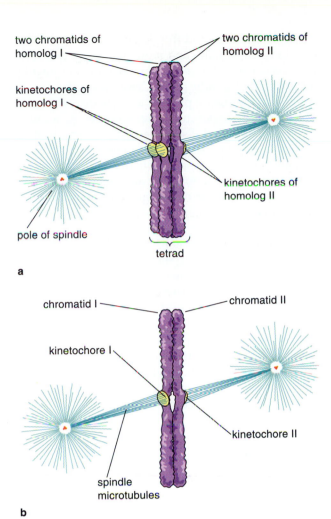

Figure 25-6 Kinetochore connections in meiosis and mitosis. **(a)** Spindle connections made by the chromosomes of a tetrad at metaphase I of meiosis (equivalent to Fig. 25-2f). Both kinetochores of one chromosome of the pair make connections leading to the same spindle pole; both kinetochores of the other chromosome of the pair connect to the opposite pole. **(b)** Spindle connections made by chromosomes at metaphase of mitosis or at metaphase II of meiosis. The two kinetochores of the chromosome connect to opposite spindle poles.

recognizable telophase I and interkinesis stages (see Figs. 25-1h and and 25-2h) varies widely among different species. In most animals, telophase I and the interkinesis between the two divisions are so transitory that meiotic cells essentially move directly from anaphase I to prophase II. In monocotyledonous plants, in contrast, the chromosomes decondense completely and nuclear envelopes temporarily surround the polar masses of chromatin during telophase I and interkinesis. All possible gradations between these extremes are found in nature. No DNA replication occurs at this time in any known organism.

During this period the single spindle of the first division disassembles and reorganizes into two spindles that form in the regions previously occupied by the anaphase I spindle poles. Centrioles and asters, or spindle pole bodies, if present, also divide at this time. The centrioles do not duplicate, however. As a result, centriole and aster division places a single centriole at each pole of the two spindles in cells in which centrioles are present. (This centriole divides later, during sperm development, in many animals.) These events complete the cellular rearrangements leading to the second meiotic division.

The Second Meiotic Division　After a brief prophase II (see Figs. 25-1j and 25-2i) the chromosomes left at the two poles of the first meiotic division move to the midpoints of the two newly formed spindles. If decondensation has occurred during telophase I, the chromosomes condense again into tightly coiled rodlets as they move to the metaphase II spindle midpoint. Attachment to the spindle takes place in the same pattern as mitosis: The two kinetochores of each chromosome make connections to spindle microtubules leading to opposite poles (Fig. 25-2j).

As a consequence of the opposite kinetochore connections, sister chromatids separate and move to opposite spindle poles at anaphase II (Figs. 25-1k and 25-2k). Each pole therefore receives the haploid number of chromatids. Sex chromosomes, when present, are distributed among the products of meiosis in patterns that depend on whether they occur in a pair or singly in the diploid cells entering meiosis and on whether they contain homologous segments that can synapse and undergo recombination (see Information Box 25-1 for details).

At telophase II (see Figs. 25-1l and 25-2l) the chromatids decondense and nuclear envelopes form around the four division products. These four nuclei have widely divergent fates in various species of animals and plants. In male animals, each nucleus is enclosed in a separate cell by cytoplasmic division, and each cell differentiates into a functional sperm cell. In female animals, only one of the four nuclei becomes functional as the egg nucleus. The other three are compartmented by unequal division of the egg cytoplasm into small, nonfunctional cells called *polar bodies* at one side of the oocyte (Fig. 25-7). This unequal division concentrates most of the cytoplasm into a single large cell that develops into the oocyte. Higher plants undergo a somewhat similar developmental pattern, in which all four products of meiosis give rise to functional sperm nuclei. Functional egg nuclei, however, arise from only one of the four meiotic products. (Figure 25-8 [page 740] summarizes the events of meiotic cell division; Figure 25-9 [page 741] outlines the time and place of meiosis in different major taxonomic groups.)

Meiosis may enter arrest at various points in different animal oocytes. In human females, as noted, an initial arrest takes place at diplotene of prophase I and persists until the time of ovulation. Meiosis then resumes and proceeds to metaphase II, when the human egg enters a second arrest that persists until fertilization. A similar pattern is followed in the mouse and rat. Molluscan eggs arrest in prophase I and remain arrested at this stage until fertilization; frog eggs arrest at metaphase II.

In higher vertebrates, meiotic arrest at metaphase II is imposed by *cytostatic factor* (*CSF*), a protein complex that in some manner maintains activity of the cyclin B/CDK2 or MPF complex until the arrest is broken at fertilization. As long as cyclin B/CDK2 remains active, neither mitotic cells nor meiotic cells can make the transition from metaphase to anaphase (see p. 641 and below). CSF contains *Mos*, a protein kinase capable of adding phosphate groups to cyclin B. As cells break arrest and enter anaphase, the cyclin B/CDK2 complex is inactivated and CSF is degraded by a Ca^{2+}-dependent proteinase. The proteinase is probably activated by Ca^{2+} released into the cytoplasm as part of the egg's response to fertilization.

RNA Transcription and Protein Synthesis During Meiosis

RNA transcription can be detected during prophase I in all organisms. This has been demonstrated, for example, by autoradiographs of cells exposed to ^{3}H-uridine, a radioactive precursor of RNA. This transcription, which includes both mRNA and rRNA, peaks in many species during the diplotene stage. In some animal oocytes, enormous quantities of ribosomes may be synthesized and packed into the cytoplasm at diplotene. As cells reach metaphase I, RNA transcription drops to undetectable levels. A brief period of RNA transcription, primarily of mRNA, may occur again after telophase II during sperm cell development in some animal species.

Incorporation of radioactive amino acids into proteins can be detected throughout meiosis. In many female animals, particularly those with large, yolky eggs, protein synthesis reaches its peak during the diplotene stage of meiotic prophase I. Most of this protein is stored in the egg cytoplasm and remains inactive until fertilization.

Many parts of meiosis are not completely understood and are currently under investigation. Among the most important are the mechanism of recombination, the role of the synaptonemal complex in this mechanism, and the genes regulating entry into meiosis and the progress of the meiotic stages. The following sections of this chapter take up these subjects in greater detail.

THE MECHANISM
OF RECOMBINATION

Classical Recombination

Early in this century, classical geneticists discovered the overall characteristics of recombination (Fig. 25-10). Prior to recombination the two members of a homologous pair might contain one gene with the alleles A and a at one site and another gene with the alleles B and b at a different site (Fig. 25-10a). After replication (Fig. 25-10b), pairing during meiotic prophase I brings the homologs together and places the genes in side-by-side register (Fig. 25-10c). Recombination between these genes (Fig. 25-10d and e) produces two chromatids with the new allelic combinations A-b and a-B (Fig. 25-10f), while two of the four chromatids remain unchanged, with the original A-B and a-b combinations. The two chromatids with the new combination of alleles are termed *recombinants*, and the two that remain unchanged are *parentals*. Because the exchange is equal and all the alleles under study are retained, this pattern of change is termed *reciprocal recombination*.

With the techniques available to classical geneticists, sites over most of the chromosome arms, with the exception of a few regions such as centromeres and telomeres, seemed equally likely to undergo recombination. Therefore the positions at which recombination occurs were considered to be random.

The frequency of recombination between two genes was found to depend on the distance between them on a chromosome. As the distance increases or decreases, the space available for a random "hit" by the recombination mechanism increases or decreases in direct proportion. To classical geneticists the smallest detectable distance for recombination seemed to be the separation between adjacent genes. The gene was therefore thought to be the unit of recombination, and it was considered unlikely that recombination could occur within the boundaries of a gene. These conclusions of classical genetics, that recombination (1) is reciprocal, (2) is random except at certain locations, and (3) involves the gene as the unit of exchange, each required revision when the techniques of molecular genetics allowed very rare recombination events to be studied.

The recombination events detectable by the techniques of classical genetics were shown, with a fair degree of certainty, to correspond to the crossovers that become visible between chromosome arms late in prophase I of meiosis. In general, the number of detectable recombination events on single chromosome arms averages about the same as the number of crossovers, about one per arm. In addition, mutations that reduce the recombination rate cause a proportionate reduction in the number of visible chiasmata.

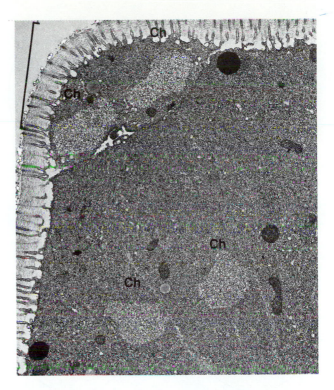

Figure 25-7 An egg of the surf clam *Spisula* with a polar body visible at one side (bracket in main figure; arrow in inset). Ch, chromatin. Main figure × 8,000; inset × 175. (Courtesy of F. J. Longo, from *J. Ultrastr. Res.* 33:495 [1970].)

The structure of crossovers, in which chromatids appear to switch between homologs, led many classical investigators to assume that recombination takes place by a process of physical breakage and exchange of chromatid segments. Because crossovers appear during meiotic prophase, recombination was assumed to take place at this time. These conclusions, that recombination takes place by breakage and exchange of chromatid segments and occurs during meiotic prophase, have been fully supported by subsequent research.

Experiments Establishing That Recombination Occurs by Breakage and Exchange During Prophase I of Meiosis

The definitive experiment establishing that chromatids break and exchange segments during recombination was carried out by J. H. Taylor in 1965, using meiotic cells of the grasshopper *Romalea*. Taylor injected grasshoppers with a radioactive DNA precursor (tritiated thymidine) and followed the distribution of label in meiotic cells by autoradiography (see Appendix p. 791). The cells of interest in the experiment

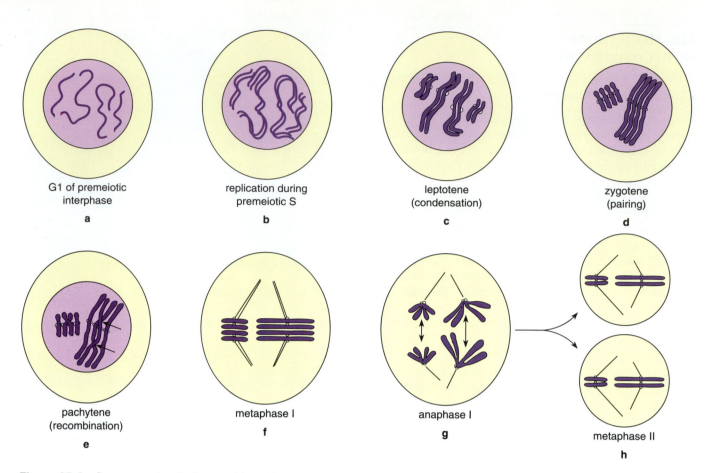

G1 of premeiotic interphase	replication during premeiotic S	leptotene (condensation)	zygotene (pairing)
a	b	c	d

pachytene (recombination)	metaphase I	anaphase I	metaphase II
e	f	g	h

Figure 25-8 Summary of meiosis, considered in a hypothetical cell containing only two pairs of chromosomes, one long and one short. At premeiotic G1 **(a)** the chromosomes of these pairs are unassociated in the nucleus. During premeiotic S **(b)**, each chromosome replicates, becoming double at all points, and now contains two chromatids. After G2, as leptotene begins **(c)**, the chromosomes condense into threads that become visible in the light microscope. Synapsis at zygotene **(d)** brings homologous chromosomes together, producing two tetrads in the nucleus, each containing two chromosomes (four chromatids). During pachytene **(e)**, recombination occurs by the exchange of segments between the chromatids of the homologs (arrows). At the next stage, diplotene (not shown), at least some decondensation with RNA transcription takes place in most species. The chromosomes condense again at diakinesis (not shown). At metaphase I **(f)** the two chromatids of each homologous chromosome connect to microtubules leading to the same spindle pole. Anaphase I **(g)** separates the two homologs of each pair and moves the haploid number of chromosomes to each spindle pole. Thus only one long chromosome and one short chromosome are present at each pole; pairs no longer exist, but the chromosomes at the poles still contain two chromatids. During interphase II and prophase II (not shown), two new spindles form at the metaphase I division poles. At metaphase II **(h)** the chromosomes move to the spindle midpoints and make microtubule attachments as in mitosis, in such a way that the two chromatids of each chromosome connect to microtubules leading to opposite spindle poles. Separation of the chromatids at anaphase II **(i)** delivers the haploid number of chromatids to the poles. The four division products at telophase II **(j)** each now contain only two chromatids (now technically called chromosomes), one long and one short. If, in this hypothetical organism, meiosis and gamete formation produce eggs and sperm, the end products will each contain two chromosomes, one long and one short. Fusion of an egg and sperm nucleus in fertilization **(k)** rejoins the pairs and returns the chromosome complement to the G1 level of premitotic cells.

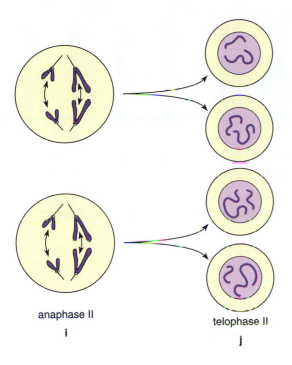

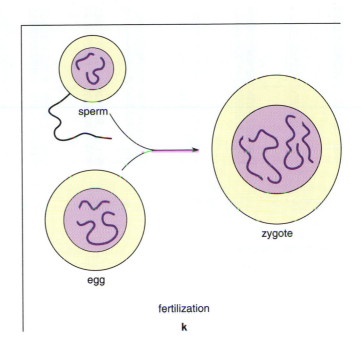

anaphase II

i

telophase II

j

sperm

egg

zygote

fertilization

k

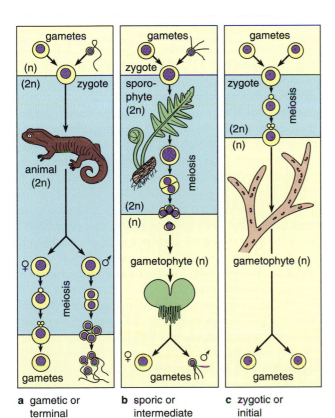

a gametic or terminal

b sporic or intermediate

c zygotic or initial

Figure 25-9 Three major patterns in the time and place of meiosis in eukaryotes (see text). The blue portions mark the diploid phase of the life cycle. n = haploid number of chromosomes; 2n = diploid number. **(a)** The gametic or terminal pattern, in which the meiotic divisions occur immediately before gamete formation. **(b)** The sporic or intermediate pattern, which takes place in higher plants and in some algae and fungi. In these organisms, fertilization produces a diploid *sporophyte* generation. Meiosis occurs in this generation, producing spores that germinate and grow by mitotic divisions into haploid individuals of the alternate *gametophyte* generation. The gametophyte generation produces eggs and sperm by differentiation of cells following mitotic divisions. Fusion of the gametes returns the cycle to the diploid sporophyte generation. **(c)** The *zygotic* or *initial* pattern typical of many fungi, in which meiosis takes place immediately after fertilization. Eggs and sperm (or simply + and − gametes) are produced by differentiation of cells following ordinary mitotic divisions. Gametes fuse to form a zygote, which immediately enters meiosis, producing four haploid cells that develop by mitotic divisions into haploid spores. The spores germinate into haploid individuals; the diploid condition is limited to the zygote. (Redrawn with permission of the Macmillan Company, Inc., from *The Cell in Heredity and Development*, by E. B. Wilson. Copyright © 1925 by Macmillan Publishing Company, Inc., renewed 1953 by Anne M. K. Wilson.)

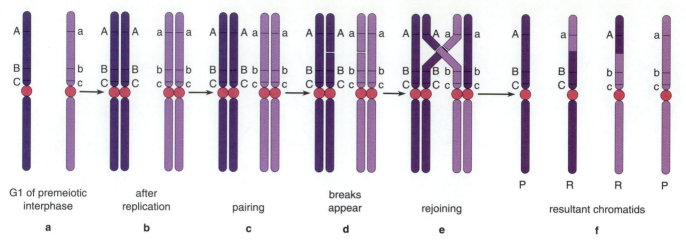

G1 of premeiotic interphase	after replication	pairing	breaks appear	rejoining	resultant chromatids
a	**b**	**c**	**d**	**e**	**f**

Figure 25-10 Recombination as determined by the techniques of classical genetics. **(a)** Before recombination the two members of a homologous pair contain a gene with the alleles A and a at one site and another gene with the alleles B and b at a different site. After replication **(b)**, pairing during meiotic prophase I brings the homologs together and places the genes in side-by-side register **(c)**. Recombination by **(d)** breakage and **(e)** exchange between these genes produces two chromatids with the new allelic combinations A-b and a-B **(f)**; these are the recombinant chromatids (R). Two of the four chromatids remain unchanged and retain the A-B and a-b combination; these are parental chromatids (P).

were those replicating their DNA in the interphase just before the last premeiotic mitosis (Fig. 25-11). Before replication began, none of the DNA was labeled (Fig. 25-11a). After replicating in the presence of label, cells entered mitotic metaphase with one nucleotide chain of each DNA helix labeled and one unlabeled (Fig. 25-11b). As a consequence, all chromatids showed the presence of radioactive label. Following this last premeiotic division the cells entered the interphase before meiosis. At this point, excess label was washed from the tissues and replication took place in unlabeled medium. Semiconservative replication of each chromosome produced two chromatids, one showing the presence of label and one completely unlabeled (Fig. 25-11c).

If no physical exchanges occurred between chromatids as part of recombination during meiotic prophase I, any single chromatid at the subsequent metaphase would be either completely labeled or unlabeled (as in Fig. 25-11d). If breakage and exchange did take place, chromatids with labeled and unlabeled segments would be detectable at the subsequent meiotic divisions (Fig. 25-11e). Taylor actually observed chromatids with labeled and unlabeled segments (Fig. 25-12). Because these could arise only by physical breakage and exchange between labeled and unlabeled chromatids, Taylor's experiment and subsequent experiments of the same type established that this mechanism underlies recombination in higher organisms. Later experiments using equivalent methods showed that recombination also occurs by breakage and exchange in prokaryotes.

The demonstration that recombination takes place during prophase I of meiosis came from the research of J. M. Rossen and M. Westergaard in 1966. The demonstration was significant because many investigators assumed at the time that recombination takes place during replication and involves a modification of the replication mechanism called *copy choice* in which the template used for assembly of a copy chain switches back and forth between homologs. For their experiment, Rossen and Westergaard used the fungus *Neottiella*, which follows the initial pattern of meiosis (see Fig. 25-9c). The gametes, derived from haploid body cells after an ordinary mitosis, fuse to form a diploid zygote. Meiosis then takes place immediately; the diploid condition persists for only the brief period between gamete fusion and anaphase I. Since the only time both chromosomes of each homologous pair are present in the same nucleus is during prophase I of meiosis, this is the only time recombination could take place.

Rossen and Westergaard found by measuring the quantity of DNA per nucleus that replication occurs while the gamete nuclei are still haploid, before they fuse to form the zygote. This result established that replication and recombination are separate events and eliminated the possibility that recombination might occur as a part of replication.

Intragenic Recombination

Geneticists began to discover exceptions to the conclusions of classical genetics when they took up the

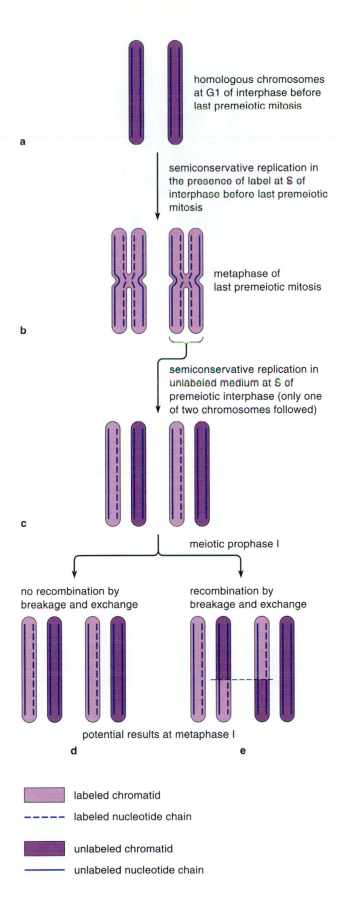

a — homologous chromosomes at G1 of interphase before last premeiotic mitosis

semiconservative replication in the presence of label at S of interphase before last premeiotic mitosis

b — metaphase of last premeiotic mitosis

semiconservative replication in unlabeled medium at S of premeiotic interphase (only one of two chromosomes followed)

c

meiotic prophase I

no recombination by breakage and exchange

recombination by breakage and exchange

potential results at metaphase I

d   e

▮ labeled chromatid

- - - - - labeled nucleotide chain

▮ unlabeled chromatid

——— unlabeled nucleotide chain

Figure 25-11 J. H. Taylor's experiment demonstrating that physical breakage and exchange of chromatid segments take place during meiosis in the grasshopper *Romalea* (see text). Chromosomes with reciprocally exchanged segments as in Fig. 25-10f were actually observed in the experiment, showing that recombination took place by physical breakage and exchange.

study of inheritance in bacteria, viruses, and fungi. In these forms the number of offspring is potentially very large, and generation times are on the order of hours or minutes, allowing offspring with traits resulting from very rare recombinational events to be detected. Detection of rare events is aided by the fact that recombination occurs at a much higher frequency in bacteria and viruses than in eukaryotes.

Although viruses and bacteria do not undergo meiosis, recombination takes place readily, by molecular mechanisms that appear to be the same or similar to those of eukaryotes. In viruses, recombination occurs in host cells, between DNA molecules originating from different viruses infecting the same cell.

In bacteria, recombination occurs between the main DNA circle and plasmids or between the main circle and segments of DNA originating from outside the cell. There are three primary routes by which DNA enters bacterial cells from outside. In *conjugation* a cytoplasmic bridge forms between two cells of the same species, and part or all of the DNA circle or plasmids of one cell pass through the bridge into the other cell. In *transformation*, bacteria take up DNA fragments directly from the surrounding medium. This route of entry is the primary route by which DNA is introduced into bacterial cells in cloning techniques (see Supplement 14-1). In the third route, *transduction*, fragments of DNA from a bacterial cell infected by a

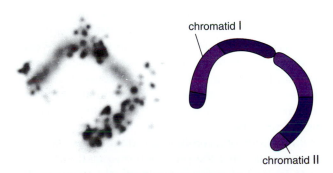

chromatid I

chromatid II

Figure 25-12 Two chromatids recovered at metaphase II of meiosis in Taylor's *Romalea* experiment. The darker color in the tracing shows the labeled segments. Reciprocal exchange of labeled segments in this manner could take place only by physical breakage and exchange. × 4,000. (Courtesy of J. H. Taylor, from *J. Cell Biol.* 25:67 [1965], by copyright permission of the Rockefeller University Press.)

Figure 25-13 Evaluation of the results of genetic crosses in *Neurospora*. (a) An ascus containing eight spores, produced by one mitotic division following meiosis. (b) A cluster of asci containing spores. The transparent walls of the asci are not visible in this light micrograph. The alleles under study in the cross produce light and dark colors in the spores. Reciprocal recombination, producing a 4:4 segregation of alleles, can be seen in the asci marked by arrows; 6:2 segregation, resulting from loss of an allele due to gene conversion, is visible in the asci marked by asterisks. (Courtesy of H. L. K. Whitehouse, from *Genetic Recombination*. Copyright © 1982 John Wiley and Sons, Inc.)

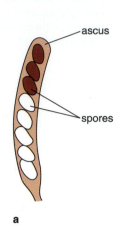

ascus

spores

a

b

virus are incorporated into viral particles and introduced into another cell during a subsequent infection. If the DNA fragments introduced by conjugation, transformation, or transduction are sufficiently homologous with the recipient cell DNA, a diploid condition is set up over the region of homology. Recombination may occur within the diploid region.

Genetic crosses with bacteria and viruses revealed very quickly that recombination can occur within the boundaries of a gene, or even within a codon. This type of genetic exchange, called *intragenic recombination*, may appear in only one out of hundreds of thousands or even millions of offspring. However, since billions of offspring can easily be produced in bacterial and viral crosses, these rare events can readily be detected.

Once intragenic recombination was established as a fact in viruses, prokaryotes, and some lower eukaryotes such as the fungus *Neurospora*, laborious observations in maize and *Drosophila*, requiring many years to complete, demonstrated that intragenic recombination also occurs in higher eukaryotes. This work made it clear that the unit of recombination in both prokaryotes and eukaryotes is potentially as small as a single base pair.

The study of recombination at the molecular level also revealed that the process is frequently *nonreciprocal*. Instead of appearing in perfectly reciprocal 2:2 ratios, as in classical genetic crosses, intragenic recombination sometimes produces a 1:3 ratio of alleles in offspring, as if one of the alleles has been converted into the opposite type. In general, the pattern of nonreciprocal recombination in which an expected allele apparently disappears and is replaced by the opposite allele is termed *gene conversion*.

Other unexpected distributions of alleles are also noted in intragenic exchanges. Some of the most interesting have been detected in *Neurospora* and other ascomycete fungi, in which the four cellular products of meiosis are enclosed in a sac called the *ascus* (Fig. 25-13a). These four cells undergo two mitotic divisions following meiosis, producing a mature ascus containing 16 haploid spore cells. Because the walls of the ascus are too narrow for the spores to slip past each other, the nucleus of each spore can be traced back to a particular cellular product of meiosis. This allows each of the four chromatids of a tetrad to be followed and assigned to an individual spore nucleus.

In *Neurospora*, genetic analysis is usually carried out after the first postmeiotic mitosis, using asci containing eight spores (as in Fig. 25-13a and b). At this eight-spore stage, a perfect 2:2 reciprocal exchange produces four spores containing one allele and four containing a different allele (as in asci marked with an arrow in Fig. 25-13b). A nonreciprocal recombination giving a 3:1 segregation after meiosis can be directly observed as a 6:2 combination at the eight-spore stage (asci marked with an asterisk in Fig. 25-13b).

Asci are also found with a 5:3 combination of alleles (as diagramed in Fig. 25-13a), indicating that one of the nuclei resulting from meiosis, even though haploid, contained both alleles at the four-cell stage. For this to happen, a DNA molecule in the region containing the alleles must be a hybrid, or *heteroduplex*, molecule—that is, must contain nucleotide chains originating from opposite homologs (Fig. 25-14a). During the first mitotic interphase following meiosis, these chains would separate (Fig. 25-14b) and serve as templates for replication. Because the templates contain noncomplementary sequences over short lengths, the

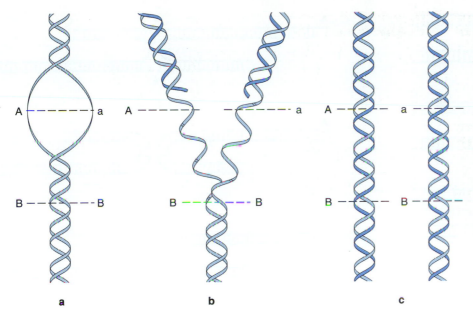

Figure 25-14 How a segment of heteroduplex DNA can account for postmeiotic segregation of alleles. **(a)** A DNA molecule containing a short heteroduplex region in which the opposite chains encode two different alleles A and a. **(b)** Unwinding of the chains during the first postmeiotic replication. **(c)** Completion of replication. The noncomplementary regions containing the A and a alleles have served as templates, producing two separate DNA molecules with different alleles in the region corresponding to the heteroduplex segment in **(a)**.

two DNA molecules resulting from replication would have different sequences in these regions (Fig. 25-14c). If the noncomplementary regions contain a genetic marker included in a cross, these differences would appear as opposite alleles following the first postmeiotic mitosis, producing the 5:3 distribution noted at the eight-spore stage in *Neurospora*. This pattern of recombination, producing differences in alleles that appear in the mitotic division following meiosis, is called *postmeiotic segregation*.

Another novel characteristic of recombination at the molecular level involves *nonexchange of flanking markers*. When a breakage and exchange event occurs, genetic markers located close to an allele under study are expected to be transferred with it to the opposite chromatid, as long as they lie within the segment being exchanged. For example, because they lie close together, allele C in the chromosome on the left in Figure 25-10a would most often be transferred with allele B to the opposite chromatid by the depicted breakage and exchange (as shown in the second chromatid from the left in Fig. 25-10f). The only event expected to prevent this transfer is a second breakage and exchange between the B and C genes, which would return allele C to its original chromatid. Second crossovers of this type are expected to be rare, however, and become rarer as the distance between the alleles under study becomes smaller. Within the boundaries of a gene, second crossovers are expected to be very rare indeed. However, in intragenic recombination, gene conversion takes place without exchange of flanking alleles at an unexpectedly high frequency, about 50% of the time.

The rare events detected by molecular genetics thus revealed three major and unexpected character-

istics of the recombination mechanism at the molecular level: (1) nonreciprocal recombination involving gene conversion, (2) formation of heteroduplex DNA leading to postmeiotic segregation, and (3) nonexchange of flanking alleles at unexpectedly high frequencies. Although these characteristics present special problems to molecular biologists attempting to understand the molecular basis of recombination, they also provide clues as to how the mechanism might work.

Molecular Models for Recombination

The rare events observed in intragenic recombination stimulated development of a series of molecular models for recombination. The model making has been more than just an intellectual exercise, because the hypotheses prompted searches for enzymes capable of carrying out steps proposed in the models. In many cases the searches were successful, and the likely progress of many steps in molecular recombination, complete with the required enzymes, has been worked out.

Although the molecular models differ in detail, all assume that "nicks" are opened in the sugar-phosphate backbones of the two DNA molecules taking part in an exchange. Nucleotide chains are then considered to unwind from the nicks and "invade" the DNA of the opposite homolog by rewinding with one of its DNA chains. The rewinding is commonly assumed to lead to formation of a crossed structure called a *Holliday intermediate*, named after R. Holliday, one of the foremost theorists in molecular recombination.

Figure 25-15 shows one of these models in simplified form, proposed originally by Holliday and modified by M. S. Meselson and C. M. Radding and others.

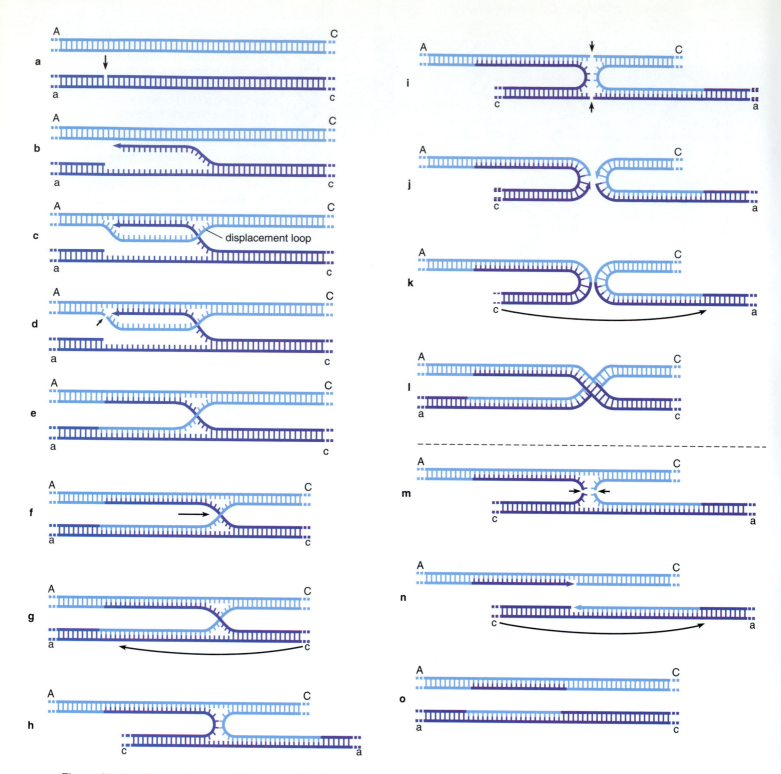

Figure 25-15 The Holliday model for recombination, as modified by M. S. Meselson and C. M. Radding (see text). A,a and C,c are flanking marker alleles.

Recombination begins as a single-chain nick is introduced by a DNA endonuclease in one of two DNA molecules entering the exchange (arrow in Fig. 25-15a). The nicked molecule is the *donor* in the exchange. The free end generated by the nick then unwinds from the donor molecule (Fig. 25-15b) and invades the opposite molecule (Fig. 25-15c). The initial invasion is considered to be a random search by the invading chain for a region of complementary sequence. If a homologous region is encountered, the invading chain rewinds with its complement in the receiving molecule (as in Fig. 25-15c). The invasion by a single chain is termed *strand transfer*. The winding forces the opposite chain of the receiving molecule away from its normal complement as a single-chain *displacement loop*.

The displacement loop is then broken by a nick (arrow in Fig. 25-15d), creating a second free end that can invade the donor molecule to set up the crossed structure shown in Figure 25-15e. Perfect complementarity is not required for the rewinding in either molecule; small noncomplementary differences in sequence, as might be expected if the rewinding DNA originates from different alleles of the same gene, will produce regions of heteroduplex DNA. The heteroduplex regions may persist or may be "corrected" by mismatch repair (see p. 678). If the mismatch is corrected, either of the two chains may be used as template. No matter which is used as template, an allele present in the chain that is removed in the correction is lost and is replaced by the complementary copy of the allele present in the template chain, accomplishing gene conversion. The mechanism proposed in the model can thus produce gene conversion as well as heteroduplex DNA.

At this point, gaps are filled in by DNA polymerase and all remaining nicks are sealed by DNA ligase (as described in Chapter 23). These reactions generate the closed, crossed structure shown in Figure 25-15e; this is the Holliday intermediate. While the two molecules are in this configuration, the DNA of both may unwind and rewind simultaneously through the crossing point. The effect of this simultaneous unwinding and rewinding, depending on the direction, will be to push the crossover point to the right or left. (The movement is shown to the right in Fig. 25-15f.) The movement, called *branch migration*, may generate additional regions of heteroduplex DNA as nucleotide chains of opposite alleles wind together.

Subsequent steps in the model free or *resolve* the crossed region of the Holliday intermediate. The mechanism proposed for this resolution, originally advanced by N. Sigal and B. M. Alberts, is easier to understand if two legs of the crossed structure are rotated through 180° (Fig. 25-15g), producing the arrangement shown in Figure 25-15h. Single-chain cuts are now made to free the crossed structure. Vertical cuts (arrows in Fig. 25-15i) sever the two DNA chains not involved in the original invasion and rewinding. This cutting generates free ends (Fig. 25-15j) that when reciprocally sealed (Fig. 25-15k) complete the crossover. Rotating the lower half back to its original position (broken arrow, Fig. 25-15l) shows that the cutting and rejoining produce the new combinations of flanking alleles A-c and a-C (Fig. 25-15l). Flanking alleles thus recombine if the resolving cuts are made in the pattern shown in Figure 25-15i.

If horizontal cuts follow the rotation shown in Figure 25-15g, the two DNA chains involved in the initial

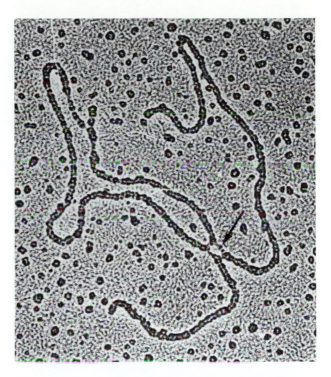

Figure 25-16   A Holliday intermediate (arrow) in DNA molecules undergoing recombination. (Courtesy of C. M. Radding, from *Cell* 25:507 [1981]. Copyright Cell Press.)

invasion and rewinding are severed (Fig. 25-15m). Sealing these free ends (Fig. 25-15n and o) resolves the DNA molecules in their original form, effectively eliminating the crossover so that no recombination of flanking alleles takes place. (Note that the flanking alleles are in their original A-C and a-c arrangement in Fig. 25-15o.) However, within the DNA region involved in the recombination event, heteroduplex regions and gene conversion due to mismatch correction can still lead to nonreciprocal exchanges and postmeiotic segregation. Because vertical or horizontal cuts are equally probable, gene conversion or postmeiotic segregation without recombination of flanking alleles would be expected about half the time, as observed in nature.

Recombining DNA molecules with the structure expected for the Holliday intermediate have been detected in DNA extracted from bacterial, yeast, and human systems and photographed under the electron microscope (Fig. 25-16). These observations give strong support to the conclusion that this configuration is an intermediate step in both prokaryotic and eukaryotic recombination, as proposed in the model.

Enzymes and Factors of Recombination

Many of the steps proposed in the recombination model can be carried out by enzymes active in DNA

Table 25-1 Enzymes and Factors Probably Active in Recombination

Enzyme or Factor	Probable Role in Recombination
DNA endonuclease	Opens single- or double-chain breaks in DNA molecules.
Unwinding enzymes (helicases)	Promote unwinding of nucleotide chains.
DNA binding proteins	Stabilize unwound DNA in single-chain form.
DNA topoisomerase I	Promotes DNA winding or unwinding.
DNA topoisomerase II	Resolves interlocks and tangles resulting from recombination.
DNA exonuclease	Removes nucleotides from exposed ends of DNA nucleotide chains.
DNA polymerase	Extends chains and fills gaps.
DNA ligase	Seals nicks after rewinding and gap filling.
Recombination enzymes (RecBCD, RecA)	Promote strand invasion, displacement loop formation, and branch migration.
RuvC	Resolves Holliday intermediates.

replication, including DNA endonucleases, unwinding enzymes, DNA polymerases, and DNA ligase (Table 25-1). A growing body of evidence indicates that many of these enzymes are actually active and required in cells undergoing recombination. A number of additional enzymes catalyzing steps unique to recombination, such as strand invasion and branch migration, have been identified in both prokaryotes and eukaryotes.

Temperature-sensitive mutants of *E. coli* (see p. 638) have been particularly valuable in the research identifying enzymes unique to recombination. One recombination enzyme identified in this way is the *RecBCD* enzyme, composed of three polypeptides encoded in the *recB*, *recC*, and *recD* genes. This enzyme, which is required for recombination in *E. coli*, can make a single-chain nick and unwind a single nucleotide chain, in the pattern proposed in Figure 25-15a and b.

A second enzyme necessary for recombination to proceed in *E. coli* is *RecA*, which can catalyze invasion of an intact DNA helix by a single nucleotide chain (as in Fig. 25-15c). The RecA enzyme, originally identified in *E. coli* by A. J. Clark and A. Margulies, can bind to an exposed single chain and catalyze rewinding of the chain with its complement in a receiving molecule. As part of this process, RecA polymerizes into a filament and combines with the receiving DNA double helix as well as the invading strand. The combination stretches the double helix to 150% of its B-conformation, setting up a structure in which base-pair spacing extends from 3.4 to 5.1 A. The stretched conformation promotes DNA unwinding and single strand invasion, which may include formation of a triple helix. (In a triple helix a third nucleotide chain winds into one of the grooves of a double helix, forming hydrogen bonds with the edges of bases exposed in the groove; see p. 372.) The invasion catalyzed by RecA eventually forms a displacement loop, creating the configuration shown in Figure 25-15c. The RecA enzyme is highly efficient in its search for homologous regions and can rapidly locate and rewind regions of homology. Once a Holliday intermediate is formed, the same versatile enzyme can catalyze branch migration in a $5' \rightarrow 3'$ direction. Both invasion and branch migration occur at the expense of ATP hydrolyzed by the enzyme. Holliday intermediates can be generated in the test tube by adding the RecBCD and RecA proteins to homologous DNA molecules. A third required enzyme recently identified in *E. coli* by S. C. West and his colleagues, RuvC, can catalyze the next step in the molecular models, resolution of the Holliday intermediate, when added to the test-tube system. Another protein, RuvB, is highly efficient in promoting branch migration.

Enzymes with properties similar to RecA have been detected in eukaryotes. D. K. Bishop and his coworkers found that yeast contain *DMC1*, a protein homologous to RecA that catalyzes recombination. In the lily, Stern and Hotta found a 45,000-dalton protein, *m-rec*, that can substitute for RecA in *E. coli* test-tube systems. (Stern's Experimental Process essay on p. 750 describes his experiments identifying m-rec activity and its significance for the recombination mechanism in eukaryotes.) An enzymatic activity promoting strand transfer was also detected in human cells.

Some of the other enzymes and factors proposed in the models have also been detected in the lily by Stern and Hotta. These investigators found an unwinding enzyme with properties similar to RecBCD, as well as DNA endonuclease, DNA ligase, and DNA polymerase activities in lily meiotic cells during prophase I. The activities of most of these enzymes appear or increase during leptotene and zygotene, peak during pachytene, and fall to low or unmeasurable levels thereafter, in agreement with their expected roles in the recombination mechanism. Thus many of the enzymes, factors, and processes predicted by the molecular models have been detected in the lily.

THE SYNAPTONEMAL COMPLEX AND GENETIC RECOMBINATION

The synaptonemal complex probably takes part in eukaryotic recombination, possibly as a framework organizing the enzymes and factors involved in breakage and exchange. Some of the evidence supporting this role is circumstantial—the synaptonemal complex is present in the right time and place to take part in recombination. It appears between homologs as they pair, persists throughout recombination, and disappears from most segments of the chromosomes when recombination is complete. The complex lies in the space between the homologs, where close molecular pairing and recombination probably occur.

There is also a correlation between presence of the complex and the normal progress of recombination. In male *Drosophila*, pairing occurs, but synaptonemal complexes do not form between homologs. No visible crossovers appear in chromosomes, and recombination cannot be detected in genetic crosses. In normal *Drosophila* females the synaptonemal complex is present, chiasmata appear, and recombination can be detected. *Drosophila* females homozygous for a mutation eliminating recombination (the *c(3)G* mutation) have no crossovers or evidence of recombination; in these females the synaptonemal complex fails to develop between homologs. Similarly, in *S. cerevisiae*, recombination in the temperature-sensitive *cdc7* mutant is absent at elevated temperatures; examination of meiotic cells shows that the synaptonemal complex is also absent. At permissive temperatures in which the mutation is inactive, both the synaptonemal complex and recombination occur. In a few species, including some insects, crossovers are restricted to certain regions of the chromosomes. The synaptonemal complex species is found only in these regions. Additional examples reinforcing the correlation between presence of the synaptonemal complex and recombination have been observed in other species, including humans.

Structure of the Synaptonemal Complex

The synaptonemal complex, first identified by M. J. Moses in 1956, is remarkably similar in appearance in different eukaryotes. It consists of a longitudinal *central element* bounded on either side by *lateral elements* (see Figs. 25-4, 25-17, and 25-18). The central element in most species appears simply as an aggregation of dense material running through the middle of the complex (as in Figs. 25-4 and 25-18). The two lateral elements usually appear to be made up of denser, more granular material than the central element. In cross section the synaptonemal complex appears flat and ribbonlike (Fig. 25-19). The side elements thicken where chromosomes attach to the nuclear envelope, forming dense, caplike structures closely fused to the nuclear envelope membranes (Fig. 25-20). In all forms of the synaptonemal complex, very thin *transverse fibers* cross the central space and connect the lateral elements with the central element. The entire complex, including lateral elements, central elements and transverse fibers, is digested by proteinases such as trypsin and pronase, indicating that proteins form a major part of its framework.

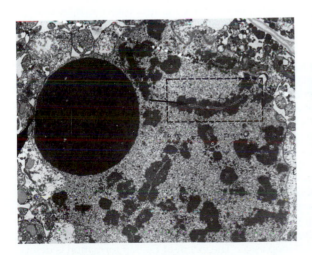

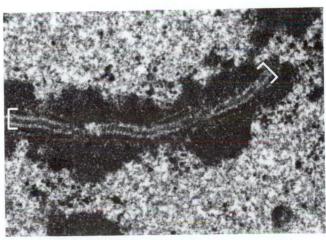

a b

Figure 25-17 Synaptonemal complex in the lily. **(a)** The paired homologs at low magnification. The synaptonemal complex lies in the narrow, regular space separating the homologs (arrow). × 5,500. **(b)** The boxed region in **(a)** at higher magnification, showing the synaptonemal complex in the narrow space separating the homologs (brackets). × 41,000. (Courtesy of P. B. Moens and Springer-Verlag, from *Chromosoma* 23:418 [1968].)

m-rec: An Enzyme That Effects Genetic Recombination in Meiotic Cells

Herbert Stern

HERBERT STERN received his Ph.D. in Botany at McGill University, Montreal, Canada, in 1945. He spent two years as a Royal Society of Canada fellow in the Department of Biochemistry at the University of California, Berkeley, five years at the Rockefeller Institute for Medical Research (now Rockefeller University), and five years at the Research Branch of Canada Department of Agriculture. He served as Professor of Botany at the University of Illinois, Urbana, from 1960–1965. Since then, he has been Professor of Biology at the University of California, San Diego. A former president of the Society for Development Biology, his research has centered on biochemical and molecular events during meiosis.

The mechanism governing crossing-over (recombination) in meiotic cells (meiocytes) has been uninterruptedly studied in a variety of ways ever since it was recognized as a prominent feature of the process about 100 years ago. Recombination itself is not unique to meiosis. It occurs in bacteria and in mitotically dividing cells of species as diverse as baker's yeast and humans (cells of the immune system in particular). The uniqueness of meiotic recombination lies in its *regularity*, which applies to its unfailing occurrence not only in every meiotic division (rare exceptions excluded), but also in every chromosome pair. One major research challenge, still unmet, is to account for the regularity just described when it is well known that the total recombinations in any particular meiotic cell are rare events relative to the number of potential recombination sites. In general, the larger the genome size the lower the number of recombinations per DNA base pair. The mouse genome, for example, is about 20 times as large as that of *Drosophila*, while the number of crossovers per DNA unit is about 6 times higher in *Drosophila*. The issue of regularity would not exist if meiotic recombinations were extremely abundant, but such abundance is clearly unwelcome as judged by the course of evolution.

We sought an answer to this issue by asking how the enzymes relevant to recombination are organized in meiosis. Mere identification of the recombinogenic enzymes present in meiocytes would be inadequate because the process requires a still undetermined number of enzymes, and the way in which their behavior is coordinated with that of the chromosomes is an essential component of the answer. One barrier to discovering the nature of that coordination is the extremely few species in which different biochemical events can be characterized and assigned to specific stages of meiosis. A satisfactory system is one in which the cells undergo meiosis synchronously and slowly enough that a sufficient number of cells at different meiotic stages can be isolated for biochemical study.

The system is even more treasured if the cells can progress through meiosis under *in vitro* conditions, thus permitting a variety of experimental manipulations. The pollen mother-cells (microsporocytes) of the Easter lily and several closely related species are a unique source of such a system. We initiated our meiotic studies using the anthers of *Lilium* and, in the course of time, it became evident that cells undergoing meiosis display a number of distinctive biochemical features that are coordinated with chromosome behavior.

We turned to the enzyme m-rec after we had found a striking change in DNA behavior in the transition from zygotene when the chromosomes are undergoing pairing, to pachytene when, as determined by cytogenetic evidence, crossing-over occurs.[1] At zygotene the chromosomes replicated select sequences of DNA that were suppressed in their replication during the S-phase. By contrast, the DNA at pachytene seemed to be in a state of turmoil. At the start of that stage a meiosis-specific endonuclease was found to introduce a large number of nicks in selected chromosomal regions that accounted for about 50% of the genome. The nicking was accompanied by intense repair synthesis. The nick-repair process did not occur in the absence of pairing and crossing-over. We were confident that recombinogenic enzymes would be active during pachytene. We selected the enzyme, later named m-rec, as one of our targets because genetic evidence unquestionably pointed to the highly purified recA protein of *E. coli* as a major component of a central recombination process. Methods have been designed for its assay although a display of its critical recombinogenic characteristics in previously untested species does not imply an identity with all other characteristics of recA protein.

The essential feature of the assay centers on the interaction between a stretch of single-stranded DNA and that of double-stranded DNA in which one strand is identical in sequence with the single-strand component. Generally, one component is circular and the other linear. If the duplex component is circular it must be supercoiled. The single strand, if circular, is necessarily intact. In an assay mixture of circular duplex and linear single strand, the enzyme catalyzes the invasion of the single strand into the duplex circle, displacing the one identical to itself by hybridizing with the complementary strand. Inasmuch as the duplex is intact, the displaced strand can neither escape the circle nor be fully displaced. The resulting structure has been named a D-loop, a product of the D-loop assay. The extent of D-loop formation is determined by using radiolabeled duplex and passing the reaction mixture through a nitrocellulose filter, which binds single-strand but not duplex DNA. Since D-loops have single-strand regions the entire complex is bound to the filter. The radioactivity bound to the filter is a measure of the number of duplexes that have undergone D-loop formation. If the substrate is a combination of linear duplex and a circular single strand, the reaction catalyzed is slightly different. In this case the invasion is effected by that strand of the duplex that is complementary to the circular DNA, and the entire strand can be transferred to form a duplex circle, a product of the strand-transfer assay. If one of the DNA components is radiolabeled, the extent of transfer can be determined by measuring the susceptibility of the reaction mixture to S1 nuclease, which does not attack double-strand DNA. If the linear duplex is radiolabeled, the transfer of one of its strands to its circular complement renders the remaining strand susceptible to nuclease digestion, so that the loss of DNA radioactivity measures the proportion of the original duplex that has been transferred to form a circular duplex. If the linear circle is labeled, the increase in DNA radioactivity measures the proportion of circular duplex formed. These recA reactions require ATP, but the requirement is not general for all recA-like proteins.

To check for the presence of a recA-like enzyme in meiotic cells, extracts of meiocytes from both lily anthers and mouse testes at unselected stages were assayed by suitable procedures. To our delight, recA-like activity was found in both groups of cells. Moreover, the requirements for D-loop and strand-transfer activities were identical to those of *E. coli* recA. We then checked vegetative tissues of various plants and somatic tissues of mice for the activity and found a 4- to 5-fold lower level in the more active nonmeiotic tissues. One distinguishing difference was found. Gel analysis of the partially purified extracts indicated the molecular weights of the rec-like proteins in meiocytes of lilies and mice to be about 43 and 45 kD (kilodaltons) respectively and those of the vegetative and somatic tissues to be about 70 and 75 kD respectively. Given the partial purity of the extracts, the significance of that finding is uncertain. A significant difference was found when the activities of the recA-like enzymes were tested at different temperatures. Lily breeders are aware that the plants are more fertile at 23–25°C than at higher temperatures at which the plants happily grow. Also, human and mouse testes prefer 33–37°C for spermatocyte fertility. We therefore compared the strand transfer activity of meiocyte and nonmeiocyte at two temperatures. Lily meiocyte recA-like enzyme was four times more active at 23°C than at 33°C. No such difference was found for the vegetative enzyme, which was somewhat lower in activity at the lower temperature. Mouse meiocyte activity was five times higher at 33°C than at 37°C, whereas that of spleen cells was slightly higher at 37°C than at 33°C. The recA-like activity of meiocytes is distinctive compared with somatic cells and the activities were designated as m-rec and s-rec respectively.

The similarities between m-rec and recA made it highly likely that m-rec functioned in meiotic recombination. The question then addressed was how its behavior during meiosis might contribute to the regularity of recombination. An answer was sought by assaying m-rec activities at different stages of meiosis, a task readily achieved with lily because of its meiotic synchrony. The result was striking. Activities were negligible during the interval of chromosome replication and remained so through the first meiotic stage, leptotene. A rise began when the cells entered the pairing stage, zygotene; activity rose sharply as the cells approached pachytene, reaching a peak value at early pachytene. A correspondingly sharp decline occurred after the cells reached midpachytene. A strong coordination exists between m-rec activity, chromosomal nick-repair activity, and cytogenetic evidence for the occurrence of crossing-over at pachytene. In a later study it was found that interference with chromosome pairing results in a much reduced rise in m-rec activity.

Two conclusions may be drawn from this study. The first of these is relevant to the largely unaccepted claim that crossing-over occurs during chromosome replication. The timing of m-rec activity validates that lack of acceptance and thus constitutes a minor contribution of the study to meiotic recombination. The second conclusion relates to the more profound consideration of the mechanism underlying the regularity of meiotic recombination. This and related studies do not fully provide a mechanism but they make possible a reasonable speculation. The latter addresses the question of how regularity might be achieved under conditions in which the frequency of recombination relative to genome size is extremely low and in which each chromosome pair is assured at least one recombination event. Clearly, one rare factor such as the recombination nodule [see p. 752] could determine the site and frequency of recombination. If so, either of two very different mechanisms might account for regularity amidst rarity. It is conceivable that the recombination system is so tightly organized that the limiting factor contains all the biochemical ingredients essential to recombination. The alternative is to consider the limiting factor as a single essential ingredient in the process, with all other ingredients present in excess. This alternative involves the siting of a few copies of a probably complex factor whose fruitful function is assured by an overwhelming excess of all the necessary supportive mechanisms. The overwhelming excess of nick-repair and m-rec activities, as well as others not described here, all coordinated with chromosome behavior at zygotene and pachytene, make the flooding of rare sites with an excess of support mechanisms a reasonable way to achieve a regularity of rare events. The more wasteful it seems, the more efficient it is.

Reference

[1]Hotta, Y.; Tabata, S.; Bouchard, R. A.; Pinon, R.; and Stern, H. General recombination mechanisms in extracts of meiotic cells. *Chromosoma* 93:140–51 (1985).

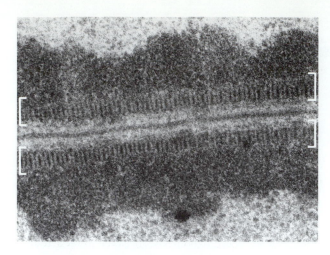

Figure 25-18 Synaptonemal complex of the fungus *Neottiella*, in which the central element is unstructured and the lateral elements (brackets) are striated. × 22,000. (Courtesy of D. von Wettstein.)

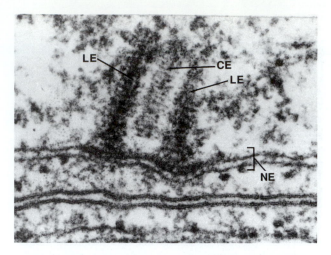

Figure 25-20 The thickened, caplike structures formed where the lateral elements (LE) of the synaptonemal complex attach to the nuclear envelope, as seen in the silkworm *Bombyx*. CE, central element; NE, nuclear envelope. (Courtesy of R. C. King.)

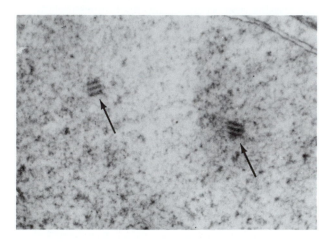

Figure 25-19 Synaptonemal complexes of *Philaenus*, a homopteran insect, in cross section (arrows). × 20,000. (Courtesy of P. L. Maillet and R. Folliot.)

Chromatin fibers are densely packed around the side elements. It is possible that DNA from the chromatin passes entirely through the lateral elements and extends into the central element, the probable locale of recombination between homologous chromatids.

Recombination Nodules in the Synaptonemal Complex

Evidence from a variety of sources suggests that sites within the synaptonemal complex in which recombination occurs are marked by *recombination nodules*, dense, egg-shaped structures embedded in the central element of the synaptonemal complex (Fig. 25-21). First discovered in *Drosophila* by A. T. C. Carpenter, the nod-

ules have also been observed in protozoa, fungi, plants, and other animals. Generally recombination nodules are present in the synaptonemal complex of organisms or types in which recombination takes place and absent in those without recombination.

Recombination nodules appear as the homologs pair, and they persist until the end of pachytene. Initially they are numerous and appear to be distributed randomly along the central element of the synaptonemal complex. Later in pachytene they are reduced to about one per chromosome arm—a number and distribution that correlates very closely with the crossovers that become visible later in prophase I.

These observations indicate that the synaptonemal complex is a framework organizing the enzymes and factors carrying out recombination, and suggest that recombination nodules are assemblies of these enzymes and factors. Carpenter noted that recombination nodules take up DNA precursors during recombination, as expected if DNA is being replicated or repaired as part of the recombination mechanism.

REGULATION OF MEIOSIS

In most organisms, cells entering meiosis are products of lines that reproduce by mitotic division. In lower eukaryotes, including protists and fungi, the change from mitosis to meiosis is frequently triggered by environmental changes, usually toward less favorable growth conditions. In *S. cerevisiae*, for example, a

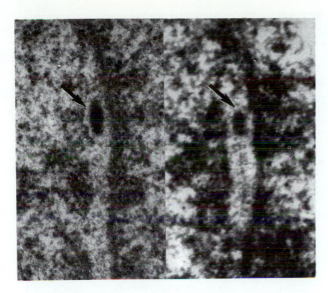

Figure 25-21 Recombination nodules (arrows) embedded in the central elements of synaptonemal complexes of rye (*Secale*). × 50,000. (Courtesy of M. Abirached-Darmency and Springer-Verlag, from *Chromosoma* 88: 299 [1983].)

switch from a full nutrient medium to one lacking carbon and nitrogen sources leads to meiosis and formation of resistant spores. In many animals, induction of meiosis involves a programmed cellular response to hormones contacting receptors at the cell surface.

The few detailed studies carried out so far in higher eukaryotes suggest that the first regulatory alterations leading toward meiosis take place in mitotic cell cycles well in advance of meiosis. The duration of DNA replication, for example, gradually increases through several mitotic cell cycles preceding meiosis in the lily and mouse.

These alterations notwithstanding, the changes regulating entry into a meiotic division probably include the usual genes controlling mitosis, with activation of a relatively few genetic sites unique to meiosis until G2. Once in G2, genes are activated that fundamentally alter division toward prophase I of meiosis and the first meiotic division. Since the second meiotic division proceeds much like a normal mitosis, the sequence at this stage is probably activated primarily by the usual mitotic genes.

The characteristics of mutations altering meiosis in *Drosophila* and *S. cerevisiae* support these conclusions (Table 25-2). Most of these mutations affect the processes of prophase I and metaphase I unique to meiosis: chromosome pairing, formation of the synaptonemal complex and recombination nodules, recombination itself, chiasma formation, and formation of the kinetochore connections leading to separation of homologous chromosomes at metaphase I. The

mutations affecting recombination are especially interesting because various mutants in this category show alterations in the rate, total amount, and location of recombination events as well as inhibition. These effects indicate that all these characteristics are under genetic control. The surprisingly large number of mutants affecting prophase I and metaphase I— more than 40 in *Drosophila* and more than 50 in yeast —reflects the complexity of these events and the multiplicity of factors, enzymes, and regulatory elements required to bring them about.

Relatively few meiosis-specific mutations affect premeiotic DNA replication or the second meiotic division. Instead, most mutations identified at these stages are the same as those altering mitosis. This supports the conclusion that the major regulatory events channeling cells toward meiosis occur after premeiotic replication, during G2 or early prophase I. It also supports the conclusion that the second meiotic division proceeds by essentially the same pathway as an ordinary mitotic division.

These observations suggest that the evolutionary adaptations that established meiosis as a separate and distinct division mechanism involved the development of pairing, recombination, and separation of homologous chromosomes during the first meiotic division. These meiotic adaptations are the most complex from the genetic standpoint and evidently are the most susceptible to disturbance through mutation.

These adaptations are not entirely unique to meiosis. Pairing of homologous chromosomes also occurs in the somatic cells of some species, as in the salivary glands of dipteran flies. Breakage and exchange of segments between the chromatids of replicated chromosomes also take place regularly during mitosis in many organisms including higher vertebrates. Because this exchange, called *sister strand exchange*, takes place between identical chromatids, it does not lead to genetic recombination. (The functional significance of these exchanges is unknown.) Even the movement of both chromatids of a chromosome to the same pole as in metaphase I of meiosis, rather than to opposite poles as in mitosis, takes place regularly in some nonmeiotic cell types.

The immediate factor controlling meiosis is the cyclin B/CDK2 complex, which rises and falls in activity during the two divisions. Activity of the complex increases at the end of G2, until the fully activated CDK2 carries out phosphorylations that trigger entry into meiotic prophase I. Activity falls at the transition to anaphase I and increases again during prophase of the second meiotic division. Cyclin B is degraded, and CDK2 activity eliminated, as cells enter anaphase II. The metaphase II meiotic arrest typical of higher vertebrates, as noted, involves delay of cyclin B/CDK2 inactivation by cytostatic factor (CSF).

Table 25-2   Some Meiotic Mutants and Their Effects in Yeast and *Drosophila*

Mutation	Effect	Mutation	Effect
Saccharomyces cerevisiae		*Drosophila melanogaster*	
cdc4	Synaptonemal complex incomplete; central elements collect in nucleolus.	*c(3)G*	No synaptonemal complex; homologs fail to separate at first meiotic division.
cdc7	No synaptonemal complex.	*Df(3)sbd*[103]	Premature disassembly of synaptonemal complex; recombination reduced.
cdc9	DNA ligase faulty.		
cdc31	Chromatids fail to separate during second meiotic division.	*mei-9*	Recombination reduced; recombination nodules in normal numbers.
hop1	No synapsis; synaptonemal complex incomplete.	*mei-41*	Recombination nodules reduced in number; recombination reduced.
dcm1	Faulty recombination; homologous to *E. coli* RecA.	*mei-218*	Recombination nodules reduced in number; intergenic, but nonintragenic recombination reduced.
ndc1	Chromatids fail to separate during second meiotic division.		
rad1	No recombination.	*mei-S282*	Recombination nodules reduced in number; recombination reduced near telomeres.
rad50	No synaptonemal complex; no recombination.		
spo10	Synaptonemal complex incomplete; central elements collect in nucleolus.	*mei-S322*	Homologs fail to separate during first meiotic division.
spo11	Synaptonemal complex present but no recombination.	*mei-332*	Recombination increased; effects more pronounced near centromeres.
spo13	First meiotic division blocked; forms diploid spores.	*mei-551*	Recombination reduced by one-half; chromosome 4 and X chromosome fail to separate during first meiotic division.
top2	Topoisomerase II faulty; chromosomes do not separate at metphase I.	*mcd*	Faulty spindle formation for first meiotic division; kinesinlike protein.
		ord	Homologs fail to separate during first meiotic division.
		pal	Chromosomes of paternal origin lost during meiosis.

As long as cyclin B/CDK2 remains active, normally until fertilization, metaphase II arrest is maintained.

Meiosis has three outcomes that are vital to sexual reproduction in eukaryotes. It reduces the chromosome number to the haploid level, so that the number does not double at fertilization. Through recombination and independent segregation of maternal and paternal chromatids, meiosis produces genetic variability in the haploid products of the division sequence. This variability is a major source of differences in the offspring of sexually reproducing organisms. Finally, through the RNA transcription and protein synthesis that occur during the diplotene stage, meiosis provides the ribosomes, enzymes, structural proteins, and raw materials needed for gamete production, fertilization, and the early stages of embryonic development.

The wide distribution of meiosis and sexual reproduction among eukaryotes reflects their importance to survival. Through these mechanisms, particularly those of recombination, an almost infinite variety of genetic types is presented for testing by the environment. Unless environmental change becomes too drastic, at least some of the variant individuals are likely to survive and reproduce.

For Further Information

Suggestions for Further Reading

Callan, H. G. 1987. Lampbrush chromosomes as seen in historical perspective. In W. Hennig, ed., *Structure and Function of Eukaryotic Chromosomes*, pp. 5–26. New York: Springer-Verlag.

Carpenter, A. T. C. 1987. Gene conversion, recombination nodules, and the initiation of meiotic synapsis. *Bioess.* 6: 232–236.

Dickinson, H. G. 1989. The physiology and biochemistry of meiosis in the anther. *Internat. Rev. Cytol.* 107:79–109.

Dunderdale, H. J., Benson, F. E., Parsons, C. A., Sharples, G. A., Lloyd, R. G., and West, S. C. 1991. Formation and resolution of recombination intermediates by *E. coli* RecA and RuvC proteins. *Nature* 354:506–510.

Hawley, R. S., McKim, K. S., and Arbel, T. 1993. Meiotic segregation in *Drosophila melanogaster* females: Molecules, mechanisms, and myths. *Ann. Rev. Genet.* 27:281–317.

Holliday, R. 1990. The history of the DNA heteroduplex. *Bioess.* 12:133–142.

John, B. 1990. *Meiosis.* New York: Cambridge University Press.

Macgregor, H. C. 1987. Lampbrush chromosomes. *J. Cell Sci.* 88:7–9.

Malone, R. E. 1990. Dual regulation of meiosis in yeast. *Cell* 61:375–378.

McLeod, M. 1989. Regulation of meiosis: From DNA binding protein to protein kinase. *Bioess.* 11:9–14.

Meyer, R. R., and Lane, P. S. 1990. The single-strand DNA-binding protein of *E. coli. Microbiol. Rev.* 54: 342–380.

Miller, R.V., and Kokjohn, T. A. 1990. General microbiology of *recA*: Environmental and evolutionary significance. *Ann. Rev. Microbiol.* 44:365–394.

Minshull, J. 1993. Cyclin synthesis: Who needs it? *Bioess.* 15:149–155.

Moens, P. B., ed. 1987. *Meiosis.* New York: Academic Press.

Moens, P. 1990. Unravelling meiotic chromosomes: Topoisomerase II and other proteins. *J. Cell Sci.* 97:1–3.

Moens, P. B., and Pearlman, R. E. 1988. Chromatin organization at meiosis. *Bioess.* 9:151–153.

Pelech, S. L., Sanghera, J. S., and Daya-Makin, M. 1990. Protein cascades in meiotic and mitotic cell cycle control. *Biochem. Cell Biol.* 68:1297–1330.

Richardson, C., and Lehman, R., eds. 1990. *Molecular Mechanisms in DNA Replication and Recombination.* New York: Liss.

Roca, A. I., and Cox, M. M. 1990. The RecA protein: Structure and function. *Crit. Rev. Biochem. Molec. Biol.* 25: 415–456.

Roeder, G. S. 1990. Chromosome synapsis and genetic recombination: Their roles in meiotic chromosome segregation. *Trends Genet.* 6:385–389.

Sawin, K. E., and Endow, S. A. 1993. Meiosis, mitosis, and microtubule motors. *Bioess.* 15:399–407.

Stahl, F. W. 1987. Genetic recombination. *Sci. Amer.* 256: 90–101 (February).

Stern, H. 1990. Meiosis. In K. W. Adolph, ed., *Chromosomes, Eukaryotic, Prokaryotic, and Viral*, pp. 3–37. Boca Raton, Fl.: CRC Press.

West, S. C. 1992. Enzymes and molecular mechanisms of genetic recombination. *Ann. Rev. Biochem.* 61:603–640.

von Wettstein, D., Rasmussen, S. W., and Holm, P. B. 1984. The synaptonemal complex in genetic segregation. *Ann. Rev. Genet.* 18:331–413.

Wickramasinghe, D., and Albertini, D. F. 1993. Cell cycle control during mammalian oogenesis. *Curr. Top. Devel. Biol.* 28:126–153.

Review Questions

1. Compare the chromosome number and DNA content per nucleus during G1, S, G2, and division in cells undergoing mitosis and meiosis.

2. In what ways does premeiotic interphase differ from a premitotic interphase?

3. Outline the stages of meiotic prophase I. What events mark the beginning and end of leptotene? What is the major event of this stage?

4. What differences are noted between the structure and arrangement of the chromosomes at the beginning of mitosis and the leptotene stage of meiotic prophase?

5. What events mark the beginning and end of zygotene? What is the major event of zygotene? What is zygDNA? zygRNA?

6. What is a chromosome pair? A homolog? An allele?

7. What events mark the beginning and end of pachytene? What major event takes place during this stage? What is rDNA amplification?

8. What is a bivalent? A tetrad?

9. What events mark the beginning and end of diplotene? What major events take place during this stage? What are lampbrush chromosomes?

10. What are chiasmata?

11. What is diakinesis? What major event takes place during this stage?

12. What are the results of meiotic prophase?

13. Outline the patterns of spindle formation during the two meiotic divisions. What patterns are followed by centrioles, if present, during the two divisions?

14. Compare the spindle connections made by the chromosomes during the first and second meiotic divisions. What is the importance of chiasmata to these connections? How do the connections lead to a reduction in the number of chromosomes? What is nondisjunction?

15. Compare the first and second meiotic divisions with a mitotic division. What are the similarities? The differences?

16. What are maternal and paternal chromosomes? How are maternal and paternal chromosomes distributed during meiosis? During mitosis?

17. Outline the steps and outcome of classical recombination. What are recombinant chromatids? Parental chromatids?

18. What is reciprocal recombination? Why was the gene thought to be the unit of recombination by classical geneticists?

19. What experiment established that recombination takes place by breakage and exchange in eukaryotes?

20. What experiment demonstrated that recombination takes place during meiotic prophase rather than replication?

21. What characteristics of fungi, bacteria, and viruses made the detection of rare recombination events practical?

22. What is intragenic recombination? Nonreciprocal recombination? Gene conversion?

23. What is postmeiotic segregation of alleles? What is heteroduplex DNA, and how does it lead to postmeiotic segregation?

24. What is meant by the nonexchange of flanking markers? What is the significance of this phenomenon to the mechanism of recombination?

25. Outline the major steps in the molecular model for recombination. What is a displacement loop? Branch migration? What is a Holliday intermediate? In what two ways can the Holliday intermediate be resolved? What is the significance of this resolution to the nonexchange of flanking markers?

26. List the major enzymes thought to participate in recombination and describe their possible roles in the process.

27. What steps are catalyzed by the RecA enzyme of *E. coli*? By RuvB and RuvC? What evidence indicates that enzymes similar to RecA occur in eukaryotes?

28. Outline the structure of the synaptonemal complex. What evidence indicates that the synaptonemal complex takes part in recombination?

29. What are recombination nodules? What evidence indicates that recombination nodules are associated with sites of recombination?

30. At what stage of the cell cycle do cells apparently become committed to a meiotic division? What stages of meiosis are most greatly affected by mutations? What is the role of the cyclin B/CDK2 complex during meiosis? During meiotic arrest? What is CSF?

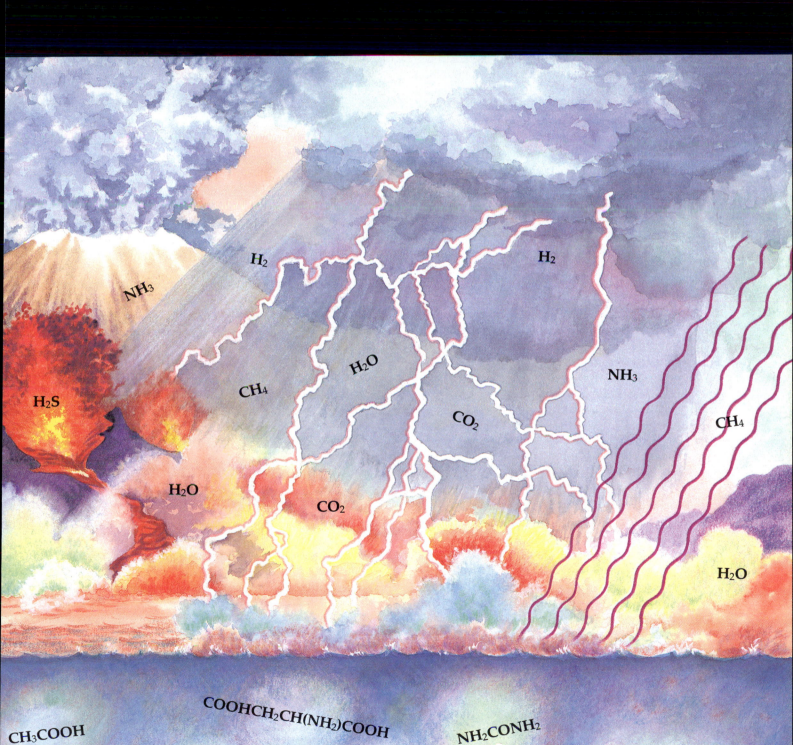

The most ancient rocky material of the solar system is estimated to be approximately 4.6 billion years old. At this distant time our planet condensed out of the primordial matter and began its long transition into the environment we know today. Microscopic, cell-like fossils resembling bacteria have been found in the Archean deposits near North Pole, Australia, laid down about 3.5 billion years ago. If the North Pole fossils are authentic, life must have appeared on earth at some time between 4.5 and 3.5 billion years ago, during the first billion years of its existence.

How did life appear on earth? A divine origin for life has been proposed by many religions. However, the possibility of a divine origin falls outside the realm of science because it cannot be tested. The same can be said for the idea that life may have come here from outer space, from a space probe launched from another planet billions of years ago. From a scientific standpoint it is necessary to hypothesize that life evolved on earth from nonliving matter through spontaneous chemical and physical processes no different from those in operation in the universe today. Hypotheses made under these assumptions are testable to the extent that the chemical and physical processes can be duplicated in the laboratory.

The scientific view of the origin of life is a more sophisticated version of a very old belief, easily as old as beliefs in divine origins, that life can arise by spontaneous generation. As late as the mid-nineteenth century, laymen and scientists alike believed that lesser forms of life could arise spontaneously and more or less constantly from rotting organic matter. A prime example is the recipe for spontaneous production of mice, reported in the 1600s by the chemist J. B. Van Helmont: "If you press a piece of underwear soiled with sweat together with some wheat in an open mouth jar, after about twenty-one days the odor changes and the ferment, coming out of the underwear and penetrating through the husks of the wheat, changes the wheat into mice."

The common belief in spontaneous generation was refuted in 1862 by Louis Pasteur's famous demonstration that nutrient fluids, sterilized and sealed against contamination, could be kept indefinitely without generation of microbial or other life. In a sense, Pasteur's refutation of spontaneous generation was too effective: scientists generally came to reject the idea that life could have arisen spontaneously at any time. As a result, even scientific hypotheses proposing chemical or molecular origins of life were not accepted as valid for nearly a hundred years after Pasteur's experiments.

This generally negative attitude remained unchanged until the 1920s, when A. I. Oparin, a Russian, and J. B. S. Haldane, an Englishman, independently developed a hypothesis that forced reconsideration of spontaneous generation. Oparin and Haldane agreed that spontaneous generation of cellular life is impossible under present-day conditions. They argued, however, that the earth's surface and atmosphere during the first millions of years of its existence were radically different from today. The primordial conditions, according to Oparin and Haldane, favored spontaneous generation of life rather than inhibiting it.

Oparin and Haldane proposed that the earth's primitive atmosphere contained primarily reduced substances such as methane (CH_4), ammonia (NH_3), and water instead of high concentrations of oxygen as in today's atmosphere. In such an atmosphere, electrons and hydrogens were readily available for conversion of inorganic substances to organic forms. Through absorption of solar energy and the effects of electrical discharges during thunderstorms, great quantities of organic molecules were supposedly produced in the atmosphere and on the earth's surface. The organic molecules would have accumulated because the two main routes by which such substances break down today, oxidation and decay by microorganisms, could not take place. As these organic compounds became more and more concentrated over many millions of years, they interacted spontaneously to produce still more complex organic substances such as nucleic acids and proteins.

These complex substances, according to the Oparin–Haldane hypothesis, constantly aggregated into random collections of molecules, some of which were able to carry out primitive living reactions. Presumably these were more successful than nonliving assemblies in competing for space and raw materials and therefore persisted. Eventually the most successful of these aggregations developed the full qualities of life, including the ability to self-reproduce.

As the chemical activities of the first collections of living matter increased, the store of organic molecules used as an energy source became depleted. Life persisted, however, through the development of photosynthesis, which took advantage of sunlight as an inexhaustible energy source for production of organic molecules from inorganic precursors.

Photosynthesis and the organisms obtaining energy by this pathway gradually became more complex until the level was achieved in which water was used as a primary source of electrons and hydrogens for photosynthesis, with oxygen released as a by-product (see p. 247). By this time, life was definitely cellular and had advanced in complexity to organisms that were the direct ancestors of present-day bacteria and cyanobacteria.

The release of oxygen in ever greater quantities by early cyanobacteria gradually changed the character of the atmosphere from reducing to oxidizing. Once this change came about, spontaneous generation of life was no longer possible because organic molecules generated outside the confines of living cells were quickly oxidized back to inorganic form. From this time on, life could arise only from preexisting life as in today's environment.

The Oparin–Haldane hypothesis was not widely accepted at first because of the general weight of opinion against spontaneous generation and the lack of an effective scientific test of the ideas. The situation remained much the same for the next 30 years, until several discoveries stimulated new interest in thinking about the origins of life. One was a successful test by S. L. Miller in 1953 of a fundamental proposal of the Oparin–Haldane hypothesis: that an energy source acting on a reducing atmosphere can spontaneously generate organic molecules. The second discovery came from analysis of the light transmitted or reflected by the atmospheres of other planets in the solar system and dust clouds in interstellar space. The outermost planets, in particular Jupiter, Saturn, and Uranus, proved to have reducing atmospheres containing some of the components proposed by Oparin and Haldane, including methane and ammonia. The interstellar dust clouds were also found to contain reduced gases and even organic molecules (see Table 26-1). Further, some meteors arriving from outer space were found to contain a variety of organic molecules including amino acids.

These discoveries showed that key assumptions of the Oparin–Haldane hypothesis were feasible and in fact appeared to have taken place. The findings set off a wave of speculation and experimentation about the conditions and reactions involved in the transition from nonliving to living matter. The contemporary conclusion from this effort is that life originated through spontaneous, inanimate processes taking place under the conditions existing on the primitive earth. The molecular assemblies harboring the spark of life advanced spontaneously through a series of intermediate stages that, although not yet cellular, were able to carry out successively more complex activities. Eventually the molecular assemblies achieved fully cellular characteristics.

STAGES IN THE EVOLUTION OF CELLULAR LIFE

Because conditions on the primitive earth were critical for evolution of cellular life, formation of the earth and its atmosphere is considered as the first stage in the long process leading from inanimate matter to living forms. This stage provided inorganic raw materials for the evolution of life and set up conditions for their interaction. The second stage produced organic molecules through interactions between inorganic substances, driven by energy sources such as lightning and ultraviolet radiation from the sun. In the third stage the organic molecules produced in the second stage assembled randomly into collections capable of chemical interaction with the environment. As the collections formed, interactions taking place within them produced still more complex organic substances, including polypeptides and nucleic acids. Some of these collections of molecules were capable of carrying out primitive living reactions.

There is little agreement on the form taken by the first spark of life in these primitive aggregates. Some investigators propose that in its most primitive form, life consisted simply of the ability of a molecular aggregate to absorb organic molecules from the environment and use them as raw materials and an energy source to grow in mass. Presumably proteins and nucleic acids were among the molecules assembled in the primitive assemblies, not yet linked by a coding system. The initial form of reproduction may simply have been breakage of larger aggregates into smaller particles still capable of growing in mass.

In the fourth stage a genetic code appeared in the primitive living aggregates. The code regulated duplication of information required for reproduction of the molecular aggregates and established the link between nucleic acids and the ordered synthesis of proteins. With these developments providing the basis for directed synthesis and reproduction, life, although primitive and precellular, was fully established in the molecular assemblies.

The genetic code, built into the structure of the nucleic acids, provided a means for introducing and recording mutations. This development established natural selection of favorable mutations as the basis for further evolutionary change. At this stage, evolution progressed from chemical to biological. The gradual accumulation of favorable mutations by natural selection accomplished the fifth and final stage in the origin of cellular life: The precellular assemblies were converted into fully organized cells, complete with a nuclear region and cytoplasm, all enclosed by an outer boundary membrane that limited and controlled the substances entering and leaving.

Figure 26-1 The Horsehead Nebula, a cloud of gas and dust particles some 1300 light years from the earth. (Courtesy of Hale Observatories.)

Table 26-1	Substances and Chemical Groups Important in the Evolution of Life Detected in Cosmic Clouds or Outer Space	
Atom, Molecule, or Radical	**Symbol**
Hydrogen atom	H
Hydroxyl radical	OH^-
Ammonia	NH_3
Water	H_2O
Formaldehyde	HCHO
Carbon monoxide	CO
Cyanogen radical	CN^-
Hydrogen cyanide	HCN
Cyanoacetylene	HC_2CN
Methyl alcohol	CH_3OH
Formic acid	HCOOH
Carbon monosulfide	CS
Formamide	$HCONH_2$
Silicon oxide	SiO
Carbonyl sulfide	OCS
Acetonitrile	CH_3CN
Isocyanic acid	HNCO
Hydrogen Isocyanide	HNC
Methylacetylene	CH_3C_2H
Acetaldehyde	CH_3CHO
Thioformaldehyde	HCHS
Hydrogen sulfide	H_2S
Methylene imine	H_2CNH

Adapted from S. W. Fox, *Mol. Cell. Biochem*, 3:1291 (1974); courtesy of S. W. Fox.

The first cells to appear on earth were probably roughly equivalent in complexity to the most primitive prokaryotes known today. Presumably this level was reached by the time the Archean deposits were laid down in Australia some 3.5 billion years ago.

Stage 1: Formation of the Earth and Its Primitive Atmosphere

According to current hypotheses, the solar system, and in fact all the stars and other bodies in space, condensed out of vast clouds of gas and dust particles. The matter in these clouds, which still persist in space (Fig. 26-1), has been identified by analysis of light or other forms of radiant energy passing through them. The interstellar clouds consist mostly of hydrogen gas at extremely low concentrations. Lesser amounts of helium and neon are also present. Other elements and compounds, including metallic iron and nickel, the silicates, oxides, sulfides, and carbides of these and other metals, inorganic carbon compounds, ammonia, and water, are also present. In addition, most notably for the evolution of life, organic molecules such as formic acid, methyl alcohol, formaldehyde, acetaldehyde, and cyanoacetylene can also be detected in the clouds (Table 26-1).

According to the condensation hypothesis, first proposed by Immanuel Kant, stars and planetary systems are continually condensing and disintegrating within interstellar dust and gas clouds. Some 4 to 5 billion years ago our solar system condensed from one of the clouds. Rapid condensation of most of the

matter around one center developed high pressure and heat in a rotating gas cloud. The intense heat and pressure set off a thermonuclear reaction, establishing the star of our solar system, the sun. The remainder of the spiraling dust and gas condensed into the planets and other bodies surrounding the sun. As the planets formed, internal pressure and heat were also generated in these bodies.

On earth, internal temperatures probably rose to levels as high as 1000° to 3000°C, causing melting and stratification of its solid matter. Metallic elements settled to form the molten core of the earth, and lighter substances, such as silicates, carbides, and sulfides of the metallic elements, floated to the surface. As the planet radiated away some of its heat, layers at the surface cooled and solidified into the rocky materials of the earth's crust. Water, initially present only as a vapor, condensed into droplets and rained down on the rocks of the crust. Eventually, after millennia of torrential rains and further cooling, water collected into the rivers, lakes, and seas of the primitive earth.

Some of the water was retained as vapor in the atmosphere. Other components of the initial atmosphere probably included H_2, N_2, CO, and CO_2. Carbon dioxide would have been steadily lost from the atmosphere through its interaction with elements at the earth's surface to form solid carbonates. However, significant quantities of CO_2 would probably have returned to the atmosphere in the gases released by erupting volcanoes. H_2S may also have been among the gases released by volcanoes. Most of the oxygen initially present would have been removed by reaction with elements of the crust and atmosphere to form oxides.

Whether ammonia and methane were present in quantity in the primordial atmosphere, as proposed by Oparin and Haldane, remains a subject of considerable controversy. These gases, according to some scientists, would have been produced by spontaneous interaction of the hydrogen, nitrogen, and carbon compounds of the primitive atmosphere and crust. This proposal is supported by spectroscopic analysis of the atmospheres of some other planets and large satellites of the solar system, in which either ammonia or methane or both are present along with water vapor, and free oxygen is absent. However, geochemists, in analyzing the oldest known sedimentary rock strata in Isua, Greenland (3.8 billion years old), find evidence of carbonates but none of the compounds expected if ammonia or methane were present in quantity on the primitive earth. According to the geochemists, the only source of hydrogen was likely to have been the H_2 condensing from the original cloud of primordial matter. In any case the early atmosphere is expected to have stabilized as moderately reducing, if only hydrogen was present, or strongly reducing, if gases such as ammonia, methane, or H_2S were present.

There is also some controversy about the presence of oxygen in the earth's primitive atmosphere. Most hypotheses developed for the spontaneous appearance of organic molecules on the primitive earth depend on a reducing atmosphere. If oxygen was present in any quantity, it becomes more difficult to propose reasonable pathways for the appearance of organic molecules. We will assume, because oxygen in quantity is generally absent from interstellar clouds and other planets or satellites with atmospheres in our solar system, that it was also absent, or at least not abundant, on the primitive earth. The time from the initial condensation of matter until the earth's environment favored the spontaneous formation of organic compounds may have been as much as half a billion years.

Stage 2: Spontaneous Formation of Organic Molecules

In the second stage in the evolution of life, organic molecules were produced by the action of various energy sources on components of the primitive atmosphere and crust. These energy sources are believed to be essentially the same as those acting on the earth and its atmosphere today: solar radiation, electrical discharges during storms, cosmic rays, radioactivity produced by atomic decay in surface rocks, shock waves from meteorites passing through the atmosphere, and heat released from the earth's core, particularly by volcanoes.

The organic molecules produced in the atmosphere through these energy sources were trapped in condensing raindrops and carried down to the crust by rainstorms. Additional material was produced at the surface and possibly at sites where vents spurted hot water under the sea. Gradually the organic molecules accumulated in lakes and seas. As their concentrations increased, interactions between organic substances and elements and compounds of the crust and atmosphere produced more varied and complex molecules.

Evidence That Organic Molecules Can Be Synthesized under Primitive Earth Conditions   Miller's experiment in 1953 was the first to demonstrate that organic molecules can be synthesized spontaneously under conditions proposed for the primitive earth. Miller, then a graduate student working with H. C. Urey, developed an apparatus to test the effects of electrical discharges on a simulated reducing atmosphere (Fig. 26-2). Miller assumed that the primitive atmosphere contained both methane and ammonia along with hydrogen and water vapor. He placed these gases in a closed apparatus and exposed them to an energy source in the form of continuously sparking electrodes. Water vapor was added to the atmosphere in the chamber from water boiled at one point in the apparatus and continually removed by cooling and condensation at another point. The condensation removed and trapped organic molecules produced by the action of the energy source on the atmosphere. Operation of the apparatus for one week produced a surprising assortment of organic compounds, including urea, biological and nonbiological amino acids, and lactic, formic, and acetic acids (Table 26-2). As much as 15% of the carbon in the simulated atmosphere was converted into organic compounds. The apparatus provided a convincing demonstration that organic compounds can actually be produced spontaneously through the action of energy sources on a reducing atmosphere.

After Miller published his results, many additional experiments were carried out using variations in the starting substances and energy input. An important substance included in many of these experiments was hydrogen cyanide (HCN), readily produced by the

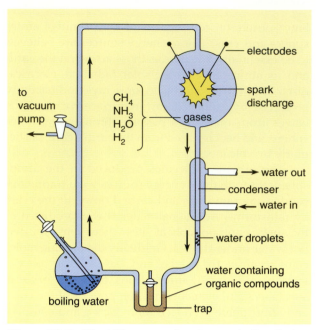

Figure 26-2 The Miller apparatus demonstrating that organic molecules can be spontaneously synthesized in a reducing atmosphere. Operation of the apparatus for one week converted 15% of the carbon in the atmosphere into a surprising variety of organic compounds. (Courtesy of S. L. Miller. Copyright 1955 by the American Chemical Society.)

action of electrical discharges on inorganic substances assumed to be present in a reducing atmosphere:

$$CO + NH_3 \longrightarrow HCN + H_2O \qquad (26\text{-}1)$$

$$2CH_4 + N_2 \longrightarrow 2HCN + 3H_2 \qquad (26\text{-}2)$$

When HCN is present, experiments readily produce additional amino acids, purines, pyrimidines, and porphyrins. HCN is also easily converted into a variety of *condensing agents* that promote assembly of amino acids into polypeptides and of nucleotides into nucleic acids. (Condensing agents are substances that promote condensation reactions, in which molecular subunits are assembled by removal of the elements of a water molecule from reactants; see p. 40.) Formaldehyde, also readily produced by interactions of the primitive gases, leads to production of sugars when included in experiments simulating primitive earth conditions.

Biological amino acids are common among the organic substances produced in these experiments. Large quantities of nonbiological amino acids are also produced, including D-forms (only the L-forms of the 20 biological amino acids are used by ribosomes in protein synthesis; see Chapter 2). Why only 20 of the many amino acids probably produced in this stage were later adapted for use in protein synthesis, and

Table 26-2	Organic Compounds Formed in Sample Runs in the Miller Apparatus (mole × 10⁵)		
	Run 1	Run 3	Run 6
Glycine	63 (2.1%)*	80 (0.46%)*	14.2 (0.48%)*
Alanine	34	9	1.0
Sarcosine	5	86	1.5
β-alanine	15	4	7.0
α-aminobutyric acid	5	1	—
Methylalanine	1	12.5	—
Aspartic acid	0.4	0.2	0.3
Glutamic acid	0.6	0.5	0.5
Iminodiacetic acid	5.5	0.3	3.9
Iminoacetic-propionic acid	0.5	—	—
Formic acid	233	149	135
Acetic acid	15.2	135	41
Propionic acid	12.6	19	22
Glycolic acid	56	28	32
Lactic acid	31	4.3	1.5
α-hydroxybutyric acid	5	1	—
Succinic acid	3.8	—	2
Urea	2	—	2
Methylurea	1.5	—	0.5
Total yields of compounds listed	15%*	3%*	8%*

Reprinted with permission from S. L. Miller, in A. I. Oparin, ed., *The Origin of Life on the Earth* (I.U.B.: Sympos. Series). New York; Pergamon Press, 1959.
* Percent yield based on carbon placed in the apparatus as methane.

these only in the L-form, are among the many unanswered questions about the origin of life.

Sugars have been produced in experiments duplicating primitive conditions by reactions in which formaldehyde units interact in alkaline solutions to produce four-, five-, and six-carbon sugars such as glyceraldehyde, ribose, deoxyribose, glucose, fructose, mannose, and xylose. Fatty acids have also been produced; in one experiment by W. R. Hargreaves, S. J. Muvihill, and D. W. Deamer, fatty acids and glycerol combined spontaneously to produce phospholipids when heated to dryness at 65°C, as they might have been in an evaporating tidepool along a primitive sea.

Adenine has proved to be the most easily produced of the nitrogenous bases occurring in nucleic acids. C. Ponnamperuma and his coworkers detected synthesis of this purine, along with small amounts of guanine, in mixtures of CH_4, NH_3, and H_2O exposed to a beam of electrons. Purines are also readily synthesized when experimental atmospheres containing HCN are irradiated with ultraviolet light. Spontaneous synthesis of pyrimidines requires more extreme conditions, such as exposure of urea to cyanoacetylene ($HC \equiv C - C \equiv N$) at elevated temperatures. Cyanoacetylene, produced spontaneously by the action of electrical discharges on CH_4 and N_2, acts as a condensing agent in these interactions.

Experiments with primitive atmospheres and energy sources have thus produced the organic building blocks for all the major biological molecules—proteins, carbohydrates, lipids, and nucleic acids. The numbers and variety of organic molecules produced in the simulation experiments depend directly on the degree to which the primitive atmosphere is reducing in character. The richest variety is obtained if methane and ammonia are among the gases being tested. If only hydrogen is present, the yield of organic material is much smaller. J. Pinto and his coworkers estimate that even if only H_2 and CO_2 were present in the primitive atmosphere, the action of solar radiation alone on these substances would have produced 3 million tons of formaldehyde each year. At this rate, formaldehyde would have reached concentrations in the oceans high enough to polymerize into more complex organic molecules within 10 million years.

One of the fundamental requirements for the synthesis and accumulation of organic molecules in any quantity in the experiments simulating primitive conditions is the absence of oxygen. If molecular oxygen is present among the reactants in quantities sufficient to convert the experimental atmosphere from a reducing to an oxidizing character, the yields of organic molecules are vanishingly small or nonexistent. If molecular oxygen was present in any quantity, we must assume that it was continually removed from the atmosphere by chemical combination into oxides, so

that the balance of the atmosphere remained predominantly reducing.

An analysis by K. A. Kvenvolden and J. Oro and their colleagues of meteorites striking the earth directly supports conclusions from the Miller-type experiments. For example, a large stony meteorite found near Murchison, Australia contained no less than 77 amino acids, including many of the 20 occurring in proteins. Purines and pyrimidines are also suspected to be among the organic molecules of the meteorite. Analysis of the tail of Halley's comet by space probes and spectroscopy also indicates that the body of the comet contains a complex mixture of organic molecules. W. F. Heubner even found evidence of a formaldehyde polymer, polyoxymethane $(H_2CO)_n$, in the comet. Interstellar clouds, as noted, also contain organic molecules including formaldehyde, HCN, acetaldehyde, and cyanoacetylene.

Some authorities have criticized the results obtained with meteorites, pointing out that the amino acids and other organic compounds could have been picked up as contaminants as these bodies impacted living organisms during passage through the atmosphere and collision with the earth's surface. This is not thought likely, however, because nonbiological amino acids predominate in the Murchison and other samples and because D- and L-forms occur in approximately equal quantities in the Murchison meteorite. If the amino acids were picked up as contaminants from living organisms, they should be primarily or exclusively in the L-form and mostly biological rather than nonbiological.

The arrival of organic molecules in meteorites and comets may have actually contributed significantly to the complement of organic molecules on the primitive earth during the first billion years of the solar system's existence. During this period, dust and asteroids were distributed throughout the solar system as remnants of the dust cloud that condensed to form the sun and planets. Much of this material was swept from space by the planets and larger satellites as they moved through their orbits during the period from about 4.5 to 3.8 billion years ago. Most of the craters of the moon were formed by collisions with interplanetary debris during this period. The collisions are estimated by C. F. Chiya to have delivered at least 1 to 10 million kilograms of organic matter per year to the primordial earth.

Spontaneous Combination of Organic Subunits into Larger Molecules Once amino acids, purines, pyrimidines, monosaccharides, and other organic subunits formed in significant quantities, they presumably assembled spontaneously into more complex molecules such as proteins and nucleic acids. A major problem presented by such assembly reactions is that they primarily involve chemical condensations. In aqueous solutions, reactions tend to run spontaneously in the opposite direction, toward hydrolysis, in which covalent bonds holding molecular subunits together are broken by the addition of the elements of a molecule of water (see p. 40). Thus it is not clear how condensations became predominant in the primitive environment.

The probability that condensations will proceed spontaneously is greatly increased if water is reduced in concentration or eliminated from the reacting systems. One means to accomplish this, already mentioned, is the presence of condensing agents among the reactants. Another is the anhydrous condition produced in ponds or pools evaporated to dryness by the sun or volcanic activity.

A third possibility is adsorption of reacting molecules on clays, originally proposed by J. C. Bernal and more recently by A. G. Cairns-Smith. Clays consist of stacked, extremely thin layers of negatively charged aluminosilicates separated by layers of water. Depending on the state of hydration, the water layers range from 1 to 5 nm in thickness. The aluminosilicate layers can adsorb and concentrate organic molecules in large quantities. Adsorption into the layers excludes water molecules and sets up anhydrous conditions promoting condensation reactions. The strongly negative charge of the aluminosilicate layers also attracts positive ions such as Na^+, K^+, Ca^{2+}, and Mg^{2+} to their surfaces. Metallic ions such as Fe^{2+}, Zn^{2+}, and Ni^{2+}, which substitute for aluminum atoms at sites in the aluminosilicate layers, can act as catalysts for condensation reactions, speeding combination of organic building blocks into more complex assemblies at moderate temperatures.

All three routes for the condensation of macromolecules have been successfully tested in experiments simulating primitive conditions. In experiments testing the spontaneous assembly of polypeptides, S. W. Fox and his colleagues melted together mixtures of anhydrous amino acids at temperatures ranging from 120° to 200°C for seven days. The procedure caused the amino acids to link into long, proteinlike polymers with molecular weights approaching 20,000, which Fox called *proteinoids*. The polymers exhibit some of the characteristics of natural proteins when dissolved in water, including hydrolysis by proteinases such as trypsin, pepsin, and chymotrypsin, and even limited ability to catalyze reactions such as ATP hydrolysis. Fox proposed that the somewhat extreme conditions required to form proteinoids might have prevailed around sites of volcanic activity, where solutions of amino acids might have been dried and heated to the required temperatures.

Polypeptides have also been produced by exposing amino acids to condensing agents under anhydrous conditions. One group of condensing agents used in these experiments, based on cyanamide ($H_2N—C≡N$) and its derivatives, is relatively stable in water. Mixtures of these compounds with amino acids in aqueous solution yield polypeptides at moderate temperatures.

Several laboratories have demonstrated that polypeptides can also be produced from amino acids by the third major route, adsorption on clays. One of the most significant of these experiments was carried out by N. Lahav and S. Chang, who detected polypeptide formation in mixtures of clay and amino acids exposed to variations in water content and temperatures fluctuating between 25° and 94°C. (The Experimental Process essay by Lahav and Chang on p. 766 describes the rationale and results of their experiments.) Clays also promote formation of many organic building blocks from more simple precursors.

Nucleotides and nucleic acids have been produced chiefly in simulated systems containing condensing agents. The condensing agents of choice have been *polyphosphates*, phosphate groups linked into chains of varying length. Besides promoting condensation reactions, polyphosphates release energy when hydrolyzed, acting as an inorganic version of ATP. Ponnamperuma and R. Mack showed that nucleotides form spontaneously in mixtures of nucleosides and polyphosphates. Similarly, A. W. Schwartz and Fox induced spontaneous assembly of nucleic acids by heating nucleotides to 65°C in the presence of polyphosphates. Polyphosphates have also been used experimentally to promote spontaneous formation of AMP from adenine, ribose, and inorganic phosphate, polysaccharides from glucose, and polypeptides from amino acids at moderate temperatures. The polyphosphates required for these reactions could have been formed through the exposure of inorganic phosphorus or phosphate groups to high temperatures (200° to 350°C), possibly in regions of volcanic activity or at hot vents in the sea floor or, as L. E. Orgel and his associates have demonstrated, at more moderate temperatures through interaction of phosphate groups with urea and ammonia.

Some of the experiments with spontaneous nucleic acid synthesis were carried out in clays. J. P. Ferris and G. E. Erten found that RNAlike molecules were produced spontaneously from nucleoside pyrophosphates suspended in clays; most of the sugarphosphate linkages in the RNAs were the 3' → 5' linkages typical of RNA.

Reactions of this type, as well as the spontaneous interactions producing simpler organic molecules, are considered to have produced significant quantities of all the major biological molecules over the millions of years following the initial formation of the earth. Accumulation of these compounds established conditions necessary for the third stage in the evolution of life, formation of organic matter into molecular aggregates capable of primitive living reactions.

Stage 3: Formation of Molecular Aggregates

The formation of aggregates requires that molecules in solution must be precipitated or brought together in arrangements that are stable enough to resist disruption. Four mechanisms have been widely considered as possibilities for spontaneous aggregation of discrete, reactive assemblies of this type: (1) *coacervation*, originally proposed by Oparin; (2) formation of *proteinoid microspheres*, proposed by Fox; (3) formation of lipid bilayers, first proposed by R. J. Goldacre; and (4) adsorption on clays, advanced as a possibility by Bernal. Each of these mechanisms has been duplicated in experiments under conditions approximating the primitive environment.

Coacervation In coacervation, polypeptides in solution assemble into discrete, stable droplets 1 to 500 μm in diameter as a result of attractions between charged groups on the surfaces of polypeptide chains. If many polar or charged groups face the surface of a coacervate formed in this way, the surrounding water molecules tend to form an ordered film or boundary layer several molecules in thickness over the surface. The charged and polar surface groups and the film of ordered water molecules give the boundary of the droplet some of the properties of a membrane. As a result, coacervates may absorb molecules from the surrounding solution and shrink or swell in response to differences in molecular concentrations inside or outside the droplets.

Oparin showed in a series of experiments that enzymes included in the solutions used to form coacervates could be taken up and concentrated inside the protein droplets. Inclusion of starch phosphorylase, an enzyme that catalyzes polymerization of glucose 1-phosphate, led to starch formation inside coacervates and to growth of the droplets if glucose 1-phosphate was present in the surrounding solution.

Although functional enzymes are not expected to have been among the polypeptides included in coacervates on the primitive earth, the experiments demonstrate that reaction systems can be spontaneously concentrated and confined in active form inside coacervate droplets. If such reactions led to extensive growth, the coacervates could have separated into smaller droplets with the same properties, thus achieving the most primitive form of growth and reproduction. By this route, according to Oparin, coacervates carrying out favorable reactions could have increased in number and displaced droplets containing less favorable reaction systems. One problem with coacervates, how-

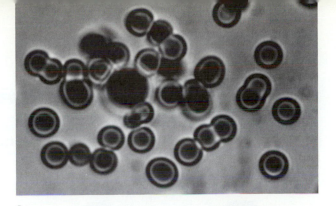

a

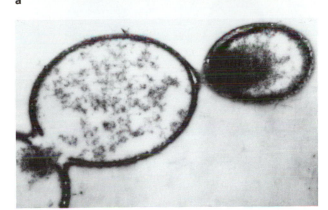

b

Figure 26-3 Fox's microspheres. (a) A collection of microspheres made by heating proteinoids in water, followed by cooling. Light micrograph, × 640. (b) Microspheres in a configuration superficially resembling cell division. Electron micrograph, × 7,800. (Courtesy of S. W. Fox, from *Molecular Evolution and the Origin of Life*, by S. W. Fox and K. Dose. San Francisco: W. H. Freeman and Company, 1972.)

ever, is stability. The droplets are highly sensitive to conditions in the surrounding solution and break down readily if conditions stray from optimum values.

Proteinoid Microspheres Fox and his colleagues showed that if a solution of proteinoids is heated in water and allowed to cool, highly stable particles 1 to 2 μm in diameter, which he termed *microspheres*, separate out of the solution (Fig. 26-3). Microspheres are surrounded by a differentiated surface layer superficially resembling a membrane and can carry out internal reactions among molecules absorbed from the surrounding solution. Some microspheres retain the weak catalytic activity of proteinoids, including the ability to increase the rate of biological reactions such as ATP hydrolysis.

Microspheres appear in electron micrographs in forms that suggest budding or fragmenting, implying that these aggregates can also carry out a primitive form of reproduction (see Fig. 26-3b). Fox proposed that microspheres formed readily in the primitive environment, producing molecular aggregates in which

How Were Peptides Formed in the Prebiotic Era?

Noam Lahav and Sherwood Chang

NOAM LAHAV is a professor at the Hebrew University of Jerusalem, Israel, who has been active in origin of life research since 1975. His major interests are the possible role of minerals in prebiotic reactions, especially peptide formation, and devising scenarios for the first prebiotic self-replication and translation reactions. SHERWOOD CHANG received his Ph.D. (in organic chemistry with a minor in physical chemistry) from the University of Wisconsin in 1967. Since then, he has been a research scientist at NASA's Ames Research Center, where he is now Chief of the Planetary Biology Branch. His interests are related to the origin of organic matter in the cosmos and the origin of life on Earth.

Many researchers in the early fifties were excited by the possibility of making contributions to the scientific understanding of the origin of life. Two scientific breakthroughs in 1953 stimulated their enthusiasm. The first, by James Watson and Francis Crick, was discovery of the double helix structure of deoxyribonucleic acid (DNA), which carries the information needed to replicate cells in all living organisms. With this knowledge the biochemistry of genetic processes could be studied in detail.

The second discovery resulted from an experiment outlined theoretically by Harold Urey and carried out by his graduate student, Stanley Miller, at the University of Chicago. This experiment [see p. 761] simulating the passage of lightning through Earth's early atmosphere over its oceans, demonstrated unambiguously that some of the biochemicals necessary for all life could have been produced by a strictly nonbiological process prior to the origin of life. Although theories about how life originated had been published earlier in the century by A. I. Oparin and J. B. S. Haldane, Miller was the first researcher to provide compelling experimental evidence in support of one of the central ideas expressed in the theories, that the interaction of energy with inorganic chemicals on the primordial Earth led to synthesis of simple organic compounds.

Soon after the publication of Miller's experiment, scientists all over the world adopted his approach and studied other energy sources such as sunlight, heat, and radioactivity for their ability to promote organic synthe-

ses. The origin of biochemical building blocks, which used to be considered a formidable barrier to understanding the origin of life, became a new area of chemical research. Moreover, it became clear that many questions about the chemical reactions leading to life's origin could be fruitfully explored through scientific research. This realization, together with the new discipline of molecular biology launched by the discovery of DNA structure, seemed to hold promise of a rapid solution to the mysteries of life's origin.

Miller's observations also paved the way for research on another key hypothesis of theories on the origin of life, namely, that simple compounds were converted by other natural processes to more complex organic compounds such as peptides and polynucleotides, the precursors of proteins and nuclei acids.

In living cells, protein synthesis occurs by chemical condensation of amino acids, a complex process catalyzed by enzymes. Enzymes also catalyze the synthesis of nucleic acids from nucleotides. Assuming that amino acids and nucleotides were readily formed on the early Earth, prebiotic processes would have been needed to convert these monomers into polymers in the absence of enzymes. Researchers who investigated experimental models for this process soon realized that the condensations were more difficult to achieve under plausible conditions than were the syntheses of the simple monomers.

As early as 1951, however, a biophysicist named J. D. Bernal suggested that minerals on the prebiotic Earth could have catalyzed the reactions of prebiotic monomers. He favored clay minerals, a family of ubiquitous aluminosilicates consisting of extremely small crystallites and possessing very chemically active surfaces. In 1966, an organic chemist, A. G. Cairns-Smith, further underscored the importance of minerals and made a case that clay crystals might have been the first life forms. Against this background in the mid-70s several scientists at NASA's Ames Research Center discussed a variety of experimental strategies to tackle the problem of peptide synthesis under prebiotic conditions. Since few experimental studies had addressed the role of clays in the origin of life, we concluded that the use of these minerals to assist the conversion of amino acids to peptides offered a promising and relatively unexplored approach.

To test our hypothesis that clay minerals could catalyze prebiotic amino acid condensation, we developed an experimental approach that also took into account natural variations in environmental conditions. Aqueous slurries of the amino acid glycine and either kaolinite or montmorillonite clay minerals were dried out by heating and held at 94°C for several days. In the second reaction cycle the mixture was cooled to room temperature, rewetted with water, and heated to dryness again. After these cycles were repeated the desired number of times, up to 27, the slurries were extracted with small amounts of ammonium hydroxide solution, and the water was analyzed for peptides. Three sets of control experiments were carried out in which the procedure was modified by lowering the temperature, omitting clays in the reaction cycles, and eliminating the rewetting step. The objectives of the control experiments were to determine how significant temperature, presence of clay, and rewetting were as independent factors in the overall results.

The rationale behind the procedures was based on the following considerations. Glycine was the major product of almost all experiments on the prebiotic synthesis of amino acids. Therefore, it seemed reasonable to use it as our starting material. Kaolinite and montmorillonite are widespread on Earth today and represent readily available model minerals known to be associated with aqueous environments. The thermal stability of glycine and its shorter polymers assure survival at the temperatures employed. Although it is known that amino acids in the solid state condense readily at temperatures above 150°C, it was considered important to carry out the reaction at temperatures below the boiling point of water. (Interestingly, microbes of ancient ancestral lineage that thrive at 105°C have been isolated recently from deep sea hydrothermal vents.) The wetting-drying procedure was intended to simulate conditions in environments where both temperature and water content fluctuate cyclically on a daily or seasonal basis. Presumably, such environments were common on the prebiotic Earth, as they are today.

Analyses of the reaction mixtures using standard chromatographic procedures showed that condensation of glycine to peptides did occur under the experimental conditions. In the absence of clays, however, negligible amounts of product were observed. The yields of peptides obtained with clays decreased with peptide length in the order diglycine > triglycine > tetraglycine > pentaglycine. The total amount of peptides increased with the number of cycles, but was always less than 1% of the initial amount of glycine. Peptide formation also took place at temperatures as low as 60°C, but at a very low rate. Heating alone was much less efficient than heating with wetting-drying cycles.

Our results showed that productivity in the condensation of glycine (and probably other amino acids) into peptides is greatly increased by the presence of clay minerals. Furthermore, the reactions are enhanced by environmental fluctuations of temperature and water content. Therefore, analogous prebiotic reactions could have occurred in common environmental settings like tidepools and seashores in the absence of implausible chemical condensing agents or enzymes.

Yet to be solved, however, are two related problems. First, how important were environmental rhythms to the origin of life? In our experiments, enhanced peptide synthesis occurred as a result of a dynamic balance between condensation and hydrolysis in a cyclically fluctuating environment. The significance of environmental fluctuations for stages of chemical evolution beyond peptide synthesis is unknown and remains to be studied. Such fluctuations might have been more important on the prebiotic Earth when day lengths were shorter and tides were more frequent and intense than now.

Second, how were peptides with non-random amino acid sequences formed and the directions for their synthesis encoded in the evolving assemblages of prebiotic polymers? Prebiotic peptides with random sequences would have limited functional capability. Eventually, a primitive mechanism would have to be "invented" by which peptides with essentially the same amino acid sequence could be produced repeatedly. A step toward such "directed" synthesis of peptides might have been provided by a process similar to that of our simulation. Given a fluctuating environment containing various amino acids, it is probable that certain peptides would be preferred over others because of the characteristics of specific monomers and short polymers. These factors include solubility, strength of adsorption to clay particles, stability to hydrolysis, and ease of incorporation into a growing peptide chain. Peptides with non-random sequences of amino acids would be of great interest. Conceivably, some could catalyze peptide bond formation itself, while others could fulfill different functions necessary in the chemical evolution of polymers into more complex systems.

the first reactions of life may have appeared and reproduced.

Lipid Bilayer Assembly Phospholipids and some other lipid molecules assemble spontaneously into bilayers when suspended in water (see p. 100). Frequently the spontaneously formed bilayers round up into closed vesicles consisting of a saclike, continuous boundary "membrane" enclosing an inner space. W. R. Hargreaves and D. W. Deamer, for example, showed that phospholipids synthesized under simulated primitive earth conditions can assemble spontaneously into lipid bilayers forming closed vesicles (Fig. 26-4).

Spontaneously formed bilayers are relatively stable and have many properties of living membranes. Various substances, including proteins or polypeptides, can be absorbed on the bilayer surfaces or into the hydrophobic bilayer interior. Other substances can be trapped inside the vesicles as they form. The collection and concentration of organic molecules in the bilayers could have had catalytic effects and promoted a variety of biochemical reactions. In the primitive environment the vesicles could have increased in size through the further uptake of lipids and proteins. Mechanical disturbances such as wave action could have separated larger vesicles into smaller ones, providing a primitive form of reproduction.

One problem with assembly of lipid bilayers from phospholipids is that the straight-chain fatty acids required for phospholipid formation have proved difficult to produce under simulated primitive earth conditions. It is therefore uncertain whether these molecules could have reached concentrations high enough to produce lipid bilayers in any quantity.

Adsorption on Clays Spontaneous assembly by adsorption of organic molecules on clays may have been common because of extensive clay deposits expected to have formed in tidal flats and river estuaries. Clays, as noted, may also have been sites of extensive formation of both simple and more complex organic substances. The organic molecules formed and aggregated by the clays could have provided large, relatively stable areas of concentrated organic matter, heated by sunlight, in which a variety of interactions could have taken place. Besides the polymerization reactions demonstrated to be enhanced by clays, hydrolytic and other reactions, including some releasing free energy, can also be promoted by clays, especially those containing metallic ions in abundance.

In addition, clays have been shown to be capable of storing and transforming energy. Thus an energy input such as heat, light, or mechanical compression can induce in clay molecular alterations that represent a form of potential energy. On return to an original molecular arrangement, some of the energy can be

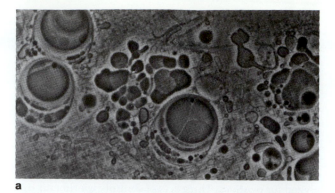

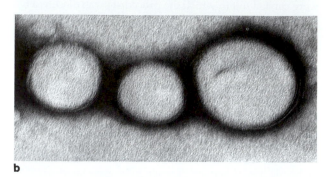

Figure 26-4 Vesicles consisting of a continuous lipid bilayer enclosing an interior space, assembled from phospholipids synthesized under simulated primitive earth conditions. **(a)** Light micrograph. (Courtesy of D. W. Deamer.) **(b)** Electron micrograph of a negatively stained preparation. (Courtesy of W. R. Hargreaves, from *Nature* 266:78 [1977].)

released, in some cases in a form different from the form absorbed. Mechanical energy absorbed by clays, for example, can be released under changing conditions as fluorescence. Clays could thus have channeled environmental energy sources into interacting systems and provided centers of aggregation of organic molecules.

Of the various mechanisms aggregating macromolecules into interacting assemblies, adsorption on clays seems to be the most likely and prevalent route. Clays circumvent a difficulty common to coacervation, microsphere formation, and bilayer assembly presented by the fact that their components—protein-like polymers or phospholipids—must be present in relatively high concentrations for spontaneous aggregation to occur. The regions reaching such concentrations might have been too limited for aggregation to progress to any extent by these mechanisms. Clays, in contrast, undoubtedly occupied wide areas and, by adsorbing substances from the environment, could have spontaneously concentrated organic molecules to levels required for interaction. Thus the admonition "from dust we came and to dust we shall return" may not be far from the truth.

All the proposed mechanisms may have contributed in regions where the special conditions required for their activity were met in the environment. And, sooner or later, the spontaneous formation of lipid bilayers would have contributed to formation of membranes in developing aggregates.

Stage 4: Development of Life in the Primitive Aggregates

Of all the processes proposed as stages in the evolution of life, the reactions accomplishing the transformation of nonliving to living matter have proved most difficult to imagine and test. However, there are some general ideas about pathways by which this difficult feat may have been accomplished. These ideas fall into three areas: (1) development of reactions harnessing an energy source to drive metabolic pathways synthesizing required molecules; (2) the appearance of enzymes, permitting these reactions to take place in ordered sequences at moderate temperatures; and (3) development of informational systems necessary for directed synthesis and reproduction in kind.

Development of Reaction Pathways Utilizing an Energy Source Oxidation-reduction reactions, in which electrons are removed from one substance (an oxidation) and delivered to another (a reduction; see p. 208) were likely to have been among the initial energy-releasing reactions of the primitive macromolecular assemblies. At first the energy from these reactions would have been utilized in single steps, through direct transfer of electrons from substances being oxidized to others being reduced as part of synthetic reactions. However, the advantages and efficiency of energy release in stepwise increments would have favored development of intermediate carriers, opening the way for the appearance of primitive electron transport systems.

Among the prime candidates for the first intermediate carriers are the *porphyrins*. (Figure 26-5 shows a pathway by which these substances may have been spontaneously produced.) Porphyrins readily gain and lose electrons; in this capacity they serve as active groups transferring electrons in the *cytochromes*, carriers that occur in the electron transport systems of both respiration and photosynthesis in present-day cells (as detailed in Chapters 8 and 9).

Porphyrins are also pigments, with electrons capable of absorbing light energy and jumping to excited orbitals at elevated energy levels (see p. 250). In this capacity they form the active centers of chlorophyll molecules in all present-day photosynthetic organisms. In photosynthesis, chlorophyll accepts electrons at relatively low energy levels, absorbs light energy, and transfers the excited electrons to acceptor substances. In primitive precells, porphyrins could also have served in both electron transport and light absorption.

Initially the developing oxidative and photosynthetic pathways probably used a common electron transport pathway, as they do in some contemporary photosynthetic bacteria (see p. 271). Electrons at elevated energy levels might enter the pathway after removal from highly reduced organic molecules absorbed from the environment (Fig. 26-6a). Alternatively, electrons might enter after being elevated to higher energy levels in a light-absorbing molecule such as a porphyrin (Fig. 26-6b and c). In the latter case the primitive photosynthetic pathway might operate either cyclically or noncyclically (see p. 272). In a cyclic pathway (Fig. 26-6b), electrons return to the light-absorbing molecule after passing through a chain of electron carriers. In a noncyclic pathway (Fig. 26-6c), electrons pass from the system containing the light-absorbing molecule to an acceptor outside the electron transport system. The electrons donated to acceptors in primitive systems could have been replaced by electrons removed through oxidation of an organic molecule or an inorganic substance such as H_2S. In some variations of the noncyclic pathways, electrons could pass from the joint photosynthesis-oxidation systems to organic molecules as final acceptors, accomplishing

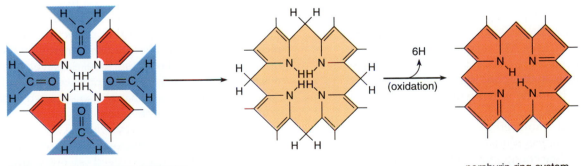

4 pyrrole rings + 4 formaldehyde molecules

porphyrin ring system

Figure 26-5 Possible route by which porphyrins may have assembled spontaneously on the primitive earth.

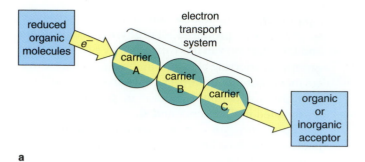

a

Figure 26-6   Possible electron transport pathways for primitive oxidative and photosynthetic reactions. **(a)** Electron flow from oxidation of highly reduced organic substances absorbed from the environment. **(b)** Cyclic electron flow around a porphyrin absorbing light and releasing electrons at elevated energy levels. **(c)** Noncyclic photosynthetic electron flow starting from highly reduced organic or inorganic donors and ending with an organic or inorganic acceptor. The same electron carriers could act in both photosynthetic and oxidative pathways. Porphyrins could have acted as electron carriers as well as photosynthetic pigments in the primitive systems, as they do in contemporary prokaryotes and eukaryotes.

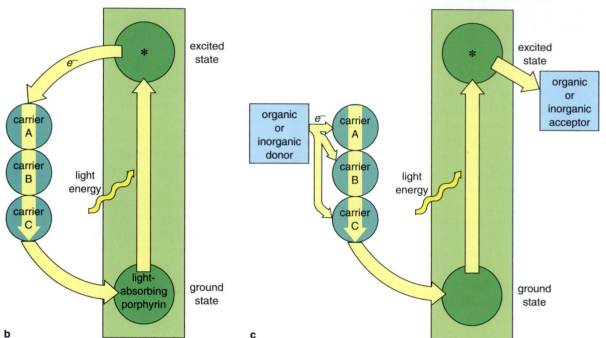

b

c

reductions in pathways synthesizing molecules required by the precells.

Development of ATP as a Coupling Agent   In contemporary cells, ATP is universally used to couple oxidative reactions releasing energy to synthetic reactions requiring energy (see p. 73). ATP is synthesized in present-day cells by two primary routes. In one the energy released by a hydrolysis or oxidation is used directly to add a phosphate group to ADP, in a process termed *substrate level phosphorylation* (see p. 210). Although relatively little ATP is produced by this mechanism in contemporary cells, it is still critical to those existing under anaerobic conditions and is considered likely to have been the initial pathway for ATP production in precells.

The second, more efficient but more complex pathway, *oxidative phosphorylation*, synthesizes ATP by indirect use of free energy released by electron transport. The free energy released as electrons cascade through a series of electron carriers is used to set up

an H^+ gradient across the membrane housing the electron transport system. ATP synthesis is then driven by the H^+ gradient, in effect by reversing an ATP-dependent H^+ pump (the F_oF_1 ATPase; see p. 232 for details).

The earliest forms of life probably utilized hydrolysis as well as oxidation to obtain free energy. ATP may first have entered cells simply as one of many organic molecules absorbed from the environment and hydrolyzed as an energy source. It is likely that energy-requiring synthetic reactions were directly coupled to the release of free energy by the hydrolysis of ATP and other substances. Reversal of the reactions hydrolyzing ATP, synthesizing ATP from ADP and phosphate, probably also occurred to some extent in the primitive environment.

Some of the free energy released during photosynthesis or electron transport by the primitive systems was probably also used in substrate-level synthesis of ATP. The likelihood of this outcome was demonstrated by W. S. Brinigar and his colleagues, who mixed a porphyrin, AMP, and inorganic phosphate with an imidazole group as a catalyst. Exposure of the mixture to ultraviolet or visible light resulted in direct synthesis of ATP.

Development of the much more efficient pathway of oxidative phosphorylation for ATP synthesis probably appeared after substrate-level phosphorylation was widely established among precells. Because oxidative phosphorylation occurs only in association with membranes, development of this pathway for ATP production was linked intimately to the appearance of lipid bilayers. Oxidative phosphorylation would also have required evolution of electron transport systems including two or more molecules capable of accepting and releasing electrons. Establishment of a concentration gradient and membrane potential may have resulted spontaneously from electron transport in some of the membrane-bound systems. This gradient provided an energy source that could be used with greater efficiency than direct substrate-level phosphorylation to drive ATP synthesis. The ATPase activity necessary for this reaction could have been present initially in weak form in the membrane or in a molecular cluster enclosed by the membrane, as in Fox's microspheres. Once established, the much greater efficiency of oxidative phosphorylation ensured that it would become the primary oxidative mechanism through which cells produce ATP.

The ability of precells to synthesize ATP in quantity would promote the use of this substance as a means to supply usable chemical energy to synthetic reactions. Because of the efficiency and versatility of energy transfer via phosphate bonds carried by ATP, this nucleotide gradually became established as the primary substance coupling reactions releasing and requiring energy.

Development of Metabolic Pathways Synthesizing Required Molecules   As these energy-releasing systems appeared, synthetic pathways also became more complex. N. H. Horowitz has suggested a pattern by which complex, sequential synthetic pathways might have appeared. Suppose that a contemporary cellular pathway makes a required substance such as an amino acid through the sequence $A \rightarrow B \rightarrow C \rightarrow D \rightarrow E$, in which A is a simple inorganic substance and E is the final organic product. Initially E was abundant in the environment and was absorbed directly by primitive aggregates. Later, as E became scarce because of its use, chemical selection favored precells that could make E from D, a slightly less complex organic substance still found in abundance in the environment. As D became exhausted, selection favored assemblies that developed the pathway $C \rightarrow D \rightarrow E$, in which the even simpler substance C could be absorbed and used as a raw material to make D. This process continued until the entire synthetic pathway, based on an essentially inexhaustible inorganic substance, was established.

Development of Polypeptide-based Enzymes   All these developing reaction systems yielding and utilizing energy were probably initially increased in rate by the relatively nonspecific catalytic activities of lipid vesicles, microspheres, coacervates, or clays. Gradually, more specialized molecules took over these catalytic functions, leading to development of polypeptide-based enzymes.

The first polypeptides included in the early catalytic systems may simply have provided a stable framework anchoring an inorganic catalyst such as a metallic ion to a primitive molecular cluster. However, since many amino acid side groups, particularly those with acidic or basic properties, also have catalytic activity, some of the catalyst-polypeptide complexes would have been more efficient in speeding reactions than an inorganic catalyst alone. The ability of polypeptide chains to undergo conformational changes in response to substrate binding probably also enhanced the effect of a bound catalytic group in speeding reaction rates.

As the structure of polypeptides became more ordered, some took on a folding conformation that held the catalyst in a position favoring certain substrates, providing the beginning of specificity in catalyzed reactions. In some cases the catalytic activity of amino acid side groups in the folded active site reduced the importance of an included inorganic group, making possible the transfer of catalytic function to the protein. Selection then singled out and perpetuated proteins providing the greatest stability, specificity, and increases in reaction rate. With the development of directed protein synthesis, catalytic polypeptides became increasingly more specific and reproducible until the first fully active and specific enzymes appeared.

Origins of the Information System   Development of any of the metabolic pathways to levels in which more than a few coordinated steps were required would not have been likely without an information system to direct synthesis and reproduction of required components, particularly enzymes. Therefore a mechanism for storing, reproducing, and translating information probably developed in parallel with the metabolic systems.

Developments leading to appearance of a coding system present the most difficult conceptual problems for theorists interested in the evolution of life. The problem is central to understanding the origins of life, because once a system evolved that could store information and direct synthesis of proteins of ordered sequence, the subsequent development of greater complexity and specificity through mutation and natural selection is relatively easy to visualize.

The major difficulty in understanding how a nucleic acid–based information system evolved is the total interdependence between nucleic acids and proteins in contemporary organisms. Without highly specific enzymatic proteins the present-day reactions replicating, transcribing, and translating the nucleic acid code could not take place. The specificity of the enzymatic proteins involved in these processes, however, depends on their amino acid sequences, which are determined by the sequences of nucleotides in nucleic acids. Thus, in contemporary systems, proteins depend on nucleic acids for their structure, and nucleic acids depend largely on proteins to catalyze their activities.

This close interdependence between nucleic acids and enzymatic proteins seemed to present an unsolvable paradox until the recent discovery that some RNA molecules can act as "ribozymes" capable of self-catalysis (see p. 86). This ability may initially have allowed RNAs to act as both informational molecules and catalysts in the primitive precells, with no requirement for enzymatic proteins to catalyze their activities. Thus a self-catalyzed "RNA world," as proposed by L. Orgel, F. H. C. Crick, C. Woese, W. Gilbert, and others, may have preceded the development of DNA as the hereditary material and provided an initial system free of close interdependence between nucleic acids and enzymatic proteins.

The RNA molecules developing as informational molecules may have been assembled initially from ribonucleotides taking part in oxidative and other metabolic reactions. Nucleotides containing a ribose sugar, such as ATP, NAD, and coenzyme A, form important parts of many metabolic pathways, including oxidation-reduction reactions in glycolysis and respiration, and protein synthesis. Their universal presence indicates that they are likely to have been prevalent among the molecules participating in metabolic reactions in precells and thus were available for chance assembly into RNAs. These RNAs could subsequently have developed into informational molecules and ribozymes.

Experiments by J. A. Doudna and J. W. Szostak showed that the RNA-based system that catalyzes splicing of RNA molecules in the protozoan *Tetrahymena* (see p. 435) can be modified to splice nucleotides together into short copies of an RNA template. According to W. Gilbert, in the primitive RNA world, self-replicative reactions of this type were the beginnings of an RNA-based informational system. Gilbert proposes that tRNA-like molecules linked to amino acids could have paired with the RNA informational molecules and influenced the assembly of polypeptides of ordered sequence. Some of the more ordered polypeptides, in turn, could have combined with the RNA informational molecules and tRNA-like molecules, stabilizing their three-dimensional form and increasing the efficiency of their interactions. Selection of the most efficient ribonucleoproteins gradually increased the specificity of the mechanism and led to gradual transfer of most of the enzymatic activities associated with nucleic acid functions from RNA to proteins. Selection also gradually restricted and perfected the coding relationship between sequences in the RNA informational molecules and the amino acid sequences of proteins.

DNA, according to the proponents of the RNA world, entered the picture some time after RNA became established as the informational molecule. At first, DNA nucleotides may have been produced simply through accidental modification of RNA nucleotides—by removal of an oxygen from the ribose subunits of the ribonucleotides. At some point, supposedly, DNA nucleotides were assembled into DNA copies of RNA informational molecules by an enzyme with reverse transcriptase activity. Once the copies were made, DNA supplanted RNA as the molecule storing information because it has greater chemical stability and can be assembled into much longer coding sequences than RNA. Transfer of information storage to DNA left RNA with its functions at intermediate steps between the stored information and protein synthesis.

By the time DNA took up its function as the informational storage molecule, interdependence between nucleic acids and proteins was fully established. The contemporary reactions self-catalyzed by RNA molecules and the viruses with RNA genomes, according to this proposal, are molecular "fossils" persisting from the days of the RNA world.

The proposal of an RNA world solves so many conceptual problems connected with the origin of life that many investigators have accepted it almost as fact. However, the proposal is not without its problems. RNA is among the most difficult molecules to produce under conditions that simulate the primitive environment. Although ribose is among the sugars produced by self-condensation of formaldehyde, the yields are very low. Further, although the purines adenine and guanine can be made relatively easily, the pyrimidines uracil and cytosine, also required to make RNA polymers, are extremely difficult to produce under simulated conditions. Thus it is uncertain whether RNA could have been assembled in quantities large enough to develop as the primary informational molecule in the primitive environment. In fact, its appearance in quantity may have had to await the development of enzymatic proteins. Further problems are presented by the fact that ribose is synthesized by the simulated systems in both D- and L-forms (see

p. 45). G. F. Joyce and his coworkers found that RNA replication stops immediately in simulated systems if mixed D- and L-forms interact.

These problems notwithstanding, there are some ideas about how the coding system may have stabilized around three-nucleotide codons specifying 20 amino acids. Systems based on one- or two-letter codons were probably eliminated because not enough amino acids could be specified. (One-letter codons can encode only four different amino acids, and two-letter codons only 16.) Wobble (see p. 478) in two-nucleotide codon-anticodon pairing would likely have reduced the practical total considerably below 16. Codon-anticodon pairing with one- or two-nucleotide codons would also be highly unstable. Although three-letter codons allow 64 different combinations, offering the possibility of encoding this many different amino acids, the problems imposed by wobble evidently limited the mutually exclusive combinations to 21—20 for different amino acids and 1 to specify "stop." Codons containing four or more nucleotides, which would have offered a much greater coding capacity, may have proved impractical simply because too many combinations were generated for any one group to be fixed by selection into a uniform genetic code. Thus, after considerable evolutionary hits and misses, a system based on three-letter codons, specifying only 20 amino acids and "stop," may have been the only likely outcome of the selection process.

Crick has suggested that the assignment of codons to particular amino acids was simply a historical accident. According to this idea, there are no special reasons why a particular codon stands for a given amino acid. The codons were simply established by random matching to amino acids very early in the evolution of cellular life. Once established, the codon assignments became fixed because a mutation that switched a coding assignment could potentially alter the amino acid sequence of almost every polypeptide in an individual and would very likely be lethal.

Another possibility is that there are definite but as yet unknown chemical or physical relationships between codons or anticodons and amino acids that limit coding assignments to the present ones. The fact that substitutions in the genetic code occur in specialized and restricted systems such as mitochondria makes this possibility seem less likely. Crick's historical accident hypothesis may therefore be correct.

Once the coding system appeared and the relationships between codons and amino acids became fixed, the evolutionary pathway leading to fully directed synthesis and reproduction was open. At this stage the primitive molecular assemblies took on the full qualities of life. At the same time, the coding system provided the basis for heritable changes caused by random mutations, and with it, the change from chemical evolution to biological evolution by natural selection.

Stage 5: Evolution of Cells from Precells

Natural selection of favorable mutations subsequently led to development of all the features of fully cellular life. One of these features was a nuclear region containing the DNA of the coding system and mechanisms for precisely replicating the DNA and transcribing it into RNA. Another feature was a cytoplasmic system, including ribosomes and the required enzymes, for translating the RNA into proteins of specified sequence. The cytoplasm also contained a photosynthetic or oxidative system supplying chemical energy for protein synthesis and assembly of other required molecules. All these systems were enclosed by a membrane consisting of a lipid bilayer, in which were embedded transport proteins controlling the molecules and ions entering and leaving the cell. Finally, the first cells included a mechanism precisely separating replicated DNA molecules and dividing the nuclear and cytoplasmic regions among daughter cells. The evolutionary pathways from precells to this level may have occupied millions of years and undoubtedly included many molecular trials that ended in failure.

LATER METABOLIC DEVELOPMENTS AND THE ORIGINS OF EUKARYOTIC CELLS

Photosynthesis, Oxygen Evolution, and Respiration

Further development of metabolic pathways in the newly emerging cells led to three sequential events of supreme importance for the appearance of eukaryotic cells. One was development of an extended electron transport system allowing water to be used as the electron donor for photosynthesis. Splitting water molecules to release electrons for this newly developed pathway resulted in the second event critical for the evolution of more advanced cells: release of large quantities of oxygen into the atmosphere. This release changed the atmosphere from a reducing to an oxidizing character, making possible the third event, the evolution of respiratory electron pathways using oxygen as final electron acceptor.

According to R. E. Dickerson and his associates and others, these developments may have taken place by the following sequence of events. Natural selection of electron transport pathways in photosynthetic cells led to both cyclic and noncyclic systems. Among these systems were some operating in the sequences shown in Figures 26-6b and c, roughly equivalent to some of the pathways of photosynthesis in the modern green

Figure 26-7 Noncyclic pathway of electron flow from water through two photosystems to an electron acceptor. The photosystems contain chlorophyll in ground (Chl) and excited (Chl*) states. The ability to use water as an electron donor freed the photosynthetic organisms evolving this pathway from the use of less abundant substances such as H_2S as electron donors. In addition, the oxygen evolved as a by-product of the two-photosystem pathway altered the atmosphere from a reducing to oxidizing character. The system may also operate cyclically, by the flow of electrons from photosystem I to the electron transport system connecting the two photosystems (dotted arrow).

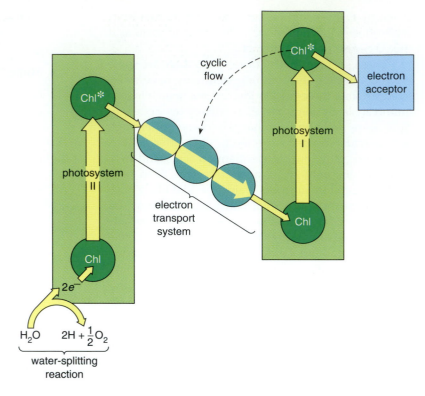

and purple sulfur bacteria (see p. 271). In these systems, porphyrin-based light-absorbing pigments had evolved fully into bacteriochlorophylls, contained in photosystems similar to those of modern photosynthetic prokaryotes. However, the primitive systems were limited to donors, such as H_2S, releasing electrons at relatively high energy levels. Such donors were not plentiful enough in the environment for these photosynthetic cells to become dominant forms.

The first evolutionary event of fundamental importance to the later evolution of eukaryotes resulted from the appearance of a modified form of one of the photosystems, capable of accepting low-energy electrons removed from water. This established the two-step pathway for light absorption and electron excitation in photosynthesis (Fig. 26-7) and allowed photosynthetic cells possessing this pathway to use the most abundant molecule of the environment as the electron donor for the photosynthetic pathway. A critical part of this development was the evolution of a mechanism catalyzing the oxidation of water:

$$H_2O \longrightarrow 2H^+ + 2e^- + 1/2O_2 \qquad (26\text{-}3)$$

As a consequence of this evolutionary development, which appeared in ancestral prokaryotes giving rise to cyanobacteria, oxygen was released to the atmosphere in ever increasing quantities.

The gradual increase in atmospheric oxygen set the stage for the development of respiratory electron transport systems. At some point a group of cells de-

veloped an extra step in their electron transport systems involving addition of cytochromes able to deliver electrons at very low energy levels to oxygen. This provided the advantage of tapping the greatest possible amount of energy from the electrons before releasing them from electron transport. Just as the ability to use water as an electron donor made groups with this adaptation dominant among photosynthetic organisms, the much greater efficiency of cells able to use oxygen as final electron acceptor made these forms dominant among those living by oxidations.

Some indication of the time scale of these events can be taken from the geological and fossil record. Using the oldest prokaryotic cells yet discovered as the beginning point some 3.5 billion years ago, evolution of oxygen-evolving photosynthetic cells from these early prokaryotes may have required another 500 million years or more. This conclusion is based on discoveries of limestone-containing deposits, similar to stromatolytes formed today by certain types of cyanobacteria, in rocks laid down 1.6 to 2.7 billion years ago (Fig. 26-8). The length of time required for the atmosphere to change from a reducing to oxidizing character has been estimated from the relative degree of oxidation of iron-containing compounds in rock layers. Rocks in relatively reduced form occur in layers deposited as recently as 1.8 billion years ago. These give way to relatively oxidized "red beds" that are about 1.5 billion years old. Therefore the transition to a predominantly oxidizing atmosphere, begun by

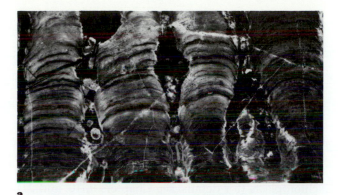

a

b

Figure 26-8 Fossil and recently formed stromatolytes. **(a)** Fossil stromatolytes from western Australia that are more than 3 million years old. (Courtesy of S. M. Awramik.) **(b)** Stromatolytes along a shoreline in western Australia. These stromatolytes were formed relatively recently, between 1000 and 2000 years ago. (Courtesy of J. W. Schopf.)

cyanobacteria nearly 3 billion years ago, may have occupied a period of 1 to 1.5 billion years (Fig. 26-9). Oxygen reached concentrations of about 15% in the atmosphere between 2.2 to 1.9 billion years ago.

Appearance of Eukaryotic Cells

Eukaryotic cells evolved from one or more prokaryotic groups in a series of developments that may have begun as long ago as 3 billion years. The process culminated in the many adaptations that characterize present-day eukaryotic cells: separation of nucleus and cytoplasm by a nuclear envelope; structural and regulatory chromosomal proteins allowing long-term differentiation and specialization; cytoplasm containing membrane-bound compartments specializing in metabolic and synthetic functions—mitochondria, chloroplasts, ER, and the Golgi complex, among others; movement by systems based on microtubules and microfilaments; endocytosis; and reproduction by mitotic cell division.

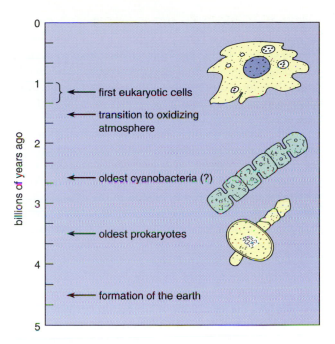

Figure 26-9 A time scale for the evolution of cellular life.

An initial development believed to underlie the appearance of eukaryotic characteristics is the appearance of motile proteins, the forerunners of contemporary microfilament- and microtubule-based motile systems. The motile proteins enabled the plasma membrane to take in materials by endocytosis, an adaptation thought to have led to the appearance of mitochondria, chloroplasts, and other eukaryotic cell structures including the ER, Golgi complex, and nuclear envelope.

The origin of mitochondria and chloroplasts through membrane activity due to motile proteins has been proposed as the *endosymbiont hypothesis*, developed in greatest detail by L. Margulis. This hypothesis proposes that mitochondria and chloroplasts originated from prokaryotic cells that were ingested by endocytosis in cell lines destined to evolve into eukaryotes (see Fig. 20-14). Instead of breaking down, the ingested cells persisted as symbionts in the cytoplasm of the feeding cells. Mutations increasing the interdependence between the symbiotic prokaryotes and their host cells eventually converted the symbionts into mitochondria and chloroplasts.

Margulis proposes that in the development of mitochondria, prokaryotes evolved to a point at which complete photosynthesis was common and oxygen was present in large quantities in the atmosphere. Among the prokaryotes were some nonphotosynthetic types that developed the capacity to engulf other cells through invaginations of the plasma membrane; these cells lived by oxidizing organic molecules ingested by this means. Some of these nonphotosynthetic cells possessed a complete respiratory system, including an electron transport system that used oxygen as final electron acceptor. These cells were therefore fully aerobic. Others were capable only of glycolytic fermentations and thus were limited to a less efficient anaerobic existence.

Interactions among these cells set the stage for evolution of mitochondria, which began when groups of nonphotosynthetic, anaerobic cells ingested nonphotosynthetic aerobic cells in large numbers. Some of the ingested aerobes persisted in the cytoplasm of the capturing cells and continued to respire aerobically in their new location. As a result, the cytoplasm of the host anaerobe, formerly limited to the use of organic molecules as final electron acceptors, became the place of residence of an aerobe capable of carrying out the much more efficient transfer of electrons to oxygen. As the symbiotic relationship became established, many functions duplicated in the aerobe were taken over by the host cell. As part of this transfer of function, most of the genes of the aerobe moved to the cell nucleus and became integrated into the host cell DNA. At the same time, the host anaerobe became dependent for survival on the respiratory capacity of the symbiotic aerobe. This gradual process of mutual adaptation culminated in transformation of the cytoplasmic aerobes into mitochondria during a period some 2.4 to 2.8 billion years ago.

This new cell type, a former anaerobe containing a symbiotic aerobe in its cytoplasm, gave rise to the eukaryotic cell line. Other features of eukaryotic cells, including the ER and Golgi complex, separation of the nucleus from the cytoplasm by the nuclear envelope, and regulatory and structural chromosomal proteins, gradually appeared in the developing eukaryotes.

Development of the nuclear envelope and ER may also have depended on the ability of the plasma membrane to invaginate as part of endocytosis (Fig. 26-10). The invaginations carried enzymatic mechanisms associated with the plasma membrane—including those controlling the attachment of ribosomes and secretion of proteins—to the cell interior. The invaginations extended inward and surrounded the nucleus and left vesicles giving rise to the ER in the cytoplasm. The Golgi complex may also have developed by the same pathway or as a secondary development from vesicles budding off from the ER.

Some time after the ancestral eukaryotic cells developed to a level including mitochondria, nuclear envelope, ER, and Golgi complex, a second major symbiotic relationship led to development of chloroplasts. Members of a nonphotosynthetic eukaryotic subgroup ingested photosynthetic prokaryotes resembling present-day cyanobacteria, capable of splitting water and evolving oxygen. The ingested photosyn-

thetic prokaryotes gradually transformed into chloroplasts by the same evolutionary processes establishing mitochondria as permanent symbionts. These early eukaryotic groups containing developing chloroplasts founded cell lines giving rise to modern algae and plants.

Several major lines of evidence support the endosymbiotic hypothesis for the origins of mitochondria and chloroplasts. One is the similarity in structure and biochemistry between aerobic bacteria and mitochondria and between cyanobacteria and chloroplasts (for details, see Chapters 9, 10, and 21). Mitochondria and chloroplasts contain circular DNA molecules, with no associated histone or nonhistone proteins, that closely resemble prokaryotic DNA in structure. The organelle DNA codes for rRNA and tRNA molecules that in sequence resemble their prokaryotic counterparts more closely than the equivalent RNAs of the eukaryotic cytoplasm. The two organelles also possess ribosomes that structurally resemble prokaryotic ribosomes and carry out protein synthesis by mechanisms that most closely resemble those of prokaryotes. For example, mitochondria, chloroplasts, and contemporary prokaryotes use formylmethionine (see p. 486) as the first amino acid in a polypeptide chain instead of the unmodified methionine used in the eukaryotic cytoplasm outside the organelles. Similarities between present-day prokaryotes and the two organelles are also noted in their reactions to inhibitors of protein synthesis. For example, chloramphenicol inhibits protein synthesis in bacteria, chloroplasts, and mitochondria but has no effects on ribosomes in eukaryotic cytoplasm. Conversely, cycloheximide inhibits protein synthesis in the cytoplasm of eukaryotic cells but has no effect on bacterial or organelle protein synthesis. Another striking similarity is in the F_oF_1 ATPase carrying out ATP synthesis in association with electron transport. This complex enzyme occurs nowhere in living systems except in the membranes of prokaryotes and in mitochondria and chloroplasts.

One of the most interesting characteristics of chloroplasts and mitochondria with reference to the endosymbiont hypothesis is the signal system directing cytoplasmically synthesized proteins into the organelle interiors (see p. 588 for details). Many proteins entering mitochondria and chloroplasts have two signals at their N-terminal ends. The first signal directs newly synthesized proteins into or through the outer boundary membrane of the organelles, which, according to the endosymbiont hypothesis, was derived from the plasma membrane of the host cells taking in the organelle precursors by endocytosis. This signal is unique to the organelle-directed proteins and does not resemble prokaryotic distribution signals. The second signal, when present, directs proteins into or through

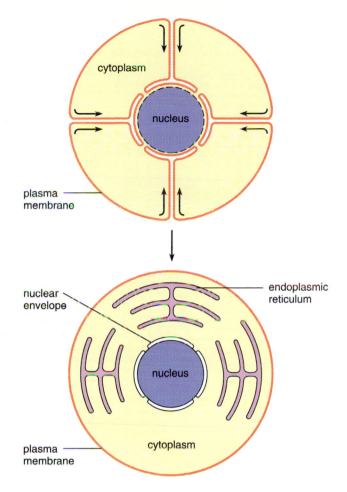

Figure 26-10 A hypothetical route for formation of the nuclear envelope and ER through invaginations of the plasma membrane.

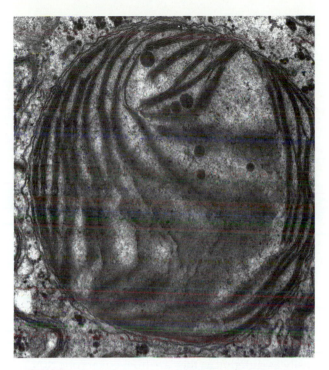

Figure 26-11 A chloroplast in a cell lining the gut of the snail *Elysia viridis*. (Courtesy of R. K. Trench, R. W. Greene, B. G. Bystrom, and the Rockefeller University Press.)

the inner boundary membranes of the organelles, supposedly derived from the plasma membrane of the prokaryotic ancestors of mitochondria and chloroplasts. This signal, as expected from the endosymbiont hypothesis, is typically prokaryotic.

Of the two organelles, chloroplasts have the most pronounced biochemical and structural similarities to their prokaryotic counterparts, the cyanobacteria. Mitochondria and present-day bacteria appear to be less closely related. This difference probably reflects the more recent origins of chloroplasts from their prokaryotic ancestors.

Another line of evidence supporting the endosymbiont hypothesis is that present-day organisms of 11 phyla, distributed among more than 150 genera, contain eukaryotic algae or cyanobacteria as photosynthetic symbionts. Cyanobacteria, which occur as symbionts in protozoa and diatoms, can in some cases hardly be distinguished from true chloroplasts. One of the most interesting examples, described by R. K. Trench and his associates, involves a group of marine snails that utilizes chloroplasts derived from algae. Initially the snails develop as larvae containing no chloroplasts. As the larvae begin feeding on algae, chloroplasts from the algal cells are taken up by endocytosis in cells lining the gut. This process continues as the larvae develop into adult snails. The chloroplasts persist in the gut cells (Fig. 26-11) and continue to carry out photosynthesis in their new location. Trench used radioactive tracers to show that organic molecules synthesized in the ingested chloroplasts diffuse into the cytoplasm of the host cells and are used as fuel substances by the snails. Individual chloroplasts may remain active in the gut cells for months. Thus the route envisioned by Margulis and others for the origins of mitochondria and chloroplasts—ingestion by endocytosis and retention in the cytoplasm—can be demonstrated to take place in the cells of contemporary organisms.

There is also ample demonstration of the ability of DNA sequences to move from organelles into the nucleus. In one experiment by P. E. Thorsness and T. D. Fox, for example, genes carried on DNA initially inserted into mitochondria appeared later in the nucleus (see p. 623 and the Experimental Process essay by Thorsness on p. 624 for further details). In many species of higher plants, fragments of the entire chloroplast DNA circle can be found inserted at scattered locations in the nuclear DNA.

Reconstruction of Ancient Lineages Leading to Contemporary Prokaryotes and Eukaryotes

There is much current interest in tracing the lineages leading from the first true cells to contemporary prokaryotes and eukaryotes. C. R. Woese and G. E. Fox and their colleagues and others have approached this problem by comparing rRNA sequences and the characteristics of biochemical systems operating among various living cells. Their work indicates that contemporary cells fall into three groups that share certain characteristics but differ enough to set them clearly apart. One group consists of present-day eukaryotes. The second consists of the bacteria and cyanobacteria (called the *eubacteria*, from *eu* = true or typical). The third group, the *archaebacteria*, are bacterialike organisms that include the *extreme halophiles*, which live in highly saline environments, the *extreme thermophiles*, which inhabit hot springs, and the *methanogens*, which release methane as a final metabolic product. Although the archaebacteria are bacterialike, their rRNAs are distinctly different from those of eubacteria, easily as different as they are from eukaryotic rRNAs.

Archaebacteria also have other features that are either unique, typically prokaryotic, or typically eukaryotic. For example, introns, a primarily eukaryotic characteristic, occur in tRNA genes in archaebacteria. The introns in these organisms occur in exactly the same position in the tRNA coding sequences as eukaryotic tRNA introns—at a site corresponding to one nucleotide to the 3' side of the anticodon—and include a sequence that can pair with the anticodon (see Fig. 15-16a). G. Olsen and S. Jackson and their coworkers found that archaebacteria also have transcription factors resembling those of eukaryotes, including TBP, a polypeptide forming part of several eukaryotic transcription factors, and TFIIB (see p. 412). The TBP factor recognizes a typically eukaryotic TATA box in the 5'-flanking regions of archaebacterial mRNA genes. Archaebacteria also initiate protein synthesis with unformylated methionine as do eukaryotes. On the other hand, archaebacteria lack nuclear envelopes and contain circular DNA molecules organized similarly to those of prokaryotes. Although the ribosomes of archaebacteria resemble those of eubacteria in dimensions, S. Altamura and his coworkers showed that the ribosomal subunits of archaebacteria form functional ribosomes when combined with eukaryotic subunits but not with eubacterial subunits.

Comparisons among archaebacteria, eubacteria, and eukaryotes suggest that a series of very early evolutionary splits may have given rise to three separate lineages leading to these contemporary forms (Fig. 26-12). At some very early point the evolutionary line split into two main branches leading to eubacteria and eukaryotes. At some time after this early split a

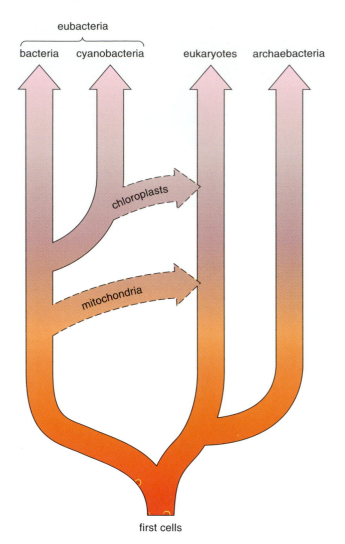

Figure 26-12 A pathway by which eubacteria, cyanobacteria, archaebacteria, and eukaryotes may have evolved from the first true cells.

side branch off the line leading to eukaryotes gave rise to the archaebacteria. Introns, which may have been present in the common ancestor of all three lineages, persisted in archaebacteria and the line leading to eukaryotes but were almost completely lost in the eubacteria in response to selection pressures for maintenance of the smallest possible genome. In the line leading to eukaryotes the advantages of the intron/exon structure of genes to protein evolution by exon shuffling (see p. 411) led to the proliferation of introns in the genome. Thus, according to this proposal, advanced by Gilbert in particular, introns are very ancient sequence elements that were present in the genomes of the first cells, largely lost in the evolution of eubacteria, and retained and elaborated in eukaryotes. It should be mentioned, however, that there is no solid evidence indicating whether eukaryotic introns are evolutionary survivors from an ancient group or an innovation that appeared separately in eukaryotes and archae-bacteria. Similarly, it is unknown whether the very few introns found in eubacteria are remnants from an earlier time or sequence elements introduced in some way from eukaryotes.

The total time required for evolution of the first fully eukaryotic cells may have extended over a billion years or more. Fossils in rocks as old as 1.4 billion years have been claimed by J. W. Schopf and D. Z. Oehler to be eukaryotic cells on the basis of apparent nuclei and division figures resembling mitosis. These conclusions are controversial, however; others maintain that the apparent nuclei result from precipitation of prokaryotic cell structures during fossilization. According to the dissenting investigators, eukaryotic cells are not likely to have been on the earth much longer than about 1 billion years. Assuming that the first fully eukaryotic cells appeared between the limits proposed by these investigators—1.4 and 1 billion years ago—and that the first cells appeared 3.5 billion years ago, evolution of fully eukaryotic cells from the first cells required 2 billion years or more. If so, this interval is twice as long as the interval from the time of the earth's origins to the appearance of the first prokaryotes (see Fig. 26-9). The amount of time required undoubtedly reflects the complexity of the adaptations in eukaryotic cells.

The events outlined in this chapter, leading from the earth's origins to the appearance of eukaryotic cells, are admittedly hypothetical and so tenuous that they may seem impossible. But given the total time span of these events, 3.5 billion years, "the impossible becomes possible, the possible probable, and the probable virtually certain. One has only to wait; time itself performs the miracles."[1] There are even those who say that, given the chemical conditions and environment of the primitive earth, the evolution of prokaryotic and eukaryotic cellular life, no matter how complex it has become, was inevitable—it was a predictable outcome of the starting conditions and the spontaneous chemical reactions taking place since the earth's origin. Therefore, according to this idea, the origin of life was chemically predetermined by the initial conditions. It may well be that a similar sequence of chemical and organic evolutionary events has occurred or is under way on other planets in the universe.

[1]From G. Wald, The origin of life, *Sci. Amer.* 191:45(1954).

For Further Information

ATP synthesis, *Chs. 8 and 9*
Chloroplasts, structure and biochemistry, *Ch. 9*
   DNA, transcription, and protein synthesis in, *Ch. 21*
Electron transport in mitochondria and bacteria, *Ch. 8*
   in chloroplasts and cyanobacteria, Ch. 9
Endocytosis, *Ch. 20*
Eukaryotic cell structure, *Ch. 1*
Genetic code, *Ch. 16*
Glycolytic fermentation, *Ch. 8*
Mitochondria, structure and biochemistry, *Ch. 8*
   DNA, transcription, and protein synthesis in, *Ch. 21*
Oxidative phosphorylation, *Ch. 8*
Photosynthesis, *Ch. 9*
Prokaryotic cell structure, *Ch. 1*
Protein synthesis, *Ch. 16*
Respiration, *Ch. 8*
Substrate-level phosphorylation, *Ch. 8*

Suggestions for Further Reading

Benner, S. A., Ellington, A. D., and Tauer, A. 1989. Modern metabolism as a palimpsest of the RNA world. *Proc. Nat. Acad. Sci.* 86:7054–7058.

Bernal, J. D. 1967. *The Origin of Life*. New York: World.

Cairns-Smith, A. G. 1985. *Seven Clues to the Origin of Life*. New York: Cambridge University Press.

Calvin, M. 1969. *Chemical Evolution*. New York: Oxford University Press.

Cech, T. R. 1986. RNA as an enzyme. *Sci. Amer.* 255:64–75 (November).

Chyba, C. F., and Sagan, C. 1992. Endogenous production, exogenous delivery, and impact-shock synthesis of organic molecules: An inventory for the origins of life. *Nature* 355:125–132.

Crick, F. H. C. 1968. The origin of the genetic code. *J. Molec. Biol.* 38:367–379.

Day, W. 1984. *Genesis on Planet Earth: The Search for Life's Beginnings*. New Haven, Conn.: Yale University Press.

DeDuve, C. 1991. *Blueprint for a Cell: The Nature and Origin of Life*. Burlington, N.C.: Patterson.

Dickerson, R. E. 1978. Chemical evolution and the origin of life. *Sci. Amer.* 239:70–86 (September).

Dickerson, R. E. 1980. Cytochrome c and the evolution of energy metabolism. *Sci. Amer.* 242:136–153 (March).

Doolittle, R. F., and Bork, P. 1993. Evolutionarily mobile modules in proteins. *Sci. Amer.* 269:50–56 (October).

Dyson, F. 1985. *Origins of Life*. New York: Cambridge University Press.

Eigen, M. 1992. *Steps Toward Life: A Perspective on Evolution*. New York: Oxford Press, 1992.

Ferris, J. P. 1987. Prebiotic synthesis: Problems and challenges. *Cold Spring Harbor Symp. Quant. Biol.* 52:29–35.

Gray, M. W. 1989. Origin and evolution of mitochondrial DNA. *Ann. Rev. Cell Biol.* 5:25–50.

Gray, M. W. 1992. The endosymbiotic hypothesis revisited. *Internat. Rev. Cytol.* 141:233–357.

Horgan, J. 1991. In the beginning. . . *Sci. Amer.* 264: 116–125 (February).

Joyce, G. F. 1989. RNA evolution and the origins of life. *Nature* 338:217–224.

Kasting, J. F. 1993. Earth's early atmosphere. *Science* 259: 920–926.

Kenyon, D. H., and Steinman, G. 1969. *Biochemical Predestination.* New York: McGraw-Hill.

Keosian, J. 1964. *The Origin of Life.* New York: Reinhold.

Knoll, A. H. 1991. End of the proterozoic era. *Sci. Amer.* 265:64–73 (October).

Knoll, A. H. 1992. The early evolution of eukaryotes: A geological perspective. *Science* 256:622–627.

Lake, J. A. 1991. Tracing origins with molecular sequences: Metazoan and eukaryotic beginnings. *Trends Biochem. Sci.* 16:46–50.

Laszlo, P. 1987. Chemical reactions on clays. *Science* 235: 1473–1477.

Loomis, W. J. 1988. *Four Billion Years.* Sunderland, Mass.: Sinauer.

Mann, A. P. C., and Williams, D. A. 1980. A list of interstellar molecules. *Nature* 283:271–725.

Margulis, L. 1970. *Origins of Eukaryotic Cells.* New Haven, Conn.: Yale University Press.

Margulis, L. 1981. *Symbiosis in Cell Evolution.* New York: Freeman.

Miller, S. L. 1987. Which organic compounds could have occurred on the prebiotic earth? *Cold Spring Harbor Symp. Quant. Biol.* 52:17–27.

Nei, M., and Koehn, R. K. 1983. *Evolution of Genes and Proteins.* Sunderland, Mass.: Sinauer.

Orgel, L. E. 1989. Evolution of the genetic apparatus. *Cold Spring Harbor Symp. Quant. Biol.* 52:9–16.

Pace, N. R. 1991. Origins of life—facing up to the physical setting. *Cell* 65:531–533.

Rebek, J. 1994. Synthetic self-replicating molecules. *Sci. Amer.* 271:48–55 (July).

Sharp, P. A., and Eisenberg, D. 1987. The evolution of catalytic function. *Science* 238:729–730, 807.

Waldrup, M. M. 1989. Did life really start out in an RNA world? *Science* 246:1248–1249.

Whatley, J. M. 1993. The endosymbiotic origin of chloroplasts. *Internat. Rev. Cytol.* 144:259–299.

Woese, C. R. 1987. Bacterial evolution. *Microbiol. Rev.* 51: 221–271.

Review Questions

1. What evidence indicates that life appeared on the earth between 4.5 and 3.5 billion years ago?

2. Why are the views that life had a divine origin or arrived on earth as contamination on a spaceship considered unscientific? Why is it necessary for scientific purposes to assume that life originated through inanimate chemical processes?

3. What is the distinction between the common and scientific attitudes about spontaneous generation? Why may spontaneous generation have been possible on the primitive earth but not now?

4. Outline the basic tenets of the Oparin–Haldane hypothesis. What gases were proposed to be present in the earth's primitive atmosphere? What is the significance of the reducing quality proposed for the primitive atmosphere?

5. What evidence supports the idea that the primitive atmosphere was mildly or strongly reducing?

6. List the stages in the evolution of cellular life. How are the earth and its atmosphere believed to have originated? What interactions would probably have removed any gaseous oxygen from the atmosphere? What is the significance of this removal for the evolution of life?

7. Outline the structure and operation of the Miller apparatus. What biological compounds were produced in the apparatus?

8. What is the significance of HCN as a component of the early atmosphere? Of formaldehyde?

9. What major classes of organic molecules have been produced in experiments simulating prebiotic conditions? What energy sources may have acted on the inorganic chemicals of the primitive environment to produce organic molecules?

10. What observations of natural systems support the results of experiments indicating that organic molecules could have been synthesized on the primitive earth?

11. What is a condensation? What conditions may have promoted condensations on the primitive earth?

12. What are clays? How may clays have figured in condensation reactions?

13. What experiments indicate that condensations producing polypeptides and nucleic acids may actually have taken place?

14. What minimum characteristics would you consider to be required for the presence of life?

15. What are coacervates? Proteinoid microspheres? What lifelike reactions have been demonstrated in coacervates and proteinoid microspheres?

16. How may lipid bilayers have been important in molecular aggregation? What major problem is connected with the possible involvement of lipid bilayers in aggregation of molecules in the primitive earth?

17. What characteristics of clays may have made these substances ideal as sites of molecular aggregation and interaction?

18. What is the difference between substrate level phosphorylation and oxidative phosphorylation in the synthesis of ATP? Does oxidative phosphorylation require oxygen?

19. How may substrate-level phosphorylation and oxidative phosphorylation have arisen in the primitive environment? How may porphyrins have been important in these developments?

20. What are the possible relationships between respiration and photosynthesis in the development of electron transport pathways?

21. What advantages does RNA offer over DNA as a primitive informational molecule? What experiments and observations support the idea that RNA may have been the first informational molecule? What problems limit this possibility?

22. What characteristics may have favored the selection of three-letter codons as opposed to one-, two-, or four-letter codons? What is meant by the "historical accident" that may have fixed the codons in their present assignments?

23. How did the appearance of a coding system alter evolution from chemical to organic?

24. What adaptations were probably important in the conversion of precells to cells?

25. Trace the developments that may have led to the use of water as an electron donor for photosynthesis.

26. What is the significance of photosynthesis to the evolution of cellular life?

27. Outline the primary differences between prokaryotic and eukaryotic cells. What environmental conditions were necessary for the appearance of eukaryotic cells?

28. Outline the possible steps in the origin of mitochondria and chloroplasts according to the endosymbiont hypothesis. What evidence supports the idea that mitochondria and chloroplasts originated from ancient prokaryotes?

29. How may the ER, Golgi complex, and nuclear envelope have originated?

30. What characteristics of archaebacteria are similar to those of eubacteria? Of eukaryotes? What relationship may ancestors of archaebacteria have had to present-day prokaryotes and eukaryotes?

Methods Described in This Appendix

Methods Described Elsewhere in the Text

LIGHT AND ELECTRON MICROSCOPY

Light and electron microscopes have been highly important tools in many of the discoveries of cell and molecular biology. Until the early 1960s, microscopy was among the primary methods used to study cells. Although the importance of microscopy to research has been overtaken by molecular methodology, many techniques still employ light or electron microscopy to analyze cell structures and functions.

The Light Microscope and Its Applications

Light microscopes are applied in ways that use different light sources and patterns of image formation. The most common application, and the one most familiar to students, is the *brightfield light microscope*, in which a beam of light is transmitted directly through the specimen and the background appears bright. Other applications commonly used to meet particular viewing or specimen requirements are the *darkfield*, *phase contrast*, *differential interference contrast*, and *fluorescence microscopes*.

The Brightfield Light Microscope In a brightfield light microscope (Fig. A-1), light rays from a source such as an incandescent bulb are focused on the specimen by a *condenser* lens. After passing through the specimen, the rays are focused into a magnified image by two lenses placed at either end of a tube. Nearest the specimen is the *objective lens;* at the opposite end of the tube is the *ocular lens*. To correct for faults inherent to glass lenses, the objective and ocular lenses are constructed from a series of lenses placed close together, usually as many as 8 to 10 lenses for the objective and 2 to 3 for the ocular. The individual lenses in the objective or ocular are mounted so close together that they act in effect as a single, highly corrected lens.

The image is observed by looking directly into the ocular lens. Coarse and fine controls move the specimen stage or lens tube to place the specimen in the

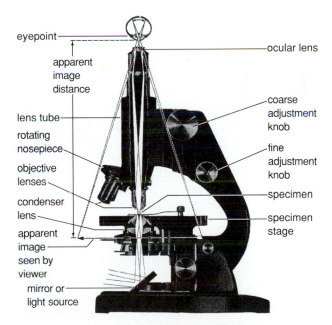

Figure A-1 The brightfield light microscope (see text). Objective lenses of different magnifying power can be selected by rotating the nosepiece. (Redrawn from an original courtesy of the American Optical Company.)

correct position for focusing by the objective lens. The position of the condenser can be adjusted so that light focused by this lens converges on the specimen and spreads into a cone of light that exactly fills the objective lens.

The extent to which a microscope can distinguish fine details in the specimen as separate, distinct image points is termed its *resolution*. In light microscopy this property depends on the interaction of several factors:

$$\text{resolution} = d = \frac{0.61\lambda}{n \sin \alpha} \qquad \text{(A-1)}$$

In this equation, λ is the wavelength of the light illuminating the specimen, n is the refractive index[1] of the medium surrounding the specimen and filling the space between the condenser and objective lens, and α is the half-angle of the cone of light entering the objective lens (Fig. A-2). The quantity 0.61 is a constant describing the degree to which image points can overlap and still be recognized as separate points by an observer. Since resolution is a measure of the ability of a microscope to image fine details, the quantity d becomes smaller as resolution improves. Therefore, for best resolution, 0.61λ should be numerically as small as possible, and $n \sin \alpha$ as large as possible.

In a high-quality light microscope operated under optimal conditions the quantities in Equation A-1 take on limits at fixed values. The value for n is pushed to its maximum by placing a drop of immersion oil (refractive index = about 1.5) in the space between the objective lens and the specimen and, in some microscopes, between the condenser lens and specimen. The half-angle of the cone of light entering the objective lens in the best microscopes is about 70°, giving a maximum value for $\sin \alpha$ of about 0.94. The quantity $n \sin \alpha$, called the *numerical aperture* of the objective lens, therefore takes on a maximum value of about 1.4.

This leaves λ, the wavelength of the light used as an illumination source, as a variable in the equation. To keep d as small as possible, λ must take on a minimum value. The lower limit of this value is fixed by the shortest wavelength of visible light usable for illumination: blue light at a wavelength of about 450

---

[1] The *refractive index* is the ratio between the velocity at which light is transmitted through a vacuum and a transmitting medium such as glass:

$$\text{refractive index} = n = \frac{\text{velocity of light in a vacuum}}{\text{velocity in a transmitting medium}}$$

The velocity of light is at its maximum in a vacuum; this velocity is given a value of 1. The velocity of light in air is almost equivalent to the velocity in a vacuum (refractive index of air = 1.0029). Glass transmits light significantly more slowly (refractive index = 1.5).

nm. Substituting this value and $n \sin \alpha = 1.4$ in Equation A-1 shows that the best resolution possible with microscopes using visible light approaches 0.2 μm, or about 200 nm. Using ultraviolet light as an illumination source ($\lambda = 250$ nm) pushes this value down to about 0.1 μm. At this level, cell organelles such as nuclei, chloroplasts, and mitochondria are clearly resolved, but smaller details such as microtubules and ribosomes remain invisible.

Objects resolved at 0.2 μm must be magnified about a thousand times to be distinguished by the human eye. This is accomplished in most light microscopes by the combination of a 100× oil-immersion objective lens with a 10× ocular. Magnification beyond this level does not improve the resolution of small objects viewed in the light microscope. For this reason, enlargement of the image beyond 1000× is often termed "empty" magnification in light microscopy.

The Darkfield Light Microscope Darkfield illumination improves the visibility and apparent resolution of very fine specimen points in the light microscope. In this application an opaque disc is placed in the center of the condenser so that light can pass only around its edges. As a result, a hollow cone of light illuminates the specimen. The cone is adjusted by the condenser control to an angle wide enough so that no light transmitted directly through the specimen can enter the objective lens (Fig. A-3). As a result, only light scattered to smaller angles by specimen points enters the objective lens, making these points appear bright against a dark background. The method improves apparent resolution because objects too small to be resolved directly can still scatter and divert the paths of light waves. The criterion for visibility is simply the amount of scattering by the point. Darkfield microscopy allows objects as small as individual microtubules, with a diameter of about 25 nm, to be visualized (as, for example, in Fig. 10-17).

The Phase Contrast Light Microscope Most of the structures in living cells are equally transparent and colorless and thus show little contrast. However, many structures differ in refractive index. The phase contrast microscope, in which differences in refractive index in the specimen are changed to differences in brightness in the image, greatly improves the visibility of these structures.

The phase contrast microscope takes advantage of the fact that light rays are bent from their path (refracted) as well as slowed as they pass through transparent objects of differing refractive index. On the average the light waves passing through structures of higher refractive index in cells are delayed by about one-fourth wavelength and are bent to a greater de-

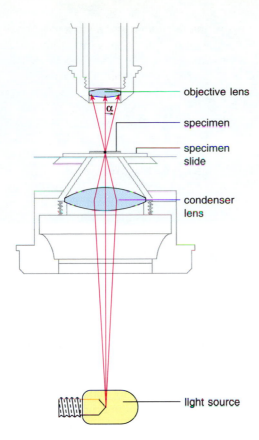

Figure A-2 Light path from the condenser to the objective lens. For maximum resolution the half-angle (α) of the cone of light entering the objective lens should be as large as possible. The maximum half-angle obtainable is about 70°.

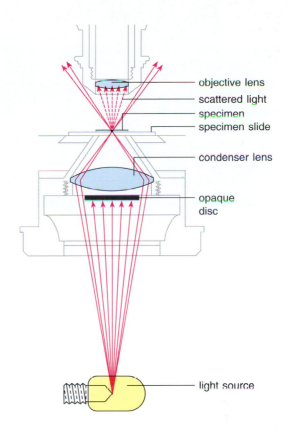

Figure A-3 Light path from the condenser to the objective lens in darkfield illumination. An opaque disc blocks the central region of the condenser so that rays transmitted through the specimen without scattering cannot directly enter the objective lens. As a result, only scattered rays contribute to image formation. Specimen points that scatter light into the objective lens therefore appear bright against a dark background.

gree than the paths followed by waves passing through less refractive structures. The rays refracted to greater and lesser degrees follow different paths through the objective lens but are focused back into the same image points. Since the rays refracted to greater and lesser degrees differ in phase by only one-fourth wavelength on the average, they do not interact strongly enough in an ordinary light microscope to alter the brightness of image points.

The phase contrast microscope makes such points visible by increasing the difference in wavelength between strongly and weakly refracted rays. This is done by introducing a *phase plate* into the paths followed by the more greatly refracted rays. The phase plate is coated with a layer of transparent material of sufficient refractive index and thickness to delay the more greatly refracted rays by an additional one-fourth wavelength. The phase plate is uncoated in the central region passing the less refracted rays, so that these rays are transmitted without additional delay.

The objective lens focuses all rays into the same image points. At image points corresponding to regions of higher refractive index in the specimen the crests of the more greatly refracted rays, since they now differ in phase from the less refracted rays by about one-half wavelength, coincide in the image with the troughs of the unrefracted waves. The two sets of waves therefore cancel each other, producing an *interference* that reduces the brightness of the image at these points. The effect over the entire image is to create differences in brightness corresponding to differences in the refractive index of the specimen. The phase contrast image makes many structures in living cells clearly visible without staining (Fig. A-4).

Differential Interference Contrast (DIC or Nomarski) Microscopy Another technique used to increase the apparent contrast of transparent, unstained tissues in the light microscope is *differential interference contrast* (*DIC*) microscopy, often termed *Nomarski interference*

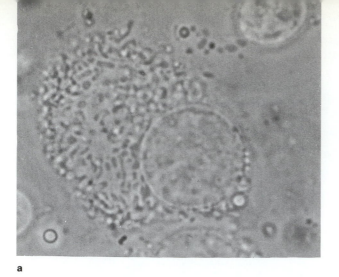
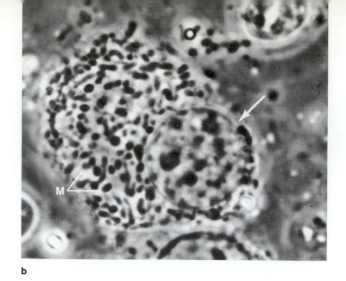

a

b

Figure A-4 The effects of phase contrast on the light microscope image. **(a)** Brightfield image of a living cell from frog liver. **(b)** The same living cell in the phase contrast microscope. The nucleus is marked by an arrow; numerous bodies, including mitochondria (M), are visible in the cytoplasm. × 2000. (Courtesy of B. R. Zirkin.)

microscopy after its inventor. DIC microscopy has several advantages over the phase contrast microscope, among them the ability to make cell structures appear three-dimensional, as if a strong shadowing light is directed over the specimen from one side (Fig. A-5).

In the DIC microscope (Fig. A-6) the illuminating beam leaving the light source is *polarized* by a filter that passes only light waves vibrating in a single plane. The polarized beam then passes through a *beam splitter* placed just below the condenser lens. The beam splitter is a prismatic crystal that separates the entering beam of light into two beams that vibrate in planes at 90° to each other. The waves in one beam follow paths that are separated by a very small lateral distance from the waves in the other beam. The beam splitter is constructed so that this separation, called *shear*, is equal to the resolution of the objective lens. After focusing by the condenser lens, the two beams pass through the specimen. In the specimen, waves in the two beams follow slightly different paths through closely adjacent regions. If the adjacent regions differ in refractive index or thickness, as they might along the edges of a cell structure, the waves are delayed or refracted differently.

After passing through the specimen, the beams are combined into a single beam by a *recombining prism* placed just below the objective lens. When recombined by the prism, the two beams interfere to a greater or lesser extent, producing differences in brightness in the image corresponding to differences in refractive index or thickness in the specimen. The brightness differences are particularly pronounced along the edges of structures of differing thickness or refractive index in the specimen, giving a strong, finely delineated highlight to structural borders such

as the boundaries of organelles or cell surfaces. In most DIC microscopes the position of the recombining prism can be shifted to add a uniform delay to one of the recombining beams to maximize interference and the resulting differences in brightness. If adjusted so that one of the beams is delayed by one-fourth wavelength, the interference patterns make one edge of an object light and the other dark, with a progressively gradated change in intensity between them. The effect in the image is as if a strong light shadows the object from one side (as in Fig. A-5). The direction of the shadow, which makes the specimen appear three-dimensional, can be moved around the object by rotating the polarizing filter below the condenser lens.

Fluorescence Microscopy The fluorescence microscope detects light emitted by fluorescent dye molecules linked to cell structures of interest. One of the most used applications of the method is *immunofluorescence*, in which a fluorescent dye is attached to an antibody molecule that can recognize and bind specifically to a cell structure.

Fluorescence results from absorption and release of light energy by electrons in the dye molecules. Light is absorbed at a certain wavelength by an electron in a dye molecule, which rises from a lower to a higher energy level when it absorbs the energy. At the elevated level an electron is highly unstable and tends to return quickly to its original energy level. As the electron returns to its original level, it releases the energy of the absorbed wavelength as light and heat. The energy released as light is termed *fluorescence*. Because some of the energy is also lost as heat, the light released as fluorescence contains less energy than the

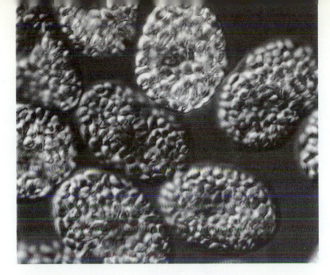

Figure A-5 Light micrograph taken with the differential interference contrast (DIC or Nomarski) microscope. The micrograph shows blood cells of the horseshoe crab *Limulus*, which function in defense against invading pathogens and in blood clotting. The granules in the cytoplasm contain the antimicrobial and clotting agents. × 1,600. Courtesy of P. B. Armstrong, from *Exptl. Cell Res.* 107: 127 (1977).

light originally absorbed and is therefore at a longer wavelength (see p. 250).

In fluorescence microscopy a specimen is stained with a fluorescent dye and illuminated with light filtered to remove all wavelengths except the one absorbed by the dye. In response to light absorption the dye fluoresces, releasing light at a different but characteristic wavelength. The fluorescent light leaving the specimen passes through another filter, the *barrier filter*, which eliminates all wavelengths except the one fluoresced by the dye. Since the illuminating wavelength is different from the wavelength of the fluorescence, no light passes through the barrier filter except for the fluoresced light. As a result, regions marked by the fluorescent dye glow strongly against a dark background (as, for example, in Fig. 1-16).

Electron Microscopy

Electron microscopes use a beam of electrons rather than light as an illumination source. The electron beam is focused by magnetic fields, generated by massive coils of wire through which precisely controlled electric currents are passed. An iron *pole piece* (see Fig. A-7) inserted into the center of the wire coil shapes the magnetic field into the three-dimensional form required for focusing electrons. Because electron lenses can be varied in focus by altering the current applied to their wire coils, electron microscopes are focused by changing lens current rather than by moving lenses as in light microscopes. For the same reason, magnification is adjusted by altering lens current rather than by substituting lenses.

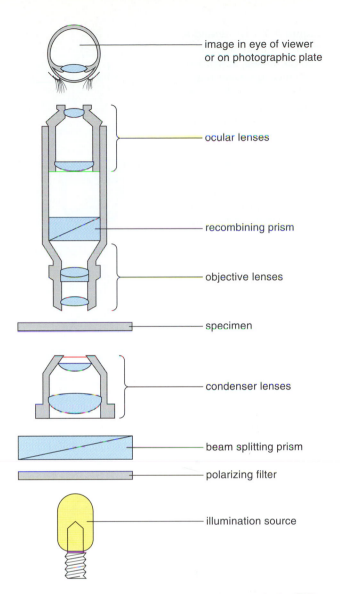

image in eye of viewer or on photographic plate

ocular lenses

recombining prism

objective lenses

specimen

condenser lenses

beam splitting prism

polarizing filter

illumination source

Figure A-6 Arrangement of optical elements in the DIC or Nomarski microscope (see text).

The pathways followed by electrons inside an electron microscope must be kept under a high vacuum because electrons are easily scattered and absorbed by gas molecules. This places special requirements on the specimen, which must be dry and nonvolatile.

Electron microscopes are employed in several different applications. Of these, two—the *transmission electron microscope (TEM)* and *scanning electron microscope (SEM)*—are most common.

The Transmission Electron Microscope (TEM) The TEM is so called because electrons forming the image pass through the specimen. In construction a TEM resembles an inverted light microscope (Fig. A-7). At the top of the central column is the illumination source, the *electron gun*, consisting of a *filament* and

Figure A-7 The arrangement of the illumination source, lenses, specimen, and viewing screen in a light microscope (left) and a transmission electron microscope (right). Each electron lens consists of a massive coil of wire wound around an iron pole piece. An electric current passed through the coil generates a magnetic field; the pole piece shapes the field into a form that focuses electrons.

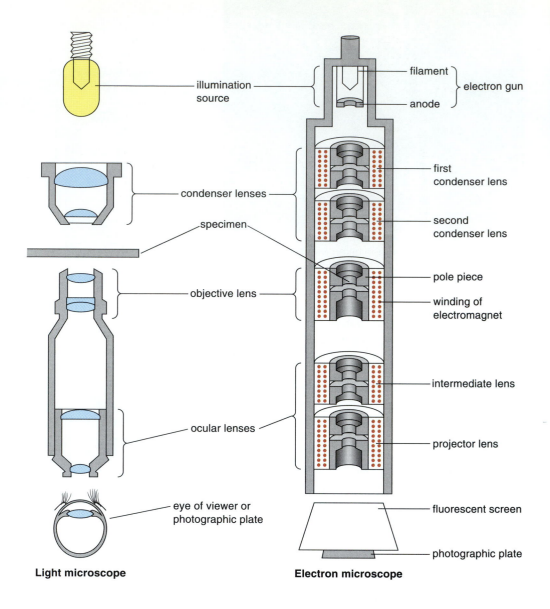

illumination source

condenser lenses

specimen

objective lens

ocular lenses

eye of viewer or photographic plate

Light microscope

filament } electron gun
anode

first condenser lens

second condenser lens

pole piece

winding of electromagnet

intermediate lens

projector lens

fluorescent screen

photographic plate

Electron microscope

anode. The filament, a thin tungsten wire, is heated to a high temperature by an electric current, which drives electrons from its surface. The filament and its holder, which are well insulated from the rest of the column, are maintained at a high negative voltage, −50,000 to −100,000 volts (v) in most microscopes. The anode is grounded and is thus positive with respect to the filament. Consequently, electrons leaving the filament are strongly attracted to the anode. As they travel from the filament to the anode, the electrons are accelerated to a velocity that depends on the voltage difference between the two locations. The wavelength of the electrons in the beam is inversely proportional to the velocity—the higher the velocity, the shorter the wavelength. Voltage differences of −50,000 to −100,000 v between the filament and anode produce wavelengths from about 0.005 to 0.003 nm.

The beam leaving the gun is focused by a series of two condenser lenses into a small, intense spot of electrons on the specimen. The specimen can be moved, allowing different regions to be illuminated by the spot. Another series of lenses, the objective, intermediate, and projector lenses, focuses the electrons passing through the specimen onto a fluorescent viewing screen at the bottom of the microscope. Each lens in the train magnifies the image, giving a total magnification varying from a few thousand to 300,000 times or more depending on the current applied to the lenses.

The viewing screen, similar to the screen of a television tube, is coated with crystals that respond to electron bombardment by emitting visible light. This process converts the electron image into a visual image. Permanent records of the image can be made by exposing a photographic plate to the electron beam at the level of the screen. Because photographic emulsions respond to electrons and light in essentially the same way, ordinary films and plates can be used for

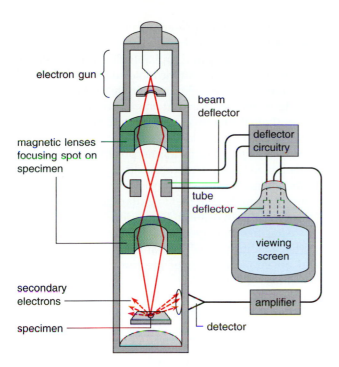

Figure A-8 The scanning electron microscope (SEM; see text).

this purpose. Airlocks are provided so that specimens and photographic plates can be introduced or removed without disturbing the microscope vacuum.

The short wavelengths of the electron beam greatly improve the resolution of the TEM as compared to light microscopes, allowing objects separated by as little as 0.5 nm to be resolved. Resolution in this range is about 200 times better than the resolving power of the best light microscopes. At these levels the TEM can easily resolve structures such as ribosomes, microtubules, microfilaments, and larger molecules such as proteins.

The Scanning Electron Microscope (SEM) The SEM, widely used to examine the surfaces of cells or isolated cellular structures, differs extensively in its construction and operation from the TEM (Fig. A-8). Only the illumination source and condenser lenses are similar. An electron gun produces in the SEM an electron beam, which is focused into an intense spot on the specimen surface by a magnetic lens system analogous to the condenser lens of a TEM. Rather than being stationary, however, the focused spot moves rapidly, or *scans*, back and forth over the specimen. Scanning is accomplished by beam deflectors, charged plates placed between the condenser lenses and the specimen.

The intense spot of electrons scanning the specimen surface excites specimen molecules to high energy levels. The excited molecules release this energy in several forms, including high-energy electrons

called *secondary electrons*. The image in an SEM is formed from the secondary electrons rather than the illuminating beam.

There are no further lens systems in the SEM. The secondary electrons leaving a particular spot on the specimen surface are picked up by a *detector* at one side of the specimen (see Fig. A-8). The detector consists of three elements: a small fluorescent screen, a photomultiplier, and a photoelectric cell. The screen emits visible light when bombarded by secondary electrons leaving the specimen surface. The photomultiplier amplifies the emitted light and projects it into the photoelectric cell. The photoelectric cell, which works on the same principle as the light meter on a camera, emits a current proportional to the amount of light striking it.

The current leaving the photoelectric cell is used to create an image in a television tube. In the tube an electron gun produces a narrow beam of electrons that is focused into a spot at the front of the tube. Deflectors inside the tube, connected to the same electronic circuits scanning the beam in the microscope, move the spot back and forth across the screen at the same rate and direction as the spot scanning the specimen. As a result, the spots on the screen and specimen are at corresponding points at any instant.

When the scanning spot in the microscope strikes a point on the specimen surface that responds by emitting large numbers of secondary electrons, the detector adjusts the scanning spot on the television screen at that point and instant to a high level of brightness. At specimen points emitting fewer secondary electrons the detector current is reduced and the spot scanning the screen at the same point and instant is dimmed. Each point on the screen therefore corresponds in brightness to the number of electrons emitted by the same relative points on the specimen. The scanning is so rapid that an apparently instantaneous image of the specimen surface is produced on the TV screen, with areas of brightness and darkness on the screen corresponding to ridges and valleys on the specimen surface. The result is an apparent three-dimensional reconstruction of the specimen surface (Fig. A-9).

Resolution in the SEM depends primarily on the size of the spot scanning the specimen inside the microscope. As the spot becomes smaller, the number of secondary electrons emitted is determined by smaller details on the specimen surface, allowing these details to be recognized as separate points with different degrees of brightness on the viewing screen. A scanning spot 5 to 10 nanometers in diameter, routinely achieved with contemporary SEMs, resolves surface details of about the same dimensions.

The fact that the SEM uses electrons emitted from surfaces rather than transmitted electrons for image

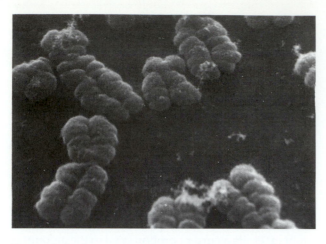

Figure A-9 A scanning electron micrograph of isolated metaphase chromosomes prepared by the G-banding technique (see p. 696). (Courtesy of C. J. Harrison, from *Cytogenet. Cell Genet.* 35:21 [1983].)

formation means that relatively thick specimens can be observed in the microscope. However, as in the TEM, the lens system, specimen, and detector must be kept at high vacuum to avoid scattering and absorption of the secondary electrons emitted by the specimen surface. The specimen must therefore be dry and nonvolatile.

SPECIMEN PREPARATION FOR MICROSCOPY

Living specimens are observed routinely in the light microscope, particularly with phase contrast microscopy. However, for many applications of light microscopy, and for essentially all uses of the electron microscope, operating limitations of the instruments make it impractical to examine living specimens. To circumvent these operating limitations, specimens are prepared for examination in various ways.

Basic Preparation Methods

Most commonly, specimens are prepared for microscopy by exposing them to chemical *fixatives*, chemicals that coagulate or crosslink specimen structures to hold them in place. The fixatives used in light microscopy include substances such as acetic acid, alcohol, formaldehyde, glutaraldehyde, and osmium tetroxide. Acetic acid and alcohol work primarily by precipitating or coagulating cell structures. Formaldehyde and glutaraldehyde introduce covalent linkages that stabilize specimen molecules with little or no coagu-

lation. The linkages preserve cell structures in very fine detail, often to molecular levels. Osmium tetroxide works similarly and probably also precipitates in and around structures at the molecular level. The ability of formaldehyde, glutaraldehyde, and osmium tetroxide to fix structures at the molecular level also makes them the fixatives of choice for electron microscopy. Osmium tetroxide acts as a stain as well as a fixative because osmium atoms, precipitated or chemically linked to cell structures, add significant contrast to the image. For most electron microscopy preparations, tissues are fixed first in glutaraldehyde and then "postfixed" with osmium tetroxide.

After fixation, specimens are *embedded* in a supportive medium, usually paraffin for light microscopy and an epoxy plastic for electron microscopy. The specimens are then *sectioned* by cutting them into slices about 10 μm thick for light microscopy and 60 to 70 nm thick for electron microscopy. Before or after sectioning, specimens are usually *stained* to improve the visibility of structures inside cells. Staining involves the addition of colored pigments for light microscopy or heavy metal ions for electron microscopy. For light microscopy, dyes can be chosen that give different colors to organelles such as nuclei or mitochondria or structures containing specific molecules such as RNA, DNA, or various proteins. For electron microscopy, specimens are stained with solutions of heavy metal salts, most often osmium tetroxide, uranyl acetate, or lead citrate, which all produce similar results in the electron image.

For some applications, specimens are squashed, smeared, or broken rather than sectioned. Smearing and squashing are used routinely to prepare materials for light microscopy—blood samples are smeared, for example, and cells squashed by pressing them between glass slides. In equivalent techniques for electron microscopy, cells are broken by various means (see p. 794) and dried on a thin plastic film. Such specimens are often *shadowed* by coating them with a thin layer of metal evaporated from a source located to one side. The metal deposits on raised surfaces of the specimen facing the source; depressions in the specimen, located in the shadow of higher points, are not coated. The effect in the electron microscope is as if a strong light has been directed toward the specimen from one side, highlighting ridges and placing depressions in deep shadow (as, for example, in Fig. 10-1).

Alternatively, particulate specimens may be *negative-stained* for electron microscopy by drying a staining solution around them. The stain outlines surface details by depositing in depressions and surface crevices during the drying process, typically producing a "ghost" image in which the specimen appears light against a dark background (as in Fig. 11-3).

In another much used technique for electron microscopy, *freeze-fracture*, cells are quick frozen in liquid nitrogen. The freezing, which is almost instantaneous, locks specimen molecules in place with little or no disturbance. The specimen is then broken while still frozen by striking it with a sharp knife edge. The fracture travels through the specimen, primarily following the hydrophobic interiors of membranes. The fractured surface is then shadowed with metal evaporated from one side, either immediately after fracturing or after a delay of some seconds or minutes. During the period between fracturing and shadowing, water escapes from the fractured surface, "etching" the specimen to produce greater surface relief and more clearly outlined details (as in Fig. 4-11a).

Analytical Methods

Many of the preparative techniques for light and electron microscopy have been modified so that they stain or mark only certain structures or molecular types for identification. Among the most important of these methods are *autoradiography*, in which specimen details are marked by combination with radioactive atoms, and *immunochemistry*, in which antibodies are used to attach pigments or heavy atoms to specific structures or molecules in the specimen. These approaches are often combined with *enzymatic digestion* to remove molecules of specific types.

Autoradiography   Autoradiography involves tagging a specimen molecule with a radioactive isotope, usually ^3H, ^{14}C, ^{32}P, ^{35}S, ^{125}I, or ^{131}I. In the most common application (Fig. A-10), living cells are "labeled" by exposing them to a substance containing one of the radioactive isotopes. The labeling substance is usually a precursor molecule used by the cell to make only the molecules of interest. For example, most cells use thymidine only as a precursor for synthesis of DNA. By exposing cells to thymidine made radioactive by the inclusion of ^3H (tritium) atoms, the DNA can be specifically labeled by radioactivity. Similarly, RNA can be specifically labeled by exposing cells to radioactive uridine; proteins can be labeled by exposure to amino acids made radioactive by the inclusion of ^3H or ^{14}C atoms.

After exposure to the radioactive precursor, cells are washed in an unlabeled medium to remove any excess label and then fixed, embedded, sectioned, and placed on microscope slides. The slides are coated in a darkroom with a photographic emulsion that has been liquified by heating. After coating, the slides are cooled and stored in the dark. During the storage period, radioactive decay exposes crystals in the photographic emulsion lying over labeled sites. The emulsion is then developed by standard photographic techniques and the sections are examined in the microscope. Developed silver grains mark radioactive areas in the preparation. The developed pattern of grains, called an *autoradiograph*, marks the sites containing the molecules of interest.

Immunochemistry   Several preparative techniques take advantage of the high specificity of antibodies, which usually react and bind only to a single antigen (see p. 545 for details). The technique is so sensitive that in many cases, proteins with sequences differing by only one or two amino acids can be separately detected and identified.

Antibodies are prepared by injecting a suitable animal, such as a rabbit or mouse, with the purified antigen of interest. After a period of one to two months an antibody specific in its reaction to the antigen can be detected in the bloodstream of the injected animal. Extraction of a relatively small quantity of blood usually yields large quantities of the antibody, which is easily separated from other blood proteins.

For light microscopy, antibodies are frequently marked by radioactivity or with fluorescent dyes to make them visible (Figure 1-16 shows cells prepared for light microscopy by the fluorescent antibody technique). For electron microscopy, antibodies are combined with an atom that scatters electrons, usually a metal atom. Gold particles or an iron-containing protein, *ferritin*, are used most often for this purpose. In either case, small, dense spots that are readily visible in the electron microscope mark sites reacting with the antibody.

Enzymatic Digestion   Enzymes are used to detect the presence of specific molecules by testing the susceptibility of cellular structures to enzymatic breakdown. If a cell structure is destroyed or extensively altered after exposure to ribonuclease, for example, it can be assumed that RNA forms a part of the affected structure. Alternatively, the presence of enzymes as components of cell structures can be tested by adding a substrate for the enzyme. Activity of the enzyme in converting the substrate to products is tested by adding a substance that reacts with the products to produce an identifiable color or precipitate. For example, in electron microscope preparations, ATP is added in combination with a soluble lead salt such as lead citrate to test for the presence of an ATPase enzyme. In sites containing an ATPase the ATP is hydrolyzed, releasing inorganic phosphate. The phosphate reacts with the lead salt, producing a dense, insoluble precipitate of lead phosphate that is directly visible in the

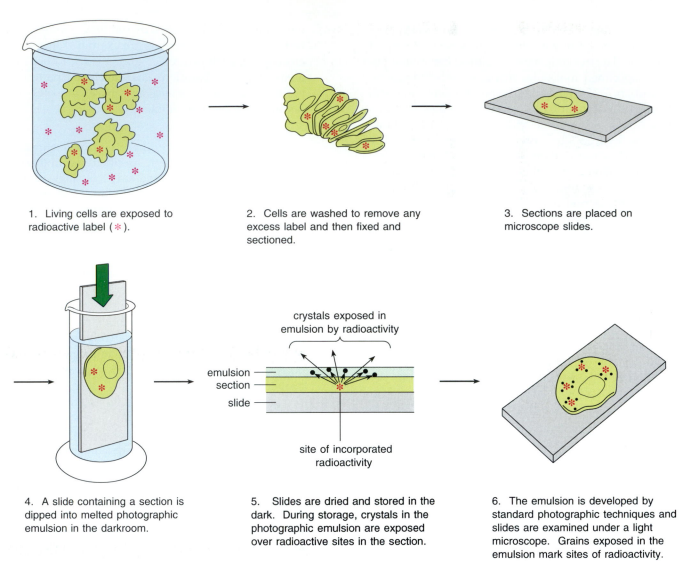

1. Living cells are exposed to radioactive label (✳).

2. Cells are washed to remove any excess label and then fixed and sectioned.

3. Sections are placed on microscope slides.

crystals exposed in emulsion by radioactivity

emulsion
section
slide

site of incorporated radioactivity

4. A slide containing a section is dipped into melted photographic emulsion in the darkroom.

5. Slides are dried and stored in the dark. During storage, crystals in the photographic emulsion are exposed over radioactive sites in the section.

6. The emulsion is developed by standard photographic techniques and slides are examined under a light microscope. Grains exposed in the emulsion mark sites of radioactivity.

Figure A-10 The techniques used to prepare cells for light microscope autoradiography.

electron microscope and marks the sites of enzymatic activity.

CELL CULTURE

Culturing Microorganisms

The conditions required for growing microorganisms such as yeast and bacteria in pure laboratory cultures are relatively simple. The much cultured human intestinal bacterium *E. coli* and the yeast *Saccharomyces cerevisiae* (brewer's yeast), for example, require only a carbon source such as glucose, a source of nitrogen, and inorganic salts. Under optimum conditions, *E. coli* completes a cycle of cell growth and division every 20 minutes, allowing cultures to reach very large numbers in a relatively short time. The cells may be grown in liquid suspension or on the surface of a solid

growth medium such as an agar gel (agar is a polysaccharide extracted from an alga). In either case, sterile techniques must be used to ensure that foreign microorganisms do not contaminate the culture.

For many experimental purposes it is desirable to work with a genetically uniform culture of bacterial or yeast cells. For these purposes a culture may be started from a single individual. Such cultures are called *clones*, and the process used to establish the cultures is called *cloning*. The techniques used for cloning microorganisms are readily adapted to the selection of mutants that are resistant to antibiotics or that require substances which normal or "wild-type" strains can make for themselves. (Figures A-11 and A-12 show how these selections are made.) The many thousands of bacterial mutants selected by these and similar techniques are widely used in genetic, biochemical, and molecular studies.

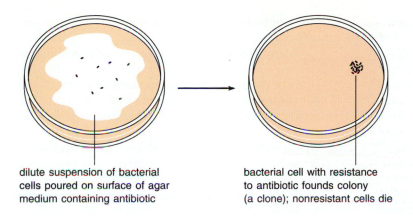

Figure A-11 Selection of a mutant bacterial strain resistant to antibiotics. A diluted sample of bacterial cells is poured over an agar medium containing the antibiotic (alternatively, the antibiotic may be mixed with the medium used to culture the bacteria). Any colonies growing on the agar surface may be assumed to be mutant strains with resistance to the antibiotic.

dilute suspension of bacterial cells poured on surface of agar medium containing antibiotic

bacterial cell with resistance to antibiotic founds colony (a clone); nonresistant cells die

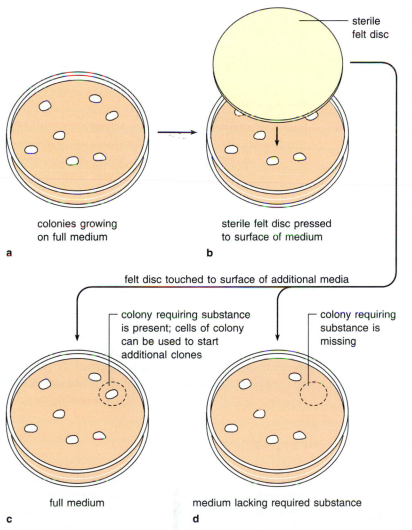

sterile felt disc

colonies growing on full medium

a

sterile felt disc pressed to surface of medium

b

felt disc touched to surface of additional media

colony requiring substance is present; cells of colony can be used to start additional clones

colony requiring substance is missing

full medium

c

medium lacking required substance

d

Figure A-12 Selection of a mutant strain requiring a substance that wild-type cells can make for themselves. **(a)** Cells are first grown on agar containing a full complement of nutrients, including amino acids and other substances that wild-type individuals can synthesize. **(b)** After colonies have grown on the agar surface, a replica of the colonies is made by pressing a piece of sterile felt against the agar. The felt is then pressed against the surface of two agar preparations, one with full nutrients **(c)** and one lacking only the substance of interest **(d)**. Any colonies absent in the preparation lacking the substance but present on the agar with the full complement can be assumed to contain mutant cells that have lost the capacity to make the substance. The colony growing on the full medium can be used to start clones of the mutants.

Culturing Higher Eukaryotic Cells

Cells from higher animals are usually isolated for culturing by treating dissected tissues with a protein-digesting enzyme such as trypsin in combination with a chemical that removes (chelates) calcium ions from the cell surfaces. The proteinase digests away extracellular proteins holding cells together in masses; Ca^{2+} removal breaks linkages binding animal cells to each other or to extracellular materials. (Many such linkages depend on Ca^{2+} for their formation and maintenance; see p. 166.) After these treatments, gentle shaking separates the cells.

The cells obtained by these methods are usually of mixed types. Separating individual cell types for culture is accomplished by a variety of methods. Some use antibodies that specifically recognize and bind surface molecules of a particular cell type. The antibodies may be attached to a supporting surface, so that the cells of interest are linked to the surface and immobilized by the antibodies. Repeated washings then remove the unwanted cell types from the surface. In another technique, antibodies developed against surface molecules of the cell type of interest are combined with a fluorescent dye, so that the desired cells glow brightly when irradiated with light at the correct wavelength. Automated separators can extract the fluorescent cells from a mixed culture at the rate of thousands per minute.

Once separated, animal cells are placed in growth media that are considerably more complex than those used for microorganisms. The media must contain the amino acids animal cells cannot synthesize for themselves (arginine, histidine, leucine, isoleucine, lysine, methionine, phenylalanine, threonine, tryptophan, and valine), plus cysteine, glutamine, and tyrosine, which are usually included for best growth. Also required are several vitamins (see p. 63), salts, and glucose. Blood serum (the fluid part of blood with the blood cells removed) must often be added to ensure growth. Many of the components supplied by serum have not been identified. Among them are probably proteins acting as *growth factors*—molecules that stimulate cell growth and division—and a variety of unknown substances required by the cells in trace quantities. Only a relatively few animal cells can be grown in fully defined media in which every required chemical is known.

Even with the addition of blood serum, most cells removed from normal tissues cannot grow indefinitely in culture. After 50 to 100 cycles of growth and division, most normal cell lines become dormant and eventually die. Cells removed from tumors, however, often have the capacity for unlimited growth. Frequently cells from normal tissues can be stimulated to grow indefinitely by exposing them to cancer-causing viruses or to *carcinogens* (see p. 654), chemicals that can induce some of the characteristics of cancer cells. Of these characteristics the most significant for cell cultures is continued growth and cell division. Cultured cells with the capacity to grow and divide indefinitely are said to be "immortalized."

Cells from higher plants are cultured after their cell walls have been removed by exposure to *cellulase*, an enzyme that digests away cell wall material. The naked plant cells, called *protoplasts*, can then be grown in chemically defined media. These techniques have been applied with greatest success to the cells of carrots and a few other higher plants.

Many differentiated eukaryotic cells retain specialized functions when grown in culture, allowing study of their activities and patterns of synthesis under controlled chemical and physical conditions. Because cultures can be cloned from single cells, sources of error resulting from the presence of mixed cell types can be avoided. Cell cultures are also invaluable for testing the toxicity of substances and evaluating the ability of chemicals to cause or cure cancer. In addition, cell cultures are much used for *genetic engineering* studies in which DNA containing genes of interest is introduced into the cells by various means. Plant cell cultures provide an advantage for many of these studies because, with the addition of plant growth hormones, complete plants can often be grown from treated and cultured cells. As a consequence, the results of experimental treatments can be observed in adult plants as well as protoplasts in culture.

Fusing Cultured Cells

Some cultured cells can be fused together to produce composite cells containing the nucleus and cytoplasm of both cell types. Exposing the cells to certain enveloped viruses or to polyethylene glycol induces the fusion (see p. 110). Once fused, the composite cells can be maintained in culture as clones. In animal cells the nuclei fuse as well as the cytoplasm, so that a single composite nucleus is formed.

Fused animal cells have proved invaluable in studies of factors regulating genetic activity, particularly in dividing cells. Cell fusion also underlies the production of *monoclonal antibodies*, which are much used in science and medicine (see p. 557). The uniform and precise reactivity of monoclonal antibodies provides distinct advantages over the antibodies obtained by injecting an antigen into a test animal, which usually consist of a large family of antibody types that collectively react with many sites on the antigen.

CELL FRACTIONATION

Once obtained in quantity from organisms or cultures, cells can be *fractionated* to yield cell organelles, structures such as ribosomes or microtubules, or individual molecular types in highly purified form.

Cells are disrupted for fractionation by one or more of several methods, including sonication (exposure to high frequency sound waves), grinding by fine glass beads or other abrasive materials, increases in internal osmotic pressure, or exposure to detergents. The cell organelles, structures, and molecules released by the breakage are suspended in a buffered salt solution and separated into fractions by spinning in a centrifuge. Centrifugation can separate organelles as large as nuclei or mitochondria, or molecules as small as proteins into separate groups. Molecules are also purified by *gel electrophoresis*, a technique in which an electrical field rather than centrifugal force is used as the driving force for separation.

Centrifugation

A centrifuge consists of a *rotor* driven at high speed by an electric motor (Fig. A-13). The rotor holds tubes containing solutions or suspensions of the materials to be centrifuged. For centrifugation at very high speeds the centrifuge is enclosed in an armored chamber that can be cooled and held at a vacuum to reduce heat and friction caused by air resistance. The centrifugal forces generated by the spinning rotor in the most powerful instruments can reach 500,000 times the force of gravity.

Movement in response to centrifugal force is related to the mass, density (density = mass per unit volume), and shape of the structures or molecules being centrifuged and to the mass, density, and viscosity of the surrounding solution. If a structure or molecule has greater mass or is denser than the surrounding solution, it moves downward in the centrifuge tube. If it has lesser mass or is less dense than the surrounding solution, it remains at the top of the tube or, if present at lower levels, is displaced by molecules of the solution and moves upward in the tube. The velocity of movement is modified by the shape of the particle, with rodlike or elliptical forms moving more slowly than spherical particles. Increasing the viscosity of the suspending solution also slows the rate of movement of structures or molecules, particularly those with nonspherical forms.

Because many discrete cell structures, such as the nucleus, mitochondria, chloroplasts, and ribosomes, differ in density and are reasonably resistant to mechanical disruption, they can be separated by centrifugation. Several centrifugations at successively higher speeds separate and purify the structures. At lower speeds, large, dense bodies such as nuclei are driven down and concentrated. Low-speed centrifugation can also be used to remove debris from the preparation. Centrifugation of the remaining solution at higher speeds drives down and concentrates structures of intermediate size and density, such as mitochondria and chloroplasts. Final centrifugation at very high speeds concentrates structures such as microtubules, microfilaments, ribosomes, and larger molecules such as nucleic acids and proteins (Table A-1).

Depending on conditions such as the ionic composition and pH of the surrounding medium, the organelles and smaller structures isolated by fraction-

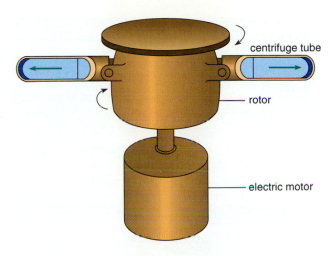

Figure A-13   The arrangement of tubes and rotor in a centrifuge. In some centrifuges the tubes are held solidly at an angle within the head instead of hinged as this diagram shows.

ation and centrifugation may retain some or all of their biological activity. Active preparations of this type, in which no whole cells are present, are called *cell-free systems*. These systems are much used in biochemical studies, particularly those investigating DNA replication, RNA transcription, and protein synthesis.

Molecules more dense than the suspending medium move downward or *sediment* at a constant rate when the speed of the centrifuge is held uniform. This constant is usually expressed in the form of *Svedberg* or *S* units, named in honor of one of the pioneers of the centrifugation technique. Larger molecules generally have higher S values. Often the S value is used rather than molecular weights to designate molecular size, particularly for nucleic acid molecules. Eukaryotic ribosomal RNAs, for example, are typically designated as 5S, 5.8S, 18S, and 28S rRNAs in order of increasing molecular size, rather than in molecular weights.

Density-Gradient Centrifugation   For more precise separation of molecules, *density-gradient centrifugation* is used. In the most used application a plastic centrifuge tube is filled with a solution of sucrose or cesium chloride (CsCl) mixed in gradually denser concentrations toward the bottom of the tube. The concentration gradient, made by laboratory equipment designed for this purpose, is adjusted so that its most dense region at the bottom of the tube is less dense than any of the molecules under study. A mixture of the molecules to be separated is carefully layered at the top of the tube.

During centrifugation the gradient separates the molecules into bands according to density. The

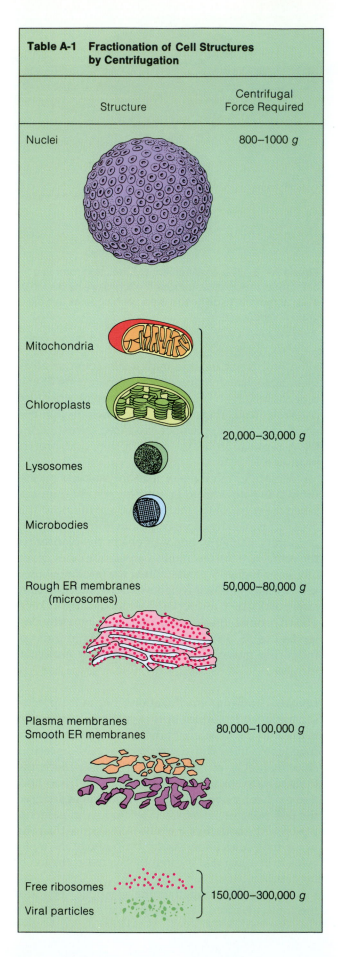

Table A-1 Fractionation of Cell Structures by Centrifugation

Structure	Centrifugal Force Required
Nuclei	800–1000 g
Mitochondria	
Chloroplasts	20,000–30,000 g
Lysosomes	
Microbodies	
Rough ER membranes (microsomes)	50,000–80,000 g
Plasma membranes Smooth ER membranes	80,000–100,000 g
Free ribosomes	150,000–300,000 g
Viral particles	

molecules collect in bands because, as they travel downward, the molecules at the leading edge of a band constantly encounter a denser solution than the molecules in the trailing edge and are slowed slightly in their progress down the tube. The molecules at the trailing edge of a band move slightly faster because they encounter a less dense medium. The tendencies of the leading molecules to slow slightly and the trailing molecules to speed maintains them in a sharply defined band as they migrate downward. Centrifugation is stopped before any of the bands has moved to the bottom of the tube. After centrifugation a hole is punched in the bottom of the tube and the bands are collected in separate fractions as they drain off.

Buoyant Density Centrifugation For highly critical work, molecules may be separated by *buoyant density centrifugation*. The technique usually employs a CsCl solution, adjusted initially to approximately the same density as the molecules under study. The sample is layered on top of the solution and the tube is spun at very high speed, sometimes for several days. As the tube spins, the CsCl molecules gradually pack more densely toward the bottom of the tube, producing an even gradient in density from top to bottom. Because the initial, uniformly dispersed CsCl solution approximately matched the density of the sample, the final CsCl gradient is less dense than the sample at the top of the tube and more dense at the bottom. The molecules in the sample descend until they reach the level at which their density and the density of the CsCl gradient are the same, where they form a sharp band. The method is so sensitive that molecules of the same type but containing different atomic isotopes can be separated into different bands. For example, this method was used in a classic study of DNA replication in which DNA molecules containing the ^{14}N and ^{15}N isotopes of nitrogen were clearly and sharply separated (see p. 660).

Gel Electrophoresis

The centrifuges used for molecular separation become more expensive, required speeds become higher, and centrifuge times become longer as the size of the molecules under study become smaller. Even with centrifuges capable of the highest speeds, molecules with molecular weights smaller than about 10,000 cannot be sedimented because of the opposing effects of diffusion. Gel electrophoresis circumvents these problems by using an electrical field instead of centrifugal force to move molecules through a gel-like medium. The technique, using relatively inexpensive equipment, rapidly and precisely separates molecules from the size of large proteins downward to very short polypeptide and nucleic acid chains or even single nucleotides.

In a gel electrophoresis apparatus (Fig. A-14) the gel is cast in a glass tube or as a slab between glass or plastic plates. For separation of proteins, gels are usually cast from polyacrylamide plastic; for nucleic acids, gels are cast from *agarose*, a complex carbohydrate molecule extracted from seaweed. The tube or slab gel is suspended so that its ends are in contact with separate buffered salt solutions. The salts used are electrolytes, which conduct an electric current in water solutions. The conducting solutions at either end of the gel are connected to the positive and negative electrodes of an electrical power source so that current flows from one solution to the other through the gel. Any molecules added to the gel migrate toward the positive or negative electrode in accordance with their own electrical charge.

The gel forms an open molecular network, with spaces just large enough to allow passage of the molecules under study along with the buffered salt solution used in the apparatus. The consistency of the gel effectively eliminates diffusion and convection, which tend constantly to disturb bands of molecules sedimenting in a centrifuge. When the current is held constant, the rate at which molecules move through the gel depends primarily on three factors: molecular size, molecular shape, and charge density.

The rate of migration with respect to size reflects the ability of different molecules to thread through the gel network. Larger molecules, because they meet more resistance in passing through the gel, move more slowly than smaller molecules. The second factor, molecular shape, has a similar effect. Molecules with extended fibrillar or elliptical forms move more slowly than spherical ones. The third factor, charge density, refers to the number of positive or negative charges per unit of surface area of the molecules under study. The higher the charge density, the more rapidly a molecule moves toward the end of the gel with opposite charge. Because the charge of many biological molecules varies with pH, the adjustment of pH is critical to the rate of migration.

In application of the technique a mixture of the molecules to be separated is layered onto the gel at the top of the tube or slab. The buffer solutions at the ends of the tube are then connected to the power source, with the electrodes placed so the charge at the bottom of the tube is opposite that of the molecules under study. The layered molecules move downward through the gel in response to the attracting charge. Because the relative rates at which the molecules move depends on size, shape, and charge density, the initial mixture of molecules gradually separates into a series of distinct bands moving at different rates through the gel. The fastest moving bands contain the

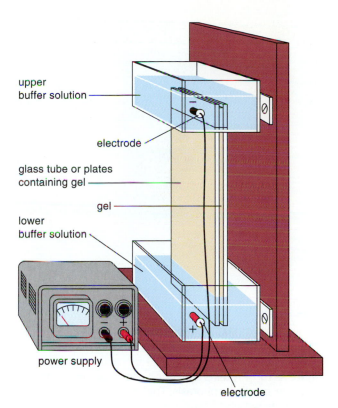

Figure A-14 The apparatus used for gel electrophoresis. The gel, confined in a glass tube or between glass or plastic plates, is suspended between electrolyte solutions. The electrolyte solutions at the two ends of the gel are connected to the positive and negative electrodes of a power source.

smallest, most compact molecules with the highest charge density. Because most or all of the molecules in the mixture form invisible bands, a dye molecule slightly smaller than any of the molecules under study is usually added to enable the operator to keep track of the separation. The apparatus is turned off as the dye molecule reaches the bottom of the gel.

After separation into distinct bands the molecules in the gels may be examined in various ways. Gels containing proteins are frequently stained with an organic dye to reveal the positions of the bands (Figure A-15 shows a tube gel prepared in this way). Protein bands are also identified by radioactive labeling or reaction with fluorescent antibodies. Bands containing nucleic acids are identified by radioactive labels or by reacting them with ethidium bromide, which fluoresces under ultraviolet light. For further analysis the molecules in individual bands can be isolated by slicing the gel into pieces or by making what is known as a *blot* (see below).

Determining Relative Molecular Weight by Gel Electrophoresis Since the rate of migration through the gels reflects molecular size, the relative positions

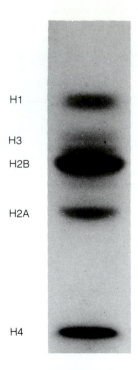

Figure A-15 Proteins separated into separate bands in a tube gel by electrophoresis. The proteins shown are the five histone proteins associated with DNA in eukaryotic chromatin. (Courtesy of W. F. Reynolds.)

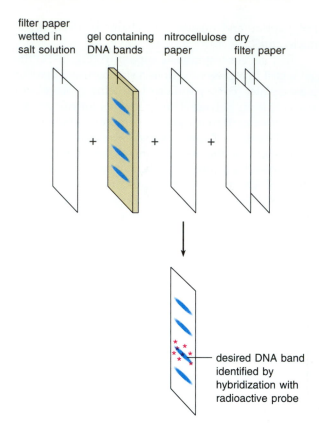

Figure A-16 The Southern blotting technique for identifying DNA bands separated by gel electrophoresis. As the salt solution is wicked through the gel and into the dry filter paper, the DNA bands are carried into the nitrocellulose paper, where they are trapped as blots. Individual bands are then identified by hybridization with a radioactive probe consisting of a DNA or RNA molecule that is complementary to the sequences of interest.

of migrating bands can provide an approximation of molecular weight. The approximation becomes more exact if the charge density and shape of the molecules under study are uniform. Nucleic acids satisfy this requirement nicely because of their uniform structure and even distribution of negatively charged phosphate groups.

Proteins, in contrast, vary widely in charge density and shape. This problem is circumvented by reacting proteins with the detergent *sodium dodecyl sulfate (SDS)*. The negatively charged SDS molecules have hydrocarbon chains that interact with the hydrophobic portions of protein molecules. Because most proteins contain roughly the same proportion of hydrophobic amino acid residues, they bind equivalent quantities of SDS, in a ratio of SDS to protein of about 1:4 by weight. The bound SDS effectively swamps out the native charge of the proteins and gives them a more or less uniform negative charge density.

The SDS coat causes most proteins to lose their native secondary structures and unwind into a random coil. The unwinding is aided if proteins are treated with a reducing agent such as mercaptoethanol to break disulfide linkages. The size of the random coil primarily reflects the length of the amino acid chain. As a consequence of the unwinding and the uniform charge density imposed by the SDS coat, different proteins move toward the positive end of the gel primarily according to the length of their amino acid chains—that is, according to molecular weight.

In practice, molecular weights are usually estimated by mixing unknowns with several readily identifiable molecules of known molecular weight. The knowns and unknowns move through the gel in such a way that the relative distance traveled, plotted against the logarithm of the molecular weight, produces a straight line. Placing the knowns and unknowns in position in the plot gives a reasonably accurate estimate of the molecular weights of the unknowns. Molecular weights determined in this way are frequently designated M_r, indicating that the weight is an estimate made relative to a known standard.

Blotting Techniques Blots are made by pressing electrophoretic gels against a piece of filter paper selected or treated to bind the proteins or nucleic acids in the bands. For a gel containing single, unpaired DNA nucleotide chains separated into bands, a piece of filter paper soaked in a salt solution is placed on one side

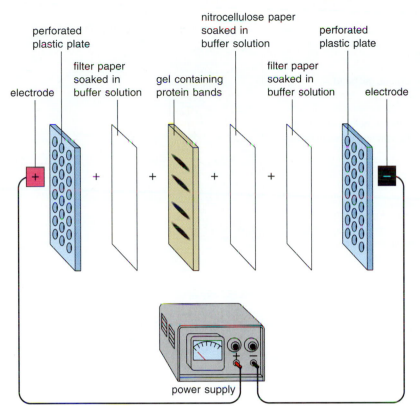

perforated plastic plate

filter paper soaked in buffer solution

electrode

nitrocellulose paper soaked in buffer solution

gel containing protein bands

filter paper soaked in buffer solution

perforated plastic plate

electrode

power supply

Figure A-17 The western blotting technique for removing protein bands from an electrophoretic gel. A gel containing protein bands is placed on a piece of nitrocellulose paper soaked in a buffer solution. The gel and nitrocellulose paper are sandwiched between additional pieces of filter paper soaked in buffer and two perforated plastic plates. Current applied across the plates drives the proteins from the gel onto the nitrocellulose paper, producing a pattern of blots corresponding to the bands.

of the gel (Fig. A-16). A piece of nitrocellulose filter paper, on which the blot will form, is placed on the other side, backed by several sheets of dry filter paper (nitrocellulose is used because it can directly bind unpaired DNA nucleotide chains). As the salt solution is drawn through the gel to the dry filter paper, it carries the molecules in the bands into the nitrocellulose paper, where the bands are deposited and tightly bound as blots. Individual blots containing DNA molecules of interest can then be identified by a technique known as *hybridization* (see below). A blot of DNA bands produced and identified in this way is called a *Southern blot*, after E. M. Southern, the investigator who originated the DNA blotting technique.

Essentially the same technique is used to make blots from a gel containing RNA bands. An RNA blot produced in this way, in which individual RNA molecules are transferred to filter paper and identified by hybridization, is termed a *northern blot*.

A more extensive modification (Fig. A-17) is necessary to blot protein bands because proteins diffuse only very slowly from gels. A gel containing protein bands is placed on a piece of nitrocellulose paper soaked in a buffer solution. (Nitrocellulose paper binds proteins as well as single-chain DNA.) The gel and nitrocellulose paper are sandwiched between soaked pieces of ordinary filter paper and two perforated plastic plates. When current is applied across the plates, the proteins are driven from the gel onto the nitrocellulose paper, producing a pattern of blots

corresponding to the bands. A protein blot formed in this way is called a *western blot*. Once in the paper, individual proteins are usually identified by means of antibodies.

Identifying Nucleic Acid Blots by Hybridization Blots produced by the Southern and northern techniques are routinely identified by *nucleic acid hybridization* (see Information Box 13-1), in which a sequence of interest is identified among other nucleotide chains by pairing it with a complementary sequence used as a "probe." If the sequence of interest is a gene, the probe may consist of part or all of the messenger RNA copied from the gene or an artificially synthesized piece of complementary DNA. A DNA copy of an mRNA made by reverse transcriptase, an enzyme that can use RNA as a template for making a DNA copy (see p. 523), is also frequently used. The DNA or RNA molecules used as probes may contain as few as 15 or so to many thousands of nucleotides.

For identification of Southern blots a solution containing a radioactive probe is poured over single-chain DNA bound in blots to the nitrocellulose paper. The probe hybridizes only with a blot containing a complementary DNA chain. The paper is washed to remove unbound probe molecules, leaving the DNA molecule of interest as an individual blot marked with radioactivity. For northern blots a radioactive single-chain DNA molecule complementary to the RNA molecule of interest is used as a probe.

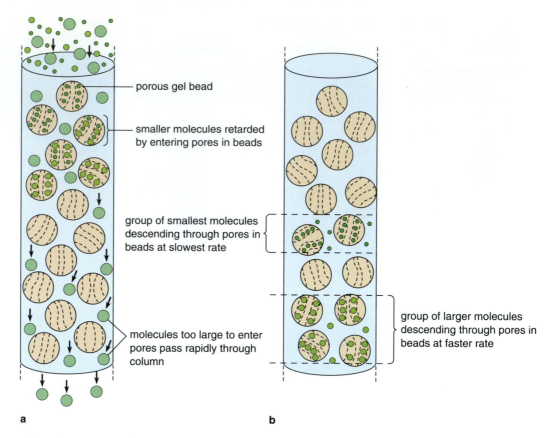

- porous gel bead

- smaller molecules retarded by entering pores in beads

group of smallest molecules descending through pores in beads at slowest rate

molecules too large to enter pores pass rapidly through column

group of larger molecules descending through pores in beads at faster rate

a   b

Figure A-18 Filtration chromatography. **(a)** A mixture of proteins of different sizes is poured into a filtration column. Proteins too large to enter pores in the beads or gel are excluded and pass rapidly through the column. Smaller molecules enter pores in the beads and are retarded. **(b)** After traveling some distance through the column, the molecules able to fit into the pores are retarded differentially according to size, with the smallest molecules traveling downward most slowly. The retardation separates the proteins into distinct bands descending through the column at different rates.

Chromatography

Chromatography is another method widely used to separate proteins according to size, shape, or charge. The technique involves passing a solution of proteins through a tube or column packed with beads or a gel matrix. Flow through the column depends on gravity or pressure; electrical currents are not used. The beads or gel retards the passage of some proteins to a greater or lesser extent and allows others to pass freely.

Chromatography frequently allows proteins to be purified in their undenatured, native state. Three different applications of the technique are commonly used: gel filtration, ion exchange, and affinity chromatography.

Gel Filtration Chromatography Filtration chromatography separates proteins from mixtures according to size and shape. For most applications the filtration

columns are packed with porous plastic beads that have openings similar in dimensions to the molecules under study (Fig. A-18).

When a buffered solution containing proteins passes through the column, proteins larger than the pores move rapidly through the spaces between the beads and leave the column without further separation (Fig. A-18a). Proteins with dimensions smaller than the pores enter the beads and are retarded. In general the smaller the molecule, the more space available inside the beads and the greater the retardation. To move the retarded proteins through the beads, additional buffer solution is added or pumped through the column. As the retarded proteins pass from bead to bead downward through the column, they separate into bands according to size and shape, with the largest and most compact proteins reaching the bottom of the column first (Fig. A-18b). The proteins are collected in separate fractions as they exit the bottom of the column.

Filtration beads can be chosen with pore sizes that retard proteins or other molecules of interest and exclude larger molecules. Thus if the proteins of interest have molecular weights below 50,000, for example, filtering beads can be chosen that exclude proteins above this size. The larger proteins do not enter the pores in the beads and quickly pass through the column.

Ion Exchange Chromatography   The ion exchange technique separates proteins according to charge rather than size or molecular weight. For this method the packing material consists of minute plastic beads that contain groups opposite in charge to the molecules of interest.

A buffered solution containing a mixture of proteins to be separated is added to a column packed with the charged beads. As the solution flows through the column, the proteins carrying an opposite charge attach to the beads. Those with no charge or the same charge as the beads pass through the column without retardation. The column is then washed with salt solutions of gradually increasing concentration. The salt ions displace the proteins by competing for binding sites on the beads. The first proteins to be displaced are those with the fewest charged groups. These leave the column first; as the salt concentration of the wash is increased, additional proteins are displaced in order of increasing charge. The proteins are collected in individual fractions as they pass in succession from the bottom of the column.

Affinity Chromatography   Affinity chromatography is a highly effective and much used method for separating individual proteins that can recognize and bind specific molecules. In this technique a molecule recognized and bound by the protein of interest is attached covalently to the beads in the column, which are otherwise inert (Fig. A-19a). A mixture of proteins is then passed through the column. Among the proteins in the mixture, only the protein recognizing and binding the substance attaches to the beads; the remainder pass through without hindrance. After the unbound proteins are washed from the column, the protein of interest can be removed from the beads by adding a solution in which the substance bound by the protein is suspended in free, unattached form. As the solution passes through the column, the freely suspended molecules of the substance compete with the molecules attached to the beads for the binding site on the protein. The competition eventually releases most or all of the protein molecules from the

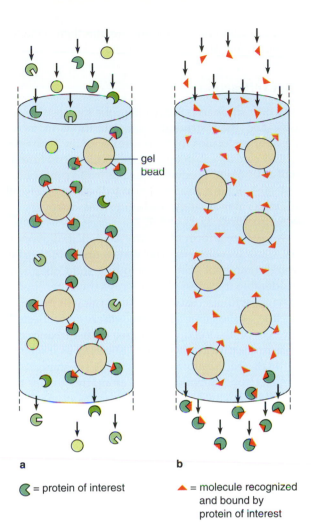

= protein of interest

▲ = molecule recognized and bound by protein of interest

Figure A-19   Affinity chromatography. A molecule recognized and bound by the protein of interest is covalently linked to the beads in the column. **(a)** A mixture of proteins is added to the column. Only the protein binding the molecule attached to the beads is trapped by the beads; other proteins pass through the column without hindrance. **(b)** The attached proteins are released by adding a solution of the molecules recognized and bound by the protein of interest. The added molecules compete for the binding site on the protein, releasing the protein from the beads.

beads for collection as a highly purified sample at the bottom of the column (Fig. A-19b).

Affinity chromatography can rapidly separate a single protein from crude cell extracts containing hundreds or thousands of proteins. For example, if the receptor for the insulin hormone is the protein of interest, the hormone is attached covalently to the beads. An insulin receptor present in a preparation of cellular proteins will attach specifically to the bound insulin when the mixture passes through the column; all other proteins exit the column without delay. Adding a solution containing unbound insulin then removes the insulin receptor from the beads.

X-Ray Diffraction

X-ray diffraction uses X rays to probe the positions of atoms and their structural configurations in molecules. For molecules that can form crystals, X-ray diffraction can reveal details of molecular structure as small as 0.15 nm. The technique has led to some of the most far-reaching and significant discoveries of cell and molecular biology, including the discovery of DNA structure.

X rays are a form of electromagnetic radiation of very short wavelength. The waves are able to pass through many substances, including solid objects of considerable thickness. Some X rays travel through a substance without disturbance, emerging in the same direction as the entering beam. Others are deflected, scattered, or absorbed. The degree of disturbance depends primarily on the number of electrons traveling in orbitals around atomic nuclei in the substance. The more electrons surrounding an atom, the greater the possibility that an X ray passing near the atom will be deflected from its path. Heavy atoms, particularly those of the heavy metals, are most effective in deflecting X rays.

X rays passing through relatively unstructured materials in which individual atoms are arranged in nonrepeating patterns are deflected in random directions. It is possible to obtain an image of such materials by placing a photographic plate or viewing screen on the side from which the beam emerges. Such images are essentially shadows, with areas of greater or lesser density depending on the relative degree of absorption or deflection of X rays by different regions. The X-ray images used in medicine and luggage inspection at airports are shadow images of this type.

X rays passing through a substance with highly ordered atoms, such as a crystal, are deflected in a significantly different pattern. As the beam enters a crystal, the X rays encounter atoms arranged in regularly repeated planes and positions. Each atom located in a specific position and plane in the crystal lattice deflects X rays to the same angle and direction (Fig. A-20). As a result, the entering beam is split into a number of smaller beams that leave the crystal at different angles. The troughs and crests of the waves in a beam leaving the crystal at the same angle may *interfere*, that is, reinforce or cancel each other, depending primarily on the distance separating layers of atoms that reflect a beam to the same angle (see Fig. A-20). The degree of reinforcement or cancellation produces differences in the intensity of the beams leaving the crystal. The combined effect of the deflection and interference, called *diffraction*, is most pro-

nounced if the X rays entering the crystal are all at the same wavelength and travel in parallel rather than diverging paths.

The number of emerging beams depends on the number of atoms in each repeating unit of the crystal. Simple crystals such as those of sodium chloride, with a repeating lattice unit containing only a few atoms, split the beam into a only a few smaller beams. Crystals with more complex lattices, such as those cast from proteins, diffract the entering beam into proportionally greater numbers of smaller beams, as many as hundreds or thousands in some cases. The number, positions, and intensities of the beams leaving a crystal are recorded on a photographic plate placed on the opposite side of the crystal from the X-ray source. The resulting distribution of spots is called an *X-ray diffraction pattern* (Fig. A-21).

Because the angles and intensities of the emerging beams depend on the locations of atoms in the repeating crystal lattice, it is possible to use the diffraction pattern to determine the positions of individual atoms in the crystal. The determination is relatively easy if the crystal is a simple one that forms a diffraction pattern with only a few spots. But unraveling the diffraction patterns produced by highly complex atomic lattices, such as those of proteins, may take literally months of calculations by the most advanced computer systems.

In many cases the diffraction patterns produced by complex crystals are compatible with several different possible lattices. The possibilities are narrowed by the addition of metal atoms to the crystal lattice. The positions taken by the metal atoms, reflected in definite alterations in the pattern of spots, are frequently compatible with only one of the several possible lattices.

Many proteins and other biological molecules such as larger nucleic acid molecules are difficult or impossible to cast into crystals. X-ray diffraction can still provide information about the molecular structure of such substances if the molecules can be roughly ordered in some fashion. In the studies leading to the discovery of DNA structure, for example, DNA molecules were roughly aligned in parallel by pulling a partially dried DNA sample into a fiber. The X-ray diffraction pattern produced by such substances blurs into partial or complete concentric rings rather than forming discrete spots. However, the spacing of the rings and their relative intensities still provided information about the dimensions of regularly repeating major patterns in the specimen. More recently, true crystals have been cast of small DNA and RNA molecules, allowing more precise analysis of these molecules with X-ray diffraction (see, for example, p. 367).

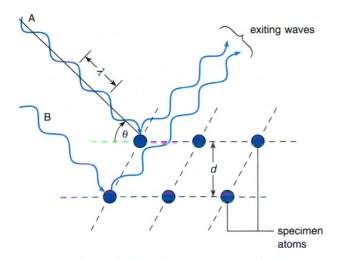

Figure A-20 X-ray diffraction of a beam of parallel X rays traveling at the same wavelength. Wave A in an entering beam of X rays is deflected to an angle by an atom in the top layer of the crystal lattice. Wave B is deflected by an atom in a corresponding location in the second layer to the same angle as wave A. However, wave B must travel a longer distance to reach the atom in the second layer. Depending on the wavelength, angle of incidence θ, and the distance *d* between the layers of atoms in the crystal lattice, the deflected waves may leave the crystal in or out of phase. If *d* is equivalent to the wavelength, the deflected rays leave the crystal more or less in phase—the crests and troughs coincide and the waves reinforce, producing a reflected beam of higher intensity. If *d* is equivalent to one-half wavelength, the deflected waves leave the crystal more or less out of phase—the troughs of one wave coincide with the crests of the other, interfering to reduce the intensity of the exiting beam. Maximally interfering beams are shown exiting the crystal in the diagram.

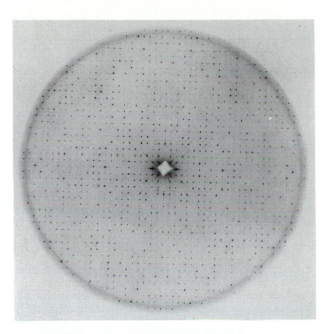

Figure A-21 The X-ray diffraction pattern produced by a crystal of the protein myosin. (Courtesy of H. M. Holden and I. Rayment, from *Prot. Struct. Funct. Genet.* 9:135 [1991]. Copyright © 1991 Wiley-Liss, a division of John Wiley and Sons, Inc.)

Suggestions for Further Reading

Andrews, A. T. 1986. *Electrophoresis,* 2nd ed. Oxford, UK: Clarendon Press.

Barnes, D. W., Sirbasky, D. A., and Sato, G. H., eds. 1984. *Cell Culture Methods for Molecular and Cell Biology.* New York: Liss.

Boatman, E. S., Berns, M. W., Walter, R. J., and Foster, J. S. 1987. Today's microscopy. *Biosci.* 37:384–394.

Bradbury, S. 1967. *The Evolution of the Microscope.* New York: Pergamon Press.

Bradbury, S. 1984. *An Introduction to the Optical Microscope.* New York: Oxford University Press.

Claude, A. 1975. The coming of age of the cell. *Science* 189:433–435.

Crewe, A. V. 1983. High resolution transmission scanning electron microscopy. *Science* 221:325–330.

de Duve, C. 1975. Exploring cells with a centrifuge. *Science* 189:186–194.

de Duve, C., and Beaufay, H. 1981. A short history of tissue fractionation. *J. Cell Biol.* 91:293s–299s.

Everhart, T. E., and Hayes, T. C. 1972. The scanning electron microscope. *Sci. Amer.* 226:54–69 (January).

Freshney, R. I. 1987. *Culture of Animal Cells: A Manual of Basic Technique.* New York: Liss.

Harris, R. 1991. *Electron Microscopy in Biology: A Practical Approach.* New York: Oxford University Press.

Meek, G. A. 1976. *Practical Electron Microscopy for Biologists,* 2nd ed. New York: Wiley.

Rogers, A. W. 1979. *Techniques of Autoradiography,* 3rd ed. New York: Elsevier/North Holland.

Scheeler, P. 1981. *Centrifugation in Biology and Medical Science*. New York: Wiley.

Spencer, M. 1982. *Fundamentals of Light Microscopy*. New York: Cambridge University Press.

Wilson, M. B. 1976. *The Science and Art of Basic Microscopy*. Bellaire, Tex.: The American Society for Medical Technology.

Review Questions

1. Diagram a light microscope, showing the positions of lenses and the specimen. What are the functions of the condenser, objective, and ocular lenses? How is a light microscope focused?

2. What is resolution? The refractive index? The numerical aperture of an objective lens? What are the effects of wavelength and the numerical aperture on resolution? Why does the use of immersion oil improve resolution?

3. Outline the operating principles of the brightfield and darkfield light microscopes. Why are objects smaller than the resolving power of the light microscope visible in the darkfield microscope?

4. Outline the operating principle of the phase contrast light microscope. What is a phase plate? How does introduction of the phase plate change differences in refractive index in the specimen to differences in contrast in the image?

5. What is fluorescence? What filters are employed to make fluorescence visible in the fluorescence microscope?

6. Compare the construction of a light microscope and a transmission electron microscope. How are the magnetic lenses used in an electron microscope focused? Why must specimens viewed in the TEM be thin and nonvolatile?

7. What are fixatives? What kinds of fixatives are used in specimen preparation for light and electron microscopy?

8. What is autoradiography? Outline the steps used in preparing tissue for autoradiography.

9. What is immunofluorescence? How is immunofluorescence used in light microscopy?

10. Outline the negative staining, shadowing, and freeze-fracture techniques used for preparing specimens for electron microscopy.

11. Compare the components of culture media for bacterial and animal cells. What is a clone? How is a clone prepared? What steps are followed in selecting a bacterial strain resistant to an antibiotic? What steps are followed in selecting a bacterial strain that requires a substance that bacteria can normally make for themselves?

12. How are animal cells obtained for culturing? How are animal cells "immortalized" for culturing? Why is this necessary?

13. What is cell fusion? How is it accomplished? In what ways have fused cells proved valuable in research? What is a monoclonal antibody?

14. What is cell fractionation? How is centrifugation used in cell fractionation? What is a cell-free system? What factors affect the rate at which molecules move downward or upward in a centrifuge tube? What is a Svedberg unit? How are Svedberg units used to identify molecules?

15. What is density-gradient centrifugation? Why do groups of molecules form highly distinct bands in density gradients? What is buoyant density centrifugation? How is the density gradient established in this technique?

16. Diagram a gel electrophoresis apparatus. What factors determine the rate at which molecules move through electrophoretic gels? What advantages does gel electrophoresis provide over centrifugation as a technique for isolating and purifying molecules from mixed samples?

17. What is the difference between a Southern, northern, and western blot?

18. What is the difference between filtration, ion exchange, and affinity chromatography? Outline the principles underlying molecular separation by each of these chromatographic methods.

19. What features of molecular structure split an X-ray beam into separate subbeams? What determines the locations and intensity of spots in an X-ray diffraction pattern?

INDEX

Steitz, T., 482, 495
stereoisomer, 44, 45*IB*
Stern, H., 729–730, 733, 748, 750–751*IB*
steroid, 46, 51, **52**
steroid hormone receptor, 499–501, 646–647
sterol, 51, 95, 98
Stillman, B., 663, 671, 681
Stoeckenius, W., 233, 276
Stöffler, G., 475
stop-transfer signal, 580, 582
STOPs. *See* stable tubule only proteins (STOPs)
storage vesicle, 10
strand transfer, 746–747
Strasburger, E., 25
stratum corneum, 344
stress fibers, 356–357, 360
stretch-gated channel, 128–129
striated muscle. *See* muscle, striated
strict anaerobe, 216
stroma, of chloroplast, 16, 249, 250
stromal lamellae, 249, 256
substitution-deletion analysis, 404
substrate, of enzyme, 77
substrate concentration, of enzyme, 81, 91–92
substrate level phosphorylation, 207, 210, 770
succinate, 220–221, **224**
succinate dehydrogenase, 221, 228
SulA gene, 645*IB*
sulfhydryl group, 39*t*, 42
Sulitzeanu, D., 559
Summers, K., 294–295
superfamilies, of genes, 522
surface receptors. *See* receptor, surface
surroundings, of defined system, 70
Sutherland, E., 157
Sutton, W., 25
SV40 system, 663
SV40 T antigen, 652, 663
Svedberg unit, 429, 795
Svoboda, K., 298–299
Swenson, K., 643
symport, 135–136
synapse, nerve, 141–143
synapsis, 730–731
synaptic cleft, 142
synaptic vesicle, 142
synaptobrevin, 593
synaptonemal complex, 730, **733**, 749, 752, **753**
synemin, 326
Szent-Gyorgyi, A., 222, 324
Szostak, J., 674, 676, 699–700, 772

T

T antigen, 652, 663
T cell, 151–152
T cells, 545, 649
   interaction of B cells and helper T cells, 562–563
   killer T cells, 563–564
   MHC genes and, 557, 560–562
T state, of allosteric enzyme, 84–85
T tubule, 329
TψC arm, 443–444
t-SNARE (target SNAP REceptor), 578
Takeichi, M., 166
talin, 187, 353, 356
Talmadge, D., 558
TATA box, 409, 447, 499, 778
TATA-binding protein (TBP), 412
TATA-less gene, 409
Tatum, E., 28–29
taxol, 288–289
Tay-Sachs disease, 97
Taylor, D., 333
Taylor, E., 319
Taylor, J., 739, 742, **743**
TBP, 778
teichoic acid, 200–202
telomerase, 674–678, 699
telomere, 383, 521, 674, 695, 699–700
telomere replication, 674–675, 678
telomere terminal transferase. *See* telomerase
telophase
   of meiosis, 729
   of mitosis, 692–693, 716–718
temperature, catalysis and, 81–82
temperature-sensitive mutants, 638–641
template, in DNA replication, 366, 367*t*
terminal web, of microvilli, 354
termination
   of transcription, 400–401
   of translation, 469
terminator codon, 454
tertiary structure
   of proteins, 58–60
   of RNA, 397
tetrad, 731
Tetrahymena, 86, **87**
tetramer, 61
   of intermediate filament, 348
tetraploid genome, 522
tetrapyrrole ring, 225, 271
TFIIB, 778
TFIID, 499, 504
TFIIIA, 438, 445, 496
TFIIIB, 438, 445
TFIIIC, 438, 445
Thach, R., 461
thalassemia, 409, 412
Theriot, J., 336
thermodynamics, laws of, 70–72
thick filament
   of sarcomere, 326, **327–328**
   in smooth muscle cell, 330–331
thin filament
   of sarcomere, 326, **327–328**
   in smooth muscle cell, 330–331
Thorsness, P., 623–625, 777
3'---> 5' exonuclease activity, 400, 459*fn*, 668

3'---> 5' phosphodiester linkage, 662
3'-end, of RNA molecule, 398, **399**, 403, 412
3'-flanking sequence, 408, **409**
3' untranslated region (3' UTR), 404, 405, **408**, 454, 508
3-ketoacyl coenzyme A, 243
3-phosphoglyceraldehyde (3PGAL), 213–215, 222–223, 261, 262–263, 264–265, 268–269
3-phosphoglycerate, 214, 261–262
threonine deaminase, regulation of, 85, **86**
thylakoid, 16, **17**, 249, **257**
thylakoid membranes, 234–235, 256
   in light reactions, 259
thymidine triphosphate (TTP), 662*ff*
thymine, 63, **64**
thyroid hormone receptor, 646–647
tight junction, 111–112, 167, 170–171, **354–355**
Tilney, L., 335–336
Tjian, R., 410
Tn element, 543
Todd, J., 564
tonofilament, 167, **168**, 344
top2, 723
topoisomerase, 373, 401
topoisomerase II, 383, 723
Toxoplasma, 568
trans element, 490–491, 491*t*, 503, 506
trans face, of Golgi complex, 576–578
trans form of carbon chain, 47, **48**
trans splicing, 415
transcription, 9, 397
   bacterial, 451–457
   5S rRNA, 438
   activators, 453
   alterations in chromatin during, 416–418
   changes in nucleosomes during, 379–380
   in chloroplast, 620
   direction, 398, **399**
   during meiosis, 738
   elongation, 400–401
   initiation, 400
   large pre-rRNA, 434–437, **440, 442**
   mechanism, 397–400
   in meiosis, 738
   in mitochondrion, 618–620
   mRNA, 403–412, 454
   pre-mRNA, 412–413
   prokaryotic, 451–457
   proofreading in, 400
   regulation, 489*ff*
   repressors, 453
   RNA, 372, 382, 397–402, 451–457
   RNA polymerases in, 402
   rRNA, 431–439, **442**, 454–455, **456**
   snRNA, 447
   termination, 401–402
   tRNA, 445–447

transcription factors, 400
   TFIIIA, 438
   TFIIIB, 438
   TFIIIC, 438
transcriptional regulation, 489–495
   DNA methylation in, 504–506
   systems controlling specific genes, 497
   See also posttranscriptional regulation; regulatory proteins
transduction, 743–744
transfection, 644
transfer RNA (tRNA), 9, 88, 369*IB*, 397
   aminoacyl, 459
   bacterial, 455–457
   in chloroplast, 616–617
   cognate, 482
   genes, 444–445
   initiator, 460, 486
   isoaccepting, 482
   in mitochondrion, 610, 612, **613**
   modifications, 444, 446
   in polypeptide assembly, 459*ff*
   processing, 446–447
   promoter, 444
   recognition by synthetases, 482–484
   structure, 442–445
   transcription, 445–447
transformation, 743
transgene, 347
transition state, 78–80
transition state analog, 79
transition vesicle, 113, 576
translation, 9, 459
   accuracy, 467–469
   amino acid activation, 459, 480–484
   chloroplast, 620–622
   elongation in, 460, 463, 466–469
   eukaryotic, 461*ff*, **471**
   genetic code and, 476–480
   initiation in, 460–463
   in meiosis, 738
   mitochondrial, 620, 622
   polypeptide assembly in, 459–463, 466–472
   in prokaryotes, 486–487
   regulation, 489, 508–510
   ribosomal structure in, 472–476
   termination in, 460, 469, **470**
translational regulation, 489, 508–510
translocation, of ribosomes, 467
   translocation, of chromosomes, 533, 538–539, 649
transmembrane complexes, 226–227
transmission electron microscope (TEM), 787–789
transplantation antigens, 557
transport, 118
   action potential in, 137–140